30th International Conference

Coastal Engineering 2006

vol. 1

Proceedings of the 30th International Conference

Coastal Engineering 2006

San Diego, California, USA 3–8 September 2006

editor

Jane McKee Smith

U.S. Army Engineer Research and Development Center
Coastal & Hydraulics Laboratory, USA

World Scientific

NEW JERSEY • LONDON • SINGAPORE • BEIJING • SHANGHAI • HONG KONG • TAIPEI • CHENNAI

Published by

World Scientific Publishing Co. Pte. Ltd.
5 Toh Tuck Link, Singapore 596224
USA office: 27 Warren Street, Suite 401-402, Hackensack, NJ 07601
UK office: 57 Shelton Street, Covent Garden, London WC2H 9HE

British Library Cataloguing-in-Publication Data
A catalogue record for this book is available from the British Library.

Cover photos: San Diego Bay (US Navy)
Mission Bay (US Corps of Engineers, Los Angeles District)
San Diego Skyline (© Joanne DiBona, San Diego Convention & Visitors Bureau)

COASTAL ENGINEERING (In 5 Volumes)
Proceedings of the 30th International Conference

ISBN-13 978-981-270-636-2 (pbk) (Set)
ISBN-13 978-981-270-989-9 (pbk) (Vol. 1)
ISBN-13 978-981-270-990-5 (pbk) (Vol. 2)
ISBN-13 978-981-270-991-2 (pbk) (Vol. 3)
ISBN-13 978-981-270-992-9 (pbk) (Vol. 4)
ISBN-13 978-981-270-993-6 (pbk) (Vol. 5)

Printed in Singapore by Mainland Press

Foreword

The 30th International Conference on Coastal Engineering (ICCE 2006) was held in San Diego, California, USA, in September of 2006. The Local Organizing Committee, led by Ronald M. Noble, is acknowledged for their dedicated preparation over many years that led to a successful conference with broad participation. A record 895 attendees from 35 countries gathered at the Manchester Grand Hyatt in San Diego to discuss research and applications in coastal engineering. The papers contained in this *Proceedings* cover a wide range of topics including waves; swash, nearshore currents, and long waves; coastal management, risk, and environmental restoration; sediment transport and morphology; and coastal structures. The authors have provided state-of-the-art contributions, and this volume could not be produced without their commitment to solving coastal engineering challenges. The members of the ASCE/COPRI Coastal Engineering Research Council (CERC) and the ICCE 2006 Technical Review Committee reviewed 851 abstracts and selected the 514 that were presented at the conference. The dedication of the Council members has led to the continued high quality and popularity of the International Conference on Coastal Engineering.

Preparation of these proceedings would not be possible with the assistance of many colleagues. Thank you to Prof. Robert A. Dalrymple, CERC Chairman, and Prof. Billy L. Edge, CERC Vice Chairman, for their guidance and encouragement. Thanks to Ron Noble and Jessica Latham, Noble Consultants, for answering my many requests for information and for their wonderful hospitality in San Diego. I would also like to thank Alison Sleath, Julie Roasti, Bill Seabergh, Tahirih Lackey, and Kyle Gordon, all of the Engineering Research and Development Center, for assistance in organizing the papers. Finally, I would like to acknowledge Chelsea Chin, World Scientific Publishing Company for bringing it all together.

Jane McKee Smith
U.S. Army Corps of Engineers
Engineer Research and Development Center
Coastal and Hydraulics Laboratory
Jane.M.Smith@erdc.usace.army.mil

Proceedings Dedication

Photo credit: Dave Werren

This ***Proceedings*** is dedicated to Orville T. Magoon who has served as Vice Chair and member of the Coastal Engineering Research Council 1979-2004. His dedication, devotion to coastal engineering and hard work has contributed to the success of the ICCEs held between 1980 and 2006.

Orville Magoon received his under-graduate degree in Civil Engineering from the University of Hawaii in 1951 and an M.S. degree in Civil Egineering from Stanford University in 1952. After thirty years of experience in the field of coastal planning, design, construction and rehabilitation of coastal structures, Mr. Magoon retired as Chief of the Coastal Engineering Branch of the Planning Division of the US Army Corps of Engineers, South Pacific Division in July, 1983. Subsequently, he balanced his time between being owner of Guenoc Winery in California and coastal engineering. He continues to be active in coastal zone management and coastal engineering through lectures and writing.

During his years of government service, Orville Magoon served on several committees and panels, including the Permanent International Association of Navigation Congresses (PIANC) Committee on Waves, a number of National Aeronautics and Space Administration panels involved with remote sensing and satellite imagery of coastlines and the National Academy of Engineering.

Mr. Magoon conceived and chaired the first ten major international specialty conferences on coastal zone management and coastal processes beginning with Coastal Zone 78 which was held in San Francisco. This successful conference series continues on a biannual basis. He organized seven international, focused conferences on coastal structures through the Rubble Mound Structures Committee of COPRI. He continues to serve on several committees of COPRI of ASCE and is an active member of the PIANC.

Orville Magoon has received several national and international awards during his career including the Jim Purpura National Coastal Engineering Award from FSBPA and the Murrough P. O'Brien Award from ASBPA. He also received the William Wisely, Arthur M. Wellington Prize, International Coastal Engineering Award and the Moffat-Nichol Awards from ASCE in recognition of his service to the coastal engineering and coastal zone management. Of special note, the ASCE and the JSCE have elevated him to Honorary Member in recognition of his contributions to the profession.

The Council thanks him on behalf of the coastal engineering community for his untiring service to the profession and leadership to the ICCE and the CERC.

Acknowledgments

Local Organizing Committee

Ronald M. Noble, Chairman, Noble Consultants, Inc.
Orville T. Magoon, Coastal Zone Foundation
Dan Muslin, TEC Infrastructure Consultants
Young C. Kim, California State University, Los Angles
Dick Seymour, Scripps Institution of Oceanography
Lesley Ewing , California Coastal Commission
Art Shak, US Army Corps of Engineers, Los Angels District
Bud Carroll, Chairman Emeritus, Montgomery Watson Harza
Donald Treadwell, Consulting Engineer

Coastal Engineering Research Council (ASCE/COPRI)

Robert A. Dalrymple, Chairman, Johns Hopkins University
Billy L. Edge, Vice Chairman, Texas A&M University
Jane McKee Smith, Secretary, U.S. Army Corps of Engineers, Engineer Research & Development Center
Robert G. Dean, University of Florida
Yoshimi Goda, Yokohama National University and ECOH Corporation
James R. Houston, U.S. Army Corps of Engineers, Engineer Research & Development Center
J. William Kamphuis, Queen's University
Orville T. Magoon, Coastal Zone Foundation
Ronald M. Noble, Noble Consultants, Inc.
Marcel Stive, Delft Hydraulics

Technical Abstract Review Committee

Ian H. Townend, Coastal Environment
Lesley Ewing, Coastal Hazards
Jurjen A. Battjes, Coastal Meteorology and Waves
Peter Nielsen, Sediment Processes
Shigeo Takahashi, Coastal Structures
Miguel A. Losada, Ports and Harbors
Charitha Pattiaratchi, Shore Protection
Masahiko Isobe, Sustainable Development

Sponsorship

The ICCE 2006 Local Organizing Committee is pleased to acknowledge the significant financial support of the following organizations:

Noble Consultants, Inc.
Federal Emergency Management Agency
Tetra Tech, Inc.
US Geological Survey
Moffatt & Nichol Engineers
H. Morgan & Cherry Noble
W.F. Baird & Associates Ltd.
Bosch Rexroth
Port of San Diego
Culbertson, Adams & Associates, Inc.
CH2M-Hill
SAIC
Parsons Brinckerhoff Quade & Douglas, Inc.
The Dutra Group
Great Lakes Dredge & Dock Company
Olsen Associates, Inc.
William J. Carroll
Weston Solutions, Inc.
Halcrow Group Ltd./HPA, Inc.
Bellingham Marine
Northwest Hydraulic Consultants
County of San Diego
DMJM Harris
Power Engineering Contractors
Manson Construction Co.
DHI Water & Environment
Donald D. Treadwell
Everest International Consultants, Inc.
Taylor Engineering, Inc.
Applied Coastal Research and Engineering, Inc.
California Coastal Coalition
Coastal Environments
Coastal Frontiers Corporation
Coastal Technology Corporation
GeoSoils, Inc.
Marathon Construction Company
TerraCosta Consulting Group, Inc.
Anchor Environmental, L.L.C.
AMEC Earth & Environmental, Inc.
International Construction Service, Inc.
John S. Meek Company, Inc.

The ICCE 2006 Local Organizing Committee also expresses sincere gratitude to all the organizations represented in the technical exhibition.

CONTENTS

VOLUME ONE

VOLUME TWO

VOLUME THREE

VOLUME FOUR

VOLUME FIVE

Waves

La Jolla, California
(Photo courtesy of Lesley Ewing)

A NOTE ON WAVE CELERITIES ON A COMPRESSIBLE FLUID

Robert A. Dalrymple[1] and Benedict D. Rogers[2]

Smoothed particle hydrodynamics (SPH) is being used to model fluid flows, including free surface waves; however, the method assumes an compressible fluid. Most water wave theories are based on the assumption of incompressibility, which is reasonable for water. Here we develop a linear wave theory for waves on a compressible fluid. The effect of the fluid compressibility on wave number is examined to determine how much error is induced by an assumption of compressibility and also how these effects might be avoided. The dispersion relationship for a free surface wave on a compressible fluid is determined and the results are compared to incompressible free surface wave theory.

INTRODUCTION

Smoothed Particle Hydrodynamics (SPH), originally developed in astrophysics, has been used to model the behavior of water waves since Monaghan first applied it in 1994. Here we examine the ability of this compressible fluid methodology to correctly propagate water waves over large distances. While various viscous and turbulence closure models will affect the details of this argument, the basic conclusion will hold, regardless of the extra terms.

Examples of the use of SPH for free surface waves include Monaghan and Kos (1999, 2000) for long waves, and Gomez-Gesteira and Dalrymple (2004) for 3-D waves and Dalrymple and Rogers (2006) for plunging breakers and surf zone turbulence. For all of these cases, following the rule-of-thumb of Monaghan (1992), the fluid compressibility is taken as small, with the Mach number is assumed to be on the order of 0.1, such that changes in density are on the order of 1 to 2 percent.

Water waves on a compressible fluid was examined first by Pidduck (1910) and later by Longuet-Higgins (1950) for microseisms. They both point out that one solution of the governing equations is similar to waves on a incompressible fluid.

Madsen and Schäffer (2006) show that the use of an artificial fluid compressibility for the Navier-Stokes equations changes the free surface wave phase speed from that on an incompressible fluid. (They examine the use of Chorin's method (1967) to solve steady-state Navier-Stokes equations with a fluid of constant density.) This raises a question about the effect of compressibility assumptions within SPH on water wave problems and whether Monaghan's 1992 rule-of-thumb is valid. This paper addresses that issue.

[1] Department of Civil Engineering, Johns Hopkins University, 3400 N. Charles Street, Baltimore, MD, 21218, USA

[2] School of Mechanical, Aerospace & Mechanical Engineering, University of Manchester, Manchester, M60 1QD, U.K.

THEORETICAL ARGUMENT

The SPH method is characterised by several important features: it is a Lagrangian method that is meshfree and the pressure is directly related to density through an equation of state. This latter feature saves computational time as it is not necessary to solve a (Poisson) equation for the pressure. The questions to be addressed here is what is the the effect of this equation of state on water wave celerities?

Following Monaghan (1994), the equation of state, relating pressure p to density ρ is given by

$$p(x,y,t) = B\rho_s\left(\left(\frac{\rho}{\rho_s}\right)^{\gamma} - 1\right), \tag{1}$$

as suggested by Batchelor (1967), where γ = 7 and ρ_s is a reference density, and B is a constant that is chosen to reduce the speed of sound so that the time step in a numerical model is not dictated by the actual (more rapid) speed of sound. Monaghan recommends that B be chosen such at that the artificial sound speed is ten times faster than any fluid flows in the model. Morris (1996) examining the stability of SPH techniques uses $\gamma = 1$, and Morris *et al.* (1997) suggest that γ = 1 is more appropriate for low Reynolds number flows.

The conservation of mass and momentum are given by

$$\begin{aligned}
\frac{\partial \rho}{\partial t} + u\frac{\partial \rho}{\partial x} + v\frac{\partial \rho}{\partial y} &= -\rho\left(\frac{\partial u}{\partial x} + \frac{\partial v}{\partial y}\right) && (2)\\
\frac{\partial u}{\partial t} + u\frac{\partial u}{\partial x} + v\frac{\partial u}{\partial y} &= -\frac{1}{\rho}\frac{\partial p}{\partial x} = -\frac{1}{\rho}\frac{\partial p}{\partial \rho}\frac{\partial \rho}{\partial x} && (3)\\
\frac{\partial v}{\partial t} + u\frac{\partial v}{\partial x} + v\frac{\partial v}{\partial y} &= -\frac{1}{\rho}\frac{\partial p}{\partial y} - g = -\frac{1}{\rho}\frac{\partial p}{\partial \rho}\frac{\partial \rho}{\partial y} - g && (4)
\end{aligned}$$

where u, v are the horizontal and vertical velocities in the x, y coordinate directions, and g is the acceleration of gravity.

Since we are examining wave motion and we have neglected frictional terms in the equations of motion, we will assume that the fluid motion is irrotational, or $\frac{\partial u}{\partial z} - \frac{\partial v}{\partial x} = 0$. This allows us to define a velocity potential, $\phi(x, y, t)$, such that the velocities are defined as

$$u = -\frac{\partial \phi}{\partial x}, \quad \text{and} \quad v = -\frac{\partial \phi}{\partial y} \tag{5}$$

At the free surface of the compressible fluid, denoted as $y = \eta(x, t)$, the motion of the surface satisfies a kinematic condition, that states the fluid follows the surface. This can be expressed as

$$\frac{\partial \eta}{\partial t} + u\frac{\partial \eta}{\partial x} = v \quad \text{on } y = \eta(x,t) \tag{6}$$

The second condition at the surface is that the pressure is constant along the surface; in Eq. 1 we have implicitly taken the constant to be zero (gage) pressure. The compressible Bernoulli equation, found by integrating the equations of motion, using the irrotationality condition, and introducing the velocity potential, is

$$-\frac{\partial \phi}{\partial t} + \frac{u^2+v^2}{2} + \frac{B\gamma}{\gamma-1}\left(\frac{\rho}{\rho_s}\right)^{\gamma-1} + gy = \text{constant} \tag{7}$$

For the case that $\gamma = 1$, then the equation is

$$-\frac{\partial \phi}{\partial t} + \frac{u^2+v^2}{2} + B\ln\rho + gy = \text{constant} \tag{8}$$

At the free surface, $\rho = \rho_s$, and both of these equations reduce to

$$-\frac{\partial \phi}{\partial t} + \frac{u^2+v^2}{2} + gy = \text{constant on } y = \eta(x,t) \tag{9}$$

The last boundary condition to be applied is that the velocity into the bottom is zero. This will be expressed as

$$-\frac{\partial \phi}{\partial y} = 0 \text{ on } y = -h \tag{10}$$

For the first part of this paper, we will assume a linear equation of state, corresponding to $\gamma = 1$. This yields a constant speed of sound, C_s, which follows from $\partial p/\partial\rho = B = C_s^2$.

All of the above conservation equations and the boundary conditions involve nonlinear terms and a perturbation approach will be employed to linearize the problem for an analytic solution. We expand the dependent variables in a power series of a small parameter, ϵ.

$$\rho(x,y,t) = \rho_0 + \epsilon\rho_1 + \epsilon^2\rho_2 + \ldots, \tag{11}$$

$$u(x,y,t) = \epsilon u_1 + \epsilon^2 u_2 + \ldots;\quad v(x,y,t) = \epsilon v_1 + \epsilon^2 v_2 + \ldots \tag{12}$$

$$p(x,y,t) = p_0 + \epsilon p_1 + \epsilon^2 p_2 + \ldots;\quad \phi(x,y,t) = \epsilon\phi_1 \tag{13}$$

$$\eta(x,t) = \epsilon\eta_1 \tag{14}$$

where $\eta(x,t)$ is the free surface displacement about the x axis. Note that η and the velocities are assumed to be zero at the lowest order.

Substituting these expansions into the conservation equations and grouping by order we obtain the zeroth order equations:

$$\frac{\partial \rho_0}{\partial t} = 0;\quad \frac{\partial \rho_0}{\partial x} = 0;\quad \frac{1}{\rho_0}\frac{\partial \rho_0}{\partial y} = -g/C_s^2 \tag{15}$$

Integrating these first order differential equations, with the condition that $\rho_0 = \rho_s$ at $y = 0$ (the zeroth order water surface) yields

$$\rho_0 = \rho_s e^{-gy/C_s^2} \tag{16}$$

At the next order the equations are:

$$\frac{\partial \rho_1}{\partial t} + v_1 \frac{\partial \rho_0}{\partial y} = -\rho_0 \left(\frac{\partial u_1}{\partial x} + \frac{\partial v_1}{\partial y} \right) \tag{17}$$

$$\frac{\partial u_1}{\partial t} = -\frac{C_s^2}{\rho_0} \frac{\partial \rho_1}{\partial x} \tag{18}$$

$$\frac{\partial v_1}{\partial t} = -\frac{C_s^2}{\rho_0} \frac{\partial \rho_1}{\partial y} - g \frac{\rho_1}{\rho_0} \tag{19}$$

These equations can be combined into one equation by taking the time derivative of Eq. 18 and substituting in Eqns. 17 and 19 and the velocity potential. Integrating with respect to x yields

$$\frac{\partial^2 \phi_1}{\partial t^2} = C_s^2 \left(\frac{\partial^2 \phi_1}{\partial x^2} + \frac{\partial^2 \phi_1}{\partial y^2} \right) - g \frac{\partial \phi_1}{\partial y} \tag{20}$$

(The constant of integration can be shown to be a true constant by developing the governing equation via Eq. 19.) This equation is an extended form of the canonical wave equation. This equation is the same as that of Pidduck (1910), except for a - sign in the last term as his y axis pointed downwards, rather than upwards as here).

The free surface boundary conditions at this order are:

$$\frac{\partial \eta_1}{\partial t} = -\frac{\partial \phi_1}{\partial y} \quad \text{on } y = 0 \text{ from Eq.6} \tag{21}$$

$$-\frac{\partial \phi_1}{\partial t} + g\eta_1 = 0 \quad \text{on } y = 0 \text{ from Eq. 9} \tag{22}$$

(23)

Combining these two linear equations yields the linear combined free surface boundary condition

$$\frac{\partial^2 \phi_1}{\partial t^2} + g \frac{\partial \phi_1}{\partial y} = 0 \quad \text{on } y = 0 \tag{24}$$

A progressive wave solution to the governing equation 20, found by separation of variables, is

$$\phi_1(x, y, t) = A\, e^{\alpha y} \cos(kx - \omega t) \tag{25}$$

$$\text{where } \alpha = \frac{g}{2C_s^2} \pm \sqrt{(k^2 - k_s^2) - \frac{g^2}{4C_s^4}} = \gamma \pm \mathcal{K} \tag{26}$$

$$\text{and } k_s = \frac{\omega}{C_s}, \tag{27}$$

and where k is the water wave number, and ω is the angular frequency of the wave. The above solution can be rewritten as

$$\phi_1(x,y,t) = A_1\, e^{\gamma y}\, e^{\mathcal{K} y}\, \cos(kx-\omega t) + A_2\, e^{\gamma y}\, e^{-\mathcal{K} y} \sin(kx-\omega t)$$

$$\text{where, again, } \mathcal{K} = \sqrt{(k^2-k_s^2)-\gamma^2} \text{ and } \gamma = \frac{g}{2C_s^2} \tag{28}$$

Next we apply the boundary conditions, beginning with the no-flow condition at the bottom, which leads to a condition relating the constants A_1 and A_2. Using this, we rewrite the velocity potential:

$$\phi_1(x,y,t) = D\, e^{\gamma y}\, (\gamma\, \sinh \mathcal{K}(h+y) - \mathcal{K}\, \cosh \mathcal{K}(h+y)) \sin(kx-\omega t) \tag{29}$$

where D is a new constant.

Substituting into the combined free surface boundary condition, Eq. 24,

$$\omega^2 = g\frac{(\gamma^2-\mathcal{K}^2)\tanh \mathcal{K}h}{\gamma \tanh \mathcal{K}h - \mathcal{K}} \tag{30}$$

This dispersion relationship relates the wave number k to the angular frequency ω. It is the same as that found by Pidduck (1910, his equations 20 and 21; note: our $\mathcal{K}$ is his μ, and γ is coincidently defined the same way).

For the case where the sound speed C_s is significantly greater than ω/k (the phase speed of the surface wave), then $\gamma << \mathcal{K}$, and $\mathcal{K} \rightarrow k$, which then reduces the dispersion relationship goes to that for incompressible fluids:

$$\omega^2 = gk_\ell \tanh k_\ell h \tag{31}$$

In deep water, defined as $kh > \pi$, the dispersion relationship (Eq. 30) reduces to $\omega^2 = g(\gamma + \mathcal{K})$. The associated velocity potential is

$$\phi_1 = Ae^{\gamma y}\, e^{\mathcal{K} y} \sin(kx-\omega t) \tag{32}$$

which satisfies the condition that the velocities go to zero as $y \rightarrow -\infty$.

The water surface displacement associated with the velocity potential (29 is found from the linear dynamic free surface boundary condition (22), which, when we assume that the free surface maximum displacement is a, yields a more detailed description of the constant D:

$$\eta_1(x,t) = a\cos(kx-\omega t) \text{ with } D = \frac{ga}{\omega}\frac{1}{(\gamma \sinh \mathcal{K}h - \mathcal{K}\cosh \mathcal{K}h)} \tag{33}$$

The associated fluid density at first order can be also obtained from our solution for the velocity potential. Again using Eq. 18 and substituting for the velocity potential, we obtain,

$$\rho_1 = \frac{\rho_s ga}{C_s^2}\, e^{-\left(\frac{gy}{2C_s^2}\right)} \left(\frac{\gamma \sinh \mathcal{K}(h+y) - \mathcal{K}\cosh \mathcal{K}(h+y)}{\gamma \sinh \mathcal{K}h - \mathcal{K}\cosh \mathcal{K}h}\right) \cos(kx-\omega t) \tag{34}$$

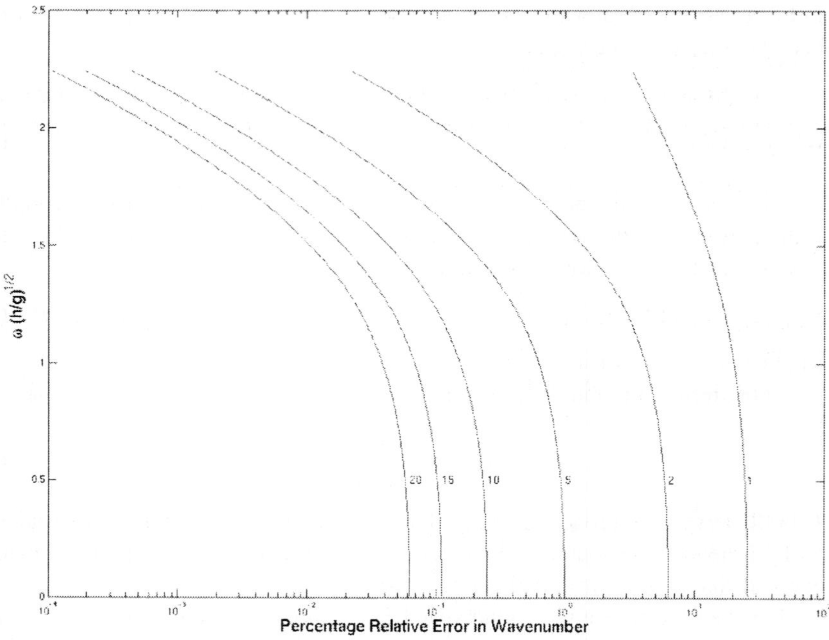

Figure 1. Dimensionless angular frequency, $\omega\sqrt{\frac{h}{g}}$, versus relative wave number difference, $100*(k-k_\ell)/k_\ell$ for C_s/C_w varying from 1 to 20.

FIRST ORDER RESULTS

The dispersion relationship (Eq. 30) for a compressible fluid is used to calculate the free surface wave number k by iteration. It is compared to the wave number k_ℓ for the same wave on an incompressible fluid (Eq. 31), with the difference between the two then being the error that results in using a compressible fluid theory to model an incompressible fluid. The only variable in the problem is the speed of sound, C_s. For these comparisons, for each given angular frequency, the ratio of the sound speed in the compressible fluid C_s to the free surface wave celerity, $C_w = \omega/k_\ell$ will be fixed at various values.

Figure 1 shows the dimensionless angular frequency $\omega\sqrt{\frac{h}{g}}$ versus the percentage difference between compressible and incompressible wave numbers, $100*(k-k_\ell)/k_\ell$ for C_s/C_w varying from 1 to 20. Clearly the errors increase dramatically as the speed of sound in the fluid approaches that of the free surface wave celerity. Also, the errors are larger in shallow water than for deep water. However, it is clear that the errors are relatively small when the ratio of speeds is around ten or more. For example, for $C_s/C_w = 10$, the maximum percentage error in wave number using the compressible model is only 0.25%.

MORE COMPLETE MODEL, $\gamma \neq 1$

The previous analytical model was developed for the equation of state: $p = B\rho_s(\rho/\rho_s) - 1)$. The next step is to examine the influence of the factor γ in Eq. 1.

Taking the derivative of the pressure, we obtain the speed of sound,

$$\frac{\partial p}{\partial \rho} = B\gamma(\frac{\rho}{\rho_s})^{\gamma-1} = C^2(\rho) \tag{35}$$

From this, we can define $C_s^2 = C^2(\rho_s) = B\gamma$, which is the sound speed at the surface.

To proceed, we follow the same perturbation expansion as before. From

$$-\frac{1}{\rho_0}(\frac{\partial p}{\partial \rho})_0\frac{\partial \rho_0}{\partial y} - g = 0 \tag{36}$$

we obtain $\frac{\partial \rho_0}{\partial y} = -\rho_0 g/C_0^2$, where $C_0^2 = C^2(\rho_0)$. Integrating by separation of variables and using the condition that $\rho = \rho_s$ at the surface ($y = 0$), we obtain

$$\rho_0 = \rho_s\left(1 - (\frac{\gamma-1}{\gamma})\frac{gy}{B}\right)^{\frac{1}{\gamma-1}} \tag{37}$$

This equation gives the zeroth order density distribution in the fluid with depth.

Solving for the velocities and substituting for the first order velocity potential yield the governing equation for the surface wave motion

$$\frac{\partial^2\phi_1}{\partial t^2} = C_0^2\left(\frac{\partial^2\phi_1}{\partial x^2} + \frac{\partial^2\phi_1}{\partial y^2}\right) - g\frac{\partial\phi_1}{\partial y} \tag{38}$$

This equation only differs from the linear equation of state governing equation (20) by the sound speed that appears in the first term on the right hand side. While before the sound speed was the constant value C_s^2, here it is the depth dependent $C_0^2(\rho_0)$, which does have a depth dependency.

The same free surface and bottom boundary conditions pertain as before. However, an analytical solution to this problem eludes us and a numerical solution is used.

Eigenvalue Approach

To compare with the analytical solution, we will assume that the solution to () can be obtained by separation of variables. Therefore, the first order velocity potential will be written as $\phi_1(x, y, t) = Y(y)\cos(kx - \omega t)$. Introducing the velocity potential and definition for $C^2(\rho_0)$ into the governing equation gives us an equation in $Y(y)$ only:

$$\frac{d^2Y}{dy^2} - \frac{g}{B\gamma(1-\alpha y)}\frac{dY}{dy} + \frac{\omega^2}{B\gamma(1-\alpha y)}Y = k^2Y \quad \text{for } -h \leq y \leq 0 \tag{39}$$

where $\alpha = (\gamma - 1)/\gamma(g/B)$. As before, for given angular frequency ω and water depth h, the wave number k is an eigenvalue. The associated boundary conditions are $\frac{\partial Y}{\partial y} = 0$ at $y = -h$ and $\frac{\partial Y}{\partial y} - (\omega^2/g)Y = 0$ at $y = 0$.

To solve, we discretize the elevation y as $y_i = -h + i\Delta y$, $i = 0, 1, 2, \ldots, N+1$, where $\Delta y = h/(N+1)$. Using second order central differences, the finite difference representation of the governing equation for $Y(y)$ is

$$Y_{i-1}\left(1 + \frac{g\Delta y}{2B\gamma(1-\alpha y_i)}\right) + Y_i\left(-2 + \frac{\omega^2 \Delta y^2}{B\gamma(1-\alpha Y_i)}\right) +$$

$$Y_{i+1}\left(1 - \frac{g\Delta y}{2B\gamma(1-\alpha y_i)}\right) = k^2 \Delta y^2\, Y_i \tag{40}$$

This equation is valid for $i = 2, 3 \ldots, N$ The equations for $i = 1$, just above the bottom, and $i = N + 1$, just below the free surface are derived by using the boundary condition as bottom and top as shown in Appendix 7 for brevity.

All of the finite difference equations then can be assembled into a matrix equation of the form: $Ay = k^2(\Delta y)^2\, y$, where A is a tridiagonal square matrix of size (NxN). The equation is an eigenvalue problem that can be solved for the real eigenvalues $k^2\Delta y^2$ and the associated eigenvectors.

For this case, there is only one positive eigenvalue that is associated with a progressive wave, the other eigenvalues are associated with non-propagating evanescent modes (Ursell, Dean and Yu, 1960).

Figures 2 and 3 show the dimensionless wave number (kh) as a function of the value of γ in the Equation of State (1) for six values of the ratio C_s/C_w. While there is a large variation in computed wave number as a function of gamma for small values of C_s/C_w, it is clear from the figures that for values of sound speed much larger than the water wave phase speed, say, ten times larger, there is no difference in wave number with increasing γ.

For values of $C_s/C_w > 5$, there is no appreciable variation in computed wave number due to the value of γ, which means that the choice of γ in the equation of state is not that important, given that the compressiblity is small.

CONCLUSIONS

The major conclusion of this study is that waves in a compressible medium propagate at nearly the same speed as that wave in an incompressible fluid when the speed of sound of the compressible fluid is at least five to ten times that of the celerity of the free surface wave. However, errors are much larger for low frequency waves, say with angular frequencies less than 0.01 rad/s. Further, the choice of γ in the equation of state is not that important which the fluid is just slightly compressible.

ACKNOWLEDGEMENTS

This work was supported by the Office of Naval Research, Coastal Geosciences Program, under Award Number N00014-04-0089.

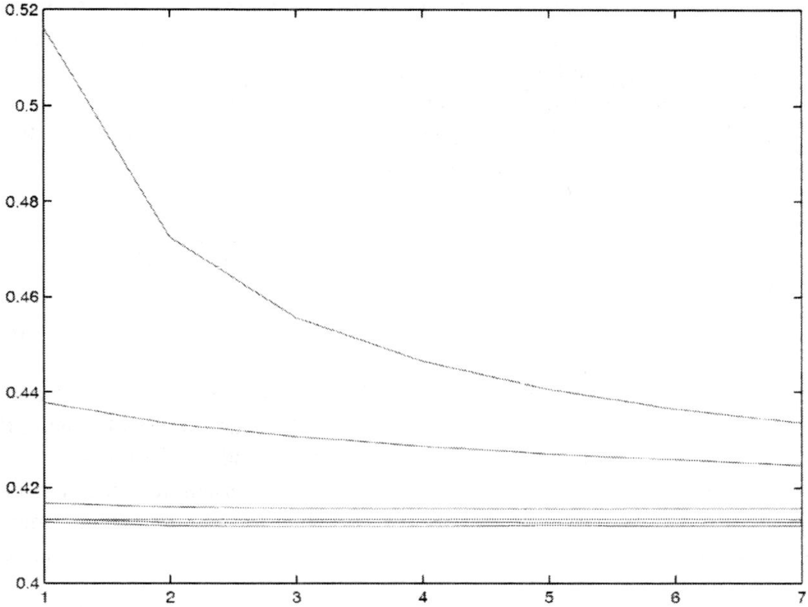

Figure 2. kh versus γ for C_s/C_w ranging from 1 to 20. Here $\omega\sqrt{h/g} = 0.4012$, corresponding to a shallow water case.

APPENDIX

At the bottom, corresponding to $y = -h$, and y_o, and Y_0, the no-flow boundary condition applies, $\frac{dY}{dy} = 0$. Writing this in second order accurate forward difference form as

$$\frac{dY}{dy} \approx \frac{-Y_2 + 4Y_1 - 3Y_0}{2\Delta y} = 0 \tag{41}$$

gives us

$$Y_0 = \frac{4Y_1 - Y_2}{3} \tag{42}$$

Substituting this value for Y_0 into the $(i = 1)$ differential equation (40) yields

$$Y_1\left(-2 + \frac{\omega^2 \Delta y^2}{B\gamma(1-\alpha y_1)} + \frac{4}{3}(1 - \frac{g\Delta y}{2B\gamma(1-\alpha y_1)}\right) + Y_2\left(\frac{2}{3} - \frac{4}{3}\frac{g\Delta y}{2B\gamma(1-\alpha y_1)}\right)$$
$$= k^2 \Delta y^2 Y_1 \tag{43}$$

At the free surface, correponding to $y = 0$, and Y_{N+1}, the linear combined free surface boundary condition is

$$\frac{dY}{dy} - \frac{\omega^2}{g} Y = 0 \tag{44}$$

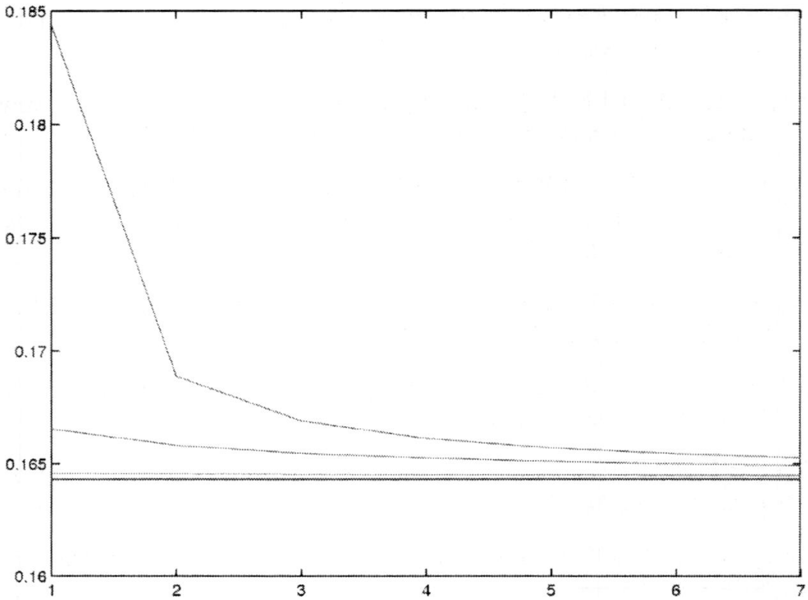

Figure 3. kh **versus** γ **for** C_s/C_w **ranging from 1 to 20. Here** $\omega\sqrt{h/g} = 1.5012$**, which is in intermediate water depth.**

Using second order accurate backwards difference, we can rewrite this as

$$Y_{N+1} = \frac{Y_{N-1} - 4Y_N}{\left(\frac{2\omega^2 \Delta y}{g} - 3\right)} \tag{45}$$

Substituting into the finite difference equation at i=N gives the form:

$$Y_{N-1}\left(1 + \frac{g\Delta y}{2B\gamma(1-\alpha y_N)} + \frac{(1 - g\Delta y/(2B\gamma(1-\alpha y_N)))}{\omega^2 2\Delta y/g - 3)}\right) +$$
$$Y_N\left(-2 + \frac{\omega^2 \Delta y^2}{B\gamma(1-\alpha y_N)} - 4\frac{(1 - g\Delta y/(2B\gamma(1-\alpha y_N)))}{\omega^2 2\Delta y/g - 3)}\right) = k^2 \Delta y^2\, Y_N \tag{46}$$

Using now these two finite difference equations for Y_1 and Y_N, and the equation 40 for Y_i, $i = 2, 3 \ldots, N-1$ results in N equations for the N unknowns Y. Given these solutions, Y_0 and Y_{N+1} can then be determined from Eqns. 42 and 45.

REFERENCES

Batchelor, G.K. 1974. **Introduction to Fluid Dynamics**, Cambridge University Press, Cambridge.

Chorin, A.J., 1967. A numerical method for solving incompressible viscous flow problems, *J. Comput. Phys.*, 2, 12-26.

Dalrymple, R.A. and B.D. Rogers. 2004. Numerical modeling of water waves with the SPH method, *Coastal Engineering*, 53/2-3, 141-147.

Gomez-Gesteira, M. and R.A. Dalrymple. 2004. Using a 3-D SPH method for wave impact on a tall structure, *J. Waterways, Ports, Coastal, and Ocean Engineering,* 130, 2, 63-69.

Longuet-Higgins, M.S. 1950. A theory of the origin of microseisms, *Phil. Trans. Roy. Soc. London, A.* 243, 867, 1-35.

Madsen, P. and H.A. Schäffer. 2006. A discussion of artificial compressibility, *Coastal Engineering*, 53, 93-98.

Monaghan, J.J. 1992. Smoothed particle hydrodynamics, *Ann. Rev. Astron. Astrophys.*, 30, 543-574.

Monaghan, J.J. 1994. Simulating free surface flows with SPH, *J. Comput. Phys.*, 110, 399-406.

Monaghan, J.J. and A. Kos, 1999. Solitary waves on a Cretan beach, *J. Waterways, Ports, Coastal, and Ocean Engineering,* 125, 145.

Monaghan, J.J. and A. Kos. 2000. Scott Russell's wave generator, *Phys. of Fluids.*, 12, 622.

Morris, J.P. 1996. A study of the stability properties of smoothed particle hydrodynamics, *Astron. Astrophys.,* 30, 543-574.

Morris, J.P., P.J. Fox, and Y. Zhu. 1997. Modeling low Reynolds Number incompressible flows using SPH, *J. Comput. Phys.*, 136, 214-226.

Pidduck, F.B. 1910. On the propagation of a disturbance in a fluid under gravity, *Proc. Roy. Society*, A, 83, 347-356.

Ursell, F. , R.G. Dean and Y. S. Yu, 1960. Forced small amplitude water waves: a comparison of theory and experiment, *J. Fluid Mech.*, 7, 33-52.

SIMPLIFIED NONLINEAR BOUSSINESQ MODELLING OF WAVES -AN ENHANCEMENT OF HIGHER ORDER HARMONICS

Y. Yao[1] and P.H. Taylor and A.G.L. Borthwick

By introducing a tunable coefficient, Zou (1999) derived simplified higher-order Boussinesq equations that improve the prediction of nonlinear wave harmonics in water of intermediate depth. The present study describes a parameter-adaptive model based on Zou's (1999) equations, with enhanced accuracy of the second order harmonics. Thus, the modification to the equation set becomes part of the solution. The model will provide better prediction of wave heights and crest elevations using modest computer resources.

INTRODUCTIONS

Historically, Boussinesq wave theory has been widely recognized (see e.g. Madsen and Mei 1969; Peregrine 1967) as a shallow water approximation to the fully dispersive and nonlinear water wave problem. The original Boussinesq equations incorporate a balance between only the lowest-order dispersion and lowest-order nonlinearity. From the 1980s, coastal and ocean engineers have shown a growing interest in modelling the propagation of regular cnoidal waves and irregular waves from intermediate to shallow water depths. As a result, advances in two aspects of the Boussinesq equations have received considerable recent attention. These aspects are the linear dispersion accuracy and the nonlinearity embedded in formulations. The analytical and computational accuracy of linear and nonlinear properties have been improved by including further higher order terms in the Taylor series expansion about the long wave limit. However, the resulting equations are extremely complex, (see e.g. Madsen and Schaffer 1998; Gobbi and Kirby 1999), and so their implementation and analysis are very difficult to achieve in practice. Furthermore, these equations contain either fifth or sixth derivatives, which are awkward and time consuming to compute numerically in the presence of noise. These twin concerns have limited the general acceptance of higher order Boussinesq equations for coastal engineering applications.

Several authors (Zou 1999; Kennedy et al. 2002) have made attempts to simplify higher order Boussinesq equations. Usually the simplified Boussinesq formulations bear a strong resemblance to the $O(\mu^2)$ equations previously in wide use. (those of Madsen and Sø rensen 1992). The new equations are linearly identical to the conventional $O(\mu^2)$ equations on a flat bottom, but not on a sloping bed. Thus, linear dispersion is identical, but the shoaling properties are much improved.

One problem that Zou (1999) and Kennedy et al. (2002) did not fully resolve is to provide an accurate estimate of the second order harmonics over a large range

[1] Engineering Science Department, University of Oxford, Parks Road, Oxford, UK. OX1 3PJ

of water depths and time scales. As shown originally for waves on constant water depth (Dean and Sharma 1981), the interaction of wave components of different frequencies contribute significantly to modification of crest-trough asymmetry (crest and troughs are obviously different in shape) and to improvement in accuracy of wave height prediction compared with linear models. The important question is whether a simple model can be embued with close to correct second order harmonics over a large range of water depths and time scales.

Hence, the prime motivation for this work is to develop a highly-efficient numerical Boussinesq wave model with accurate second order harmonics structure over the whole wave shoaling process. Two major concerns are:

1. computational efficiency in order to achieve long-time, large scale 2-D random wave simulations to be undertaken in practice;
2. accurate wave shape prediction and crest/wave height distribution statistics.

A successful numerical model of this kind would be of great value for full-scale engineering projects for which the water depth is intermediate between the depth usually associated with offshore and coastal engineering.

ZOU'S TWO PARAMETER FORMULATION

By introducing an extra source term (an additional volume flux) during the perturbation analysis, Zou (1999) proposed a new higher order Boussinesq-type equation set, with improved shoaling and nonlinear characteristics, that retain the reasonable simplicity of the Madsen and Sø rensen (1992)-type formulations. Zou's (1999) equation set reads:

$$\eta_t + \nabla \cdot [H\mathbf{U}] = 0, \tag{1}$$

and

$$\begin{aligned} \mathbf{U}_t + g\nabla\eta + (\mathbf{U}\cdot\nabla)\mathbf{U} + \Omega \; &= \; \frac{1}{2}H\nabla[\nabla\cdot(h\mathbf{U}_t)] - \frac{1}{6}H^2\nabla(\nabla\cdot\mathbf{U}_t) \\ &\quad + B_1 h^2\nabla[\nabla\cdot(\mathbf{U}_t + g\nabla\eta)] \\ &\quad + B_2\nabla[\nabla\cdot h^2(\mathbf{U}_t + g\nabla\eta)], \end{aligned} \tag{2}$$

where

$$\begin{aligned} \Omega \; &= \; \nabla\left[h^2\left(\frac{1}{3}(\nabla\cdot\mathbf{U})^2 - \frac{1}{3}\mathbf{U}\cdot\nabla^2\mathbf{U} - \frac{1}{30}\nabla^2(\mathbf{U}\cdot\mathbf{U})\right)\right] \\ &\quad - \frac{1}{2}\eta\nabla[\nabla\cdot h\mathbf{U}_t] - h\nabla\eta\nabla\cdot\mathbf{U}_t \end{aligned} \tag{3}$$

This formulation is defined in a Cartesian coordinate system (x, y, z). The fluid domain is bounded by the sea bed at $z = -h(x, y)$ and the free surface at $z = \eta(x, y, t)$, where t is time. $\nabla \equiv (\frac{\partial}{\partial x}, \frac{\partial}{\partial y})$ is the horizontal gradient operator.

$H = \eta + h$ is the total water depth. $\eta(x, y, t)$ and $U(x, y, t)$ are the two dependent variables, where U is the depth-averaged velocity.

In order to match closely the Stokes shoaling gradient over a larger wave depth range, Zou (1999) proposed the two coefficients should be specified as

$$B_1 = \frac{29}{885}, \qquad B_2 = \frac{2}{59}, \qquad B_1 + B_2 = \frac{1}{15} \tag{4}$$

Furthermore, Zou (1999) attempted to match the quadratic transfer function $G_{\pm}^{\text{Stokes}}(\omega_1, \omega_2)$ to the solution from Stokes' second order constant depth theory. This leads to a change to the Ω term in Eqn.(3), into the following form,

$$\begin{aligned} \Omega &= \nabla\left[h^2\left(\frac{1}{3}(\nabla\cdot\mathbf{U})^2 - \frac{1}{3}\mathbf{U}\cdot\nabla^2\mathbf{U} - \frac{B_3}{30}\nabla^2(\mathbf{U}\cdot\mathbf{U})\right)\right] \\ & -\frac{1}{2}\eta\nabla[\nabla\cdot h\mathbf{U}_t] - h\nabla\eta\nabla\cdot\mathbf{U}_t \end{aligned} \tag{5}$$

where $B_3 = 5.3$.

Except for certain extra third order nonlinear terms, Zou's (1999) formulations retain the structure of the widely used Madsen and Sø rensen (1992) equation set, given by Eqn.(6) and (7). This gives the advantage of being easily extendable from our previous numerical Boussinesq model (Weston et al. 2003), developed in Oxford.

$$\eta_t + \nabla[(h+\eta)\mathbf{U}] = 0, \tag{6}$$

and

$$\mathbf{U}_t + g\nabla\eta + (\mathbf{U}\cdot\nabla)\mathbf{U} = \frac{1}{3}h\nabla^2(h\mathbf{U}_t) + Bh\nabla^2[h(\mathbf{U}_t + g\nabla\eta)] \tag{7}$$

FOURIER ANALYSIS FOR SECOND ORDER SUB- AND SUPER-HARMONICS

The accuracy of second order transfer is the most important and desirable nonlinear characteristic in intermediate and shallow water wave modelling where the energy transfer across the frequency spectrum is substantial. The transfer functions for the second order bound sub- and super-harmonics are derived on the basis of the Zou (1999) 1-D equations, constant depth equations. The standard method of Fourier analysis undertaken herein can be referred to Madsen and Schaffer (1998). The results are compared to the exact expressions derived from the nonlinear boundary value problem for the Laplace equation (Sand and Mansard 1986).

For the simplest second order harmonics case, we consider a simple first order wave group made up of just two frequencies ω_n and ω_m, plus their second order interactions. The first order superpositions are given by:

$$\eta^{(1)}(x,t) = \eta_n + \eta_m, \qquad U^{(1)}(x,t) = \left(\frac{\omega_n}{k_n h}\eta_n + \frac{\omega_m}{k_m h}\eta_m\right)$$

where

$$\eta_n = a_n \cos(\omega_n t - k_n x) + b_n \sin(\omega_n t - k_n x)$$

and

$$\eta_m = a_m \cos(\omega_m t - k_m x) + b_m \sin(\omega_m t - k_m x)$$

Each of the two wave components are considered to satisfy the linear dispersion relation.

The next step is to look for second order solutions to the Zou (1999) equations. The bound second order wave train consists of four new frequencies: one sub-harmonic $\omega_p = \omega_n - \omega_m$ and three super-harmonics $\omega_p = \omega_n + \omega_m, \quad \omega_p = 2\omega_n, \quad \omega_p = 2\omega_m$. The four second order wave numbers are determined by $k_p = k_n - k_m$ and three super-harmonics $k_p = k_n + k_m, \quad k_p = 2k_n, \quad k_p = 2k_m$, which come from combinations of k_n, k_m and do not satisfy the linear dispersion relation. The resulting second order solution can be expressed by

$$\eta^{(2)}(x,t) = \eta_{nm}^{-} + \eta_{nm}^{+} + \eta_{nn}^{+} + \eta_{mm}^{+}, \qquad U^{(2)}(x,t) = U_{nm}^{-} + U_{nm}^{+} + U_{nn}^{+} + U_{mm}^{+}$$

where

$$\begin{aligned} \eta_{nm}^{\pm} &= \epsilon\delta G_{\eta}^{\pm}(a_p \cos(\omega_p t - k_p x) + b_p \sin(\omega_p t - k_p x)) \\ U_{nm}^{\pm} &= \epsilon\delta G_{U}^{\pm}(a_p \cos(\omega_p t - k_p x) + b_p \sin(\omega_p t - k_p x)) \end{aligned}$$

and where

$$a_p = \frac{1}{h}(a_n a_m \mp b_n b_m), \quad b_p = \frac{1}{h}(a_m b_n \pm a_n b_m)$$

$$\omega_p = \omega_n \pm \omega_m, \quad k_p = k_n \pm k_m$$

and

$$\delta = \begin{cases} \frac{1}{2} & \text{for } n = m \\ 1 & \text{for } n \neq m \end{cases}$$

Next we derive the $G_{\eta}^{\pm}$ functions on the basis of Eqn. (1), and (2). The full solution of wave motion variables are the sum of first order and second order components.

$$\eta(x,t) = \eta^{(1)} + \eta^{(2)}, \qquad U(x,t) = U^{(1)} + U^{(2)}$$

By substituting the combination of the above expressions of η, U into (1) and (2), and collecting together terms of $O(\epsilon)$ we obtain the following algebraic system, with unknowns of second order transfer functions:

$$\begin{pmatrix} m_{11}^{(2)} & m_{12}^{(2)} \\ m_{21}^{(2)} & m_{22}^{(2)} \end{pmatrix} \begin{pmatrix} G_{\eta}^{\pm} \\ G_{U}^{\pm} \end{pmatrix} = \begin{pmatrix} F_1^{\pm} \\ F_2^{\pm} \end{pmatrix} \tag{8}$$

leading to the solution

$$G_{\eta}^{\pm} = \frac{F_1^{\pm} m_{22}^{(2)} - F_2^{\pm} m_{12}^{(2)}}{m_{11}^{(2)} m_{22}^{(2)} - m_{21}^{(2)} m_{12}^{(2)}} \tag{9}$$

For the case of the bound superharmonic $\omega_p = \omega_n + \omega_m$, the coefficients in (9) read

$$m_{11}^{(2)} = \omega_p, \quad m_{12}^{(2)} = -k_p(1 + \frac{1}{15}\kappa_p^2), \quad m_{21}^{(2)} = -k_p h, \quad m_{22}^{(2)} = \omega_p(1 + \frac{2}{5}\kappa_p^2)$$

$$F_1^+ = \frac{1}{2}k_p\left(\frac{\omega_n}{k_n} + \frac{\omega_m}{k_m}\right), \tag{10}$$

and

$$F_2^+ = \frac{1}{2}k_p\frac{\omega_n\omega_m}{hk_nk_m}\left[1 + \frac{1}{3}\left((\kappa_n - \kappa_m)^2 + \frac{B_3}{5}(\kappa_n + \kappa_m)^2\right.\right.$$
$$\left.\left.-\kappa_n^2\frac{\omega_n k_m}{\omega_m k_n}(\frac{2\kappa_n + 3\kappa_m}{\kappa_n + \kappa_m}) - \kappa_m^2\frac{\omega_m k_n}{\omega_n k_m}(\frac{2\kappa_m + 3\kappa_n}{\kappa_n + \kappa_m})\right)\right] \tag{11}$$

where $k_p = k_n + k_m$ and $\kappa_p = \kappa_n + \kappa_m$. The subharmonic transfer function is obtained simply by changing the sign of the quantities $(\omega_m, \kappa_m, k_m)$ in the above solution. For the case of $\omega_m = \omega_n$ the problem simpifies to the second-harmonic transfer analysed by Zou (1999).

The target second-order transfer function G_η^{Stokes}, determined directly from the nonlinear boundary value problem for the Laplace equation, was given by, for instance, Ottesen-Hansen (1978) for the subharmonics and Sand and Mansard (1986) for the superharmonics.

THE RESULTS

Fig.(1) shows the ratio of the present formulation to the Stokes theoretical super-harmonic transfer function. Different B_3 settings are considered. Fig.(1a), (1b), (1c), and (1d) are cases with $B_3 = 1.0, 2.0, 3.0, 5.3$ respectively. Fig.(2a), (2b), (2c), and (2d) present the equivalent sub-harmonics plots. Attention are drawn to the two major features of these contour series:

- the 10% or 5% error zone bounded by corresponding contour lines;
- the exact (unit ratio) harmonics accuracy represented by green lines in super-harmonics contours.

First, the 5% error zone moves upwards along the kh axis as B_3 increases. This means that, generally speaking, a larger value of B_3 gives an improved super-harmonics over a larger range of kh. Second, for every B_3 setting, the green line (along which the super-harmonics is exact) intersects with the diagonal at a certain kh. And this intersection moves continuously up the diagonal as B_3 increases. It suggest an B_3 adaptive scheme can be imposed in order to maintain good second order harmonics as the wave structure evolves from deep to shallow water.

ADAPTIVE PARAMETER SCHEME

Given the above Fourier analysis, one of the most important features of Zou's (1999) equation set is that the accuracy of the second order harmonics (compared

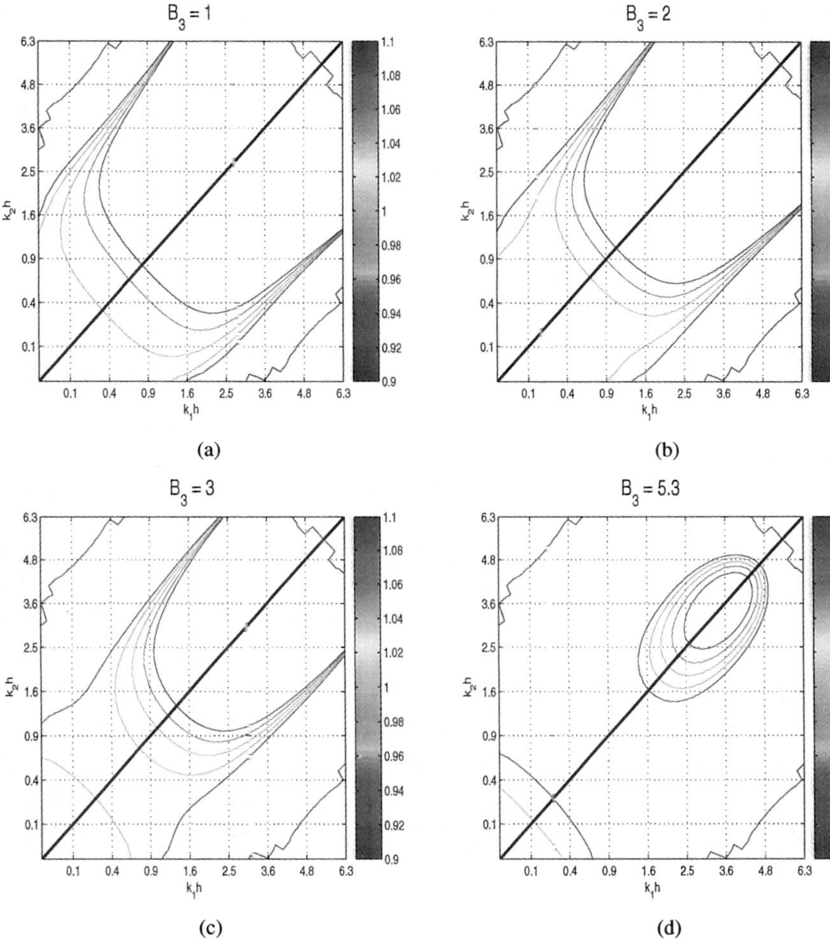

FIG. 1: Super-harmonics transfer function accuracy variation with B_3

to the Stokes theoretical transfer function), can be modified to a specified higher accuracy simply by changing the parameter, B_3. It is possible to get exact super-harmonics for same-frequency wave interactions, at all kh up to 2.5, if B_3 is varied with kh.

Therefore, Fourier analysis at second order shows that B_3 should be a function of the local kh value. This results in a match to the exact Stokes solution for the same frequency wave-wave interactions. This function implies that B_3 should be continuously adapted in the evolving wave structure for shoaling. A number of super-harmonics accuracy calculations with the settings of $1 < B_3 < 5.5$ have been carried out. In each calculation, we collect B_3 value and the

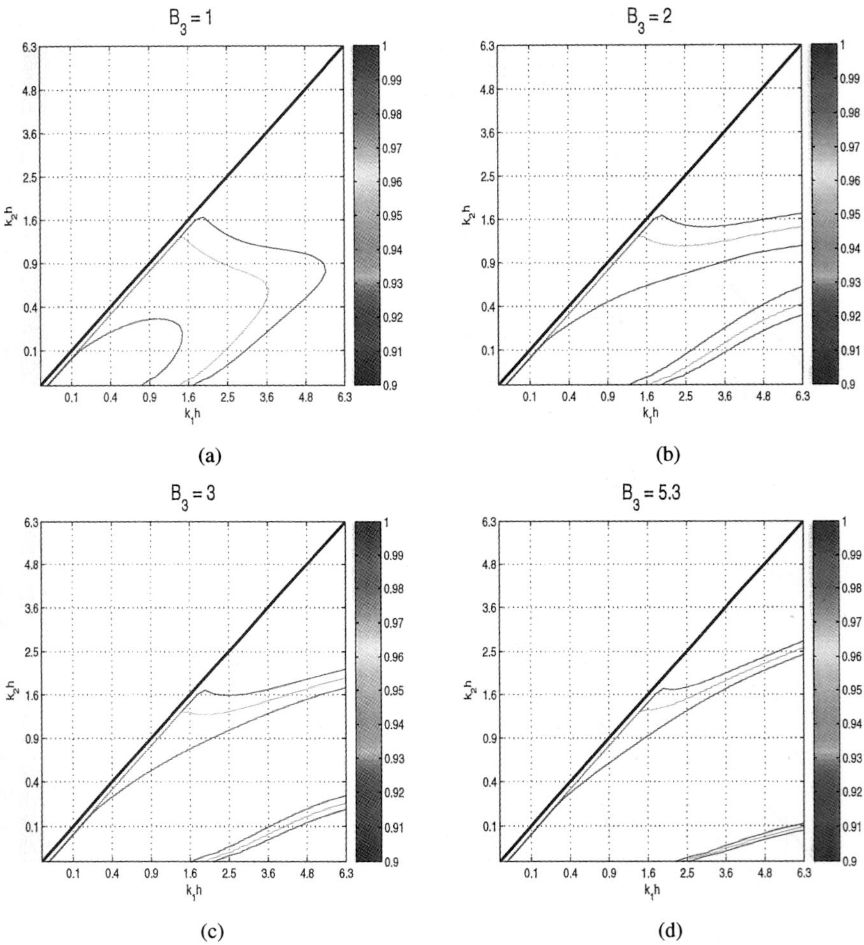

FIG. 2: Sub-harmonics transfer function accuracy variation with B_3

corresponding kh value at which the accuracy achieves its exact Stokes prediction. Then a continuous function representing B_3 is obtained through a least squares polynomial fit to the collected data. The formulation is given by Eqn.(12). Fig.(3) shows this optimized B_3 variation with kh.

$$B_3 = -0.257(kh)^3 + 0.41(kh)^2 + 2.8kh + 0.035, 1.0 < B_3 < 5.5 \qquad (12)$$

If we make an adaptive modification to the B_3 coefficient as a function of kh, we can require the 1:1 diagonal line in Fig.(4) to be the exact solution for the super-harmonic self-interaction term. When we do this, the question is how good is the accuracy when we move away from the diagonal?

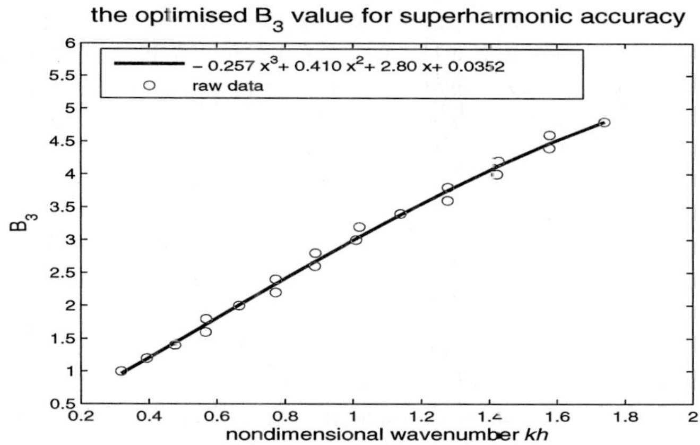

FIG. 3: B_3 Parameter adaptive scheme

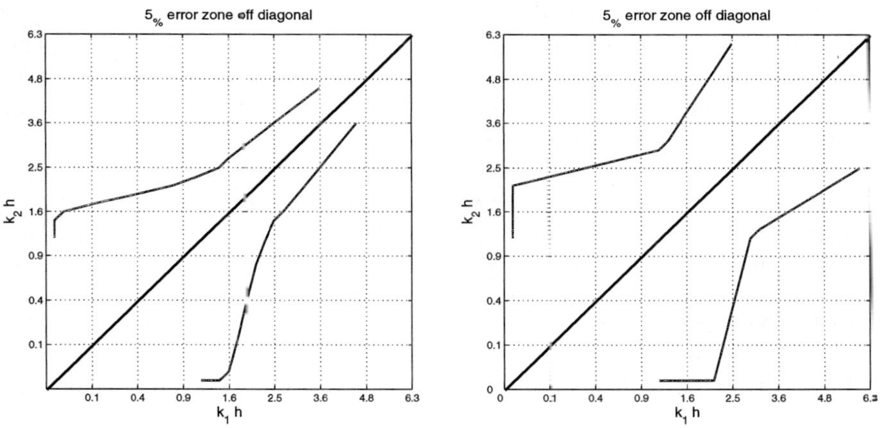

(a) Super-harmonics (b) Sub-harmonics

FIG. 4: Overall error zone of 2nd order harmonics transfer function after B_3 is adapted

The answer is that it is excellent. Fig.(4) shows the overall 5% error zone, after we have adapted B_3 to obtain the exact solution along the leading diagonal. Fig.(4.a) shows a large region bounded by two red lines within which the error in the super harmonics is less than 5%. The equivalent contour for sub-harmonics is equally encouraging and is shown in Fig.(4.b). Therefore, considerable overall enhancement of second order harmonics is demonstrated.

Thus, we conclude that for a realistic wave sea-state (such as that defined as a

JONSWAP spectrum), the errors associated with the Boussinesq approximations to both second order sum and difference bound components can be made very small. This is one of the main advantages of the present adaptive optimization of the (Zou 1999) equation set. The second and equally important desirable feature associated with Zou's equation set is its computational efficiency.

CONCLUSION

A simplified high-order Boussinesq equation (Zou 1999) is analysed and enhanced with emphasis on nonlinear properties over a large range of wavenumber. The Fourier type analysis concentrates on second- and third-order transfer functions including amplitude dispersion for nonlinear regular waves. One of the most important features of Zou (1999) equation set is that the accuracy of the second order harmonics (compared to the Stokes theoretical transfer function), can be modified to a specified higher accuracy simply by changing the parameter, B_3. It is possible to get exact super-harmonics for same-frequency wave interactions, at all kh up to 2.5, if B_3 is varied with kh. We come to the remarkable conclusion that a simple Boussinesq model can be developed with almost exactly correct second order harmonics over a large range of kh, simply by adapting a single parameter, B_3. This adaption is given in terms of the function of $B_3 = -0.257\,(kh)^3 + 0.41\,(kh)^2 + 2.8\,kh + 0.035$, satisfying $1.0 < B_3 < 5.5$. Thus $B_3(x, t)$ as a function of space and time locally optimizes the second order solutions everywhere in the simulation domain.

The method of coupling a parameter in the equation set to the local structure of the evolving wave system is novel. This adaptive model is believed to greatly applicable in the engineering context of long-distance gently sloped seabed, shallow water wave modelling.

ACKNOWLEDGMENTS

The first author is thankful to the Clarendon Fund sponsored by Oxford University Press. Without the financial support from this Fund, this research could not have been accomplished.

REFERENCES

Dean, R. G. and J. N. Sharma. 1981, 30 June-4 July. Simulation of wave systems due to nonlinear directional spectra, *International symposium on hydrodynamics in coastal engineering*, Trondheim, Norway, ASCE, 1211C1222.

Gobbi, M. F. and J. T. Kirby. 1999. Wave evolution over submerged sills: tests of a high-order Boussinesq model, *Coastal Eng.,* 40(277).

Kennedy, A. B., J. T. Kirby, and M. F. Gobbi. 2002. Simplified higher-order Boussinesq equations I. linear simplifications, *Coastal Engineering, 44*, 205C 229.

Madsen, O. S. and C. C. Mei. 1969. The transformation of a solitary wave over an uneven bottom., *J. Fluid Mech., 39*, 781–791.

Madsen, P. and H. Schaffer. 1998. Higher-order Boussinesq-type equations for surface gravity waves: Derivations and analysis, *Phil. Trans. R. Soc., Ser., 356*, 1–59.

Madsen, P. and Q. Sø rensen. 1992. A new form of the Boussinesq equations with improved linear dispersion characteristics. Part 2: A slowly varying bathymetry, *Coastal Engineering, 18*, 183–204.

Ottesen-Hansen, N. E.. 1978. Long period waves in natural wave trains. Progress report 46, Institiute Hydrodyn. and Hydraulic Engineering, Technical University of Denmark.

Peregrine, D. H.. 1967. Long waves on a beach, *J. Fluid Mech.,* 27(4), 815–827.

Sand, S. E. and E. P. D. Mansard. 1986. Reproduction of higher harmonics in irregular waves, *Ocean Engineering, 13*, 57–83.

Weston, B. P., P. H. Taylor, A. G. L. Borthwick, and A. C. Hunt. 2003. A Godunov-type modelling of extreme wave run-up, *Proc.International Symposium on Shallow Flows, 1*, 227–234.

Zou, Z. L.. 1999. A simplifed higher order Boussinesq equations, *Ocean Engineering, 3*, 21–26.

DEVELOPMENT OF A BOUSSINESQ-RANS VOF HYBRID WAVE MODEL

Khairil I Sitanggang[1] , Patrick J. Lynett[2], and Philip L.-F Liu[3]

Efficient and accurate large scale numerical modeling of wave propagation from deep water to the shoreline is difficult. The difficulty arises from the fact that such a simulation requires a large number of computational grid points and has to account for both the nonbreaking and breaking wave processes. In this study, two widely used wave models: the inviscid Boussinesq-equation-type- and the RANS-based- wave models are two-way-coupled to create a hybrid-wave model that can perform accurate and efficient large scale, nonbreaking and breaking, wave simulation.

INTRODUCTION

Large scale simulation of wave propagation from deep water to shoreline is difficult. Such a simulation requires a huge number of computational grids and includes both the non-breaking and breaking wave processes. With the current state-of-the-art wave model, it is hard at the same time to meet both the efficiency and accuracy in a single model for that purpose. For modeling the non-breaking wave propagation in the prebreaking zone, it is usually sufficient to employ the potential wave model with very good accuracy and efficiency. Although attempts have been made to include the turbulence mechanism in this model, the applicability of the model for simulating the breaking wave processes is still limited. To better understand the breaking wave process, engineers and scientists have approached it through the RANS-based wave model. Not only can this type of wave model accurately calculate the breaking wave model due to the bathymetry change, with the VOF fluid-tracking technique it can also calculate the breaking wave due to coastal structures. To drive this model from the deep water is still limited by excessive computational time. With the current computing technology, it is still not feasible to apply the model for large scale simulation.

To overcome this dilemmatic situation, in this study we coupled two widely used wave models: the Boussinesq equation and RANS wave models. Brief descriptions of both models follow shortly. In implementing this coupled model, the former will be applied on the prebreaking zone and the latter on the breaking zone. During the calculation, both models are exchanging pertinent variables. Since both models are both accurate and efficient in said domains, the coupled model will be expectedly accurate and efficient.

[1] Coastal and Ocean Engineering Division, Civil Engineering Department, Texas A&M University, College Station, TX 77843-3136, USA

[2] Coastal and Ocean Engineering Division, Civil Engineering Department, Texas A&M University, College Station TX 77843-3136, USA

[3] School of Civil and Environmental Engineering, Cornell Unversity, Ithaca, NY 14853, USA

The Boussinesq-type equation wave model has been successfully used in the past decade for simulating wave propagation in the ocean of relatively great depth to breaking zone. The pioneering work on the Boussinesq wave model is that of Peregrine (1967) who introduced the weakly Boussinesq equation model over variable bathymetry; its implementation was limited to the shallow water regime, $kh \sim 0.3$. Enhancement of the model which pushed its accuracy further off-shore and improved the dispersive properties followed thereafter. Madsen and Sørensen (1992), for instance, through manipulation of the dispersive term of the depth-averaged Boussinesq equation, managed to push the model accuracy to intermediate depth. Nwogu (1993) used the velocity at arbitrary depth, instead of the depth-averaged velocity, in his derivation to get an excellent dispersive properties and good accuracy up to $kh \sim 3$.

In deriving the weakly nonlinear Boussinesq equation, higher order derivative terms had been truncated. Hence its implementation to simulate highly nonlinear wave, i.e. ratio of wave height to wave length is big, is not accurate. To better capture the high nonlinearity properties in the model, Wei et al. (1995) retained higher order derivative terms and derived a more accurate highly nonlinear Boussinesq equation model with excellent dispersive properties up to $kh \sim 3$. Similar attempts were made by a number of researchers, e.g. Gobbi et al. (2000), Agnon et al. (1999), and Lynett and Liu (2004). As a result, the current version of the Boussinesq model can be implemented to an extremely deep water with a very good accuracy.

Despite being accurate on the nonbreaking zone, the Boussinesq model still uses an approximation method in calculating breaking wave nearshore (Kennedy et al. (2000) and Chen et al. (2000)). Moreover, the inviscid assumption made in the model will incur irrotational velocity field over the entire simulation domain. In the nearshore zone where the wave breaks, the velocity field is very rotational due to intense turbulence. Hence, using an inviscid model to calculate breaking wave cannot, in general, produce accurate velocity field. Since breaking wave to some extent is responsible for the nearshore sediment transport processes, accurate calculation of velocity field and turbulence in the breaking zone is very important, if we, for instance, want to study the dynamics of nearshore zone.

The RANS-VOF (will be simply termed RANS hereafter) wave model has been widely used to simulate breaking wave in the nearshore zone and wave breaking due to wave interaction with structure. Lin and Liu (1998), for instance, employing the nonlinear eddy viscosity model, performed a numerical study of breaking wave in the surf zone. Comparison with the experimental data showed that the numerical calculation of the breaking wave was in good agreement with the data. The model was also tested for nonbreaking wave simulation. Perfect match with the analytical solution was found. Employing Lin and Liu's model, Lara et al. (2006) studied the wave interaction with breakwater. This study showed that the model could correctly calculate the wave reflection, shoaling, and breaking near the structure and hence could be used as a tool for design purposes. Using the same type of model, Liu and Cheng (2001) investigated the behavior of a solitary wave propagating over a step. Both the nonbreaking and breaking cases were considered. Here the author also found that the model was able to compute the turbulence generated by the breaking and flow separation at the corner of the step. The application of the model extended to the study of waves generated by nearshore submarine mass-movement, Yuk et al. (2006). The numerical simulation result was compared with laboratory experiment and analytical solution. Very good agreement was obtained for the submarine and aerial mass movements. Other researchers, Zhao et al. (2004), also found success in studying the breaking wave using RANS model with a multi-scale turbulence model.

Typical simulation of wave breaking using RANS model is usually computationally expensive. This is due to the iterative solver employed in the model to solve the pressure equation. This expensive computation hinders users from performing large scale simulation. Hence, the implementation of the RANS model is limited to a small domain size with relatively small number of computational grids.

Noting the strengths of the two models as briefly described above, it is very advantageous to couple the two to create a 'single' hybrid wave model that can accurately and efficiently simulate wave propagation over large domain. *This idea motivates the present study.*

In the next section the authors will briefly describe the governing equations and the solution techniques of the models. Further details of the models can be found in the cited references. After that the coupling method is presented and followed by the model test which will validate the hybrid wave model.

BOUSSINESQ EQUATION MODEL

Following Lynett and Liu (2002), the governing equation of the Boussinesq equation model consists of the depth-averaged continuity equation:

$$\frac{\partial H}{\partial t} + \nabla \cdot (H\mathbf{u}_\alpha) - \nabla \cdot \left\{ H \left[\left(\frac{1}{6}\left(\eta^2 - \eta h + h^2\right) - \frac{1}{2} z_\alpha^2 \right) \nabla S + \left(\frac{1}{2}(\eta - h) - z_\alpha \right) \nabla T \right] \right\} = 0 \quad (1)$$

and the momentum equation:

$$\frac{\partial \mathbf{u}_\alpha}{\partial t} + \frac{1}{2}\nabla\left(\mathbf{u}_\alpha \cdot \mathbf{u}_\alpha\right) + g\nabla\eta + \frac{\partial}{\partial t}\left\{\frac{1}{2}z_\alpha^2\nabla S + z_\alpha\nabla T - \nabla\left(\frac{1}{2}\eta^2 S + \eta T\right)\right\} + \nabla\left\{\frac{\partial\eta}{\partial t}\left(T+\eta S\right) + \left(z_\alpha - \eta\right)\left(\mathbf{u}_\alpha\cdot\nabla\right)T + \frac{1}{2}\left(z_\alpha^2-\eta^2\right)\left(\mathbf{u}_\alpha\cdot\nabla\right)S + \frac{1}{2}\left(T+\eta S\right)^2\right\} = 0 \quad (2)$$

where $S = \nabla\cdot\mathbf{u}_\alpha$, $T = \nabla\cdot(h\mathbf{u}_\alpha) + \partial h/\partial t$, h = depth, η = free surface elevation, $H = h+\eta$, $\mathbf{u}_\alpha$ = horizontal velocity at a reference level, z_α (Figure 1), and t = time. The reference level is so chosen that the resulting frequency dispersion characteristic of the Boussinesq model agrees well with linear theory (Nwogu (1993)). In this model, weak frequency dispersion is assumed. The temporal derivative of depth, $\partial h/\partial t$ in (1) and (2), allows the simulation of wave generated by bathymetry movement such as landslide which in many instances has caused tsunami.

The higher-order predictor-corrector Adams-Bashford finite difference scheme is used to solve (1) and (2). In the predictor level, the free surface elevation and horizontal velocities are calculated using the explicit scheme:

$$\eta_{i,j}^{n+1} = \eta_{i,j}^{n} + \tfrac{1}{12}\Delta t\left(23E_{i,j}^{n} - 16E_{i,j}^{n-1} + 5E_{i,j}^{n-2}\right), \quad (3)$$

$$U_{i,j}^{n+1} = U_{i,j}^{n} + \tfrac{1}{12}\Delta t\left(23F_{i,j}^{n} - 16F_{i,j}^{n-1} + 5F_{i,j}^{n-2}\right) + 2\left(F_1\right)_{i,j}^{n} - 3\left(F_1\right)_{i,j}^{n-1} + \left(F_1\right)_{i,j}^{n-2}, \quad (4)$$

$$V_{i,j}^{n+1} = V_{i,j}^{n} + \tfrac{1}{12}\Delta t\left(23G_{i,j}^{n} - 16G_{i,j}^{n-1} + 5G_{i,j}^{n-2}\right) + 2\left(G_1\right)_{i,j}^{n} - 3\left(G_1\right)_{i,j}^{n-1} + \left(G_1\right)_{i,j}^{n-2}, \quad (5)$$

where E, F, F_1, G, and G_1 depend on u, v, and η from the previous time steps, t^n, t^{n-1}, and t^{n-2}. The velocity field, u and v, are obtained from U and V through a tridiagonal linear system of linear equations. Since explicit scheme is used, the solution is in general not accurate.

To refine the predictor's solution, the implicit and iterative corrector step is invoked:

$$\eta_{i,j}^{n+1} = \eta_{i,j}^{n} + \tfrac{1}{24}\Delta t\left(9E_{i,j}^{n+1} + 19E_{i,j}^{n} - 5E_{i,j}^{n-1} + E_{i,j}^{n-2}\right), \quad (6)$$

$$U_{i,j}^{n+1} = U_{i,j}^{n} + \tfrac{1}{24}\Delta t\left(9F_{i,j}^{n+1} + 19F_{i,j}^{n} - 5F_{i,j}^{n-1} + F_{i,j}^{n-2}\right) + \left(F_1\right)_{i,j}^{n+1} - \left(F_1\right)_{i,j}^{n}, \quad (7)$$

$$V_{i,j}^{n+1} = V_{i,j}^{n} + \tfrac{1}{24}\Delta t \left(9G_{i,j}^{n+1} + 19G_{i,j}^{n} - 5G_{i,j}^{n-1} + G_{i,j}^{n-2}\right) + (G_1)_{i,j}^{n+1} - (G_1)_{i,j}^{n}. \quad (8)$$

The corrector is iterated until the following criteria:

$$\frac{\sum \left|f^{n+1} - f^{n}\right|}{\sum \left|f^{n}\right|}, \quad (9)$$

is met for $f \equiv u$, v, and η.

RANS-VOF MODEL

RANS equations consist of the mean continuity equation:

$$\frac{\partial \langle u_i \rangle}{\partial x_i} = 0 \quad (10)$$

and the momentum equation:

$$\frac{\partial \langle u_i \rangle}{\partial t} + \langle u_j \rangle \frac{\partial \langle u_i \rangle}{\partial x_j} = -\frac{1}{\langle \rho \rangle}\frac{\partial \langle p \rangle}{\partial x_i} + g_i + \frac{1}{\langle \rho \rangle}\frac{\partial \langle \tau_{ij} \rangle}{\partial x_j} - \frac{\partial \langle u_i' u_j' \rangle}{\partial x_j}. \quad (11)$$

The product of the density and the correlation of the velocity fluctuation, $\rho\langle u_i' u_j' \rangle$, is called the Reynolds stress. To close (11), the nonlinear eddy viscosity model was used in the RANS model. Detailed description can be found in Lin and Liu (1998).

Similar to its predecessor, Kothe et al. (1994), the RANS model uses the two-step projection method to solve (10) and (11). The first step of the method is to solve (11) using an explicit scheme to get the provisional velocity, $\tilde{u}_i^{n+1}$, at the next time step, t^{n+1}:

$$\frac{\tilde{u}_i^{n+1} - u_i^{n}}{\Delta t} = -u_j^{n}\frac{\partial u_i^{n}}{\partial x_j} + g_i + \frac{\partial \tau_{ij}^{n}}{\partial x_j}. \quad (12)$$

In this step the pressure effect is not yet included and hence the velocity field is only an approximation to the correct velocity. In the second step, the provisional velocities are updated to get the correct velocities, u_i^{n+1}:

$$\frac{u_i^{n+1} - \tilde{u}_i^{n+1}}{\Delta t} = -\frac{1}{\rho^{n}}\frac{\partial p^{n+1}}{\partial x_i}. \quad (13)$$

The new pressure information, p^{n+1}, is obtained from the Poisson pressure equation:

$$\frac{\partial}{\partial x_i}\left(\frac{1}{\rho^{n}}\frac{\partial p^{n+1}}{\partial x_i}\right) = \frac{1}{\Delta t}\frac{\partial \tilde{u}_i^{n+1}}{\partial x_i} \quad (14)$$

which is a combination of (13) and the continuity equation (10). In both steps, staggered grid system, i.e. pressure variable calculated at the center of the mesh and velocities on its face (Figure 1), had been employed.

In solving (14) using the second order centered finite difference scheme, a penta-diagonal system of linear equations arises. To solve this system, an iterative preconditioned conjugate gradient method is employed. In this case, the Incomplete Cholesky factorization is used as preconditioner. For typical wave simulation, most of the CPU time is spent to carry out the iterative process in the solver. Quite often the nearshore breaking wave simulation requires large number of computational grids to better capture the turbulence details. In such cases, the computation time can be too expensive to perform, and parallelization of the model is an option to speedup this computation.

HYBRID WAVE METHOD

Clearly the two models are different in physics and solution techniques. These two will pose difficulties in the effort to couple them.

Figure 1 depicts the grid systems used in the two models. Here, we consider the coupling of the 1D-Boussinesq model and the 2D-RANS model. The superscripts 'B' and 'R' stand for Boussinesq and RANS respectively, F is the volume of fluid in the cell, and p pressure. The first and the last 3 grids, i.e. -2, -1, 0 and $N+1$, $N+2$, $N+3$, in the Boussinesq's grid sytem serve as the left and right boundary conditions for this model. For the RANS model, the first grid, 1, and last grid, $M+1$, serve as the left and right boundary conditions. Hence, the first and last computational grids of the Boussinesq model is grid 1 and N and grid 2 and M for RANS-VOF.

The idea behind the hydrodynamic coupling is that one model serves as the boundary condition for the other. Refering to Figure 1, grid N of the Boussinesq model provides F, u, and v for grid 1 of the RANS model to be used as boundary conditions. Similarly, u and F in grids 2, 3, and 4 of the RANS model are used in grids $N+1$, $N+2$, and $N+3$ of the Boussinesq model as boundary conditions.

Now, for coupling the two models the authors follow the algorithm below. Note that it is a 2-way coupling, in which the model passes information to each other.

2-way coupling of the Boussinesq and RANS-VOF model.

1. Calculate the velocity, u_{α}^{n+1}, and free surface elevation, η^{n+1}, in the predictor (3), (4), and (5).
2. Do the iteration in the corrector, (6), (7), (8), until both u_{α}^{n+1} and η^{n+1} converge. In both the predictor and corrector steps, the boundary conditions are the pertinent RANS' variables at time step t^n.
3. Calculate the provisional velocities, $\tilde{u}^{n+1}$ and $\tilde{v}^{n+1}$, in (12).
4. Solve (14) for p^{n+1} and with this newly calculated pressure, update the velocity $\tilde{u}^{n+1}$ and $\tilde{v}^{n+1}$ via (13) to get the correct velocity u^{n+1} and v^{n+1}. Both steps (3) and (4) use the pertinent Bossinesq's variables from time step t^{n+1} as boundary conditions.
5. Using the correct velocities, distribute the fluid volume, F, over computational cells.

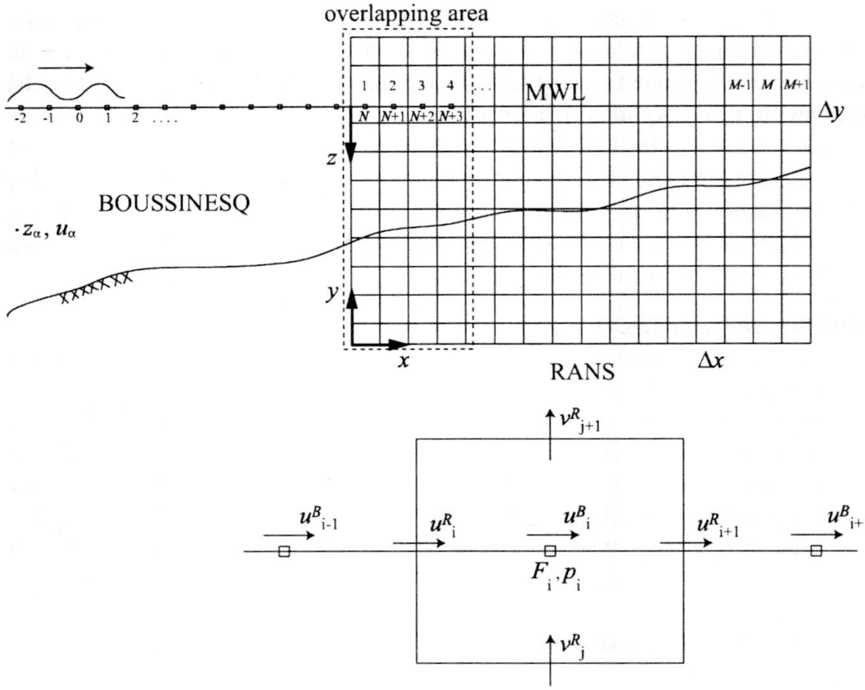

Figure 1. Hybrid wave model grid system.

In determining the boundary conditions for step 1 and 2, the RANS model must provide the Boussinesq model with free surface elevation and the velocity at the reference level in each column $N+1$, $N+2$, and $N+3$. This free surface elevation is simply the difference between the fluid height in the column and the mean water level (MWL). To determine the reference velocity, a second order polynomial of the form $P = \beta + \gamma z + \theta z^2$ is fitted to three arbitrary RANS' horizontal velocities from three consecutive depths z_{j-1}, z_j, and z_{j+1} in each column. Although any three consecutive depths may arbitrarily be picked, the authors found that the depths near the reference level provide the best simulation result. The reason to use a second order polynomial to fit the three velocities will become obvious shortly. The computed coefficient β of the fitted quadratic polynomial is the value sought for reference velocity and is used as boundary condition in step 1 and 2.

The fluid distribution and the vertical structure of the velocities in column 1 of RANS mesh are required in step 3 and 4 as boundary conditions. The fluid distribution is determined by first identifying the cells below, at, and above the free surface. Those cells below the Boussinesq's free surface are full cells. These cells are flagged with identifier $F = 1$. The cell at the free surface is a partially full cell and has identifier $0 < F < 1$. The cells above the free surface are void cells with $F = 0$ identifier. The vertical structures of the Boussinesq's horizontal and vertical velocities are calculated via:

$$u = u_\alpha - \frac{1}{2}\left(z - z_\alpha^2\right)\left(u_\alpha\right)_{xx} - \left(z - z_\alpha\right)\left(hu_\alpha\right)_{xx} \tag{15}$$

$$v = -z\left(u_\alpha\right)_x - \left(hu_\alpha\right)_x . \tag{16}$$

Note that the form (15) was previously used for quadratic polynomial fitting to determine the reference velocity.

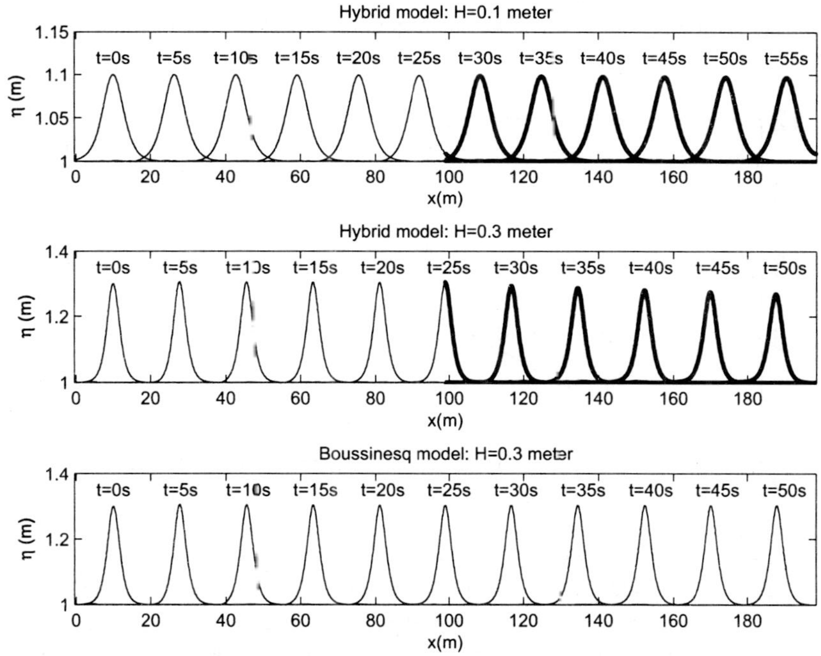

Figure 2. Solitary wave propagation along a channel of constant depth d=1 meter. The calculation is carried out using the Boussinesq and Hybrid models as indicated on top of each figure. The thin line is Boussinesq model and thick line RANS model.

MODEL TEST

Solitary wave propagation

As the first test of the hybrid model, the propagation of a solitary wave along a channel of constant depth was considered. The analytical solution of the weakly-nonlinear Boussinesq equation, Wei and Kirby (1995), served as the initial condition for this simulation. The velocity and the free surface elevation of this wave take the form:

$$u = A\,\mathrm{sech}^2\left[B\left(x - Ct\right)\right] \tag{17}$$

$$\eta = A_1\,\mathrm{sech}^2\left[B\left(x - Ct\right)\right] + A_2\,\mathrm{sech}^4\left[B\left(x - Ct\right)\right], \tag{18}$$

A, B, C, A_1, and A_2 are constants that depend on the channel depth and wave height. The depth of the channel was 1 meter and two simulations with wave heights 0.1 and 0.3 meter were considered. In this case, the molecular and eddy viscosities were set to zero, hence there was no molecular nor Reynold stress acting on the fluid. Although the weakly-nonlinear wave drove the model, in this test the highly-nonlinear form of the Boussinesq model was employed. For comparison, the 0.3 meter wave was also generated using the Boussinesq model alone.

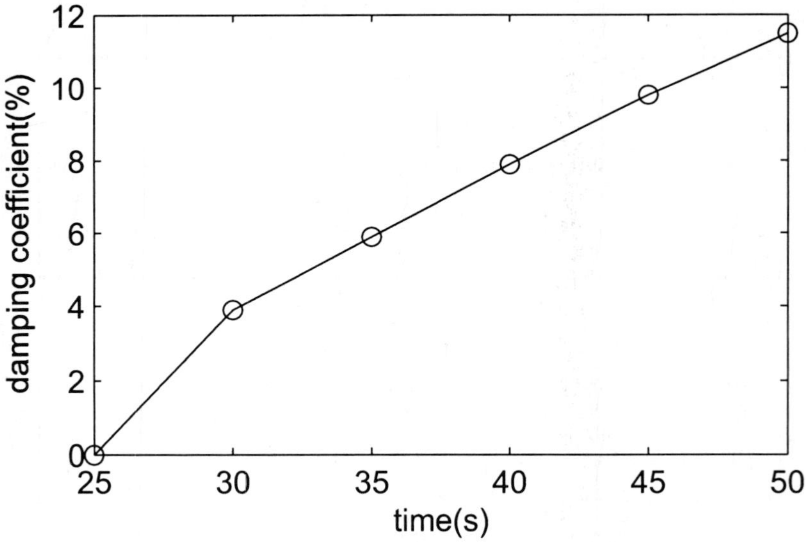

Figure 3. Wave height reduction of a solitary wave, H=0.3 meter, propagating along a channel of constant depth, d=1 meter.

Figure 2 depicts the soliton as it moved from its initial location (x=10 meter) to the right. Apparent from the first figure, the 0.1 meter wave did not change in wave height as it propagated from its initial location to the right. For higher wave, $H = 0.3$ meter, the wave height was damped. The damping coefficient is given in Figure 3. Comparing both the Hybrid's and Boussinesq's simulations for wave height, $H = 0.3$ meter, it is apparent that the solitons traveled with the same constant speed ≈ 3.56 m/s (compared with ≈ 3.28 m/s for $H = 0.1$ meter) in the two models. The fact that the 0.3 meter wave in the hybrid wave model traveled with constant speed and its height diminished indicates that numerical dissipation occurred in the RANS model particularly for highly nonlinear wave. In a larger domain, the wave height, after traveling over a large distance, may significantly change in height, as suggested by Figure 3.

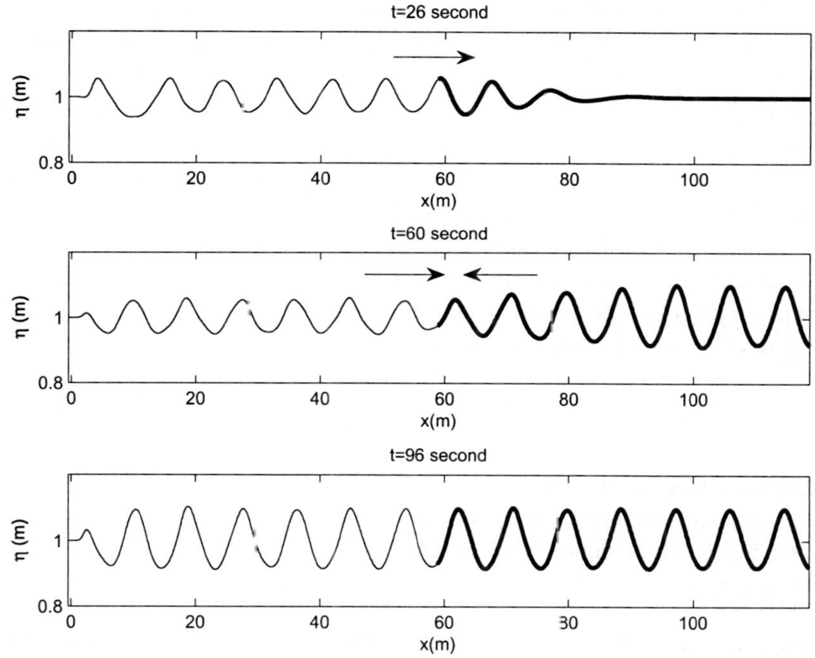

Figure 4. Sinusoidal wave, $H = 0.1$ meter, $T = 3$ seconds, propagated ($t = 26$ seconds)in a semi-closed channel, $d = 1$ meter. The wave was reflected by the right impermeable wall and traveled back ($t = 60$ seconds) to the source and eventually both the incoming- and reflected-waves superposed to create a perfect standing wave ($t = 96$ second).

Standing wave simulation

In the solitary wave simulation, the wave traveled in one direction only. To test the model for wave traveling in both directions, the creation of a standing wave in a channel (closed at one end and open at the other) was considered. At the channel's mouth, a sinusoidal wave was generated. This wave propagated to the closed end and was reflected by the wall. The reflected wave traveled back toward the source. The two identical opposing waves would create a standing wave inside the channel. For this simulation, the channel length and depth were 100 meters and 1 meter respectively and the wave height was 0.1 meter and period 3 seconds. Similar to the previous test, the grid sizes were 0.1 meter in the x-direction and

0.03 meter in the y-direction.

The result of this simulation is given in Figure 4. The snapshot of the wave at $t = 26$ seconds shows that the wave was traveling to the right and just entered the RANS domain (thick line). After being reflected by the impermeable wall on the right side, the wave traveled back to the source, as indicated by the arrow in the snapshot at $t = 60$ seconds. The superposition of the two waves eventually created a 0.2 meter standing wave at $t = 96$ seconds. This simulation demonstrates the ability of the current hybrid wave model to simulate wave propagation in opposite direction.

CONCLUSION

Two widely used wave models, the Boussinesq and RANS wave models, are coupled to perform large scale wave simulation. Since both models were proven accurate, the coupled hybrid model with proper implementation will be expectedly accurate and efficient. The results of the preliminary simulations with uni- and bi-directional waves indicates that the hybrid algorithm can correctly handle the information transfer between the models on their interface.

For future work, breaking wave simulation will be performed to investigate the validity of the hybrid model. To increase the efficiency of the model, the authors also plan to parallelize the model so that it can be run on a distributed memory system.

ACKNOWLEDGMENTS

The research presented here was partially supported by grants from the National Science Foundation (CTS-0427014, CTS-0427115).

REFERENCES

Agnon, Y., Madsen, P. A., and Schaffer, H. 1999. A new approach to higher order Boussinesq models, *Journal of Fluid Mechanics,* 399, 319–333.

Chen, Q., Kirby, J. T., Dalrymple, R. A., Kennedy, A. B., and Chawla, A. 2000. Boussinesq modeling of wave transformation, breaking, and runup. II: 2D, *Journal of Waterway, Port, Coastal, and Ocean Engineering,* 126(1), 48–56.

Gobbi, M. F., Kirby, J. T., and Wei, G. 2000. A fully nonlinear Boussinesq model for surface waves. Part II, *Journal of Fluid Mechanics,* 405, 182–210.

Kennedy, A. B., Chen, Q., Kirby, J. T., and Dalrymple R. A. 2000. Boussinesq modeling of wave transformation, breaking, and runup. I: 1D, *Journal of Waterway, Port, Coastal, and Ocean Engineering,* 126(1), 39–47.

Kothe, D. B., Mjolness, R. C., and Torrey, M. D. 1965. *A Computer Program For Incompressible Flows With Free Surfaces*, Los Alamos National Laboratory, LA-12007-MS, 110 pp.

Lara, J. L., Garcia, N., and Losada, I. J. 2006. RANS modelling applied to random wave interaction with submerged permeable structures, *Coastal Engineering,* 53(5–6), 395–417.

Lin, P. and Liu, P. L.-F. 1998. A numerical study of breaking waves in the surf zone, *Journal of Fluid Mechanics,* 359, 239–264.

Liu, P. L.-F. and Cheng, Y. 2001. A numerical study of the evolution of a solitary wave over a shelf, *Physics of Fluids,* 13, 1660–1667.

Lynett, P. and Liu, P. L.-F. 2002. A numerical study of submarine-landslide-generated waves and run-up, *Proceedings of the Royal Society of London A,* 458, 2885–2910.

Lynett, P. and Liu, P. L.-F. 2004. A two-layer approach to water wave modeling, *Proceedings of the Royal Society of London A,* 460, 2637–2669.

Madsen, P. A. and Sørensen, O. R. 1992. A new form of the Boussinesq equations with improved linear dispersion characteristics. Part 2, *Coastal Engineering,* 18, 183–204.

Madsen, P. A., Bingham, H. B., and Liu, H. A. 2002. A new Boussinesq model for fully nonlinear waves from shallow to deep water, *Journal of Fluid Mechanics,* 462, 1–30.

Nwogu, O. 1993. Alternative form of Boussinesq equations for nearshore wave propagation, *Journal of Waterway, Port, Coastal, and Ocean Engineering,* 119(6), 39–47.

Peregrine, D. H. 1967. Long waves on a beach, *Journal of Fluid Mechanics,* 27(4), 815–827.

Ward, S. N. 2001. Landslide tsnunami, *Journal of Geophysical Research,* 106, 11201–11215.

Wei, G. and Kirby, J. T. 1995. A time-dependent numerical code for extended Boussinesq equations, *Journal of Waterway, Port, Coastal, and Ocean Engineering,* 126(5), 251–261.

Wei, G., and Kirby, J. T., Grilli, S. T., and Subramanya, R. 1995. A fully nonlinear Boussinesq model for surface waves. Part 1. Highly nonlinear unsteady waves, *Journal of Fluid Mechanics,* 294, 71–92.

Yuk, D., Yim, S. C., and Liu, P. L.-F. 2006. Numerical modeling of submarine mass-movement generated waves using RANS model, *Computers & Geosciences,* 32(7), 927–935.

Zhao, Q., Armfield, S., and Tanimoto, K. 2004. Numerical simulation of breaking waves by a multi-scale turbulence model, *Coastal Engineering,* 53(1), 395–417.

INITIATION OF THE BREAKING PROCESS IN BOUSSINESQ-TYPE WAVE MODELS

Giuseppe R. Tomasicchio[1], Felice D'Alessandro[2], and Paolo Veltri[2]

The Breaking Celerity Index (BCI) is proposed as a new wave breaking criterion for phase resolving wave propagation models. The BCI effectiveness in determining the breaking initiation location has been verified for Boussinesq-type wave models against data from different experimental investigations conducted with incident regular waves propagating on uniform beach profiles (Utku 1999; Gonsalves Veloso dos Reis 1992). Moreover, in one case, the comparison has considered the numerical results from the COBRAS model based upon the Reynolds Averaged Navier Stokes (RANS) equations (Liu et al. 2000). Numerical simulations have been performed with the 1D-FUNWAVE model (Kirby et al. 1998). With regard to the adopted experimental conditions, the breaking location has been calculated for different trigger mechanisms (Zelt 1991; Kennedy et al. 2000; Utku 1999). The proposed BCI gives a better agreement with the physical data respect to the other trigger criteria, both for spilling and plunging breaking events. A second paper, in preparation, extends the confirmation of the BCI with further available data for uniform beach slopes (Ting and Kirby 1995; Lara et al. 2005) and for a barred beach (Tomasicchio and Sancho 2002) under regular and irregular incident wave conditions.

INTRODUCTION

In the last decades the coastal research community has experienced a renewal of activity for developing more accurate predictions of the effects of wave and currents on beaches and coastal structures. In particular, the shallow water steepening has been identified as the primary cause of breaking for waves on a beach. As the waves reach a limiting steepness, a portion of the wave front becomes vertical and then overturns and forms a jet of water with generation of turbulence (Peregrine and Svendsen 1978). Several studies addressing wave-induced kinematics over laboratory planar beaches (Nadaoka and Kondoh 1982; Cox et al. 1995; Brunone and Tomasicchio 1997), and field monotonic profiles (George et al. 1994) have been made. A number of studies on the initiation of the wave breaking process have been conducted for over a century. Results from early studies (Miche 1944) are classified as the phase-averaged type. The breaking formulations are described by water depth and wave characteristics, such as wave height, wave period, and wave length, at wave breaking location.

Efforts have also been spent to extend the validity of the Boussinesq type equations (BTE) to the shallow water conditions (Madsen et al. 1991; Nwogu 1993; Wei et al. 1995). The advances in computer sciences have allowed to use the phase resolving BTE models for near-shore wave transformations. The BTE

---

[1] Engineering Dept., University of Lecce, Via per Monteroni, corpo O - 73100 Lecce, Italy
[2] Soil Conservation Dept., University of Calabria, Ponte Bucci 42/B - 87036 Rende, Italy

do not consider the physics of wave breaking and thus become invalid in the surf zone. Accordingly, there have been attempts to introduce wave breaking into BTE. A wave breaking model for phase-resolving type numerical models requires two mechanisms to simulate the wave breaking numerically. One is a trigger mechanism and the other one is an energy dissipation mechanism (extra momentum dissipation model). In the BTE models the initiation of the breaking process is simulated when the trigger mechanism is activated. There are two types of phase-resolving wave breaking trigger mechanisms to determine the initiation/cessation of the process. Both triggers are controlled by the free surface kinematics: the slope angle variation (Schäffer et al. 1993), or the vertical acceleration of the free surface elevation (Kennedy et al. 2000). They adopt three parameters to determine the critical conditions, which are the initiation condition, the termination condition, and the duration time. However, the three coefficients are somewhat hypothetical to match numerical results to experimental data, therefore, they have to be changed by the type of breaker and the bathymetry.

THE RTFN CRITERION

An improvement respect to the three parameters trigger mechanisms has been introduced with the Relative Trough Froude Number (RTFN) theory for BTE models (Utku 1999; Okamoto and Basco 2006). The most important concept for the RTFN theory is the analogy with a moving hydraulic jump in a one-dimensional open channel flow. The similarity between wave breaking and a moving hydraulic jump as a propagating bore has been suggested for a long time (e.g. Peregrine and Svendsen 1978; Basco and Svendsen 1984). The energy dissipation calculation of the surface roller model (Schäffer et al. 1993) is based on this concept. An hydraulic jump in open channel flow changes from an undular jump to a fully developed surface roller jump as the Froude number attains larger values. By analogy between the undular jump and a non-breaking wave, and between the surface roller jump and breaking wave, the transition from the undular jump to a surface roller jump is assumed to be the critical condition of the wave breaking in the RTFN theory. Therefore, breaking of wave and breaking of roller are considered as similar phenomena and it can be estimated that the critical RTFN condition for the wave breaking initiation falls in the range between 1.36 and 1.60 according to Reinnauer and Hager (1995). The Froude number in open channel hydraulics is the ratio of the inertia forces and the gravity forces in the fluid. It can be also assumed as the ratio of the flow velocity in the fluid and the propagation celerity of a surface disturbance. In accordance with this assumption, the RTFN has been defined as (Utku 1999):

$$RTFN = F_{r_t} = \frac{C_{crest} - u_{trough}}{C_{trough}} \tag{1}$$

where C_{crest} is the wave celerity at crest, C_{trough} and u_{trough} are the wave celerity and the horizontal component of the water particle velocity at trough, respectively. In shallow water conditions, $C_{trough} = \sqrt{gh}$ where h = still water depth and g = gravity acceleration. The theoretical study has proved that the RTFN theory provides a constant value, CTFN, as critical condition for the initiation of wave breaking. CTFN has been calibrated with physical data (Utku 1999). Trials have been taken with CTFN between 1.3 and 1.6 and it was found that for a 1:20 slope, CTFN = 1.47 determines the best fit line to the experimental data (Utku 1999; Okamoto and Basco 2006).

The BTE model computes RTFN at each grid point and time. If the RTFN exceeds 1.47, wave breaking begins; when the RTFN drops below the critical value, the wave breaking process is turned off.

THE BREAKING CELERITY INDEX

The Breaking Celerity Index (BCI) is proposed as a new wave breaking criterion for phase resolving wave propagation models. It has been obtained by coupling two different existing trigger mechanisms (Utku 1999; Kennedy et al. 2000).

The criterion proposed by Kennedy et al. (2000) considers the initiation of the wave breaking when the vertical acceleration of the free surface elevation, η_t, exceeds a controlling parameter, $\eta_t^{(I)}$, resembling the shallow water celerity. In accordance with this criterion, the onset of wave breaking is assumed to occur when $\eta_t > \eta_t^{(I)} = b\sqrt{gh}$ with $b = 0.35 \div 0.65$. This condition determines the addition of the extra momentum term to the momentum equation at the calculation node.

As detailed by Kennedy et al. (2000), the value of coefficient b was selected to obtain the agreement between the results from Wei et al. (1995) and Nwogu (1993) equations and the experimental measurements: b = 0.35 is adopted for a barred beach; b = 0.65 is considered for a monotone sloping beach (Kirby et al. 1998). However, the relatively large range of variation of b (0.35÷0.65) increases the uncertainties on the calculated results.

The second criterion that the BCI considers is the RTFN (Utku 1999) which is detailed in the previous section.

The BCI is originated from the two criteria mentioned above. In particular, following Kennedy et al. (2000), η_t is computed and compared with a new controlling parameter with b = 1. The threshold value to be compared with η_t is called BCI (Breaking Celerity Index) and is obtained from Eq. 1 assuming that the wave breaking may be described by means of the solitary wave theory. As a consequence, the BCI is valid at highly nonlinear wave conditions. A solitary wave is normally considered as a wave with infinite wave length, so that it is a special form of a very long, shallow water wave (kh <<1, with k=wave number). The solitary wave celerity at the wave crest is given as (Russel 1844):

$$C_{crest} = \sqrt{g(H+h)} \tag{2}$$

The value of C_{crest} is calculated by Eq. 2 with the assumption $H = \eta_{br} = a$, where a = solitary wave amplitude and η_{br} = free surface elevation at breaking (Figure 1).

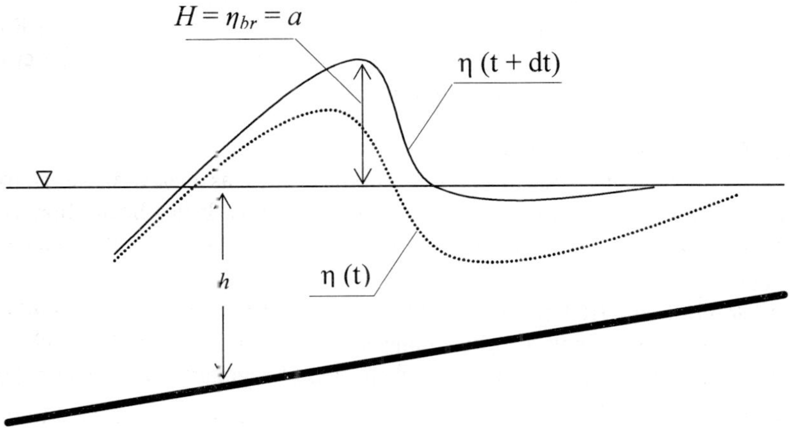

Figure 1. Definition sketch of η (t) in proximity of breaking over a slope.

Theoretically, there is no trough for a solitary wave. A point, infinitely far ahead from the crest location, is considered as the trough. The trough celerity may be written as:

$$C_{trough} = \sqrt{gh} \tag{3}$$

Introducing the obtained expressions for C_{crest} and C_{trough} and the critical value CTFN = 1.47 into Eq. 1 the BCI may be written as:

$$BCI = b\sqrt{gh} = \frac{\sqrt{g(h+a)} - u_\alpha}{1.47} \qquad b = 1 \tag{4}$$

where u_α= horizontal velocity component at a reference water depth $z = z_\alpha = -0.53h$ (Nwogu 1993). At each grid node and for each time step, the BTE model solves the depth integrated equations for the free surface elevation, η, and u_α. Consequently, η_t is computed and compared with BCI. The breaking event starts when/where η_t > BCI.

MODEL VERIFICATION

The wave breaking initiation position, x_b, is defined as the horizontal distance from the toe of the beach slope to the wave crest location where the wave breaking event occurs. The aim of the study is to determine x_b for a shoaling wave on a monotone sloping beach. Numerical simulations have been performed with 1D-FUNWAVE (FUlly Nonlinear WAVE) model (Kirby et al. 1998) referring to the fully nonlinear BTE of Wei et al. (1995), after altering the subroutine for breaking initiation with the BCI trigger mechanism. The extra momentum dissipation term due to wave breaking has not been changed from the original 1D-FUNWAVE eddy viscosity model (Kennedy et al. 2000).

Two types of numerical simulations have been conducted to evaluate the effectiveness of the BCI in determining x_b. The verification has considered data from different experimental investigations (Utku 1999; Gonsalves Veloso dos Reis 1992; Ting and Kirby 1995; Lara et al. 2005; Okamoto et al. 2006; Tomasicchio and Sancho 2002). At first, only incident regular wave conditions acting on uniform slope beaches (Utku 1999; Gonsalves Veloso dos Reis 1992) have been considered. For one case the BCI verification has been extended to the numerical results from COBRAS model based upon the RANS equations (Liu et al. 2000).

In addition, the BCI confirmation (not presented herein) has been conducted for the case of large scale facilities considering both a monotone sloping beach (Lara et al. 2005) and a barred beach under regular and irregular wave conditions (Tomasicchio and Sancho 2002).

Plan slope beach

Two different plane slopes have been considered: 1:20, 1:50 (Utku 1999; Gonsalves Veloso dos Reis 1992). In both cases the experimental longitudinal shore profile presented a flat horizontal bottom with still water depth, h_0, followed by the uniform slope. Table 1 summarizes the experimental geometrical characteristics and the corresponding incident regular wave conditions.

Table 1. Experimental characteristics and incident wave conditions.

Experimental settings	Uniform slope	h_0 (m)	H_i (m)	T (s)
University of Liverpool (*G. V. dos Reis, 1992*)	1:20 1:50	0.2	0.015 ÷ 0.070 0.015 ÷ 0.066	0.7 ÷ 1.5 0.7 ÷ 1.5
Old Dominion University (*Utku, 1999*)	1:20	0.3 0.4 0.5	0.016 - 0.035 - 0.052	1 - 1.5 - 2

For all the considered experimental conditions, the value of x_b has been calculated with the different trigger mechanisms (Zelt 1991; Kennedy et al. 2000; Utku 1999), including the proposed BCI.

In the simulations, the grid size, Δx, was 0.025 m, and the adopted time increments, Δt, were 0.01 s and 0.02 s. Figures 2, 4 show the measured *vs* the calculated values of x_b. With regard to the 1:20 plan slope case (Fig. 2), the inclination of the least squares fit line does not change very much for the considered criteria. As a consequence, the accuracy has been determined by finding the interception of the least squares fit line for each case. It was found that, although the different considered criteria predict x_b accurately, the BCI and Kennedy (b = 0.65) offer the best fit lines to the experimental observations; in fact, as shown in Figures 2 (d) and (e), the corresponding fit lines almost match the "$x_{b_{calc}} = x_{b\,meas}$" condition, where $x_{b_{meas}}$ = measured x_b and $x_{b_{calc}}$ = calculated x_b. The calculations with RTFN have been carried out at Old Dominion University (Okamoto and Basco 2006). The obtained least squares fit lines and the correlation coefficient, R^2, for the 1:20 case, are given in Table 2.

Table 2. Least squares fit lines and R^2 for the different criteria (1:20 slope).

Criterion	Least squares fit line	R^2
Zelt	y = 1.08x – 0.70	0.98
RTFN	y = x – 0.19	0.98
Kennedy (b = 0.35)	y = 1.04x – 0.48	0.98
Kennedy (b = 0.65)	y = 1.06x – 0.17	0.99
BCI	y = 1.05x – 0.16	0.99

In order to extend the BCI verification, for one case only, the x_b value calculated with the BCI has been compared with the corresponding x_b determined with COBRAS wave model (Liu et al. 2000). The case (H_i = 0.02 m, T = 1.2 s, h_0 = 0.20 m, slope 1:20) has been selected among the experimental data from University of Liverpool (Gonsalves Veloso dos Reis 1992). The simulation time was 150 s. The cell size in x-direction varies from 0.02 m at the generation region down to 0.005 m at the toe of the slope: it has been kept constant onshore along the slope and in y-direction (0.005 m). This mesh has been shown suitable to define surf zone hydrodynamics at the breaking point. During the analysis of the calculated time series at each node, two ways of determining x_b with the numerical simulations have been considered. One of them was to individuate the maximum of the upper envelope of η along the beach. When breaking occurs, large values of Turbulence Kinetic Energy (TKE) are observed. For this reason, the second way consisted in the identification of the peaks of maximum TKE at the crest of the wave. The difference between both methodologies is negligible, and any of them is reliable. In this paper, the first method has been considered. Figure 3 shows the upper and lower envelope of η (full line) and mean water level (dashed line) along the wave propagation domain.

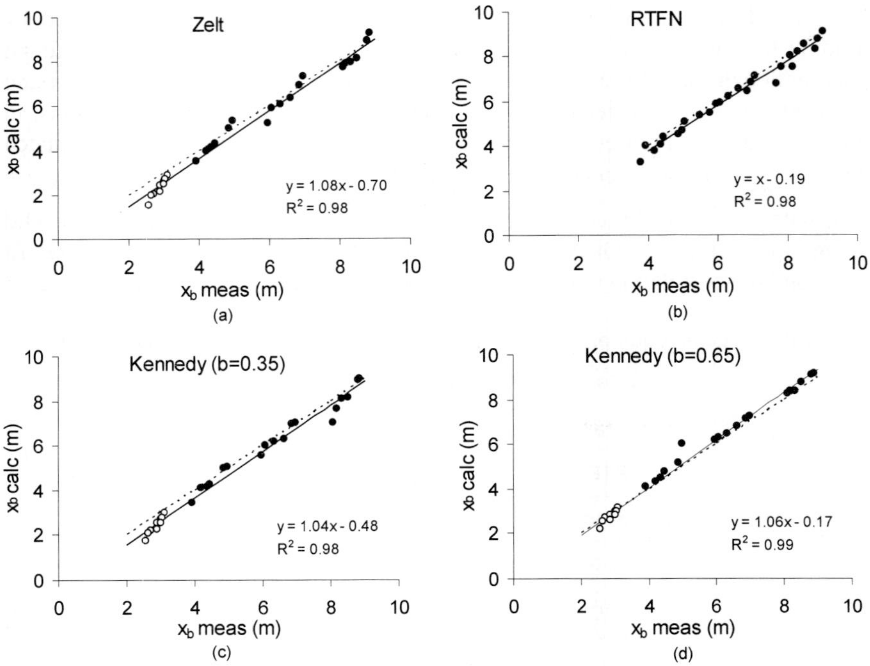

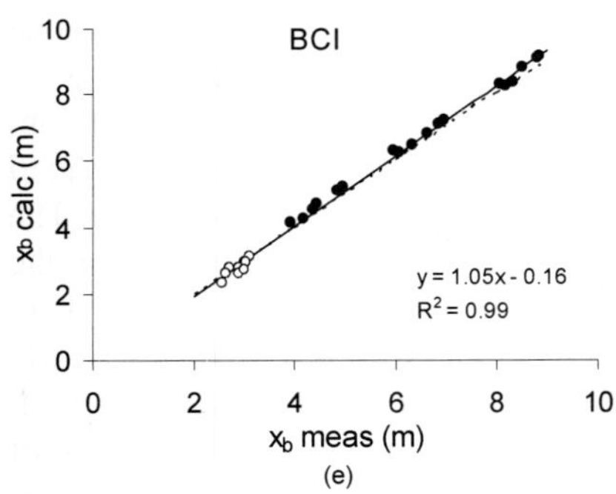

Figure 2.
○ University of Liverpool data
● Old Dominion University data
Slope 1:20
Experimental vs calculated values of x_b for the different breaking criteria
—— least squares fit line
- - - - "$x_b$calc = $x_b$meas " line

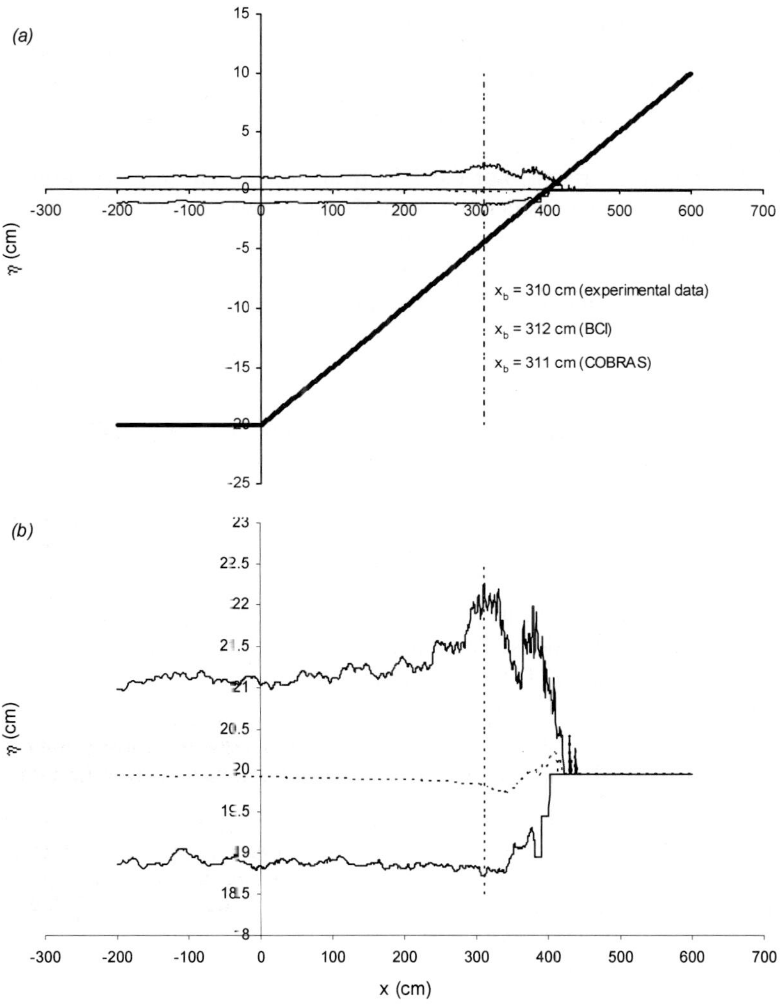

Figure 3. Upper and lower envelope of η (full line), mean water level (dashed line).

The x_b is located where the upper envelope attains the maximum. COBRAS value (vertical dashed line) and the experimental data have a good agreement. For the considered case, the BCI has showed a good accuracy in predicting x_b in comparison with the corresponding experimental data and COBRAS.

The verification of the BCI capabilities in determining x_b has been extended to the 1:50 slope case with experimental data from University of Liverpool (Gonsalves Veloso dos Reis 1992). In this case, the RTFN criterion has not been included in the comparisons. For the 1:50 plan slope beach (Fig. 4), differently from the 1:20 case, the considered criteria shown a lower accuracy in predicting x_b, with higher values of the interception of the least squares fit line for the cases in Figures 4 (a) and (b). With regard to the BCI, a comment is necessary: by substitution of Eq.s 2 and 3 into Eq.1, the following relationship is obtained:

$$RTFN = \sqrt{1 + H / h} \tag{5}$$

For $\gamma = H_b/h_b$ in the range 0.78 ÷ 1.56 with H_b and h_b wave height and water depth at breaking, respectively (Weggel 1972), CTFN falls in the range 1.33 ÷ 1.60. This range resembles that for the breaking of the surface roller in the transition of undular jump given by Reinnauer and Hager (1995). The slightly lower accuracy of the BCI in predicting x_b for the 1:50 case, makes evident the influence of the bottom slope on the CTFN which is assumed 1.47 in Eq. 4 and suggests that the critical value of the RTFN for initiation of wave breaking requires further investigations. Therefore, a deeper discussion on the sensitivity analysis of the adopted CTFN = 1.47 in the BCI formula is needed. However, also for the 1:50 case, the BCI and Kennedy (b = 0.65) give a better response. As shown in Figures 4 (c) and (d), the least squares fit lines almost match the "$x_{b_{calc}} = x_{b\,meas}$" lines and the scatter is negligible. With reference to the criterion in Kennedy et al. (2000), in accordance with Kirby et al. (1998), the coefficient $b = 0.65$, respect to $b = 0.35$, has been found to be more suitable for waves breaking on monotone sloping beaches for both the 1:20 and 1:50 conditions. The least squares fit lines and the corresponding values of R^2, for the 1:50 slope case, are given in Table 3.

Table 3. Least squares fit lines and R^2 for different criteria (slope 1:50).

Criterion	Least squares fit line	R^2
Zelt	y = 1.52x – 4.47	0.98
Kennedy (b = 0.35)	y = 1.33x – 2.89	0.98
Kennedy (b = 0.65)	y = 1.05x – 0.21	0.96
BCI	y = 0.99x + 0.24	0.96

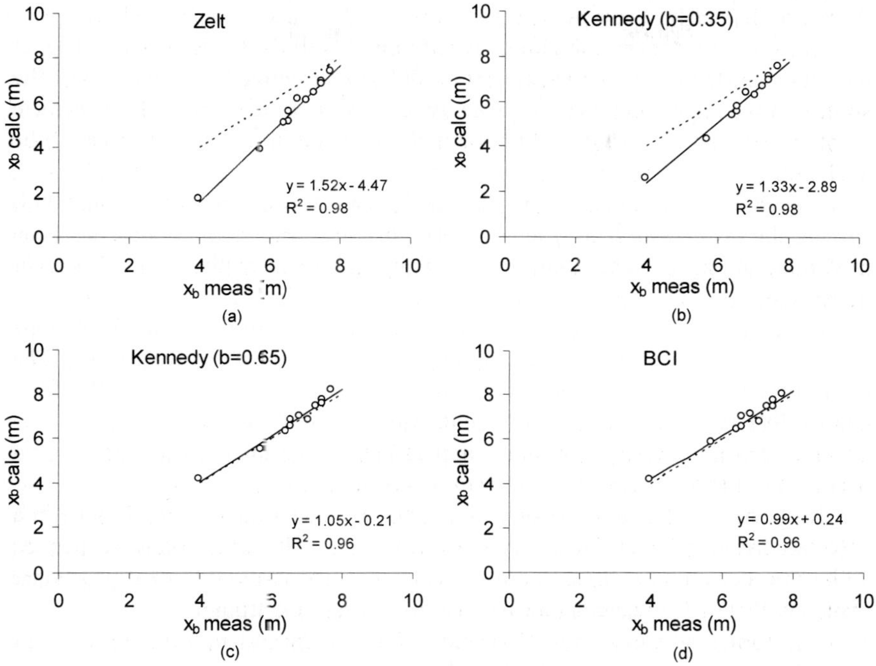

Figure 4. ○ **University of Liverpool data. Slope 1:50.**
Experimental vs calculated values of x_b for the different breaking criteria.
(—— least squares fit line - - - - "$x_b$calc = $x_b$meas " line).

CONCLUSIONS

The Breaking Celerity Index (BCI) is proposed as a wave breaking criterion for phase-dependent wave models.

The BCI has been obtained by coupling two different existing trigger mechanisms (Utku 1999; Kennedy et al. 2000) and defined with the use of the solitary wave theory. Compared to the two coupled criteria, the BCI reduces the number of parameters to be calibrated and the uncertainties on the values to be assigned.

The BCI has been verified for the 1D-FUNWAVE model (Kirby et al. 1998) solving the BTE respect to η and u_α. The original wave model code has been modified adding a subroutine for breaking initiation with the BCI trigger mechanism.

Experimental data of x_b for regular waves propagating over a uniform slope beach have been considered (Utku 1999; Gonsalves Veloso dos Reis 1992). The capability of the different breaking criteria (Zelt 1991; Kennedy et al. 2000; Utku 1999) in determining x_b has been investigated.

The BCI gives a larger accuracy in predicting x_b against the other considered criteria, both for 1:20 and 1:50 uniform beach slopes.

Respect to the RTFN criterion, use of the BCI decreases the CPU time in a 1/20 ratio resulting in a practical relevant advantage. The reduction in the needed CPU time occurs since BCI doesn't need to consider any additional procedure (Misra et al. 2003) to calculate wave celerity at crest and trough.

The numerical simulations have showed that, independently from spilling or plunging breaker type, the BCI presents a larger numerical stability and robustness and does not produce any failure in the calculation run. When not using the BCI, a number of cases have been experienced where the numerical model went unstable crashing during the simulation.

A second paper, in preparation, extends the verification of the BCI with more data for uniform beach slopes (Ting and Kirby 1995; Lara et al. 2005) and barred beaches (Tomasicchio and Sancho 2002; Okamoto et al. 2006). The preliminary results confirm that the use of BCI can be extended to more complicated beach profiles. The case of irregular incident wave conditions has also been considered.

Further verification of the BCI will need to include the case of a beach with the extension limited by a coastal structure.

AKNOWLEDGMENTS

The authors thank Prof. David R. Basco from Old Dominion University (USA) for the fruitful discussions and for providing the experimental data.

They also gratefully acknowledge Prof. Inigo J. Losada and Dr. Javier Lara from University of Cantabria (Spain) for providing the simulation results from the COBRAS wave model.

REFERENCES

Basco, D.R., and Svendsen, I.A. 1984. Modeling turbulent bore propagation in the surfzone, *Proceedings of 19th Int. Conf. on Coastal Engineering*, ASCE, 99-114.

Brunone, B., and Tomasicchio, G.R. 1997. Wave kinematics at steep slopes: second-order model, *J. of Waterway, Port, Coastal and Ocean Engineering*, 123, 223-232.

Cox, D.T., Kobayashi, N., and Okayasu, A. 1995. Experimental and numerical modeling of surf zone hydrodynamics, *Research Report N_0 CACR-95*-07, *(Ph.D. Dissertation of 1st author)*, Center for Applied Coastal Research, University of Delaware, Newark.

George, R.A., Flick, R.E., and Guza, R.T. 1994. Observations of turbulence in the surf zone, *Journal of Geophysical Research*, 99 (C1), 801-810.

Gonsalves Veloso dos Reis, M.T.L. 1992. Characteristics of waves in the surf zone, *MS Thesis*, Department of Civil Engineering, University of Liverpool.

Kennedy, A.B., Chen, Q., Kirby, J.T., and Dalrymple, R.A. 2000. Boussinesq modeling of wave transformation, breaking and run-up. I: 1D, *Journal of Waterway, Port, Coastal and Ocean Engineering*, 126, 39-47.

Kirby, J.T., Wei, G., Chen, Q., Kennedy, A.B., and Dalrymple, R.A. 1998. FUNWAVE 1.0 Fully Nonlinear Boussinesq Wave Model Documentation and User's Manual, *Research Report N_0 CACR-98-06*, Center for Applied Coastal Research, University of Delaware, Newark.

Lara, J.L., Losada, I.J., and Liu, P.L.-F. 2005. Breaking waves over a mild gravel slope: experimental and numerical analysis, submitted for revision to *Journal of Geophysical Research.*

Liu, P.L.-F., Lin, P., Hsu, T., Chang, K., Losada, I.J., Vidal, C., and Sakakiyama, T. 2000. A Reynolds averaged Navier-Stokes equation model for nonlinear water wave and structure interactions, *Proceedings of Coastal Structures '99*, Balkema, 169-174.

Madsen, P.A., Murray, R., and Sørensen, O.R. 1991. A new form of the Boussinesq equations with improved linear dispersion characteristics, *Coastal Engineering*, 15, 371-388.

Miche, R. 1944. Movements ondulatories des mers en profondeur constante ou decroissante, *Annales des Ponts et Chaussees.*

Misra, S.K., Kennedy, A.B., and Kirby, J.T. 2003. An approach to determining nearshore bathymetry used remotely sensed ocean surface dynamics, *Coastal Engineering*, ASCE, 47, 265-293.

Nadaoka, K. and Kondoh, T. 1982. Laboratory measurements of velocity field structure in the surf zone by LDV, *Coastal Engineering*, 25, 125-145.

Nwogu, O. 1993. Alternative form of Boussinesq equations for nearshore wave propagation, *J. of Waterway, Port, Coastal and Ocean Engineering*, 119, 618-638.

Okamoto, T., and Basco, D.R. 2006. The Relative Trough Froude Number for initiation of wave breaking: theory, experiments and numerical model confirmation, *Coastal Engineering*, in press.

Okamoto, T., Basco, D.R., and Fortes, C.J. 2006. The Relative Trough Froude Number for termination of wave breaking, *Proceedings of 30th International Conference on Coastal Engineering*, in press.

Peregrine, D.H., and Svendsen, I.A. 1978. Spilling breakers, bores and hydraulic jumps, *Proceedings of 16th Int. Conf. on Coastal Engineering*, ASCE, 540-550.

Reinnauer, R.R., and Hager, W.H. 1995. Non breaking undular hydraulic jump, *Journal of Hydraulic Research*, 33, 683-704.

Russel, J.S. 1844. Report on Waves, *14th Meeting of the British Association for the Advancement of Science*, 311-390.

Schäffer, H.A., Madsen, P.A., and Deigaard, R. 1993. A Boussinesq model for waves breaking in shallow water, *Coastal Engineering*, 20, 185-202.

Ting, F.C.K., and Kirby, J.T. 1995. Dynamics of surf-zone turbulence in a strong plunging breaker, *Coastal Engineering*, 24, 177-204.

Tomasicchio, G.R., and Sancho, F. 2002. On wave induced undertow at a barred beach, *Proceedings of 28th Int. Conference on Coastal Engineering*, ASCE, 557-569.

Utku, M. 1999. The Relative Trough Froude Number. A new criteria for wave breaking, *Ph.D. Dissertation*, Dept. of Civil and Enviromental Engineering, Old Dominion University, Norfolk, VA.

Weggel, J.R. 1972. Maximum breaker height, *Journal of Waterways, Harbours, and Coastal Engineering Division*, ASCE 98, 529-548.

Wei, G., Kirby, J.T., Grilli, S.T., and Subramanya, R. 1995. A fully non linear Boussinesq model for surface waves. I: highly non linear, unsteady waves, *Journal of Fluid Mechanics*, 294, 71-92.

Zelt, J.A. 1991. The run-up of nonbreaking and breaking solitary waves, *Coastal Engineering,* 15, 205-246.

AN EFFICIENT CURVILINEAR NON-HYDROSTATIC MODEL FOR FREE SURFACE WAVES

Chin H. Wu, Doo-Yong Choi, and Hengliang Yuan

A curvilinear non-hydrostatic model based upon the Navier-Stokes equations is developed to simulate wave propagation and transformation. An implicit method with a domain-decomposition strategy is employed. At the top-layer, an integral approach is used to takes the full effects of non-hydrostatic pressure into account. Through a series of numerical experiments it is demonstrated that the model using a small number of vertical layers is capable of effectively simulating the deep-water and shall-water wave propagation. The model is applied to simulate wave propagation in a gradually converging-diverging and a circular channel. Efficiency and accuracy of the model is discussed.

Introduction

Accurate prediction of wave climate in deep-water and shallow-water is a prime subject in coastal engineering and physical oceanography (Mei and Liu 1993; Battjes 2006). Over years many efforts have been paid to develop a depth-integrated model (DIM), as reviewed by Liu and Losada (2002), since the reduction of the vertical dimension effectively saves computational costs. The essence of the DIM is to account for the non-hydrostatic pressure effects (attributed to the vertical acceleration of fluids) to a certain degree while eliminating the vertical coordinate. For example, the conventional Boussinesq model was derived by accounting for the equal importance of weak nonlinearity and dispersion (Peregrine 1967). Further efforts have been made to remove the original constraints by including higher-order effects (Madsen et al. 1991; Nwogu 1993; Wei et al. 1995), a high-order method that consists of fifth-derivative operators (Madsen et al. 2002), or a multi-layer idea that trades fewer unknowns and higher derivatives for more unknowns and lower derivatives (Lynett and Liu 2004; Hsiao et al. 2005). While the latest Boussinesq-type models have been shown to be capable of simulating highly nonlinear and dispersive waves, continuously developing efficient models for resolving nonlinear and dispersive wave propagation and unified models that can predict wave transformation including shoaling, refraction, diffraction, and reflection from deep-water to shallow-water zone remains a challenging task.

In recent years, many non-hydrostatic models that track the free-surface motion using a single-valued function of the horizontal plane have been developed (Mahadevan et al. 1996; Casulli and Stelling 1998; Mayer et al. 1998;

Environmental Fluid Mechanics Lab, Department of Civil and Environmental Engineering, University of Wisconsin-Madison Madison, WI 53706, USA

Casulli 1999; Namin et al. 2001; Li and Fleming 2001; Kocyigit et al. 2002; Lin and Li 2002, Stelling and Zijlema 2003; Chen 2003; Yuan and Wu 2004; Bradford 2005; Lee et al. 2006; Choi and Wu 2006). Unlike hydrostatic models (HM) that neglect the vertical acceleration or the depth-integrated formulation like Boussinesq-type models that are more applicable to relatively shallow-water conditions, models based upon a full non-hydrostatic pressure distribution can effectively simulate both coastal and deep-water surface wave motions (Lin and Liu 1998; Stelling and Zijlema 2003; Yuan and Wu 2004). In addition the effects of currents on wave evolutions (Yuan and Wu 2006) and wave induced current modulations (Groeneweg and Battjes 2003) can be easily included in non-hydrostatic models. Other examples include internal wave evolution, upwelling and downwelling motion, and flow over abruptly changing bathymetry (Marshall et al. 1997).

While the development of non-hydrostatic models continues, two challenges remain. First, the computation of non-hydrostatic models is more expensive than that of the HM. In particular, three-dimensional calculation is demanding even in the absence of viscous and turbulence effects. Second, the implementation of the free-surface pressure boundary condition is difficult, especially for a staggered grid (Mayer et al. 1998). Some models (Casulli and Stelling 1998; Namin et al. 2001; Kocyigit et al 2002) assume a hydrostatic pressure at the top layer, and others (Casulli 1999; Casulli and Zanolli 2002) use a hydrostatic relation. Hodges and Street (1999) and Bradford (2005) employ a ghost cell approach at the top layer, which still leads to a need of relatively large number of vertical layers to simulate dispersive waves.

To address this issue, Stelling and Zijlema (2003) propose a Keller-box method, based upon an explicit project method, to arrange the pressure at the cell faces rather than the cell centers. The pressure boundary condition at the free surface can therefore be assigned to zero without any approximation. Different from the Keller-box scheme, Yuan and Wu (2004, 2006) develop an integral approach, based upon an implicit method, to remove the top-layer hydrostatic assumption under a staggered grid framework. The top-layer pressure is implicitly expressed as a function of the free-surface elevation (hydrostatic pressure component) and the vertical acceleration (non-hydrostatic pressure component). The top-layer integral approach is also successfully employed to an explicit project non-hydrostatic model under a staggered grid (Choi and Wu 2006). All these models are shown to have the capability of simulating surface-wave propagation using a small number of vertical layers (2 ~ 4 layers). While their results are promising, this type of non-hydrostatic models has yet to be applied to simulate wave propagation under irregular boundaries.

In this paper, we expand the existing Cartesian non-hydrostatic free surface model of Yuan and Wu (2006) to an orthogonal curvilinear system, and investigate the accuracy and efficiency of curvilinear non-hydrostatic model through dispersive wave conditions.

Mathematical Formulation

The non-hydrostatic model is based upon the three dimensional incompressible Reynolds-averaged Navier-Stokes equations (RANS), which describes the conservations of mass and momentum. The equations in the Cartesian coordinates along the x, y, and z directions for the velocity u, v, and w the velocity components can refer to Yuan and Wu (2006). By maintaining the vertical coordinate z, the governing equations under the orthogonal curvilinear coordinate system (ξ,η,z) are

$$\frac{1}{h_1 h_2}\frac{\partial}{\partial \xi}(h_2 U)+\frac{1}{h_1 h_2}\frac{\partial}{\partial \eta}(h_1 V)+\frac{\partial w}{\partial z}=0, \tag{1}$$

$$\begin{aligned}&\frac{\partial U}{\partial t}+\frac{U}{h_1}\frac{\partial U}{\partial \xi}+\frac{V}{h_2}\frac{\partial U}{\partial \eta}+w\frac{\partial U}{\partial z}+\frac{UV}{h_1 h_2}\frac{\partial h_1}{\partial \eta}-\frac{V^2}{h_1 h_2}\frac{\partial h_2}{\partial \xi}\\ &=-\frac{1}{h_1}\frac{\partial p}{\partial \xi}+\frac{1}{h_1}g_\xi+\frac{1}{h_1 h_2}\frac{\partial}{\partial \xi}\left(h_2\tau_{\xi\xi}\right)+\frac{1}{h_1 h_2}\frac{\partial}{\partial \eta}\left(h_1\tau_{\xi\eta}\right)+\frac{\partial \tau_{\xi z}}{\partial z}+\frac{\tau_{\xi\eta}}{h_1 h_2}\frac{\partial h_1}{\partial \eta}-\frac{\tau_{\eta\eta}}{h_1 h_2}\frac{\partial h_2}{\partial \xi}\end{aligned} \tag{2}$$

$$\begin{aligned}&\frac{\partial V}{\partial t}+\frac{U}{h_1}\frac{\partial V}{\partial \xi}+\frac{V}{h_2}\frac{\partial V}{\partial \eta}+w\frac{\partial V}{\partial z}-\frac{U^2}{h_1 h_2}\frac{\partial h_1}{\partial \eta}+\frac{UV}{h_1 h_2}\frac{\partial h_2}{\partial \xi}\\ &=-\frac{1}{h_2}\frac{\partial p}{\partial \eta}+\frac{1}{h_2}g_\eta+\frac{1}{h_1 h_2}\frac{\partial}{\partial \xi}\left(h_2\tau_{\xi\eta}\right)+\frac{1}{h_1 h_2}\frac{\partial}{\partial \eta}\left(h_1\tau_{\eta\eta}\right)+\frac{\partial \tau_{\eta z}}{\partial z}+\frac{\tau_{\xi\eta}}{h_1 h_2}\frac{\partial h_2}{\partial \xi}-\frac{\tau_{\xi\xi}}{h_1 h_2}\frac{\partial h_1}{\partial \eta}\end{aligned} \tag{3}$$

$$\frac{\partial w}{\partial t}+\frac{U}{h_1}\frac{\partial w}{\partial \xi}+\frac{V}{h_2}\frac{\partial w}{\partial \eta}+w\frac{\partial w}{\partial z}=-\frac{\partial p}{\partial z}+g_z+\frac{1}{h_1 h_2}\frac{\partial}{\partial \xi}\left(h_2\tau_{\xi z}\right)+\frac{1}{h_1 h_2}\frac{\partial}{\partial \eta}\left(h_1\tau_{\eta z}\right)+\frac{\partial \tau_{zz}}{\partial z} \tag{4}$$

where $h_1=\left(\left(\frac{\partial x}{\partial \xi}\right)^2+\left(\frac{\partial y}{\partial \xi}\right)^2\right)^{0.5}$ and $h_2=\left(\left(\frac{\partial x}{\partial \eta}\right)^2+\left(\frac{\partial y}{\partial \eta}\right)^2\right)^{0.5}$ are the scale metric factors of the $\xi-$, and $\eta-$direction, respectively; $U(\xi,\eta,z,t)$ and $V(\xi,\eta,z,t)$ are the transformed horizontal velocity components in the $\xi-$, and $\eta-$direction, respectively, which are related to the Cartesian horizontal velocities as $U=\frac{1}{h_1 h_2}(uy_\eta - vx_\eta)$ and $V=\frac{1}{h_1 h_2}(-uy_\xi + vx_\xi)$; and g_ξ, and g_η are the gravitational acceleration in $\xi-$, and $\eta-$direction, respectively. Under the Boussinesq hypothesis and isotropic turbulence assumption, the normalized shear stresses, divided by a constant reference density, are described in terms of mean velocities as follows

$$\begin{aligned}&\tau_{\xi\xi}=2\nu_e\left(\frac{1}{h_1}\frac{\partial U}{\partial \xi}+\frac{V}{h_1 h_2}\frac{\partial h_1}{\partial \eta}\right),\quad \tau_{\eta\eta}=2\nu_e\left(\frac{1}{h_2}\frac{\partial V}{\partial \eta}+\frac{U}{h_1 h_2}\frac{\partial h_2}{\partial \xi}\right),\quad \tau_{zz}=2\nu_e\frac{\partial w}{\partial z},\\ &\tau_{\xi\eta}=\nu_e\left[\frac{h_2}{h_1}\frac{\partial}{\partial \xi}\left(\frac{V}{h_2}\right)+\frac{h_1}{h_2}\frac{\partial}{\partial \eta}\left(\frac{U}{h_1}\right)\right],\quad \tau_{\xi z}=\nu_e\left(\frac{\partial U}{\partial z}+\frac{1}{h_1}\frac{\partial w}{\partial \xi}\right),\quad \tau_{\eta z}=\nu_e\left(\frac{\partial V}{\partial z}+\frac{1}{h_2}\frac{\partial w}{\partial \eta}\right),\end{aligned} \tag{5}$$

where $\nu_e=\nu_t$ (turbulent eddy viscosity)$+\,\nu$ (kinematic viscosity).

Four types of boundary conditions are used. First for wall boundaries, a no-slip boundary condition is employed to impose zero surface velocity or a standard wall function approach for viscous flows. If inviscid flows, a null gradient condition for the tangential velocity components, *i.e.*, free-slip, is imposed. An impermeability condition is specified for the normal component of velocity to the wall. Second, at the free surface $z = \varsigma(\xi,\eta,t)$, the kinematic boundary condition

$$w\big|_{z=\varsigma} = \frac{\partial \varsigma}{\partial t} + \frac{U}{h_1}\frac{\partial \varsigma}{\partial \xi} + \frac{V}{h_2}\frac{\partial \varsigma}{\partial \eta} \tag{6}$$

is used. Integrating the continuity equation (6) over the water depth and using Eq. (6) gives a conservative form of free surface equation

$$\frac{\partial \varsigma}{\partial t} + \frac{1}{h_1 h_2}\frac{\partial}{\partial \xi}\left(\int_{-h}^{\varsigma} h_2 U dz\right) + \frac{1}{h_1 h_2}\frac{\partial}{\partial \eta}\left(\int_{-h}^{\varsigma} h_1 V dz\right) = 0 . \tag{7}$$

In addition by integrating the vertical momentum equation without the viscous terms from the center of free-surface cell, $z = z^* = [\varsigma(\xi,\eta,t) - \Delta z]/2$, to the free surface, $z = \varsigma(\xi,\eta,t)$, the resulting equation gives

$$p_{z=z^*} = g\left[\varsigma - z^*\right] + \frac{\partial}{\partial t}\int_{z=z^*}^{z=\varsigma} w dz + \frac{\partial}{\partial \xi}\left(\int_{z=z^*}^{z=\varsigma} \frac{U}{h_1} w dz\right) + \frac{\partial}{\partial \eta}\left(\int_{z=z^*}^{z=\varsigma} \frac{V}{h_2} w dz\right) + \frac{U}{h_1} w\Big|_{z^*}\frac{\partial z^*}{\partial \xi} + \frac{V}{h_2} w\Big|_{z^*}\frac{\partial z^*}{\partial \eta}\Bigg|$$

$$+ w\Big|_{z^*}\frac{\partial z^*}{\partial t} - w^2\Big|_{z=z^*}, \tag{8}$$

where the first term at the right-hand side denotes the hydrostatic pressure, and the rest of right-hand side terms contribute to the non-hydrostatic pressure. Third, velocity components at the inflow are specified by either analytical solutions or laboratory conditions. Finally at outflow boundaries, the Sommerfeld radiation boundary condition

$$\frac{\partial \phi}{\partial t} + \frac{c\cos\theta}{h_1}\frac{\partial \phi}{\partial \xi} + \frac{c\sin\theta}{h_2}\frac{\partial \phi}{\partial \eta} = 0, \tag{9}$$

is used, where θ is the angle between the normal direction of the boundary and the direction of the propagating of wave, c is the wave celerity, and ϕ can be U, V, or ζ. In addition a sponge layer technique (Yuan and Wu 2004) are incorporated into the computational domain.

Numerical Method

The model is based upon a staggered finite difference method following the Crank Nicholson scheme in time and the central difference scheme in space. Details discretization can refer to Yuan and Wu (2004). All flow-field components (i.e., the pressure, velocities, and free-surface elevation) are simultaneously solved at each time step by the following procedure. First, for layers below the top-layer, using discretized continuity equation (1) and vertical momentum equation (4), the vertical velocity component and pressure are expressed as functions of horizontal velocities. For the top-layer, discretization forms of equation (6) and (9) would give the vertical velocity component and pressure as functions of horizontal velocities by use of the discretized form of equation (7). Substituting the above two relationships of vertical velocity component and pressure into the horizontal momentum equations (2) and (3) to form a tri-block diagonal matrix system with the unknown of horizontal velocity components. Second, the resulting matrix system is solved by a domain decomposition method (Wu and Yuan 2007), which decomposes the full 3D matrix into a series of vertical two-dimensional (2D) problems. A direct solver (Tannehill et al. 1997) can be used to solve the solution for each vertical 2D problem. The solution to the original 3D solution is achieved by using an iteration procedure. Finally, the vertical velocity component and pressure can be updated by back substituting the solved horizontal velocity component.

Model characteristics and applications

The 3D non-hydrostatic model has been tested previously with linear 3D standing waves in a closed basin, finite amplitude sloshing motions, wave propagation over a submerged bar, and 3D wave transformation over an elliptic shoal on a sloped bottom (Yuan and Wu, 2004). Detailed discussion of accuracy and efficiency of the model can refer to Yuan and Wu (2006). We only summarize the general characteristics of the model here. Figure 1 shows the CPU time, normalized with respect to that of the reference resolution, versus the resolution-increasing factor, defined as the ratio of the increasing resolution of the reference resolution. The CPU time increases linearly as the horizontal resolution (squares) increases. However the CPU times increase e-fold as the vertical resolution (circles) increases. This result overall indicates that reducing vertical resolution is more effective in terms of saving the computational cost.

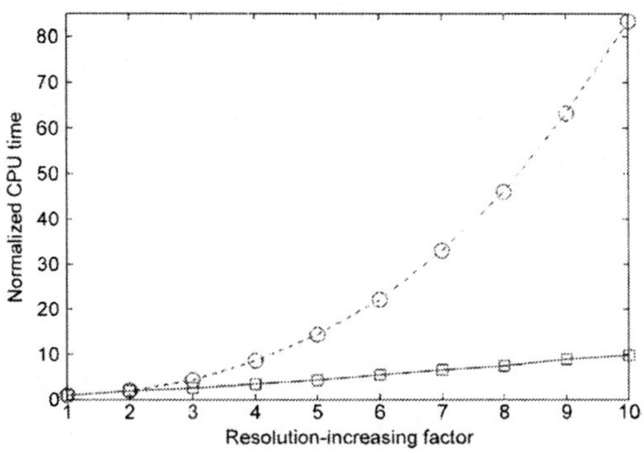

Figure 1. Normalized CPU time versus the horizontal resolution-increasing factor (squares) and the vertical resolution-increasing factor (circles).

Figure 2 summarizes the accuracy of model results for simulating a two-dimensional numerical h = 1m deep wave tank with a total length of 12 incident wavelengths L, with the last two wavelengths serving as a sponge layer under a range of Kh, where K is the wave number. Accuracy of the model is assessed by a normalized percentage error $\varepsilon = |c_{\text{linear}} - c_{\text{model}}| / c_{\text{linear}}$, where c_{model} is the wave speed from the model calculation and $c_{\text{linear}} = (g \tanh(Kh)/K)^{0.5}$ is the liner wave speed. For $Kh = 1$ the model using only two layers match well the linear wave speed and has only an error of ε = 0.05 %, indicating the accuracy of the model. The five-layer model is sufficient to accurately calculate the linear dispersion up to $Kh = 5$. Overall increasing vertical number of layers allows the model to accurately simulate deeper water wave propagation.

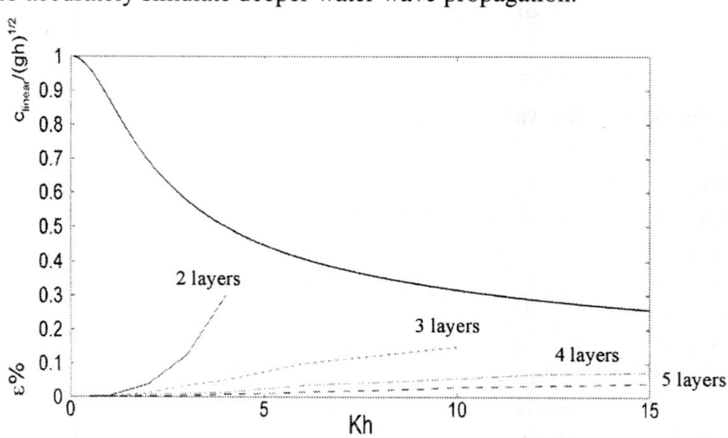

Figure 2. The upper line shows the normalized wave speed versus relative depth Kh for linear dispersion. The bottom lines represent the relative errors of the non-hydrostatic model based upon different layers.

In this paper, the non-hydrostatic model is extended to a curvilinear coordinate system. We apply the extended model to two examples: (i) linear waves in a gradually converging-diverging (CD) channel, and (ii) wave propagation in a circular channel.

To show the conservation of energy of the non-hydrostatic model in a boundary fitted domain, a linear progressive wave is simulated over a channel with gradually varied width while water depth is kept constant. Since the change of the channel width is much less than the wave length, waves can propagate without significant reflection due to varying-width channels. Under this condition, the envelope of free surface elevation along a channel can be analytically predicted by the theoretical Green's formula (Lamb 1932), i.e.,

$$H(x) = H_0 \sqrt{\frac{B_0}{B(x)}} \tag{10}$$

where $H(x)$ and $B(x)$ are the wave height and the channel's width, respectively, at the location x along a channel. The subscripts "0" of $H(x)$ and $B(x)$ represent the inlet of the channel location, i.e. $x = 0$. In this test, linear monochromatic sinusoidal waves with a wave period T = 5 s and water depth h = 1 m, are imposed at the inflow boundary. The channel width begins with 30 m, converges to the half of inflow width after 10 wave lengths, and then diverges back to the original width after 20 wave lengths. A cosine function is used to describe the variation of channel width

$$B(x) = \frac{B_0}{4}\cos\left(\frac{2\pi}{20L}x\right) + \frac{3B_0}{4}, \tag{11}$$

where $B_0 = 30\,\text{m}$ and L = wave length.

In the model, the varying-width channel on the physical x – y domain is transformed to the horizontal ξ - η computational domain, which is discretized into 800 × 40 uniform grids. Only two vertical layers are employed since $Kh = 0.412$. The time step Δt = T/80 s is chosen to ensure the Courant number less than unity. Fig. 3 shows a perspective view of wave propagation in the converging and diverging channel. As the width of the channel gradually reduces, the wave height in the channel increases. At the end of the channel the wave height returns to the original. Uniformity of the wave crest across the channel width also indicates negligible reflection. Fig. 4 shows the comparison of surface displacements at the mid-section of the channel predicted by the model and the envelopes of analytical solutions calculated by Eq. (11). The excellent agreement verifies the extended model under an orthogonal curvilinear coordinate system. In addition the efficiency of the new model is demonstrated by the use of two vertical layers.

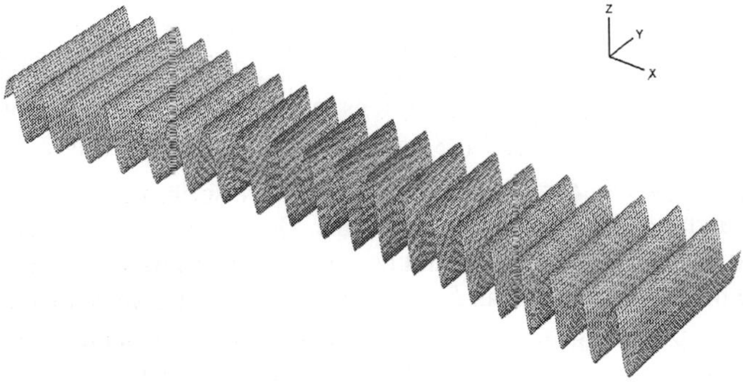

Figure 3. A perspective view of wave propagation in a CD channel

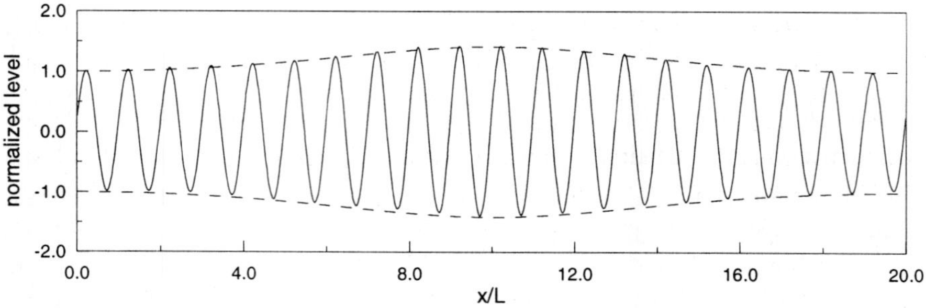

Figure 4. Comparison of normalized surface displacements between the model results (solid lines) and analytical solutions (dotted lines).

The second application of the model is to simulate wave transformation through a circular channel. This is a classical example to examine model's capability in resolving reflection and diffraction of waves due to the wall. This example was first explored by Rostafinski (1976) in the context of acoustics to study sound propagation in a curved duct. In recent years this has been extended to water wave application to examine the effects of curvilinear terms caused by grid stretching in a numerical sense (Dalrymple et al. 1994; Kirby et al. 1994; Beji and Barlas 2004; Shi and Kirby 2005). Interestingly all these studies used unified depth-integrated models, either the mild slope or Boussinesq equations. Here, this example is used to verify our newly developed orthogonal curvilinear non-hydrostatic model.

In this test the channel has a constant water depth d = 1 m. The radii of inner and outer channels are 20 m and 30 m, respectively. To find the corresponding analytical solution, we superimpose different modes of the Bessel functions of the first kind J_{ν_n} and the second kind Y_{ν_n} , i.e.,

$$\varsigma(r,\theta,t)=\sum_{n=1}^{N}\left[a_n J_{\nu_n}(Kr)+b_n Y_{\nu_n}(Kr)\right]e^{i\nu_n\theta}e^{i\omega t} \tag{12}$$

where r and θ are the radius and angle in a polar coordinate; ω is the angular frequency; N is the total number of modes. The first four modes ν_n, the order of the Bessel function, are 18.18, 24.23, 27.74, and 33.50, which can be obtained by solving no flux boundary conditions at the inner and outer walls. The coefficients a_n and b_n in Eq. (28) are determined by imposing a linear wave with the wave period T = 2 s, yielding $kd = 1.20$, at the inlet boundary $\theta = 0$ and the boundary conditions at the two walls. Details of this procedure can refer to Beji and Barlas (2004) and will not be given here for brevity.

In the model the horizontal ξ - η computational domain is discretized by a set of 350 × 30 grids, equivalent to a constant angular-grid spacing $\Delta\theta = \pi/350$ rad and a constant radial distance $\Delta r = 1/3$ m on the physical domain. In the vertical direction, 2 and 4 layers are employed. At the outflow boundary, a sponge layer with a radiation boundary is employed. The time step

$\Delta t = T/50$ s is chosen, giving the maximum Courant number at the inner wall less than 1. The model is run for 50 sec = 25 T to ensure a periodic steady-state condition. Figure 5 shows a perspective view of the computed surface elevation in the channel. The wave initially propagates uniformly at the inlet. At the bend of channel, the wave starts to diffract and energy starts to pile up at the outer sidewall, where the reflection takes place. Afterwards, the combined effects of diffraction, reflection, and incident wave result in higher harmonics at the outer wall than the inner wall. Overall the wave transformation phenomena are consistent with those predicted by the unified depth-integrated models (Dalrymple et al. 1994; Kirby et al. 1994; Shi and Kirby 2005). Figure 6 compares the normalized surface elevation between analytical solutions and the model results at the inner wall, center line, and outer wall of the channel. For $0 \le \theta \le \pi/2$, the beginning half of the channel, the model results using both 2-layer and 4-layer model are in good agreements with the analytical solutions. For $\theta \ge \pi/2$, the overall comparison is reasonable but noticeable phase differences are observed in the 2-layer model results. On the other hand, the 4-layer model results remarkably match with the analytical solutions, consistent with the characteristics of the non-hydrostatic model on the Cartesian coordinate in Figure. 2.

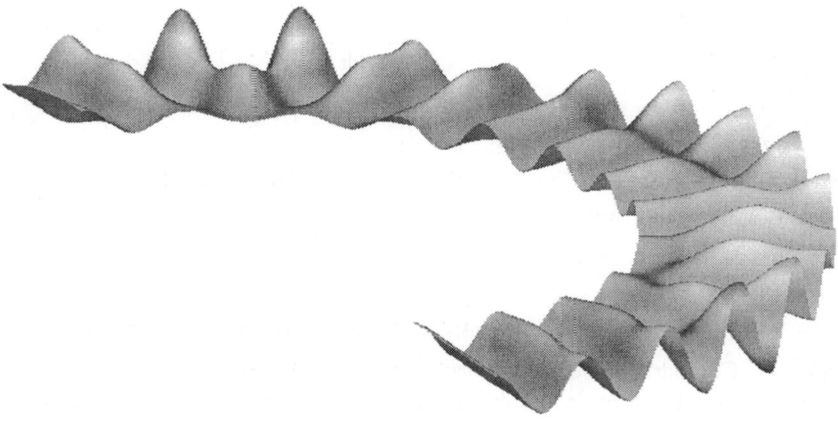

Figure 5. A perspective view of wave propagation in a circular channel

Summary and conclusions

In this paper, we extended the existing non-hydrostatic model in a Cartesian coordinate (Yuan and Wu 2006) to a curvilinear coordinate system. Accuracy and efficiency of the model is examined. It is found that the model using a small number of vertical layers (2 ~5) can resolve relatively deep water waves. The implement of an integral form of top-layer pressure equation dramatically improves the efficiency of traditional non-hydrostatic models. The model is

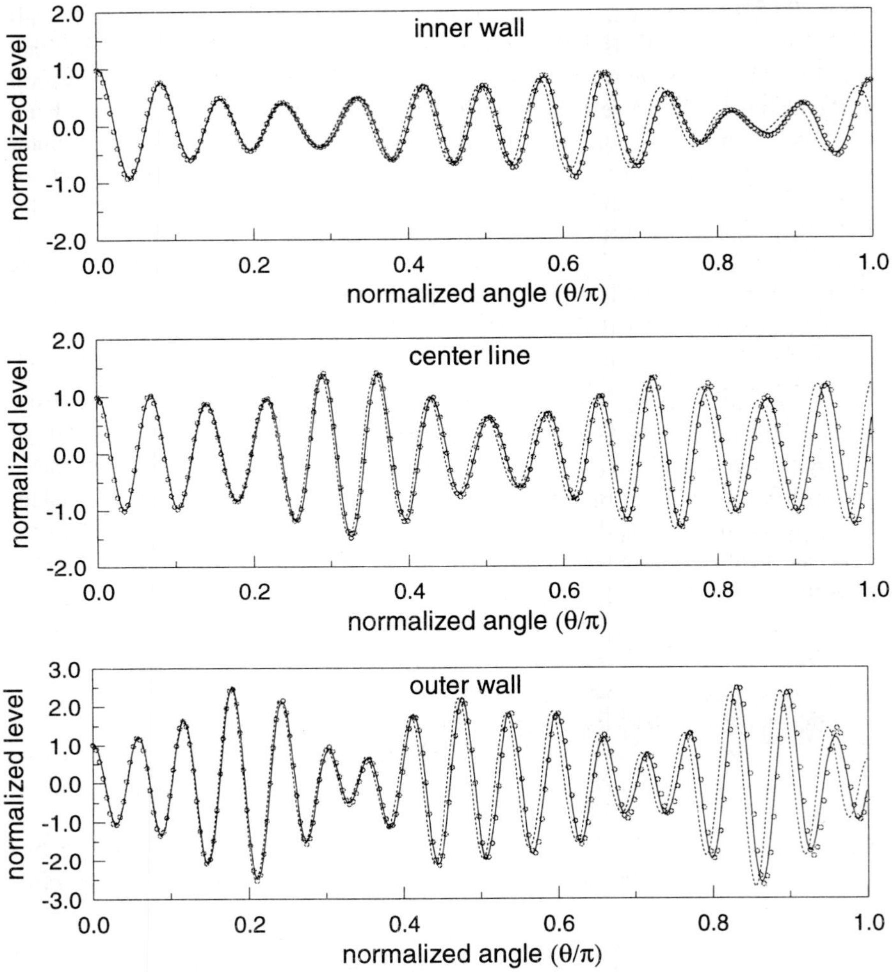

Figure 6. Comparisons of the normalized free surface level of analytical solutions (circles), 2-layer model results (dotted lines), and 4-layer model results (solid lines).

applied to simulate two benchmark examples: linear waves in a gradually converging-diverging channel and wave propagation in a circular channel. Excellent agreement between model results using small number vertical layers and analytical solutions further indicate the efficiency of the model. In addition, the non-hydrostatic model can be used to irregular wave propagation under different water depth condition. The result will be reported shortly.

Acknowledgements

This study was supported in part by the University of Wisconsin Sea Grant (NA16RG2257), the North Temperate Lakes Long-Term Ecological Research (NSF 0217533), and the Wisconsin Alumni Research Foundation (WARF 040061).

References

Battjes, J.A. 2006. Developments in coastal engineering research, *Coastal Engineering*, 23, 121-132.

Beji, S. and B. Barlas. 2004. Boundary-fitted non-linear dispersive wave model for regions of arbitrary geometry, *Int. J. Numerical Method in Fluids*, 45, 643-657

Bradford, S.F. 2005. Godunov-based model for nonhydrostatic wave dynamics. *J. of Waterway, Port Coastal and Ocean Engineering-ASCE*, 131, 226-238.

Casulli, V. 1999. A semi-implicit finite difference method for non-hydrostatic, free-surface flows, *Int. J. Numerical Method in Fluids,* 30, 425-440.

Casulli, V. and G.S. Stelling. 1998. Numerical simulation of 3D quasi-hydrostatic, free-surface flows, *J. of Hydraulic Engineering-ASCE*, 124, 678-686.

Casulli, V. and P. Zanolli. 2002. Semi-implicit numerical modeling of nonhydrostatic free-surface flows for environmental problems, *Mathematical and Computer Modeling,* 36, 1131-1149.

Chen, X. 2003. A fully hydrodynamic model for three-dimensional, free-surface flows, *Int. J. Numerical Method in Fluids,* 42, 929-952.

Choi, D.Y. and C.H., Wu. 2006. A new efficient 3D non-hydrostatic free-surface flow model for simulating water wave motions, *Ocean Engineering*, 33 587-609.

Dalrymple, R.A., J.T. Kirby, and P.A. Martin. 1994. Spectral methods for forward-propagating water waves in conformally-mapped channels, *Applied Ocean Research*, 16, 249-266.

Groeneweg, J. and J.A. Battjes. 2003. Three dimensional wave effects on a steady current, *J. Fluid Mech.*, 478, 325–343.

Hodges, B.R. and R.L. Street. 1999. On simulation of turbulent nonlinear free-surface flows, *Journal of Computational Physics,* 151, 425-457.

Hsiao, S.-C., P. Lynett, H.-H. Hwung, and P. L.-F. Liu. 2005. Numerical Simulations of Nonlinear Short Waves Using a Multilayer Model, *Journal of Engineering Mechanics*, 131, 231-243.

Kirby, J.T., R.A. Dalrymple, and H. Kaku. 1994. Parabolic approximations for water waves in conformal coordinate systems, *Coastal Engineering*, 23, 185-213.

Kocyigit, M.B., R.A. Falconer, and B. Lin. 2002. Three-dimensional numerical modeling of free surface flows with non-hydrostatic pressure, *Int. J. Numerical Method in Fluids,* 40, 1145-1162.

Lamb, H. 1932. *Hydrodynamics*, Dover Publications, New York.

Lee, J.W., M.D. Teubner, J.B. Nixon, and P.M. Gill. 2006. A 3-D non-hydrostatic pressure model for small amplitude free surface flows. *Int. J. Numerical Method in Fluids*, 50, 649-672

Li, B. and C. Fleming. 2001. Three-dimensional model of Navier-Stokes equations for water waves. ASCE *Journal of Waterway, Port, Coastal, and Ocean Engineering*, 6-25.

Lin, P.Z. and P.L.-F. Liu. 1999. Internal wave-maker for Navier-Stokes equations models. *Journal of Waterway, Port, Coastal, and Ocean Engineering*, 125, 207-212.

Lin, P.Z. and C.W. Li 2002. A σ-coordinate three-dimensional numerical model for surface wave propagation, *Int. J. Numerical Method in Fluids*, 38, 1045-1068.

Liu, P.L.F., I.J. Losada. 2002. Wave propagation modeling in coastal engineering, *J. of Hydraulic Research*, 40, 229-240.

Lynett, P. and P.L.-F. Liu. 2004. A two-layer approach to wave modeling. *Proceedings of the royal society of London Series A*, 460, 2637-2669.

Madsen, P.A., R. Murray, and O.R. Sorensen. 1991. A new form of the Boussinesq equations with improved linear dispersion characteristics, *Coastal Engineering*, 15, 371-388.

Madsen, P.A., H.B. Bingham, and H. Liu. 2002. A new Boussinesq method for fully nonlinear waves from shallow to deep water, *J. of Fluid Mechanics*, 462, 1-30.

Mahadevan, A., J. Oliger, and R.L. Street. 1996. A non-hydrostatic mesoscale ocean model. Part 1: well posedness and scaling, *J. Phys. Oceanogr.* 26, 1868-1880.

Marshall, J., C. Hill, L. Perelman, A. Adcroft. 1997. Hydrostatic, quasi-hydrostatic, and nonhydrostatic ocean modeling, *Journal of Geophysical Research*, 102, 5733-5752.

Mayer, S., A. Garapon, and L.S. Sorensen. 1998. A fractional step method for unsteady free-surface flow with applications to non-linear wave dynamics. *Int. J. Numer. Meth. Fluids*, 28, 293-315.

Mei, C.C., P.L.F. Liu, 1993. Surface waves and coastal dynamics, *Annual Review of Fluid Mechanics* 25, 215-240.

Namin, M., B. Lin, and R. Falconer. 2001. An implicit numerical algorithm for solving non-hydrostatic free-surface flow problems, *Int. J. Numerical Method in Fluids,* 35, 341-356.

Nwogu, O. 1993. Alternative form of Boussinesq equations for nearshore wave propagation. *J. Waterway, Port, Coastal, Ocean Eng.,* 119, 618-638.

Peregrine, D.H. 1967. Long waves on a beach, *J. Fluid Mech.*, 27, 815-827.

Rostanski W. 1976. Acoustic systems containing curved duct sections, *J. of Acoustic Society of America*, 60, 23–28.

Shi, F., and J.T. Kirby. 2005. Curvilinear parabolic approximation for surface wave transformation with wave-current interaction. *Journal of Computational Physics*, 204, 562-586.

Stelling, G.S. and M. Zijlema. 2003. An accurate and efficient finite-difference algorithm for non-hydrostatic free-surface flow with application to wave propagation. *Int. J. Numerical Method in Fluids,* 43, 1-23.

Tannehill, J.C., D.A. Anderson, and R.H. Pletcher. 1997. *Computational Fluid Mechanics and Heat Transfer* (2nd ed.), in: W.J. Minkowycz, E.M. Sparrow eds., Series in Computational and Physical Processes in Mechanics and Thermal Sciences (Taylor & Francis: Hemisphere Publishing Corporation, Bristol, PA, 1997) 717-724.

Wei, G., J.T. Kirby, S.T. Grilli, and R. Subramanya. 1995. A fully nonlinear Boussinesq model for surface waves. Part 1 Highly nonlinear unsteady waves, *J. Fluid Mech.*, 294, 71-92.

Wu, C.H. and Yuan, H.L. 2007. Efficient non-hydrostatic modelling of surface waves interacting with structures, *Applied Mathematical Modelling*, 31, In Press.

Yuan, H.L. and C.H. Wu. 2004. An implicit three-dimensional fully non-hydrostatic model for free-surface flows, *Int. J. Numerical Method in Fluids,* 46, 709-733

Yuan, H.L. and C.H. Wu. 2006. Fully non-hydrostatic modeling of surface waves. *J. of Engineering Mechanics-ASCE*, 132, 447-456.

IRROTATIONAL, ROTATIONAL AND TURBULENT FLOWS IN BREAKING WAVES

Yasunori Watanabe[1] and Ayumi Saruwatari[2]

This paper presents a reasonable method for decomposing instantaneous velocity field in breaking waves, into irrotational, rotational and turbulent velocity components. In this method, Nowgu's approach (1993) for Boussinesq type equation, applying slip bottom velocity computed in a large eddy simulation, is used for estimating the irrotational velocity in breaking waves, and the rotational velocity is obtained on the basis of Helmholtz decomposition. A transition from potential wave energy in incident waves to rotational flows within breaking-induced large-scale vortices and small-scale turbulence through a wave breaking process can be expressed using the proposing decomposition method.

INTRODUCTION

When incident waves, approximated as potential irrotational flows, break in a shallow water, the potential wave energy is rapidly transfered to rotational kinetic energy due to production of large-scale vortices by splashing jets and also turbulent energy owned by small-scale random velocity fluctuations. The formations of the large-scale vortices and evolution of the turbulence during wave breaking process can be computed using turbulence models (Lin and Liu 1998; Watanabe and Saeki 1999; Christensen and Deigaard 2001), and the detailed features of organized vortex structures under consecutively splashing jets of breaking waves have been identified in a large eddy simulation (LES) (Watanabe et al. 2005). These primitive turbulence computations that need to be carried out in a very fine grid system generally require expensive computational costs although they give us details of flow field at high resolution.

In terms of computational cost, a depth-averaged wave equation model is advantageous for computing potential wave field outside of a surf zone. In order to extend the applicable regime of the model to the surf zone, a wave breaking model to dissipate the wave energy needs to be given in computations.

Many depth-averaged wave equation models assuming irrotational potential flow under waves and the corresponding wave breaking models have been proposed. As mentioned, the potential energy is first transfered to the rotational kinetic energy to produce the large-scale vortices and then they are decomposed into numbers of smaller vortices to finally be turbulence that can be statistically determined through a cascade-down process (See figure 1). Finally, the turbulent energy is dissipated as heat when the turbulence scale becomes less than the Kolmogorov length-scale. However, in most of the wave breaking models,

[1] School of Engineering, Hokkaido University, North 13 West 8, Sapporo, 060 8628, Japan
[2] School of Engineering, Hokkaido University, North 13 West 8, Sapporo, 060 8628, Japan

sudden production of turbulence after wave breaking is implicitly assumed to directory dissipate the irrotational wave energy on the basis of a bore analogy or a mixing distance turbulence model. There is no effect of the rotational flows predominating the surf zone flow in such conventional wave breaking models. While such models are designed for reproducing the attenuation of wave heights with the propagation of breaking waves because the irrotational component of the flow affects pressure gradients or surface deformations, they are inappropriate for estimating the velocity field involving large-scale vortices and transient turbulence which become predominant during the wave breaking process.

This paper presents transitions from irrotational to rotational flows in a surf zone and also to small-scale turbulence intensified in rotational velocity gradients under breaking waves. New approach to decompose the instantaneous velocity field into these irrotational and rotational velocities in breaking waves is proposed in this paper. Typical features of the velocity distributions of these decomposed components through the wave breaking process for spilling, plunging and intermediate breakers and the evolution of depth-averaged kinetic energy for them are also investigated, which are needed for developing a reasonable breaking wave model that should be consistent with the local velocity field in a depth-averaged type of wave equation.

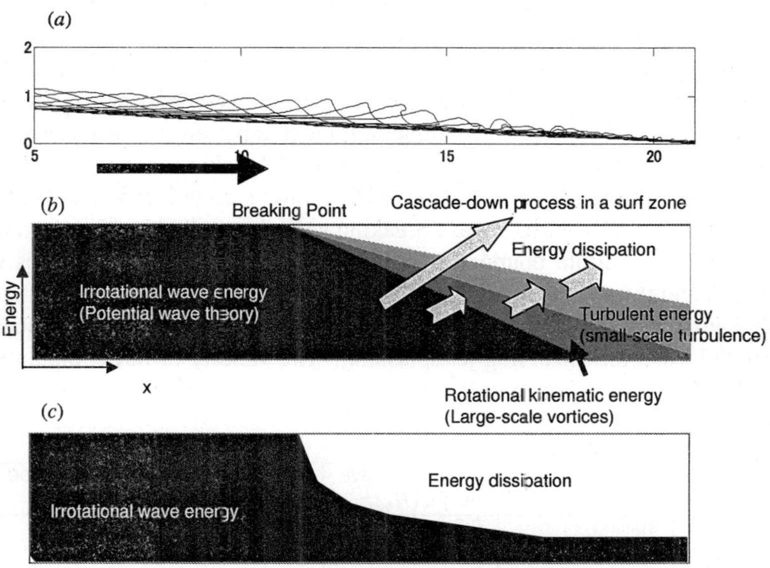

Figure 1. Wave energy transfer to rotational and turbulent kinetic energy. (*a*) free-surface evolution, (*b*) kinetic energy transfer through a cascade-down process, (*c*) attenuation of wave energy by a conventional wave breaking model based on a bore analogy.

VELOCITY DECOMPOSITION

Large Eddy Simulation

Performing filtering operations to Navier-Stokes equation, the following grid scale(GS) momentum equation can be obtained.

$$\frac{\overline{D\boldsymbol{u}}}{Dt} = -\nabla\overline{p} + \nu\nabla^2\overline{\boldsymbol{u}} - \nabla\boldsymbol{\tau} + \boldsymbol{g} \tag{1}$$

Here, ν is the kinetic viscosity, $\boldsymbol{g}$ is the gravity acceleration. When the instantaneous velocity $\boldsymbol{u}$ is expressed by the filtered velocity $\overline{\boldsymbol{u}}$ and fluctuation $\boldsymbol{u}'$,

$$\boldsymbol{u} = \overline{\boldsymbol{u}} + \boldsymbol{u}', \tag{2}$$

the sub-grid scale (SGS) stress $\boldsymbol{\tau}$ is defined by

$$\boldsymbol{\tau} = \boldsymbol{L} + \boldsymbol{C} + \boldsymbol{R}, \tag{3}$$

where the Leonard tensor $\boldsymbol{L} = \overline{\overline{\boldsymbol{u}} \otimes \overline{\boldsymbol{u}}} - \overline{\boldsymbol{u}} \otimes \overline{\boldsymbol{u}}$, cross-stress tensor $\boldsymbol{C} = \overline{\overline{\boldsymbol{u}} \otimes \boldsymbol{u}'} + \overline{\boldsymbol{u}' \otimes \overline{\boldsymbol{u}}}$ and Reynolds SGS tensor $\boldsymbol{R} = \overline{\boldsymbol{u}' \otimes \boldsymbol{u}'}$. In this study, $\boldsymbol{\tau}$ is modeled by nonlinear SGS viscosity model (Yakhot and Orszag 1986). The governing equation (1) is computed in a same manner as Watanabe and Saeki (1999) and also Watanabe et al. (2005). The computations were carried out in a three-dimensional numerical wave flume shown in figure 2. Second-order Cnoidal waves were given at the offshore boundary, and a periodic boundary condition was imposed at side walls of the wave flume. Although a bottom boundary layer may affect the rotational flow field via interactions with breaking-induced vortices, in this study, the bottom boundary effects have been ignored to focus on the velocity transitions due to only surface wave breaking such that a slip boundary condition was given at the bottom boundary. Spilling and plunging breaking waves were computed, and the proposing decomposition method was tested for non-breaking waves (see Table 2). Figure 2 shows the typical computational result of sequential surface deformations involving jet splashing. The validity of our computational method have been confirmed in Watanabe et al. (2005).

Table 1. Wave conditions.

	Breaker type	Relative wave height (H_b/h_b)	Wave period (s)	Bottom slope
run 1	spilling	0.8	1.15	1:20
run 2	plunging	1.01	2.75	1:20
run 3	non-breaking	—	1.15	—

Helmholtz decomposition

According to Helmholtz decomposition, any vector field can be split into a sum of gradient of scalar potential ($\nabla\phi$) and curl of vector potential ($\nabla \times \boldsymbol{\psi}$).

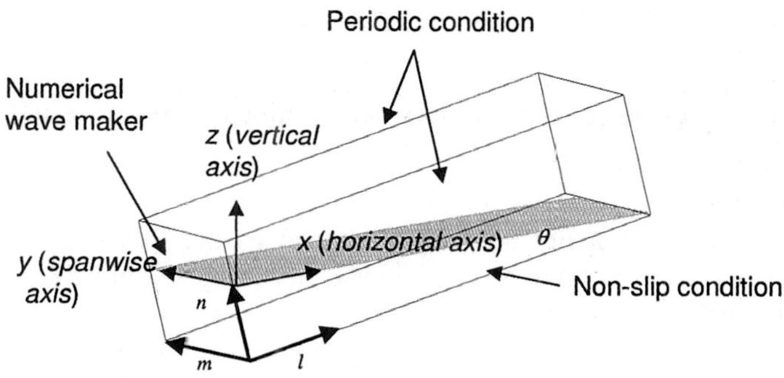

Figure 2. Computation domain.

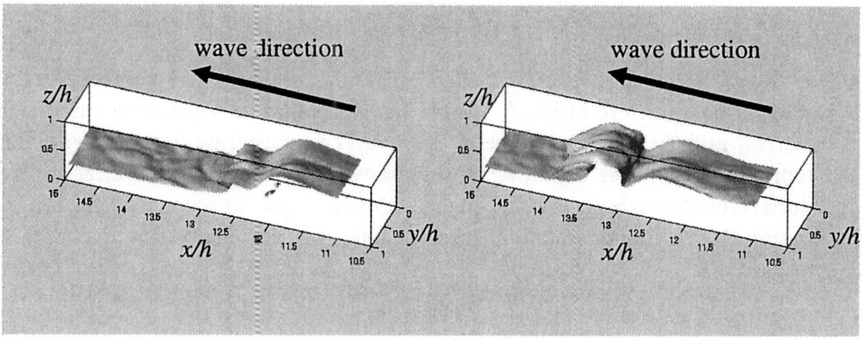

Figure 3. Examples of the computed surface shapes of breaking waves.

In another word, arbitrary velocity $\boldsymbol{u}$ can be decomposed into rotational $\boldsymbol{u}_r$ and irrotational $\boldsymbol{u}_i$ velocities.

$$\boldsymbol{u} = \boldsymbol{u}_i + \boldsymbol{u}_r \tag{4}$$

Filtering equation (4), we have

$$\overline{\boldsymbol{u}} = \overline{\boldsymbol{u}_i} + \overline{\boldsymbol{u}_r}. \tag{5}$$

If the filtered irrotational velocity $\overline{\boldsymbol{u}_i}$ is obtained on the basis of the Nowgu's approach for Boussinesq type equation model (Nowgu 1993), the filtered rotational velocity $\overline{\boldsymbol{u}_r}$ can be estimated by subtracting $\overline{\boldsymbol{u}_i}$ from the filtered instantaneous velocity $\overline{\boldsymbol{u}}$ that can be computed by using LES. The fluctuation velocity $\boldsymbol{u}'$ in equation (2) is also decomposed into irrotational and rotational parts:

$$\boldsymbol{u}' = \boldsymbol{u}_i' + \boldsymbol{u}_r'. \tag{6}$$

It may be a good assumption to neglect $\boldsymbol{u}_i'$ for small-scale turbulence which is developed though a scale-down process of vortices. Equation (2) can be written by

$$\boldsymbol{u} = \overline{\boldsymbol{u}_i} + \overline{\boldsymbol{u}_r} + \boldsymbol{u}_r'. \tag{7}$$

Using equation (7) instead of equation (2), rotational SGS stress and turbulent energy can also be computed via LES.

Irrotational velocity

The irrotational velocity field is determined by the Boussinesq-type wave equation model for which the bottom slip velocity computed in LES is given as a boundary condition. The method to estimate the filtered irrotational velocity $\overline{\boldsymbol{u}_i}$ based on the Nwogu's approach is outlined bellow. The slip boundary condition at a bottom for filtered velocities is

$$\overline{w} = -\epsilon\mu^2\overline{\boldsymbol{u}} \cdot \boldsymbol{\nabla}h, \tag{8}$$

$$\boldsymbol{\nabla} \times \overline{\boldsymbol{u}} = 0, \tag{9}$$

where $\boldsymbol{\nabla} = (\partial/\partial x, \partial/\partial y)$, $\overline{\boldsymbol{u}} = (\overline{u}, \overline{v})$, $\overline{u}$ is the filtered cross-shore velocity, $\overline{v}$ is the along-shore velocity, h is the water depth, the nonlinear parameter $\varepsilon = a_0/h_0$ and dispersion parameter $\mu = h_0/l$ (a_0: representative amplitude, h_0: representative water depth, l: representative wavelength). All variables are non-dimensionalized by a_0, h_0, l, water density ρ and gravity acceleration g. Integrating the continuity equation from the bottom to z and substituting equation (8), we have

$$\overline{w} = -\mu^2(\frac{\partial}{\partial x}\int_{-h}^{z}\overline{u}dz + \frac{\partial}{\partial y}\int_{-h}^{z}\overline{v}dz). \tag{10}$$

A Taylor expansion for $\overline{\boldsymbol{u}_i}$ in z with respect to the bottom level is

$$\overline{\boldsymbol{u}_i} = \overline{\boldsymbol{u}_{ib}} + (z+h)\frac{\partial\overline{\boldsymbol{u}_{ib}}}{\partial z} + \frac{(z+h)^2}{2}\frac{\partial^2}{\partial z^2}\overline{\boldsymbol{u}_{ib}} + \cdots, \tag{11}$$

where $\overline{\boldsymbol{u}_{ib}}$ is the slip velocity at a bottom. The bottom boundary conditions lead

$$\frac{\partial\overline{\boldsymbol{u}_{ib}}}{\partial z} = -\mu^2[\boldsymbol{\nabla}(\overline{\boldsymbol{u}_{ib}} \cdot \boldsymbol{\nabla}h) + \boldsymbol{\nabla} \cdot \overline{\boldsymbol{u}_{ib}}\boldsymbol{\nabla}h]. \tag{12}$$

Substituting equations (11) and (12) into equation (10),

$$\begin{aligned} \overline{w} = & - \mu^2\boldsymbol{\nabla}[(z+h)\overline{\boldsymbol{u}_{ib}}] + \mu^4\boldsymbol{\nabla} \cdot [\frac{(z+h)^2}{2}[\boldsymbol{\nabla}(\overline{\boldsymbol{u}_{ib}} \cdot h) + (\boldsymbol{\nabla} \cdot \overline{\boldsymbol{u}_{ib}})\boldsymbol{\nabla}h]] \\ & - \mu^2(\frac{\partial}{\partial x}\int_{-h}^{z}\overline{u_r}dz + \frac{\partial}{\partial y}\int_{-h}^{z}\overline{v_r}dz) + O(\mu^6). \end{aligned} \tag{13}$$

Ignoring higher order terms than $O(\mu^4)$, the first and second terms of the right hand side of equation (13) represent the irrotational vertical velocity $\overline{w_i}$ and the third term corresponds to the rotational vertical velocity $\overline{w_r}$. Integrating an irrotational condition for $\overline{\boldsymbol{u}_i}$ from the bottom to z,

$$\overline{\boldsymbol{u}_i} - \overline{\boldsymbol{u}_{ib}} = \int_{-h}^{z} \nabla \overline{w_i} dz. \tag{14}$$

Substituting the irrotational part of equation (13) and keeping the terms up to $O(\mu^2)$, equation (14) becomes

$$\overline{\boldsymbol{u}_i} = \overline{\boldsymbol{u}_{ib}} + \mu^2 (\frac{h^2}{2} - \frac{z^2}{2}) \nabla (\nabla \cdot \overline{\boldsymbol{u}_{ib}}) - \mu^2 (h+z) \nabla [\nabla \cdot (h \overline{\boldsymbol{u}_{ib}})] + O(\mu^4). \tag{15}$$

Since the bottom slip velocity, obtained by using LES of breaking wave field where the slip boundary condition is imposed at the bottom, must be irrotational; $\overline{\boldsymbol{u}_b} = \overline{\boldsymbol{u}_{ib}}$, the irrotational horizontal velocity (15) and vertical velocity $\overline{w_i} = -\mu^2 \nabla [(z+h) \overline{\boldsymbol{u}_{ib}}]$ at arbitrary z can be obtained if the computed slip horizontal velocity $\overline{\boldsymbol{u}_b}$ is given. The rotational velocities can also be estimated by

$$\overline{\boldsymbol{u}_r} = \overline{\boldsymbol{u}} - \overline{\boldsymbol{u}_b} - \mu^2 (\frac{h^2}{2} - \frac{z^2}{2}) \nabla (\nabla \cdot \overline{\boldsymbol{u}_b}) + \mu^2 (h+z) \nabla [\nabla \cdot (h \overline{\boldsymbol{u}_b})], \tag{16}$$

$$\overline{w_r} = \overline{w} + \mu^2 \nabla [(z+h) \overline{\boldsymbol{u}_b}]. \tag{17}$$

RESULTS

Although breaking waves typically produce three-dimensional velocity field (Watanabe et al. 2005), a two-dimensional velocity distribution on only vertical-horizontal plane is discussed in this study for simplification.

Validation

Velocity field in non-breaking waves is well approximated by irrotational potential flows except in very thin surface and bottom boundary layers. Therefore, the computed instantaneous velocity $\overline{\boldsymbol{u}}$ in non-breaking wave filed, where the slip boundary condition is imposed, should coincide with the irrotational velocity $\overline{\boldsymbol{u}_i}$ except in the vicinity of the free-surface.

Figure 3 shows the comparison of the instantaneous velocity distributions obtained in the LES computation with the irrotational velocity estimated by the method described in the previous section. The agreements in both of horizontal and vertical velocity distributions between $\overline{\boldsymbol{u}}$ and $\overline{\boldsymbol{u}_i}$ ensure appropriate estimations of irrotational velocity by our proposing method using the computed slip bottom velocity.

Irrotational, rotational and turbulent flows in breaking waves

Figure 3 shows typical vertical distributions of the irrotational velocity at successive wave phases in run 1 (spilling breaking wave). The irrotational velocity profiles described by a power series in the vertical coordinate (equation 15) vary in time and space under breaking waves. On the other hand, the high rotational velocity is found to concentrate at the wave crest where is a horizontal roller

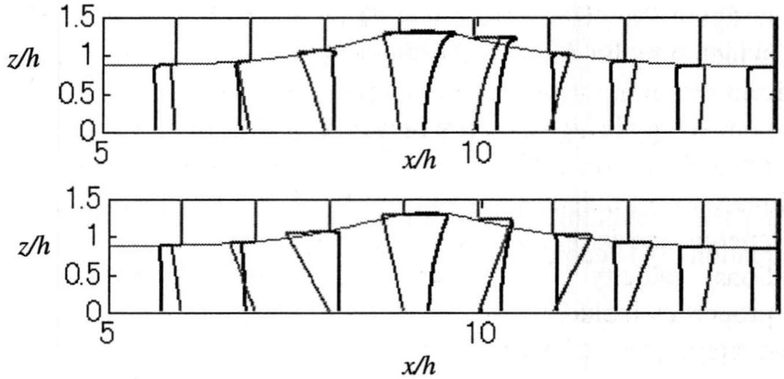

Figure 4. Computed velocity ($\overline{u}$ and $\overline{w}$) profiles (top) and estimated irrotational velocity ($\overline{u_i}$ and $\overline{w_i}$) profiles (bottom). black: horizontal velocity; gray: vertical velocity.

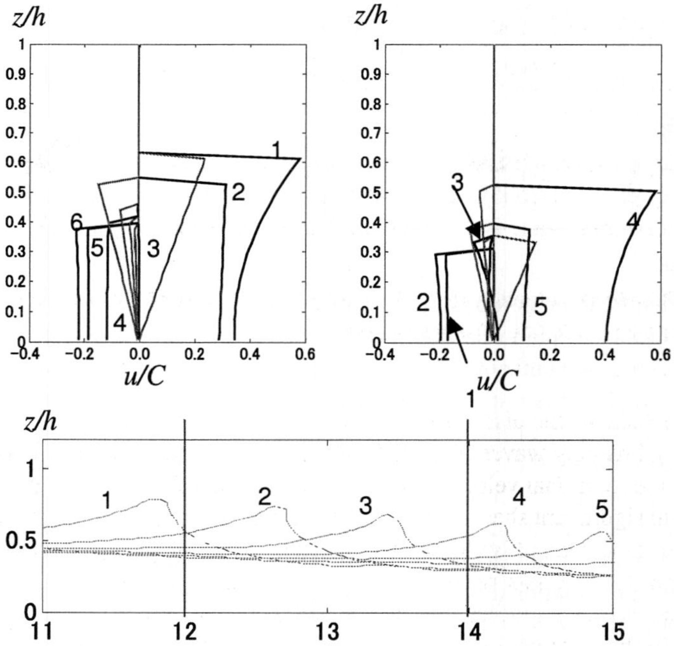

Figure 5. Typical horizontal (black) and vertical (gray) irrotational velocity distributions at x/h = 12 (top left) and x/h = 14 (top right) in run1 (spilling breaker).

vortex (see figure 3). There are also rotational flows near the surface behind the crest, which may be due to the roller vortex left from the wave crest, while no significant rotational flows can be observed bellow a wave trough level. The SGS turbulent energy is highly intensified within the crest region where rotational flows significantly occur, indicating a high correlation between distributions of the rotational and turbulent flows. It is also found there is no correlation of turbulent energy and irrotational velocity distribution (figure 3), and therefore the irrotational velocity is unacceptable for being a parameter to determine turbulent properties including energy dissipations in a wave breaking model for the depth-averaged type of wave equation.

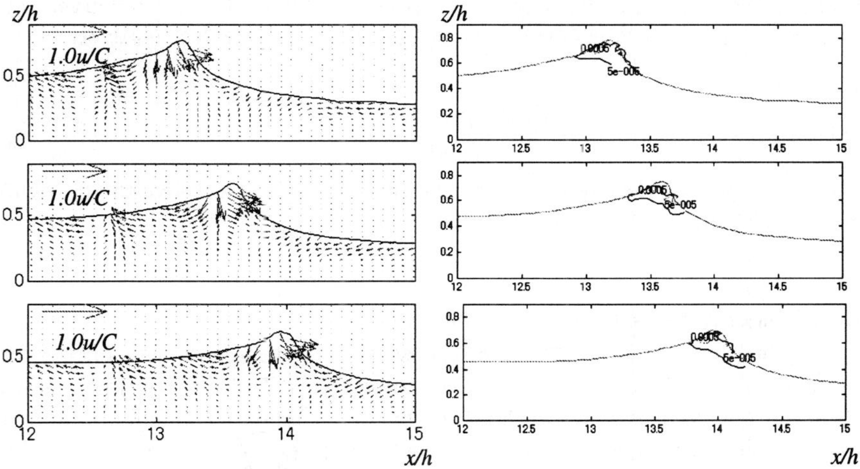

Figure 6. Rotational velocity vector plots (left) and contours of turbulent energy (right) in run1. time interval: T/16 (T: wave period).

Figure 3 shows the distributions of the rotational vectors and turbulent energy in plunging breaking waves (run 2). While there is no rotational flow before wave breaking, the rotational velocity quickly appears to have strong velocity gradients and to form significant shear field near the wave crest and jet after breaking. Since turbulence is developed by local shears, strong turbulent energy simultaneously occurs also near the jets where the rotational velocity gradients are significant. Since local velocity gradients in the irrotational component are much smaller than that in the rotational one, the irrotational velocity may has no contribution to produce turbulence. These results suggest that the turbulent production and dissipation should be modeled by features of the rotational flow as a model parameter in a wave breaking model. It should be noted that estimations of the rotational velocity within overturning jets are inaccurate since the irrotational velocity is unable to be estimated by equation (15) when a free-surface becomes a double function with respect to the vertical coordinate. In this study, the overturning jets are assumed to possess only the rotational velocity.

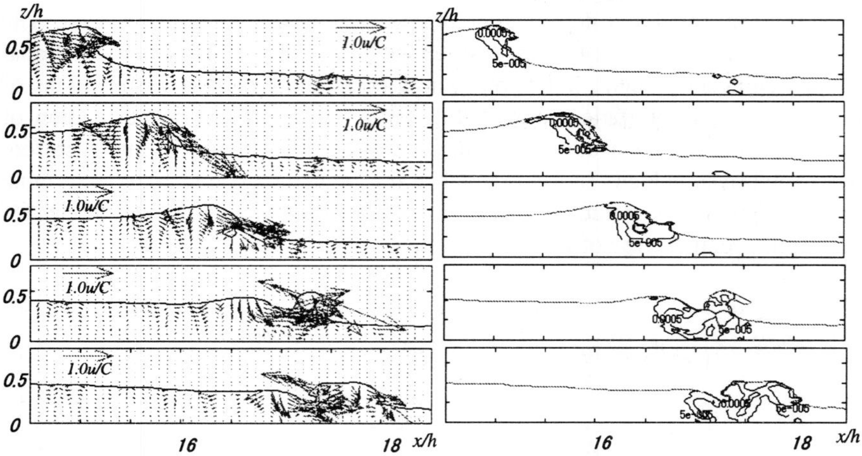

Figure 7. Rotational velocity vector plots (left) and contours of turbulent energy (right) in run2. time interval: T/16 (T: wave period).

Kinetic energy transfer in a surf zone

Figure 3 shows the spacial distributions of depth-averaged kinetic energy for irrotational, rotational and turbulent flow components at successive wave phases in the spilling breaker (run 1). The rotational kinetic energy is rapidly transfered from the potential wave energy after wave breaking, and it tends to gradually decrease as the breaking wave propagates while almost same amount of turbulent energy is consecutively produced. The rotational energy is always less than 10 % of the irrotational energy thoughout the wave propagation, and the turbulenent energy is much less than the rotational energy, indicating the irrotational potential flow governs flow field in the spilling breaking waves and the resulting rotational flows consecutively induce weak quasi-steady turbulence.

In a plunging breaker, the rotational energy begins to be slowly produced after wave breaking, and then it rapidly increases after the wave plunging phase. The cross-shore distributions of the irrotational energy become steep to concentrate at the wave face with the propagation of breaking waves. The intensity of the rotational energy becomes comparable to the irrotational one, unlike a spilling breaker, indicating significant effects of rotational (vortex) flows in the breaking process for plunging waves. The cross-shore distribution of the turbulent energy shows analogous features with one of the rotational kinetic energy, also indicating the rotational energy transfered from the potential wave energy is a major parameter to determine the turbulent energy. An appropriate estimation of the rotational velocity component is necessary for modeling the turbulent energy and also dissipation via the wave breaking event.

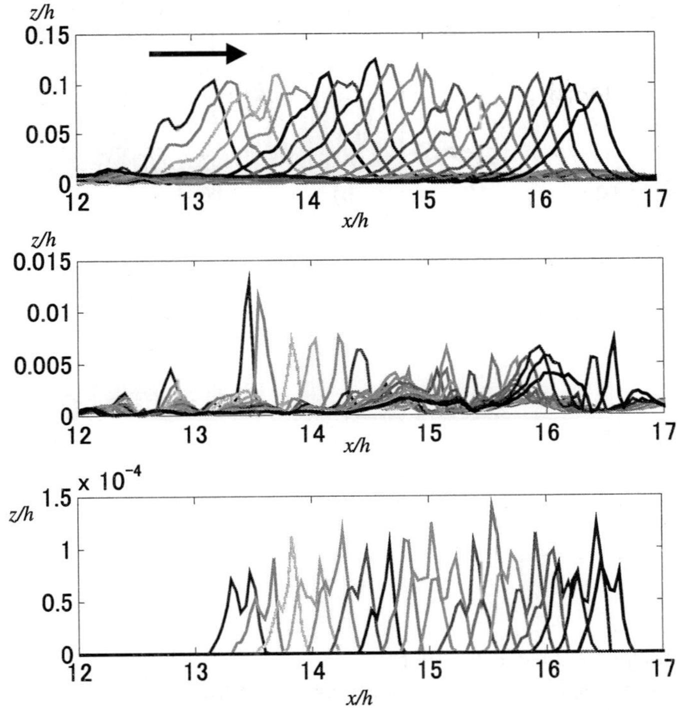

Figure 8. Depth-averaged kinetic energies of irrotational (top), rotational (middle) and turbulent flows(bottom) in run1. time interval: T/16.

CONCLUSIONS

A method to decompose instantaneous velocity field into irrotational, rotational and turbulent components, based on Nowgu's approach for Boussinesq type equation model and computational approach for turbulent flows, has been presented in this study. Transitions from irrotational flows in incident potential waves to rotational (vortex) flows and also turbulence after wave breaking in spilling and plunging breakers have been investigated. It was found there are analogies between rotational and turbulent flows in terms of their distributions and depth-averaged kinetic energy while no correlation between the irrotational and turbulent flows has not been observed. Therefore, the turbulent properties such as turbulent production and dissipation should be modeled by parameters featuring the rotational flows not irrotational flows. An appropriate model for the rotational flows in breaking waves needs to be developed.

ACKNOWLEDGMENTS

Financial support for this study is provided by JSPS Grants-in-Aid Scientific Research (18760368) and a part of this study was done thorough JSPS bilateral

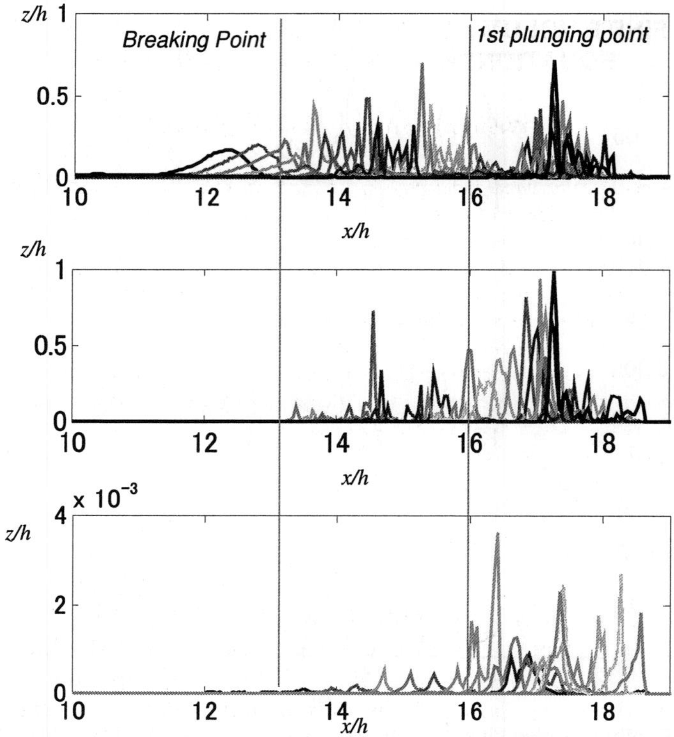

Figure 9. Depth-averaged kinetic energies of irrotational (top), rotational (middle) and

research program.

REFERENCES

Christensen, E. D. and R. Deigaard. 2001. Large eddy simulation of breaking waves, *Coastal Engineering, 42*, 53–86.

Lin, P. and P. L.-F. Liu. 1998. A numerical study of breaking waves in the surf zone, *Journal of Fluid Mechanics, 359*, 239–264.

Nowgu, O.. 1993. Alternative form of boussinesq equations for nearshore wave propagation, *Journal of Waterway, Port, Coastal and Ocean Engineering, 119*, 618–638.

Watanabe, Y. and H. Saeki. 1999. Three-dimensional large eddy simulation of breaking waves, *Coastal Engineering Journal,* 41(3), 281–301.

Watanabe, Y., H. Saeki, and R. J. Hosking. 2005. Three-dimensional vortex structures under breaking waves, *Journal of Fluid Mechanics, 545*, 291–328.

Yakhot, V. and S. A. Orszag. 1986. Renormalization group analysis of turbulence (i. basic theory), *Journal Sci. Comp., 1*, 3–51.

FINITE VOLUME SOLUTION OF BOUSSINESQ-TYPE EQUATIONS ON AN UNSTRUCTURED GRID

Walid El Asmar[1] and Okey Nwogu[1]

A new numerical method is developed to solve a set of two-dimensional Boussinesq water wave evolution equations over an unstructured grid. The governing mass and momentum conservation equations are discretized over an irregular triangular grid, with a staggered placement of the variables. The free surface elevation is defined at the centroid of the triangles, while the normal component of the velocity is defined at the mid-point of the triangle edges. The mass conservation equation is then integrated over a control volume defined over each triangle while the momentum equations are integrated over a control volume formed from two adjacent triangles. A modified Crank-Nicolson scheme is used to integrate the equations in time. Two numerical experiments are used to evaluate the conservation properties and accuracy of the numerical method: solitary wave propagation in a curved channel, and interaction of solitary waves with a vertical circular cylinder.

1. INTRODUCTION

Numerical models that solve Boussinesq-type water wave evolution equations are commonly used to investigate surface wave propagation and transformation in coastal regions. Most of the models use finite difference schemes to discretize the equations over uniformly-spaced rectangular grids (e.g. Abbott *et al.*, 1978; Wei and Kirby, 1995; Nwogu and Demirbilek, 2001). The popularity of finite difference schemes is largely based on their simplicity and ease of implementation. However, the use of structured grids can severely restrict the potential application of such models to complex boundary problems such as coastal flooding over complex topography, wave propagation in curved channels, wave interaction with coastal structures of arbitrary shape, and wave agitation in harbors of arbitrary shape. Unstructured grids provide users the flexibility of modeling complex geometries. In addition, the grid resolution can be refined where needed such as near structures or in shallow regions.

Several techniques have been developed to solve Boussinesq-type equations over irregular shaped domains. Wang *et al.* (1992) and Shi *et al.* (2001) developed curvilinear-grid based finite difference schemes. The curvilinear grid approach offers the simplicity of structured grid schemes while improving their flexibility. However, curvilinear methods might have mapping problems for complex grid geometries. Hence, their application remains limited to computational domains with relatively simple boundaries.

[1] Dept. of Naval Arch. and Marine Eng., University of Michigan, Ann Arbor, MI, 48109-2145, USA

Finite element solution schemes for Boussinesq equations have been developed by Antunes do Carmo *et al.* (1993), Ambrosi and Quartapelle (1998), Walkley and Berzins (2002), Woo and Liu (2004) and others. Compared to structured finite difference schemes, the adoption of an unstructured grid can add considerable difficulty to the discretization of the equations, treatment of the third-order spatial derivative terms that govern the dispersive properties of the equations, and change the conservation properties of the discretized equations. Walkley and Berzins (2002), for example, showed that large local variations in mesh size, or the existence of dominant orientations of the cell faces could have a noticeable impact on the numerical solution.

Classical finite volume schemes that are commonly used to solve the shallow water equations have also been modified to solve Boussinesq-type equations. Soares-Frazão and Zech (2002) proposed a "hybrid finite-volume scheme", where the dispersive terms are considered to be forcing terms and placed on the right-hand of the equations. The finite volume method is used to solve the shallow water part of the equations with the dispersive forcing terms evaluated using a finite difference scheme. Bradford and Sanders (2002) proposed a two-step approach. They initially neglect the dispersive terms, thus solving the shallow water equations. The temporary solutions are then used to estimate the dispersive effects and obtain new values for the main variables. Hybrid finite volume schemes that do not include the dispersive terms as an integral part of the solution are bound to have convergence problems in deeper water as the effects of frequency dispersion become relatively more important.

An important consideration in the development of unstructured grid schemes for Boussinesq-type equations with either finite element or finite volume methods is the stability of the numerical scheme. Placement of the water surface elevation and velocity variables at the element nodes can lead to spurious numerical oscillations that would require filtering and/or other stabilization measures. Spurious numerical oscillations can be avoided by using the staggered mesh concept of Harlow and Welch (1965). Although staggered meshes locally conserve mass, momentum and energy on structured Cartesian grids, local conservation is not necessarily guaranteed for unstructured grids. Perot (2000), however, demonstrated that it is feasible to achieve local conservation of mass, momentum, kinetic energy and circulation with certain staggered-mesh unstructured grids. His approach relies on the concept of the dual orthogonal mesh (Delaunay/Voronoi tessellation). A staggered-grid finite volume method for solving the Euler equations that does not rely on creating a dual Delaunay-Voronoi mesh was proposed by Wenneker *et al.* (2002). Wenneker *et al.* (2003) later demonstrated that the scheme also conserves mass, momentum and energy.

In this paper, a consistent finite volume method is developed to solve a set of two-dimensional Boussinesq-type water wave evolution equations on an irregular triangular grid using an approach similar to that of Wenneker *et al.* (2002). In contrast to previous finite element and finite volume Boussinesq solvers where both the free surface elevation and two components of the horizontal velocities are co-located at the nodes of the elements, we adopt a staggered placement of the variables. The mass conservation equation is integrated over a control volume defined over each triangle while the momentum equations are integrated over a control volume formed from two adjacent triangles. A reconstruction scheme is used to retrieve the full velocity vector at the vertices of the triangles. The numerical model is used to investigate the propagation of solitary waves in a curved channel, and interaction of solitary waves with a vertical circular cylinder.

2. NUMERICAL METHOD

Governing Equations

The governing equations are the weakly-nonlinear set of Boussinesq-type mass and momentum conservation equations derived by Nwogu (1993). The depth-integrated mass conservation equation can be written as:

$$\eta_t + \nabla \cdot \boldsymbol{Q} = 0 \tag{1}$$

where $\nabla = (\partial/\partial x, \partial/\partial y)$, $\eta(\boldsymbol{x},t)$ is the water surface elevation and $\boldsymbol{Q}(\boldsymbol{x},t)$ is the volume flux density given by:

$$\boldsymbol{Q} = \int_{-h}^{\eta} \boldsymbol{u}\, dz = (h+\eta)\boldsymbol{u}_\alpha + h\left\{ \left[\frac{(z_\alpha + h)^2}{2} - \frac{h^2}{6} \right] \nabla(\nabla \cdot \boldsymbol{u}_\alpha) + \left[(z_\alpha + h) - \frac{h}{2} \right] \left[\nabla(\boldsymbol{u}_\alpha \cdot \nabla h) + (\nabla \cdot \boldsymbol{u}_\alpha)\nabla h \right] \right\} \tag{2}$$

where h is the water depth, $\boldsymbol{u}_\alpha$ is the horizontal velocity at a reference elevation $z = z_\alpha$ in the water column. The momentum equation is given by:

$$\boldsymbol{u}_{\alpha,t} + g\nabla\eta + (\boldsymbol{u}_\alpha \cdot \nabla)\boldsymbol{u}_\alpha + z_\alpha\left[\nabla(\boldsymbol{u}_{\alpha,t} \cdot \nabla h) + (\nabla \cdot \boldsymbol{u}_{\alpha,t})\nabla h\right] + \frac{1}{2}\left[(z_\alpha + h)^2 - h^2\right]\nabla(\nabla \cdot \boldsymbol{u}_{\alpha,t}) = 0 \tag{3}$$

where g is the gravitational acceleration. The elevation of the velocity variable z_α is chosen to minimize differences between the linear dispersion characteristics of the model and the exact dispersion relation for small amplitude waves and is given by $z_\alpha = -0.535h$ (Nwogu, 1993).

Grid Topology and Placement of Variables

A staggered placement of the variables is adopted as shown in Fig. 1. The water surface elevation is defined at the centroids of the triangles, while the normal component of the velocity is defined at the midpoints of the triangle edges. The velocities at the cell faces are assumed to be continuous in contrast to standard shallow water finite volume schemes that allow for discontinuous velocities at cell faces.

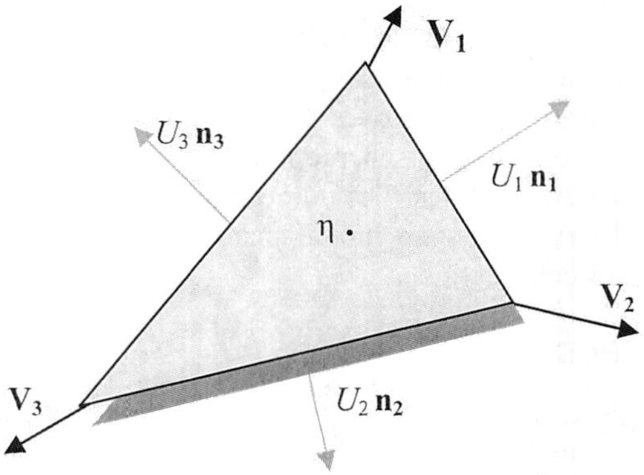

Figure 1. Staggered placement of variables on a triangular grid.

Control Volume Integrated Equations

The control volume used for the mass conservation equation consists of a single triangular cell. The mass equation (Eq. 1) is integrated over the control volume to yield:

$$\frac{\partial}{\partial t}\int_{\Omega_\beta}\eta\, d\Omega = -\int_{\Omega_\beta}\nabla\cdot \boldsymbol{Q}\, d\Omega \tag{4}$$

where Ω_β is the surface area of cell β. The control volume for the momentum equation is composed of two adjacent grid cells (see Fig. 2). The momentum equation is initially projected in the direction normal to the common edge, and then integrated over the combined control volume area. The control volume-integrated momentum equation for water of constant depth is given by:

$$\begin{aligned}&\frac{\partial}{\partial t}\left(\int_{\Omega_E}\boldsymbol{u}_\alpha\cdot \boldsymbol{N}_{\mathrm{E}}d\Omega + \alpha h^2\int_{\Omega_E}\nabla(\nabla\cdot\boldsymbol{u}_\alpha)\cdot\boldsymbol{N}_{\mathrm{E}}d\Omega\right)\\ &-g\int_{\Omega_E}\nabla\eta\cdot\boldsymbol{N}_{\mathrm{E}}d\Omega - \int_{\Omega_E}\nabla(\boldsymbol{u}_\alpha\cdot\boldsymbol{u}_\alpha)\cdot\boldsymbol{N}_{\mathrm{E}}d\Omega \;=\; 0\end{aligned} \tag{5}$$

where β and γ denote adjacent cells, α = -0.392, $\boldsymbol{N}_{\mathrm{E}}$ is the unit normal vector to common edge, and $\Omega_E = \Omega_\beta + \Omega_\gamma$ is the surface area of the control volume. The advection term in Eq. (5) has been modified by assuming the flow to be irrotational in the horizontal plane, i.e. $(\boldsymbol{u}_\alpha\cdot\nabla)\boldsymbol{u}_\alpha = \nabla(\boldsymbol{u}_\alpha\cdot\boldsymbol{u}_\alpha)$. The projected momentum equation is independent of the choice of direction for $\boldsymbol{N}_{\mathrm{E}}$. A simple convention used in this paper is to define a unique direction for $\boldsymbol{N}_{\mathrm{E}}$ by orienting all unit normal vectors away from the origin of the coordinate system.

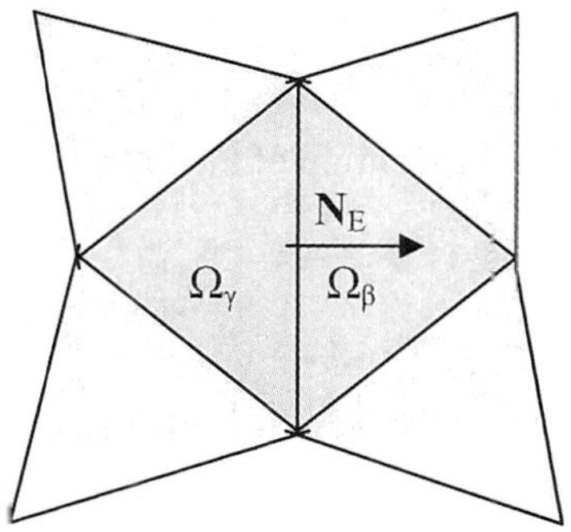

Figure 2. Control volume of integrated momentum equation.

Evaluation of Area Integrals

In order to numerically evaluate the mass conservation equation (Eq. 4), the divergence theorem is used to convert the area integral of the volume flux term into a line integral along the triangle edges. The water surface elevation is assumed to be constant over the area of the cell and defined at the centroid of the cell, while the flux terms are assumed to be constant along the cell edges and defined at the mid-point of the cell edges. The mass conservation equation then becomes:

$$\Omega_\beta \frac{\partial \eta_\beta}{\partial t} = -\sum_{i=1}^{3} l_i \left((h+\eta)_i U_i - \left(\alpha + \frac{1}{3} \right) h_i^3 \left[\nabla(\nabla \cdot \boldsymbol{u}_\alpha) \cdot \boldsymbol{n}_i \right] \right) \tag{6}$$

where η_β is the water surface elevation at the centroid of cell β, l_i are the lengths of the sides of cell i, $\boldsymbol{n}_i$ are their respective outward unit normal vectors, and $U_i = \boldsymbol{u}_\alpha \cdot \boldsymbol{n}_i$ is the outward normal component of the velocity $\boldsymbol{u}_\alpha$ at the midpoint of cell edge i.

The values of η at the cell edges are estimated using a weighted average of the values of η at the centroids of neighboring cells:

$$\eta_i = \psi_\beta \eta_\beta + \psi_\gamma \eta_\gamma \tag{7}$$

The weighting coefficients ψ_β, ψ_γ are calculated based on the relative distance of the centroids $[(x_\beta, y_\beta), (x_\gamma, y_\gamma)]$ from the midpoint of the cell edge:

$$\begin{aligned} \psi_\beta &= \frac{x_\gamma (x_\gamma - x_\beta) + y_\gamma (y_\gamma - y_\beta)}{(x_\gamma - x_\beta)^2 + (y_\gamma - y_\beta)^2 + (x_\beta y_\gamma - x_\gamma y_\beta)^2} \\ \psi_\gamma &= \frac{x_\beta (x_\beta - x_\gamma) + y_1 (y_1 - y_\gamma)}{(x_\beta - x_\gamma)^2 + (y_\beta - y_\gamma)^2 + (x_\gamma y_3 - x_\beta y_\gamma)^2} \end{aligned} \tag{8}$$

The second term on the right hand side of Eq. (6) represents the dispersive term in the mass conservation equation. It requires the evaluation of the projected value of the gradient of the divergence of the velocity vector $(\nabla(\nabla \cdot \boldsymbol{u}_\alpha))$ along the cell edges. The divergence theorem is initially used to approximate $\nabla \cdot \boldsymbol{u}_\alpha$ as a scalar quantity defined at the centroid of the cells:

$$\nabla \cdot \boldsymbol{u}_\alpha = \frac{1}{\Omega}\sum_{i=1}^{3} l_i U_i \tag{9}$$

By making use of the vector identity $\nabla f \cdot \boldsymbol{A} = \nabla \cdot (f\boldsymbol{A}) - f\,\nabla \cdot \boldsymbol{A}$ and the divergence theorem, the dispersive term is then evaluated as:

$$\nabla(\nabla \cdot \boldsymbol{u}_\alpha) \cdot \boldsymbol{n}_i = \frac{1}{\Omega}\sum_{j=1}^{4} l_j (\nabla \cdot \boldsymbol{u}_\alpha)_j (\boldsymbol{n}_j \cdot \boldsymbol{n}_i) \tag{10}$$

The values of $\nabla \cdot \boldsymbol{u}_\alpha$ along the cell edges are obtained from the centroidal values using the scalar reconstruction scheme given in Eq. (7).

The divergence theorem is also used to transform the momentum equation:

$$\Omega_E \frac{\partial}{\partial t} U_E + \alpha h^2 \sum_{i=1}^{4} l_i \frac{\partial}{\partial t}(\nabla \cdot \boldsymbol{u}_\alpha)_i (\boldsymbol{N}_E \cdot \boldsymbol{n}_i) = -\sum_{i=1}^{4} l_i \left[g\eta_i + (\boldsymbol{u}_\alpha \cdot \boldsymbol{u}_\alpha)_i\right](\boldsymbol{N}_E \cdot \boldsymbol{n}_i) \tag{11}$$

It should be noted that the dispersive term in the momentum equation (second term on the left hand side of Eq. 11) can yield up to thirteen normal velocity components corresponding to five edges of the two adjacent cells of the surface of integration, and eight additional edges from cells adjoining the surface of integration.

The velocity-squared term at the mid-point of the cell edges $(\boldsymbol{u}_\alpha \cdot \boldsymbol{u}_\alpha)$ is approximated as the average of the values at the neighboring nodes. The full velocity vectors at the nodes are reconstructed from the normal components of the velocity at the midpoints of all the edges that connect to the node using the least-squares minimization scheme of Dukowicz and Meltz (1992). The scheme minimizes the difference, *f*, between the projected nodal velocity $\boldsymbol{u}$ in the normal direction to the face and the actual normal velocity at the face:

$$f = \sum_{i=1}^{k} \psi_i \left[\boldsymbol{u} \cdot \boldsymbol{n}_i - U_i\right]^2 \tag{12}$$

where ψ_i are empirically assigned weighting coefficients and k is the number of edges that connect to the node. If all weighting coefficients are assigned a value of one, the least-squares solution is given by:

$$\boldsymbol{u} = \frac{1}{4ab - c^2}\begin{bmatrix} ce - 2bd \\ cd - 2ae \end{bmatrix} \tag{13}$$

where $a = \sum_{i=1}^{k} \psi_i (n_i^x)^2 \quad b = \sum_{i=1}^{k} \psi_i (n_i^y)^2 \quad c = \sum_{i=1}^{k} 2\psi_i n_i^x n_i^y$

$$d = -\sum_{i=1}^{k} 2\psi_i n_i^x U_i \quad e = -\sum_{i=1}^{k} 2\psi_i n_i^y U_i$$

The ψ coefficients can also be used to enforce Dirichlet boundary conditions at boundary nodes. This is done, for example, by assigning the coefficients of the inner faces a value of one, and the coefficients of the boundary faces a value much greater than one.

Time Integration

The mass and momentum equations are integrated in time using an iterative Crank-Nicolson scheme with a predictor-corrector scheme used to provide an initial guess. The scheme consists of three stages. At every time step $t = n\Delta t$, a predictor scheme is used to calculate the values of the variables at $t = (n+1/2)\Delta t$. These values are used in the following corrector stage to calculate the values of the variables at $t = (n+1)\Delta t$; which are then used as an initial guess in an iterative Crank-Nicholson scheme. A summary of the approach is provided in Table 1 below.

Table 1. Summary of Crank-Nicolson scheme.

	Water Surface Elevation	Normal Component of Velocity
Predictor	$\underline{\eta}^{n+\frac{1}{2}} = \underline{\eta}^{n} + \frac{\Delta t}{2} f_c\left(\underline{\eta}^{n}, \underline{U}^{n}\right)$	$\underline{U}^{n+\frac{1}{2}} = \underline{U}^{n} + \frac{\Delta t}{2} A^{-1} f_m\left(\underline{\eta}^{n}, \underline{U}^{n}\right)$
Corrector	$\underline{\eta}^{n+1} = \underline{\eta}^{n} + \Delta t f_c\left(\underline{\eta}^{n+\frac{1}{2}}, \underline{U}^{n+\frac{1}{2}}\right)$	$\underline{U}^{n+1} = \underline{U}^{n} + \Delta t A^{-1} f_m\left(\underline{\eta}^{n+\frac{1}{2}}, \underline{U}^{n+\frac{1}{2}}\right)$
Iterative	$\underline{\eta}^{n+1} = \underline{\eta}^{n} + \Delta t f_c\left(\underline{\eta}^{n+\frac{1}{2}}, \underline{U}^{n+\frac{1}{2}}\right)$	$\underline{U}^{n+1} = \underline{U}^{n} + \Delta t A^{-1} f_m\left(\underline{\eta}^{n+\frac{1}{2}}, \underline{U}^{n+\frac{1}{2}}\right)$
	where $\underline{\eta}^{n+\frac{1}{2}} = \frac{1}{2}\left(\underline{\eta}^{n} + \underline{\eta}^{n+1}\right)$,	$\underline{U}^{n+\frac{1}{2}} = \frac{1}{2}\left(\underline{U}^{n} + \underline{U}^{n+1}\right)$

The matrix A is constructed from the left hand side of the discretized momentum equation. The matrix is relatively sparse with each row containing a maximum of 13 non-zero entries. Two sparse matrix solvers were implemented, the Generalized Minimum Residual (GMRES) method and the Successive Overrelaxation (SOR) scheme.

3. NUMERICAL RESULTS

Propagation of Solitary Waves in Curved Channels

Shi *et al.* (1998) investigated the propagation of solitary waves through curved channels using a curvilinear finite difference method. Their work focused on the effect of channel width and bending sharpness on the reflected and transmitted waves. We consider one of Shi *et al.*'s numerical experiments involving a curved channel with a smooth 90° bend. The channel is $5h$ wide and has an inner radius of $10h$ at the bend. The channel's upstream and downstream legs are both $100h$ long.

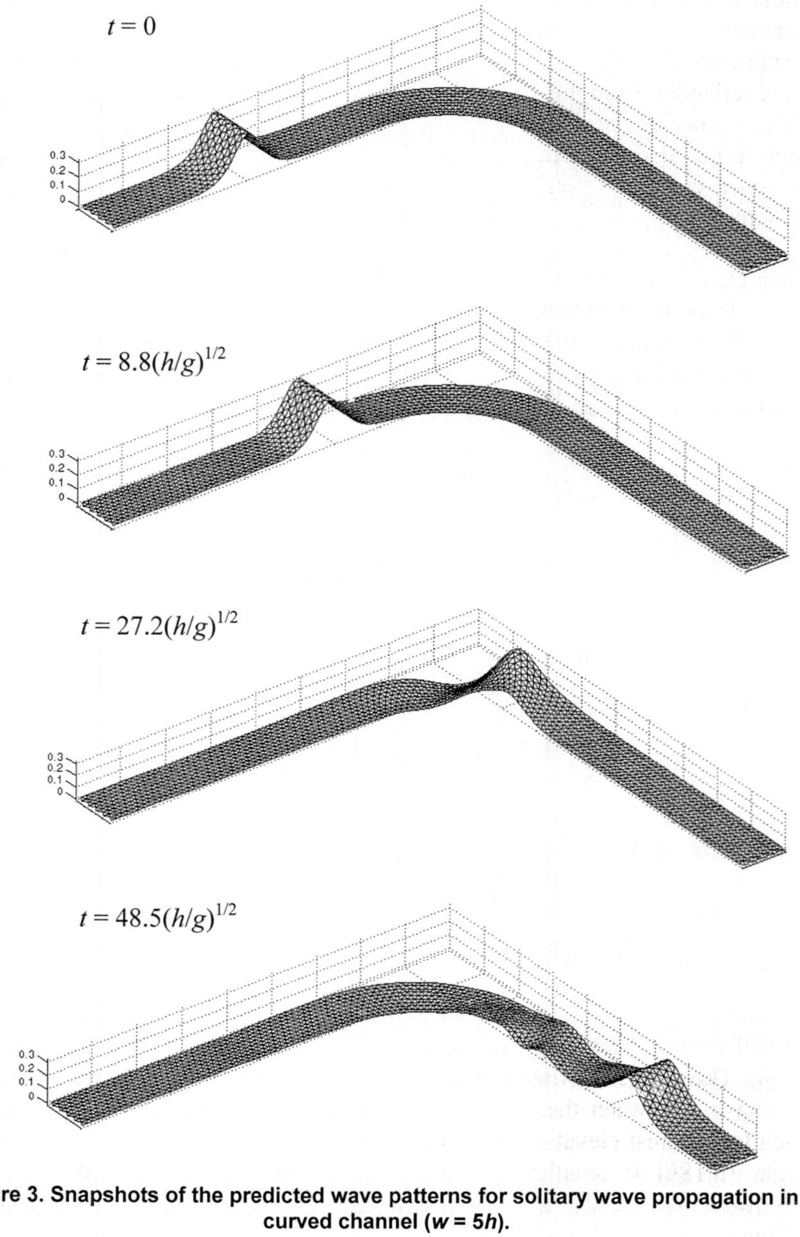

Figure 3. Snapshots of the predicted wave patterns for solitary wave propagation in a curved channel ($w = 5h$).

The computational domain was discretized into irregular triangular cells with side lengths $\Delta l \approx h$. A solitary wave of height H=0.3h was initially placed at a distance x = 85h from the upstream edge of the channel. The solitary wave profile was obtained from a Fourier series solution of the Boussinesq equations. Numerical simulations were performed with time step size $\Delta t = 0.05(h/g)^{1/2}$ for a duration of $48.5(h/g)^{1/2}$. The predicted wave patterns are shown in Figure 3. Wave reflection from the outer wall of the bend leads to a larger wave elevation in that corner as the waves propagate around the bend. The wave front in the outer corner also accelerates, eventually catching up with the slower waves in the inner corner to form a uniform crest in the downstream section. The predicted wave patterns are similar to those presented in Shi *et al.* (1998).

The transmitted wave profile at $t = 48.5(h/g)^{1/2}$ is shown in Figure 4. The transmitted wave had an amplitude of 0.259h with its crest is located at x = 133h. Shi *et al.* predicted an amplitude of 0.276h with the crest located at x = 136h. Both results are comparable, given the different form of Boussinesq equations employed by Shi *et al.* (1998), as well as differences in the numerical scheme.

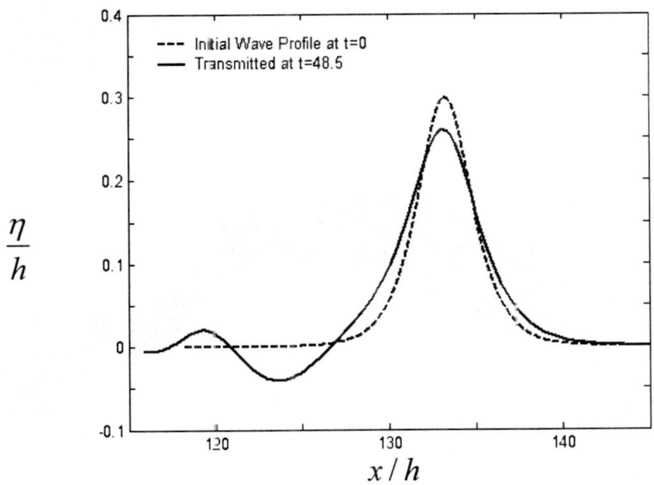

Figure 4. Initial and transmitted wave profile at $t = 48.5(h/g)^{1/2}$.

The simulations were repeated for a wider channel with a width of 10h. The channel width is now comparable to the solitary wavelength ($\lambda = 2\pi/(3H/4h^3)^{1/2} \approx 13h$). The predicted wave patterns are shown in Fig. 5. For the wider channel, the phasing between the incident and reflected waves at the bend leads to a much larger crest elevation (~ 1.5H) at the outer corner. The transmitted wave height (0.18h) is smaller compared to the narrower channel results. The transmitted wave profile is also no longer uniform across the crest and has larger trailing waves.

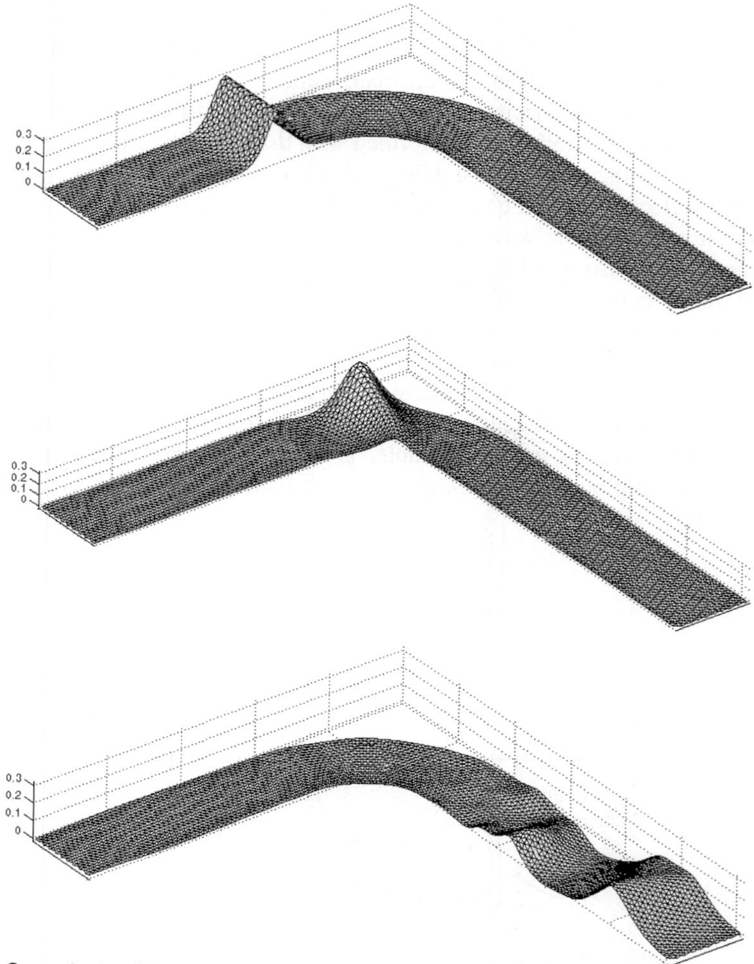

Figure 5. Snapshots of the predicted wave patterns for solitary wave propagation in a curved channel (w = 10h).

Solitary Wave Interaction with a Vertical Circular Cylinder

Antunes do Carmo *et al.* (1993) conducted laboratory experiments to investigate the interaction of a solitary wave with a vertical cylinder. The wave flume was 9 m long, 0.55 m wide and had a water depth of 0.15m. A 0.16m-diameter circular cylinder was installed in the middle of the flume. The center of the cylinder was located at x = 4.5m, y = 0.275m relative to an origin defined at the lower corner of the upstream end of the flume. Six wave probes were used to measure the water surface elevation at the following locations: Gauge 1 = (4.400, 0.275), Gauge 2 = (4.500, 0.170), Gauge 3 = (4.500, 0.045), Gauge 4 = (4.600, 0.275), Gauge 5 = (4.975, 0.275), and Gauge 6 = (5.375, 0.275).

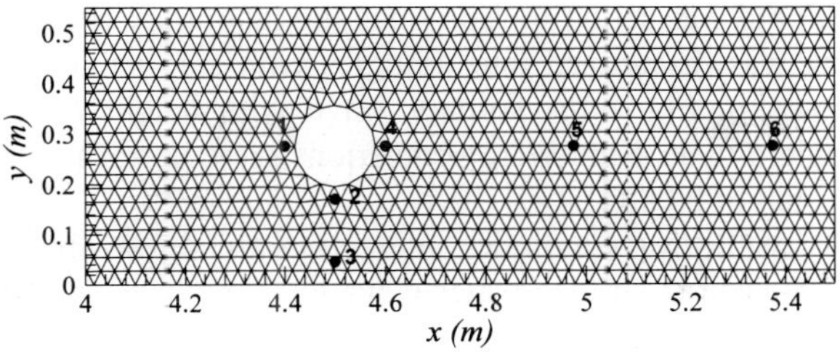

Figure 6. Triangular mesh used for solitary wave-cylinder interaction test.

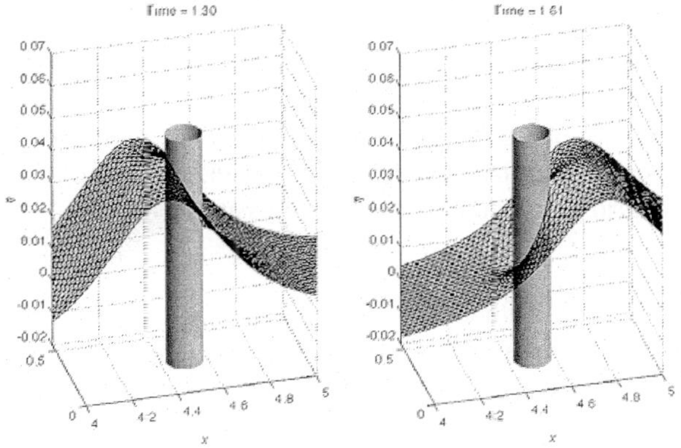

Figure 7. 3-D view of solitary wave interaction with cylinder.

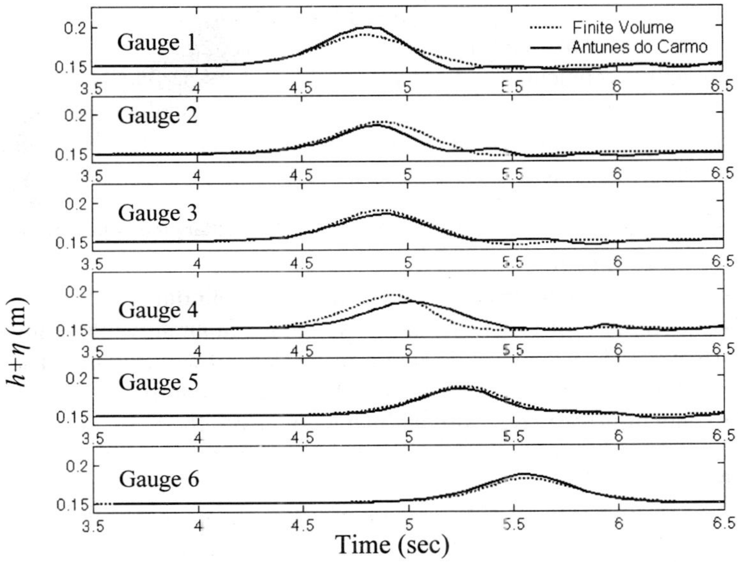

Figure 8. Measured and predicted time histories for solitary wave-cylinder interaction test.

A numerical flume was set up to reproduce the laboratory experiments. The 9m by 0.55m tank was discretized into triangles with side lengths of the order of 0.03m as shown in Fig. 6. A solitary wave with height H = 0.0375m was generated inside a tank. Numerical simulations were performed for duration of 6.5s at a time step of 0.008s.

Snapshots of the wave-cylinder interaction are shown in Fig. 7 while the predicted and measured surface water elevation time histories are compared in Fig. 8. Reasonable agreement is observed between the numerical and experimental results at Gauges 3, 5 and 6, which are located farthest from the cylinder. Greater discrepancies are seen at Gauges 1, 2 and 4, which are in the immediate vicinity of the cylinder. Antunes do Carmo *et al.* (1993) obtained similar amplitude and phase shifts with their finite element Boussinesq model. Wu (2004) obtained better model-data comparisons with a Navier-Stokes solver. We suspect that the phase differences are related to numerical and analytical errors in the weakly nonlinear model phase speeds and plan to repeat the tests on a finer mesh grid with a fully nonlinear Boussinesq solver.

4. CONCLUSIONS

A new numerical method has been developed to solve Boussinesq-type water wave evolution equations over an unstructured grid. The scheme is based on the finite volume concept where the governing mass and momentum equations are integrated over local control volumes in order to locally satisfy the conservation laws. A staggered placement of the surface elevation and velocity variables is adopted to avoid spurious numerical oscillations and ensure stability. The scheme solves for the normal velocities along cell edges, as opposed to the two velocity components, making it relatively more computationally efficient. Preliminary numerical experiments have been conducted for solitary wave propagation in a curved channel and solitary wave interaction with a circular cylinder. The tests have demonstrated the viability of the staggered unstructured grid approach. Additional work is currently being performed to extend the model to the fully nonlinear Boussinesq equations.

ACKNOWLEDGMENTS

This work was supported in part by Michigan Sea Grant College Program under Grant No. NA16RG1145 from the National Sea Grant College program, and the Coastal Inlets Research Program of the U.S. Army Engineer Research and Development Center. The authors would like to thank Tso-Ren Wu for providing data from the wave-cylinder interaction experiment.

REFERENCES

Abbott, M. B., Petersen, H. M., and O. Skovgaard. 1978. On the numerical modelling of short waves in shallow water, *J. Hyd. Res.*, **16**, 173-203.

Ambrosi D., and L. Quartapelle. 1998. A Taylor-Galerkin method for simulating nonlinear dispersive water waves, *J. Comp. Phys.*, **146**, 546-569.

Antunes do Carmo, J. S., Seabra-Santos, F. J. and E. Barthélemy. 1993. Surface waves propagation in shallow-water: A finite element model, *Int. J. Num. Methods Fluids*, **16**, 447 – 459.

Bradford, S. F., and B. F. Sanders. 2002. Finite-volume models for unidirectional nonlinear dispersive waves, *ASCE Journal of Waterway, Port, Coastal and Ocean Engineering*, **128**, 173-182.

Dukowicz, J. K., and B. J. A. Meltz. 1992, Vorticity errors in multidimensional Lagrangian codes, *J Comp. Phys.*, **99**, 115-134.

Harlow, F. H., and J. E. Welch. 1965. Numerical calculations of time dependent viscous incompressible flow of fluid with a free surface, *Physics of Fluids*, **8**, 2182-2189.

Nwogu, O. 1993. Alternative form of Boussinesq equations for nearshore wave propagation, *ASCE Journal of Waterway, Port, Coastal and Ocean Engineering*, **119**, 618-638.

Nwogu, O.G., and Demirbilek, Z. 2001. BOUSS-2D: A Boussinesq Wave Model for Coastal Regions and Harbors. ERDC/CHL Technical Report TR-01-25, U.S. Army Engineer Res. and Dev. Center, Vicksburg, MS.

Perot, B. 2000. Conservation properties of unstructured staggered mesh schemes, *J. Comp. Phys*, **159**, 58–89.

Shi, A., Teng, M.H. and T.Y. Wu. 1998. Propagation of solitary waves through significantly curved shallow water channels, *J. Fluid Mech.*, **362**, 157-176.

Shi, F., Dalrymple, R.A., Kirby, J.T., Chen, Q., and A. Kennedy. 2001. A fully nonlinear Boussinesq model in generalized curvilinear coordinates, *Coastal Engineering*, **42**, 337-358.

Soares-Frazão, S., and Y. Zech. 2002. Undular bores and secondary waves, experiments and hybrid finite-volume modeling, *J. Hyd. Res.*, **40**, 33-43.

Walkley, M., and M. Berzins. 2002. A finite element model for the two-dimensional extended Boussinesq equations, *Int. J. Num. Methods Fluids*, **39**, 865-885.

Wang, K. H., Wu, T. Y., and G. T. Yates. 1992. Three-dimensional scattering of solitary waves by vertical cylinder, *J. Waterway, Port, Coastal and Ocean Eng.*, **118**, 551-566.

Wei, G., and J.T. Kirby. 1995. Time-dependent numerical code for extended Boussinesq equations, *J. Waterway, Port, Coastal and Ocean Eng.*, **121**, 251-261.

Wenneker, I., Segal, A., and P. Wesseling. 2002. A Mach-uniform unstructured staggered grid method, *Int. J. Num. Methods Fluids*, **39**, 1209-1235.

Wenneker, I., Segal, A., and P. Wesseling. 2003. Conservation properties of a new unstructured staggered scheme, *Computers and Fluids,* **32**, 139-147.

Woo, S.-B., and P.L.-F. Liu. 2001. A Petrov-Galerkin finite element model for one-dimensional fully nonlinear and weakly dispersive wave propagation, *Int. J. Num. Methods Fluid*, **37**, 541-575.

Wu, T.-S. 2004. A numerical study of three-dimensional breaking waves and turbulence effects, Ph.D. Dissertation, Cornell University, 265 pp.

IMPROVEMENTS TO SHALLOW WATER FREQUENCY DOMAIN MODELS

J. Veeramony[1], J. M. Kaihatu[2], and K. L. Edwards[3]

Frequency domain representations of the non-linear Boussinesq equations are being increasingly used to predict wave transformation in the nearshore region. Even after implementing improved dispersion and shoaling characteristics in these models, the prediction of third moments such as skewness and asymmetry are not significantly improved. Improvement of these quantities are especially important when the wave information is used to drive sediment transport models. In this article, other options to improve the third moments are explored. Specifically, three different options are studied. The first option is to retain all the non-linear terms, the second option is to solve both the continuity and momentum equation separately, and the third option is to use a different wave breaking mechanism. The first two options improved predictions in different parts of the domain whereas the third option did not significantly improve predictions in the surf zone.

INTRODUCTION

In recent years, much effort has been focused towards improving frequency domain models in shallow water, especially those derived from Boussinesq equations. In the time domain, significant improvements have been made to the classical Boussinesq equations of Peregrine (1967). The equations which were valid for only a fairly narrow region of shoaling were extended to deeper water by improving the linear dispersion characteristics (Nwogu, 1993) as well as by retaining higher order dispersive terms (Madsen & Schäffer, 1998) and closer to the breaking region by retaining higher order non-linear terms (Wei et al., 1995; Madsen & Schäffer, 1998).

Correspondingly, the frequency domain equations of Freilich & Guza (1984), derived from the equations of Peregrine (1967) have undergone significant improvements. The frequency domain model derived directly from the time-domain equations exhibited significantly poorer dispersion properties compared to the equivalent time-domain formulations. Chen & Liu (1995) showed that the dispersion properties can be improved by using the velocity potential formulation of the equations as the starting point. On the other hand, the linear shoaling properties are shown to be worse. Kaihatu & Kirby (1998) added a linear shoaling correction factor to alleviate this problem, but it did not address the poor predictions of third-order moments such as skewness and asymmetry.

In this article, we investigate three different options to improve predictions from frequency domain formulations of the Boussinesq equation which are described in the following section. In the next section, we present the results from the model and compare the relative merits of the three options. In the last section, we present our conclusions and recommendations.

[1]Jacobs Engineering Inc., C125, Building 1009 Stennis Space Center, MS 39529
[2]Zachry Dept. of Civil Engng, Texas A&M University, 3136 TAMU, College Station, TX 77843
[3]Naval Research Laboratory, C126, Building 1009, Stennis Space Center, MS 39529

MODEL FORMULATIONS

Frequency domain formulations of the Boussinesq model in one-dimension starts with the weakly non-linear two-equation model given by the continuity and momentum equations

$$\eta_t + [(h+\eta)U]_x = 0 \tag{1}$$

$$U_t + g\eta_x + UU_x + B_1 gh^2\eta_{xxx} + \left(B_1 - \frac{1}{3}\right)h^2 U_{xxt} - hh_x U_{xt} = O(\delta^2, \delta\mu^2, \mu^4) \tag{2}$$

where η is the water surface elevation measured from the still water line, U depth-averaged velocity, g is the acceleration due to gravity, h is the local water depth, $\delta = a/h$ is the non-dimensional parameter that describes the relative amplitude and is a measure of the non-linearity, $\mu = kh$ is the non-dimensional parameter that describes the relative wave length and is a measure of wave dispersion, a is the wave amplitude, k is the wave number, and B_1 is the free parameter used to improve the dispersion of the equations. Following the procedure outlined in Kaihatu & Kirby (1998), we get the one-dimensional wave equation in η

$$\begin{aligned}\eta_{tt} - g(h\eta_x)_x - (hUU_x)_x - B_1 gh^3\eta_{xxxx} - 3B_1 gh^2 h_x\eta_{xxx} \\ + \left(B_1 - \frac{1}{3}\right)h^2\eta_{xxtt} - hh_x\eta_{xtt} + (\eta U)_{xt} = 0\end{aligned} \tag{3}$$

where the velocity is eliminated in all the linear terms. Now, we assume that η can be represented by a Fourier series

$$\eta = \sum_{n=1}^{N} A_n e^{i(\int k_n \, \mathrm{dx} - \omega_n t)} \tag{4}$$

which when substituted in the linear terms of the wave equation gives

$$\sum_{n=1}^{N} (i\beta_1 A_{nx} + \beta_2 A_n + i\beta_3 h_x A_n + i\beta_4 k_x A_n)\, e^{i\phi_n} = \Lambda, \tag{5}$$

where

$$\phi_n = \int k_n \mathrm{dx} - \omega_n t \tag{6}$$

$$\beta_1 = 2ghk_n - 4Bgh^3k_n^3 + 2\left(B_1 - \frac{1}{3}\right)\omega_n^2 k_n h^2, \tag{7}$$

$$\beta_2 = ghk_n^2 - \omega_n^2 + \left(B_1 - \frac{1}{3}\right)\omega_n^2 k_n^2 h^2 - B_1 gh^3 k_n^4, \tag{8}$$

$$\beta_3 = -\omega_n^2 k_n h + gk_n - 2B_1 gh^2 k_n^3, \tag{9}$$

$$\beta_4 = \left(B_1 - \frac{1}{3}\right)\omega_n^2 h^2 + gh - 6B_1 gh^3 k_n^2. \tag{10}$$

The right hand side of (5) contain the non-linear terms. For the weakly non-linear model, these terms are usually reduced assuming triad interactions dominate the energy transfer between harmonics. An equivalent method to calculate the terms was presented by (Bredmose, Schäffer & Madsen, 2004) wherein the non-linear terms are expanded using the chain rule and the derivatives calculated in frequency space. For example, to calculate $\eta_x U$, we have

$$\eta_x = \sum_{n=1}^{N} -ikA_n e^{i(\int k\,\mathrm{dx}-\omega t)} \tag{11}$$

The spatial derivative can then be transformed into time series applying the inverse FFT to the right hand side of (11), and similarly the time series of U can be calculated and the term $\eta_x U$ is obtained in the time domain by simple multiplication of the time series. The Fourier amplitudes of the product is then calculated using the forward FFT. As shown by (Bredmose et al., 2004), this procedure also yields considerable savings in computational time compared to the traditional method of reducing the double sums using triad interactions.

If, on the other hand, we start from the fully non-linear (up to $O(\mu^2)$) one-dimensional equations given by (Madsen & Schäffer, 1998), then the resulting wave equation for η is similar to (5), only now Λ represents all terms of higher non-linearity ($O(\delta)$ and higher). Using the traditional method of triad interactions to calculate the non-linear terms becomes computationally prohibitive. However, the technique demonstrated by Bredmose et al. (2004) can easily be extended to calculate the additional non-linear terms without excessive computational load. Because the additional non-linearity has been shown to significantly improve predictions of wave height, asymmetry and skewness (Wei et al., 1995; Madsen & Schäffer, 1998) in the near-breaking region for the time-domain model, it is expected that the frequency domain model will yield similar improvements.

Going back to the derivation of the weakly non-linear model, we see that there is considerable ambiguity when combining the continuity and momentum equation to get the wave equation for η. This ambiguity results in decreased accuracy for amplitude dispersion and linear shoaling. The linear shoaling is improved by adding the linear wave equation factored by an arbitrary parameter which is then calibrated to provide the best match to linear shoaling. These manipulations can be avoided by keeping the two equations separate and solving for η and U separately. This means that the velocity U is also expanded as a Fourier series

$$U = \sum_{n=1}^{N} B_n e^{i(\int k\,\mathrm{dx}-\omega t)} \tag{12}$$

and substituted into the continuity and momentum equation along with the expression for η given by (4) and we get two equations in the two unknowns A_n and B_n

$$A_{nx} = G_1(A_n, B_n, h, k) \tag{13}$$
$$B_{nx} = G_2(A_n, B_n, h, k) \tag{14}$$

which are used to solve for A_n and B_n.

The last option investigated in this article to improve the predictions of third moments uses a different wave breaking formulation. In the model of Kaihatu & Kirby (1998), the implementation of energy dissipation due to wave breaking follows that of Mase & Kirby (1992), which is based on the bulk energy dissipation model of Thornton & Guza (1983). In this form, the breaking term is added to the wave equation as

$$i\beta_1 A_{nx} + \beta_2 A_n + +i\beta_3 h_x A_n + i\beta_4 k_x A_n + (F\alpha_0 + (1-F)\alpha_1)A_n = O(\delta) \tag{15}$$

where the value of F, which ranges from 0 to 1, determines whether the dissipation is frequency independent ($F = 0$) or has only frequency-squared dependence ($F = 1$). We will compare the results using this formulation to that of the eddy viscosity form described by Kennedy et al. (2000). This breaking term appears as an additional term in the momentum equation in the time domain as

$$U_t + UU_x + g\eta_x + (\nu_b U_x)_x = O(\mu^2, \delta\mu^2). \tag{16}$$

MODEL COMPARISONS

First, the fully non-linear model and the two-equation model are compared to the weakly non-linear form for the case of harmonic generation. The comparison between the one- and two-equation models for the weakly non-linear form was described in detail by Bredmose (2002) and is not repeated here. The models are tested against the data of Chapalain et al. (1992), Case A. A sinusoidal wave of period $T = 2.5s$, amplitude $A_1 = 0.042m$ was generated at the wave maker in a tank with constant depth of $h = 0.4m$. Energy is transferred between the primary wave component and its harmonics. The amplitudes of the waves were measured at different locations in the tank. Figure 1 shows the comparison between the models and the data. Since the models do not include frictional dissipation along the channel, we concentrate on the comparison in the initial section of the wave tank. The non-linear parameter $\delta(= 0.1)$ is small. Even so, the fully non-linear model shows improvement over the weakly non-linear model. In particular, the extended model tracks the increase in energy of the second harmonic very closely and doubles the increase in energy of the third harmonic, bringing the predictions much closer to the measured value.

To test prediction of wave transformation on a constant slope, we use two sets of data. The first data set is Case 2 of Mase & Kirby (1992). A Pierson-Moskowitz spectrum with peak frequency $f_p = 1.0Hz$, root-mean-square wave height $H_{rms} = 0.04m$ at a water depth of $h_0 = 0.47m$ is input at the wave maker

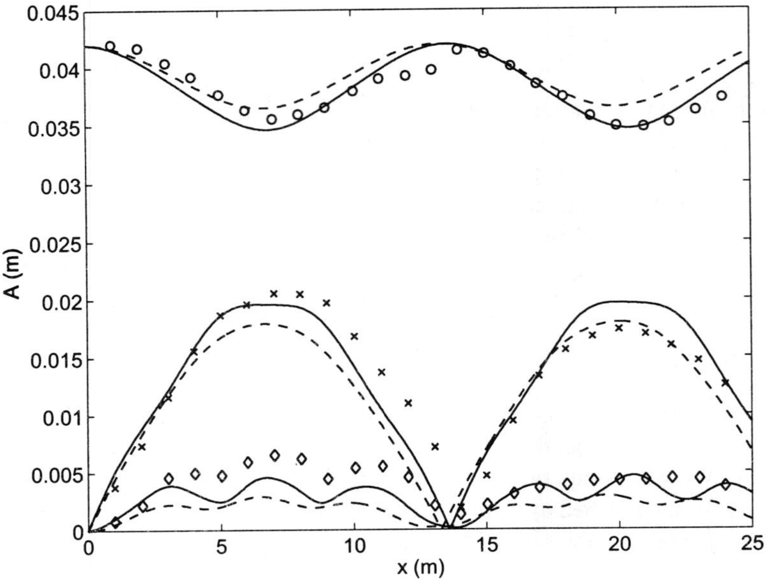

Figure 1: Comparison of weakly non-linear and fully non-linear models to data from Chapalain et al. (1992), Case A. Discrete points are data, solid lines are the fully non-linear model, and dashed lines are the weakly non-linear model.

and the waves are allowed to propagate up a constant impermeable slope of 1 : 20. Thus the waves over most of the domain replicates deep-water conditions which is a rigorous test for the energy transfer between the harmonics. Also, the surf zone is small and therefore the wave breaking is not dominant. Figure 2 shows the comparison of frequency spectra at six different locations between the various models and data.

All the models are initialized using the Fourier amplitudes from the data at the toe of the slope. Comparing the weakly and fully non-linear model, we see that in deep water there is no significant difference between the two models over most of the frequency range. However, in the low-frequency range the fully non-linear model does not compare to the data as well as the weakly non-linear model. The fully non-linear form transfers more energy to the low frequency component compared to the measurements. The amplitudes at the low-frequency remains large as the waves are propagated into shallower water. In the high-frequency region both models perform similarly through most of the domain. However, as the depth gets shallower, the amplitudes predicted by the weakly non-linear model stay small whereas prediction from the fully non-linear model is more aligned with that of the observations. The two-equation models, on the other hand, perform significantly better in the deeper water for all frequencies. The high-frequency tail is still predicted to have smaller amplitudes, but is not significantly different from

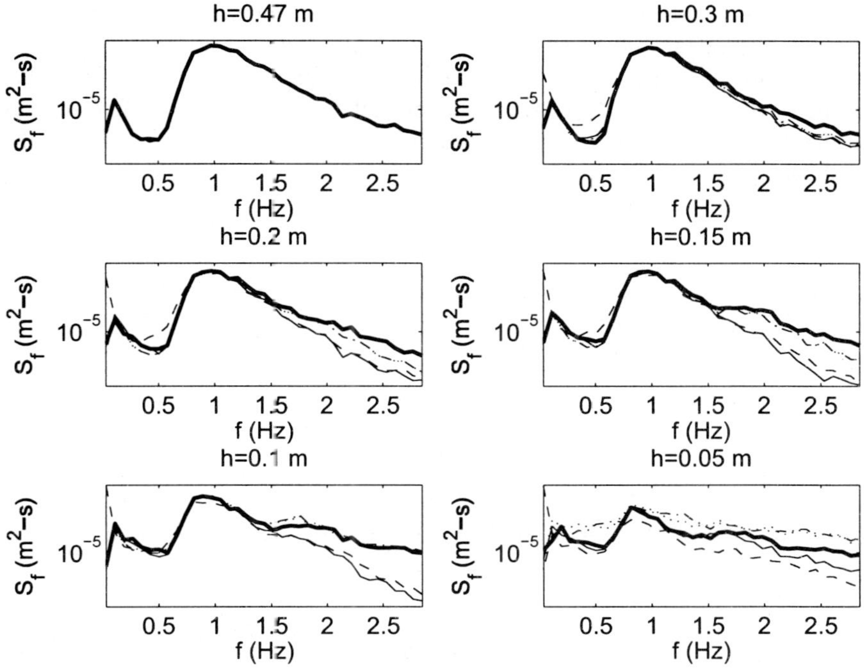

Figure 2: Comparison of the frequency spectra predicted by the models compared to the data for Mase & Kirby (1992). Thick solid line is data, thin solid is weakly non-linear model, dashed line is the fully non-linear model, dash-dot is the two-equation model with bulk dissipation and the dotted line is the two-equation model with eddy viscosity breaking.

the data. On the other hand, as we get into the breaking region the predictions significantly worsen with much larger amplitudes at all frequencies. That seems the case regardless of the breaking mechanism used. It seems to indicate that not enough energy is taken out of the system by the breaking mechanism for the two-equation model. Looking at the overall development of the spectra, it is obvious that the two-equation model preserves the energy in the waves better. However, the energy taken out by the breaking mechanism is the same as that in the one-equation model. Thus, the two-equation model over-predicts the energy once the waves start breaking. The purpose of this study is not to tune the breaking parameters and therefore we did not investigate what the optimal breaking parameters are.

Figure 3 compares the root-mean-square wave height H_{rms}, the skewness and asymmetry given by the models to the data. The water depths are shown in the x-axis. The wave height predictions by all models are very similar up to the location where waves start to break. The two-equation models overestimate the wave heights from there to the minimum depth. The skewness estimate in deeper water is excellent for the two-equation model. The fully non-linear model gives results closer to the data than the weakly non-linear model, but does not predict the maximum value of the skewness. The trend for the asymmetry estimates follows that of the wave skewness with the two-equation models providing excellent estimates at the deeper water depths and the weakly non-linear model doing better in the shallower regions.

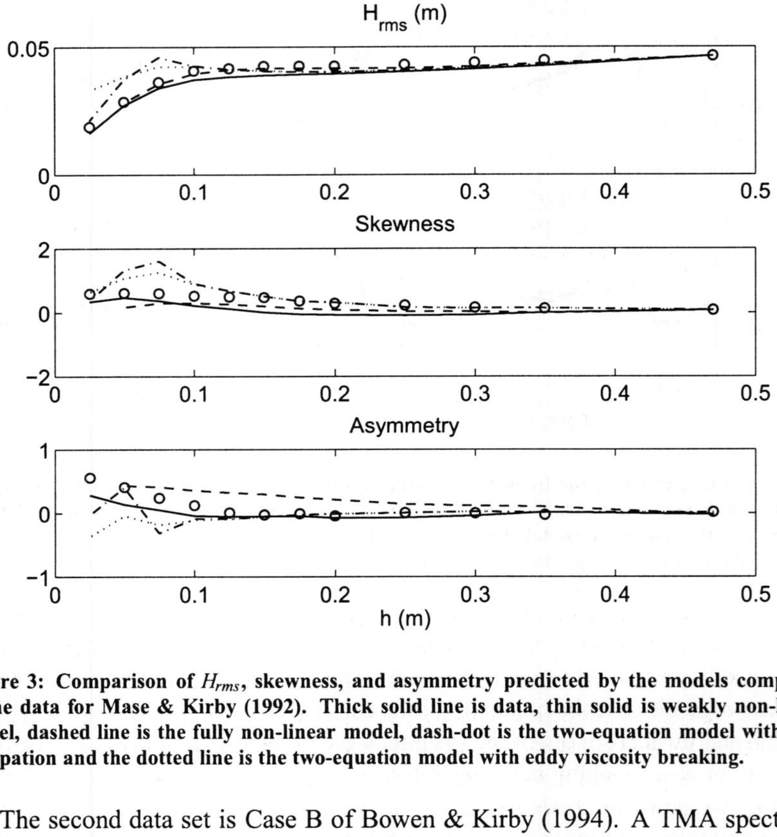

Figure 3: Comparison of H_{rms}, skewness, and asymmetry predicted by the models compared to the data for Mase & Kirby (1992). Thick solid line is data, thin solid is weakly non-linear model, dashed line is the fully non-linear model, dash-dot is the two-equation model with bulk dissipation and the dotted line is the two-equation model with eddy viscosity breaking.

The second data set is Case B of Bowen & Kirby (1994). A TMA spectrum with peak frequency $f_p = 0.225Hz$ and a root-mean-square wave height $H_{rms} = 0.08m$ at a water depth of $h_0 = 0.44m$ is input at the wave maker and the waves are allowed to propagate up a constant impermeable slope of 1 : 35. Thus, the waves are in shallow water over most of the domain, and because the non-linear parameter is fairly large to begin with ($\delta \simeq 0.182$), the influence of the additional non-linear terms is expected to be significant. Figure 4 shows the comparison of the spectra from the data to the weakly and fully non-linear models at six different depths. We see that the spectral evolution is similar for both models. However, as before, the fully non-linear model tends to amplify the energy in the low frequency region.

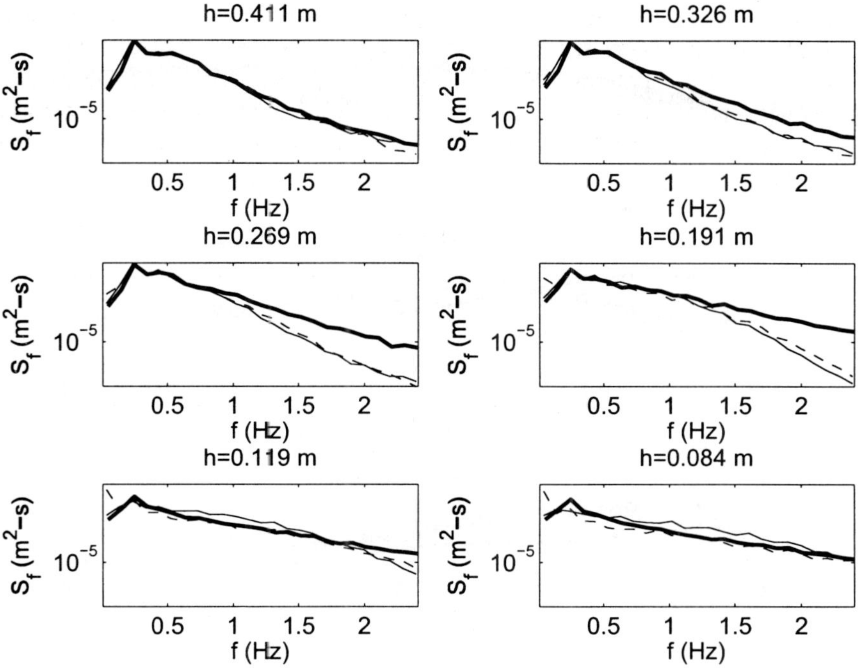

Figure 4: Comparison of the frequency spectra predicted by the models compared to the data for Bowen & Kirby (1994). Thick solid line is data, thin solid is weakly non-linear model, and dashed line is the fully non-linear model.

If we look at the estimates of the wave statistics shown in Figure 5, we see that the prediction of the H_{rms} is very similar for both models and they reproduce the data quite well. However, for the skewness and asymmetry we see a substantial improvement when we use additional non-linear terms. As mentioned before, the waves have fairly high amplitude even at the wave maker for this experiment and as such, we expect that the additional terms are needed to get a better estimate of the wave shapes. This is borne out by the results.

CONCLUSIONS

In an attempt to improve the estimates of wave statistics, we looked at different options for improving the shallow water frequency domain model based on the time domain Boussinesq equations. We tried three approaches. The first approach was to incorporate additional non-linearity in the frequency domain equations. The second approach was to solve for the water surface and velocity simultaneously, and the third approach was to use a different wave breaking mechanism to improve predictions inside the surf zone.

It was found that incorporating additional non-linearity in the model showed

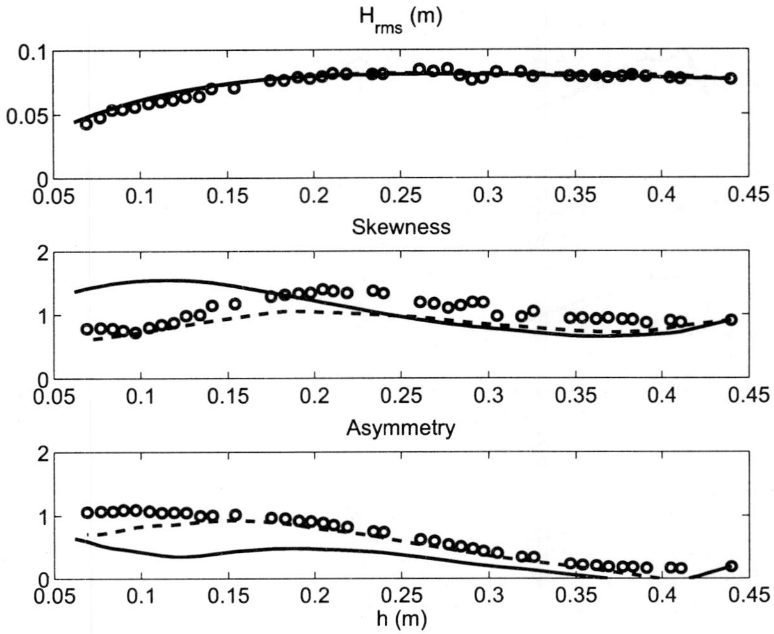

Figure 5: Comparison of H_{rms}, skewness, and asymmetry predicted by the models compared to the data for Bowen & Kirby (1994). Thick solid line is data, thin solid is weakly non-linear model, dashed line is the fully non-linear model.

better results when the waves were in intermediate to shallow water in the whole domain. However, the model overestimates the energy at the lower frequencies. This can probably be attributed to the fact that the models currently do not solve for the wave set-up, so adding this equation to the model will likely improve predictions. Solving for the water surface elevation and velocity simultaneously meant that first-order substitutions were not used to eliminate the velocity from the linear terms of the model. Thus, the deep water characteristics of the time domain model was preserved in the frequency domain which resulted in improved predictions in the deep water region.

Using a breaking mechanism different from that based on the bulk dissipation model of Thornton & Guza (1983) did not improve the estimates in the surf zone. Chen et al. (1997) points out that predicted spectra is insensitive to whether the dissipation function is dependent on f_n^4, f_n^2 or f_n. However, the third moments can be affected by the shape of the dissipation function. Furthermore, Kirby & Kaihatu (1996) show that the eddy-viscosity model to simulate wave breaking has an f_n^2 dependence. Thus, in the final analysis, the eddy-viscosity model is not substantially different than the formulation of Mase & Kirby (1992).

In conclusion, there is merit in using the additional non-linear terms in the frequency domain version as well as keeping the continuity and momentum equation separate. The model would most likely show the most improvement if these two options were combined.

ACKNOWLEDGMENTS

This work was supported by the Office of Naval Research through the Naval Research Laboratory Base Program 6.1 Project "Nonlinear Interactions in Nearshore Environments (NINE)." This is NRL Publication NRL/JA/7320-06-???? and has been approved for public release; distribution unlimited.

REFERENCES

Bowen, G. D. & Kirby, J. T. (1994). Shoaling and breaking random waves on a 1:35 laboratory beach. Technical Report No. CACR-94-14, Center for Applied Coastal Research, Department of Civil Engineering, University of Delaware.

Bredmose, H. (2002). *Deterministic modelling of water waves in the frequency domain.* PhD thesis, Maritime Engineering, Department of Mechanical Engineering, Technical University of Denmark.

Bredmose, H., Schäffer, H. A., & Madsen, P. A. (2004). Boussinesq evolution equations: numerical efficiency, breaking and amplitude dispersion. *Coastal Engineering, 51*, 1117–1142.

Chapalain, G., Cointe, R., & Temperville, A. (1992). Observed and modeled resonantly interacting progressive water-waves. *Coastal Engineering, 16*, 267–300.

Chen, Y., Guza, R. T., & Elgar, S. (1997). Modeling spectra of breaking surface waves in shallow water. *Journal of Geophysical Research, 102*(C11), 25035–25046.

Chen, Y. & Liu, P. L.-F. (1995). Modified Boussinesq equations and associated parabolic models for water wave propagation. *Journal of Fluid Mechanics, 288*, 351–381.

Freilich, M. H. & Guza, R. T. (1984). Nonlinear effects on shoaling surface gravity waves. *Philosophical Transactions of the Royal Society of London, A311*, 1–41.

Kaihatu, J. M. & Kirby, J. T. (1998). Two-dimensional parabolic modeling of extended Boussinesq equations. *Journal of Waterway, Port, Coastal, and Ocean Engineering, 124*(2), 57–67.

Kennedy, A. B., Chen, Q., Kirby, J. T., & Dalrymple, R. A. (2000). Boussinesq modeling of wave transformation, breaking, and run-up. I: 1D. *Journal of Waterway, Port, Coastal, and Ocean Engineering, 126*(1), 39–47.

Kirby, J. T. & Kaihatu, J. M. (1996). Structure of frequency domain models for random wave breaking. In *Proceedings of the 25th International Conference on Coastal Engineering*, (pp. 1144–1155). ASCE.

Madsen, P. A. & Schäffer, H. A. (1998). Higher order boussinesq-type equations

- derivation and analysis. *Phil. Trans. R. Soc. Lond., Series A,*, *356*, 3123–3184.

Mase, H. & Kirby, J. T. (1992). Hybrid frequency-domain KdV equation for random wave transformation. In *Proceedings of the 23rd International Conference on Coastal Engineering*, (pp. 474–487)., Venice.

Nwogu, O. (1993). An alternative form of the Boussinesq equations for nearshore wave propagation. *ASCE Journal of Waterway, Port, Coastal, and Ocean Engineering*, *119*, 618–638.

Peregrine, D. H. (1967). Long waves on beaches. *Journal of Fluid Mechanics*, *27*, 815–827.

Thornton, E. B. & Guza, R. T. (1983). Transformation of wave height distribution. *Journal of Geophysical Research*, *88*(C10), 5925–5938.

Wei, G., Kirby, J. T., Grilli, S. T., & Subramanya, R. (1995). A fully nonlinear Boussinesq model for surface waves. I. Highly nonlinear, unsteady waves. *Journal of Fluid Mechanics*, *294*, 71–92.

A HYBRID MODEL FOR NEARSHORE NONLINEAR WAVE EVOLUTION

Kacey L. Edwards[1] Jayaram Veeramony[2], and James M. Kaihatu[3]

Field and laboratory data show the presence of an equilibrium in the high frequencies of surf zone frequency spectra. In this region, dissipation of high frequency energy equilibrates with nonlinear interactions that transfer energy primarily from low frequencies to high frequencies. The nonlinear interactions that describe nearshore wave evolution increase model complexity, which results in an increase in computational expense. To reduce the computational burden of a nonlinear wave model, it is combined with a parameterization for surf zone wave spectra, and ifferent implementations of the parameterizations are tested and the results show that the predicted spectra and H_{rms} compare well with data and the full model nonlinear parameters skewness and asymmetry are not accurately predicted. Several ideas for future improvements are suggested.

INTRODUCTION

It is well known that as waves propagate through the nearshore environment, they change shape. In the shoaling zone, simple sinusoidal waves transform into waves with narrow, peaked crests and shallow, broad troughs. Furthermore, near the area of breaking the waves have a "saw-toothed" (Kirby, 1997) shape where the front face of the wave is steep and the back of the wave is more gentle sloping.

Wave transformation in the nearshore environment occurs as a result of energy transfers between different frequency components of the wave. Although energy is transferred between all frequencies, the dominant transfer is from low to high frequency. In addition, dissipation preferentially removes energy from the high frequency components. In the surf zone, the transfer of energy to high frequencies and the dissipation of energy from higher frequencies reaches an equilibrium; spectra evolve toward a broad, featureless shape for a range of high frequencies.

The transfer of energy and resulting equilibrium range of high frequency waves is easily seen in frequency and wavenumber spectra. As depth decreases, energy at the peak frequency (f_p) or peak wavenumber (k_p) of a

[1]Oceanography Division, Naval Research Laboratory, 1009 Balch Blvd., Stennis Space Center, MS 39529, USA

[2]Jacobs Engineering, 1009 Balch Blvd., Stennis Space Center, MS 39529, USA

[3]Zachry Department of Civil Engineering, Texas A&M University, 3136 TAMU, College Station, TX 77843-3136, USA

spectrum decreases and the tail of the spectrum obtains a gentler slope. In the surf zone, the slope or shape of the spectral tail remains relatively constant, although the amount of energy in the spectral tail may decrease. The transformation of spectral shape in the nearshore environment has been observed in field and laboratory data, as well as model results (Smith & Vincent, 2003; Herbers & Burton, 1997).

Laboratory Experiments

Two such laboratory experiments are the experiments of Mase & Kirby (1992) (referred to as MK92) and Bowen & Kirby (1994) (referred to as BK94). MK92 measured random waves propagating up a 1 : 20 bottom slope. At the wavemaker, the Pierson-Moskowitz spectrum was characterized by $H_{rms} = 0.04$ m and the $f_p = 1.0$ Hz. Therefore, MK92 is an example of mostly deep water waves, and the data are characterized by weak nonlinear interactions and a narrow surf zone. BK94 measured random waves propagating up a 1 : 35 bottom slope. At the wavemaker, $H_{rms} = 0.08$ m and $f_p = 0.225$ Hz. Therefore, their case is an example of mostly shallow water waves. Nonlinear interactions have more of an effect on the transformation of wave spectra, and the surf zone is wider, relative to MK92.

Despite weak nonlinear interactions in the MK92 data, changes in the spectral shape are apparent (Figure 1). In deeper water, where nonlinear interactions are the weakest, there is little change in the spectral shape. However, as depth decreases and the waves shoal, energy at the low frequencies decreases, and energy at higher frequencies increases. Therefore, the slope of the spectral tail decreases. In the surf zone, significant energy is lost from the spectral peak, and the spectrum is much broader. The slope of the spectral tail is flatter.

Modeling Nonlinear Waves

Because waves do not retain a simple, sinusoidal shape throughout the shoaling and surf zones, modeling the evolution of waves through the nearshore environment requires a more complex approach than linear theory. The transfer of energy that causes the changes in the shapes of waves and spectra is facilitated by nonlinear, triad interactions. These three-wave interactions must be included in a model of nearshore wave evolution to accurately predict wave shape statistics, which in turn are necessary for sediment transport problems. Unfortunately, increasing the complexity of a nearshore wave model by including triad interactions significantly increases the computational expense of the wave model.

To reduce the computational burden of a nonlinear wave model, we present a hybrid model that combines a weakly nonlinear, dispersive wave model (Kaihatu & Kirby, 1995) with a parameterization of the surf zone equilibrium range (Smith & Vincent, 2003). In addition to reducing the

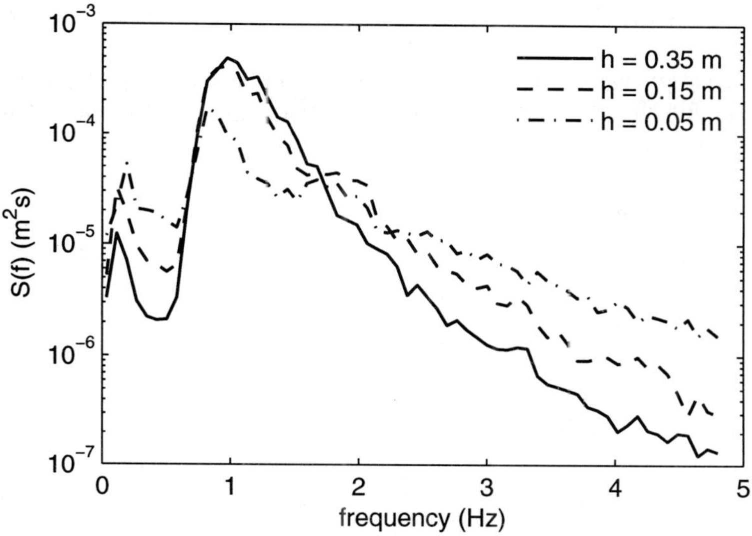

Figure 1: Frequency Spectra determined from measurments collected during the MK92 laboratory experiment.

computational burden, we anticipate that the hybrid model's estimation of spectral tails is an improvement over a full implementation of the weakly nonlinear model. We will examine model results in the form of frequency spectra, as well as wave properties computed from the frequency spectra including H_{rms}, skewness, and asymmetry.

HYBRID MODEL

Nonlinear Wave Model

Many types of models are capable of evolving waves through the nearshore environment, and each model has its advantages and disadvantages. Described herein is a frequency domain, weakly nonlinear, dispersive, one-dimensional wave model (Kaihatu & Kirby, 1995) (referred to as NLMSE). Derivation of the model begins with the boundary value problem expressed in terms of the velocity potential (ϕ) in only one horizontal dimension. The governing equation and boundary conditions are

$$\phi_{xx} + \phi_{zz} = 0; \qquad -h \le z \le 0, \qquad (1)$$

$$\phi_z = -h_x \phi_x; \qquad z = -h, \qquad (2)$$

$$g\eta + \phi_t + \frac{1}{2}(\phi_x^2 + \phi_z^2) + \eta\phi_{zt} = O(\epsilon^3); \qquad z = 0, \qquad (3)$$

$$\eta_t - \phi_z + \eta_x\phi_x - \eta\phi_{zz} = O(\epsilon^3); \qquad z = 0, \qquad (4)$$

where ϕ is the velocity potential, ϵ ($= ka$, $\ll 1$) scales the nonlinear terms terms, h is the local water depth, η is the instantaneous free surface elevation, and the subscripts refer to derivatives in space (x and z) and time (t) respectively.

The boundary value problem, governed by the Laplace equation (1), describes the movement of an inviscid, irrotational fluid. Furthermore, the problem is restricted to progressive waves

$$\phi = \sum_{n=1}^{N} -\frac{ig}{\omega_n} f_n A_n(x,y) e^{i(\int k_n \, \mathrm{dx} - \omega_n t)} + \text{c.c.}, \tag{5}$$

where N is the number of frequency components.

The evolution equation for the amplitudes A_n is

$$\begin{aligned} A_{n_x} &+ \frac{(kCC_g)_{n_x}}{2(kCC_g)_n} A_n + \alpha_n A_n = \\ &- \frac{i}{8(kCC_g)_n} \left[\sum_{l=1}^{n-1} R A_l A_{n-l} e^{i \int (k_l + k_{n-l} - k_n) dx} \right. \\ &\left. + 2 \sum_{l=1}^{N-n} S A_l A_{n+l} e^{i \int (k_{n+l} - k_l - k_n) dx} \right]. \end{aligned} \tag{6}$$

$C = \omega_n / k_n$ is phase speed, and $C_g = \partial \omega_n / \partial k_n$ is group velocity. The coefficients R and S are the subharmonic and superharmonic triad interactions given by

$$\begin{aligned} R = &\frac{g}{\omega_l \omega_{n-l}} \left[\omega_n^2 k_l k_{n-l} + (k_l + k_{n-l})(\omega_{n-l} k_l + \omega_l k_{n-l}) \omega_n \right] \\ &- \frac{\omega_n^2}{g} (\omega_l^2 + \omega_l \omega_{n-l} + \omega_{n-l}^2), \end{aligned} \tag{7}$$

$$\begin{aligned} S = &\frac{g}{\omega_l \omega_{n+l}} \left[\omega_n^2 k_l k_{n+l} + (k_{n+l} - k_l)(\omega_{n+l} k_l + \omega_l k_{n+l}) \omega_n \right] \\ &- \frac{\omega_n^2}{g} (\omega_l^2 - \omega_l \omega_{n+l} + \omega_{n+l}^2). \end{aligned} \tag{8}$$

The one dimensional model is initialized offshore with a wave spectrum. It moves forward in space determining complex amplitudes for a given number of frequency components (N).

Because the evolution equation is derived assuming weak nonlinearity and mild slope, the model retains the linear depth dependence and linear dispersion relation given by

$$f_n = \frac{\cosh k_n (h+z)}{\cosh k_n h} \quad \text{and} \tag{9}$$

$$\omega_n^2 = g k_n \tanh k_n h. \tag{10}$$

Although the NLMSE includes terms on the left hand side that combine shoaling, refraction and diffraction, it does not account for energy lost to friction or breaking. Therefore, Kaihatu & Kirby (1995) add the third term (dissipation) to the left hand side of the evolution equation. The dissipation term is based on the dissipation model of Thornton & Guza (1983). They define the total energy dissipation as

$$\epsilon_b = \frac{3\sqrt{\pi}}{16}\rho g \frac{B^3 f_p}{\gamma^4 h^5} H_{rms}^7, \tag{11}$$

where B and γ are free parameters with values 1.0 and 0.6, respectively, and f_p is peak frequency.

Using the approach of Mase & Kirby (1992) the dissipation model (11) is distributed between frequency independent and frequency dependent components.

$$\alpha_n = \alpha_{n_0} + \left(\frac{f_n}{f_p}\right)^2 \alpha_{n1}, \tag{12}$$

$$\alpha_{n_0} = F\epsilon_b', \tag{13}$$

$$\alpha_{n_1} = [\epsilon_b' - \alpha_{n_0}] \frac{f_p^2 \sum_{n=1}^{N} |A_n|^2}{\sum_{n=1}^{N} f_n^2 |A_n|^2}, \tag{14}$$

$$\epsilon_b' = \frac{\epsilon_b}{\rho g \sum_{n=1}^{N} C_g |A_n|^2}. \tag{15}$$

N is the number of frequency components, and F is a free parameter that determines what percentage of the dissipation model (11) is distributed to the frequency independent function (13).

The right hand side of the evolution equation contains nonlinear sums, which replicate the nonlinear interactions between waves of different frequencies. Unfortunately, the complexity of the nonlinear sums makes the NLMSE model computationally expensive. The nonlinear sums require a number of computations on the order of N^2. In addition, the NLMSE model can potentially overpredict energy in very high frequencies (Figure 2).

Surf Zone Parameterization

To reach our goals of reducing the computational expense of modeling wave evolution in the nearshore environment and improving the ability to replicate high frequency energy, we employ the surf zone parameterization of Smith & Vincent (2003).

Until 2003, little effort was put into parameterizing surf zone waves. Following the development of deep water wave parameterizations (Resio, Pihl, Tracy & Vincent, 2001) and previous work by Thornton (1977) and Zakharov (1999), Smith & Vincent (2003) described a parameterization for surf zone($H_{rms}/h > 0.4$) wave spectra.

Relying on wavenumber spectra from many laboratory and field experiments, Smith and Vincent (2003) found that the range of frequencies for which energy transfer and dissipation equilibrated was composed of two subranges. They call the two subranges the Zakaharov range and the Toba range. The Zakharov range is valid for low wavenumbers ($k_p < k < 1/h$), and the Toba range is valid for high wavenumbers ($k \geq 1/h$). Therefore, both ranges exist only if $k_p h << 1$.

Each subrange has its own shape and energy level. The ranges are defined mathematically as power laws in wavenumber space and are written

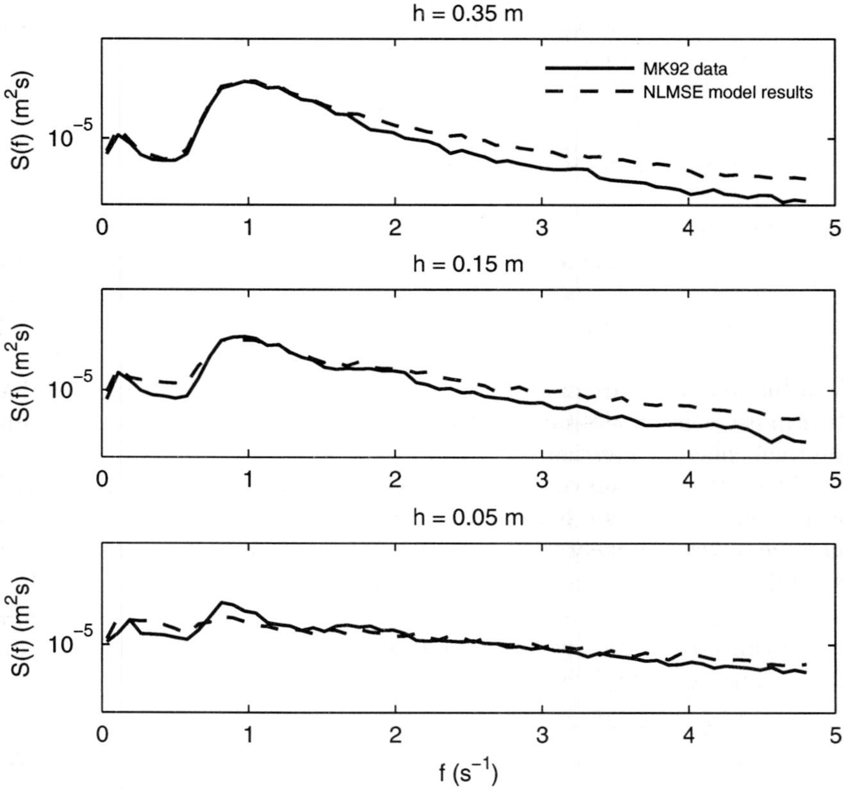

Figure 2: Results from the NLMSE model.

$$S(k) = \beta_{Toba} k^{-5/2}; \qquad kh > 1 \tag{16}$$

$$S(k) = \beta_{Zak} k^{-4/3}; \qquad k_p < k \leq 1/h \tag{17}$$

$$\beta_{Toba} = \alpha_{Toba} h^{0.5} \tag{18}$$

$$\beta_{Zak} = \alpha_{Zak} h^{1.67} \tag{19}$$

Because the subranges intersect at $k = 1/h$, $\alpha_{Toba} = \alpha_{Zak} = 0.0103 = \alpha$. In addition, the dependence on depth requires that the energy level decreases as depth decreases.

Combining the Two Components

Use of the parameterization in conjuction with the NLMSE model insures a reduction in computation expense because the number of frequency components explicitly determined by the NLMSE model decreases. To combine the two, use the relation

$$S(f_n)df = S(k_n)dk_n. \tag{20}$$

In addition, complex amplitudes are determined for the equilibrium range using the relation

$$A_n = \sqrt{\frac{S(f_n)df}{2}} e^{i\theta_n} \tag{21}$$

where θ_n is a random phase produced by a random number generator.

ALTERNATIVE β_{Toba}

Because the parameterization was developed for surf zone wave spectra, it does not represent offshore wave spectra well. At offshore locations, the slope of the parameterization compares reasonably to data; however, the β coefficients overestimate the energy of the equilibrium range (Figure 3). Limiting use of the parameterization to the surf zone region works well, but hinders any computational savings. This is especially true for cases similar to MK92 where the surf zone is relatively narrow.

An alternative approach to determine the beta coefficient was explored. β_{Toba} was written not only as a function of water depth but also as a function of the nonlinearity parameter H_{rms}/h. This was accomplished by writing

$$\alpha = C \frac{H_{rms}}{h}. \tag{22}$$

C is a constant that defines α as a fraction of the nonlinearity parameter. It is given by the relation

$$C = \alpha_0 \frac{h_0}{H_{rms,0}} \tag{23}$$

where α_0, h_0, and $H_{rms,0}$ represent offshore values, and

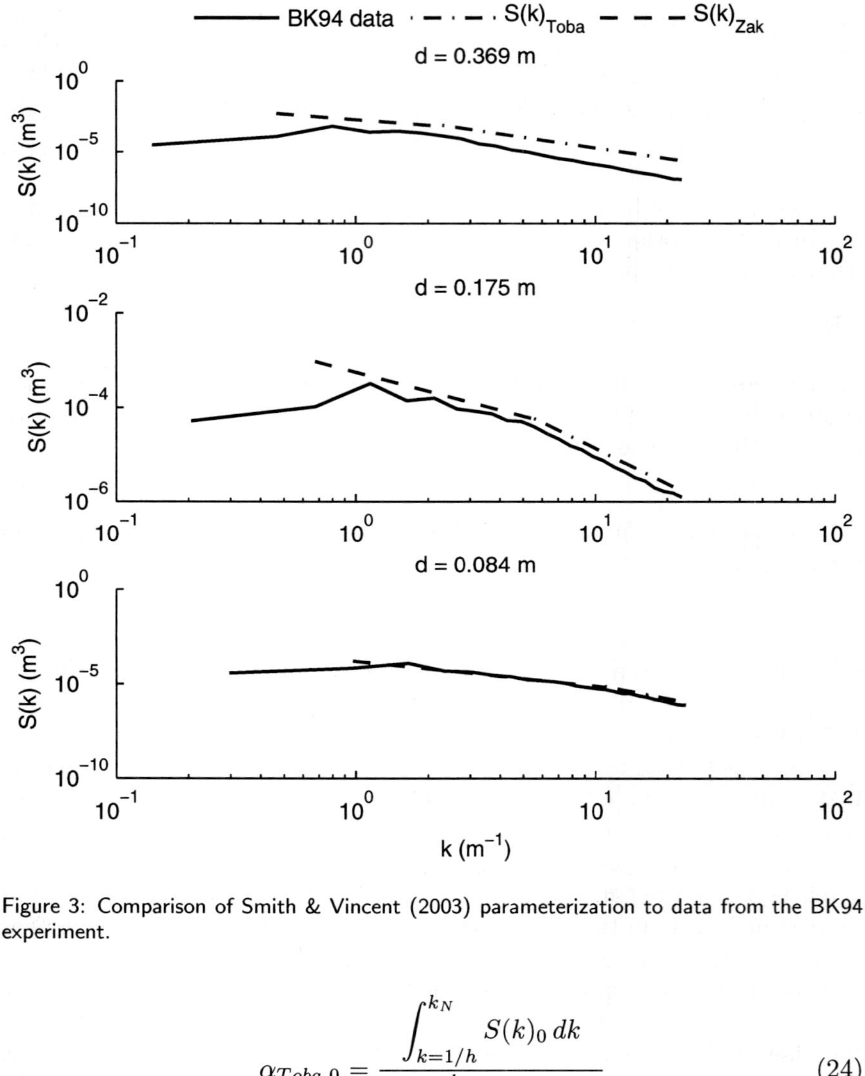

Figure 3: Comparison of Smith & Vincent (2003) parameterization to data from the BK94 experiment.

$$\alpha_{Toba,0} = \frac{\displaystyle\int_{k=1/h}^{k_N} S(k)_0 \, dk}{h_0^{0.5} \displaystyle\int_{k=1/h}^{k_N} k^{-5/2} \, dk} \tag{24}$$

where $S(k)_0$ is the wavenumber spectrum at the offshore boundary.

Frequency spectra resulting from the NLMSE model, the hybrid model using $\alpha = 0.0103$ (referred to as SNV) and the hybrid model using (22) (referred to as NALF) are compared in Figure 4. As expected, offshore the SNV hybrid model provides the same results as the NLMSE model. In addition, the NALF hybrid model compares well; the results overestimate energy at the higher frequencies slightly less than the NLMSE and SNV hybrid models do. However, as depth decreases, the NALF hybrid model underestimates the frequency spectrum. The SNV hybrid model compares to the data well in the parameterized range, butoverestimates low frequency energy. Well into the surf zone, both hybrid models compare well.

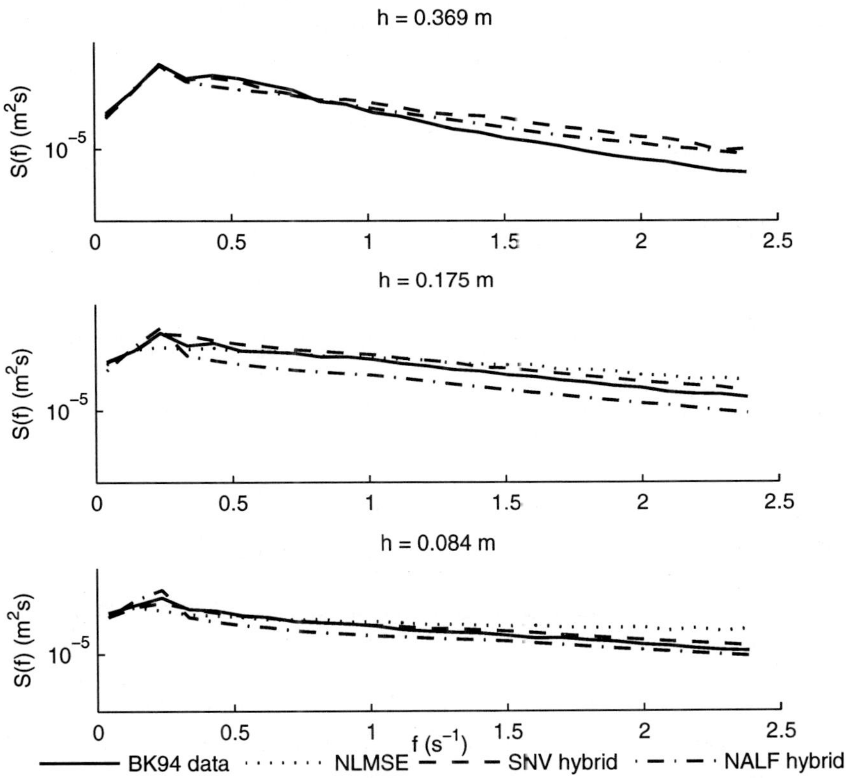

Figure 4: Frequency spectra resulting from the hybrid models and the NLMSE model.

The SNV hybrid model compares well through the modeled domain, but the computational expense is not lessened. The NALF hybrid model required only fifteen percent of the NLMSE model's computational time. The NALF hybrid model compares well to the data at the offshore location because the constant, C, depends on offshore parameters. This fraction, however, is not constant through the entire domain.

Comparisons of wave parameters determined from the data and the various model results reflect what is observed in the frequency spectra comparisons (Figure 5). The SNV hybrid model replicates H_{rms} outside the surf zone, but once the surf zone is reached and the parameterization is employed, the SNV hybrid model overestimates the H_{rms} values before settling into agreement with the data. The NALF hybrid model significantly underestimates H_{rms} until well into the surf zone.

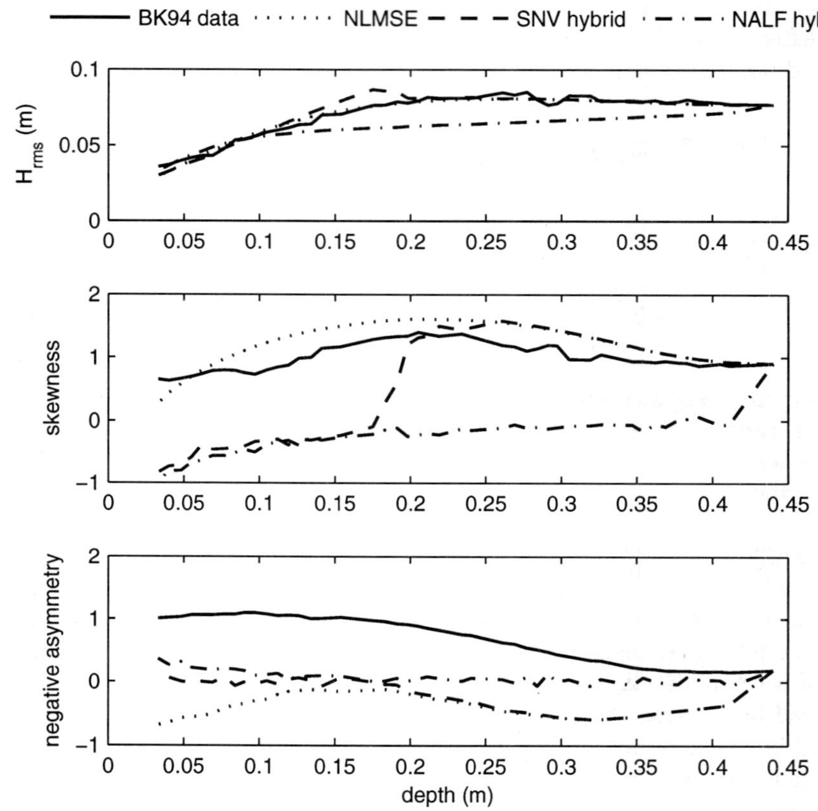

Figure 5: Wave parameters determined from NLMSE and hybrid model results. Top: H_{rms}, Middle: skewness, Bottom: asymmetry.

Comparisons of the parameters skewness and asymmetry show that neither of the hybrid models reproduce the nonlinear properties. Accurate energy transfers require correct phase locking. The random phases, θ_n, used to determine complex amplitudes in the equilibrium range do not ensure the correct phase locking.

PHASE REPRESENTATION

In an effort to improve the nonlinear properties, two additional phase representations were explored. First, the offshore phases were held constant through the domain. Then, the evolution equation was used to evolve the phase through the domain.

To obtain the evolution equation for the phases, we began by assuming that

$$A_n = a_n e^{i\phi_n(x)}. \tag{25}$$

After substituting (25) into (6), we solved for the partial derivative of ϕ_n and used the result to specify phase for the equilibrium range so that

$$\begin{aligned}\phi_{n_x} = &-\frac{1}{8\omega_n C_{g,n} a_n} \\ &\cdot \left[\sum_{l=1}^{n-1} R a_l a_{n-l} e^{i\int(k_l+k_{n-l}-k_n+\phi_l+\phi_{n-1}-\phi_n)dx}\right. \\ &\left.+ 2\sum_{l=1}^{N-n} S a_l a_{n+l} e^{i\int(k_{n+l}-k_l-k_n+\phi_{n+l}-\phi_l-\phi_n)dx}\right].\end{aligned} \tag{26}$$

Figure 6 shows skewness and asymmetry results from the SNV hybrid model using random, constant, and evolved phases. Neither the constant phase nor the evolved phase appears to improve the hybrid model's ability to produce the nonlinear parameters or H_{rms}.

SUMMARY

Modeling wave evolution through the shoaling and surf zones requires a complex model that is computationally expensive. In this study, the complex model is a frequency domain, weakly nonlinear, dispersive one dimensional wave model (Kaihatu & Kirby, 1995). The computational expense is reduced by limiting the number of frequency components explicitly determined by the nonlinear model and determining the complex amplitudes at the excluded frequencies with a parameterization.

The surf zone parameterization as described by Smith & Vincent (2003) was combined with the NLMSE model (Kaihatu & Kirby, 1995) to form a hybrid model. Because the parameterization applies to only the surf zone, no computational savings was acheived. However, the hybrid model produced a better replication of high frequency energy.

Using an alternative descripton of β_{Toba} allows the parameterization to be used outside the surf zone. Therefore, the computational burden of evolving waves through a domain is lessened. However, β_{Toba}'s depth dependence dominates the parameterization, and the hybrid model does not compare well in the surf zone.

Model results are reiterated in the computation of H_{rms}. The SNV hybrid model provides H_{rms} estimates that compare well offshore and inside the surf zone. However, there is a region of overestimation because for an area of the domain, the SNV hybrid overestimates the low frequency energy. For most of the BK94 domain, the NALF hybrid underestimates H_{rms}.

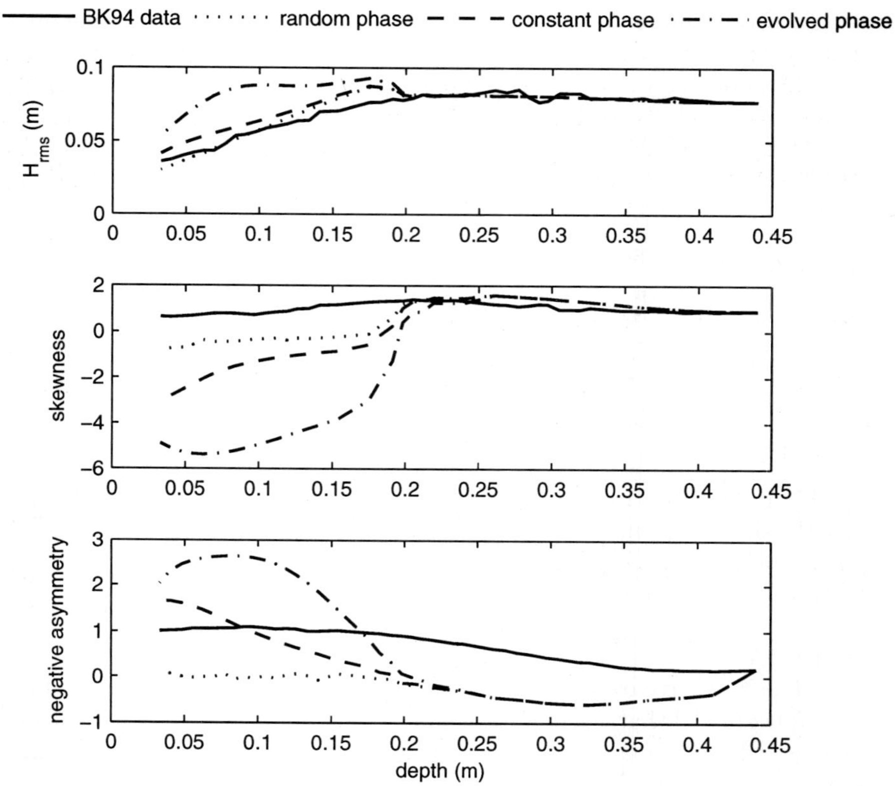

Figure 6: Wave parameters determined from the SNV hybrid model results using different phase representations. Top: H_{rms}, Middle: skewness, Bottom: asymmetry.

Although the hybrid models give reasonable H_{rms} values, the skewness and asymmetry are clearly inaccurate. Using a random phase to determine complex amplitudes in the equilibrium range interferes with the phase locking of the triad interactions. However, neither of the two alternative approaches of determining phase for the parameterized range provided any improvement over the use of the random phase.

Both aspects of this study (the energy level and the phase representation) require additional investigation. To improve the β coefficient, the $k = 1/h$ limit will be explored as well as a more reliable method to determine the constant C (23). Furthermore, perhaps a gentler approach to employing the parameterization will improve the results and reduce the computational expense. Rather than forcing the spectrum to the parameterization, maybe it would be better to guide it toward the parameteriztion.

ACKNOWLEDGMENTS

Special thanks to Dr. Jane Smith of the US Army Engineer Research and Development Center's Coastal and Hydraulics Laboratory and Dr. James Kirby at the University of Delaware for providing data needed for this study. The work was supported by the Office of Naval Research through the 6.1 NRL Core Project "Nonlinear Interactions in Nearshore Environments," Program Element 61153N.

REFERENCES

Bowen, G. D. & Kirby, J. T. (1994). Shoaling and breaking random waves on a 1:35 laboratory beach. Technical report, Center for Applied Coastal Research, Department of Civil Engineering, University of Delaware, Delaware.

Herbers, T. H. C. & Burton, M. C. (1997). Nonlinear shoaling of directionally spread waves on a beach. *J. Geophys. Res.*, *102*, 21101–21114.

Kaihatu, J. M. & Kirby, J. T. (1995). Nonlinear transformation of waves in finite water depth. *Physics of Fluids*, *7*, 1903–1914.

Kirby, J. T. (1997). Nonlinear, dispersive long waves in water of variable depth. In J. N. Hunt (Ed.), *Gravity Waves in Water of Finite Depth*, volume 10 of *Advances in Fluid Mechanics* (pp. 55–126). Great Britain: Computational Mechanics Publications.

Mase, H. & Kirby, J. T. (1992). Hybrid frequency-domain kdv equation for random wave transformation. In Edge, B. L. (Ed.), *Proceedings for the 23rd International Conference of Coastal Engineering*, (pp. 474–478)., Reston, VA.

Resio, D. T., Pihl, J. H., Tracy, B. A., & Vincent, C. L. (2001). Nonlinear energy fluxes and the finite depth equilibrium range in wave spectra. *J. Fluid Mech.*, *95*, 159–176.

Smith, J. M. & Vincent, L. V. (2003). Equilibrium ranges in surf zone wave spectra. *J. Geophys. Res.*, *108*, 3366–3376.

Thornton, E. B. (1977). Rederivation of the saturation range in the frequency spectrum of wind-generated gravity waves. *J. Phys. Oceanogr.*, *7*, 137–140.

Thornton, E. B. & Guza, R. T. (1983). Transformation of wave height distribution. *J. Geophys. Res.*, *88*, 5925–5938.

Zakharov, V. (1999). Statistical theory of gravity and capillary waves on the surface of a finite-depth fluid. *J. Geophys. Res.*, *88*, 5925–5938.

A STOCHASTIC MODEL FOR NONLINEAR SURFACE WAVES OVER TOPOGRAPHY

T. T. Janssen[†], T. H. C. Herbers[‡] and J. A. Battjes[§]

Based on a deterministic angular-spectrum model, we derive a stochastic model for nonlinear, inhomogeneous random wave evolution over nearshore topography. The stochastic model is a weak-refraction, wide-angle diffraction approximation, and accounts for quadratic nonlinearity and depth-induced breaking. Comparison to observations confirms that 1) inhomogeneity in a topography-induced refractive focus, and 2) nonlinear dynamics in shoaling waves on a natural beach, are well represented in the stochastic model.

1. Introduction

Ocean waves propagating into coastal areas refract on submerged topography and – through nonlinearity – transform from near-sinusoidal waves to the characteristic skewed, pitched-forward shapes at the onset of breaking; eventually – with further decreasing of the water depth – the ordered wave motion loses its energy to small-scale turbulent motions in the surf. The nearshore propagation and transformation of ocean waves profoundly affects coastal hydrodynamics and is clearly of great interest to coastal engineering practice.

Many deterministic, phase-resolving modeling approaches for nearshore ocean waves exist. However, these models require detailed, phase-resolving boundary conditions, which in practice are generally not available. Furthermore, most engineering applications (e.g. the modeling of morpho-dynamic evolution, structural fatigue & design criteria) utilize the wave field statistics (e.g. average measures of wave height/period), which requires averaging over large ensembles (in particular for higher-order statistics such as surface skewness). Alternatively, the wave motion can be regarded as a realization of an underlying stochastic process – a paradigm-shift which has lead to great advances in our theoretical understanding of ocean wave physics [e.g. Hasselmann, 1962], and in the

[†] Department of Oceanography, Naval Postgraduate School, Monterey, CA 93940, USA. Corresponding author: tjanssen@nps.edu

[‡] Department of Oceanography, Naval Postgraduate School, Monterey, CA 93940, USA.

[§] Environmental Fluid Mechanics, Faculty of Civil Engineering & Geosciences, Delft University of Technology, Stevinweg 1, 2628 CN, Delft, The Netherlands operational wave modeling capability [e.g. Komen et al., 1994]. However, since the mathematical description for water wave motion is nonlinear, a stochastic description invariably results in an open set of equations for the statistical moments, for which some sort of approximation, or closure hypothesis, is required – reminiscent of the closure issues in turbulence [see e.g. Orszag, 1970; Salmon, 1998, and many others].

In the present work we derive a stochastic model for nonlinear wave propagation in shallow coastal regions, including the surf zone, without inherent limitations on (linear) frequency dispersion, and suitable for inhomogeneous wave fields over two-dimensional topography. The stochastic model development is based on the deterministic angular-spectrum model derived in Janssen et al. [2004, 2006]. To extend modeling capability to the surf zone we include a parametric wave breaking model [Janssen and Battjes, 2006], and a relaxation to Gaussian statistics in the nonlinear closure [see e.g. Herbers et al., 2003]. Additional forcing terms that account for the processes of e.g. wind forcing and dissipation due to steepness instability can be added, but are not considered here. Instead, we focus on the linear and nonlinear wave processes over two-dimensional topography, in particular the spatial structure and correlations in heterogeous wave fields.

The paper is organized as follows. In §2 we derive a stochastic model based on a deterministic angular-spectrum model [Janssen et al., 2006]. The stochastic representation of spatial inhomogeneity and its relevance to nearshore wave modeling is discussed in §3. In §4 we compare predicted and observed spectral evolution across a cylindrical beach. Our main findings are summarized in §5.

2. Stochastic transport equations

We consider the evolution of weakly nonlinear, random surface waves over submerged topography. We refer to the horizontal coordinates x and y as the principal and lateral direction and since we are interested in wave propagation in coastal areas, we let the principal and lateral direction coincide with the cross-shore and alongshore direction respectively. The bottom topography is assumed to vary gently – but arbitrarily – in the cross-shore direction, whereas in the alongshore direction we assume that topographical variations are weak. Accordingly, we decompose the bottom as

$$h(\mathbf{x}) = \overline{h}(x) - \widetilde{h}(\mathbf{x}) \tag{1}$$

where $\overline{h}$ is a one-dimensional reference depth, and $\widetilde{h}$ the two-dimensional residue. The two-dimensionality is assumed weak such that $\widetilde{h}_0/\overline{h} \ll 1$, with $\widetilde{h}_0$ a representative amplitude of $\widetilde{h}$.

Further, assuming the wave field stationary in time, we decompose the surface elevation function $\eta(\mathbf{x}, t)$ in a multi-frequency angular-spectrum [see e.g. Stamnes, 1986; Janssen et al., 2006]

$$\eta(\mathbf{x}, t) = \sum_{q_1, p_1 = -\infty}^{\infty} \widehat{\eta}_1^1(x) \exp\left[i\left(\lambda_1 y - \omega_1 t\right)\right]. \quad (2)$$

Here $\omega_1 = p_1 \Delta\omega$ and $\lambda_1 = q_1 \Delta\lambda$, with $\Delta\omega$ and $\Delta\lambda$ the discrete angular frequency and lateral wavenumber spacing respectively. Numerical sub- and superscripts on wave-related variables are associated with frequency and lateral wavenumber components respectively; for instance, the $\widehat{\eta}_1^1$ is shorthand notation for $\widehat{\eta}_{\omega_1}^{\lambda_1}$ and thus represents the spectral amplitude for component (ω_1, λ_1) of the surface elevation function.

The spatial evolution of the wave field in (2), including the effects of quadratic nonlinear effects, can be described by a set of transport equations [Janssen et al., 2006]

$$\left[\frac{d}{dx} - i\varkappa_1^1\right] A_1^1 = \widetilde{\mathcal{G}}_{1;2}^{\lambda}\{M_1^1(y) + D_1(y), A_1^2\} + i \sum_{\mathbf{v}_2} \mathcal{W}_{(1-2)2}^{(1-2)2} A_{(1-2)}^{(1-2)} A_2^2, \quad (3)$$

where $A_1^1 = \sqrt{V_1^1}\widehat{\eta}_1^1$. The M_1^1 and D_1 denote bottom topography forcing and (frequency weighted) dissipation; the nonlinear term in (3) accounts for quadratic finite-amplitude effects. Explicit expressions for M_1^1, D_1 and the nonlinear coupling coefficient $\mathcal{W}_{12}^{12}$ are given in Appendix A. Further, the $V_1^1 = C_{g,1}(\overline{h})\varkappa_1^1(\overline{h})/k_1(\overline{h})$ is the principal component of the group speed vector, with the principal wavenumber $\varkappa_1^1 = \sqrt{(k_1(\overline{h}))^2 - (\lambda_1)^2}$ for $k_1 \geq \lambda_1$. In (3) the operator $\widetilde{\mathcal{G}}$ symbolically denotes the back and forth transformation between the wavenumber and physical domain (pseudo-spectral operation), written here as

$$\widetilde{\mathcal{G}}_{n;m}^{\lambda}\{a(y), b^{\lambda_m}\} = \frac{1}{L_y}\int_{-L_y/2}^{L_y/2} \sum_{q_m} a(y) b^{\lambda_m} \exp\left[i\left(\lambda_m - \lambda_n\right) y\right] dy, \quad (4)$$

such that the scatter terms are treated as products in the spatial domain rather than convolutions in the wavenumber domain. In (4) L_y is the lateral extent of the domain.

The transport equation (3) represents a deterministic model for the evolution of weakly nonlinear waves over quasi-2D topography where the topography scatter term M_1^1 redistributes the spectral amplitudes across the lateral wavenumber components to correct wave ray geometry relative to Snel's law to account for the refractive effects over the two-dimensional topography [e.g. Janssen et al., 2006]. (Extension of the angular-spectrum modeling approach to fully two-dimensional topographies is discussed in Janssen et al. [2007].)

From equation (3) we derive a set of transport equations for the statistical moments. Multiplication of (3) with $(A_1^2)^*$, where $*$ denotes the complex conjugate, adding the conjugate equation, and ensemble averaging the result yields (upon letting $\Delta\lambda, \Delta\omega \to 0$)

$$\left[\frac{d}{dx} - i\Lambda_1^{1;2}\right] \mathcal{E}_1^{12} = \mathcal{G}_{1;3}^{\lambda}\{N_1^1(y), \mathcal{E}_1^{31}\} + \left(\mathcal{G}_{2;3}^{\lambda}\{N_1^2(y), \mathcal{E}_1^{32}\}\right)^* + i \iint d\omega_3 d\lambda_3 \left(\mathcal{W}_{(1-3)3}^{(1-3)3} \mathcal{C}_{(1-3)3}^{(1-3)32} - \mathcal{W}_{(1-3)3}^{(2-3)3} \left(\mathcal{C}_{(1-3)3}^{(2-3)31}\right)^*\right), \quad (5)$$

where – for notational convenience – we combined the linear terms into a single symbol $N_1^1(y) = M_1^1(y) + D_1(y)$. The operator $\mathcal{G}$ is defined as

$$\mathcal{G}_{n;m}^{\lambda}\{a(y), b(\lambda_m)\} = \frac{1}{2\pi} \iint a(y) b(\lambda_m) \exp\left[i\left(\lambda_m - \lambda_n\right) y\right] dy d\lambda_m, \quad (6)$$

with integration limits from $-\infty$ to ∞. Further, in (5), $\Lambda_{1..N}^{1..N;M} = \varkappa_1^1 + \ldots + \varkappa_N^N - \varkappa_{1+\ldots+N}^M$, and the statistical moments are given by

$$\mathcal{E}_1^{12}(x) = \mathcal{E}(\omega_1, \lambda_1, \lambda_2, x) = \lim_{\Delta\lambda, \Delta\omega \to 0} \frac{\langle A_1^1 (A_1^2)^* \rangle}{\Delta\lambda^2 \Delta\omega}, \quad (7)$$

$$\mathcal{C}_{12}^{123}(x) = \mathcal{C}(\omega_1, \omega_2, \lambda_1, \lambda_2, \lambda_3, x) = \lim_{\Delta\lambda, \Delta\omega \to 0} \frac{\langle A_1^1 A_2^2 (A_{(1+2)}^3)^* \rangle}{\Delta\lambda^3 \Delta\omega^2}. \quad (8)$$

In what follows we refer to the moments $\mathcal{E}$ and $\mathcal{C}$ as the spectrum and bispectrum respectively. In a similar fashion we obtain an evolution equation for the bispectrum that reads

$$\left[\frac{d}{dx} - i\left(\Lambda_{12}^{12;3} + i\mu_{12}^{123}\right)\right] \mathcal{C}_{12}^{123} = \mathcal{G}_{1;4}^{\lambda}\{N_1^1(y), \mathcal{C}_{12}^{423}\} + \mathcal{G}_{2;4}^{\lambda}\{N_2^2(y), \mathcal{C}_{12}^{143}\} + \left(\mathcal{G}_{3;4}^{\lambda}\{N_{(1+2)}^3(y), (\mathcal{C}_{12}^{124})^*\}\right)^* + 2i \int d\lambda_4 \left[\mathcal{W}_{(1+2)(-2)}^{(1+4)(-4)} \mathcal{E}_2^{24} \mathcal{E}_{(1+2)}^{(1+4)3} + \mathcal{W}_{(1+2)(-1)}^{(2+4)(-4)} \mathcal{E}_1^{14} \mathcal{E}_{(1+2)}^{(2+4)3} - \mathcal{W}_{12}^{(3-4)(4)} \mathcal{E}_1^{14} \mathcal{E}_2^{2(3-4)}\right] \quad (9)$$

where – following Holloway and Hendershott [1977], and Herbers et al. [2003] – we close the nonlinear hierarchy by tentatively writing the fourth cumulant contribution to the bispectrum evolution as a linear damping term $\mathcal{C}^{(4)} = -\mu_{12}^{123} C_{12}^{123}$. This closure approximation allows nonlinearity to develop locally but acts as a decorrelator to attenuate the memory of the wave system. In the presence of strong nonlinearity this prevents the (unphysical) overshoot behavior typical for the quasi-normal closure ($\mathcal{C}^{(4)} = 0$) [see e.g. Salmon, 1998]. This closure relaxation toward Gaussian statistics is heuristic but provides the model with a means to de-couple spectral triads over long distances [e.g. Orszag, 1970]. Explicit expressions for the damping term μ_{12}^{123} are given in Appendix B.

3. Refraction-diffraction of inhomogeneous wave fields

The spectrum $\mathcal{E}_1^{12}$ and bispectrum $\mathcal{C}_{12}^{123}$ are two- and three dimensional in lateral wavenumber space respectively, which is a dimensional extension by one (with respect to the laterally homogeneous case). This reflects the dimensional extension of the wave covariance function (of which the spectrum is the Fourier transform) in a heterogeneous wave field. After all, the covariance function for an inhomogeneous (or non-stationary, if the heterogeneity is in time) process depends on the separation *and* location, whereas for homogeneous/stationary processes it is merely a function of the separation (or lag). Physically, the dimensional extension of the spectrum accounts for the phase coupling (or directional coherence) between pairs of non-collinear wave components and the topography, thus resolving the interference patterns of e.g. crossing wave trains in topography-induced caustic regions. In that respect the present approach differs fundamentally from most conventional stochastic models based on the energy balance equation (such as SWAN or WAVEWATCH) which assume spatial homogeneity of the wave field from the outset and thus omit phase coupling between non-collinear components.

However, the $\mathcal{E}$ and $\mathcal{C}$ are not 'local' in the lateral sense, which hampers their geometrical interpretation. To clarify this, consider the surface elevation variance at location (x, y)

$$\begin{aligned}\langle \eta^2(x,y) \rangle &= \iint d\lambda\, d\omega \left(\int \frac{\mathcal{E}(\omega, \lambda + \lambda'/2, \lambda - \lambda'/2, x)}{\sqrt{V_\omega^{\lambda+\lambda'/2} V_\omega^{\lambda-\lambda'/2}}} \exp\left[i\lambda' y\right] d\lambda' \right) \\ &= \iint \mathcal{S}(\omega, \lambda, x, y)\, d\lambda d\omega. \end{aligned} \tag{10}$$

Thus from (10) it follows that the quantity

$$\mathcal{S}(\omega, \lambda, x, y) = \int \frac{\mathcal{E}(\omega, \lambda + \lambda'/2, \lambda - \lambda'/2, x)}{\sqrt{V_\omega^{\lambda+\lambda'/2} V_\omega^{\lambda-\lambda'/2}}} \exp\left[i\lambda' y\right] d\lambda', \tag{11}$$

which essentially is the inverse Fourier transform of the $\mathcal{E}_1^{12}/\sqrt{V_1^1 V_1^2}$ with respect to the difference wavenumber, can be interpreted as a two-dimensional frequency-directional spectrum. In other words, $\mathcal{S}(\omega, \lambda, x, y)\, d\lambda d\omega^\dagger$ represents the contribution to the sea surface elevation variance associated with a 'wave-ray' tangent to $(\sqrt{k(\omega)^2 - \lambda^2}, \lambda)$ passing through location (x, y). From S, bulk statistics such as e.g. mean wave height, period, and direction can be estimated with conventional techniques.

† $\mathcal{S}(\omega, \lambda, x, y)$ is also sometimes referred to as a Wigner distribution [e.g. Bastiaans, 1979]

3.1. *Wave evolution over a topographical lens*

To verify the model's ability to capture topography-induced wave field inhomogeneity and correlations in crossing wave trains, we compare simulations with the linear stochastic model (eq. (5) without the nonlinear term involving the bispectrum $\mathcal{C}$) to observations of wave evolution over a submerged shoal in the laboratory [Vincent and Briggs, 1989]. The experimental bottom topography consists of an elliptic shoal, with its crest 15.24 cm below still-water level, placed on an otherwise uniform depth of $h = 45.72$ cm [Figure 1]. Further details are found in Vincent and Briggs [1989].

The computational domain measures $L_y = 30$ m by $L_x = 20$ m in lateral and principal direction respectively. Consequently, the alongshore wavenumber spacing is $\Delta\lambda = 2\pi/L_y = 1/15\pi$ rad/m. The spatial domain is discretized with $\Delta x = 0.1$ m, $\Delta y = 0.24$ m. Comparison is made to wave heights observed along the instrumented lateral transects 5 and 6 that are indicated in Figure 1.

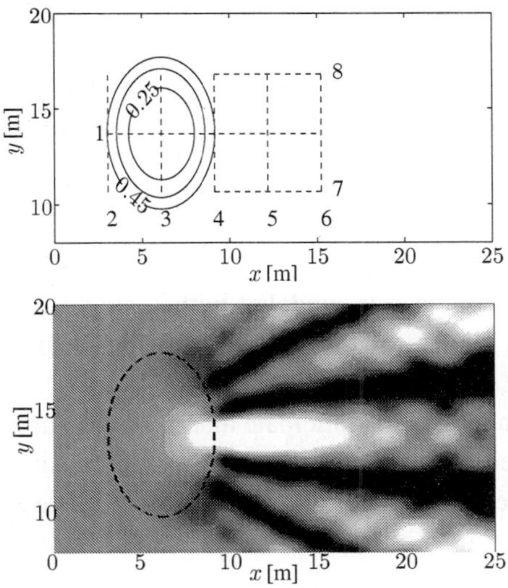

Figure 1. Top panel shows bottom profile experimental set-up Vincent and Briggs [1989]. Depth is in meters; instrumented transects are indicated by dashed lines. Bottom panel shows predicted normalized wave height for monochromatic, unidirectional incident wave train ($\omega = 1.45\pi$ rad) for the same area as shown in top panel; linear color scaling with dark(light) associated with lower(higher) values.

In our comparison we consider two cases: 1) a monochromatic, unidirectional incident wave field, with angular frequency $\omega = 1.45\pi$ rad/s [see top panels Figure 2], and 2) a random wave field (TMA spectrum) with moderate directional spreading [case N1 of Vincent and Briggs, 1989] and peak frequency $\omega_p = 1.54\pi$ rad/s [see bottom panels Figure 2].

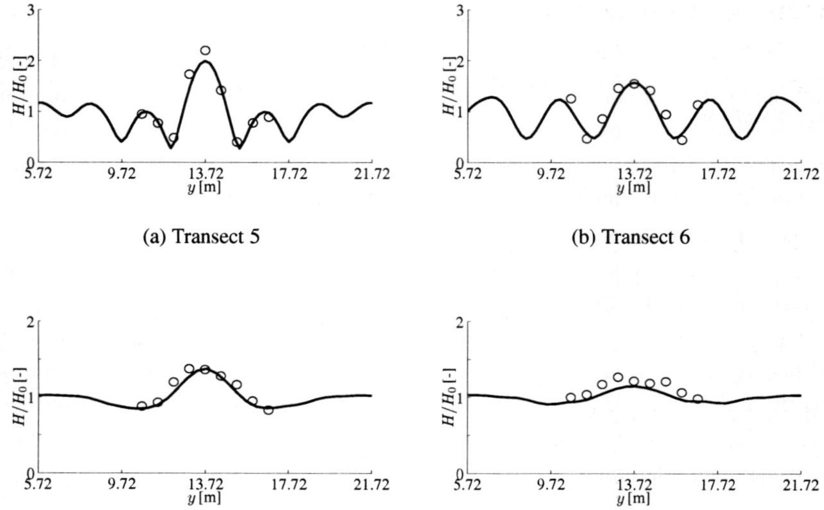

(a) Transect 5 (b) Transect 6

(c) Transect 5 (d) Transect 6

Figure 2. Comparison of observed (circles) [Vincent and Briggs, 1989] and predicted (solid line) wave heights. Top panels (a & b): monochromatic, normally incident waves ($\omega = 1.45\pi$ rad/s). Bottom panels (c & d): directionally-spread, random incident waves (TMA spectrum), $\omega_p = 1.54\pi$ rad/s [case N1 Vincent and Briggs]. Note differences in wave height scales for upper and lower panels.

The monochromatic case is modeled with a single frequency, whereas the frequency spectrum for the random wave case is discretized into 20 equidistant positive frequencies, with $\Delta\omega = 0.2\pi$ rad/s.

From geometrical optics we anticipate that this topography results in wave-ray convergence in the lee of the mound – reminiscent of focusing of light by a burning lens. In the caustic region, geometrical optics is incomplete and diffraction effects (included in the present model) must be accounted for [see e.g. Mach, 1926]. Moreover, in a statistical sense, the crossing seas in the refractive focus are generally correlated since – for a coherent or narrow-band incident wave field – they originate from the same source, scattered in different directions over the topography. For the monochromatic case [top panels Figure 2], which is the archetype of a fully coherent incident wave field, the wave heights exhibit a standing wave height pattern along the lateral transects, indicating phase coupling between the crossing wave trains. The stochastic model accurately resolves these lateral variations [top panels Figure 2]. For random, directionally-spread incident waves [lower panels Figure 2], the wave height variations along the lateral transects in the refractive focus are much more gradual. In effect, the decrease in coherency of the incident waves 'smoothes out' the standing waves in the convergence region due to phase mixing [as also discussed in Vincent and Briggs, 1989]. Also for this case, predicted wave heights are in good agreement with observations.

4. Nonlinear wave evolution across a cylindrical beach

To verify the model's representation of the nonlinear physics in shoaling waves, we compare model predictions to observations made during Sandyduck (1997) near Duck, North Carolina [see Herbers et al., 2003, who discuss the same case]. The model is initialized with the observed spectrum at the FRF array [located roughly at the 8 m depth contour, see Figure 3], which is discretized into 60 frequencies ($\Delta\omega = 0.0137\pi$ rad/s) and 128 alongshore wavenumbers ($\Delta\lambda = 0.0018\pi$ rad/m). The spectral evolution is integrated with a relatively fine, cross-shore step size $\Delta x = 0.5$ m to resolve the anticipated rapid bi-spectral modulations for off-resonant quadratic interactions. Predicted spectra are compared to observations at the nearshore arrays [indicated by B, C, E and F in Figure 3]. Since the bottom topography at Duck is laterally nearly uniform, we consider the evolution of a laterally homogeneous wave field over the laterally averaged topography. The model thus consists of eq.'s (5) and (9) with $\mathcal{E}_1^{ij} = \mathcal{E}_1^{ij}\delta_{ij}$ and $\mathcal{C}_{12}^{ijm} = \mathcal{C}_{12}^{ijm}\delta_{(i+j)m}$, where δ_{ij} denotes a Kronecker delta.

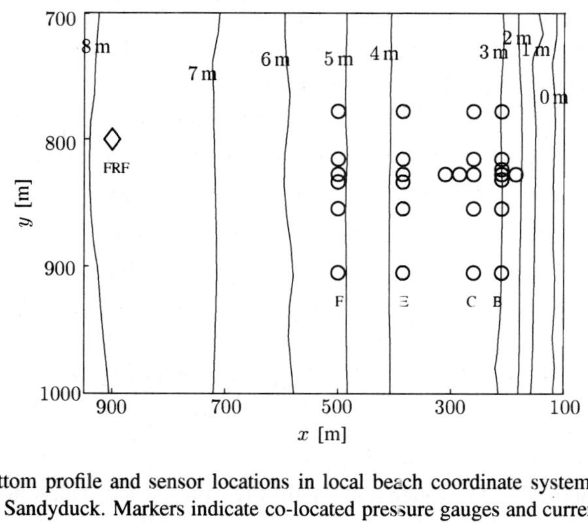

Figure 3. Bottom profile and sensor locations in local beach coordinate system of Field Research Facility during Sandyduck. Markers indicate co-located pressure gauges and current meters.

On August 10 a narrow-band wave field was observed at the FRF array, arriving from a slight Southerly direction ($\theta_m \approx -5^0$) with respect to the

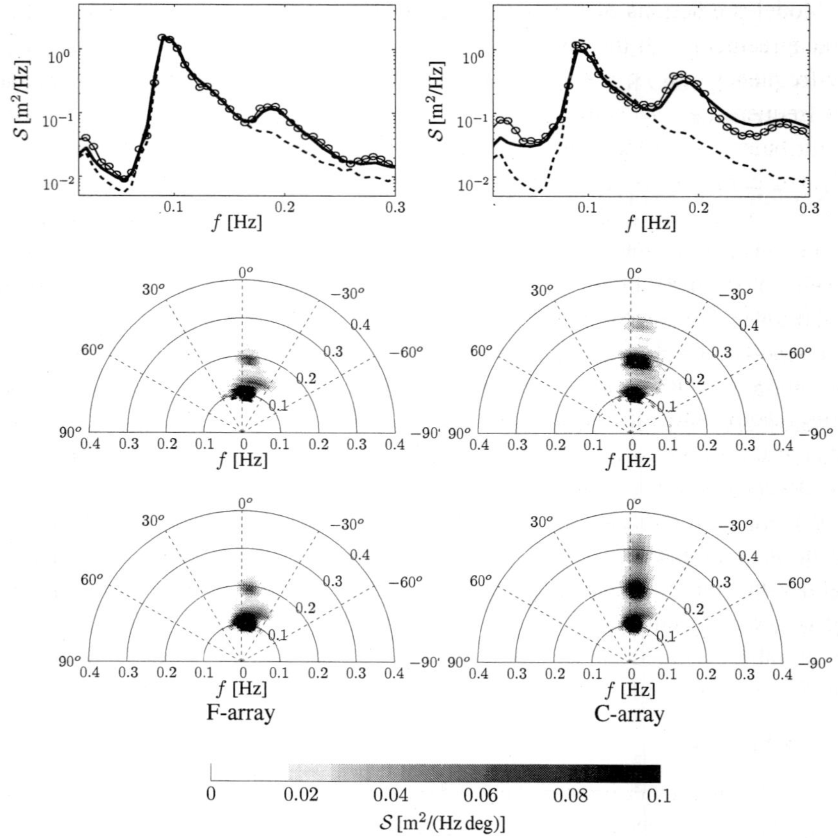

Figure 4. Comparison of observed and predicted spectra for August 10, 1997, at F- (left panels) and C-array (right panels). Top panels: observed (circles) and predicted (solid line) frequency spectra; dashed line represents offshore spectrum (FRF). Middle and lower panels are observed and predicted directional spectra respectively.

shore-normal, with $f_p = 0.09$ Hz and $H_{m0} = 0.85$ m. Positive(negative) angles, θ, are associated with waves approaching from Northerly(Southerly) directions with respect to the shore-normal (corresponding to 70^0 clockwise from true North).

The propagation from the offshore boundary (FRF-array) to the F-array [Figure 4] is accompanied by enhancement of energy at super-harmonic frequencies and a considerable amplification of low-frequency motion (difference interactions), qualitatively consistent with the interaction rules for quadratic nonlinear interactions. Over the bar and upon reaching array C, energy levels at harmonic frequencies are further amplified, at the expense of the energy around the main peak.

Model predictions of the two-dimensional spectra [Figure 4] are in overall good agreement with the observations, both for harmonic frequencies and at the low-frequency end of the spectrum. We note that generally the agreement at low frequencies, say below 0.03 Hz, is somewhat less good. In part this can be attributed to the fact that these long-wave components are likely to include reflected components, for which the model does not account. Moreover, for very low frequencies, the assumed slow variation of the topography, relative to the wavelength, may be violated. Notably, at the C- and B-array (B is not shown), model-predicted spectra are somewhat narrower in direction, and more turned toward shore-normal than is observed. These discrepancies are not well understood at present but could be associated with anomalous refraction by shear wave motion in the nearshore current field [Henderson et al., 2006]. Also, at the nearshore locations, the energy at higher frequencies appears somewhat more diffused across the spectrum than observed (right panel Figure 4), which may be an artifact of the relaxation term.

We found good agreement between predicted and observed evolution of the third-order bulk statistics of skewness and asymmetry (not shown) that characterize the 'peakedness' and 'pitching forward/backward' of the waves [Elgar and Guza, 1985].

5. Conclusions

Based on a deterministic angular-spectrum model [Janssen et al., 2004, 2006] we have derived a new stochastic model for wave evolution over two-dimensional topography. The model is closed in the linear sense and inherits the weak-refraction, wide-angle diffraction approximation of the underlying angular spectrum model [Dalrymple and Kirby, 1988; Suh et al., 1990; Janssen et al., 2006]. Two-dimensionality of forcing terms (e.g. topography and dissipation) is accounted for through a scattering approach, which governs the evolution of the off-diagonal or cross-spectrum terms ($\mathcal{E}_1^{12}$, with $\lambda_1 \neq \lambda_2$). These cross-components represent the correlation between non-collinear wave components and capture the wave field's spatial inhomogeneity.

We have given a brief summary of the stochastic formalism and physical interpretation. Preliminary verification with laboratory and field observations confirm 1) the stochastic modeling capability of combined refraction-diffraction, and 2) the accurate representation of nonlinear evolution of a random, directionally-spread wave field across a natural beach. Further verification of the representation of inhomogeneous wave evolution, and discussion and validation of the nonlinear closure relaxation is presented in Janssen [2006] and Janssen et al. [2007].

Acknowledgements

This research was supported by the Technology Foundation STW, applied science division of NWO, in the Netherlands, under project number DCB.6025. Part of this work was done while TTJ held an NRC Research Associateship at the Naval Postgraduate School, funded by the National Science Foundation in the USA. We thank Michael Briggs and Charles Vincent for providing us with their laboratory observations and documentation. Further, we thank Robert Guza, Steve Elgar and the staff of the Field Research Facility of the U.S. Army Engineer Waterways Experiment Station's Coastal Engineering Center, for sharing Sandyduck field data.

Appendix A. Deterministic forcing terms

The topography scattering term M_1^1 in (3) can be written as

$$M_1^1 = i\frac{g}{2\omega_1 V_1^1}\left(1-T_1^2\right)\left[k_1^2\widetilde{h} + i\varkappa_1^1\mathcal{P}_1^1\widetilde{h}_x + i\lambda_1\left(q_1T_1-1\right)\widetilde{h}_y + k_1^3\frac{C_1}{2C_{g,1}}\frac{(1-T_1^2)^2}{T_1}\left[\frac{1}{(1-T_1^2)}-\mathcal{P}_1^1\right]\widetilde{h}^2\right] \quad \text{(A.1)}$$

where C_1 and $C_{g,1}$ are phase and group speed respectively,

$$q_1 = k_1\overline{h}, \qquad T_1 = \tanh q_1,$$

and

$$\mathcal{P}_1^1 = \frac{1}{2}\left[\left(\frac{k_1}{\varkappa_1^1}\right)^2 - 1\right] - \frac{C_1}{2C_{g,1}}\left(1-q_1T_1\right). \quad \text{(A.2)}$$

The dissipation term in (3) is derived from the expressions in Janssen and Battjes [2006] and is written here as

$$D_1 = -\left(r_0 + r_2\frac{\widetilde{m}_0}{\widetilde{m}_2}|\omega_1|^2\right)\frac{m_1\sqrt{2m_0}}{\pi h\widetilde{m}_0} \left[\left(R^3+\frac{3}{2}R\right)\exp\left[-R^2\right] + \frac{3}{4}\sqrt{\pi}\left(1-\text{erf}\left(R\right)\right)\right]. \quad \text{(A.3)}$$

In (A.3) $R = \gamma\overline{h}/H_{\text{rms}}$ with – after Battjes and Stive [1985] – $\gamma = 0.5 + 0.4\tanh\left(33S_0\right)$, where S_0 denotes the offshore wave steepness. The r_0 and r_2 govern the frequency weighting of the dissipation function. We have used $r_0 = 0.1$ and $r_2 = 0.9$. Further in (A.3) we have used the spectral moments

$$\widetilde{m}_n(x,y) = \iint |\omega_1|^n\widehat{\mathcal{E}}_1^1(x,y)\,d\lambda_1 d\omega_1, \quad \text{(A.4)}$$

$$m_n(x,y) = \iint |\omega_1|^n\mathcal{S}_1^1(x,y)\,d\lambda_1 d\omega_1. \quad \text{(A.5)}$$

where $S_1^1(x,y) = S(\omega_1, \lambda_1, x, y)$ defined in (11), and

$$\widehat{\mathcal{E}}_1^1(x,y) = \widehat{\mathcal{E}}(\omega_1, \lambda_1, x, y) \\ = \int \mathcal{E}(\omega_1, \lambda_1 + \lambda'/2, \lambda_1 - \lambda'/2, x, y) \exp[i\lambda' y] \, d\lambda'. \quad \text{(A.6)}$$

The quadratic nonlinear coupling coefficient in (3) reads

$$\mathcal{W}_{12}^{12} = \sqrt{\frac{V_1^1}{V_2^2 V_{(1+2)}^{(1+2)}}} \frac{\omega_{12} \Lambda_{12}^{12;(1+2)}}{\omega_1 \omega_2 \left(k_{12} T_{12} - \omega_{1+2}^2/g\right)} \left[\mathcal{D}_{12}^{12} + \frac{g k_{12} T_{12}}{\omega_{12}} \mathcal{R}_{12}^{12}\right] \quad \text{(A.7)}$$

where $k_{12} = |\mathbf{k}_1^1 + \mathbf{k}_2^2|$, $T_{12} = \tanh k_{12}\overline{h}$, $\Lambda_{ij}^{ij;n}$ is defined in the text following (5), and

$$\mathcal{D}_{12}^{12} = -\frac{1}{2} \Big[\omega_1 k_2^2 \left(1 - T_2^2\right) + \omega_2 k_1^2 \left(1 - T_1^2\right) \\ -2\left(\omega_1 + \omega_2\right)\left(k_1 k_2 T_1 T_2 - \mathbf{k}_1^1 \mathbf{k}_2^2\right)\Big], \quad \text{(A.8)}$$

$$\mathcal{R}_{12}^{12} = \frac{1}{2g} \left[\omega_1 \omega_2 \left(k_1 T_1 + k_2 T_2\right) + g\left(k_1 k_2 T_1 T_2 - \mathbf{k}_1^1 \mathbf{k}_2^2\right)\right]. \quad \text{(A.9)}$$

Appendix B. Nonlinear closure relaxation

The relaxation term in the bispectrum evolution equation (9) is written here as

$$\mu_{12}^{123} = \beta \left[\frac{|\mathcal{N}_1^{11,(D)} + \mathcal{N}_2^{22,(D)} + \mathcal{N}_{(1+2)}^{33,(D)}| + |\mathcal{N}_1^{11,(NL)} + \mathcal{N}_2^{22,(NL)} + \mathcal{N}_{(1+2)}^{33,(NL)}|}{\mathcal{E}_1^{11} + \mathcal{E}_2^{22} + \mathcal{E}_{(1+2)}^{33}}\right] \quad \text{(B.1)}$$

with

$$\mathcal{N}_i^{jj,(D)} = 2\Re\{\mathcal{G}_{j;n}^{\lambda}\{D_i(y), \mathcal{E}_i^{nj}\}\} \\ \mathcal{N}_i^{jj,(NL)} = -2 \iint d\lambda_n \, d\omega_m \, \mathcal{W}_{(i-m)m}^{(j-n)n} \Im\{\mathcal{C}_{(i-m)m}^{(j-n)nj}\}, \quad \text{(B.2)}$$

Relaxation to Gaussian statistics is thus governed by the intensity with which energy is exchanged with other triads or lost through dissipation. Based on a few trial runs the (dimensionless) tuning constant is set to $\beta = 1.5$.

References

Bastiaans, M. J.. 1979. Transport equations for the Wigner distribution function, *Optica Acta,* 26(10), 1265–1272.

Battjes, J. A. and M. J. F. Stive. 1985. Calibration and verification of a dissipation model for random breaking waves, *J. Geophys. Res.,* 90(C5), 9159–9167.

Dalrymple, R. A. and J. T. Kirby. 1988. Models for very wide-angle water waves and wave diffraction, *J. Fluid Mech., 192*, 33–50.

Elgar, S. and R. T. Guza. 1985. Observations of bispectra of shoaling waves, *J. Fluid Mech., 161*, 425–448.

Hasselmann, K.. 1962. On the non-linear energy transfer in a gravity-wave spectrum, I, *J. Fluid Mech., 12*, 481–500.

Henderson, S. M., R. T. Guza, S. Elgar, and T. H. C. Herbers. 2006. Refraction of surface gravity waves by shear waves., *J. Phys. Oceanogr., 36*, 629–635.

Herbers, T. H. C., M. Orzech, S. Elgar, and R. T. Guza. 2003. Shoaling transformation of wave frequency-directional spectra, *J. Geophys. Res., 108*, doi:10.1029/2001JC001304.

Holloway, G. and M. C. Hendershott. 1977. Stochastic closure for nonlinear rossby waves, *J. Fluid Mech., 82*, 747–765.

Janssen, T. T.. 2006. *Nonlinear surface waves over topography*. PhD dissertation, Delft University of Technology, Delft, The Netherlands.

Janssen, T. T. and J. A. Battjes. 2006. A note on wave energy dissipation over steep beaches., *Submitted to Coast. Engn.,*.

Janssen, T. T., T. H. C. Herbers, and J. A. Battjes. 2004. A discrete spectral evolution model for nonlinear waves over 2D topography, *Proc. 29th Int. Conf. Coastal Eng.*, ASCE, 119–131.

Janssen, T. T., T. H. C. Herbers, and J. A. Battjes. 2006. Generalized evolution equations for nonlinear surface gravity waves over two-dimensional topography, *J. Fluid Mech., 552*, 393–418.

Janssen, T. T., T. H. C. Herbers, and J. A. Battjes. 2007. Stochastic modeling of surface waves in coastal regions: topography-induced inhomogeneity, *To be submitted,*.

Komen, G. J., M. Cavaleri, M. Donelan, K. Hasselmann, S. Hasselmann, and P. A. E. M. Janssen. 1994. *Dynamics and modelling of ocean waves*, Cambridge University Press, London.

Mach, E.. 1926. *The principles of physical optics*, Dover.

Orszag, S. A.. 1970. Analytical theories of turbulence, *J. Fluid Mech., 41*, 363–386.

Salmon, R.. 1998. *Lectures on Geophysical Fluid Dynamics* (1st ed.),, Oxford University Press, New York, 378 p.

Stamnes, J. J.. 1986. *Waves in focal regions*, Adam Hilger.

Suh, K. D., R. A. Dalrymple, and J. T. Kirby. 1990. An angular spectrum model for propagation of Stokes waves, *J. Fluid Mech., 221*, 205–232.

Vincent, C. L. and M. J. Briggs. 1989. Refraction-diffraction of irregular waves over a mound, *J. Wat., Port, Coast. Ocean Eng.,* 115(2), 269–284.

NON-LINEAR RANDOM WAVE GROUPS IN FINITE WATER DEPTH

Felice Arena[1], Vincenzo Nava[1,2], Diego Pavone[1,3], Alessandra Romolo[4]

Some effects of non-linearity are investigated for sea wave groups at a finite water depth. For this purpose, the Boccotti's quasi-determinism theory is firstly applied to describe the linear wave groups. Therefore, the second-order solution is derived for the more general condition of three-dimensional wave groups, at an arbitrary water depth, when a high crest occurs. Finally, some effects of finite bandwidth of the spectrum and of finite water depth are analyzed.

INTRODUCTION

This paper deals with the effects of finite water depth for the space-time propagation of non-linear random wave groups with a large wave.
Firstly, the wave groups are investigated by applying the Boccotti's Quasi-Determinism (QD) theory (1981-2000). This theory gives both the linear free surface displacement and the velocity potential in the space-time domain when a large crest height (or a large wave height) occurs at a fixed time and location.
The predicted structure of the linear three-dimensional wave groups was verified with some small-scale field experiments (see Boccotti, 2000, for a complete review). An alternative approach for the derivation of the Quasi-Determinism theory, as well as field verification in the Atlantic Ocean, was given by Phillips et al. (1993*a*, 1993*b*).
The second-order analytical solution of the QD theory was derived recently for long-crested waves: both the mechanics of non-linear wave groups (Arena, 2005; Arena & Fedele, 2005) and the probability of exceedance of non-linear crest height (Fedele & Arena, 2005) were investigated.
In this paper, the Quasi-Determinism theory is generalized for the more general condition of second-order three-dimensional waves on a finite water depth.
For this purpose, the classical Sharma & Dean (1979) solution for modelling the second-order sea states is considered. Therefore the analytical expression of the second-order three-dimensional free surface displacement (as well as of the velocity potential) is given when a large wave occurs at a fixed time and location. Finally both the effects of finite bandwidth and of directional spreading

---

[1] Department of Mechanics & Materials, 'Mediterranea' University, Loc. Feo di Vito, Reggio Calabria, 89100, Italy – E-mail: arena@unirc.it
[2] Current affiliation: Rice University, Houston, Texas, USA – Civil Engineering Dept. – E-mail: vincenzo.nava@unirc.it
[3] Current affiliation: Snamprogetti S.p.A. - via Toniolo 1 – Fano (PU) 61032, Italy – Email: diego.pavone@snamprogetti.eni.it
[4] Department of Mechanics & Materials, 'Mediterranea' University, Loc. Feo di Vito, Reggio Calabria, 89100, Italy – E-mail: aromolo@unirc.it

function are investigated at a varying water depth, to obtain the modification of the free surface displacement.

LINEAR THEORY OF WIND-GENERATED WAVES

Following the theory of wind-generated waves, given by Longuet-Higgins (1963) and Phillips (1967), a random sea state is defined as a sum of a very large number N of periodic components, with infinitesimal amplitudes a_i, frequencies ω_i different from each other and phase angles ε_i uniformly distributed on (0, 2π) and stochastically independent of each other.

Consequently, the linear free-surface displacement and the velocity potential are stationary Gaussian random processes in time domain and their expressions (for the more general condition of three-dimensional random waves) are respectively provided by:

$$\eta(x,y,t)=\sum_{i=1}^{N} a_i \cos(k_i x\sin\theta_i + k_i y\cos\theta_i - \omega_i t + \varepsilon_i) \tag{1}$$

$$\phi(x,y,z,t)=g\sum_{i=1}^{N} a_i\omega_i^{-1}\frac{\cosh[k_i(d+z)]}{\cosh(k_i d)}\sin(k_i x\sin\theta_i + k_i y\cos\theta_i - \omega_i t+\varepsilon_i) \tag{2}$$

where, for the ith component, the direction of advance makes the angle θ_i with the y-axis, and the wave number, k_i, is related to the angular frequency, ω_i, for any fixed depth, d, by the well-known *linear dispersion rule*:

$$k_i \tanh(k_i d) = \omega_i^2 / g\,. \tag{3}$$

Furthermore, for three-dimensional sea waves, the amplitudes a_i, the angular frequency, ω_i, and the directions of advance θ_i of every elementary component, are such as to form a directional wave spectrum $S(\omega, \theta)$.

By defining $S(\omega)$ the frequency spectrum and $D(\theta;\omega)$ the spreading directional function, we have

$$S(\omega,\theta)=S(\omega)D(\theta;\omega). \tag{4}$$

A typical directional spectrum is the JONSWAP-Mitsuyasu spectrum (Hasselmann et al., 1973; Mitsuyasu et al., 1975) whose mathematical form is:

$$S(\omega,\theta)=\alpha\, g^2\omega^{-5} \cdot\exp\left[-\frac{5}{4}\left(\frac{\omega_p}{\omega}\right)^4\right]\exp\left\{\ln\gamma\exp\left[-\frac{(\omega-\omega_p)^2}{2\sigma^2\omega_p^2}\right]\right\}K(n)\left|\cos\left[\frac{1}{2}(\theta-\bar{\theta})\right]\right|^{2n} \tag{5}$$

where ω_p is the peak angular frequency, $\bar{\theta}$ is the angle between the y-axis and the dominant direction of the spectrum, α is the Phillips' parameter, γ and σ are shape parameters (for the mean JONSWAP they are equal to 3.3 and 0.08 respectively), $K(n)$ is the normalizing factor

$$K(n) = \left[\int_0^{2\pi} \cos^{2n}\left(\frac{1}{2}\theta\right) d\theta \right]^{-1}, \tag{6}$$

and n is a parameter depending on the frequency

$$n = n_p(\omega/\omega_p)^5 \text{ if } \omega \le \omega_p; \quad n = n_p(\omega_p/\omega)^{2.5} \text{ if } \omega > \omega_p, \text{ with } n_p{=}20. \tag{7}$$

'QUASI-DETERMINISM' THEORY FOR RANDOM LINEAR WAVE GROUPS WHEN A VERY HIGH CREST OCCURS

The theory of 'Quasi-Determinism' was derived by Boccotti for the mechanics of linear three dimensional wave groups with either a very high crest (first formulation [1981-1982]) or a very large crest-to-trough wave height (second formulation [1983, 1989, 1997, 2000]).

In this paper the theory is applied in its first formulation, which enables us to investigate the evolution in the space-time domain of random 3D linear wave groups when a crest occurs. The QD theory starts from the assumption that in a Gaussian sea a local maximum, generated by a wave group at the apex of its development stage (Boccotti, 2000), occurs at some time and location.

In more detail, if a very large wave crest of given elevation H_C occurs at point x_0, y_0 at the time instant t_0, with $H_C/\sigma \to \infty$ [this means that the crest height is very large with respect to the standard deviation of the free surface displacement], the free surface displacement (at point x_0,$+X$, y_0+Y) and the velocity potential tend asymptotically, with probability approaching 1, to the deterministic expressions:

$$\overline{\eta}(x_0+X, y_0+Y, t_0+T) = \frac{\Psi(X,Y,T;x_0,y_0)}{\Psi(0,0,0;x_0,y_0)} H_C \tag{8}$$

and

$$\overline{\phi}(x_0+X, y_0+Y, z, t_0+T) = \frac{\Phi(X,Y,z,T;x_0,y_0)}{\Psi(0,0,0;x_0,y_0)} H_C \tag{9}$$

where $\Psi(X,Y,T;x_0,y_0)$ and $\Phi(X,Y,z,T;x_0,y_0)$ are the space-time covariance functions, which are respectively defined as

$$\Psi(X,Y,T;x_0,y_0) \equiv < \eta(x_0,y_0,t)\eta(x_0+X,y_0+Y,t+T) > \tag{10}$$

$$\Phi(X,Y,z,T;x_0,y_0) \equiv < \eta(x_0,y_0,t)\phi(x_0+X,y_0+Y,z,t+T) > \tag{11}$$

η and ϕ being stationary Gaussian processes such as respectively the free surface displacement and the velocity potential of the wind wave field [i.e. Equations (1) and (2)].

Long-crested waves

For long-crested waves propagating in an undisturbed field (in the direction concordant with the Y-axis), when an exceptional crest height of given elevation, H_C, occurs at the point y_0 and at the time t_0, the linear deterministic free surface displacement at the point y_0+Y, and the linear deterministic velocity potential at the point $(y_0+Y,\ z)$, at the time instant t_0+T, are derived by making explicit respectively expressions (8) and (9). The free surface displacement is:

$$\overline{\eta}_1(y_0+Y,t_0+T)=\frac{H_C}{\sigma^2}\int_0^\infty \cos(kY-\omega T)S(\omega)\mathrm{d}\omega \tag{12}$$

where σ^2 is the variance of the free surface displacement.

Short-crested waves

The first formulation of the Quasi-Determinism theory [Expressions (8) and (9)] may be also applied to achieve the linear deterministic functions of water surface and velocity potential for three-dimensional wave groups. In this condition, the free surface displacement (8) may be rewritten as

$$\overline{\eta}_1(x_0+X,y_0+Y,t_0+T)=\frac{H_C}{\sigma^2}\int_0^\infty\int_0^{2\pi} \cos(kX\sin\theta+kY\cos\theta-\omega T)S(\omega,\theta)\mathrm{d}\theta\,\mathrm{d}\omega\,. \tag{13}$$

EXTENSION TO SECOND-ORDER OF THE QUASI-DETERMINISM THEORY FOR RANDOM WAVE GROUPS WITH A VERY HIGH CREST

The second-order solution is obtained by considering the free surface displacement and the velocity potential which satisfy the set of partial differential equations for an irrotational flow, with an incompressible and inviscid flow.

In more detail, by applying the perturbation procedure, the surface displacement and the velocity potential are approximated by a truncated perturbation series

$$\overline{\eta}(X,Y,z,T)\equiv\sum_{i=1}^{2}\varepsilon^i\overline{\eta}_i(X,Y,z,T)+\mathrm{o}(\varepsilon^2)\qquad \overline{\phi}(X,Y,z,T)\equiv\sum_{i=1}^{2}\varepsilon^i\overline{\phi}_i(X,Y,z,T)+\mathrm{o}(\varepsilon^2)$$

in the nonlinearity parameter $\varepsilon=kH_C<<1$, k being the wave number and H_C the crest amplitude, where $\overline{\eta}_1$ and $\overline{\phi}_1$ are the known first-order solutions.

The final expressions of the second-order deterministic components both of the free surface displacement $\bar{\eta}_2$ and of the velocity potential $\bar{\phi}_2$ for long-crested wave groups are

$$\bar{\eta}_2(y_0+Y,t_0+T)=\frac{H_C^2}{4\sigma^4}\int_0^\infty [S(\omega)]^2\frac{2k}{\sinh(2kd)}\,\mathrm{d}\omega$$
$$+\frac{H_C^2}{4\sigma^4}\int_0^\infty\int_0^\infty S(\omega_i)S(\omega_j)\left\{A_{ij}^-\cos(\varphi_i-\varphi_j)+A_{ij}^-\cos(\varphi_i+\varphi_j)\right\}\mathrm{d}\omega_i\,\mathrm{d}\omega_j \tag{14}$$

$$\bar{\phi}_2(y_0+Y,t_0+T)=$$
$$-g\frac{H_C^2}{2\sigma^4}\left\{\int_0^\infty [S(\omega)]^2\frac{k}{\omega d}\,\mathrm{d}\omega\right\}Y-g\frac{H_C^2}{4\sigma^4}\left\{\int_0^\infty [S(\omega)]^2\frac{2k}{\sinh(2kd)}\,\mathrm{d}\omega\right\}T$$
$$+g^2\frac{H_C^2}{4\sigma^4}\int_0^\infty\int_0^\infty S(\omega_i)S(\omega_j)\,\omega_i^{-1}\omega_j^{-1}\left\{\frac{B_{ij}^-}{(\omega_i-\omega_j)}\frac{\cosh[k_{ij}^-(d+z)]}{\cosh(k_{ij}^-d)}\sin(\varphi_i-\varphi_j)\right. \tag{15}$$
$$\left.+\frac{B_{ij}^+}{(\omega_i+\omega_j)}\cdot\frac{\cosh[k_{ij}^+(d+z)]}{\cosh(k_{ij}^+d)}\sin(\varphi_i+\varphi_j)\right\}\mathrm{d}\omega_i\,\mathrm{d}\omega_j$$

where

$$\varphi(\omega_j,Y,T)\equiv k_jY-\omega_jT\,. \tag{16}$$

Note that the main part on both expressions (14) and (15), relatively to the double integrals, was achieved by Fedele & Arena (2005) and Arena & Fedele (2005) for the first formulation and by Arena (2005) for the more general second formulation of the QD theory. In those papers the well-known Sharma & Dean solution (1979) for wind-generated waves, to the second-order in a Stokes expansion, was used.

Here, in the velocity potential, a term proportional to T is considered, which derives from the Whitham (1974) discussion on the second-order problem of Stokes waves in finite depth, where is highlighted as the free surface and the potential cannot be mean zero process at the same time [see also Dalzell, 1999; Boccotti 2000; Sections 1.7.2 and 2.6.3]. In addition, by taking into account equilibrium laws concerning the effects on the second-order free surface displacement, it have to be considered also a term proportional to Y for the complete solution of the nonlinear velocity potential.

In the proposed solution of the second-order components of water profile (14) and velocity potential (15), the $A_{ij}^\pm$ and $B_{ij}^\pm$ interaction kernels are given by

$$A_{ij}^\pm=(B_{ij}^\pm-k_ik_j\pm\rho_i\rho_j)/\sqrt{\rho_i\rho_j}+\rho_i+\rho_j \tag{17}$$

$$B_{ij}^{\pm} = \Big\{ (\sqrt{\rho_i} \pm \sqrt{\rho_j}) \Big[\sqrt{\rho_j}(k_i^2 - \rho_i^2) \pm \sqrt{\rho_i}(k_j^2 - \rho_j^2) \Big] + 2\left(\sqrt{\rho_i} \pm \sqrt{\rho_j}\right)^2 (k_i k_j \mp \rho_i \rho_j) \Big\} \Big/ \left(\left(\sqrt{\rho_i} \pm \sqrt{\rho_j}\right)^2 - k_{ij}^{\pm} \tanh\left(k_{ij}^{\pm} d\right)\right) \tag{18}$$

with

$$k_{ij}^{\pm} = | k_i \pm k_j | \quad \text{and} \quad \rho_i = k_i \tanh(k_i d) = \omega_i^2 / g \, . \tag{19}$$

THREE-DIMENSIONAL NON-LINEAR RANDOM WAVE GROUPS

The generalization of the second-order solution for the more general condition of three-dimensional waves was proposed by Pavone & Arena (2004), by generalizing the approach given by Arena (2005) and Fedele & Arena (2005). They obtained that, when a high crest occurs at point X=0, Y=0 at time T=0, the free surface displacement exact up to the second order in the Stokes expansion, is given by:

$$\begin{aligned} \bar{\eta}(X,Y,T) = {} & \frac{H_C}{\sigma^2} \int_0^{\infty} \int_0^{2\pi} S(\omega,\vartheta) \cos(kX \sin\vartheta + kY \cos\vartheta - \omega T) \, \mathrm{d}\vartheta \, \mathrm{d}\omega \\ & + \frac{H_C^2}{4\sigma^4} \int_{\omega_1=0}^{\infty} \int_{\omega_2=0}^{\infty} \int_{\vartheta_1=0}^{2\pi} \int_{\vartheta_2=0}^{2\pi} S(\omega_1,\vartheta_1) S(\omega_2,\vartheta_2) \\ & \cdot \left[C_{12}^{-} \cos(\tilde{\varphi}_1 - \tilde{\varphi}_2) + C_{12}^{+} \cos(\tilde{\varphi}_1 + \tilde{\varphi}_2) \right] \mathrm{d}\vartheta_2 \, \mathrm{d}\vartheta_1 \, \mathrm{d}\omega_2 \, \mathrm{d}\omega_1 \end{aligned} \tag{20}$$

where:

$$\begin{aligned} \tilde{\varphi}_1 &= k_1 X \sin\vartheta_1 + k_1 Y \cos\vartheta_1 - \omega_1 T, \\ \tilde{\varphi}_2 &= k_2 X \sin\vartheta_2 + k_2 Y \cos\vartheta_2 - \omega_2 T, \end{aligned} \tag{21}$$

$$C_{nm}^{\pm} \equiv \frac{ \dfrac{2\left(\sqrt{k_n} \pm \sqrt{k_m}\right)^2 \left(\vec{k}_n \cdot \vec{k}_m \mp k_n k_m\right)}{\left(\sqrt{k_n} \pm \sqrt{k_m}\right)^2 - k_{nm}^{\pm}} - \left(\vec{k}_n \cdot \vec{k}_m \mp k_n k_m\right) }{\sqrt{k_n k_m}} + \left(k_n + k_m\right) \tag{22}$$

$$k_{nm}^{\pm} \equiv \left| \vec{k}_n \pm \vec{k}_m \right| . \tag{23}$$

The expression (20) enables us to determine the second-order structure of three-dimensional wave group, when a high crest of given elevation H_C occurs.

APPLICATION: NON-LINEAR WAVE GROUPS IN DEEP WATER

Long-crested (2D) waves

The structure of non-linear 2D wave groups in deep water, when a high crest occurs, is obtained from Eq.(14). The nonlinear free surface displacement, for the mean JONSWAP spectrum, is shown in Figure 1: the crest is higher and sharper, the trough is lower and flatter, with respect to the linear profile. The nonlinear effects depend upon the wave steepness (or, equivalently, upon the Phillips' parameter of the spectrum): if we define the steepness as $S=H_s/L_{p0}$, with $L_{p0}=gT_p^2/(2\pi)$, it is easy to verify that for the mean JONSWAP spectrum

$$S = 0.35\sqrt{\alpha}\,. \tag{24}$$

Really, this term is not so important for the calculation of second-order effects. For α ranging between 0.008 and 0.02, the steepness S is between $3.1\cdot 10^{-2}$ and $5.0\cdot 10^{-2}$ and the maximum amplitude of the second order term $\overline{\eta}_2$ ranges between 0.10 and 0.12 times the linear crest height, respectively (see Figure 2). By varying α the most important variation in the structure of the wave groups is in the wave period of the individual waves, as we may appreciate in figure 2. Anyway, this effect is well predicted by linear theory too.

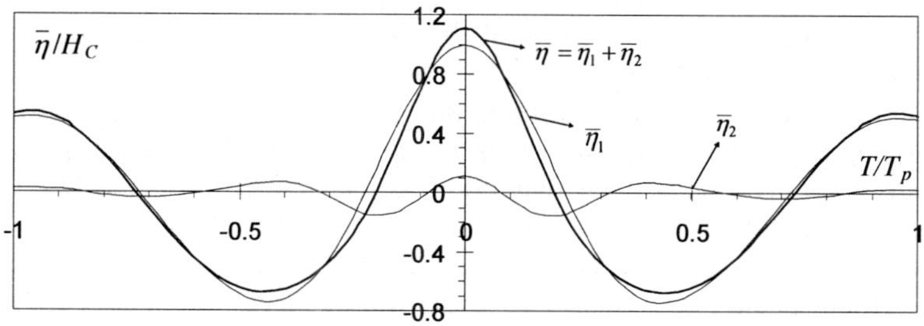

Figure 1. The free surface displacement when a large crest height H_C occurs in deep water: linear profile $\overline{\eta}_1$, second-order term $\overline{\eta}_2$, total second-order profile $\overline{\eta}$.

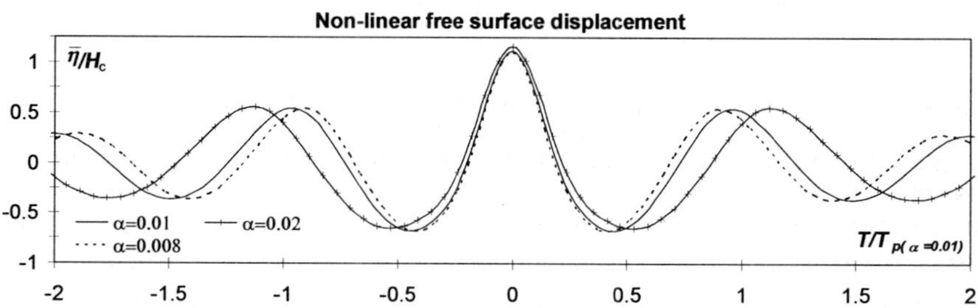

Figure 2. The second-order total free surface displacement ($\overline{\eta}_1+\overline{\eta}_2$) when a large crest height H_C occurs in deep water, for different values of the Phillips' parameter.

A characteristic value of the Phillips' parameter α is 0.01. In this condition the second-order wave group for long-crested random waves is shown in Figure 1 (see Arena, 2005 and Arena & Fedele, 2005) and the total second-order amplitude of the highest crest is 1.11 times the linear one.

Short-crested (3D) waves

In this section Eq. (20) has been applied in order to investigate the second-order effects for three-dimensional wave groups.

Figure 3 shows the profile of the wave fronts along a direction perpendicular to the wave propagation. In this figure it can be noted that the second order effects do not involve a modification in the transversal extent of the waves: indeed both to the first and to the second order, the wave front length is about 2.5 times the wavelength L_{p0}.

By examining the above mentioned figure it could be also noted that along all the wave front extent, the second order free surface displacement is always higher than the linear one and the quotient between these two quantities increases as we move towards the centre; i.e. at time t_0 it is equal to 1.02 for X=0.5 L_{P0} and to 1.10 for X=0.

Figure 4 shows the front profile of the wave that, at different times, is at the centre of the group envelope. In this figure it can be observed the group evolution: there is the first developing stage during which the central wave increases its height, the apex stage at the time t_0, then the decay stage during which the group bring down.

The quotient between the second order and the linear crest increases during the developing stage (for T<0) and decreases after the time T=0, when the highest crest occurs. In particular the above ratio is equal to 1.06 at time t_0-4T_P, 1.07 at time t_0-2T_P and finally reaches a maximum equal to 1.10 at time t_0.

Therefore we can conclude that the second order effects are influenced by the wave group evolution. Finally, in order to evaluate the influence of the directional spreading function on the wave groups, in Figure 5 are shown three different wave fronts, each corresponding to a different value of the parameter n_p in Eq. (7). It can be observed that as the directional spreading decreases (larger value of n_p) the wave front becomes longer.

APPLICATION: NON-LINEAR WAVE GROUPS IN FINITE WATER DEPTH

Finite-depth influence on free surface displacement for long-crested waves

In this section, the influence of finite depth on the non-linear free surface displacement for random sea wave groups will be examined. In particular, there will be pointed out the second-order corrections to the wave-group profile at different relative depths d/L_{p0}; the spectrum is the mean JONSWAP.

In order to appreciate just the finite-depth effects, the free surface displacement

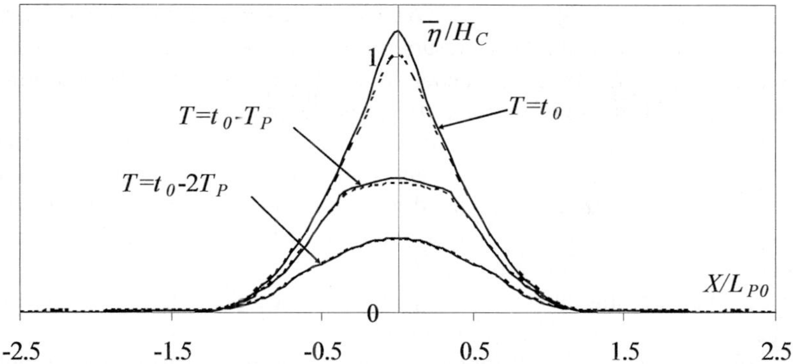

Figure 3. Wave-front profiles along x axis, $y=y_0$. Continuous line represents the second order displacement, the dashed one the linear displacement.

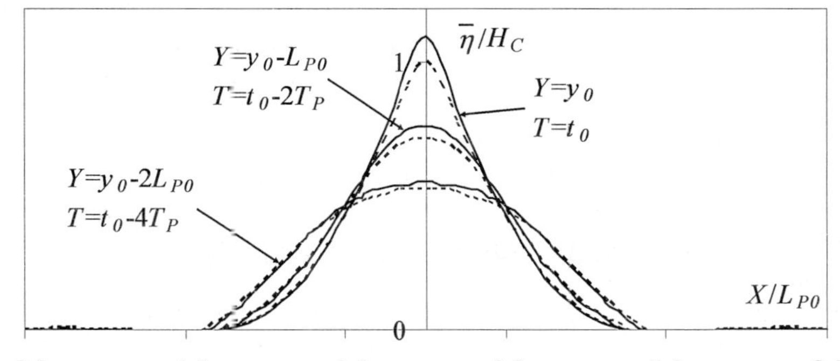

Figure 4. Front profiles of the wave that at different times is at the centre of the group envelope. Continuous line represents the second order displacement, the dashed one the linear displacement.

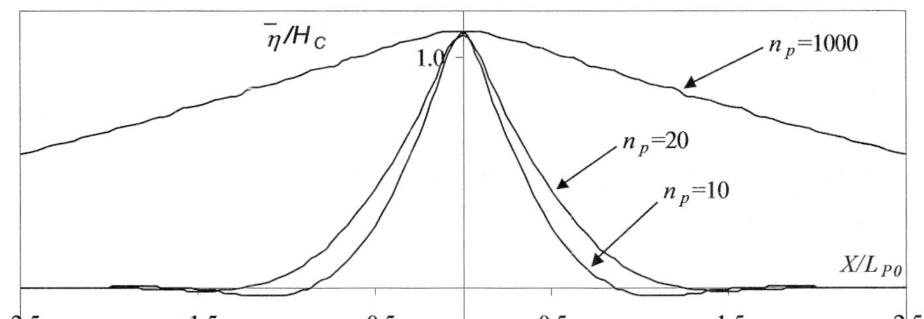

Figure 5. Second order wave-fronts at $t=t_0$, $y=y_0$, corresponding to different values of the directional spreading.

is plotted in the time domain (see Figure 6). It has been considered a wave group when a crest with height H_c=8.0m occurs, in a sea state with significant wave height H_s of 8.0m. In the upper panel of Figure 6, the linear free surface displacement time-history is reported. It represents the analytical QD solution (first formulation); it can be understood easily that all the linear time histories are the same in the time domain, at any value of the water depth *d*. Some modifications may be instead appreciated in the second panel and third panel of Figure 6: in that picture we may notice the time-histories of the second-order component at three different relative depths.

We can see that the amplitude of the crest of the second-order component is almost the same along the different depths (from d/L_{p0}=0.50 up to d/L_{p0}=0.15 it increases of just 1.9%); the difference is greater if we consider the trough amplitudes, which increase of about 59%.

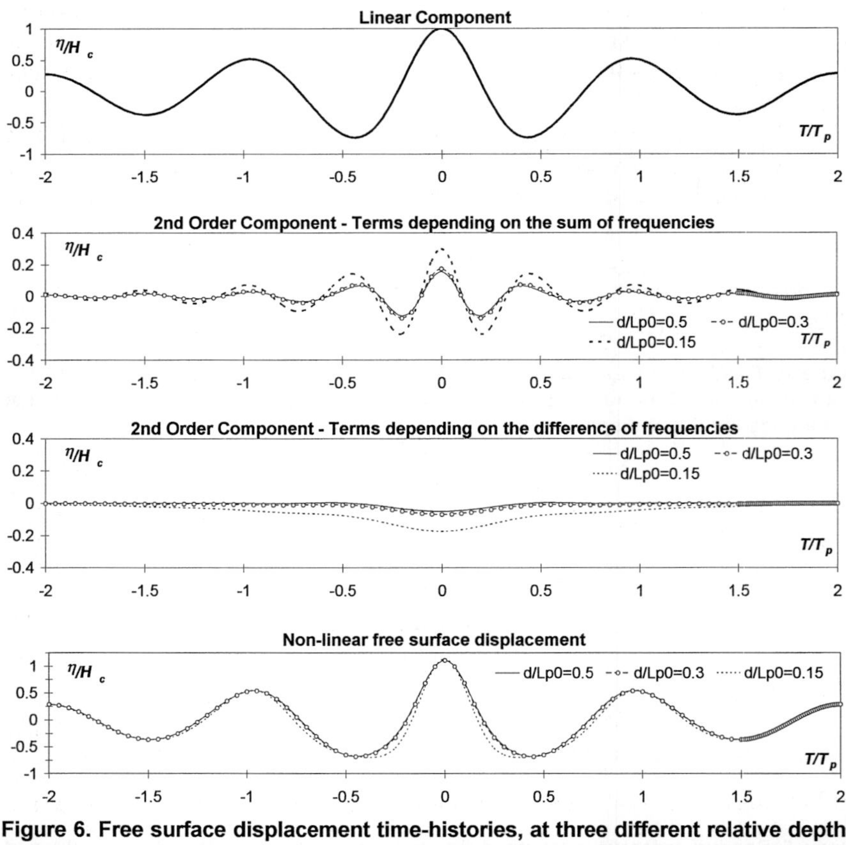

Figure 6. Free surface displacement time-histories, at three different relative depths, when a large crest height H_C occurs: (i) linear prediction; (ii-iii) second-order terms; (iii) total non-linear profile.

The second-order effect is weakened when considering the total free surface displacement as a sum of the linear component and the second-order one. For the sake of clearness in the lower picture of Figure 6 are reported the free surface displacement profile at the same three different relative depths. At d/L_{p0}=0.3, the global wave height increases of just 0.2% and at d/L_{p0}=0.15 of about 2% with respect to the deep water value 8.95 m.

Furthermore, what is important to notice is not only the entity of the second-order component, but also the shape of it. In fact, it is really strongly influenced by the set-down term in the second-order QD solution, represented by the term dependant on the difference of frequencies (third panel of Figure 6): this term makes the difference between random wave groups with either narrow-band or finite bandwidth of the spectrum.

To appreciate the effects of finite bandwidth of the spectrum, in Figure 7-middle panel the second-order component for random wave groups at d/L_{p0}=0.15 is reported, calculated for the narrow-band and the mean JONSWAP spectrum. As it can be seen, as a consequence of the set-down term the second-order correction is different at a finite water depth: for finite bandwidth a lowering on the maximum crest on the global solution (lower panel of Figure 7) greater than 12% is noticed.

This different behavior between random wave groups with narrow-band and finite bandwidth of the spectrum is well reported in Figure 8, in which it is plotted the quotient between heights of non-linear and linear largest wave crest,

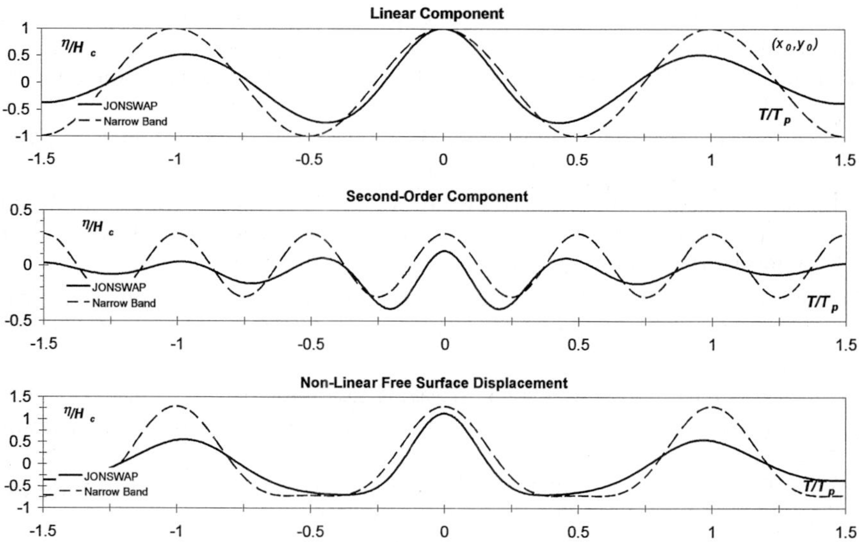

Figure 7. Random waves when a large wave crest of given height H_c occurs at the relative depth d/L_{p0}=0.15: comparison between infinitely narrow spectrum and mean JONSWAP. Upper panel: linear components; middle panel: second-order components; lower panel: total second-order free surface displacement.

as a function of bottom depth *d*. In deep water the two curves, calculated for the narrow-band and mean JONSWAP spectra, tend to be very close to each other; for decreasing water depth it can be noticed that the ratio is much greater for the infinitely narrow spectrum rather than for finite bandwidth of the spectrum. Finally, Figure 9 shows the space-time evolution of wave groups both in deep water and at a depth d/L_{p0}=0.15.

REFERENCES

Arena, F. 2005. On non-linear very large sea wave groups, *Ocean Engineering*, 32, 1311-1331.

Arena, F., Fedele, F., 2005. Non-linear space-time evolution of wave groups with a high crest, *ASME Journal of Offshore Mechanics and Ocean Engineering*, 127, 1, 46-51.

Boccotti, P. 1981. On the Highest Waves in a Stationary Gaussian Process, *Atti Acc Ligure di Scienze e Lettere*, 38, 271–302.

Boccotti, P. 1982. On Ocean Waves with High Crests, *Meccanica*, 17, 16–19.

Boccotti, P. 1983. Some New Results on Statistical Properties of Wind Waves, *Applied Ocean Research*, 5, 134–140.

Boccotti, P. 1989. On Mechanics of Irregular Gravity Waves, *Atti Acc Naz Lincei, Memorie*, 19, 11–170.

Boccotti, P. 1997. A General Theory of Three-dimensional Wave Groups, *Ocean Engineering*, 24, 265–300.

Boccotti, P. 2000, *Wave mechanics for ocean engineering*, Elsevier Science, Oxford, 1–496.

Dalzell, J. F., 1999. A note on finite depth second order wave-wave interactions, *Applied Ocean Research*, 21, 105-111.

Fedele, F. and F. Arena. 2005. Weakly nonlinear statistics of high random waves, *Physics of Fluids*, 17, 2, paper 026601, 1-10.

Hasselmann K., Barnett T. P., Bouws E., et al., 1973. Measurements of Wind Wave Growth and Swell Decay During the Joint North Sea Wave Project (JONSWAP). *Deutsche Hydrogr Zeitsch*, A8, 1–95.

Longuet-Higgins M.S. 1963. The effects of non-linearities on statistical distributions in the theory of sea waves. *Journal of Fluid Mechanics*, 17,459-480.

Mitsuyasu H., Tasai F. and Suara T., 1975. Observation of Directional Spectrum of Ocean Waves Using a Clover-Leaf Buoy. *Journal of Physical Oceanography*, 5, 750-760.

Pavone D. and Arena F., 2004. On the nonlinear structure of sea wave groups. *Proc. 1st Workshop MODECI* 2004, Rende (CS) Italy, 439-450.

Phillips, O.M. 1967. "The theory of wind-generated waves". *Advances in Hydroscience*, 4, 119-149.

Phillips, O.M., Gu, D., et al. (1993), On the expected structure of extreme waves in a Gaussian sea. *Journal of Physical Oceanography*, vol. 23, pp. 992–1000 & pp. 2297–2309.

Sharma J. N., and R. G. Dean, 1979. Development and Evaluation of a Procedure for Simulating a Random Directional Second Order Sea Surface and Associated Wave Forces. *Ocean Engineering Rep. 20, University of Delaware*, 112 pp.

Whitham G. B. 1974. Linear and Nonlinear Waves. *New York: Wiley*.

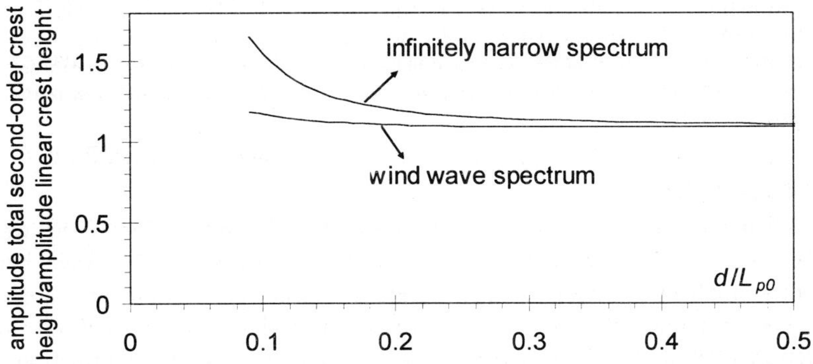

Figure 8. The quotient between heights of non-linear and linear largest wave crest, as a function of bottom depth *d*: comparison between narrow-band and wind-wave spectra.

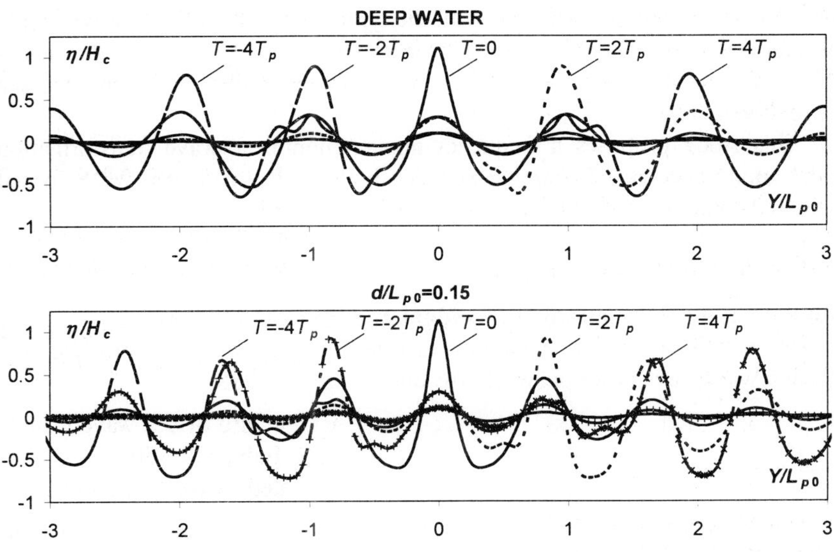

Figure 9. Space time evolution of wave groups in deep water and at d/L_{p0}=0.15.

FREQUENCY-DOMAIN NONLINEAR WAVE TRANSFORMATION MODEL INCORPORATING INCIDENT AND REFLECTED WAVE COUPLING

Kazuhiko Honda[1] and Hajime Mase[2]

This study proposed a frequency-domain nonlinear wave transformation model taking account of coupling between incident and reflected waves. The present wave model is easy to solve numerically even for the case where very shallow water area exists like reef bathymetry. The present wave model was applied to the analysis for wave transformations on reefs of laboratory and field experiments. Concerning with significant wave heights and mean water level changes, the wave model was able to estimate the observations reasonably.

INTRODUCTION

When high incident waves come into reef area from offshore, the waves break on the reef. After that, the wave energy reduces and the wave height becomes small in nearshore area. In terms of coastal protection, reefs are very effective to reduce waves. In order to understand effects of reef on wave reduction quantitatively, it is necessary to simulate wave transformations on reefs accurately.

There are various wave transformation models; among them, the Boussinesq equations are applied to solve water surface elevations and flow velocities at each point and time. In this case, the time interval and grid size should be small enough compared with the typical wave period and wavelength. Therefore, it is not suitable for a wide area calculation. Moreover, the model tends to be unstable for a very shallow water suddenly changed from deep sea like reef bathymetry.

This study proposes a frequency-domain nonlinear wave transformation model taking account of coupling between incident and reflected waves based on the KP equation. This model is easy to solve numerically.

FORMULATION OF WAVE MODEL

Simultaneous KP Equations for Incident and Reflected Waves

The simultaneous KP equations incorporating coupling of incident and reflected waves are shown below (Kirby and Vengayil, 1988):

[1] Tsunami Research Center, Port and Airport Research Institute, Nagase 3-1-1, Yokosuka, Kanagawa, 239-0826, Japan

[2] Disaster Prevention Research Institute, Kyoto University, Gokasho, Uji, Kyoto, 611-0011, Japan

$$\left\{\frac{1}{c}\zeta_t^+ + \zeta_x^+ + \frac{h_x}{4h}\left(\zeta^+ + \zeta^-\right) + \frac{3}{2h}\zeta^+\zeta_x^+ + \frac{h^2}{6}\zeta_{xxx}^+\right\}_x + \frac{1}{2h}\left(h\zeta_y^+\right)_y = 0 \quad (1)$$

$$\left\{\frac{1}{c}\zeta_t^- - \zeta_x^- + \frac{h_x}{4h}\left(\zeta^+ + \zeta^-\right) - \frac{3}{2h}\zeta^-\zeta_x^- - \frac{h^2}{6}\zeta_{xxx}^-\right\}_x - \frac{1}{2h}\left(h\zeta_y^-\right)_y = 0 \quad (2)$$

where ζ^+ and ζ^- are the incident and reflected wave components, h is the water depth, c is the phase celerity, subscripts of t, x and y mean the partial differentiation with respect to their variables. Kirby and Vengayil (1988) assumed that the wave propagates predominantly in the x-direction, while there is a weak dependence in the y-direction. In other words, the wave-propagation direction is assumed to be within a small angle to the x-axis.

Frequency-Domain Simultaneous KP Equations

We employ the Fourier series representation for wave components with complex amplitudes as follows:

$$\zeta^+ = \sum_{n=1}^{N}\left[\frac{1}{2}A_n \exp\left\{in\left(\hat{k}_1 x - \omega_1 t\right)\right\} + c.c.\right] \quad (3)$$

$$\zeta^- = \sum_{n=1}^{N}\left[\frac{1}{2}B_n \exp\left\{in\left(-\hat{k}_1 x - \omega_1 t\right)\right\} + c.c.\right] \quad (4)$$

where A_n and B_n are the complex amplitudes of wave components, $\hat{k}_1$ is the characteristic fundamental wave number, ω_1 is the characteristic fundamental angular velocity, $c.c.$ indicates the conjugate complex.

Eqs. (3) and (4) are substituted into the simultaneous KP equations represented by Eqs. (1) and (2) (Mase and Kirby, 1992). Then, we obtain the frequency-domain simultaneous KP equations as follows:

$$\frac{\partial A_n}{\partial x} + \frac{(Cg_n)_x}{2(Cg_n)}A_n - i\left(k_n - n\hat{k}_1\right)A_n - i\frac{\left\{(CCg)_n A_{ny}\right\}_y}{2(kCCg)_n}$$

$$+\frac{3ink_1}{8h}\left\{\sum_{l=1}^{n-1}A_l A_{n-l} + 2\sum_{l=1}^{N-n}A_l^* A_{n+l}\right\} - \frac{(Cg_n)_x}{2(Cg_n)}B_n \exp\left(-2in\hat{k}_1 x\right) + \alpha_n A_n = 0 \quad (5)$$

$$\frac{\partial B_n}{\partial x'} + \frac{(Cg_n)_{x'}}{2(Cg_n)}B_n - i\left(k_n - n\hat{k}_1\right)B_n - i\frac{\left\{(CCg)_n B_{ny}\right\}_y}{2(kCCg)_n}$$

$$+\frac{3ink_1}{8h}\left\{\sum_{l=1}^{n-1}B_l B_{n-l} + 2\sum_{l=1}^{N-n}B_l^* B_{n+l}\right\} - \frac{(Cg_n)_{x'}}{2(Cg_n)}A_n \exp\left(-2in\hat{k}_1 x'\right) + \alpha_n B_n = 0 \quad (6)$$

where $\alpha_n A_n$ and $\alpha_n B_n$ are the energy damping terms, Cg is the group velocity, $*$ indicates the conjugate complex, x' is the negative coordination of x-axis as $x' = -x$.

Damping Coefficient

In order to take account of energy dissipation, the damping terms were added to Eqs. (5) and (6). The damping coefficient α_n should be determined theoretically or experimentally. When the damping coefficient is real, only energy change occurs. On the other hand, when the damping coefficient is imaginary, the damping term makes the phase change. When the damping coefficient is complex, the damping term changes both energy and phase. In this study, we take the damping coefficient to be real.

Using the inverse FFT for the calculated complex amplitudes, A_n and B_n, we can obtain water surface variations in the time domain and can calculate wave characteristics such as representative wave heights, periods, crest heights, and so on.

In this study, we use the damping coefficient proposed by Mase and Kirby (1992). Their formula is based on the expected value of energy dissipation rate by Thornton and Guza (1983) and the formula of damping coefficient depends on frequencies.

Thornton and Guza (1983) formulated the expected value of energy dissipation rate based on a bore model and a distribution of broken wave heights as follows:

$$\langle \varepsilon_b \rangle = \frac{3\sqrt{\pi}}{16} \rho g B^3 \bar{f} \frac{H_{rms}^5}{\gamma^2 h^3} \left[1 - 1 \Big/ \left(1 + \left(H_{rms} / \gamma h \right)^2 \right)^{5/2} \right] \tag{7}$$

where ρ is the density of sea water, g is the gravity acceleration, B is the wave breaking coefficient, $\bar{f}$ is the characteristic frequency, H_{rms} is the root mean square wave height, γ is the coefficient of relationship between h and H_{rms}.

In order to represent the experimental tendency of energy dissipation, Mase and Kirby (1992) determined the damping coefficient considering frequency dependence:

$$\alpha_n = \alpha_0 + \left(f_n / \bar{f} \right)^2 \alpha_1 \tag{8}$$

$$\alpha_0 = F \beta_{(x)} \tag{9}$$

$$\alpha_1 = (1 - F) \beta_{(x)} \bar{f}^2 \sum_n^N |A_n|^2 \Big/ \sum_n^N f_n^2 |A_n|^2 \tag{10}$$

$$\beta_{(x)} = \langle \varepsilon_b \rangle \Big/ \sum_n^N |A_n|^2 \tag{11}$$

In this study, we use the values employed by Mase and Kirby (1992):

$$B = 1.0, \quad \gamma = 0.6, \quad F = 0.5 \tag{12}$$

When the water depth is very shallow, we introduce the following threshold of wave height H_{lim}:

$$H_{\lim} = Kh \tag{13}$$

$$\alpha_n = \begin{cases} 0 & (H_{rms} \le H_{\lim}) \\ \alpha_0 + (f_n/\bar{f})^2 \alpha_1 & (H_{rms} \ge H_{\lim}) \end{cases} \tag{14}$$

When H_{rms} is smaller than H_{lim}, the effect of energy dissipation is ignored; this means a kind of reforming.

Mean Water Level Change

In one-dimensional case, we used the formula derived by Longuet-Higgins and Stewart (1962) as

$$\frac{\partial A_0}{\partial x} + \frac{1}{2h}\frac{\partial S_{xx}}{\partial x} = 0 \tag{15}$$

$$S_{xx} = \sum_{n=1}^{N}\left(\frac{2k_n h}{\sinh(2k_n h)} + 1\right)|A_n|^2 \tag{16}$$

where A_0 is the water surface elevation and S_{xx} is the radiation stress.

In two-dimensional case, we use the following equation:

$$\frac{\partial}{\partial x}\left[\frac{\partial A_0}{\partial x} + \frac{1}{2h}\frac{\partial S_{xx}}{\partial x}\right] + \frac{1}{2h}\frac{\partial}{\partial y}\left(h\frac{\partial A_0}{\partial y}\right) = 0 \tag{17}$$

COMPARISON BETWEEN OBSERVATIONS AND PREDICTIONS

Comparison with Laboratory Observations

Takayama et al. (1977) carried out the laboratory experiments for wave transformations on reef. Table 1 shows the experimental condition of the model tests. In these experiments, two kinds of offshore water depth were used. Four kinds of random waves were generated for each offshore water depth. H_0' and T_0' are the incident wave height and wave period respectively.

Figure 1 shows the bathymetry of the model reef. The slope of the reef is 1/10, and the height of the model reef is 0.5m.

In the calculation, we used the same wave height, wave period and water depth as those of the experiments. We employed the Bretschneider-Mitsuyasu spectrum for the incident waves, since it is difficult to represent the same spectrum as the experimental waves.

Table 1. Experimental Condition

h [m]	WAVE-1		WAVE-2		WAVE-3		WAVE-4	
	H_0'	T_0'	H_0'	T_0'	H_0'	T_0'	H_0'	T_0'
0.52	0.190	2.16	0.097	1.64	0.038	1.08	0.093	1.99
0.55	0.199	2.14	0.101	1.63	0.040	1.09	0.098	1.99

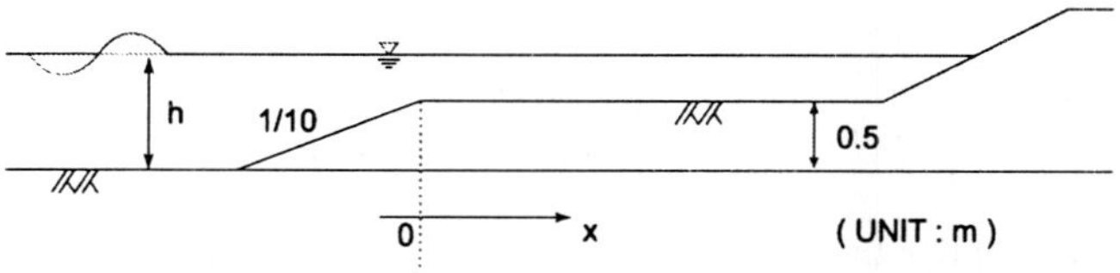

Figure 1. Bathymetry of the model reef

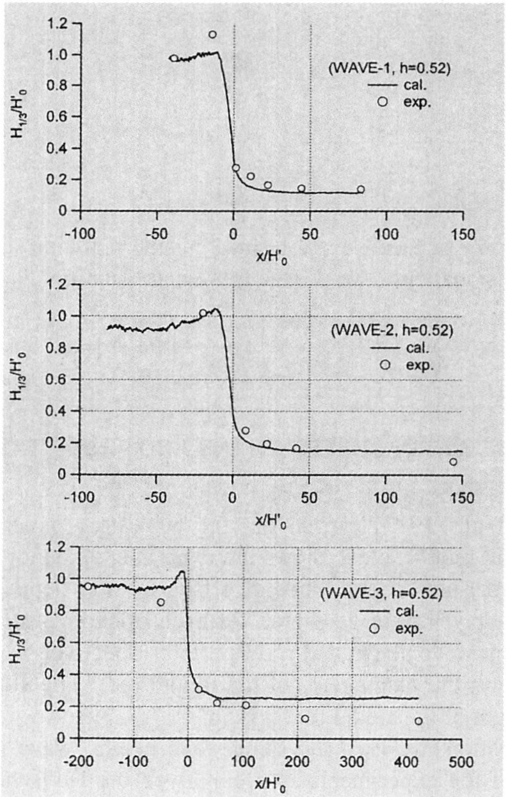

Figure 2. Spatial change of significant wave heights

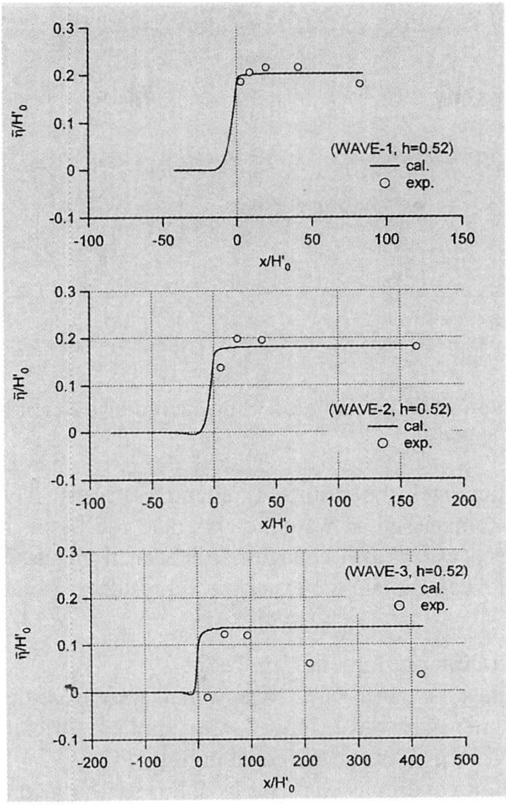

Figure 3. Spatial change of mean water level changes

Figure 2 shows the comparison of the calculated wave heights with the measured ones. In the case of WAVE-1 and WAVE-2, the measured and calculated wave heights show good agreement. On the other hand, in the case of WAVE-3 where the incident waves are small, the calculated wave heights become a little larger than the measured ones.

Figure 3 shows the spatial change of mean water level changes obtained by the experiments and calculations. The measured mean water level changes and calculated ones agree well in the case of WAVE-1 and WAVE-2. On the other hand, in the case of WAVE-3, the measured mean water level changes become smaller shoreward. It was reported that the measured wave heights and mean water level changes were not accurate in the case of WAVE-3 because of water leakage between the model reef and the wall of water basin.

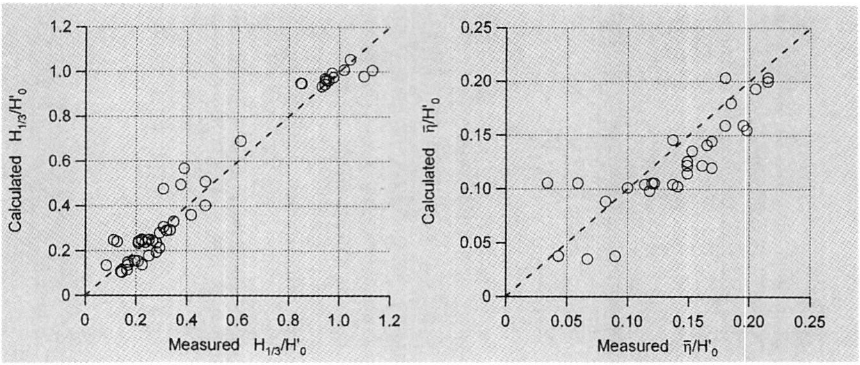

Figure 4. Comparison of all calculated and measured significant wave heights and mean water level changes

Figure 4 summarizes the results of all comparisons. The left hand side figure shows the comparison of wave heights, and the figure in the right hand side is for the mean water level changes. It is seen from these figures that the present wave model can explain the experimental results reasonably.

Comparison with Observations at Uza Coast

The Uza Coast is quasi one dimensional without little lateral change. Therefore, one-dimension wave model was applied to calculate the wave transformations over the coastal reef shown in Fig. 5.

The calculation conditions were set as follows. The grid size is 1.0m, the time step is 1.0s, and the frequency spectrum at St-1 is the Bretshneider-Mitsuyasu spectrum.

Figure 6 shows the comparison between the calculated and measured wave heights and mean water level changes for one of cases. The incident waves break at the edge of reef, and the mean water level change is elevated (wave set-up). The measured and calculated wave heights and mean water level changes agree very well.

Figure 7 shows all results of comparisons between observations and calculations at each observation station. It is recognized from these results that the present wave model is useful to estimate wave transformations for various field wave conditions.

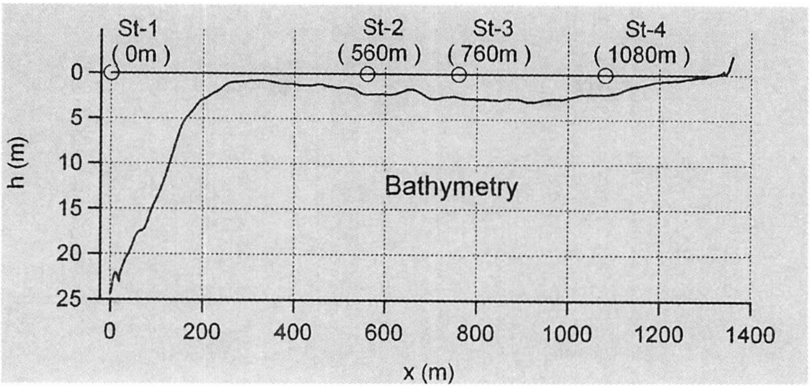

Figure 5. Bathymetry of the Uza Coast

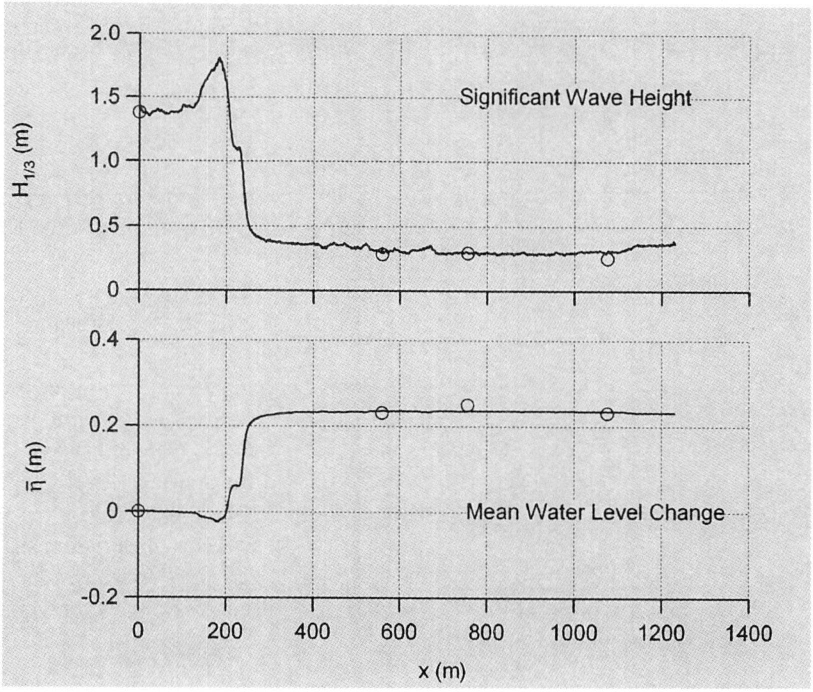

Figure 6. Spatial distribution of significant wave heights and mean water level changes

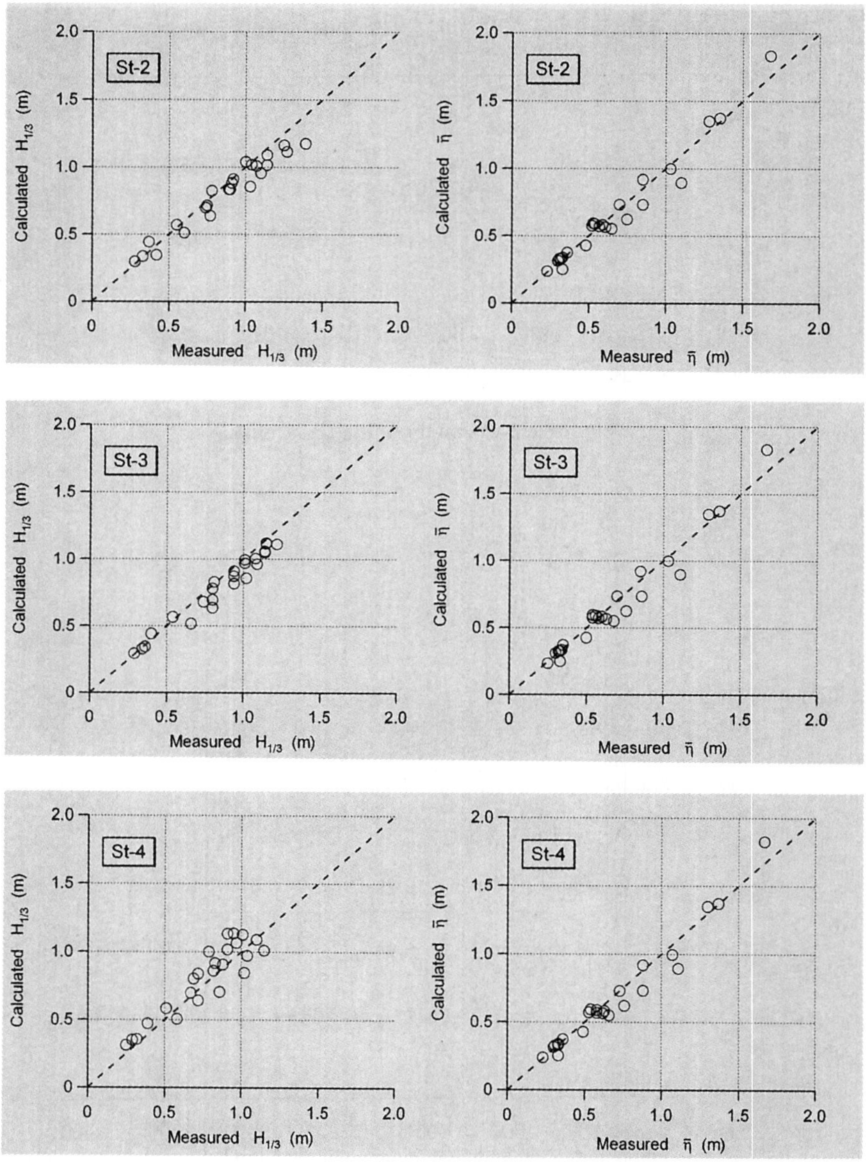

Figure 7. Comparison of all observations and calculations

Comparison with Observations at Mizukama Coast

The Mizukama coast has complex bathymetry with lateral change as Fig. 8. The coastal line is straight and has seawalls at x=265m. There are three sets of observed data, but without information of wave direction and directional spreading.

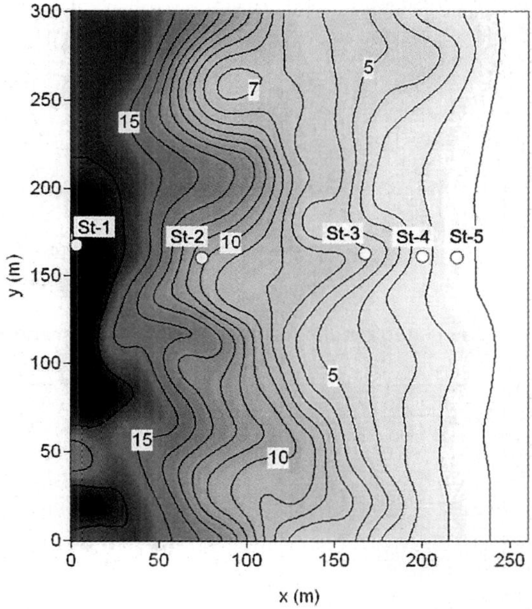

Figure 8. Bathymetry of the Mizukama Coast

The calculations were carried out under the condition that the time step is 1.0s and the grid size is 5.0m. We used the wave height and wave period of incident waves observed at the station St-1, and assumed the Bretschneider-Mitsuyasu spectrum as the incident wave spectrum. We used four case of directional spreading parameter (S_{max} = ∞, 75, 25, 10) and two case of incident angle to x-axis (θ = 0, 20 deg.).

Figure 9 shows the calculated wave height and mean water level change for the case that the wave is one-directional and the angle is 0 degree. Figure 10 shows the calculated results for S_{max} = 25 and θ = 20 degree. The wave height increases and decreases due to wave shoaling, refraction and breaking. In the calculation using S_{max} = 25, the concentration of wave energy by refraction becomes smooth compared to that using S_{max} = ∞. In the nearshore area, the wave height decreases and the mean water level change increases.

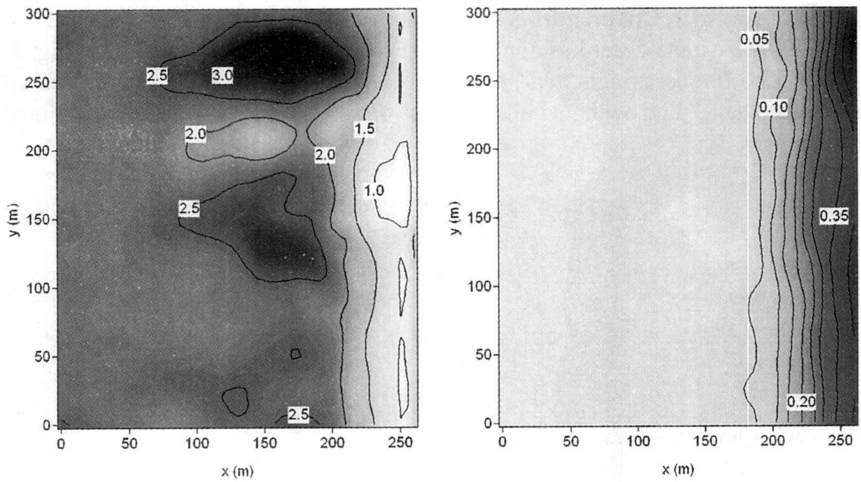

Figure 9. Calculated results for the case of S_{max} = ∞ and θ = 0

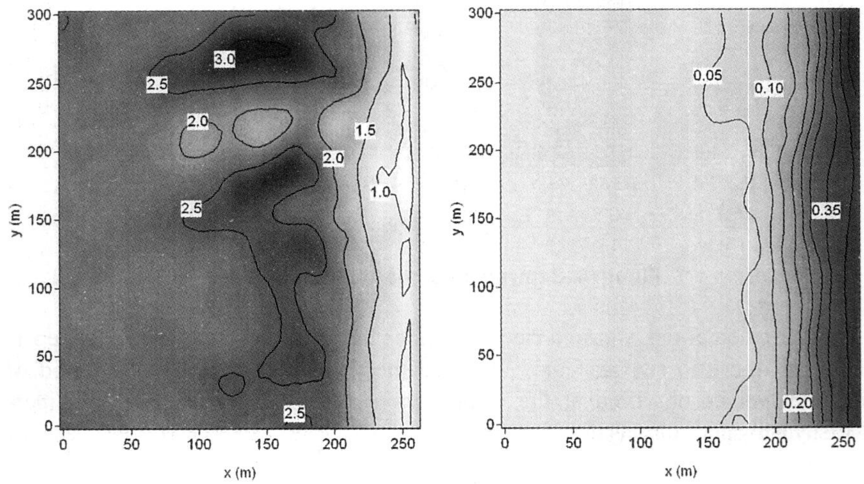

Figure 10. Calculated results for the case of S_{max} = 25 and θ = 20

Figure 11 summarizes all results of comparison between all calculated and measured significant wave heights and mean water level changes at each observation station. The measured wave heights agree well with the calculated ones. On the other hand, the calculated mean water level changes become larger than the measured ones at St-4 and St-5.

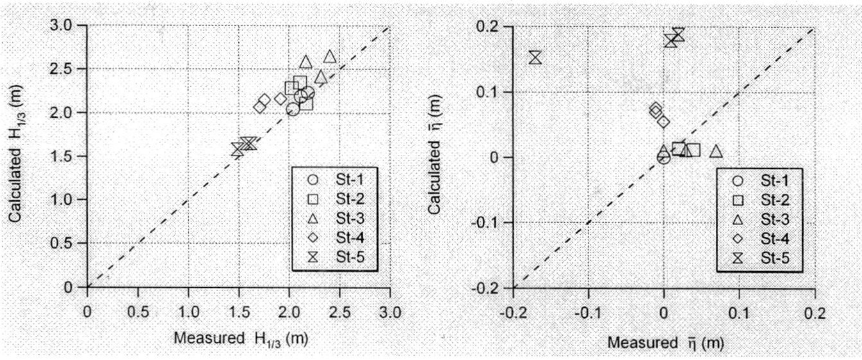

Figure 11. Comparison of all calculated and measured significant wave heights and mean water level changes

The present two-dimensional wave model could not predict the mean water level change in the area near shoreline; however, we doubt accuracy of data of mean water level changes at the station St-4 and St-5, since wave set-up should be observed in the surf zone.

CONCLUSIONS

This study proposed a frequency-domain nonlinear wave transformation model taking account of coupling between incident and reflected waves, and it was found that the present wave model is useful even for a very shallow water area like reef bathymetry. Horizontally two-dimensional wave model and one-dimensional wave model were checked against laboratory and field observations.

Concerning with significant wave heights, the horizontally two-dimensional wave model and one-dimensional model were able to predict observations. For prediction of mean water level changes, the present one-dimensional mean water level change model can predict observations. On the other hand, the two-dimensional mean water level change model should be more applied to experiments and observations.

REFERENCES

Kirby, J.T. and P. Vengayil. 1988. Nonresonant and resonant reflection of long waves in varying channels, *Jour. Geophysical Res.*, Vol.93, pp.10782-10796.

Longuet-Higgins, M.S. and R.W. Stewart. 1962. Radiation stress and mass transport in gravity waves, with application to "surf beats", *Jour. Fluid Mech.*, Vo.13, pp.481-504

Mase, H. and J.T. Kirby. 1992. Hybrid frequency-domain KdV equation for random wave transformation, *Proc. 23rd Coastal Eng. Conf.*, ASCE, pp.474-487

Takayama, T., Y. Kamiyama and O. Kikuchi. 1977. Wave Transformation on a reef, *Tech. Note of the Port and Harbour Res. Inst.*, No.278, 32p

Thornton, E.B. and J.T. Guza. 1983. Transformation of wave height distribution, *Jour. Geophysical Res.*, Vol.88, No.C10, pp.5925-5938

PROGRESS ON A CONVOLUTION METHOD FOR NONLINEAR DISPERSIVE WAVE TRANSFORMATION OVER GENTLY VARYING BATHYMETRY

Hemming A. Schäffer [1]

Developments toward a nonlinear and fully dispersive irregular-wave model in two horizontal dimensions are presented. The model relies on time stepping the two free surface boundary conditions with a closure between the horizontal and vertical velocity expressed in terms of a convolution integral. The formulation is fully explicit in space and thus no equations need to be solved. The convolution integral involves an exponentially decaying impulse response function. In practice, this reduces the infinite limits of the integral to horizontal distances of only several water depths. Generalization from constant to variable bathymetry is investigated by introducing a slight distortion of the local horizontal coordinates.

INTRODUCTION

Boussinesq-type models have evolved tremendously over the past decade (Kirby, 2003). The latest developments practically eliminate the usual water depth to wavelength restriction and treat highly nonlinear waves with great accuracy (Madsen et al. 2002, 2003,...). However, the increase in accuracy and range of application has been accompanied by a significant increase in computational effort. The present paper uses an alternative approach by which the high accuracy and large application range is attainable at lesser computational cost. The method is fully dispersive and thus there is no restriction on the water depth. For constant depth a similar approach to the linear part of the model has previously been used for computing the kinematics field from the kinematics at the still water level as assumed to be provided by a numerical wave model (Schäffer, 2003). The perturbation approach of West et al. (1987) is applied for inclusion of nonlinearity to any order. The linear part of the approach may be regarded as an alternative to the linear mild-slope equation, but valid for irregular waves with a spectral bandwidth limited only by resolution. The present work shows progress toward an extension to two horizontal dimensions of the channel model of Schäffer (2004, 2005). The theoretical developments are preliminary and need further checking. Lengthy calculations are sometimes omitted in the transition between the equations.

---

[1] SchäfferWaves, Sortedam Dossering 59H st., DK-2100 Copenhagen Ø, Denmark. Hemming@SchafferWaves.dk

THE NONLINEAR TIME-STEPPING PROBLEM AND THE LINK TO LINEAR THEORY

The model is based on the time-stepping of the fully nonlinear free surface boundary conditions which may be expressed as

$$\eta_t = \tilde{w}(1+\nabla\eta\cdot\nabla\eta) - \tilde{\mathbf{V}}\cdot\nabla\eta \tag{1}$$

$$\tilde{\mathbf{V}}_t = -g\nabla\eta - \tfrac{1}{2}\nabla\left(\tilde{\mathbf{V}}\cdot\tilde{\mathbf{V}} - \tilde{w}^2(1+\nabla\eta\cdot\nabla\eta)\right) \tag{2}$$

Here tilde denotes variables at the free surface and the velocity $\tilde{\mathbf{V}}$ is defined by $\tilde{\mathbf{V}} \equiv \tilde{\mathbf{u}} + \nabla\eta\tilde{w}$, where $(\mathbf{u}, w)$ is the particle velocity vector and η is the surface elevation. In terms of the velocity potential $\mathbf{\Phi}$, $\tilde{\mathbf{V}}$ equals the gradient of the surface potential, $\tilde{\mathbf{V}} = \nabla\tilde{\mathbf{\Phi}}$, and (2) may equivalently be expressed as

$$\tilde{\mathbf{\Phi}}_t = -g\eta - \tfrac{1}{2}\left((\nabla\tilde{\mathbf{\Phi}})^2 - \tilde{w}^2(1+\nabla\eta\cdot\nabla\eta)\right) \tag{3}$$

Either form of the surface conditions have been widely used by several authors, e.g. West et al. (1987).

To advance (1) and (2) in time, a relation between the dependent variables is needed by which $\tilde{w}$ can be determined at every time step, given $\tilde{\mathbf{V}}$ and η. As for the high-order spectral method of West et al. an explicit form is used, but instead of spectral computations in Fourier space, the final model is expressed in physical-space. The desired closure may be written

$$\tilde{w} = f\,\tilde{\mathbf{V}}; \quad f = f(h;\eta) \tag{4}$$

where f is a suitable operator and h is the water depth. Using a combination of Taylor expansion and perturbation theory, West et al.[2] derived an arbitrary-order procedure to be used at every time step by which the nonlinear closure can be expressed in terms of the linearized version

$$w_0 = f_0\mathbf{u}_0; \quad f_0 = f_0(h) \equiv f(h;0) \tag{5}$$

where subscript "0" on the velocities refers to variables at the still water level. Although the approach of West et al. is based on $\tilde{\mathbf{\Phi}}$, it easily adapts to a $\tilde{\mathbf{V}}$ formulation and the remaining challenge is to derive the a physical-space operator f_0 applicable to mild-slope bathymetry. The remainder of this paper is dedicated to this task. Eventually, complex shaped domains and associated lateral boundary conditions also need to be addressed.

---

[2] Several techniques like that of West et. al (1987) were compared analytically by Schäffer (2006) and found to be essentially the same.

THE BASIC MILD-SLOPE KINEMATICS RELATION

From the impermeable-bottom boundary condition and the continuity equation Madsen and Schäffer (1998) derived an implicit relation between $\mathbf{u}_0$ and w_0 in terms of infinite series differential operators. In the formulation of Agnon et al. (1999), this is given by

$$\cos(h\nabla)w_0 + \sin(h\nabla)\mathbf{u}_0 + \nabla h \cdot \left(-\sin(h\nabla)w_0 + \cos(h\nabla)\mathbf{u}_0\right) = 0 \tag{6}$$

Here

$$\cos(h\nabla) \equiv \sum_{n=0}^{\infty}(-1)^n \frac{h^{2n}}{(2n)!}\nabla^{2n} \tag{7}$$

$$\sin(h\nabla) \equiv \sum_{n=0}^{\infty}(-1)^n \frac{h^{2n+1}}{(2n+1)!}\nabla^{2n+1} \tag{8}$$

with the understanding that powers of nabla should be interpreted as

$$\begin{aligned} &\nabla^{2n}\mathbf{u} \equiv \nabla(\nabla^{2n-2})(\nabla\cdot\mathbf{u}); \quad &&\nabla^{2n+1}\mathbf{u} \equiv \nabla^{2n}(\nabla\cdot\mathbf{u}) \\ &\nabla^{2n}w \equiv \nabla^{2n}w; &&\nabla^{2n+1}w \equiv \nabla(\nabla^{2n}w) \end{aligned} \tag{9}$$

depending on whether the power is even or odd and whether the operand is a vector or a scalar function. Although the goal of Madsen and Schäffer was to derive higher-order Boussinesq-type equations, (6) is generally valid and needs neither restriction on the water depth nor on the bottom slope. The mild-slope assumption will be made throughout the remainder of this paper, but no limits will be imposed on the water depth. Thus, depth changes may well accumulate to order-of magnitude depth differences provided they occur gradually.

A further mild-slope simplification of (6) can be obtained as

$$\cos(h\nabla)hw_0 + \sin(h\nabla)h\mathbf{u}_0 = 0 \tag{10}$$

assuming

$$|\nabla\times\mathbf{u}_0| = O(|\nabla h|) \tag{11}$$

by which a distinction between $\nabla(\nabla\cdot\mathbf{u}_0)$ and $\nabla^2\mathbf{u}_0$ is no longer needed in the bottom slope term.

THE MILD-SLOPE SOLUTION TO THE KINEMATICS RELATION

The challenge is now to solve (10) for w_0 introducing no further assumptions[3]. This requires inversion of the cosine operator. For constant depth the result is simply the secant operator, but for variable depth additional depth-gradient terms appear. These may well be derived in the framework of ordinary physical space. However, a simpler way goes via a scaled and slightly distorted physical space. This scaling and distortion is needed at a later stage in the derivation but is useful already now. The idea is to absorb the water depth in the operators. For constant depth, this merely requires a scaling of the horizontal distances by the depth. Defining nabla in scaled space and denoting it by $\tilde{\nabla}$, then $h^n \nabla^n$ turns into $\tilde{\nabla}^n$. The variable-depth problem in on horizontal dimension was successfully treated by introducing a new dependent variable, $s = s(x)$, defined by $\partial s / \partial x = 1/h$. As a generalization to two horizontal dimensions, we thus introduce $s = s(x, y)$ by

$$|\nabla s| = \frac{1}{h} \qquad (12)$$

Taking (r, θ) as local polar coordinates in the original physical space, we get to mild-slope accuracy

$$s = \frac{r}{h}\left(1 + \frac{1}{2}\frac{r}{h}\nabla h \cdot \nabla r\right) \qquad (13)$$

An appropriate orthogonal coordinate may be shown to be

$$\gamma = \theta + \frac{1}{2}\frac{r}{h}\nabla h \cdot r\nabla\theta \qquad (14)$$

We now define a distorted physical space with the property that (s, γ) are the polar coordinates. The equivalent rectangular coordinates are denoted $(\tilde{x}, \tilde{y})$. Defining vector and gradient operators in distorted space and expressing velocities in basis vectors of the $(\tilde{x}, \tilde{y})$-system, we obtain the relations

---

[3] Had this been a Boussinesq-type approach, the infinite series operators would have been truncated, thus restricting the water depth/dispersion.

$$h^n\nabla^n w_0 = \left(\tilde{\nabla}^n - \frac{1}{2}\nabla h\cdot\left(n(n-1)\tilde{\nabla}^{n-2}\tilde{\nabla} - n\tilde{\nabla}^{n-1}\right)\right)\tilde{w}_0 \qquad \text{for even } n$$

$$h^n\nabla^{n-1}\nabla\cdot\mathbf{u}_0 = \left(\tilde{\nabla}^{n-1} - \frac{1}{2}\nabla h\cdot\left(n(n-1)\tilde{\nabla}^{n-2} - n\tilde{\nabla}^{n-2} - \tilde{\nabla}^{n-2}\right)\right)\tilde{\nabla}\cdot\tilde{\mathbf{u}}_0 \quad \text{for odd } n$$

(15)

Regarding each of these as the n 'th term of a Taylor series for an even or odd operator function, α , this further gives

$$\alpha(h\nabla) = \left(\alpha(\tilde{\nabla}) - \frac{1}{2}\nabla h\cdot\left(\alpha''(\tilde{\nabla})\tilde{\nabla} - \alpha'(\tilde{\nabla})\right)\right) \qquad \text{for even } \alpha$$
$$\alpha(h\nabla) = \alpha(\tilde{\nabla}) - \frac{1}{2}\nabla h\cdot\left(\alpha''(\tilde{\nabla})\tilde{\nabla} - \alpha'(\tilde{\nabla}) - \frac{\alpha(\tilde{\nabla})}{\tilde{\nabla}}\right) \quad \text{for odd } \alpha \tag{16}$$

by which

$$\cos(h\nabla) = \cos(\tilde{\nabla}) - \frac{1}{2}\nabla h\cdot\left(-\cos(\tilde{\nabla})\tilde{\nabla} + \sin(\tilde{\nabla})\right)$$
$$\sin(h\nabla) = \sin(\tilde{\nabla}) - \frac{1}{2}\nabla h\cdot\left(-\sin(\tilde{\nabla})\tilde{\nabla} - \cos(\tilde{\nabla}) - \frac{\sin(\tilde{\nabla})}{\tilde{\nabla}}\right) \tag{17}$$

Substituting these in (10) and solving for $\tilde{w}_0$ gives

$$\tilde{h}\tilde{w}_0 = -\frac{\tan(\tilde{\nabla})}{\tilde{\nabla}}\tilde{\nabla}\cdot\tilde{h}\tilde{\mathbf{u}}_0 - \frac{1}{2}\nabla h\cdot\left(1 + \tan^2(\tilde{\nabla}) + \frac{\tan(\tilde{\nabla})}{\tilde{\nabla}}\right)\tilde{h}\tilde{\mathbf{u}}_0 \tag{18}$$

Fourier transformation with respect $(\tilde{x}, \tilde{y})$ yields

$$\mathcal{F}\left(\tilde{h}\tilde{w}_0\right) = -\frac{\tanh\kappa}{\kappa}\mathcal{F}\left(\tilde{\nabla}\cdot\tilde{h}\tilde{\mathbf{u}}_0\right) - \frac{1}{2}\nabla h\cdot\left(1 - \tanh^2\kappa + \frac{\tanh\kappa}{\kappa}\right)\mathcal{F}\left(\tilde{h}\tilde{\mathbf{u}}_0\right) \tag{19}$$

and returning to distorted physical space using the convolution theorem finally gives

$$\tilde{h}\tilde{w}_0 = -\int_{-\infty}^{\infty}\int_{-\infty}^{\infty}\left(R_a(s')\tilde{\nabla}\cdot\tilde{h}\tilde{\mathbf{u}}_0(\tilde{\mathbf{x}}') + \frac{1}{2}R_b(s')\nabla h\cdot\tilde{h}\tilde{\mathbf{u}}_0(\tilde{\mathbf{x}}')\right)d\tilde{\mathbf{x}}' \tag{20}$$

where

$$R_a(s) \equiv \mathcal{F}^{-1}\left(\frac{\tanh\kappa}{\kappa}\right) \tag{21}$$

and

$$R_b(s) \equiv \mathcal{F}^{-1}\left(1-\tanh^2\kappa+\frac{\tanh\kappa}{\kappa}\right) \tag{22}$$

This concludes the present stage of the theoretical mild-slope development. It would be attractive if the two terms of (20) could be merged while eliminating the need for an explicit depth gradient term. This was accomplished for the case of one horizontal dimension (Schäffer, 2005).

THE CONSTANT-DEPTH SOLUTION TO THE KINEMATICS RELATION

For constant depth the explicit solution reduces to

$$w_0(x,y) = -\frac{1}{h}\int_{-\infty}^{\infty}\int_{-\infty}^{\infty}\nabla u_0(x',y')R_a(s')\,dy'dx' \tag{23}$$

where

$$s' = \frac{1}{h}\sqrt{(x-x')^2+(y-y')^2} \tag{24}$$

The impulse response function can be expressed in the two different forms

$$R_a(s) = \frac{1}{\pi}\sum_{n=0}^{\infty}\mathrm{K}_0\left((2n+1)\frac{\pi}{2}s\right) \qquad \text{(fine for far field)} \tag{25}$$

and

$$R_a(s) = \frac{1}{2\pi}\left(\frac{1}{s}+\sum_{n=1}^{\infty}\frac{(-1)^n}{\sqrt{\left(\frac{s}{2}\right)^2+n^2}}\right) \qquad \text{(neat for near field)} \tag{26}$$

where the latter has the advantage of an alternating sign. This allows for fast summation by a range of different techniques, and it may well prove more efficient then (25) even for the farther parts of the field. Figure 1 shows the variation of R_a. The exponential decay (as seen analytically in (25)) eliminates the need for the infinite limits of the convolution integral. Thus, computing the volume under the impulse response function

$$\begin{aligned} V(s_{\max}) &\equiv \iint\limits_{\substack{s<s_{\max}\\0<\theta\le 2\pi}} R_a\,dA \\ &= 2\pi\int_0^{s_{\max}} R_a(s)\,s\,ds \end{aligned} \tag{27}$$

gives

$$
\begin{aligned}
V(2) &= 0.9 \\
V(4) &= 0.99 \\
V(5) &= 0.999 \\
V(\infty) &= 1
\end{aligned}
\tag{28}
$$

showing that e.g. 99.9% of the impulse response is accounted for within a horizontal neighborhood of only five water depths.

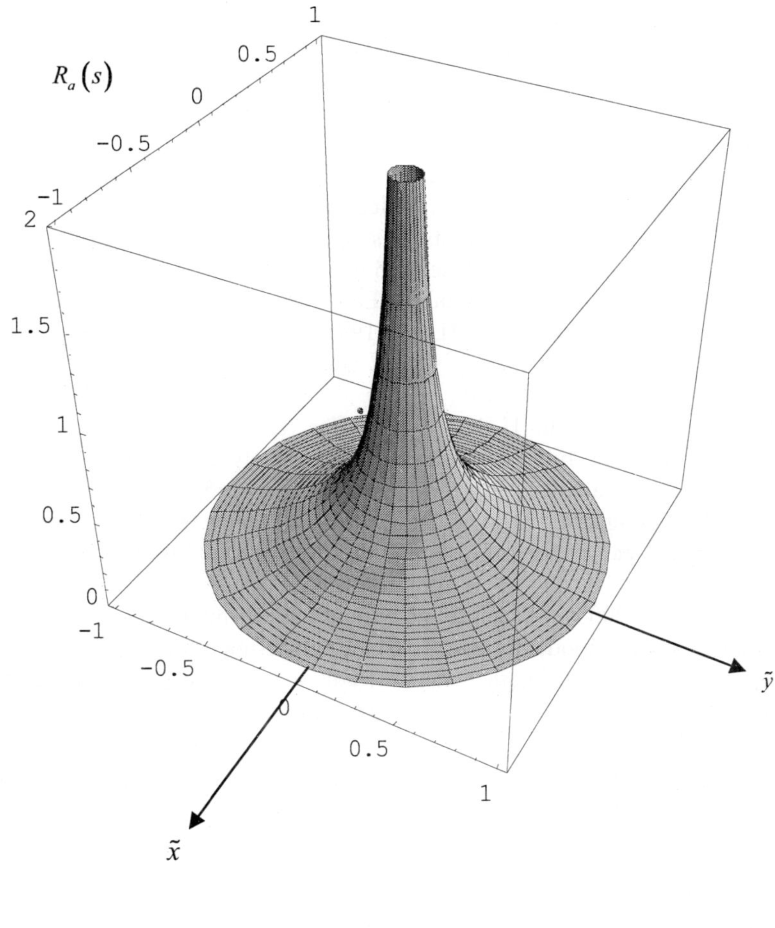

Figure 1. The constant-depth impulse response function.

CONCLUDING REMARKS

This work in progress aims at developing an efficient deterministic wave model applicable for problems with horizontal scales of several kilometers while including all potential-flow wave phenomena (except those of extreme nonlinearity). Compared with the classical mild-slope equation the approach includes irregularity and nonlinearity. Compared with Boussinesq models it eliminates the depth limitation and may include higher-order nonlinearity. Compared to higher-order spectral methods, it allows for order-of-magnitude depth variations and eventually also for arbitrarily shaped lateral boundaries. This paper has outlined the theoretical progress in the mild-slope development of the underlying linear model in two horizontal dimensions.

ACKNOWLEDGMENTS

Support from the Danish Research Council for Technology and Production Sciences (FTP grant No 26-04-0191) is greatly appreciated.

REFERENCES

Agnon, Y. Madsen, P.A. and Schäffer, H.A. (1999). A new approach to high order Boussinesq Models. J Fluid Mechanics, 399, 319-333.

Kirby, J.T. (2003). Boussinesq models and applications to nearshore wave propagation, surf zone processes and wave-induced currents. In Advances in Coastal Engineering, V. C. Lakhan (Ed.), Elsevier.

Madsen, P.A. and Schäffer, H.A. (1998). Higher-order Boussinesq-type equations for surface gravity waves: derivation and analysis. Phil Trans R Soc Lond A, 356, 3123-3184.

Madsen, P.A., Bingham, H.B. and Schäffer, H.A. (2003). Boussinesq-type formulations for fully non-linear and extremely dispersive water waves: Derivation and analysis. Proc. Roy. Soc. Lond. A. 459, 1075-1104.

Madsen, P.A., Bingham, H.B. and Liu, H. (2002). A new Boussinesq method for fully non-linear waves from shallow to deep water. J. Fluid Mech. 462, 1-30.

Schäffer, H.A. (2003). Accurate determination of internal kinematics from numerical wave model results. Coastal Engineering 50, 199-211.

Schäffer, H.A. (2004). Another step towards a post-Boussinesq wave model. Proc 29th Int. Conf. on Coastal Engrg., Lisbon, Portugal, 19-24 September 2004.

Schäffer, H.A. (2005). A convolution method for 1-D nonlinear dispersive wave transformation over a mild-slope bottom. Waves '05: Proc. 5th Int. Symp. On Ocean Wave Measurement and Analysis, Madrid, Spain, 3-7 July 2005.

Schäffer, H.A. (2006). Comparison of Dirichlet-Neumann operator expansions for nonlinear surface gravity waves. Submitted.

West B.J., Brueckner, K.A. Janda, R.S., Milder, D.M. and Milton, R.L. (1987). A new numerical method for surface hydrodynamics. J. Geophys. Res. 92(C11), 11803-11824.

THE IMPORTANCE OF DIRECTIONALITY IN THE EVOLUTION OF LARGE WAVES IN INTERMEDIATE AND SHALLOW WATER

Vasiliki Katsardi[1] and Chris Swan[1]

Earlier work by Katsardi & Swan (2004) showed that the evolution of large waves in intermediate and shallow water depths may be fundamentally different to that which occurs in deep water. In the latter case linear dispersion is the dominant process. In contrast, in intermediate and shallow water depths, the evolution of uni-directional wave fields is dominated by nonlinear wave modulation rather than linear focusing. In the present paper, numerical calculations based upon the fully nonlinear, three-dimensional wave model proposed by Bateman, Swan & Taylor (2001, 2003) are presented. These consider a number of realistic frequency spectra of varying bandwidth and contrast the evolution of large waves with a variety of directional spreads. The results of these calculations show that the evolution of large waves in intermediate and shallow water depths is critically dependent upon the directionality of the wavefield. In particular, it raises questions over the appropriateness of adopting uni-directional wave trains in both physical model studies and numerical simulations.

INTRODUCTION

This paper concerns the formation of large waves in intermediate and shallow water depths and, more specifically, considers whether recent advances in our understanding of how large waves evolve in uni-directional seas in both deep and shallow water are equally valid in multi-directional seas. Particularly, this paper builds upon earlier uni-directional work presented at ICCE2004 (Katsardi & Swan (2004)). This work showed that the evolution of large waves in intermediate and shallow water depths is dominated by nonlinear wave modulation; the largest events occurring as a modular instability in front of a truncated quasi-regular wave train.

In contrast, in deep water it is well established that the evolution of a wave field is governed by linear dispersion; the largest waves occurring as transient events in random and directionally spread wave fields due to the focusing of freely propagating wave components. This is the process whereby a large number of wave crests, corresponding to different frequencies and having different directions of propagation, are superimposed at one point. This produces a large wave crest that rapidly disperses in both space and time.

---

[1]Department of Civil & Environmental Engineering, Imperial College, London, SW7 2AZ, UK. E-mail: vasiliki.katsardi@imperial.ac.uk & c.swan@imperial.ac.uk

Historically, these events were investigated using long time-domain simulations, where the input is based upon either linear or second-order random wave theory. However, having appreciated the nature of these focusing events, Tromans *et al.* (1991) built upon earlier work of Lindgren (1970) and Bocotti (1983) and proposed the so-called *NewWave* model. This describes the expected or most probable shape of a large linear wave of given height as the scaled autocorrelation function of the underlying spectrum and thereby eliminates the need for long time-domain simulations.

Johannessen & Swan (2003) showed that the nonlinear increase in crest elevation in a deep-water focused wave-group was the result of the transfer of energy to higher frequencies. Furthermore, they showed that the extent of these energy transfers were strongly dependent upon the directionality of the sea-state. This work was extended by Gibson & Swan (2006) in which they showed that the deep-water energy transfers were dominated by third-order or four-wave resonant interactions and that the magnitude of these were indeed dependent on directionality. More specifically, they showed that very large nonlinear increases in the maximum crest elevation were only observed in very long-crested sea-states. Nevertheless, nonlinearity remains important leading to significant changes in the crest elevation, the wave slope and the underlying directionality. The purpose of the present paper is to determine whether the effects of directionality are equally important in intermediate and shallow water depths.

WAVE MODELLING

Recent advances in wave modelling allow fully nonlinear predictions of the evolution of large waves in realistic wave fields, involving a significant spread of wave energy in both frequency and direction. In particular, Bateman, Swan & Taylor (2001) provide a highly efficient wave model in which a spatial description of the water surface elevation, $\eta(x,y)$, and the velocity potential on that surface, $\phi(x,y,\eta)$, are time-marched using the fully nonlinear free-surface boundary conditions coupled with a Taylor series expansion of the Dirichlet – Neumann operator. In a follow-up paper, Bateman, Swan & Taylor (2003) apply a related approach to accurately describe the internal water particle kinematics based on the previous solution for $\eta(x,y)$ and $\phi(x,y,\eta)$. An essential element of both of these wave models lies in their computational efficiency. This is not sought for its own sake, but is an absolute requirement necessary to achieve the very high resolution in both the wave number and the directional domains, without which realistic wave fields cannot be successfully modelled. Taken together these two wave models, hereafter referred to as *BST*, provide a complete solution of highly nonlinear transient wave-events arising in irregular or random wave fields, involving a significant directional spread. A brief review of this model is given in Katsardi & Swan (2004); with full details provided in the original papers (Bateman, Swan & Taylor, 2001 & 2003).

DISCUSSION OF RESULTS

The numerical calculations presented within this paper concern the evolution of a *JONSWAP* spectrum defined by

$$S_{\eta\eta}(\omega) = \frac{\alpha g^2}{\omega^5}\exp(-\beta\frac{\omega_p^4}{\omega^4})\gamma^{\exp\left[-\frac{(\omega-\omega_p)^2}{2\omega_p^2\sigma^2}\right]} \tag{1}$$

where α=0.0081, β=1.25, σ=0.07 for $\omega \leq \omega_p$ and 0.09 for $\omega > \omega_p$. In each of the cases presented the peak spectral frequency is given by ω_p=0.628rad/s, corresponding to a peak period of T_p=10s, and the peak enhancement factor is given by γ=2.5.

A common approach adopted in descriptions of directional seas is to apply a constant directional distribution to all the frequency components within a uni-directional sea-state (or spectrum); the former acting to spread the wave energy relative to the mean wave direction. One commonly adopted directional distribution is due to Mitsuyasu (1975):

$$a(\theta) = \lambda \cos^{2s}\left(\frac{\theta}{2}\right), \tag{2}$$

where s is the so-called Mitsuyasu directional spreading parameter, $a(\theta)$ is the proportion of the total amplitude that propagates at the angle θ from the mean direction and λ is a normalising coefficient. Field observations (Jonathan *et al.* (1994)) suggest that severe storms in the northern North-Sea may be characterised by s=7.

Within this paper we will revisit the limiting uni-directional wave case (Case A9.5UD corresponding to uni-directional seas with an input amplitude sum of A=9.5m) first investigated in Katsardi & Swan (2004), and will present new numerical calculations involving a range of directional spreads. These will be based upon the Mitsuyasu spreading parameter (Eq.(2)), covering values in the range of $7 \leq s \leq 500$. At the lower limit (s=7), this corresponds to a very short-crested wave event; whilst the upper limit (s=500) represents a very long-crested (near uni-directional) wave field. In all other respects the definition of the wave spectrum is held constant (T_p=10s, γ=2.5 and A=9.5m).

Figure 1 shows a sequence of spatial wave profiles, $\eta(x)$, describing the evolution of two nonlinear wave fields corresponding to cases A9.5UD and A9.5s7; the left-hand column describing the limiting uni-directional (UD) and the right-hand column the short-crested (s=7) wave case. In the latter case the

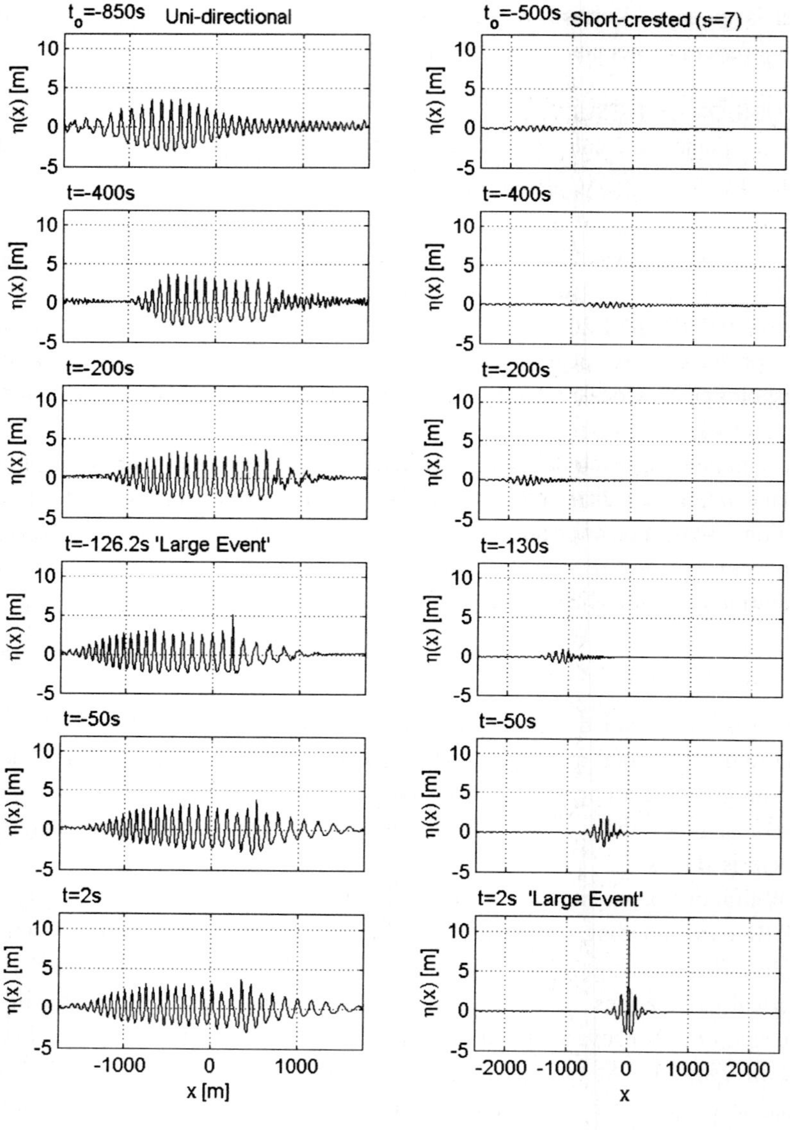

Figure 1. Spatial and temporal evolution of the water surface elevation: the role of directionality in the evolution of Case A9.5. Left-hand figures correspond to the uni-directional case (A9.5UD); whilst the right-hand figures concern the centerline profiles of the short-crested (s=7) sea-state (A9.5s7).

profiles are presented along the centerline cross-section defining the mean wave direction and about which the evolution is symmetric (i.e. $0<x<L_x$ and $y=L_y/2$, where L_x is the domain length and L_y the domain width). At each point in time the maximum surface elevation is located along this line. As shown previously in Katsardi & Swan (2004), the evolution of the uni-directional wave case is not governed by linear dispersion, but by the modulation of a quasi-regular wave train that can be clearly identified from t=-400s. In contrast, the short-crested wave-group is controlled by linear dispersion, with the largest wave appearing as an almost perfectly focused wave-event. In particular, Case A9.5UD shows almost no evolution of the surface profile up to the occurrence of the large event at t=-126.2s, and certainly no evidence of the group becoming more compact even after the large event. In contrast, Case A9.5s7 is highly dispersed for $t \leq$-200s, and becomes rapidly more compact with a near perfectly focused wave-group occurring at t=2.0s; the latter being very close to the linearly predicted focus time (t=0s). This evolution is consistent with a *NewWave*-type event and has been observed in deep water for both uni-directional and directionally spread wave fields.

In Fig. 2 the evolved amplitude spectrum in the uni-directional wave case is compared with the corresponding amplitude spectrum along the centerline of the short-crested wavefield. In both cases, two spectra are presented. The first describes all of the amplitude components (parts (a) and (b)); while the second describes only the *odd*-order[2] components (parts (c) and (d)). The latter is of interest because it provides an approximation of the freely propagating wave components; a full explanation of this being provided by Johannessen & Swan (2003) and, more recently, Gibson & Swan (2006). In Case A9.5UD the spectrum becomes narrower around its peak, explaining the quasi-regular form of the corresponding surface profile (Fig. 1). However, it becomes broader overall, with side-band instabilities growing either side of the spectral peak. This is entirely consistent with the fact that the large wave-event occurs as a modular instability at the front of the elongated wave group.

In the short-crested wavefield, Figs. 2(b) and 2(d) show that the changes in the wave spectrum are much less significant, particularly in respect of the freely-propagating wave spectrum (Fig. 2(d)). Indeed, whilst the dominant changes in the uni-directional wave case (Fig. 2(a) and 2(c)) appear to involve changes in the free-wave spectrum, hence the changing nature of the wave- group, the opposite appears to be true in the short-crested (s=7) wave-group. Evidence of

---

[2] For example, interactions occurring at first, third, fifth, etc. orders.

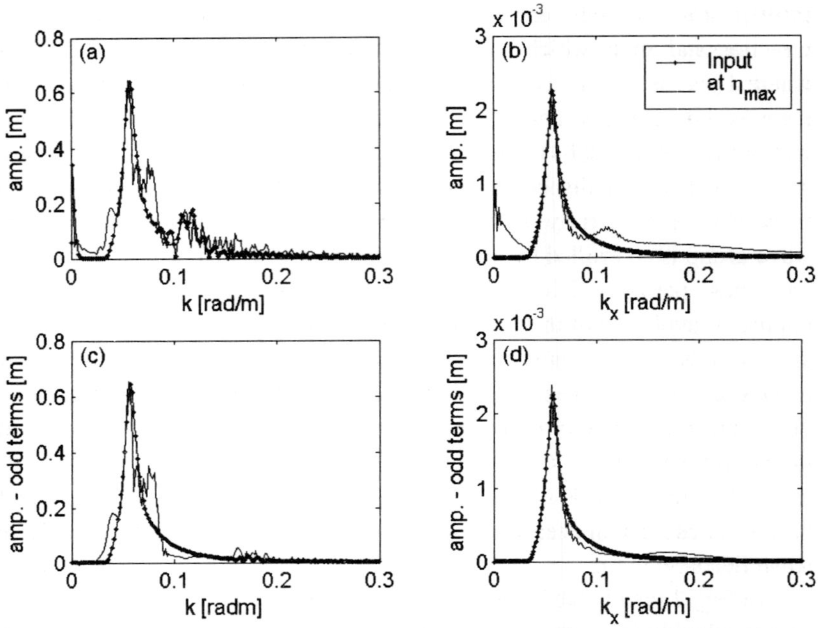

Figure 2. Spectral evolution for a uni-directional wave-group (Case A9.5UD) and a directionally spread wave-group (Case A9.5s7) at the time of the largest wave-event. (a) Uni-directional, all terms; (b) Multi-directional, all terms; (c) Uni-directional, odd-terms; (d) Multi-directional, odd-terms.

this is provided by the fact that the only significant changes in the spectral shape correspond to the development of the associated bound waves; the latter involving both 'sum' and 'difference' terms as would be expected of a shallow water environment. This result has two important implications. First, it stresses the growing importance of the bound wave interactions as the water depth reduces, particularly the low frequency or low wavenumber components corresponding to the set-down beneath a wave-group. With an increase in the magnitude of these components, the maximum crest elevation will undoubtedly reduce. Second, the growth of appreciable bound waves cannot itself alter the nature of the wave group, particularly the general form of the wave group envelope. As a result, one would not expect to see the effects of wave modulation dominate this directionally spread wave-group.

As waves propagate into shallow water the effects of wave refraction typically lead to a reduction in the directional spread; individual waves becoming more long-crested. To investigate the importance of this change, several other cases have been examined. In all other respects the definition of the input wave spectrum is held constant (T_p=10s, γ=2.5 and A=9.5m). Two cases of gradually reducing directional spread are considered in addition to the very short-crested wave case (A9.5s7) considered earlier. These cases are denoted by A9.5s45 and A9.5s500; where the final digits define the directional spread, s=45 and 500 respectively. These cases may be characterized as being relatively short-crested (s=45) and very long-crested (s=500).

In both cases the phasing of the wave components was adjusted according to linear theory so as to produce a focal event and the evolution of the wave field calculated fully nonlinearly. The resulting large wave-events are presented in Fig. 3; the A9.5s7 and A9.5UD cases investigated previously added to facilitate comparisons. The left-hand column shows the centerline cross-section of the spatial water surface profile, $\eta(x)$, corresponding to the occurrence of the largest wave-event. In contrast, the right-hand column shows the corresponding centerline cross section of both the input and the evolved odd-term spectrum at the instant of the large wave-event.

We have already noted that the large wave-event corresponding to the most short-crested wavefield (Case A9.5s7) was well focused. However, with only a small decrease in the directional spread, the *focal quality* of the large wave-event reduces significantly. In this context the term focal quality merely indicates how close a wave-event is to the idealized or perfectly focused case in which all the wave components come into phase at one point in space and time. When the latter event arises the wave group will be symmetric about the largest wave crest. Evidence of a reduction in the focal quality with relatively small changes in the directional spread is provided by the large wave event corresponding to s=45. In this case there is clear asymmetry in the depth of the adjacent wave troughs, with consequent reduction in the maximum crest elevation. At this point it is important to note that this asymmetry is not merely a function of the initial or input phasing of the wave components. Indeed, a systematic attempt at iterating this phasing to improve the focal quality proved totally unsuccessful. These observations may be explained by the fact that although *directional focusing* continues to dominate the evolution of the wavefield, the competitive mechanism of *wave modulation* plays a small but non-negligible role.

It is also interesting to note that, the focal quality does not continue to decrease with a further reduction in directionality. Evidence of this is provided by the large wave event arising in the most long-crested sea-state (A9.5s500); the latter appearing to have a similar focal quality to that arising in case A9.5s45. However, the most important result lies in the fact that although there may be some variation in the focal quality, there is no evidence that any of the directionally spread wave cases may evolve into a quasi-regular wave profile; the latter indicating the dominance of wave modulation and only occurring in the equivalent uni-directional wavefield (Case A9.5UD on Fig. 3(a)). It is particular-

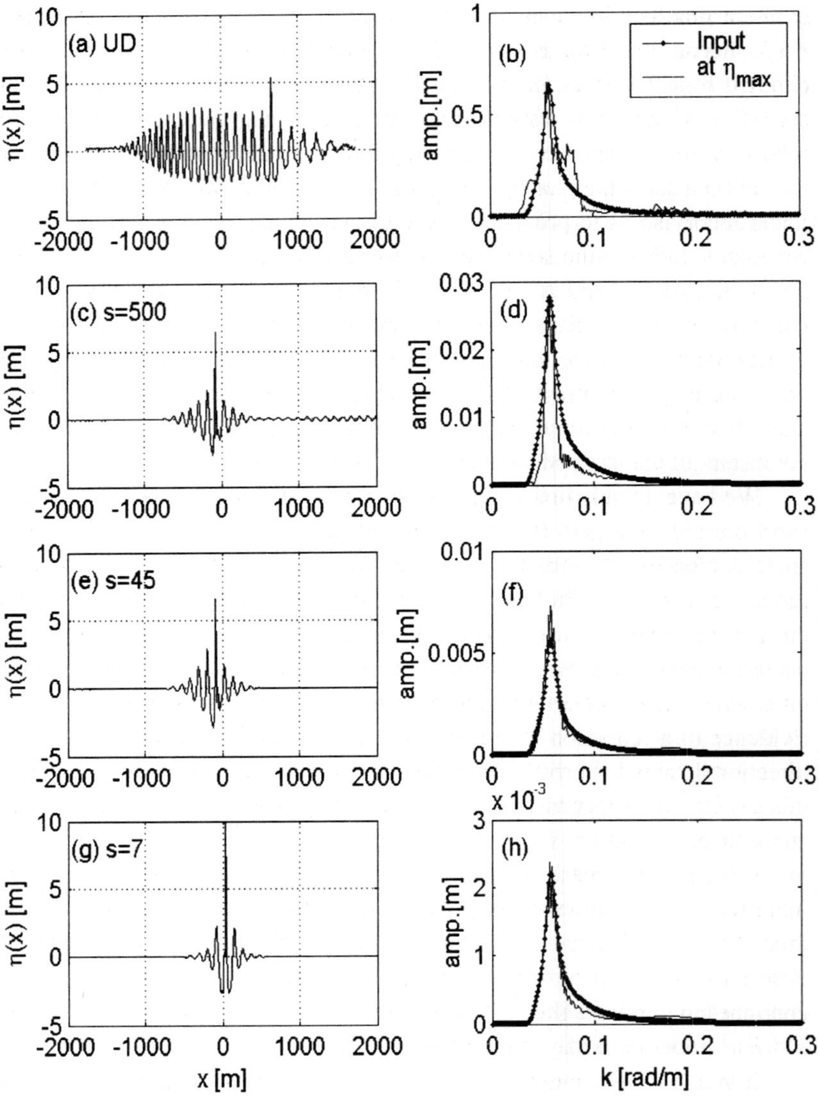

Figure 3. Extreme wave-events arising in various directionally spread sea-states; Left-hand column: water surface elevation, *η(x)*; Right-hand column: initial and evolved odd-spectra in the mean wave direction.

ly interesting that this remains valid in even the very long-crested sea-state, Case A9.5s500 on Fig. 3(c), involving the introduction of a small directional spread. Nevertheless, the maximum crest elevations for all cases, with the notable exception of the very short-crested case (s=7), is significantly smaller than that predicted by linear theory. The explanation for this is two-fold. First, the increasing significance of the low frequency or difference-terms, which are out of phase at the point of the large wave-event, and thus cause reduced crest elevations. Second, the large wave-events are not perfectly focused. This latter effect is, at least in part, related to a reduction in directionality. Since the focusing of wave components is related to both frequency dispersion and directionality, a reduction in the latter will weaken the focusing process. Evidence of this is provided by the surface elevation given on Fig. 3 in which only the most long-crested case (s=7 on Fig. 3(g)) appears fully focused; the latter being demonstrated by the depth of the adjacent wave troughs.

However, in all of the cases noted above, important changes to the wave spectrum are identified. In Fig. 3 the right-hand column describes the centerline cross-section of the odd-order spectra. For each directional spread it contrasts the input spectrum with the local spectrum corresponding to the occurrence of the largest wave event; the difference between the two describing the spectral evolution associated with an extreme wave-event. For all cases involving a directional spread ($7 \leq s \leq 500$) these results show that the spectrum defining the freely propagating wave components becomes narrower during the evolution of a large wave. This is in marked contrast to the uni-directional wave case (A9.5UD in Figs. 3(a) & 3(b)) which becomes narrower in the vicinity of the spectral peak but broader overall.

Although the changes in the spectral shape with the directionality of the wavefield are undoubtedly important, accounting in part for the reduction in crest elevation, the most significant change arises with the introduction of directionality; even in those circumstances where the wavefield remains very long-crested. The marked difference between Figs. 3(a) & 3(b) and Figs. 3(c) and 3(d) providing a very clear example of this.

Given that the evolution of these waves is strongly influenced by the underlying directionality, it is instructive to contrast the evolution of a short-crested wave group in deep and intermediate water depths. These comparisons are provided on Fig. 4. Setting aside subtle changes in the shape of the envelope of the wave group, which may be accounted for in terms a reduction in the effective bandwidth and hence reduced frequency dispersion, the wave events look remarkably similar. Both are highly focused, primarily due to directional effects and neither display any evidence that wave modulation is significant. This

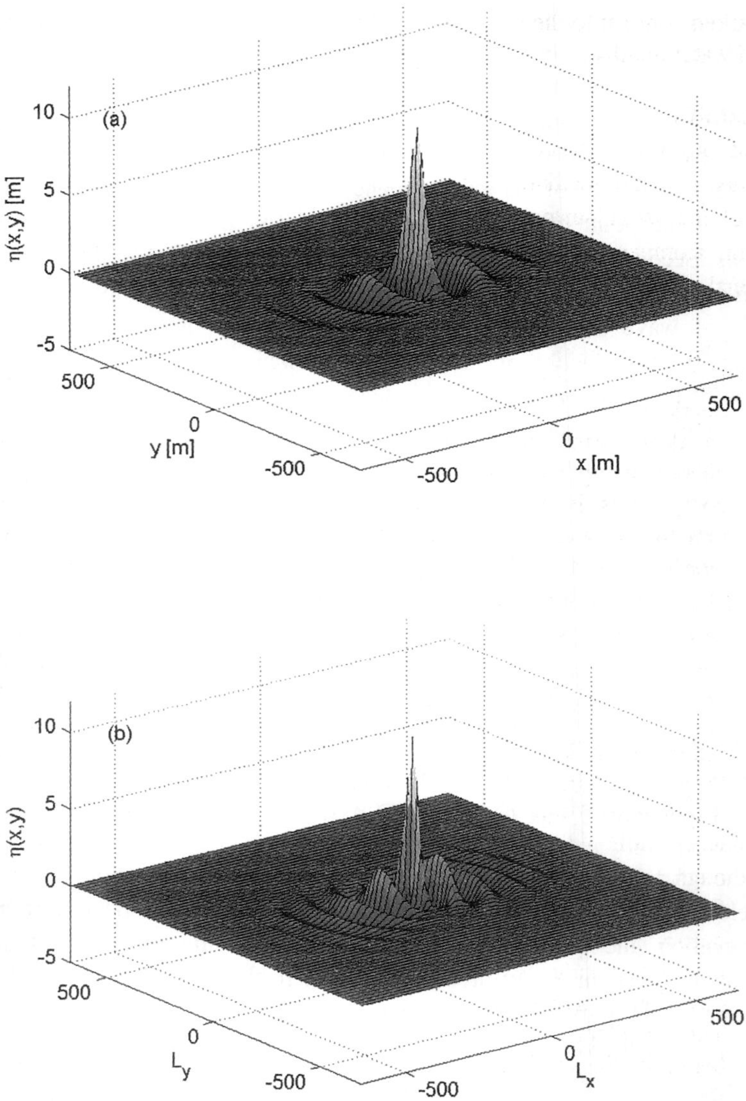

Figure 4. Surface profile at the extreme event in (a) deep water (Case A9.5s7d) and (b) intermediate water (Case A9.5s7).

is in marked contrast to the uni-directional results, particularly those occurring in reduced water depths.

CONCLUSIONS

This paper has considered the effect of directionality on the evolution of large waves in intermediate water depths; specific attention being paid to the nature of the water surface elevation and to changes in the underlying wave spectrum. Comparisons between a number of wave cases involving varying directional spreads have led us to the following conclusions.

1. The significance of directionality in the evolution of large waves in finite water depths cannot be over-estimated, particularly as the water depth reduces.
2. In deep water the evolution of large waves in any sea state encompassing a realistically based broad range of frequency components is governed by frequency dispersion in which a large number of wave components superimpose to produce a focused or nearly focused NewWave-type event. This is not to say that nonlinearity is not important, merely that it modifies rather than replaces the fundamental processes governing the evolution of the largest waves.
3. As the water depth reduces so does the effectiveness of frequency dispersion; the resulting wave group behaving as if it were progressively more narrow-banded. In terms of the evolution of a large wave event a reduction in water depth is, in part, equivalent to a reduction in frequency bandwidth; both weakening the process of frequency dispersion. If this is weakened sufficiently, the process of nonlinear wave modulation replaces frequency dispersion as the controlling mechanism responsible for the evolution of the largest waves and fundamentally different type of extreme wave events arise. The latter occurs as a rapidly developing local instability at the front of a truncated regular wave train and has previously been observed in uni-directional wave trains.
4. The introduction of directionality fundamentally changes the balance between linear focusing and nonlinear modulation. In effect, the focusing process now arises due to the combined effects of frequency dispersion and directional focusing; the latter relatively unaffected by a reduction in water depth. As a result, the introduction of even a small directional spread, resulting in a relatively long-crested wavefield, leads to the continued dominance of the focusing process. Indeed, in directional wave fields the present numerical calculations show no evidence that nonlinear wave modulation can become a controlling mechanism.

5. However, this is not to say that the waves are not highly nonlinear with marked differences with the linearly predicted solutions. Indeed, the present results show that as the water depth reduces the departures from linear predictions increase. In particular, it is shown that the spectral broadening that accounts for the increased crest elevations observed during the evolution of large deep water waves is reversed such that the spectrum of the freely propagating wave components becomes more narrow-banded. Furthermore, the present calculations also show that the bound wave components, particularly the frequency difference (or wavenumber difference) terms that account for the set-down beneath a wave group, become progressively more important as the water depth reduces. These effects demonstrate the importance of incorporating the full nonlinearity when describing extreme waves in intermediate and shallow water depths; both leading to a reduction in the maximum crest elevation associated with an extreme wave event. Indeed, these results contribute to the growing body of evidence that suggests extreme waves in intermediate water depths may be overestimated by the traditional limiting criteria commonly applied in the engineering design.

REFERENCES

Bateman, W.J.D., Swan, C. and Taylor, P.H., 2001. On the efficient numerical simulation of directionally spread surface water waves. J. Comp. Physics, 174, 277-305.

Bateman, W.J.D., Swan, C. and Taylor, P.H., 2003. On the calculation of the water particle kinematics arising in a directionally spread wavefield. J. Comp. Physics, 186, 70-92.

Bateman, W.J.D. and Swan, C., 2006. Realistic Descriptions of Extreme Ocean Waves. Part I: The application of recent advances in wave modelling. Submitted to: Applied Ocean Research.

Bocotti, P., 1983. Some new results on statistical properties of wind waves. Appl. Ocean Res., 5, 134-140.

Craig, W. and Sulem, C., 1993. Numerical simulation of gravity waves. J. Comp. Phys. 108, 73-83.

Gibson, R. and Swan, C., 2006. The evolution of large ocean waves: the role of local and rapid spectral changes. Proc. R. Soc. London., In press.

Lindgren, G., 1970. Some properties of a normal process near a local maximum. Ann. Math. Stat., 41, No.6, 1870-1883.

Johannessen, T.B. and Swan, C., 2001. A laboratory study of the focusing of transient and directionally spread water surface. Proc. R. Soc. London, Ser. A, 457, 971-1006.

Johannessen, T.B. and Swan, C., 2003. On the nonlinear dynamics of wave groups produced by the focusing of surface-water waves, Proc. R. Soc. London, Ser. A, 459, 1021-1052.

Jonathan, P., Taylor, P. H. and Tromans, P.S., 1994. Storm waves in the northern sea. Proc. of Conf. on the Behaviour of Offshore Structures, BOSS94.

Katsardi, V. and Swan, C., 2004. The evolution of large waves in shallow water. Proc. 29th Int. Conf. of Coastal Eng., Lisbon, 69-82.

Lindgren, G., 1970, Some properties of a normal process near a local maximum. Ann. Math. Stat., 41, No.6, 1870-1883.

Longuet-Higgings, M.S., and Stewart, R.W., 1960. Changes in the form of short gravity waves on long waves and tidal currents. J. Fluid Mech., 8, 565-583.

Mitsuyasu, H., 1975. Observations of directional spectrum of ocean waves using a cloverleaf buoy. J. Phys. Ocean., 16, 750-760.

Sharma, J.N. and Dean, R.G., 1981. Second-order directional seas and associated wave forces. Society of Petroleum Eng. J., 4, 129-140.

Tromans, P.S., Anaturk, A.R. and Hagemeijer, P., 1991. A new model for the kinematics of large ocean waves - application as a design wave, Proc. 1st Int. Conf. Offshore & Polar Eng., UK. 3, 64-71.

INCORPORATION OF WAVE TRANSMISSION AND OVERTOPPING EFFECTS INTO A MULTIDIRECTIONAL RANDOM WAVE TRANSFORMATION MODEL

Kazuya Oki[1], Hajime Mase[2] and Tetsuo Sakai[3]

In order to improve the ExEBED wave model (a high-order differential wave model based on an energy balance equation with a diffraction term and an energy dissipation term), the effects of wave transmission and overtopping are incorporated into the ExEBED wave model. The numerical results by the present wave model are compared with the experimental measurements by Pan et al. (1998) to validate the model.

INTRODUCTION

The progress of computer technology makes it available to analyze complicated wave fields with high accuracy. Of course, the more complicated the computational model becomes, the heavier the computation load becomes and the more time the calculation needs.

In the meanwhile, phase-averaged models need less computational power and it is easy to handle the wave forecast in large sea area. Therefore, this kind of models is suitable for a practical use. SWAN (Booij et al. 1996) is also a phase-averaged model.

Authors have developed the wave model based on an energy balance equation with a diffraction effect. For instance, Oki et al. (2004) adapted the QUICK scheme and suppressed the numerical diffusion. The model is named the 'ExEBED' wave model. Mase et al. (2005) improved the calculation method of wave reflection in back-marching.

The objective of this study is an incorporation of wave transmission and overtopping effects into the ExEBED wave model. These effects are important in practical applications. Each transmitted wave or wave overtopping is verified by comparing with a typical experimental result. In order to investigate the appropriateness of the present wave model in a practical case, the results by the model are compared with the experimental measurement by Pan et al. (1998), where wave height distributions of multidirectional random waves around the detached breakwaters in a wave flume are reported.

[1] Department of Urban and Environmental Engineering, Kyoto University, Kyotodaigaku-katsura, Nishikyo-ku, Kyoto, 615-8540, Japan
[2] Disaster Prevention Research Institute, Kyoto University, Gokasho, Uji, Kyoto, 611-0011, Japan
[3] Department of Urban and Environmental Engineering, Kyoto University, Kyotodaigaku-katsura, Nishikyo-ku, Kyoto, 615-8540, Japan

ExEBED WAVE MODEL

The energy balance equation which has a diffraction term and an energy dissipation term for the steady-state conditions is shown as

$$\frac{\partial(v_x S)}{\partial x}+\frac{\partial(v_y S)}{\partial y}+\frac{\partial(v_\theta S)}{\partial \theta} \\ =\frac{\kappa}{2\omega}\left[\left(CC_g\cos^2\theta S_y\right)_y-\frac{1}{2}CC_g\cos^2\theta S_{yy}\right]-\varepsilon_b S \tag{1}$$

where (x, y) is the horizontal coordinates and θ is the wave direction measured counterclockwise from the x-axis, S is the directional wave spectral density, κ is the coefficient of a diffraction effect, ω is an angular frequency, C is a wave celerity, C_g is a group velocity and ε_b is an energy dissipation coefficient. The subscript, y, in the wave diffraction term in Eq. (1) represents partial differentiation with respect to y. The wave diffraction term is introduced by Mase (2001). The characteristic velocities, (v_x, v_y, v_θ), are respectively defined as

$$\left(v_x, v_y, v_\theta\right) \\ =\left[C_g\cos\theta, C_g\sin\theta, \frac{C_g}{C}\left(\sin\theta\frac{\partial C}{\partial x}-\cos\theta\frac{\partial C}{\partial y}\right)\right] \tag{2}$$

The numerical diffusion term represented by the second derivative is similar to the wave diffraction term which is also denoted by the second derivative. In order to suppress the numerical diffusion separately from the effect of wave diffraction, Oki et al. (2004) employed a quadratic upstream interpolation for convective kinematics (QUICK), proposed by Leonard (1979), for discretization of Eq. (1). This wave model is called the "ExEBED", which stands for an **Ex**tended **E**nergy **B**alance **E**quation with a **D**iffraction term.

Consequently, an algebraic equation after discretization is shown as follows:

$$A_1S_n^{ijk}+A_2S_n^{i(j-2)k}+A_3S_n^{i(j-1)k} \\ +A_4S_n^{i(j+1)k}+A_5S_n^{i(j+2)k}+A_6S_n^{ij(k-2)} \\ +A_7S_n^{ij(k-1)}+A_8S_n^{ij(k+1)}+A_9S_n^{ij(k+2)}=-BS_n^{(i-1)jk} \tag{3}$$

where $i = 1\sim I$ and $j = 1\sim J$ are the grid numbers related to the x-axis and y-axis respectively, $n = 1\sim N$ is the number of frequency components, $k = 1\sim K$ is the identification number for the wave direction. The grid system of the directional spectrum S_n and the characteristic velocity employ a staggered scheme for the arrangement of their definition point. The coefficient $A_1\sim A_9$ and B are the functions of the characteristic velocities, the differential grid size of i, j and k, the wave velocity, the group-velocity, the diffraction influence coefficient and the energy dissipation coefficient. The formulation of these coefficients in Eq. (3) is described in detail by Mase et al. (2005). The parameterization of wave breaking energy dissipation, ε_b, is determined from Goda's wave breaking formula (1975). The significant wave height, $H_{1/3}$, the significant wave period, $T_{1/3}$, and the mean wave direction can be computed from the calculated value of S_n^{ijk}.

In the computation of the energy balance equation model, the wave reflection can not be solved at the same time to calculate in the direction of the incident wave propagation. But the reflected wave can be computed in the backward direction after the forward calculation by setting a calculated value as the incident condition at the boundary in front of the reflective structures. When computing reflection from the structures which was arranged at the arbitrary angle with crossing the Cartesian grid, the reflection angle can be properly treated by using the method of Mase et al. (2005).

TRANSMITTED WAVE MODEL

Since the calculation proceeds in the direction of wave propagation, the wave energy behind the impermeable structures is set to be zero. When the incident wave is transmitted through a permeable structure, the wave energy behind the structure is newly set as an incident condition from the calculated value in front of the structure. The energy to be given as a transmitted condition is the product of the calculated value in front of the structure and the square of transmission coefficient, K_{tr}. The spectral density, $S(x, y, f)$, in the cell just behind the permeable structure is set as

$$S(x, y, f) = K_{tr}^2 S(x_{in}, y_{in}, f) \tag{4}$$

K_{tr} is given a priori. (x_{in}, y_{in}) is the position of the cell which is the origin of the transmitted wave. The angle of a structure in the Cartesian grid and the transmission coefficient at the cell just behind the structure are set by the same way of Mase et al.'s reflection condition (2005).

In this study, the energy of transmitted wave is calculated from the one at the cell in front of the structure. The angle of the structure at the cell is just perpendicular to the direction of the transmitted wave propagation as shown in Fig. 1. In this formulation, the width of the structure is not considered but its effect is included into the transmission coefficient which is set at each point where the transmitted condition should be given.

Figure 2 shows the sketch of normalized spectral density distribution. Dotted line represents incident wave spectrum and solid line represents those of a transmitted wave. The spectral density distribution has the same shape of the incident wave spectrum.

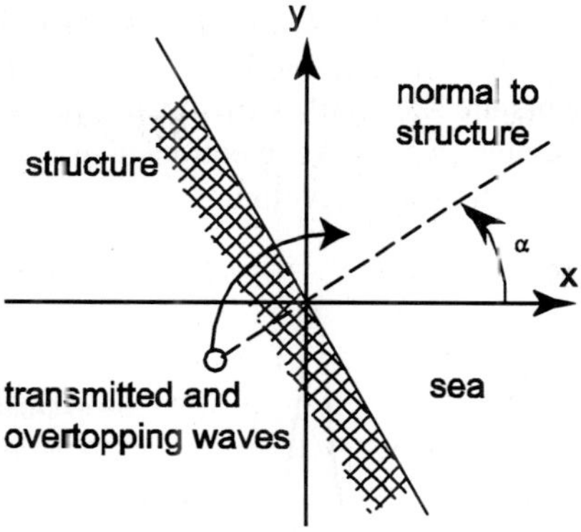

Figure 1. Modeling of transmitted and overtopped waves

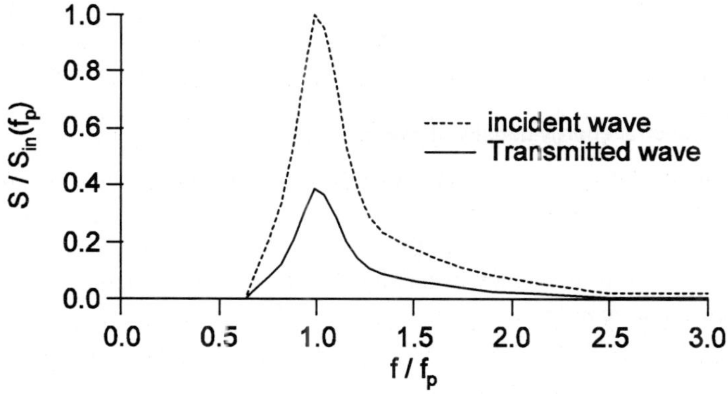

Figure 2. Spectral energy density of transmitted wave

WAVE OVERTOPPING MODEL

At a cell in back of the structure where wave overtopping occurs, the spectral density is set as follows:

$$S_{ot}(x,y,f)=K_{ot}^2\left[(1-K_r)S_{in}(x_{in},y_{in},f)+K_rS_{in}(x_{in},y_{in},f/2)\right]. \tag{5}$$

Here, S_{ot} is the frequency spectrum of wave overtopping, K_r is the rate of energy transfer to the double frequency components and S_{in} is the spectrum in front of the structure. K_{ot} is the coefficient of overtopping. This formulation is based on the literatures (e.g. J.W. van der Meer et al. 2000) showing that high frequency components of transmitted wave are generated in comparison with those of the incident wave. A certain part of the energy spectrum in the original distribution is transferred to high frequency components in the new distribution given behind the structure.

Figure 3 represents the sketch of normalized spectral density distributions of incident and overtopping wave. As mentioned above, the double frequency components of overtopping wave develop.

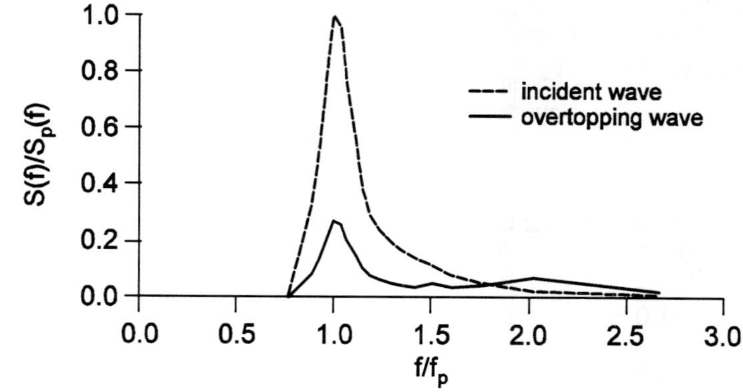

Figure 3. Spectral energy density of overtopped wave

As a similar model, van der Meer (2000) proposed the spectral density distribution of overtopping wave, in which 40% of the total-energy of the overtopping wave is allocated uniformly between the components of $1.5*f_p$ and $3.5*f_p$ (f_p : the peak frequency of incident wave) and the components higher than $3.5*f_p$ are set to be zero.

K_{ot} is determined by Goda's wave overtopping criteria (1969) given by

$$K_{ot} = 0.5\left[1 - \sin\frac{\pi}{2\alpha}\left(\frac{h_c}{H_i} + \beta\right)\right] \tag{6}$$

where $\alpha = 2.2$ and $\beta = 0.25$. Since the incident wave height H_i is a known value during the calculation, only crown height of a structure, h_c, is necessary for estimation of K_{ot}. It is confirmed by a preliminary calculation that the wave height behind a breakwater calculated by this model agrees with the Goda's empirical formula. The predominant direction of the overtopping wave is also taken into consideration as the cell condition at the back of structure.

For the directional distribution characteristic, it is shown that the angular spreading of overtopping waves generally becomes broader than that of the incident wave according to the preceding literature; however it is difficult to generalize. Therefore, in the present ExEBED wave model, the characteristic of the directional distribution of the wave overtopping is set as same as that of the incident wave.

To verify this wave overtopping model, the overtopping wave heights are calculated against various height of structure. K_r=0.4 is adopted with reference to the past studies. Figure 4 shows the calculated result of the wave height rate against the crown height of structure. When the crown height of structure is small, the calculated result is a little small comparing with Goda's wave overtopping coefficient but agrees generally.

The spectrum of overtopping wave, S_{ot}, changes accompanying the variation of the coefficient, K_r, around 0.4, but the sensitivity of S_{ot} against the parameter, K_r, is found to be small by trial calculations.

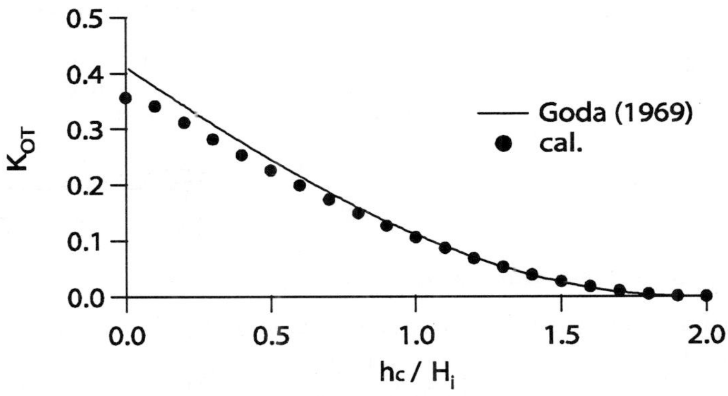

Figure 4. Wave transmission coefficient due to wave overtopping

APPLICATION TO PAN ET AL.'S EXPERIMENT

In order to validate the models discussed above, the results by the present ExEBED wave model are compared with the Pan et al.'s (1998) experimental data of wave transformation around the detached breakwaters. Pan et al. compared the numerical results by their own wave model with the experimental measurements and mentioned that it was important to take the effects of reflected and transmitted waves into consideration. Because the wave overtopping doesn't occur in this experiment, the influence of the wave overtopping is examined by computation.

Figure 5 shows the experimental setup by Pan et al. (1998) where five detached breakwaters were arranged in shallow water area. Multidirectional random wave was used and the significant incident wave height was 0.05m, the wave period was 1.2sec, the predominant wave direction measured counterclockwise from shore-normal was 20 degrees. In the present calculation, the grid size was set to 0.1m. The calculation method of reflection derived by Mase et al. (2005) was adopted.

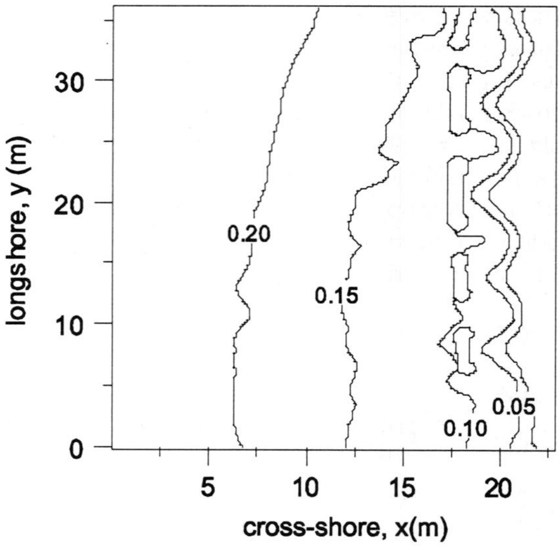

Figure 5. Experimental setup carried by Pan et al. (1998)

Figure 6 shows the wave height distribution in front of the breakwaters (x = 16.5m). The solid circles denote experimental results. The solid line and the broken line represent the result by the present ExEBED wave model and that by the wave model without reflection and transmission effects, respectively. Since the reflection and transmitted coefficients are not mentioned in the experiment, they are set to 0.4 and 0.3 respectively in the calculation. In the results of no reflection calculation, wave height distribution is almost constant in the y-direction. In the results of the present ExEBED wave model, the wave height is large in front of the breakwater and small among breakwaters. In qualitatively, the tendency shown by the solid line provides a better fit to the experimental data than the result shown by the broken line.

Figure 7 shows the wave height distribution along the line of y = 17.3m which passes among the breakwaters. As for both calculated results, there is not a remarkable difference and the wave height distributions agree with the experimental results quantitatively, because the influences of reflection and transmission are small around there.

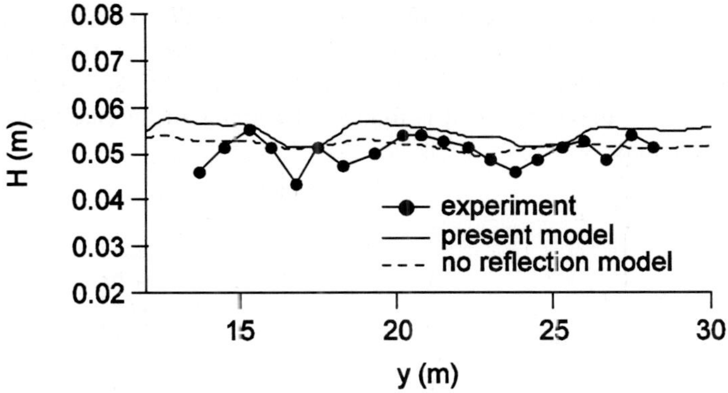

Figure 6. Wave height distribution in front of breakwaters (x = 16.5m)

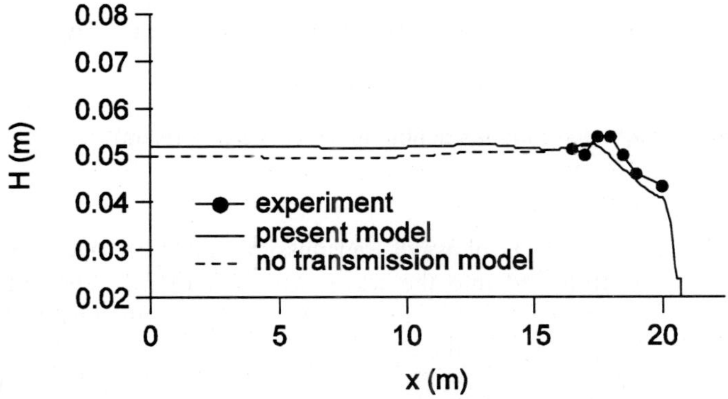

Figure 7. Wave height distribution in the cross-shore direction (y = 17.3m)

Furthermore, the effects of transmitted and overtopping waves on the wave height distribution behind the breakwaters are investigated numerically. Four cases as shown in Table 1 are tested. The conditions of the incident wave are the same as above.

Table 1. Calculation conditions

Case 1	no wave transmission, no wave overtopping
Case 2	only wave transmission
Case 3	only wave overtopping
Case 4	wave transmission, wave overtopping

Figure 8 shows the wave height distribution behind the breakwaters along the line of x = 19.2m. In the case of wave overtopping, the crown height of breakwater is set as h_c = 0m. There are little differences behind the gap of breakwaters for all cases. On the other hand, as is easily predicted, the wave height of Case 1 which does not consider either effects of transmission or overtopping is small. The difference between the results of Case 1 and Case 4 is remarkable. From these results, it is found to be important to take these effects into considerations.

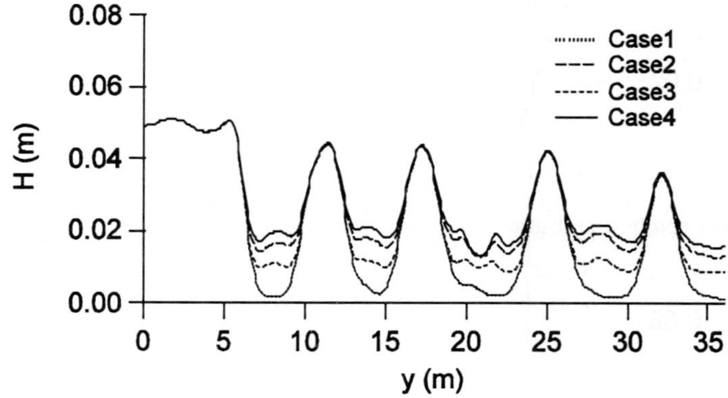

Figure 8. Wave height distributions behind breakwaters (x = 19.2m)

CONCLUSIONS

In this study, the effects of wave transmission and wave overtopping are formulated and incorporated into the wave model, ExEBED. The validity of these modeling for transmission and overtopping was confirmed by performing the calculation for the fundamental experimental conditions.

Furthermore, in order to verify the ExEBED with the effects of wave transmission and overtopping for a practical application, the results by the present model were compared with the experimental data (Pan et al. 1998) of multidirectional random waves around the detached breakwaters in a wave basin. The tendency of wave height distribution in front of breakwaters calculated by the present ExEBED wave model fit to the experimental results better than those calculated by no-reflection wave model.

In addition, the influences of transmitted and overtopping waves on the wave height distribution behind the breakwaters were investigated numerically. The result by the present ExEBED wave model produced a remarkable difference from that by the wave model which does not considers ether effects of wave transmission or wave overtopping. From these results, it was found to be important to take these effects into consideration.

REFERENCES

Booij, N., Holthuijsen, L.H. and Ris, R.C. 1996. The "SWAN" wave model for shallow water, *Proceedings of 25th International Conference on Coastal Engineering*, ASCE, 668-676.

Goda, Y. 1696. Re-analysis of laboragory data on wave transmission over breakwaters, *Report of the port and harbor research institute*, 8, 1-18.

Goda, Y. 1975. Deformation of irregular waves due to depth-controlled wave breaking, *Report of the port and harbor research institute*, 14, 59-106. (in Japanese)

J.W van der Meer, E. Regeling and J.P. de Waal. 2000. Wave transmission : spectral changes and its effects on run-up and overtopping, *Proceedings of 27th International Conference on Coastal Engineering*, ASCE, 2156-2168.

Mase, H. 2001. Multidirectional random wave transformation model based on energy balance equation, *Coastal Engineering Journal*, JSCE, 43, 317-337.

Mase, H., Oki, K., T.S. Hedges and H.J. Li. 2005. Extended energy-balance equation wave model for multidirectional random wave transformation, *Ocean Engineering*, 32, 961-985.

Oki, K., Mase, H. and Hedges, T.S. 2004. Energy balance equation model with diffraction effect for random waves, *Proceedings of 29th International Conference on Coastal Engineering*, ASCE, 881-893.

Pan, S., N. MacDonald, J. Nicholson and B.A. O'Cornor. 1998. Hydrodynamics Associated with Offshore Structures, *Department of Civil Engineering*, University of Liverpool. Report No. CE/06/98, 14p.

THE RELATIVE TROUGH FROUDE NUMBER FOR TERMINATION OF WAVE BREAKING

Takashi Okamoto[1], David R. Basco[2], and Conceição Juana Fortes[1]

Wave breaking termination is an important boundary for the determination of the wave breaking area especially when termination occurs in the middle of the domain, such as on a bar-trough shaped beach. However, this subject has not been closely examined. The development of a wave breaking index based on more experimental and theoretical evidence is needed. The Relative Trough Froude Number (RTFN) is a wave breaking index based on the moving hydraulic jump concept. The critical condition for wave breaking termination was determined from wave tank experiments on a horizontal bottom as 1.15 (Okamoto and Basco, 2004). In this paper, we examine two issues related to the termination condition of wave breaking (1) how the RTFN and other wave breaking indices approach the termination condition on a horizontal bottom, and (2) the bottom slope influence on the termination condition. The RTFN approaches the termination condition at a slower rate than the other indices based on free surface geometry so that the index value stays noticeably higher than the critical condition during wave breaking. Wave properties in the trough area show different patterns from the one obtained on a horizontal bottom. By using two hypothesizes, the RTFN at the "termination" location comes close to the theoretical value, but further investigation is required.

1. Introduction

Determination of the wave breaking region is an important factor in wave transformation studies because the mechanism of wave breaking is completely different from potential flow wave motion. It is also very important for nearshore hydrodynamics because the dissipation energy through wave breaking is the driving force for near-shore phenomena such as wave set-up, longshore currents, etc.

In most cases, wave breaking occurs very close to the shoreline and continues until the wave dissipates almost all the energy at the shoreline. The termination of wave breaking occurs so close to the shore that the wave breaking area can be defined as the area between where wave breaking begins and the shoreline. However, in some cases, e.g. on a bar-trough shaped beach, wave breaking ceases in the middle of the domain so that it has to be determined explicitly by certain criteria. Under these circumstances, determining the termination location by certain criteria is as important as determining the initiation point.

Phase-resolving wave transformation models, such as the Boussinesq equation model, are now commonly used in near shore hydrodynamics studies.

[1] Hydraulics and Environment Dept., National Laboratory of Civil Engineering, Av. do Brasil 101, Lisbon, 1700-066, Portugal
[2] Coastal Engineering Centre, Old Dominion University, Norfolk, VA, 23508, USA

The Boussinesq equation itself does not contain the physics of wave breaking so extra momentum terms have to be added to the governing equation to reproduce wave breaking numerically. And, the decision as to which wave the extra momentum terms should be added is made by some external criteria. Different models exist but the decision mechanism is essentially the same. It contains three control values (initiation condition, termination condition, and duration) and the critical condition changes in time using these three parameters. These parameters however, have not been determined by either physical experiments or theoretical analysis. They are coefficients adjusting the numerical results to some real world data, therefore these parameters have to be calibrated and changed according to the breaker type and bathymetrical setting.

More detailed studies regarding the behavior of waves after wave breaking and development of a wave breaking index based on more physical evidence are needed. The Relative Trough Froude Number (RTFN) is an index number for wave breaking introduced by Utku and Basco (2002). They conducted wave tank experiments on a 1:20 plane slope beach and concluded that the critical RTFN for the wave breaking initiation is stable for different types of input wave conditions and the presence of an opposing current. For the critical RTFN at wave breaking termination, Okamoto and Basco (2004) conducted wave tank experiments on a horizontal beach section and confirmed that the experimental results at the termination location were very close to the theoretical expectation determined by using linear and nonlinear wave theories

In numerical models, the index behavior during wave breaking is as important as the termination condition because the numerical model checks values continually to determine whether the wave is in the breaking state. The bottom slope also affects the termination condition.

It is known that the critical condition for the wave breaking initiation varies with input wave conditions and bottom slope conditions (Goda, 1970; Weggel, 1972). Similarly this could apply to the termination condition. In this paper, we focus on input wave conditions and bottom slope on breaking wave termination.

2. Wave decay and the termination condition

Many studies on the post-wave-breaking phenomena have been undertaken on plane sloping beaches or monotonically decreasing water depth cases. We know how the wave height changes after wave breaking initiation (wave decay). The wave height decreases rapidly over a short distance after initiation, then wave breaking continues while the wave height decreases more slowly. Svendsen et al. (1978) defined two regions; (1) the radically changing "outer breaking region", and (2) a more stabilized zone as an "inner breaking region". The important fact is that wave breaking continues even when the wave height becomes smaller than the initiation height and stabilized with that smaller height.

Dally et al. (1985) derived a wave decay equation using the concept of a stable wave height. An analytical expression of wave decay on a horizontal bottom is given as;

$$\frac{H}{d}=\left\{\left[\left(\left(\frac{H}{d}\right)_b^2-\Gamma^2\right)\exp\left(-K\frac{x}{d}\right)\right]+\Gamma^2\right\}^{1/2} \tag{1}$$

where: H is wave height, d is still water depth, K is a decay coefficient, and Γ is a normalized, stable wave height. As the equation shows, the relative wave height (H/d) decreases exponentially and converges asymptotically to the stable wave height.

The stable wave height is very similar to the termination condition of wave breaking. It can be treated as the wave height at the termination location if no shoaling or nonlinear wave transformation is involved. For horizontal bottom cases, Horikawa and Kuo (1966) found H/d=0.35~0.4 as the termination condition based on physical experiments. This value for the termination condition was used for the theoretical analysis of the RTFN (Okamoto and Basco, 2004). One problem with the stable wave height condition is that the relative wave height does not change very quickly or already becomes stable in the inner breaking region. We will discuss this later.

Different types of wave breaking indices are required for phase-resolving, wave transformation models, because they provide different information from traditional, phase-averaged wave properties, such as wave height and wave length. A few indices have been introduced for this purpose. Front face angle of the wave is an index used in the wave breaking module introduced by Schäffer et al. (1993). The only reference related to the value of a termination condition is an estimated front face angle in the inner surf zone, which is given as *in the order of 10 degrees.* The important point here is that the reference value was taken from the breaking wave, not at the moment of termination. Therefore, the termination condition given in their paper (10 degrees), was not determined by any experimental data, or theoretical estimation. The given value, 10 degrees, however, must be close to the actual termination condition because of the decay pattern. Madsen et al. (1997) modified it to 7 degrees, for cases with a bar-trough profiled beach and spilling breaker as a result of calibration of the model. Note that this result does not mean that the experimental result of the front surface angle is seven degrees. The numerical front face angle was seven degrees at the termination location determined by some experimental data.

A different index is employed in the FUNWAVE numerical model (Kirby et al., 1998) where vertical speed of the free surface ($\partial\eta/\partial t$) is employed. There is no explanation of how the termination condition was determined in Kirby et al. (1998) or related papers (Kennedy et al., 1999; Chen et al., 1999). The vertical speed of free surface and front face angle seem to be completely different quantities but they are closely related when the relation between spatial and time derivatives of the free surface profile given by Sørensen (1998) is employed:

$$C=-\eta_t/\eta_x \tag{2}$$

where, C is wave celerity and subscripts t and x denote time and spatial derivatives, respectively. The initiation and termination conditions of vertical free surface speed are given as $\alpha\sqrt{gd}$. If the wave is classified as a shallow water wave with wave celerity defined as $\sqrt{gd}$, the relation between the vertical speed of free surface and front face angle becomes very simple. Under this condition, the termination condition, α=0.15, is equivalent to 8.5 degrees.

The Relative Trough Froude Number (RTFN) is a wave breaking index introduced by Utku (1999). The definition equation of the RTFN was modified by Okamoto and Basco (2006) as follows;

$$Fr_t = \frac{C_{crest} - u_{trough}}{C_{trough}} \tag{3}$$

where, C is wave celerity and u is particle velocity. Subscripts denote the location in the wave profile.

The critical value of the RTFN at the termination location was calculated by using linear wave theory and the stable wave height on horizontal bottom cases (Okamoto and Basco, 2004) Using H/d=0.35~0.4, the theoretical expectation for the value at the termination location was RTFN=1.175~1.20. Results of wave tank experiments were in the range of 1.1 to 1.2 and the averaged value of fourteen test cases was 1.15. Fig. 1 displays the experimental result of the behavior of the RTFN on a horizontal bottom. The X-axis is a relative distance scale in which x=0 denotes the beginning of the horizontal section (B.P.) and x=1 denotes the termination location (T.P.).

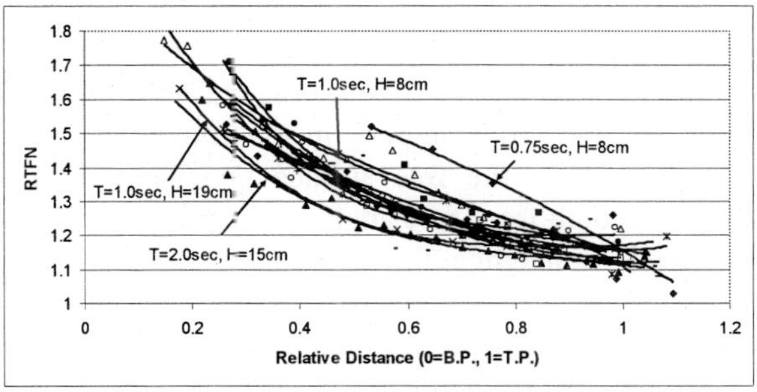

Figure 1. RTFN evolution on horizontal bottom (from Okamoto and Basco, 2004)

3. Behavior of wave breaking indices on a horizontal bottom

Behavior of each wave breaking index is important for numerical modeling because the wave breaking module compares the index value to the termination condition continuously during breaking to decide when breaking ends. Most of the wave height change occurs in the outer breaking region and does not change as much in the inner breaking region. If the index value becomes very close to the termination condition during the outer breaking region and does not decreases much in the inner breaking region, then the decision for termination becomes a very difficult task for the wave breaking module.

To see how the RTFN changes during wave breaking compared to other indices, the relative wave height (H/d), front face angle, and vertical speed of free surface were calculated from the experimental data on the horizontal bottom (Okamoto and Basco, 2004). The front face angle was calculated from the vertical speed of the free surface by using a relation derived from Eq. 2. The original data for three wave periods (1.0, 1.5, and 2.0sec) and four wave heights was tested for each wave period so that twelve cases were compared. For more details regarding the experimental settings, see Okamoto and Basco (2004).

The data contains some scatter so that the comparison was made with the best-fit curves. Since the wave decay equation by Dally et al. (1985) has been widely accepted for the evolution of wave height after breaking, the exponentially decreasing curve given by Eq. 1 was used for the comparison. To adopt it to the other indices and the relative distance scale, Eq. 1 was modified to:

$$f(x) = \left\{\left(f_0^2 - f_\infty^2\right)\exp\left(-Kx\right) + f_\infty^2\right\}^{1/2} \tag{4}$$

where f is index quantity and K is decay coefficient, and subscript 0 denotes x=0, i.e. the beginning of the horizontal section, and ∞ denotes the stabilized condition. The decay coefficient, the stabilized condition, and the value at x=0, i.e. f_0 were determined by using trial and error methods to minimize the least-squared error.

The averaged relative wave height of fourteen cases was 0.43 at the termination location. This is slightly larger than the stabilized wave height condition based on the experiments by Horikawa and Kuo (1966). The averaged stabilized angle was 13.7 degrees and the averaged stabilized vertical speed was 0.22, which corresponds to 14.0 degrees in the converted free surface angle. These indices are based on the free surface geometry. Therefore, showing slightly larger values in both front face angle and vertical speed can be explained by the slightly higher wave height at the termination location.

The decay coefficient, K is compared to see how each index decreases towards the stabilized condition. Together with the decay coefficient, the value at the middle of the horizontal section (at x = 0.5), R, was calculate to see how much decay occurs from the initiation value by the mid-point of the horizontal section.

$$R = \frac{f_{ini} - f_{x=0.5}}{f_{ini} - f_{x=1.0}} \times 100 \tag{5}$$

Because wave breaking initiation occurs before the horizontal section and we do not have data at the initiation location, theoretical initiation conditions were used for the calculation. The initiation conditions are 1.45 for the RTFN and 0.78 for the relative wave height. For the front face angle, there are various choices for the initiation condition, but we choose 27 degrees because it is the maximum free surface steepness of a cnoidal wave, the reference value given by Schäffer et al. (1993). The vertical speed for wave breaking initiation, 0.51 was converted from the initial condition of front face angle, for the same reason.

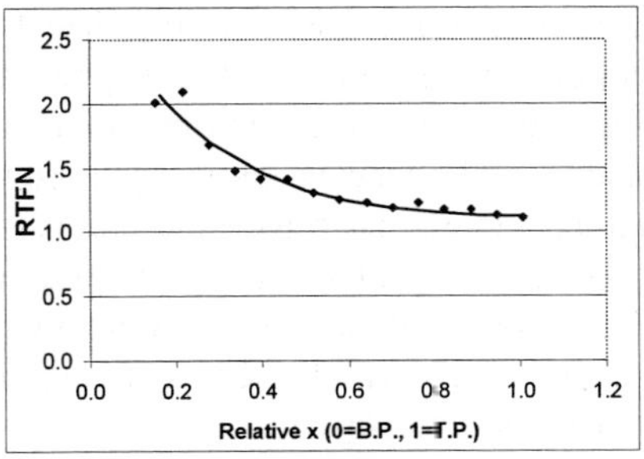

(a) RTFN

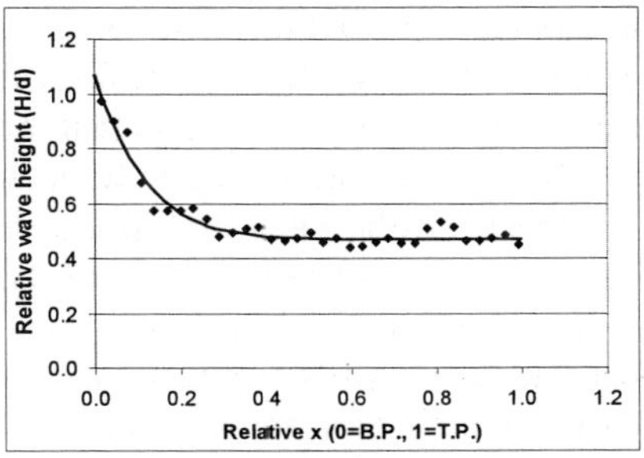

(b) Relative Wave Height

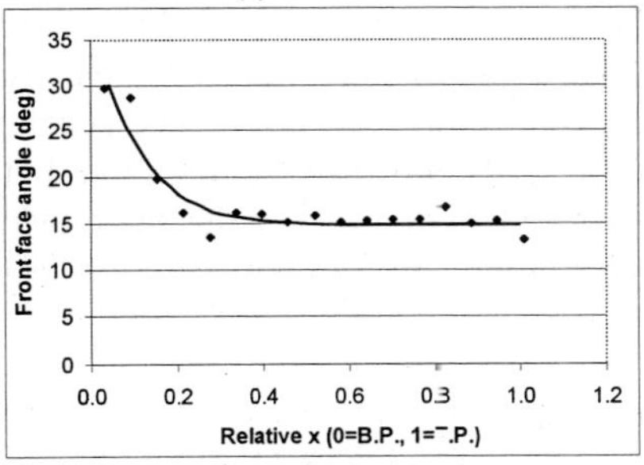

(c) Front Face Angle

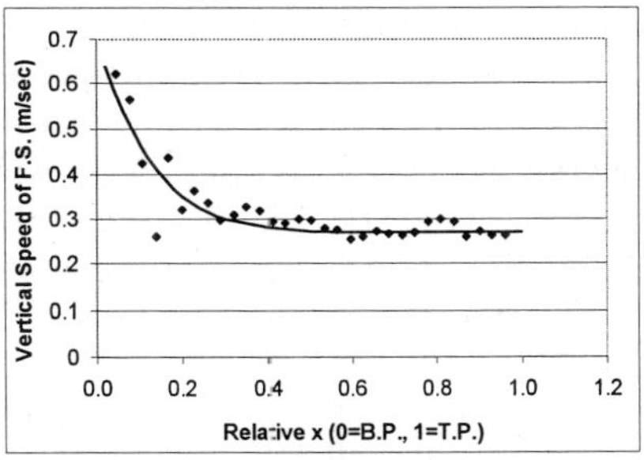

(d) Vertical Speed of Free Surface

Figure 2. Comparison of index evolution on horizontal section (T=2.0sec, H=6cm)

Fig. 2 displays index evolutions for the case with T=2.0sec and H=6cm on the horizontal section. Wave breaking begins very close to the beginning of horizontal section (about 0.1m before) with this wave input condition so that most of the life of wave breaking happens on the horizontal section. In this case, the termination location (x=1 on the relative distance scale) is about 3.3m. As shown, all indices based on free surface geometry have similar decay patterns. The values quickly decrease in the first half and it becomes almost constant by the middle of horizontal section. Compared to the free surface geometry indices, the RTFN more gradually decreases toward the termination location.

Table 1. Comparison of decay characteristics among wave breaking indices (averages of 12 cases tested)

Index	Decay Coef. (K)	% of decay (R)
RTFN	3.5	41.9
H/d	9.3	95.7
Front face angle	9.2	96.4
Vertical speed	12.3	95.1

Table 1 compares the averaged value of decay coefficient, K, and amount of decay at the middle of the horizontal section R, for the twelve cases tested. The decay coefficient of the RTFN is smaller than the other indices, which indicates that the RTFN decreases more slowly than the other indices. The amount of decay means that the value of the RTFN is still about half-way between the initiation condition and the value at the termination location at the middle of horizontal section, but the other indices have already become very near to the termination value. Note that this calculation was based on the exponentially decreasing regression curve so that R never becomes 100 in the finite range.

The decay patterns of indices based on free surface geometry follow the description of the outer and inner breaking regions. However, the index values are almost indistinguishable from the termination values far before the actual termination occurs. Therefore, the slightest error in the measurement or numerical calculation might give a false decision on the wave breaking termination location and the entire area of wave breaking.

The main reason why the RTFN exhibits a gently sloping curve and the value in the middle is still distinguishably larger than the value at the termination is wave celerity (change in bore propagation speed). As reported in Svendsen et al., (2003), the wave celerity is accelerated by wave breaking. The accelerated celerity decreases to the linear wave celerity towards termination but the transition is gentle. Fig. 3 displays the celerity evolution for the same test case shown above. The shape of the celerity evolution curve is similar to the shape of the RTFN curve.

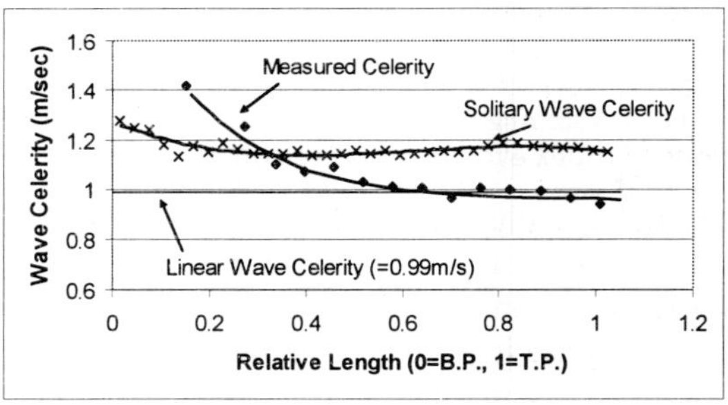

Figure 3. Measured wave celerity evolution (T=2.0sec, H=6cm) and comparison to linear and solitary wave theories

4. The critical RTFN on a bar-trough shaped beach

4.1 Overview and experimental settings

The critical condition for wave breaking initiation depends on the input wave conditions and bottom slope angle (Goda, 1970; Weggel, 1972). Calibration results of the stabilized wave height condition for plane slope beaches also varied by the slope angle (Dally et al., 1985). Therefore, it is possible that the termination condition of the RTFN is affected by the bottom slope angle. An experiment was conducted to determine if there is a bottom slope effect on the termination condition of the RTFN.

The experiment was carried out at the National Laboratory of Civil Engineering (LNEC) in Lisbon, Portugal. The wave flume is 32m long from the wave maker to the end of the flume. A simple, bar-trough shaped beach profile was constructed as shown in Fig. 4. Water depth at the bar-top was set 10cm and 15cm.

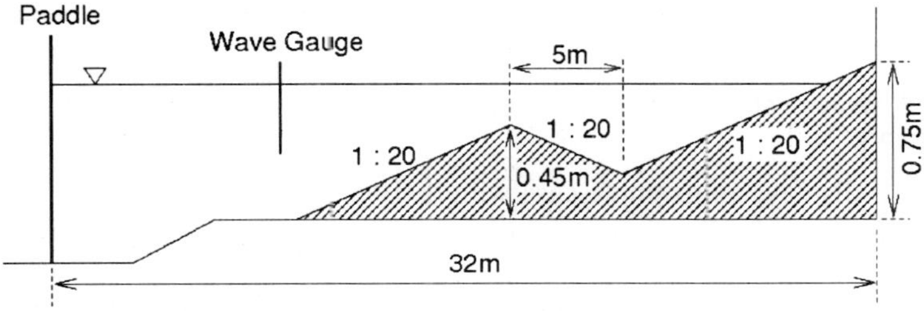

Figure 4. Bathymetrical setting of the wave tank at LNEC

Since the original bottom configuration of the tank is not flat, a wave gauge was installed at the toe of the front slope and input conditions were calibrated by using data obtained from this wave gauge. Four different wave periods (T=1.1, 1.5, 2.0, and 2.5sec) were used in this experiment. Four wave heights (H=8, 10, 15 and 20cm) were tested for d=10cm at the bar crest and two wave height conditions (10 and 15cm) were tested for d=15cm at the bar crest.

Wave celerity was calculated by taking the time difference of the wave crest between two wave gauge records separated with a known distance (20cm). Seven wave gauges were installed at each time for the wave celerity calculation and moved as a set. Particle velocity was measured by an Acoustic Doppler Velocimeter (ADV). The probe was placed roughly in the middle of the depth of the water column. The sampling rate of wave gauges was 100Hz and that of the ADV was 25Hz.

Termination locations were determined by eye-observation. The crest location when bubbles disappear was considered as the termination location. About 50 samples were collected in this way and the averaged value was obtained for each case tested.

4.2 Evolution of wave celerity and height in the trough region

Wave celerity was calculated from the obtained data. The evolution of wave celerity shows similar patterns as on the horizontal bottom. The wave celerity accelerates at the start of wave breaking, then decreases. However, the minimum celerity was observed before the termination location.

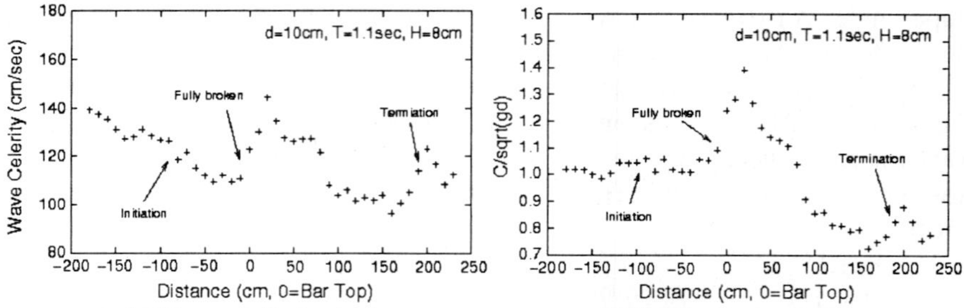

(a) Wave celerity (cm/sec) (b) Relative wave celerity

Figure 5. Wave celerity evolution; (a) absolute number (cm/sec), (b) ratio to linear wave theory

Fig. 5 displays an example case with d=10cm, T=1.1sec, and H=8cm. Note that the "Fully broken" location was estimated from the shape of the wave gauge record. As shown, the wave celerity accelerates after the wave becomes fully broken, then it decreases. But it becomes minimum at about 100cm/sec, at x=150cm. When measured celerity is normalized by the theoretical value for shallow water celerity, it corresponds to about 0.7. It is low compared to the theory. Then it increases for a short distance towards the termination location (x=191cm), but the celerity is still less than the theory. Fig. 6a displays the wave celerity evolution after the crest of the bar for all 23 cases. The X-axis is again the relative distance scale, in which x=0 is the bar-top and x=1 is the termination location. The same tendency can be seen in all cases. The data scatter in the area close to the bar-top is due to the difference of wave breaking age (difference of the initiation location) so it is not important. But the minimum celerity appears around x=0.9 for all the cases and the ratio to the linear wave theory celerity are between 0.68 and 0.83, with 0.74 on average.

The relative wave height was also calculated and plotted on the relative distance scale in Fig 6b. The decay pattern is very similar to the horizontal bottom case until about x=0.8. The H/d ratio asymptotically decreases to near 0.4 by the middle and then becomes very close to 0.4 at x=0.8. Then, it decreases again towards the termination location and the value at the termination location becomes 0.3 on average. Note that the water depth increases as the wave propagates, so that wave height has to increase to keep the H/d ratio nearly constant. The wave height actually increases between x=0.4 and 0.8, then decreases.

The results for both wave celerity and relative wave height have an unexpected result near the end of the wave breaking region. In this region, the evolution pattern is very different than the results on a horizontal section. This may mean that the turbulence in the wave breaking has actually terminated before the bubbles disappear from the water surface and our observations are slightly in error.

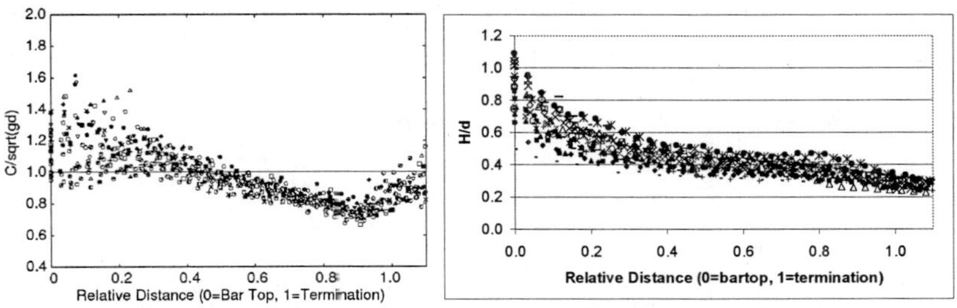

(a) Relative wave celerity (b) Relative wave height

Figure 6. Wave celerity and relative wave height evolution

The minimum celerity is much slower than the theory suggests, but as shown in Fig. 6a, in a relatively small range. Table 2 provides the minimum wave celerity for all cases tested. There is some scatter especially with wave period, but celerities with the same water depth at the bar-top show similar values in each group (except one case with d=15cm, T=1.1sec, and H=15cm).

Table 2. Minimum wave celerity (cm/sec) (a) : d=10cm

	1.1sec	1.5sec	2.0sec	2.5sec
8cm	101.1	106.8	105.5	109.3
10cm	103.8	100.7	106.5	112.8
15cm	98.0	102.5	108.3	112.9
20cm	N/A	101.8	110.2	115.6
(b) : d=15cm				
	1.1sec	1.5sec	2.0sec	2.5sec
10cm	115.6	117.8	123.7	128.4
15cm	101.4	114.5	123.9	130.6

The average minimum celerity for d=10cm is 106.4cm/sec and the average becomes 119.5cm/sec for d=15cm by excluding T=1.1sec and H=15cm. Linear wave celerity's for d=10 and 15cm are 99cm/sec and 121cm/sec, respectively. The measure values are close to the celerity at the shallowest water depth at the bar-top rather than the celerity calculated with the local water depth. Therefore, our hypothesis introduced here is that the breaking wave "remembers" the shallowest water depth experienced during the wave breaking and the celerity once accelerated by the wave breaking decreases towards a celerity associated with that minimum water depth.

To investigate the influence of set-up at the termination location, set-up was calculated from the wave gauge record by taking time-average. The wave celerity calculated with fixed water depth (d=10 or 15cm) and set-up were compared with the measured wave celerity, such as $C_{measured} / \sqrt{g\left(d_{bt} + \bar{\eta}\right)}$, where d_{bt} is depth at the bar-top and $\bar{\eta}$ is the set-up. The ratio of the celerity with set-up becomes very close to one, except one case with d=15cm, T=1.1sec, and H=15cm. The maximum value, 1.085 appeared for d=10cm, T=2.5sec, and H=15cm and the minimum value is 0.939 for d=15cm, T=1.5sec, and H=15cm.

The average of 22 cases (excluding one case) becomes 1.004. Fig. 7 displays the comparison between the measured celerity and $\sqrt{g(d_{bt}+\bar{\eta})}$.

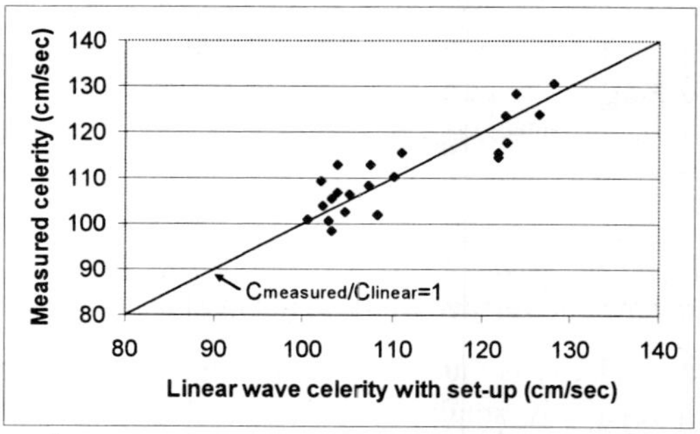

Figure 7. Comparison between the measured wave celerity and linear wave celerity calculated with fixed water depth and set-up

4.3 The minimum RTFN

As discussed in the previous section, the trend of the evolution pattern for wave celerity and wave height changes before the visual determination of the termination location. Considering the fact that H/d becomes nearly stable before the location where the celerity becomes the minimum, it appears that changes near the end of wave breaking may be governed by different criteria. The reason for this change is unknown. It might suggest that the turbulence generation ceases at the location where the trend of evolution changes and only the advection of pre-generated turbulence (i.e. the bubbles) exists for the last 20 or 30cm (in this scale) of wave breaking. Therefore, the RTFN was calculated by using the following hypothesizes:

1. Wave breaking "eventually" ceases where the wave celerity becomes minimum.
2. The wave celerity decreases towards a value related to the shallowest water depth experienced during wave breaking.

In other words, the breaking wave travels thru the trough region as if it propagates on an imaginary horizontal bottom beginning from the bar-top.

Table 3. Minimum RTFN (a) : d=10cm

	1.1sec	1.5sec	2.0sec	2.5sec
8cm	1.13	1.20	1.19	1.16
10cm	1.12	1.13	1.17	1.21
15cm	1.13	1.13	1.19	1.22
20cm	N/A	1.13	1.22	1.24
(b) : d=15cm				
	1.1sec	1.5sec	2.0sec	2.5sec
10cm	1.08	1.12	1.09	1.15
15cm	0.96	1.09	1.11	1.20

For the RTFN calculation, the set-up was excluded from the trough celerity. The spatial relation between the wave crest and trough was determined by the time difference in the wave gauge record for the particle velocity. Other wave gauge record taken at the same time was also referred for the determination of trough position. Table 3 shows the minimum RTFN for 23 test cases. The results become very similar to those for a horizontal bottom. The average (excluding a case with d=15cm, T=1.1sec, and H=15cm) is 1.16.

5. Summary and recommenced further investigations

Two topics related to wave breaking termination were discussed in this paper. The behavior of wave breaking indices on a horizontal bottom reveals that the RTFN decreases more gently than the other indices which are based on the free surface geometry. Therefore, the RTFN stays noticeably larger than the critical condition until the wave reaches the termination location. The other index values approach the critical condition far before the actual termination location. Index values based on free surface geometry in the inner breaking region are soon close to the critical condition so that the slightest error in the calculation could bring the wrong location for termination when used in numerical models.

The termination condition on a bar-trough profile was tested experimentally. The measured quantities showed some unexpected behavior from the theoretical expectation. Therefore, two hypothesizes were introduced to keep the problem simple. As a result, the RTFN at the location where the celerity becomes the slowest is very close to the theoretical expectation and results found on the horizontal bottom. Further experiments are needed to make further conclusions regarding a bottom slope effect on the termination condition of wave breaking.

ACKNOWLEDGMENTS

This work is funded by the Science and Technology Foundation of the Ministry of Science, Technology, and Higher Education, Portugal (SFRH/BPD/20508/2004 and POCTI/CTA/45431/2002). The authors gratefully acknowledge this financial support.

REFERENCES

Chen, Q., J.T. Kirby, R.A. Dalrymple, A.B. Kennedy, and A. Chawla. 2000. Boussinesq Modeling of Wave Transformation, Breaking, and Run-up. II:2D. *Journal of Waterway, Port, Coastal, and Ocean Engineering*, Vol. 126, No. 1, January/February, 44-56.

Dally, R.D, R.G. Dean, and R.A. Dalrymple. 1985. Wave Height Variation Across Beaches of Arbitrary Profile, *Journal of Geophysical Research*, Vol. 90, No. C6, 11,917-11,927.

Horikawa, K. and C.T. Kuo. 1966. A study of wave transformation inside surf zone, *Proceedings of 10th ICCE*, ASCE, 217-233

Goda, Y.. 1970. A Synthesis of Breaking Indices, *Transactions of JSCE*, Vol.2, Part 2, 39-49. (in Japanese)

Kennedy, A.B., Q. Chen, J.T. Kirby, and R.A. Dalrymple. 2000. Boussinesq Modeling of Wave Transformation, Breaking and Runup I: 1D. *Journal of Waterway, Port, Coastal and Ocean Engineering*, ASCE, Vol. 126, No. 1, January/February, 39-47.

Kirby, J.T., G. Wei, Q. Chen, A.B. Kennedy, and R.A. Dalrymple. 1998. FUNWAVE 1.0 Fully Nonlinear Boussinesq Wave Model Documentation and User's Manual, Research Report No. CACR-98-06, Center for applied coastal research, University of Delaware, Newark, Delaware.

Okamoto, T. and D.R. Basco. 2004. Experimental Verification of the Relative Trough Froude Number (RTFN) Method for Wave breaking for Boussinesq Models, *Proceedings of 29th ICCE (Lisbon),* ASCE, 357-369.

Okamoto, T. and D.R. Basco. 2006. The Relative Trough Froude Number for Initiation of Wave Breaking : Theory, Experiments and Numerical Model Confirmation, *Coastal Engineering*, 53, 675-690.

Madsen, P.A., H.A. Schäffer, and O.R. Sørensen. 1997. Surf Zone Dynamics Simulated by a Boussinesq Type Model. Part I: Model Description and Cross-Shore Motion of Regular Waves, *Coastal Engineering*, 32, 255-288.

Schäffer, H.A., P.A. Madsen, and R.A. Deigaard. 1993. A Boussinesq Model for Wave Breaking in Shallow Water, *Coastal Engineering*, 20, 185-202.

Sørensen, O.R., P.A. Madsen, and H.A. Schäffer. 1998. Nearshore wave dynamics simulated by Boussinesq type models, Proceedings of 26th ICCE, ASCE, 272-285.

Svendsen, I.A., P.A. Madsen, and J.B. Hansen. 1978. Wave characteristics in the surf zone, *Proceedings of 16th ICCE*, ASCE, 520-539.

Svendsen, I.A., W. Qin, and B.A. Ebersole. 2003. Modeling waves and currents at the LSTF and other laboratory facilities, *Coastal Engineering*, 50, 19-45.

Utku, M.. 1999. The Relative Trough Froude Number: A New Wave Breaking Criteria, Ph.D. Dissertation, Dept. of Civil and Environmental Eng., Old Dominion University, Norfolk, Virginia.

Utku, M. and D. R. Basco. 2002. A New Criteria for Wave Breaking Based on the Relative Trough Froude Number, *Proceedings of 28th ICCE (Wales)* ASCE, 258-268.

ESTIMATION OF INCIDENT AND REFLECTED WAVES USING MULTIPLE SENSORS

Miguel Figueres and Josep R. Medina[1]

An upgrade of the time-domain method for separating incident and reflected waves, LASA-FN, is presented. This upgrade involves the use of multiple sensors to perform the separation of the incident and reflected waves. The equations of a fully nonlinear wave model are used to calculate pressure and velocity at any point of the fluid, and the surface elevation. This method can be used in different water conditions, including shallow water conditions where nonlinear terms are significant.

INTRODUCTION

Incident and reflected waves are superimposed in the measurements taken in front of a structure, in both laboratory and field experiments. The reflection is a fundamental phenomenon that influences the behavior of the structure, such as stability or overtopping. The separation of incident and reflected waves is necessary to analyze and predict the response to incident wave conditions of most maritime structures. In addition, most models and formulas for maritime structural design depend on the incident waves, and sometimes, the reflection conditions are essential for the design.

There are several methods for separating incident and reflected waves, but most are based on the linear dispersion equation and the linear superposition of waves, especially the methods developed from the two-point method popularized by Goda and Suzuki (1977), useful to analyze noise-free numerical simulations. Frequency-domain methods, like the methods developed by Mansard and Funke (1981) or Baquerizo et al. (1997), are less sensitive to the noise of the signal but they still assume linear dispersion, linear superposition of waves, stationary wave components and periodic waves. These assumptions are implicit in the methods that use discrete Fourier transform (DFT) algorithms to calculate the wave components of irregular waves.

The LASA-FN method proposed by Figueres and Medina (2004) has proved to be more accurate in highly nonlinear conditions, similar to shallow water conditions, due to the use of the fully nonlinear local wave model as described by Fenton (1999). Moreover, this wave model makes it possible to obtain the values of pressure or velocity at any point of the fluid, so that multiple sensors can be used to perform the analysis.

The use of pressure sensors has several advantages: the high frequency energy (noise) is dissipated if the sensor is located at the bottom of the wave flume, and the pressure sensors don't require as much calibration as the wave

[1] Laboratory of Ports and Coasts, *Universidad Politécnica de Valencia*, Camino de Vera s\n, Valencia, 46022, Spain

gauges. In this paper, a method that uses a combination of water surface and pressure sensors in the separation process is described.

LASA METHOD

In this time-domain method, the optimization of the parameters of the wave model in each small local segment of the measured record is performed using simulated annealing algorithms. The results of the optimization in each time window are overlapped to obtain the water surface elevation corresponding to the incident and reflected wave trains at any point of the record.

Different wave models have been used, from the first version of the "LASA" method (Medina 2001), where a Stokes-II model was included, to the "LASA-FN" (Figueres and Medina 2004), where a fully nonlinear wave model was used, to improve the performance in nonlinear wave conditions.

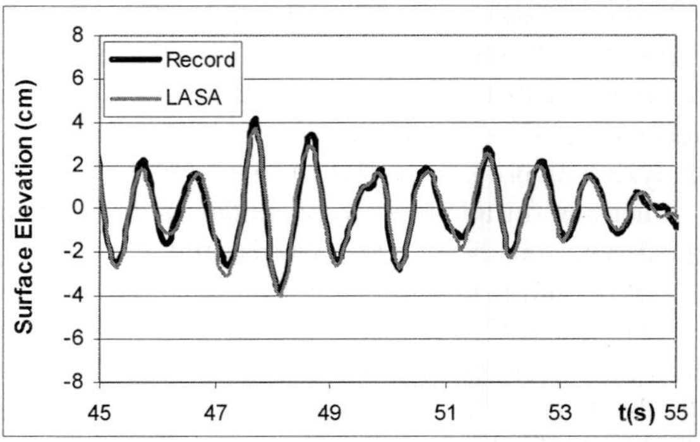

Figure 1. Comparison of measured and calculated waves with highly nonlinear conditions in a random wave test.

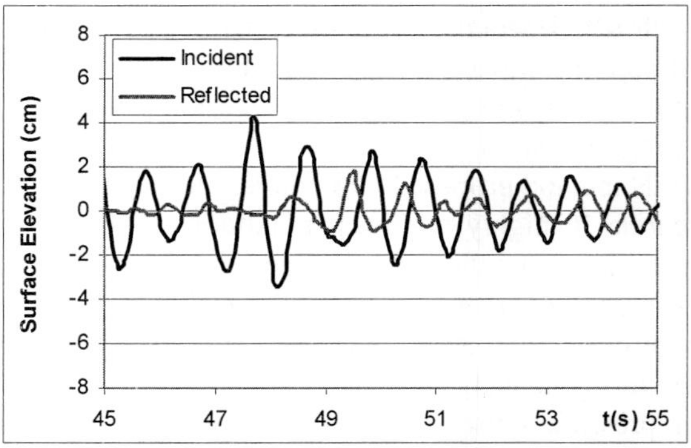

Figure 2. Separation of incident and reflected waves with highly nonlinear conditions in a random wave test.

The parameters of the wave model are optimized using a cost function that compares the measurements with the sum of the wave model components, so as to perform the local approximation. In this case, both water surface and pressure measurements are included in the cost function:

$$Cost = \frac{\sum_{n=1}^{N}\sum_{i=1}^{p} w1_p\left(\eta_{measured}(x_i,n\Delta t)-\eta_{calculated}(x_i,n\Delta t)\right)^2}{\sum_{n=1}^{N}\sum_{i=1}^{p}\left(\eta_{measured}(x_i,n\Delta t)-\bar{\eta}_{measured}\right)^2} + \frac{\sum_{n=1}^{N}\sum_{j=1}^{q} w2_q\left(P_{measured}(x_j,n\Delta t)-P_{calculated}(x_j,n\Delta t)\right)^2}{\sum_{n=1}^{N}\sum_{j=1}^{q}\left(P_{measured}(x_j,n\Delta t)-\bar{P}_{measured}\right)^2} \tag{1}$$

where 'N' is the number of measurements in a time window, 'p' and 'q' are the number of wave gauges and pressure sensores respectively, '$\eta(x,n\Delta t)$' and '$P(x,n\Delta t)$' are the measurements.

Although the pressure can be expressed as water column height, so that every term dimension of the equation is 'length', it is necessary to use weights to ponderate surface elevation and pressure measurements, '$w1$', '$w2$', in cases where pressure sensors and wave gauges are in the same position or very close. If these cases, the values of '$w1$', '$w2$' corresponding to the sensors that are close or in the same position, are lower than 1, while the other sensor's weights are equal to 1.

FULLY NONLINEAR WAVE MODEL

Fenton (1999) solved the problem of numerical fully nonlinear waves for irrotational and incompressible fluids in a two dimensional wave field. The solution gives the exact value of the free surface $\eta(x,t)$, horizontal velocity $u(x,z,t)$, vertical velocity $v(x,z,t)$ and pressure $p(x,z,t)$:

$$\eta(x,t) = 2\sum_{j=0}^{N}{}'' E_j \cos jk(x-ct) \tag{2}$$

$$u(x,z,t) = \sqrt{\frac{g}{k}}\sum_{j=1}^{N} jB_j \frac{\cosh jkz}{\cosh jkd}\cos jk(x-ct) \tag{3}$$

$$v(x,z,t) = \sqrt{\frac{g}{k}}\sum_{j=1}^{N} jB_j \frac{\sinh jkz}{\cosh jkd}\sin jk(x-ct) \tag{4}$$

$$\frac{p(x,z,t)}{\rho} = R - gz - \frac{1}{2}\left(\left(u(x,z,t)-c\right)^2 + v^2(x,z,t)\right) \tag{5}$$

where 'k' is the wave number, 'c' is the wave celerity and 'R' is a value related to the mean pressure at the bottom.

The solution only exists in a specific area inside the rectangle defined by $H/L\in[0,0.13]$ and $H/d\in[0,0.75]$, ('H': wave height, 'L': wave length, 'd' depth), as shown in Figure 3. This area is wide enough to include the wave conditions usually given in field and laboratory experiments.

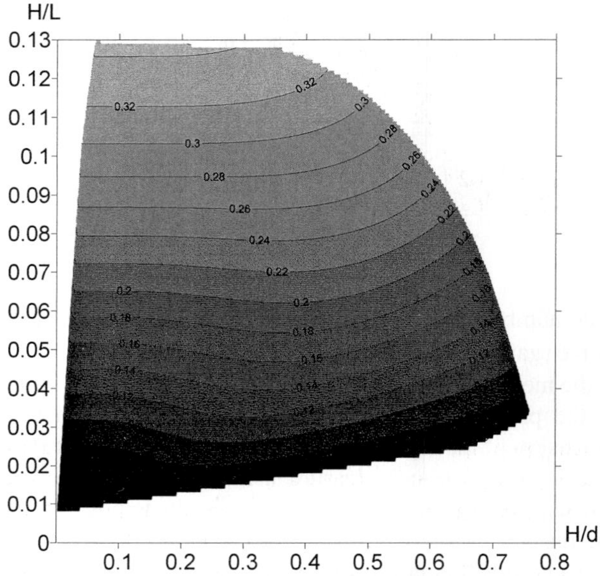

Figure 3. Existing solution area and values of B_1.

Inside this area, the dimensionless parameters $R{\cdot}k/g$, $c\sqrt{k/g}$, $T\sqrt{g/d}$, B_j, and E_j (j=1,…,N) can be expressed as a function of H/d and H/L. To obtain these parameters, a linear system of $2N+5$ equations needs to be solved in an iterative algorithm, whose initial solution is the linear approximation and the value of N is at least 20 to obtain an accurate solution in highly nonlinear conditions. The computational cost of each of the dimensionless parameters for each pair of values (*H/d*, *H/L*) is too high to use in the simulated annealing process of the LASA method; thus, a faster approximation is needed.

ARTIFICIAL NEURAL NETWORK

Artificial Neural Networks (ANN) have proved to be very efficient to solve function approximation problems (see Medina et al. 2002). In this case, the solutions of Fenton equations in a high resolution grid of more than 16300 values of (H/d, H/L) are the training data set to optimize the parameters of a feed-forward ANN, obtaining a value of 'MSE/var' of the ANNs approximations lower than 10^{-5} using 7 nonlinear neurons in the hidden layer (Figure 4). One ANN is required to perform the interpolation of each variable $R{\cdot}k/g$, $c\sqrt{k/g}$, $T\sqrt{g/d}$, B_j, and E_j (j=1,…,N) as a function of the pair of values

(H/d, H/L), and an additional ANN to calculate 'H/L' as a function of (H/d, $T\sqrt{g/d}$).

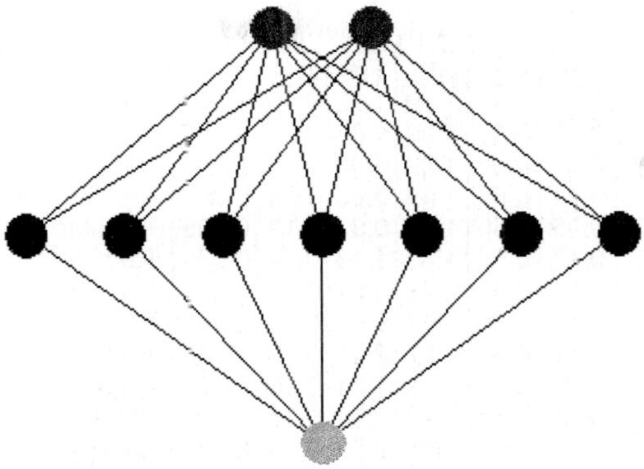

Figure 4. Neural Network configuration.

The approximation of the exact solution by using ANNs fits very well (Figure 5) and the computational cost of each variable is very low; therefore, this efficient approximation can be used in the simulated annealing procedure.

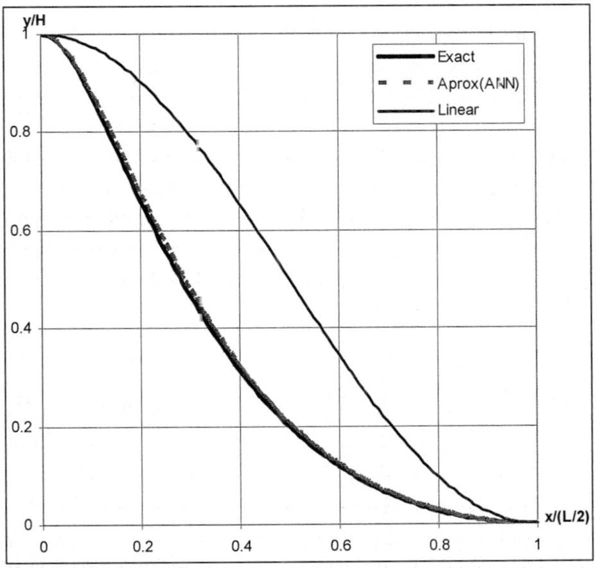

Figure 5. Example of wave model approximation by using ANN (H/L=0.083,H/d=0.5).

EXPERIMENTAL TESTS

The experimental tests done at the LPC-UPV wave flume, described in Figueres and Medina (2004), included the measurements of pressure and free-surface elevation sensors. The wave test conditions were: 0.04<H/L<0.11 and 0.2<H/d<0.6. A high reflective vertical wall, a low reflective mound breakwater and a very low reflective parabolic structure were tested.

Table 1. Experimental tests

Test	H(m)	d(m)	L(m)	T(s)	H/d	H/L	Comments	
FN01_016	0.08	0.50	1.93	1.32	0.16	0.041	Regular	Breakwater
FN01_027	0.13	0.50	1.50	1.03	0.26	0.087	Regular	Breakwater
FN03_011	0.05	0.50	1.22	0.95	0.10	0.041	Random	Breakwater
FN03_018	0.14	0.58	1.49	0.95	0.24	0.094	Random	Breakwater
FN04_015	0.05	0.50	1.27	0.97	0.10	0.039	Regular	Ramp
FN04_027	0.13	0.58	1.83	1.15	0.22	0.071	Regular	Ramp
FN05_011	0.05	0.50	1.22	0.94	0.10	0.041	Random	Ramp
FN05_018	0.14	0.58	1.49	0.95	0.24	0.094	Random	Ramp
FN06_015	0.05	0.50	1.27	0.97	0.10	0.039	Regular	Wall
FN06_017	0.10	0.50	1.41	1.01	0.20	0.071	Regular	Wall
FN07_011	0.05	0.50	1.22	0.94	0.10	0.041	Random	Wall
FN07_018	0.14	0.58	1.49	0.95	0.24	0.094	Random	Wall

Results

Four different combinations of multiple sensors were analyzed for each of the experimental tests selected. These combinations are:

1. Two pressure sensors (PP).
2. Two wave gauges (SS).
3. Two pressure sensors and a wave gauge (PPS)
4. Three wave gauges (SSS).

The pressure sensors are located at the same position as the first pair of wave gauges, so that results can be compared. The results of the tests are shown in Table 2.

Table 2. Results of the experimental tests for different combinations of sensors

	C1 - PP			C2 - SS			C3-PPS			C4 - SSS		
Test	RMSE	Hi(m)	CR	RMSE	Hi(m)	CR	RMSE	Hi(m)	CR	RMSE	Hi(m)	CR
FN01_016	1.63%	0.064	0.29	0.73%	0.067	0.27	0.73%	0.080	0.25	0.39%	0.088	0.28
FN01_027	1.87%	0.131	0.20	1.02%	0.131	0.24	0.78%	0.141	0.19	0.46%	0.151	0.22
FN03_011	4.71%	0.050	0.31	3.83%	0.048	0.30	3.71%	0.059	0.27	3.46%	0.069	0.26
FN03_018	5.19%	0.125	0.24	4.06%	0.130	0.27	3.98%	0.137	0.23	3.60%	0.148	0.27
FN04_015	1.65%	0.050	0.13	0.76%	0.048	0.16	0.90%	0.057	0.14	0.41%	0.069	0.14
FN04_027	1.63%	0.113	0.16	0.96%	0.117	0.18	0.85%	0.124	0.17	0.67%	0.139	0.17
FN05_011	4.25%	0.054	0.14	3.49%	0.052	0.17	3.58%	0.061	0.14	3.11%	0.071	0.16
FN05_018	4.56%	0.143	0.12	3.77%	0.133	0.13	3.73%	0.147	0.12	3.27%	0.156	0.14
FN06_015	2.26%	0.046	0.77	1.46%	0.045	0.83	1.75%	0.055	0.80	1.28%	0.066	0.80
FN06_017	2.81%	0.098	0.83	1.69%	0.093	0.77	2.08%	0.119	0.76	1.44%	0.120	0.76
FN07_011	5.70%	0.043	0.69	4.45%	0.038	0.74	4.70%	0.046	0.69	4.20%	0.062	0.72
FN07_018	5.33%	0.133	0.75	4.53%	0.135	0.76	4.66%	0.143	0.70	3.98%	0.156	0.74

The plot of the relative mean square errors (RMSE) is shown in Figure 6. The results for the sensors combinations SS, PPS, and SSS are quite similar. However, the RMSE of the PP combination is higher.

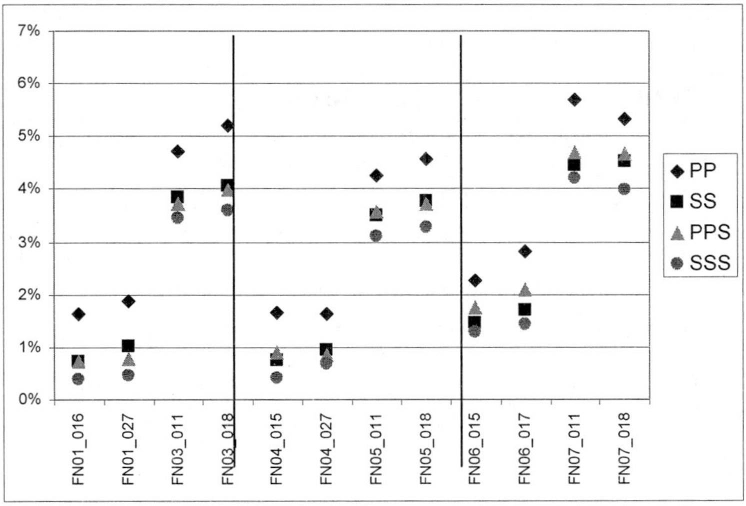

Figure 6. RMSE of the analysis with different sensor combinations.

CONCLUSIONS

An upgrade of the LASA-FN method is presented. The fully nonlinear wave model allows for the analysis of a wide range of experimental conditions in the time-domain, with highly nonlinear and non stationary wave trains, as well as the use of multiple sensors. Several combinations of pressure sensors and wave gauges were tested. The errors are small and mostly caused by the noise of the signal. When only pressure sensors are involved, the analysis becomes more sensitive to small errors in the position of the sensor.

The error introduced by the superposition of a nonlinear incident and reflected wave train is small, although it increases in highly reflective conditions.

The LASA method has proved to be very flexible and other wave models could be used in the optimization process.

ACKNOWLEDGEMENTS

This research was funded by the Ministry of Science and Technology through grant MODUMAR (BIA2005-03197). Miguel Figueres is financially supported by the Spanish *Secretaría de Estado de Educación y Universidades*. The authors wish to thank Dr. Debra Westall (*Universidad Politécnica de Valencia*) for her revision of this English text.

REFERENCES

Baquerizo, A., Losada, M.A., Smith, J.M. and Kobayashi, N. (1997). Cross-shore variations of wave reflection from beaches. *J. Wtrwy., Port., Coast. and Oc.Engrg.*, ASCE, 123(5), 274-279.

Fenton, J.D. (1999). Numerical Methods for Nonlinear Waves, in Advances in Coastal and Ocean Engineering, P.L.-F. Liu (Ed). World Scientific. pp. 241-324.

Figueres, M. and Medina, J.R. (2004). Estimating incident and reflected waves using a fully nonlinear wave model. *International Conference on Coastal Engineering*, Lisbon. ASCE, ICCE'2006: 594-603.

Mansard, R.P.D. and Funke, E.R. (1981). The measurement of incident and reflected spectra using a least squares method. Proc., 17th ICCE, ASCE, 154-172.

Medina, J.R. (2001). Estimation of Incident and Reflected Waves Using Simulated Annealing. J. Wtrwy., Port., Coast. and Oc. Engrg., ASCE, 127(4), 213-221.

Medina, J.R., González-Escrivá, J.A, Garrido, J. and De Rouck, J. (2002). Overtopping análisis using neural networks. ASCE, ICCE'2002: 2165-2177.

ON THE RUN-UP AND DRAW-DOWN OF BREAKING SOLITARY WAVES

Kao-Shu Hwang[1] Yu-Hsuan Chang[1] Hwung-Hweng Hwung[2]

Hsin-Hung Chen[1]

The flow characteristics of breaking solitary waves on plane beaches are investigated with emphasis on the run-up and draw-down processes. A series of experiments were conducted in the SUPER TANK of Tainan Hydraulics Laboratory with three plane beaches of 1:20, 1:40 and 1:60. Solitary waves of which relative wave heights ranged from 0.031 to 0.231 were generated. The surface elevation changes were recorded by capacitance type wave gages distributed at 96 stations. The velocities were measured by acoustic Doppler velocimeters and electric-magnetic velocimeters. The surface wave profiles evolutions and velocity time series are displayed and discussed. In addition, the run-up heights and the breaking index obtained from this large-scale experiment are also shown and are compared to results of previous studies.

INTRODUCTION

A tsunami is a very long wave that could be generated by a quick displacement of huge water volume in the ocean. A solitary wave propagate on a plane beach has been used to simulate the behavior of a tsunami approaching onshore. Regarding coastal hazard caused by a tsunami, the run-up height has been the most important index to evaluate its impact. Accordingly there are a number of studies have contributed on this topic. However, the run-up and draw down process of a solitary wave on a plane beach is still a challenging issue that involves the complex interactions of fluid, air and the solid boundary from the view point of hydrodynamics, especial for a breaking solitary wave.

For analytical approaches, the non-linear shallow water equations have been widely employed to investigate the run-up and draw down on a uniform plane beach for non-breaking cases. The pioneer work by Carrier & Greenspan (1958), for example, proposed a nonlinear transformation to convert the nonlinear shallow water equations to a single linear equation. And this made analytical solutions possible for a few special initial-value problems. Carrier & Greenspan (1958)'s solutions were further simplified by Synolakis (1987), in which the solution was applied to a solitary wave propagates from a constant water depth then running up on a plane beach. Based on the Green function representation of fully nonlinear shallow water equation, Carrier, Wu and Yeh (2003) developed an analytical-numerical hybrid method that is able to compute both incoming and

[1] Tainan Hydraulics Laboratory, National Cheng-Kung University, Tainan, 709, TAIWAN.
[2] Department of Hydraulic and Ocean Engineering, National Cheng-Kung University, No. 1, Da-Sheh Road, Tainan, 701, TAIWAN.

outgoing waves with arbitrary initial wave form. While for breaking cases, due to the complex flow evolution during the breaking process, it has been necessary to use numerical models so far. Based on Boussinesq equations, Zelt (1991) develop a model on Lagrangian coordinate to investigate both the non-breaking and breaking solitary wave run-up on plane beaches. The wave breaking was modeled by an eddy viscosity model with empirical turbulent length scale coefficient. Li and Raichlen (2002) used the weighted essentially non-oscillatory (WENO) scheme to solve the nonlinear shallow water equation. Using this scheme, the solitary wave breaking can be modeled without ad-hoc terms.

As for laboratory experiments, most of the early works have investigated the evolution of the shoaling wave heights, the breaking criterion and the run-up heights. The breaking index of solitary waves propagating on different slopes of plane beaches was reported by Ippen and Kulin (1954). Papanicolaou and Raichlen (1987) investigated the surf zone characteristics of solitary waves on five gentle beaches by using high-speed camera. To verify the numerical results, Grilli et. al. (1997) had collected a few experimental data on the breaking index and other characteristics of solitary waves on slopes. Synolakis (1987) conducted a series of experiments on a 1:19.85. Equations to predict the run-up heights for the non-breaking and the breaking solitary waves were proposed. The shoreline movement was reported in Li and Raichlen (2002), who used high-speed video camera to detect the wave front locations vary with time. In comparison with the surface wave information, there were limited literatures that had revealed the internal flow fields of a breaking solitary wave. Among these, Li and Raichlen (2001) measured the velocity near the original shoreline by laser Doppler velocimeters (LDV). Their measurements were obstructed by the presence of air bubbles after breaking. Hence there is still no data available beyond the breaking point, let alone the run-up region. Jensen, Pederson & Wood (2003) had tried to reveal the detailed distribution of velocity in the run-up region by particle image velocimetry (PIV). Their results finally focused on the early stages in the run-up of nearly breaking waves on a relatively much steeper slope, 1:5.4 (10.54°). In spite of all these efforts, there were rare papers reported on the run-up and draw-down processes of breaking solitary waves. Therefore, attempts to model this phenomenon have been hampered. For example, Lin, Chang & Liu (1999) had tried to simulate this process based on RANS equation and an improved k-ε model. The computed surface wave profile evolutions, both the breaking and the non-breaking cases, had been compared with experiments and found good agreement. However, for the internal flow fields, only the non-breaking cases had been verified.

There are two major difficulties to be overcome in swash zone velocity measurements. The first one is the usually very thin layer of the run-up flow, especial when the experiments were conducted in a smaller scale or on a milder slope. The second one is the highly air bubbles entrained during the breaking process that always cause the velocimeters lose their function. To cope with the first problem, it is practical to use a larger scale flume that is, supposedly, able to produce a run-up flow with thicker water body. Then some of the popularly used instruments may be possible to be applied. In present study, experiments in a large-scale wave flume were performed to acquire much more data in depth in the run-up region. Although the state-of-the-art non-intruded velocimeters such as PIV and LDV are not applicable in such a large flume, comprehensive acoustic Doppler velocimeters and electro-magnetic velocimeters were used instead. The variations of water surface and velocity at different locations were presented and discussed.

EXPERIMENTS

The experiments were conducted in the Super Tank, of which length, 300.0 m, width, 5.0 m and depth, 5.2 m, in Tainan Hydraulics Laboratory (*THL*), National Cheng Kung University. The tank was divided into a constant water depth region and an inclined bottom. Three plane beaches of slopes 1:20, 1:40 and 1:60 were constructed with a smooth layer of concrete. For tests on the 1:20 sloping bottom, the water depth in the constant depth region, h_0, was 3.2 m. For cases in the 1:40 and the 1:60 sloping bottoms, it was 2.2 m (Figure 1). Solitary waves of which relative incident wave heights, H/h_0, ranging from 0.031 to 0.231 were generated, where H is the wave height in the constant depth region.

The definitions of normalized parameters are shown in Figure 2. The normalized time, t^*, the normalized horizontal distance to the initial shoreline, x^*, and the normalized horizontal velocity, u^*, are defined as:

$$t^*=t\sqrt{(g/h_0)},\quad x^*=x/h_0,\quad u^*=u/C_0 \tag{1}$$

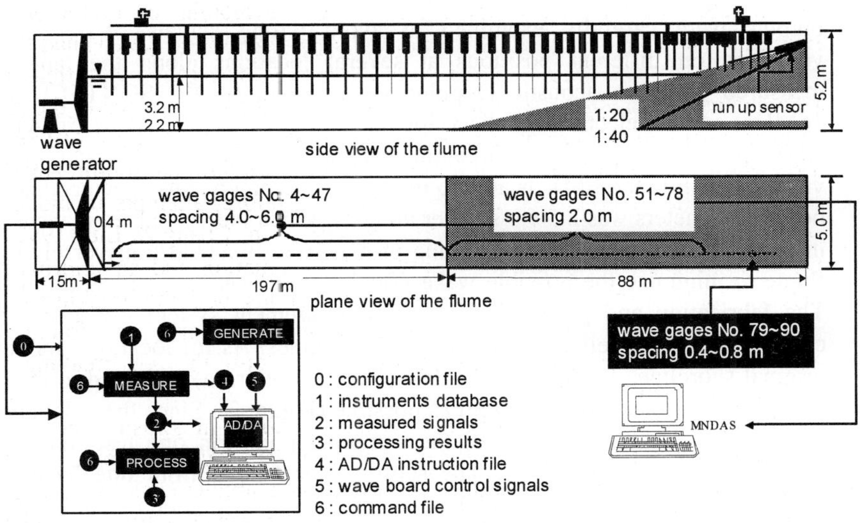

Figure 1. Experimental set-up.

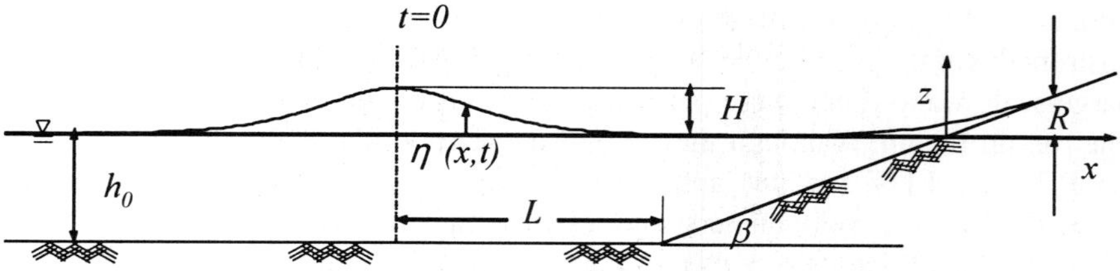

Figure 2. Definition sketch.

where C_0 is the phase velocity of the solitary wave in the constant depth region. The half length of the horizontal extent of the wave, L, is defined as the horizontal length from the wave crest to the point that the amplitude is 5% of the wave height. Based on the Boussinesq equation, it thus gives as:

$$L=\frac{1}{\sqrt{3H/4h_0}}\ COSH^{-1}\left(\sqrt{1/0.05}\ \right) \tag{2}$$

The referred time base $t=0$ was counted when the wave crest passed a distance L away from the toe of the slopping bottom.

Time series of local water surface elevation were recorded by capacitance-type wave gages distributed at 96 stations along the tank. Each sensor of the gages was mounted at 0.4 m away from the sidewall. The resolution of the most gages is 12 bits on a dynamic range of 1.0 m. Some longer (1.2 m) gages were installed in the surf zone, where the wave heights sometimes exceed 1.0 m. Among the 96 gages, there were 14 gages located beyond the initial shoreline. As these gages were entirely out of water before the wave passing by, their outputs were carefully examined to avoid wrong signals. Water particle velocities on different positions at several locations along the tank were measured by three acoustic Doppler velocimeters (ADV, SONTECH ADV 16MHz) and six electromagnetic velocimeters (EMV, ALEC ACM250-A). The ADV is able to measure three-dimensional components of water particle velocities, while the EMV is able to measure two components. The accuracy for both velocimeters was 2.5 cm/s under all tests. When the velocimeters were used in the surf zone, all the down-looking probes were mounted transversely to the flume to minimize the possible wake effects by the sensors' bodies (Figure 3). The fabrication and positioning of this measurement system was under very careful control to well align all measuring volumes. As for locations above the original shoreline, a side-looking ADV (Nortek, Vectrino) was used to measure the velocity as close to the bottom as possible (Figure 3). Since the velocimeters were not enough to measure all the sampling points in one test run, it was necessary to repeat the tests. The sensors were move from one location to another and the signals were referred to the wave gages, which were located at fixed places for all test runs. The maximum run-up was determined by visual observation.

Figure 3. The Super Tank and the instrumentation.

SOLITARY WAVE GENERATION

The waves were generated by a hydraulically driven, dry back, piston type wave-maker with a maximum stroke of 2.0 meter. The wave-board was driven one way forward with a constant speed to simulate an offshore water volume displacement. One example of the wave board displacement time history and wave profiles evolution in constant depth region are shown in Figure 4. As indicated by Goring (1979), subsequent trail waves are inevitable under this generation method. The surface elevation record was compared with the 1st order theory. It shows that the shapes of the wave leading faces of the experiment and the theory are very close (Figure 5).

RESULTS AND DISCUSSION

Breaking index and run-up heights

Although there have been a number of literatures reported on the breaking index and run-up heights of solitary waves propagating on plane beaches, some more data acquired in this large scale flume are added here. Table 1 shows the test conditions and the observed breaking wave height to depth ratio, H_b/h_b. As had indicated by previous studies, the breaking wave height to depth ratio of a solitary wave propagating on a plane beach is apparently larger than the commonly recognized value, 0.78, which was obtained for a solitary wave traveling in constant water depth. For the same beach slope, the breaking index becomes smaller as the relative incident wave height becomes larger. Meanwhile, such a negative relation between the two parameters becomes less obvious as the beach slope becomes milder.

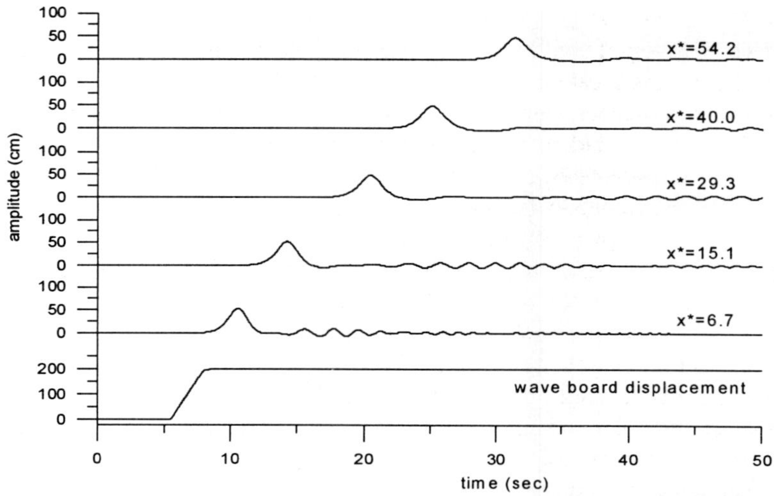

Figure 4. wave board displacement time histories and wave profiles evolution in constant depth region.

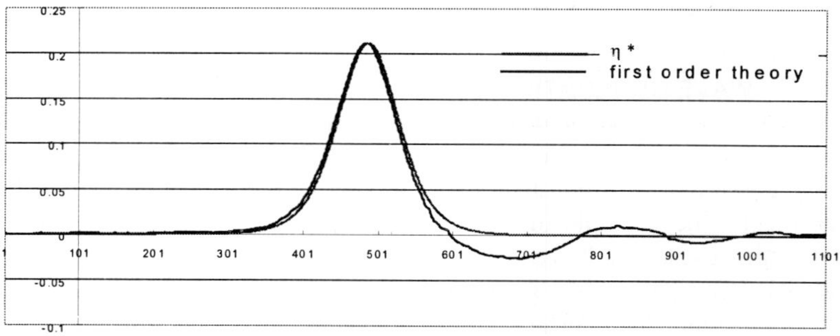

Figure 5. Comparison of surface wave profiles in constant depth region.

Table 1 Test condition and results							
$\cot\beta$	h_o (m)	H/h_o	H_b/h_b	$\cot\beta$	h_o (m)	H/h_o	H_b/h_b
20	3.2	0.031	-	40	2.2	0.161	1.429
20	3.2	0.086	3.883	40	2.2	0.195	1.456
20	3.2	0.095	4.083	40	2.2	0.215	1.400
20	3.2	0.100	3.823	60	2.2	0.063	1.300
20	3.2	0.106	3.313	60	2.2	0.117	1.374
20	3.2	0.132	2.812	60	2.2	0.136	1.147
20	3.2	0.140	2.948	60	2.2	0.161	1.274
40	2.2	0.053	1.667	60	2.2	0.188	1.184
40	2.2	0.108	1.412	60	2.2	0.202	1.219
40	2.2	0.112	1.368	60	2.2	0.231	1.228
40	2.2	0.133	1.422				

Ippen and Kulin (1954) conducted a series of experiment to investigate the shoaling and breaking of solitary waves on three plane beaches: 1:15.4, 1:20 and 1:43.5. It showed the breaking index kept around 1.2 for the 1:43.5 slope. While for the other two steeper slopes, the breaking index had a negative relation varied with relative incident wave heights. Grilli, Svendsen & Subramanya (1997) proposed a parameter, S_0, which considered both the beach slope and the relative incident wave height. Based on 17 experimental data quoted from four researches and their own numerical results, an empirical formula that the breaking index grows exponentially with S_0 was obtained. The data acquired in present study were compared with this empirical formula, as plotted in Figure 6. The observed H_b/h_b is in a good agreement with the predicted value, except for the point of which S_0 is around 0.16. This case was the wave with smallest relative incident wave height carried out on our 1:40 slope. A common feature for those of smaller relative incident wave height was that their trail waves were relatively much more pronounced than those of larger relative incident wave height under current wave generation method. Yet it is presumed that this deviation may because of the solitary waves generated in present study were not as pure as previous studies did.

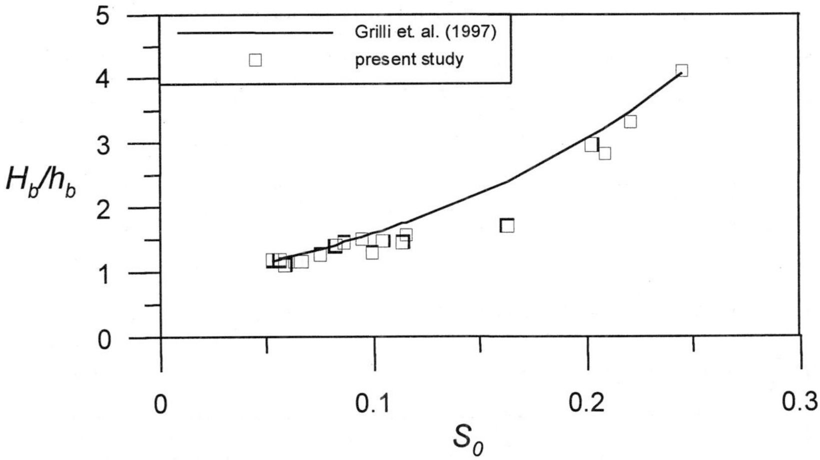

Figure 6 The variation of observed and predicted breaking index with S_0 (S_0 was defined in Grilli et. al. 1997).

The normalized run-up heights of different relative incident wave heights under the three slopes are shown in Figure 7. For solitary waves propagating on plane beaches of the same slope, the run-up heights obviously increase as the relative incident wave heights increase. However, the increasing rate becomes smaller as the plane beach becomes milder. On the other hand, for solitary waves of the same relative incident wave height, the steeper the slope, the higher it could run up.

There were a few empirical formula proposed to predict the run-up heights of breaking solitary waves. Among them, results obtained from the formula proposed by Synolakis (1987) and the "energy balance model (EBM)" proposed by Li and Raichlen (2003) were compared. The experiments being conducted under only one beach slope, Synolakis's empirical formula evaluates the breaking solitary wave run-up height with a simple function of H/h_0. The EBM, on the other hand, had taken both H/h_0 and the beach slope into account. For the slope of 1:60 and 1:40 cases, the EBM seems to predict the run-up heights well, though there are a little bit over-estimated. Since the EBM assumed the bottom and side-wall friction loss to be very small, so it is not surprising that most of our observed data are smaller than their prediction. As for the 1:20 beach slope, Synolakis's formula was also employed. For this case, the both formula and the experimental results are close for smaller wave heights. However, when the relative wave height larger than 0.1, the both predictions become under-estimated. And this deviation seems to getting larger as H/h_0 increases. Actually, both Synolakis's empirical formula and the EBM were based on a series of experimental data. While on the other hand, each of our data shown was obtained from several times repeated runs. In addition to the mentioned "not pure" solitary wave, the relatively large water depth used (3.2 m) in the 1:20 beach may be another source that cause this difference.

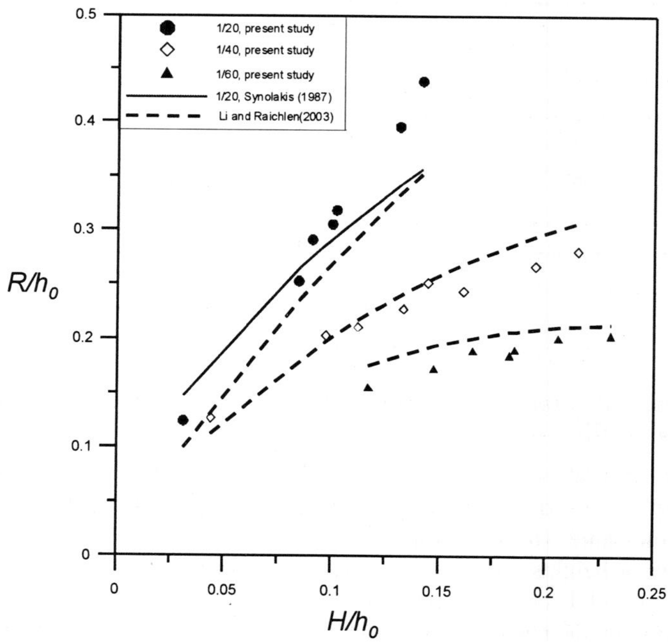

Figure 7. The variation of the normalized run-up with relative wave height.

Surface elevation and velocity time series

The water surface and horizontal velocity time histories of a solitary wave (H/h_0=0.096) traveling on the 1:20 slope are displayed in Figure 8. It is clear to see that the wave height increases gradually and the front face becomes steeper due to non-linear effects before breaking. The front face then becomes nearly vertical at breaking point (x^*=-1.0). Additionally, an interesting phenomenon is that the water surface time histories after x^*=-1.4 apparently appears a narrow peak at the leading edge attaching on an extended triangular wedge-shaped region. For locations in the surf zone, the velocities were measured at least on three levels: the position 2 cm below the still water level, the middle level and 3 cm above the bottom. It was found that the water column under the still water level moved uniformly with time, so that the data detected on the middle level are presented as a typical record to each section. The variation trend of the velocity is almost the same with that of the water surface changes. It increases rapidly as the surface elevation rises instantly, and then it reaches the maximum as the local surface elevation reaches its summit. After the wave crest passes, the water body still moving onshore but with a decreasing changing rate and finally turns into offshore. As for those above the still water level, theoretically, the peak horizontal velocity would be significantly larger than those under the still water level. However, since both the ADVs and the EMVs were not able to respond well in this region, the data were abandoned.

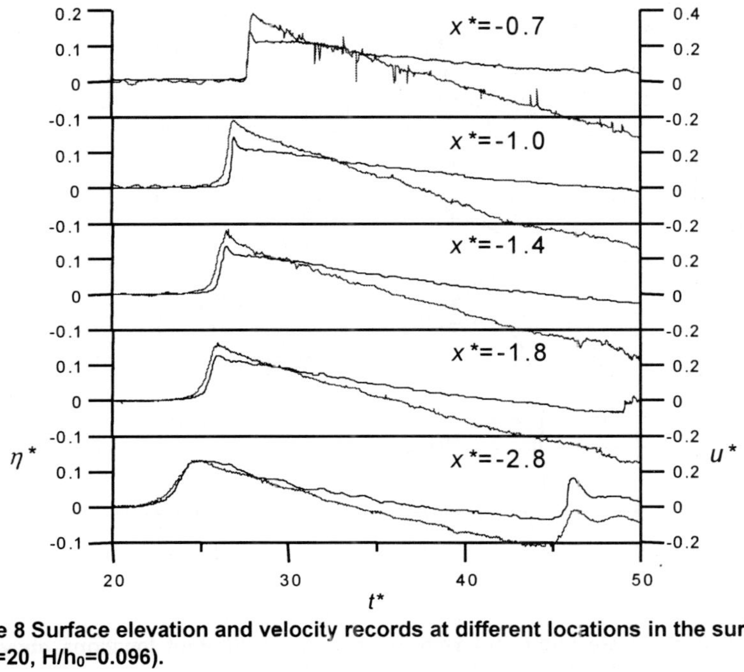

Figure 8 Surface elevation and velocity records at different locations in the surf zone (cotβ=20, H/h₀=0.096).

The wave front turns into a turbulent bore and climbs up the beach after breaking (Figure 9). Time series of the local water level above the initial shoreline exhibits a rather sharp front but with a very gentle rear face. As the tip of the run-up tongue passes the measuring volume, drop out signals (set to be zero) come out because of high-density air bubbles. Better and qualified signals were not available until the highly air bubbles contained front totally pass away. In this case, another bore formed during the draw-down process and it also causes drop out signals. The available data show that the local velocity in the run-up region decreases almost linearly with time. After the local water level reaches it's maximum, the flow continuously move upward with decelerated speed. Then at a certain moment, with almost the same decelerated rate, it stops and starts drawing down. By locating the time when the local velocity of each monitoring section becomes zero (had been marked with a small circle in Figure 9), it can be seen that the water body in the lower part of the run-up region starts drawing down earlier than that of the upper part. That is, not all the run-up water move together upward before the run-up front reaches the highest point. In fact, during a certain moment when the run-up tongue front still climbs up, the water in the lower part has started flowing down. And this phenomenon can also be witnessed from the visual observation on the wave flume.

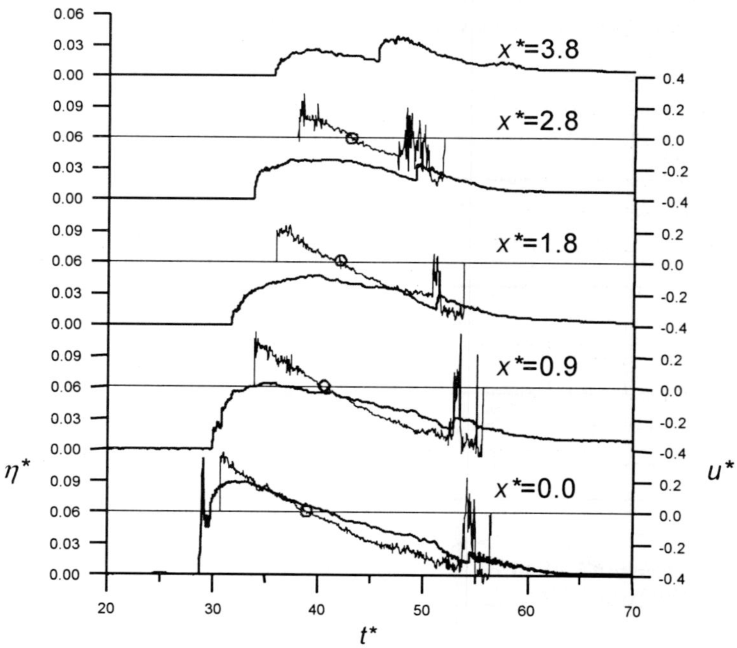

Figure 9 Surface elevation and velocity records at different locations above the initial shoreline (the lower figure) (cotβ=20, H/h$_0$=0.096).

Figure 10 and Figure 11 show the case of which H/h_0=0.215 on the 1:40 slopping bottom. The most striking feature of the water surface time histories is that the narrow peak at the leading edge becomes much more protruded (Figure 10). A similar wave profile that had also been reported by Zelt (1991) can be found when it was exhibited in the space domain (Figure 12). After breaking, the wave front propagates toward the shoreline, possessing more or less the same shape with a narrow peak still at the leading edge but with dramatically decaying wave amplitude. Being with a steeper incident wave propagating on a milder slope of beach, it breaks at a distance notably farther away from the initial shoreline (x^*=11.8), as compared to the previous case. It means that the post breaking wave front travels a longer distance to the initial shoreline than the previous case. Interacting with the initially quiescent water body, there should be much more energy consumed before it climbing up the "dry" beach. The velocity observed on the initial shoreline, though not completed, is smaller in this case. As a consequence, even the H/h_0 of this case is double more than the previous case, its relative run-up height is still smaller.

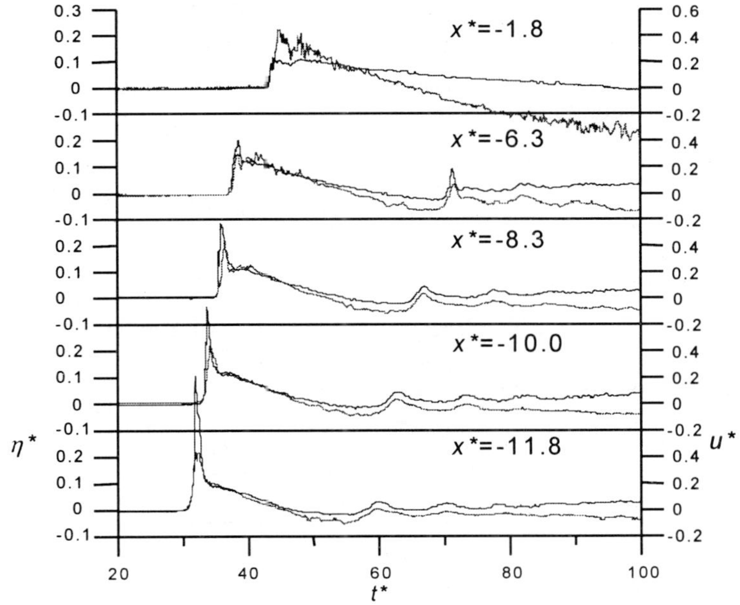

Figure 10 Surface elevation and velocity records at different locations in the surf zone (cotβ=40, H/h_0=0.215).

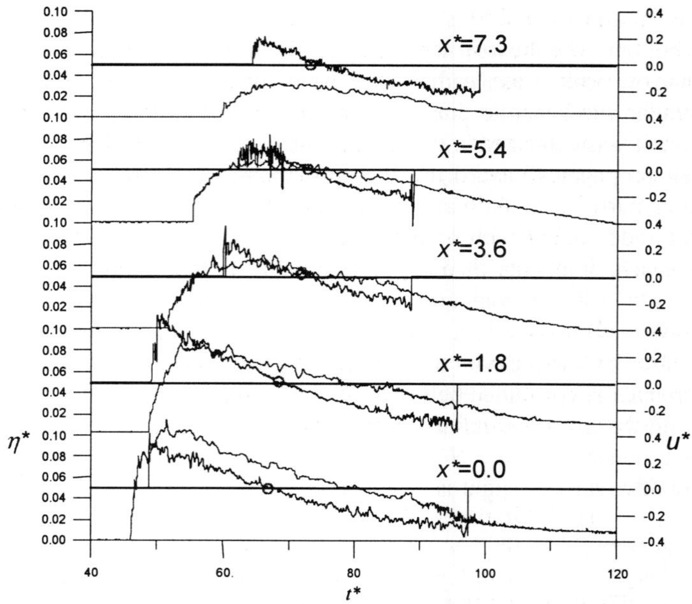

Figure 11 Surface elevation and velocity records at different locations above the initial shoreline (cotβ=40, H/h$_0$=0.215).

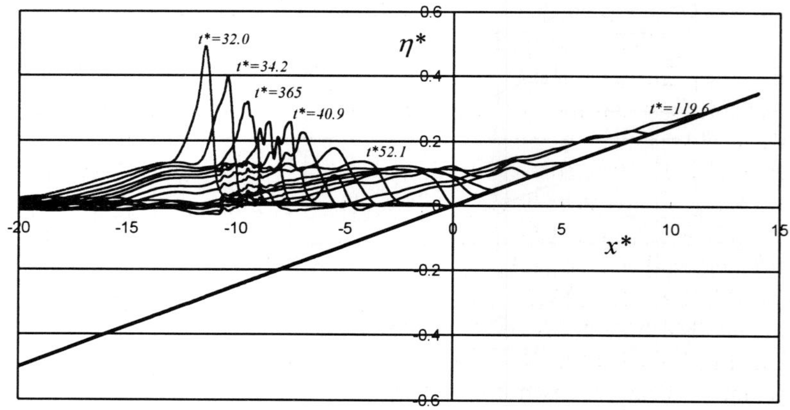

Figure 12 Surface wave evolutions from breaking to run-up (cotβ=40, H/h$_0$=0.215).

CONCLUDING REMARKS

A large scale experimental study of solitary waves breaking on plane beaches was conducted, particularly focused on the run-up and draw down process. Attempts to measure the velocities in the run-up region although not completely successful, some valuable data were acquired. Nevertheless, how to reveal the velocity within the air bubble contained water body is the most essential future work. For plane beaches of the same slope, the run-up heights of breaking solitary waves increase as the relative incident wave heights increase. However, the increasing rate becomes smaller as the plane beach becomes milder. While for the same relative incident wave height, the steeper the beach slope, the higher it runs up. For the 1:40 and the 1:60 beaches, the EBM can predict the run-up height well. For the 1:20 beach, both Synolakis's formula and the EBM can predict well when H/h_0 are smaller than 0.1, but the deviations between our observed data and the empirical formulas become larger when H/h_0 increases.

ACKNOWLEDGMENTS

The authors would like to express sincerely gratitude to the "NCKU Project of Promoting Academic Excellence & Developing World Class Research Centers" which was supported by the Ministry of Education, TAIWAN, R.O.C.

REFERENCES

Carrier, G. F. & H. P. Greenspan (1958) "Water waves of finite amplitude on a sloping beach" , J. Fluid Mech., vol. 17, pp. 97-110.

Carrier, G. F., T. T. Wu and H. Yeh (2003) "Tsunami run-up and draw-down on a plane beach" , J. Fluid Mech., vol. 475, pp. 79-99.

Goring, D. G. (1979) "Tsunami: The propagation of long waves on a shelf", Ph. D. thesis, California Institute of Technology.

Grilli, S. I. A. Svendsen and R. Subramanya(1997) " Breaking criterion and characteristics for solitary waves on slope, J. Waterway, Port, Coastal and Ocean Eng., vol. 123, pp. 102-112.

Ippen, A. T. and G. Kulin (1954) "The shoaling and breaking of the solitary waves", Proc. 5th International Conference on Coastal Engineering, pp. 24-47.

Jensen A., G. K. Pederson and D. J. Wood (2003), An experimental study of wave run-up at a steep beach, J. Fluid Mech., vol. 486, pp. 161-188.

Li Y. and F. Raichlen (2001), Solitary wave runup on plane slopes, J. Waterway, Port, Coastal and Ocean Eng., vol. 127, pp. 33-44.

Li Y. and F. Raichlen (2002), Non-breaking and breaking solitary wave run-up, J. Fluid Mech., vol. 456, pp. 295-318.

Li Y. and F. Raichlen (2003), Energy balance model for breaking solitary wave runup, J. Waterway, Port, Coastal and Ocean Eng., vol. 129, pp. 47-59.

Lin, P., K. A. Chang & P. L.-F. Liu (1999) " Runup and rundown of solitary waves on slopping beaches", J. Waterway, Port, Coastal and Ocean Eng., vol. 125, pp. 247-255.

Synolakis, C. E. (1987) "The run-up of solitary waves", J. Fluid Mech., vol. 185, pp. 523-545.

Zelt, J. A. (1991) " The run-up of nonbreaking and breaking solitary waves" , Coastal Engineering, vol. 15, pp. 205-246.

TURBULENT INTERFACIAL BOUNDARY CONDITIONS FOR SPILLING BREAKERS

Shubhra K. Misra [1], Maurizio Brocchini [2] and James T. Kirby [3]

This paper presents experimental observations as well as theoretical formulations of the flow structure in the turbulent shear layer in spilling breakers. The theoretical model is developed within the framework of a three-layer system, with the top layer being the two-phase (air/water) turbulent surface layer, and the bottom layer comprising the single phase (water) irrotational flow. The model development in the single phase (water) turbulent shear layer is guided by and validated with PIV flow measurements of a turbulent hydraulic jump. The breaker shear layer is seen to resemble a classical mixing layer with entrainment of fluid into the reverse flow region localized about the mean surface. Preliminary results are reported on interfacial boundary conditions which account for the contributions from the surface layer as well as the evolution of turbulent and mean quantities in the breaker shear layer. Such interfacial boundary conditions can be incorporated into Reynolds-averaged Navier-Stokes numerical models for simulating the spatio-temporal evolution of the turbulent and mean flow in spilling breakers.

1. INTRODUCTION

An accurate modeling of turbulence generated by breaking waves is essential in fully understanding and simulating the complicated mechanisms of mass, momentum, and energy transfer from organized wave motion to nearshore current and sediment suspension and transport processes in the surf-zone. There are few theoretical and experimental studies which provide a detailed spatial description or a fundamental understanding of the temporal evolution of the mean and turbulence flow in the breaker (Duncan, 2001). Non-intrusive experimental methods such as particle image velocimetry (PIV) and video based imaging techniques have contributed significantly to a better qualitative and quantitative understanding of the instantaneous and mean flow structure (Lin and Rockwell, 1995; Govender *et al.*, 2002). Motivated by these observations, simplified theoretical models have been formulated to describe the flow in the breaker (Longuet-Higgins and Turner, 1974; Cointe and Tulin, 1994). In the hydraulic jump/bore model of Madsen and Svendsen (1983), which provided motivation for the

[1]HALCROW HPA, 22 Cortlandt Street, 33rd Floor, New York, NY 10036, USA
[2]Istituto di Idraulica e Infrastrutture, Viarie, Universit Politecnica delle Marche, Via Brecce Bianche, 60131 Ancona, Italy
[3]CACR, University of Delaware, Newark, DE 19716, USA

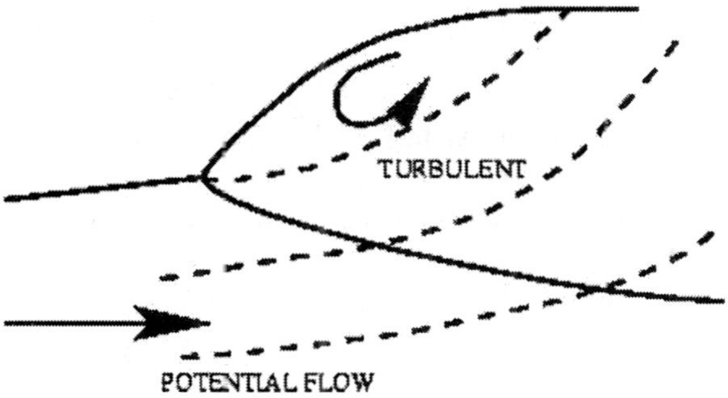

Figure 1: Schematic representation of the hydraulic jump/bore model. Adapted from Madsen and Svendsen (1983).

development of the present theoretical model, the turbulence was assumed to spread downstream in a wedge originating at the toe of the breaker front (Figure 1). The turbulence was modeled by a simplified $k-\epsilon$ model allowing for non-equilibrium in the turbulent kinetic energy.

The theoretical models mentioned above have primarily been "bulk" formulations, neglecting various details of the flow structure. The underlying assumptions in these studies are that of a generally accepted conceptual description of the breaker shear layer as a mixing layer with or without a recirculating roller which, however, remain unquantified and unformulated in the details of their structure and interaction with each other or the rest of the flow. In particular, the effects of the two-phase surface layer have not been paid much attention. In recent years, with a steady increase in computational resources facilitating a more accurate numerical modeling of the two-phase surface turbulence, heterogenous domain decomposition techniques have been used to separately model the viscous breaker region and the irrotational flow underneath, where a critical component of the coupling is the mean pressure at the interface (Rhee and Stern, 2002; Iafrati and Campana, 2005). This paper is aimed at formulating such interfacial boundary conditions from a rigorous theoretical description of the flow aided by high-resolution measurements of the flow in a spilling breaker.

In section 2, laboratory PIV experiments for an air-entraining weakly turbulent hydraulic jump are described. The underlying framework and assumptions of the theoretical formulation are discussed and validated with the PIV measurements in section 3 and the interfacial boundary conditions are highlighted. Conclusions and discussions are presented at the end of the paper in section 4.

2. LABORATORY MEASUREMENTS OF A TURBULENT HYDRAULIC JUMP

Turbulent hydraulic jumps fall under the general category of steady spilling breakers such as those generated by submerged hydrofoils (Duncan, 2001). In similtude with saturated surf-zone breakers, the upstream Froude number (Fr) is found to be in the range 1.15 to 1.26. For such low Froude number turbulent jumps, the energy dissipation in the transition region from supercritical to subcritical flow comes primarily from sheared turbulence accompanied by free surface air-entrainment. The most comprehensive experimental study to date of low Froude number, weakly turbulent jumps with breaking and air entrainment is that of Svendsen *et al.* (2000). They found that, contrary to the traditionally simplified formulations of a hydraulic jump, the effects of non-uniform velocities and non-hydrostatic pressure are important for mass and momentum conservation. Little attention was focused on the intermittency of the surface and its influence on the flow structure. A single roller was found to reside in the region bounded from below by, what the authors referred to as, a dividing streamline (above which the net volume flux was zero), and from above by the mean surface.

The PIV measurements discussed in this paper are of a laboratory generated turbulent hydraulic jump with breaking and air entrainment. The upstream to downstream depth ratio was 1.25, and the upstream Froude number was 1.2. Details of the experimental set-up and flow parameters can be found in Misra (2005). A non-intrusive detection of the instantaneous air-water interface from the raw PIV images was achieved through sophisticated image processing algorithms (Misra *et al.*, 2005) and the instantaneous velocities were computed using a phase-correlation based hierarchial, multi-pass motion estimation algorithm that has been extensively tested with ground-truth measurements (Thomas *et al.*, 2005). 1020 image pairs were collected and mean and root mean square turbulent quantities were calculated using suitable conditional averaging techniques (Misra *et al.*, 2005).

Figure 2 shows that the surface-normal velocity is nearly zero in the reverse-flow region. The flow is from left to right, and the toe of the breaker is at $\tilde{x} = \frac{\mathrm{x}}{h_0} \sim 0.47$, where x is the streamwise coordinate and h_0 is the upstream water depth. y is the vertical coordinate and the origin of the coordinate system is defined at the flume bottom at an arbitrary upstream (of the toe) location. There are large, positive values of surface normal velocity, $\hat{V}_n$, below the mean surface and are an indication of the entrainment of water from below. It is also clear from the present data that, though most

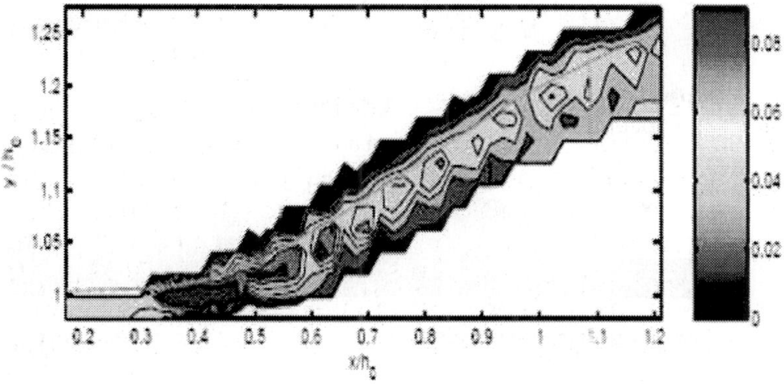

Figure 2: Mean surface-normal velocity $\hat{V}_n$ (m/s). Solid line is the ensemble-averaged free surface.

of the entrainment occurs near the toe, inflow from below occurs all over the reverse-flow region with approximately equidistant peaks of upwelling. As opposed to the assumption made by IS00, the mean shear is finite and negative at the mean surface. The present topology of the reverse-flow region is different from the concept of an isolated recirculating eddy/'roller' bounded on the top by a mean surface and on the bottom by a dividing streamline, since there clearly is reverse flow above the mean surface. As noted by Banner and Phillips (1974), we believe it essential to replace the idea of a mean surface in the reverse-flow region by an intermittency zone, which in the present measurements, is taken into account through the conditional averaging technique applied to the fluctuating air-water interface; the fluctuations follow a Gaussian distribution (Misra *et al.*, 2005).

It is fairly well-established that there is an intense shear layer formed below a fully-formed, quasi-steady spilling breaker, which spreads downstream from the foot of the breaker, and qualititave observations have suggested this to be a "mixing layer" (Peregrine and Svendsen, 1978; Hoyt and Sellin, 1989). Lin and Rockwell (1995) had found coherent vortical motions, and concentrated vorticity along a "mixing layer" originating from the toe. However, there has been no detailed experimental evidence for the suggested classification of the shear layer in a fully-formed breaker as a mixing layer. Figure 3 shows the Reynolds shear stress in the breaker shear layer of the hydraulic jump. The Reynolds shear stress is negative throughout the shear layer, implying an upward diffusion of momentum. The Reynolds shear stress has its largest value at $\tilde{x} \sim 0.59$, $\tilde{y} \sim 1$, downstream of the toe.

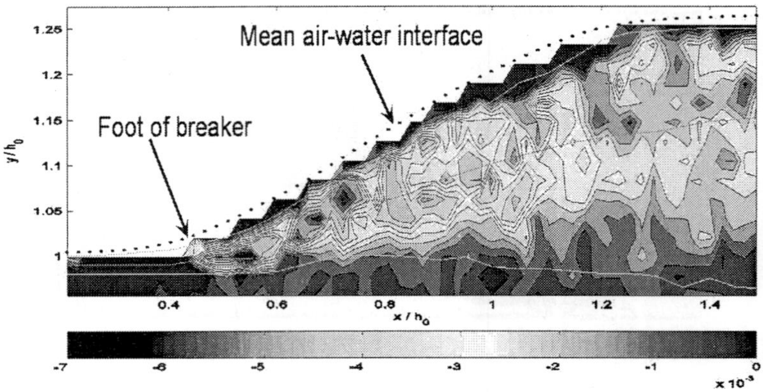

Figure 3: Reynolds shear stress, τ' (m^2/s^2) in the mixing layer. The dotted line is the mean surface and the thin solid white lines in the mixing layer represent 10 % of the maxima of the Reynolds stress.

The width of the mixing layer, W, is calculated and made dimensionless with h_0 and is shown in Figure 4 along with the linear least squares regression fit. In terms of the absolute coordinate $\tilde{x}$, there is a linear increase in the width till about $\tilde{x} \sim 1.2$, after which it becomes approximately constant and the flow resembles a wake. This is in good qualitative agreement with the spreading rate of mixing layers where the "wedge" is known to spread in a linear fashion and indicates that the breaker shear layer can be classified as a mixing layer with its origin downstream of the toe and extending downstream till it changes to a wake-type flow. This transition occurs at approximately the location of the peak negative curvature of the mean free surface. Therefore, the breaker shear layer is seen to resemble a plane mixing layer. The mean value of the non-dimensional growth rate in the mixing layer ($\approx$0.29) is found to be in good agreement with values found earlier by Brown and Roshko (1974).

3. THEORETICAL MODEL FOR SPILLING BREAKERS

Partially motivated by the measurements, a comprehensive theoretical model has been developed for the description of turbulence in the shear layer of unsteady spilling breaking waves by extending the formulations in Brocchini and Peregrine (2001a,b) and Brocchini (2002). The governing equations have been formulated within the framework of a three-layer system (see Figure 5) in which the top layer represents the intermittent effects of two-phase free-surface turbulence (Brocchini, 2002) and the bottom layer is assumed to be irrotational. The middle layer is the single phase

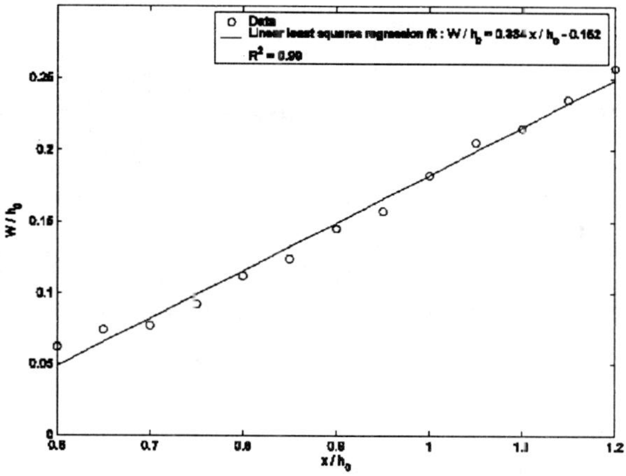

Figure 4: Streamwise variation of the width of the mixing layer. Measurements (circles) and linear regression fit (solid line).

mixing layer in which turbulence is generated due to the continuous shearing action of the flow downstream of the toe of the breaker. In a mobile coordinate system (s, n), equations of mass and linear momentum in the turbulent thin layer in covariant form are derived including a correct representation of two-phase flow near the surface, streamline and geometric curvature, unsteadiness, non-hydrostatic pressure, and local rotation which are all inherent to the dynamics of the layer of turbulent flow. Turbulent averaging leads to Reynolds-type equations. A k-ϵ closure for Reynolds stresses based on Boussinesq eddy viscosity concept is used and a transport equation allows for non-equilibrium of the turbulent kinetic energy.

These contributions give a complete account of the motion of the unsteady spilling breaker. They also make these equations very complex and the solution of the complete equations a very difficult task. Scaling arguments are adopted to retain only the most important terms in the basic equations and to help compare relative weights of terms appearing in different equations. As is well known, mixing layers (or shear layers) are characterized by a scale velocity $\tilde{U}$ for the cross-stream (n) variation of the mean velocity component in the streamwise direction $(\hat{\mathbf{s}})$. A first useful relationship which characterizes the flow in the turbulent layer expresses the observation that the layer is thin, i.e., the thickness, b, is much smaller than the streamwise length scale, L:

$$b \ll L \quad \text{or} \quad \frac{b}{L} = \epsilon \ll 1. \tag{1}$$

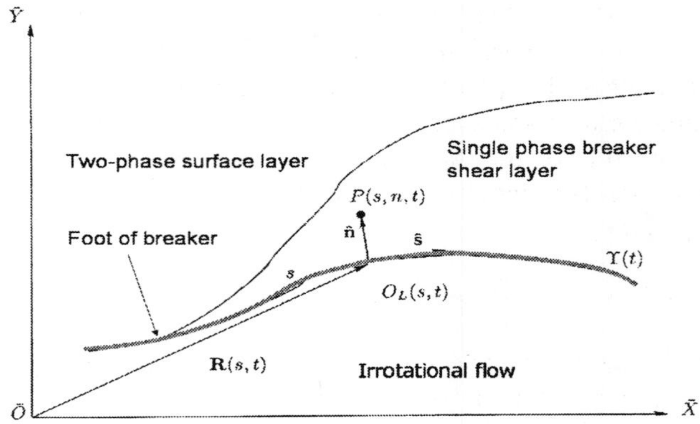

Figure 5: Global geometry adopted in the model for the system wave - turbulent thin layer - surface layer.

L represents the size of the region of the wave with the highest curvature. From the present measurements, considering the variation of the calculated length scale of the mixing layer, values of ϵ, the geometric parameter, averaged over the breaker shear layer are of the order of 0.03, which is slightly smaller than typically observed values of order 0.06 for jets and mixing layers (Tennekes and Lumley, 1972). These values are reasonable enough to satisfy the 'thin layer' approximation.

We also introduce a dimensionless kinematic parameter μ which represents the ratio of the scales for turbulent velocities and mean velocities, and use it to implicitly estimate the size of the different contributions in the following analysis:

$$\mu = \frac{\tilde{u}}{\tilde{U}}, \quad \mathcal{O}(\epsilon) < \mu < \mathcal{O}(1). \tag{2}$$

In the following, when referring to the leading order or $\mathcal{O}(1)$ model, the contributions of terms of $\mathcal{O}(\epsilon)$ and higher are neglected. Also, the particular flow regime analyzed is made explicit depending on the size of μ. For example, at the leading edge of the breaker, where intense turbulence is generated by the shear induced by the flow separation, turbulent velocities are of the same order as the mean velocities, and $\mu \sim \mathcal{O}(1)$, whereas in the far wake region, $\mu \sim \mathcal{O}(\epsilon)$.

Similar to Madsen and Svendsen (1983), a cubic polynomial is used for the cross-flow profile for the mean horizontal velocity. In a significant difference from their model, the two-phase surface layer contributions are taken into account by enforcing the kinematic free surface boundary condition at the top of the single phase shear layer, and using the results obtained in Brocchini and Peregrine (2001b) and Brocchini (2002). The other matching conditions used to solve for the flow are continuity of mean velocity at the top and bottom of the layer and vanishing turbulent shear stresses at the interface with the irrotational flow underneath. The mean streamwise velocity is given by

$$U(s,\sigma,t) = \hat{U} + (U_b - \hat{U})\sigma^2 + A\left[U_b - \hat{U} - b(s)\Omega\right]\sigma^2(1-\sigma), \qquad (3)$$

where U_b and $\hat{U}$ are the mean velocities at the bottom and top of the layer respectively, $\sigma = \frac{n}{b}$ is a non-dimensional vertical coordinate, Ω denotes the rotation and A is the coefficient of the cubic term and represents the mean shear at the interface with the surface layer. For $\Omega = 0$, this reduces to

$$U(s,\sigma,t) = \hat{U} + (U_b - \hat{U})\sigma^2[1 + A(1-\sigma)]. \qquad (4)$$

The analytical solution for the mean horizontal velocity can be checked using the present measurements. A cubic polynomial was fit to the measured values for U. This is shown as the blue solid line in Figure 6. The coefficient of the cubic term (A) was calculated from the fit and is shown at each horizontal location. Once A had been calculated, by approximating $\hat{U}$ and U_b with the mean horizontal velocity values at the bottom and top (at the contours for 10 % of the maximum Reynolds shear stress; see Figure 3) of the mixing layer, the mean horizontal value at each vertical location was calculated from equation (4), and this is shown as the red line in the figure. Because of the small thickness of the shear layer near the toe, few data points were available for a robust fit to the profile. Therefore, the streamwise locations were chosen in the middle part of the shear layer. Equation (4) is seen to provide a good approximation to the vertical variation of the mean horizontal velocity in the shear layer.

Using the profiles for the mean and turbulent quantities, the leading order equations are analytically integrated across the width of the layer. The leading-order contributions from the normal component of the momentum equation are essentially a balance between the normal turbulent stress and the normal mean pressure gradient. Using the profiles for the normal turbulent stresses and the turbulent kinetic energy, this can be integrated across the layer from $n = 0$ to $n = b$ to give

$$P_0 = P_b + 2\rho C_3 \Lambda^2 f^2 \left[(-A+2)\left(-A + 2 + \frac{2b\Omega}{f}\right)\right]. \qquad (5)$$

Λ denotes the ratio of turbulence production and dissipation, f denotes the streamwise variation of the flow, ρ the density, and C_3 is a constant of the

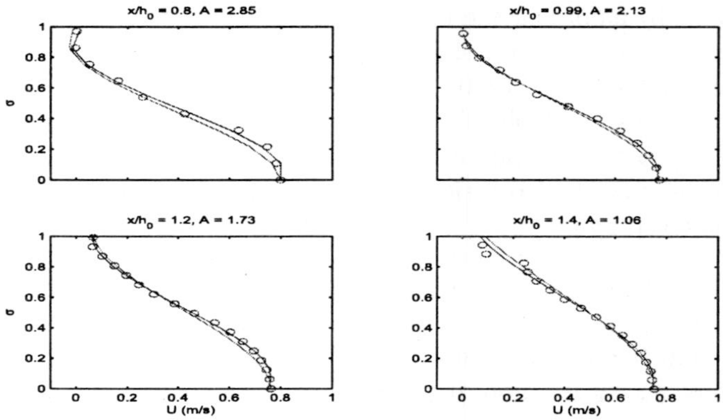

Figure 6: Cubic profile fit (blue solid line) for mean horizontal velocity data (circles). Red line is the estimated mean velocity from equation (4). σ is the non-dimensional vertical coordinate.

turbulence model. P_0 and P_b are respectively the pressure at the bottom and top of the layer, and P_b can be obtained from the dynamic boundary condition derived in Brocchini and Peregrine (2001b) (see their equation (5.19)) which includes the surface layer contributions. Note the presence of extra pressure with respect to P_b due to rotation and non-zero mean shear stress (represented by $A \neq 2$) at the top of the layer. Note that the effect of gravity is absent at leading order and is only felt at $(O(\epsilon))$. The normal turbulent stresses make a dominant contribution to the non-hydrostatic component of the pressure. In the absence of rotation and mean shear at the top of the layer ($\Omega = 0$ and $A = 2$), $P_0 = P_b$. In other words, there is uniform pressure across the layer. On integrating the $\mathcal{O}(1)$ contributions from the streamwise momentum equation,

$$\hat{\mathbf{n}}\cdot\frac{\partial \mathbf{R}}{\partial t} = \frac{C_t \Lambda f^2 (2-A)\left[A - \left(2 + \frac{2b\Omega}{f}\right)\right]^2}{\left(U_b - \hat{U}\right)}, \tag{6}$$

which describes the unsteady vertical motion of the layer. Note that the steady state is recovered for $A = 2$ which represents vanishing mean shear stress at the surface.

The normal integration of the $\mathcal{O}(1)$ turbulent kinetic energy equation results in an expression for $\Lambda(s)$ describing the local imbalance between production and dissipation is found. In the absence of rotation,

$$\Lambda^2(s) = \frac{-6A^3 + 63A^2 - 126A + 210}{105(2-A)^3\left[(2-A)^2 - 4\frac{(1-2A)}{(2-A)} + \frac{C_\varepsilon}{C_t}\right]}. \tag{7}$$

C_ε and C_t are constants in the turbulence model (Launder and Spalding, 1972). The first thing to note in the above expression is the singularity at $A = 2$. Since Λ^2 is the dissipation divided by the production, near the leading edge of the breaker, $\Lambda^2 < 1$. The singularity appears when the production vanishes, and this is expected to occur in the downstream direction. In contrast to Madsen and Svendsen (1983), where A is a constant (chosen as a best-fit to data), $A(s)$ in the present model varies in the streamwise direction, and its variation can be modeled in terms of the dynamics of the turbulence near the foot of the breaker. The streamwise dependence follows from the more complex flow situation analyzed here (e.g. non-hydrostatic pressure, vertical accelerations, etc.) with respect to those of Madsen and Svendsen (1983).

Instead of using the profiles for the mean and turbulent quantities, the integration of the governing equations across the layer can also be done by defining layer-averaged variables with the aim of developing a model which, apart from its governing variables, is determined entirely by the information provided from the underlying flow so that the integrated equations also serve as boundary conditions for the irrotational flow below. The variables for the model are the layer thickness, b, the layer-averaged mean horizontal velocity, $\overline{U} \equiv \frac{\int_0^b U dn}{b}$, and the layer-averaged turbulent kinetic energy, $\overline{k} \equiv \frac{\int_0^b k dn}{b}$. The analysis assumes $\Omega = 0$. The layer-averaged equations for conservation of mass, momentum and turbulent kinetic energy are given as

$$\frac{\partial b}{\partial t} + C_B\frac{\partial b}{\partial s} - C_U\frac{\partial \overline{U}}{\partial s} + C_K\frac{\partial \overline{k}}{\partial s} = C_R, \tag{8}$$

$$\frac{\partial \overline{U}}{\partial t} + M_B\frac{\partial b}{\partial s} + M_U\frac{\partial \overline{U}}{\partial s} + M_K\frac{\partial \overline{k}}{\partial s} = M_R, \quad \text{and} \tag{9}$$

$$\frac{\partial \overline{k}}{\partial t} + K_B\frac{\partial b}{\partial s} + K_U\frac{\partial \overline{U}}{\partial s} + K_K\frac{\partial \overline{k}}{\partial s} = K_R, \tag{10}$$

or in matrix form as

$$\left\{\begin{array}{c} b \\ \overline{U} \\ \overline{k} \end{array}\right\}_t + \left[\begin{array}{ccc} C_B & C_U & C_K \\ M_B & M_U & M_K \\ K_B & K_U & K_K \end{array}\right]\left\{\begin{array}{c} b \\ \overline{U} \\ \overline{k} \end{array}\right\}_s = \left[\begin{array}{c} C_R \\ M_R \\ K_R \end{array}\right],$$

where the subscripts denote partial differentiation with respect to the variables, and the coefficients are functions of the dependent variables. The above system of equations can either be regarded as a description for the turbulent dynamics in the thin layer, or as turbulent free surface boundary conditions for the irrotational flow below. The two flows are coupled through the mass exchange and pressure at the boundary. The mass exchange is represented by the mean velocities, V_0 and $\hat{U}$. The layer-averaged pressure ($\overline{P}$) and the pressure at the top of the layer (P_b) can easily be expressed in terms of the layer-averaged turbulent kinetic energy, $\overline{k}$. Since a consistent set of equations for the $O(\epsilon)$ model has contributions from the $O(1)$ normal momentum equation, using the $O(1)$ relationship between the normal turbulent stress and the turbulent kinetic energy, the pressure at the base of the layer, for the $\mathcal{O}(\epsilon)$ model, is given by

$$P_0 = P_b + 2\rho C_3(k_b - k_0), \tag{11}$$

where k_b and k_0 represent the turbulent kinetic energies at the bottom and top of the shear layer. Since the diffusion of turbulent kinetic energy at the bottom has been neglected, in the absence of rotation, the profile for the turbulent kinetic energy gives $k_0 = 0$. Also, since k_b is related to $\overline{k}$ through $k_b = A_k\overline{k}$ (A_k represents the coupling between the surface layer and the single phase turbulent layer), and $P_b = \rho C_p\overline{k}$, the pressure at the base of the layer, and as a boundary condition for the irrotational flow is found to be

$$P_0 = \overline{k}\rho(C_p + 2C_3A_k). \tag{12}$$

Therefore, solving the evolution equation for the layer-averaged turbulent kinetic energy, yields the solution for the pressure at the base of the layer, which is used in the dynamic boundary condition (the Bernoulli equation) to solve for the potential flow underneath (Iafrati and Campana, 2005).

4. CONCLUSIONS

PIV measurements were used to provide a detailed picture of the mean and turbulent structure of an air-entraining hydraulic jump set up in Froude similitude with surf-zone breakers. The traditional concept of a roller in the form of a recirculating region is seen to be only a partial description of the mean flow near the interface, since along with a weak downslope flow above the mean surface, there is mass exchange with the underlying flow across the lower boundary of the roller. The turbulence structure of the breaker shear layer is seen to resemble that of a mixing layer originating at the foot of the breaker followed by a wake further downstream. The average growth rate is in good agreement with observed values in typical mixing layers. Model equations for a comprehensive description of the flow in the breaker shear layer have been developed, including non-hydrostatic and unsteady effects, streamline and geometric curvature, rotation and the effects of surface turbulence. The geometric and kinematic scaling assumptions are justified from the present measurements. The vertically integrated model provides evolution equations for the mean and turbulent flow, as well as the motion of the interface. The mean shear at the top of the layer significantly affects the mean and turbulent flow in the layer. Experimental results validate the assumed classification of the breaker shear layer as a classical mixing layer and the vertical profiles of the mean velocity in the mixing layer are found to be in good agreement. The layer-averaged equations are coupled to the underlying flow through mass entrainment and excess pressure at the interface, and they serve as boundary conditions for the flow below and can be used in Reynolds-averaged Navier-Stokes numerical models to describe the evolution of the flow in the shear layer of a spilling breaker.

ACKNOWLEDGEMENTS

A substantial part of this research has been funded by the National Oceanographic Partnership Program (NOPP) grant number N00014-99-1-1051. Part of this research was funded by the University of Genoa and the College of Marine Studies, University of Delaware. Prof. D. H. Peregrine's contributions in the theoretical development of the model are gratefully acknowledged. SKM acknowledges the travel and conference registration support provided by HALCROW HPA.

References

Banner, M. L. and Phillips, O. M. (1974). On the incipient breaking of small scale waves. *J. Fluid Mech.*, **65**, 647–656.

Brocchini, M. (2002). Free surface boundary conditions at a bubbly/weakly-splashing air-water interface. *Phys. Fluids*, **14**(6), 1834–1840.

Brocchini, M. and Peregrine, D. H. (2001a). The dynamics of turbulent free surfaces. part 1. Description. *J. Fluid Mech.*, **449**, 225–254.

Brocchini, M. and Peregrine, D. H. (2001b). The dynamics of strong turbulence at free surfaces. part 2. Free-surface boundary conditions. *J. Fluid Mech.*, **449**, 255–290.

Brown, G. L. and Roshko, A. (1974). On density effects and large structures in turbulent mixing layers. *J. Fluid Mech.*, **64**, 775–816.

Cointe, R. and Tulin, M. P. (1994). A theory of steady breakers. *J. Fluid Mech.*, **276**, 1–20.

Duncan, J. H. (2001). Spilling breakers. *Ann. Rev. Fluid Mech.*, **33**, 519–547.

Govender, K., Mocke, G., and Alport, M. (2002). Video imaged surf zone wave and roller structures and flow fields. *J. Geophys. Res.*, **107**(C10), 3177.

Hoyt, J. W. and Sellin, R. H. J. (1989). Hydraulic jump as "Mixing layer". *J. Hyd. Engrng.*, **115**(12), 1607–1614.

Iafrati, A. and Campana, E. F. (2005). Free-surface fluctuations behind microbreakers: space-time behaviour and subsurface flow field. *J. Fluid Mech.*, **529**, 311–347.

Launder, B. E. and Spalding, D. B. (1972). *Mathematical models of turbulence*. Academic Press, London and New York.

Lin, J. C. and Rockwell, D. (1995). Evolution of a quasi-steady breaking wave. *J. Fluid Mech.*, **302**, 29–44.

Longuet-Higgins, M. S. and Turner, J. S. (1974). An entraining plume model of a spilling breaker. *J. Fluid Mech.*, **63**, 1–20.

Madsen, P. A. and Svendsen, I. A. (1983). Turbulent bores and hydraulic jumps. *J. Fluid Mech.*, **129**, 1–25.

Misra, S. K. (2005). *The turbulent dynamics of quasi-steady spilling breakers - Theory and experiments*. Ph.D. thesis, University of Delaware.

Misra, S. K., Thomas, M., Kambhamettu, C., Kirby, J. T., Veron, F., and Brocchini, M. (2005). Estimation of complex air-water interfaces from PIV images. *Exp. Fluids*, **40**(5), 764–775.

Peregrine, D. H. and Svendsen, I. A. (1978). Spilling breakers, bores and hydraulic jumps. In *Proceedings of the 16th ICCE, Hamburg*, pages 540–555. ASCE.

Rhee, S. H. and Stern, F. (2002). RANS model for spilling breaking waves. *J. Fluids Engrng.*, **124**, 424–432.

Svendsen, I. A., Veeramony, J., Bakunin, J., and Kirby, J. T. (2000). The flow in weak turbulent hydraulic jumps. *J. Fluid Mech.*, **418**, 25–57.

Tennekes, H. and Lumley, J. L. (1972). *A first course in turbulence*. MIT Press.

Thomas, M., Misra, S. K., Kambhamettu, C., and Kirby, J. T. (2005). A robust motion estimation algorithm for PIV. *Meas. Sci. and Tech.*, **16**(3), 865–77.

WHITECAPS: EFFECTS OF WAVE AGE AND AIR STABILITY

Dag Myrhaug[1], Lars Erik Holmedal[1] and Olav H. Slaattelid[2]

The paper considers the effects of wave age and air stability on whitecaps at sea. This is made by using the logarithmic mean wind velocity profile including a stability function as well as adopting a recent wave age dependent sea surface roughness formula. The results are valid for wind waves in local equilibrium with the steady wind. An example of results demonstrates clear effects of wave age and air stability on whitecaps.

INTRODUCTION

Breaking of wind waves plays an important role in air-sea exchange processes due to the enhancement of turbulence and the entrainment of air. To quantify the occurrence of breaking wind waves is difficult; the whitecap coverage, which is defined as the area of whitecaps per a unit sea surface, has often been used. Existing data on whitecaps show large scatter when they are plotted versus the wind speed only; other important factors are the near-water air stability and the state of development of surface waves, see e.g. Sugihara et al. (2004). Thus it is important to examine how whitecaps depend on meteorological and wave conditions.

The boundary layer flow over the sea surface is complicated by the sea surface roughness over waves, which depends on air-sea interactions. Reviews of the subject are given in, e.g., Smith et al. (1996) and Jones and Toba (2001).

For strong winds the effect of temperature stratification is minimal due to mixing of the air. However, for weaker and moderate winds, i.e. for wind velocities (at the 10 m elevation) up to about 25 m/s, the presence of stratification effects due to temperature gradients has been documented (Smith, 1980; Andersen and Løvseth, 1995). In spring the stratification is mainly stable, and in autumn it is mainly unstable, since the seawater temperature is lower and higher than the air temperature. By analyzing wind speed observations over the Southern North Sea covering a period of about seven years, Coelingh et al. (1996) quantified this. Their results are summarized in Table 1 showing that stable and unstable conditions (caused by stratification) cover 85-90 percent of all hours in the various seasons of the year. Spillane et al. (1986) analyzed whitecaps observations (262 cases) with wind speed, water temperature and thermal stability. They documented that for wind speeds up to 18 m/s both stable, unstable and near-neutral conditions were present in the observations. This justifies the relevance of the present study.

---

[1] Department of Marine Technology, Norwegian University of Science and Technology, NO-7491 Trondheim, Norway

[2] Norwegian Marine Technology Research Institute, NO-7450, Trondheim, Norway

Table 1. Percentage of atmospheric stability versus season of the year.

Condition	Season				
	Spring	Summer	Fall	Winter	Annual
Stable	40-50	30-40	10-25	20-30	30-35
Unstable	40-45	55-60	65-80	60-75	55-60
Neutral	10-15	~10	~10	10-15	~10

Sugihara et al. (2004) analyzed field observations of whitecap coverage and proposed the formula

$$W_c = \left[a\left(u_* - b\right)\right]^3 \ ; \ (a, b) = (2.20, 0.109) \tag{1}$$

where W_c is the whitecap coverage given in percent, u_* (m/s) is the friction velocity equal to the square root of the vertical flux of horizontal momentum at the sea surface. The basis for Sugihara et al. to use Eq. (1), was the results by Phillips (1985) who found the dissipation rate of wave energy to be proportional to u_*^3 based on the equilibrium spectrum of wind waves. Moreover, there is a direct connection between the dissipation rate and the wave breaking; thus $W_c \sim u_*^3$ by assuming a linear relationship between W_c and the dissipation rate, and the consequence is Eq. (1) (Sugihara et al., 2004). Sugihara et al. showed figures plotting $W_c^{1/3}$ versus u_* and $W_c^{1/3}$ versus the mean wind speed at the 10 m elevation, U_{10}. The dependence of $W_c^{1/3}$ on u_* supported the validity of Eq. (1), but the data scattering was larger than for $W_c^{1/3}$ versus U_{10}. This was attributed to the larger uncertainties in estimating u_* than measuring U_{10}. Their data covered both near-neutral and unstable stratification, and they found that air instability promotes wave breaking. They also distinguished between the state of development of the waves; whether the sea surface was in the state of developing waves or decaying waves. This was classified on the basis of time series of the significant wave height H_s; developing waves if the time variation rate of H_s was positive, and decaying if the time variation rate of H_s was negative. For a given value of U_{10} they found that the whitecap coverage was larger for decaying than for developing waves. Moreover, from directional wave spectra they divided their data into groups of pure wind sea, swell-dominated sea as well as combined sea, finding that whitecaps are mainly produced by wind waves and suppressed by the presence of swell. More details are given in Sugihara et al. (2004).

Another quantity of interest is the breaking frequency f_{br}. Phillips (1985) used the whitecap residence time to determine f_{br}. By assuming the residence time to be constant and denoting it as $t_{disrupt}$ (as the whitecaps is understood as disruption of the surface), the breaking frequency is given as (Eifler, 2005)

$$f_{br} = \frac{W_c}{t_{disrupt}} \quad ; \quad t_{disrupt} = 0.5s \tag{2}$$

This value of $t_{disrupt}$ is based on Snyder et al.'s (1983) results from open-ocean measurements, although other investigations have indicated somewhat lower values (~0.3 s). More details, as well as a review of the topic, are given in Eifler (2005).

The purpose of the present paper is to investigate the effects of wave age and air stability on the whitecap coverage at sea. This is achieved by using Eq. (1) and to determine u_* by using the logarithmic mean wind velocity profile including a stability function according to Kraus and Businger (1994) accounting for the air stability, as well as adopting a recent sea surface roughness formulation including the wave age according to Volkov (2001). The results are valid for the development of wind waves which are in local equilibrium with the wind, and for a given steady wind speed that starts to blow over flat water without waves until the waves are fully developed. Thus, the approach does not cover situations with unsteady wind, or for wind over the sea surface in sea states of developing or decaying waves as defined by Sugihara et al. (2004).

THEORETICAL BACKGROUND

Sea surface shear stress

The effect of air stability on the sea surface boundary layer structure usually scales with the dimensionless stability parameter $\zeta = z/L$. Here z is the height above the sea surface and L is buoyancy length scale known as the Monin-Obukhov length defined as $L = -u_*^3 T_0 / g\kappa Q_0$ (Arya, 1982); κ is the von Karman's constant (=0.4), T_0 is the sea surface temperature, and Q_0 is the sea surface kinematic heat flux. Physically ζ expresses the ratio between the potential energy required at a particular level to mix the potential temperature gradient and the turbulent kinetic energy supplied by the wind shear at the level. For $\zeta < 0, \zeta = 0, \zeta > 0$ the stratification is referred to as unstable, neutral and stable, respectively. For large values of $|L|$, mechanical mixing dominates buoyancy in the turbulent intensive production in the boundary layer. In general $|\zeta| \ll 1$ close to the surface, regardless of the magnitude. Arya (1982) gives a review of the dimensionless parameters and their relative importance for the structure of stratified boundary layers.

Close to the surface Kraus and Businger (1994) give the mean wind velocity profile

$$U(z) = \frac{u_*}{\kappa}\left[\ell n\frac{z}{z_0} - \psi(\zeta)\right] \tag{3}$$

where $\psi(\zeta)$ is often referred to as the stability function taking into account stratification effects, and z_0 is the sea surface roughness which generally depends on u_* and will be discussed later. Following Kraus and Businger (1994), the stability function is parameterized as

$$\psi = \ell n\frac{(1+x^2)(1+x)^2}{8} - 2arctgx + \frac{\pi}{2},\ x = (1-16\zeta)^{1/4};\text{unstable}(\zeta < 0) \tag{4}$$

$$\psi = -5\zeta;\ \text{stable}\ (\zeta > 0) \tag{5}$$

One should note that Eqs. (4) and (5) are derived from observations over land, but a number of indirect results suggest that these equations are also valid over water (Kraus and Businger, 1994). According to Panofsky and Dutton (1984), Eq. (4) is the most widely used parameterization for unstable air, although other parameterizations exist. For stable air all measurements suggest Eq. (5), with estimates of the constant in the range 4.7-5.2 (instead of using 5). More details are given in Panofsky and Dutton (1984). However, one should note that recent analysis of Norwegian coastal wind measurements by Heggem (1997) gave slightly different values of the stability function, i.e. Eq. (4) with $x = (1-17\zeta)^{1/4}$ and $\psi = -3.6\zeta$ for unstable and stable conditions, respectively, while the Charnock constant was found to be 0.0172 (see the next section). The shape of the atmospheric boundary layer by using Eqs. (3) to (5) for different values of L is shown in e.g. Young (1998, Fig. 1).

For $z = z_{10} = 10\,\text{m}$, Eq. (3) can be rearranged to

$$u_* = \frac{\kappa U_{10}}{\ell n\frac{z_{10}}{z_0} - \psi\left(\frac{z_{10}}{L}\right)} \tag{6}$$

where z_0 depends on u_*. Now u_* can be determined from Eq. (6) by iteration for given values of U_{10} and L by substituting an appropriate model of z_0 (see below), and the sea surface drag coefficient can be expressed as

$$C_D = \left(\frac{u_*}{U_{10}} \right)^2 \tag{7}$$

The sea surface drag coefficient for neutral stability, C_{Dn}, is then given from Eqs. (6) and (7) as

$$C_{Dn} = \kappa^2 \left(\ell n \frac{z_{10}}{z_0} \right)^{-2} \tag{8}$$

Sea surface roughness parameter

The roughness of the sea surface depends on air-sea interaction conditions and is difficult to estimate; no consistent theory exists on the relation between z_0 and the roughness of the sea surface. Since Charnock (1955) proposed his well-known formula from a dimensional argument, many different formulas have been proposed, see e.g. Smith et al. (1996). Essentially it is a discussion of to what extent laboratory and ocean-wave systems actually involve precisely the same physics, i.e. if extrapolation of the laboratory data to the field using non-dimensional quantities such as wave age or dimensionless fetch is feasible. Hopefully, fundamental studies in the laboratory (see e.g. Banner and Peirson, 1998) together with field investigations will contribute to clarify the matter.

Volkov (2001) provided the following expression for the sea surface roughness:

$$\begin{aligned} z_0^* &= 0.03x\exp(-0.14x) \ \text{for} \ 0.35 < x < 35 \\ z_0^* &= 0.008 \ \text{for} \ 35 \le x \end{aligned} \tag{9}$$

where $z_0^* = gz_0 / u_*^2$ is the dimensionless roughness, and $x = c_p / u_*$ is the wave age. Moreover, g is the acceleration of gravity, and c_p is the phase speed associated with wind waves with peak frequency σ_p. Eq. (9) is obtained as a fit to existing data (see Fig. 10.6, Volkov (2001)). The non-dimensional roughness has a maximum value at c_p / u_* around 10 (see Fig. 1.15, Jones et al. (2001)), and for $c_p / u_* > 35$ (which can also represent light wind over swell) the non-dimensional roughness is near 0.01. Volkov's expression seems to be a reasonable compromise between simplicity and accuracy based on the state of knowledge at that time. It should be noted that Eq. (9) is valid for steady wind and wind waves in local equilibrium with the wind in a laterally homogeneous flow. Further background and details are given in Volkov (2001) and Jones et al. (2001). Moreover, the influence of unsteady wind speed on the wave response is discussed in e.g. Toba and Jones (2001).

Here the Volkov (2001) model will be used to serve the purpose of taking into account the effect of wave age, and the wave age independent Charnock (1955) formula will be used as a reference; the latter formula is given by

$$z_0^* = \beta \quad ; \quad \beta = 0.0185 \tag{10}$$

where the given β-value, i.e. the Charnock constant, is often used. It should be noted that this β-value is not considered conclusive as the values of β are given within a range; 0.015 to 0.025, see Toba et al. (1990). By using the dispersion relationship for linear waves in deep water, the phase speed associated with waves with peak period $T_p = 2\pi / \sigma_p$ is given by $c_p = gT_p / 2\pi$, and thus the wave age can be expressed as $(g/2\pi)(T_p / u_*)$. By replacing u_* by U_{10}, c_p / U_{10} is also referred to as wave age. Realistic wave age limits representing wind waves are often taken as $0.03 \le c_p / U_{10} \le 1.0$ (Toba et al., 1990).

RESULTS AND DISCUSSION

Myrhaug and Slaattelid (2003) considered the effects of sea roughness and air stability on the sea surface wind stress over waves, which are in local equilibrium with the wind, by using Eqs. (6) and (8) for the Volkov (Eq. (9)) and the Charnock (Eq. (10)) models of z_0. By the awareness of the Volkov roughness, that was a follow-up study of Myrhaug and Slaattelid (2002) where other roughness formulas were used; one with z_0 increasing with wave age (Toba et al., 1990); the other with z_0 decreasing with wave age (Donelan et al., 1993). Figures 1 to 3 give a summary of some of these results which are relevant to this study.

Figure 1 shows the neutral sea surface drag coefficient C_{Dn} versus the wave age $c_p / U_{10} = gT_p / 2\pi U_{10}$ in the range 0.03 to 1.0 according to Eqs. (6) and (8) for the wave age dependent Volkov model of z_0 (Eq. (9)) and for the wave age independent Charnock model of z_0 (Eq. (10)). It appears that C_{Dn} increases as the wave age increases for young waves; then C_{Dn} reaches a maximum at the wave age of about 0.4, before it decreases approaching the Charnock results for fully developed waves.

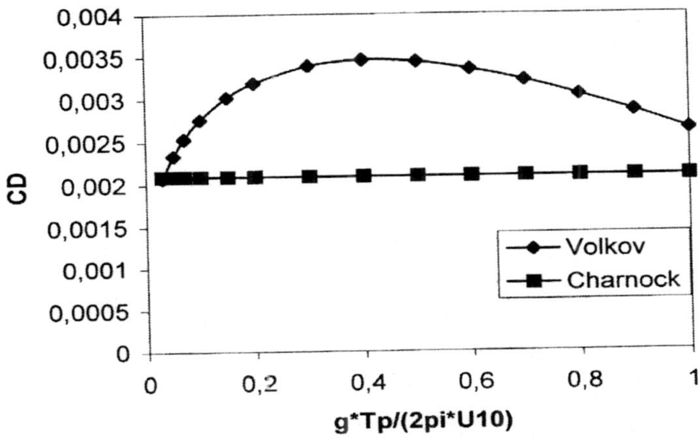

Figure 1. Neutral drag coefficient C_{Dn} versus wave age for Volkov and Charnock (reproduced from Myrhaug and Slaattelid (2003, Fig. 1)).

As a reference case C_D versus the air stability z_{10}/L according to Eqs. (6), (7), (4) and (5) using the Charnock roughness in Eq. (10) is shown in Fig. 2. Compared with neutral stability, C_D increases as the flow becomes more and more unstable, while C_D decreases as the air stability increases.

Figure 3 shows C_D versus the air stability z_{10}/L for different wave ages c_p/U_{10} according to Eqs. (6), (7), (4) and (5) using the Volkov roughness in Eq. (9) for $0.35 < c_p/u_* < 35$. For a given wave age the results are qualitatively the same as those given in Fig. 2 using the Charnock-formula. For a given air stability for unstable conditions $(z_{10}/L < 0)$, C_D increases as c_p/U_{10} increases for the lower to moderate wave ages $(0.03 \le c_p/U_{10} \le 0.5)$. However, for the higher wave ages $(0.7 \le c_p/U_{10} \le 1.0)$ the results show another behaviour, in particular as $|z_{10}/L|$ decreases. It is observed that the curves representing the two higher wave ages cross the curves representing the lower wave ages. For stable conditions $(z_{10}/L > 0)$ it is observed that the curves representing some of the wave ages cross each other. Moreover, it is observed that the C_D-dependence on the air stability is generally an order of magnitude larger than the dependence on the wave age. In particular this is evident for stable conditions $(z_{10}/L > 0)$ in Fig. 3 where the results only vary slightly with wave age.

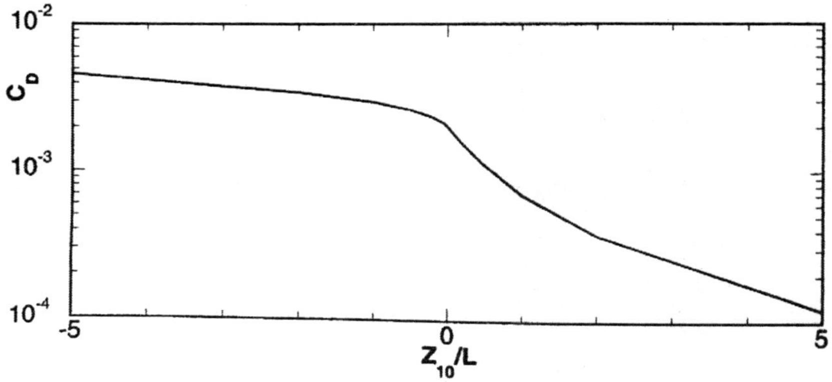

Figure 2. Drag coefficient C_D versus z_{10}/L for Charnock (taken from Myrhaug and Slaattelid (2002)).

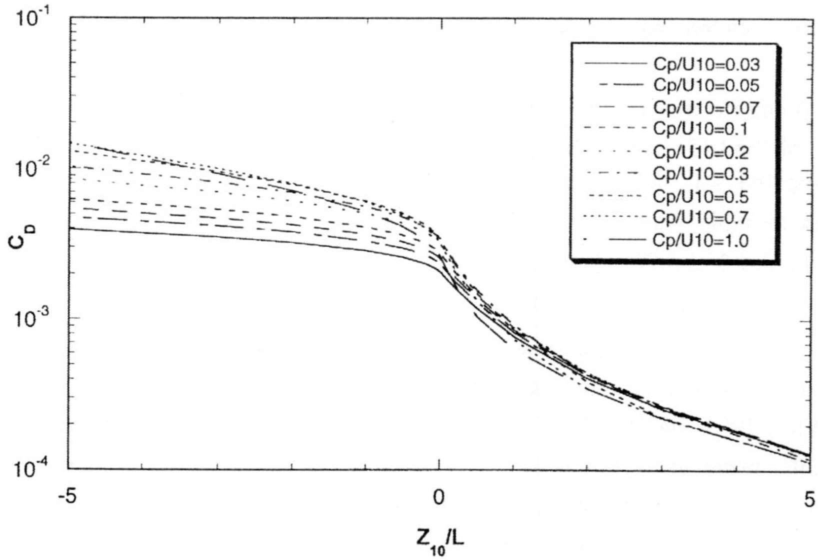

Figure 3. C_D versus z_{10}/L and wave age c_p/U_{10} for Volkov for $0.35<c_p/u_*<35$ (taken from Myrhaug and Slaattelid (2003)).

Table 2. Example of results with U_{10} = 20 m/s for neutral (*N*), stable (*S*) (*L*=10 m), and unstable (*U*) (*L*= -10m) stratification. The results are given using the wave age independent Charnock (1955) roughness; the wave age dependent Volkov (2001) roughness for wave ages c_p/U_{10}=0.03 (very young waves), 0.4 (intermediate wave age), 1.0 (fully developed waves).

Variables	Stability	Charnock	Volkov c_p/U_{10}=0.03	Volkov c_p/U_{10}=0.4	Volkov c_p/U_{10}=1.0
T_p (s)		--	0.38	5.12	12.8
u_* (m/s)	N	0.91	0.91	1.18	1.03
	S	0.54	0.56	0.60	0.51
	U	1.10	1.07	1.54	1.44
z_0 (cm)	N	0.16	0.15	1.11	0.41
	S	0.055	0.058	0.22	0.021
	U	0.23	0.18	1.82	1.25
$C_D \cdot 10^{-3}$	N	2.1	2.1	3.5	2.6
	S	0.73	0.78	0.89	0.64
	U	3.0	2.8	5.9	5.2
W_c (%)	N	5.5	5.5	13.1	8.3
	S	0.85	0.98	1.3	0.69
	U	10.4	9.5	31.2	25.1

Table 2 gives an example of results with $U_{10} = 20$ m/s for neutral, stable and unstable stratification. The results are obtained by using Volkov for the wave ages $c_p / U_{10} = 0.03$ (very young waves), 0.4 (intermediate wave age), 1.0 (fully developed waves), as well as Charnock as a reference case. For Volkov the T_p values corresponding to the wave ages are given. The results in Table 2 show the correct qualitative behaviour, i.e. the effect of stable and unstable stratification is to reduce and to increase the turbulence in the flow as compared to neutral flow, respectively, which is well known. More specifically, the values of u_*, z_0 and C_D are reduced and increased as compared to neutral flow when the flow becomes stable and unstable, respectively. Consequently, W_c (and the breaking frequency f_{br} given in Eq. (2)) have the same qualitative behaviour as these flow variables; air stability and air instability suppresses and enhances wave breaking as compared to neutral flow, respectively; the latter agrees qualitatively with the results in Sugihara et al. (2004). This qualitative behaviour is the same for Charnock and Volkov for given wave ages. Moreover, it appears that the results for Charnock and Volkov, wave age = 0.03 are close.

For a given air stability the Volkov results suggest that the whitecaps increase as wave age increases from very young waves $(c_p / U_{10} = 0.03)$ to an intermediate wave age $(c_p / U_{10} = 0.4)$, and then decrease as the wave age increases towards the fully developed wave age $(c_p / U_{10} = 1.0)$. The results in Table 2 also suggest that the whitecaps dependence on the stability generally is an order of magnitude larger than the dependence on the wave age, which is consistent with the results for C_D as shown in Fig. 3. Overall, the results in Table 2 demonstrate clear effects of wave age and air stability on whitecaps.

SUMMARY

An example of results has demonstrated:

1. Clear effects of wave age and air stability on whitecaps.
2. For a given air stability, whitecaps reach a maximum for intermediate wave age.
3. The whitecaps dependence on air stability is generally an order of magnitude larger than the dependence on wave age.
4. Air stability and air instability suppresses and enhances wave breaking as compared to neutral flow, respectively; the latter agrees qualitatively with the results in Sugihara et al. (2004).

REFERENCES

Andersen, O.J., and J. Løvseth. 1995. Gale force maritime wind. The Frøya data base. Part 1: Sites and instrumentation. Review of the data base, *J. Wind Eng. Ind. Aerodyn.*, 57, 97-109.

Arya, S. P. 1982. Atmospheric boundary layers over homogeneous terrain. *Engineering Meteorology*, E. Plate, ed., Elsevier, New York, 233-267.

Banner, M.L. and W.L. Peirson. 1998. Tangential stress beneath wind-driven air-water interface, *J. Fluid Mech.,* 364, 115-145.

Donelan, M.A. et al. 1993. On the dependence of sea surface roughness on wave development, *J. Phys. Oceanogr.,* 23, 2143-2149.

Charnock, H. 1955. Wind stress on a water surface, *Q. J. R. Meteorol. Soc.*, 81, 639-640.

Coelingh, J. P. et al. 1996. Analysis of wind speed observations over the North Sea, *J. Wind Eng. Ind. Aerodyn.*, 61, 51-69.

Eifler, W. 2005. The near-surface boundary layer, *Marine Turbulence*, Baumert, H. Z. et al. eds., Cambridge University Press, Cambridge, UK, 250-272.

Heggem, T. 1997. Measurements of coastal wind and temperature. Sensor evaluation, data quality and wind structures, Dr.Sci. thesis, Dept. of Physics, NTNU, Trondheim, Norway.

Jones, I.S.F. et al. 2001. Overview. *Wind Stress over the Ocean*, Jones, I. S. F. and Y.Toba, eds., Cambridge University Press, Cambridge, UK, 1-33.

Jones, I.S.F. and Y.Toba. 2001. *Wind Stress over the Ocean*, Cambridge University Press, Cambridge, UK.

Kraus, E.B. and J.A. Businger. 1994. *Atmosphere-Ocean Interaction*, Oxford University Press, New York.

Myrhaug, D. and O.H. Slaattelid. 2002. Wind stress over waves: Effects of sea roughness and atmospheric stability, *ASME J. Offshore Mech. Arct. Eng.*, 124, 169-172.

Myrhaug, D. and O.H. Slaattelid. 2003. Addendum to "Wind stress over waves: Effects of sea roughness and atmospheric stability" *ASME J. Offshore Mech. Arct. Eng.*, 125, 219-220.

Panofsky, H.A. and J.A. Dutton. 1984. *Atmospheric Turbulence*, John Wiley & Sons.

Phillips, O.M. 1985. Spectral and statistical properties of the equilibrium range in wind-generated gravity waves, *J. Fluid Mech* , 156, 505-531.

Smith, S.D. 1980. Wind stress and heat flux over the ocean in gale force winds, *J. Phys. Oceanogr.,* 10, 709-726.

Smith, S.D. et al. 1996. Air-sea fluxes: 25 years of progress, *Boundary Layer Meteorol.,* 78, 247-290.

Snyder, R.L. et al. 1983. On the formation of whitecaps by a threshold mechanism, Part III: field experiments and comparison with theory, *J. Phys. Oceanogr.,* 13, 1505-1518.

Spillane, M.C. et al. 1986. Whitecaps and global fluxes, *Oceanic Whitecaps*, E.C. Monahan and G. MacNiocaill, eds., D. Reidel Publ. Company, Dordrecht, The Netherlands, 209-218.

Toba, Y. et al. 1990. Wave dependence on sea-surface wind stress, *J. Phys. Oceanogr.*, 20, 705-721.

Toba, Y. and I.S.F. Jones. 2001. The influence of unsteadiness, *Wind Stress over the Ocean*, Jones, I.S.F. and Y. Toba, eds., Cambridge University Press, Cambridge, UK, 190-205.

Sugihara, Y. et al. 2004. Imaging measurement of whitecaps at sea observation tower. *Proc. 29th Conf. on Coastal Eng., ASCE,* Lisbon, Portugal, Vol. 1, 1082-1092.

Volkov, Y. 2001. The dependence on wave age, *Wind Stress over the Ocean*, Jones, I.S.F. and Y. Toba, eds., Cambridge University Press, Cambridge, UK, 206-217.

Young, I.R. 1998. An experimental investigation of the role of atmospheric stability in wind wave growth, *Coastal Engineering,* 34, 23-33.

LARGE EDDY SIMULATION OF REGULAR WAVES BREAKING OVER A SLOPING BEACH

Pierre Lubin[1], Hubert Branger[2], Olivier Kimmoun[3]

The scope of this paper is to show the results obtained for the Large Eddy Simulations (LES) of three-dimensional regular waves shoaling and breaking over a sloping beach, by solving the Navier-Stokes equations in air and water. The interface tracking is achieved by a Lax-Wendroff TVD scheme, which is able to handle with interface reconnections. The breaking processes including shoaling, overturning, splash-up and air entrainment, will be presented and discussed.

INTRODUCTION

Great improvements have been brought to the knowledge of the hydrodynamics and the general processes occurring in the surf zone, widely affected by the breaking of the waves (Peregrine 1983, Christensen et al. 2002). Nevertheless, the turbulent flow structure is still very complicated to investigate. The interest of the numerical approach is to provide a complete and accurate description of free surface and velocity evolutions in both air and water media during the breaking of the waves, which must lead to the understanding of energy dissipation and turbulent flow structures generation processes. On the basis of the published numerical works (Lubin et al. 2006) and the experiments conducted by Kimmoun et al. (2004, 2006), a numerical study of regular waves breaking over a sloping beach is presented. Large Eddy Simulation have already been found to be a very reliable method for the unsteady simulation of the Navier-Stokes equations, as already shown in the recent works of Watanabe et al. (2005) and Christensen (2006).

The numerical methods are first introduced. We implemented up to date subgrid scale models to describe turbulence generated in both air and water media. Then, the experimental wave tank is presented, with the visualizations of the plunging breaking waves over the sloping beach. The results of two-dimensional (2D) and three-dimensional (3D) numerical simulations of the regular waves are finally compared to the experimental data. A discussion about the general behavior of the flow in both media is held. We aim at illustrating the capacity of the numerical tool to give an accurate description of the flow, in terms of instantaneous quantities, and the improvement in the numerical modelling of breaking waves since it includes the air entrainment process neglected in most previous existing models.

---

[1] Université Bordeaux 1, TREFLE-ENSCPB UMR CNRS 8508, 16 avenue Pey-Berland, Pessac, France
[2] IRPHE UMR CNRS 6594, Aix-Marseille Université, 49 rue Joliot Curie, Marseille, France
[3] LRD Ecole Centrale de Marseille, 38 rue Joliot Curie, Marseille, France

MODEL AND NUMERICAL METHODS

The numerical methods have already been fully described by Lubin et al. (2006). We solve the Navier-Stokes equations in air and water, coupled with a subgrid scale turbulence model (Large Eddy Simulation – LES). The numerical tool is well suited to deal with strong interface deformations occurring during wave breaking, for example, and with turbulence modelling in the presence of a free surface in a more general way. The numerical tool has already been shown to give accurate results for coastal applications (Lubin et al. 2004). For example, comparisons with experimental and numerical results have been shown by Lubin (2004) and Helluy et al. (2005) for the practical application of the propagation and the breaking of a solitary wave over a submerged reef.

Governing equations

On a fixed orthogonal curvilinear grid, an incompressible multiphase phase flow between non-miscible fluids can be described by the Navier-Stokes equations in their multiphase form. For each medium, a phase function C, or "color" function, is used to locate the different fluids standing $C = 0$ in the outer media, $C = 1$ in the considered medium. The interface between a medium and the entire domain is repaired by the discontinuity of C between 0 and 1. In practice, $C = 0.5$ is used to characterize this surface. The governing equations for the Large Eddy Simulation of an incompressible fluid flow are classically derived by applying a convolution filter to the unsteady Navier-Stokes equations. The resulting set of equations reads:

$$\nabla.\boldsymbol{u} = 0 \tag{1}$$

$$\rho\left(\frac{\partial \boldsymbol{u}}{\partial t} + (\boldsymbol{u}\cdot\nabla)\boldsymbol{u}\right) = \rho\boldsymbol{g} - \nabla p + \nabla\cdot\left((\mu+\mu_T)(\nabla\boldsymbol{u}+\nabla'\boldsymbol{u})\right) + \boldsymbol{F} - \frac{\mu}{K}\boldsymbol{u} \tag{2}$$

$$\frac{\partial C}{\partial t} + \boldsymbol{u}.\nabla C = 0 \tag{3}$$

$$\begin{aligned} \rho = \rho_1 \text{ and } \mu = \mu_1 \text{ if } C \geq 0.5 \\ \rho = \rho_2 \text{ and } \mu = \mu_2 \text{ if } C < 0.5 \end{aligned} \tag{4}$$

with velocity $\boldsymbol{u}$ and pressure p, assuming $\boldsymbol{g}$ as the gravity vector, ρ as the density, μ as the viscosity, μ_T as the turbulent viscosity, t as the time and $\boldsymbol{F}$ as the superficial tension volume force. To deal with solid obstacles within the numerical domain, it is possible to use multi-grid domains, but it is often much simpler to implement the Brinkman theory: the numerical domain is then considered as a unique porous medium. The permeability coefficient K defines the capability of a porous medium to let pass the fluids more or less freely through it. If this permeability coefficient is great ($K \rightarrow +\infty$), the medium is equivalent to a fluid. If it is nil, we can model an impermeable solid. A real porous medium is modelled with intermediate values of K. It is then possible to model moving rigid boundaries or complex geometries. To take this coefficient K into account in our system of equations, we thus add an extra term, called Darcy term, $(\mu / K)\, \boldsymbol{u}$. Model (Eqs. 1 to 4) describes the entire hydrodynamics and geometrical processes involved in the motion of multiphase media.

Large scale turbulence is described by solving the flow equations (Eqs. 1 and 2), the small scale turbulence, which is not resolved by the flow model, is taken into account through a subgrid scale model. To represent the dissipative effect of the small turbulent structures a turbulent viscosity μ_T is calculated with the Smagorinsky model or the Mixed Scale model. Both models are based on the resolved scales through the strain rate tensor, but the Mixed Scale model takes also into account some subgrid information. This model has been checked to be much more efficient. We also implemented a selective function to check that the velocity field is turbulent and requires a surgrid scale model to be turned on. All the details can be found in Sagaut (1998).

Numerical methods

A Finite-Volume method on a staggered mesh is carried out to discretize the Navier-Stokes equations and an adaptive augmented Lagrangian technique is investigated to solve the coupling between pressure and velocity in the equations of motion (Eqs. 1 and 2). The interface tracking is achieved by a Lax-Wendroff TVD scheme, which is able to handle with interface reconnections.

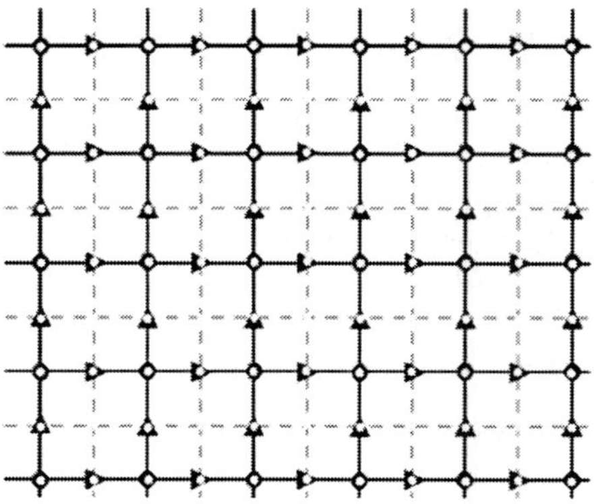

Figure 1. Unknown grid discretization. Black dots: pressure; black arrowheads : velocity components ; light gray dots : color function.

A dual grid, or underlying grid (Liovic et al. 2006), is used to discretize the advection equation (Eq. 3). This technique allows to have an improved accuracy for the interface description. It can be seen in Fig. 1 that the color function is defined on a twice finer mesh than the staggered mesh used for the Navier-Stokes equations. All the references and details concerning the numerical methods can be found in Vincent et al. (2003) and Lubin (2004).

EXPERIMENTAL CONFIGURATION

The experiments were performed in the Ecole Centrale wave tank in Marseille. A side view of the experimental apparatus is given in Fig. 2. The glass-windowed tank is 17 m long and 0.65 m wide. The water depth was set at $d = 0.735$ m. The 1/15 sloping beach was about 13 m long, starting at about 4 m away from the wavemaker. The length of the surf zone was about 3 m. Camera PIV measurements were done in twelve different locations from the incipient breaking location up to the swash zone. Fifth order Stokes waves were generated, corresponding to the analytical solution developed by Fenton (1985). The wave period was T = 1.3 s and wave amplitude before the sloping beach was a = 0.07 m. The wavelength was L = 2.5 m and the measured height at breaking was H_b = 0.137 m. The waves are observed to start breaking at about 2.65 m away from the shoreline, or 12.375 m away from the wavemaker.

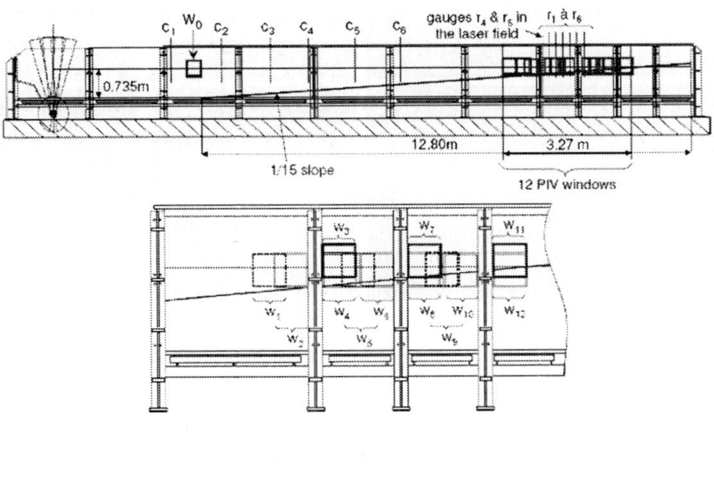

Figure 2. Experimental wave tank.

A sketch of a wave breaking event is displayed in Fig. 3. The wave starts breaking between windows 1 and 2 showing a short spilling phase (W2 and W3). Then a jet of liquid is rapidly ejected from the wave crest (W4) and the overturning wave front curls forward. A first splash-up is generated when the jet of liquid hits the front face of the wave (W5). We can then see a large amount of air entrained with foam and bubbles. Some other splash-ups are then generated

(W6). A roller propagates towards the shoreline, with a great air-water mixing area (W7 to W11). It can be seen in windows 7 and 8 that the bubbles are generated in the upper part of the water column, and are advected towards the bottom with a slight slanting axis (W9 and W10). The volume of entrained bubbles decreases gradually till the wave crosses the shoreline (W12), runs up before coming back. This is in agreement with the general description of Peregrine (1983), for example. More details of the experiments can be found in Kimmoun et al. (2004) and Kimmoun and Branger (2006).

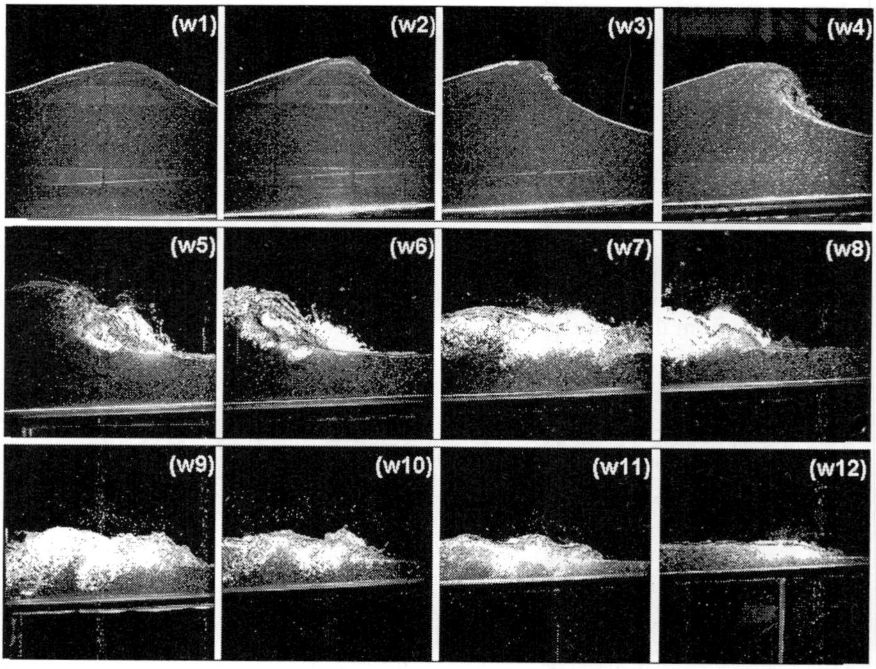

Figure 3. Wave breaking in the twelve PIV interrogation windows.

LARGE EDDY SIMULATION OF PLUNGING BREAKING WAVES

Wave generation procedure

Prior to the simulation of the laboratory tests, an effort has been made to implement and validate the procedure of regular and irregular wave generation developed by Lin and Liu (1999). The method consists in introducing an internal mass source function in the continuity equation for a chosen group of cells defining the source region. We adapted the analytical developments of Fenton (1989) for the fifth order Stokes wave theory, corresponding to the experimental conditions, to the source function method. In Fig. 4, we show the velocity field around the source function, acting like a pump. Two wave trains of surface gravity waves are thus generated, as the free surface responds to a pressure increment defined within the source region cells. The two wave trains propagate in opposite directions towards the both ends of the numerical domain.

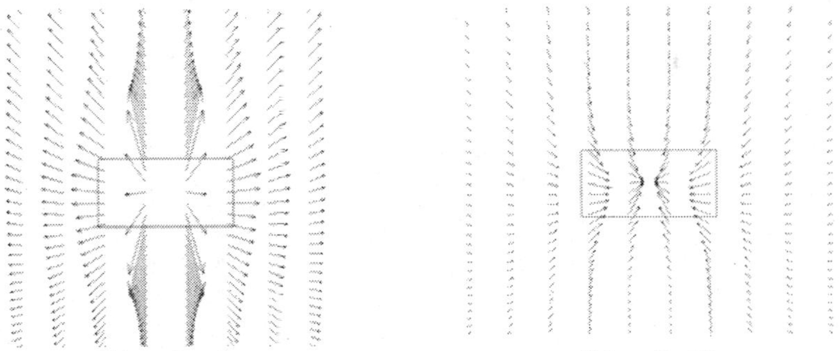

Figure 4. Velocity fields generated by the presence of the source function.

Computational configuration

The computational domain is 20 m long and 1.2 m high (Fig. 5). The sloping beach starts at $x = 7$ m, the source function being located at $x = 3.5$ m and $z = 0.3675$ m (the center of the source region is at $d/2$). The numerical beach is considered as an impermeable solid obstacle, the permeability coefficient K being initialized at zero (Eq. 4). The source region is 0.06 m wide and 0.0735 m high. The area and the location of the source function have been designed applying the rules described by Lin and Liu (1999). 520 000 mesh grid points are used to discretize the numerical domain (2000 x 260). The grid system is uniform along the wave propagation direction, with a constant cell length $\Delta x = 1.10^{-2}$ m, and non-uniform in the vertical direction, the minimum cell height being $\Delta z = 2.5\ .10^{-3}$ m. These values have to be divided by two, in both directions, for the free surface description thanks to the dual grid. The time step is chosen to ensure a Courant-Friedrichs-Levy condition less than 1, necessary for the explicit advection of the free surface.

As already said, two wave trains are generated and propagate in opposite directions towards the both ends of the numerical domain. An open boundary condition is thus set at the left side of the numerical domain to let the outgoing wave exit the numerical domain. In order to ensure that no numerical reflection occurs at the left side of the numerical domain, a sponge layer is set in addition to the open boundary condition. It consists in a region where the permeability coefficient K is chosen such that the outgoing wave train is properly attenuated before reaching the open boundary. The simulations have been carried out on fast PC's. About twenty breaking waves have been simulated. Only instantaneous quantities are presented and discussed.

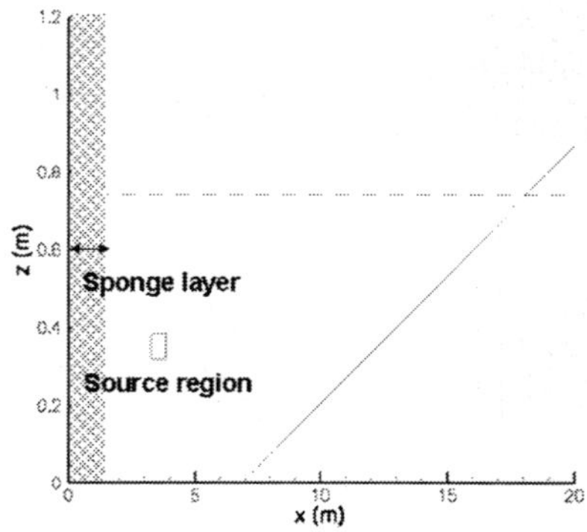

Figure 5. Sketch of the numerical configuration for the regular wave breaking simulation. Dashed line: mean water depth *d* = 0.735 m. The toe of the 1/15 sloping beach starts at *x* = 7 m.

2D numerical results

In Fig. 6, we present the numerical results corresponding to the 10th breaking wave. The figures correspond to the twelve experimental pictures taken at the different locations in the surf zone (Figs. 3). A general good agreement has been found in term of wave propagation. The main observed difference is that we miss the first short spilling event.

Some elements can be brought to discuss this discrepancy. On the basis of the experimental measured waveheight at breaking H_b, the surf similarity parameter has been evaluated to be $\xi = 0.285$, which usually characterizes a spilling breaker (Battjes 1988). Spilling breaking waves involve a complex combination between the wave propagation and the vorticity generation based on perturbations appearing on the crest of the steepening face of the wave. Moreover, the measured height of the white cap is about 1 mm (Fig. 3 – windows 2 and 3). So, numerically, the mesh grid refinement we use is obviously not sufficient enough to be able to capture this feature. The numerical diffusion is also probably responsible, even if it has been checked to be very low (the numerical diffusion of the free-surface is contained in three mesh grid cells).

Once the front face of the crest steepens and becomes vertical (Fig. 6 - windows 1 to 4), a jet of water is projected (Fig. 6 - between windows 4 and 5), and free falls down forward into a characteristic overturning motion. The plunging breaking wave is then responsible for the generation of jet-splash cycles, this, in turn, being responsible for the generation of a sequence of large scale coherent vortices.

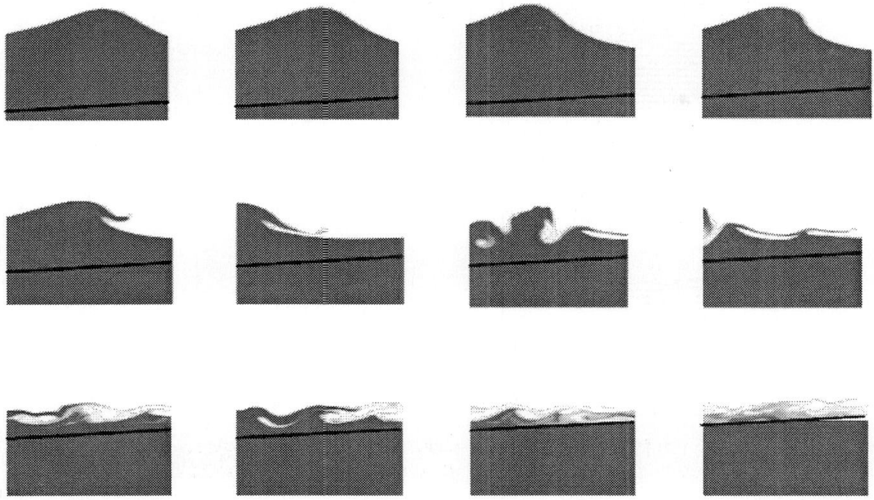

Figure 6. Numerical results for a wave breaking at the twelve PIV interrogation windows locations (10th wave).

A first high splash-up is generated (Fig. 3, window 6 - Fig. 6 window 7) and the successive splash-ups are also simulated. We have air entrapped and entrained in the water. And once the wave moves into a roller propagating towards the shoreline (Fig. 6 – windows 8 to 12), we have very similar results for the free surface description. It is distorted and very dynamic (Fig. 3 – windows 8 to 12). Nevertheless, some discrepancies appear. Due to the coarse mesh grid resolution, the dislocation of the gas pockets into small bubbles cannot be simulated, even if, in the numerical results, the gas pockets correspond to some air-water mixing zones observed in the experimental pictures. Turbulence is associated with air entrainment, which is responsible for wave energy damping in the surf zone. In the experiments, it appears that the entrained air bubbles are contained mostly in the large structures and diffused towards the bottom due to the eddies. The rate of energy dissipation is increased with the bubble penetration depth and strong vertical motions are induced by the rising air bubbles. These mechanisms are mostly 3D, which cannot be taken into account in a 2D numerical simulation.

In Fig. 7, we present the numerical velocity fields corresponding to the 10th breaking wave. The figures correspond to three experimental pictures taken at the different locations in the surf zone.

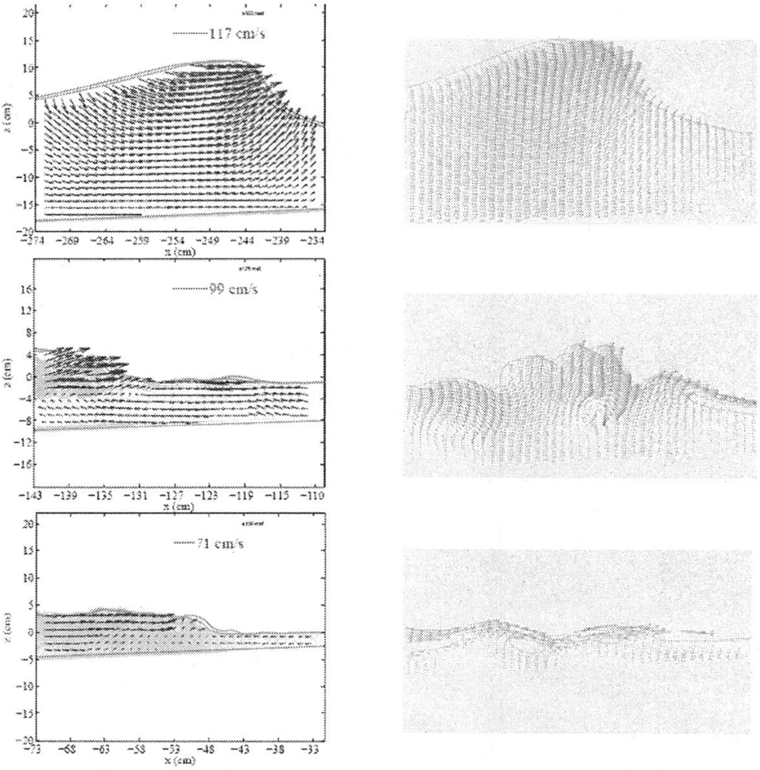

Experimental Numerical

Figure 7. Comparison between instantaneous PIV and numerical velocity fields at different locations.

A high velocity region is located at the crest of the wave. A high splash-up is rising, with high velocities directed upward and towards the wave propagation direction, which generates counter-rotative vortices, as observed by Bonmarin (1989). This process is observed to repeat, each successive splash-up being weaker than the preceding one. Some large volumes of air and water are put into rotation. Co-rotative vortices are then generated by these successive splash-ups, as observed by Miller (1976) or Sakaï et al. (1986). High velocities are located near the free-surface, due the jet-splash cycles, during the bore propagation.

Considering these simulations as a validation step, our numerical model gives very satisfactory and encouraging results for this 2D configuration. Nevertheless, wave breaking is a 3D two-phase turbulent problem, so the first numerical results will be presented hereafter, consisting in a first attempt.

3D numerical results

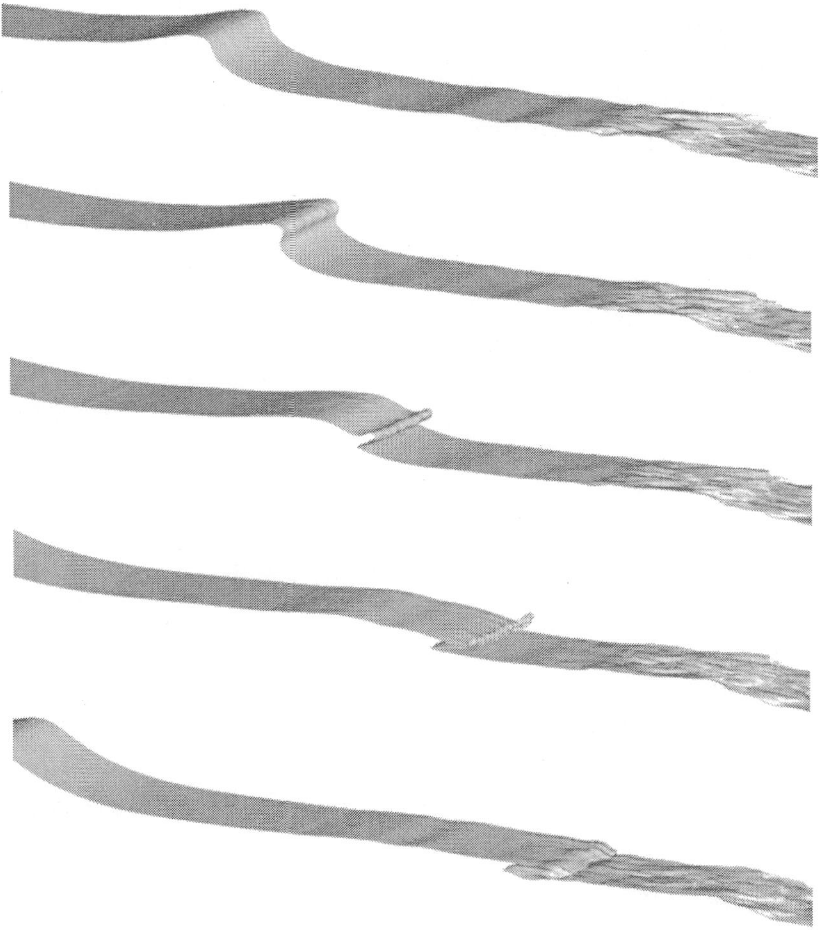

Figure 8. Numerical results for 3D waves breaking (14th wave). Color function C = 0.5.

The same initial condition than previously detailed is used, spread along the transverse direction. The numerical wave tank is 30 cm wide. But the numerical domain has been shortened and is discretized with 660 000 mesh grid points (550 x 60 x 20) to be still able to run the simulation on a fast PC with a reasonable accuracy. The numerical domain is now 15.5 m long. The grid system is uniform along the wave propagation direction, with a constant cell length $\Delta x = 2.8\ .10^{-2}$ m, and non-uniform in the vertical direction, the minimum cell height being $\Delta z = 8.4\ .10^{-3}$ m. A constant cell width $\Delta y = 1.5\ .10^{-2}$ m is used.

These values have to be divided by two, for all directions, for the free surface description thanks to the dual grid.

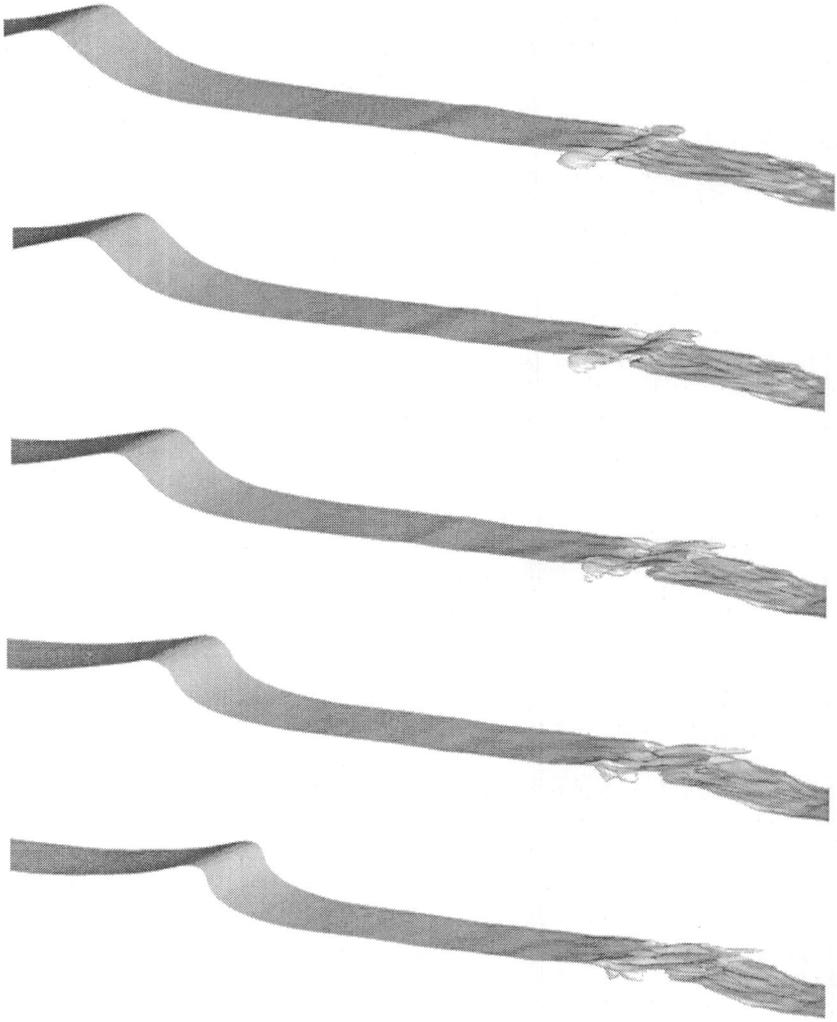

Figure 9. Numerical results for 3D waves breaking (end of the 14th wave). Color function C = 0.5.

About fifteen breaking waves have been simulated. The number of grid points is obviously not sufficient enough again to take into account the dislocation of the pockets of gas into small bubbles, but the flow dynamics is correctly described in the surf zone with air entrainment. A plunging breaker type is still observed. Figs. 8 and 9 present the numerical results corresponding to the 14th wave breaking. The amount of air entrained in the water is less important than previously observed in the 2D numerical results due to the less accurate description of the plunging jet, responsible for a great quantity of air to be entrapped.

CONCLUSION

The numerical results presented in this paper concerns instantaneous quantities, simulating 2D, and then 3D, regular waves breaking over a sloping beach. Numerical simulations are now undertaken to calculate phase averaged and fluctuating quantities. Turbulence will then be studied quantitatively and compared with the experiments. Our model was found to be reliable to describe correctly and accurately the complicated two-phase flow interactions that happens when waves break. The breaking process, in terms of wave overturning and splash-up occurrence, is in accordance with the general observations given in the literature. The air entrainment is described, which is important as it plays a great role in the energy dissipation process.

But 3D numerical simulations remain very demanding and time consuming. Two-phase flows simulations and turbulence modelling require fine mesh cells to be very accurate, even if Large Eddy Simulation is supposed to allow us to save some mesh points. Some theoretical work is also led to model what happens in the vicinity of the color function C = 0.5 (Lubin 2004, Labourasse et al. 2006).

ACKNOWLEDGMENTS

The authors would like to acknowledge the financial and scientific support of the French INSU - CNRS (Institut National des Sciences de l'Univers - Centre National de la Recherche Scientifique) program IDAO (« Interactions et Dynamique de l'Atmosphère et de l'Océan » and the IDRIS - CNRS (« Institut du Développement et des Ressources en Informatique Scientifique ») and CINES (« Centre Informatique National de l'Enseignement Supérieur ») for the material support (project ter 2237).

REFERENCES

Battjes, A. 1988. Surf-zone dynamics. *Ann. Rev. Fluid Mech.*, 20, 257-293.

Bonmarin, P. 1989. Geometric properties of deep-water breaking waves, *J. Fluid Mech.*, 209, 405-433.

Christensen, E. D., D. J. Walstra, and N. Emarat. 2002. Vertical variation of the flow across the surf zone, *Coastal Engineering*, 45, 169-198.

Christensen, E. D. 2006. Large eddy simulation of spilling and plunging breakers, *Coastal Engineering*, 53, 463- 485.

Fenton, J. D. 1985. A fifth-order Stokes theory for steady waves. *Journal of Waterway, Port, Coastal and Ocean Engineering*, 111, 216-234.

Helluy, P., F. Gollay, S. T. Grilli, N. Seguin, P. Lubin, J.-P. Caltagirone, S. Vincent, D. Drevard and R. Marcer. 2005. Numerical simulations of wave breaking. *Mathematical Modelling and Numerical Analysis Mathematical Modelling and Numerical Analysis*, 39 (3), 591-608.

Kimmoun O., H. Branger and B. Zucchini. 2004. Laboratory PIV measurements of wave breaking on a beach. *Proceedings of 14th International Offshore and Polar Engineering Conference*, 3, 293-298.

Kimmoun K. and H. Branger. 2006. A PIV investigation on laboratory surf-zone breaking waves over a sloping beach, *corrected and submitted to J. Fluid Mech.*

Labourasse, E., D. Lacanette, A. Toutant, P. Lubin, S. Vincent, O. Lebaigue, J.-P. Caltagirone, P. Sagaut. 2006. Towards Large Eddy Simulation of isothermal two-phase flows: governing equations and a priori tests, *under press for International Journal of Multiphase Flow.*

Lin P. and P. L.-F. Liu. 1999. Internal wave-maker for Navier-Stokes equations models, *Journal of Waterway, Port, Coastal and Ocean Engineering*, 125, 322-330.

Liovic P., M. Rudmanb, J.-L. Liowc, D. Lakehal and D. Kothe. 2006. A 3D unsplit-advection volume tracking algorithm with planarity-preserving interface reconstruction, *Computers and Fluids*, 35 (10), 1011-1032.

Lubin, P. 2004. *Large eddy simulation of plunging breaking waves*, PhD thesis (in English), University of Bordeaux.

Lubin, P., S. Vincent, J.-P. Caltagirone and S. Abadie. 2004. Three-dimensional large eddy simulation of vortices induced by plunging breaking waves, *Proceedings of 29th International Conference on Coastal Engineering*, ASCE, 344-356.

Lubin, P., S. Vincent, S. Abadie and J.-P. Caltagirone. 2006. Three-dimensional Large Eddy Simulation of air entrainment under plunging breaking waves, *Coastal Engineering*, 53, 631-655.

Miller, R. L. 1976. Role of vortices in surf zone prediction: sedimentation and wave forces, Soc. Econ. Paleontol. Mineral. Spec. Publ., 24, Ed. R. A. Davis, R. L. Ethington, 92-114.

Peregrine, D. H. 1983. Breaking waves on beaches, *Ann. Rev. Fluid Mech.*, 15, 149-178.

Sagaut, P. 1998. Large eddy simulation for Incompressible Flows, Springer, Verlag.

Sakaï, T., T. Mizutani, H. Tanaka and Y. Tada. 1986. Vortex formation in plunging breaker, *Proceedings of 20th International Conference on Coastal Engineering*, ASCE, 711-723.

Vincent, S., J.-P. Caltagirone, P. Lubin and T. N. Randrianarivelo. 2003. An adaptative augmented Lagrangian method for three-dimensional multi-material flows, *Computers and Fluids*, 33, 1273-1289.

Watanabe, Y., H. Saeki and R. J. Hosking. 2005. Three-dimensional vortex structures under breaking waves, *J. Fluid Mech.*, 545, 291-328.

INVESTIGATION OF DECOMPOSITION METHODS OF TURBULENT FLOW FIELD BENEATH WIND WAVES

Sang-Ho Oh[1], Natsuki Mizutani[2] and Kyung-Duck Suh[3]

The performance of two flow field decomposition techniques, the linear spatial filter and the proper orthogonal decomposition, in turbulence decomposition from a two-dimensional velocity field was investigated by using experimental data of velocity field beneath laboratory-generated wind waves. The main interest was in the identification of turbulent eddy structures embedded in the unsteady velocity field by the decomposition techniques that are used. The linear spatial filter presented good results showing almost the same vorticity map from the decomposed velocity field as the one from the original velocity field. As long as the size of the window function implemented in the linear spatial filter is adequately determined, turbulent flow field is well extracted by the filter in terms of preservation of turbulent eddy structures. In contrast, somewhat distorted vorticity map was obtained from the turbulent velocity field decomposed by the proper orthogonal decomposition, which implies poor performance of the method in turbulence decomposition.

INTRODUCTION

Recent advance of flow visualization techniques allows the measurement of instantaneous subsurface flow fields of wind waves, where turbulence influences in mixing and transfers of momentum, heat, and gases across the air-sea interface. In order to investigate such turbulence properties in this subsurface boundary layer, it is required to decompose the original velocity field into turbulent and non-turbulent parts. However, the unsteadiness of the wind waves precludes the use of common turbulence decomposition methods, such as ensemble or phase averging-based techniques, in which turbulence is defined as the fluctuation around the mean, but requests a different technique. At the present, there is no decomposition method that fully accounts for the unsteadiness induced by the wave motions.

Traditionally, understanding of turbulence in the top layers beneath the wavy free surface relied on point-based measurements, such as laser Doppler velocimetry (LDV) or acoustic Doppler velocimetry (ADV). Hence, turbulence decomposition techniques have also been developed to be suitable for use of these measuring instruments. Benilov and Filyushkin (1970) developed linear filtration technique, which is based on the assumption that wave orbital motion is

---

[1] Coastal Engineering Research Department, Korea Ocean Research & Development Institute, 1270 Sa2-dong, Sangrok-gu, Ansan-si, Kyunggi-do, 426-744, Korea

[2] Department of Civil Engineering, Osaka Sangyo University, 3-1-1 Nakagaito, Daito, Osaka, 574-8530, Japan

[3] School of Civil, Urban & Geosystem Engineering, Seoul National University, San 56-1, Shillim-dong, Gwanak-gu, Seoul, 151-742, Korea

linearly related to the surface elevation whereas not correlated with turbulent motion. Nonlinear least squares method proposed by Jiang et al. (1990) estimates turbulent motion from the measured velocity signal by assuming that the wave orbital motion is two-dimensional and potential. Thais and Magnaudet (1995) introduced a more general separation technique called as triple decomposition method. In this method, orbital rotational motion beneath wind waves is also estimated as well as the orbital potential motion and the turbulent components. A common feature of these turbulence decomposition methods is that they correlate the measured velocity with the surface displacement in order to separate orbital contribution from the velocity and then estimate turbulent motion as a remnant of the velocity.

Since these methods are only suitable for turbulence decomposition of single-point velocity time history data, it is desirable to develop a different technique suitable for two-dimensional velocity data. One way of such development is to extend the point-based decomposition method to spatially resolved data fields (*e.g.* Cowen and Monismith 2004). Another way is to use some of the decomposition techniques that are widely used in researches on other unsteady flows, which is the main interest of the present study. In this paper, two different types of such decomposition techniques, the linear spatial filter and the proper orthogonal decomposition, were applied to experimental data of two-dimensional velocity field of laboratory-generated wind waves that were obtained by digital particle image velocimetry (DPIV) and their adequacy was investigated.

TURBULENCE ESTIMATION FROM 2-D VELOCITY DATA

Linear Spatial Filter

Decomposition of turbulent flow field can be accomplished by using the linear spatial filter, which separates turbulent fluctuations $u_t(x,t)$ by extracting temporal mean velocity vectors $\overline{u}(x,t)$ and large-scale components $u_l(x,t)$ from the original velocity field $u(x,t)$:

$$u_t(x,t) = u(x,t) - \overline{u}(x,t) - u_l(x,t) \tag{1}$$

where $u_l(x,t)$ is computed using a low-pass two-dimensional spatial filter based on the assumption that the wave-related motions are represented by the low-frequency velocity components that govern large-scale motions of the velocity field. This method can be considered as the two-dimensional extension of the linear filtration technique (Benilov and Filyushkin 1970), as pointed out by Siddiqui et al. (2004). In this study, the linear spatial filter is adopted because it is very simple to be applied and the linear filtration technique, its one-dimensional counterpart, yields reasonably good results in turbulence decomposition of single-point velocity time series data.

Proper Orthogonal Decomposition

Recent researches on turbulence reveal that large-scale eddy structures in turbulent flows play an important role in overall dynamics of the flow and affect mixing, noise, vibrations, heat transfer, and etc. As a result, many techniques have been developed in order to properly identify such eddy structures embedded in the background turbulence. Proper orthogonal decomposition (POD), one of such techniques that was first proposed by Lumley (1967), is a procedure for extracting a basis for a modal decomposition from an ensemble of signals. The decomposition technique is efficiently used in many practical problems as a method of identifying vortical structures in turbulent flows. Although there are several other vortex identification methods and their performance in detecting large-scale eddy structures is generally similar (Bonnet and Delville 2001), POD is selected in this study because it is known to be an unbiased method for deducing the eddy structures.

The fundamental idea of POD is the projection of the velocity field u into an orthogonal coordinate system ϕ optimal in terms of energy, which is accomplished by maximizing $<(u,\phi)^2>/(\phi,\phi)$. This maximization is equivalent to obtaining a solution of the integral eigenvalue problem

$$\int < u(x)u(x')) > \phi(x')dx' = \lambda\phi(x) \tag{2}$$

The solution of Eq. (2) is a series of eigenvalues and eigenmodes, which are sorted in decreasing order so that the first mode contains most of the energy. According to Lumley (1967), the most energy-containing structures are related to the primary eigenmode that makes the most significant contribution to the flow field. If we assume that the most energy-containing structures in the original flow field are associated with the wave-related motions, turbulent fluctuations can be deduced by extracting the contribution due to the primary eigenmode

$$u_t(x,t) = u(x,t) - \lambda_1\phi_1 \tag{3}$$

EXPERIMENTAL DATA

The performance of the two turbulence decomposition methods was examined using experimental data of velocity fields beneath laboratory-generated wind waves. The subsurface velocity fields were observed using a digital particle image velocimetry (DPIV) system that is synchronized with a capacitance wave gauge to measure surface displacement simultaneously. The experiment was carried out in a wind-wave flume whose uniform test section is 1.5 m wide, 1.3 m high, and 23.5 m long. The sidewalls of the flume consist of glass plates and the top is covered with wood plates. The water depth was kept constant at 53 cm for all the tests, in which the spectral peak wave component always fell under the deepwater condition.

Figure 1 shows the schematic diagram of the experimental setup. In order to conduct the DPIV experiment in daytime, a dark room was constructed. The laser sheet was projected vertically upward along the centerline of the flume, parallel with the sidewalls. In order to obtain synchronized wave and image data of almost identical wave properties, the wave gauge was setup at the center of the field of view longitudinally and 3 mm away from the edge of the laser sheet laterally. By this approach, the air-water interface could be obtained for each DPIV image. The CCD camera was positioned below the wave trough level with its angle of 20° upward. The field of view of the camera was 23.3 cm wide and 21.8 cm high for all the test cases, and the corresponding image resolution was 0.455 mm/px. The sampling rate of the wave gauge was 200 Hz for all the test cases, whereas the frame transfer rate of the CCD camera varied depending on the experimental conditions. The camera registered wave images at a constant frame transfer rate and shutter speed. A total of 2184 DPIV images was captured continuously and stored in a built-in memory temporarily and then transferred to the host computer.

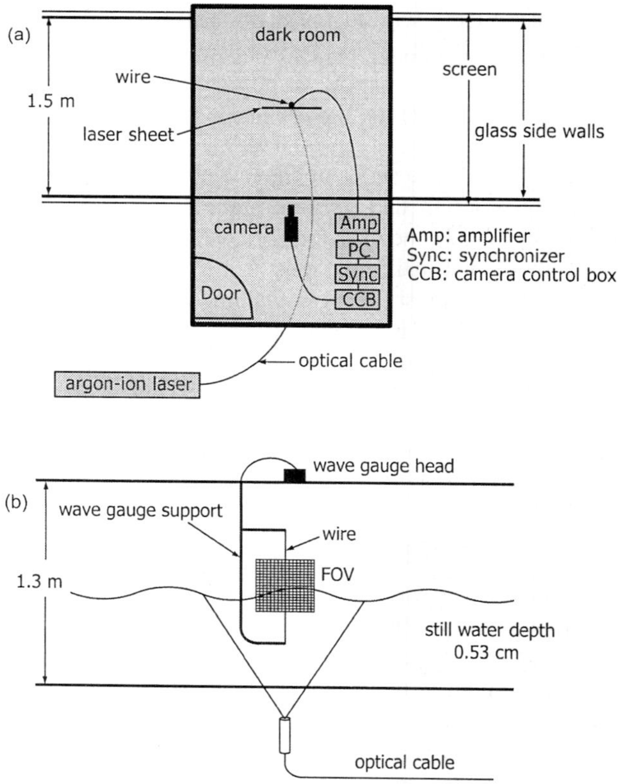

Figure 1. Experimental setup: (a) plan view and (b) side view.

The acquired images were analyzed by a PIV analysis software, INSIGHT (TSI Inc.) after denoising the background noise of the raw DPIV images. The size of an interrogation window was 32 × 32 pixels and the window was overlapped by 50 % with neighboring ones. This produces a 31 × 29 array of velocity vector maps for each DPIV image, the resolution of which was 7.3 × 7.3 mm^2. The raw velocity vector matrix calculated by the software was further subjected to post-processing in which spurious vectors were detected and replaced by alternative values using weighted average of the eight surrounding vectors of 3 × 3 window. The post-processing was done by using a computer program developed by the authors. In general, the number of corrected spurious vectors was less than 3% of the total number of velocity vectors.

The measurements of subsurface velocity field were made for three different wind speeds (approximately 7, 10, and 14 m/s) at two fetch lengths (11.4 and 23.2 m) along the flume. Among these test conditions, the performance of the two turbulence decomposition techniques was examined using the test condition in which fetch length is 11.4 m and wind speed is 14 m/s. Further details about the DPIV measurement and the subsequent procedure of data processing are described in Oh et al. (2005) who investigated breaking criteria of deepwater wind waves.

RESULTS

It is well known that the appearance of eddy structures is closely related to the local turbulent motion in a flow field. This implies that a good turbulence decomposition method would yield the turbulent vorticity field that well contains the vortical structures appearing in the original vorticity field. Figure 2 shows the results of turbulence decomposition by the linear spatial filter and the proper orthogonal decomposition, respectively. In the vorticity fields by both methods, only large energy-containing eddy structures are retained by applying the vorticity-based conditional sampling method (Hussain and Hayakawa, 1987) to the vorticity fields. The performance of turbulence decomposition by both methods is well compared in the figure, especially in the turbulent vorticity fields.

As shown in Figure 2(b), most of the vortical structures appearing in the original vorticity field remain in the turbulent vorticity field subjected to the linear spatial filter. In contrast, Figure 2(c) shows a slightly distorted turbulent vorticity map obtained by the proper orthogonal decomposition, particularly in the near-surface region. This implies that the proper orthogonal decomposition over-estimates the turbulent velocity components around this region, which is confirmed in the turbulent velocity field. Hence, the linear spatial filter provides better results of turbulence decomposition than the proper orthogonal decomposition in terms of preservation of turbulent eddy structures. In the following, the effects of the window function on the performance of the linear spatial filter are discussed.

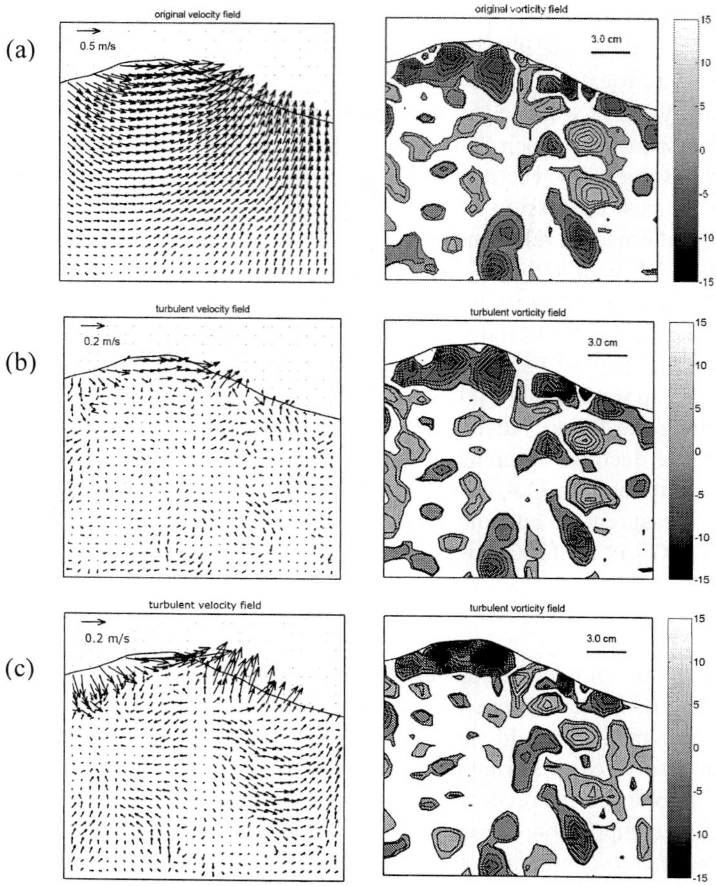

Figure 2. Instantaneous velocity and vorticity fields of (a) original and turbulent fields obtained by (b) the linear spatial filter and (c) the proper orthogonal decomposition.

Effects of Size of Window Function

In the implementation of the linear spatial filter, the large-scale components $u_l(x,t)$ were computed by the low-pass two-dimensional spatial filter as shown in Eq. (4)

$$u_l(x,t) = \frac{1}{N}\sum_{i=1}^{N} W(x_i)u(x_i,t) \tag{4}$$

where $W(x_i)$ is the rectangular window function of unit magnitude and N is the number of effective vectors within the window. In the neighborhood of the water surface, N was smaller than the total number of the grid points enclosed by the window.

It was found that size of the window function affected the performance of the linear spatial filter. In this study, therefore, the size of the window function was determined so as to guarantee the survival of the large-scale vortical structures in the turbulent velocity field after it was subjected to the linear spatial filter. Figure 3 shows the difference in the wave-related and turbulence vorticity fields depending on the size of the window function. As shown in the figure, many of the vortical structures appearing in the original vorticity field were filtered out in the turbulence vorticity field as the size of the window function decreases. On the contrary, the intensity of wave-related vorticity increases with the decrease of the window size. A wider window than the size of 9×9 was also tested but it made little difference from the results obtained by using the 9×9 window.

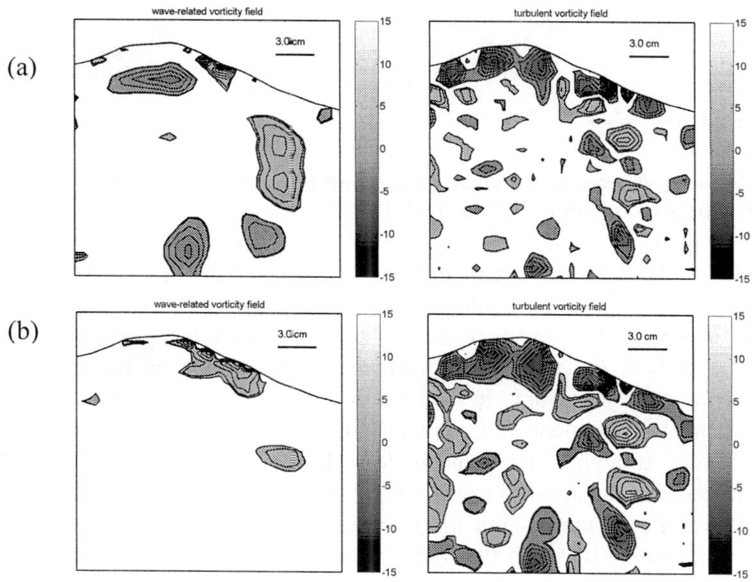

Figure 3. Wave-related and turbulent vorticity fields when the size of the window function is (a) 5×5 and (b) 9×9.

Shown in Figure 4 are the effects of window size on the spectral energy levels of the wave-related and turbulent velocity components. The vertical velocity spectra shown in the figure are estimated from the velocity time history at 5 cm below the still water level. It is clearly seen that the spectral energy of the turbulent velocity component increases with the window size over the whole frequency range with the increase of the window size. In contrast, the wave-related components show an opposite trend, while their energy level at the spectral peak frequency is not influenced by the window size. This implies that a better results of turbulence decomposition is obtained when the 9×9 window was used. Similarly to Figure 3, difference in the velocity spectra of the wave-related and turbulent components was insignificant when a wider window than the 9×9 window is used. Hence, the size of the window function was determined as 9×9 in this study.

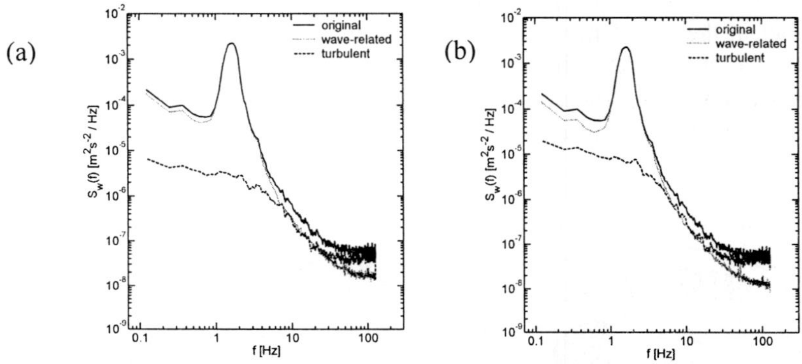

Figure 4. Velocity spectra of the wave-related and turbulent components when the size of the window function is (a) 5×5 and (b) 9×9.

In order to quantitatively estimate the performance of the linear spatial filter, the velocity spectra of the wave-related and turbulent components are normalized as shown in Figure 5. In the left figure, it is clearly seen that the filter effectively separates the wave-related and turbulent motion in the frequency ranges near the peak frequency. In the high frequency ranges, however, some of the wave-related component still remains, implying the contamination of the turbulent field, although its contribution decreases with the frequency. Meanwhile, most of the spectral energy in the low frequency range is associated with the wave-related motion but the turbulent contribution is not completely zero, approximately 10% of the total value. This contribution would be misidentified turbulence due to the spatial filtering similar to the one reported by Melville et al. (1985) in their application of the linear filtration technique of Benilov and Filyushkin (1970). The right figure shows that in the frequency range lower than the peak frequency, the sum of the energy by the two components is less than that of the original spectrum. The energy level in the low frequency range is about one order of magnitude smaller than the energy at peak frequency so that the total loss of energy is insignificant, approximately 2% in this study.

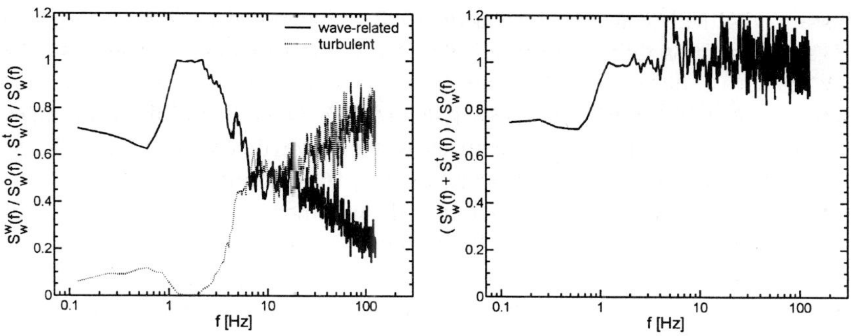

Figure 5. Relative magnitudes of the wave-related and turbulent spectra to the original spectrum.

CONCLUSIONS

The performance of two turbulence decomposition techniques, the linear spatial filter and the proper orthogonal decomposition, was examined in terms of preservation of turbulent eddy structures when velocity fields beneath wind-generated waves are subjected to each turbulence decomposition technique. The eddy structures appearing in the original velocity field were well retained by the linear spatial filter. In contrast, a distorted vortical field that is different from the original vorticity field, especially near surface region, was obtained by the proper orthogonal decomposition. This implies that the linear spatial filter is better in identifying vortical structures in the velocity field than the proper orthogonal decomposition. It was found that the performance of the linear spatial filter was influenced by the size of the window function. When the size of the window function was not sufficiently large, some turbulent eddy structures were filtered out in the decomposed turbulent vorticity field. If the size of window function is suitably determined according to a velocity field subjected to the filter, the use of linear spatial filter is expected to be acceptable for the investigation of vortical structures embedded in the turbulent flow field of wind waves, while further examination is requested for other respect of turbulent properties.

REFERENCES

Benilov, A.Y., and B.N. Filyushkin. 1970. Application of methods of linear filtration to an analysis of fluctuations in the surface layer of the sea, *Izv. Acad. Sci. USSR Atmos. Oceanic Phys., Engl. Trasnsl.,* 6, 477-482.

Bonnet, J.-P. and J. Delville. 2001. Review of coherent structures in turbulent free shear flows and their possible influence on computational methods, *Flow, Turbulence and Combustion*, 66, 333-353.

Cowen, E.A. and S.G. Monismtih. 2004. The turbulent boundary layer structure beneath a gravity wave and its proper decomposition, *PIV and Water Waves* (Ed: Grue, J., P. L.-F. Liu, and G.K. Pedersen), 311-313.

Hussain, A.K.M.F. and M. Hayakawa. 1987. Eduction of large-scale organized structures in a turbulent plane wake, *Journal of Fluid Mechanics*, 180, 193-229.

Jiang, J.Y., R.L. Street, and S.P. Klotz. 1990. A study of wave-turbulence interaction by use of a nonlinear water wave decomposition technique, *Journal of Geophysical Research*, 95(C9), 16037-16054.

Lumley, J.L. 1967. The structure of inhomogeneous turbulent flows, *Atmospheric Turbulence and Radio Wave Propagation* (Ed: Yaglom, A.M., and V.I. Tatarsky), 166-178.

Melville, W.K., R.J. Rapp, and E.S. Chan. 1985. Wave breaking, currents and mixing, *The Ocean Surface* (Ed: Toba, Y, and H. Mitsuyasu), 413-418.

Oh, S.-H., N. Mizutani, K.-D. Suh, and N. Hashimoto. 2005. Experimental investigation of breaking criteria of deepwater wind waves under strong wind action. *Applied Ocean Research*, 27, 235-250.

Thais, L., and J. Magnaudet. 1995. A triple decomposition of the fluctuating motion below laboratory wind water waves, *Journal of Geophysical Research*, 100(C1), 741-755.

Siddiqui, M.H.K., M.R. Loewen, W.E. Asher, and A.T. Jessup. 2004. Coherent structures beneath wind waves and their influence on air-water gas transfer, *Journal of Geophysical Research*, 109, C03024.

NUMERICAL SIMULATION OF 3-D WAVE BREAKING ON BREAKWATER

Nguyen Danh Thao[1] and Tomoya Shibayama[2]

This paper presents a three-dimensional breaking wave model to investigate the spatial and temporal variation of fluid motions under breaking waves. A three-dimensional Large Eddy Simulation (LES) is used to solve the governing equations for the flow parameters. Fluid motions in the area near the breakwater and its turbulences as well as wave overtopping on an inclined breakwater are investigated. Through the model, the spatial and temporal variations of the hydrodynamic problems of breaking waves on a breakwater are examined.

1. INTRODUCTION

It is known that turbulence generated by breaking waves has important effects on most of the fluid dynamical processes within the surf zone such as wave transformation, generation of near-shore currents, diffusion of materials, and sediment transport. Under the effect of breaker-generated turbulence, the surf zone is an area where the hydrodynamics are very complex with extremely complicated fluid motion. The turbulent flow field has been the subject of intensive experimental study. Most investigations have been carried out in the laboratory, where the well controlled conditions make a systematic study possible, but field measurements have also been made of turbulence in breaking waves.

The most direct way to investigate the flow in the surf zone numerically is to solve the basic equations for Newtonian fluids, namely the Navier-Stokes equations, since this approach can provide time-dependant detail of the flow characteristics. It is possible to calculate the complex flow to give detailed information about the velocities, motion properties and others. This direct solution of the dynamic equations is called Direct Numerical Simulation (DNS). Duy and Shibayama (1997) carried out a detailed investigation and evaluation of breaking wave kinematics and turbulence in the surf zone, using time-averaged technique to solve averaged forms of Navier-Stokes equations and show that the algebra Reynolds stress model can give good results in the inner surf zone. Lin and Liu (1998), Bradford (2000) also got very good results using this technique. Although DNS is a good method for understanding turbulence, the applicable phenomena are limited because the scale of the

[1] Grad. Student, Dept. of Civil Eng., Yokohama National University, 79-5, Tokiwadai, Hodogaya-ku, Yokohama, 240-8501 JAPAN
[2] Professor, Dept. of Civil Eng., Yokohama National University, 79-5, Tokiwadai, Hodogaya-ku, Yokohama, 240-8501 JAPAN

turbulence is confined to the grid scale used in the calculation. In this case, DNS requires the use of very smooth meshes to resolve turbulence.

One approach to solve this problem is called Large Eddy Simulation (LES), which simulates only the larger-scale eddy motions which are resolvable by the numerical methods, while the effect of smaller-scale eddy motions is simulated indirectly. The fundamental premise of this approach is that the large-scale eddies are affected by boundary conditions, causing most of the Reynolds stress and must be computed. The small-scale turbulences are weaker, therefore less critical since they contribute less to the Reynolds stress. The use of LES models gives very detailed information about the flow structures with good evaluating anisotropic turbulence, and it is possible to achieve much higher Reynolds numbers than with DNS.

Wijayaratna et al. (2000) presented a numerical simulation of wave transformation, breaking and run-up on sloping beds. He used a full three-dimensional Navier–Stokes solver to study the three-dimensional turbulent flow structures in the surf and swash zones. Watanabe and Saeki (1999), Christensen and Deigaard (2001), Suzuki et al. (2004) and Christensen (2006) also found similar results from three-dimensional simulations of breaking waves. Thao et al. (2006) investigated breaking wave motions around a slope-type breakwater using LES and got reasonable results although they have not directly verified the model with experimental data.

This paper presents a three-dimensional breaking wave model to investigate the spatial and temporal variation of fluid motions under breaking waves in front of and around a breakwater. A three-dimensional Large Eddy Simulation (LES) is used to solve the governing equations for the flow parameters. Through the model, the spatial and temporal variations of the hydrodynamic problems of breaking waves in front of a breakwater are examined.

2. NUMERICAL MODEL

2.1. Governing Equations and Sub-grid Scale Model

After Wijayaratna et al. (2000), the flow field is calculated based on the three-dimensional Large Eddy Simulation (LES) in which the governing equations are three-dimensional Navier-Stokes equations. The simulation is carried out directly only for the larger scale turbulences, where the effect of the smaller scale turbulences, i.e. smaller than the grid scale, are simulated indirectly by taking into account a sub-grid scale (SGS) model.

The Navier-Stokes equations consist of a continuity and momentum equations as described below:

$$\frac{\partial \overline{u_i}}{\partial x_i} = 0 \tag{1}$$

$$\frac{\partial \overline{u_i}}{\partial t} + \overline{u_j}\frac{\partial \overline{u_i}}{\partial x_j} = -\frac{1}{\rho}\frac{\partial \overline{p}}{\partial x_i} + \nu\frac{\partial^2 \overline{u_i}}{\partial x_j^2} - \frac{\partial}{\partial x_j}\left(\tau_{ij}\right) + g_i \tag{2}$$

where $\overline{u_i}$ is the mean (spatially averaged) velocity in the ith component, t is time, ρ is the fluid density, $\overline{p}$ is the mean dynamic pressure, g_i is gravitational acceleration, and τ_{ij} is the sub-grid scale (SGS) Reynolds stress:

$$\tau_{ij} = \overline{u_i^{'}u_j^{'}} = \left(\overline{u_i u_j} - \overline{u}_i\overline{u}_j\right) \tag{3}$$

The sub-grid scale Reynolds stress contains local averages of the small scale field, so the model for it should be based on the local velocity field or on the past history of the local fluid in order to obtain a closure. So far, several models for determining the SGS Reynolds stress have been proposed, such as Smagorinsky model (Smagorinsky, 1963), dynamic model (Germano, 1991), k-equation model (see Christensen, 2006), deconvolution model (Domaradzki and Saiki, 1997). Smagorinski (1963) was the first to postulate a model for the SGS stresses. At present, Smagorinsky model is the most widely used model, due to two main reasons: the model yields sufficient diffusion and dissipation to stabilize the numerical computation, and low-order statistics of the larger eddies are usually insensitive to the SGS motions. In the model, τ_{ij} is modeled by using an isotropic SGS eddy viscosity to present the momentum exchange.

$$\tau_{ij} = -2\nu_t S_{ij} \tag{4}$$

where ν_t is the eddy viscosity (Smagorinski eddy viscosity) and S_{ij} is the deformation tensor.

$$\nu_t = \left(C_s\Delta\right)^2\sqrt{2S_{ij}S_{ij}} \tag{5}$$

$$S_{ij} = \frac{1}{2}\left(\frac{\partial u_i}{\partial x_j} + \frac{\partial u_j}{\partial x_i}\right) \tag{6}$$

Δ is the length scale, which depends on the grid size. If the steps in the three coordinate directions are different, the overall scale of the SGS motion is:

$$\Delta = \sqrt[3]{\Delta x\Delta y\Delta z} \tag{7}$$

C_s is the Smagorinsky coefficient in the order of 0.07 to 0.21 (Schumann, 1987). $C_s = 0.1$ is used for the present simulations.

2.2. Boundary conditions

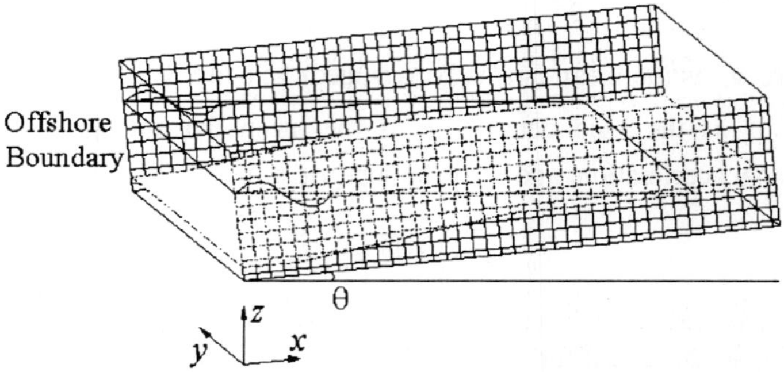

Fig.1. Computational domain and the 3-D grid system

For the 3-D, the boundary conditions are specified at the following boundaries:

- Offshore Boundary

The boundary conditions for the incoming wave at the offshore boundary are obtained by applying 3rd order cnoidal wave theory or 5th order Stokes wave theory, depending on the calculated Ursell parameter U_r (Nishimura et al., 1977).

$$U_r = \frac{HL^2}{h^3} \begin{cases} \geq 25\text{: cnoidal waves} \\ < 25\text{: Stokes waves} \end{cases} \tag{8}$$

where H is the incident wave height, L is the incident wave length, and h is the wave depth.

- Onshore Boundary

There is no out flow at the onshore boundary. An impermeable slope (breakwater) is used as the onshore limit of the computational domain.

- Bottom Boundary

Non-slip condition is applied at the bottom boundary, where the velocity components on this boundary are set to be zero and the pressure is approximated to be hydrostatic.

- Lateral Boundaries

Lateral boundaries in the three-dimensional models are treated in the same way as the bottom boundary. The main flow is assumed to be restricted between two parallel walls.

- Free Surface Boundary

Volume of Fluid (VOF) method is used to detect the free surface. In the computational domain, each cell is categorized as being a surface cell, full cell or empty cell. The density function f indicates 1 at any grid cell fully occupied by fluid and 0 if it contains no fluid. For simplicity, the free surface is assumed to be located on a surface where f is equal to 0.5. When f is greater or equal to 0.5, the cell is considered to be a water cell while it is an air cell otherwise. The status of each cell is checked and corrected at each time step by using the following equation:

$$\frac{\partial f}{\partial t}+u\frac{\partial f}{\partial x}+v\frac{\partial f}{\partial y}+w\frac{\partial f}{\partial z}=0 \tag{9}$$

The Cubic Interpolated Pseudo-particle (CIP, Yabe, 1991) method is used to update the VOF function f in time. The free surface is considered as a curve linking the partially filled cells. The velocity components in fictitious points outside the free surface boundary (if required) will be calculated to satisfy the mass conservation criteria. The irregular star method (Chan and Street, 1970) is used to solve the pressure at the grid cells neighboring the free surface.

2.3. Numerical procedures

Free surface is updated at the beginning of each time step. The pressure for this new free surface is computed next, using a Poisson equation, which is derived through the divergence form of Navier-Stokes equations and the mass conservation condition at the next time step. Successive Over-Relaxation (SOR) method is used for an iterative solution. For this new free surface and pressure, the velocities are computed by solving the governing equations. Since non-linear terms must be solved indirectly in order to maintain a high numerical accuracy, Cubic Interpolated Pseudoparticle (CIP) method is used to give an accurate solution with less numerical diffusion for a substantial differential equation. With this method, an indirect explicit way is carried out through both the non-advection and advection phases (Wijayaratna et al., 2000).

3. NUMERICAL RESULTS

The simulations of breaking wave computation are carried out with two specific cases: a central section of a detached breakwater (Case 1) and entire detached breakwater (Case 2). Regular wave generated at the offshore boundary moves along a sloping bottom to the inclined breakwater. The numerical set-up and incident wave conditions are shown in Table 1.

Table 1. Numerical set-up and incident wave conditions

Case	Type of Breakwater	Wave height (cm)	Wave period (s)	Water depth (cm)	Bottom slope	Breakwater slope	Computational domain (cm)
1	A central section	5.5	1.6	64	1/20	1/2	800 x 120 x 112
2	Detached breakwater	5.5	1.6	36	1/20	1/2	400 x 60 x 56

Figure 2 shows the computational results of flow velocity vectors for a breaking wave in a cross-shore vertical section at the middle of the wave domain in Case 1. The plan-view of the LES velocity field for this case (z = 44cm) is shown in Figure 3. The wave crest moves forward and increases in height due to the bottom slope. It achieves maximum energy and breaks just before attacking the breakwater.

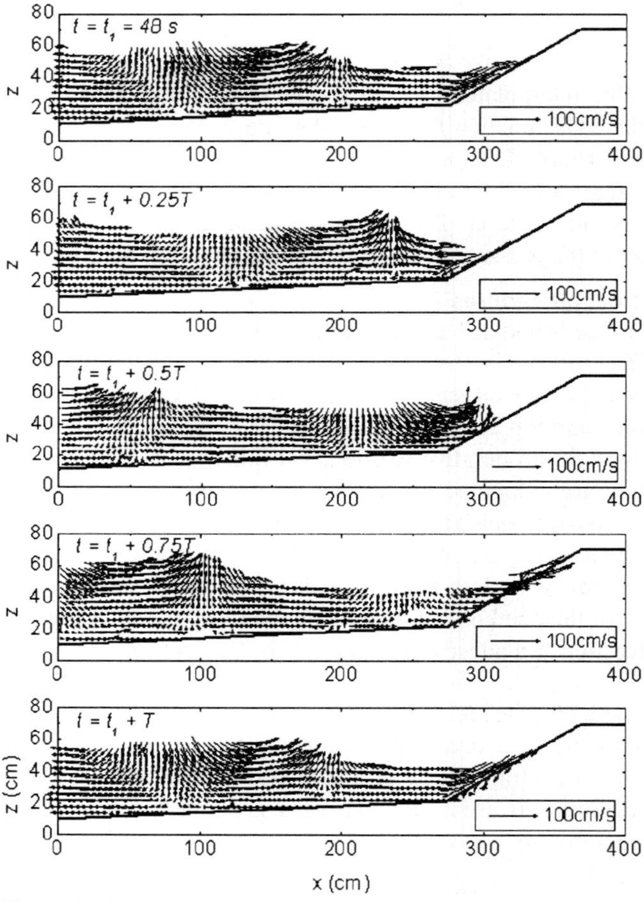

Fig. 2. Flow velocity vectors of wave breaking in a cross-shore vertical section (Case 1, y = 60cm)

In the breaking zone, the horizontal and vertical velocity components are relatively small before wave breaking. However, these components increase promptly just after wave breaking, especially in the horizontal direction. The horizontal velocity increases when the wave breaks but then gradually attenuates. The wave continues to run-up and then runs down along the surface of the breakwater, just before the following wave arrives.

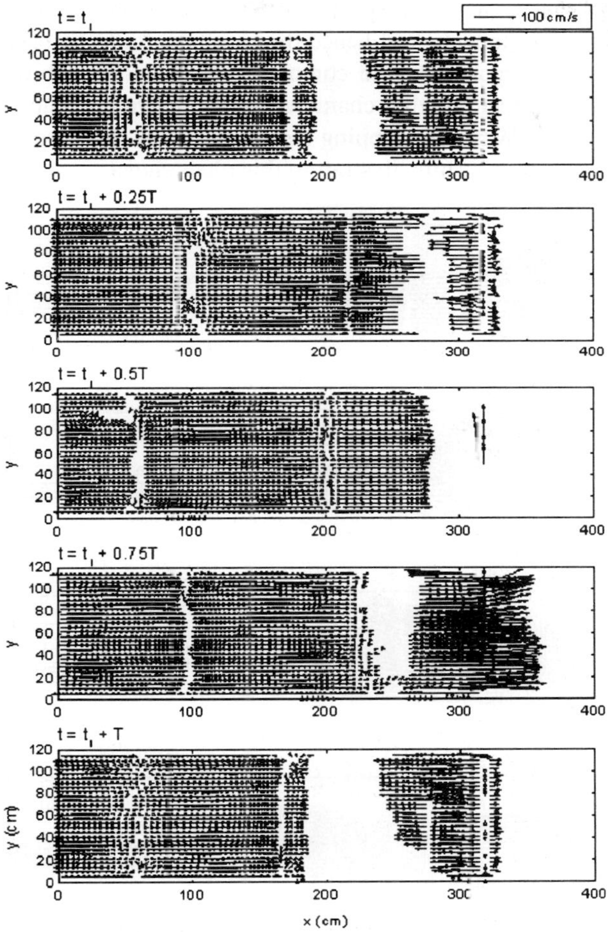

Fig. 3. LES velocity field on a horizontal section (Case 1, z = 44cm)

Figure 4 and Figure 5 show the flow velocity field of the interaction between wave and inclined breakwater at a cross-shore vertical section at the middle of the domain (y = 30cm) and at a horizontal section (z = 22cm). Main flow patterns of velocity vectors show the general trend of wave breaking, including turbulence areas around the breakwater, especially around the edges of the breakwater. When the wave crest arrives at the breakwater and breaks, velocity becomes very high. At that time, turbulences appear in the areas in front of it. As it is blocked by the structure, the wave is divided into two main flows to pass over the obstacle. Turbulences occur when the wave motion changes rapidly. When the wave pass over the structure, turbulences are very clear behind the breakwater. Even though experimental data is not available to verify the model and there are a few discrepancies in the post breaking zone, the main flow pattern and subsequent waves all follow the same trend proving the reliability of the model. And hence the results seem to be reasonable. This model is able to demonstrate the complex motion of wave breaking on the breakwater with most of the characteristics of hydrodynamic motion as observed in reality. Wave overtopping also occurred in this case. However, the images just show some results illustrating this phenomena.

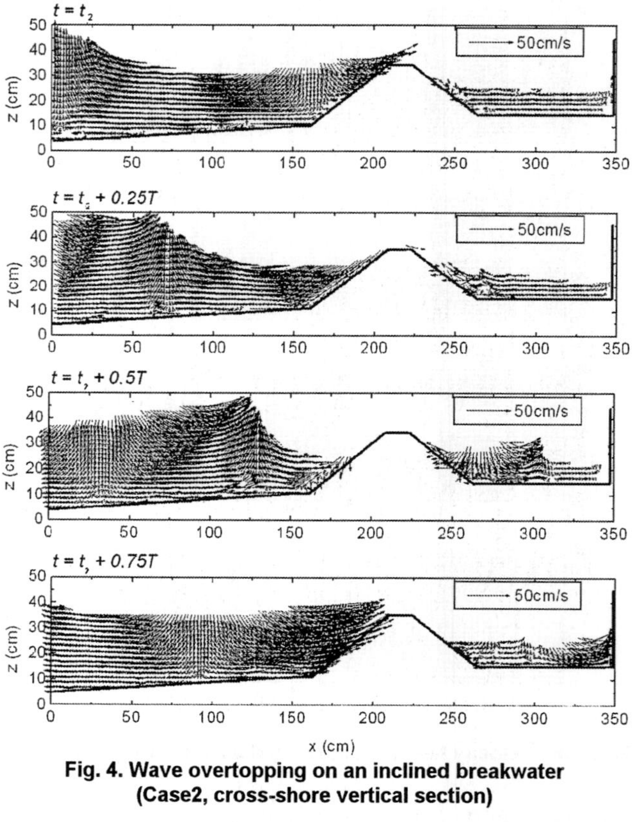

Fig. 4. Wave overtopping on an inclined breakwater (Case2, cross-shore vertical section)

Figure 6 and Figure 7 show the flow velocity field at a lateral cross section in front of (z = 160cm) and just behind the breakwater (x = 260cm). At the time $t = t_2 + 0.25T$, very little turbulence is present. Turbulences in lateral direction are weak in the front areas, corresponding to the case when the wave is far from the breakwater; but they are strong after passing the breakwater. The two panels indicating the time $t = t_2 + 0.5T$ show how the turbulence is generated by the breaker when it reaches the breakwater and how the turbulence is spreading over the vertical. The figure illustrates how the local variation in the turbulence can be affected clearly by the coastal structure.

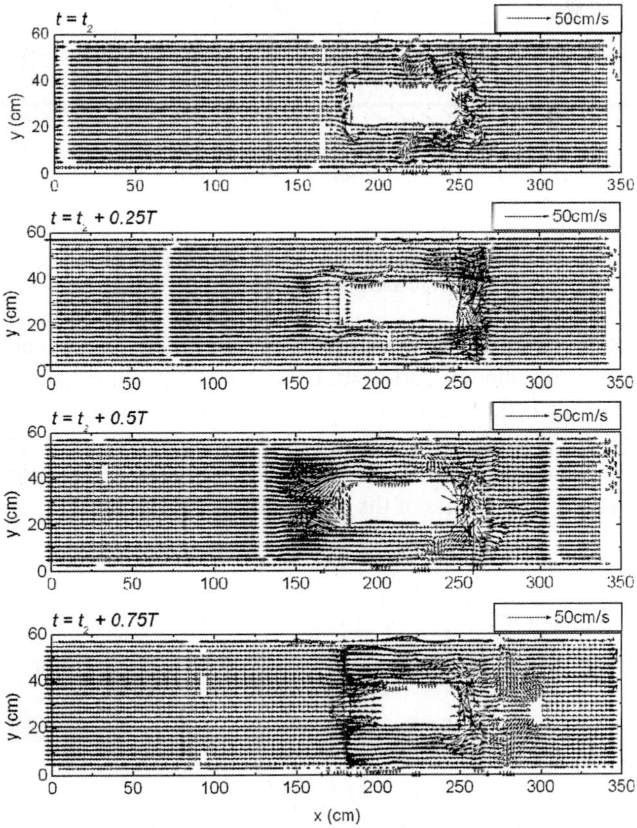

Fig. 5. Wave overtopping on an inclined breakwater (Case2, horizontal section, z = 22cm)

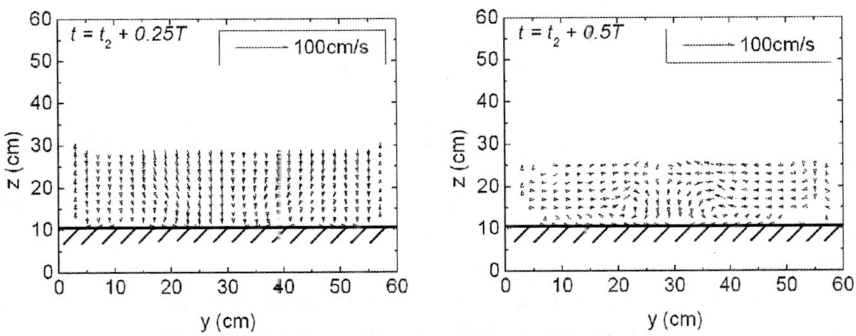

Fig. 6. A lateral cross section in front of the breakwater (Case 2, x = 160cm)

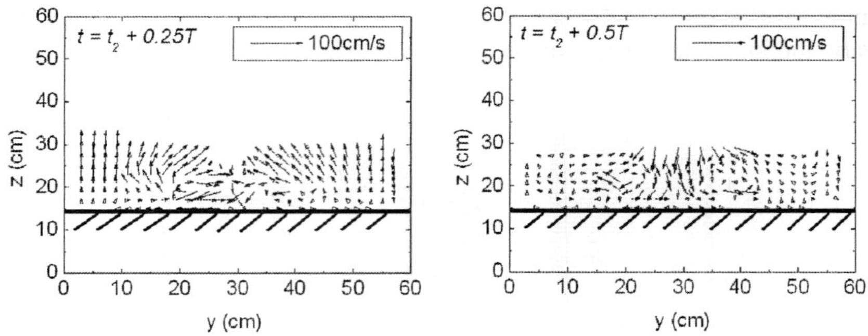

Fig. 7. A lateral cross section just behind the breakwater (Case 2, x = 260cm).

Figures 8, 9 and 10 show contour plots of the vorticity components about y, z and x axes due to temporal and spatial changes. They show that the vorticity generated in vertical, transverse as well as longitudinal axes is very strong at the edge of the breakwater. The reason is that the water flows move towards or pass over the structure and the change in direction will generate turbulent motion, influenced by the sides of the breakwater.

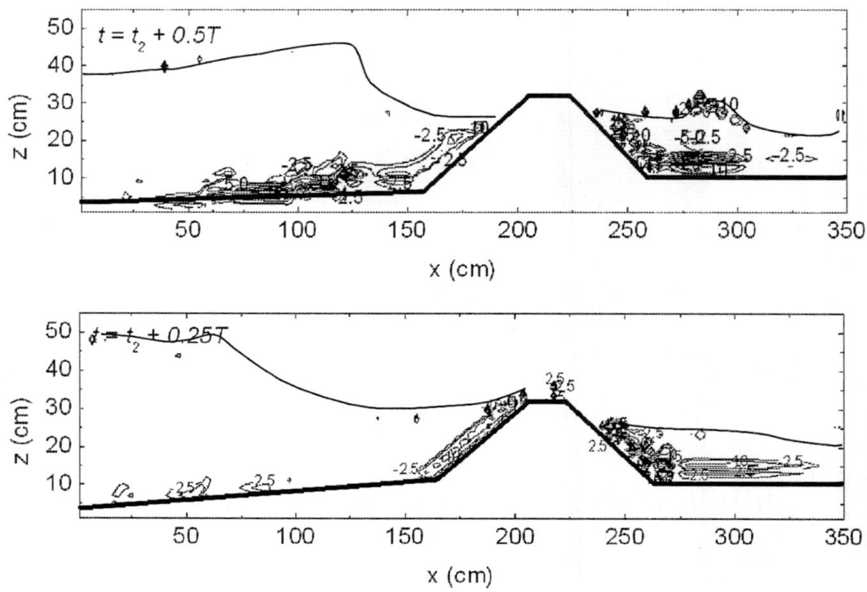

Fig. 8. Vorticity plot for a cross-shore vertical section (Case2, y = 30cm)

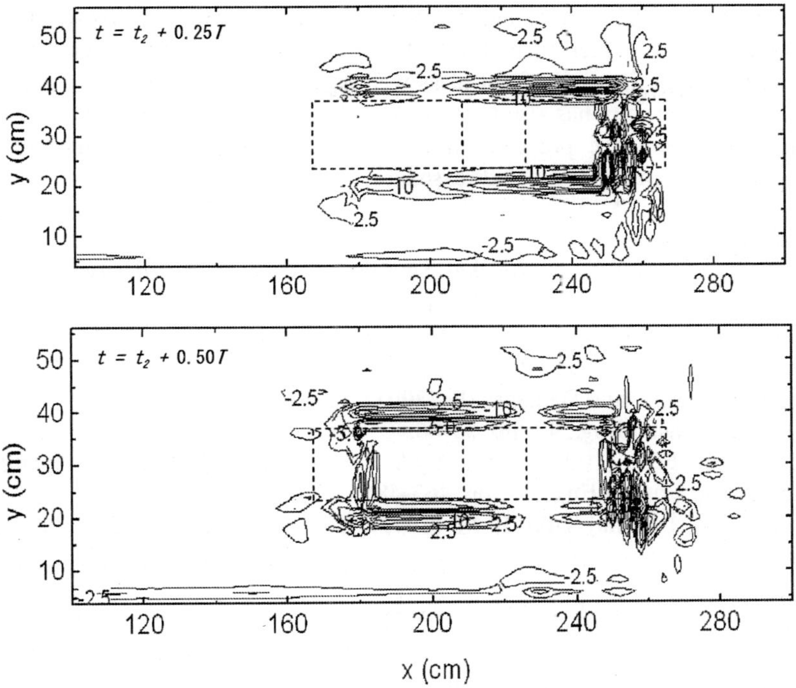

Fig. 9. Vorticity plot for a horizontal section (Case2, z = 22cm)

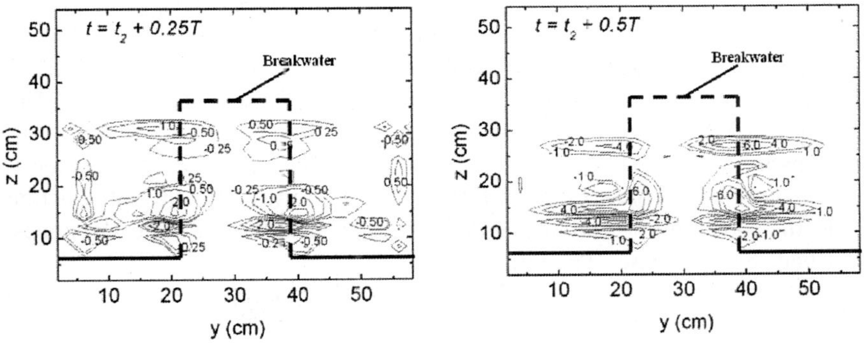

Fig. 10. Vorticity plot for a lateral cross section in front of the breakwater (Case 2, x = 160cm)

4. CONCLUSIONS

A 3-D LES model is used to investigate the complex fluid motions in the area near a coastal structure as well as wave overtopping on an inclined breakwater. Although laboratory and field measurement data are not available, the model is capable of simulating reasonably well some characteristics of wave motion, such as water surface elevations, water particle velocities, and the turbulences that appear in front of as well as around the coastal structure. The evolution of vorticity has also been studied. Through the numerical results, it is found that the wave processes and highly nonlinear characteristics of the breaking wave can be simulated reasonably.

5. REFERENCES

Bradford, S.F. (2000): *Numerical Simulation of Surf Zone Dynamics*, J. of Waterway, Port, Coastal and Ocean Eng., Vol. 126, No.1, pp. 1-13.

Chan, R. K. C. and Street, R. L. (1970): *A computer study of finite-amplitude water waves*, J. Comp. Phys., Vol. 6, pp. 68-94.

Christensen, E.D. and Deigaard, R. (2001): *Large Eddy Simulation of Breaking Waves*, Coastal Eng., 42, pp. 53-86.

Christensen, E.D. (2006): *Large Eddy Simulation of Spilling and Plunging Breakers*, Coastal Eng., 53, pp. 463-485.

Domaradzki, J.A. and Saiki, E. M. (1997): *A subgrid-scale model based on the estimation of unresolved scales of turbulence*, Phys. Fluids, Vol. 9, pp. 2148-2164.

Duy, N.T. and Shibayama, T. (1997): *A Convection-diffusion model for Suspended Sediment in the Surf Zone*, Journal of Geophysical Research, Vol. 102, No.C10, pp. 23,169-23,186.

Germano, M., Piomelli, U., Moin, P., Cabot, W.H. (1991): *A dynamic subgrid-scale eddy viscosity model*, Phys. Fluids A, Vol. 3, pp. 1760-1765.

Lin, P. and Liu, P.L.-F. (1998): *A Numerical Study of Breaking Waves in the Surf Zone*, J. Fluid Mech. Vol. 359, pp. 239-264.

Nishimura, H., M. Isobe and K. Horikawa (1977): *Higher Order Solutions of the Stokes and the Cnoidal Waves*, J. Fac. Eng., Univ. of Tokyo, B-34, No. 2, pp. 267-293.

Schumann, U. (1987): *Direct and Large Eddy Simulation of Turbulence – Summary of the State of the Art 1997*, Lecture series 1987-06, Introduction to the Modeling of Turbulence, Von Karman Institute for Fluid Dynamics, Rhode Saint Genese, Belgium, pp. 1-36.

Smagorinsky, J. (1963): *General Circulation Experiments with the Primitive Equations*, Mon. Weath. Rev., Vol. 91, No. 3, pp. 99-164.

Suzuki, T. and Okayasu, A. (2004): *Numerical Simulation of 3-D Intermittent Sediment Suspension in the Surf Zone*, Proc. 29th Int. Conf. Coastal Eng., ICCE, Vol. 2, pp. 1793-1804.

Thao N.D., Takagi, H., Shibayama, T. and Okayasu, A. (2006): *3-D Large Eddy Simulation of Breaking Wave Motions around a Slope-Type Breakwater*, Annual Journal of Coastal Eng., JSCE, Vol. 53, pp.751-755 *(in Japanese)*.

Watanabe, Y. and Saeki, H. (1999): *Three Dimensional Large Eddy Simulation of Breaking Waves*, J. of Coastal Eng., JSCE, Vol. 41, pp 281-301.

Wijayaratna, N. and Okayasu, A. (2000): *DNS of Wave Transformation, Breaking and Run-up on Sloping Beds*, Proc. 4th Int. Conf. on Hydrodynamics, Vol. 2, Yokohama, Japan, pp. 527-532.

Yabe, T. and Aoki, A. (1991): *A Universal Solver for Hyperbolic Equations by Cubic-Polynomial Interpolation I. One-dimensional solver*, Comp. Phys. Comm., Vol. 66, pp.219-232.

SPH-LES TWO-PHASE SIMULATION OF WAVE BREAKING AND WAVE-STRUCTURE INTERACTION

Giovanni Cuomo[1], Andrea Panizzo[2] & Robert Dalrymple[3]

This paper presents a Smoothed Particle Hydrodynamics (SPH) model for the simulation of two-phase flows. The model solves the Navier-Stokes Equations for compressible fluids and is able to handle fluids with high density and compressibility ratio such in the air-water case. Viscosity in the fluids is evaluated using a large eddy simulation approach. The model is applied to the simulation of some example cases including dam break, wave breaking and wave-structure interaction.

MODELLING COMPRESSIBLE TWO PHASE FLOWS

The modeling of wave breaking at shores represents one of the most fascinating challenges to date for numerical experimentalists, as efficiently describing the complex phenomenology involved with the dramatic modification of waves at breaking requires numerical models that are able to follow the development of the wave field and to correctly represent the physics of the overall process. Similarly, when waves break curling against a seawall or slam underneath a jetty deck, trapping air between the water surface and the structure, correctly describing the nature of the flow requires modeling of several complex phenomena, including compressibility of air-water mixture.

Several numerical techniques have been proposed to model complex and distorted free surface flows, following both Eulerian and Lagrangian approaches. Among the latter, Smoothed Particle Hydrodynamics (SPH) belongs to the so-called mesh-free methods and is therefore able to deal naturally with multiply connected computational domains. During the last decade SPH has been demonstrated to be able to deal with the description of a large number of hydrodynamic systems, including free surface flows (Monaghan, 1994; Dalrymple et al., 2001), impact generated waves (Monaghan & Kos, 2000), impacts of water waves with structures (Gesteira & Dalrymple, 2004), underwater landslide generated waves (Panizzo & Dalrymple, 2004) and breaking waves (Gotoh & Sakai, 1999, Rogers & Dalrymple 2004).

Works by Gotoh et al (2001), Lo & Shao (2002) and Shao & Gotoh (2004) demonstrated the ability of SPH in efficiently modelling turbulent flows by mean of Large Eddy Simulation based on a Subparticle Scaling (SPS). Rogers & Dalrymple (2005) have recently presented an application of the SPS to a compressible SPH model for the simulation of breaking waves.

---

[1] University of Roma TRE, V. Volterra 62, 00146, Roma, Italy, email: cuomo@uniroma3.it
[2] University of L'Aquila, Monteluco di Roio, 67040 ,AQ, Italy, email: panizzo@ing.univaq.it
[3] Johns Hopkins University, Baltimore, 21211, MD, USA, email: rad@jhu.edu

Figure 1. Aerated wave impact against a breakwater and curling wave breaking

Air entrapment plays an important role in the overall process. Relative importance of aeration on wave loading is confirmed by experimental evidences (Obray et al. 2004, Cuomo 2005) and field measurements (Bullock et al. 2004). Interest in numerical modeling of compressible flows has significantly increased in the last decade. Recently Peregrine et al. (2004, 2006) extended previous work by Peregrine and Thais (1996) to model compressible flows by means of a mixture of incompressible water with an ideal gas.

The most significant attempt to model two-phase flows with SPH is that of Colagrossi & Landrini (2003). The model succeeds in describing the overall features of the flow, but is not able to account for relative compressibility of the two phases. This paper presents a new SPH formulation for two-phase turbulent flows (Cuomo, 2005) that allows accounting for the effective compressibility of both fluid and gas phase.

SPH FORMULATION FOR TWO-PHASE FLOWS

In the Smoothed Particle Hydrodynamics (SPH) formulation, the fluid is represented as a series of elementary particles whose kinematics and pressure evolve as a function of the corresponding properties of their neighboring particles. The relative influence of each particle on the evolution of its neighboring elements is a function of the inter-particle distance by means of a weighting function or "kernel" W(x) so that:

$$\int_{\Omega} A(\mathbf{x}_j) \cdot W(\mathbf{x}_i - \mathbf{x}_j, h) dx = A(\mathbf{x}_i) \tag{1}$$

where A is the generic field variable. Other than the one expressed by Equation 1, the kernel function (Fig. 2) has the following properties:

$$\begin{aligned} &\int_{\Omega} W(\mathbf{x}_i - \mathbf{x}_j, h) dx = 1 \\ &\lim_{h \to 0} W(\mathbf{x}_i - \mathbf{x}_j, h) = \delta(\mathbf{x}_i - \mathbf{x}_j) \end{aligned} \tag{2}$$

where δ is the Dirac delta, *h* is the maximum interparticle distance within which each particle feels the influence of its neighbors and Ω is a spatial domain including the i-th particle and having dimensions O(h). The value of h is theoretically infinite, but since the relative importance of particles on each other motion rapidly decreases with increasing interparticle distance, h is usually less then 3 times the initial particle spacing, depending on the shape of $W(\mathbf{x}_i-\mathbf{x}_j,h)$ (W_{ij} hereinafter).

In the simulations presented in this paper, we use the expression for W_{ij} first suggested by Johnson and Beissel (1996), that is:

$$W(\mathbf{x}_i-\mathbf{x}_j,h)=\frac{3}{2\pi h^2}\begin{cases}\frac{1}{4}\left(\frac{\mathbf{x}_i-\mathbf{x}_j}{h}\right)^2+\left(\frac{\mathbf{x}_i-\mathbf{x}_j}{h}\right)+1 & 0\le\frac{\mathbf{x}_i-\mathbf{x}_j}{h}\le 2\\ 0 & \frac{\mathbf{x}_i-\mathbf{x}_j}{h}>2\end{cases} \tag{3}$$

The main advantage of introducing the kernel function is given by the simplicity with which the differential operators can be expressed as a function of the spatial derivatives of the kernel function only as follows:

$$\nabla\cdot\mathbf{u}_i=\sum_j(\mathbf{u}_j-\mathbf{u}_i)\cdot\nabla W_{ij}\frac{m_j}{\rho_j} \tag{4}$$

$$\nabla\Phi_i=\sum_j(\Phi_j-\Phi_i)\nabla W_{ij}\frac{m_j}{\rho_j} \tag{5}$$

This leads to the following "canonical" expression for the evolution equations of the i-th fluid particle:

$$\frac{d\rho_i}{dt}=-\rho_i\cdot\sum_j(\mathbf{u}_j-\mathbf{u}_i)\cdot\nabla W_{ij}\frac{m_j}{\rho_i} \tag{6}$$

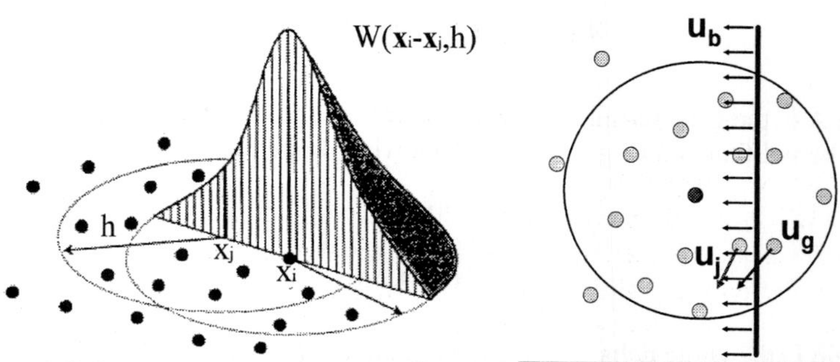

Figure 2 Schematic representation of kernel function (left) and imposition of fully reflective boundary conditions by means of ghost particles (right)

$$\frac{d\mathbf{u}_i}{dt} = -\frac{1}{\rho_i} \cdot \sum_j (p_j + p_i + \Pi_{ij}) \nabla W_{ij} \frac{m_j}{\rho_i} + \mathbf{f}_i \tag{7}$$

$$\frac{d\mathbf{x}_i}{dt} = \mathbf{u}_i \tag{8}$$

in Equations 6-8: t is time, ρ_i is the density of the i-th particle at position $\mathbf{x}_i$, $\mathbf{u}_i$ its velocity vector, m_i its mass, p_i its pressure, and $\mathbf{f}_i$ and Π_{ij} are respectively the resulting body forces vector felt by the particle and an artificial viscosity term (Monaghan, 1994).

Equations 6 and 7 are coupled by means of the following state equation (Batchelor, 1967):

$$p(\rho) = P_0 \left[\left(\frac{\rho}{\rho_0} \right)^{\gamma} - 1 \right] \tag{9}$$

In which ρ_0 is the reference density of the fluid, γ is its polytropic index ($\gamma = \gamma_a = 1.4$ in air and $\gamma = \gamma_w = 7$ in water) and P_0 is function of the speed of sound (c_s) in the fluid so that:

$$c_s = \sqrt{\gamma \cdot P_0 / \rho_0} \tag{10}$$

In order to both improve the stability of the model and deal with the simulation of interfacial flows of fluids characterised by small density ratio, the standard SPH formulation for divergence and gradient operator has been rewritten in the form (Colagrossi & Landrini, 2003):

$$\nabla \cdot \mathbf{u}_i = \sum_j (\mathbf{u}_j - \mathbf{u}_i) \cdot \nabla W_{ij} \frac{m_j}{\rho_j} \tag{11}$$

$$\nabla \Phi_i = \sum_j (\Phi_j - \Phi_i) \nabla W_{ij} \frac{m_j}{\rho_j} \tag{12}$$

in which m_j/ρ_j is used instead of m_j/ρ_i.

Boundary conditions

Reflective boundary conditions are imposed on solid boundaries by means of ghost particles generated as the fluid particle approaches the boundary (Fig 2). For each i-th particle, ghost particles are generated by mirroring (with respect to the solid boundary) fluid particles in Ω_i that are close enough to the boundary for their ghost to fall within Ω_i. Ghost particle have positions $\mathbf{x}$, velocity $\mathbf{u}$ and pressure p given by:

$$\begin{aligned} \mathbf{x}_g &= \mathbf{x}_b + \mathbf{x}_{ij} & \mathbf{x}_{ij} &= \mathbf{x}_i - \mathbf{x}_j \\ \mathbf{u}_g &= 2 \cdot \mathbf{u}_b + \mathbf{u}_{ij} & \mathbf{u}_{ij} &= \mathbf{u}_i - \mathbf{u}_j \\ p_g &= p_j \end{aligned} \tag{13}$$

Using Equations 13 ensure satisfaction of free slip-boundary conditions.

To speed up the computation, each particle carries a list with the information of each neighboring particle (Allen and Tildesley, 1987). In this approach, ghost particles only exist within the list of the particle they belong to.

Density reinizialization

To ensure exactly the consistency between mass, density and occupied area we re-initialize density by means of the following:

$$\rho = \sum_j m_j \cdot W_{ij}^1 \tag{14}$$

where W_{ij}^1 is the normalized kernel at first order (Belytchko et al., 1998) which can be written as:

$$W_{ij}^1 = \left[\beta_0(\mathbf{x}_i) + \beta_1(\mathbf{x}_i)\cdot(x_i - x_j) + \beta_2(\mathbf{x}_i)\cdot(y_i - y_j)\right]\cdot W_{ij} \tag{15}$$

Coefficients β_0, β_1 and β_2 are defined as:

$$\begin{pmatrix} \beta_0 \\ \beta_1 \\ \beta_2 \end{pmatrix} = \left(\sum_j W_{ij} \cdot \begin{bmatrix} 1 & x_i - x_j & y_i - y_j \\ x_i - x_j & (x_i - x_j)^2 & (x_i - x_j)\cdot(y_i - y_j) \\ y_i - y_j & (x_i - x_j)\cdot(y_i - y_j) & (y_i - y_j)^2 \end{bmatrix} \right)^{-1} \cdot \begin{pmatrix} 1 \\ 0 \\ 0 \end{pmatrix} \tag{16}$$

Using a normalized kernel function ensures mass conservation close to boundary and at the interface between fluids with a high density ratio.

Two phase flows: compressibility and interface sharpness

In simulations involving fluids with high density ratio, as in the case of air and water (ρ_a/ρ_w = 1/1000) a series of instabilities will rise at the interface. To improve the stability of the model, Colagrossi and Landrini (2003) enforce $P_{0w} = P_{0a}$ in Equation 9. For Equation 10, this involves: $c_{sw} = \sqrt{1.4/1} \cdot \sqrt{P_0}$ and $c_{sw} = \sqrt{7/1000} \cdot \sqrt{P_0}$, leading to, $c_{sa} \cong 14 \cdot c_{sw}$ or, which is the same, to air being less compressible than water.

Since the main reason for this research is to investigate effect of air entrapment at breaking and during violent wave-structure interaction, we need to correctly represent the relative compressibility of both fluids as well as that of the air-water mixture. This has been achieved here by removing such constraint. Results shown in this paper have been obtained assuming $c_{sa} = c_{sw}$, (that is

$P_{0w} \cong 200 \cdot P_{0a}$) which ensures water to be less compressible then air; further new formulation allows the use of any compressibility ratio.

To improve interface sharpness, the following inter-particle force has been introduced:

$$I_{ij} = \alpha \left[\left(\frac{\Delta}{\mathbf{x}_{ij}} \right)^{b_1} - \left(\frac{\Delta}{\mathbf{x}_{ij}} \right)^{b2} \right] \tag{17}$$

where $b_1 > b_2$, α and Δ are empirical parameters. Equation 17 describes a short-range repulsive, long-range attractive inter-particle force similar to that given by Lennard-Jones potential. I_{ij} only acts between particles of different densities and vanishes quickly away from the interface.

SPS turbulence method for compressible fluids

The model solves the Navier-Stokes equations for compressible fluids in which the Reynolds stress is formulated by an eddy viscosity model by means of the SPS stress tensor $\boldsymbol{\tau}$ with elements given by:

$$\frac{\tau_{ij}}{\rho} = (\overline{u}_i \overline{u}_j - \overline{u_i u_j}) = 2\nu_t \cdot \overline{S}_{ij} - \frac{2}{3} k \delta_{ij} \tag{18}$$

where k is the SPS turbulent kinetic energy, δ_{ij} is the Kronecker's delta, $\boldsymbol{S}$ is the SPS strain-rate tensor with elements given by:

$$\overline{S}_{ij} = \frac{1}{2} \left(\frac{\partial u_i}{\partial x_j} + \frac{\partial u_j}{\partial x_i} \right) \tag{19}$$

and ν_t is the turbulence eddy viscosity, modelled here according to what proposed by Smagorinsky (1963), in figures:

$$\nu_{ts} = (C_s \cdot \Delta l)^2 \cdot \left| \overline{S}_{ij} \right| \tag{20}$$

where C_s is the Smagorinsky constant and Δl is the particle-particle spacing.

TWO PHASE DAM-BREAK

In this section we present the application of the model to some example two-phase flows and discuss some key features of the new formulation.

The first case study is the numerical simulation of the dam-break flow described in Buchner and van Ballegoyen (1997). The problem geometry is sketched in Fig. 3. The water, initially at rest, evolves freely on a dry bottom until it reaches a vertical wall, water flows up the wall until the motion inverts downwards and the water tongue collapses backwards trapping an air pocket.

For this simulation we use N_w = 1152 water particles and N_a = 9770 air particles. The evolution of the system is reproduced in Fig 4, showing pressure contour plots. Super-imposed quiver plots represent the velocity field.

Energy conservation

The total energy of the system can be expressed as the sum of the kinetic, potential and internal energy of each particle:

$$E_{TOT} = \sum_i \rho_i \cdot \left(\frac{|\mathbf{u}_i|^2}{2} + \mathbf{g} \cdot y_i + e_i \right) \cdot dV \tag{21}$$

where **g** is the gravitational acceleration and *e* is the specific internal energy given by:

$$\frac{De_i}{Dt} = -\frac{p}{\rho} \cdot div\mathbf{u}_i \tag{22}$$

The evolution in time of the energy of the system is plotted on the right hand side of Fig. 3, showing very little energy loss during simulation.

Air and water compressibility

It might be worth noticing a fundamental difference between this model and that of Colagrossi and Landrini (2003). Details of frames from simulation are given in Fig. 5, showing different pressure contours in air and water at instant when water tongue closes the loop trapping the air bubble. When compared to results by Colagrossi and Landrini (2003), the new formulation appears to give a more realistic description of pressure distribution within the two fluids, with little compressible water particles having pressure significantly higher then the more compressible air particles, in agreement with continuum mechanics.

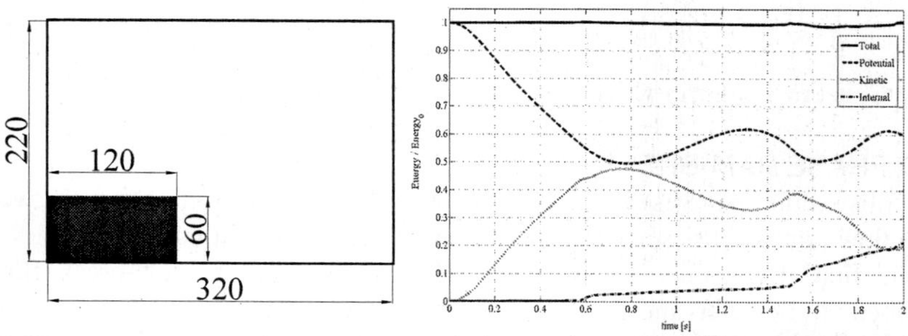

Figure 3 Schematic representation of dam-break problem (left, all dimensions [cm]) and evolution in time of system energy (right)

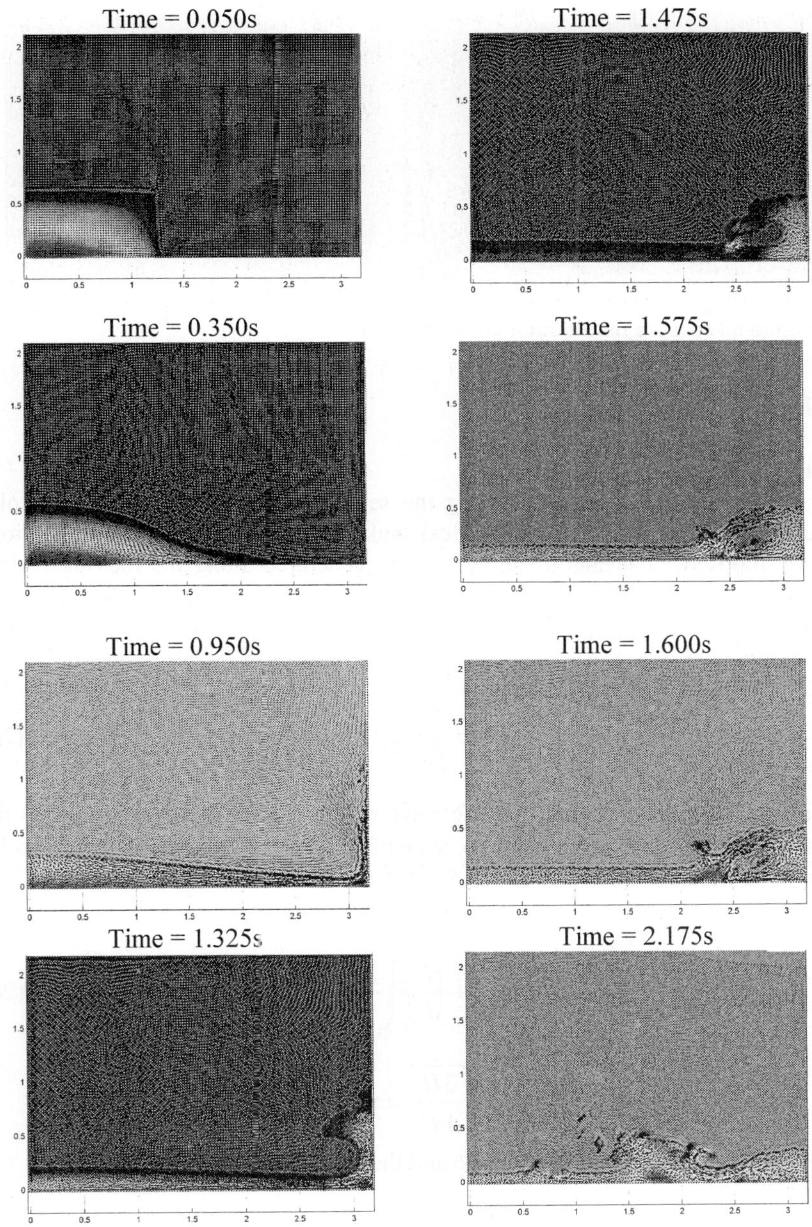

Figure 4 Snapshots from dam-break flow simulation, frames ordered from top to bottom, left to right. All dimensions in meters.

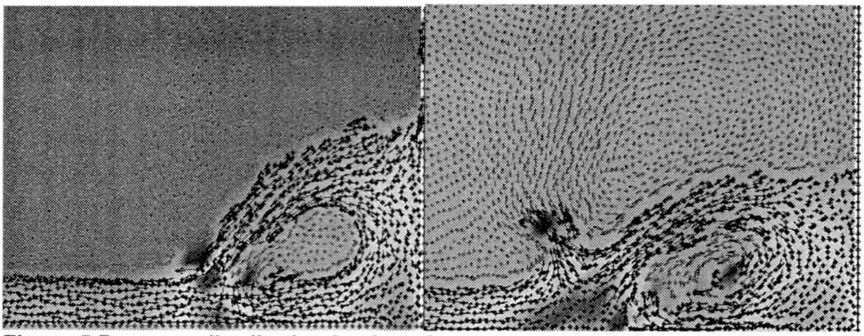

Figure 5 Pressure distribution in air and water

NUMERICAL WAVE TANK

Wave generation

Waves are generated by moving the set of particles representing the solid boundary at one end of the numerical tank, the effective boundary conditions being imposed by means of ghost particle generated along the moving paddle according to Equations 13. The same approach has been used to generate both solitary and sinusoidal periodic waves. In general, the position ξ of the paddle is given by:

$$\xi(t) = \frac{H}{kd} \cdot \int_0^{\bar{\theta}} f(\theta) d\theta \tag{23}$$

where H is the wave height, k is the wave number at water depth d and θ is the phase angle. Given a general $f(\theta)$, equation 23 is an implicit equation to be solved numerically.

For solitary wave $f(\theta) = sech^2\theta$ so that at i-th time instant:

$$\xi(t^i) = \frac{H}{kd} \cdot \left\{ 1 + \tanh\left[\left(3.8 + \frac{H}{d}\right) \cdot \left(\frac{2t^i}{\tau} - 1\right) - \left(k \cdot \xi(t^{i-1}) - \frac{H}{d}\right) \right] \right\} \tag{24}$$

where $\tau = \frac{2}{kc}\left(3.8 + \frac{H}{d}\right)$, $k = \sqrt{\frac{3H}{4d^3}}$ and $c = \sqrt{g(d+H)}$.

For sinusoidal waves $f(\theta) = \sin\theta$ and thus:

$$\xi(t^i) = c \cdot t^i - \theta^i / k \tag{25}$$

where $c = L/T$ is the wave celerity for a wave of period T at water depth d (according to linear theory) and:

$$\theta^i = \theta^{i-1} - \frac{\theta^{i-1} - kc\cdot t + \dfrac{H}{2d}\left(1-\cos\theta^{i-1}\right)}{1+\dfrac{H}{2d}\sin\theta^{i-1}} \tag{26}$$

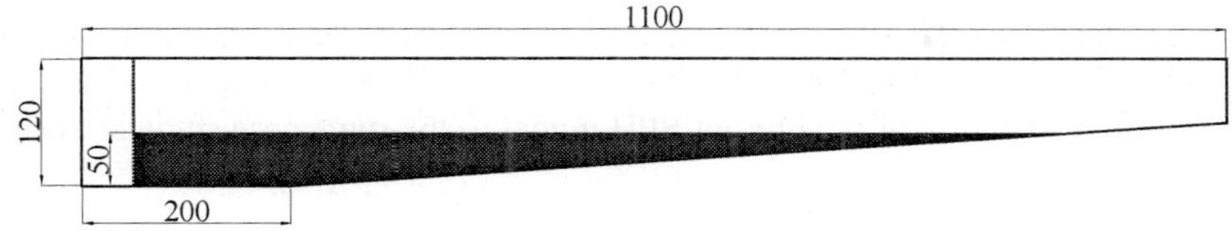

Figure 6 Numerical wave tank used in the simulation of solitary wave breaking

Wave breaking on a shore

The model is now applied to the simulation of wave breaking on a plane beach (Fig. 6). Both solitary and periodic waves are used in the experiments. Waves are generated at one end of the numerical wave tank by means of a piston type wave-maker and forced to break on a 1:15 foreshore at the other end of the tank. This case study refers to physical model tests by Li and Raichlen (2003), water depth at the paddle is h = 0.5m and H/h = 0.4.

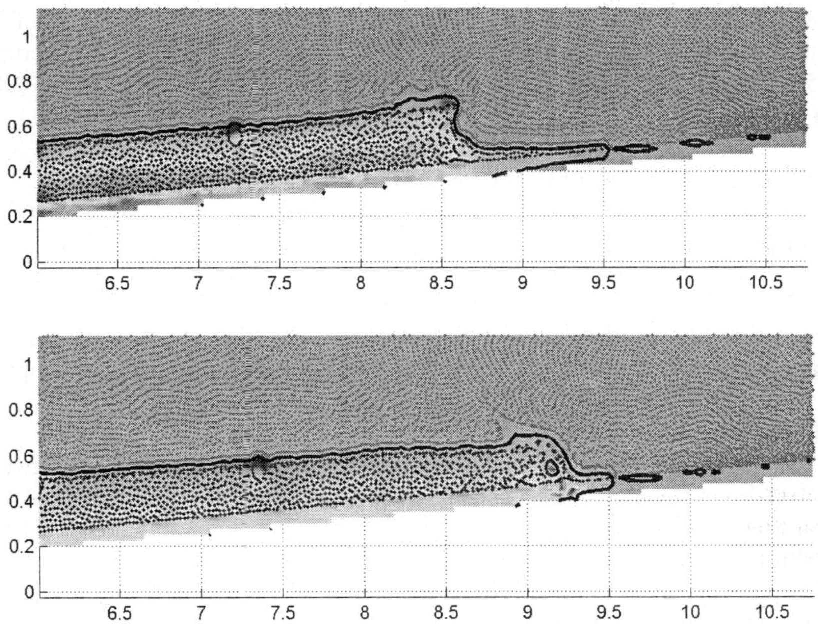

Figure 7 Snapshots from simulation of solitary wave breaking. Incipient breaking (top) and plunging (bottom), all dimensions in meters.

For this simulation we use N_w = 3600 water particles and N_a = 9960 air particles. Two snapshots from the simulation are shown in Fig. 7, showing pressure contours in the two phases. The wave progressively increases its steepness up to reach breaking condition, then curls trapping an air pocket.

Wave-impact on a vertical wall

Takahashi et al. (1983) performed physical experiments on impulsive wave forces on caisson breakwater with step (1:10) foreshore. Here, qualitative results from the application of the present SPH model to the same case study are briefly

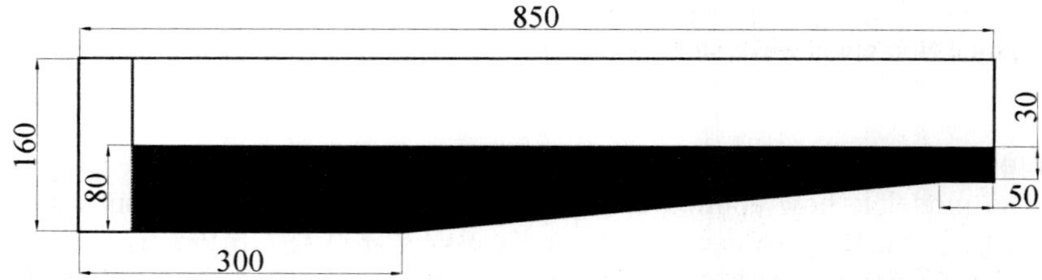

Figure 8 Numerical wave tank used to simulate wave impacts on a vertical wall

presented.

The numerical wave tank used in this simulation is represented in Fig. 8, regular sinusoidal waves (H=0.3m, T=3.04s) are generated at a water depth of 0.8m and forced to break in front of a vertical wall placed at the other end of the flume. The water depth in front of the wall is 0.3m and the length of the berm is 0.5m. For this simulation the we use N_w = 8480 water particles and N_a = 10880 air particles.

CONCLUSIONS AND FURTHER WORK

A new two-phase SPH model has been presented that is able to simulate complex flows involving high density and comressibility ratio. The model has been successflully applied to the simulation of complex flows including dam break and wave breaking. The model allows accounting for relative compressibility of the liquid and gas phase, resulting in a more realistic representation of the flows and significant improvement over previous attempts.

ACKNOWLEDGEMENTS

This work presents part of the results from a joined research project between University of Rome, L'Aquila and Johns Hopkins. The authors wish to thank Dr. Ben Rogers, Muthukumar Narayanaswamy and Shan Zou for extremely kind and useful collaboration. First author studentship is funded by Japanese Society for Promotion of Science through its Post-Doctoral fellowship program. Dr. Shigeo Takahashi is grateful acknowledged for many useful discussion and for continuously encouraging this research.

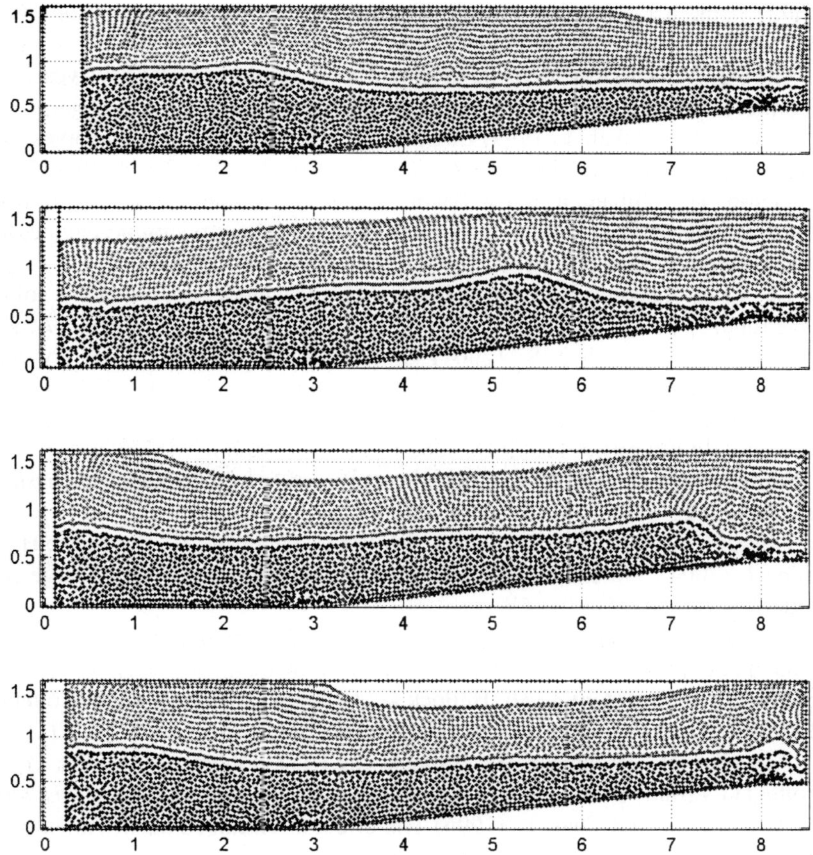

Figure 9 Snapshots from simulation of wave impacts on vertical wall. From top to bottom: Wave generation, shoaling, breaking and air entrapment at impact. All dimensions in meters.

REFERENCES

Allen M. and Tildesley D.(1987) *Computer Simulations of Liquids*, Claredon Press, Oxford

Belytchko, T., Y. Krongauz, J. Dolbow, and C. Gerlach. (1998) On the completeness of meshfree particle methods. *Int. J. Num. Meth. Eng.* 43,785-819.

Buchner, B., and van Ballegoyen, G., (1997) FPSO Green Water Loading, Vol. A1, Discussion Report, *MARIN Report* No. 1.13644, Dec.

Colagrossi & Landrini (2003) Numerical simulation of interfacial flows by smoothed particle hydrodynamics. *J. of Comp. Phys.* 191 (2003) 448–475

Cuomo (2005) "Dynamics of wave-induced loads and their effects on coastal structures", PhD Thesis.

Dalrymple, R. A., Knio, O., Cox, D. T., Gomez-Gesteira, M. & Zou, S.

(2002) Using a Lagrangian particle method for deck overtopping. *Proc. Waves 2001*, ASCE.

Goring, Y. and Raichlen, F. (1980) The generation of long waves in the laboratory. *Proc. of 17th ICCE*, ASCE, New York, 763-783.

Gotoh, H. & Sakai, T.(1999) Lagrangian simulation of breaking waves using particle method. *Coastal Engineering Journal*, Vol. 41, Nos. 3 & 4, 303-326

Gotoh, H., Shibahara, T. & Sakai, T.(2001) Sub-particle-scale turbulence model for the MPS method – lagrangian flow model for hydraulic engineering. *Computational Fluid Dynamics Journal*, Vol. 9, No. 4, 339-347

Li, Y. & Raichlen, F. (2003) Energy balance model for breaking solitary wave runup. *Journal of Waterw., Port, Coastal and Oc. Eng.* Mar/Apr, 47-59

Lo, E.Y.M. & Shao S. (2001) Simulation of near-shore solitary wave mechanics by an incompressible SPH method *Appl. Oc. Res.*, 24, 275-286.

Monaghan, J. J. (1994). Simulating free surface flows with SPH *Journal Computational Physics*, 110, 399- 406.

Monaghan, J. J. & Kos, A. (2000) Scott Russell's wave generator. *Physics of Fluids*, 12, 622-630.

Obhrai C., Bullock G., Wolters G., Muller G. Peregrine H., Bredmose H. and Grune J. (2004) Violent wave impact on vertical and inclined walls: large scale model tests. *Proc. of 29th ICCE.*

Panizzo A. & Dalrymple, R. A. (2004). SPH modelling of underwater landslide generated waves. *Proc. of 29th ICCE*, ASCE.

Peregrine H., Bredmose H., Bullock G., Obhrai C., Muller G. and Wolters G. (2004) Water wave impact on walls and the role of air. *Proc. of 29th ICCE.*

Peregrine H., Bredmose H., Bullock G., Hunt, Obhrai C., Muller G. and Wolters G. (2006) Water wave impact on walls and the role of air. (*in this book)*

Peregrine H. and Thais L. (1996) The effect of entrained air in violent water wave impacts, *J. Fluid Mech.* 325 pp.377-397.

Rogers, B. & Dalrymple, R. A. (2004) Three-dimensional modelling of wave breaking *Proc. of 29th ICCE*, ASCE.

Rogers, B. & Dalrymple, R. A. (2005) Three dimensional SPH-SPS modelling of wave breaking *Proc. of WAVES 2005 Conference*, Madrid.

Shao, S & Gotoh, H. (2004) Simulating coupled motion of progressive wave and floating curtain wall by SPH-LES model. *Coastal Engineering Journal*, Vol. 46, No. 2, 171-202

Smagorinsky J. (1963) General circulation experiment with the primitive equations. I. The basic experiment. *Mon. Weath. Rev.*, 91, 99-164

Takahashi Shigeo, Tanimoto Katsutoshi and Suzumura Satoshi (1983) Generation Mechanism of Impulsive Pressure by Breaking Wave on a Vertical Wall, *PARI Internal Report*, Vol.22 No.4 1983.12 (in Japanese)

NONLINEAR WAVE INTERACTION WITH SUBMERGED AND SURFACE-PIERCING POROUS STRUCTURES

Okey Nwogu[1] and Zeki Demirbilek[2]

A coupled Boussinesq-boundary integral method is developed to simulate nonlinear water wave interaction with structures consisting of multiple layers with different physical and hydraulic characteristics. The flow field in the water region is modeled with a modified set of Boussinesq-type equations, with additional terms to account for the flow of water into/out of the porous region. The equations of motion for the porous regions include an empirical Forchheimer-type term for laminar and turbulent frictional losses, and an inertial term for acceleration effects. A boundary integral formulation based on Green's third identity is used to close the problem for the porous region. The coupled equations for the evolution of the free surface and boundary values of the tangential velocities are integrated in time using an iterative Crank-Nicolson scheme. At each time step, the Boussinesq problem is solved for the water region to determine the pressure at porous interface. The boundary integral problem for the porous region is then solved to determine the normal velocities along porous interface. The model is used to investigate wave interaction with a vertical surface-piercing porous structure and wave transmission over submerged breakwaters. Comparisons between the numerical model predictions and laboratory data show generally good agreement for both the wave field inside the structure and the reflection/transmission coefficients.

1. INTRODUCTION

Porous structures such as rubblemound breakwaters typically consist of multiple dissipative layers with different physical and hydraulic characteristics. A better understanding of the dissipation, reflection and transmission characteristics of porous structures is required to accurately predict wave conditions in navigation channels and harbors, sediment dynamics behind submerged breakwaters, and erosion and sedimentation processes near coastal inlets.

Early theoretical studies of wave-porous bed interaction were concerned with the attenuation of surface waves propagating over permeable horizontal seabeds (e.g. Putnam, 1949; Reid and Kajiura, 1957). Linear wave theory was used to describe the overlying wave field while Darcy's law was used to describe the flow inside the porous layer. Darcy's law is valid for flows with low Reynolds number. As the flow velocity or particle size increases, Darcy's linear relationship between the discharge velocity and pressure gradient breaks down and the quadratic drag law proposed by Forchheimer (1901) becomes more applicable. Sollitt and Cross (1972) presented a modified set of equations

[1] Dept. of Naval Architecture and Marine Eng., University of Michigan, Ann Arbor, MI, 48109-2145, U.S.A.
[2] Coastal and Hydraulics Lab , U.S. Army Engineer R & D Center, Vicksburg, MS,

for fluid motion within a porous medium that included both a Forchheimer-type drag term and inertial effects. An eigenfunction expansion method was used to solve the linearized form of the equations for waves normally incident on vertical surface-piercing porous structures.

In many applications, it is important to investigate wave propagation over non-horizontal seabeds or wave interaction with non-rectangular breakwaters. The porous medium could also consist of multiple layers with different porosities and dissipative properties. Several investigators (e.g. Sulisz 1985; Mallayachari and Sundar 1994) have used linear boundary integral methods to simulate wave interaction with multi-layer porous structures. Linear models, however, cannot accurately describe the nonlinear wave steepening and breaking processes that occur on the slopes of submerged and surface-piercing trapezoidal structures. These complicated processes can be better addressed by numerically solving either the full Navier-Stokes equations (e.g. Liu *et al.* 1999) or asymptotic theories for weakly nonlinear/weakly dispersive waves.

Flaten and Rygg (1991) developed an asymptotic Boussinesq-type theory for shallow water wave propagation over a permeable bed. Darcy's law was used to model the flow inside the porous layer. Cruz *et al.* (1997) developed a two-layer weakly nonlinear Boussinesq model that included both fluid acceleration effects and quadratic drag losses inside the porous layer. The dispersive and damping properties of the equations were also improved by introducing additional dispersive terms to the momentum equations using the approach of Madsen and Sorensen (1992). Hsiao *et al.* (2002) extended the two-layer approach to large amplitude waves by using a fully nonlinear set of Boussinesq equations for the water region. While the two-layer Boussinesq approach is suitable for modeling wave propagation over permeable seabeds, it is not ideal for studying wave interaction with multi-layer porous structures of arbitrary cross-section. The equations assume that the thicknesses of the porous layers are much smaller than the incident wavelength and the interfacial/seabed slopes are mildly varying. For structures that are a fraction of a wavelength, the spatial discretization schemes used for the open water region are also too coarse to resolve details of the flow field inside the porous structure.

In this paper, a numerical model is presented for simulating nonlinear wave interaction with multi-layer porous structures of arbitrary shape. The coupled model consists of a set of fully nonlinear Boussinesq equations for the water region and boundary integral equations for each of the porous layers. This approach does not make any assumption as to the profile of the flow field inside the porous medium. The numerical model is initially calibrated with data from laboratory experiments on wave interaction with a vertical porous breakwater (Requejo *et al.* 2002). The numerical model is then used to investigate wave transmission over submerged breakwaters and compared with the laboratory data of Seabrook and Hall (1998).

2. THEORETICAL BACKGROUND

Governing Equations for Water Region

Consider the interaction of weakly dispersive surface waves with a multi-layered surface-piercing breakwater as shown in Figure 1. A fully nonlinear set of Boussinesq equations is used to describe the fluid motion in the water region. The evolution of the free surface $\eta(\boldsymbol{x},t)$ is governed by the depth-integrated mass conservation equation:

$$\eta_t + \nabla \cdot \boldsymbol{Q} = q_n^{(p)} \tag{1}$$

where $\nabla = (\partial/\partial x, \partial/\partial y)$, $\eta(\boldsymbol{x},t)$ is the water surface elevation, $q_n^{(p)}$ represents the normal flow through the water-porous layer interface and $\boldsymbol{Q}(\boldsymbol{x},t)$ is the volume flux density given by:

$$\boldsymbol{Q} = \int_{-h}^{\eta} \boldsymbol{u}^{(w)}\, dz \tag{2}$$

$\boldsymbol{u}^{(w)}$ is the horizontal fluid velocity in the water region and $h(\boldsymbol{x})$ is the water depth.

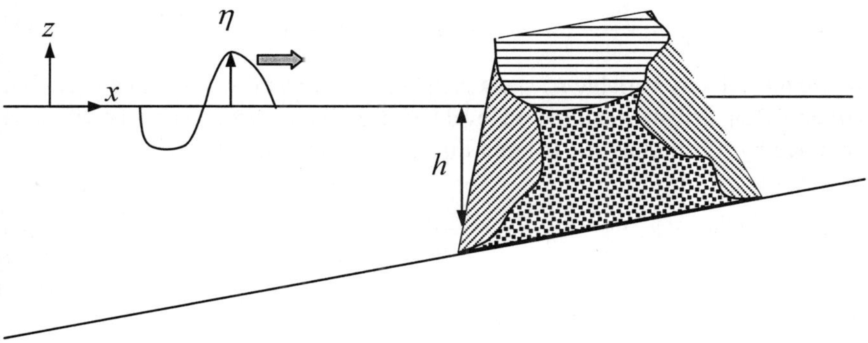

Figure 1. Definition Sketch.

The momentum equation is derived by evaluating the Euler equations at the free surface to obtain:

$$\boldsymbol{u}_t^{(w)}\Big|_\eta + w_t^{(w)}\Big|_\eta \nabla\eta + g\nabla\eta + \left(\boldsymbol{u}_\eta^{(w)}\cdot\nabla\right)\boldsymbol{u}_\eta^{(w)} + w_\eta^{(w)}\nabla w_\eta^{(w)} = 0 \tag{3}$$

where $w^{(w)}$ is the vertical velocity. The system of equations (1) to (3) requires knowledge of the vertical profile of the flow field in order to evaluate the horizontal and vertical velocities at the free surface and the volume flux density. For shallow water waves, the vertical profile of the flow field is obtained by expanding the velocity potential, Φ, as a Taylor series about an arbitrary elevation, z_α, in the water column (Nwogu 1993):

$$\Phi^{(w)}(\boldsymbol{x},z,t) \quad = \quad \phi_\alpha + (z - z_\alpha)\left.\frac{\partial \phi^{(w)}}{\partial z}\right|_{z=-h} + \frac{1}{2}\left[(z_\alpha + h)^2 - (z + h)^2\right]\nabla^2\phi_\alpha \tag{4}$$

where $\phi_\alpha = \Phi^{(w)}(\boldsymbol{x},z,t)$. The vertical velocity at the bottom of the water region depends on the inflow/outflow from the porous region and is obtained by enforcing the continuity of normal velocity at the water-porous layer interface:

$$\left.\frac{\partial \phi^{(w)}}{\partial z}\right|_{z=-h} + \boldsymbol{u}^{(w)}\cdot\nabla h \quad = \quad \left.\frac{\partial \phi^{(p)}}{\partial n}\right|_{z=-h}\sqrt{1+\nabla h\cdot\nabla h} \quad = \quad q_n^{(p)} \tag{5}$$

By substituting Eq. (5) into Eq. (4) and differentiating, we obtain the vertical distributions of the horizontal and vertical velocities as:

$$\begin{aligned} \boldsymbol{u}^{(w)}(\boldsymbol{x},z,t) \quad = \quad \nabla\Phi^{(w)} \quad &= \quad \boldsymbol{u}_\alpha + (z_\alpha - z)\left[\nabla(\boldsymbol{u}_\alpha\cdot\nabla h) + (\nabla\cdot\boldsymbol{u}_\alpha)\nabla h - \nabla q_n^{(p)}\right] \\ &+ \quad \frac{1}{2}\left[(z_\alpha + h)^2 - (z+h)^2\right]\nabla(\nabla\cdot\boldsymbol{u}_\alpha) \end{aligned} \tag{6}$$

$$w^{(w)}(\boldsymbol{x},z,t) \quad = \quad \frac{\partial \Phi^{(w)}}{\partial z} \quad = \quad q_n^{(p)} - \left[\boldsymbol{u}_\alpha\cdot\nabla h + (z+h)\nabla\cdot\boldsymbol{u}_\alpha\right] \tag{7}$$

where the velocity variable $\boldsymbol{u}_\alpha = \boldsymbol{u}(\boldsymbol{x},z_\alpha,t)$ is the horizontal velocity at $z = z_\alpha$. The volume flux density required to evaluate the mass conservation equation is obtained by substituting Eq. (6) into Eq. (2) and integrating to yield:

$$\begin{aligned} \boldsymbol{Q} \quad = \quad (h+\eta)\Bigg\{\boldsymbol{u}_\alpha + &\left[\frac{(z_\alpha + h)^2}{2} - \frac{(h+\eta)^2}{6}\right]\nabla(\nabla\cdot\boldsymbol{u}_\alpha) + \\ &\left[(z_\alpha + h) - \frac{(h+\eta)}{2}\right]\left[\nabla(\boldsymbol{u}_\alpha\cdot\nabla h) + (\nabla\cdot\boldsymbol{u}_\alpha)\nabla h - \nabla q_n^{(p)}\right]\Bigg\} \end{aligned} \tag{8}$$

The momentum equation can be written in terms of the velocity variable $\boldsymbol{u}_\alpha$ by substituting the expressions for $\boldsymbol{u}^{(w)}$ and $w^{(w)}$ (Eqs. 6 and 7) into Eq. 3 to obtain:

$$\begin{aligned} &\boldsymbol{u}_{\alpha,t} + g\nabla\eta + \left(\boldsymbol{u}_\eta^{(w)}\cdot\nabla\right)\boldsymbol{u}_\eta^{(w)} + w_\eta^{(w)}\nabla w_\eta^{(w)} + (z_\alpha - \eta)\left[\nabla(\boldsymbol{u}_{\alpha,t}\cdot\nabla h) + (\nabla\cdot\boldsymbol{u}_{\alpha,t})\nabla h - q_{n,t}^{(p)}\right] \\ &+ \frac{1}{2}\left[(z_\alpha + h)^2 - (h+\eta)^2\right]\nabla(\nabla\cdot\boldsymbol{u}_{\alpha,t}) - \left[(\boldsymbol{u}_{\alpha,t}\cdot\nabla h) + (h+\eta)\nabla\cdot\boldsymbol{u}_{\alpha,t} - q_{n,t}^{(p)}\right]\nabla\eta \\ &+ \frac{1}{h+\eta} f_w \boldsymbol{u}_b^{(w)}\left|\boldsymbol{u}_b^{(w)}\right| + \frac{1}{h+\eta}\nabla\left(\nu(h+\eta)\nabla\cdot\boldsymbol{u}_\alpha\right) \quad = \quad 0 \end{aligned} \tag{9}$$

where v and f_w are the eddy viscosity and bottom friction coefficient respectively intended to account for wave energy dissipation due to wave breaking and bottom friction. The wave breaking model is based on the eddy viscosity model of Nwogu (1996).

Equations (1), (8) and (9) represent a modified set of fully nonlinear Boussinesq equations for the water region. Compared to the fully nonlinear equations of Nwogu (1996), Eqs. (1) and (9) contain additional terms involving $q_n^{(p)}$ to account for flow into and out of the porous structure. The normal flow through the water-porous layer interface will be determined by solving a boundary value problem for the porous region.

Governing Equations for Porous Region

The asymptotic Taylor series expansion used to describe the flow field in the water region could also be applied to shallow porous layers (e.g. Hsiao *et al.*, 2002). However, this approach is cumbersome to apply to structures that consist of multiple porous layers of arbitrary cross-sections. Since the flow inside the porous layer satisfies the Laplace equation, Green's third identity may be applied over any closed boundary S to relate the values of the potential inside the bounded region with the boundary values of the velocity potential $\Phi^{(p)}(\xi)$ and its normal derivative $\partial\Phi^{(p)}(\xi)/\partial n$. For field points $\boldsymbol{x}$ that lie along the boundary, the velocity potential is given by:

$$\Phi^{(p)}(\boldsymbol{x}) = \frac{1}{\pi}\oint_S \left[G(\boldsymbol{x};\xi)\frac{\partial\Phi^{(p)}}{\partial n}(\xi) - \Phi^{(p)}(\xi)\frac{\partial G}{\partial n}(\boldsymbol{x};\xi) \right] dS \qquad (10)$$

where $G = -\ln|\boldsymbol{x}-\xi|$ is the Greens function and ξ represents the source point on the boundary S over which the integration is performed. The integral equation is evaluated numerically by dividing the porous layer boundaries into a finite number of segments with the values of $\Phi^{(p)}$ and $\partial\Phi^{(p)}/\partial n$ assumed to be constant over each segment and defined at the midpoints of the segments. The integral equation for each porous region can thus be rewritten as a matrix equation in terms of discrete values of $\Phi^{(p)}$ and $\partial\Phi^{(p)}/\partial n$:

$$\sum_{j=1}^{N} \left\{ \left[A_{ij} + \delta_{ij} \right] \Phi_j^{(p)} + B_{ij} \frac{\partial\Phi_j^{(p)}}{\partial n} \right\} = 0 \qquad (11)$$

where N is the number of segments on the boundary and the matrix coefficients A_{ij}, B_{ij} correspond to the integral of the Green's function or its normal derivative over the individual segments. When $i \neq j$, the integrals are approximated by assuming the Green's function (or normal derivative) to be constant over each segment and evaluated at the midpoint of the segment. When $i = j$, the integrals become singular and are evaluated analytically.

Each porous layer may have up to four different types of boundary conditions corresponding to: 1) matching boundary with water region, 2) free surface, 3) matching boundary with another porous layer, and 4) impermeable boundary.

Along matching boundaries with the water region, we enforce the continuity of pressure and normal velocities. The solution of the problem for the water region provides pressure boundary conditions for the porous layer while the solution for the porous layer provides normal velocity boundary conditions for the water region. Following Sollitt and Cross (1972), we assume that the flow in the porous region is governed by a Forchheimer-type resistance term plus an inertial term to account for fluid acceleration around solid particles. The continuity of pressure boundary condition is obtained by evaluating the modified Euler equations for the porous region along the interface $z = z_i$. This can be expressed in terms of the tangential velocity along the interface $q_s^{(p)} = \nabla\Phi^{(p)}$ as:

$$\begin{aligned} & c_m \boldsymbol{q}_{s,t}^{(p)} + g\nabla z_i + \frac{c_m}{\lambda}\left(\boldsymbol{u}_i^{(p)} \cdot \nabla\right)\boldsymbol{u}_i^{(p)} + \frac{c_m}{\lambda} w_i^{(p)} \nabla w_i^{(p)} \\ & + f_l \lambda \boldsymbol{q}_s^{(p)} + f_t \lambda \boldsymbol{q}_s^{(p)} \sqrt{\boldsymbol{u}_i^{(p)} \cdot \boldsymbol{u}_i^{(p)} + w_i^{(p)2}} = -\frac{1}{\rho} \lambda \nabla p_i^{(w)} \end{aligned} \tag{12}$$

where c_m is an inertial coefficient, which can be written in terms of the added mass coefficient, c_a, using $c_m = 1 + (1-\lambda)\, c_a$, λ is the effective porosity of the medium, $(\boldsymbol{u}_i^{(p)}, w_i^{(p)})$ are the discharge velocities on the porous layer side of the interface, and f_l and f_t are empirical laminar and turbulent friction factors. The friction factors can be related to the permeability of the medium (Sollitt and Cross, 1972) or the characteristic stone size, d, and porosity by (e.g. Van Gent 1995):

$$\begin{aligned} f_l &= a_0 \frac{(1-\lambda)^2}{\lambda^3} \frac{\nu}{d^2} \\ f_t &= b_0 \left(1 + \frac{7.5}{K_C}\right) \frac{(1-\lambda)}{\lambda^3} \frac{1}{d} \end{aligned} \tag{13}$$

a_0, and b_0 are empirical constants, and $K_C = \pi H/(\lambda d)$ is the Keulegan-Carpenter number with H the characteristic wave height. The empirical coefficients depend on the particle shape, surface roughness and Reynolds number of the flow.

The pressure boundary condition (Eq. 12) is also applied along phreatic surface boundaries in the porous region with $z_i = \eta$ and $p_i = 0$. The location of the phreatic surface is obtained by solving the kinematic free surface boundary condition:

$$\eta_t = \frac{1}{\lambda} \frac{\partial \Phi^{(p)}}{\partial n} \sqrt{1 + \nabla\eta \cdot \nabla\eta} \tag{14}$$

The normal velocities $\partial\Phi^{(p)}/\partial n$ are obtained as part of the solution of the boundary integral equation.

Along matching boundaries between porous layers, we enforce the continuity of $\Phi^{(p)}$ and $\partial\Phi^{(p)}/\partial n$ while the zero normal flow boundary condition ($\partial\Phi^{(p)}/\partial n = 0$) is enforced along impermeable boundaries.

Numerical Solution

The governing mass and momentum equations for the evolution of the free/phreatic surfaces and velocity variables are integrated in time using a modified Crank-Nicolson scheme. At each time step, we initially solve the Boussinesq problem for water region to determine the pressure forcing along all porous interfacial boundaries. The porous-region momentum equation is then integrated to determine the velocity potential along the interface while the kinematic boundary condition is integrated to update the location of the phreatic surface. The boundary integral problem for the porous region is then solved to determine normal velocities along the interfacial boundaries. Finally, the Boussinesq problem is re-solved with updated values of the normal velocity along porous interface. This procedure is repeated at each time step until convergence.

3. NUMERICAL RESULTS

Wave Interaction with a Vertical Surface-Piercing Breakwater

The performance of the numerical model was initially evaluated with data from the laboratory experiments of Requejo *et al.* (2002), who investigated the interaction of periodic waves with a crib-style rectangular breakwater. The experiments were conducted in a 12-m long and 0.56-m wide wave flume using two structure widths (B = 0.56m, 1.12m), two characteristic stone sizes (d = 0.0105m, 0.0378m), three wave periods (T = 1.2s, 1.7s, 2.7s), and five wave heights (H = 0.02m to 0.09m). Five wave probes were used to measure the wave field in front of the structure and two or three wave probes were used to measure inside the structure. The experimental layout is shown in Fig. 2.

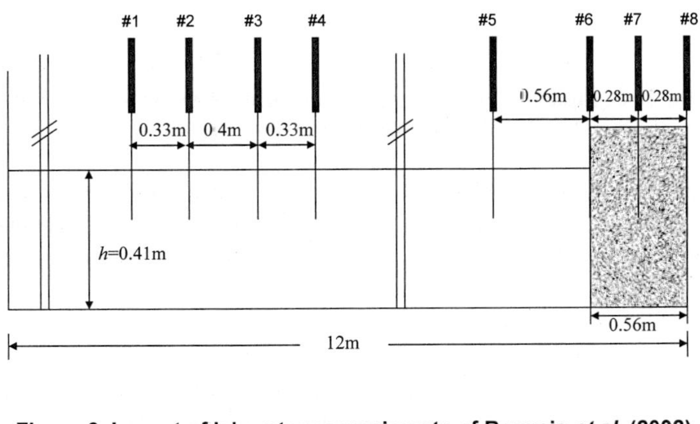

Figure 2. Layout of laboratory experiments of Requejo *et al.* (2002).

Table 1. Summary of Best-Fit Empirical Constants for H=0.034m, T=1.7s.

Structure	B(m)	d(m)	a_o	b_o
#1	1.12	0.0105	100	1.5
#2	0.56	0.0105	100	1.5
#3	1.12	0.0378	100	0.6
#4	0.56	0.0378	100	0.6

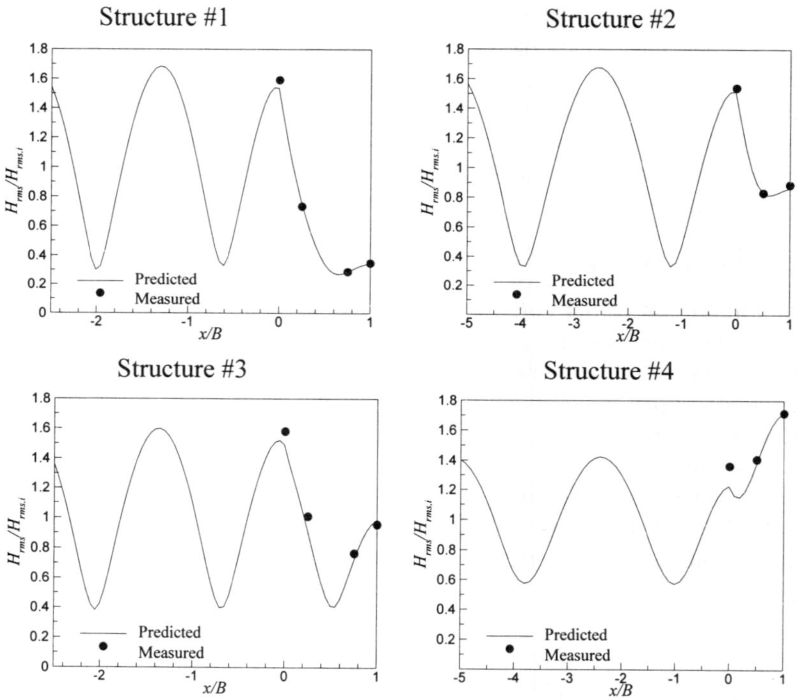

Figure 3. Comparison of measured and predicted wave heights for H=0.034m, T=1.7s.

We investigated the sensitivity of the numerical model predictions of the wave heights inside the porous structure to the values of the empirical constants a_o, b_o. Numerical simulations were carried out for one of the test conditions (H = 0.034m, T = 1.7s) for different combinations of a_o and b_o, with a_o varying from 0 to 300, and b_o from 0 to 3. These range of values were selected to cover the optimal values a_o = 100 and b_o = 1.1 recommended by Requejo *et al.* (2002). The value c_a = 0.4 was used for added mass coefficient as suggested by Van Gent (1995). The values of a_o and b_o that yielded the least square error between the measured and predicted wave heights inside the structure are summarized in Table 1. The empirical constant b_o associated with turbulent losses was smaller for Structures #3 and #4 that have the larger stone size.

The numerically simulated wave height distributions with the best-fit coefficients are compared to the measured wave heights in Figure 3. Although the reflection coefficients were similar for some of the test conditions, the different structures have fairly different wave height variations inside the structure. Excellent agreement is observed between the measured and predicted wave heights inside the structure for Structures #1, #2, and #3.

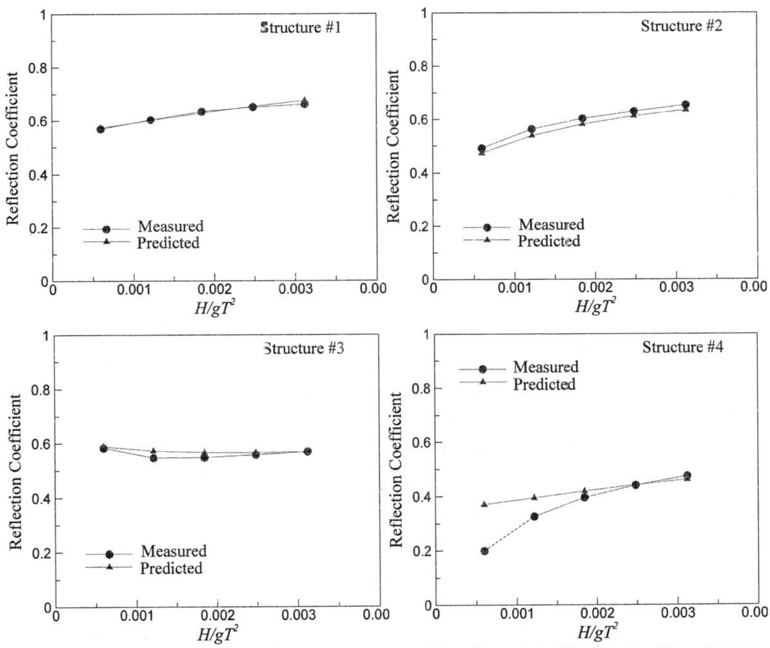

Figure 4. Measured and predicted reflection coefficients (T = 1.7s).

The best-fit empirical constants for Structures #3 and #4 (a_o = 100, b_o = 0.6) were used in numerical simulations for other wave conditions and structure configurations to investigate the potential relative error in using a universal set of constants to estimate the reflection properties of porous structures. The measured and computed reflection coefficients are shown in Figure 4 for T =1.7s, and Figure 5 for T = 1.2s. Excellent agreement is observed between the predicted and measured reflection coefficients for Structures #1, #2 and #3 for T = 1.7s. The numerical model also reproduces the observed trends in the variation of reflection coefficient with wave steepness. For the T = 1.2s test conditions, the numerical model predicts larger values of the reflection coefficient (5-25%) but generally reproduces the correct trends in the variation of reflection coefficient with wave steepness.

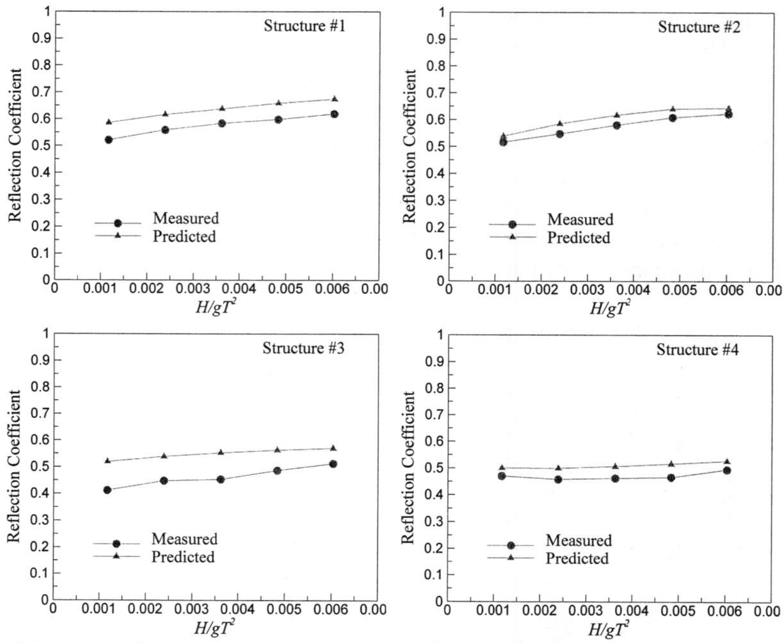

Figure 5. Measured and predicted reflection coefficients (*T* = 1.2s).

Wave Propagation over a Submerged Breakwater

Seabrook and Hall (1998) carried out an extensive series of tests to investigate wave transmission over submerged rubblemound breakwaters. The tests were conducted in a 47-m long and 1-m wide wave flume at Queens University, Canada. The layout for the experiments is shown in Figure 6. The breakwater cross-section had a crest height of h_s = 0.25m and 1V:3H side slopes. The breakwater consisted of a relatively coarse inner core layer with d_{50} = 0.017m and two layers of primary armor with d = 0.059m. Over 800 irregular wave tests were performed with 13 breakwater geometries, 5 water levels, 3 spectral peak periods, and 4 significant wave heights.

Since the transmission coefficient for submerged breakwaters depends on parameters such as the relative submergence depth and relative crest width, we carried out two sets of numerical simulations. The first set of simulations were conducted for a constant crest width B = 0.6 m and varying submergence depths d_s = 0.05, 0.1, 0.15 and 0.2m. For the second set of simulations, we varied the crest width B = 0.6, 1.5, 2.5 and 3.5m and used a submergence depth of 0.05m. The numerical simulations were performed for an incident sea state characterized by a JONSWAP spectrum with H_{mo} = 0.1 m, T_p = 1.5 s and γ = 3.3. A porosity of 0.5 and empirical constants a_o = 100, b_o = 0.6, and c_a = 0.4 were used for the model simulations.

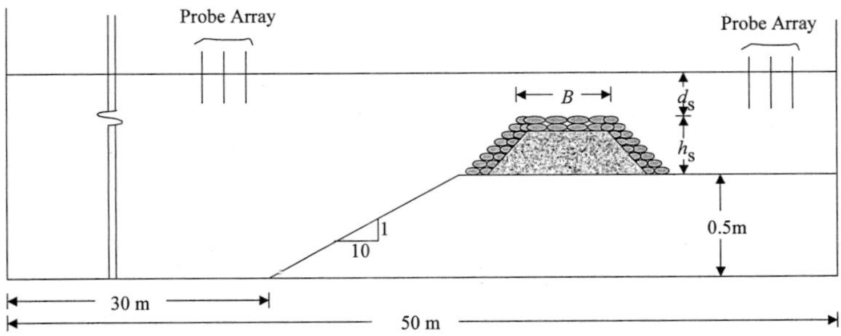

Figure 6. Layout for the Seabrook and Hall (1998) experiments.

Figure 7 shows a comparison of the variation of the measured and predicted transmission coefficients with relative submergence depth, d_s. For non-breaking waves with $d_s/H_{mo} = 2$, the transmitted wave height is about 80% of the incident wave height. As d_s decreases, the transmitted height decreases to about 40% of the incident wave height. Wave breaking over the breakwater also plays an important role in the wave energy dissipation process. Figure 9 shows a comparison of the wave transmission coefficient as a function of the breakwater crest width for a submergence depth of 0.05m. The results show a decrease in transmission coefficient with increasing crest width due to both wave breaking and energy dissipation inside the porous structure.

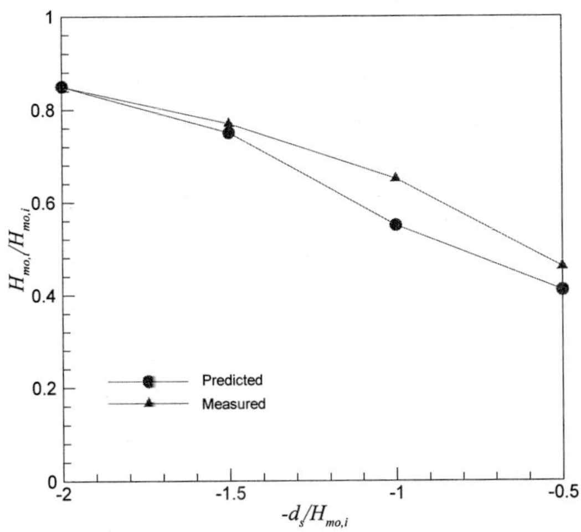

Figure 7. Variation of transmission coefficient with relative submergence depth.

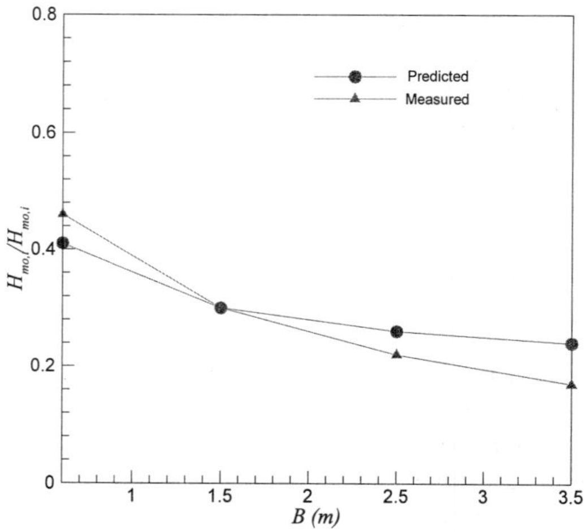

Figure 8. Variation of transmission coefficient with structure crest width.

CONCLUDING REMARKS

A coupled Boussinesq-boundary integral model has been developed to simulate nonlinear wave interaction with surface-piercing and submerged porous structures. The model solves vertically integrated Boussinesq-type equations for the water region and evolution equations for the phreatic surface and boundary values of the tangential velocities in the porous region. A boundary integral method is used to close the problem for porous region and determine the relationship between the normal and tangential velocities along the boundaries. Since the energy dissipation within porous structures is governed by empirical laminar and turbulent frictional factors, the applicability of the model formulation and determination of the friction factors have been investigated by comparing numerical model predictions of wave reflection and transmission coefficients through porous breakwaters with data from two laboratory experiments. Overall, the model performed reasonably well and was able to reproduce the observed variation of reflection coefficient with wave steepness with one set of empirical constants. Good agreement was also observed for the wave field inside the structure, provided the empirical coefficients are suitably chosen.

ACKNOWLEDGMENTS

This work was supported by the Coastal Inlets Research Program of the U.S. Army Engineer Research and Development Center. Permission to publish this paper was granted by the Chief of Engineers, U.S. Army Corps of Engineers. The authors would also like to thank C. Vidal and S.R. Seabrook for providing their laboratory data.

REFERENCES

Cruz, E. C., Isobe, M. and A. Watanabe. 1997. Boussinesq equations for wave transformation on porous beds. *Coastal Eng.*, **30**, 125–156.

Flaten G., and O.B. Rygg. 1991. Dispersive shallow water waves over a porous seabed, *Coastal Eng.*, **15**, 347-369.

Forchheimer, P., 1901. Wasserbewegung durch boden. *Z.* Ver. Deutsch. Ing., **45**, 1782-1788.

Hsiao, S.-C., Liu, P.L.-F., and Y. Chen. 2002. Nonlinear water waves propagating over a permeable bed, *Proc. R. Soc. Lond. A*, **458**, 1291–1322.

Liu, P. L.-F., Lin, P., Chang, K.-A. and T. Sakakiyama. 1999. Numerical modeling of wave interaction with porous structures, *J. Waterw. Port Coastal Ocean Eng.*, **125**, 322–330.

Madsen, P.A., and O.R. Sørensen. 1992. A new form of the Boussinesq equations with improved linear dispersion characteristics. Part 2. A slowly varying bathymetry, *Coastal Eng.*, **18**, 183–204.

Mallayachari, V., and V. Sundar. 1994. Reflection characteristics of permeable seawalls, *Coastal Eng.*, **23**, 135-150.

Nwogu, O., 1993. Alternative form of Boussinesq equations for nearshore wave propagation, *J. Waterw. Port Coastal Ocean Eng.*, **119**, 618-638.

Nwogu, O.G., 1996. Numerical Prediction of breaking waves and currents with a Boussinesq model. *Proc. 25th Int. Conf. on Coastal Eng.*, ASCE, 4807-4820.

Putnam, J.A., 1949. Loss of wave energy due to percolation in a permeable sea bottom. *Trans. Am. Geophys. Union*, **30**, 349-356.

Reid, R.O. and K. Kajiura. 1957. On the damping of gravity waves over a permeable sea bed, *Trans. Am. Geophys. Union*, **38**, 662-666.

Requejo, S., Vidal, C., and I.J. Losada. 2002. Modelling of wave loads and hydraulic performance of vertical permeable structures. *Coastal Eng.*, **46**, 249-276.

Seabrook, S.R., Hall, K.R., 1998. Wave transmission at submerged rubblemound structures, *Proc. 26th Int. Conf. on Coastal Eng.*, ASCE, Copenhagen, Denmark. pp. 2000-2013.

Sollitt, C.K., Cross, R.H., 1972. Wave transmission through permeable breakwaters. *Proc. 15th Int. Conf. on Coastal Eng.*, ASCE, 1827-1846.

Sulisz, W., 1985. Wave reflection and transmission at permeable breakwaters of arbitrary cross-section. *Coastal Eng.*, 9, 371-386.

Van Gent, M.R.A., 1995. Wave interaction with permeable coastal structures. Ph.D. Thesis, Delft University of Technology, Delft, The Netherlands.

DEVELOPMENT OF A STRICTLY COMBINED BEM-VOF METHOD AND ITS APPLICATION TO WAVE PROPAGATION THROUGH 3 DOMAINS

Sang-Ho KIM[1], Masaru YAMASHIRO[2], Akinori YOSHIDA[3], Noriaki HASHIMOTO[4] and Shuguang YAN[5]

A strictly combined BEM-VOF model has been developed by making the best use of both method's merits. To carry out stable and accurate computations, the water surfaces obtained from BEM and VOF are smoothly connected in the transmission domain at each time step by using weighted means, and the velocity potential and its normal derivative in BEM and the value of the function of the volume of fluid in VOF are also corrected. The validity of the model was confirmed by comparing numerical results with theoretical values for Stokes' 5th order wave propagation through three domain BEM-VOF-BEM calculation as well as two domain BEM-VOF calculation. Wave propagation for a random waves through two domains was also checked. The model makes it practically possible to calculate wave-structure interactions under random waves by reducing the computational time and minimizing the numerical dissipation which is inevitable especially in single VOF calculations.

INTRODUCTION

Several numerical methods to simulate interactions between waves and structures have been developed and have produced numerous useful results. However, because each numerical method has both the advantage and the disadvantage owing to respective characteristic of the method, the development of more versatile and efficient numerical method has been required. Recently, several hybrid numerical methods which combine different numerical methods have come to be studied. Maeno et al.(2002) combined the Volume Of Fluid (VOF) method and Finite Element Method (FEM) to investigate the dynamic behavior of the seabed around a seawall under wave motion. Itoh et al.(2002) have combined the Discrete Element Method (DEM) and VOF method to study the deformation of rubble structures. Sitanggang et al.(2006) have shown a method which couples a Boussinesq model with a Reynolds Averaged Navier-Stokes (RANS) VOF model.

Among the models available for the simulation of wave motions, VOF method and Boundary Element Method (BEM) are seemingly the most popular ones. VOF method can be used for numerical simulations of wave motions including wave breakings. VOF method, however, generally needs enormous

[1] Dep. of Civil Engineering, Kyushu Univ., 6-10-1 Hakozaki, Fukuoka, 812-8581, Japan
[2] Dep. of Civil Engineering, Kyushu Univ., 6-10-1 Hakozaki, Fukuoka, 812-8581, Japan
[3] Dep. of Civil Engineering, Kyushu Univ., 6-10-1 Hakozaki, Fukuoka, 812-8581, Japan
[4] Dep. of Civil Engineering, Kyushu Univ., 6-10-1 Hakozaki, Fukuoka, 812-8581, Japan
[5] Dep. of Civil Engineering, Kyushukyoritsu Univ., 1-8 Jiyugaoka Kitakyushu, Japan

computational time and large computational storage memories; thus practically it is difficult to use VOF method for random wave calculations because long and stable computation (e.g. for more than 100 significant wave periods) is required to obtain statistically meaningful results. On the other hand if the wave motion is assumed to be a potential motion, BEM, which is a much faster and more accurate method than VOF method, can be effectively used.

When we consider a rather wide range of wave field, which has a structure as shown in Fig. 1, intense wave motion occurs around the structure but gentle wave motions are regenerated in the area a little far from the structure and it seems that potential theory is sufficiently valid. Consequently, the combined BEM-VOF method which uses BEM to the gentle wave motion in the region far from the structure and VOF to the drastic wave motion around the structure can be a good candidate. Based on this idea, We have developed a combined method of BEM and VOF for more efficient and accurate calculation of interactions between random waves and structures (Yan et al. 2003 a,b : 2004).

As similar studies, Guignard et al.(1999) and Lanchaume et al.(2003) have introduced a method coupling BEM and VOF for the calculation of wave breakings on a sloped beach. Biausser et al.(2003) and Drevard et al.(2005) have extended the method to 3-D calculations. Grilli et al.(2004) have compared the experiments and the coupled BEM-VOF model for solitary wave shoaling and breaking. These studies use one-way coupling in which the results of the BEM calculation are used as just initial data for VOF calculation, and thus the reflected waves from the VOF domain to the BEM domain cannot be treated.

In this study, we have proposed a strictly combined BEM-VOF method which can deal with the two-way wave propagations between the VOF and BEM domain. Then we have verified the method by comparing the numerical results with theoretical ones for Stokes 5th order waves and random waves. In addition to 2-domain configurations which connect BEM and VOF domain, 3-domain configurations (BEM-VOF-BEM) were conducted to further reduce the time consuming VOF domain.

BEM AND VOF MODEL

BEM model

Let us consider the wave field closed by S1~S4 boundaries (Fig. 2). The water depth is constant. Fluid motion is assumed to be incompressible and

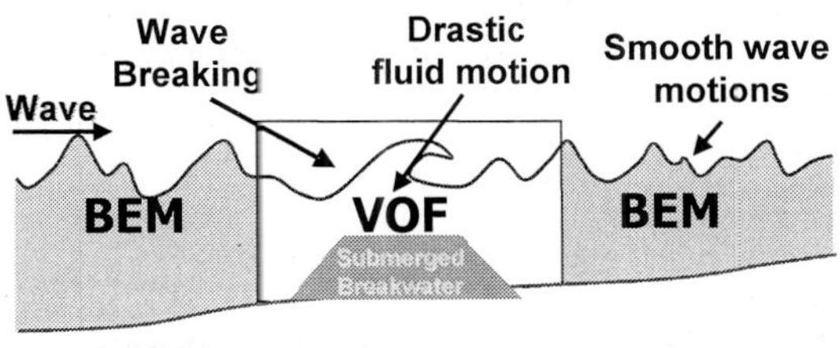

Figure 1. Calculation using a combined BEM-VOF method

inviscid wave motions; and thus the velocity potential $\phi(x,z,t)$ exists, and must satisfy the following Poisson equation in the whole fluid region.

$$\nabla^2\phi(x,z,t) = q^*(x,z,t) \quad (q^*(x,z,t) = U^*(z,t)\delta(x-x_1)) \tag{1}$$

where $U^*(z,t)$ is the flux density at the wave-making source, and is zero elsewhere. δ is the dirac delta function, and $x1$ is the x-position of the source.

Potential theory shows harmonic function in a closed domain, $\phi(X)$, which is a solution of the Poisson equation, can be expressed with Green's second identity.

$$\phi(X) = \frac{-1}{\alpha}\int_\Gamma \left\{\phi(X_b)\frac{\partial G(r)}{\partial n} - G(r)\frac{\partial \phi(X_b)}{\partial n}\right\}ds + \int_\Gamma q^* G(r)ds \tag{2}$$

where $G(r) = \log r + \log r^*$; r is the length between an arbitrary point $X = (x,z)$ in the domain and a point $X_b = (x_b, z_b)$ on the boundary; r^* is the length between the point X and its mirror image X_b^* with respect to the uniform bottom boundary; Γ is the whole boundary except the bottom boundary. α is defined to take the angle of the element when X is on the boundary, and to take 2π when X is inside the boundary. n denotes inward normal to the boundary of the region, and the direction of the integral with respect to s is taken counterclockwise along the boundary Γ.

Equation 3 expresses the time variation of the velocity potential ϕ along with the node point on the water surface $\eta(x,t)$; the node point is assumed to move only in the vertical direction. Eq. 4 is the kinemetic boundary condition, and Eq.5 is the combined boundary condition of Eq. 3 , Eq. 4 and the Bernoulli equation.

$$\frac{d\phi}{dt} = \frac{\partial\phi}{\partial t} + \frac{\partial\eta}{\partial t}\cdot\frac{\partial\phi}{\partial z} \quad (\text{on } S_2) \tag{3}$$

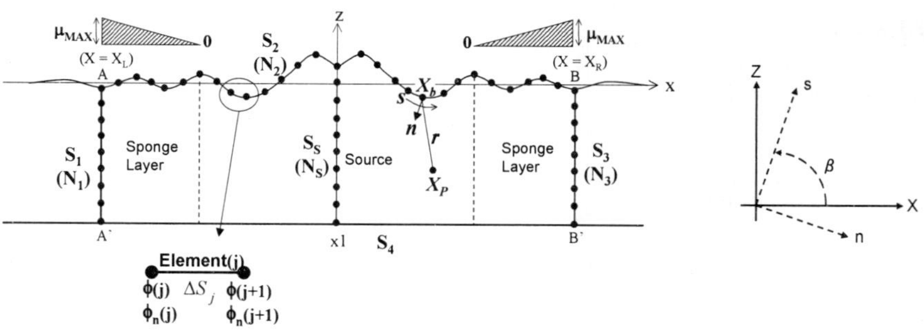

Figure 2. Definition sketch and coordinate system

$$\frac{\partial \eta}{\partial t} = \frac{-1}{\cos\beta} \cdot \frac{\partial \phi}{\partial n} \ (\text{on } S_2) \tag{4}$$

$$\frac{d\phi}{dt} + g \cdot \eta + \frac{1}{2}\left\{ \left(\frac{\partial \phi}{\partial s}\right)^2 + \left(\frac{\partial \phi}{\partial n}\right)^2 + 2\tan\beta \cdot \frac{\partial \phi}{\partial n} \cdot \frac{\partial \phi}{\partial s} \right\} + \mu\phi - \int_{x1}^{x} \frac{\partial \mu}{\partial x}\phi dx = 0(\text{on } S_2) \tag{5}$$

where β is the angle between (x, z) and (s, n) coordinate system, and μ is the decay coefficient in the sponge layer. g is the acceleration of gravity.

The open boundary condition is expressed as follows :

$$\begin{aligned} \frac{\partial \phi}{\partial t} &= C_0 \frac{\partial \phi}{\partial n} - \mu\phi + \int_{x1}^{x} \frac{\partial \mu}{\partial x}\phi \Big|_{onS_2} dx \ (\text{on } S_1) \\ \frac{\partial \phi}{\partial t} &= -C_0 \frac{\partial \phi}{\partial n} + \mu\phi - \int_{x1}^{x} \frac{\partial \mu}{\partial x}\phi \Big|_{onS_2} dx \ (\text{on } S_3) \end{aligned} \tag{6}$$

where C_0 is the phase velocity of the incident wave for regular waves and is the long wave velocity for random waves. More details can be found in Ohyama et al.(1991).

Γ is divided into a number of small elements $\Delta S_j \ (j = 1,2,\cdots,N)$, and ϕ and $\phi_n (= \partial\phi / \partial n)$ on each element is interpolated by their values at the node points as follows :

$$\begin{aligned} \phi(X_b) &= \left(1 - \frac{s}{\Delta S_i}\right)\phi(j) + \left(\frac{s}{\Delta S_i}\right)\phi(j+1) \\ \phi_n(X_b) &= \left(1 - \frac{s}{\Delta S_i}\right)\phi_n(j) + \left(\frac{s}{\Delta S_i}\right)\phi_n(j+1), \quad (s: 0 \to \Delta S_i) \end{aligned} \tag{7}$$

The finite-difference approximation of Eq.2 with respect to the elements yields the simultaneous linear equations as follows :

$$\begin{aligned} &\sum_{j=1}^{N} \left\{ (\overline{Q}_{ij} - \delta_{ij})\phi(j) - Q_{ij}\phi_n(j) + \overline{E_{ij}^*}\phi(j+1) - E_{ij}^*\phi_n(j+1) \right\} \\ &= \sum_{j=1}^{N_s} \left[Q_{ij} U^*(j) + E_{ij}^* U^*(j+1) \right] \quad (i = 1,2,\cdots,N) \end{aligned} \tag{8}$$

$$Q_{ij} = E_{ij} - E_{ij}^*, \ \overline{Q_{ij}} = \overline{E_{ij}} - \overline{E_{ij}^*} \tag{9}$$

where $E_{ij} = \frac{1}{\alpha_i}\int_{\Delta S_i}(\log r_i + \log r_i^*)ds$, $\overline{E}_{ij} = \frac{1}{\alpha_i}\int_{\Delta S_i}\frac{\partial(\log r_i + \log r_i^*)}{\partial n}ds$,

$$E_{ij}^* = \frac{1}{\alpha_i \Delta S_j}\int_{\Delta S_i} s(\log r_i + \log r_i^*)ds \ , \ \overline{E_{ij}^*} = \frac{1}{\alpha_i \Delta S_j}\int_{\Delta S_i}\frac{\partial(\log r_i + \log r_i^*)}{\partial n}ds$$

In the calculation of the integral of $\log r$, not a numerical estimation but an analytical solution based on the integral of complex function $\log z$ and $\log z^*$ $(z = x + iy,\ z^* = x - iy)$ is used.

The procedure of BEM calculation is as follows:

1. The water surface of the present time-step (m) is obtained from Eq.4 by using ϕ_n of previous time-step (m-1).
2. Q_{ij} and $\overline{Q_{ij}}$ are calculated along with the water surface.
3. ϕ of the m time-step is obtained from the boundary condition (Eq. 5, 6).
4. ϕ_n on the boundary in the m time-step are obtained by solving simultaneous linear equations.
5. Repetition of these step will advance a solution.

The velocity potential at any point inside the boundary is obtained through Eq.2 using the calculated ϕ and ϕ_n on the boundary, and consequently the velocity and pressure at any point are also obtained.

VOF model

Several modified VOF models from the original SOLA-VOF model are proposed but we use the original SOLA-VOF model because the primary aim of this study is to develop the way to combine BEM and VOF. The whole calculation domain is divided into small cells. The continuity equation, Navier-Stokes equations and advection equation, which are described below, are solved by using the staggered mesh in which the velocity is located on the boundary of each cell and the pressure and the value of function F are located at the center of the cell. The basic procedure for advancing a solution through one increment in time consists of the following three steps; Firstly, the velocity of new time step is calculated by solving the N-S equations with boundary condition. The calculated velocity usually cannot satisfy the continuity equation, so that secondly the pressure is adjusted in each cell and the velocity is changed to fit this pressure adjustment. This process is iteratively continued until the modified velocity satisfies the continuity equation in each cell. Finally, the F function is updated to give the new fluid configuration.

$$\frac{\partial u}{\partial x} + \frac{\partial w}{\partial z} = 0 \tag{10}$$

$$\frac{\partial u}{\partial t}+u\frac{\partial u}{\partial x}+w\frac{\partial u}{\partial z}=-\frac{1}{\rho}\frac{\partial P}{\partial x}+\upsilon\left(\frac{\partial^2 u}{\partial x^2}+\frac{\partial^2 u}{\partial z^2}\right)$$
$$\frac{\partial w}{\partial t}+u\frac{\partial w}{\partial x}+w\frac{\partial w}{\partial z}=-\frac{1}{\rho}\frac{\partial P}{\partial z}-g+\upsilon\left(\frac{\partial^2 w}{\partial x^2}+\frac{\partial^2 w}{\partial z^2}\right) \tag{11}$$

$$\frac{\partial F}{\partial t}+\frac{\partial Fu}{\partial t}+\frac{\partial Fw}{\partial z}=0 \tag{12}$$

where ρ is the water density; p is the pressure; υ is the coefficient of kinematic viscosity; F is the VOF function.

COMBINED BEM-VOF MODEL

The way of combination

The BEM and VOF domains are located to overlap some width as transmission domain (T.D.), and the boundary of the BEM domain is put exactly on the boundary of the cells in the VOF domain as shown in Fig. 3.

In the case that the data are given from the BEM domain to VOF domain, the value of ϕ at the center of VOF's boundary cells and their adjacent cells are calculated by Eq.2 with the values of ϕ and ϕ_n on the boundary of the BEM domain. The horizontal and vertical velocities on the boundary cells are estimated by the central difference scheme of ϕ. The vertical velocity for the surface cell is calculated by the extrapolation using the vertical velocity of two consecutive cells under the surface cell. The vertical velocity of the bottom is always set zero. These velocities are used as initial condition for VOF calculation.

On the other hand, when the data are given from the VOF domain to BEM domain, the value of ϕ on the right boundary of the BEM domain, which is used

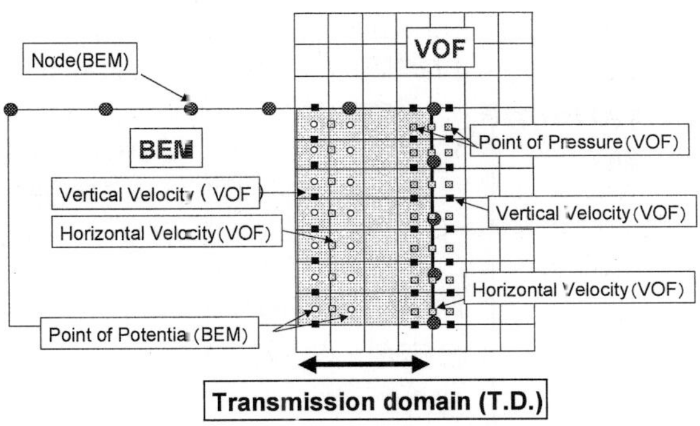

Figure 3. Locations of domains, velocity, pressure and node points.

as a boundary condition for BEM calculations, is calculated through Bernoulli equation using obtained velocities and pressures in the VOF calculation; First, consider two cell columns adjacent to either side of the right boundary of the BEM domain. The vertical velocity and the pressure on the boundary of the BEM domain are obtained by averaging the adjacent vertical velocities and adjacent pressures in the adjacent two cell columns. Because the horizontal velocity in the VOF calculation is obtained on the same vertical line as the right boundary of the BEM domain, it is directly used. However, the vertical locations of the obtained horizontal and vertical velocities and pressures deviate from the node positions on the boundary of the BEM domain, so that the velocities and pressures are adjusted by using the spline interpolation. Then, using these adjusted velocities and pressures, ϕ is obtained from Eq. 13, which is led by using Bernoulli equation for the first term in the right-hand of Eq. 3,.

$$\frac{d\phi}{dt} = -\frac{P_{VOF}}{\rho} - g \cdot z - \frac{1}{2} \cdot \left\{ (U_{VOF})^2 + (V_{VOF})^2 \right\} + \frac{\partial Z}{\partial t} \cdot V_{VOF} \tag{13}$$

where the forth term of the right side represents the variation of the velocity potential due to the vertical movement of node point ($Z(z)_{on\,S_2}$). The value of ϕ_n is directly given from the horizontal velocity. The value of ϕ and ϕ_n at the edge are estimated by the extrapolation.

The connection of water surfaces

The water surface in the T.D. is calculated by both BEM and VOF, and both the water surfaces do not necessarily agree because the calculation accuracy is naturally different each other. This difference causes the instability and eventually the breakdown of computation. To avoid this instability, the water surfaces are smoothly connected in the T.D. at each time-step by using weighted means as seen in Fig. 4. Along with this correction of the water surface, the values of ϕ, ϕ_n, Q_{ij} and $\overline{Q_{ij}}$ are corrected in BEM as follows:

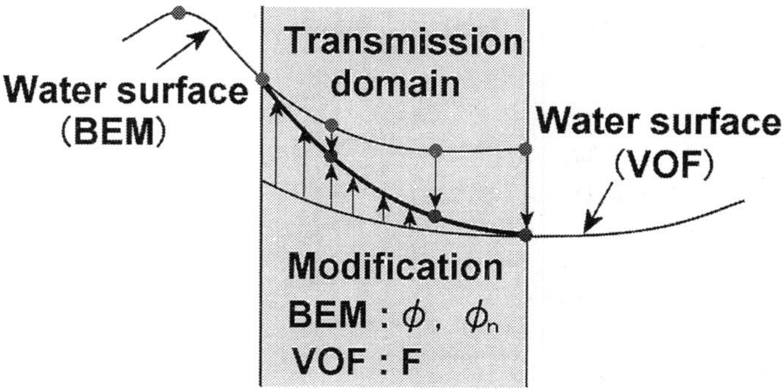

Figure 4. Connection of water surfaces in transmission domain (T.D.)

The finite-difference approximation of Eq. 4 is given by

$$\eta(m)=\eta(m-2)-\frac{2\Delta t}{\cos\beta(m-1)}\cdot\phi_n(m-1) \tag{14}$$

where Δt is the time increment.

The value of ϕ_n at the previous time step(m-1) is corrected through Eq. 14. The finite-difference approximation of Eq 5 is given below (Eq. 15), and the value of ϕ at (m) time step is corrected through this Eq. 15 using the corrected value of ϕ_n.

$$\phi(m)=\phi(m-2)-2g\cdot\Delta t\cdot\eta(m-1)-NL(m-1)-D(m-1) \tag{15}$$

in which

$$NL(m-1)=\Delta t\left\{\phi_s(m-1)^2-\phi_n(m-1)^2+2\tan\beta(m-1)\cdot\phi_s(m-1)\cdot\phi_n(m-1)\right\}$$

$$D(m-1)=2\Delta t\left\{\mu\phi(m-1)-\frac{\mu_{MAX}}{B}\sum_{k=x1}^{x}\phi(m-1)\Delta S_k\right\},\quad \phi_s=\frac{\partial\phi}{\partial s}$$

where B is the width of the sponge layer. Finally, Q_{ij} and $\overline{Q_{ij}}$ are also recalculated for the corrected water surface boundary.

On the other hand, the value of F function is corrected in VOF model along with the correction of the water surface. If the value of F function is modified artificially, the case that the previous air cell changes to the fluid cell or surface cell (or its inverse case) occurs generally, so that, strictly speaking, the correction of velocities is also needed. However, because the velocities on the water surface are updated at the next time-step, and moreover, the correction of F function is carried out just one time in 10~20 VOF time steps owing to (1/10~1/20) smaller VOF time step than the BEM time step, the correction of velocities is not performed.

The flow of calculation

Because the wave-making source is placed in the BEM domain, BEM calculation is carried out one time step before VOF calculation; for BEM calculation, the boundary values given from VOF are one time step behind. Therefore, the calculation about the reflected wave from the VOF domain to the BEM domain generally causes a time lag. To prevent this time lag, the double calculations of BEM are performed.

The left sketch of Fig. 5 shows the procedure of the calculation. The BEM calculation at N time-step is first performed by initial data of N time-step and N-1 time-step data of BEM and VOF. Then velocities obtained from BEM is used to determine the increment of velocities on VOF time step at the boundary cells.

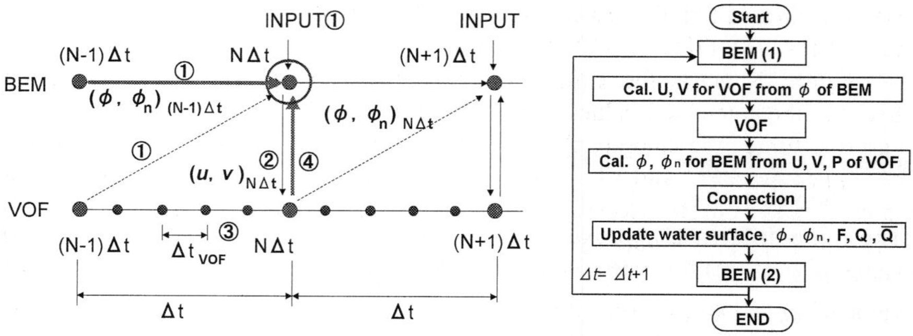

Figure 5. Procedure of the combined BEM-VOF model

After VOF calculation with a number of short time steps reaches to N-step, BEM calculation for the same N-step, not for N+1 step, is performed again, under the boundary condition obtained by using VOF's results at N-step. And using this second results, the calculation proceeds to N+1 step. This double calculations of BEM, enable us to calculate the reflected wave into BEM domain. The flowchart of this procedure of the double calculation is illustrated in Fig. 5. In the BEM calculation, the corrected ϕ, ϕ_n, Q_{ij} and $\overline{Q_{ij}}$ along with the correction of the water surface are used in the second BEM calculation. In the VOF calculation, the corrected value of F function is reflected on next time step.

The verification of the combined BEM-VOF model

The verification was carried out with regular and random waves in the area as shown in Fig. 6. The right boundary condition of the VOF domain was an open or wall condition. The water depth was constant 1m and the width of the T.D. was set 0.3L or 0.15L (L means the wave length for the regular wave, and the significant wave length for random waves).

Input wave was set 5th order stokes wave of T (wave period) =3.0sec., H (wave height) =0.18m in the case of regular wave calculation and the wave of Bretschneider-Mitsuyasu spectrum of $T_{1/3}$ (significant wave period) =3.0sec., $H_{1/3}$ (significant wave height) =0.18m in the case of random wave calculation. The running time of the regular wave calculation and random wave calculation

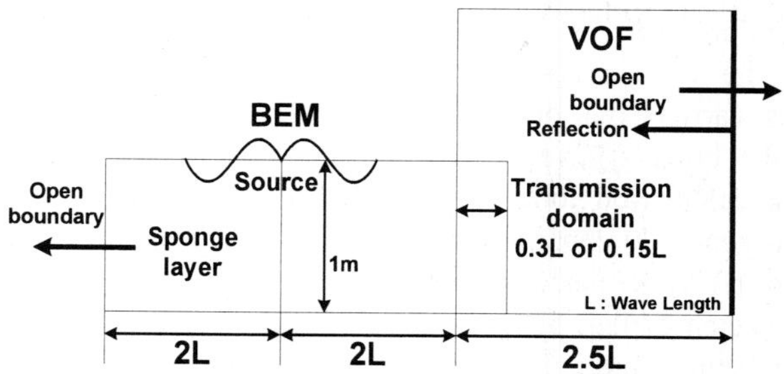

Figure 6. Domain for the verification of model

was set 30T and $100T_{1/3}$ respectively. The BEM time step was 0.01sec. and VOF time step was 0.001sec. (1/10 of BEM).

Figure 7 shows the comparison for the spatial profile of wave motion between the calculated values and theoretical ones. It is seen that the numerical values estimate the theoretical ones very well and this indicates that the process of connecting BEM and VOF domain is successful for two way propagations of incident and reflected waves. The water surface in the T.D., calculated from BEM and VOF, are compared in Fig. 8 (regular wave, at 20T) and in Fig. 9 (random wave, at $30T_{1/3}$) for the T.D. lengths of 0.3L and 0.15L. the comparisons also give very good agreement regardless of the T.D. length. Fig. 10 shows the spatial distribution of wave spectra in the cases that the T.D. is set 0.3L and 0.15L. Although a certain amount of variation along the flume is seen, the overall spectral shapes are very similar and thus the results can be practically acceptable.

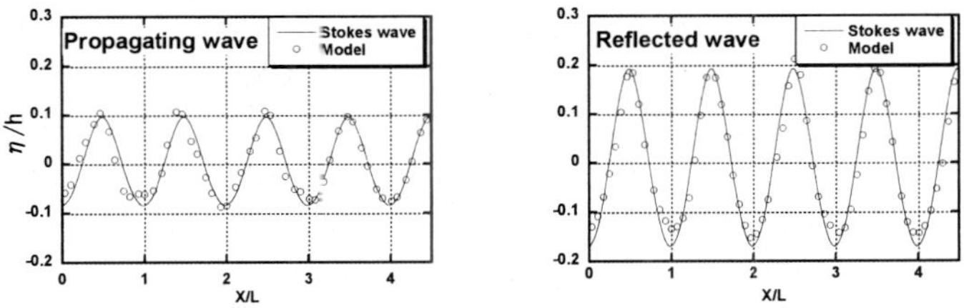

Figure 7. Comparison of wave profile between model and theoretical Stokes wave

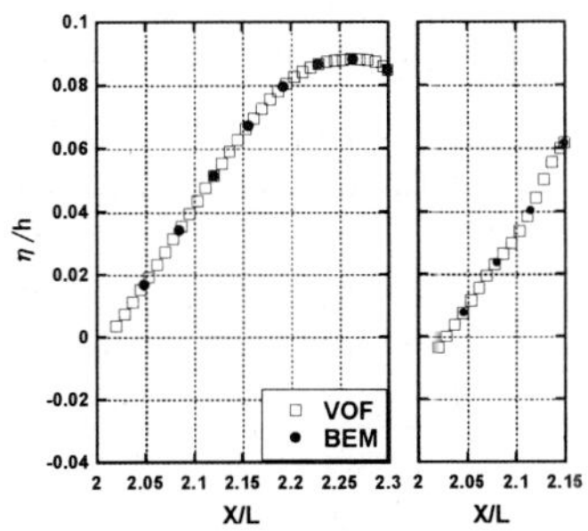

Figure 8. Water surfaces in T.D. (regular wave)

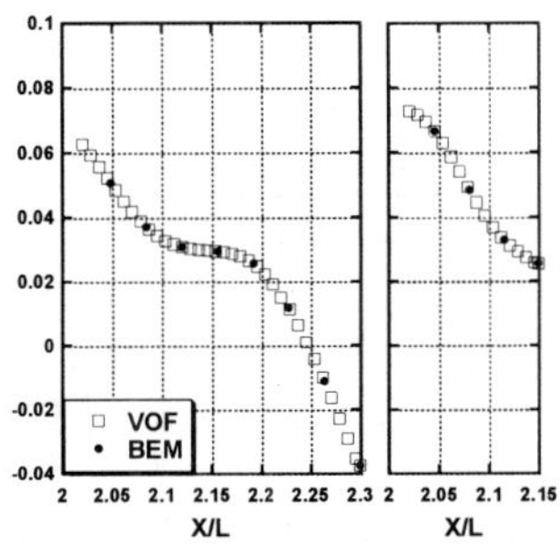

Figure 9. Water surfaces in T.D. (random waves)

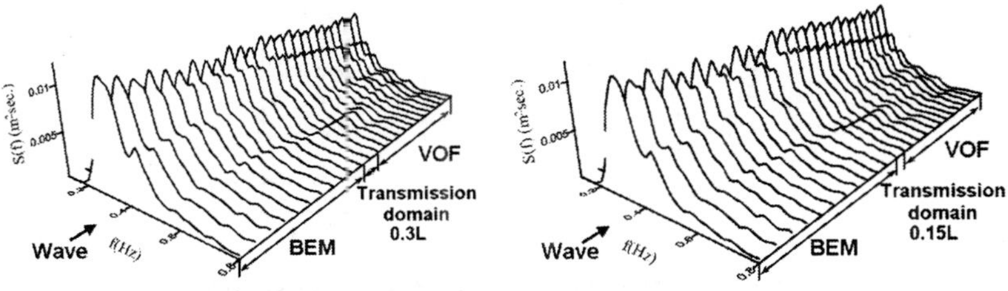

Figure 10. Spatial distribution of wave spectra

THREE DOMAIN BEM-VOF-BEM CALCULATION

When we are interested in wave fields leeward of the structure, it is often needed to calculate wave breakings and regenerated wave propagations. Considering the use of the two-domain calculation of BEM-VOF for this situation, the wave breakings and following wave propagations must be treated in VOF domain. This necessitates to use large VOF domain and leads to huge computational time and less accuracy. Since the wave field far from the structure can be expressed by BEM, it is well expected to reduce computational labor by reducing VOF domain as much as possible by further connecting BEM domain after VOF domain. Hence, the 3-domain BEM-VOF-BEM calculation was carried out for domains as shown in Fig. 11.

The above-mentioned connecting method which achieves two-way wave propagations was used in the left-hand side BEM-VOF connection. In the right-hand side VOF-BEM connection, one-way connecting method (single BEM calculation at each time step) was used. The initial wave was set 5th order stokes

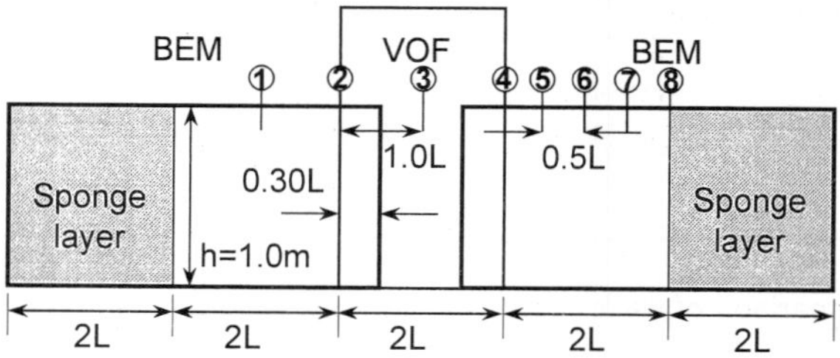

Figure 11. Domain for 3-domain calculation

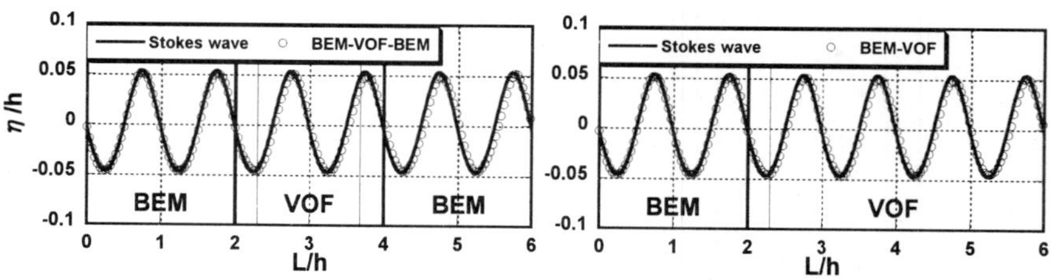

Figure 12. Comparison of wave profile between model and theoretical Stokes wave

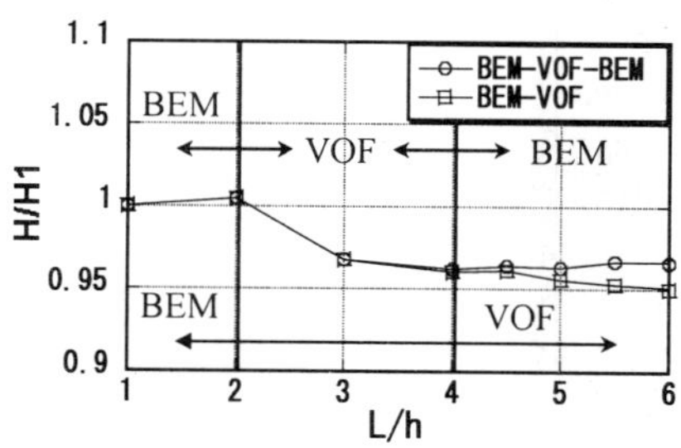

Figure 13. Comparison of the wave heights between 2-domain and 3-domain calculation

wave of H=0.1m, T=2.0sec. The running time of the calculation was set 30T and the time spacing of BEM and VOF was 0.01 and 0.005 sec. respectively. The two-domain calculation, in which the VOF-BEM domain in Fig. 11 was replaced with a single VOF domain, was also carried out for comparison.

Figure 12 shows the comparison of the spatial wave profile between the calculated values and theoretical ones at the moment of 30T in the case of BEM-VOF-BEM and BEM-VOF. It is seen that the numerical values agree with the theoretical ones very well in all the cases. Figure 13 shows the spatial change of the wave height ratio normalized with the wave height at the observing point No1. The decay in wave height along with wave propagation in VOF calculation has been well noticed. The 3-domain calculation shows the more uniform wave height comparing with the 2-domain calculation. In other words, the 3-domain calculation is more useful than the 2-domain calculation in the case that a wide range of regenerated wave field after wave breakings is involved.

Moreover, the running time of the 3-domain calculation is overwhelmingly faster than 2-domain one. The computational time ratio between the 3-domain and the 2-domain calculation is 1:3.1 (CPU : AMD-Athron 2700+, memories : 1GB). Hence, the reasonable reduction of VOF domain by using BEM-VOF-BEM calculation allows us an efficient computation, which also makes it practically possible to calculate the interaction between random waves and structures.

CONCLUSION

The strictly combined BEM-VOF method, which is much faster in computation and has a better accuracy in comparison with single VOF method, was developed. The method makes use of the excellent accuracy and time saving characters of BEM, and of VOF's ability to calculate intense wave motion such as wave breakings. The model can treat propagating and reflecting waves between BEM and VOF domains successfully. Since the combined BEM-VOF method can considerably reduce the domain size of the time consuming VOF method, the method is particularly useful for random wave calculations in which long time computation for large numbers of waves are required to obtain statistically meaningful results.

ACKNOWLEDGMENTS

A part of this study was accomplished with the support of the Grant-in-Aid for Young Scientists (B) No.16760413 by Ministry of Education, Culture, Sports, Science and Technology (MEXT), for which the authors express their appreciation.

REFERENCES

Biausser, B., Grilli, S.T. and Fraunié, P. 2003. Numerical simulations of three-dimensional wave breaking by coupling of a VOF method and a Boundary

Element method, *Proceedings of the 13th international Offshore and Polar Engineering Conference*, ISOPE, 333-339.

Biausser, B., Grilli, S.T. and Fraunié, P. 2004. Numerical analysis of the internal kinematics and dynamics of three-dimensional breaking waves on slopes, *Proceedings of the 14th international Offshore and Polar Engineering Conference*, ISOPE, 247-256.

Drevard, D., Marcer, R. Grilli, S.T., Fraunié, P. and Rey, V. 2005. Experimental Validation of a coupled BEM-Navier-stokes model for solitary wave shoaling and breaking, *Proceedings of the 5th international Ocean Wave Measurement and Analysis*,-.

Grilli, S.T., Gilbert, R.W., Lubin, P., Vincent, S., Astruc, D., Legendre, D., Duval, M., Kimmoun, O., Branger, H., and Devrard, D. 2004. Numerical Modeling and experiments for Solitary wave shoaling and breaking over a sloping beach, *Proceedings of the 5th international Ocean Wave Measurement and Analysis*, 306-312.

Guignard, S., Grilli, S.T., Marcer, R., and Rey, V. 1999. Computation of shoaling and breaking waves in nearshore areas by the coupling of BEM and VOF methods, *Proceedings of the 9th international Offshore and Polar Engineering Conference*, ISOPE, 304-309.

Itoh, K., Higuchi, Y., Toue, T., and Katsui, H. 2002. Numerical simulation of deformation of rubble structures by DEM and VOF. *Proceedings of 12th international Offshore and Polar Engineering Conference*, 714–721.

Lachaume, C., Biausser, B., Grilli, S.T., Fraunié, P., and Guignard, S. 1999. Modeling of Breaking and Post-breaking Waves on Slopes by Coupling of BEM and VOF Methods, *Proceedings of the 13th international Offshore and Polar Engineering Conference*, ISOPE, 353-359.

Maeno, S., Bierawski, L.G., and Fujita, S. 2002. A Study on dynamic behavior of the seabed around a seawall under wave motion Using the VOF-FEM Model, *Proceedings of the 12th International Offshore and Polar Engineering Conference*, VOL. 3, 579-603.

Sitanggang, K.I., Lynett, P., and Liu, P.L.-F. 2006. Development of a boussinesq-rans vof hybrid wave model, *Proceedings of 30th International Conference on Coastal Engineering*, ASCE, to be published.

Yan, S., Yamashiro, M., Yoshida, A., Ayukawa, K., and Irie, I. 2003a. Development of a combined method of BEM and VOF for analyzing wave motion, *Annual Journal of Civil Engineering in the Ocean,* JSCE, 83-88(*in Japanese*).

Yan, S., Yamashiro, M., Yoshida, and Irie, I. 2003b. A combined method of BEM and VOF for analyzing wave motion, *Proceedings of coastal engineering*, JSCE, 11-15(*in Japanese*).

Yan, S., Yamashiro, M., Yoshida, Inokuchi, Y., and Irie, I. 2004. An application of a combined method of BEM and VOF for solving non-linear wave-structure interactions, *Annual Journal of Civil Engineering in the Ocean,* JSCE, 689-694(*in Japanese*).

VELOCITY MEASUREMENTS IN AERATED REGION OF PLUNGING BREAKING WAVE

Ho-Joon Lim[1] and Kuang-An Chang[2]

Bubble velocity and turbulence intensity in the aerated region of a plunging breaker generated in a two-dimensional wave tank were investigated using a novel measurement technique - bubble image velocimetry (BIV). The mean velocity was calculated from ensemble averaging 20 instantaneous velocity fields. It was found that the maximum velocities right before and right after the first wave impinging are 1.89 m/s (1.46*C*) and 2.12 m/s (1.63*C*), respectively, with *C* being the phase speed. The maximum turbulence intensity is around 0.85 m/s that is approximately 45% of the maximum velocity.

INTRODUCTION

Breaking waves have been of great interest to researchers for several decades. The understanding of velocity and void fraction in breaking waves is essential in the study of coastal and offshore structures, scalar mixing, air entrainment, energy dissipation, and sediment transport. However, reliable quantitative measurement technique was not available until the late 1970s. Non-intrusive optical techniques for velocity measurement, such as laser Doppler velocimetry (LDV) and particle image velocimetry (PIV), have become available to obtain quantitative data during the last two decades. These techniques have been successfully applied in numerous laboratory experiments (Nadaoka *et al.*, 1989; Ting and Kirby, 1995; Chang and Liu, 1999). However, these methods have had great difficulties due to the signal drop out when applied to the study of multiphase flows, such as the aerated region in the upper part of the breaking waves. Only few experimental results in the aerated region were reported, including Jansen (1986) and more recently, Govendor *et al.* (2002). Jansen employed flow visualization technique with the use of ultraviolet light to illuminate fluorescent tracer particles fed into the air bubble region. Govender *et al.* obtained velocity and turbulence flow fields over the full depth by analyzing image structures composed of tracers and bubbles using direct digital correlation with high-speed strobing for illumination. However, these methods lack of a defined measurement plane (i.e., no light sheet) so the displacements of the particles and bubbles are not well defined.

The objective of this study is to investigate the flow property in the aerated region of plunging breaking waves with a newly developed technique called bubble image velocimetry (BIV) (Ryu *et al.*, 2005). The BIV technique directly

---

[1] Department of Civil Engineering, Texas A&M University, College Station. TX 77843, U.S.A.
[2] Department of Civil Engineering, Texas A&M University, College Station. TX 77843, U.S.A.

uses bubbles and gas-liquid interfaces as the tracer and measures the fluid velocity by correlating the 'texture' of bubbles and gas-liquid interfaces in the images. No laser light sheet is needed while the depth of field is limited to minimize the error. The BIV technique measures velocity mainly in the multiphase region where the PIV technique does not work well or does not work at all.

EXPERIMENTAL SETUP

The experiments were performed in a 2D glass-walled wave tank, which is 35 m long, 0.91 m wide, and 1.2 m deep equipped with a flap-type wavemaker. A 1:5.5 sloping beach with a layer of horsehair is at the other end of the tank to absorb the wave energy and reduce reflection. The water depth was maintained at 0.80 m throughout the experiments. Wave focusing technique was used to generate the desired plunging breaking waves in the deep water.

The BIV technique was used to obtain the velocity field in the aerated region after the wave impinges its front surface and breaks. The images were captured by a high speed camera mounted with a 105 mm focal lens. The camera has a resolution of 1024×1024 pixels, a maximum frame rate of 1200 frames per second (fps), and a 10 bit dynamic range. The camera framing rate was set at 500 fps and the aperture was set at f/1.8 through out the experiments. Regular 600 W light bulbs with reflecting mounts were used to illuminate the flow. No lasers are needed in the measurements.

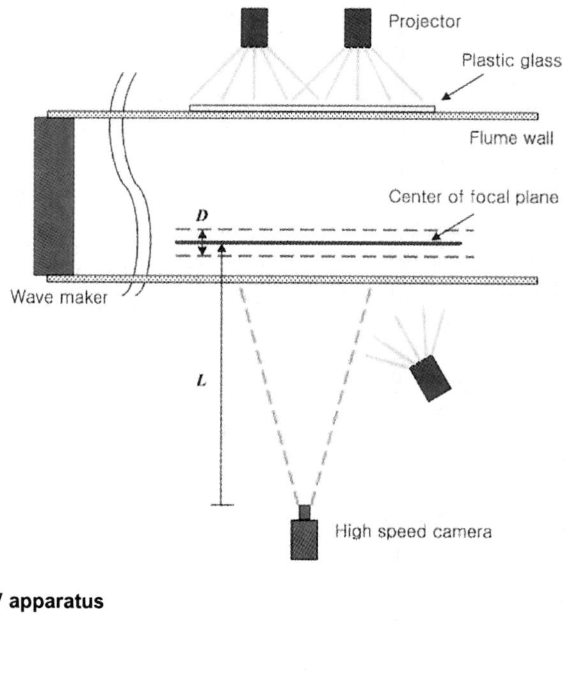

Figure 1. BIV apparatus

The apparatus of BIV is shown in Fig. 1. The depth of field (DOF) for the captured images is 21 cm with the camera located at 4.7 m in front of the center of the DOF, resulting in an uncertainty of 2.2% in the acquired images for later velocity determination. The principle and validation of BIV are described in details by Ryu *et al* (2005).

Three fields of view (FOV) were used to cover the entire aerated region of the plunging wave in the BIV measurements. Note that the origin $(x, z) = (0, 0)$ is at the impinging point and the stationary water level, as shown in Fig. 2. The size of three FOVs is fixed as 66.0 cm × 49.5 cm. The locations of the FOVs starts at $(x, z) = (-5.0, -30.1)$ cm, $(x, z) = (54.3, -32.5)$ cm, and $(x, z) = (141.0, -32.6)$ cm. Note that there is an overlap of 6.7 cm between FOVs 1 and 2 whereas there is a gap of 21 cm in between FOVs 2 and 3 due to the steel column of the wave tank.

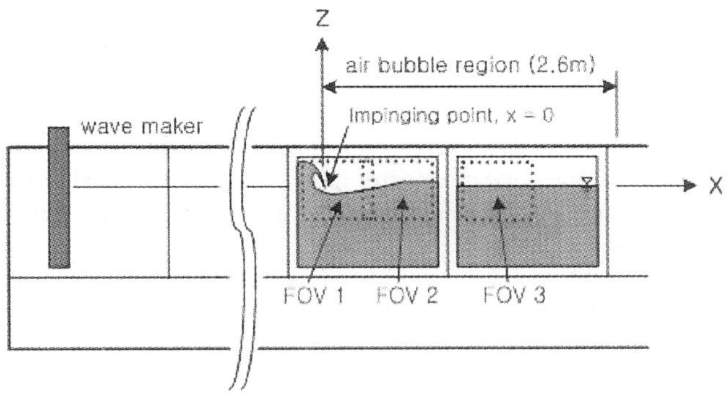

Figure 2. BIV fields of view

The time interval between the recorded images is 2.0 ms that is equal to the time separation between the consecutive frames captured by the high speed camera. The images were processed using commercial software from LaVision Inc. The velocity fields were calculated using an adaptive multi-pass algorithm with an initial interrogation window size of 32×32 pixels and a final window size of 16×16 pixels with a 50 % overlap. A median filter was subsequently applied to eliminate the spurious vectors in the calculated velocity maps. The experiments were repeated for 20 times with the same test condition when taking BIV measurements at each FOV. The mean velocity and turbulence intensity were calculated from ensemble averaging the 20 instantaneous velocity fields.

RESULTS AND DISCUSSION

The free surface elevation was measured using a wave gauge located right in front of the impinging point at $x = -0.2$ m. Note that the impinging point is located at $x = 0$ and $t = 0$ indicates the moment of impingement. The measured waves profile is shown in figure 3; the wave height of the plunging breaker is 20.4 cm.

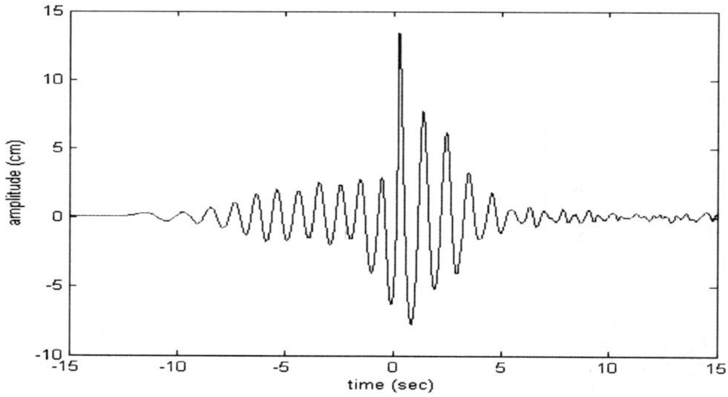

Figure 3. Free surface elevation measured at *x* = -20 cm

The ensemble-averaged mean velocity fields at FOVs 1, 2, and 3 are shown in figure 4. The mean velocity was extracted from the measured instantaneous velocity as $u_i = \langle u_i \rangle + u'_i$ with u being the instantaneous velocity, subscript i the direction, $\langle u_i \rangle$ the mean velocity, and prime the turbulent fluctuations. Note that the impinging point is at (x, z) = (0, 0.5) cm. The time separation between the consecutive panels in the figure is 0.06 s. Figures 4(a1) to 4(d1) show the velocity field in FOV 1 right after the wave impinging onto its front surface and created the first splash. The maximum velocity (hereafter "velocity" indicates the ensemble-averaged *mean* velocity) right before the impinging is 1.89 m/s that is equivalent to 1.46C with C = 1.30 m/s being the phase speed of the primary wave (the wave that led to the breaking event). The maximum velocity occurred after the impinging is 2.12 m/s (1.63C) at the location of (x, z) = (23.2, 2.7) cm at t = 0.12 s, as shown in figure 4(b1). This velocity is, surprisingly, greater than the maximum velocity prior to the wave impingement. It may be caused by the splash-up that reacts due to the impingement is accelerated by the forward motion of wave surface.

By examining the velocity field in figure 4 we see all the fluid particles are in forward motion, not as indicated in some roller models (e.g., Schaffer et al., 1993). However, if we are able to mount the camera moving with the wave, the pattern can be revealed. We thus subtract the phase speed from the measured velocity in figure 4 to examine the velocity pattern.

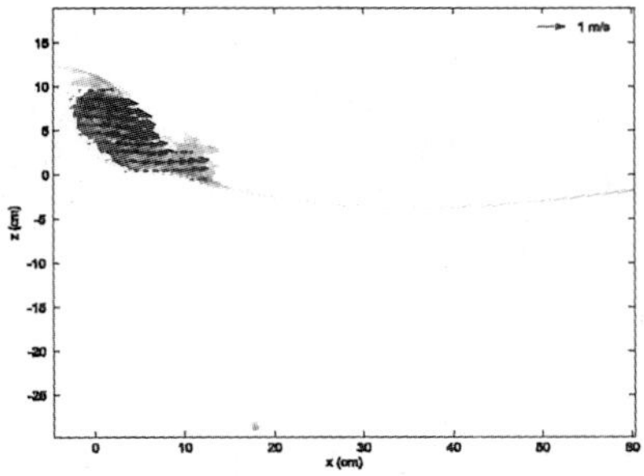

(a1) $t = 0.06$ s

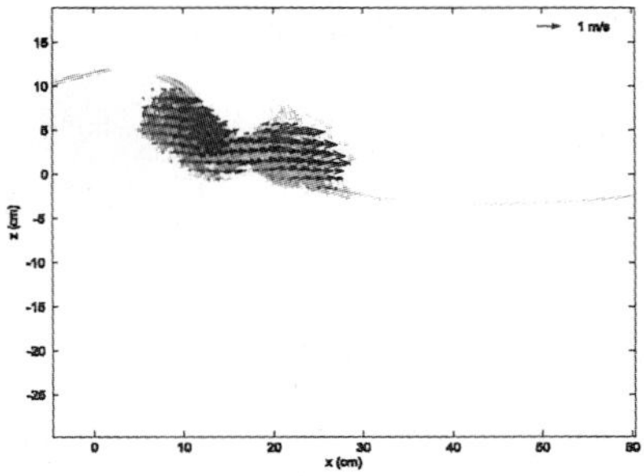

(b1) $t = 0.12$ s

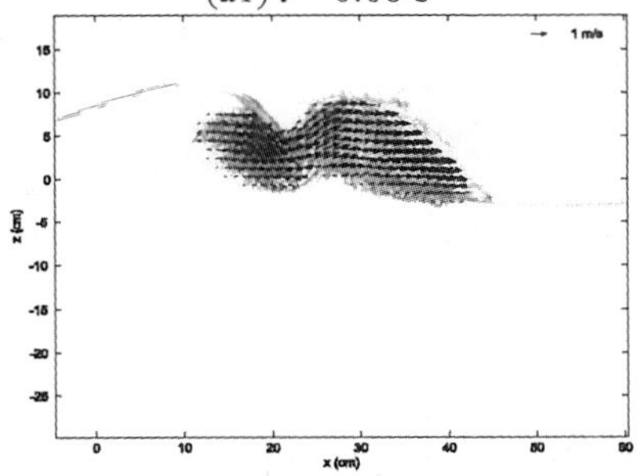

(c1) $t = 0.18$ s

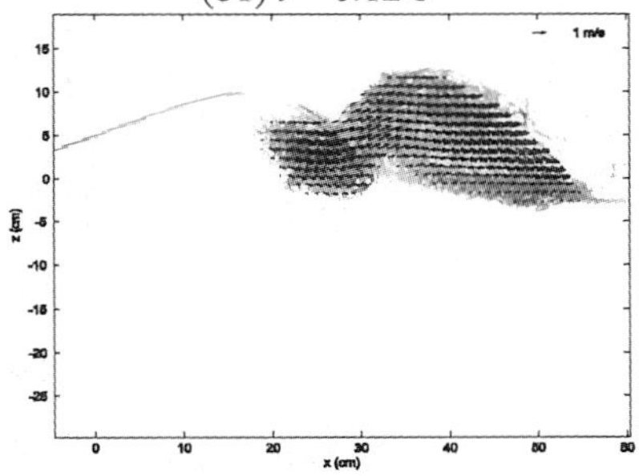

(d1) $t = 0.24$ s

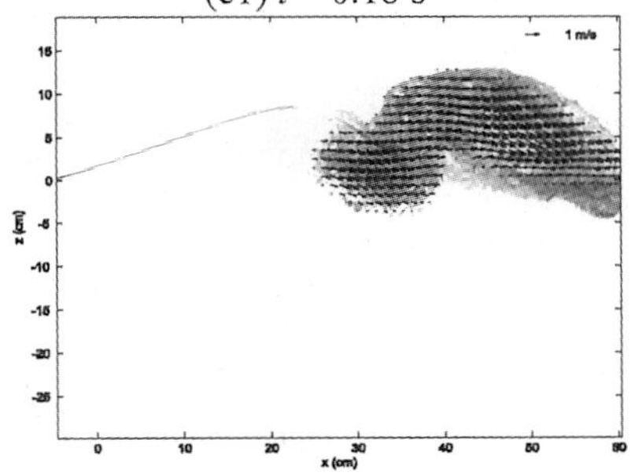

(e1) $t = 0.30$ s

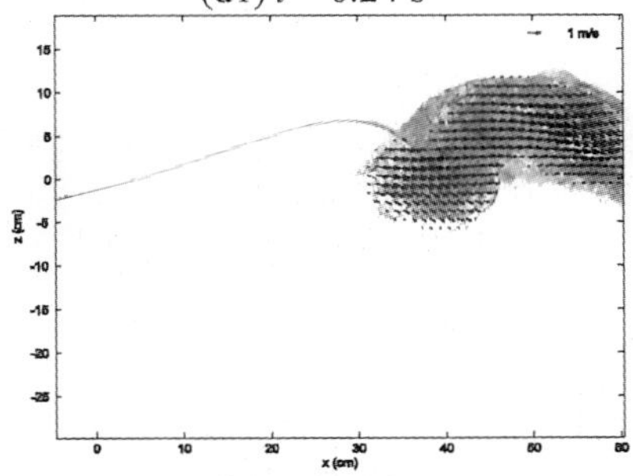

(f1) $t = 0.36$ s

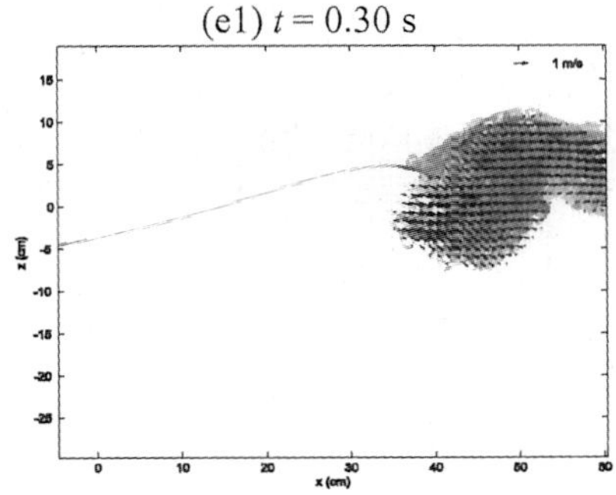

(g1) $t = 0.42$ s

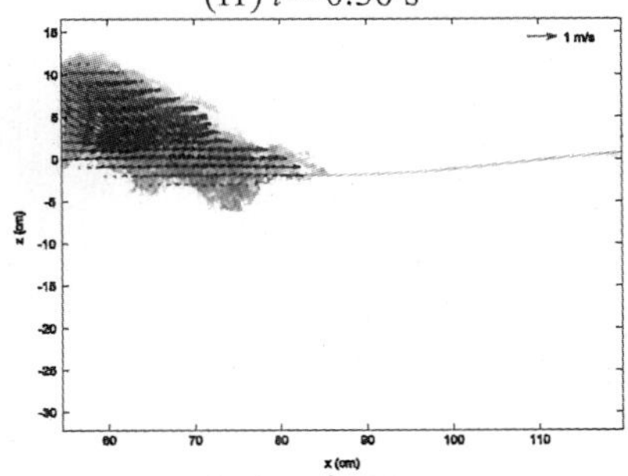

(g2) $t = 0.42$ s

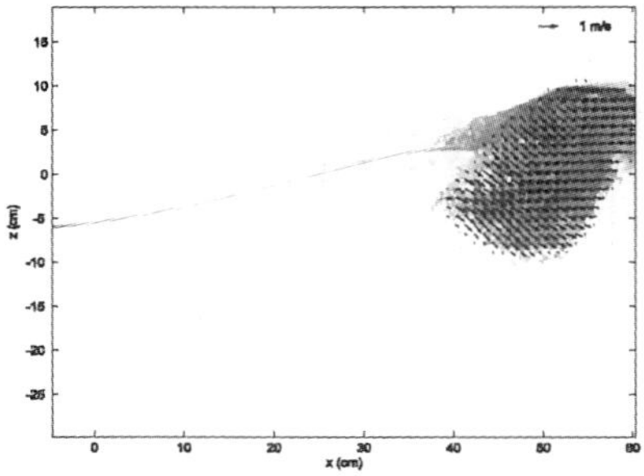

(h1) $t = 0.48$ s

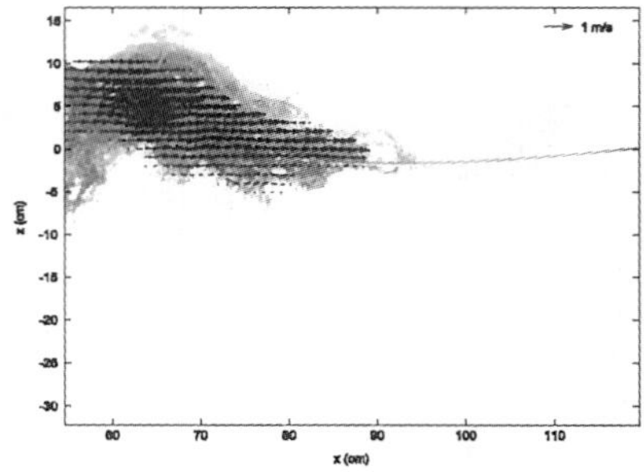

(h2) $t = 0.48$ s

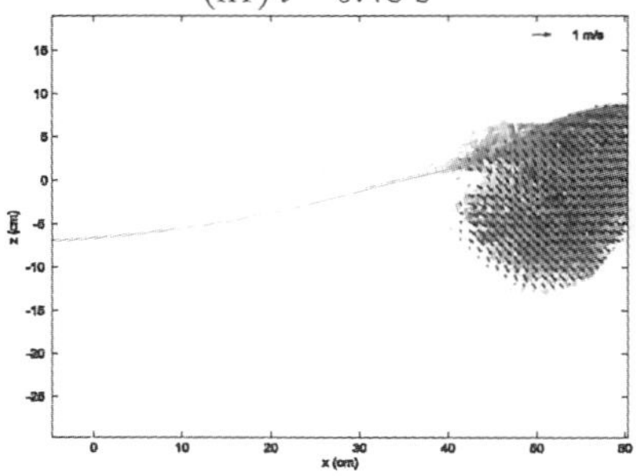

(i1) $t = 0.54$ s

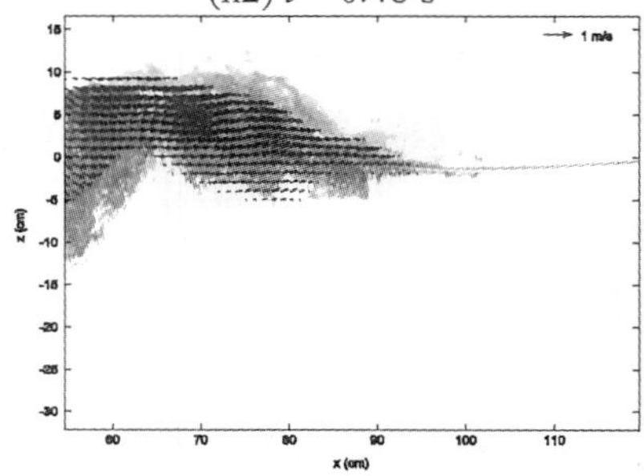

(i2) $t = 0.54$ s

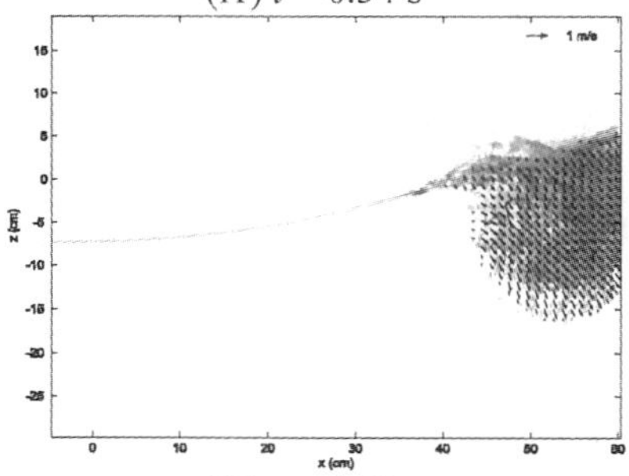

(j1) $t = 0.60$ s

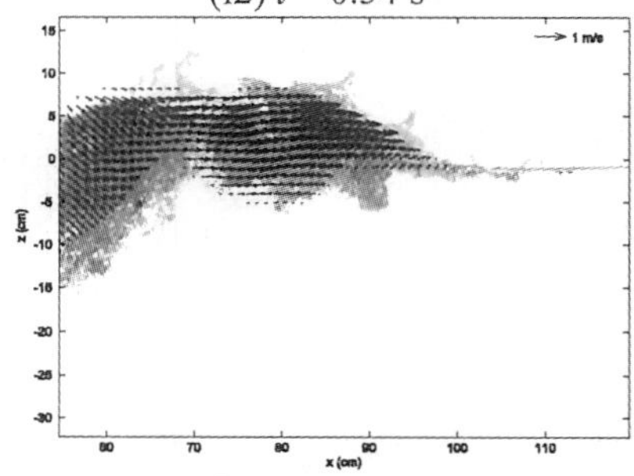

(j2) $t = 0.60$ s

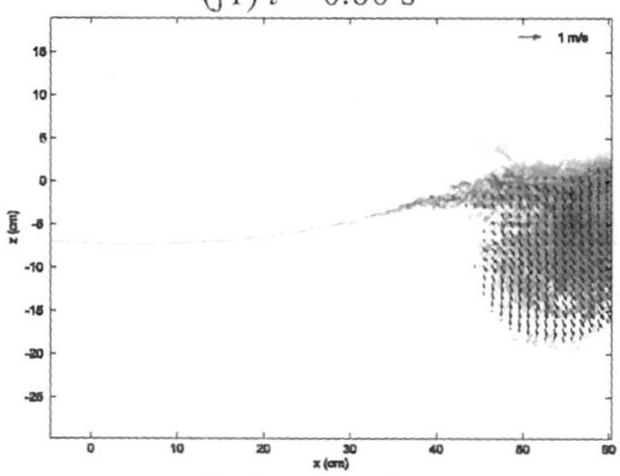

(k1) $t = 0.66$ s

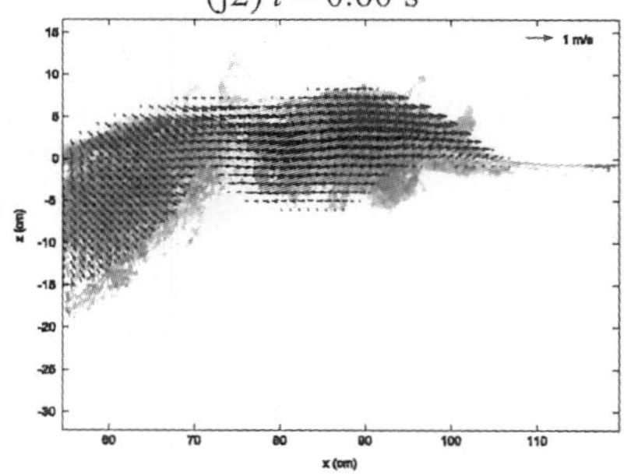

(k2) $t = 0.66$ s

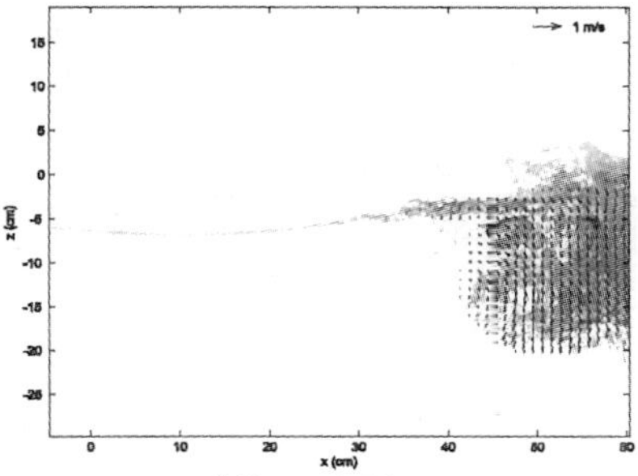

(l1) $t = 0.72$ s

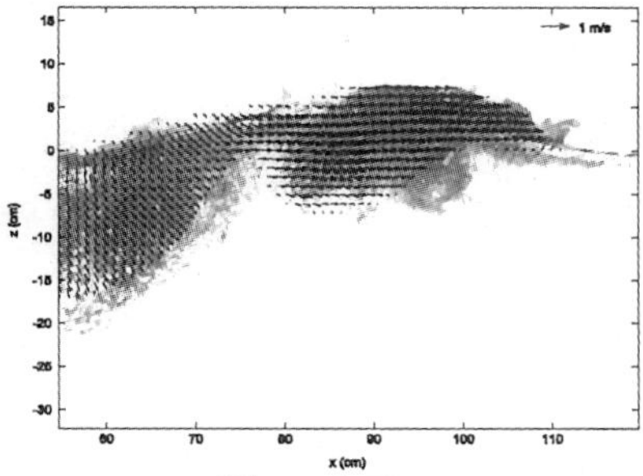

(l2) $t = 0.72$ s

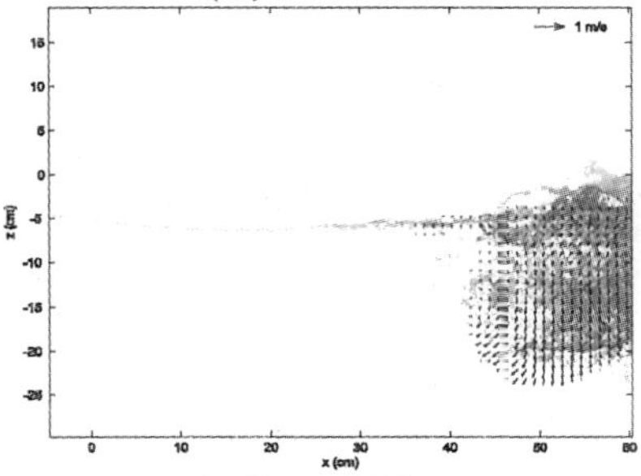

(m1) $t = 0.78$ s

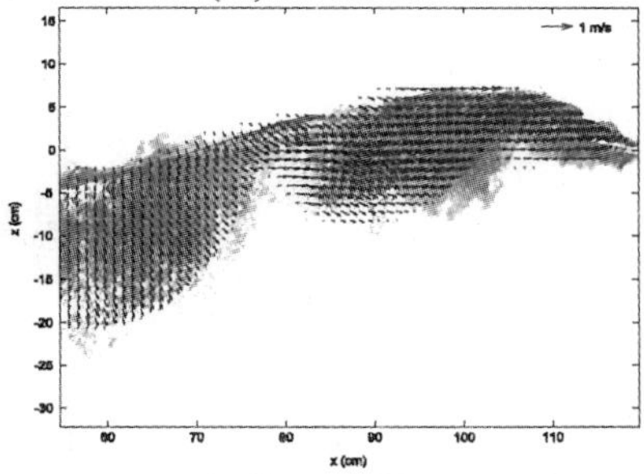

(m2) $t = 0.78$ s

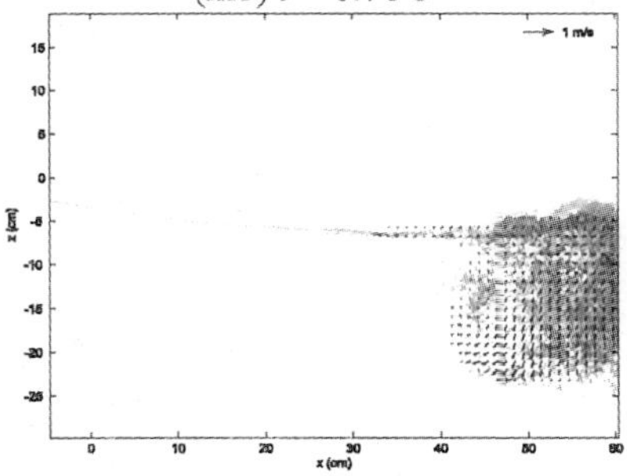

(n1) $t = 0.84$ s

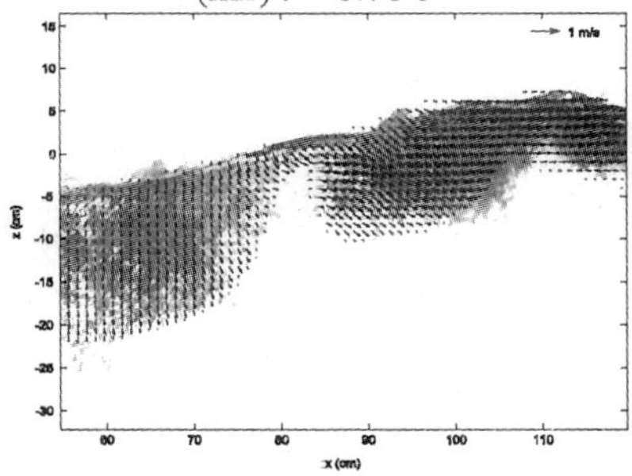

(n2) $t = 0.84$ s

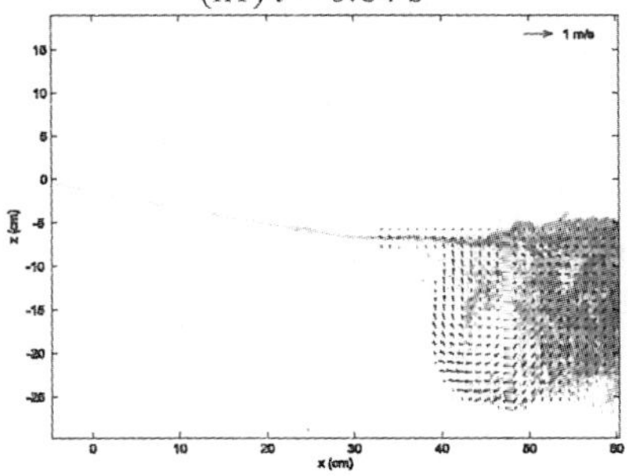

(o1) $t = 0.90$ s

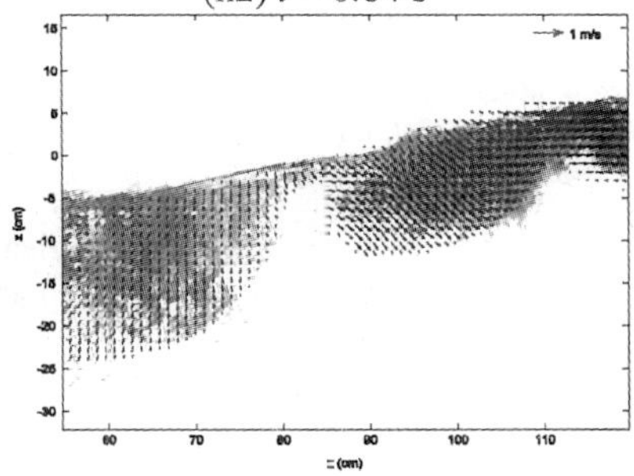

(o2) $t = 0.90$ s

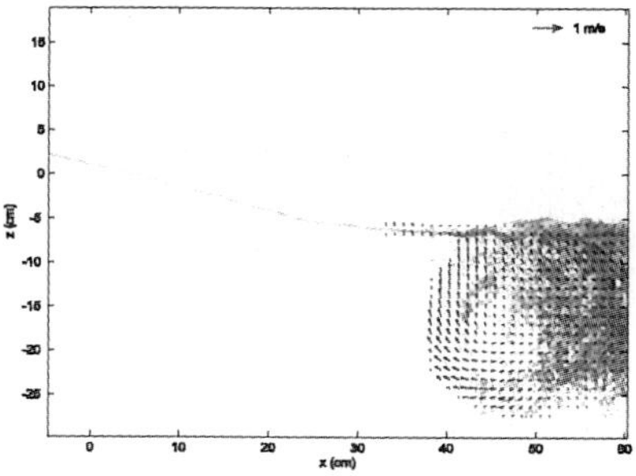

(p1) $t = 0.96$ s

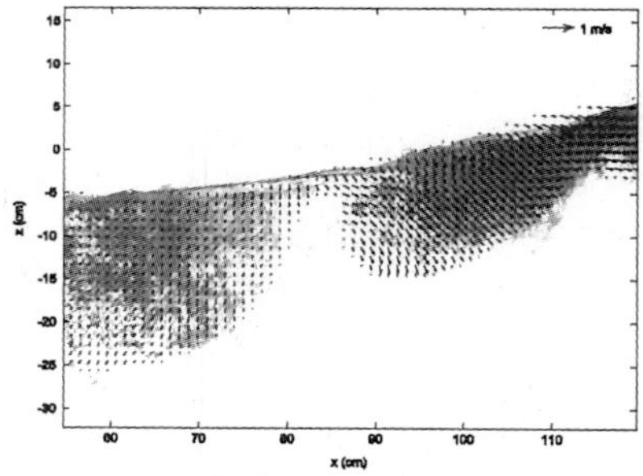

(p2) $t = 0.96$ s

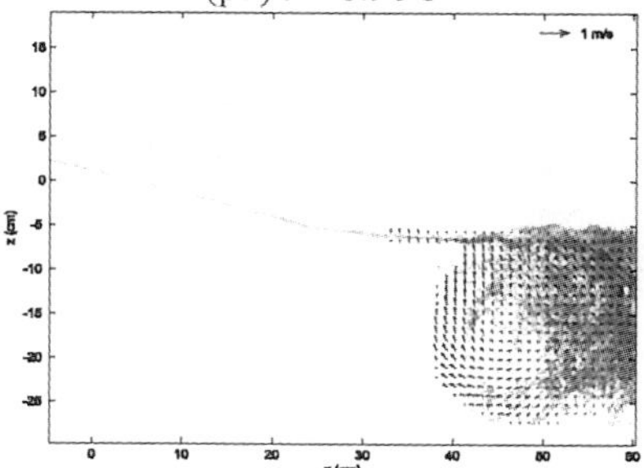

(q1) $t = 1.02$ s

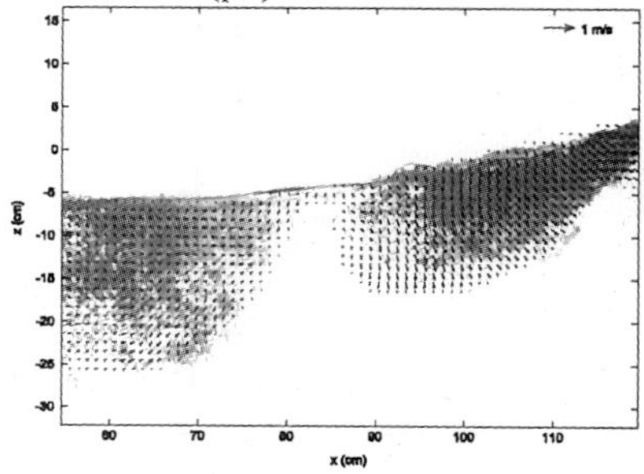

(q2) $t = 1.02$ s

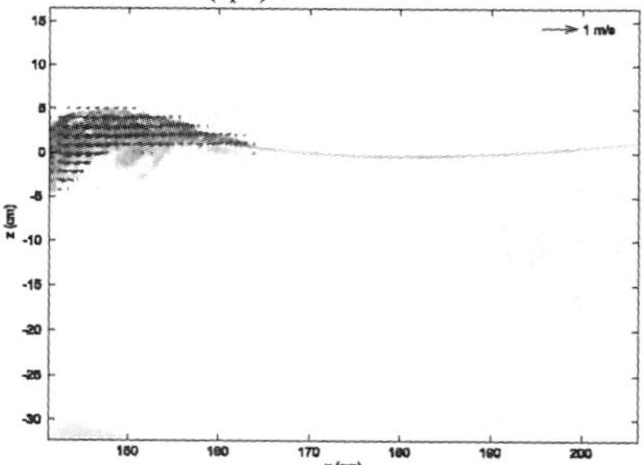

(r3) $t = 1.02$ s

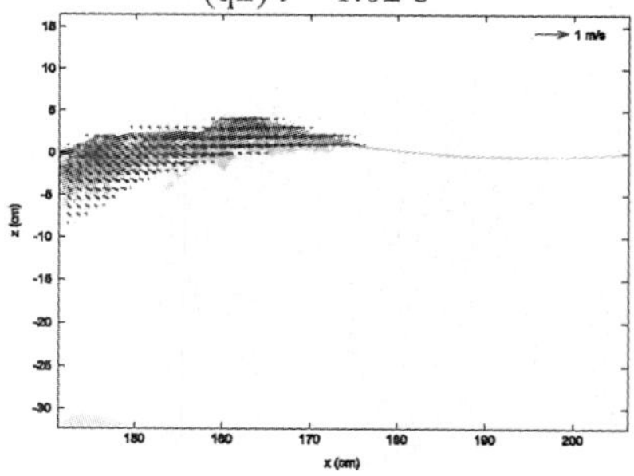

(s3) $t = 1.14$ s

Figure 4. Mean velocities in the aerated region. The subfigure number indicates the corresponding FOV, i.e., "a1" is the subfigure "a" from FOV "1".

Figure 5 shows the result with clear circulation of the rolling pattern, corresponding to Fig. 4(c1). The roller pattern is somewhat different from what the researchers predicted. It shows two rollers with one formed by the curving down of the overturning jet of the breaking wave while the other from the impingement of the first splash.

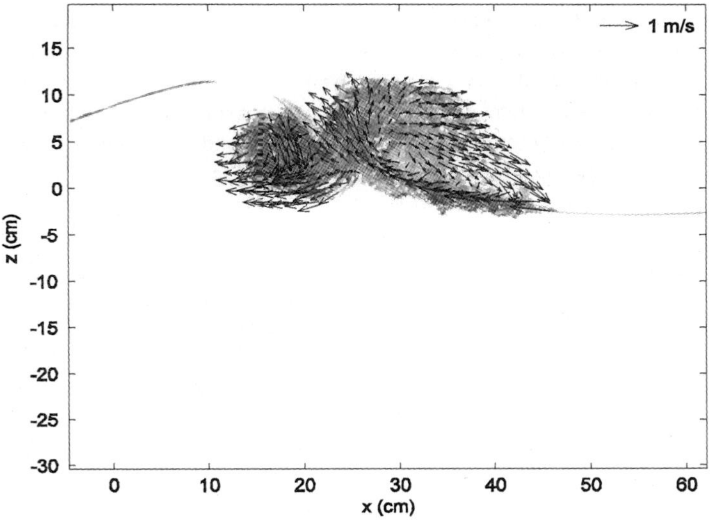

Figure 5. Example of roller pattern revealed by moving with the phase speed at t = 0.18 s (corresponding to Fig. 4(c1)).

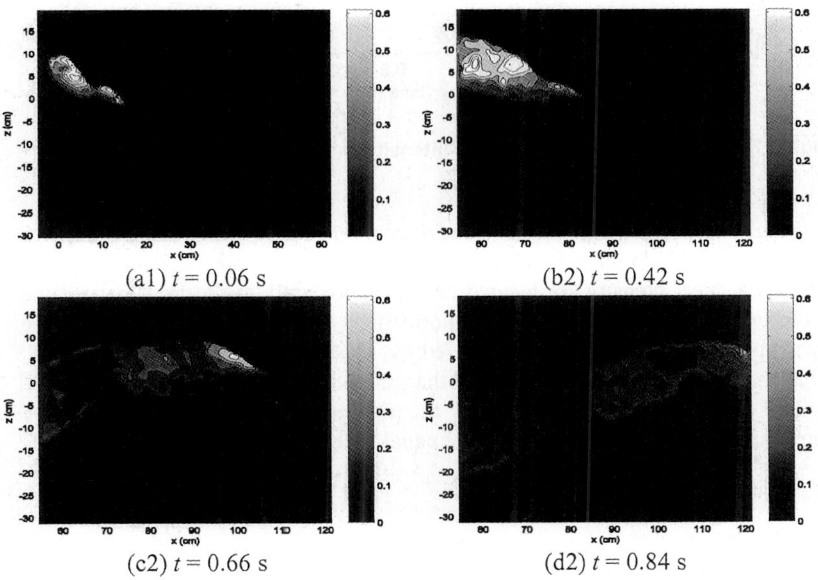

Figure 6. Turbulence intensity

The turbulent intensity, defined as $I=\sqrt{u'u'+w'w'}$, is shown in figure 6. The generation of turbulence starts from the impinging of the wave. The maximum turbulent intensity is around 0.85 m/s that is approximately 0.65C or 45% of the maximum velocity. Turbulence intensity has local maxima at each impingement in a breaking event (first impingement: $t \approx 0.06$ s; second impingement: $t \approx 0.42$ s; third impingement: $t \approx 0.66$ s ~ 0.84 s) as shown in Fig 7.

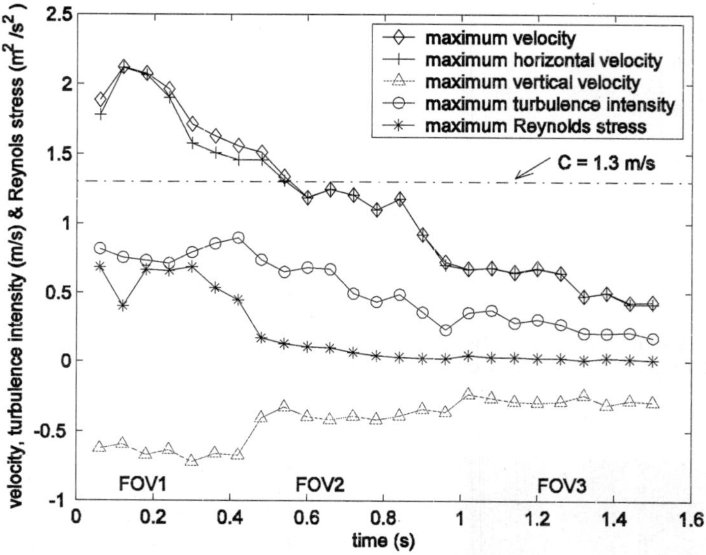

Figure 7. Maximum velocity, turbulence intensity and Reynolds stress

Fig. 7 shows the time series of the maximum velocity, maximum turbulence intensity and maximum Reynolds stress. Note that the local maximum velocity and turbulence intensity are located nearly coincidently at the front part of the wave crest. The time series of the normalized maximum velocity, turbulence intensity and Reynolds stress normalized by, V_{max} and V_{max}^2 are shown in Fig. 8. That the maximum velocity is larger than the wave celerity lasted almost 0.7T with T being the wave period. In the figure the dimensionless turbulence intensity is approximately constant and equal to 0.4 to 0.5.

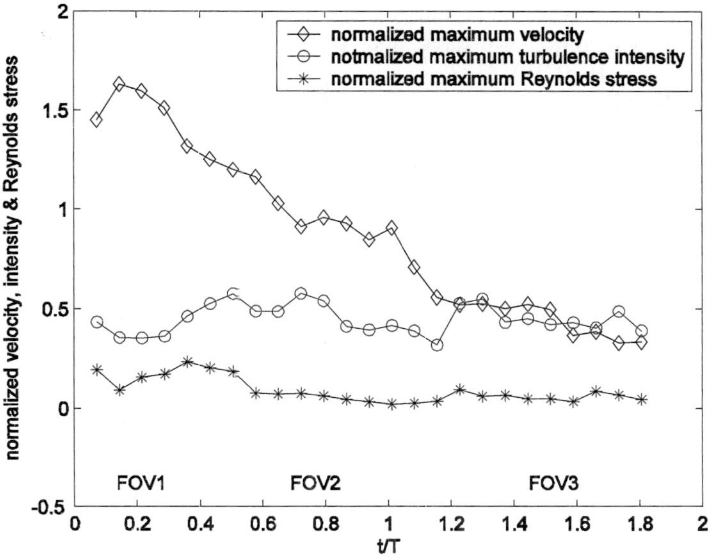

Figure 8. Dimensionless maximum velocity, turbulence intensity and Reynolds stress normalized by *C*, V_{max} and V_{max}^2, respectively.

SUMMARY AND CONCLUSIONS

BIV can be successfully applied to measure velocities in the aerated region of a breaking wave by directly tracing air bubbles and air-water interfaces. When waves break, the maximum velocity in the plunging breaker reaches $1.46C$, while the maximum velocity reaches $1.63C$ at the initial splash-up rebounded after the first impingement. It may be caused by the splash-up that reacts due to the impingement is accelerated by the forward motion of wave surface. The duration when the maximum velocity is greater than the wave celerity lasts almost $0.7T$ with T being the wave period. The maximum turbulent intensity is approximately $0.5V_{max}$ over the entire breaking processes.

The BIV technique is been successfully applied in wave breaking measurements. It could also be applied to many other multiphase flows in hydraulic and ocean engineering, such as hydraulic jump, dam-break flow, bubble plume, and wave overtopping.

REFERENCES

Chang, K.-A. and Liu, P. L.-F. 1999. Experimental investigation of turbulence generated by breaking waves in water of intermediate depth, *Physics of Fluids*, 11, 3390-3400.

Govender, K., Mocke, G. P. and Alport, M. J. 2002. Video-imaged surf zone wave and roller structures and flow fields, *Journal of Geophysical Research*, 107(C10), 3177.

Jansen, P. C. M. 1986. Laboratory observations of the kinematics in the aerated region of breaking waves, *Coastal Engineering*, 9, 463-477.

Nadaoka, K., Hino, M. and Koyano, Y. 1989. Struc-ture of turbulent flow field under breaking waves in the surf zone, *Journal of Fluid Mechanics*, 204, 359-387.

Schaffer, H.A., Madsen, P.A. and Deigaard, R. 1993. A Boussinesq model for waves breaking in shallow water, *Coastal Engineering*, 20, 185-202.

Ting, F.C.K and Kirby, J.T. 1996. Dynamics of surf-zone turbulence in a spilling breaker, *Coastal Engineering*, 27, 131-160.

Ryu, Y, Chang, K.-A. and Lim, H.-J. 2005. Use of bubble image velocimetry for measurement of plunging wave impinging on structure and associated greenwater, *Measurement Science and Technology*, 16, 1945-1953.

LABORATORY STUDY OF TURBULENT FLOW STRUCTURES IN BREAKING WAVES BY PARTICLE IMAGE VELOCIMETRY

Francis C. K. Ting[1]

The instantaneous turbulent velocity field associated with a broken solitary wave propagating on a plane slope was measured in the longitudinal transverse plane in the middle part of the water column and near the bottom using particle image velocimetry. These measurements showed that large-scale turbulence first arrived at the bottom in the form of a downburst of turbulent fluid. The downburst diverged at the bed and quickly spread into the surrounding non-turbulent flow. The transverse spacing between downbursts was up to several times the local still water depth. In the middle part of the water column, downbursts arrived shortly after the wave crest passed. Each downburst was followed immediately by two counter-rotating vortices. The latter must extend to the free surface since vortices cannot end in the interior of the fluid. Counter-rotating vortices were a prominent feature of the flow field at mid-depth and each vortex had a typical length scale of ½ to one water depth. It was suggested that these vortices were produced by bending and stretching of the primary spanwise vorticity generated in the wave breaking process, possibly as a result of three-dimensional deformation in the water surface.

INTRODUCTION

The turbulent flow field under breaking waves is characterized by organized flow structures which occur intermittently in space and time. These coherent structures play an important role in sediment suspension in the surf zone. Nevertheless, our present understanding of the characteristics of the large eddies in breaking waves remains largely qualitative. The main reason is that there are few detailed measurements of these large-scale flow structures. In a recent study, we have measured the instantaneous turbulent velocity field in the near-bed region of a broken solitary wave propagating on a plane slope using the stereoscopic particle image velocimetry (PIV) technique. The advantage of studying a single breaking wave is that the wave breaking process and the evolution of the associated turbulent velocity field can be studied separately from the effects of return flow and residual turbulence. Those measurements have confirmed that large-scale organized flow structures were the sources of most of the turbulence energy and turbulent shear stresses. The type of organized flow structures most frequently observed was a downburst of turbulence descending from above and diverging at the bed (Ting, 2006). In this present paper, we present the results from a second set of PIV measurements taken in the middle part of the water column at about the same cross-shore location and with the same incident wave conditions. These measurements have provided new

[1] Department of Civil and Environmental Engineering, South Dakota State University, Brookings, South Dakota, 57007, USA

information on the structure of the downbursts at an earlier stage of development.

Several previous studies that are closely related to the present work are briefly mentioned here. In a laboratory study of regular breaking waves on a plane beach, Nadaoka et al. (1989) observed that the wave breaking process generates spanwise vortices with axis of rotation parallel to the wave crest. Behind the wave crest, entrained air bubbles revealed the presence of obliquely descending vortices with counter-rotating vorticity. Nadaoka et al. (1989) suggested that the formation and evolution of the obliquely descending vortices are related to the stretched velocity field around the saddle point of streamlines between two adjacent spanwise vortices.

Watanabe et al. (2005) computed the three-dimensional vortex structures under spilling and plunging regular waves using large-eddy simulation (LES). They found that perturbation and stretching of vorticity in the saddle region between the rebounding jet and the primary spanwise vortex lead to the formation of stretched vortex loops with counter-rotating vorticity, which subsequently envelop the primary spanwise vortices. They suggested that these stretched vortex loops were the obliquely descending vortices previously observed by Nadaoka et al. (1989) and others.

In a laboratory study of spilling regular waves, Kubo and Sunamura (2001) observed obliquely descending vortices and a new type of coherent structures which they called downbursts. A downburst is formed when water spilling down the front face of the breaking wave produces a turbulent, aerated water mass, and the collapse of the subsequent splash-up in reverse breaking causes the aerated water mass to descend toward the bottom without marked rotation. Kubo and Sunamura (2001) observed that downbursts produced more sediment movement than obliquely descending vortices.

In two numerical studies, obliquely descending vortices in breaking waves were reproduced in large-eddy simulations by Watanabe and Saeki (1999), and Christensen and Diegaard (2001). More recently, Rogers and Dalyrmple (2004) have modeled the formation of downbursts using smoothed particle hydrodynamics (SPH). In the author's opinion, the forms and characteristics of organized flow structures found in numerical simulations may depend critically on how the free surface deformation is modeled. At present, it is only possible to assess these models by comparing the predicted and measured flow fields. If accurate measurements can be obtained close to the free surface, the measured data may also shed light on the generation mechanism(s) of the large eddies.

EXPERIMENTAL PROCEDURE

Figure 1 shows a schematic drawing of the experimental arrangement. The experiment was conducted in a tilting flume, 0.9 m wide, 0.75 m deep and 25 m long. A solitary wave was generated with a piston type wave maker installed over a horizontal false bottom. The generated wave immediately propagated onto a 1 in 50 plane slope (the bottom of the tilted flume). The still water depth and wave height at the wave maker were 0.30 m and 0.22 m, respectively. The solitary wave broke as a spilling/weakly plunging wave. The maximum wave height measured just before breaking was 0.229 m, and the corresponding still water depth was 0.236 m. Stereoscopic PIV measurements were taken in a horizontal plane at a height of 70 mm above the flume bed where the still water depth was 152 mm. At this cross-shore location, the wave-height-to-water-depth ratio was approximately constant and equaled to 0.60 (Ting, 2006). In this study, the wave conditions and experimental set-up were the same as those used in Ting (2006) unless stated otherwise.

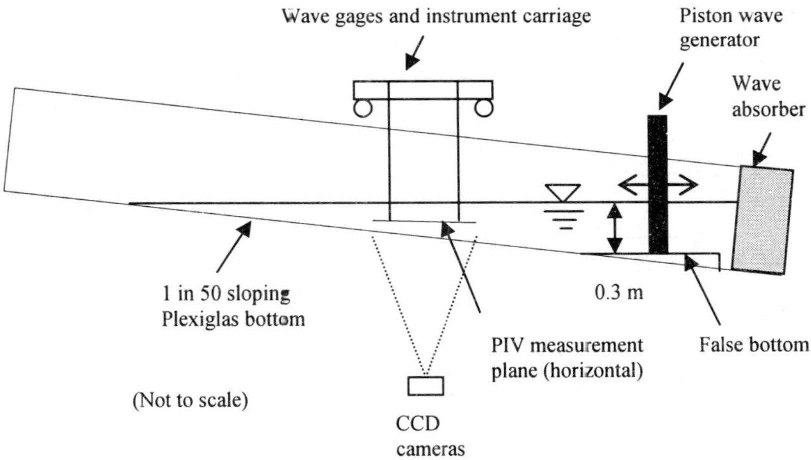

Figure 1. Experimental arrangement.

In the PIV measurements, two CCD cameras were positioned underneath the flume to view the horizontal light sheet through a Plexiglas bottom. Silver-coated hollow glass spheres with a mean particle diameter of 13 microns and density of 1.3 g/cc were used as seeding particles. The field of view (FOV) was approximately 250 mm × 250 mm. A total of 38 test runs were completed under identical experimental conditions. In each test run, 100 instantaneous velocity vector fields (frames) were obtained. The sampling interval was 1/15 s. An ensemble-averaged velocity field was computed for each of the 100 time steps. The ensemble-averaged velocity field for each time step was subtracted from the corresponding measured velocity field to yield the instantaneous turbulent velocity field. Figure 2 shows a comparison of the measured ensemble-averaged velocities obtained from the present PIV measurements and the ADV measurements previously reported in Ting (2006). Note that the PIV velocities have been averaged over the entire FOV, whereas the ADV velocities are "point" measurements. Because of spatial averaging, the PIV velocities are generally smaller in magnitude. This effect is most pronounced in the vertical velocity under the back face of the wave (bottom plot), where the longitudinal gradient in vertical velocity is large. Inspection of the ensemble-averaged velocity fields shows that the vertical velocity from the PIV indeed reaches a minimum value of about -0.05 m/s behind the wave crest consistent with the ADV measurements.

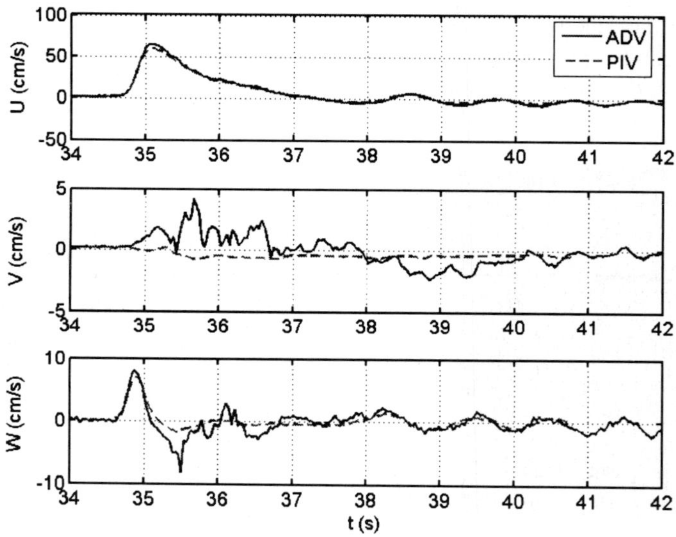

Figure 2. Comparison of PIV and ADV velocities. These measurements were taken at a height of 70 mm (center of FOV) above the flume bottom. The PIV velocities have been averaged over the entire FOV.

RESULTS

Among the 38 test runs conducted in this study, two distinct types of flow structures were observed: a downburst of turbulence descending into the FOV from above (~40%), and counter-rotating vortices (~80%). Downbursts were typically observed shortly after the wave crest passed, when the organized wave-induced velocity was large. Because of this, these structures were quickly advected out of the FOV and could be observed for only a short period of time (usually 2 to 5 frames). Downbursts were often followed by one or more pairs of counter-rotating vortices, which passed through the FOV at a later time. In about 40% of the test runs, counter-rotating vortices were the only type of large, organized flow structures observed. No organized flow structures were observed in two test runs.

Figure 3 shows an example of a downburst from test 36, frames 15 and 16. From the ensemble-averaged velocity field, we have determined that the wave crest arrived at frame 15 and the main body of the solitary wave passed through the FOV completely after frame 48 (15 frames = 1 s). The solitary wave propagated in the X direction, and the Y direction is across the flume. Frame 15 shows a downburst of turbulence descending into the FOV from above. The turbulence velocities inside the downburst are directed downward and outward. The corresponding PIV image on the left shows the presence of many bright spots in the region occupied by the downburst, which we believe to be reflected light from entrained air bubbles. In some PIV images, we could see light shadow behind some air bubbles which indicates that those air bubbles were in the light sheet plane.

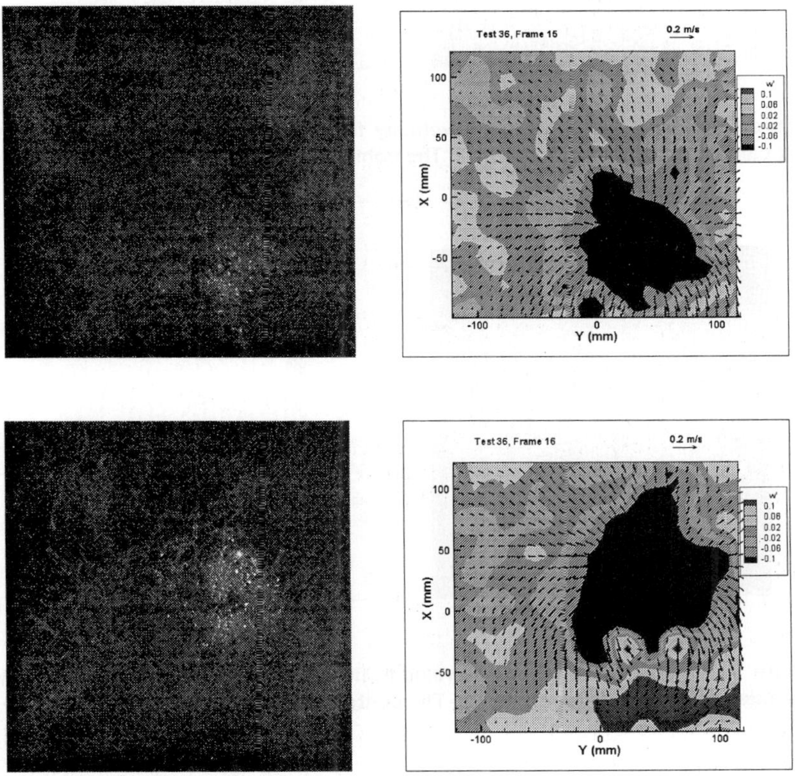

Figure 3. Instantaneous turbulent velocity fields and the corresponding PIV images (left camera) from test 36, frames 15 and 16. The contour variable is *w'* in m/s.

Frame 16 shows the instantaneous turbulent velocity field 1/15 s later. Two counter-rotating vortices have emerged from behind the downburst. These vortices continued to grow in size until they were advected out of the FOV (after frame 18). The corresponding PIV image on the left again indicates the presence of entrained air bubbles in the region occupied by the downburst.

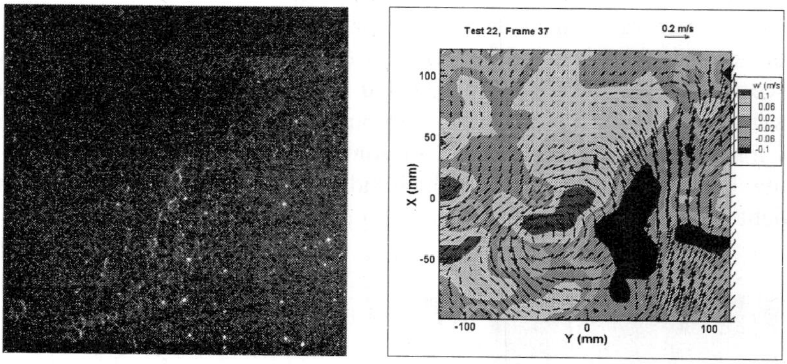

Figure 4. Instantaneous turbulent velocity field and the corresponding PIV image (left camera) from test 22, frame 37. The contour variable is *w′* in m/s.

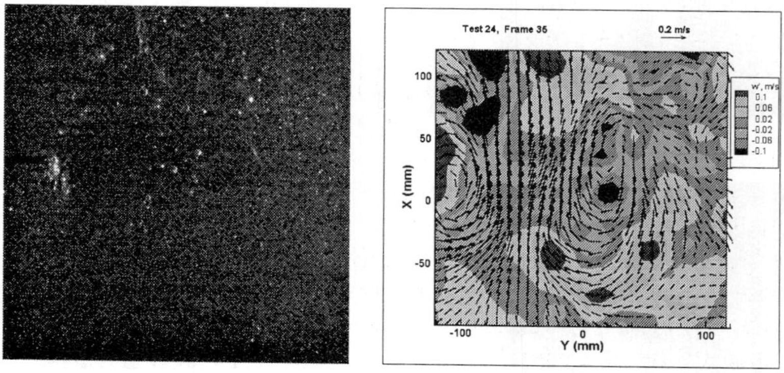

Figure 5. Instantaneous turbulent velocity field and the corresponding PIV image (left camera) from test 24, frame 35. The contour variable is *w′* in m/s.

In most test runs, one or more pairs of counter-rotating vortices were observed, usually later in time. Two typical examples are shown in Figures 4 and 5. Because of their large sizes, only part of the vortex structures was captured. These measurements showed that a region of large horizontal velocities existed between the vortices. The size of each vortex is comparable to the local still water depth, that is, around 150 mm. The corresponding PIV images indicate that air bubbles are entrained in the vortices. Some of the vortices captured in this study have velocity distributions that are reminiscent of an obliquely descending eddy. For example, the vertical velocity for the counter-clockwise vortex shown in Figure 4 is generally downward in the first and fourth quadrants and upward in the second and third quadrants.

The PIV measurements shown in Figure 3 indicate that downbursts and counter-rotating vortices are connected; two counter-rotating vortices are attached to each downburst of turbulence. Thus, the vortices seen in Figures 4 and 5 were most likely following behind downbursts that had reached the measuring depth upstream from the FOV. The vortices were then advected through the FOV. In Ting (2006), it was found that downbursts were more numerous than vortices in the near-bed region. Together with the results of this present study, those measurements suggest that transport of breaking-wave-generated turbulence to the bottom is initiated by downbursts.

An important question is what mechanisms produced the downbursts and counter-rotating vortices observed in these experiments. In the model proposed by Watanabe et al. (2005), the counter-rotating vortices are vortex loops produced by stretching and bending of perturbed vorticity in the saddle region between the rebounding jet and primary spanwise vortex. As these streamwise vortices develop, they wrap around the adjacent spanwise vortices to form a rib structure, and a complex vorticity field is developed. This process is analogous to the formation of streamwise vortex tubes in a plane free shear layer described in Lasheras and Choi (1988). There are, however, some important differences between the flow field in breaking waves and in a mixing layer. In a mixing layer, the spanwise vortices are produced through Kelvin-Helmholz instability. As the spanwise vortices are advected downstream, they grow by entraining irrotational fluid from the exterior flow. Stretching of weak perturbed vorticity near the stagnation point between adjacent spanwise vortices in the direction of maximum positive strain then leads to the formation of secondary streamwise vortices with counter-rotating vorticity. Lasheras and Choi (1988) found that the spanwise vortices largely maintain their two-dimensionality during this stretching process. The streamwise vortices then wrap over the top of the downstream vortex and under the bottom of the upstream vortex to create a more complex three-dimensional vorticity field. In contrast to a mixing layer, the breaking wave flow field is strongly influenced by gravity. Furthermore, the free surface is highly irregular once the wave has broken. Therefore, the question arises whether the formation and evolution of vortex structures in breaking waves are controlled by the same mechanisms found in mixing layers, where gravity forces are of minor or no importance. We speculate that the downward momentum associated with the falling water mass in the wave breaking process initiates the downbursts of turbulence observed in the PIV measurements. As seen, these downbursts can reach the bottom in the case of a weakly plunging wave. Thus, the spanwise vortices would also be subjected to large downward forces. We have observed that even in a two-dimensional wave flume, the initial wave breaking is not exactly uniform along the wave crest. The spanwise perturbation may be the result of uncontrolled upstream disturbance and/or side wall effects present in the flume. Figure 6 shows the development of the primary spanwise vortex at two different instance of time during the initial breaking of a weakly plunging regular wave on a plane slope. The picture on the left shows a local disturbance emerging from somewhere along the primary spanwise vortex, which otherwise has a rather smooth surface. Observation of many repeated waves showed that this disturbance quickly grew in a cloud of air bubbles extending obliquely downward. On both sides of this aerated column, entrained air bubbles were seen to rotate in opposite directions suggesting the existence of counter-rotating vorticity. After the wave had broken, similar bubble cloud structures emerged intermittently in space and time in a denser and denser pattern as the broken wave propagated onshore.

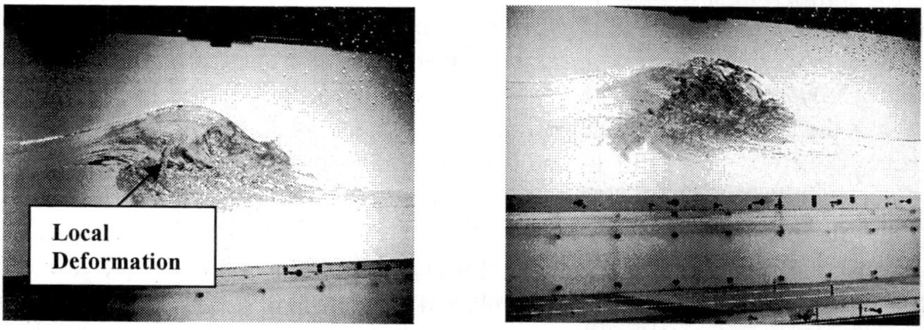

Figure 6. Initial breaking of a weakly plunging regular wave on a 1 in 50 plane slope. The wave period was 4 s. These pictures show the evolution of the primary spanwise vortex. Note that the pictures were not taken in the same wave cycle.

So far, our PIV measurements have captured only a small part of the flow field. To investigate the spatial distribution of the downbursts, a third set of PIV measurements was taken with a smaller solitary wave and a larger FOV. In these experiments, the laser light sheet was parallel to the plane slope and separated from the bottom by a distance of 2 mm. The wave height of the solitary wave and the still water depth in the constant-depth region in front of the wave maker were reduced to 0.08 m and 0.20 m, respectively. These wave conditions were chosen by trial and error so that a smaller broken solitary wave that was geometrically similar to the one described in the early part of this paper was produced in the same measuring area. The wave height (and water depth) ratio in the two sets of experiments was approximately 2.9 to 1. In addition, the two

PIV cameras were positioned to image two different flow regions side by side. The viewing angle of the cameras was 90° (i.e., perpendicular to the light sheet plane). With this experimental set-up, the out-of-plane velocity component was lost, but the larger FOV (approximately 200 mm in the longitudinal direction and 400 mm in the transverse direction) together with the smaller wave height have allowed us to capture entire flow structures as well as their spatial distributions.

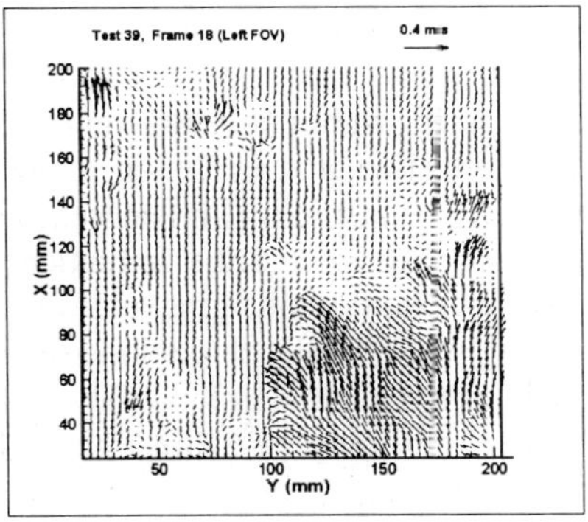

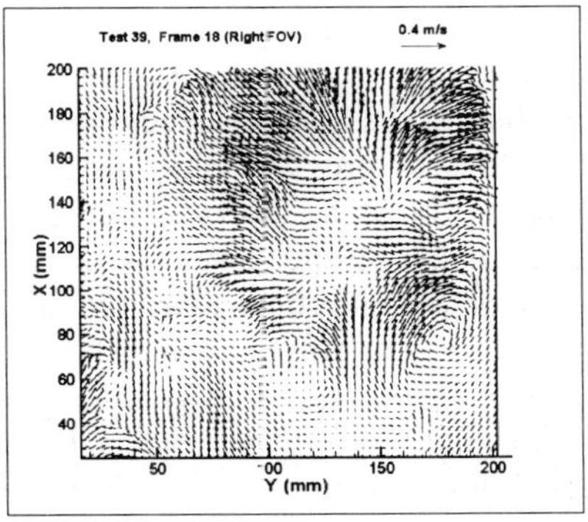

Figure 7. Instantaneous turbulent velocity fields from the left and right cameras from test 39, frame 18. The left and right fields of view have an overlap of 14 mm in the Y (transverse) direction. The wave crest arrived at frame 8 and the main body of the solitary wave passed through the FOV completely after frame 39 (15 frames = 1 s). Thirty six test runs were used to construct an ensemble-averaged velocity field, which was then subtracted from the measured velocity field to yield the turbulent velocity field. The laser light sheet was parallel to the bottom and the two planes were separated by a distance of 2 mm. The still water depth at the center of FOV was 52 mm.

Figure 7 shows a typical example from test 39, frame 18. A complete flow structure impinging on the bed was captured in the FOV of the right camera. This flow structure first appeared in the FOV as a down-rush of turbulent fluid diverging at the bed. It can be inferred from conservation of mass that the vertical velocity is downward inside the downburst. Two counter-rotating vortices soon emerged at the upstream end of this downburst of turbulence. The "diameter" of each vortex is around 50 mm, which is comparable to the local still water depth. This result is also consistent with the size of the vortices shown in Figures 4 and 5.

A second flow structure was captured in the FOV of the left camera. The flow field outside the two flow structures has much smaller turbulence velocities. Although a detailed analysis is still underway, inspection of all 36 test runs shows a similar picture. Up to three (completely or partially captured) structures could be seen across the FOV of the two cameras. These structures usually appeared in the FOV at different time and thus they were not aligned in the transverse direction (their distributions were more like that shown in Figure 7), which suggests that they were individual structures rather than part of a vortex loop or filament. The center-to-center spacing between adjacent structures in the transverse direction varied among test runs but was typically around 150 to 200 mm, which is equivalent to 3 to 4 water depths. Hence, up to four downbursts might be distributed across the width of the flume (0.9 m) around one cross-shore location.

DISCUSSION

The results of this study may be compared with published numerical simulations to provide some insight on the formation and evolution of large eddies in breaking waves. Using large eddy simulation, Christensen and Diegaard (2001) and Watanabe et al. (2005) have predicted the formation of counter-rotating vortices extending obliquely downward in plunging and weakly plunging regular waves. Watanabe et al. (2005) used a fully three-dimensional model. Christensen and Diegaard (2001) modeled a three-dimensional flow field, but specified the free surface to be uniform in the transverse direction. The numerical simulations from Christensen and Diegaard (2001) for a weakly plunging breaker show that the streamwise vortices are closely packed in the transverse direction, and have a typical length scale of ½ of the water depth. Since the free surface is two-dimensional in their numerical model, these results suggest that three-dimensional water surface deformation is not critical to the formation of the counter-rotating vortices. This is consistent with the development of streamwise vortices in a mixing layer (see Lasheras and Choi, 1988), where the base-flow is two-dimensional. Then, the question arises whether the formation and evolution of vortex structures are still governed by the same instability mechanisms found in a mixing layer after the wave has broken, when the wave has developed a highly irregular three-dimensional free surface geometry, or some other mechanisms would predominate. For example, the numerical simulations of Watanabe et al. (2005) predict that the interaction of the secondary, streamwise vortices with the primary, spanwise vortices would produce a rib structure. The rib vortices would lie longitudinally along the bottom after the wave passes. So far, our PIV measurements have not been able to confirm this. Instead, the measurements show that the near-bed flow field is scattered with individual downbursts that are separated from one another in space and time.

Currently, we are investigating what other mechanisms can produce the downbursts and counter-rotating vortices seen in the PIV measurements. Numerical models of plunging breakers typically assume that the water surface is uniform in the transverse direction until the plunging point. We suspect that this artificial constraint on the water surface geometry might lead to certain preferred flow structures, for example more stable quasi-two-dimensional spanwise vortices. Thus, we are focusing our attention on the strong, spanwise vorticity generated by wave breaking. An important question is whether local disturbance in the spanwise vortices due to non-uniform wave breaking along the crest line may precipitate three-dimensional instability of these vortices. One form of vortex structures consistent with the PIV measurements is a vortex loop with counter-rotating vorticity that is stretched out over the water column. The two legs of the vortex loop are tilted in the wave direction, connected beneath the water surface upstream by a transverse vortex and open at the free surface downstream. The three-dimensional structure would be similar to that of a hairpin vortex in a wall boundary layer, but inverted. A vortex loop like this might be produced by stretching and bending of the primary, spanwise vortices as a result of non-uniform wave breaking in the transverse direction. If the falling water mass causes one part of the vortex to descend faster than the rest of the vortex tube, strong velocity gradient in the vertical direction might produce enough stretching and bending to deform the spanwise vortex into a vortex loop with counter-rotating vortices. If this mechanism proves correct, accurate prediction of the breaking wave flow field may lie in the correct treatment of the complex free surface deformation occurring in three-dimensional wave breaking.

CONCLUSIONS

The instantaneous turbulent velocity field under a broken solitary wave propagating on a plane slope was measured in the longitudinal transverse plane in the near-bed region and in the middle part of the water column using the PIV technique. These measurements showed that a typical large-scale organized flow structure under the broken wave had the form of two counter-rotating vortices attached to a downburst of turbulence. The downburst arrived in the middle part of the water column shortly after the wave crest passed, and thus could be observed for only a short period of time. The attached counter-rotating vortices grew rapidly with time and became a prominent feature of the flow field. In the near-bed region, downbursts of turbulence diverged at the bed and covered a large surface area compared to the counter-rotating vortices. The length scale of each vortex was comparable to the local water depth, and the transverse spacing between downbursts was up to 3 or 4 times the water depth. Significant discrepancies were found between the measured flow field and published results from numerical simulations, which suggests that the vortex generation mechanisms in breaking waves may be different from those found in simple shear flows. Future investigation would determine whether the counter-rotating vortices are generated by instability in the primary spanwise vortices and what role three-dimensional wave breaking plays in the formation and evolution of these large eddies.

ACKNOWLEDGMENTS

This study was funded by the National Science Foundation (NSF) through Grant OCE-0525676 and the Office of Naval Research (ONR) through Grant N00014-00-1-0461. The support of NSF and ONR is gratefully acknowledged.

REFERENCES

Christensen, E. D., and Deigaard, R. 2001. Large eddy simulation of breaking waves, *Coastal Engineering*, 42, 53-86.

Kubo, H., and Sunamura, T. 2001. Large-scale turbulence to facilitate sediment motion under spilling breakers, *Proceedings of the Fourth Conference on Coastal Dynamics*, Lund, Sweden, 212-221.

Lasheras, J. C., and Choi, H. 1988. Three-dimensional instability.of a plane free shear layer: an experimental study of formation and evolution of streamwise vortices, *Journal of Fluid Mechanics*, 189, 53-86.

Nadaoka, K., Hino, M., and Koyano, Y. 1989. Structure of turbulent flow field under breaking waves in the surf zone, *Journal of Fluid Mechanics*, 204, 359-387.

Rogers, B. D., and Dalrymple, R. A. 2004. SPH modeling of breaking waves, *Proceedings of the 29th International Conference on Coastal Engineering*. World Scientific Press, 415-427.

Ting, F. C. K. 2006. Large-scale turbulence under a solitary wave, *Coastal Engineering*, 53, 441-462.

Watanabe, Y., and Saeki, H. 1999.Three-dimensional large eddy simulation of breaking waves, *Coastal Engineering Journal*, 41 (3 & 4), 281-301.

Watanabe, Y., Saeki, H., and Hosking, R. J. 2005. Three-dimensional vortex structures under breaking waves, *Journal of Fluid Mechanics*, 545, 291-328.

EXPERIMENTAL STUDY OF THE PLUNGING WAVE PROCESS

Sin Hwei Lydia Lim[1] and Eng Soon Chan[2]

By means of flow visualization techniques, the key features of a two-dimensional plunging wave process had been identified. Previous literature was synthesized and new observations utilized to develop a qualitative picture of the breaking process. In the overturning stage, visual evidence suggested that the undulations on the underside of the plunging jet contributes to the roughness and instability of the air tube, affecting the length scales of its breakdown in the form of the sprayback. It was also realized that the shape of the sprayback can be attributed to the kinematics of individual particles forming the streamline. Scaling the sprayback suggested that they are at an order of five to ten centimeters apart. It was also found that the "wail" phenomenon previously noted in Kway (2000) is actually part of the front face of the elliptical air tube submerged into the water. The length scale of the bubble penetration visually suggested that the direct effects of breaking were felt to a depth of the order of the wave height, which was approximately fifteen centimeters in this experiment.

INTRODUCTION

The physical complexities of wave breaking in deep water are widely recognized but still are not well understood after many years of study. (Griffin et al. 1996) Current theoretical and numerical treatments of the problem fail at the point where the surface impacts upon itself, and models of the later stages of breaking are dependent on experimental measurements and observations for guidance and testing. Field measurements alone cannot fill this need and therefore well-controlled laboratory experiments are necessary to advance our knowledge (Rapp and Melville 1990). In this research, two-dimensional breaking waves are simulated under laboratory conditions to investigate the fluid dynamical features associated with deep water wave breaking.

The many important roles that wave breaking plays warrant an extensive research in this topic in recent years. However, hindered by theoretical and experimental difficulties, advance towards understanding the geometric, kinematics and dynamics of deepwater wave breaking has been relatively slow. From an experimental point of view, observations and measurements are not easy for various reasons: breaking is an unsteady phenomenon occurring intermittently, starting suddenly and evolving rapidly; in addition, it presents three-dimensional and multiphase aspects (Bonmarin 1989). Compared with the laboratory studies, the advances of numerical computations for the breaking wave in the transitional stage after impingement are even slower. Because of the

[1] Surbana International Consultants Pte. Ltd., Reclamation & Infrastructure Division, 168 Jalan Bukit Merah #01-01 Surbana One Singapore 150168

[2] Professor, Dept. of Civil Engineering, National University of Singapore, 21 Lower Kent Ridge Road, Singapore 119077

high turbulence intensity and rapid change of free surface geometry during this period, direct computation becomes very difficult (Lin and Liu 1998). The present study aims to elucidate the features and mechanisms associated with deep water plunging breaker so as to contribute to future theoretical modeling efforts.

WAVE FLUME EXPERIMENTS

The plunging waves were generated in a laboratory ferrocement flume capable of generating two-dimensional waves, spanning 35.22m long, 2m wide and 1.3 metres deep. The wave tank was filled with fresh water to a depth of 0.8 metres. See Fig. 1. A wave absorber in the form of a beach with a slope of 0.29 at the opposite end of the tank served to dampen the effect of reflections. Because a finite length packet was generated in these experiments, image acquisition can be made on the transmitted packet before reflection travel back to the image acquisition point. Observations and image capture were acquired through the glass panels provided on both sides of the flume extending 6 metres. Flow visualization techniques using highspeed videography (frame interval of 0.04 seconds) and photography are employed to capture the features of this rapidly changing event.

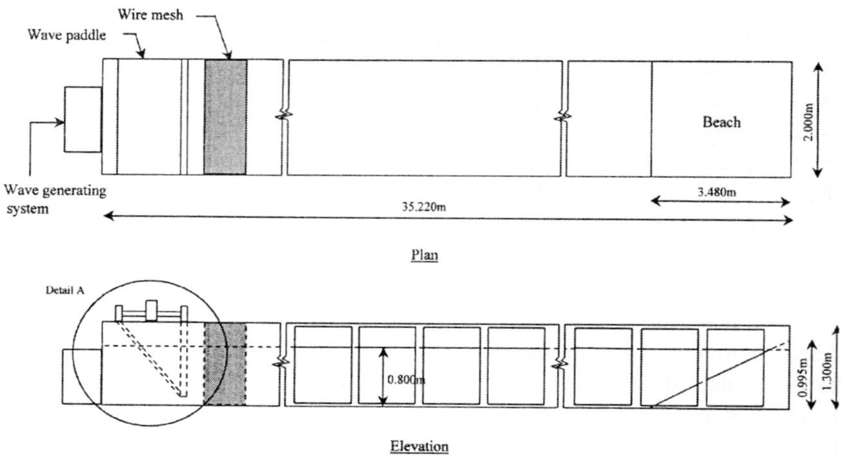

Figure 1 Layout and Dimensions of Wave Flume

The following Table 1. shows the parameters used as input data to the algorithm used for generating the constant steepness wave packet.

OVERVIEW OF THE WAVE BREAKING PROCESS

In order to have a general overview of the breaking phenomenon, a schematic sequence of the plunging breaker is depicted in Fig. 2 and a description of each stage is outlined as follows, based on the photographs and video images obtained.

Table 1. Wave Parameters and Input Variables to Algorithm	
Wave parameters	
Characteristic wave period, T (s)	1.205
Characteristic wave length, L (m)	2.217
Characteristic phase speed, C=L/T (m/s)	1.84
Characteristic wave number k_c (m^{-1})	2.834
Wave Variables input to algorithm	
Δf, frequency bandwidth of components(Hz)	0.54
f_{max}, highest frequency (Hz)	1.1
N, number of components	28
d, water depth (m)	0.8
x_b, Breaking distance from the paddle (m)	15.7
t_b, Breaking time (s)	23

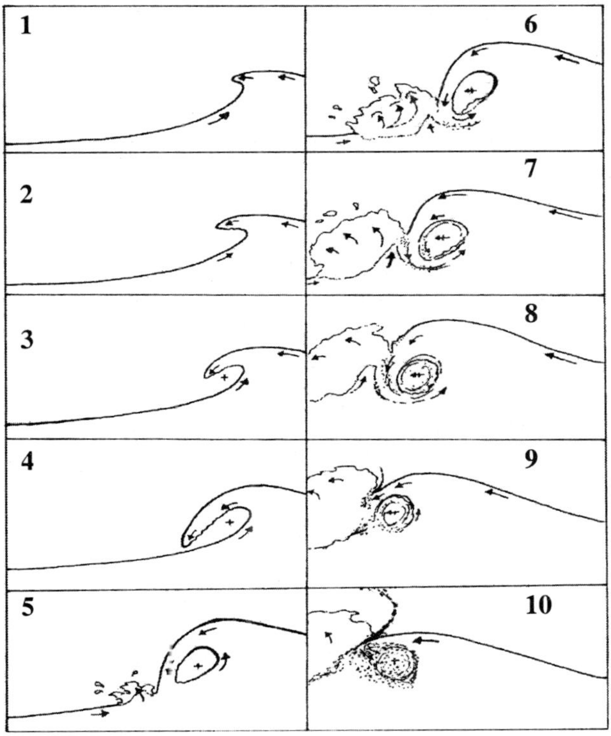

Figure 2. Schematic Sequence of breaking Plunger

Description of each stage of the wave breaking process

1. The wave steepening stage is marked by the asymmetric profile of the front surface of the wave.
2. At the beginning of wave overturning, a jet of fluid ejects itself forward towards the trough in front, trapping air in the process.
3. The underside of the overturning jet develops a certain roughness that seems to occur at regular intervals, resembling those of capillary waves. This feature will be discussed in detail in the section under *overturning*.
4. When the overturning jet is just about to strike the oncoming trough at a plunge point, a well-formed elliptical-shaped air tube can be observed.
5. The impact of the falling jet with the undisturbed water surface creates a small initial *splash-up*. This is the beginning of vortices and turbulence generation.
6. The momentum and weight of the overturning jet forces it to penetrate the oncoming trough, pushing down the trapped air core in the process. The region between the original jet and the splash-up is very much aerated and acts as a source of turbulence due to the high shear developed between streams of fluid moving in opposite directions.
7. As the air tube is being pushed further into the water, a "wall" can be observed from below the water surface, similar to recent observations (Kway, 2000). This feature will be discussed in detail in the section under *air tube formation*.
8. The crest continues its horizontal translation and the splash-up reaches the same surface elevation as the wave crest.
9. The crest at this point "widens" up and pushes down on the air tube. The unstable air tube has a tendency to float up due to its buoyancy and its filmsy nature. This causes an overall compressive force on the air tube and squeezes part of the trapped air in the air tube out in the form of a sudden squirt. This is termed *sprayback* in this paper.
10. This is when the sprayback attains its maximum height and intensity. The shape of the sprayback resembles the curve of a crescent. This feature will be discussed later under *Sprayback Formation*. Simultaneously, the air tube is observed to lose its intact shape and structure and begins to break down into infinitesimal bubbles.

OVERTURNING

Wave overturning marks the beginning of the plunging wave process. To study this stage in detail, a series of photographs were taken with a NIKON high-speed camera. In Fig. 3, the wave crest has overturned and the plunging jet is projecting forward. The underside of the plunging jet has already developed some form of wavy texture as it traps a cylindrical column of air. The roughness of the jet can be shown by the strong light reflections thrown off by the undulations. These undulations, resembling vertical capillary waves, formed as a result of surface tension due to decrease in radius of curvature, occurred at certain intervals. Results from the numerical modeling (Fig. 4) of the plunging breaker based on similar wave parameters showed that the velocity vectors on the inner side of the plunging jet have higher magnitudes than those on the upper surface of the jet. This velocity difference can also possibly constitute the reason for the uneveness observed in Fig.3.

Figure 3 Capillary waves on the underside of the plunging jet

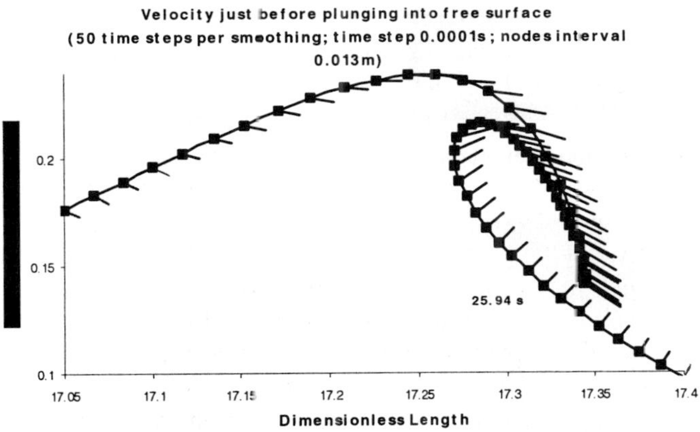

Figure 4 Numerical Modelling Results of Velocity Vectors just before plunging

AIR TUBE FORMATION

Upon impingement, the air tube starts to develop roughness due to turbulence from the plunge point. The rotating momentum of the air tube also causes the plunging jet with the underside roughness observed in the overturning stage to rotate around the air tube. As the wave propagates forward, the air tube continues rotating, maintaining an intact structure until the sudden emergence of a spray (termed "Sprayback" in this paper), see Fig. 5. Simultaneously, the air tube disintegrates into a mass of white entrained bubbles. Synchronized photographs taken using two video cameras at two different angles (Fig. 6) clearly showed that the air tube disintegrates (see red arrow in Fig. 6) at the point when the sprayback emerges and attains its maximum intensity.

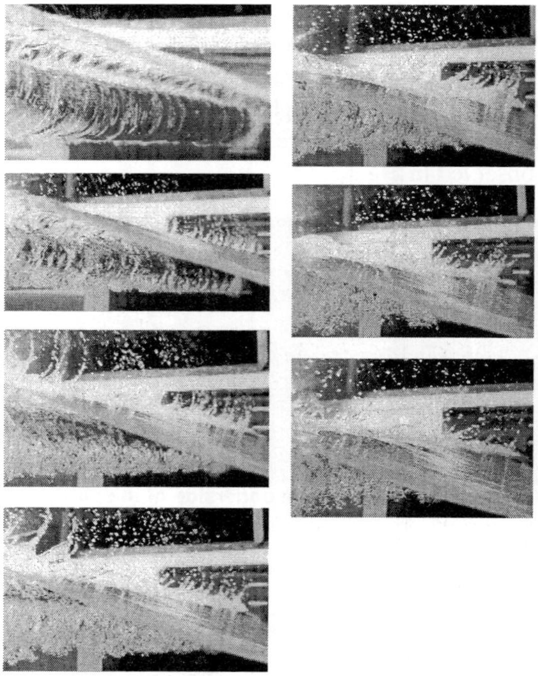

Figure 5. Air tube development (series photos at 0.02s interval)

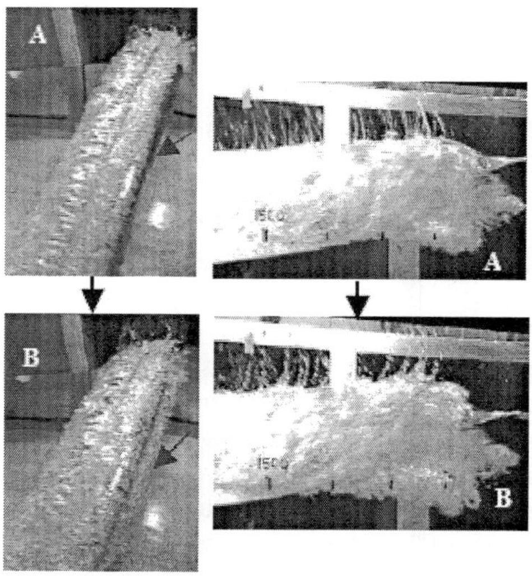

Figure 6 Synchronized Video Images showing air tube breakdown coincides with sprayback at maximum intensity (Set A and B are taken 0.04 s apart, all four images are from the same run)

When the video camera is placed at a downstream location below the water surface, it can be observed that a "wall" (Fig. 7) is present at after impact of the plunging jet with the trough surface. This "wall" phenomenon is noted in Kway (2000) which is described as a well-defined wall formed by the strong plunging jet as it penetrated the water mass.

Figure 7 "Wall" phenomenon

THE SPRAYBACK FORMATION

The sprayback develops an increasing curvature as the wave propagates forward. The water droplets forming the sprayback follows a crescent shape as more and more water droplets emerged. However, upon detailed analysis, it is realized that the curvature formed by the water droplets is not a path taken by all the droplets that join the path. The changing curvature at the wave crest splash-up interface causes new emerging water droplets to follow different stream lines (Fig. 8). Water droplets following different streamlines together form the curvature of the sprayback at any instant.

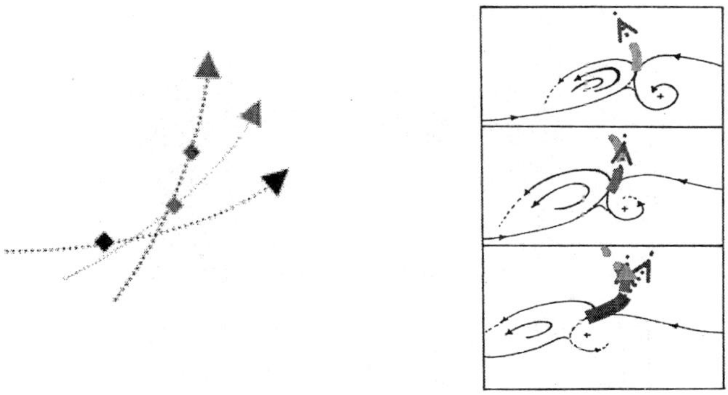

Figure 8 water droplets on different streamlines making up the curvature of sprayback

As shown in Fig. 9, the sprayback occur at certain interval across the length of the entire two-dimensional air tube. To have an idea of the order of magnitude of the spacing between each spray of water droplets, a length scale (at 5 cm interval) is located at the point of where sprayback emerge and images of the entire breaking process are recorded. Apparently, from Fig. 9 below, the spraybacks are regularly spaced at about 5-10 cm interval.

Figure 9 Sprayback Intervals

THE SPLASH-UPS

After the first splash-up is generated, it plunges at the next plunging point in front of it, and in turn generates the second splash-up. This splash-plunge cycle, similar to a domino effect, is observed to repeat several times with decreasing plunging intensity. The height and bubble-entraining capacity of the splash-ups decreases from the first to the last due to eventual loss of energy through turbulence and shear. It is observed from the series of video images (at 0.04s interval) from Fig. 10 that the splash-ups becomes more disorganized from the first to the third and the breakdown of the plunging air tube itself constitutes the first streak of bubbles left on the water surface.

Figure 10 Formation of steaks of bubbles on water surface due to splash-plunge cycles

BUBBLE DISSIPATION

Each splash-plunge cycle will result in air entrainment, trapping air into the water and later the air dissipates as tiny air bubbles. The plunging action of each splash-up serves as a source of momentum which pushes more air into the water. These bubbles will initially penetrate deeper into the water with the downward momentum. However, as the momentum reduces due to the frictional resistance between the air-water interface of the bubbles, buoyancy forces dominate and the bubbles embark on an upward direction towards the water surface instead. An attempt was made to scale the bubble penetration depth by putting marks at equal intervals on the glass walls. From Fig. 11, it is realized that the first bubble zone, which is due to the breakdown of the air tube enclosed by the plunging jet, penetrated to a maximum depth of about 20cm from the Still Water Level (SWL) and about 16cm from the water surface. This suggests that the direct effects of breaking are typically felt to a depth of the order of the wave height, which is about 15 cm in this experiment.

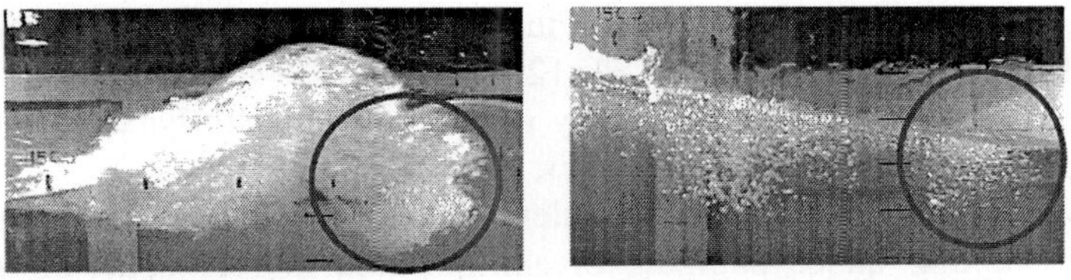

Maximum depth from water surface — Maximum depth from SWL

Figure 11 Penetration depth of First Bubble Zone

CONCLUSIONS

A laboratory study has been conducted to investigate the key features of the two-dimensional plunging wave before and after breaking through the use of flow visualization techniques. Six key elements emerge for the plunging type breaker.

1. The underside roughness of the plunging jet contributes to the uneveness and instability of the air tube, affecting the length scales of its breakdown and sprayback.
2. The emergence of the sprayback is due to the breakdown of the air tube.
3. The shape of the sprayback can be attributed to the kinematics of individual particles forming the streamline.
4. The "wall" phenomenon previously noted in Kway (2000) is actually part of the front face of the elliptical air tube submerged into the water.
5. Scaling the backspray reveals that the sprays are at an order of 5 to 10 cm apart, consistent with the scales of the air tube surface undulations.
6. The length scale of the bubble penetration depth visually suggests that the direct effects of breaking are typically felt to a depth of the order of the wave height, which is about 15 cm in this experiment.

ACKNOWLEDGMENTS

Sin Hwei Lydia Lim wishes to express her gratitude to Mr Loh Yan Hui and Mr Chia Way Seng from Surbana International Consultants Pte. Ltd. (Reclamation & Infrastructure Division) for supporting her attendence at ICCE2006. She is thankful to her supervisor Professor Chan Eng Soon for his patient guidance throughout the research.

REFERENCES

Basco, D.R. 1985. A qualitative description of wave breaking. *Journal of Waterway, Port, Coastal and Ocean Engineering*, Vol. 111, No. 2, 171-188.

Chanson, H., and Lee, J-F. 1997. Plunging jet characteristics of plunging breakers. *Coastal Engineering* 31, 125-141.

Kway, J. H. L. 2000 Laboratory study of the kinetics and dynamics of deep-water wave breaking." *P.H.D. thesis*, NUS, Singapore

Lin, P., and Liu, P.L.-F. 1998. Turbulence transport, vorticity dynamics, and solute mixing under plunging breaking waves in surf zone. *Journal of Geophysical Research*, Vol. 103, No. C8, 15,677-15,694.

Melville, W. K. 1996. The role of surface-wave breaking in air-sea interaction, *Annu. Rev. Fluid. Mech.*, 28, 279-321.

Mizuguchi, M. 1986. Experimental study on kinematics and dynamics of wave breaking. *Coastal Engineering*, Chapter 45, 589-603.

Nadaoka, K., Hino, M., and Koyano,Y. 1989. Structure of the turbulent flow field under breaking waves in the surf zone. *J. Fluid Mech*, vol. 204, 359-387.

Peregrine, D.H., and Banner, M.L. 1993. Wave breaking in deep water. *Annu. Rev. Fluid Mech*, 25: 373-97.

Yasuda, T., Mutsuda, H., and Mizutani, N., 1999 Relationships of plunging jet size to kinematics of breaking waves with spray and entrained air bubbles. *Coastal Engineering Journal*, Vol. 41, Nos 3 & 4 269-280

TURBULENT FEATURES BENEATH BREAKING WAVES

Stefan Schimmels[1], Zhu Zhang[1] and Torsten Schlurmann[2]

Transient breaking waves exceeding a critical steepness are investigated by means of numerical simulations based on the Reynolds-averaged Navier-Stokes equations. For the approximation of Reynolds stresses four different closure models are used: the first order standard k-ε-model, RNG k-ε-model and SST k-ω-model and a second order Reynolds stress model. Up to the breaking point the influence of turbulence models is insignificant and the numerical results agree very well with laboratory PIV measurements. The turbulence quantities like turbulent kinetic energy and turbulent viscosity that develop during the breaking and post breaking process are investigated and the influence of the different turbulence models is analyzed.

INTRODUCTION

The process of wave breaking is one of the most complex and difficult subjects in fluid mechanics and by far not been fully understood yet. Many results of laboratory experiments or numerical simulations investigating breaking waves have been presented in the past. However, almost all of them concentrated on depth induced breaking when a wave approaches a beach. In a series of papers e.g., Ting and Kirby (1995, 1996) analyzed spilling and plunging breaking waves on a 1:35 slop in the laboratory. Some of these results were later used by Lin and Liu (1998) to validate their numerical model.

Harms and Schlurmann (2004) presented the results of PIV measurements under breaking transient wave packages. Due to the relatively coarse resolution of 1.598 x 1.598 mm per pixel, the PIV data precluded the analysis of turbulence and only mass transport velocities have been investigated. To further complete the studies of breaking waves exceeding a critical steepness the current paper presents numerical simulations of the turbulent features beneath these waves using the Reynolds-averaged Navier-Stokes (RANS) equations with different turbulence closure models and the Volume-of-Fluid (VOF) method.

NUMERICAL MODEL

Governing Equations

The governing equations to solve for the wave breaking problem are the incompressible Reynolds-averaged Navier-Stokes (RANS) equations, with the underlying assumption that the flow field described by velocities and pressure can be divided in a mean and a fluctuating part. Using Einstein's summation

---

[1] Institute of Fluid Mechanics, Leibniz Universität Hannover, Appelstr. 9a, 30167 Hannover, Germany

[2] Institute for Environment and Human Security, United Nations University, Hermann-Ehlers-Str. 6, 53113 Bonn, Germany

convention, i.e. summation over identical indices, mass and momentum conservation of the mean flow field can be written in Cartesian tensor form as:

$$\frac{\partial u_i}{\partial x_i} = 0, \tag{1}$$

$$\frac{\partial u_i}{\partial t} + \frac{\partial u_i u_j}{\partial x_j} = \frac{1}{\rho}\frac{\partial}{\partial x_j}\left(\mu\left(\frac{\partial u_i}{\partial x_j} + \frac{\partial u_j}{\partial x_i}\right) - \rho u_i' u_j'\right) - \frac{1}{\rho}\frac{\partial p}{\partial x_i} + g_i, \tag{2}$$

where δ_{ij} is the Kronecker delta, ρ is the fluid density, u_i are the mean velocity components, p is the mean pressure, g_i are the components of gravitational acceleration, μ is the molecular viscosity of the fluid and $u_i'u_j'$ are correlations of the velocity fluctuations. In (2) we already made the assumption of a Newtonian fluid, where the viscous stresses are linearly related to the rate-of-strain tensor of the mean flow field by the molecular viscosity of the fluid. However, to close the formulation the unknown correlations of the velocity fluctuations have to be determined. Their influence on the mean flow field can be viewed as a form of stress, wherefore these terms are often referred to as Reynolds stresses.

The most straightforward way to close the problem is to solve additional transport equations for the Reynolds stresses which can directly be derived from the Navier-Stokes equations. However, the resulting equations contain unknown higher order correlations that have to be modeled. Nevertheless, the assumptions made in modeling the corresponding terms are on a higher order level so that this way of calculating the effect of turbulence on the mean flow field in the Reynolds averaging sense seems to be the most sophisticated. It can be regarded as a second order closure model and will be described later.

Another and more common approach to determine the Reynolds stresses is the Boussinesq hypothesis, relating the unknown second order correlations of turbulent fluctuations linearly to the mean flow field by a turbulent viscosity μ_t:

$$-\rho u_i' u_j' = \mu_t\left(\frac{\partial u_i}{\partial x_j} + \frac{\partial u_j}{\partial x_i}\right) - \frac{2}{3}\rho k \delta_{ij}, \tag{3}$$

where k is the turbulent kinetic energy.

There are also more complicated nonlinear formulations that have been applied to determine the Reynolds stresses (e.g. Lin and Liu 1998). However, the major task in modeling the effects of turbulence on the mean flow field in the Boussinesq sense remains the same. It is reduced to finding an appropriate turbulent viscosity, which in contrast to its molecular counterpart is space and time dependent and a property of the flow, not of the fluid. Even if a nonlinear model is used the modeling assumptions are directly made on the Reynolds stresses and hence, this approach is regarded as first order closure model and will be described next.

First order closure models

The most common approach to determine the turbulent viscosity is to relate two turbulent quantities for which additional transport equations have to be solved. The most obvious choice is the turbulent kinetic energy which is defined as the trace of the Reynolds stress tensor:

$$k = \frac{1}{2} u_i' u_i' . \tag{4}$$

Its transport equation can directly be derived from the Navier-Stokes equations and can be written as:

$$\frac{\partial k}{\partial t} + \frac{\partial k u_i}{\partial x_i} = \frac{1}{\rho} \frac{\partial}{\partial x_i} \left(\left(\mu + \frac{\mu_t}{\sigma_k} \right) \frac{\partial k}{\partial x_i} \right) + \frac{1}{\rho} P_k - \varepsilon . \tag{5}$$

Two major assumptions are made in (5):

- Higher order correlations of velocity and pressure fluctuations are modeled by a turbulent diffusivity that is related to the turbulent viscosity by a Prandtl number for turbulent kinetic energy σ_k, which has a value of 1.0.
- The second term on the right hand side represents the production of turbulent kinetic energy and is modeled in a manner consistent with the Boussinesq hypothesis:

$$P_k = -\rho u_i' u_j' \frac{\partial u_i}{\partial x_j} = \mu_t \left(\frac{\partial u_i}{\partial x_j} + \frac{\partial u_j}{\partial x_i} \right)_j \frac{\partial u_i}{\partial x_j} . \tag{6}$$

The third term on the right hand side of (5) is the dissipation rate of turbulent kinetic energy. As it is a priori not known it is straightforward to make it the second turbulent quantity for which an additional transport equation is solved, leading to the popular class of *k*-ε-models. For the standard *k*-ε-model (Launder and Spalding, 1972) the transport equation for turbulent dissipation rate is:

$$\frac{\partial \varepsilon}{\partial t} + \frac{\partial \varepsilon u_i}{\partial x_i} = \frac{1}{\rho} \frac{\partial}{\partial x_i} \left(\left(\mu + \frac{\mu_t}{\sigma_\varepsilon} \right) \frac{\partial \varepsilon}{\partial x_i} \right) + \frac{1}{\rho} C_1 \frac{\varepsilon}{k} P_k - C_2 \frac{\varepsilon^2}{k} . \tag{7}$$

In this equation even more assumptions have to be made than for the transport equation of k in (5). Among the turbulent Prandtl number σ_ε, with a typical value of 1.3, the second and third term on the right hand side of (7), representing production and dissipation of ε, contain further empirical parameters C_1 and C_2 with values of 1.44 and 1.92, respectively.

The turbulent viscosity finally results from k and ε with the additional empirical constant $C_\mu = 0.09$

$$\mu_t = \rho C_\mu \frac{k^2}{\varepsilon}. \tag{8}$$

If the transport equations for k and ε are derived from the Navier-Stokes equations by a mathematical technique called 'renormalization group' (RNG) one arrives at a very similar result compared to (5), (7) and (8). The major differences between the standard and the RNG k-ε-model are:

- The diffusion of k and ε is modeled with an effective viscosity and analytically derived turbulent Prandtl numbers $\sigma_k = \sigma_\varepsilon \approx 0.718$.
- The constants C_1, C_2 and C_μ are also derived analytically and have values of $C_1 = 1.42$, $C_2 = 1.68$ and $C_\mu = 0.0845$, respectively.
- The main difference is an additional term in the transport equation for ε depending on the rate-of-strain tensor of the mean flow field which reduces the destruction of ε in areas of rapid strain. Hence, in these areas the turbulent kinetic energy is less, yielding a lower turbulent viscosity.

Instead of solving an additional transport equation for the turbulence dissipation rate ε, Wilcox (1988) proposed to solve an additional transport equation for the specific dissipation rate ω, which can also be thought of as the ratio of ε to k. In the standard k-ω-model the transport equations for k and ω are very similar to equations (5) and (7):

$$\frac{\partial k}{\partial t} + \frac{\partial k u_i}{\partial x_i} = \frac{1}{\rho}\frac{\partial}{\partial x_i}\left(\left(\mu + \frac{\mu_t}{\sigma_k}\right)\frac{\partial k}{\partial x_i}\right) + \frac{1}{\rho}P_k - \beta^* k\omega, \tag{9}$$

$$\frac{\partial \omega}{\partial t} + \frac{\partial \omega u_i}{\partial x_i} = \frac{1}{\rho}\frac{\partial}{\partial x_i}\left(\left(\mu + \frac{\mu_t}{\sigma_\omega}\right)\frac{\partial \omega}{\partial x_i}\right) + \frac{1}{\rho}\alpha\frac{\omega}{k}P_k - \beta\omega^2. \tag{10}$$

Also the turbulent viscosity is defined in a comparable way to (8) by:

$$\mu_t = \rho\alpha^* \frac{k}{\omega}. \tag{11}$$

However, a major difference lies in the coefficients which almost all depend on the Reynolds number. In the high Reynolds number limit all coefficients approach constant values which are in detail: $\sigma_k = 2.0$, $\sigma_\omega = 2.0$, $\alpha^* = 1.0$, $\beta^* = 0.09$, $\alpha = 1.0$ and $\beta = 0.072$.

As was shown above the formulations for the k-ε-models and the k-ω-model are quite similar. Therefore Menter (1994) proposed the shear-stress transport (SST) k-ω-model which incorporates a blending function between the standard k-ω-model and the standard k-ε-model transformed into a k-ω formulation. The transport equations for k and ω are basically the same as in equations (9) and (10). The major differences between the standard and the SST k-ω-model are:

- A blending function to smoothly switch between the robust and accurate formulation of the k-ω-model in the near-wall region and the free-stream independence of the k-ε-model in the far field.
- A modified definition of the coefficient α^* to account for the transport effects of the principal turbulent shear stress.
- An additional cross-diffusion term in the transport equation for ω.

For further details about the SST k-ω-model the reader is referred to the literature.

All first order closure models assume turbulence to be isotropic, which in case of breaking waves might not strictly be true. To circumvent this constraint the closure of the RANS equations (1) and (2) has to be made on a higher order level, which will be described in the next section.

Second order closure model

In order to make no first order assumptions on the turbulent velocity field and the distribution of Reynolds stresses, each of these is transported individually. The according transport-equations can directly derived from the Navier-Stokes equations and have the following general form:

$$\frac{\partial u_i'u_j'}{\partial t}+\frac{\partial u_k u_i'u_j'}{\partial x_k}=\frac{1}{\rho}\frac{\partial}{\partial x_k}\left(\mu\frac{\partial u_i'u_j'}{\partial x_k}\right)-\frac{1}{\rho}\frac{\partial}{\partial x_k}\left(\rho u_i'u_j'u_k'+p\left(\delta_{kj}u_i'+\delta_{ik}u_j'\right)\right)$$
$$-\left(u_i'u_k'\frac{\partial u_j}{\partial x_k}+u_j'u_k'\frac{\partial u_i}{\partial x_k}\right)+\frac{1}{\rho}p\left(\frac{\partial u_i'}{\partial x_j}+\frac{\partial u_j'}{\partial x_i}\right)-\frac{1}{\rho}2\mu\frac{\partial u_i'}{\partial x_k}\frac{\partial u_j'}{\partial x_k} \quad (12)$$

The first and third term on the right hand side represent molecular diffusion and turbulence production and have not to be modeled. However, the second, fourth and fifth term, representing turbulent diffusion, pressure strain and dissipation have new unknowns and need some model assumptions.

The details of the Reynolds stress model (RSM) and the assumptions made in modeling the unknown terms can be found in the literature (e.g. Launder 1989). The basic approaches can be summarized as follows:

- The last term in (12) is evaluated by solving an additional transport equation for the scalar turbulence dissipation rate ε – like equation (7) – and distributing the total dissipation rate equally to the main diagonal elements of the Reynolds stress tensor.
- The pressure strain term is classically modeled by decomposing it in a slow and a rapid pressure strain part involving turbulent kinetic energy, dissipation rate and turbulence production terms.
- Turbulent diffusion is modeled with a turbulent diffusivity just like in the first order closure models. The turbulent Prandtl number is 0.82 and the turbulent viscosity is computed by equation (8) where the turbulent kinetic energy is obtained from the trace of the Reynolds stress tensor (eq. (4)).

The fidelity of RSM predictions is still limited by the closure assumptions employed to model the various terms in the exact transport equations for the Reynolds stresses. Particularly the modeling of the pressure-strain and dissipation-rate terms is challenging, and often considered to be responsible for compromising the accuracy of RSM predictions. Nevertheless, since the RSM accounts for the effects of streamline curvature, swirl, rotation and rapid changes in strain rate in a more rigorous manner than first order closure models, it has greater potential to give accurate predictions for wave breaking. However, the alleged higher accuracy is paid by computational effort as for 2D simulations five and for 3D simulations even seven additional equations have to be solved.

Numerical schemes

The governing equations (1) and (2) and the transport equations for turbulent quantities are discritized using the Finite Volume Method on a Cartesian grid. For the spatial discretization of advective terms a third order accurate QUICK scheme is used while diffusion terms are central-differenced giving second order accuracy. Temporal discretization is first order accurate using an implicit formulation. The resulting nonlinear equation system is implicitly linearized and sequentially solved with pressure and velocities coupled by a PISO scheme.

The free water surface is tracked by a Volume-of-Fluid (VOF) method, where the governing equations are solved for two non-interpenetrating phases (water and air) that are identified by an additional variable defining their volume fraction. The actual interface is reconstructed assuming a piecewise linear approach where the calculation of the position of the linear interface in each computational cell is based on information about the volume fraction and its derivatives inside the cell.

Model Setup

Harms and Schlurmann (2004) carried out their laboratory experiments on transient breaking waves in a 24 m long and 30 cm wide wave flume. They investigated breaking events in 30 cm water depth by focusing the components of a JONSWAP spectrum with a peak frequency of 0.8 Hz and a total amplitude of 12 cm. The concentration point was about 12.5 m behind the wave maker and breaking set in as the wave steepness exceeded the critical value.

To reproduce the experiments and further investigate the results of turbulence modeling numerical simulations were performed solving the above set of equations in a 2D vertical domain. This simplification is justified since lateral variations of the mean flow field were not observed in the laboratory experiments and could be proved to be negligible in full 3D test runs. The numerical wave tank used here is sketched in Figure 1.

The total length of the computational domain is 30 m with the first 15 m making up the actual area of interest and the last 15 m used to damp out waves by smoothly stretching the computational grid in this area. The total height is 45 cm with the lower 30 cm filled up by water and the remaining 15 cm above filled up by air. To accurately resolve the free water surface and the velocity fields in the breaking region the initial grid size of $dx = dy = 1.50$ cm was once refined 6 cm above and below still water level to be $dx = dy = 0.75$ cm in this area. Although the treatment of time derivatives is implicit, making the solution unconditionally stable, the time step was chosen small to avoid spurious numerical diffusion.

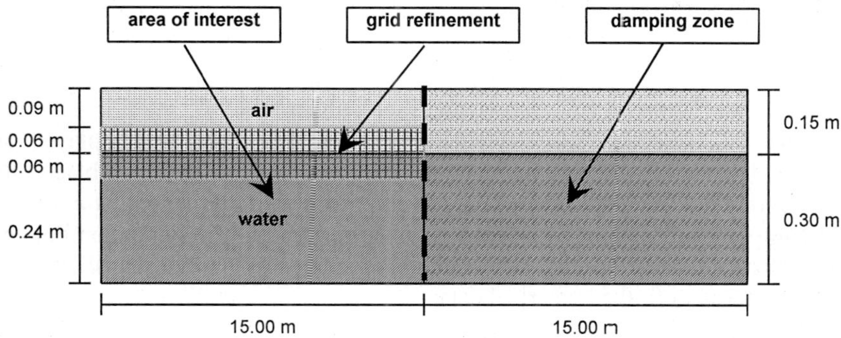

Figure 1: Definition sketch of computational domain

The bottom and right boundary are assumed to be frictionless closed walls while the top boundary is open allowing air to leave and enter the computational domain. At the left boundary the waves will enter the computational domain. For the present investigation there are no reflections inside the wave tank to be expected and it is not necessary to allow waves to leave the domain over the left boundary. Hence, the most straightforward and most accurate way to generate waves is to prescribe the water surface elevation and the corresponding particle velocities at the boundary.

During the laboratory experiments the water surface elevation was measured by a capacitive wave gauge placed 8 m behind the wave maker. This data can now be used as a boundary condition for the numerical simulations. However, the velocities under the waves are not known and must be determined theoretically. The best way to calculate the velocity field under irregular waves is by means of a Local Fourier Approximation (Sobey 1992) in which the nonlinear wave boundary value problem is solved locally.

The initial conditions for the mean quantities are simply zero velocities and a hydrostatic pressure distribution. The initial settings for turbulent quantities, however, have to be treated with care. Even if the quiescent initial state will actually have no turbulent fluctuations, the turbulent kinetic energy k for instance must have a finite value for numerical reasons. Otherwise the production terms for the turbulence dissipation rate ε or the specific dissipation rate ω, would become singular. Furthermore the production of k would always be zero as it is proportional to the turbulent viscosity which in turn is dependent on k. Hence, the initial conditions for turbulence quantities are chosen in such a way that the turbulence level is very small but finite.

RESULTS

Comparison with PIV data

The analysis of the PIV data was done in terms of mean flow quantities, which can now serve as a validation of the present numerical results. Although several different turbulence models as described above have been used in the present numerical study their influence on the mean flow field and the water surface elevation up to the onset of breaking is rather small. Hence, it seems sufficient if only one turbulence model is considered here for reference.

Figure 2 shows six snapshots of PIV raw data with a time step of 0.1 s as presented in Figure 4 of Harms and Schlurmann (2004). Additionally the instantaneous velocity fields obtained from the PIV data are displayed as vectors. The result of the numerical simulation is shown as the thick black line representing the water surface which was computed with the standard *k*-ε-model. The time steps at the bottom of each panel correspond to the simulation time of the numerical model. Therefore the original time steps of the PIV snapshots were slightly shifted to fit the numerical data.

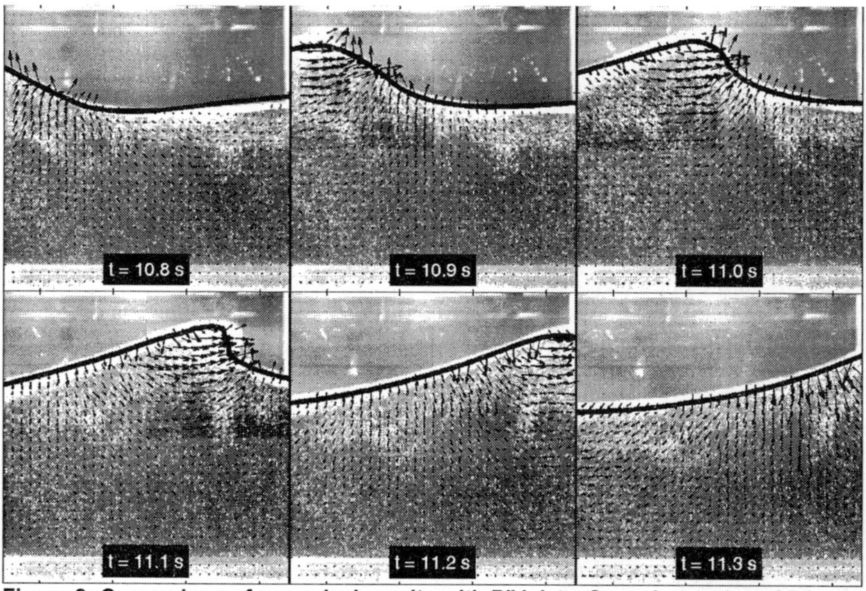

Figure 2: Comparison of numerical results with PIV data. Snapshot series of PIV raw data with underlying velocity field and water surface elevation from numerical simulation (black line).

The water surface in the laboratory experiment can either be deduced from the PIV raw data itself or from the upper boundary of the velocity vectors. The overall agreement between the laboratory data and the numerical results is excellent concerning the water surface elevation. The steepening of the front face of the wave and the onset of breaking are as well predicted by the numerical model as the evolution of the following wave trough.

A direct comparison of the flow field under the wave is not useful as the laboratory measurements represent instantaneous velocities whereas the numerical results are ensemble mean values. However, Harms and Schlurmann (2004) also presented depth averaged horizontal mass-transport velocities for which random fluctuations are canceled out somehow. Figure 3 shows a time series of the depth averaged horizontal velocity at the focusing point of the wave just before breaking occurs. The numerical results (full line) were obtained with the standard k-ε-model and the laboratory data (open circles) correspond to location #19 in Figure 7 of Harms and Schlurmann (2004).

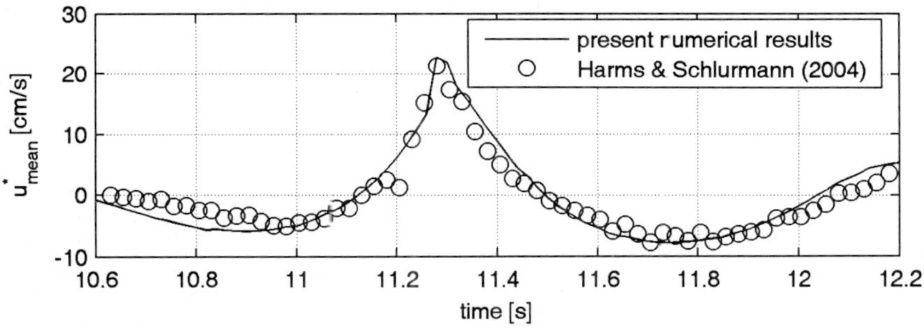

Figure 3: Comparison of numerical results with PIV data. Time series of depth averaged horizontal mass-transport velocity at the focusing point just before breaking sets in.

As for the water surface elevation above the agreement of the numerically determined velocities with the laboratory measurements is also very good. Considering the uncertainties in the measurements as seen by the slight variations in the data in Figure 3 the small deviations with the numerical result are within the accuracy of the measurements.

Analysis of Turbulence during and after Breaking

Up to the focusing point no breaking occurs so the flow field is irrotational and the intensity of turbulence is very small. Therefore the kind of turbulence closure is negligible and as shown above the numerical results agree almost perfectly with the laboratory data. After the onset of breaking, however, turbulence is produced and its amount and influence on the mean flow field depends on the assumptions of the turbulence model.

Figures 4 and 5 present the evolution of turbulent kinetic energy after the wave starts to break. The water surface is marked by the line and the amount of turbulent kinetic energy is indicated by the contours. The first time step was chosen to be the onset of breaking and corresponds to the lower middle panel of Figure 2. As illustrated above the SST k-ω-model is based on a transformed k-ε-model which becomes most pronounced away from a wall. Figure 4 shows the results for the standard k-ε-model on the left and for the SST k-ω-model on the right and as could be expected both have very similar characteristics.

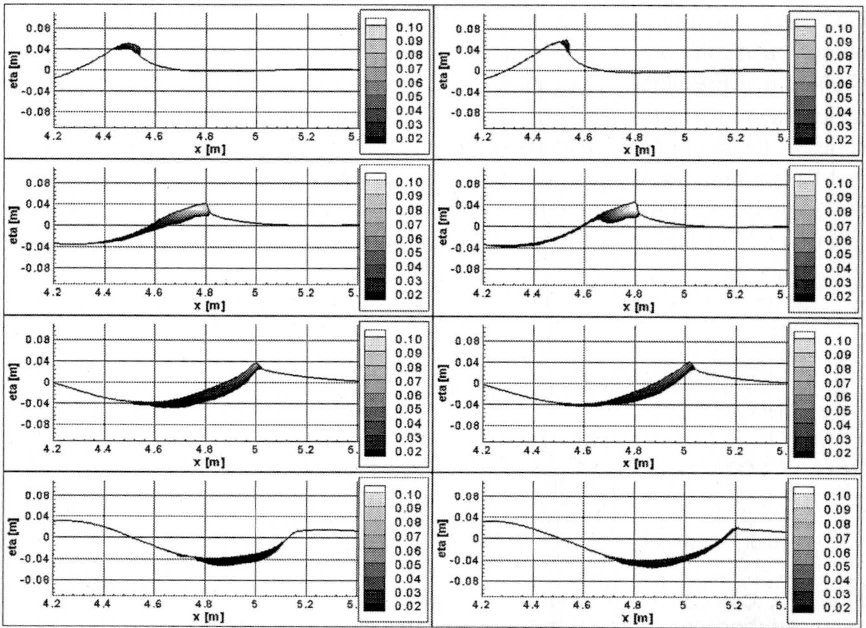

Figure 4: Evolution of turbulent kinetic energy for standard k-ε-model (left) and SST k-ω-model (right). Time steps from top to bottom: 11.2 s, 11.44 s, 11.68 s, 11.92 s.

At the onset of breaking in the top panels of the figure the increased strain rates at the front face of the wave produces turbulent kinetic energy which is then transported with the wave celerity and further amplified during the breaking process. As the wave propagates the turbulence dissipation rate starts to outweigh the production and the amount of turbulent kinetic energy decreases until it is almost completely gone in the lower panels of the figure.

The RNG formulation of the k-ε-model inhibits turbulent kinetic energy by reducing the destruction of its dissipation rate. The results of the RNG k-ε-model are presented in Figure 5 on the left and interestingly have a very comparable behavior to those of the second order RSM which are shown on the right.

As the wave starts to break neither of the models shows any production of turbulent kinetic energy. The first noticeable amount can only be seen almost half a second later when the actual breaking process is almost finished. Due to the different evolution of turbulence compared to the other two models above also the water surface elevation does not look the same as in Figure 4. The celerity of the wave crest is slightly larger and some oscillations occur at the back face of the wave. This can be attributed to the very low level of turbulent kinetic energy during the actual breaking event which keeps turbulent viscosity and thus, the balancing effect on mean flow variables small.

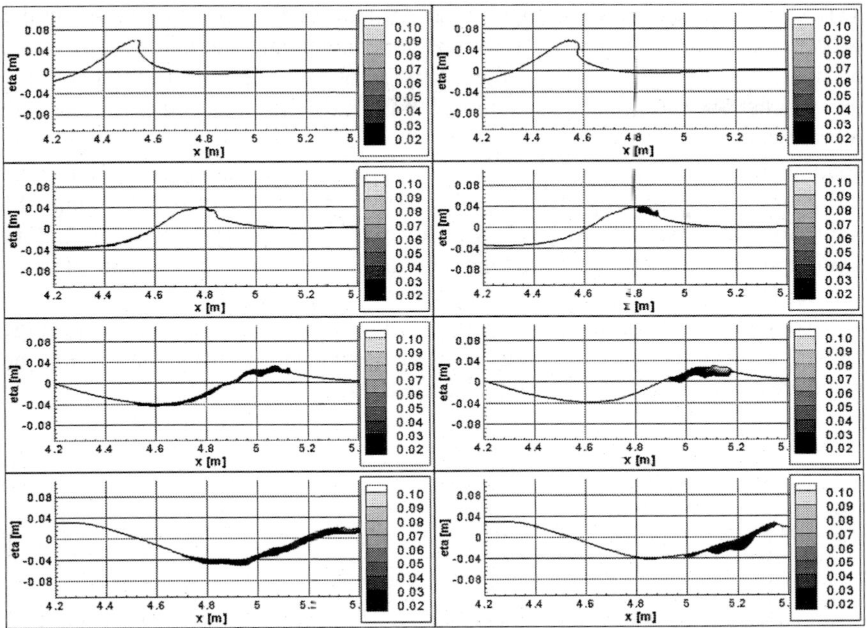

Figure 5: Evolution of turbulent kinetic energy for RNG k-ε-model (left) and RSM (right). Time steps from top to bottom: 11.2 s, 11.44 s, 11.68 s, 11.92 s.

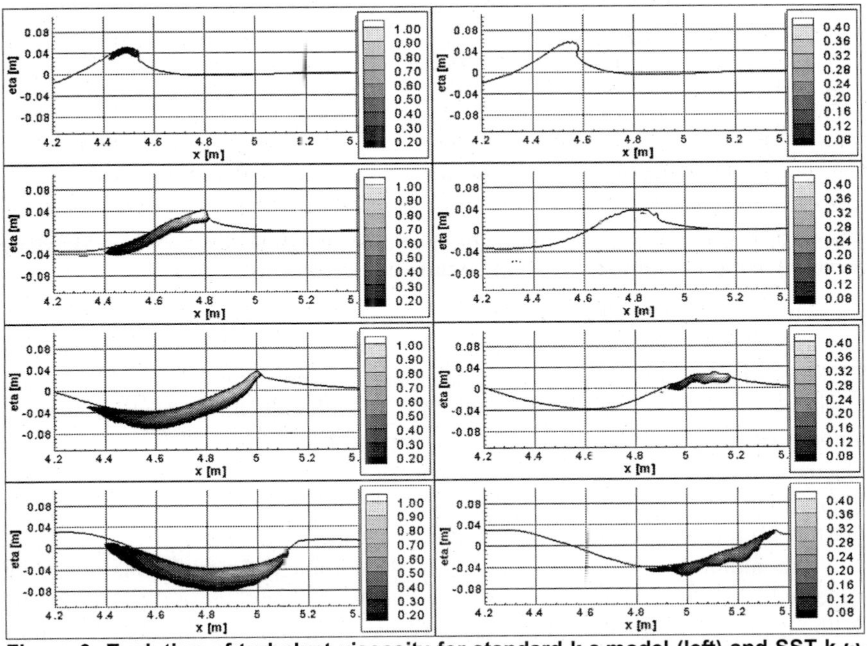

Figure 6: Evolution of turbulent viscosity for standard k-ε-model (left) and SST k-ω-model (right). Time steps from top to bottom: 11.2 s, 11.44 s, 11.68 s, 11.92 s.

Turbulent viscosity does not only smooth the mean flow field but is also a major source for the production of turbulent kinetic energy and a measure of its dissipation rate. Figure 6 shows the evolution of turbulent viscosity where only the results for the standard *k*-ε-model and RSM are presented as they are representative for the k-ω-model and the RNG *k*-ε-model, respectively. A comparison with Figures 4 and 5 shows a similar behavior as for the evolution of turbulent kinetic energy which might have been expected regarding the definitions in equations (8) and (11). However, also note the different contour scaling in Figure 6 indicating that the total amount of turbulent viscosity is much smaller for the RSM.

To conclude the analysis Figure 7 shows the evolution of the average amount of turbulent kinetic energy (top) and turbulent viscosity (bottom) in the wave crest which is defined to be the part 1 cm above mean water level.

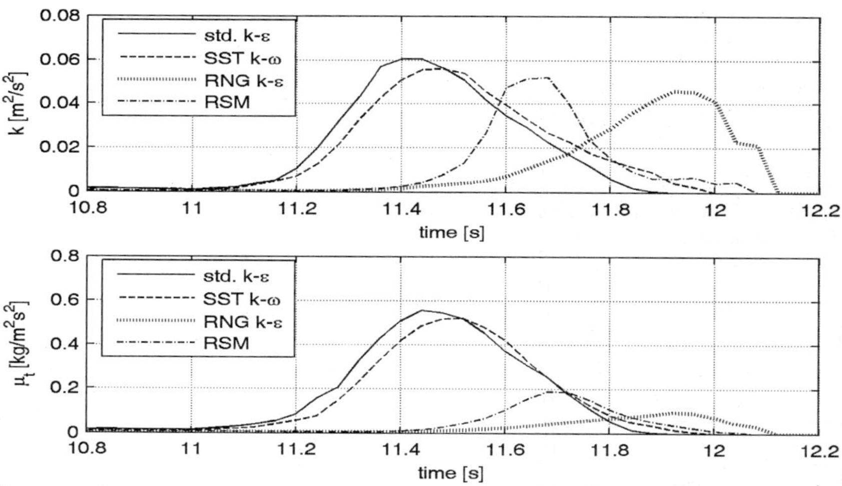

Figure 7: Average amount of turbulent kinetic energy (top) and turbulent viscosity (bottom) in the wave crest (z > 1 cm) for different turbulence models.

Turbulent kinetic energy and turbulent viscosity are in phase for each of the four models. However, as shown above the peak amounts for the RSM and the RNG *k*-ε-model occur significantly later and are not coupled with the actual breaking event which starts at about 11.2 s. Although the maximum average amount of turbulent kinetic energy is comparable for all models the turbulent viscosity for the standard k-ε-model and the SST k-ω-model is significantly higher than for the other two models which obviously predict a larger turbulent dissipation rate as could be expected at least for the RNG model.

The completely different behavior of the turbulence models was not to be expected and is a little unfortunate as the results somehow loose their reliability. Without further verification it might not be said which of the models is the most reasonable but, there is at least a tendency away from the results of the RSM and

RNG k-ε-model as it seems unlikely that almost no turbulence is produced when the wave starts to break. However, why these models behave like this remains an open question.

CONCLUSIONS

The present paper was a logical continuation of the experimental work by Harms and Schlurmann (2004). Turbulence quantities under a transient breaking wave were investigated by numerical simulations based on the RANS equations. Four different turbulence models were used to close the equations and the characteristic features of each model were explained.

To accurately reproduce the laboratory experiment the water surface elevation and the velocities of the incoming waves were directly prescribed at the boundary. The former was taken from gauge measurements and the flow field under the waves was calculated by means of a Local Fourier Approximation. A comparison with PIV data showed almost perfect agreement and proved the accuracy of the numerical model at least up to the onset of breaking.

The development of turbulence under the breaking wave was analyzed by means of turbulent kinetic energy and turbulent viscosity. The standard k-ε-model and the SST k-ω-model showed very similar reasonable results. Although the RNG k-ε-model and even more the RSM have a reputed higher fidelity their results seem questionable compared to the other two models. However, without a more thorough analysis and a comparison with measurements this remains an open question which will be addressed in future work.

REFERENCES

Harms, J. and T. Schlurmann. 2004. Time Variant Mass Transport beneath Breaking Waves, *Proceedings of 29th International Conference on Coastal Engineering*, ASCE, 441-453.

Launder, B.E. and D.B. Spalding. 1972. *Lectures in Mathematical Models of Turbulence*, Academic Press, London, England, 1972.

Launder, B.E. 1989. Second-Moment Closure: Present... and Future?, *Inter. J. Heat Fluid Flow*, 10(4), 282-300.

Lin, P. and P.L.-F. Liu. 1998. A numerical study of breaking waves in the surf zone, *Journal of Fluid Mechanics*, 359, 239-264.

Menter, F.R. 1994. Two-Equation Eddy-Viscosity Turbulence Models for Engineering Applications. *AIAA Journal*, 32(8), 1598-1605.

Sobey, R.J. 1992. A local Fourier approximation method for irregular wave kinematics, *Applied Ocean Research*, 14, 93-105.

Ting, F.C.K. and J.T. Kirby. 1995. Dynamics of surf-zone turbulence in a strong plunging breaker, *Coastal Engineering*, 24, 177-204.

Ting, F.C.K. and J.T. Kirby. 1996. Dynamics of surf-zone turbulence in a strong spilling breaker, *Coastal Engineering*, 27, 131-160.

Wilcox, D.C. 1988. Reassessment of the Scale Determining Equation of Advanced Turbulence Models, *AIAA Journal*, 26(11), 1299–1310.

NONLINEAR SURF ZONE WAVE PROPERTIES AS ESTIMATED FROM BOUSSINESQ MODELLING : RANDOM WAVES AND COMPLEX BATHYMETRIES

Rodrigo Cienfuegos[1], Eric Barthélemy[2], Philippe Bonneton[3] and Xavier Gondran[2]

The present work aims at investigating the ability of Boussinesq-type equations and breaking-wave parameterizations to reproduce nonlinear properties of surf zone waves. We compare results produced by two different breaking models : those proposed by Kennedy et al. (2000) and by Cienfuegos et al. (2005). Both breaking strategies are implemented in a fully nonlinear and weakly dispersive Boussinesq code (Cienfuegos et al., 2006a,b). In the first part we calibrate model parameters on the spilling regular wave experiment conducted by Ting and Kirby (1994). In the 2nd part, we apply the breaking Boussinesq models on a new laboratory experiment on random waves propagating over uneven bathymetries.

INTRODUCTION

Over the last 15 years important practical improvements for Boussinesq-type models have been introduced extending their applicability range into deeper waters and into the surf and swash zones. The breaking processes have been modelled by means of an *ad-hoc* extra term, usually written with a diffusive mathematical form, added in the momentum conservation equation of the inviscid (non dissipative) set of Boussinesq equations. Essentially, three different approaches to parameterize breaking terms can be found in the literature : i) roller-based models (e.g. Brocchini et al., 1992; Schäffer et al., 1993), the turbulent eddy viscosity analogy (e.g. Zelt, 1991; Kennedy et al., 2000), and iii) models accounting for the vorticity generation under breakers (e.g. Veeramony and Svendsen, 2000; Musumeci et al., 2005). An alternative approach was recently proposed by Cienfuegos et al. (2005) where a diffusive-type term is also introduced in the mass conservation equation. These authors use experimental and theoretical results on shallow water breakers to obtain a physical interpretation for mass and momentum mixing coefficients and show good agreement between computed and measured free surface profiles in a regular wave surf zone.

Eventough the available wave breaking formulations have been extensively calibrated and validated on regular wave experiments and uniform beach slopes, in nature, propagation and breaking is a random process. Consequently, there is still a need for testing these models on realistic situations where random waves

[1] Departamento de Ingeniería Hidráulica y Ambiental, Pontificia Universidad Católica de Chile, Vicuña Mackenna 4860, casilla 306, correo 221, Santiago de Chile. Email : racienfu@ing.puc.cl

[2] Laboratoire des Écoulements Géophysiques et Industriels, BP 53, 38041 Grenoble Cedex 9, France.

[3] Département de Géologie et d'Océanographie, Université de Bordeaux I, Av. des Facultés 33405 Talence, France.

propagate over uneven bathymetries. Ozanne et al. (2000) and Bayram and Larson (2000) reported some of the few examples where computed free surface profiles were compared against field measurements. Nevertheless, their numerical comparisons were somehow mistrusted by the poor nonlinear characteristics that were embedded in the chosen low order version of Boussinesq-type equations. It is worth noting that those authors didn't show results on the nonlinear energy transfer produced by their models.

The present work aims at investigating the ability of Boussinesq equations and breaking-wave parameterizations to reproduce nonlinear properties of surf zone waves and nonlinear energy transfer in the shoaling and surf zones. We focus in particular on wave height, skewness and asymmetry since it appears that these properties are paramount for sediment transport prediction (see for instance Drake and Calantoni, 2001). We investigate the application of two breaking models, those proposed by Kennedy et al. (2000) (KENN hereafter) and the one developed by Cienfuegos et al. (2005) (CIEN in the following). Both parameterizations are implemented in the fully nonlinear finite volume code SERR-1D (see Cienfuegos et al., 2006a,b, for details).

In the first part of the paper we calibrate breaking models and compare their results for the regular spilling breaking experiment reported by Ting and Kirby (1994). In the second part, we move on to an experimental random wave test keeping the same parameter values. We investigate model predictions in terms of free surface profiles in time domain and the nonlinear energy evolution and transfer in the frequency domain.

MODEL CALIBRATION ON A REGULAR WAVE EXPERIMENT

Breaking-wave models

Depth-averaged mass and momentum conservation equations can be written in the following generic form,

$$\frac{\partial h}{\partial t} + \frac{\partial}{\partial x}(hu) - D_h = 0, \tag{1}$$

$$\frac{\partial u}{\partial t} + \frac{1}{2}\frac{\partial u^2}{\partial x} + g\frac{\partial h}{\partial x} + \Gamma_d - \frac{1}{h}D_{hu} = 0, \tag{2}$$

where h is the water depth, u is the depth-averaged horizontal velocity, D_h and D_{hu} represent extra terms accounting for energy dissipation by breaking, Γ_d groups all high-order dispersive (non-hydrostatic) terms, and g is the gravitational acceleration; variables x and t denote space and time coordinates. In surf zone Boussinesq models, a breaking criterion must be adopted to turn on/off breaking dissipation, and wave crests need to be followed since a wave-by-wave approach must be considered (see for instance Kirby, 2003).

KENN model introduces breaking effects in the momentum equation only, i.e. $D_h = 0$. The remaining breaking term reads,

$$D_{hu} = \frac{\partial}{\partial x}\left(\nu_{hu}\frac{\partial hu}{\partial x}\right), \tag{3}$$

with the eddy viscosity coefficient, ν_{hu}, given as,

$$\nu_{hu} = B\,\delta_b^2\,h\,\frac{\partial h}{\partial t}, \tag{4}$$

where δ_b is a constant mixing length coefficient, and B is the parameter which activates breaking terms avoiding an impulsive behaviour. The breaking initiation/cessation criterion is based on threshold values for the time derivative of the free surface (see Kennedy et al., 2000, for details).

On the other hand, CIEN breaking model includes a mass conservative diffusivity term in the continuity equation intended to mimic local mass redistribution inside the roller and mixing layer effects. While the breaking induced momentum term is written as in (3), the additional mixing term in eq. (1) reads,

$$D_h = \frac{\partial}{\partial x}\left(\nu_h\frac{\partial h}{\partial x}\right). \tag{5}$$

In this model, mass and momentum diffusivity coefficients, ν_h and ν_{hu}, are written as,

$$\nu_h = -\delta_h\frac{c\,d}{\tan\Phi}\,\Lambda(X/l_r), \quad \text{for } 0 \leq X \leq l_r, \tag{6}$$

$$\nu_{hu} = -\delta_{hu}\frac{c\,d}{\tan\Phi}\,\Lambda(X/l_r), \quad \text{for } \ 0 \leq X \leq l_r, \tag{7}$$

where δ_h and δ_{hu} are order-one coefficients, c is the wave celerity, d is the mean water depth and Φ is the breaker angle. Λ is a self-similar shape function that locally distributes diffusivity over the breaker front ensuring that extra terms are conservative over a breaking event. This form relies on experimental results produced by Svendsen et al. (2000) and is written as,

$$\Lambda(X/l_r) = \exp\left(\frac{X}{l_r} - 1\right)\left[\left(\frac{X}{l_r} - 1\right) + \left(\frac{X}{l_r} - 1\right)^2\right],$$

with X a local coordinate system moving with the wave ($X = 0$ is located at the crest), and l_r the roller horizontal length. The different parameters have been scaled using empirical and theoretical knowledge on quasi-steady shallow water breakers and do have a physical interpretation (see Cienfuegos et al., 2005, for details).

In the present work we simplify the model taking δ_{hu} and Φ as constants and assuming that $\delta_h = 0.1\delta_{hu}$. Finally, the roller length is related to local wave properties using Cointe and Tulin (1994) theory of steady breakers calibrated on hydraulic jumps in similarity with surf zone waves (see Cienfuegos et al., 2004). It follows that $l_r/d = \frac{0.97}{\tan\Phi}(1-\gamma)$ introducing the breaker index $\gamma = H/d$, with H being the local wave height.

Optimal parameter values

The numerical values of model parameters are calibrated on the regular wave spilling breaking data reported by Ting and Kirby (1994). Cnoidal waves with incident height $H_0 = 0.127$ m and period $T = 2.0$ seconds propagate towards a uniform beach of 1:35 slope. The still water level is fixed at $h_0 = 0.4$ m in the horizontal part of the flume (see Fig. 1). Phase-averaged time series of free surface elevation are vailable at 21 locations, before and after breaking.

To test and calibrate the breaking models we define an error index based on measured and computed time-domain wave properties over the whole domain. Following Kennedy et al. (2000) we introduce left-right asymmetry as,

$$As = \frac{\langle \mathcal{H}^3(\eta - \overline{\eta})\rangle}{\langle (\eta - \overline{\eta})^2\rangle^{3/2}}, \tag{8}$$

where $\langle\cdot\rangle$ is the mean operator, $\mathcal{H}$ is the Hilbert transform, η is the free surface elevation relative to the still water level and overbar denotes time averaging. Similarly, crest-trough asymmetry or wave skewness is estimated as,

$$Sk = \frac{\langle (\eta - \overline{\eta})^3\rangle}{\langle (\eta - \overline{\eta})^2\rangle^{3/2}}. \tag{9}$$

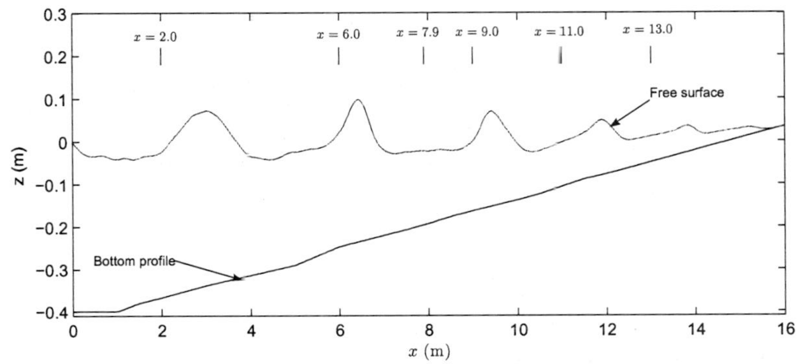

FIG. 1. Definition sketch for Ting and Kirby (1994) spilling breaking experiment.

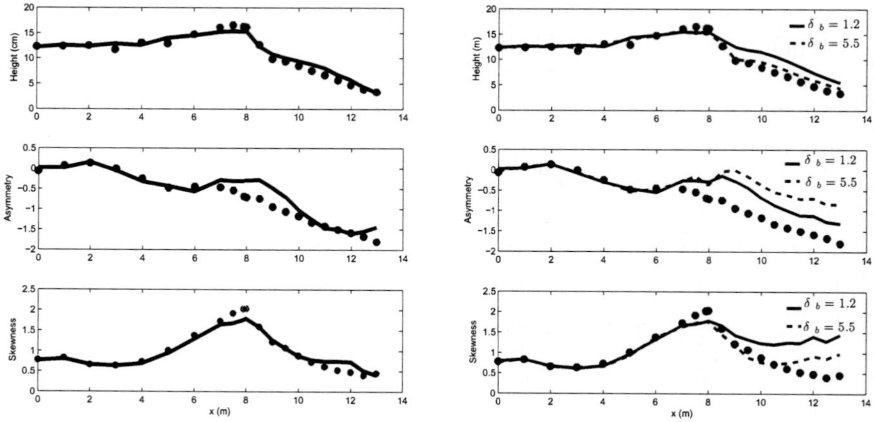

FIG. 2. Comparison between measured (•) and computed (–, – –) wave properties for Ting and Kirby (1994) experiment. Left panel: CIEN breaking model. Right panel: KENN breaking model.

A root mean square error (RMSE) can be computed for wave height, asymmetry and skewness; the calibration process consists in minimizing the average error for these three properties.

In the left panel of Fig. 2 we show optimal results obtained using CIEN model. The best agreement between computed and measured properties is produced for $\Phi = 17.0°$, $\gamma = 0.7$ and $\delta_{hu} = 5.0$ with an average RMSE of 14.2 % (RMSE for H, As and Sk being respectively 7.3 %, 24.1 % and 11.3 %). The biggest error occurs for wave asymmetry in the vicinity of the breaking point in the transition zone. Nevertheless, the model succeeds in accurately reproducing nonlinear wave evolution in the inner surf zone. It is worth noting that breaking terms are activated when the local breaker angle reaches 30°, conversely, breaking stops if it falls below an angle of 8°.

Results obtained using KENN breaking model are shown in the right panel of Fig. 2. Even though we closely follow the methodology detailed in Kennedy et al. (2000), the numerical values for parameter δ_b in eq. (4) and for the iniciation criterion of breaking are to be adjusted in the calibration. Indeed, in the present example the critical value for the free surface time derivative had to be lowered to $\eta_t^I = 0.55\sqrt{g\,d_0}$ in order to match the breaking point (d_0 is the local still water depth).

We conducted several computations using different values for δ_b because this coefficient sets the intensity of breaking energy dissipation. In Fig. 2 we show results for $\delta_b = 1.2$, which is the value recommended by Kennedy et al. (2000), and for $\delta_b = 5.5$.

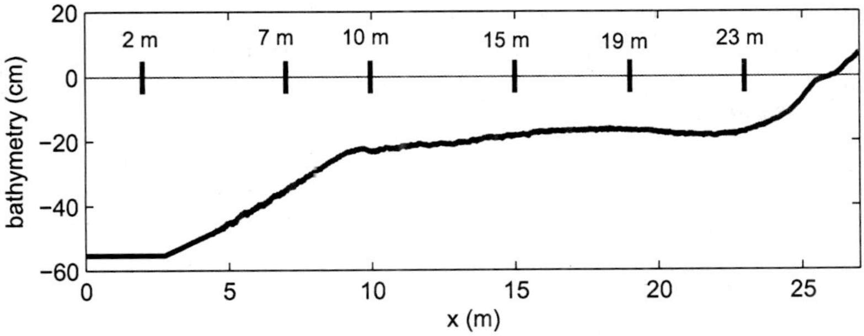

FIG. 3. Bottom profile and wave gauge locations for the random wave experiment.

The default value, $\delta_b = 1.2$, produces an over estimation of wave height in the inner surf zone in agreement with results published in Kennedy et al. (2000). On the other hand, even though the slope of the wave asymmetry curve is in agreement with the measured one, there is a vertical shift in the absolute value of this property all over the surf zone. Finally, it is seen that wave skewness in the inner surf zone is heavily overestimated by KENN model. The average RMSE for H, As and Sk is nearly 31.6 % (The RMSE for each one is respectively 16.6 %, 39.9 % and 38.3 %).

In order to improve wave height estimates in the inner surf zone, we set $\delta_b = 5.5$ to force higher energy dissipation. This parameter value allows to reduce the RMSE in wave height to 7.6 %, which is similar to the error we got using CIEN model. The RMSE in Sk is also smaller than before being now of nearly 19.2 %. Unfortunately, the RMSE in wave asymmetry, As, is considerably larger than before (61.1 %) which means that by forcing the model with such a high δ_b, the characteristic surf zone saw-tooth wave profile is completely lost. The average RMSE error for the three properties is now of nearly 29.3 %, i.e. not so different from the one we obtained by applying the recommended value $\delta_b = 1.2$. Hence, it seems that the default value for this parameter represents a reasonable compromise. Indeed, improving the estimate of wave height in the inner surf zone by forcing δ_b will inevitably increase the error in wave asymmetry predictions.

In what follows we will use $\delta_b = 1.2$ as suggested by Kennedy et al. (2000) and we freeze the calibrated parameter values for CIEN model. In the following we move on to a more realistic situation where random waves propagate over uneven bathymetries.

APPLICATION TO RANDOM WAVES

Experimental set-up

Our experiments were carried out in a flume 36 m long and 55 cm wide equipped with a piston wave generator. The water depth at rest is 55.3 cm. The mean overall slope is approximately 1/40 (see Fig. 3). The sloping bottom consists of a loose material for which the low density ($1.19\,\mathrm{g\,cm^{-3}}$) and the median diameter $d_{50} = 0.6$ mm allow representative sheet-flow and suspension regimes. The experiments were designed in order that the Froude number is of the same magnitude as those of natural environments under mild wave forcing. Length scales are roughly 1/10 and thus time scales roughly 1/3. The Shields number in the shoaling zone and the Rouse number in the breaking zone (ratio of turbulent agitation to the settling velocity of the sediment) are also of the same magnitude as those of natural environments. Irregular waves are generated according to a JONSWAP spectrum (peak enhancement factor =3.3 and peak frequency at $f_p = 0.33$ Hz). We checked that these waves conform to the expected spectrum and that they follow a Rayleigh distribution at 2 m downstream of the wave maker. Starting from a strongly disrupted initial condition, the beach profile was formed by the breaking waves in about 15 hours. During the following 150 hours of the experiment, the beach profile did not change in the mean. This confirms that equilibrium state between the bed morphology and its hydrodynamical forcing can be reached in laboratory conditions. The form of the equilibrium profile is at the transition between a barred and a terraced beach, similar to the one obtained in a wave basin for spilling waves by Wang et al. (2003). The large waves tend to break where the bottom also breaks ($x = 10$m) from steep to mild slope. These large waves were observed to break up to the beach. Small waves tend to break very close to the shoreline. The large terrace creates a very challenging situation for numerical models. Waves that are at the threshold of breaking stop and start breaking all along this terrace.

Model results

The numerical model is now applied to the random wave experiment described above for the quasi-equilibrium beach configuration. The location of wave gauges where measurements are available is depicted in Fig. 3. The time serie of free surface elevation at the wave gauge located at $x = 2.0$ m is used to prescribe in the numerical model the incident random wave field at the seaward boundary. A moving shoreline condition, adapted from Lynett et al. (2002), is implemented on the right boundary (see Cienfuegos et al., 2006b, for details).

We perform computations over 1200 s using CIEN and KENN breaking models with fixed parameter values as calibrated in Subsection 2.2. Comparisons between computed and measured free surface elevations at different locations are presented in Fig. 4. It can be seen that wave deformation in the shoaling zone ($x = 2 - 10$ m) is accurately represented by the model thus confirming the good nonlinear properties of the chosen set of Boussinesq equations. Numerical predictions for waves propagating in the terrace-like part of the beach ($x = 10-19$ m) where the biggest waves continue breaking are also in good agreement with the experimental data. In the upper part of the beach, after the bar, most of the waves break and only minor differences (essentially phase-lag errors) can be observed between CIEN and KENN breaking models in this time window. In addition, H_{rms}, set-up and intra-phase statistics (aymmetry and skewness) are plotted in Fig. 5. The overall agreement between computed and measured free surface statistics is considered adequate for this challenging test. In particular, H_{rms} and set-up evolution in the shoaling and surf zone is very well reproduced by both models. However, the computed set-up starts shorewards probably because the breaking iniciation criterion cannot accurately reproduce all types of breaking mechanisms. It is worth noting that CIEN model produces a slight underprediction of the set-up level in the inner zone. On the other hand, the numerical estimates of asymmetry and skewness produced by both breaking models are very similar unlike the situation we had in the regular wave experiment. There is a slight tendency to underestimate horizontal asymmetry in the inner surf zone while skewness prediction is quite reasonable. The randomness of the incident wave field and the averaging process used to compute wave statistics probably contribute to smear out model results in this case. In addition, it seems that wave reformation over the terrace-like bathymetry is over predicted probably because breaking terms are turned off too early. This could explain why wave asymmetry is not accurately reproduced inside the surf zone neither in CIEN nor in KENN models.

Finally, we perform a spectral analysis in order to study the nonlinear energy transfer inside the domain and results are presented in Fig. 6. As expected, for the first wave gauge located at $x = 2$ m the prescribed narrow band JONSWAP spectrum is recovered centered at the peak frequency $f_p = 0.33$ Hz. However, a low frequecy peak at $f_p = 0.027$ Hz is also evident. Further analysis suggest

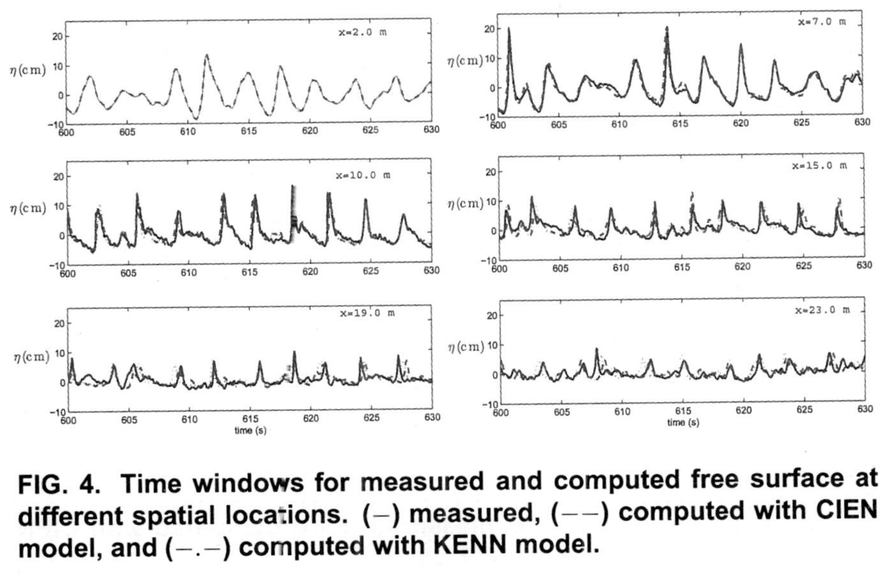

FIG. 4. Time windows for measured and computed free surface at different spatial locations. (–) measured, (– –) computed with CIEN model, and (–.–) computed with KENN model.

that this low frequency motion corresponds to a nearly standing wave for which the wave-length is nearly equal to the flume's length. This is also sustained by the spatial evolution of the energy content at this frequency since maximum values are reached near the two edges of the flume while at $x = 15$ m the energy is much smaller. Antinodes of the standing wave are located near the channel edges, while the node is near its center. The numerical model underpredicts the energy content at the antinodes but overpredicts its value at the node. However it succeeds in reproducing the nonlinear harmonic generation for both, the infragravity wave and the incident narrow band Jonswap spectrum. Energy dissipation by breaking is also well reproduced and differences between predictions produced by CIEN and KENN models are mild.

CONCLUSIONS

In this work we implemented the wave-breaking parameterizations proposed by Kennedy et al. (2000) and Cienfuegos et al. (2005) on a fully nonlinear and weakly dispersive Boussinesq-type model. We calibrated model parameters on the regular wave experiment conducted by Ting and Kirby (1994) trying to minimize the joint RMSE on wave height, horizontal asymmetry and skewness. For this particular experiment, Cienfuegos et al. (2005) breaking model was able to provide an accurate prediction of these three quantities (joint RMSE of 14%). On the other hand, the error produced by Kennedy et al. (2000) model could not be reduced below 29 % mainly because decreasing the RMSE in wave heights in the inner surf zone produced inaccurate estimates of horizontal asymmetries.

In the second part, surf zone Boussinesq models were applied to a random wave experiment carried out in the LEGI's wave flume (36 m). A narrow

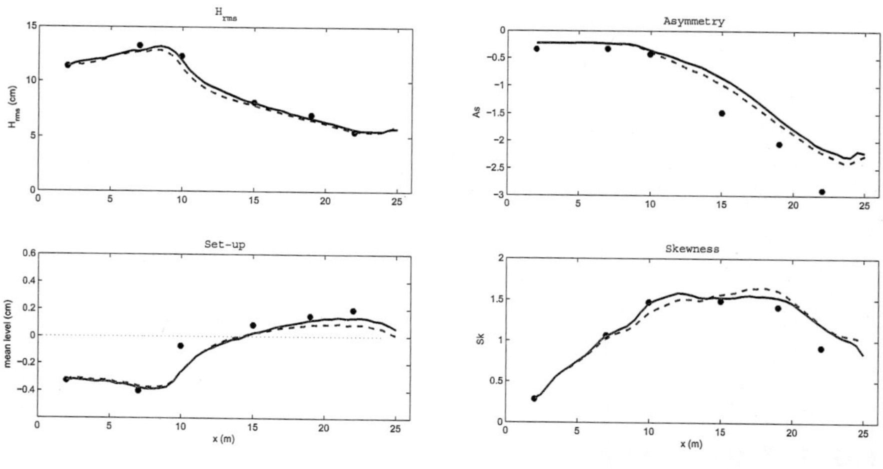

FIG. 5. Wave properties. (•) measured, (−−) computed with CIEN model, and (−) computed with KENN model.

band Jonswap spectrum propagated towards a terrace-like beach bathymetry. Comparisons between measured and predicted results showed an overall good agreement in the shoaling and surf zones. Unlike the regular wave case, the application of the two breaking models only produced minor differences. This could be attributed to the randomness of wave propagation since at one particular location inside the surf zone, wave statistics are computed over breaking and non breaking waves. On the other hand, over the quasi-horizontal terrace-like part of the beach, breaking seems to be prematurely turned off in both models (both model uses a similar breaking criterion). Consequently, wave reformation is over predicted by the numerical model and wave statistics is computed on a larger number of non breaking waves. This situation could also explain why different breaking models are producing rather similar results in this experiment.

Finally, a spectral analysis showed that breaking Boussinesq models were able to correctly reproduce nonlinear energy transfer in the shoaling and surf zones. Harmonic generation in the shoal and energy dissipation by breaking was accurately represented using both breaking parameterizations. On the other hand, a low-frequency nearly standing wave was generated in the wave flume. This infragravity motion was also captured by the numerical models showing only slight amplitude differences between measurements and computations. The question of infragravity wave generation under random wave fields needs to be investigated in more detail since long term beach morphology seems to respond to these low frequency forcing. For instance, in LEGI's random wave experiment beach bathymetry is in quasi-equilibrium and the observed infragravity wave is a nearly standing one.

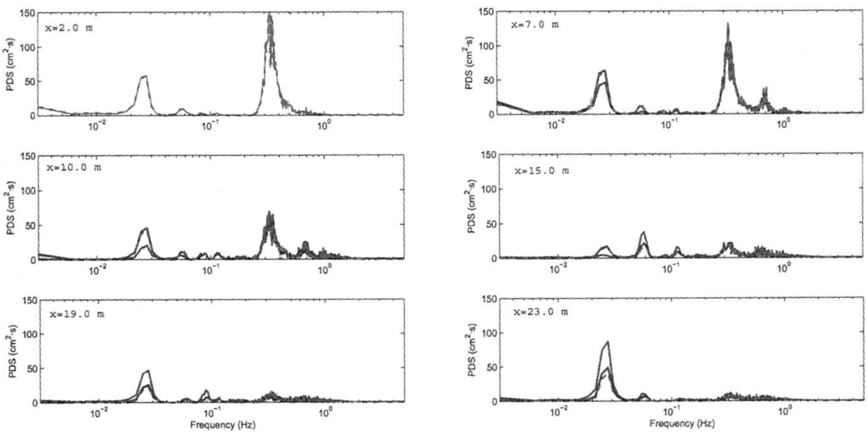

FIG. 6. Spectral analysis : power density spectra. (–) measured, (– –) computed with CIEN model, and (–.–) computed with KENN model.

ACKNOWLEDGMENTS

This work was partially funded by French research program PATOM. The first author gratefully acknowledges the financial support from CONICYT and The French Embassy in Chile through a PhD scholarship.

REFERENCES

Bayram, A. and M. Larson. 2000. Wave transformation in the nearshore zone : comparison between a Boussinesq model and field data, *Coastal Eng., 39*, 149–171.

Brocchini, M., M. Drago, and L. Iovenitti. 1992. The modelling of short waves in shallow waters. Comparisons of numerical models based on Boussinesq and Serre equations, *23th. Int. Conf. Coastal Eng.*, Venice, Italy, 76–88.

Cienfuegos, R., E. Barthélemy, and P. Bonneton. 2004. Roller modelling in the context of undertow prediction, *29th. Int. Conf. Coastal Eng.*, Lisbon, Portugal, 318–330.

Cienfuegos, R., E. Barthélemy, and P. Bonneton. 2005. A new wave-breaking parametrization for Boussinesq-type equations, *5th Int. Symp. Oc. Wave Meas. Anal.*, Madrid, Spain. CD-ROM.

Cienfuegos, R., E. Barthélemy, and P. Bonneton. 2006a. A fourth order compact finite volume scheme for fully nonlinear and weakly dispersive Boussinesq-type equations. Part I : model development and analysis, *Int. J. Num. Meth. Fluids,* 51(11), 1217–1253.

Cienfuegos, R., E. Barthélemy, and P. Bonneton. 2006b. A fourth order compact finite volume scheme for fully nonlinear and weakly dispersive Boussinesq-type equations. Part II : boundary conditions and validation, *Int. J. Num. Meth. Fluids,*. in press.

Cointe, R. and M. Tulin. 1994. A theory of steady breakers, *J. Fluid Mech., 276*, 1–20.

Drake, T. G. and J. Calantoni. 2001. Discrete particle model for sheet flow sediment transport in the nearshore, *J. Geophys. Res.,* 106(15), 19859–19868.

Kennedy, A., Q. Chen, J. Kirby, and R. Dalrymple. 2000. Boussinesq modelling of wave transformation, breaking and runup. I : 1D, *J. Waterw. Port Coastal and Oc. Eng.,* 126(1), 39–48.

Kirby, J.. 2003. Boussinesq models and applications to nearshore wave propagation, surfzone processes and wave-induced currents, V. Lakhan (Ed.), *Advances in Coastal Modelling*, 1–41. Elsevier.

Lynett, P., T. Wu, and P. Liu. 2002. Modeling wave runup with depth-integrated equations, *Coastal Eng., 46*, 89–107.

Musumeci, R., I. Svendsen, and J. Veeramony. 2005. The flow in the surf zone : a fully nonlinear Boussinesq-type approach, *Coastal Eng., 52*, 565–598.

Ozanne, F., A. Chadwick, D. Huntley, D. Simmonds, and J. Lawrence. 2000. Velocity predictions for shoaling and breaking waves with a Boussinesq-type model, *Coastal Eng., 44*, 361–397.

Schäffer, H., P. Madsen, and R. Deigaard. 1993. A Boussinesq model for waves breaking in shallow water, *Coastal Eng., 20*, 185–202.

Svendsen, I., J. Veeramony, J. Bakunin, and J. Kirby. 2000. The flow in weak turbulent hydraulic jumps, *J. Fluid Mech., 418*, 25–57.

Ting, F. and J. Kirby. 1994. Observation of undertow and turbulence in a laboratory surf zone, *Coastal Eng., 24*, 51–80.

Veeramony, J. and I. Svendsen. 2000. The flow in surf-zone waves, *Coastal Eng., 39*, 93–122.

Wang, P., B. Ebersole, and E. Smith. 2003. Beach-profile evolution under spilling and plunging breakers, *J. Waterw. Port Coastal Oc. Eng.,* 129(1), 41–46.

Zelt, J.. 1991. The run-up of nonbreaking and breaking solitary waves, *Coastal Eng., 15*, 205–246.

WAVE HEIGHT DISTRIBUTIONS ALONG A BEACH PROFILE

Carmen Castillo[1] and Miguel A. Losada[2]

Forecasting probabilities associated to local values of wave height in a beach profile is a very important subject in coastal engineering. The proposed model provides local probabilities for wave height, consecutive wave heights and maximum wave height at any point of a given profile and for a given sea state.

INTRODUCTION

One of the most necessary subjects in coastal engineering is the forecasting of wave height evolution across a beach profile and its associated probability.

Many authors have dealt with this problem proposing different CDFs and statistical descriptors evolution. Two of the simplest models were proposed by Battjes (1974) and Goda (1975).

Battjes (1974) suggested that the probability of non-breaking wave height exceeding the local limiting wave height H_b can be added to the probabilty of nonbreaking H_b (see figure 1). This model is only valid for saturated beaches.

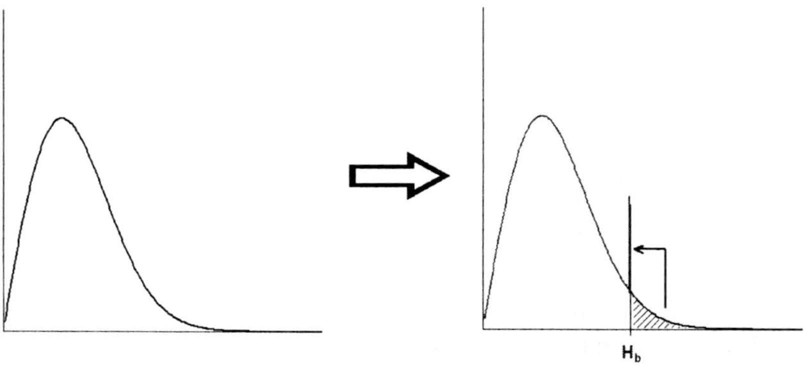

Figure 1. Battjes's model for redistributing the probability.

Goda (1975) redistributed uniformly the probability of the Rayleighan non-breaking waves in order to define the probability of breaking waves (see figure 2). This model does not agree with the physics of the problem.

---

[1] Civil Engineering Department, University of Castilla-La Mancha, Edif. Politécnico, Avda. Camilo José Cela s/n, 13071, Ciudad Real, Spain
[2] Coastal Engineering Group, Centro Andaluz de Medio Ambiente, Avda. Mediterráneo s/n, 18006 Granada, Spain

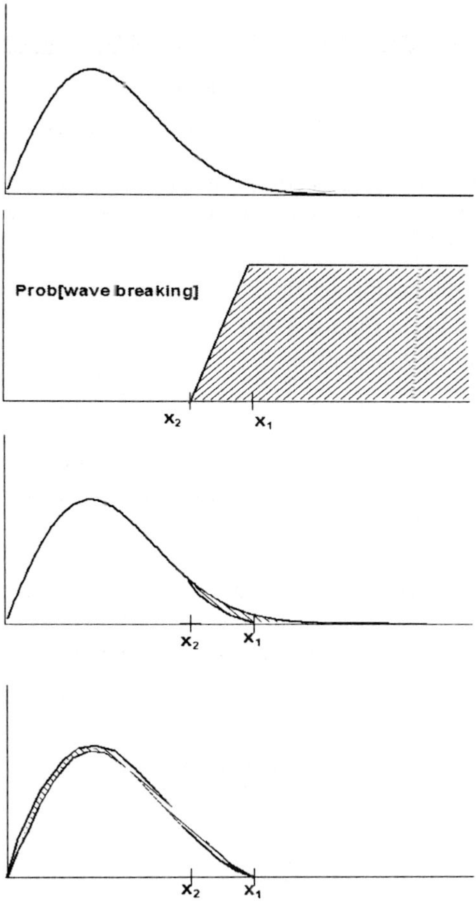

Figure 2. Goda's model for redistributing the probability.

Castillo (2004) proposed a model for local CDFs of different variables (water column components) where the parameters depended on the local value of kh.

This model was applied to wave height (see Castillo et al., 2004) and it was found that wave height was a Weibull for minima with parameters also depending on kh.

This last model is used to obtain the distributions for other variables related to wave height as conditional CDF of a wave height provided the previous one is known or the maximum wave height $H_{\max,N}$ provided the number of waves N is known. The model also shows that the statistical descriptors evolution in the profile depends on local *kh*.

WAVE HEIGHT DISTRIBUTION

Castillo (2004), Castillo et al (2004) and Castillo and Losada (2005) defined a Weibull for minima $H-W_m(\lambda,\delta,\beta)$ distribution for local wave height in a given sea state with the location parameter $\lambda=0$ and the scale and shape parameters, δ,β depending on the local values of *kh* and the offshore wave statistical descriptors (see figures 3 and 4) and used the model to analyse other quantities such as energy disipation.

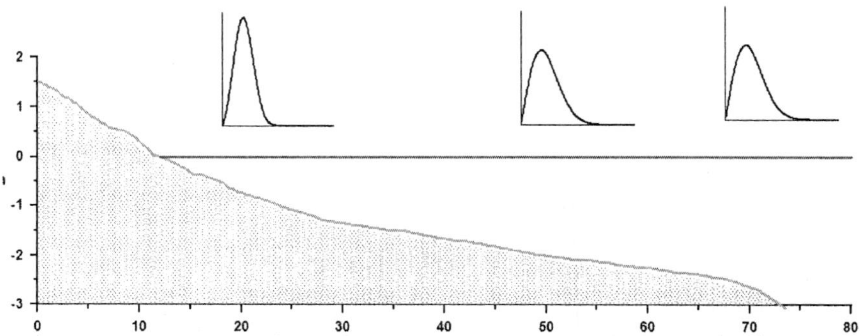

Figure 3. Castillo's model for wave height distribution.

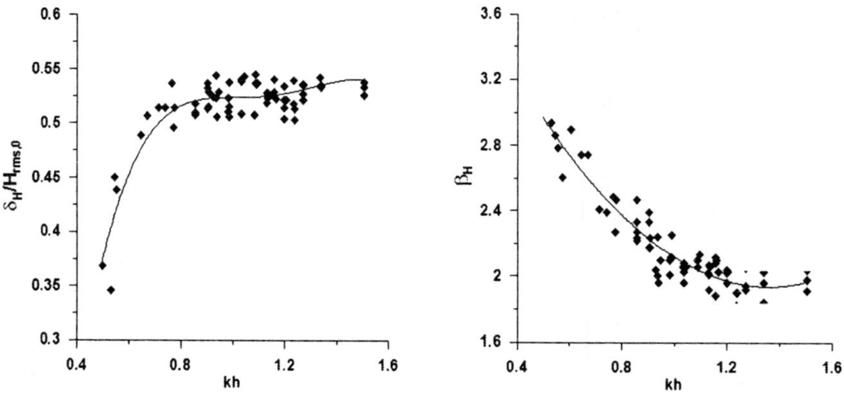

Figure 4. Statistical descriptors of wave height depending on kh.

The model has been validated with SUPERTANK data (Krauss and McKee, 1994). There are many ways of showing the validation of the model. One of them is comparing different empirical and theoretical values of wave height as the significant wave height, the root mean squared wave height or the mean wave height (see Figure 5 for H_{rms} comparison). For a Weibull for minima distribution, the theoretical values are:

$$
\begin{aligned}
H_S &= \delta\left[(\log 3)^{1/\beta} + \frac{3}{\beta}\Gamma\left(\frac{1}{\beta}, \log 3\right)\right] \\
H_{rms}^2 &= \delta^2 \Gamma\left(1 + \frac{2}{\beta}\right) \\
\overline{H} &= \delta\,\Gamma\left(1 + \frac{1}{\beta}\right)
\end{aligned}
\qquad (1)
$$

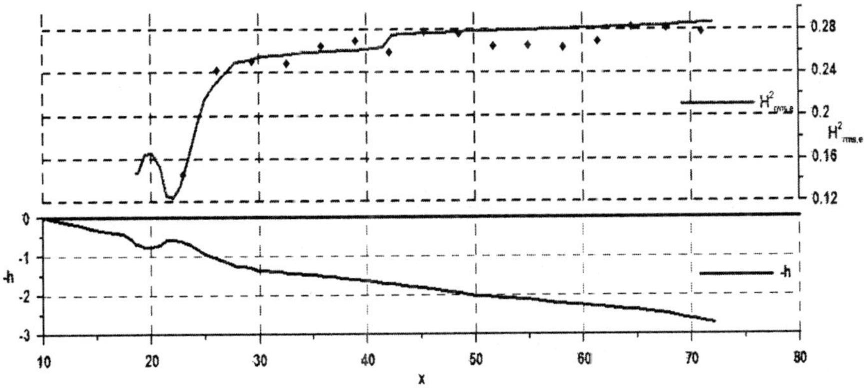

Figure 5. Validation of the model. Comparison between theoretical and empirical Hrms

It is important to compare how the model fits the experiments for the greatest values due to the fact that the model will be used for estimating the maximum wave height distribution. Figure 6 shows the goodness of fit of the proposed distribution for the maximum values, that is for some of the quantiles. The figure shows very good agreement far from the shoreline and the values are overestimated as the shoreline is approached but the agreement is still quite good.

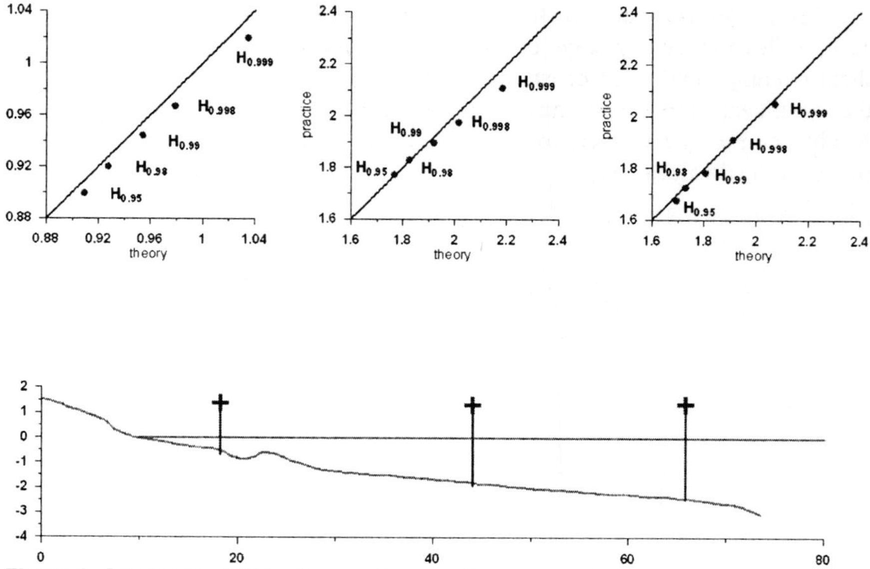

Figure 6. Goodness of fit for the maximum values.

CONSECUTIVE WAVE HEIGHT DISTRIBUTIONS

Proposing a model for probabilities of a wave height provided the previous ones are known implies an independence test which is shown in figure 7. The figure shows that consecutive wave heights are highly dependent (left hand side) for almost all cases except some near the shoreline. Some dependence is also shown between a wave height and the two previous ones (center figure) and almost no dependence at all with the one three waves before/after. Therefore, it is assumed that a wave height probability depends only on the value of the previous wave.

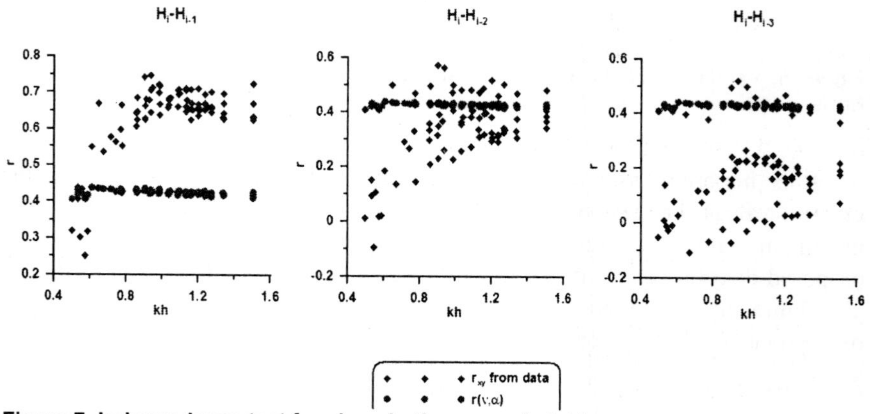

Figure 7. Independence test for close-in-time wave heights.

Castillo (2004) used a bivariate weibull for minima distribution (see Arnold et al, 1999) for the correlation between short and long wave oscillations. The same model is applied to consecutive wave height probability. The conditional probability model is shown in equation (2) so the probability of non-exceedance of a wave height y provided the previous one was x depends on these values and the wave height parameters, δ and β, and a new correlation parameter θ which depends on kh again as shown in figure 8.

$$F_{H_i|H_{i-1}=x}(y) = 1 - \exp\left\{-\left[1+\theta\left(\frac{x}{\delta}\right)^{\beta}\right]\left(\frac{y}{\delta}\right)^{\beta}\right\}\left[1+\theta\left(\frac{y}{\delta}\right)^{\beta}\right], \quad x, y > 0 \qquad (2)$$

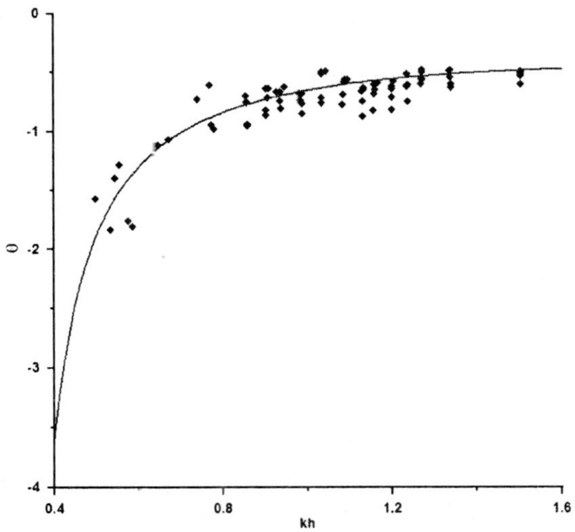

Figure 8. θ **dependence on kh.**

The validation of this model is shown in figure 9 where a comparison is made between the empirical and theoretical survival functions.

MAXIMUM WAVE HEIGHT DISTRIBUTION

The previous wave height distributions are used for estimating the probabilities for the maximum wave height in a sea state with N waves.

$$F_{H_{\max,N}}(x) = \text{Prob}\left[H_{\max,N} \le x\right] = \text{Prob}\left[H_1 \le x, H_2 \le x, \ldots, H_N \le x\right] \quad (3)$$

Two cases have been analysed: the first one is the simplest case where the wave heights are independent and the second one is based on the hypothesis that a wave height only depends on the previous one.

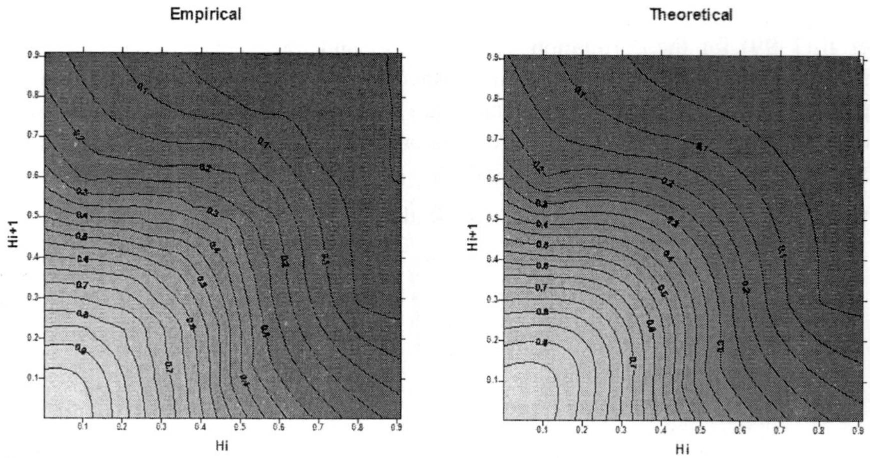

Figure 9. Validation of the model for consecutive wave heights.

Uncorrelated wave heights

In this case, the joint CDF of all the wave height for the value x is equal to the product of all the CDFs of each wave height, that is:

$$\mathrm{Prob}[H_1 \le x,\ldots,H_N \le x] = \mathrm{Prob}[H_1 \le x] \times \mathrm{Prob}[H_2 \le x] \times \ldots \times \mathrm{Prob}[H_N \le x] \quad (4)$$

so the CDF for the maximum wave height becomes:

$$\mathrm{F}_{H_{\max,N}}(x) = [F_H(x)]^N = \left[1 - \exp\left\{-\left(\frac{x}{\delta}\right)^{\beta}\right\}\right]^N \quad (5)$$

Figure 10 shows the evolution of the dimensionless maximum wave height $H_{\max,N} / H_{rms,off}$ with $H_{rms,off}$ the offshore root mean squared wave height which is an input data of the model. As can be seen, the probable values for the maximum wave height decrease as the shoreline is approached and the range of values also becomes smaller.

Correlated wave heights

In this case, the CDF of the maximum wave height is attained from the conditional probabilities for consecutive waves, that is:

$$\begin{aligned} &\mathrm{Prob}[H_1 \le x,\ldots,H_N \le x] = \mathrm{Prob}[H_1 \le x] \times \mathrm{Prob}[(H_2 \le x,\ldots,H_N \le x) \,|\, H_1 \le x] \\ &= \mathrm{Prob}[H_1 \le x] \times \mathrm{Prob}[H_2 \le x \,|\, H_1 \le x] \times \ldots \times \mathrm{Prob}[H_N \le x \,|\, H_{N-1} \le x] \end{aligned} \quad (6)$$

$$\mathrm{F}_{H_{\max,N}}(x) = \frac{\left[F_{(H,H)}(x,x)\right]^{N-1}}{\left[F_H(x)\right]^{N-2}} \quad (7)$$

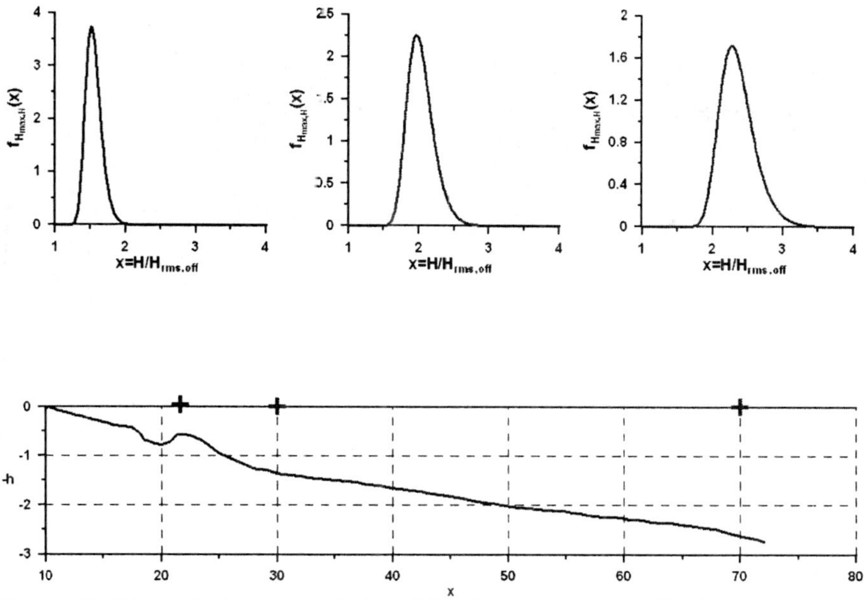

Figure 10. Dimensionless wave height PDF along a beach profile for uncorrelated waves.

$$F_{H_{\max,N}}(x)=\frac{\left[1-2\exp\left\{-\left(\frac{x}{\delta}\right)^{\beta}\right\}+\exp\left\{-2\left(\frac{x}{\delta}\right)^{\beta}+\theta\left(\frac{x}{\delta}\right)^{2\beta}\right\}\right]^{N-1}}{\left[1-\exp\left\{-\left(\frac{x}{\delta}\right)^{\beta}\right\}\right]^{N-2}} \tag{8}$$

Figure 11 shows the evolution of the CDF of the dimensionless maximum wave height $H_{\max,N}/H_{rms,off}$. As can be seen, the probable values for the maximum wave height decrease as the shoreline is approached and the range of values also becomes smaller.

When comparing both models, for the same probability, the wave height in the correlated model is a little bit smaller than the one in the uncorrelated model.

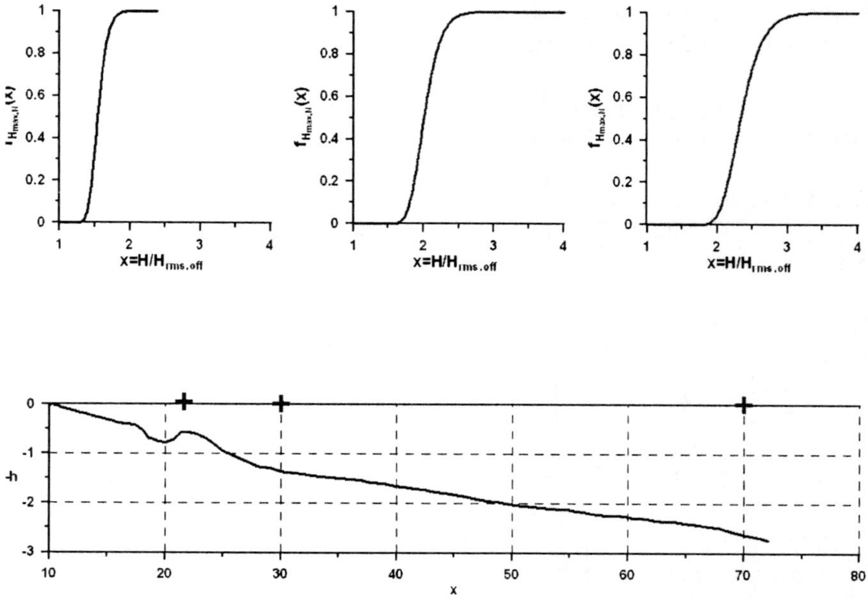

Figure 11. Dimensionless wave height CDF along a beach profile for correlated waves.

ABSOLUTE PROBABILITIES

Previous results are conditional probabilities to a given sea state. Absolute probabilities are defined by:

$$F_X^*(x) = \text{Prob}[\mathrm{X} \le \mathrm{x}] = \sum_j \text{Prob}\left[\mathrm{X} \le \mathrm{x} \middle| sea\ state_j\right] \times \text{Prob}\left[sea\ state_j\right]$$
$$F_X^*(x) = \sum_j F_X(x)\text{Prob}\left[sea\ state_j\right] \tag{9}$$

where $F_X^*(x)$ is the absolute probability, $F_X(x)$ is any of the CDFs proposed in previous sections and sea state j is characterized by wave height and wave period parameters.

CONCLUSIONS

The proposed method allows obtaining the conditional local probabilities of wave heights, consecutive wave heights and maximum wave heights for a given sea state and initial profile. The parameters of these distributions depend on the local value of kh.

Therefore, it is possible to describe the evolution of the probability distributions related to wave heights across a beach profile whose geometry is known as well as the statistical descriptors of wave oscillation approaching the beach profile (offshore the depth of closure).

The absolute probabilities can be easily defined with the application of an offshore wave climate as described in Castillo and Solares (2006).

ACKNOWLEDGMENTS

Carmen Castillo is indebted to the University of Castilla-La Mancha for partial funding.

REFERENCES

Arnold, B. C., Castillo, E. and Sarabia, J. M. 1999. Conditional Specification of Statistical Models. Springer, New York.

Battjes, J. A. 1974. Computation of set-up, longshore currents, run-up and overtopping due to wind-generated waves. Ph.D Thesis.

Castillo, M. C. 2004. Fiabilidad y optimización en ingeniería marítima. PhD. Thesis. (in Spanish) University of Granada.

Castillo, M. C., Baquerizo, A. and Losada, M. A. 2004. Temporal and spatial statistical variability of the wave height in the surf zone. *Proc. 29th Int. Conf. on Coastal Engineering*, Vol. 1, pp 997-1008.

Castillo, M. C., Losada, M. A. 2005. Evolución espacial de descriptores estadísticos del estado de mar. *VIII Jornadas Españolas de Ingeniería de Costas y Puertos*, resumen 38.(in Spanish)

Castillo, M. C., Solares, C. 2006. Statistics applied to wave climate on a beach profile. *Proc. Int. Conf. on Mathematical and Statistical Modeling (ICMSM)*, CD-R.

Goda, Y. 1975. Deformation of irregular waves due to depth-controlled wave breaking. Rept. Port & Harbour Res. Inst., Vol. 14, No. 3, pp 59-106. (in Japanese).

Kraus, N. C. and McKee Smith, J. 1994. Supertank Laboratory Data Collection Project. Vol. I: Main text. Tech. Report, CERC-94-3, 274 pp, US Army Eng. Wat. Exp., St. Vicks, Mss.

A MODIFIED MODEL FOR WAVE DECAY ON BEACHES

Ching-Piao Tsai[1] and Hong-Bin Chen[2]

This paper presents a modified model for the prediction of the wave decay after wave breaking on beaches. The wave energy-flux equation involving the energy dissipation was adopted in the model. The energy dissipation rate presented in this study was assumed to be nonlinear proportion to the difference between the local energy flux and the stable energy flux. The best values of the relative coefficients in the energy-flux equation were obtained from the laboratory data based on the least-squares-error procedure. The present model performs well for the prediction of the wave decay in the surf zone based on the comparisons to the laboratory data. The results show that the predictions are available for a fairly wide range of beach slopes from very gentle to very steep.

INTRODUCTION

The wave energy dissipation due to the wave breaking is mainly transformed into the turbulent energy. It is a very complex hydrodynamic process, and is one of the most fascinating phenomena of wave transformations. The wave breaking does not only produce large force to the structures, but also generate the nearshore current circulation to affect the shore topography. Thus the estimation of the wave height transformation across the surf zone is an essential part for the nearshore hydrodynamics as well as the design of coastal structures.

It is most convenient to employ the wave energy balance equation to estimate the wave height decay due to the wave breaking. The steady state wave energy-flux equation for one-dimensional monochromatic wave is given by

$$\frac{\partial(EC_g)}{\partial x} = -\Phi_d \tag{1}$$

in which E is the wave energy per unit surface area, C_g is the group velocity, and Φ_d is the energy dissipation rate per unit surface area. For the estimation of the energy dissipation rate, Le Méhauté (1962) approximated the wave breaking as a propagating bore. Horikawa and Kou (1966) represented it in terms of turbulent velocity fluctuations which were assumed to decay exponentially with distance from the wave breaking point. Mizuguchi (1980) introduced the eddy viscosity into the energy dissipation rate based on the wave and beach profile characteristics. Later, Dally et al. (1985) developed a wave-breaking model

---

[1] Department of Civil Engineering, National Chung Hsing University, Taichung 402, Taiwan
[2] Department of Tourism, Restaurant and Hotel Management, Kainan University, Taoyaun 338, Taiwan

based on the concept of a stable wave height within the surf zone for a given water depth.

There were two assignable parameters included in the model of Dally et al. (1984), namely the wave decay factor, K, and the stable wave factor, Γ. The values of these two factors are varied with the beach slopes, and it was preferable to choose single values $K = 0.15$ and $\Gamma = 0.40$ for all beach slopes. It gives satisfactory results for the beach slopes less than 1/20, but it does not perform well on the steep slopes.

For the case of coasts in the east of Taiwan, some beach slopes are very steep (to 1/5), where an industrial harbor and many fishery harbors located there. Thus the prediction of the wave height transformation for this area was needed. For extending the model of Dally et al. (1984) to a fairly wide range of the beach slopes, the previous experimental data involving those of the steep slopes are collected in this study to develop a modified model for the estimation of the wave decay in the surf zone.

MODEL DEVELOPMENT

In Dally et al. (1985), the rate of energy dissipation per unit plan area Φ_d in Eq. 1 was assumed to be proportional to the difference between the local energy flux and the stable energy flux. In the present modified model, the energy dissipation rate is also assumed to be related with the local energy flux and the stable energy flux but it is given by a nonlinear form as

$$\frac{\partial\left(EC_g\right)}{\partial x}=\frac{-EC_g}{h}a\left[\frac{EC_g-\left(EC_g\right)_s}{EC_g}\right]^b \tag{2}$$

in which $(EC_g)_s$ is the energy flux associated with the stable wave, h is the still water depth, a and b are dimensionless coefficients to be determined. It is identical to the assumption of Dally et al. (1985) if $a = K$ and $b = 1$.

Based on the laboratory data of Horikawa and Kuo (1966), the height of stable wave, H_s, is given by

$$H_s = \Gamma h \tag{3}$$

The stable wave factor, Γ, is a dimensionless coefficient whose value is between 0.35 and 0.40. The group velocity is taken as $C_g = \sqrt{gh}$ based on the shallow wave theory, and $\partial/\partial x = (\partial h/\partial x)(\partial/\partial h) = -\tan\beta(\partial/\partial h)$, $\tan\beta$ is the bottom slope, Eq. 2 can then be written as

$$\frac{\partial\left(H^2\sqrt{h}\right)}{\partial h}=\frac{H^2}{\tan\beta\sqrt{h}}a\left(1-\frac{\Gamma^2 h^2}{H^2}\right)^b \tag{4}$$

This equation can easily be solved by the explicit finite-difference numerical scheme.

CALIBRATION

Referring to Dally et al. (1985), a least squares procedure is adopted to determine the best values for the dimensionless coefficients *a* and *b*. The error function to be minimized is defined by

$$\varepsilon = \left[\sum_{j=1}^{N} \left(H_{pj}^2 - H_{mj}^2 \right)^2 / \sum_{j=1}^{N} H_{mj}^2 \right]^{1/2} \tag{5}$$

where ε is the least square error, H_{mj} is the measured wave height, H_{pj} is the wave height at that location as predicted by the numerical scheme for given incipient conditions and value of Γ, and N is the number of data points analyzed.

Besides the laboratory data obtained in Horikawa and Kuo (1966) for slopes of 1/20, 1/30, 1/65 and 1/80, the data obtained in Sakai et al. (1986) for slope of 1/15 and Tsai et al. (1997) for slopes of 1/10 and 1/5 are also included in the analysis. As indicated in Dally et al. (1985) that the value of K is between 0.35 and 0.40. For simplifying the determination of the best values of a and b, the recommended value 0.40 of K in Dally et al. (1985) is specified in the present model for all slopes.

Figure 1 displays the errors for the data taken on slopes of 1/10 and 1/5 which vary with the value of b. The similar variation of error could be found for all slopes and the minimum error occurs at about $b = 2$. For the specified values of $b = 2$ and $K = 0.40$ for all slopes, the best fit values for a are presented in Table 1. Based on the best fit results, we have a regression line for the value of the coefficient a as shown in Figure 2, and it is given by:

$$a = 0.178 + 0.88 \tan\beta \tag{6}$$

It can be seen that the a value tends to a constant of 0.178 for gentle beaches.

RESULTS AND COMPARISONS

Based on the best fit results, the model of Dally et al. (1985) now is modified and extended to calculate the wave decay on the steep slopes. Figure 3 illustrates the wave height decay in the surf zone as a function of the incipient condition, H_b/h_b, where the H_b and h_b represent the wave height and still water depth at breaking point. The result shows the faster decay of the wave height is associated with the higher value of H_b/h_b.

Table 1. Best fit values for *a*, *b* and Γ.

Slope	*a*	*b*	Γ	Minimum error
1/80	0.198	2.0	0.40	0.0882
1/65	0.188	2.0	0.40	0.0770
1/30	0.194	2.0	0.40	0.1070
1/20	0.169	2.0	0.40	0.1066
1/15	0.271	2.0	0.40	0.0516
1/10	0.242	2.0	0.40	0.0901
1/5	0.412	2.0	0.40	0.0847

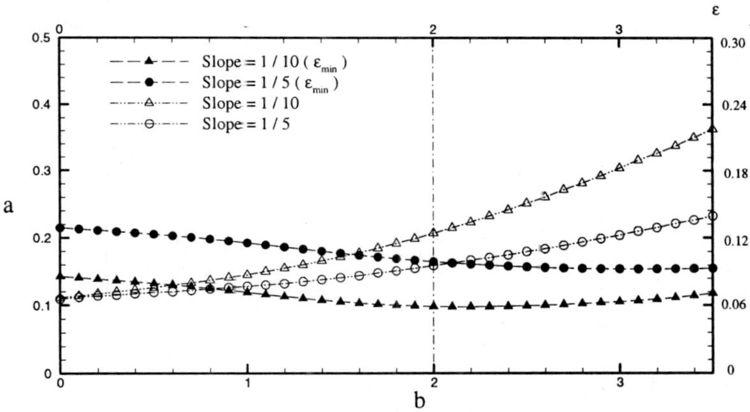

Figure 1. Variations of the error with *b*, and the corresponding fit value of *a*.

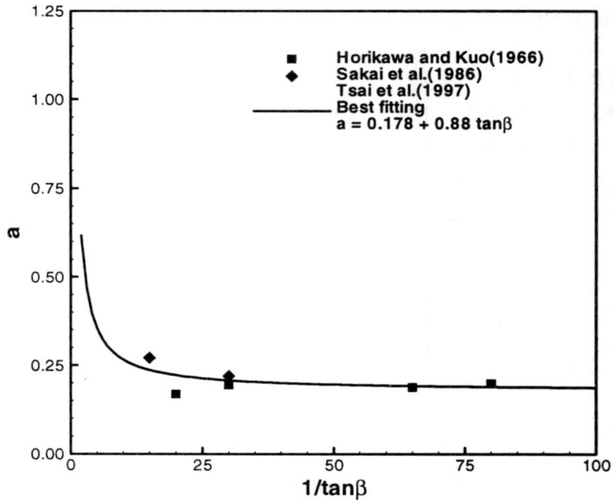

Figure 2. The regression line of the best fit value of *a* vs. slope.

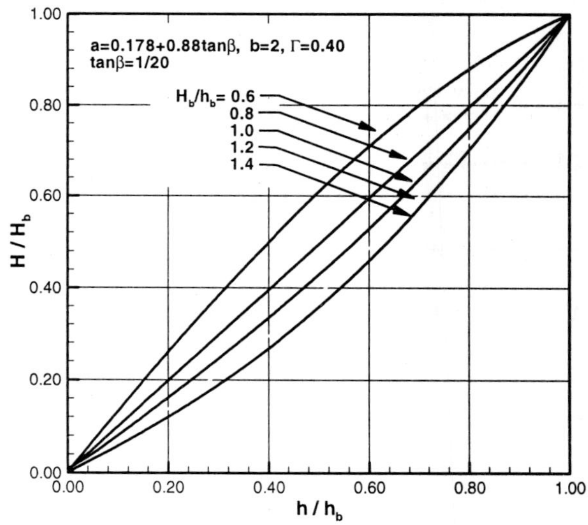

Figure 3. Wave height decay as function of H_b/h_b on a beach slope of 1/20.

The comparisons among the present modified model, the model of Dally el al. (1985) and the laboratory data of the wave height decay are shown in Figures 4-7 for the beach slopes of 1/30, 1/15, 1/10 and 1/5. It is noted that the laboratory data for the case of 1/30 in Horikawa and Kuo (1966) was used in the calibration of the model, but the data of Sakai et al. (1986) is verified in Figure 4. The values of the wave decay factor and the stable wave factor, $K = 0.15$ and $\Gamma = 0.40$, are set in the model of Dally et al. (1985). As indicated in Dally et al. (1985), the figures show that the use of $K = 0.15$ and $\Gamma = 0.40$ does not perform well on the steep slope. In Dally's model, it needs to give different values of K and Γ for the cases of steeper slope. But they only showed the values of K and Γ for the case of slope of 1/12. While in the present model, the coefficients in Eq. 4 have been presented in a function of slope. The comparisons in Figures 4-7 show that the present predictions are in good agreement with the laboratory data for a wide range of slopes from very gentle to 1/5.

The set-up and set-down of the mean water level in the surf zone were not included in the present analysis because of no available measured laboratory data in the aforementioned literature. The wave set-up will affect the wave height transformation after the wave breaking, especially in case of steep slope. However, the measured wave height in the experiments included the outcome of the effect of the wave set-up. The laboratory data of steep slopes were included in the present analysis, thus the effect of the set-up to the wave height decay was covered in the present empirical coefficients. At least, the present model performs well if only the wave height decay is to be calculated.

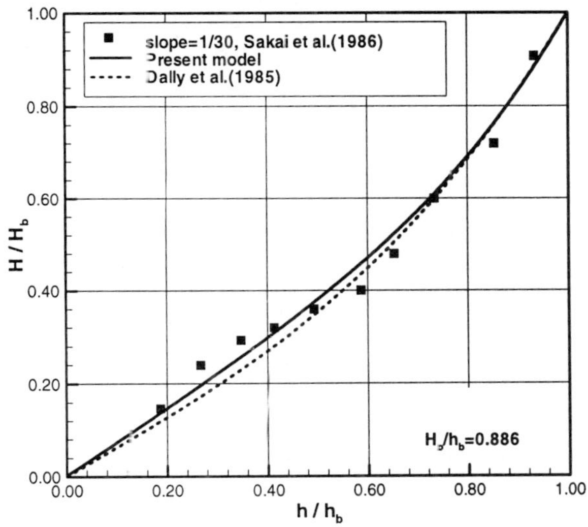

Figure 4. Comparisons of wave height decay on a beach slope of 1/30.

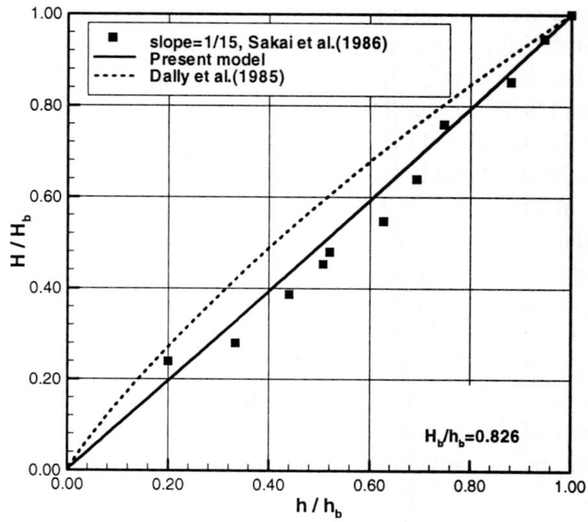

Figure 5. Comparisons of wave height decay on a beach slope of 1/15.

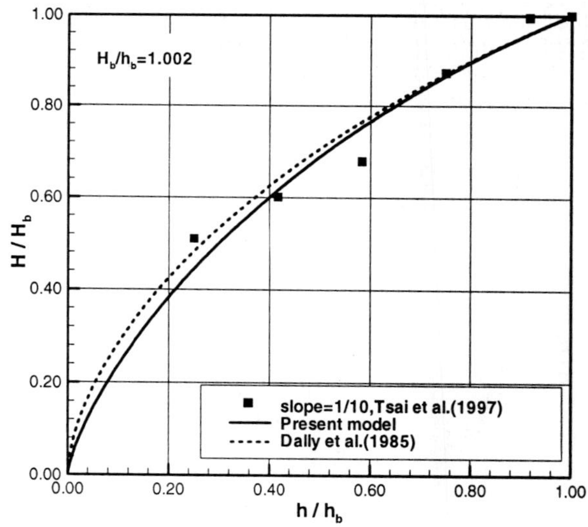

Figure 6. Comparisons of wave height decay on a beach slope of 1/10.

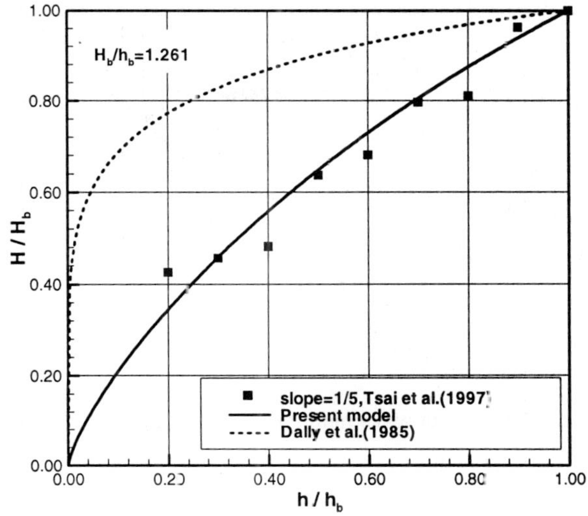

Figure 7. Comparisons of wave height decay on a beach slope of 1/5.

Figures 8 and 9 display the samples of the prediction of the wave decay after the wave breaking as compared to the measured data for beaches of 1/10 and 1/5, in which x_b is the location of the wave breaking and x-axis is directed to the onshore. The results show that the present model performs well in the prediction of the wave decay even for the steep slopes.

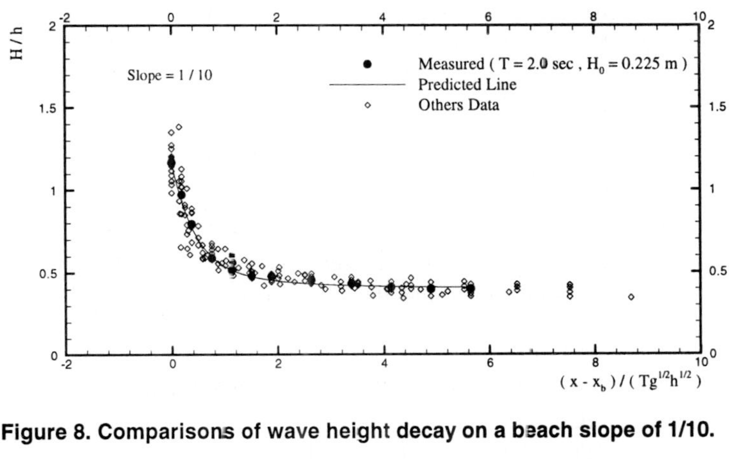

Figure 8. Comparisons of wave height decay on a beach slope of 1/10.

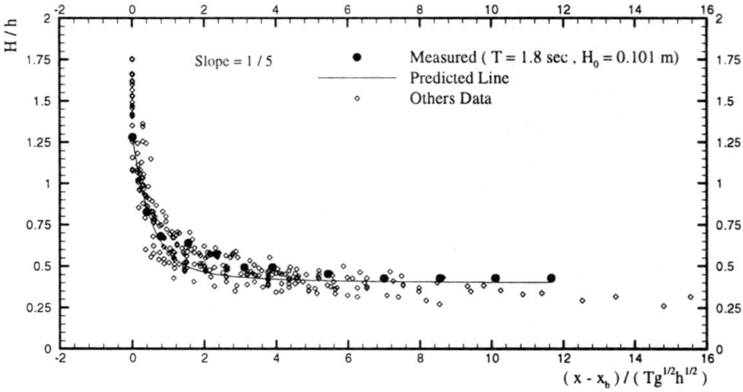

Figure 8. Comparisons of wave height decay on a beach slope of 1/5.

CONCLUSIONS

Based on the analysis method of Dally et al. (1985) and the laboratory data, the model for describing the wave transformation in the surf zone was modified and developed for a fairly wide range of beach slopes from 1/80 to 1/5. In the model, the steady state wave energy-flux equation for one-dimensional monochromatic wave involved the energy dissipation was adopted. The energy dissipation rate presented in this study was assumed to be nonlinear proportion to the difference between the local energy flux and the stable energy flux. The best values of the relative coefficients in the energy-flux equation were obtained from the laboratory data based on the least squares error procedure. Based on the comparisons with the laboratory data, it shows the present model performs well for the prediction of wave decay in the surf zone.

ACKNOWLEDGMENTS

The authors are grateful to National Science Council, Taiwan, for the financial support to this study.

REFERENCES

Dally, W.R., R.G. Dean, and R.A. Dalrymple. 1985. Wave height variation across beaches of arbitrary profile, *Journal of Geophysical Research*, 90(C6), 11917-11927.

Horikawa, K. , and C.T. Kuo. 1966. A study of wave transformation inside surf zone, *Proceedings of 10th International Conference on Coastal Engineering*, ASCE, 217-233.

Le Méhauté, B. 1962. On non-saturated breakers and the wave run-up, *Proceedings of 8th International Conference on Coastal Engineering*, ASCE, 77-92.

Mizuguchi, M. 1980. An heuristic model of wave height distribution in surf zone, *Proceedings of 17th International Conference on Coastal Engineering*, ASCE, 278-289.

Sakai, S., K. Hiyamizu, and H. Saeki. 1986. Wave height decay model with a surf zone, *Proceedings of 20th International Conference on Coastal Engineering*, ASCE, 686-696.

Tsai, C.P., Y.Y. Lee, and H.B. Chen. 1997. Experiments on wave deformations over steep slope seabed, *Proceedings of 19th Ocean Engineering Conference in Taiwan*, 137-144 (In Chinese).

EXPERIMENTAL WAVE BREAKING IN SWAN

L.H. Holthuijsen[1] and N. Booij[2]

The dissipation of waves in the surf zone is generally well predicted by spectral wave models but the triad wave-wave interactions and the fraction of breaking waves are not. The present study shows that the estimate of the fraction of breakers can be improved by correlating one coefficient of a conventional model to the average dissipation rate in the surf zone. In addition, it is suggested that the basic characteristics of triad wave-wave interactions in the surf zone (initial generation of higher harmonics, followed by a flattening of the spectral tail) can be simulated efficiently using an analogy with quadruplet wave-wave interactions.

INTRODUCTION

Spectral wave models are reasonably successful in predicting the evolution of the significant wave height through the surf zone. This success is due to the formulation that represents depth-induced breaking. Since the fraction of breakers is an essential parameter in this formulation, one would expect this fraction to be predicted with equal success. This is surprisingly not the case. The error is typically one or two orders of magnitude. The success in predicting the significant wave height is also not matched by a similar success in predicting the mean wave period. This is due to the fact that the theory and the numerical techniques for triad wave-wave interactions in spectral models are not well developed.

METHOD (FRACTION OF BREAKERS)

One of the best known formulations for wave breaking in the surf zone is due to Battjes and Janssen (1978). The rate of dissipation in this model is controlled by an assumed maximum possible wave height for a given water depth. In this formulation the probability density function of the wave heights is assumed to be a *clipped* Rayleigh distribution with all breaking wave heights concentrated at the maximum wave height. This model also predicts the fraction of breaking waves with

$$\frac{1-Q_b}{\ln Q_b} = -\left(\frac{H_{rms}}{H_{\max}}\right)^2 \tag{1}$$

[1] Faculty of Civil Engineering and Geosciences, Delft University of Technology, Stevinweg 1, Delft, 2628 CN, the Netherlands

[2] Digital Hydraulics, Willemsplein 535, Rotterdam, 3016 DR, the Netherlands

where Q_b is the fraction of breakers, H_{rms} is the root-mean-square (rms) wave height and H_{max} is the wave height of all breaking waves at a given position and local water depth. Quantitative observations of breakers are few and this estimate of Q_b has been barely verified.

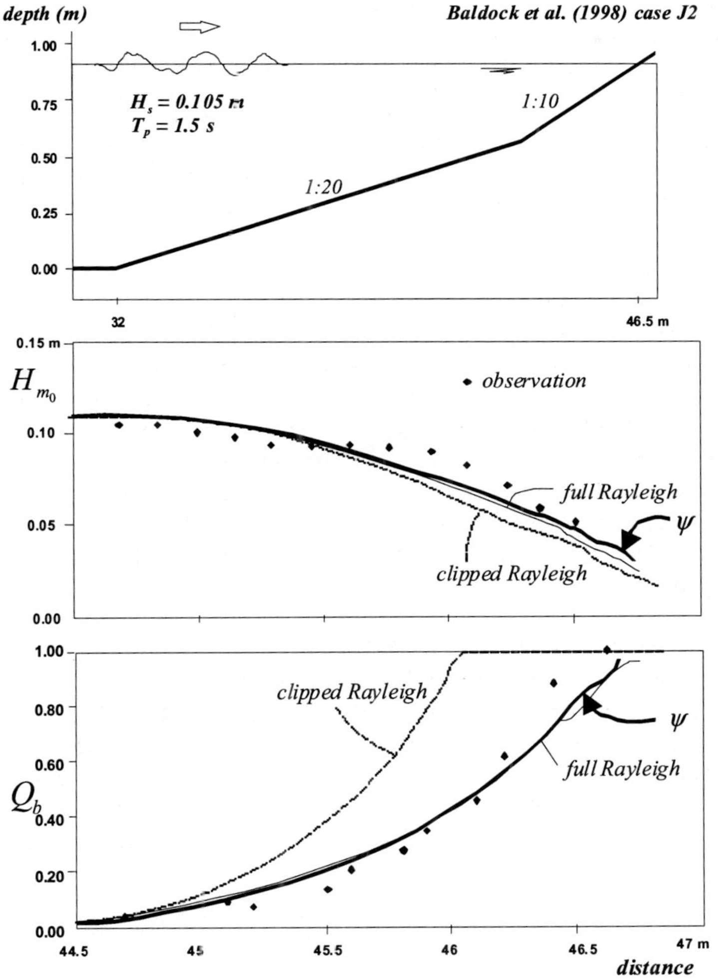

Fig. 1 An observation of the fraction of breakers by Baldock et al. (1998) and the predictions of Battjes and Janssen (1978; clipped Rayleigh), Baldock et al. (1998; full Rayleigh) and the present study (ψ).

One of the few observations, over a rather steep bottom, is due to Baldock et al. (1998). It is given in Fig. 1, together with the above prediction of Q_b and of the significant wave height $H_{m_0} = 4\sqrt{m_0}$ (where m_0 the integral is over the variance density spectrum; all computed with the SWAN wave model, Booij et al. 1999). The prediction of the significant wave height in this case is reasonable in view of the steepness of the bottom, but the quality of the prediction of Q_b is obviously poor.

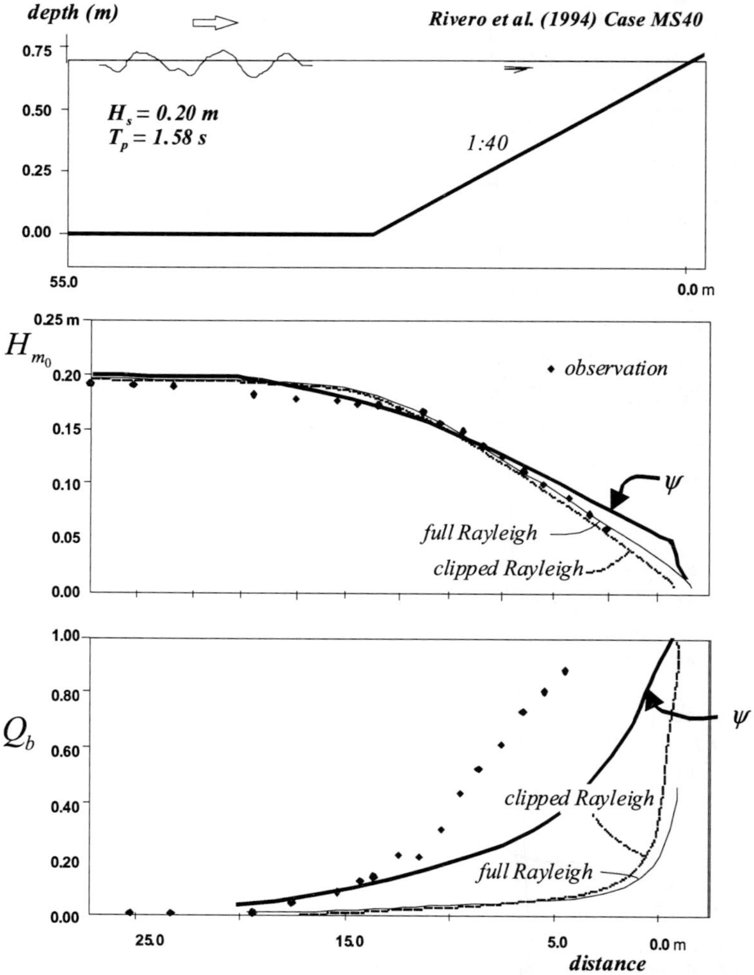

Fig. 2 Same as Fig. 1 for Rivero et al. (1994).

This was also noted by Baldock et al. (1998) who proposed using a variation on the formulation of Battjes and Janssen (1978), based on the *full* Rayleigh distribution for the wave heights:

$$Q_b = \exp\left[-\left(H_{cr}/H_{rms}\right)^2\right] \quad (2)$$

where H_{cr} is the wave height above which all waves break. This improved the prediction of Q_b for these observations significantly, while it only slightly affected (improved) the prediction of the significant wave height (see Fig. 1). However, we find that the performance of this prediction of Q_b is again rather poor for other observations: those of Rivero et al. (1994), see Fig. 2. In an attempt to improve on this, we found with trial computations, that the dissipation in the formulations of both Battjes and Janssen (1978) and Baldock et al. (1998) is rather insensitive to the exact values of Q_b and α as long as $\alpha = (\gamma/0.73)^4$, where α is the proportionality coefficient in these dissipation formulations. This leaves γ as a convenient tuning parameter. We subsequently fitted this parameter to the 8 cases of Baldock et al. (1998), Rivero et al. (1994), Cai et al. (1992) and Boers (2005). These fitted values were plotted as a function of various parameters such as H_{rms}/d (where d is local depth), incident and local Ursell numbers, the incident and local wave steepness, the Iribarren number and the normalized overall dissipation rate (see below).

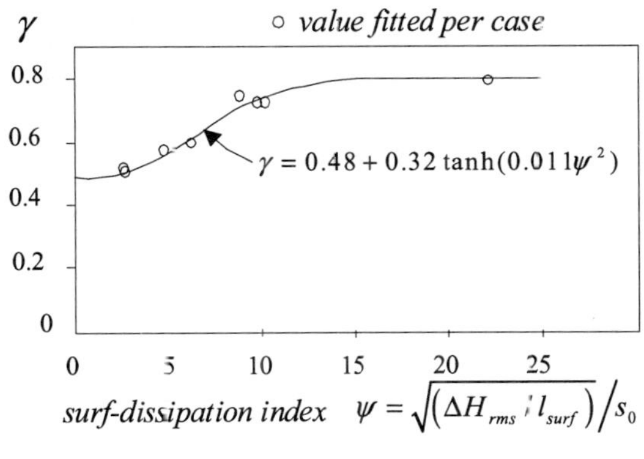

Fig. 3 The fitted values of γ as a function of the normalized dissipation rate.

We found the highest correlation with this normalized dissipation rate defined as $\psi = \sqrt{\left(\Delta H_{rms}/l_{surf}\right)}/s_0$, where ΔH_{rms} is the drop of H_{rms} over the surf zone (not necessarily ending at the beach but possibly in deeper water behind a shoal), l_{surf} is the width of the surf zone and s_0 is the incident wave steepness (defined as H_{rms}/L_{peak}, where L_{peak} is the peak wave length). The best fit through these values is obtained with (see Fig. 3):

$$\gamma = 0.48 + 0.32\tanh(0.011\psi^2) \tag{3}$$

In SWAN, the width of the surf zone l_{surf} is determined between the location most off-shore where the rate of decrease of H_{rms} is more than 5% per wave length (the incident $H_{rms,i}$ is defined at this point), and the then nearest down-wave location where H_{rms} has dropped to 10% of $H_{rms,i}$ or increases again.

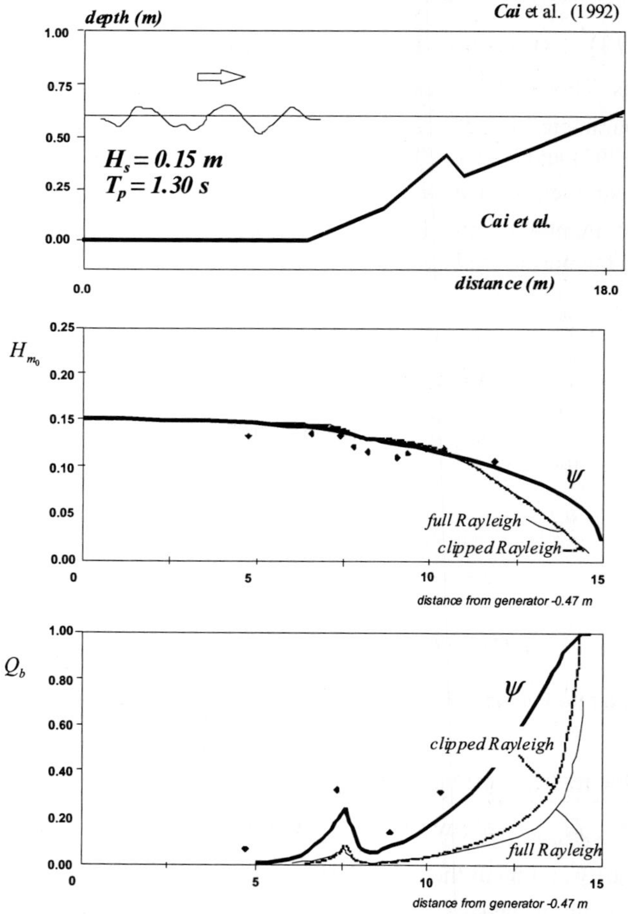

Fig. 4 Same as Fig. 1 for Cai et al. (1992).

RESULTS (FRACTION OF BREAKERS)

The computational results with SWAN for the 8 cases considered show considerable improvement in estimating the fraction of breakers (see Figs. 2 and 3 for the above cases of Baldock et al. and Rivero et al. and Figs. 4 and 5 for one case of Cai et al. and Boers each), except behind the two bars in the observations of Boers (2005). Behind these bars, the model does not reproduce the observed doubling of the fraction of breakers. This may be due to (a) differences in definition (Boers emphasizes that he counted waves with a roller rather than breaking waves) or (b) counting splash-ups of the breakers as

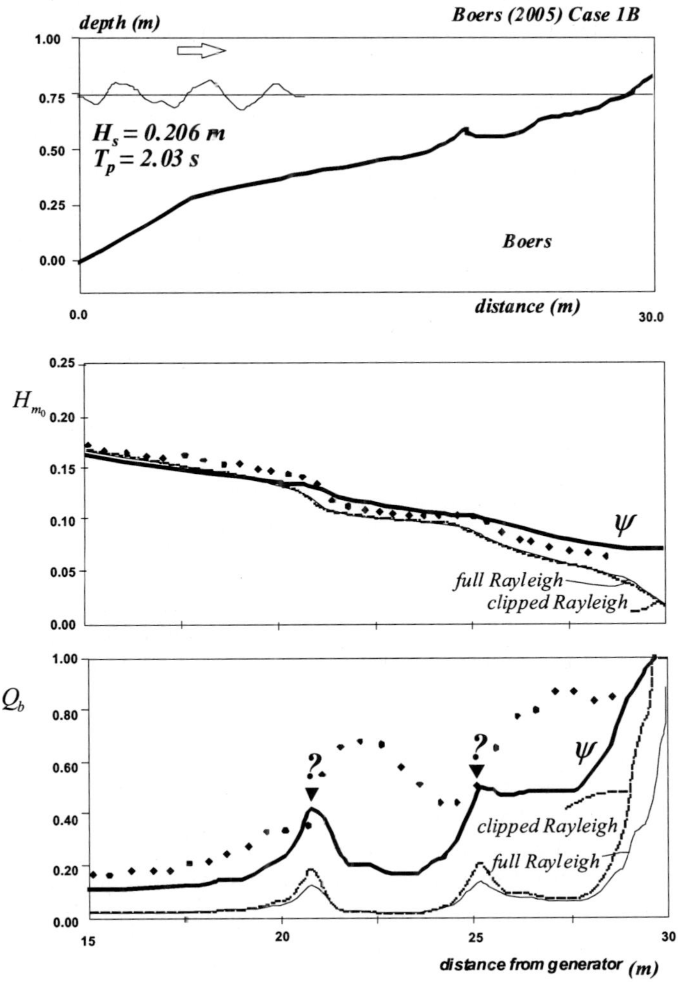

Fig. 5 Same as Fig. 1 for Boers (2005) with unresolved doubling of the fraction of breakers behind the bars (initiation indicated with a question mark ?).

independent breakers (this does not seem to be the case in these observations, nor in the similar case of Cai et al., 1994; D. Basco and M. Boers, personal communications, 2006) or (c) the merging of waves while the number of breakers remains more-or-less constant, thus increasing the fraction of breakers (D.H. Peregrine, personal communication, 2006, supported by Boers, 1996 who comments on his observations that 'Often two or more waves are united to one wave …').

METHOD (TRIAD WAVE-WAVE INTERACTIONS)

For waves entering the surf zone with a unimodal spectrum, triad wave-wave interactions tend to generate second and higher harmonics of the peak frequency, at least initially, i.e., at the deep-water edge of the surf zone (Boers, 1996, his Figs. 3.3a-i; see Fig. 6). Further inside the surf zone, the same interactions remove these secondary peaks and create a smooth $f^{-4/3}$ tail for $kd < 1$ (where k is wave number and d is the local depth; Vincent and Smith, 2002).

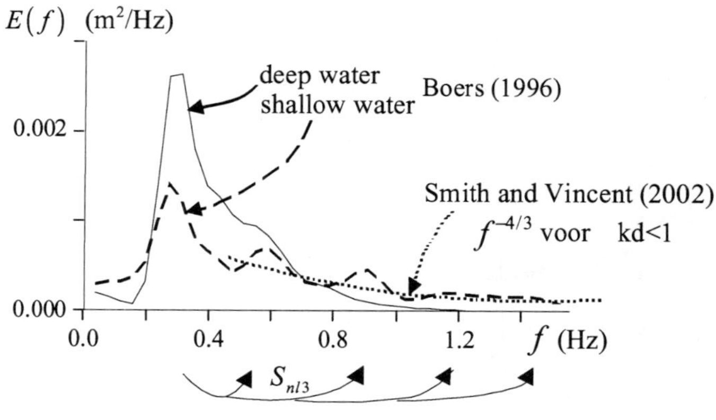

Fig. 6 The evolution of a unimodal spectrum from outside the surf zone to a multimodal spectrum inside the surf zone (dashed line) and the converged shape of the spectral tail (dotted line).

If these interactions are accounted for in an operational wave model at all, they are severely parameterized, e.g., the LTA approximation of Eldeberky (1996) in SWAN, which only generates the 2nd, 4th, 8th etc. harmonics of the spectral peak (instead of all harmonics, at least initially) and does not reproduce the final smooth tail. To improve such parameterization, we consider an analogy with the quadruplet wave-wave interactions (e.g., Hasselmann, 1962), the general expression of which can be written as:

$$S_{nl4}(\mathbf{k}_4) = \iint\iint T_a N(\mathbf{k}_1)\, N(\mathbf{k}_2)\, N(\mathbf{k}_3)\, d\mathbf{k}_1 d\mathbf{k}_2 - N(\mathbf{k}_4) \iint\iint T_b\, N(\mathbf{k}_1)\, N(\mathbf{k}_2)\, d\mathbf{k}_1 d\mathbf{k}_2 \qquad (4)$$

where $S_{nl4}(\mathbf{k}_4)$ is the source term in the spectral action balance for wave component $\mathbf{k}_4$ (where $N(\mathbf{k}_4)$ is its action density). The other wave components that are involved are $\mathbf{k}_1$, $\mathbf{k}_2$ and $\mathbf{k}_3$ subject to the resonance condition $\mathbf{k}_1+\mathbf{k}_2=\mathbf{k}_3+\mathbf{k}_4$ and $\sigma_1+\sigma_2=\sigma_3+\sigma_4$ (the four frequencies corresponding to the four wave numbers). The coefficients T_a and T_b are complicated functions of the wave numbers involved. We removed one component, reformulated the result in terms of frequency for collinear interactions and renumbered the result, with the following result:

$$S_{nl3}(\sigma_1,\theta)=\int\delta S_{nl3,\sigma_1,\sigma_3}(\sigma_1,\sigma_2,\sigma_3,\theta)\,d\sigma_2$$

$$\delta S_{nl3,\sigma_1,\sigma_3}(\sigma_1,\sigma_2,\sigma_3,\theta)=N(\sigma_2,\theta)\left[\beta_1 N(\sigma_3,\theta)-\beta_2 N(\sigma_1,\theta)\right] \tag{5}$$

subject to $\sigma_3=\sigma_1+\sigma_2$, and S_{nl3} is the source term of component (σ_1,θ) and $\delta S_{nl3,\sigma_1,\sigma_3}(\sigma_1,\sigma_2,\sigma_3,\theta)$ represents the transfer of action between component (σ_1,θ) and component (σ_3,θ), as affected by component (σ_2,θ). The coefficients β_1 and β_2 are coefficients that we formulated on the basis of dimensions, symmetry and the condition that the spectral tail of the variance density spectrum shall converge towards $\sigma^{-4/3}$.

RESULTS (TRIAD WAVE-WAVE INTERACTIONS)

To demonstrate the behavior of the above formulation, we consider a situation with a flat beach (slope 1:100) and short-crested, random incident waves with 1 m significant wave height and 10 s peak period (JONSWAP spectrum with $\cos^4(\theta)$ directional distribution). We show the spectra in Fig. 7 in f - space, multiplied with $f^{4/3}$, so that an $f^{-4/3}$ tail of the spectrum would be represented by a constant level. It is obvious from the upper panels with spectra in this illustration, that without triad wave-wave interactions no higher harmonics are generated. With the LTA approximation only second harmonics are generated (evident from the second harmonic of the spectral peak). The spectral feature at the highest frequencies in these LTA computations (it looks like a third harmonic but it isn't) is a spurious effect due to the default high-frequency cut-off of the LTA computations at 2.2 times the mean frequency (we used SWAN version 40.41). The above quadruplet analogy initially generates harmonics at multiples of the peak frequency and also forces the spectral tail to converge towards the

$f^{-4/3}$ limit. These are the basic effects of triad wave-wave interactions and they can apparently be simulated with the above simple expressions.

The improvement in predicting the mean wave period by using the quadruplet analogy is illustrated for the observations of Boers (1996) in Fig. 8. It also shows that this analogy does not reduce the quality of the prediction of the significant wave height (which is dominated by the Battjes-Janssen dissipation model).

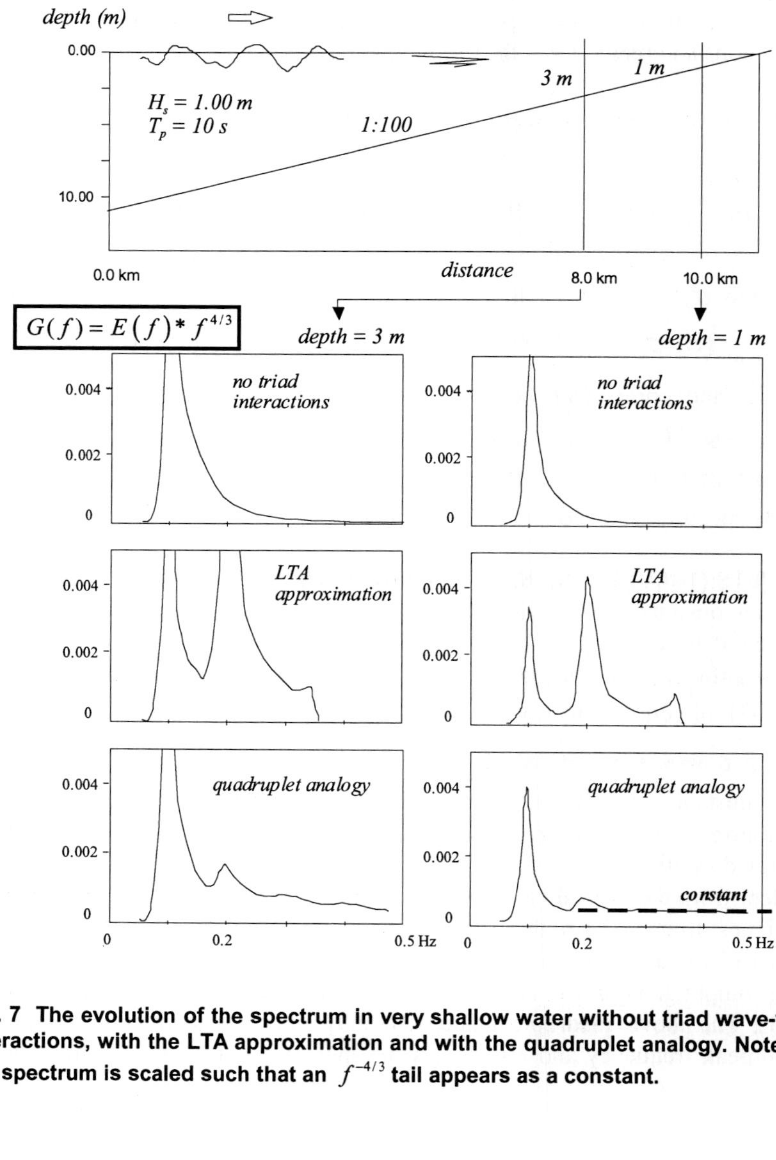

Fig. 7 The evolution of the spectrum in very shallow water without triad wave-wave interactions, with the LTA approximation and with the quadruplet analogy. Note that the spectrum is scaled such that an $f^{-4/3}$ tail appears as a constant.

CONCLUSIONS

The fraction of breaking waves and the maximum wave height in the Battjes-Janssen model for depth-induced breaking are tunable model parameters rather than physical realities. By using an empirical relationship between the maximum wave height (normalized with the local depth) on the one hand and the averaged dissipation rate across the surf zone (normalized with the incident wave steepness) on the other, the fraction of breaking waves can be estimated more realistically. In addition, it seems possible to parameterize the triad wave-wave interactions using an analogy with the quadruplet wave-wave interactions. Preliminary tests show the desired behavior of this parameterization of initially generating all higher harmonics (rather than only the 2^n-th harmonics as in the LTA) and converging towards the required smooth spectral tail. The prediction of the mean frequency of the waves in the surf zone seems to improve accordingly.

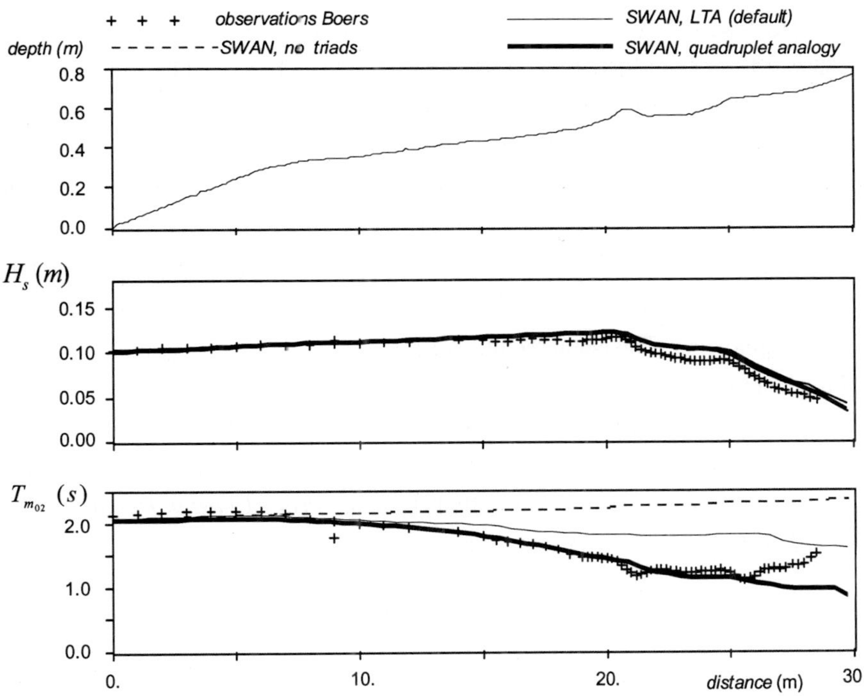

Fig. 8 The prediction of the significant wave height and the mean wave period without triad wave-wave interactions, with the LTA approximation and with the quadruplet analogy.

REFERENCES

Booij, N., R.C. Ris and L.H. Holthuijsen, 1999, A third-generation wave model for coastal regions, Part I, Model description and validation, *J. Geophys. Res.*, 104, C4, 7649-7666

Battjes, J.A. and J.P.F.M. Janssen, 1978, Energy loss and set-up due to breaking of random waves, *Proc. 16th Conf. Coastal Engineering*, Hamburg, ASCE, New York, 569-587

Baldock, T.E., P. Holmes, S. Bunker and P. van Weert, 1998, Cross-shore hydrodynamics within an unsaturated surf zone, *Coastal Engineering*, 34, 173-196

Boers, M., 1996, Simulation of a surf zone with a barred beach; Part 1: Wave heights and wave breaking, Delft University of Technology, Faculty of Civil Engineering, Communications on Hydraulic and Geotechnical, Report No. 96-5, 116 p.

Boers, M., 2005, Surf Zone Turbulence, Ph.D. Thesis, Delft University of Technology, Faculty of Civil Engineering and Geosciences, 171 p.

Cai, M., D. Basco and J. Baumer, 1992, Bar / trough effects on wave height probability distributions and energy losses in surf zones, *Proc. 23rd Int. Conf. Coastal Engineering*, Venice, ASCE, New York, 103-115

Eldeberky, Y., 1996, *Nonlinear transformation of wave spectra in the nearshore zone*, Ph.D. thesis, Delft University of Technology, Department of Civil Engineering, The Netherlands, 203 p.

Hasselmann, K., 1962, On the nonlinear energy transfer in a gravity-wave spectrum. Part 1.General theory, *J. Fluid Mech.*, 12, 481-500

Rivero, F.J., A.S.-Arcilla and D. Beyer, 1994, Comparison of a wave transformation model with LIP-11D data, *Proc. Coastal Dynamics '94*, 518-532

ON VISUALLY OBSERVED WAVES AT SAN CLEMENTE, CALIFORNIA, USA

Chuck Mesa[1]

The intent of this paper is to investigate the efficacy of visual observations as a means to obtain engineering accurate information in the surf zone. This paper analyzes a data set of visual observations in the surf zone by comparison with numerical predictions of the same waves. Comparison of wave period histograms indicates generally good agreement. Comparison of wave direction histograms indicates generally poor agreement. After appropriate transformation of the numerical predictions, comparison and analysis of wave height histograms indicate generally good agreement. The results further indicate the visual observer is estimating the H_{10} wave heights in the surf zone. Human nature introduces some influences on the visual observations.

INTRODUCTION

Quantitative information of the nearshore wave climatology is of vital importance to the field of coastal zone management. Coastal engineers and scientists, government, harbor officials, civic planners, and mariners require accurate characterization of the site specific conditions for rational decision making in any endeavor in the coastal zone. Through increasingly widespread dissemination, this information is also vital to promote public safety and the responsible use and enjoyment of coastal resources by recreational boaters, surfers and all marine enthusiasts.

The need for nearshore wave climatology information is worldwide. Contrastingly, there typically is a dearth of measured field data available. Obstacles to field data collection are primarily and foremost the monetary cost of conventional long term field data collection programs. Conventional field data collection methods typically include bottom mounted pressure sensors, surface riding buoys, and increasingly LIDAR and satellite based systems. Coastal zone projects are often regional in nature and it is typically cost prohibitive to obtain measured wave climatology throughout the entire region. Individual site conditions often vary widely in any given region, yet due to cost considerations a single location is often assumed representative of the entire region. Visual observations have not traditionally been used as a source of engineering accurate wave climatology information. Visual observations have some human induced influences, but are simple to conduct and are very cost effective relative to other means of field data collection.

This paper describes the results of a field data collection program of visual observations at San Clemente, California, USA. Surf zone measurements were

---

[1] U.S. Army Corps of Engineers, Los Angeles District, 915 Wilshire Blvd, Los Angeles, CA, 90017-3401, United States

obtained by highly trained visual observers. The data was compiled and compared to numerical model predictions at the same location to determine the accuracy of the observed data.

VISUAL OBSERVATIONS

Visual observations of the surf zone conditions at San Clemente were obtained by Surfline, Inc of Huntington Beach, California. Surfline employs a network of trained personnel who visually estimate and report wave climate twice daily at the shore in a variety of locations along the California coast. The wave climate observations are forwarded to a central office where they are collected and made available to the public through a subscription service.

The wave climate was developed from data collected for a 14 month period from Feb2002 – Mar2003. A trained observer visually estimated the breaking wave climate at the shoreline twice daily, typically at 0600 (first light) and at 1300; the times were approximate. Waves were observed for a period of 5-10 minutes and characteristics measured included height, period, and direction. There were 782 measurements over the observation period. Specific details of the observations include:

1. Wave heights (crest to trough) were observed and the minimum, average, and maximum were visually estimated to the nearest 1 ft. Visual wave height observations in the surf zone are typically calibrated based on other surfers in surf zone. The standing body of a person board surfing provides a quasi fixed frame of reference to obtain estimates. It is usual and customary to estimate wave heights based on the following scale: ankle high = 1 ft; knee high = 2 ft; waist high = 3 ft; chest high = 4 ft; shoulder high = 5 ft; head high = 6 ft. Conditions where no surfers are present are dependent on the skill and experience of the observer to obtain an accurate estimate.
2. Wave period is based on an average of 30 waves over the 5-10 minute observation period and was reported to the nearest 1 second.
3. Wave directions were reported relative to the beach normal and estimated to the nearest 5 degrees. Wave directions were recorded as normal (0-10 degrees); slightly from the left (or right) (10-25 degrees); significantly from the left (or right) (greater than 25 degrees).

NUMERICAL PREDICTIONS

The Coastal Data Information Program (CDIP) operated by the Scripps Institution of Oceanography is an extensive network for monitoring waves along the coastlines of the United States, with a strong emphasis on the Pacific coast. The numerical model utilized in this study is an integral part of the CDIP program and is used for detailed climatic predictions throughout the southern California region. Hourly nearshore model predictions were made based on a nearby, deep water directional wave buoy and a linear wave model. Directional spectra were estimated from a Datawell Directional Waverider Buoy (7 km offshore, 370 m depth) using the Maximum Entropy Method and transformed to the 25 ft (7.5 m) depth contour using a spectral refraction wave model. The numerical transformation model used is valid for wave periods within the 4-25 second range. There were 10,151 predictions over the observation period.

WAVE PERIOD COMPARISON

The comparison of wave period histograms is shown in Figure 1. The double peaked histogram in this figure illustrates the dual sea/swell nature of the wave climate which is very common in southern California. Shorter period waves are typically associated with sea conditions; longer period waves are typically associated with swell conditions. The cut-off between sea and swell is somewhat subjective but in southern California is generally accepted between 8-10 seconds.

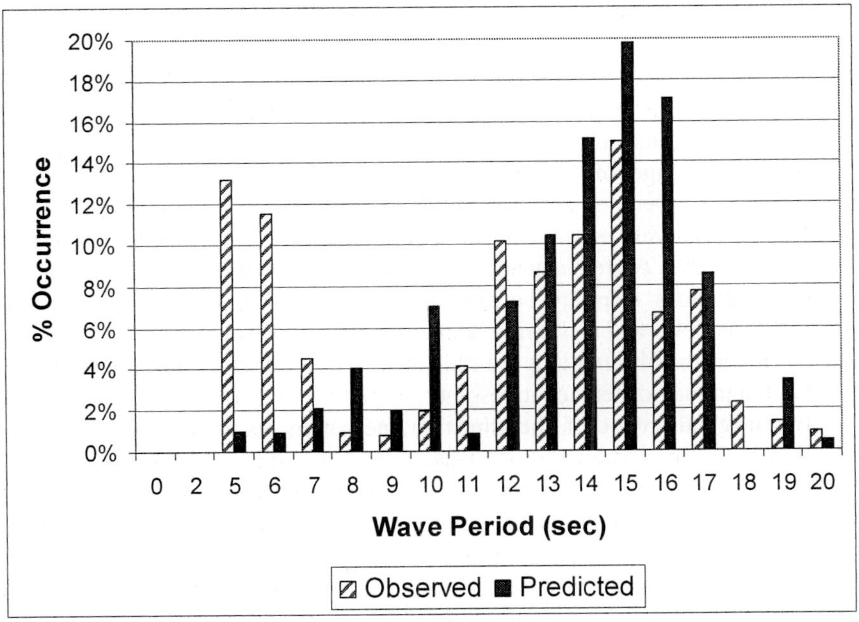

Figure 1. Comparison of wave period histograms of visual observations and numerical predictions.

The visual observations and the numerical predictions both indicate very good agreement for the dominant swell peak at 15 seconds. There is less agreement in the secondary sea peak. The numerical predictions indicate the peak between 8-10 seconds whereas the visual observations indicate a very well defined peak at 5-6 seconds. The exact reasons for this discrepancy are uncertain although it is theorized that some of this difference can be attributed to the manner these values are determined. The visual observer counts the time required for the passage of 30 waves to determine an average wave period value, whereas the numerical model computes spectral peak period. In monochromatic swell conditions it is expected that this difference is minimal. In clean, long period swell conditions a visual observer should have little difficulty determining the average wave period and should be consistent with the spectral peak period. In southern California it is common for local onshore winds to dominate in the late afternoon creating chaotic, high energy, short period seas. It is typical for the spectral peak period to shift from the swell frequencies into the sea frequencies. The model predictions utilize a cut-off frequency of 4 seconds and the spectral peak period could deviate substantially within the 4-10 second range of sea periods. A visual observer has no pre-established cut-off frequency and would not discriminate between very short period waves. Therefore, it is conceivable that counting waves and the resultant average value would result in the preponderance of 5 second observations.

WAVE DIRECTION COMPARISON

The comparison of wave direction histograms is shown in Figure 2. This comparison indicates a poor agreement. The observed wave directions are spread over a range of 125 degrees. The predictions are depicted in two direction bands. Detailed analyses of the numerical predictions indicate that the histogram is relatively narrow banded approximating a log-normal shape, with a sharp peak at 225 degrees and decreasing to 250 degrees. The results shown illustrate the gross disparity between the two measurement techniques. The numerical model computes wave direction with a resolution of 0.1 degree whereas the visual observations are estimated within 5 degrees; this represents an order of magnitude difference in resolution.

The original intent of the direction measurements was for purposes of sediment transport. Although the visual observation direction data as shown is hardly suitable for sediment transport calculations, visual observations can yet be beneficially used. The visual observer measured absolute direction relative to shore normal and expressed in terms of angular degrees. The data, however, indicates that the human observer is unable to discriminate with the resolution necessary for coastal engineering applications. The human observer is able to express direction to the nearest 5 degrees whereas coastal engineering applications typically require resolution on order of 1 degree or less. However, it may instead be useful for the human observer to express wave direction in terms of "drift". A visual observer can typically express longshore drift more easily than wave direction in angular degrees. Longshore drift can be estimated by any object floating in the surf zone such as surfers or natural or man-made debris. Longshore drift can be directly correlated to sediment transport direction. The direction of longshore drift is approximately equivalent to the sign of sediment transport, i.e. left or right of shore normal is equivalent to the + or – in sediment transport calculations. Thus, it could be very valuable to sediment transport studies for the visual observer to record longshore drift in addition to or in lieu of absolute wave direction.

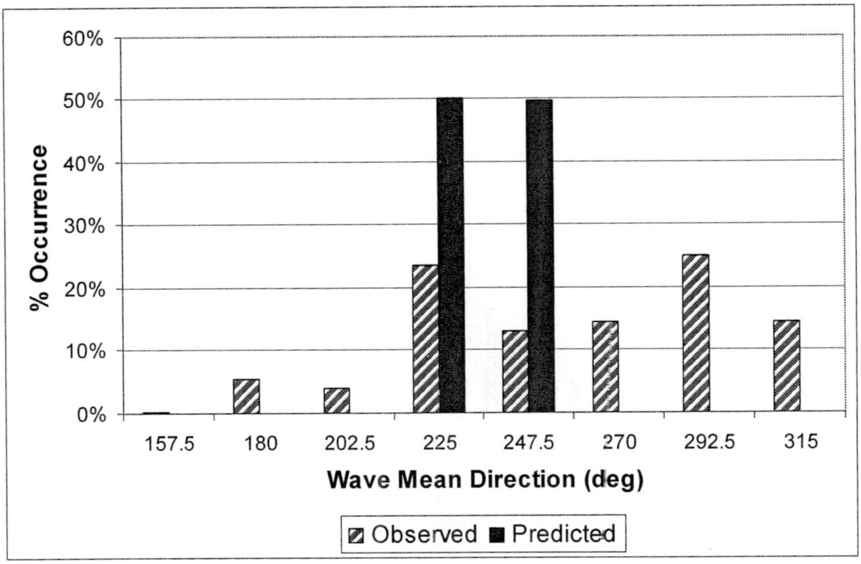

Figure 2. Comparison of wave direction histograms of visual observations and numerical predictions.

WAVE HEIGHT COMPARISON

The comparison of wave height histograms is shown in Figure 3. Some general observations can be made regarding the wave height climatology. The observations and predictions both illustrate a relatively benign wave climate during the 14 month measurement period, all wave heights are less than 7.0 ft. The maximum predicted significant wave height is 5.8 ft; the maximum observed wave height is 7 ft which was recorded three times during the 14 month measurement period. The most commonly occurring predicted significant wave height is in the range 1.6-1.8 ft; the most commonly occurring observed wave height is 3 ft and includes approximately 60% of all observations. This relatively benign wave climate is unexpected given the somewhat severe winter wave climate that can occur in southern California.

These histograms are not directly comparable. The numerical predictions are distributed continuously; the visual observations are distributed discreetly in 1 foot bins that are a result of the manner (surfer body scale) in which the wave heights are estimated. Furthermore, the numerical predictions are significant wave heights H_s made at the 25 ft depth (7.5 m) while the visual observations are breaking wave heights H_b made in the surf zone at the breaking depth. As part of the analysis described hereinafter the numerical predictions are transformed to the surf zone for direct comparison with the visual observations.

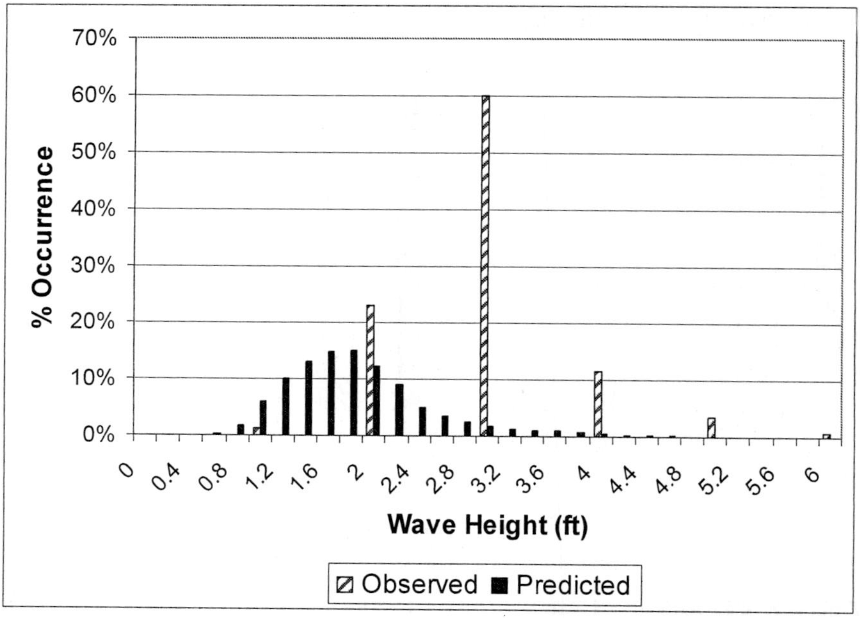

Figure 3. Comparison of wave height histograms of visual observations and original numerical predictions.

WAVE TRANSFORMATION

The comparison of wave height histograms previously described indicates that the histograms are not directly comparable. In order to directly compare the numerical predictions to the visual observations, it is necessary to transform the numerical predictions to the same location as the observations. Each numerical H_s prediction was first converted to H_{10}, H_5, and H_1 using standard statistical parameter multipliers, resulting in four data sets of numerical predictions. Each resultant numerical prediction was then shoaled to breaking using breaking index γ_b = 0.5, 0.6, 0.7, and 0.78, and then rounded to the nearest 1 foot, resulting in sixteen combinations of data. The data is compiled into histograms which can then be directly compared to the observed data.

WAVE HEIGHT COMPARISON (TRANSFORMED)

The sixteen histograms were arranged into four groups based on their statistical parameter heritage (H_s, H_{10}, H_5, and H_1) and qualitatively evaluated for goodness of fit against the observed data. This qualitative analysis indicates that the H_{10} group is the overall best fit with the observed data among the group of four statistical parameters. Figure 4 shows the results of the H_{10} group, and also the H_s group for illustration purposes. It is clear in the figure that the H_{10} group better represents the overall shape of the measured histogram than the H_s group.

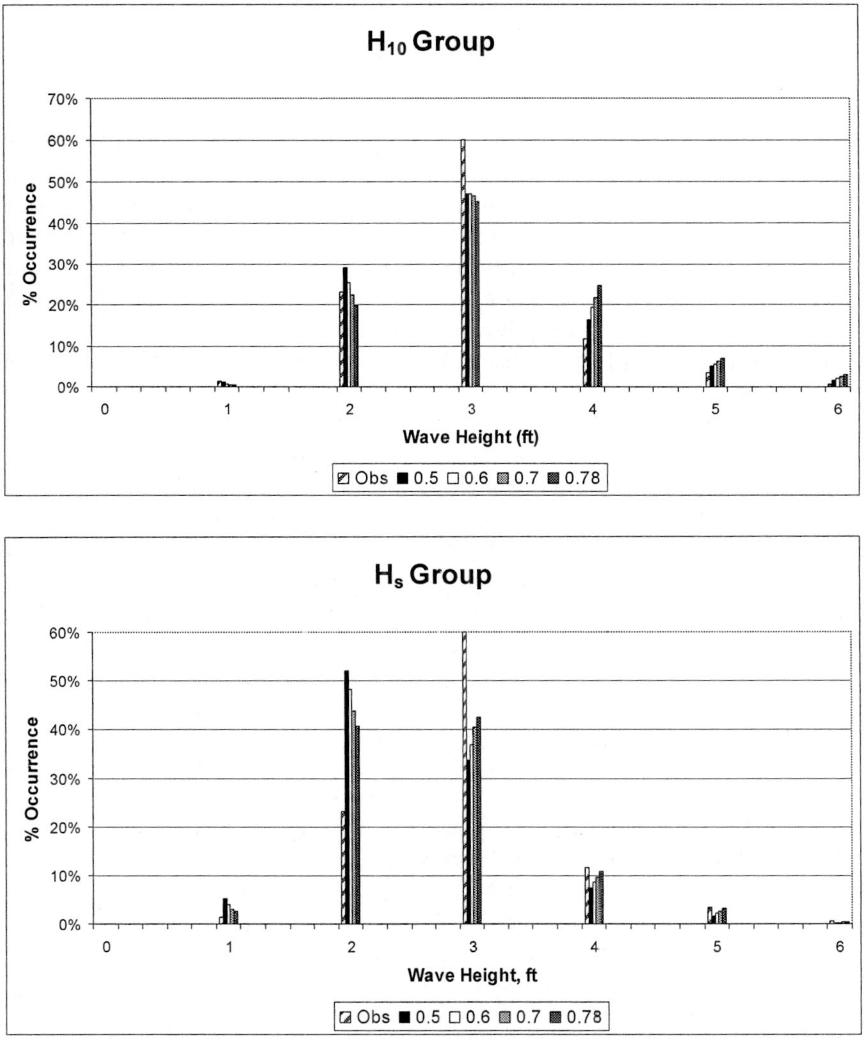

Figure 4. Comparison of wave height histograms of visual observations and transformed numerical predictions.

In particular the observed data histogram peak wave heights at 2 ft and 3 ft are substantially better represented by H_{10} than their H_s counterparts. The low percent occurrence of the distribution tails renders a meaningful qualitative comparison difficult.

A quantitative analysis is desired to determine the best overall fit as the aforementioned qualitative analysis is subjective and largely based on the eye of the beholder. The degree of skill is defined as

$$Skill = 1 - \varepsilon$$

where ε is a sum-of-squares measure of discrepancy between the numerical predictions and the observations. Thus, a perfect match between model predictions and observations would result in *Skill* = 1 and inaccurate predictions would result in deviations from unity. The discrepancies are not computed constantly across the range of observed data. The discrepancies are weighted directly proportional to the percent occurrences of the individual bins of the observed data. Weighting based on the percent occurrence of the individual bins emphasizes the peak of the observed data distribution which accounts for approximately 82% of the observations (wave height range 2-3 ft), and minimizes any undue influence by the tails of the observed data distribution which occur substantially less frequent.

The results of the quantitative skill analysis for all sixteen data sets are shown in Table 1. The results are shown as an array of skill values with axes consisting of the statistical wave parameters and breaking index γ_b. These results indicate a general trend of decreasing skill with increasing breaking index for the H_{10}, H_5, and H_1 statistical parameters and the reverse for the H_s parameter. There is a distinct preference for the H_{10} group over the H_s, H_5, and H_1 groups across the range of breaking indices. Within the H_{10} group there is a similarly strong skill value between γ_b = 0.5, 0.6, and 0.7, with the γ_b = 0.5 exhibiting the highest skill by a small amount. The combination of H_5 and γ_b = 0.5 is also a strong candidate for the highest skill.

Table 1. Skill values representing fit between numerical predictions and visual observations.

	γ_b = 0.5	γ_b = 0.6	γ_b = 0.7	γ_b = 0.78
H_s	0.34	0.45	0.55	0.62
H_{10}	0.74	0.73	0.70	0.62
H_5	0.71	0.61	0.52	0.43
H_1	0.24	0.14	0.06	0.01

The results shown in Table 1 have practical implications. Significant wave height is typically the wave statistical parameter most commonly used in applications involving the characterization of wave climate. It is commonly accepted that the human observer will estimate significant wave height H_s in lieu of any other wave statistical parameter, and most reported wave heights that are visually based are assumed to be H_s unless specified otherwise. Only an observer trained in the coastal sciences would understand a priori or otherwise be able to distinguish between H_s and other wave statistical parameters. However, the results shown herein clearly indicate that H_{10} has the highest correlations. In other words, a visual observer is actually "seeing" H_{10} in the surf zone instead of other wave statistical parameters.

CONCLUSIONS

Based upon visual observation data of the surf zone collected for 14 months at San Clemente, California, USA, the analysis and results contained herein indicate that visual observation is a reasonable method to obtain engineering accurate information in the surf zone. Recognizing the limitations of the human eye to discern ocean wave properties, visual observations by a trained observer are simple to obtain and have substantial cost advantages over conventional wave measurement methods. The data collected in this study suggests that visual observers obtain very accurate wave period information. The results do not indicate high confidence in the ability of the human observer to resolve wave direction expressed in angular degrees. However, it is suggested that longshore drift direction be substituted in lieu of wave angular degrees, whereby some very valuable sediment transport related information can be obtained. The results indicate a high degree of confidence in the human observer ability to accurately describe the breaking wave heights. Furthermore, the data suggests that the human observer is "seeing" H_{10} in the surf zone in lieu of H_s or other statistical parameters.

The wave climate measured during this study was not particularly severe. In fact, the measured wave climate was relatively benign. It is acknowledged that the conclusions drawn herein may degrade as the sea severity increases. The ability of the human eye to accurately estimate very large waves degrades as the wave heights increase, and is substantially muted in the presence of extreme waves. It is recognized the utility of the conclusions developed herein are directly related to the strengths and frailties of human nature. Nonetheless, the method described herein can be extremely useful worldwide in any location where an experienced visual observer can simply go to the beach.

ACKNOWLEDGEMENTS

The author wishes to gratefully acknowledge Mr. Sean Collins of Surfline, Inc and Dr. William O'Reilly of the Scripps Institution of Oceanography for their assistance in this study.

WAVE FIELD MODIFICATIONS BY BATHYMETRIC ANOMALIES WITH GRADUAL TRANSITIONS IN DEPTH: A REVIEW

Christopher J. Bender[1] and Robert G. Dean[2]

This paper reviews several methods to determine wave field modifications resulting from bathymetric anomalies (pit or shoal) with gradual transitions in depth. The analytic (semi-numeric) method proves valid for linear waves in a uniform depth domain with an arbitrary number of forms of different, but uniform, depths combined to represent either a pit or a shoal. Possible forms include trenches (of shoals) of infinite length, of circular planform, or of elliptic planform. Sections present the theory, formulation, and results in the form of contour plots of the energy reflection or relative amplitude in the presence of the depth anomaly. The bathymetric forms induce wave transformation through processes of wave refraction, wave diffraction, and wave reflection. The elliptic solution allows asymmetry in the solution for oblique incident wave angles.

INTRODUCTION

Modifications of nearshore bathymetry have the potential to affect wave transformation processes and thus to alter the equilibrium shoreline planforms on the landward side of borrow areas. Recent cases of erosional hot spots (EHS) associated with beach nourishment projects have increased interest in the prediction of the mechanisms of borrow pit alteration on the local wave field. Currently, a need exists for a better understanding of how borrow pit characteristics such as size and depth interact with the incident wave field to determine the effect of the pits on the nourished beach and to anticipate the effects of various designs.

This paper provides a summary of recent work by the authors on wave transformation by nearshore bathymetric anomalies. Solutions progress in complexity from bathymetric anomalies represented as trenches or shoals of infinite length, to circular pits or shoals, and then to elliptic forms. Application of the general method of Takano (1960) allows an analytical (semi-numerical) solution to the two- and three-dimensional pit/shoal problem. The method allows gradual transitions in depth as a series of uniform steps.

The analytic models developed within this study do not intend to replace or supersede the more complex numerical schemes that allow arbitrary bathymetry and more wave-related processes, but to develop alternative models that provide a semi-numerical (analytic) solution.

---

[1] Taylor Engineering, 9000 Cypress Green Dr Suite 200, Jacksonville, FL, 32259, USA
[2] Civil and Coastal Engineering Dept, Univ. of Florida, Gainesville, FL, 32611. USA

PREVIOUS STUDIES

Many authors have applied analytic methods to investigate wave field modifications by bathymetric changes. Early efforts centered on the effects of a step of infinite length (2-Dimensional Model) on normally incident long waves, trench, or sill of uniform depth in an otherwise uniform depth domain. Many of the solutions divide the fluid domain into two regions: one region comprising the fluid within the change in bathymetry and the other region consisting of the remainder of the fluid domain. Many studies have employed this technique with a progression from normally incident long waves over a step to solutions that account for oblique incidence and currents along a trench. The paragraphs below highlight some of the previous work in this area.

The analytic study of the linear long wave transformation caused by a bathymetric anomaly began with Lamb (1932) for the case of an infinite step (2-D form). Many others have extended the research for a 2-D trench or shoal to include arbitrary water depth [Takano (1960), Newman (1965), Mei and Black (1969)], oblique wave incidence [Miles (1982), Kirby and Dalrymple (1983)], and currents flowing along a trench [Kirby et al. (1987)]. The practical application of the 2-D models is limited to domains such as navigation channels or extended underwater breakwaters that allow limited longshore variation.

The addition of the variation in the longshore direction allowed 3-Dimensional analytic models to investigate the impact of a pit or shoal of finite extent in shallow water [Black and Mei (1970), Williams (1990), McDougal et al. (1996)] and arbitrary water depth (Williams and Vazquez (1991)]. Recently, complex numerical models have been developed to determine wave propagation in domains of arbitrary bathymetry. The numerical models account for many wave related processes, such as bottom friction, non-linear effects, and wave breaking, among others. These models include REF/DIF-1 (Kirby and Dalrymple, 1994), FUNWAVE 1.0 (Kirby et al., 1998), STWAVE (Smith et al., 2001) and SWAN (Holthuijsen et al., 2000).

Little previous work in coastal engineering literature employs Mathieu functions or elliptic coordinates. Chen and Mei (1973) investigate exciting forces and moments of incident waves on vertical cylinders with a finite draft and elliptical cross section. A recent study by Gutiérrez-Vega et al. (2003) provides a visual analysis of general Mathieu function behavior with several applications that include water waves traveling in elliptic domains. The study highlights similarities between Mathieu function behavior and Bessel function solutions in cylindrical domains.

FORMULATION AND THEORY

Trench of Infinite Length

Bender and Dean (2003) present a method to evaluate wave transformation and energy reflection in arbitrary water depth by pits or shoals of infinite length — 2-D model. The method applies previous work by Takano (1960), and Kirby and Dalrymple, (1983). The solution allows a series of steps of uniform depth to approximate gradual transitions in depth for the bathymetric anomalies. Figure 1 sketches a domain with a stepped trench with three steps. In this method, a domain with J regions will contain J-1 steps and boundaries. Each region will have a specified depth and each boundary between regions will have a specified x location that requires application of matching conditions.

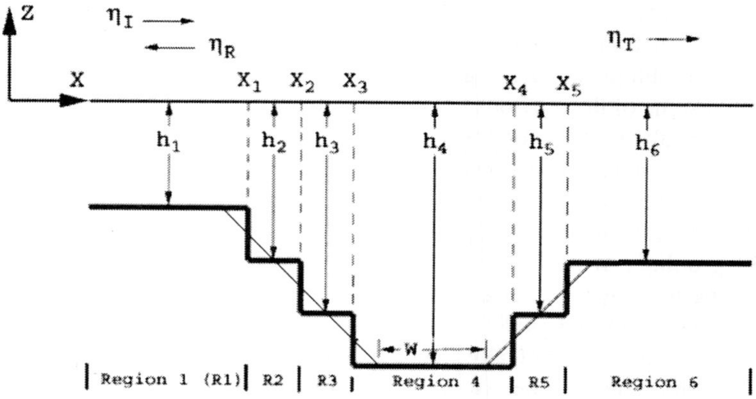

Figure 1. Definition sketch for trench with stepped transitions.

The boundary value problem entails specification of the boundary conditions at the free surface and bottom and specifies a radiation condition to develop a solution in each region of the form:

$$\phi_j(x,z) = A_j^{\pm}\cosh\left[k_j\left(h_j+z\right)\right]e^{\pm ik_j(x)} + \sum_{n=1}^{\infty} B_{j,n}^{\pm}\cos\left[\kappa_{j,n}\left(h_j+z\right)\right]e^{\pm\kappa_{j,n}\left(x-x_j\right)}$$

$$\left(j=1\to J\right),\ \left(n=1\to N\right) \qquad (1)$$

In Eq. 1, k and κ equal the wave number of the propagating modes and evanescent modes.

At each boundary the application of matching conditions depends on whether the boundary represents a step up or a step down. With the incident wave specified, the matching conditions develop a set of equations with 2(J-1)N+2(J-1) unknown coefficients. At each boundary (x_j) the matching conditions must include the appropriate evanescent mode contributions from the adjacent boundaries (x_{j-1}, x_{j+1}). Solution of the resulting set of simultaneous equations requires a linear matrix equation with the value of N large enough to ensure convergence of the solution.

Pit of Circular Planform

Bender and Dean (2005) present a method to calculate the wave transformation caused by a 3-D axisymmetric bathymetric anomaly (pit or shoal). This model extends previous work by allowing domains in arbitrary water depth and gradual transitions in depth with sidewall slopes approximated by a series of steps of uniform depth. The solution uses a cylindrical coordinate system defined in Fig. 2, which shows the case of a pit with abrupt transitions.

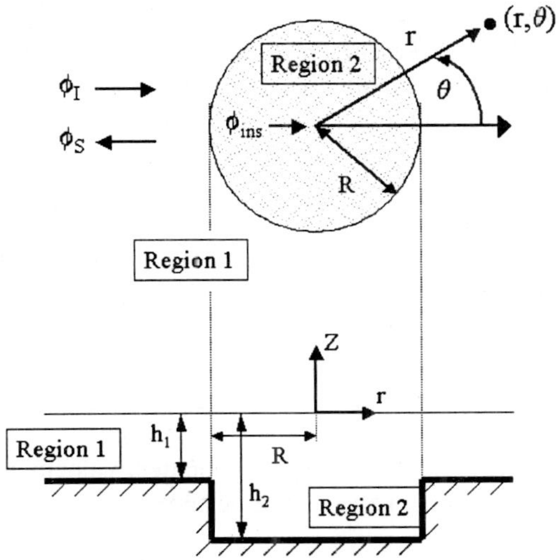

Figure 2. Definition sketch for circular pit with abrupt depth transitions.

The equations of linear wave propagation in cylindrical coordinates contain Bessel functions, which are expressed in a truncated series $(m = 1 \rightarrow M)$. The solution contains evanescent modes that decay with distance from the interface between regions to extend the solution into the arbitrary depth region. The solution requires matching conditions at each interface between regions of constant depth. Black and Mei (1970) formulated this problem for the case of a cylinder floating at the surface or fixed on the bottom. The solution starts with the definition of an incident velocity potential in the form of a plane progressive wave:

$$\phi_I(r,\theta,z,t) = M_I\left[\sum_{m=0}^{\infty}\gamma_m \cos(m\theta)J_m(k_1 r)\right]\frac{\cosh[k_1(h_1+z)]}{\cosh(k_1 h_1)}e^{-i\omega t} \quad (2)$$

where $\gamma_m = 1$ for m = 0 and $2i^m$ otherwise, $M_I = \frac{-igH}{2\sigma}$ where H equals the incident wave height, r is the radial distance from the center of the bathymetric anomaly to the point in the fluid domain and θ is the angle to the point measured counter-clockwise from the θ origin as shown in Fig. 2, k_1 represents the wave number in the region outside the anomaly of depth, h_1, ω equals the angular frequency, g equals the acceleration due to gravity, and m equals the number of the Bessel function modes with total number M included in the summation.

Pit of Elliptic Planform

Bender and Dean (2005b) present a method to calculate wave transformation in the presence of bathymetric anomalies of elliptic form. The elliptic model contains many similarities to the axisymmetric model presented in the previous section. The elliptic method requires application of the cylindrical coordinate system where (u,v) replace (r,θ) in the axisymmetric (cylindrical) system. Each region has a specified depth and each interface between regions has a specified u location where the matching conditions are applied. At each boundary, the matching conditions depend on whether the boundary represents a step up or a step down. Specifying the incident wave allows the formation of a set of equations with $2N_sN_e(M+1)$ unknown coefficients where N_e equals the number of evanescent (non-propagating) modes and M equals the number of Mathieu function modes in the solution.

The solution starts with the definition of an incident velocity potential in the form of a plane progressive wave valid in elliptic coordinates:

$$\phi_I(u,v,z,t) = M_I\left[\begin{array}{l}\sum_{m=0}^{\infty}\gamma_m ce_m(v,q_1)ce_m(\theta,q_1)Je_m(u,q_1)+\\ \quad\sum_{m=1}^{\infty}\gamma_m se_m(v,q_1)se_m(\theta,q_1)Jo_m(u,q_1)\end{array}\right]\frac{\cosh[k_1(h_1+z)]}{\cosh(k_1h_1)}e^{-i\omega t} \quad (3)$$

where the previous sections have defined γ_m, g, m, ω, k, h, z, and M_I. In Eq. 3, θ equals the angle of incidence on the ellipse, v equals the radial distance from the center of the bathymetric anomaly to the point in the fluid domain, u equals the angle to the point measured counter-clockwise from the origin.

RESULTS AND DISCUSSION

Trench of Infinite Length

Solution of the 2-D model allows investigation of the energy reflection caused by long pits or shoals with gradual transitions in depth (side slopes). Figure 3 presents the reflection caused by a Gaussian shoal of infinite length. The inset plot shows the domain, which places the steps approximating the Gaussian form at 0.05 m depth intervals. Similar to the case for a Gaussian trench, a single major peak in the reflection coefficient, K_R, occurs followed by minimal reflection at larger values of k_1h_1.

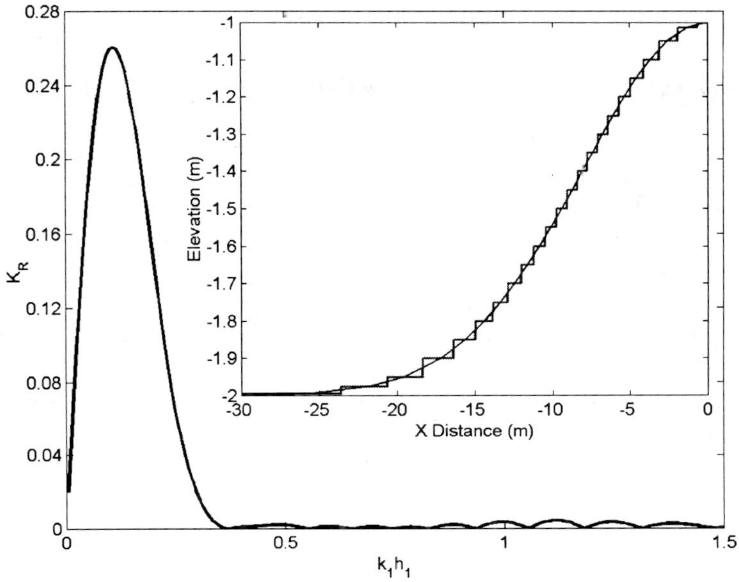

Figure 3. Reflection coefficients versus k_1h_1 for Gaussian shoal; only one-half of the symmetric shoal cross-section shown.

Figure 4 presents an asymmetric trench case where the upwave and downwave depths and transition slopes equal different values — larger upwave depth. In this figure two asymmetric trenches with the same cross-sectional area are compared: a trench with an abrupt transition (solid line) and a trench with an upwave transition slope (s_1) equal to 1 and a downwave slope (s_2) of 0.2 (dashed line). The figure also includes a third trench with an upwave transition slope (s_1) equal to 0.2 and a downwave slope (s_2) of 1.0 (dotted line). The trenches with the gradual side slopes produce the same reflection coefficients; therefore, Figure 4 only shows two values of reflection coefficient versus dimensionless wave number. The figure shows the sloped transitions significantly reduce the value of K_R, especially as k_1h_1 increases.

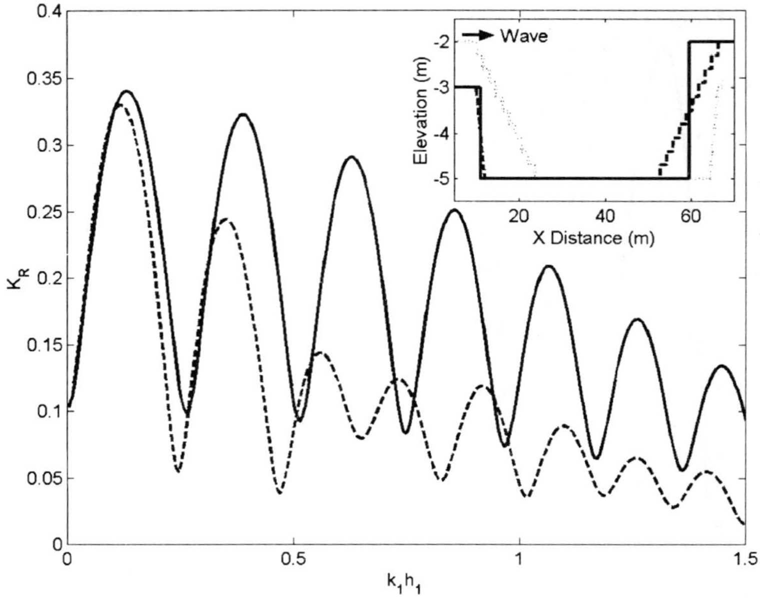

Figure 4. Reflection coefficients versus k_1h_1 for an asymmetric abrupt transition trench, an asymmetric trench with different h_1 and h_5 values and s_1 not equal to s_2.

Pit of Circular Planform

Solution of the 3-D axisymmetric step-model establishes the complex velocity potential anywhere in the fluid domain. From the complex velocity potential, the model calculates and applies the wave height, wave angle, and energy flux to show the wave field modification caused by a bathymetric anomaly. The axisymmetric model allows investigation of the effect of the depth transition slope on the wave field. Differences between the incident wave field with and without the presence of the bathymetric anomaly indicate changes in the wave height and wave direction caused by the change in depth. Bender and Dean (2004) present a method to apply the wave parameters landward of the bathymetric anomaly in a nearshore shoaling and refraction model that calculates potential shoreline response through a one-line shoreline change model.

The axisymmetric step-model was compared to REF/DIF-1 for a pit with linear transitions in depth with a slope equal to 0.1. The analysis applies a bathymetry that causes significant reflection and compares the results of the axisymmetric step-model, which accounts for reflection, to REF/DIF-1, which does not allow reflection in the upwave direction. Figure 5 shows the relative amplitude for a transect taken in the cross-shore direction with a cross-section of the pit taken through the centerline shown in the inset. In this example, 10 steps approximate the linear slope in the axisymmetric step-model and REF/DIF employed a grid spacing of 4.0 m in the X and Y directions. The axisymmetric step-model results (solid line) indicate significant reflection in the upwave region with wave sheltering in the downwave region. The relative amplitude values for REF/DIF-1 employing linear wave theory show good agreement with the 3-D Step Model results in the downwave region with a slight divergence inside and just downwave of the pit. Thus, the reflection captured by the axisymmetric step-model does not result in significant differences in the relative amplitude in the downwave region for the two models for this cross-shore transect. With REF/DIF-1 employing non-linear wave theory, the relative amplitude was generally unchanged near the pit, but with significantly less wave sheltering at large distances shoreward of the pit.

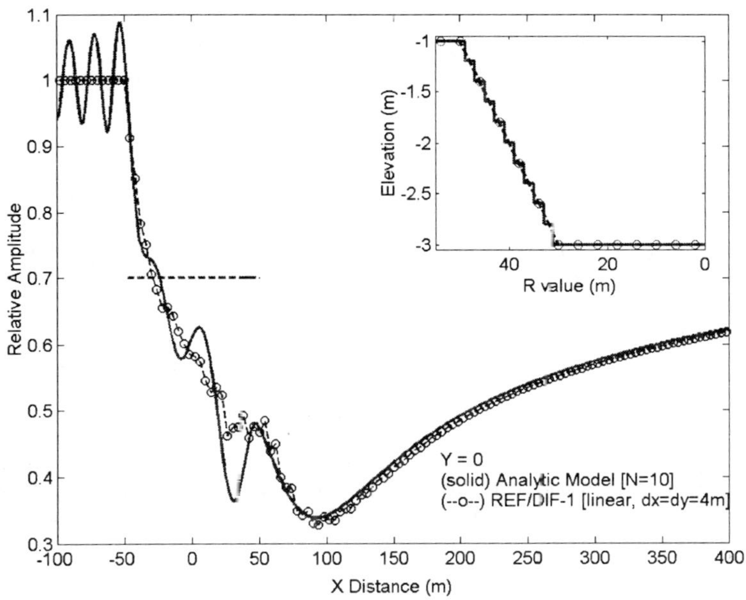

Figure 5. Relative amplitude using axisymmetric step-model and REF/DIF-1 for cross-shore transect at Y = 0 for pit with linear transitions in depth; cross-section of pit bathymetry through centerline included.

Unlike the 2-D case of waves over a trench or shoal of infinite length, determining the energy reflection from a bathymetric anomaly in 3-D proves difficult. To determine the energy reflection caused by an axisymmetric bathymetric anomaly Bender and Dean (2004) developed method using far-field approximations of the Bessel functions along a half-circular arc bounding the region of reflected energy. Summing the reflected energy through this arc and comparing it to the amount of energy incident on the anomaly allows calculation of the reflection coefficient.

Figure 6 shows the reflection coefficients for four pits in the shape of truncated cones of equal volume with constant bottom radius (slopes = abrupt, 1, 0.2, 0.05). The smaller transition slopes reduce greatly the reflection caused by the pit for most values of k_1h_1. The figure shows K_R oscillates with k_1h_1 with a decreasing trend as k_1h_1 increases and no instances of complete transmission other than $k_1h_1 = 0$.

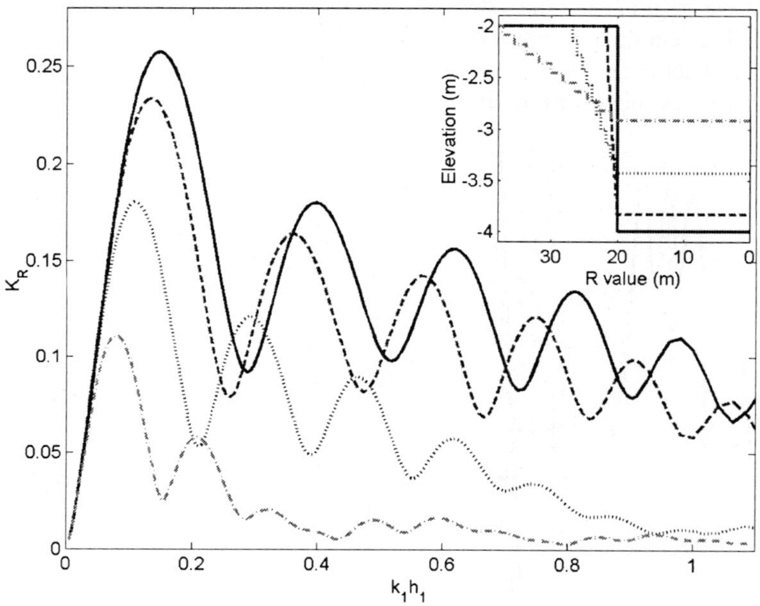

Figure 6. Reflection coefficient versus k_1h_1 based on far-field approximation and constant volume and bottom width pits; cross-section of pit bathymetries through centerline included.

Pit of Elliptic Planform

The elliptic modeling effort focused on the effect of the variation in the longshore and cross-shore extent of the elliptic form on the wave field modification. Figure 7 presents a contour plot of the relative amplitude for an elliptic shoal with sloped sidewalls (truncated cone). In this figure, the incident waves propagate with an angle of 45 degrees. The depth anomaly consists of five ellipses of uniform depth with the largest having a semi-major axis (a) length of 22.6 m and a semi-minor axis (b) length of 10.4 m. The transitions in depth equal 1:3 for the semi-major axis and 1:6 for the semi-minor axis. The domain has a uniform depth of 3.0 m outside the depth anomaly and depths of 2.7, 2.4, 2.1, 1.8 and 1.5 m for each elliptical step that comprise the shoal.

Figure 7 clearly shows the asymmetry in the wave field developed by the elliptic form and the convergence of the wave field caused by the shoal with a large area of wave focusing evident downwave of the shoal. Two bands of relative wave amplitude less than one, caused by the diverging wave field, flank this area.

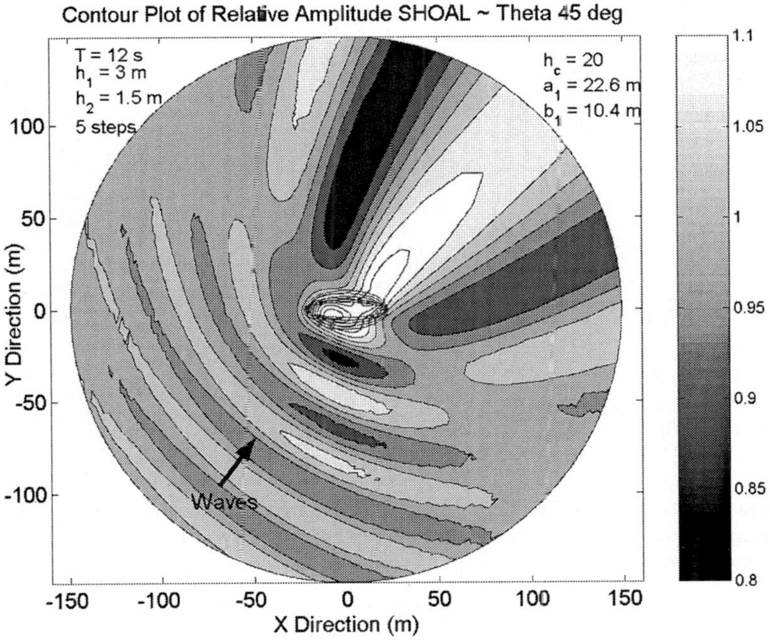

Figure 7. Contour plot of relative wave amplitude for elliptic shoal with five depth transitions and incident wave angle equal to 45 degrees; labels on figure provide model parameters.

Figure 8 presents a comparison of the relative amplitude along transects in the direction of wave propagation and through the center of the elliptical forms for three incident wave angles: 0, 45, and 90 degrees. To facilitate comparison between the three cases the distance along the x-axis is shifted relative to the semi-major axis of the largest ellipse for each case, a_1. The figure indicates a significant difference in the reflection for an incident wave angle of 90 degrees compared to 0 degrees. With the shape of the elliptic shoal, waves propagating from 90 degrees encounter an obstruction 1.5 m below the water surface and 40 m wide. In comparison, waves propagating from 0 degrees encounter an obstruction only 8.0 m wide at the same depth.

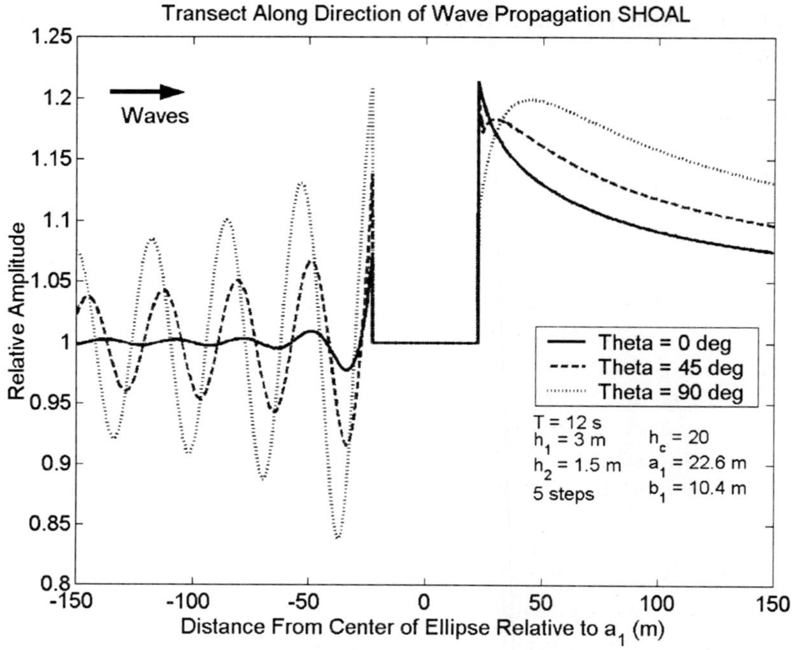

Figure 8. Relative wave amplitude along transects in the direction of wave propagation (0, 45, and 90 deg) through the center of the elliptic shoal shifted to a_1; labels on figure provide model parameters.

Additional efforts have focused on the development of a method to evaluate energy reflection (far-field energy approach) and shoreline change in the presence of elliptic forms — similar to the axisymmetric solution. Convergence problems with the Mathieu functions solutions at sufficiently high orders to resolve the solution and at the required distances away from elliptic form have limited progress. Current efforts are underway to develop solution procedures that address the convergence problems.

CONCLUSIONS

This paper reviews several semi-numeric models to evaluate wave transformation in the presence of bathymetric anomalies with gradual transitions in depth. The models apply to either trenches or shoals of infinite length (2-D model) or axisymmetric and elliptic forms (3-D models). The models apply matching conditions at depth discontinuities to solve unknown coefficients in a matrix of equations.

The results presented show how the sidewall slope of the bathymetric anomaly affects the wave transformation and/or energy reflection caused by the form. Anomalies with more gradual transitions in depth cause less reflection and reduced wave focusing or spreading at landward locations. The semi-numeric models developed provide a comparative tool for more robust fully-numeric solutions.

REFERENCES

Bender, C. J. and Dean, R. G. 2003. Wave Transformation by Two-Dimensional Bathymetric Anomalies With Sloped Transitions, *Coastal Engineering*, 50, 61-84.

Bender, C. J. and Dean, R. G. 2004. Potential Shoreline Changes Induced by Three-Dimensional Bathymetric Anomalies with Gradual Transitions in Depth, *Coastal Engineering*, 51, 1143-1163.

Bender, C. J. and Dean, R. G. 2005. Wave Transformation by Axisymmetric Three-Dimensional Bathymetric Anomalies with Gradual Transitions in Depth, *Coastal Engineering*, 52, 331-351.

Bender, C. J. and Dean, R. G. 2005b. Wave Field Modifications by Elliptic Bathymetric Anomalies with Gradual Transitions in Dept, *Proceedings of the 29th International Conference on Coastal Engineering (ICCE) 2004*, Lisbon, ASCE, 243-255.

Black, J.L. and Mei, C.C. 1970. Scattering and radiation of water waves, *Technical Report No. 121*, Water Resources and Hydrodynamics Laboratory, Massachusetts Institute of Technology.

Chen, H.S., and C.C. Mei. 1973. Wave Forces on a Stationary Platform of Elliptical Shape, *Journal of Ship Research*, 17(2), 61-71.

Gutiérrez-Vega, J. C., R. M. Rodríguez-Dagnino, M. A. Meneses-Nava, and S. Chávez-Cerda. 2003. Mathieu functions, a visual approach, *Am. J. Phys,* 71(3), 233-242.

Holthuijsen, L.H., Booij, N., Ris, R.C., Haagsma, IJ.G., Kieftenburg, A.T.M.M., and Kriezi, E.E. 2000. SWAN User Manual, SWAN Cycle III, Version 40.11, Department of Civil Engineering, Delft University of Technology. GA Delft, the Netherlands.

Lamb, H. 1932. *Hydrodynamics,* Dover, Section 176.

Mei C.C., and Black, J.L. 1969. Scattering of surface waves by rectangular obstacles in water of finite depth, *Journal of Fluid Mechanics,* 38, 499-511.

Kirby, J.T. and Dalrymple, R.A. 1983. Propagation of obliquely incident waves over a trench, *Journal of Fluid Mechanics,* 133, 47-63.

Kirby, J.T. and Dalrymple, R.A. 1994. User's manual, combined refraction / diffraction model, REF/DIF-1, Version 2.5, Center for Applied Coastal Research, Department of Civil Engineering, University of Delaware. Newark, Delaware.

Kirby, J.T., Dalrymple, R.A. and Seo, S.N. 1987. Propagation of obliquely incident waves over a trench. Part 2, Current flowing along trench. *Journal of Fluid Mechanics,* 176, 95-116.

Kirby, J. T., Wei, G., Chen, Q., Kennedy, A. B. and Dalrymple, R. A. 1998. FUNWAVE 1.0, Fully nonlinear Boussinesq wave model. Documentation and user's manual. Report CACR-98-06, Center for Applied Coastal Research, Department of Civil and Environmental Engineering, University of Delaware.

Miles, J.W. 1982. On surface-wave diffraction by a trench, *Journal of Fluid Mechanics,* 115, 315-325.

McDougal, W.G., Williams, A.N. and Furukawa, K. 1996. Multiple pit breakwaters, *Journal of Waterway, Port, Coastal, and Ocean Engineering,* 122, 27-33.

Newman, J.N. 1965a. Propagation of water waves over an infinite step, *Journal of Fluid Mechanics,* 23, 23-29.

Smith, J.K., Sherlock, A.R., and Resio, D.T. 2001. STWAVE: Steady-state spectral wave model, User's manual for STWAVE, Version 3.0, *Final Report ERDC/CHL SR-01-1*, U. S. Army Engineer Research and Development Center, Vicksburg, Mississippi.

Takano, K. 1960. Effets d'un obstacle paraléllélépipédique sur la propagation de la houle, *Houille Blanche*, 15, 247.

Williams, A.N. 1990. Diffraction of long waves by a rectangular pit, *Journal of Waterway, Port, Coastal, and Ocean Engineering*, 116, 459-469.

Williams, A.N. and Vasquez, J.H. 1991. Wave interaction with a rectangular pit, *Journal of Offshore Mechanics and Arctic Engineering*, 113, 193-198.

INVERSE METHODOLOGY FOR COASTAL WAVE DATA ASSIMILATION

Alexandre Coli[1,2], João Alfredo Santos[1] and António A. Pires-Silva[2]

The offshore sea wave spectra estimation, based on nearshore wave observations, is important to define initial conditions for some regional wave propagating models as well as for data assimilation by global wave models. Among all procedures to estimate the offshore spectrum, the inverse methodology has the advantage of doing without any prior knowledge on the spectrum shape, its minimum, maximum or integral values. This paper presents an inverse methodology for sea wave spectrum estimation and it discuss its application to three test cases with ideal bottom configuration – a sloping bottom, a headland and a bay. To simulate the wave data at the measuring points, the SWAN spectral propagation model was used while the matrix relating the offshore spectrum ordinates and the directional spectral moments at these points was built using the BACKTRACK ray tracing model. In all test cases, the good agreement between the offshore directional curve shapes, peak directions and curve area, for one given frequency, shows the capabilities of the inverse methodology in this kind of problems. In addition, the application of this technique on controlled test cases showed the influence of the smoothness constrain on the solution quality.

INTRODUCTION

To circumvent the lack of coastal wave data, the use of numerical models at different space and time scales has become a common practice. Naturally, numerical models at the regional or local scale have to be initialized with the appropriate offshore boundary conditions. In the absence of deep water directional wave measurements, a way has to be worked out to fulfill this condition. Backward ray tracing methodologies from locations where observations are at one's disposal into unsheltered deep water, and subsequent forward propagation to uninstrumented coastal spots, have been used for this purpose (Coli *et al.*, 2002).

The present wide spread availability of deep water global models simulations prompt the nesting with smaller scale numerical models. The uncertainty associated with global models results may affect the performance of these regional numerical wave models, especially when applied to coastal regions with a complex bathymetry.

O'Reilly and Guza (1998) suggest a methodology to improve the accuracy of the large scale wave models forecast spectra at the shelf break before they are used to initialize the shallow water models. The approach consists, basically, to

---

[1] Hydraulics and Environment Department, Laboratório Nacional de Engenharia Civil, Av. do Brasil, 101, Lisbon, Portugal, 1700-066

[2] DECivil / Instituto Superior Técnico, Technical University of Lisbon, Av. Rovisco Pais, Lisbon, Portugal, 1049-001

adapt the inverse methods, which has been so fruitful in solid-earth geophysics (Menke, 1984) and also in oceanography (Bennett, 1992), to assimilate nearshore wave observations into a linear wave propagation model. As in most inverse problems, a substantial underdetermination occurs. To overcome this issue the regularity or smoothness of the solution is a form of additional information that is often used.

The data assimilation methodologies deal, essentially, with the merge of information from multiple sources, including remote sensing instruments, ground-based sensors and mathematical models. In this sense it shares some conceptual similarities to the optimization and calibration problems in model evaluation and to the interpolation, smoothing and filtering in time series analysis (McLaughlin, 2002). A broad overview of the subject of data assimilation, especially in meteo-ocean related matters, can be found in Talagrand (1997).

The purpose of this study is to perform a comprehensive evaluation of the inverse methodology in idealized and controlled situations. This way, the strengths and shortcomings of this approach to the offshore directional estimation issue will be highlighted.

Generally, a fairly good agreement between the imposed spectra and the inverse estimate has been registered for different conditions and the methodology shows capability to capture the specifics of the synthetic experiment.

After this introduction, it follows a summary of the inverse methodology applied to the offshore spectral estimation, a validation of the methodology implementation with a sloping bottom case and a set of test cases with idealized geometries. The conclusions are presented on the last section.

INVERSE METHODOLOGY

Nearshore sea wave conditions are the result of the transformations endured by an offshore sea state as it propagates over the continental shelf. If these transformations could be numerically modeled and nearshore wave measurements are available, the "inverse methodology" enables the estimation of the offshore directional spectrum ordinates. The name "inverse methodology" arises because the numerical model used to propagate the sea state characteristics from offshore to nearshore is employed to estimate offshore parameters, i.e. it is used the inverse way.

For instance, if a directional wave buoy is used, the following directional moments of the wave spectrum at that point, $S(f,\theta)$, can be evaluated:

$$b_i = \int_0^{2\pi} S(f,\theta) m_i(\theta) d\theta \tag{1}$$

where $m_1(\theta) = \cos\theta$, $m_2(\theta) = \sin\theta$, $m_3(\theta) = \cos 2\theta$ and $m_4(\theta) = \sin 2\theta$. Note that the b_i moments vary with the frequency f for which they are computed.

If one knows the relation between the coastal spectrum at that frequency and the offshore spectrum at the same frequency it is possible to define a forward model that allows one to forecast the results of the measurements (the directional moments) given the ordinates of the offshore directional spectrum. This relation can be established by means of linear spectral refraction theory (Longett-Higgins 1957), which shows that $C_g\ S(f,\theta)/k$ is invariant along a wave ray, i.e.:

$$S(f,\vartheta)=\frac{k}{k_0}\frac{C_{g0}}{C_g}S_0\left[f,\Gamma(f,\theta)\right] \tag{2}$$

where $S_0(f,\theta_0)$ is the continuous offshore directional spectrum, $S(f,\theta)$ is the coastal spectrum, k is the wave number magnitude, C_g is the group velocity, the subscript "0" denotes offshore and unsubscripted variables are at coastal locations. The frequency dependent relationship $\Gamma(f,\theta)$ between offshore wave angles θ_0 and coastal wave angles θ is obtained by backward ray tracing from the point where directional moments are to be known into deep water where it is assumed that S_0, is spatially homogeneous (an assumption that restricts the use of this model to swell generated by distant storms).

Since the quantities measured at one buoy location are directional moments of the spectrum, equation (2) is combined with the equation (1) to give:

$$\int_0^{2\pi} S(f,\theta)\,m_p(\theta)\,d\theta=\frac{k}{k_0}\frac{C_{g0}}{C_g}\int_0^{2\pi} S_0\left[f,\Gamma(f,\vartheta)\right]m_p(\theta)\,d\theta \tag{3}$$

$p=1,2,3,4,5$

To make computations simpler the integral in the above equation is replaced by a summation. This means that one is discretizing the offshore spectrum in frequency directional bins, the surface of the directional spectrum in each of those regions being replaced by a constant level. Let

$$S_o=\left[S_{01},\ \ S_{02},\ \ S_{03},\ \ \ldots,\ \ S_{0N}\right]^T \tag{4}$$

designate the vector of the offshore energy densities in each of the N contiguous directional bins for a given frequency band, that is the discretized offshore spectrum. To construct the discrete forward wave model, each component of the offshore directional distribution is assigned unit energy and substituted into (3) to obtain the linear transfer coefficients between the discretized offshore spectrum and the coastal observations. The theoretical relationship between the offshore directional spectrum and the coastal measurements is thus approximated as a set of linear equations:

$$\mathbf{b}=\mathbf{A}\,\mathbf{S}_0 \tag{5}$$

where $\mathbf{b}$ is the coastal measurements vector, whose number of components (M) depends on the number of coastal points with wave measurement and on the type of it (directional or not), and $\mathbf{A}_{(MxN)}$ is the so called data kernel matrix, where each element A_{ij} of matrix $\mathbf{A}$ represents the contribution of the jth offshore directional component in the ith coastal observation.

The number of unknowns in the inverse problem, N components of $\mathbf{S}_0$ vector, greatly outnumbers the M equations and the problem is underdetermined. To get an overdetermined problem, a smoothness constrain is included in the problem definition. This is a reasonable assumption since away from the generation area the directional spectrum is not supposed to vary wildly.

A simple, smooth $\mathbf{S}_0$ consistent with coastal observations if found by including penalties in the overall least square fit for directional roughness (smoothness condition is the imposition of the differences among neighboring intervals to be not significant). These conditions are implemented as additional lines in the $\mathbf{A}$ matrix and have the general form:

$$\begin{bmatrix} w_d & -w_d & 0 & \cdots & 0 & 0 \\ 0 & w_d & -w_d & & 0 & 0 \\ \vdots & & & \ddots & & \vdots \\ 0 & 0 & 0 & & w_d & -w_d \end{bmatrix} \begin{bmatrix} S_{01} \\ S_{02} \\ \vdots \\ S_{0N} \end{bmatrix} = \begin{bmatrix} 0 \\ 0 \\ \vdots \\ 0 \end{bmatrix} \tag{6}$$

where w_d is the directional smoothness constrain. The corresponding values in the $\mathbf{b}$ column-vector are set to zero. With N directional intervals it exist $N-1$ additional equations in $\mathbf{A}$ matrix. Also, to reinforce the lateral boundaries to be equal to zero, 6 unit-elements are set at each side of the matrix $\mathbf{A}$ and they are assigned to be zero in the vector $\mathbf{b}$.

The matrix $\mathbf{A}$ and the $\mathbf{b}$ vector built in this way turn the inverse problem in an overdeterminated one and it can be solved by the Non Negative Least Squares (NNLS) technique by Lawson and Hansen (1974). This technique provides a non negative solution, i.e., the directional distribution of the offshore discrete spectrum at a frequency band that is consistent to $\mathbf{b}$ in a least squares sense (2-norm):

$$\min_{S_o>0} \left\| \mathbf{AS}_0 - \mathbf{b} \right\|_2 \tag{7}$$

VALIDATION OF THE METHODOLOGY IMPLEMENTATION

With all problem variables previously known (the offshore directional distribution, the nearshore spectrum moments and the wave transformation model), the proposed methodology can be applied. If the methodology implementation is valid then the estimated offshore directional distribution has to match the initial one when the nearshore spectrum moments are obtained directly through the kernel matrix.

The chosen situation consisted of irregular waves that propagated over a sloping bottom. This way, the kernel matrix $\mathbf{A}$ was built using the transformation of the offshore spectrum directional bins, for a given frequency band, into the nearshore directions considering solely linear wave refraction over a sloping bottom. The frequency band chosen was centered in 0.08 Hz (12.5 seconds) and the nearshore spectra were computed for two points located over the 10 m and 20 m depth contours. For the 0.08 Hz frequency and for each 0.1° directional interval between 90° and 270°, the directional transformation that depends on the frequency $\Gamma(f,\theta)$ was obtained by the application of the Snell's Law (simple refraction). Instead of the values resulting from the ray tracing computations, the offshore directions were characterized by the 5° direction bins that contained them and matrix $\mathbf{A}$ was computed as described in the previous section.

To represent the offshore directional distribution ($\mathbf{S}_0$) at 0.08 Hz, a cosine function to the power of 20, with maximum ordinate of 0.3, was adopted. Two peak directions of 180° and of 135° were tested as well as a bimodal directional spectrum with peaks at 135° and 225°.

Each $\mathbf{b}$ vector was computed by $\mathbf{AS}_o$ product to guarantee their univocally origin of the corresponding $\mathbf{S}_0$.

Figure 1 shows a comparison between the offshore directional distribution that was input to the linear refraction model and the offshore directional distributions estimated by the proposed methodology for different values of w_d. In the three Figure 1 panels, it is evident the influence of the w_d values on the inverse problem results. Nevertheless, the panels show also that a w_d value (in these cases, $w_d = 0.01$) producing a good agreement between the solution and the initial function $\mathbf{S}_0$ can always be found.

Naturally, the zero moments of the estimated directional distributions agree quite well with those of the offshore directional distributions that were input to the linear refraction model (Table 1).

Table 1. Area under the curves of the directional distribution of the reference spectrum (S_0) and the inverse methodology estimated spectrum (S_0 – IM), with $w_d = 0.01$.

Inclined Plane	Area under the curve (m^2/Hz)	
	S_0 (Reference)	S_0 - IM
Case 180°	9.695	9.519
Case 135°	9.695	9.576
Bimodal Case	19.390	19.202

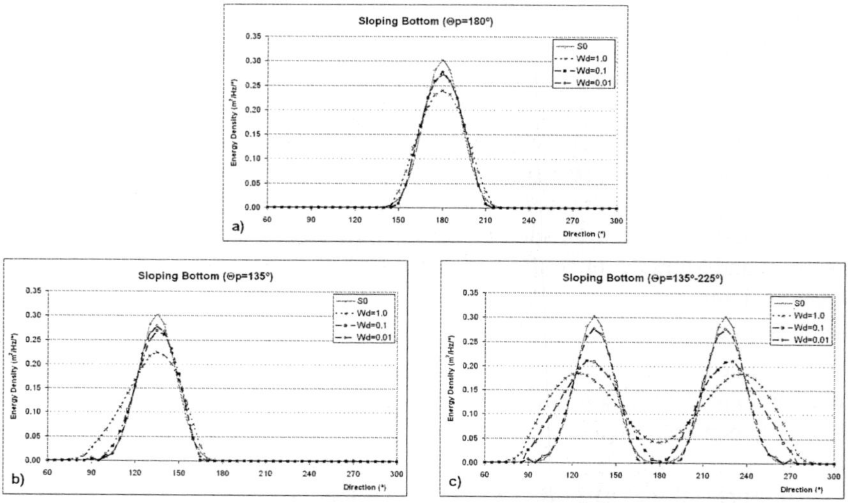

Figure 1. Directional distributions of the reference spectrum (S0) and of the estimated spectra for different values of w_d, for the 0.08 Hz frequency. Note the good agreement for the case with $w_d = 0.01$.

From the good agreement between the directional distributions of the reference spectra (S0) and the inverse methodology estimates (S0 – IM), it may be concluded that the implemented inverse methodology is valid.

TEST CASES WITH IDEALIZED CONFIGURATION

The synthetic case studies aim to clarify the inverse methodology capabilities by considering three idealized bottom configurations: a Sloping Bottom, a Bay and a Headland. While the Sloping Bottom is a simple refraction problem, the Bay case can represent a pocket beach or a sheltered area. The Headland case illustrates the importance of an obstacle and its sheltering on wave propagation.

Whatever the bottom configuration, Figure 2, the study area is rectangular 15 km long (x-axis) and 10 km wide (y-axis). Apart from a area with a flat bottom with a water depth of 130 m (which corresponds to deepwater for a 12.5 s period wave) the bottom has a slope of 1:34.615 and it is defined by a grid with rectangular elements 150 m long and 100 m wide. Both the Bay and the Headland bottom have conical shapes with 130 m height and maximum radius of 4500 m that match the Sloping Bottom (Figure 2).

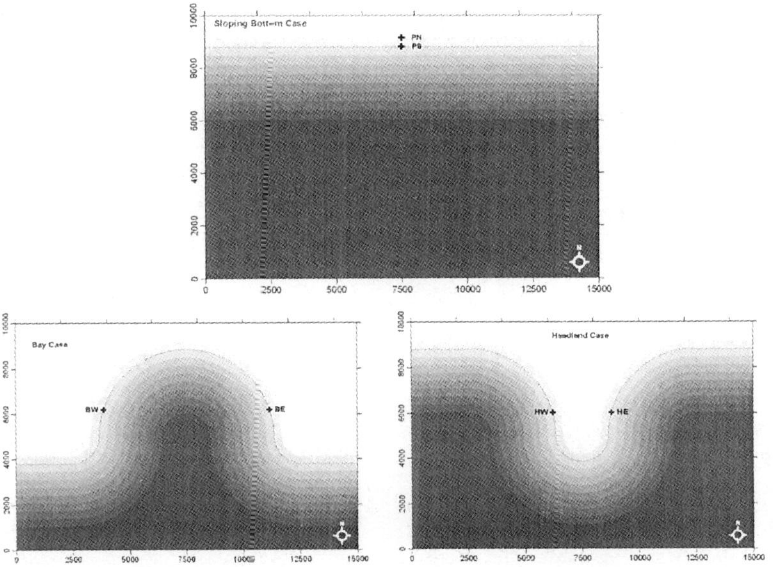

Figure 2. Bottom configurations for the synthetic cases with "buoy" places and the 20 m contour stressed.

In all test cases, there are two sea wave "measurement" points. On the Sloping Bottom these points are over the 10 m and 20m depth contours; on the Bay, they are located over the 20 m depth contour at each side of the bay entrance and, on the Headland, they are at each side of the headland also over the 20 m depth contour.

The SWAN model (Booij *et al.*, 1999) was used to simulate the sea wave data at each "measurement" point of these test cases. The SWAN is a phase average model able to propagate the sea waves up to the breaking zone. Conceptually, the SWAN model is considered a third generation model, based on the balance of the wave action equation, where the wave action is the quotient between the directional spectra and its intrinsic frequency.

The SWAN offshore boundary conditions were a parameterization of a directional spectrum. For each test case, two sea states, with significant wave height of 2m, peak period of 12.5 s and peak directions of 180° (South/normal incidence) and of 135° (Southeastern incidence) were considered.

For each set of these parameters, the directional spectrum was defined as a product of a JONSWAP frequency spectrum, with a peak enhancement factor of 3.3, by a directional distribution function of cosine to a power of 20 type. The frequency was discretized into 22 values ranging from 0.0507 to 0.25 Hz, with a logarithmic distribution. The directional range covers 360° with a resolution of 0.5°.

Figure 3 shows the vector maps corresponding to the cases with offshore peak direction of 135°. These vector maps are graphical plots with vectors on each grid node composed by the product of the peak direction versor and the significant wave height.

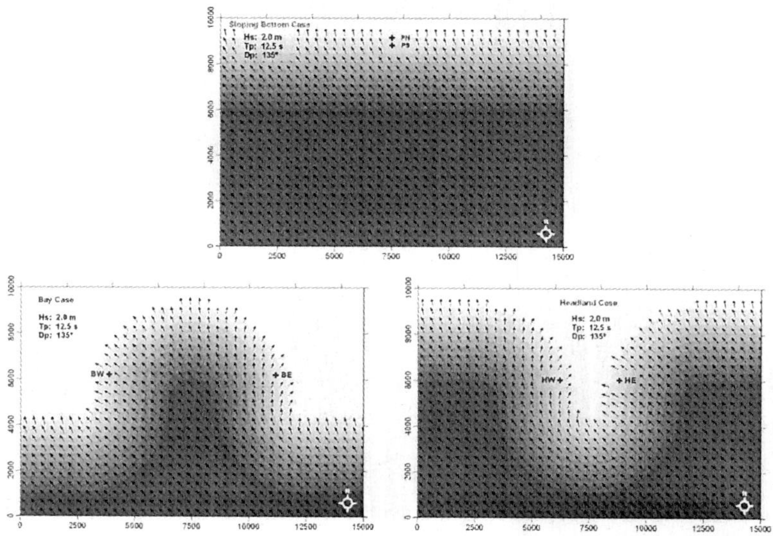

Figure 3. SWAN wave propagation over the computational domain of the three test cases, with the initial offshore spectrum defined by $H_s = 2m$ **,** $T_p = 12.5s$ **and** $\theta_p = 135°$ **.**

All computations were done in stationary mode with SWAN version 40.41, without wind or currents and with refraction, diffraction, shoaling and dissipation by bottom friction and wave breaking.

The SWAN results at each "measurement" point (directional spectrum ordinates) were combined to build the vector **b** of each case:

$$\mathbf{b} = [b_1, b_2, b_3, \ldots, b_M]^T \tag{9}$$

each component representing one of the five directional moments (Eq. 1), for the frequency of 0.08 Hz, i.e. $M = 10$.

Matrix **A** was evaluated by using a backward ray tracing model, BACKTRACK. This wave model is embedded into TRANSFER methodology (Coli *et al.*, 2002) and computes wave rays based on spectral linear refraction (Longuet-Higgins, 1957) for a given bottom configuration and for a wave period. Figure 4 shows the fan-type wave ray diagrams for one of the "measurement" points and for each of the three bottom configurations.

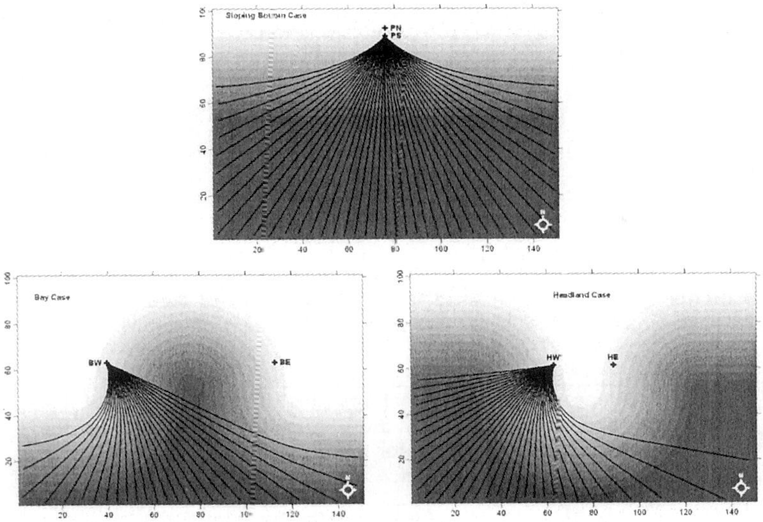

Figure 4. BACKTRACK fan wave rays results for the three test cases and for $T_p = 12.5s$ and for the directions ranging from 90° to 270° taken from one of the "measurement" point.

The BACKTRACK model was run for the frequency of 0.08 Hz, for each "measurement" point at each bottom configuration. The rays cover a 180° range (from 90° to 270°), with a 0.1° discretization at the "measurement" points. Only the rays that reach offshore were considered and their offshore directions were categorized into 5° intervals to build matrix $\mathbf{A}$.

Once matrix $\mathbf{A}$ and vector $\mathbf{b}$ have been created, the smoothness constrains, with $w_d = 1.0$ (Eq. 6), and the zero boundary conditions were added and the inverse problem is then solved.

The goodness of the inverse methodology estimates is assessed by comparing the shape, the peak position and the area of the directional distribution with those of the known SWAN boundary conditions.

Sloping Bottom

The results of the Sloping Bottom case for the frequency of 0.08 Hz, obtained from the solution of the inverse problem, are shown on Figure 5.

For both test cases the shape of the estimated directional distribution is similar to the shape of the corresponding offshore SWAN spectrum and their peaks are at the same directions.

Table 2, with the directional distribution area for the frequency of 0.08 Hz, supports the good agreement of the 180° test case, since the area difference is less than 3%. For the 135° case the area difference reaches 15%.

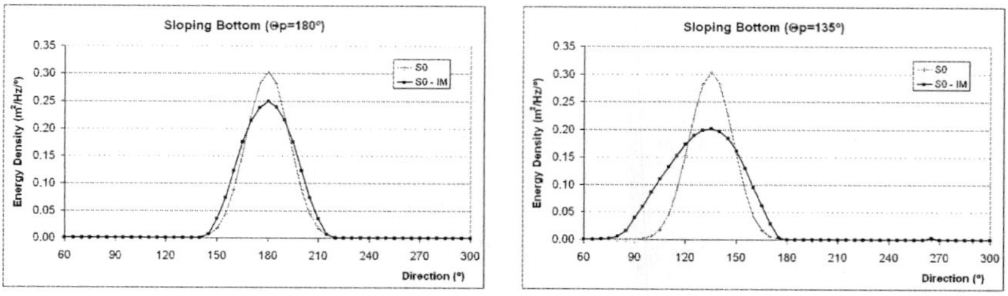

Figure 5. Directional distribution of the reference spectra (S0) and estimated spectra (S0-IM) of the Sloping Bottom case, for the frequency of 0.08 Hz. A) Case with offshore peak direction of 180°; b) Case with offshore peak direction of 135°.

Table 2. Areas under the directional distribution curves of the reference spectrum (S_0) and the estimated spectrum of the proposed inverse methodology for the Sloping Bottom (S_0-SB), Bay (S_0-B) and Headland (S_0-H) cases.

	Area under the curve (m²/Hz)			
	S_0	S_0-SB	S_0-B	S_0-H
Case 180°	9.695	9.928	10.901	10.090
Case 135°	9.695	11.182	12.046	10.661

A question that may arise is "why the results obtained here for the 135° case with "real" data are worst than those obtained in the validation procedure for the same bottom configuration?" Since it is easy to see that the BACKTRACK results in a sloping bottom that are employed to build matrix A used here are similar to those that can be obtained from the Snell law and which were employed in matrix A used in the validation, it is not difficult to hint that the differences in the results can be a consequence of using SWAN results at the "measurement" points. However, Figure 6 shows that what really matters is the w_d value employed in the smoothness constrain. In fact, the estimate at the validation case using $w_d = 1$ (S0 – validation) does not differ much from the one obtained here (S0 – test case).

Conversely, the $w_d = 1$ value is a consequence of using SWAN data since w_d values lower than 1, which produced better estimates in the validation, cannot be used here. Otherwise worst estimates would be produced. In conclusion, the use of SWAN results induces the choice of w_d which, in turn influences the offshore estimate.

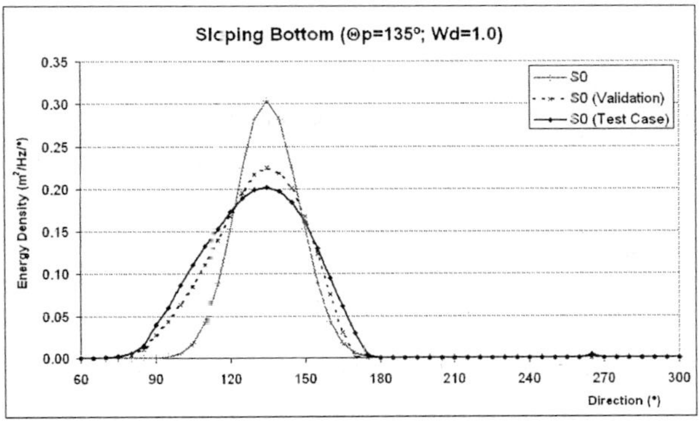

Figure 6. Directional distribution of the reference spectrum (S0) and estimated spectra of the Sloping Bottom case, for the frequency of 0.08 Hz, computed for the validation case (S0 (Validation)) and for the test case (S0 (Test Case)).

Headland

For the Headland case, the same offshore spectra with peak directions of 180° and of 135° were set at the SWAN model grid boundary. For both test cases the estimated directional distributions agree quite well with the corresponding SWAN boundary conditions in terms of shape, peak position and area of the estimated curves (Figure 7 and Table 2).

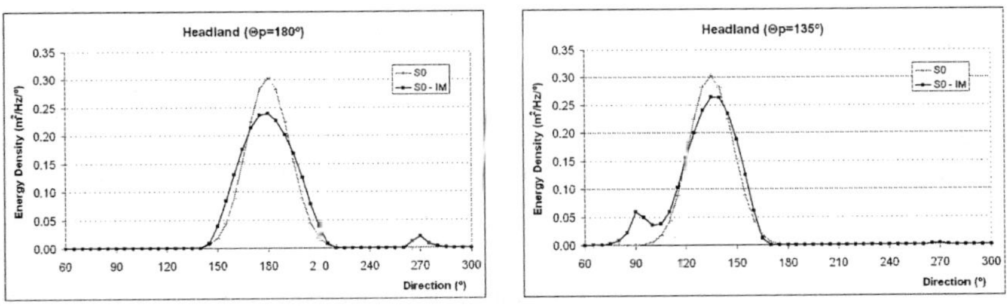

Figure 7. Directional distribution of the reference spectrum (S0) and estimated spectrum (S0-IM) of the Headland case, for the frequency of 0.08 Hz. A) Case with offshore peak direction of 180°; b) Case with offshore peak direction of 135°.

Although one of the "measurement" points can be sheltered by the headland, depending on the direction of the incident sea state, Figure 3, the same does not occur for the other point. Apart the 180° (the direction of the headland), one "measurement" point will always be directly reached by waves and, thus, a stronger measurement signal at that point will be obtained. This feature seems to be the reason for better results in this case than in the Bay case.

Bay

The proposed inverse methodology results for the Bay case are presented on Figure 8, which has the directional distribution imposed on the offshore SWAN boundary and the inverse methodology estimate.

The estimation for the 180° case was good, as in the other bottom configurations. However, the inverse methodology results for the 135° case present the major differences relative to the SWAN offshore boundary conditions, as can be seen on Figure 8 b) and Table 2. The estimated peak period, however, was 135°.

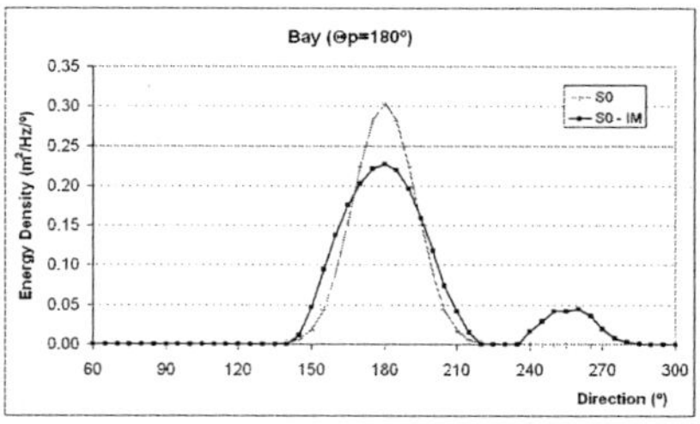

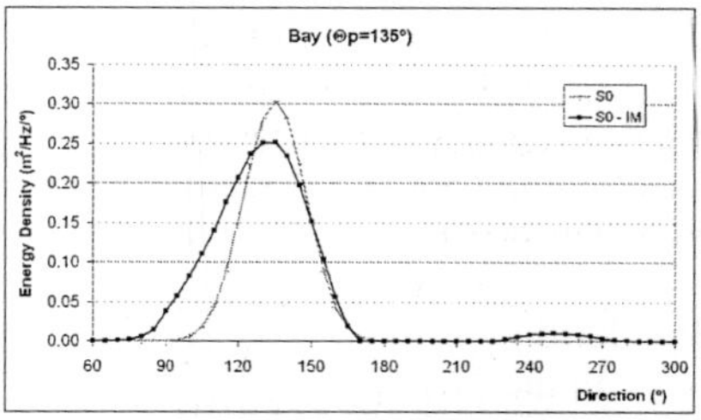

Figure 8. Directional distribution of the reference spectrum (S0) and estimated spectrum (S0-IM) of the Bay case, for the frequency of 0.08 Hz. A) Case with offshore peak direction of 180°; b) Case with offshore peak direction of 135°.

The areas under the directional distributions for the Bay cases are indicated on Table 2 stressing the good estimation for the 180° case and the difference of 24% (the worst adjustment) for the 135° case.

CONCLUSIONS

This paper discussed a methodology to estimate the offshore directional distribution energy of sea waves, for a given frequency that uses shallow water spectral moments computed at measurement sites.

The data kernel matrix is obtained by using linear refraction and by considering a discrete offshore directional distribution. The overdetermination of this problem is reached by imposing a smoothness constrain and the inverse problem is solved by the Non Negative Least Squares (NNLS) algorithm.

The implementation of the inverse methodology was validated with a problem of sea waves propagating over a sloping bottom, situation in which is trivial the use of the Snell's law. For each of the three wave propagation directions (normal to the coast, 45° to the coast and bimodal) the estimate of the offshore directional distribution was considered close to the expected solution, i.e. the imposed offshore directional distribution. It was also observed that the results are sensitive to the smoothness constrain and this value is, apparently, independent of the dominant offshore direction estimation.

This methodology was also applied to sea states propagating over three idealized bottom configurations: a Sloping Bottom, a Headland and a Bay. For these cases, the data kernel matrix was built using the results of a backward ray tracing model and the wave spectral moments at shallow water were computed using the SWAN model.

In all test cases, the good agreement between the offshore directional curve shapes, peak directions and curve area, for one given frequency, shows the capabilities of the inverse methodology in this kind of problems. In addition, the application of this technique on controlled test cases showed the influence of the smoothness constrain on the solution quality.

ACKNOWLEDGEMENTS

Alexandre Coli would like to acknowledge the financial support of the Portuguese Foundation for Science and Technology (FCT/MCTES), through the grant SFRH/BD/12488/2003.

REFERENCES

Bennett, A.F. 1992. *Inverse Methods in Physical Oceanography,* Cambridge University Press, Cambridge, 356 pp.

Booij, N., R.C. Ris, and L.H. Holtthuijsen. 1999. A Third-generation Wave Model for Coastal Regions, Part I, Model Description and Validation, *Journal of Geophysical Research*, 104 (C4): 7649-7666.

Coli, A.B., J.A. Santos, C.J. Fortes, R. Capitão and M.M. Carvalho. 2002. Extreme Wave Regime Propagation to Douro's Mouth: Analysis by the Numerical Models BACKTRACK-REFSPEC and SWAN, *Proceedings of Littoral 2002*, Oporto, 185-196.

Lawson, C.L. and R.J. Hanson. 1974. *Solving Least Squares Problems*, Prentice-Hall, New Jersey, 340 pp.

Longuet-Higgins, M.S. 1957. On the Transformation of a Continuous Spectrum by Refraction, *Proceedings of the Cambridge Philosophical Society*, 53(1): 226-229.

McLaughlin, D. 2002. An Integrated Approach to Hydrologic Data Assimilation: Interpolation, Smoothing, and Filtering. *Advances in Water Resources*, 25 (8-12), 1275-1286.

Menke, W. 1984. *Geophysical Data Analysis: Discrete Inverse Theory*, Academic Press, San Diego, 289 pp.

O'Reilly, W.C. and R.T. Guza. 1998. Assimilating Coastal Wave Observations in Regional Swell Predictions. Part I: Inverse Methods, *Journal of Physical Oceanography*, 28: 679-691.

Talagrand, O. 1997. Assimilation of Observations. An Introduction. *Journal of Meteorological Society of Japan*, 75 (1B): 191-209.

WAVE CELERITY ESTIMATION AND DEPTH INVERSION USING LINEAR FEATURE EXTRACTION FROM NEARSHORE VIDEO IMAGES

Jeseon Yoo [1], Hermann Fritz [1], Kevin Haas [1], Paul Work [1], and Christopher Barnes [2]

A remote sensing technique was developed to measure wave celerity in the nearshore and inversely estimate water depth using linear feature extraction from video imagery. Linear feature extraction of wave crest signatures involves image frame differencing and post-processing techniques to remove persistent foam in the wave breaking zone. Each wave crest track extracted from the processed image cross-shore timestack is used to calculate wave celerity and obtain the wave breaking point. For the nonlinear depth inversion, wave height and water depth at the breaking point of each wave is estimated by associating the nonlinear shallow water equation based phase speed with wave breaker formulas. In addition, the amplitude dispersion effects after the wave breaking are computed from Dally's wave dissipation model, and subtracted from the celerity profiles to calculate water depths inversely. The results of the determined beach profiles are compared to in-situ field measurements.

INTRODUCTION

In order to retrieve wave phase speeds and water depths from the remotely collected wave signals in the form of image intensity, mainly four different types of wave image signal analysis approaches have been developed: three dimensional Fourier Fast Transform (3-D FFT: Dugan et al., 2001), cross-shore image timestack in space–time domain (Stockdon and Holman, 2000), image time exposure (Aarninkhof et al., 2005), and sparse time-lapse imaging (Dalrymple et al., 1998) methods. Earlier works typically utilize image pixel intensity as a proxy of water surface elevation to compute depth induced wave properties, since planar remote sensing techniques have limitations to measure the water surface level without stereo-optics. The previous depth inversion techniques show good performances to estimate water depths in the intermediate water up to wave breaking lines, although most of them use the linear dispersion relation to compute water depth using remotely measured wave properties. However, those applications to the wave breaking region yield significant errors in depth inversion modeling compared to the intermediate water depth. The main errors are primarily originating from two sources: poor quality of wave image intensity signal and non-inclusion of nonlinear amplitude effects. The quality of

[1] Civil & Environmental Engineering, Georgia Tech - Savannah, 210 Technology Circle, Savannah, GA, 31407. U.S.A., gtg435p@mail.gatech.edu, fritz@gatech.edu, kevin.haas@gtsav,gatech.edu, paul.work@gtsav.gatech.edu.
[2] Electrical & Computer Engineering, Georgia Tech - Savannah, 210 Technology Circle, Savannah, GA, 31407. U.S.A., chris.barnes@gtsav.gatech.edu.

pixel intensity time series in the breaking zone deteriorates due to foam caused by wave breaking. Consequently, noise due to foam in the breaking zone contributes to the errors of wave parameter estimation by video remote sensing. Inclusion of nonlinear amplitude effects also can not be considered in the depth inversion modeling due to incapability of water surface elevation measurements by remote sensing platforms. However, a number of previous studies have shown that the shallow water linear dispersion relation underestimates the wave speed within the surf zone because of finite amplitude effects (Suhayda and Pettigrew, 1977; Thornton and Guza, 1982; Holland, 2001).

Thus, the purpose of this study is to improve estimation of the surf zone bathymetry using video wave image enabled wave signature extraction, involving development of image processing techniques to minimize the influence of the persistent foam on extracting wave signals from video wave images. Also, this study aims to include the nonlinear effects in the depth inversion modeling by extracting wave height information from video images.

FIELD DATA COLLECTION

Nearshore video wave images were recorded at the NTSC standard of 30 Hz at Myrtle Beach, SC, on Dec. 11, 2003 by Haas et al. (2004). Also, simultaneous measurements of cross-shore beach profiles and water surface elevations were conducted for the verification of nonlinear depth inversion. An upward-looking current profiler (Nortek Aquadopp) used for determining the mean water level measured water pressure, flow velocities at 1 Hz data sampling rate. The beach profile was surveyed by using a Total Station (Sokkia SET610) and 2 sets of single prisms on poles. Video image sequences were down-sampled to 3 Hz.

LINEAR FEATURE EXTRACTION

The nearshore wave imagery treated with the present image processing technique have strongly reformed intensity signals, even though wave crest signatures moving in the surf are obviously discernable. The processed image intensity signals are not adequate to be analyzed in the frequency domain, since the processed intensity signals corresponding to original wave crests are enhanced by this image processing technique and loose their natural intensity pattern of the original wave signals. Thus, this study analyzes the wave imagery preferably in the time domain by tracking individual wave crest signatures from the processed surf zone imagery.

Image Frame Differencing

Image frame differencing techniques to subtract the initial non-moving background from image sequence have been applied to various research fields such as visualization in robotics, automated road traffic evaluation, etc. In the video based coastal research, Lippmann and Holman (1989) introduced frame differencing to quantify sand bar morphology in the nearshore by subtracting image frames separated by a time interval (0.5 ~ 1.0 seconds) to eliminate persistent signals attributed to residual foam accumulation.

As applied and shown by Lippmann and Holman, the image frame differencing technique is an effective method to reduce or remove effects of drifting foam in the breaking zone on wave signals. Herein, the frame differencing is conducted by subtracting from a target image the average of 2 to 3 successive images in its neighborhood. Consequently, it removes both the non-moving background and the foam patches more clearly than the case of Lippmann and Holman's application, retaining moving features in the image frame. Since foam remains behind in terms of wave propagation while a wave crest tumbled down in the surf zone leads in the front, the method of backward differencing, instead of central or forward differencing, is applied:

$$I_i^{new}(x,y) = I_i(x,y) - \frac{1}{3}\sum_{n=i-2}^{n=i} I_n(x,y) + K \tag{1}$$

where I_i = pixel intensity of ith image; I_i^{new} = pixel intensity of ith image after frame differencing; *(x,y)* = pixel coordinates in an image; K = constant for intensity level adjustment (K = 127 selected in this study). After application of the frame differencing, average filtering is applied to the frame differenced images using three different sizes of the sub-windows (horizontally × vertically: 15×1, 9×1, and 1×3 pixels) in order to smooth out the residual spot noise.

Thereafter, a modified version of the rectification method by Holland et al. (1997) is applied to rectify video frame coordinates to the real world coordinates. In addition, the present video technique involves collecting time-series of pixel intensity at a pixel array taken from the rectified imagery as 1-D slice along cross-shore, yielding a cross-shore timestack (space-time image). This cross-shore timestack illustrates the variations of pixel intensity in time at each cross-shore location. Herein, a set of two cross-shore timestacks is generated from both the rectified raw and processed image sequences. Fig. 1 shows segmental cross-shore timestacks collected from Myrtle video captured on Dec. 11, 2003. Fig. 1(a) is a cross-shore timestack of raw image sequence, and Fig. 1(b) is its pairing timestack of processed image sequence. The processed imagery timestack is used to extract individual wave crest tracks. The raw imagery timestack is used for estimating wave periods and detecting wave breaking points through association with the extracted wave crest features.

Extraction of Wave Crest Tracks

Previously, Catalan (2005) generated mean removed cross-shore timestacks from video imagery captured in a 1-D wave flume. He extracted wave front trajectories from a binary image of the mean removed timestack using a zero-up crossing based edge detection method. He used only regular wave conditions in which patterns of wave front trajectories are uniform. He could easily isolate individual waves, since an incoming wave was not under influence of persistent foam created by the previous wave. In the nearshore field, the tracks of wave crests in the raw imagery timestack (Fig. 1(a)) is hardly distinguished due to their associations with the white foam patches generated by the precedent waves, whereas the processed imagery timestack (Fig. 1(b)) clearly illustrates tracks of wave crests approaching the shore line. The wave crest tracks can be detected and extracted by tracking pixels of highest intensity values within a movable interrogation sub-window of 2×2 pixel size on the processed imagery timestack in the wave propagation direction.

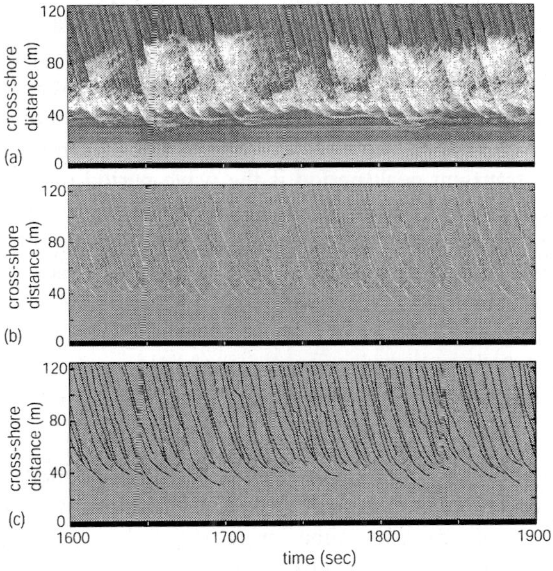

Figure 1. Segmental cross-shore timestacks from the Myrtle video captured on Dec. 11, 2003: (a) raw image timestack; (b) processed image timestack; (c) extracted wave crest features.

WAVE PROPERTY ESTIMATION

Wave Phase Speed

Wave phase speed measured by remote sensing techniques has been used as a key parameter to determine the nearshore bathymetry since World War II. In general, water level data in time-series collected at different locations are required to calculate wave celerity (c). Several ways to retrieve wave phase speed from the wave level data has been suggested for the field measurements (Suhayda and Pettigrew, 1977; Lippmann and Holman, 1991; Stockdon and Holman, 2000; Holland, 2001; Catalan, 2005). In the present study, local wave phase speed is calculated by taking the first derivative at each location of a wave trajectory extracted in the space (y: cross-shore direction) – time (t) domain.

Wave Period

Dominant wave periods can be obtained by applying a spectral analysis to water surface elevation data. The implicit assumption in the spectral analysis is that the wave field is stationary, not having temporal change of its statistical properties. However, in the random wave field, the peak period from spectral analysis can not be used as sole wave period for each wave crest. On the other hand, in the field, transitory groups of incident waves exist and are observed in the nearhsore, having the time variation of the wave breaking locations. The raw imagery cross-shore timestack in Fig. 1(a) also shows different locations of wave breaking in groups such that some wave groups break early far from shore but other groups break late on shore. Liu (2000) adopted a wavelet analysis to investigate the time localized characteristics of wind waves in intermittent groups.

Herein, similarly, a wavelet analysis adopting the Morlet wavelet as a mother wavelet is applied to pixel intensity time-series taken at each location in the raw imagery cross-shore timestack to obtain transient peak periods T_t. The definition of the transient peak period is the wave period of the peak energy at each time in the wavelet spectrum domain. The transient peak periods computed from the raw imagery timestack range approximately 3 ~ 12 seconds in the offshore region. The period profile of each wave crest track along the cross-shore shows consistent values up to the wave breaking line, meaning that each wave maintains its own wave period during its propagation to the shoreline. So, the wave period for a wave crest track is selected as the mean value of the transient peak period profile in the offshore.

Wavelength

The spatial wavelength along cross-shore can be obtained by applying the frequency domain complex empirical orthogonal function (CEOF) analysis to a cross-shore array of intensity time-series (Stockdon and Holman, 2000). However, either the raw or the processed pixel intensity data in the wave breaking region can not be input to the CEOF analysis, since the data lost shapes of natural waves. Instead, the wave length L is calculated as $L = C \times T$.

DEPTH INVERSION MODEL

Waves approaching the shore become highly nonlinear in the surf zone, and their phase speeds are strongly influenced by nonlinearity of the wave amplitudes. However, previous works for depth inversion using remotely collected wave image data (Stockdon and Holman 2000; Dugan et al., 2001) adopted the linear dispersion relation, since actual water level data is not available by the remote sensing platforms. In order to perform nonlinear depth inversion in the shallow water, Catalan (2005) sampled a hybrid data set composed of wave period and phase speed data retrieved from wave imagery and water surface level time-series generated by the combined refraction/ diffraction model (REF/DIF1). The model requires input of measured wave amplitudes at the offshore boundary which can not be provided by remote sensing.

The present depth inversion algorithm includes the nonlinear amplitude dispersion effects by detecting wave breaking points and retrieving wave amplitudes at the breaker points from nearshore video imagery without any help of in-situ measured data input for boundary conditions. Also, the algorithm includes a wave dissipation model for computing wave heights dissipated after breaking and subtracting their contributions from the shallow water wave theory based celerity. The algorithm produces beach profiles from the measured wave phase speed profiles by subtracting contributions of wave heights from the shallow water wave theory based celerity.

Detection of Wave Breaking Points

During wave shoaling, wave form asymmetry becomes greatest tending to exhibit saw-tooth shapes. At the moment of wave breaking, the front face of waves is commonly overturned, showing dark shadow in the front face. The dark shadow is observable in the raw imagery cross-shore timestack.

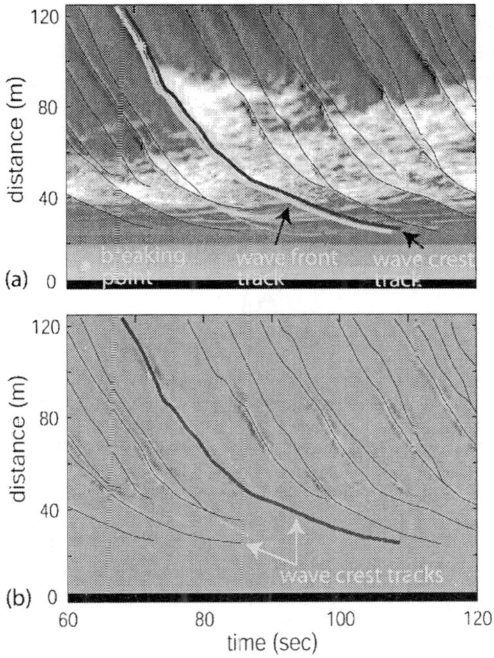

Figure 2. (a) Individual wave crest tracks and a wave front face track superimposed on top of a raw imagery timestack; (b) the crest tracks extracted from the processed imagery timestack.

The trajectory of each wave crest is obtained by extracting each wave crest track from the processed imagery cross-shore timestack as mentioned earlier. With regard to extracting the trajectory of the front face of each wave, the front face leads the crest at each location, forming a trajectory almost identical to the crest trajectory. Thus, herein, the trajectory of the front face of individual waves is taken as the same form to the crest trajectory, being shifted to the past side of the crest trajectory event (at time t) as much as $t - 2 \cdot \Delta t$ in the time axis in the cross-shore timestack. Fig. 2(a) shows a pair of extracted front face and crest trajectories (the two adjacent trajectories).

In order to obtain the pixel intensity profile of each wave crest, individual wave crest trajectories extracted from the processed imagery timestack are superimposed on top of the raw imagery timestack as shown in Fig. 2(a). Thereafter, the intensity profile of each wave crest is obtained by following its trajectory in the raw imagery cross-shore timestack. Similarly, the intensity profile of the wave front face can be obtained. Fig. 3(a) shows the intensity profile of a wave crest obtained by tracing the intensity of the crest trajectory of the pair in Fig. 2(a). The Fig. 3(b) shows the intensity profile of the wave front face obtained by tracing the intensity of the front face trajectory. The breaking point of each wave is determined as the location at the lowest value of the intensity profile of the wave front face as shown in Fig. 3(b), since the wave front face is commonly darkest at the breaking moment, particularly in plunging type breaker observed in this study.

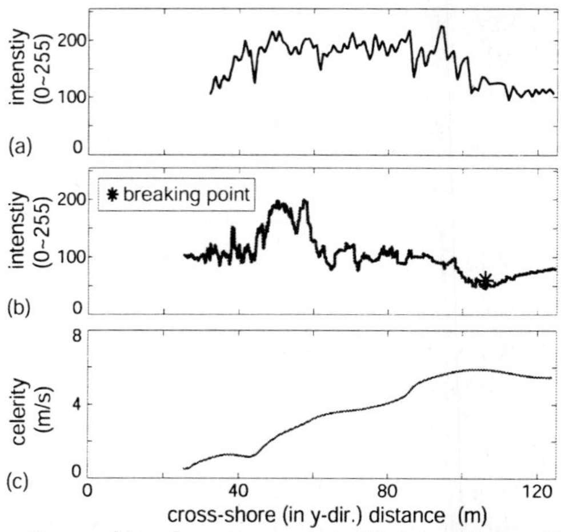

Figure 3. (a) Intensity profile of a wave crest; (b) intensity profile of the wave front face; (c) the profile of wave phase speed.

Wave Breaker Model

The wave breaker model is composed of a wave phase speed equation based on a nonlinear shallow water theory and a wave breaker formula. The breaker model calculates wave amplitudes and water depths at wave breaker points using the two equations along with input data of wave phase speed values, wave periods and wavelengths at the breaker points retrieved from video imagery.

In order to fulfill the amplitude effects on wave propagation speed in shallow water, several nonlinear shallow wave theories have been suggested (McCowan, 1891; Thornton and Guza, 1982; Holland, 2001). In the present study, the wave phase speed equation obtained from the characteristic form of the nonlinear shallow water equation is used:

$$c = \sqrt{g(h+\eta)} \tag{2}$$

where g is gravitational acceleration, and h is water depth, and η is the water surface elevation. Since the instantaneous water level is consistent with a wave-by-wave approach.

Most of the breaker formulas have been established empirically or semi-empirically through calibration from laboratory data. In the present study, Miche (1944)'s breaker formula is adopted as the primary index to calculate wave heights and water depths at breaking points along with the nonlinear shallow water equation based wave phase speed equation, since this breaker formula requires only local wave parameters which are provided from video imagery. Four different types of breaker formula are also tested to investigate the sensitivity of the depth inversion results to the breaker types.

The phase speed equation in Eq. (2) includes the wave crest elevation rather than the wave height (H). On the other hand, all of the breaker formulas are expressed with the wave height instead of the wave crest elevation (η_c). Hansen (1990) suggested an empirical formula for the ratio of η_c/H at breaking points from experimental data of periodic waves:

$$\left(\frac{\eta_c}{H}\right)_B = 1 - 0.5\tanh\left(\frac{4.85}{\sqrt{U_B}}\right) \tag{3}$$

where $U_B = 10.1 \cdot m^{0.20} \cdot (H_o / L_o)^{-1}$ (m: local beach slope, and the subscript o: offshore conditions), and the subscript B indicates breaking points.

Wave Dissipation Model

The wave dissipation model estimates dissipated wave heights and water depths starting from the breaker points by associating a wave dissipation model with the nonlinear shallow water theory based wave phase speed.

For quantifying the wave decay for breaking random waves in the breaking zone, two main approaches have been proposed: turbulent bore energy dissipation based probabilistic approach and stable energy approach. Herein, the second approach developed by Dally et al. (1985) is applied:

$$\frac{\partial\left(H^2 h^{1/2}\right)}{\partial y} = -\frac{K}{h}\left[H^2 h^{1/2} - \Gamma^2 h^{5/2}\right] \tag{4}$$

where K is a dimensionless decay coefficient, and Γ is a dimensionless stable wave factor. Dally et el. determined best fit values for K and Γ for regular waves from small scale laboratory data, and found that fixed values of $K = 0.15$ and $\Gamma = 0.40$ represent robustly all the data from different beach slopes. Dally (1992) applied Eq. (4) to random breaking waves in the nearshore field with the same fixed values of K and Γ by a wave-by-wave approach, and verified that the modeled height statistics were comparable to the field measurements. The wave dissipation algorithm is schemed by a finite backward stepping method.

At the initiation of wave breaking, the ratio of η_c/H is observed to exceed the value of 0.85. However, the ratio of η_c/H starting from the breaker points tends to approach the symmetric wave shape value of 0.5 (Hansen, 1990). Hansen (1990) proposed the ratio of η_c/H as a parabolic variation form after wave breaking by calibrating laboratory data of various beach slopes:

$$\frac{\eta_c}{H} = 0.5 + \left[\left(\frac{\eta_c}{H}\right)_B - 0.5\right]\left(\frac{h}{h_B}\right)^2 \tag{5}$$

Beach Profile Iteration

The term U_B in Eqs. (3) and (5) for the ratio η_c/H at the breaker points is dependent of local beach slope. Also, other wave breaker formulas require local beach slopes to determine breaking wave heights. However, the beach slope is not known before executing the depth inversion algorithm. Herein, initial value of the ratio η_c/H at the breaker points is set to be 0.75, since the range of η_c/H at breaking is found as 0.7 ~ 0.8 (Hansen, 1990). In addition, the Miche (1944)'s breaker formula which requires only local wave parameters is employed.

For testing different types of wave breaker formulas which require local beach slopes, the Miche's breaker formula is used at the first run to obtain and provide initial beach profile for the second run from which the selected different type of wave breaker formula is applied. The initial beach profile estimated through the first run offers local beach slopes to the term U_B and the different types of breaker formula, enabling the beach profile to be recalculated through the next run. In this way, the beach profile is updated by iteration until a certain residual limit. Herein, the residual is defined as:

$$R^k = \sqrt{\sum_{i=1}^{N}\left(h_i^k - h_i^{k-1}\right)^2} \tag{6}$$

where k indicates k^{th} iteration, and i is location of the depth estimated in y-axis, and N is the final number of i. The limit for the residual is selected as 10^{-3} m.

RESULTS AND DISCUSSION

Breaking Wave Heights and Water Depths

From the Myrtle video data collected on Dec. 11, 2003, 6,138 raw images were captured at 3 Hz frequency during a 34.1 minute time-span and processed in order to extract individual wave crest trajectories. A total of 447 individual wave crest tracks were extracted from the processed imagery cross-shore timestack. The wave crests with bad tracks were removed by limiting the celerity change rate (Δc) per Δt (= 1s) to less than 0.7 m/s. In addition, the celerity profiles which deviated beyond upper and lower boundaries of 95% confidence interval of the locally averaged celerity profile were removed. Retained 198 wave celerity profiles (crest tracks) were used to detect breaking points and estimate wave heights and water depths at the breaking points.

Largely two groups of breaking points were detected by utilizing trajectories of both the wave crest and the wave front face as well as the raw imagery cross-shore timestack: bar crest area (outer surf breaking group) and steep slope area close to the shoreline (inner surf breaking group) as shown in Fig. 4(c). In order to investigate nonlinear effects of wave amplitude dispersion on wave celerity in the surf, all of individual celerity profiles in each group were locally averaged. Fig. 4(a) shows that the nonlinear effect of the outer surf breaking group is greater than that of the inner surf breaking group in the outer surf zone where wave heights of the outer surf breaking group are expected to be greater than the other group, while vice versa in the inner surf zone. Both the outer surf breaking and the inner surf breaking celerity profiles show significant nonlinearity caused by the wave amplitude dispersion effects, compared to the linearly converted celerity profile from the surveyed bathymetry.

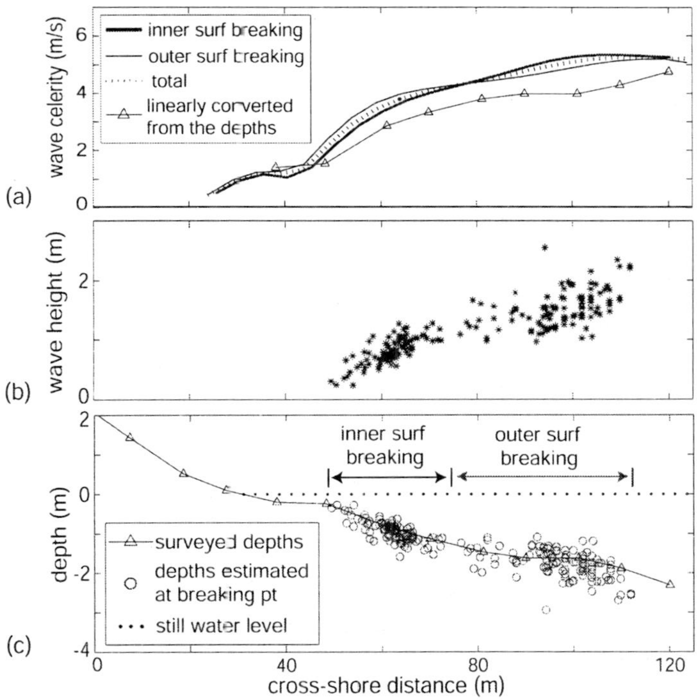

Figure 4. (a) Nonlinear amplitude dispersion effects on wave celerity profiles; (b) Breaking wave heights; (c) water depths estimated using Miche's breaker formula.

Results of wave heights and depths at the breaking points computed by using the Miche (1944)'s breaker formula and Eq. (2) are shown in Fig. 4. By introducing the nonlinear shallow wave and the breaker formula, the nonlinear effects on wave celerity could be subtracted to estimate water depths at breaking points. In general, the estimated water depths have good agreements with the measured depths, even though they have a tendency to be slightly over-estimated.

Bathymetry Inverted by the Wave Dissipation Model

Starting from each breaking point and using its breaking wave height and water depth estimated by Miche (1944)'s breaker formula, the dissipated wave height and water depth were computed by Dally's wave dissipation model and Eq. (2) from its celerity profile. Upper panel in Fig. 5(a) shows two examples of beach profiles derived by the wave dissipation model. All of beach profiles estimated from celerity profiles were locally averaged as shown in Fig. 5(b). The result of locally averaged beach profile is significantly improved, compared to the result by the linear depth inversion. The absolute normalized relative error of the depths by the wave dissipation model is around 11.8 %, whereas the linear depth inversion resulted in an error of 49.3 %.

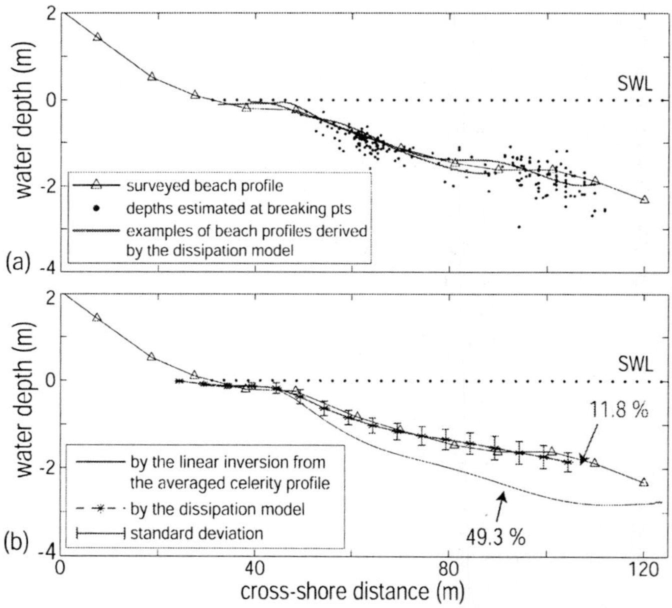

Figure 5. Bathymetry estimated by Miche's breaker formula and Dally's wave dissipation model.

In order to investigate the sensitivity of the present depth inversion techniques to different types of breaker formula, four different types other than the Miche (1944)'s breaker formula were tested: Goda (1970), Sunamura and Horikawa (1974), Kampius (1991), and Hansen (1990). Those all four breaker formulas were found in a breaker formula related study by Rattanapitikon et al. (2003). For the initial guess of beach profile required for local beach slopes and/or offshore condition parameters in the four breaker formulas, the Miche's breaker formula was used at the first run. The four formulas replaced the

Miche's formula after the first run. The inverted beach profiles by the four formulas showed similar results as shown in Fig. 6, indicating that the present depth inversion technique is not sensitive to types of breaker formula.

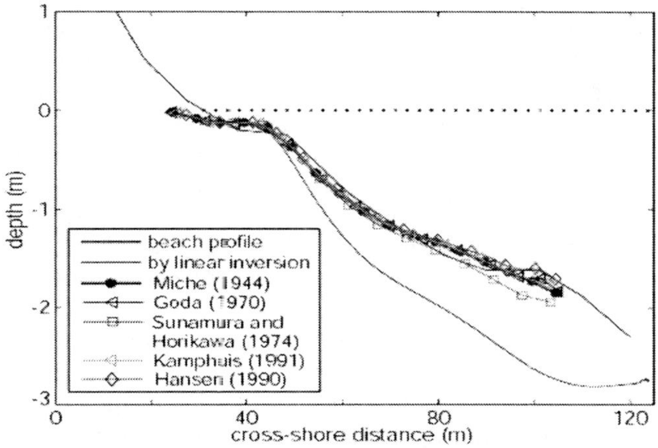

Figure 6. Bathymetry estimated using four different types of breaker formula.

CONCLUSION

Video based water depth estimates in the surf zone are limited by foam due to wave breaking and nonlinear amplitude effects. In the present study, individual wave crest trajectories were detected and extracted successfully from nearshore wave images by means of image frame differencing and image post-processing. Further, extracted wave crest tracks were used to calculate wave celerity and obtain wave breaking points. For the nonlinear depth inversion, water depths were calculated from the measured celerity profiles by applying the five wave breaker formulas and Dally's wave dissipation model. The results of the determined beach profiles are compared to in situ field measurements.

ACKNOWLEDGMENTS

This research was primarily funded by the U.S. National Geospatial-Intelligence Agency. Also, the authors would like to thank the U.S. Geological Survey, the South Carolina Sea Grant and the National Oceanographic Partnership Program for their financial support during the data collection.

REFERENCES

Aarninkhof, S.G.J., B.G. Ruessink, and J.A. Roelvink. 2005, Nearshore subtidal bathymetry from time-exposure video images, *J. Geophys. Res.*, 110, C06011, doi:10.1029/2004JC002791.

Catalan, P.A. 2005. Hybrid Approach to Estimating Nearshore Bathymetry using Remote Sensing, *M.S. Thesis, Ocean Eng., Oregon State Univ.*, US.

Dally, W.R. 1992. Random breaking waves: Field verification of a wave-by-wave algorithm for engineering application, *Coast. Eng.*, 16, 369-397.

Dally, W. R., R.G. Dean, and , R.A. Dalrymple. 1985. Wave height variation across beach of arbitrary profile, *J. Geophys. Res.,* 90, C6, 11917-11927.

Dalrymple, R.A., A.B. Kennedy, J.T. Kirby, and Q. Chen. 1998. Depth inversion for nonlinear waves shoaling over a barred-beach, *in Proc. 26th Inter. Conf. on Coast. Eng.*, 2395-2408.

Dugan, J.P., C.C. Piotrowski & J.Z. Williams. 2001. Water depth and surface current retrievals from airborne optical measurements of surface gravity wave dispersion, *J. Geophys. Res.*, 106(C8), 16903–16915.

Haas, K., H. Demir, P. Work, G. Voulgaris, and S. Obley. 2004. Myrtle Beach Nearshore Experiment, Dec 10 to Dec 15th, 2003, Part II: Morphodynamic & Remote Imagery Measurements, *Technical Report, School of Civil and Envi. Eng., Georgia Tech – Savannah*, US.

Hansen, J.B. 1990. Periodic waves in the surf zone: Analysis of experimental data, *Coast. Eng.*, 14, 14-14.

Holland, T.K. 2001. Application of the linear dispersion relation with respect to depth inversion and remotely sensed imagery, *IEEE Trans. on Geos. and Rem. Sens.*, 39 (11), 2060-2071.

Holland, K.T., R.A. Holman, and T.C. Lippmann. 1997. Practical use of video imagery in neareshore oceanographic field studies, *IEEE Journal of Oceanic Engineering*, 22(1), 81-92.

Lippmann, T.C., and R.A. Holman. 1989. Quantification of sand bar morphology: a video technique based on wave dissipation. *J. Geophys. Res.*, 94, 995–1011.

Lippmann, T.C., and R.A. Holman. 1991. Phase speed and angle of breaking waves measured with video techniques, *Coastal Sediments ' 91*, 542-556.

Liu, P.C. 2000. Wave grouping characteristics in nearshore Great Lakes, *Ocean Eng.*, 27, 1221–1230.

McCowan, J. 1891. On the solitary wave, *Philosophical Magazine*, 32, 45-58.

Miche, R. 1944. Mouvements ondulatoires de la mer, *Ann. des Ponts et Chaussees*, Chap. 114, 131-164, 270-292, and 369-406.

Rattanapitikon,W., T. Vivattanasirisak, and T. Shibayama. 2003. A proposal of new breaker height formula, *Coast. Eng. J.*, 45, 1, 29-48.

Stockdon, H.F. & R.A. Holman. 2000. Estimation of wave phase speed and nearshore bathymetry from video imagery, *J. Geophys. Res.*, 105, 22015–22033.

Suhayda, J.N., and N.R. Pettigrew. 1977. Observations of wave height and wave celerity in the surf zone, *J. Geophys. Res.*, 82 (9), 1419-1424.

Thornton, E.B., and R.T. Guza. 1982. Energy saturation and phase speeds measured on a natural beach, *J. Geophys. Res.*, 87 (C12), 9499-9508.

BATHYMETRIC ESTIMATION BASED ON WAVE REFRACTION PATTERNS

Kristen D. Splinter[1] and Robert A. Holman[1]

As waves enter and travel through the nearshore region, they refract and shoal as a response to gradients in the bathymetry. Many algorithms using the dispersion relationship have been developed to exploit this shoaling variation in the magnitude of the wave number, k, to allow remote sensing estimation of the underlying bathymetry. However, these methods typically ignore the refractive turning of waves as an exploitable signal. A two-step algorithm is developed to estimate bathymetry gradients from gradients in directional wave information. Local wavenumber and angle are extracted from images of surface waves using Hilbert Transform techniques. Bathymetry gradients are calculated from an augmented form of the refraction equation. Synthetic testing of the model shows that under ideal conditions, the model accurately determines two-dimensional bathymetry at wavelength scales with errors on the order of 5% of the true depth. Imagery from field data collections provides challenges for accurate data extraction of wavenumber and wave direction fields. New methods for preprocessing images using standard deviation and wavenumber filters are employed. Data extraction methods using wavelength scale least-squares fits to wrapped phase are developed and tested and show promise at this stage of the work.

INTRODUCTION

Motivation

As waves approach the shore, they refract and shoal as a response to gradients in the underlying bathymetry. As waves break, momentum is transferred to mean flows, driving sediment transport and changes in the beach profile. This inter-dependence, wherein the waves influence the bathymetry and the bathymetry influences wave propagation and breaking patterns is of interest to coastal scientists.

In order to successfully use nearshore models to study this dynamic region, accurate knowledge of the bathymetry is necessary. The surf zone, however, is a very dynamic area, exposed to large wave forces and daily tidal variation that make it hard to monitor the changing bathymetry with in-situ instruments. Recently, there has been a move to remote sensing, utilizing cameras and satellites as an alternative way to monitor the nearshore on both short and long-term timescales. Many of these methods rely on the depth dependence of wave propagation (the dispersion relationship) to estimate bathymetry from observed propagation characteristics. It is also recognized, however, that waves change direction due to refraction over bathymetry gradients, often producing patterns that are distinguishable in a single snapshot. Here we develop a refraction-based

---

[1] College of Oceanic and Atmospheric Sciences, Oregon State University, Corvallis, OR 97330, USA

model to exploit this directional change information for estimation of underlying bathymetry from a single aerial photograph that includes strong wave refraction patterns.

Previous remote sensing methods

Ideally, we would prefer to have the capacity to survey every beach of interest using high-resolution GPS techniques. This, however, is very costly and time consuming. Alternatively, LIDAR and the use of hyper-spectral techniques have been successful in directly mapping nearshore bathymetry under certain conditions. As well, methods have been developed for the estimation of bathymetry from surface wave propagation patterns based on the dispersion relationship, relating the magnitude of the wave number (k) and wave period (T) to depth:

$$h = \frac{\tanh^{-1}\left(4\pi^2 / gkT^2\right)}{k} \tag{1}$$

(Dalrymple, et al. 1998); (Stockdon and Holman 2000); (Misra, et al. 2003). Although all these methods provide reasonable estimates of nearshore bathymetry, they are not always practical. For instance, dispersion-based methods require a time series of image data to estimate wave period. Time series data may not be available, especially from satellites, so a method that requires only single frames is attractive.

Current method

Our method exploits the relationship between the gradients in directional wave information and the gradients in local bathymetry. The authors acknowledge that this is not meant as a replacement to the other methods discussed, but as a substitute when they are not available due to environmental conditions or lack of timeseries data.

The objectives of this research are to:

1. Develop a method for bathymetry estimation based on gradients in the local directional wavenumber.
2. Develop a method to extract the required directional wave information from optical imagery.

Our method is limited to images with clean swell and can only resolve bathymetric features with scales greater than the local wavelength.

METHODS

Refraction-based method for determining bathymetry gradients

Depth gradients are explicitly solved for using the refraction equation (Eqn. 2). For numerical stability, depth gradients are only solved for in the cross-shore direction. This is done with the use of the chain rule for the cross-shore gradient in wavenumber as given in Eqn. 3.

$$\nabla \times \vec{k} = \frac{\partial(k\sin\theta)}{\partial x} - \frac{\partial(k\cos\theta)}{dy} = 0 \tag{2}$$

$$\frac{\partial k}{\partial x} = \frac{\partial k}{\partial h}\frac{\partial h}{\partial x} \tag{3}$$

Substituting Eqn. 3 into Eqn. 2 and rearranging, we have a relationship that describes the cross-shore gradients in depth as a function of local wavenumber and wave angle and their spatial gradients (Eqn. 4).

$$\frac{\partial h}{\partial x} = \frac{k\frac{\partial h}{\partial k}\left(\frac{\partial\cos\theta}{\partial y} - \frac{\partial\sin\theta}{\partial x}\right)}{\sin\theta - \cos\theta\frac{\partial k}{\partial y}\Big/\frac{\partial k}{\partial x}} \tag{4}$$

Method for extracting wave direction data from images

Two-dimensional band-passed Hilbert transforms are used to extract local wave phase (ϕ) from images. Directional wavenumber components (k_x, k_y) are calculated from the spatial gradients in local wave phase (Eqn. 5) (Havlicek, et al. 1998) and the wave angle (θ) is determined by taking the inverse tangent of the two directional wavenumber components (Eqn. 6).

$$\left[k_x, k_y\right] = \nabla\phi(x,y) = \mathrm{Re}\left[\frac{\nabla z(x,y)}{j * z(x,y)}\right], \tag{5}$$

where z is the analytic image determined from the Hilbert transform.

$$\theta = \tan^{-1}\left(k_y/k_x\right) \tag{6}$$

SYNTHETIC TESTING

Two synthetic beach profiles are used to test the algorithms. First we use a planar beach on a 500 x 500 m grid with dx = dy = 2m grid spacing. The second beach profile includes a Gaussian shoal superimposed on the planar profile to add alongshore-variable bathymetry. We first test the bathymetry estimation algorithm using known wavenumbers and wave angles calculated from the known phase of a shoaling wave. For the beach profiles tested, maximum errors in the depth estimate are $O(10^{-2}\text{m})$.

The second set of synthetic testing includes extracting the wave information from a synthetic image (Figure 1). To avoid contamination of the Hilbert Transform process due to edge effects, the data are tapered at the cross-shore edges and a 50 m band adjacent to cross-shore edges of the domain are neglected. Maximum errors on bathymetry estimates are on the order of 5% of the actual depth for the cases tested (Figure 2). Based on the positive results of our synthetic testing, we now apply these techniques to real images.

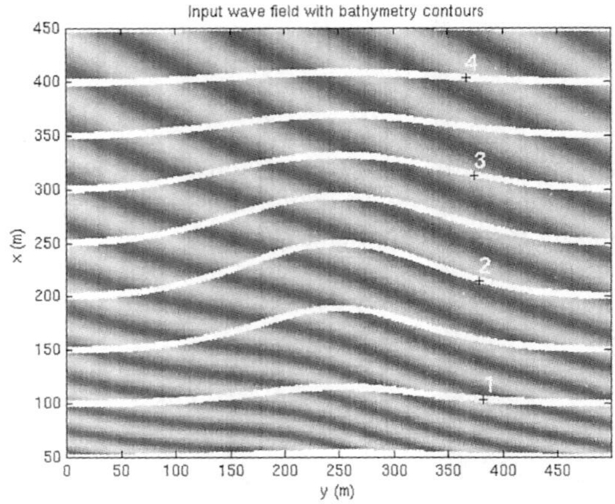

Figure 1. Input synthetic wave field with bathymetry contours labeled in white. Offshore wave angle, $\theta_0 = 30^0$, T = 8 sec.

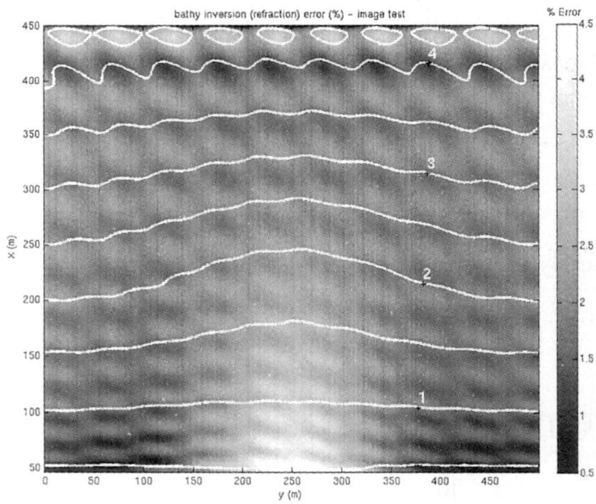

Figure 2. Bathymetry estimate from refraction-based algorithm shown in line contours (m) with % error of estimate (non-dimensional) shown in grayscale.

FIELD APPLICATION

Introduction

Testing of the algorithm is done using data from rectified video images and aerial photography at sites where bathymetry has been collected. There are several issues that need to be addressed when applying the above techniques to field data. Incoming waves can be broad-spectrum and become non-linear as they enter shallow water. As well, they may exhibit patterns of varying wave amplitude due to wave-wave interaction and bending due to wave-current interaction that are not accounted for in this model. Optically, images are contaminated with both high and low frequency signatures due to wave slope and lighting effects. The optical signature of the wave is based on its local slope so small scale features (wind chop) will be exaggerated compared to the slowly varying swell (Walker 1994). There are also complications in the region between non-breaking and breaking waves where the modulation transfer function of the optical image shifts in phase from crests being dark (non-breaking) to light (breaking). Within this region, the phase estimate will be inconsistent with the true wave phase. Accurate methods of data extraction to resolve the dominant swell are the focus of the remainder of this paper.

Pre-Filtering Images

Although our eyes are trained to extract certain features from an image there are many difficulties in using optical images as our data source with the current techniques developed. Figure 3 is an example of an image from Black's Beach, CA taken during the NCEX experiment in the fall of 2003. Wave swell is visible to the viewer's eye, but if we look at a cross-shore transect of the pixel intensity (Figure 4), we gain a very different view of the incident wave field. The transect does not show the clean swell patterns our eyes are able to focus in on, but instead shows very high-frequency noise, sharp spikes and dips at the approximate locations of wave crests/troughs, and a low-frequency trend of increasing intensity as we move onshore. It is our objective to pre-filter the pixel intensity data using robust image processing techniques in order to capture the dominant incoming swell. We currently do this in a two-step process by first applying a standard deviation filter weighting the local variance of the image to reduce the breaking wave signature and increase the signature of the offshore waves and secondly by applying a band-pass filter in the wavenumber domain to remove frequency components not expected to be related to the swell. These processes are discussed below.

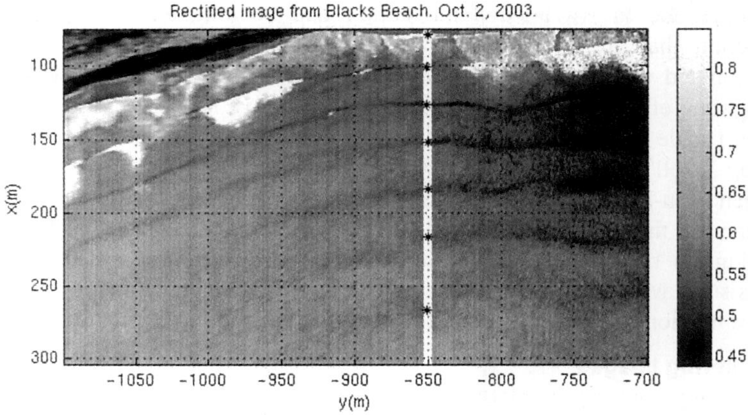

Figure 3. Image of incident waves approaching Black's Beach during NCEX. A transect of pixel intensity at y = -850m is shown in white with hand-picked locations of assumed wave crests denoted by the black stars. This transect is used in the following discussion to compare filtering techniques.

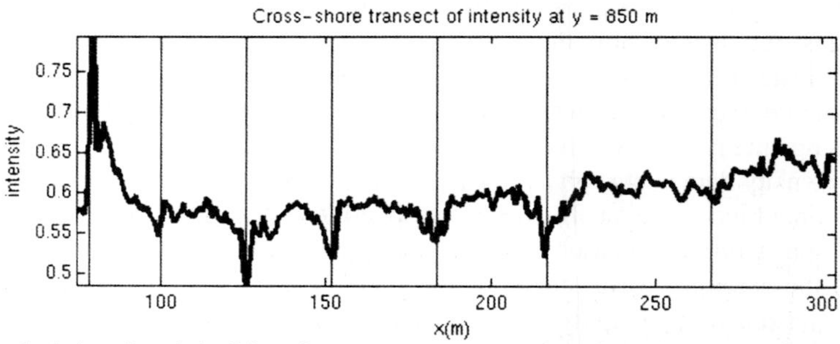

Figure 4. Intensity plot of the chosen transect as a function of cross-shore location showing the cross-shore variation in intensity. Black vertical lines represent the locations of the above mentioned hand-picked wave crest locations for comparison between visual estimates and those existing in the intensity transect.

Standard Deviation Filters. To remove trends in the image and to reduce the dominance of wave breaking optical signatures, we use a weighted standard deviation filter applied at every point in the image using a 2D convolution method. The weighting filter is a normalized 2D convolution of two orthogonal N-point Gaussian curves, with $\sigma_G=N/3$ points. For the image presented here, $N = 50$ (~50 m), approximately equal to the dominant wavelength visible in the image.

The formula for the filtered image (at every point) is:

$$I_{FILT} = \frac{(I_o - \mu_w)}{\sigma_w}, \tag{7}$$

where I_0 is the original image intensity, μ_w (Eqn. 8) is the weighted mean of the intensity image, and σ_w (Eqn 9) is the weighted standard deviation.

$$\mu_w = \frac{\sum_{i=1}^{N}\sum_{j=1}^{N} w_{ij} I_{ij}}{\sum_{i=1}^{N}\sum_{j=1}^{N} w_{ij}} \tag{8}$$

$$\sigma_w = \sqrt{\left(\frac{\sum_{i=1}^{N}\sum_{j=1}^{N} w_{ij} I_{ij}^2}{(N'-1)\frac{\sum_{i=1}^{N}\sum_{j=1}^{N} w_{ij}}{N'}} - \frac{\left(\sum_{i=1}^{N}\sum_{j=1}^{N} w_{ij} I_{ij}\right)^2}{(N'-1)\frac{\left(\sum_{i=1}^{N}\sum_{j=1}^{N} w_{ij}\right)^2}{N'}} \right)} \tag{9}$$

where N' is the number of points we are summing over, equal to N^2 in this example.

The image filtering (Figure 5) yields a stronger visual wave signature offshore and reduces the optical dominance of the breakers in the surf zone. Comparing the same cross-shore transect (Figure 4 and Figure 6), we have a more sinusoidal variation in intensity while maintaining the correct locations of the wave crests. The fundamental frequency of the incident swell becomes more apparent after this processing technique, but the image is still dominated by high-frequency signals that must be removed.

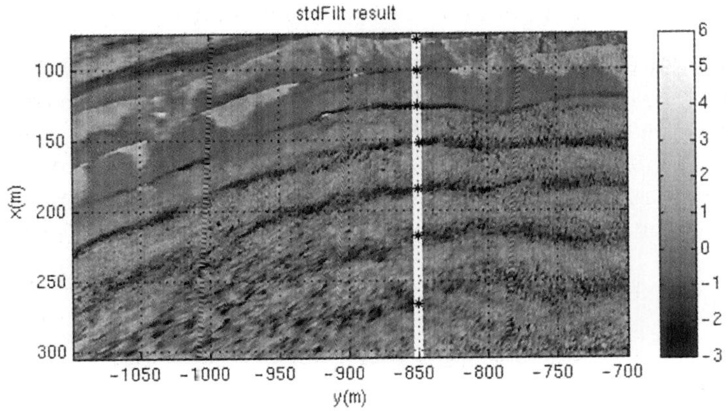

Figure 5. Intensity image after applying the STD filtering technique.

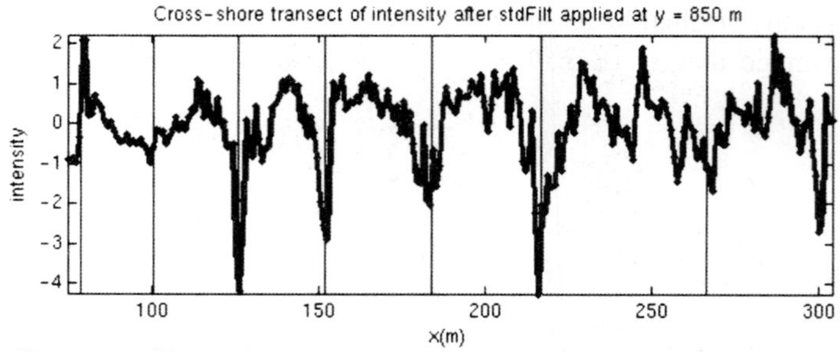

Figure 6. Transect of intensity vs. cross-shore location. Notice that the shoreline is no longer a dominant signal and the low-frequency trend has been removed. Wave patterns are more dominant while maintaining the location of the wave crests

Filtering in the wavenumber domain. The second step of the image processing technique involves band-pass filtering in the wavenumber domain to remove any energy that is not associated with the incident swell. For the current case, band-passing is done based on visual observation of the original image to determine approximate wavelengths of interest in both directions. In the cross-shore direction, we band-pass to remove any low-frequency trends left from the changes between breaking and non-breaking regions as well as the high-frequency signals that result from the wave slope dependence of optical signals. In the alongshore direction, we low-pass filter the image to remove the high-frequency signal. The band-pass filter, computed in the wavenumber domain, is confined to 30 m < L_x < 150 m in the x-direction and is tapered at both ends using a Hanning filter with half-width equal to 15 points in the wavenumber domain. The low-pass filter is cut-off at L_y = 30 m and has the same tapering at the cut-off end as the band-pass filter. The final filtered image (Figure 7) is much smoother while retaining the incident swell band seen in the original image.

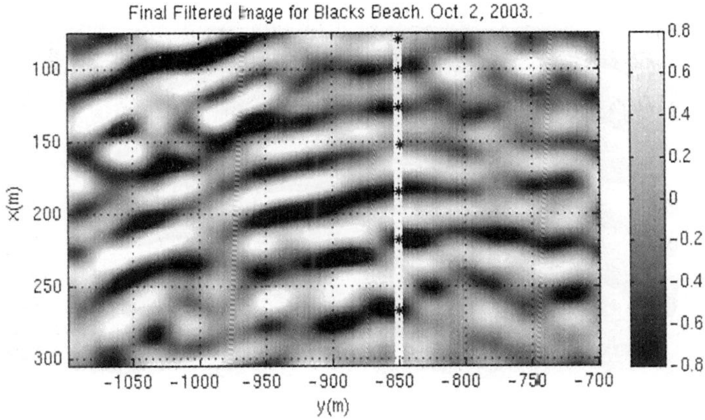

Figure 7. Final filtered image used in phase estimation. Image has been band-passed in the wavenumber domain to remove energy not associated with the incident swell. Directional wave patterns are clearly visible.

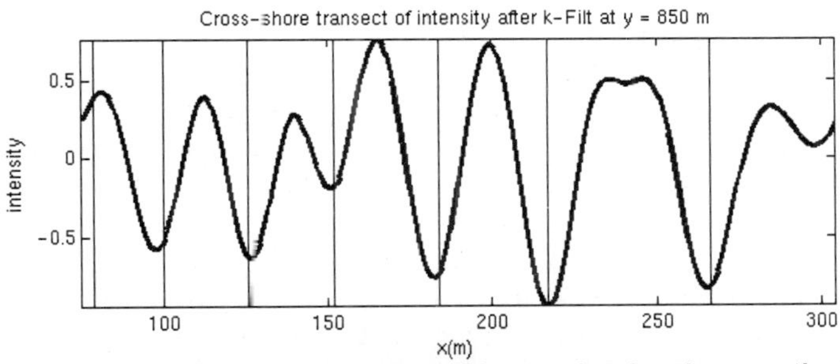

Figure 8. Intensity transect of the final filtered image showing the smoothed incident swell patterns consistent with the original wave information.

Phase Estimation

Wave phase is estimated using 2D Hilbert Transforms, where a 2D-FFT is done on the image and the direction of the Hilbert Transform is applied in the direction of dominant wave propagation (cross-shore). Phase estimates are good in areas of strong wave signal (Figure 9). In areas of initial wave breaking and in areas of low wave amplitude we observe jumps in the phase estimate that are not consistent with the original image.

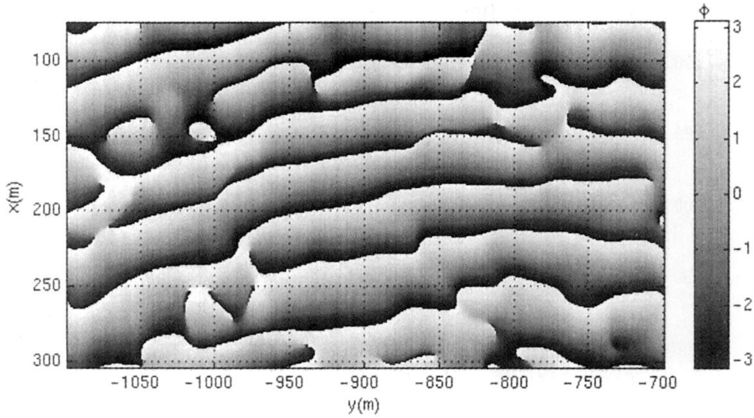

Figure 9. Estimate of phase from the 2D Hilbert Transform on the filtered intensity image for Black's Beach, CA, Oct. 2, 2003. Inconsistencies in phase estimates occur during the transition between breaking and non-breaking region and at the offshore where wave slope is minimal.

Wavenumber Estimation

As discussed above, directional wave information is calculated from the local gradients in phase (Havlicek, et al. 1998) . Figure 10 and Figure 11 show that these local estimates are very noisy and essentially unusable for our refraction-based model.

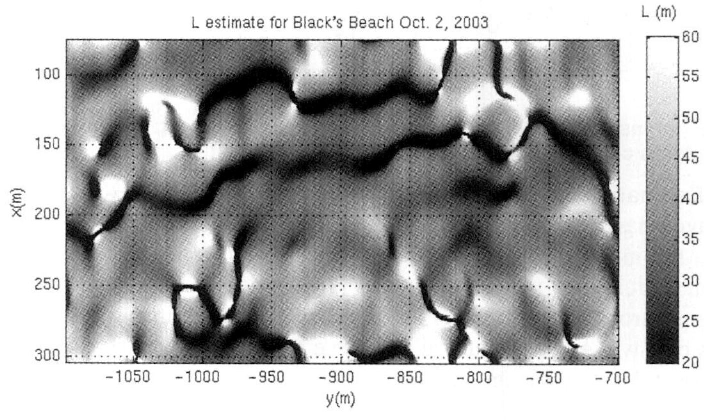

Figure 10. Wavelength estimates using method proposed by (Havlicek, et al. 1998). We see disagreement in areas where phase estimates are also inconsistent with original input. Estimates show lots of local variability.

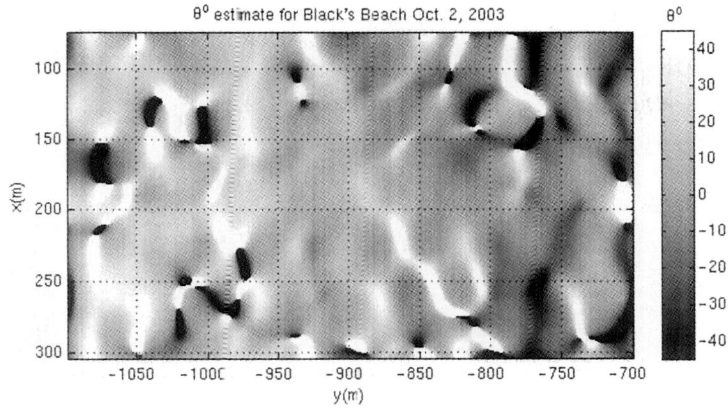

Figure 11. Wave angle estimate using local k_x and k_y. We see inconsistencies in areas where phase estimates are also inconsistent with the original input.

Since the local gradient method is a generally noisy estimator and blows up at phase jumps, we have developed a regional operator to estimate the directional wave information over larger (wavelength) scales. A global unwrapping of phase is a challenge to do in 2D because the unwrapped phase should be independent of the integration path taken. We do a 2D unwrap of the phase on a floating 50 x 50 point (~50 m x 50 m) grid and then do a least-squares fit to the plane in order to get an estimate of unwrapped phase at wavelength scales. Wavenumbers are then defined as:

$$k_\alpha = \frac{\partial \phi_{LSQ}}{\partial x_\alpha} \tag{10}$$

The definition of wave angle does not change with this new method. Estimates of wavenumber and angle are more consistent (Figure 12 and Figure 13), but still display areas of false estimates in areas of phase jumps. Research into improved solutions is ongoing.

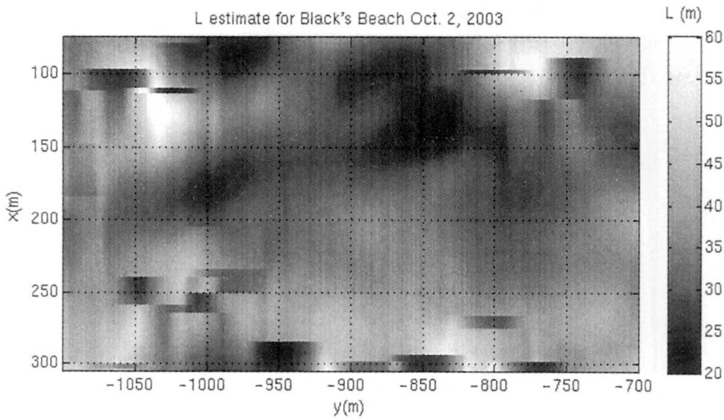

Figure 12. Estimate of wavelength using phase estimates from the least-squares fit to a 2D 50-pt unwrapped phase. General trends are improved over the local gradient method, but there are still areas where we see large jumps that are unrealistic. These areas match the locations where the transition of non-breaking to breaking waves occur and where wave amplitude is low.

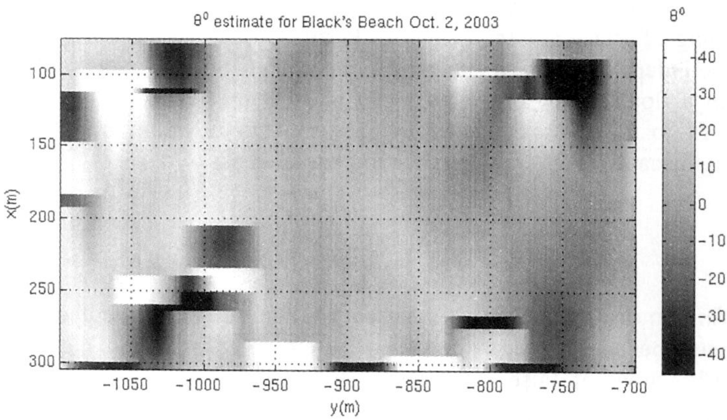

Figure 13. Estimates of wave angle using phase estimates from the gradients of a least-squares fit to a 2D 50-pt unwrapped phase. General trends are improved over previous method but there are still areas where we see large jumps that are unrealistic. These areas match the locations where the transition of non-breaking to breaking waves occur and where wave amplitude is low.

CONCLUSIONS

A two-step algorithm is developed to estimate bathymetry gradients based on directional wave information. Synthetic testing of the model shows that under ideal conditions, the method proposed can accurately determine 2D bathymetry at wavelength scales. However, imagery from field data collections provides challenges for accurate data extraction of wavenumber and wave direction fields. New methods for preprocessing images using standard deviation and wavenumber filters are employed. Data extraction methods using wavelength scale least-squares fits to wrapped phase are developed and tested. For the few cases tested it is not yet clear whether directional wave data can be estimated sufficiently well enough to allow bathymetry estimation.

AKNOWLEDGEMENTS

Funding for this research is provided by the Office of Naval Research Coastal Geosciences program, grant N00014-02-1-0154 and under the SECNAV/CNO Chair program, grant N00014-03-1-0973.

REFERENCES

Dalrymple, R. A., A. B. Kennedy, J. T. Kirby and Q. Chen. 1998. Determining depth from remotely-sensed images, *Proceedings of International Conference in Coastal Engineering*, ASCE, 2395-2408.

Havlicek, J. P., J. W. Havlicek, N. D. Mamuya and A. C. Bovik. 1998. Skewed 2D Hilbert Transforms and Computed AM-FM Models, *Proceedings of IEEE International Conference on Image Processing*, IEEE, 602-606.

Misra, S. K., A. B. Kennedy and J. T. Kirby. 2003. An approach to determining nearshore bathymetry using remotely sense ocean surface dynamics, *Coastal Engineering*,47, 265-293

Stockdon, H. F. and R. A. Holman. 2000. Estimation of wave phase speed and nearshore bathymetry from video imagery, *Journal of Geophysical Research*,105, 22,015-22,033

Walker, R. E. 1994. *Marine light field statistics*, John Wiley and Sons, Inc., New York.

WAVE PROPAGATION OVER A SHALLOW MUDDY SHELF: A FIELD EXPERIMENT

Sergio Jaramillo[1], Alexandru Sheremet[1], and Mead Allison[2]

We describe a comprehensive field experiment conducted on the muddy shelf fronting Atchafalaya Bay, Louisiana, in the vicinity of the subaqueous Atchafalaya River delta. In early spring, cold fronts pass over the experiment site with a periodicity of roughly two weeks. The rains, river floods and wave fields associated with the fronts produce episodic resuspension events which in turn affect significantly wave propagation (enhance wave dissipation). The experiment represents the first successful attempt to collect high-resolution coherent observations of wave and sediment dynamics in a cohesive sedimentary environment.

INTRODUCTION

Limited laboratory and field observations show that surface waves are strongly dissipated by muddy seabeds. Laboratory experiments of wave propagation in the presence of a highly viscous bottom fluid-mud layer (Gade, 1958) showed about 80% wave energy loss over only 2.6 wavelength, for layer thickness of about 1.3 times the wave boundary layer thickness. In the field, (Wells and Coleman, 1981) observed more than 90% of the incident energy loss as waves propagated across the 20-km wide shallow mudflats off the coast of Suriname; (Mathew et al., 1995) report 95% of the incident energy was lost as waves crossed the 1.1 km-wide mudbanks off the coast of India. Mud-enhanced long-wave damping has been observed in water depths greater than 20 m near the Mississippi Delta (Forristall and Reece, 1985). By comparison, weaker dissipation effects (50% to 75% loss) have been observed across the much wider (100 km), sandy North Carolina shelf, attributed to the interaction of waves with bottom ripples (Ardhuin et al., 2003).

A number of physical mechanisms for mud-induced wave dissipation have been proposed, based on different models of sediment rheology: poro-elastic solids (Yamamoto and Takahashi, 1985); viscous Newtonian fluids (Dalrymple and Liu, 1978); Bingham fluids (Mei and Liu, 1987); generalized Voigt solids (Jiang and Mehta, 1996; Jiang and Mehta, 2000). The diversity of these approaches mirrors the variability of field conditions. Over smooth and hard consolidated muds, wave dissipation can be similar or even weaker than over sandy bottoms (Hsiao and Shemdin, 1980; Shemdin et al., 1980; Yamamoto and Takahashi, 1985; Lee, 1995). By contrast, the fluid-mud phase, with non-Newtonian characteristics at concentrations above $\sim$5 kg/m^3, induces wave-dissipation effects that are orders of magnitude higher (Gade, 1957; Chou et al., 1993; Foda et al., 1993).

[1] Civil and Coastal Engineering, University of Florida, 365 Weil Hall, Gainesville, FL 32611, USA
[2] Earth and Environmental Sciences, Tulane University, Dinwiddie Hall, New Orleans, LA 70118, USA.

These models are based on the fundamental assumption that the sediment phase does not change over time-scales characteristic of wave evolution. There is strong evidence, however, that this is not true, at least in the (more interesting) case of energetic wave fields. In the laboratory, muds have been shown to be sensitive to wave- and current-induced turbulence (Chou et al., 1993; Foda et al., 1993; deWitt, 1995). Wave action liquefies bottom sediment by building up pore pressure and breaking the sediment matrix. Field evidence of massive submarine landslides (Sterling and Strohbeck, 1975) associated with energetic waves suggest supports this hypothesis. Near-bottom high-density suspensions (fluid-mud layers) have been observed to form due to hindered settling in the wake of storms (Wells and Kemp, 1984; Allison et al., 2000; Winterwerp, 2001; Sheremet et al., 2005).

Modeling the mud-phase transitions is also complicated by the fact that rheology changes in the bed can modify considerably the physics of wave-bottom interaction. A series of recent field experiments (Sheremet and Stone, 2003; Sheremet et al., 2005) on the Atchafalaya shelf showed unexpectedly strong short-wave dissipation in areas with soft cohesive bed sediments, when compared to normal (negligible dissipation) over "equivalent" areas with sandy bottoms. Sediment monitoring devices recorded the formation of a fluid-mud layer with sediment concentrations of over 10 kg/m^2, that coincided with spectrum-wide wave damping. These observations raise intriguing questions about the role of nonlinear surface-surface, and surface-lutocline wave interactions in coupling wave and sediment dynamics. They also highlight the strong coupling between the state of bed sediment and surface wave processes.

The physics of wave dissipation over muddy beds are not well understood, in part because of the scarcity of field observations. Previous field work on the Louisiana coast has focused either on wave propagation (e.g. Forristall and Reece, 1985; Sheremet and Stone, 2003), or on sediment dynamics (Allison et al., 2000; Draut et al., 2005; Neill and Allison, 2005), with minimal attention paid to companion processes – sediment dynamics in the former case, waves in the latter.

Here, we present what we consider to be the first systematic effort to collect high-resolution coherent sediment and wave dynamics field data. In the Section "Field Experiment" we describe the instrumentation used and field experiment sites. A brief overview of the data sets collected is presented in the Section "Observations". Section "Conclusions" summarizes the discussion.

FIELD EXPERIMENT

Fluid muds dominate the western Louisiana coast and the adjacent muddy chenier coast shoreface (Allison et al., 2000; Draut et al., 2005). Atchafalaya River discharges approximately 84 million tons/year of cohesive sediments inside Atchafalaya Bay where subaqueous mud clinoform deltas is being created (?). In winter and early spring, the area is swept periodically (about twice every month) by cold atmospheric fronts accompanied by energetics wells (characteristic period 8-10 s). In addition to an increase in river sediment discharge due to rainfall,

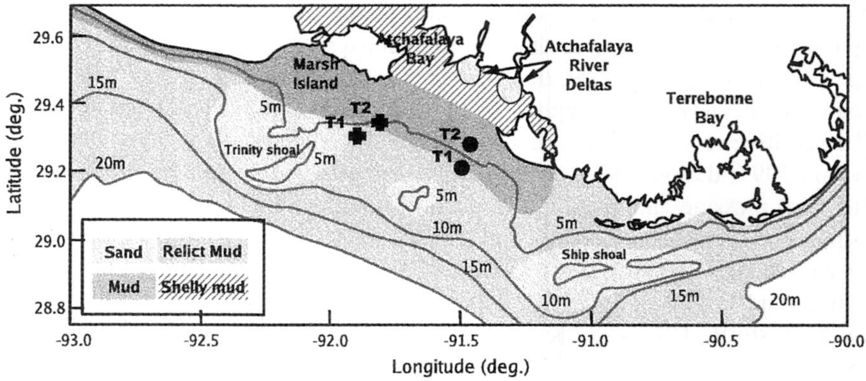

Figure 1. Qualitative facies map of surficial sediments on the Atchafalaya Shelf, Louisiana, USA (Neill and Allison, 2005). Relict mud are mantled by thin (less than 20-cm thick) ephemeral modern mud. The locations of the two instrument clusters during the two valid deployments are marked by circles (February 28 to March 13) and crosses (March 13 to April 1).

wave-induced bottom turbulence resuspends significant quantities of sediments. In the wake of a typical atmospheric perturbation event (3-10 day time scale), rapid depositional events lead the formation of episodic fluid-mud layers with a duration of about 12 hours, and thickness < 30 cm (Wells and Kemp, 1984; Allison et al., 2000).

Over the shallow shelf fronting Atchafalaya Bay, the sedimentary structure is more complex (Figure 1), with large adjacent patches of different sediment characteristics, and a large-scale gradation of grain size from clay-rich, cohesive sediments in the west to silts and fine-medium sands to the east. The width of the shallow inner Atchafalaya shelf (the 10-m isobath is in some places about 50 km offshore), and its sedimentary fabric make it a prime location for the study of wave-bottom interaction under variable mud-phase conditions.

Two instrument clusters, denoted here by "T1" and "T2", were deployed near the 5-m isobath on the Atchafalaya shelf (crosses and dots, Figure 1) over a period of about two months (February-April, 2006). To study the effects of the spatial sedimentary heterogeneity of the area the clusters were deployed in 4 runs, each approximately 14-days long, at different positions and configurations. A 1 to 2-day turnaround time was used for instrument retrieval, cleaning, and redeployment, and data downloading. For various reasons (instrument malfunction, tripod overturning, battery failure) each of the tripods yielded complete datasets in only two runs, corresponding to the locations shown in Figure 1. Overall, the valid data sets cover a period of about 4 weeks, from February 13 to March 14, which includes two storms. During redeployment

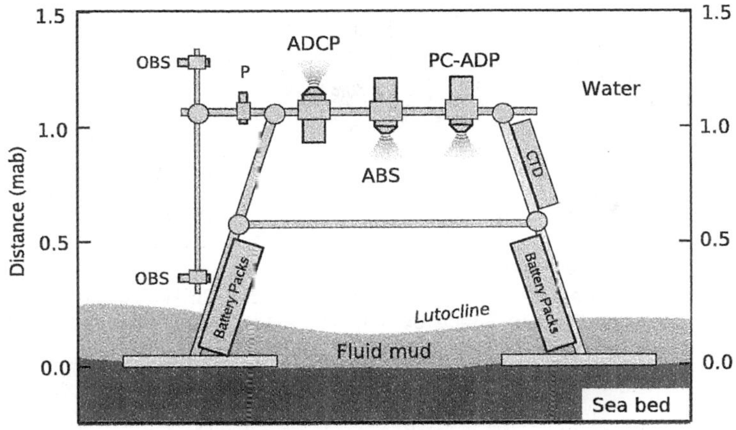

Figure 2. Basic configuration of an instrument cluster. Not all the instruments shown were available on both tripods. Abbreviations: ABS (Acoustic Backscatter Sensor, only at T2), PC-ADP (Pulse-Coherent Acoustic Doppler Profiler), ADCP (Acoustic Doppler Current Profiler), OBS (Optical Backscatter Sensor), CTD (Conductivity and temperature), and P (pressure sensor). The LISST (Laser In Situ Scattering and Transmissometry) device is not shown.

cruises, water samples and cores were also collected for subsequent sediment characterization.

The basic components of the two instrument clusters are shown in Figure 2.

The basic element of a cluster was a downward-looking Sontek 1.5-MHz Pulse Coherent Acoustic Doppler Profiler (PC-ADP) paired with an upward-looking Nortek AWAC 1-MHz Acoustic Doppler Current Profiler (ADCP), which collected current velocities throughout the entire water column, with increased resolution in the first $\sim$ 50 mab (meters above the bottom). The PC-ADP measured velocities continuously for 14 days at 2 Hz in 17 3-cm bins (0-51 cmab), with a blanking distance of about 0.6m from the instrument head (total distance to bottom was about 1.1 m). The ADCP (instrument head at approximately 1.1 mab) covered the upper water column recording 2-minute velocity averages in 50-cm cells. With a blanking distance of approximately 0.45, this yielded 5-6 measurement points, depending on the tidal phase. Both of the instruments were also outfitted with pressure sensors sampling at 2Hz, and the AWAC also provided estimates of directional surface wave spectra. Some instrument redundancy was built into the cluster by adding an independently-recording Paroscientific pressure transducer (approximately 54 cmab sampling at 2Hz in bursts of 2 hours and 50 minutes every 3 hours).

Sediment dynamics was monitored using two OBS (Optical Backscatter Sensor) located at 1.14 mab, and 0.44 mab, and recording one-minute averages of

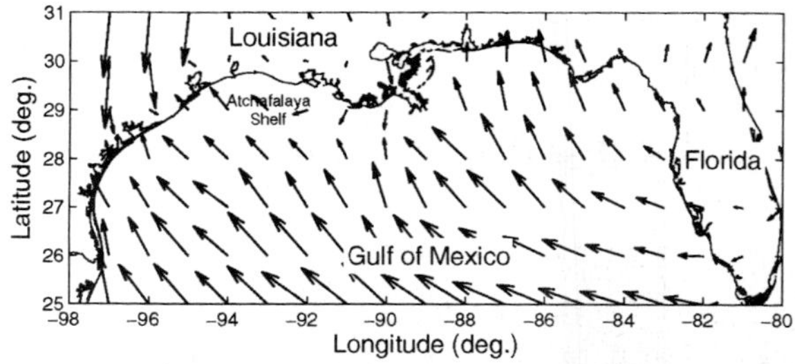

Figure 3. Surface wind field on February 17 at 12:00 hrs numerically simulated using the Coupled Ocean-Atmosphere Prediction System (COAMPS) run at the Naval Research Laboratory, Stennis Space Center, MS. Arrow length is proportional with wind speed (maximum value on this plot is approximately 8 m/s).

suspended sediment concentration (SSC) once every 3 hours. A Seabird MicroCat CT probe sampled salinity and temperature once every 20 minutes at 85 cmab.

On the onshore T2 platform, a number of additional instruments were deployed to improve sediment monitoring capabilities of the cluster. Sedimentary strata formation (e.g. lutocline) was monitored with a single-frequency (700 MHz) custom-made Acoustic Backscatter Sensor (ABS). The instrument was located approximately at 1 mab and measured the intensity of acoustic return from 100 9-mm cells covering a distance of 0.3 to 1.2 m from the instrument head. Suspended sediment characteristics (particle and floc size) were monitored using a Sequoia Scientific, Laser In Situ Scattering and Transmissometry device (LISST-100X).The PC-ADP on the T2 cluster also logged the records from two OBS and Seabird MicroCAT CT. The sampling scheme of the PC-ADP at T2 was modified to 10-min, 2Hz burst on every 30 minutes in 60 2-cm cells.

OBSERVATIONS

Two frontal passages occurred at the experimental sites over the one-month period covered by the valid data sets. The perturbation associated with the first front lasted for about three days, 15 to 18 February 2006. Figure 2 shows the map of numerically simulated 1-degree resolution surface winds for February 17 at 12:00 hrs, when the front approximately lined up with the Texas coast. The position of the front is indicated by the discontinuity in wind direction (Figure 2). Before the arrival of the front S-SE winds are typical, associated with intense swells (periods 8-10 s). In the wake of the front wind direction switches to N-NW, usually with lower intensity, and associated with short (5 s and less) seas.

Figure 3 shows a summary of wave and current observations at T1 during this storm. Locally generated short waves (here defined as having frequencies larger than 0,.2 Hz) are fairly intense throughout the storm (Figure 3A). Swell (frequencies less than 0.2 Hz) energy ramps up slowly to reach the peak of intensity before midnight February 17th. The frequency bands of swell and sea

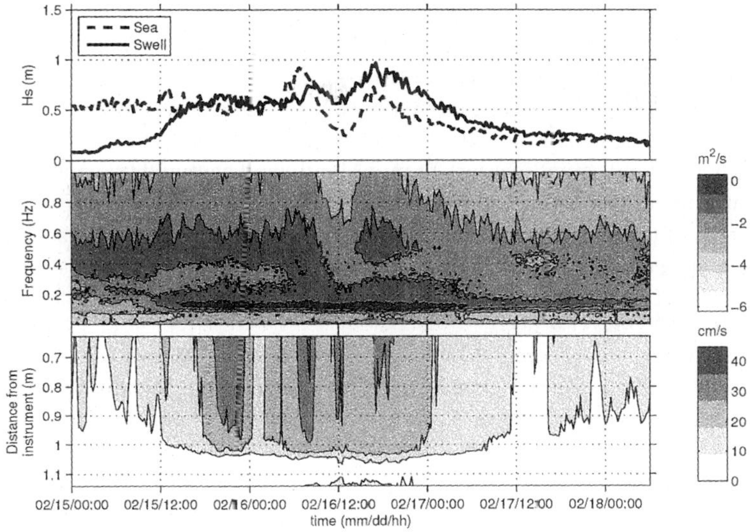

Figure 4. PC-ADP measurements at T1 of: A) Significant wave height of seas (frequency ≥ 0.2 Hz) and swell (frequency <0.2 Hz); B) Spectral density (m^2/Hz); and C) 17-minutes averages of near bottom current velocity, versus time.

are well differentiated in the spectrum (bimodal spectra, Figure 3B). Figure 3C shows 17-minute averages of current velocity. Mean velocity values are well correlated to wave activity. The local minimum seen at about 1.1 m distance from the instrument head corresponds to the hydrodynamic bottom (zero velocity). The apparent increase of velocity below the bottom, visible in this plot mainly during the peak of the storm, is associated with low pulse coherence and are likely just noise.

CONCLUSIONS

A number of quite different approaches have been proposed for wave-bed interaction in muddy environments, based on different mud rheologies, assuming a stable and unique mud state throughout the duration of the interaction. In contrast, laboratory experiments suggest that bed rheology does change on time scales comparable to those of wave evolution. Liquefaction, fluidization, resuspension, settling, fluid-mud formation, and consolidation into soft bed can all occur over the lifetime of a storm. The theory requires widely different rheological descriptions for each of these states, and suggests that different mechanisms of wave-sediment interaction are active in each case.

Field observation support for these assertions is patchy, in part due to the lack of coherent, high-resolution measurements of wave, current and sediment dynamics. The field experiment presented here is an effort to fill this gap.

The site chosen for the experiment has been previously documented as a dynamic cohesive sedimentary environment. We were able to overcome difficulties raised by the sheer dimension of the area, intensity of human activity, and weather and collected high resolution (2-Hz sampling, 3-cm vertical resolution near-bed) hydrodynamic and sediment measurements. The data sets span about a month (mid February to mid March) and captured two major storms, associated to cold fronts passing through the region.

Details of the ongoing analysis of bed sediment response to wave activity and its consequences on wave propagation will be published elsewhere.

ACKNOWLEDGMENTS

This work was supported by ONR, Awards N00014-03-1-0200 and N00014-06-1-0229. We are grateful to Dr. Holland, NRL Stennis Space Center for additional equipment support. Data was collected under the supervision of Captain Steve Dartez, head of the Field Support Group, Coastal Studies Institute, Louisiana State University.

REFERENCES

Allison, M., G. Kineke, E. Gordon, and M. Goni. 2000. Development and reworking of a seasonal flood deposit on the inner continental shelf off the atchafalaya river., *Continental Shelf Res.*, 20, 2267–2294.

Ardhuin, F., W. O'Reilly, T. Herbers, and P. Jessen. 2003. Swell transformation across the continental shelf. Part I: Attenuation and directional broadening., *J. Phys. Oceanogr.*, 33, 1921–1939.

Chou, H.-T., M. Foda, and J. Hunt. 1993. Rheological response of cohesive sediments to oscillatory forcing., A. Mehta (Ed.), *Nearshore and Estuarine Cohesive Sediment transport*, Coastal Estuarine Sci., 42, AGU, Washington DC, 126–148.

Dalrymple, R. A. and P.-F. Liu. 1978. Waves over soft muds: a two-layer fluid model., *J. Phys. Oceanogr.*, 8, 1121–1131.

deWitt, P. J.. 1995. Liquefaction of cohesive sediments caused by waves., *Delft Studies in Integrated Water Management 6*. Delft University Press.

Draut, A. E., G. Kineke, D. Velasco, M. Allison, and R. Prime. 2005. Influence of the Atchafalaya River on recent evolution of the chenier-plain inner continental shelf, northern Gulf of Mexico., *Continental Shelf Res.*, 25, 91–112.

Foda, A. M., J. Hunt, and H.-T. Chou. 1993. A nonlinear model for the fluidization of marine mud by waves., *J. Geophys. Res.*, 98(C4), 7039–7047.

Forristall, G. Z. and A. Reece. 1985. Measurements of wave attenuation due to a soft bottom: the SWAMP experiment., *J. Geophys. Res.*, 90, 3367–3380.

Gade, H.. 1957. *Effects of a non-rigid impermeable bottom on plane surface waves in shallow water*. Phd thesis, Texas A&M University.

Gade, H. G.. 1958. Effects of a nonrigid, impermeable bottom on plane surface waves in shallow water., *J. Marine Res.*, 16(2), 61–82.

Hsiao, S. V. and O. Shemdin. 1980. Interaction of ocean waves with a soft bottom, *J. Phys Oceanogr.,* 10, 605–610.

Jiang, F. and A. Mehta. 1996. Mudbanks of the Southwest coast of India V: Wave attenuation, *J. Coastal Res.,* 12, 890–897.

Jiang, F. and A. Mehta. 2000. Lutocline behavior in a high-concentration estuary, *Journal of Waterway, Port, Coastal and Ocean Engineering,* 126(6), 324–328.

Lee, S.. 1995. *Response of mudshore profiles to waves*. Phd thesis, University of Florida, Gainesville.

Mathew, J., M. Baba, and N. Kurian. 1995. Mudbanks of the Southwest coast of India I: Wave characteristics, *J. Coastal Res.,* 11, 168–178.

Mei, C. C. and P.-F. Liu. 1987. A Bingham-plastic model for a muddy seabed under long waves, *J. Geophys. Res.,* 92, 14,581–14,594.

Neill, C. F. and M. Allison. 2005. Subaqueous deltaic formation on the Atchafalaya shelf, Louisiana, *Marine Geology,* 214, 411–430.

Shemdin, O. H., S. Hsiao, H. Carlson, K. Hasselmann, and K. Schulze. 1980. Mechanisms of wave transformation in finite-depth water, *J. Geophys. Res.,* 85, 5012–5018.

Sheremet, A., A. Mehta, B. Liu, and G. Stone. 2005. Mechanisms of wave transformation in finite-depth water, *Est. Coastal Shelf Sci.,* 63, 225–233.

Sheremet, A. and G. Stone. 2003. Observations of nearshore wave dissipation over muddy sea beds, *J. Geophys. Res.,* 108(C11), doi: 10.1029/2003JC001885.

Sterling, G. H. and G. Strohbeck. 1975. Failure of South Pass 70 platform B in hurricane Camille, *J. Pet. Tech.,* 27, 263–268.

Wells, J. T. and J. M. Coleman. 1981. Physical processes and fine-grained sediment dynamics, coast of Surinam, South-America, *J. Sed. Pet.,* 51, 1053–1068.

Wells, J. T. and G. Kemp. 1984. Interaction of surface waves and cohesive sediments: Field observations and geologic significance, A. J. Mehta (Ed.), *Estuarine Cohesive Sediment Transport. Lecture Notes on Coastal and Estuarine Studies*, Volume 14, 44–65.

Winterwerp, J. C.. 2001. Stratification effects by cohesive and noncohesive sediment, *J. Geophys. Res.,* 106, 22559–22574.

Yamamoto, T. and S. Takahashi. 1985. Wave damping by soil motion, *J. Waterway, Port, Coastal, and Ocean Eng,* 3, 62–77.

NUMERICAL AND EXPERIMENTAL STUDY ON WAVE PROPAGATION OVER MUD

OVEISY A.[1], SOLTANPOUR M.[2], SHIBAYAMA T.[3], MASUYA Y.[3]

Due to high wave energy dissipation of wave-mud interaction, the common 2D horizontal hydrodynamic models for the simulation of wave field can not be applied to muddy coasts. The present study offers a multi-layered numerical model to simulate the wave transformation on 2D horizontal plane under regular waves. Time dependent mild slope equations are employed including the combined wave refraction, diffraction and breaking. A series of wave basin experiments have been performed to obtain the required data for the verification of the numerical model.

INTRODUCTION

The major aspects of wave mud interaction such as wave attenuation and mud mass transport has long been the subject of the scientific researches. In coastal areas covered with soft mud bed, traveling waves generate an inter-surface wave between the water layer and the fluid mud layer resulting to high wave attenuation. This wave energy dissipation, which generally is much more significant than wave height decay due to bottom friction, can not be ignored in the simulation of wave transformation in coastal muddy environments.

During the past decades, a number of studies have been conducted on different aspects of wave-mud interaction including two most important phenomena, i.e. wave attenuation and mud mass transport (Sakakiyama and Bijker, 1989; Ross and Mehta, 1990; Shibayama and An, 1993; Shibayama et. al., 1989). However, almost all of these studies have been conducted on horizontal beds under laboratory conditions. There have also been few efforts to consider the effect of bottom slope on wave transformation and mud mass transport (e.g., Soltanpour et al., 2003). In order to approach the real field problems, where the changes of cross-shore profiles along the shoreline can not be ignored, the influence of bottom configuration on wave height transformation in a general three-dimensional bathymetry should also be examined.

Here a hydrodynamic model is presented considering the various effects of shoaling, wave breaking and energy absorption of soft mud bed on wave height transformation. Simulating the high energy dissipation of fluid mud layers, the proposed wave model can be applied on real field conditions.

[1] *Civil engineering Department, Queen's University, Kingston, Ontario K7M 1B6, Canada*
[2] *Civil engineering Department, K.N. Toosi University of Technology, Tehran, Iran*
[3] *Civil engineering Department, Yokohama National University, Yokohama, Japan*

NUMERICAL MODEL

The proposed numerical model consists of a two dimensional vertical wave-mud interaction model and a two dimensional horizontal wave transformation model. These models are coupled to make a pseudo three-dimensional model. The multi-layered vertical model is used to calculate the wave attenuation rate on mud. The value of wave attenuation rate is introduced to wave transformation model to simulate wave transformation on fluid mud bed.

Wave-Mud Interaction Model

Following Tsuruya et al. (1987), the fluid system is divided into N layers in which the water layer is represented by N=1.

Figure 1 shows the sketch of the multi-layered model with the water layer on top.

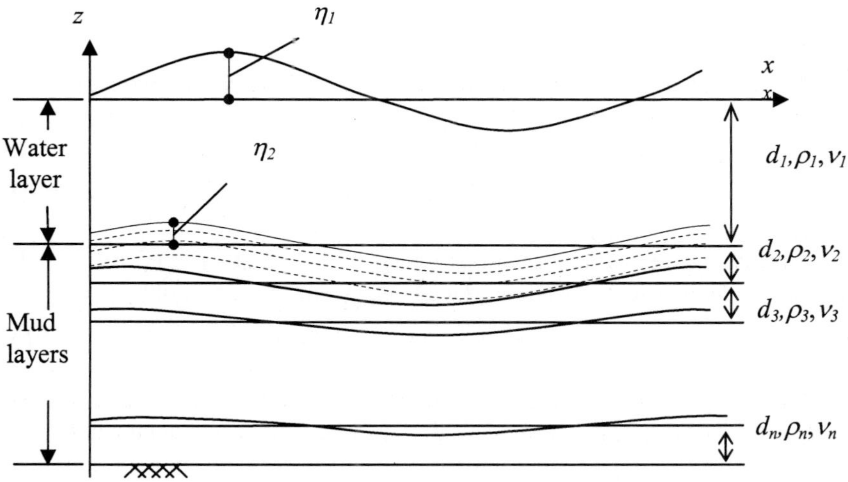

Figure 1. Definition of multi-layered model

The linearized Navier-Stokes equations, neglecting the convective accelerations and the continuity equation for an incompressible fluid can be expressed as

$$\frac{\partial u_j}{\partial t} = -\frac{1}{\rho_j}\frac{\partial p_j}{\partial x} + \nu_{e,j}\left(\frac{\partial^2 u_j}{\partial x^2} + \frac{\partial^2 u_j}{\partial z^2}\right) \tag{1}$$

$$\frac{\partial w_j}{\partial t} = -\frac{1}{\rho_j}\frac{\partial p_j}{\partial z} + \nu_{e,j}\left(\frac{\partial^2 w_j}{\partial x^2} + \frac{\partial^2 w_j}{\partial z^2}\right) \tag{2}$$

$$\frac{\partial u_j}{\partial x} + \frac{\partial w_j}{\partial z} = 0 \tag{3}$$

where x and z are the horizontal and vertical coordinates, the subscripts j indicate the layers, and the parameters t, ρ, ν_e and p represent the time, density, kinematic viscosity of mud and dynamic pressure, respectively.
The separable, periodic solutions for $\hat{u}_j$, $\hat{w}_j$ and $\hat{p}_j$ are assumed as

$$\hat{u}_j(x,z;t) = u_j(z)\exp[i(kx-\sigma t)] \tag{4}$$

$$\hat{w}_j(x,z;t) = w_j(z)\exp[i(kx-\sigma t)] \tag{5}$$

$$\hat{p}_j(x,z;t) = p_j(z)\exp[i(kx-\sigma t)] \tag{6}$$

where k is the unknown complex wave number, namely

$$k = k_r + ik_i \tag{7}$$

Displacements of water surface and interfaces, η_j, are represented by:

$$\eta_j = a_j \exp[i(kx-\sigma t)] \tag{8}$$

where a_j is the amplitude of the displacement of the j_{th} layer and the water surface is expressed as η_1.

Substituting the real and imaginary parts of wave number into Eq. (8), the expression of water surface and interfacial displacements can be obtained as

$$\eta_j = a_j \exp(-k_i x)\exp[i(k_r x-\sigma t)] \tag{9}$$

Therefore, the real part of the wave number, k_r, gives the wave length $L = 2\pi / k_r$ and its imaginary part, k_i, presents the wave attenuation rate, assuming the exponential wave height decay.
Substituting Eqs. (4) and (5) into the continuity Eq. (3) results to

$$u_j = \frac{iw_j'}{k} \tag{10}$$

where the prime represents the differential with respect to z. Introduction of Eq. (10) into Eq. (1) yields an expression for p_j

$$p_j = \frac{\rho_j \nu_{e,j}}{k^2}(w_j''' - w_j' \lambda_j^2) \tag{11}$$

where

$$\lambda_j^2 = k^2 - i\sigma \nu_{e,j}^{-1} \tag{12}$$

Substituting p_j into the vertical momentum, Eq. (2), yields the forth order differential equation for w_j

$$w_j''''-(k^2+\lambda_j^2)w_j''+k^2\lambda_j^2 w_j=0 \tag{13}$$

The solutions can be obtained as

$$w_j(z)=A_j\sinh k(\sum_{n=1}^{j}d_n+z)+B_j\cosh k(\sum_{n=1}^{j}d_n+z)+$$

$$C_j\exp[\lambda_j(\sum_{n=1}^{j-1}d_n+z)]+D_j\exp[-\lambda_j(\sum_{n=1}^{j}d_n+z)] \tag{14}$$

where d_n is the thickness of n_{th} layer. The complex constants A_j, B_j, C_j, D_j, and the unknown variables k and a_j are determined from the boundary conditions.

5N boundary conditions are required for a viscous fluid model of N layers. The unknown constants and variables are determined from the boundary conditions at the water surface, the interfaces and the rigid bottom. The details can be found in Tsuruya et al. (1987).

Rheological Model of Mud

Proper rheological model of mud should be adopted in order to investigate wave-mud interaction. Mud in general can range from being a highly rigid and weakly viscous material to one that can be approximated as a purely viscous fluid, depending on the properties of the constituent sediment and the ambient fluid. Considering the complexity of rheological behavior, the visco-elastic-plastic model has been adopted in the present study to develop a predictive behavior model for wave-mud interaction (Shibayama and An, 1993). The constitutive equations are expressed as:

$$\sigma_{ij}=2\mu_e\dot{e}_{ij} \tag{15}$$

$$\mu_e=\mu_1+\frac{iG}{\omega} \qquad \text{for} \quad (\frac{1}{2}\sigma_{ij}\sigma_{ij}\le\tau^2{}_y) \tag{16}$$

$$\mu_e=u_2+\frac{\tau_y}{\sqrt{4|\Pi_e|}} \quad \text{for} \quad (\frac{1}{2}\sigma_{ij}\sigma_{ij}>\tau^2{}_y) \tag{17}$$

where σ_{ij} is the deviator part of stress tensor, $\dot{e}_{ij}$ is the deviator part of strain rate tensor, G is the shear modulus of elasticity, μ_1 is the viscosity of mud in the viscoelastic state, μ_2 is the viscosity of mud in the viscoplastic state, τ_y is yield stress, ω is the angular frequency of wave and $4|\Pi_e|$ is expressed as

$$4|\Pi_e| = 2(\frac{\partial u}{\partial x})^2 + 2(\frac{\partial w}{\partial z})^2 + (\frac{\partial u}{\partial z} + \frac{\partial w}{\partial x})^2 \tag{18}$$

where u and w are velocity components in x and z direction, respectively. The rheological viscoelastic parameters of mud, i.e. shear modulus and viscosity, are calculated from the empirical equations offered by Shibayama and An (1993):

$$\mu_1 = 10^{(3.353-9.56\times10^{-3}W)} \times T \tag{19}$$

$$\log G = 3.761 - 1.05\times10^{-2}W + (0.147 - 3.38\times10^{-3}W) \times \log(T - 0.522 - 1.23\times10^{-3}W) \tag{20}$$

where T is the wave period and W is the water content ratio of fluid mud (%). The viscoplastic parameters, μ_2 and τ_y were adopted from the laboratory experiments of Tsuruya et al. (1987):

$$\tau_y = 1.494\times10^6 W^{-2.452} \tag{21}$$

$$\mu_2 = 8.465\times10^3 W^{-1.344} \tag{22}$$

Equivalent viscosity. In the viscoplastic state of the visco-elastic-plastic model, the objective $4|\Pi_e|$ of the representative viscosity, can be approximated by a viscoelastic field using an iterative method. The real parts of Eqs. (4) and (5) are written as

$$\hat{u}_j(z) = |u_j| \exp(-k_i x) \cos(k_r x - \sigma t + \alpha_j) \tag{23}$$

$$\hat{w}_j(z) = |w_j| \exp(-k_i x) \cos(k_r x - \sigma t + \beta_j) \tag{24}$$

where α_j and β_j are the arguments of u_j and w_j, respectively. Taking the partial derivatives of Eqs. (23) and (24) with respect to x and z

$$\frac{\partial \hat{u}_j}{\partial x} = -|k||u_j| \exp(-k_i x) \sin(k_r x - \sigma t + \alpha_j + \theta_k) \tag{25}$$

$$\frac{\partial \hat{u}_j}{\partial z} = -|u'_j| \exp(-k_i x) \cos(k_r x - \sigma t + \alpha'_j) \tag{26}$$

$$\frac{\partial \hat{w}_j}{\partial x} = -|k||w_j| \exp(-k_i x) \sin(k_r x - \sigma t + \beta_j + \theta_k) \tag{27}$$

$$\frac{\partial \hat{w}_j}{\partial z} = -\left|w'_j\right| \exp(-k_i x)\cos(k_r x - \sigma t + \beta'_j) \tag{28}$$

where α'_j and β'_j and θ_k are the arguments of u'_j, w'_j and k, respectively. Substituting Eqs. (25) ~ (28) into Eq. (3) leads to

$$4\left|\Pi_e\right| = \frac{1}{2}\{\left|k\right|^2(2\left|u_j\right|^2 + \left|w_j\right|^2) + (\left|u_j'\right|^2 + 2\left|w_j'\right|^2)$$

$$-2\left|u_j'\right|\left|w_j\right|\left|k\right|\sin(\beta_j + \theta_k - \alpha_j')\}\exp(-2k_i x) \tag{29}$$

Wave Transformation Model

Governing equations. The governing equations are time dependent mild slope equations which were introduced by Nishimura et al. (1983):

$$\frac{\partial Q_y}{\partial t} + c^2 \frac{\partial}{\partial y}(\xi) + f_D . Q_y = 0 \tag{30}$$

$$\frac{\partial Q_x}{\partial t} + c^2 \frac{\partial}{\partial x}(\xi) + f_D . Q_x = 0 \tag{31}$$

$$\frac{\partial \xi}{\partial t} + \frac{1}{n}\{\frac{\partial (nQ_x)}{\partial x} + \frac{\partial (nQ_y)}{\partial y}\} = 0 \tag{32}$$

where Q_x and Q_y are flow rates and ξ is water surface elevation. In these equations, c is wave celerity and $n=c_g/c$. f_D is the total dissipation coefficient which includes energy loss caused by wave-mud interaction (f_{DM}) and wave breaking (f_{DB}) in the surf zone. The dissipation rate of fluid mud is calculated using wave mud interaction model for each computational node.

The governing equations (30) to (32) are discretized and a staggered mesh scheme is employed. Eqs. (30) to (31) cannot be discretized into an explicit form considering the nonlinear dissipation term with respect to the flow rate. An iterative scheme is employed to reach the steady state solution. Computation is made by finite difference method, in which the details can be found in Horikawa (1988).

Breaking dissipation. Breaking dissipation coefficient is calculated as follows (Watanabe and Maruyama, 1986),

$$f_{DB} = \alpha_D \tan\beta \sqrt{\frac{g}{h}(\frac{\hat{Q}}{Q_r} - 1)} \tag{33}$$

$$\hat{Q} = \sqrt{\hat{Q}_x^2 + \hat{Q}_y^2}, \quad Q_r = \gamma' \sqrt{gh^3}$$

where α_D and γ' are set to 2.5 and 0.25, respectively and $\tan\beta$ is the average bottom slope. The breaking point is determined using the breaker index, i.e. the ratio of the wave height to water depth (h_B / H_B), as a predefined value.

Mud dissipation. The attenuation of wave height on a horizontal bed can be approximated by an exponential decay (Gade, 1958),

$$H(x) = H_0 e^{-k_i x} \tag{34}$$

where H_0 is the incident wave height at x=0, and k_i is the wave attenuation coefficient. The attenuation effect of mud is related to f_{DM} as

$$\begin{aligned} \varepsilon_{Dm} &= -\frac{d}{dx}(C_g E) = -C_g(2\rho g \frac{H}{8} \times \frac{dH}{dx}) = -C_g(\frac{2E}{H} \times \frac{dH}{dx}) \\ &= 2EC_g(-\frac{1}{H} \times \frac{dH}{dx}) = 2C_g k_i E \end{aligned} \tag{35}$$

considering

$$\nabla.(c_g E) = -f_D n E \tag{36}$$

where

$$E = \rho g a^2 / 2 \tag{37}$$

it is concluded that $f_{DM} = 2k_i.c$ assuming mild slope bottom bathymetry.

EXPERIMENTAL SETUP

Four different cases of experiments were conducted in hydraulic laboratory, Yokohama National University. Fig. 2 shows the sketch of experimental setup. Commercial kaolinite of 12 cm thickness was used as bed sediment in a 1m×1m box in the wave basin. The test conditions are presented in Table 1. Wave characteristics are measured at points 1 and 2, 30 cm from the beginning and the end of mud box. Measured wave heights are presented in Table 2.

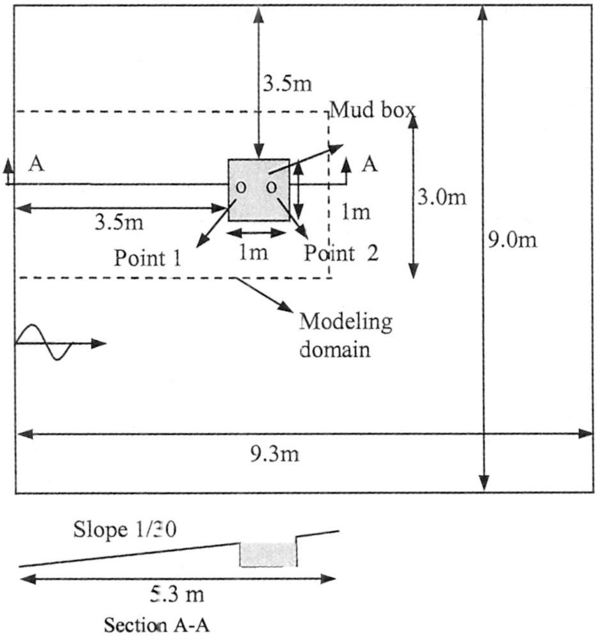

Figure 2. Experimental setup and modeling domain.

Table 1. Incident wave and fluid mud characteristics.

	Incident Wave Height	Period	Mud Water Content	Wave Angle
Case 1	6.61 cm	1.00 s	130 %	10°
Case 2	3.80 cm	1.00 s	130 %	0°
Case 3	3.78 cm	1.00 s	130 %	0°
Case 4	4.47 cm	1.00 s	130 %	0°

Table 2. Measured wave height at point 1 and 2.

	Wave Height at Point 1	Wave Height at Point 2
Case 1	1.88 cm	1.23 cm
Case 2	2.59 cm	1.53 cm
Case 3	2.66 cm	1.78 cm
Case 4	3.50 cm	2.00 cm

MODEL VRIFICATION

Table 3 shows the comparison between the numerically simulated wave heights and the laboratory measurements. An acceptable agreement is observed between the measured and simulated values. A sample of wave height contours have also been depicted for case 3 in Fig. 3.

Table 3. Measured and calculated wave height at point 1 & 2.

	Case 1	Case 2	Case 3	Case 4
Measured Point 1 (cm)	1.88	2.59	2.66	3.50
Measured Point 2 (cm)	1.23	1.53	1.78	2.00
Calculated Point 1 (cm)	2.34	3.29	3.25	3.12
Calculated Point 2 (cm)	1.19	1.88	1.84	1.84

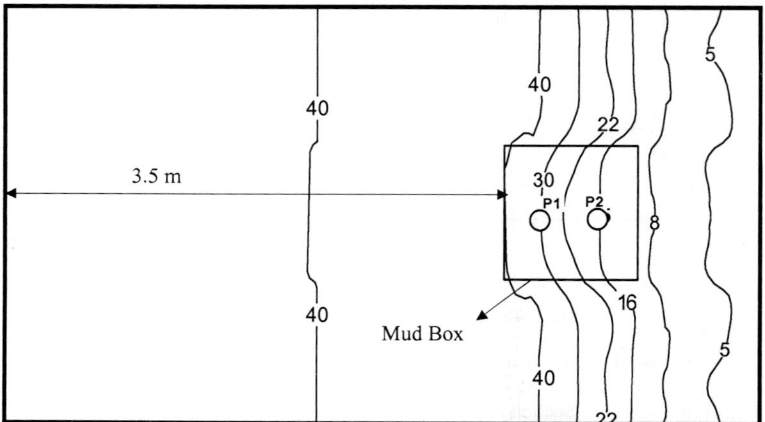

Figure 3. Wave height contours considering fluid mud and breaking dissipation for case 3 (unit: mm)

CONCLUSION

The wave attenuation on fluid mud was included as a part of dissipation terms in mild slope equations considering a multi-layer visco-elastic-plastic model. The wave transformation model includes combined wave refraction, diffraction and reflection. The model also considers the wave energy dissipations due to the existence of fluid mud layer and wave breaking. Model performance was examined comparing the numerical results with the conducted laboratory experiments. It is concluded that the numerical model is capable to predict the wave transformation on a general bathymetry.

REFERENCES

Gade, H.G. 1958. Effects of a non-rigid, impermeable bottom on plane surface waves in shallow water, *Journal Marine Research*, 16(2), 61-82.

Horikawa, K. 1988. *Nearshore dynamics and coastal processes*, University of Tokyo Press.

Nishimura, H., K. Maruyama, and K. Hiraguchi. 1983. Wave field analysis by finite difference method, *Proc. 30th Japanese Conference on Coastal Engineering*, JSCE, 123-127 (in Japanese).

Ross, M. A., and A. J. Mehta 1990. Fluidization of soft estuarine mud by waves. In: *The Microstructure of Fine Grained Sediments*: From Mud to Shale, R. H. Bennett ed., Springer-Verlag, New York, 185-191

Sakakiyama, T., E. W. Bijker. 1989. Mass transport velocity in mud layer due to progressive waves, *Report of Coastal Engineering Group, Coastal Engineering Division,* Delft University of Technology

Shibayama T., N. N. An. 1993. A visco-elastic-plastic model for wave-mud interaction, *Coastal Engineering in Japan,* 36, 67-89.

Shibayama T., M. Nakanishi, H. Ishihara and S. Sato. 1987. Mud suspension due to waves, *Proceedings of Coastal Engineering*, JSCE, 34, 271-275 (in Japanese).

Soltanpour, M., T. Shibayama, and T. Noma. 2003. Cross-shore mud transport and beach deformation model, *Coastal Engineering Journal,* 45(3), 363-386.

Tsuruya, H., S. Nakano and J. Takahama. 1987. Interaction between surface waves and a multi-layered mud bed, Rep. No. 5, *Port and Harbor Res. Inst.,* Ministry of Transport, Japan.

Watanabe, A., and K. Maruyama, 1986. Numerical modeling of nearshore wave field under combined refraction, diffraction and breaking. *Coastal Engineering in Japan*, Vol. 29, pp. 19-39.

IRREGULAR WAVE TRANSFORMATION ON MUD PROFILES INCLUDING THE FLUIDIZATION OF MUD LAYER

M. Soltanpour[1], S.A. Haghshenas[2] and T. Shibayama[3]

The present study offers a numerical cross-shore wave-mud interaction model that can be used to predict fluid mud thickness and irregular wave transformation along soft mud profiles. The rheological constitutive equations of visco-elastic (VE) material are adopted for the behavior of fluid mud layer under the wave action. Extending an analytical model of fluidization to irregular waves, the fluidized depth of mud layer under overlaying wave action is determined. Spectral calculation method is used for the simulation of irregular wave transformation. The capability of the model to predict the fluidized depth and the wave height transformation along muddy profiles is also examined.

INTRODUCTION

Although muddy coasts are not very common, there are many coastal areas, normally adjacent to big rivers, that the bottom layer has been covered with soft mud. Since Gade (1958), the interaction of waves and low density muddy beds has been studied in several ways. Two major phenomena that have received the most attentions are wave attenuation and mud mass transport. There have also been some efforts to formulate the fluidization of mud, based on the characteristics of the wave and undisturbed mud layer (Ross and Mehta, 1990; Foda et al., 1993; Soltanpour et al., 2005). Recently, the effect of irregularity has also been included in a number of studies (Zhang and Zhao, 1999; Soltanpour et al., 2002). However, the fluidization of mud layer under irregular waves and the attenuation of irregular waves on mud profiles have not been studied yet.

The present study offers a numerical cross-shore wave-mud interaction model that can be used to predict fluid mud thickness and irregular wave transformation along soft mud profiles. Spectral calculation method is used for simulation of irregular wave-mud interaction (Soltanpour et al., 2004). An integrated analytical model is applied to determine the fluid mud thickness under the combined effects of irregular wave components. The cross-shore model includes wave shoaling and the dissipation effects of fluid mud bed and wave breaking on the transformation of wave spectra along the profile.

---

[1] Civil Engineering Department, K.N. Toosi University of Technology, No.1346, Vali-Asr St., P.C. 19697, Tehran, Iran, soltanpour@kntu.ac.ir

[2] Civil Engineering Department, K.N. Toosi University of Technology, No.1346, Vali-Asr St., P.C. 19697, Tehran, Iran, sahaghshenas@sina.kntu.ac.ir

[3] Civil Engineering Department, Yokohama National University, 79-5 Tokiwadai, Hodogaya-ku, Yokohama, 240-8501 Japan, tomo@ynu.ac.jp

FLUIDIZATION OF MUD LAYER

Fluidization under A Monochromatic Wave

It is assumed that a water layer of a given thickness is sitting on top of a stratified two-layered mud bed, where the upper one is a generalized visco-elastic layer (Fig. 1). The behavior of this fluidized mud layer is simulated, using a complex viscosity coefficient, which is a function of the imposed strain in some nonlinear manner. The lower stationary layer is assumed to be elastic.

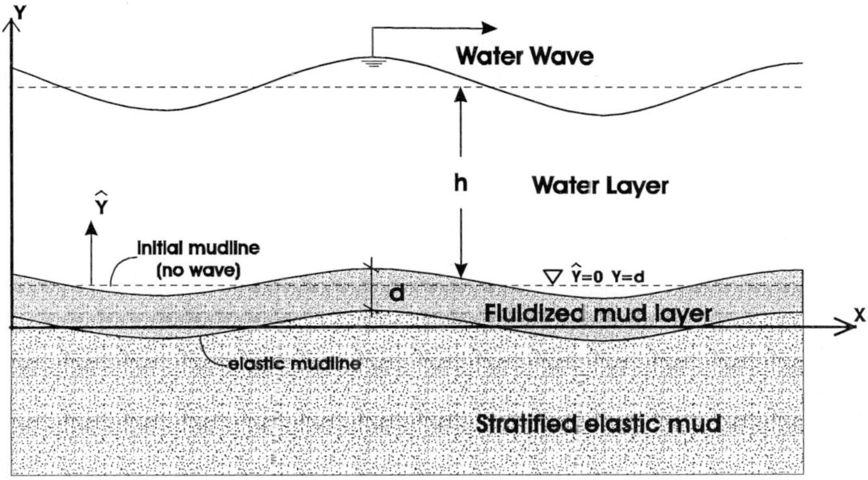

Figure 1. Definition sketch for the mud-fluidization model (After Foda et al., 1993).

The thickness of the fluid mud layer is related to the resulting strains due to the imposed wave height. The water layer response as well as the fluidized mud layer's are decomposed into an inviscid potential flow plus a visco-elastic boundary layer correction near the elastic mud line (Foda et al., 1993). The simplified boundary layer structure for considering the interaction between water wave and mud layer is adopted. Assuming a monochromatic wave, a velocity potential function, ϕ, is introduced to describe the potential flow component in the water layer and the fluidized mud layer of thicknesses h and d, respectively. The velocity potential function must satisfy Laplace equation

$$\nabla^2 \phi = 0 \tag{1}$$

The horizontal and vertical components of the inviscid velocity vector, $\hat{u}$ and $\hat{v}$, are defined as

$$\hat{u} = \partial \phi / \partial x \,, \hat{v} = \partial \phi / \partial y \tag{2}$$

The linearized kinematic and dynamic free-surface boundary conditions at $y = h + d$ are

$$\frac{\partial \phi}{\partial y} - \frac{\partial \eta}{\partial t} = 0 \tag{3}$$

$$\frac{\partial \phi}{\partial t} + g\eta = 0 \tag{4}$$

where η is the water surface relative to the mean water level. The total horizontal velocity, u is defined as a composition of the inviscid velocity $\hat{u}$ and a rotational velocity U inside the fluidized mud layer, i.e.

$$u = \frac{\partial \phi}{\partial x} + U \tag{5}$$

where U is governed by the following boundary layer equation

$$\frac{\partial U}{\partial t} = \upsilon_{ve} \frac{\partial^2 U}{\partial y^2} \tag{6}$$

On the other hand, velocities in the elastic bed, u and v, are defined as

$$u = \frac{\partial \phi}{\partial x} + \frac{\partial \psi}{\partial y}, v = \frac{\partial \phi}{\partial y} - \frac{\partial \psi}{\partial x} \tag{7}$$

where ϕ is governed by Laplace equation. ψ satisfies the elastic shear wave equation

$$c_s \nabla^2 \psi = \frac{\partial^2 \psi}{\partial t^2}, c_s = (G_e / \rho)^{1/2} \tag{8}$$

where c_s is the depth dependent speed of elastic shear wave for a stratified bed. The elastic shear wave-number s is defined as ω / c_s. The continuity of velocities and stresses across $y = 0$ are the desired boundary conditions which relate the solutions above and below the elastic mud line at $y = 0$

$$\left(\frac{\partial \phi}{\partial y}\right)^+ = \left(\frac{\partial \phi}{\partial y} - \frac{\partial \psi}{\partial x}\right)^- \tag{9}$$

$$\left(\frac{\partial \phi}{\partial x} + U\right)^+ = \left(\frac{\partial \phi}{\partial x} + \frac{\partial \psi}{\partial y}\right)^- \tag{10}$$

$$\left(\rho\frac{\partial^2\phi}{\partial t^2}\right)^+ = \left(2G\frac{\partial v}{\partial y} + \rho\frac{\partial^2\phi}{\partial t^2}\right)^- \quad (11)$$

$$\frac{\partial}{\partial t}\left(\rho\upsilon_{ve}\frac{\partial U}{\partial y}\right)^+ = \left(G\left[\frac{\partial u}{\partial y} + \frac{\partial v}{\partial x}\right]\right)^- \quad (12)$$

The plus superscript indicates that the expression belongs to a point just above the mud line, $y = 0^+$, while the minus superscript indicates to a point just below it, $y = 0^-$. Equations (9) to (12) satisfy the continuity of vertical velocity, horizontal velocity, normal stress and shear stress, respectively. Furthermore, the solution should decay to zero at infinite depth, i.e., ϕ and $\psi \to 0$ as $y \to -\infty$. The above system of equations forms the desired eigen-value problem for the wave-mud bed model, in which the fluidized thickness of mud layer is unknown. The thickness of the top fluid mud layer is determined by requiring that the elastic strain at the top of the elastic bed be equal to the limiting or yield strain for the given mud material. This will result in the desired relationship between the depth of fluidization and the imposed water wave height.

Model Extension to Irregular Wave

The fluidization model is extended to irregular waves assuming that the irregular wave consists of an infinite number of component regular waves with different wave amplitudes and frequencies. The wave spectra are defined as

$$S(f) = \frac{1}{\Delta f}\sum_{f}^{f+\Delta f}\frac{1}{2}a_n^2 \quad (13)$$

where a_n is the amplitude of the nth wave component. The waves in the range (f_{n-1}, f_n) can be treated as one component wave with the frequency of

$$\hat{f}_n = \frac{(f_n + f_{n-1})}{2} \quad (14)$$

and amplitude of

$$a_n = \sqrt{2S(\hat{f}_n)\Delta f_n} \quad (15)$$

where $S(\hat{f}_n)$ is the average value of spectra in the range (f_{n-1}, f_n).

Considering the periodic response of the imposed shear strain and its corresponding shear stress (Chou et al., 1993), the maximum shear strain and its corresponding maximum shear stress of an irregular wave is calculated by superimposing of maximum shear strains and stresses of regular wave components

$$\gamma_{imp} = \sqrt{\sum_n \left(\gamma_{\max,n}{}^2\right)} \tag{16}$$

$$\tau_{imp} = \sqrt{\sum_n \left(\tau_{\max,n}{}^2\right)} \tag{17}$$

where $\gamma_{\max,n}$ and $\tau_{\max,n}$ are maximum shear strain and maximum shear stress of each wave component, respectively. The thickness of the fluidized mud layer is determined by comparing the maximum shear strain, i.e. γ_{imp}, at the top of the elastic bed with the limiting or yield strain for the given mud material. A try and error procedure will result in the desired depth of fluid mud layer under the overlaying irregular wave.

WAVE-FLUID MUD INTERACTION MODEL

After the fluidization, the fluid system above the stationary mud, i.e. the fluidized mud layer and water layer, is divided into N sub-layers, in which the water layer is represented by $N=1$ (Fig. 2). The governing equations are linearized Navier-Stokes equations and the continuity equations which are solved considering the boundary conditions. VE rheological model is adopted for fluid mud behavior. The VE parameters, i.e. μ and G, are adopted from the laboratory experiments of Shibayama and An (1993). Using different mud characteristics in the horizontal sub-layers of the fluid mud layer, the vertical distribution of fluid mud properties can be considered in simulation. The mud mass transport in fluid mud as well as the wave attenuation rate, k_i, is calculated by this model.

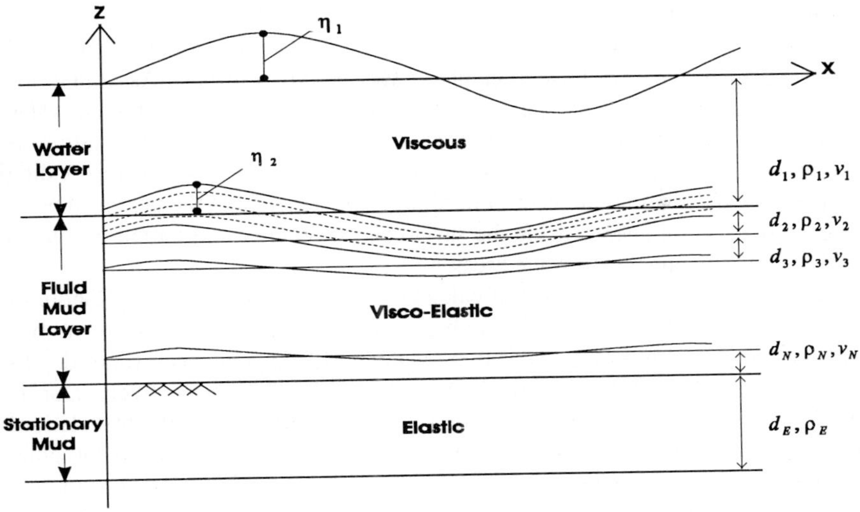

Figure 2. Definition of multi-layered wave-fluid mud interaction model.

IRREGULAR WAVE TRANSFORMATION

Assuming the exponential decay of wave height over a horizontal mud bed, the energy dissipation rate of mud bottom, ε_{D_m}, can be related to the wave attenuation rate, k_i, as (Soltanpour et al., 2003)

$$\varepsilon_{Dm} = -\frac{d}{dx}(C_g E) = 2C_g k_i E \tag{18}$$

where $E = \rho g H^2 / 8$ is the wave energy per unit surface area, ρ is the water density and C_g is the group velocity. After the breaking line, ε_D reflects both dissipation effects of wave breaking and fluid mud, i.e.

$$\varepsilon_D = \varepsilon_{Dm} + \varepsilon_{Db} \tag{19}$$

where ε_{Db} is the energy dissipation rate of wave breaking.

Computing the amplitudes of the wave components, the modeling wave spectra at $x = x_j$ can be defined as

$$\hat{S}_j(\bar{f}_n) = \frac{1}{2} a_{jn}^2 \Delta f_n \tag{20}$$

MODEL PERFORMANCE

The proposed model is used to compute the thickness of the fluidized mud and irregular wave transformation along the muddy profiles. Assuming the incident wave spectrum, the spectrum is divided into a number of regular wave components and the resultant fluidized depth on the first grid location is determined. Then the wave attenuation rate is calculated and the corresponding wave energy dissipation rate is used to find the wave height at the next grid point for each wave component. Inside the surf zone, the energy dissipation of mud is combined by the one due to wave breaking.

Fig. 3 shows the measured and calculated wave spectra at a site offshore of the town of Alleppey in Kerala, India, where the measurements are at two stations with a distance of about 300 m (Mathew, 1992). The mean water depths of at offshore site and the inshore site were 10 m and 5 m, respectively. During monsoon waves, a wave energy reduction of about 85% between the two stations was observed. As it is observed, the measured wave spectrum is in good agreement with the numerical simulations. The calculated thickness of fluid mud is also in the range of reported values in the area, i.e. up to 2 m (Kumar, 1997).

CONCLUSION

Summary and main conclusions of the present study are as follows:

1. Combining an analytical fluidization model and a wave-fluid mud interaction model, a cross-shore wave transformation model is presented.
2. Comparing the results of numerical simulation with the observed field data of Mathew (1992), it is revealed that the proposed wave transformation model is capable to predict the wave spectrum transformation along the muddy profiles. However, more field data is necessary for the verification of numerical model.

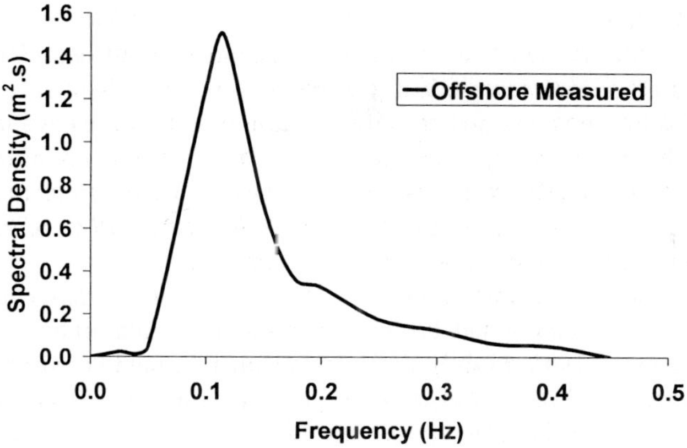

a. Wave spectrum at offshore station

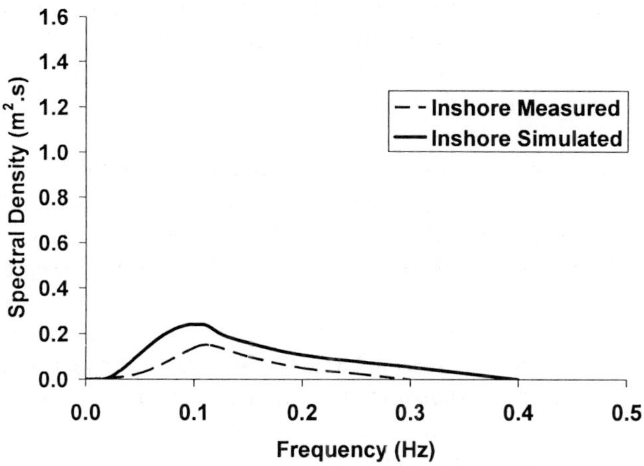

b. Measured and simulated wave spectra at inshore station

Figure 3. Measured and simulated wave spectra at Alleppey in Kerala, India (Mathew, 1992) (W=210%, G=5kPa, ρ_E=1300kg/m^3, ρ_{VE}=1250kg/m^3).

REFERENCES

Chou, H. T., M. A. Foda, and J. R. Hunt. 1993. Rheological response of cohesive sediments to oscillatory forcing, in *Nearshore and Estuarine Cohesive Sediment Transport, Coastal Estuarine Sci. Ser.*, vol.42, edited by A. J. Mehta, chap. 8, AGU, Washington, D. C., pp. 126-148.

Foda, M. A., Hunt, J. R. and Chou, H. T. 1993. A nonlinear Model for the Fluidization of Marine Mud by Waves. American Geophysical Union, *Journal of Geophysical Research*, Vol. 58, No. C4, pp. 7039-7047.

Gade, H. G. 1958. Effects of a non-rigid, impermeable bottom on plane surface waves in shallow water, *Jour. Marine Research*, Vol. 16, No. 2, pp. 61-82.

Kumar, P.K.Sassi 1997. The mud banks of south-west coast of India and its effects on the coastline, *Proceedings of Second Indian National Conference on Harbour and Ocean Engineering*, pp. 1191-1204.

Mathew, J. 1992. Wave-Mud Interaction in Mud banks, *Ph.D. dissertation*, Cochin University of Science and Technology, Cochin, Kerala, India.

Ross, M. A., Mehta, A. J. 1990. Fluidization of soft estuarine mud by waves. In: *The Microstructure of Fine Grained Sediments: From Mud to Shale*, R. H. Bennett ed., Springer-Verlag, New York, pp. 185-191.

Shibayama T., An, N. N. 1993. A visco-elastic-plastic model for wave-mud interaction, *Coastal Engineering in Japan*, Vol. 36, No. 1, pp. 67-89.

Soltanpour, M., Shibayama, T. 2002. Wave-mud interaction under irregular waves, *Proc. 28th Coastal Eng. Conf.*, ASCE, pp. 2949-2958.

Soltanpour M., Shibayama T., Noma T. 2003. Cross-shore mud transport and beach deformation model, *Coastal Engineering Journal*, Vol. 45, N0. 3, pp. 363-387.

Soltanpour, M., Shibayama, T., Masuya, Y., Sabzevari, I. 2004. Wave attenuation and mud mass transport under irregular waves, *Proc. 29th Coastal Eng. Conf.*, ASCE, pp. 1851-1860.

Soltanpour, M., Shibayama, T., Haghshenas, S.A., 2005. An integrated hydrodynamic model on muddy coasts, *Proc. 8th InterCoh Conf.* (to be published).

Zhang, Q. H., Zhao, Z. D. 1999. Wave-mud interaction: wave attenuation and mud mass transport, *Coastal Sediments "99"*, pp. 1867-1880.

DAMPING OF GRAVITY WAVES BY FIELDS OF FLEXIBLE VEGETATION

Sávio Freire Lima[1], Claudio Freitas Neves[1], Nara Maria Luzzi Rosauro[2]

This work presents a method for determining wave damping through vegetation fields which does not assume the individual plants as rigid, fixed structures. The proposed model combines empirical observations and a mathematical formulation that computes the depth integrated, time averaged force on the stem, based on a single unknown parameter, the drag coefficient. Video images of the movement of one isolated stem were used to calibrate the model. Experiments were conducted on a wave flume, using stretches of 6 mm flexible nylon ropes to simulate a particular type of vegetation, *Brachiaria subquadrípara*. The expression of the force on a single stem is introduced on the time averaged, depth integrated momentum equations for the fluid as a resistance term, multiplied by the spatial density of stems and by an ensemble coefficient. The wave height decay can thus be determined. Experimental results indicate that the interaction among stems plays an important role on total wave decay.

INTRODUCTION

The presence of aquatic vegetation in lakes, coastal wetlands and open coasts changes the local hydrodynamic circulation and plays an important role on the environment (Bruno & Kennedy 2000; Mendez & Losada 2004). The hydrodynamic behavior of the plants varies significantly, and any attempt to finding a general theory for plant-flow interaction is bound to fail. It is well accepted, though, that size, geometry and mechanical properties of the individuals, their spatial density and distribution, as well as flow pattern should be determinant on the study of wave damping (Mork 1996). Lima (2005) presents a comprehensive review of the subject and points out that all studies are restricted to specific conditions.

This study was motivated by the observation of two species, *Brachiaria subquadrípara* and *Brachiaria mutica*, two weeds commonly found in pastures in Brazil (Carbonari *et al.* 2003), which can survive flooding by higher water levels in hydroelectric power plant reservoirs. Fields of these plants actually protect the shores against wave action, as it was confirmed during exploratory field trips around Ilha Solteira Lake, a man-made reservoir. Indeed, wherever the vegetation was removed, shore erosion was likely to occur. In addition, ship waves also affected the morphological equilibrium of the margins along the waterway, thus contributing to an excess of sedimentation in the lake.

In order to quantify the degree of damping caused by vegetation, an experiment at Ilha Solteira reservoir was conducted in May 2003, as a cooperative effort of the Federal University of Rio de Janeiro (COPPE/UFRJ),

[1] Ocean Engineering Program/COPPE, Federal University of Rio de Janeiro. Caixa Postal 68508, 21945-900 Rio de Janeiro, RJ, Brazil.

[2] Hydraulic Research Institute, Federal University of Rio Grande do Sul, Av. Bento Gonçalves, 9500, 90.650-001 Porto Alegre, RS, Brazil.

the Federal University of Rio Grande do Sul (IPH/UFRGS) and the "Julio de Mesquita Filho" São Paulo State University (FEIS/UNESP), sponsored by the Brazilian Ministry of Education (CAPES). Directional spectra were measured in the middle of the vegetation, in 1 m water depth, and farther offshore, outside the vegetated area, in 5 m water depth. The wave induced underwater movement of the vegetation was filmed and its magnitude was measured using a topographic rod.

REVIEW OF EXISTING MODELS

There are various models for simulating the interaction between waves and the vegetated area. Some consider it as a high viscosity layer (e.g. Price *et al.* 1968; Mork 1996), others represent the vegetation as a friction force at the bottom (e.g. Camfield 1983), but most researchers seeks to simulate the forces induced by the waves along the body of the plant (e.g. Dalrymple *et al.* 1984; Kobayashi *et al.* 1993; Mendez & Losada 2004). In most models, though, the displacement of the plant itself is disregarded, hypothesis which is not supported by evidence, at least in the case of *B. subquadripora* at Ilha Solteira lake.

Dalrymple *et al.* (1984) considered the plants as slender rigid structures, which caused a reaction force against the wave flow. Based on energy balance arguments and assuming a quadratic law for the drag force, these authors derived an expression for wave height decay along the vegetation field

$$H(x) = H_o\left[\frac{1}{1+\beta x}\right] \tag{1}$$

where H_o is the starting wave height and β is given by the expression:

$$\beta = \frac{H_o\, N_h\, C_D\, \phi}{24\, g\, c_g\, k\pi}\,\frac{\left[\sinh^3(\lambda kh)+3\sinh(\lambda kh)\right]}{3\cosh^3(kh)}\left(\frac{gk}{\sigma}\right)^3 \tag{2}$$

where N_h is the number of specimens per unit area, h is the local water depth, λ is the ratio between the stem submerged length to water depth, c_g is the wave group velocity, k is the wave number, σ is the wave frequency, C_D is a drag coefficient for the entire field.

Kobayashi *et al.* (1993) assumed that the drag force was a linear function of the fluid particle velocity, and found a solution in terms of an exponential decay.

Which decay model better reproduces the field conditions remains an open question, intrinsically related to the *a priori* assumed expression for drag. For short propagation distances, both models are asymptotically similar, but for longer distances, their expressions would tend to differ more significantly. Because the horizontal extension of the vegetation field is usually too short compared to the wave length, it is hard to precise which model would better represent the physical processes involved.

Mendez & Losada (2004) have extended the solution of Dalrymple *et al.* (1984) to shallow water and mildly sloped bottom. This formulation is restricted to shallow water, a condition which is seldom fulfilled at lakes, where wave periods are short, wave steepness is high, and nearshore bottom slopes are comparatively higher than those at ebb shoals or ocean beaches.

All previous models have assumed the stems to be rigid, a severe restriction on the applicability to highly flexible plants. Even Mendez *et al.* (1999), who have considered the plant movement on the computation of the hydrodynamic forces, modeled the plant stems as cantilever beams, fixed at the bottom, and subject to hydrodynamical transversal loads.

This article presents an alternative approach which considers large displacements of the flexible stems and a non-linear wave theory to compute the flow dynamics. Table 1 compares four studies found in the literature about the problem to the model proposed by Lima (2005). Most of them are based upon a global description of the vegetation field and try to determine the final effect on the wave. The total resistance is just the arithmetical sum of many individual resistances, given by the density of individuals per unit area. However, this hypothesis is flawed because, when large displacements occur, it is highly probable that adjacent stems interact with each other, or get entangled, or produce vortices which affect the motion of the neighboring stems. Therefore, other types of resistance may appear.

Table 1 – Summary of most frequently used models to represent wave damping by vegetation.

	Reviewed Models				
	DALRYMPLE *et al.* (1984)	KOBAYASHI *et al.* (1993)	MENDEZ *et al.* (1999)	MENDEZ & LOSADA (2004)	Proposed Model
Plant Influence	Energy dissipation	Distributed force	Distributed force	Energy dissipation	Average force
Stem movement	No	No	Yes	No	Yes
Model of force	Quadratic drag	Linear drag	Linear drag and inertia	Quadratic drag	Quadratic drag (inertia)
Bottom bathymetry	Plane and horizontal	Plane and horizontal	Plane and horizontal	Plane, constant slope	Plane and horizontal
Wave theory (*)	Linear	Linear	Linear	Linear, shallow water	Stream Function
Stem Interaction	No	No	No	No	Yes

(*) Reference to wave theory which is used for computing hydrodynamic forces.

PROPOSED MODEL

The aquatic vegetation, such as the one found at Ilha Solteira Lake, is modeled by fixed length, buoyant, flexible stretches of nylon rope, attached at the bottom, and put in movement by the oscillatory wave motion. In this paper, the ropes will be referred to as "flexible elements" and mathematically modeled as a string of nodes, unable to resist to bending moments. As the waves propagate through a field of such elements, there is an exchange of momentum, part of the kinetic energy of the flow is dissipated, and the wave height decays.

The current work is structured in three steps. First, the depth integrated, time averaged momentum equations are derived in order to include an average resisting force due to the presence of the elements. The reasoning follows that by Longuet-Higgins, who favored the momentum instead of energy balance equation, because it is possible to include the radiation stress terms, mean level gradients and the reaction of the elements. Second, a calibrated model of the displacement of the element is developed, based on the observations of a single stem, providing the hydrodynamic resisting force which will be introduced on the momentum equations previously derived (Lima *et al.* 2005). Third, a wave decay equation is obtained, which depends on geometrical properties of the field and on an "ensemble parameter" α.

Field Model

The depth integrated, time averaged momentum equations may be written as (e.g. Dean & Dalrymple 1984):

$$\frac{\partial}{\partial t}M_i + \frac{\partial}{\partial x_j}\left(U_i M_j\right) = p_{atm} - \rho_f g\left(h+\overline{\eta}\right)\frac{\partial \overline{\eta}}{\partial x_i} - \frac{\partial}{\partial x_j}S_{ij} + \frac{1}{T}\int_0^T F_i\, d\overline{t} \qquad (3)$$

where S_{ij} is the radiation stress and the indexes i and j refer to directions x and z, respectively; M_i is the total mass transport, resulting from the superposition of an arbitrary current and the second order wave mass transport; U_i is the total depth averaged current velocity; $\overline{\eta}$ is the mean free surface elevation; $\overline{p}_{atm}$ is the atmospheric pressure; ρ_f is the fluid density; g is gravity; and F_i is the depth averaged resisting force exerted by the elements on the flow.

Assuming a steady wave regime, long crested waves propagating along the x-direction, and retaining terms up to $O(H)^2$, equation (3) reduces to:

$$H\frac{\partial}{\partial x}H = -\frac{8}{\rho_f g(4n-1)}\left[\frac{1}{4}\rho g H^2\frac{\partial}{\partial x}n + \rho_f g\left(h+\overline{\eta}\right)\frac{\partial}{\partial x}\overline{\eta} + \overline{F}_x\right] \qquad (4)$$

where n is the ratio between wave group celerity and phase velocity.

Field Resistance

The total resisting force exerted on the fluid per unit horizontal area, in case all elements are identical, is a function of the force exerted by a single element ($\overline{F}_x^1$) during one wave period, multiplied by the spatial density of stems (N_h) and an empirical coefficient α, which will be denominated as an "ensemble interaction coefficient". Hence,

$$\overline{F}_x = \alpha N_h . \overline{F}_x^1 \qquad (5)$$

Different from other models found in the literature, the total resisting force is not the simple summation of individual effects, because of the inclusion of the ensemble coefficient. Later, by means of dimensional analysis an expression for α will be determined, as a function of flow parameters and stem properties.

Resisting Force by a Single Stem

Consider a flexible stem, which is set in movement by an oscillatory flow and the only restoring force is buoyancy. By Newton's law, the resisting force of the stem against the flow equals in magnitude the force exerted by the flow on the stem (Figure 1), as a result of the relative movement between the fluid and the stem.

Assuming the stem has uniform mass density, it is divided into n segments of constant length, ℓ_n, which are acted upon by hydrodynamic forces. At a given instant of time t', the center of mass of each segment will have velocity $\vec{u}_{v(n,t')}$ while the fluid will be moving with velocity $\vec{u}_{f(n,t')}$.

Due to the small resistance of the plant to bending moments, the stem is therefore idealized as a string of nodes, equally spaced, and freely articulated on the x-z plane. The depth is constant and equal to h. A Cartesian coordinate system is fixed at the bottom, so that the free surface is positioned at $z = h+\eta$. For the nth segment, a system of three equations is established – a constitutive equation which establishes that the stem has fixed length and the momentum balance in the horizontal and vertical directions:

$$\begin{cases} \Delta x_{(n,t')} = \left[\Delta x_{(n-1,t')} + \dfrac{\Delta z_{(n-1,t')}}{\mathrm{tg}\,\theta_{(n,t')}} \right] - \dfrac{\Delta z_{(n,t')}}{\mathrm{tg}\,\theta_{(n,t')}} \\ F_{Tx(n,t')} - T_{(n,t')} \sin\theta_{(n,t')} + T_{(n+1,t')} \sin\theta_{(n+1,t')} = \dfrac{2M}{\Delta t} \left[\dfrac{\Delta x_{(n,t')}}{\Delta t} - u_{vx(n,t')} \right] \\ F_{Tz(n,t')} + F_E - T_{(n,t')} \cos\theta_{(n,t')} + T_{(n+1,t')} \cos\theta_{(n+1,t')} = \dfrac{2M}{\Delta t} \left[\dfrac{\Delta z_{(n,t')}}{\Delta t} - u_{vz(n,t')} \right] \end{cases} \tag{6}$$

where $F_{Tx(n,t')}$ and $F_{Tz(n,t')}$ are the horizontal and vertical components of the hydrodynamical force acting on nth the node, $T_{(n,t')}$ is the interaction between nodes (internal stresses on the stem), $\theta_{(n,t')}$ is the angle of the nth node with the vertical at a given time t', and M is the mass of the cylindrical element $M = \rho_v V_{el}$ whose density is ρ_v and volume V_{el}.

For the total length of the stem, with N nodes, there will be 3N equations and 3N unknowns. The solution of the system provides the connecting forces between nodes as well as the vertical and horizontal displacements of each node. Further details about the derivation of Eq. (6) and the solution of the system are presented by Lima *et al.* (2005).

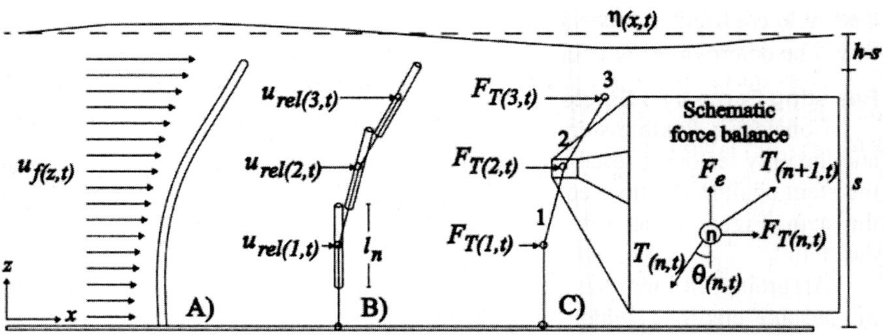

Figure 1 – Mechanical model for the displacement of one single stem. A) Original distorted form. B) Discretization into segments. C) Force balance at the nodes.

The generalized Morrison equation, accounting for the movement of the structure, is assumed to be valid (e.g. Chakrabarti 1987):

$$\vec{F}_T = \underbrace{\frac{1}{2} C_D \rho_f \vec{A}_{eq} \vec{u}_{rel} \left| \vec{u}_{rel} \right|}_{1} + \underbrace{k_m \rho_f V_{el} \vec{a}_{rel}}_{2} + \underbrace{\rho_f V_{el} \vec{a}_f}_{3} \tag{7}$$

where C_D is the drag coefficient, $\vec{A}_{eq}$ is the projection of the area of the stem on the direction normal to the relative velocity between fluid and flow, $\vec{u}_{rel} = \vec{u}_f - \vec{u}_v$, k_m is the added mass coefficient which depends on the geometry of the structure, and $\vec{a}_{rel} = \vec{a}_f - \vec{a}_v$ is the relative acceleration. The drag coefficient C_D depends both on the flow and on the structure, and is a function of the Reynolds number and Keulegan-Carpenter coefficient.

For establishing the mechanical model of the stem, its movement was filmed on video both from above and sideways, in order to certify that transversal movements could be neglected.

Dean's Stream Function Wave Theory (Dean 1968; Dean and Dalrymple 1984) was used to compute the flow kinematics, because it accounts for the asymmetry between crest and trough velocities in order to compute the hydrodynamical forces.

EXPERIMENTAL PROCEDURES

Two sets of experimental work were conducted to characterize the damping caused by a specific type of flexible element using the proposed approach. The first set of experiments was performed with a single artificial element, and the second with fields of the same elements. For the element fields, two different spacing between elements were used: 50 mm and 25 mm. Artificial stems whose hydrodynamic behavior was similar to that of *B. subquadripora*, one of the vegetal species observed at Ilha Solteira reservoir, were used in all experiments.

The choice of the artificial element was based on two criteria: (i) the stems should have small resistance to flexion and be maintained on the vertical position by buoyancy alone; (ii) under the action of waves, whose heights and periods were similar to those measured in the reservoir, the stem displacements should have the same magnitude as those observed in the native species.

The selected element consisted of a braided nylon rope with a diameter of 6 mm and density $\rho_v = 681.43$ kg/m^3, which satisfied the imposed conditions and was easy to manipulate.

All experimental activities were conducted at the Waterways Research Institute (INPH) in Rio de Janeiro. The flume was 43 m long, 0.8 m wide and 1 m deep, it had at one end a piston type, computer controlled wave generator and at the other end a 5 m long sloping beach of granulated material. All tests were conducted with monochromatic waves.

Measurements of the Behavior of One Isolated Element

For calibrating the single stem model, the movement of marked points along the stem, each one 5 cm apart from the other, was recorded on video for 1 minute. Knowing the position of these points in different times enabled the calculation of the velocity and acceleration along the stem, and the drag coefficient was thus determined as the best fit for the theoretical expression (7).

Two water levels were used – 0.75 and 0.55 m – and, for each case, wave heights were approximately 0.10 and 0.20 m with periods of 1.0, 1.5 and 2.0 s. Stem lengths were always 0.1 m smaller than the water depth, to ensure that, even for the highest wave, the stem remained submerged, thus avoiding occurrence of other phenomena caused by reduction of buoyancy. Test conditions are described in full detail in Lima (2005) and Lima *et al.* (2005).

Measurement of the Behavior of a Stem Field

Figure 2 presents the experimental set-up, indicating the location of the field of artificial vegetation and points of measurement in the flume. Each artificial stem was fixed in a metallic grid mounted at the bottom of the flume, creating a 3 m long, 0.75 m wide field. The stems were spaced 50 mm apart, resulting in a spatial density N_h= 400 stems/m^2, and 25 mm apart, giving a density N_h= 1600 stems/m^2. Figure 3 presents the assembling of the artificial stems and Figure 4 shows the field positioned in the flume.

Nine combinations of wave conditions were tested: heights of 0.05, 0.10 and 0.15 m, and periods of 1.00, 1.25, and 1.50 s; water depth was kept at 0.70 m and the stems were 0.60 m long.

The experiments were performed in three stages: (i) without the artificial element field; (ii) with the field of density 400 stems/m^2; (iii) with the field of density 1600 stems/m^2. During experiments wave heights were recorded along the whole extension of the field with 5 probes. Two 16 MHz micro-ADV measured the velocities at elevations of 70% and 50% of the water depth above the bottom, along the vertical coincident with the last wave probe. Each experiment was reproduced 3 times and the averaged decay rate was used for calibration of the model. Considering the 9 selected wave conditions, the experiments without vegetation and the two different spacings of elements, a total of 81 experiments were performed.

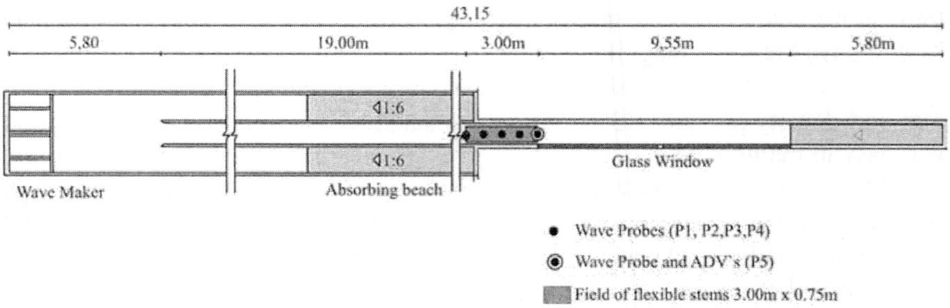

Figure 2 – Scheme of the experimental apparatus for measurement of element fields.

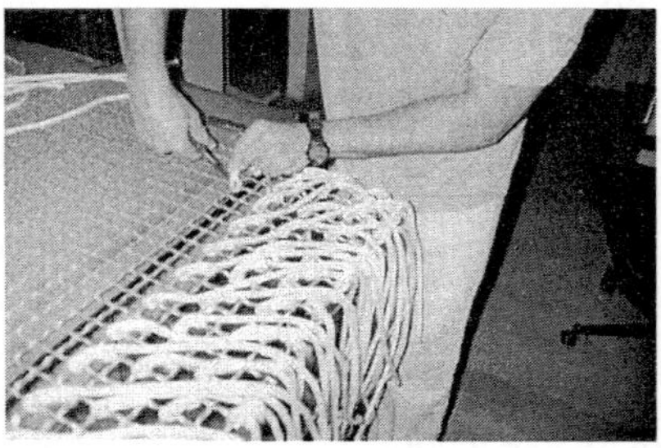

Figure 3 – Preparing of field experiments by assembling of 0.6 m artificial stems in 3 sets of 1.00 x 0.75 m metal grid.

Figure4 – Preliminary tests with field of artificial elements.

RESULTS

Behavior of a Single Stem

Twelve experimental configurations were simulated with the one-stem mathematical model, using the coefficient C_D as the fitting parameter. Terms which depended on the volume of the stem, such as inertia force, were neglected. For all tests, the sample interval used for comparison with the model consisted of three wave periods. The interval was selected after the stem movement had become harmonic, in order to neglect the initial transient wave, as well as before a reflected wave had reached the measurement site.

With the objective of obtaining a more general expression for the wave decay, it is necessary to establish an equation to represent $\overline{F}_x^1$ as a function of directly measurable variables of the flow and the stem. With the use of dimensional analysis it is possible to obtain such an expression. Assuming a functional relation between these parameters as:

$$\left(\frac{\overline{F}_x^1 T^2}{\rho_v s^4}\right) = a\left(\frac{H}{L}\right)^b \left(\frac{h}{L}\right)^c \left(\frac{\phi}{L}\right)^d \tag{8}$$

where a, b, c and d are coefficients to be determined by multivariate linear regression and $\overline{F}_x^1$ was generated by the numerical model. The values obtained were $a = 0.133$, $b = 3$, $c = -4$ and $d = 1$. Substituting these values in Eq. (8) the final equation for the drag force is obtained:

$$\overline{F}_x^1 = 0.133\frac{\rho_v \phi \lambda^4 H^3}{T^2} \tag{9}$$

The final result becomes an expression with good representativity (R^2=0.95). Equation (9) presents a cubic term on wave height, which, according to discussion that will be presented later, brings important consequences in the calculation of wave decay.

Behavior of a Stem Field

Using Eq. (8), an expression for the wave height decay is obtained:

$$H^{1-b}\frac{\partial}{\partial x}H = -a\frac{8\alpha N_h s^4}{gT^2(4n-1)L^b}\left(\frac{\rho_v}{\rho_f}\right)\left(\frac{h}{L}\right)^c\left(\frac{\phi}{L}\right)^d \tag{10}$$

Integrating this expression, considering that $H=H_0$ for $x=0$, gives:

$$\frac{H}{H_o} = \left[1 - a\frac{8\alpha N_h s^4(2-b)}{gT^2(4n-1)L^b H_o{}^{2-b}}\left(\frac{\rho_v}{\rho_f}\right)\left(\frac{h}{L}\right)^c\left(\frac{\phi}{L}\right)^d x\right]^{\frac{1}{2-b}} \tag{11}$$

and by substituting the obtained values of coefficients a, b, c e d in equation (11) one obtains an expression of the type:

$$\frac{H}{H_o} = \frac{1}{1+\tilde{\beta}x} \tag{12}$$

where

$$\tilde{\beta} = 1.064\frac{\alpha N_h \phi \lambda^4 H_0}{gT^2(4n-1)}\frac{\rho_v}{\rho_f} \tag{13}$$

This form of decay is the same as that obtained analytically by Dalrymple *et al.* (1984) when considering the vegetation structures as rigid cylinders and neglecting their movement. The fact that experimental data led to a similar solution suggests that equation (12) might have a wider use in phenomenon of wave decay for fields of slender, flexible elements, and supports the hypothesis of a quadratic law for interaction fluid-structure.

Expression (13) enhances those aspects that influence directly the final phenomenon investigated in this work, i.e., the magnitude of wave decay such as the increasing decay with buoyancy (ρ_v/ρ_f) and stem diameter (ϕ). There is a strong non-linear dependence on the portion of the water depth occupied by the stem (λ), defined as the rate between its length (s) and the water depth (h). This study did not consider the case in which the stems pierce the water surface (λ>1). Waves of larger period (longer waves) are less affected by vegetation, since the decay rate depends on the inverse of the square of the period; consequently, a vegetation field works as a filter for shorter waves, exactly those which would cause greater damage to the shore. Finally, it should be noted that the derived relation is valid for any values of relative water depth

All parameters in Eq. (13), but the coefficient that describes stem interaction (α), are determined by experimental conditions. By minimizing the average quadratic error between experimental and numerical values, this parameter was estimated for each of the experiments. Figure 5 presents a comparison between the measured curves and those predicted by the model for α=1 and for values of α equal to the fitted values. The abbreviation EMsf refers to the proposed Empirical Model for stem field.

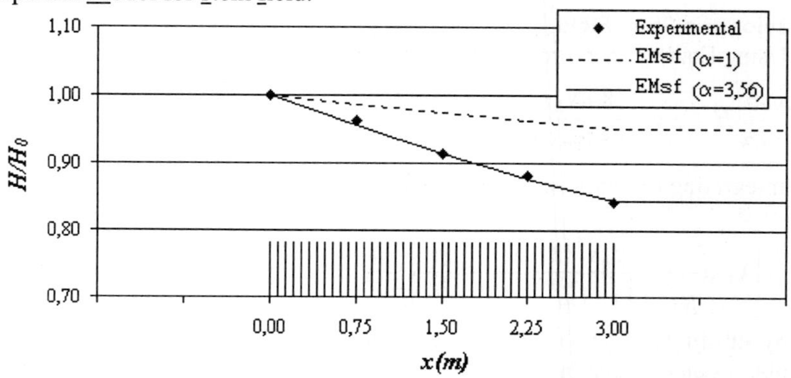

Figure 5 – Wave height decay for experiment 01-Ch (E_v=0.025m).

A significant increment in wave decay occurs when the number of stems per square meter was increased from 400 (E_v=0.050 m) to 1600 (E_v=0.025 m). In the first case, the reduction of wave height along the field length varied between 2.5 and 5%, whereas for the second case the decay rate was about 15%. This result is very significant, particularly because the field length is of the order of one to two wave lengths only.

The best fit values for the stem interaction coefficient are presented in Table 2 and, as expected, α was more significant in experiments with higher stem field density as a result of increased direct contact between adjacent elements. It is also observed that α increases with wave period (T) and decreases with the incident wave height (H_o).

The resistance to the flow imposed by the group of stems is in average four times greater than that estimated by adding up individual forces for all stems in the field. This result indicates that, contrary to what is usually proposed by some other theoretical approaches found in the literature, stem interaction is a phenomenon of greater significance. Hence, estimating the total resistance for the field by adding up individual resistances cannot be justified, particularly for flexible stems, where entanglement and/or direct contact is likely to occur.

By dimensional analysis, the dependence of α was established as.

$$\alpha = a''\left(\frac{H_0}{E_v}\right)^{b''}\left(\frac{s}{L}\right)^{c''}\left(\frac{\phi}{E_v}\right)^{d''} \tag{14}$$

Table 2 – Coefficient representing interaction between stems (α) determined for each wave configuration used in the experiments.

Wave	H_o (m)	T(s)	α (E_v=0.050)	α (E_v=0.025)
01-Ch	0.05	1.00	2.309	3.563
02-Ch	0.05	1.25	4.553	6.320
03-Ch	0.05	1.50	8.789	12.237
04-Ch	0.10	1.00	2.102	1.865
05-Ch	0.10	1.25	3.396	3.293
06-Ch	0.10	1.50	4.604	6.124
07-Ch	0.15	1.00	0.694	1.291
08-Ch	0.15	1.25	0.756	2.507
09-Ch	0.15	1.50	3.459	4.410
		Average	**3.407**	**4.623**
		Standard Deviation	**2.475**	**3.340**

By multivariate linear regression, it was finally obtained:

$$\alpha = 7.44\frac{L^2\phi^{3/2}}{H_o s^2 E_v^{1/2}} \tag{15}$$

where the wave length used in the equation was estimated by 3rd Order Stream Function Theory.

Decay Model

Discussions based on theoretical considerations and experimental evidences allows, at this point, the proposal of a decay model to be applied for gravity waves traveling along a vegetated field formed by flexible elements. The wave height can be obtained for any point using the expression:

$$\frac{H}{H_o} = \frac{1}{1+\tilde{\beta}x} \tag{16}$$

where substitution of the expression obtained for α into Eq. (13) gives:

$$\tilde{\beta} = 7.916\frac{L^2\lambda^4}{gT^2(4n-1)s^2}\frac{\rho_v}{\rho_f}\phi^{5/2}N_h^{5/4} \tag{17}$$

In this way, for the same stem, the decay rate will be a function only of water depth and wave length, and not dependent on the incident wave height. Figure 6 presents curves of wave decay for the studied conditions.

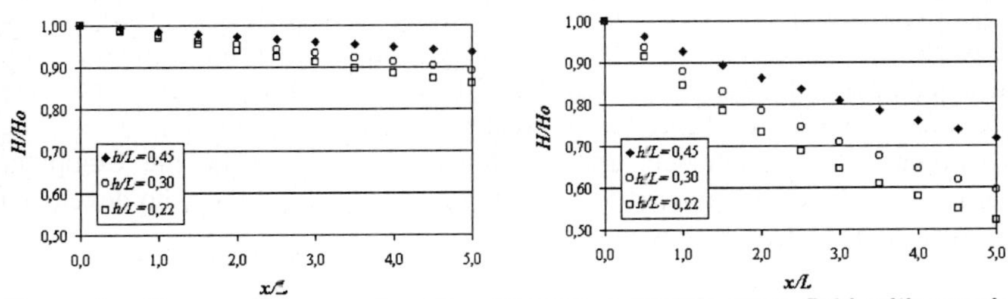

Figure 6 – Decay of a wave traveling through a flexible stem field with spacing E_v=0.050 m (left) and E_v=0.025 m (right).

CONCLUSION

This article presented a model of wave damping by a field of artificial flexible stems, situation which simulates the propagation of waves through vegetation along the shores of lakes. Two processes are modeled independently. First, the resisting force against the oscillatory flow by one single stem is simulated by a mechanical model which combines physical observation with a finite differences numerical model. Then, the ensemble resisting force and the wave height decay is modeled by an equation derived from the time averaged, depth integrated horizontal momentum equation. The total force is multiplied by a coefficient which expresses the combined effect of the ensemble and depends on the geometry of the stems, their spatial density, initial wave height and wave length (or period).

Although only drag forces have been considered in the conceptual model, the stem response was satisfactory and, after calibration, it was possible to generate time series of force, velocity and position of the stem.

For the wave conditions and mechanical properties of the stems that were tested, the ensemble resistance of the field was four times bigger than the sum of the individual forces caused by each stem. Because of the high flexibility of the stems, they contact each other or get entangled as the waves propagate through. For quantifying this effect, wave decay was measured at 5 points within the field. However, interaction among stems may also be present in case of less flexible or rigid structures, for lee vortices behind the stems may interact with each other and affect the resisting forces.

The formulations found in the literature for wave height decay usually fall in two categories: algebraic or exponential expression. The present study, based on dimensional analysis and empirical formulation, obtained an algebraic equation, similar to the one proposed by Dalrymple *et al.* (1984) for rigid structures. The empirical coefficient in the resulting expression turned out to be independent of the initial wave height but dependent of the relative depth.

The derived expressions might be an important tool in biological and environmental studies and could be used for designing low cost shore protection devices, a relevant investigation for navigation channels and artificial lakes.

ACKNOWLEDGEMENTS

The present work was funded by a grant of Brazilian Ministry of Education, Project CAPES/PROCAD 0144/01-0 – "Friends of Boussinesq – Cooperative Network for Research on Shallow Water Coastal Hydrodynamics", with participation of COPPE/UFRJ, FEIS/UNESP and IPH/UFRGS. The authors are especially thankful to Institute for Waterway Research (INPH), where the experiments were conducted, and to Moacyr de Paula, from the Computer Center of the Federal University of Rio de Janeiro (NCE/UFRJ) who developed the software for image processing.

REFERENCES

BRUNO, J. F., KENNEDY, C. W., 2000, "Patch-size dependent habitat modification and facilitation on New England cobble beaches by Spartina alterniflora". *Oecologia*, v. 122, p. 98-108.

CAMFIELD, F. E., 1983, "Wind-wave growth with high friction". *Journal of Waterways, Por., Coastal and Ocean Engineering*, v. 109, n. 1, p. 115-117.

CARBONARI, C.A., MARTINS, D. TERRA, M.A., 2003, "Controle de Brachiaria subquadripara e Brachiaria mutica através de diferentes herbicidas aplicados em pós-emergência". *Planta Daninha*, v.21, Edição Especial, p.79-84.

CHAKRABARTI, S. K., 1987, *Hydrodynamics of offshore structures*. 1 ed. New York, Springer-Verlag.

DALRYMPLE, R. A., KIRBY, J. T., HWANG, P.A., 1984, "Wave refraction due to areas of energy dissipation". *Journal of Waterways, Port., Coastal and Ocean Engineering*, v. 110, n. 1, p. 67-79.

DEAN, R. G., 1968, *Relative validities of water wave theories*. In: Technical Progress Report No. 16, University of Florida, Gainesville.

DEAN, R. G., DALRYMPLE, R. A., 1984, *Water wave mechanics for engineers and scientists*. 1 ed. Englewood Cliffs, New Jersey, Prentice-Hall Inc.

KOBAYASHI, N., RAICHLEN, A. W., ASANO, T., 1993, "Wave attenuation by vegetation". *J. Waterways, Port, Coastal and Ocean Eng. Div.*, v.199, no.1, p. 30-48.

LIMA, S. F., 2005. *Amortecimento de ondas gravitacionais por campos de vegetação simulados por elementos de grande flexibilidade*. Doctoral Dissertation, Ocean Engineering Program, Federal University of Rio de Janeiro, COPPE/UFRJ.

LIMA, S. F., VASCO, J.R., NEVES, C.F., MACIEL, G.F., ROSAURO, N.M.L., 2005, "Simulation of Flexible Vegetation Motion Induced by Water Waves". In: Proceedings, Waves 2005, ASCE, CD-ROM, Madrid.

MENDEZ, F. J., LOSADA, I. J., LOSADA, M.A., 1999, "Hydrodynamics induced by wind waves in a vegetation field". *Journal of Geophysical Research*, v. 104, n. C8, p. 18383-18396.

MENDEZ, F. J., LOSADA, I. J., 2004, "An empirical model to estimate the propagation of random breaking and nonbreaking waves over vegetation fields". *Coastal Engineering*, v. 51, p. 103-118.

MORK, M., 1996, *Wave attenuation due to bottom vegetation*. In: Waves and nonlinear processes in hydrodynamics, Kluwer Academic Publishers, p. 371-382.

PRICE, W. A., TOMLINSON, K. W., HUNT, J. N., 1968, "The effect of artificial seaweed in promoting the build-up of beaches". In: *Proceedings of the 11th Coastal Engineering Conference*, v. 1, p. 570-578.

THREE-DIMENSIONAL MODELING OF AIR-SEA INTERACTION

Jennifer L. Regis[1] and Donald N. Slinn[1]

We have modified a fully 3D, volume of fluid model, entitled TRUCHAS, to examine air-sea interactions, with particular emphasis on the deep-water realm. The ability of TRUCHAS to model multiple fluids within the same computational domain allows for a more detailed analysis of wave-breaking events. Validation of the model's predictive capabilities is accomplished through correlation with data from laboratory experiments conducted at the University of Miami's Air-Sea Interaction Salt-Water Tank (ASIST), possessing both wind and wave generating capabilities. Model results have shown good comparison with measured laboratory data for such experiments. Future work with this model will include a more comprehensive examination of the transfer of momentum from air to sea during a breaking episode, as well as in-depth studies of the generation and development of wind-waves and a detailed analysis of the impact of wave forcing on bridge deck structures.

INTRODUCTION

The breaking of water waves is crucial in the transfer of heat, gases, and momentum between the ocean and atmosphere. Observations of these events have provided qualitative support to this end, and yet accurate quantification of momentum transfer for strong winds and nonlinear waves has remained a challenge. In recent years, advances have been made in the development of algorithms to capture and describe air-sea interaction. Most, however, are idealized and only capable of modeling fluid flow within the two-dimensional approximation. While some three-dimensional studies have had success in characterizing wave breaking in shallow water, few studies have explored the phenomenon in deep-water settings. In addition, many models have their basis in potential flow, and thus approximate or ignore many of the nonlinear characteristics of deep-water breaking waves.

We employ a three-dimensional, time dependent, finite difference model, entitled TRUCHAS, to address the issue of deep-water breaking waves. This volume of fluid model accurately depicts flow characteristics in both the water and air regimes. A great many details of the flow development are available for analysis from the model output. These include wind and water velocities, pressure gradients in both the air and sea around a breaking wave, the development and evolution of wind-generated waves, and the corresponding transfer of momentum from the atmosphere to the ocean.

---

[1] Department of Civil and Coastal Engineering, University of Florida, 365 Weil Hall, Gainesville, Florida, 32611, USA

METHODOLODY

Model Dynamics

TRUCHAS is a fully 3D, finite difference, time-dependent model of great competence. A successor of such models as the SOLA-VOF Method and RIPPLE, TRUCHAS couples its algorithms with a volume tracking method to accurately evaluate the fractional volume of each fluid in every mesh cell. The numerical code includes numerous algorithms to allow for the modeling of such phenomena as heat transfer and phase changes, chemical reactions, solid mechanics, electromagnetics, and fluid dynamics. This work will focus on the fluid dynamics algorithms inherent in the code.

TRUCHAS is capable of handling multiple immiscible fluids of varying density within a single domain. This added capacity allows for more detailed studies of the characteristics of both air and water regimes during air-sea interaction events. Momentum is conserved in TRUCHAS simulations through continuous solution of the Navier Stokes Equations These equations can be represented as a single Conservation of Momentum equation, shown in Eq. 1,

$$\frac{\partial(\rho \mathbf{u})}{\partial t} + \nabla \cdot (\rho \mathbf{u}\mathbf{u}) = -\nabla p + \nabla \cdot \boldsymbol{\tau} + f_B + f_S + f_D \quad (1)$$

where **u** is the three-dimensional fluid velocity, ρ is the density, p is pressure, $\boldsymbol{\tau}$ is the shear stress tensor, f_B represents body forces, f_S captures any surface forces, and f_D is the drag force included to describe flow in the vicinity of a solid or open boundary. Conservation of mass is then maintained via Eq. 2.

$$\frac{\partial \rho}{\partial t} + \nabla \cdot (\rho \mathbf{u}) = 0 \quad (2)$$

The fluid flow algorithm present within TRUCHAS is largely dependent on the Volume of Fluid method. TRUCHAS initiates its flow algorithm with a multidimensional Piecewise Linear Interface Calculation (PLIC) method to determine the fractional volume of each fluid within every mesh cell. The fractional volumes of each fluid, f_k, are bounded within the range $0 \leq f_k \leq 1$ as depicted below.

$$f_k = \begin{cases} 1 & \text{inside fluid k} \\ >0, <1 & \text{at the fluid k interface} \\ 0 & \text{outside fluid k} \end{cases}$$

The density of each cell can thus be defined in terms of fractional volumes of fluid, shown in Eq. 3.

$$\rho = \sum f_k \rho_k \quad (3)$$

Substituting this new density definition into the Conservation of Mass equation, Eq. 2, and utilizing the constraint that ρ_k must remain constant in time, an evolution equation for fractional volume, Eq. 4, can be ascertained,

$$\frac{\partial f_k}{\partial t} + \nabla \cdot (f_k \mathbf{u}) = 0 \quad (4)$$

where **u** is again the three-dimensional fluid velocity and f_k represents the volume fraction of fluid k. As Eq. 4 is continually resolved within the flow algorithm, TRUCHAS is able to track fluid volumes and march them forward in time via a first order forward Euler time step.

Model Improvement

We have added a new inflow boundary condition to the code to accurately allow for the influx of waves into the model's computational domain, effectively acting as a wavemaker for numerical simulations. Free-surface displacement, horizontal and vertical orbital velocities are specified at the inflow (x = 0 cm) boundary using the simple linear wave equations taken from Dean and Dalrymple (1991), shown below respectively in Eq. 5. Cross-tank velocities are taken as zero at the inflow boundary as there is assumed negligible movement across the numerical mesh.

$$\eta = H\cos(kx - \sigma t)$$
$$u = \frac{H}{2}\sigma\frac{\cosh(k(h+z))}{\sinh(kh)}\cos(kx - \sigma t) \qquad (5)$$
$$w = \frac{H}{2}\sigma\frac{\sinh(k(h+z))}{\sinh(kh)}\sin(kx - \sigma t)$$

In these equations, H is representative of wave height, k is the wave number, σ represents angular wave frequency, t is time, x is our horizontal position, and z is the vertical position, taken as zero at the free-surface with increasing negative values descending from the free-surface to the bed.

A snapshot of a deep-water wave simulation utilizing this new inflow boundary condition is given in Fig. 1. A numerical beach has been included to dissipate wave energy and to decrease reflection from the tank end. An open boundary condition is imposed at the top of the domain as well as at the outflow boundary above the numerical beach. This is accomplished by setting the pressure equal to ambient pressure in these regions. The constant-pressure outflow condition above the beach allows waves to propagate across the numerical domain, run up the beach, and flow completely out of the system. Mass is thus conserved within the domain without allowing the water level of the

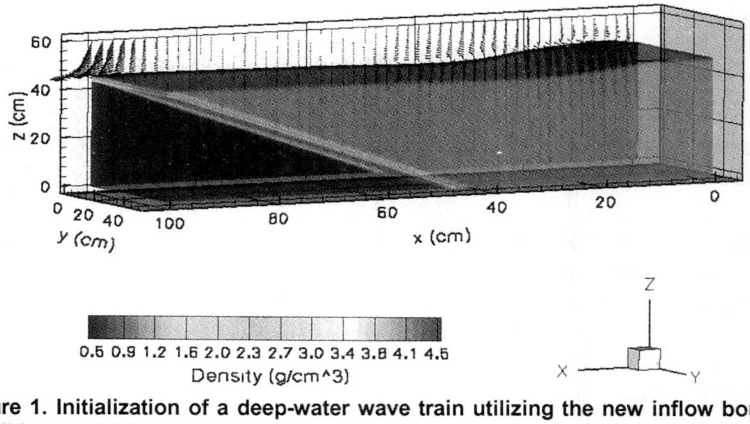

Figure 1. Initialization of a deep-water wave train utilizing the new inflow boundary condition, t = 0.75 seconds.

tank to rise. After numerous simulations were performed involving various wave train characteristics, we felt confident that we could force TRUCHAS to generate any wave climate of interest. We then moved on to validation of the model for use with deep-water spilling breakers.

ASIST EXPERIMENT

Laboratory Setup

To validate the competence of TRUCHAS to simulate deep-water breaking waves, a reliable data set was needed with which to compare model results. Mark Donelan and Brian Haus at the University of Miami provided such information using their Air-Sea Interaction Salt-Water Tank (ASIST). ASIST is a 15m by 1m by 1m stainless steel and acrylic tank fitted with a programmable wavemaker. Donelan and Haus programmed the wavemaker to produce a Gaussian wave packet in which waves were designed to coalesce and break at a certain point within the wave tank. Measurements of free-surface elevations were taken using multiple laser elevation gauges. Utilizing the free-surface elevation data and methods outlined by Donelan et al. (1992), orbital velocities were then tabulated at centimeter increments down the water column. Readings were taken every 10 milliseconds at two locations within the flow field. Free-surface elevation and orbital velocity data collected at point A, located approximately 1 meter upstream of the breaking event, served a means by which to force numerical simulations at the inflow boundary. Measurements taken at point B, approximately 1 meter downstream of the spilling breaker, provided information by which model accuracy could be evaluated.

Numerical Setup

The computational mesh shown in Fig. 2 was generated for simulations of the ASIST laboratory experiments. The numerical domain is 60 centimeters in

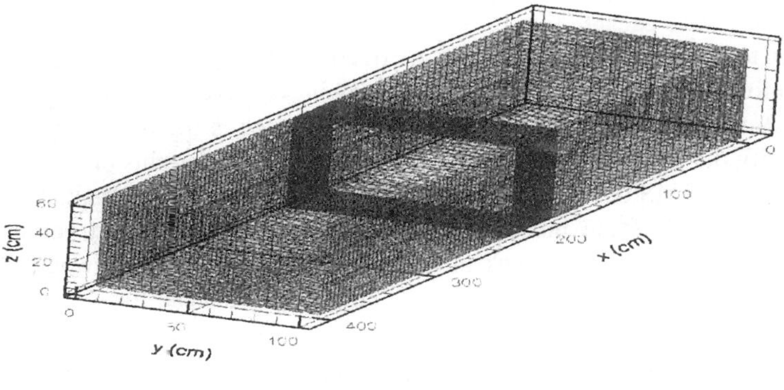

Figure 2. Numerical mesh utilized for ASIST simulations.

height and 1 meter across the tank. To reduce computational time, the numerical mesh is 4 meters in length, as the simulations focus primarily on the small region just upstream to just downstream of breaking in the ASIST experiments. The mesh uses a clustered scheme. A finely-resolved region exists in the center of the domain in which breaking is expected to occur, with increasingly coarser grid resolution towards the numerical tank walls. Such grid structuring allows for a more detailed focus on the breaking region while keeping computational time to a minimum. Free-surface elevation data as well as orbital velocity data collected at point A in the ASIST laboratory setting provided the basis for wave forcing at the inflow boundary.

RESULTS

TRUCHAS simulations of the ASIST experiment show that the model is adept at accurately propagating and focusing the Gaussian wave packet. The waves coalesce to form a steep wave event within the expected breaking region of the numerical mesh, as shown in Fig. 3. Visually, however, the simulations

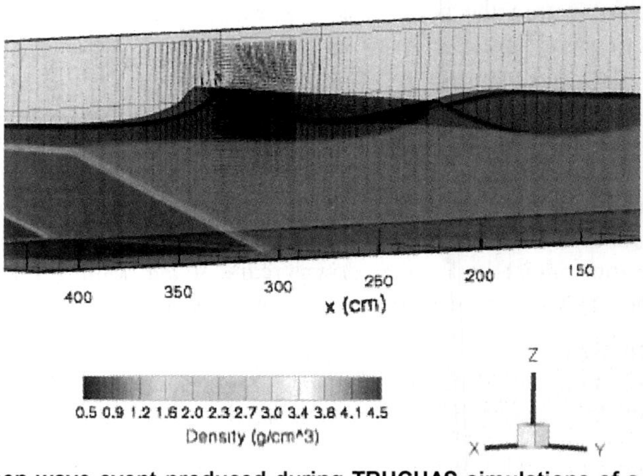

Figure 3. A steep wave event produced during TRUCHAS simulations of a focusing Gaussian wave packet.

seem to show little evidence of an overturning event. On the contrary, the waveform actually appears to devolve back into a more stable configuration.

A telltale indication of a breaking occurrence is turbulent generation. We quantify the amount of turbulence being produced during this process by looking at the total RMS deviation from the mean velocity across the tank at the point in the simulation where breaking was expected to occur. Great variance in the mean velocity across the tank in the expected breaking region would lend itself to the plausibility of wave breaking. Figure 4 shows the mean orbital velocity and RMS deviation from the mean velocity at the point in the simulation where breaking was anticipated. It can be observed from the mean velocity plot that that there exists an area of high orbital velocities in the crest of the steepened wave of approximately 90 cm/s. These increased velocities also appear to be

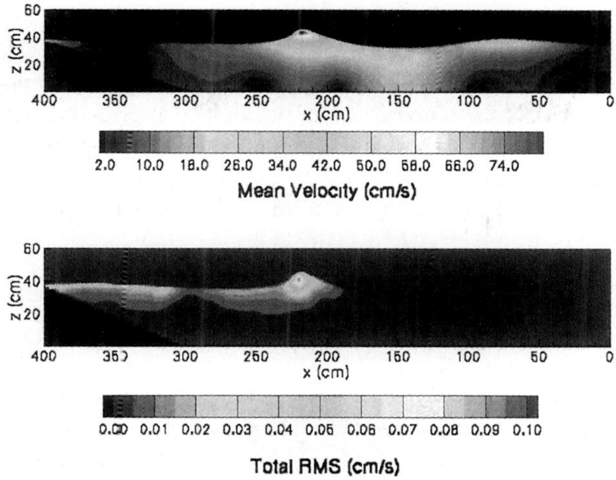

Figure 4. Mean orbital velocity and total RMS deviation from the mean velocity for the steep wave event.

impinging on the forward face of the wave crest, as one would expect in a wave on the verge of overturning. However, the RMS deviation from the mean velocity indicates that while there is some deviation in velocity across the tank at the expected breakpoint, these deviations are less than 0.1 cm/s. Such values are over an order of magnitude smaller than those indicative of a spilling event. Further analysis seems to substantiate this end.

Slight manipulation of the inflow wave data, however, can produce a spilling event similar to that experienced in ASIST settings. Multiplication of the laboratory free-surface elevations measured at point A by 1.3 and calculation of corresponding orbital velocities results in a more energetic wave packet for TRUCHAS forcing purposes. Utilizing this new data at the inflow boundary is resultant in a wave packet that evolves on towards breaking, as shown in Fig. 5. The spilling breaker produced in these simulations occurs upstream of the expected breaking region, thus existing in a more coarsely resolved region of the numerical mesh. If we again examine the mean velocities and total RMS deviations from the mean velocities for this simulation, depicted in Fig. 6, we can see high velocities of nearly 100 cm/s clustered in the wave crest. There still exists deviation in the mean velocity across the tank in the breaking region, but the values now closer to 1.5 cm/s. Such values are representative of those expected during a breaking episode. It should also be noted that these total RMS deviations are being observed in the finely resolved region of our mesh, downstream of the breaking event for this simulation. Due to the fact that the spilling breaker occurred upstream in a more coarsely resolved region of the domain, small-scale RMS deviations in velocity cannot be picked up in these regions. It is expected, however, that total RMS deviation velocity values would be even greater upstream where the breaking event actually occurred for this

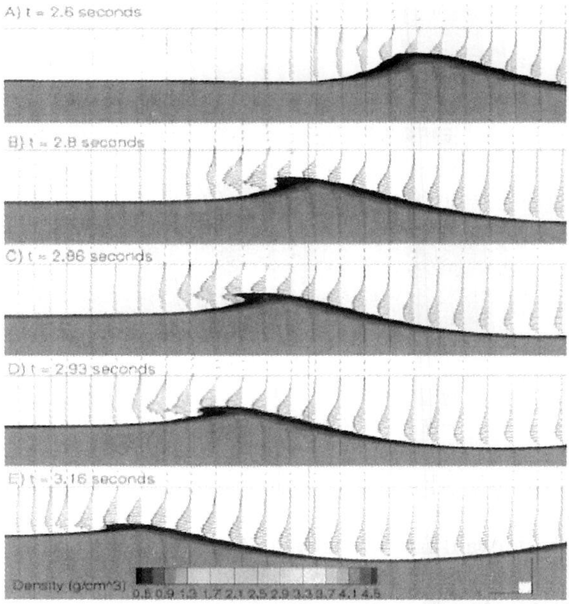

Figure 5. A series of snapshots depicting a spilling breaker, as generated by TRUCHAS.

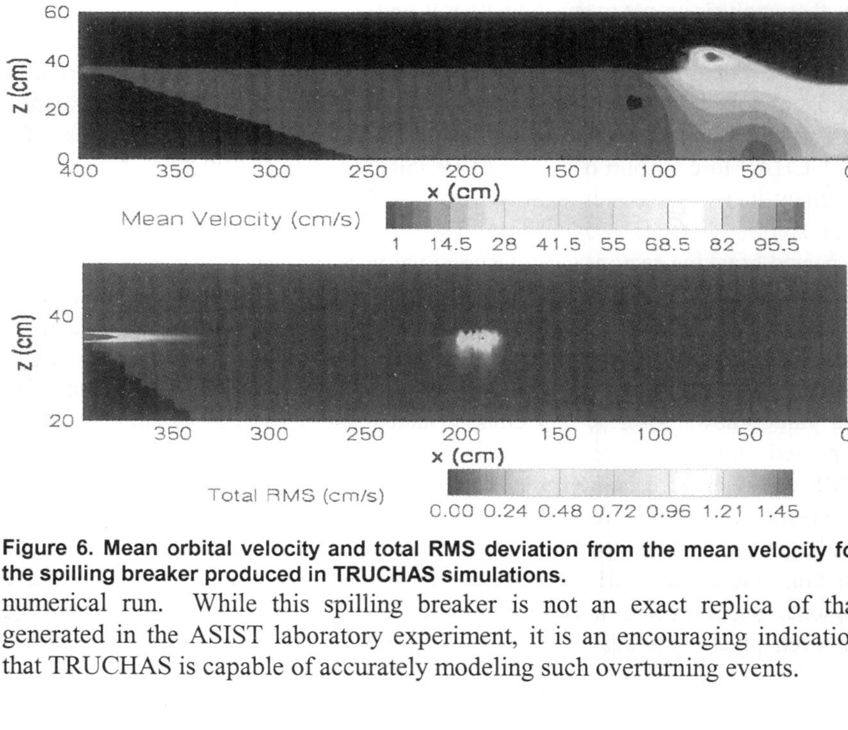

Figure 6. Mean orbital velocity and total RMS deviation from the mean velocity for the spilling breaker produced in TRUCHAS simulations.

numerical run. While this spilling breaker is not an exact replica of that generated in the ASIST laboratory experiment, it is an encouraging indication that TRUCHAS is capable of accurately modeling such overturning events.

FUTURE WORK

Wind-Waves

We would like to take full advantage of the fact that TRUCHAS can accurately model both air and sea regimes and look at the development of wind-generated waves and wind-wave interaction. A snapshot of an initial test simulation is given in Fig. 7. During this simulation, a 2 m/s wind is blown over a sinusoidal wave field. Air flow patterns can already be seen developing over the wave field with progression down the wave tank. Figure 8 shows a pressure contour on the free-surface for this same test case. We can use such techniques to examine the development and growth of waves under the influence of the wind and the extent to which the wave system is forced by the wind. We hope that this study will lead to more information regarding the transfer of momentum from air to sea as well as other details of air-sea interaction. We would like

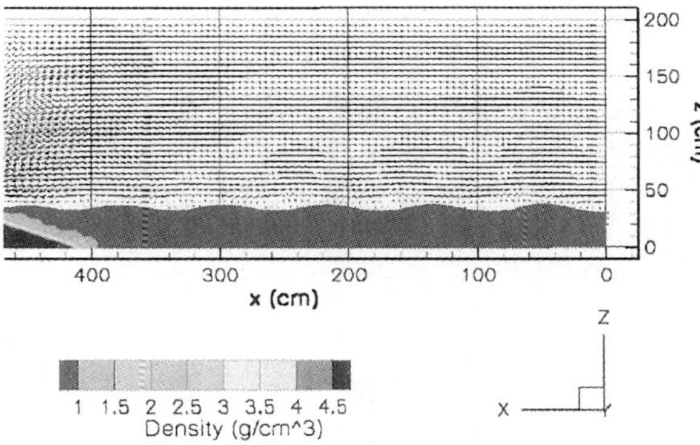

Figure 7. Snapshot of a TRUCHAS simulation involving a 2 m/s wind blowing over a sinusoidal wave field.

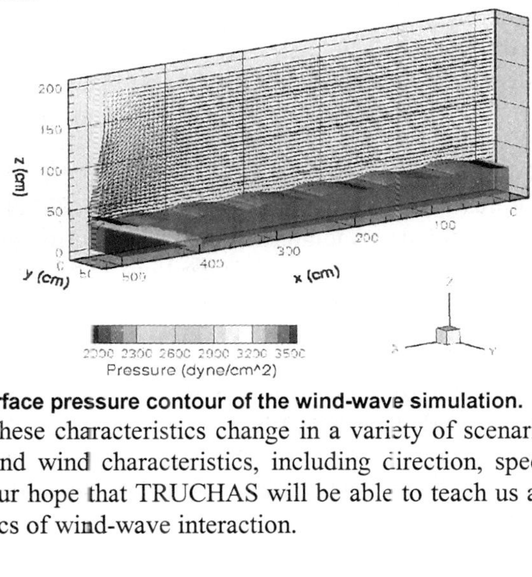

Figure 8. Free-surface pressure contour of the wind-wave simulation.

to look at how these characteristics change in a variety of scenarios of varying wave climates and wind characteristics, including direction, speed, fetch, and duration. It is our hope that TRUCHAS will be able to teach us a lot about the complex dynamics of wind-wave interaction.

Forcing on Bridge Decks

In lieu of the nation's recent hurricanes, we also have a project starting with the Florida Department of Transportation to look at wave forcing on bridge deck structures. Many bridges in hurricanes such as Katrina have failed not as a result of overwash, but because they were actually lifted right up off of their pilings by large waves. Laboratory investigations are being conducted at the University of Florida's Coastal Laboratory to investigate such forcing. It is our hope that TRUCHAS will be able to accurately reproduce these laboratory experiments, yielding good agreement between model and experimental data. Should this be the case, we can then use TRUCHAS to scale-up the laboratory experiments to more realistic scenarios, such as that shown in Fig. 9. This test case involves a 3

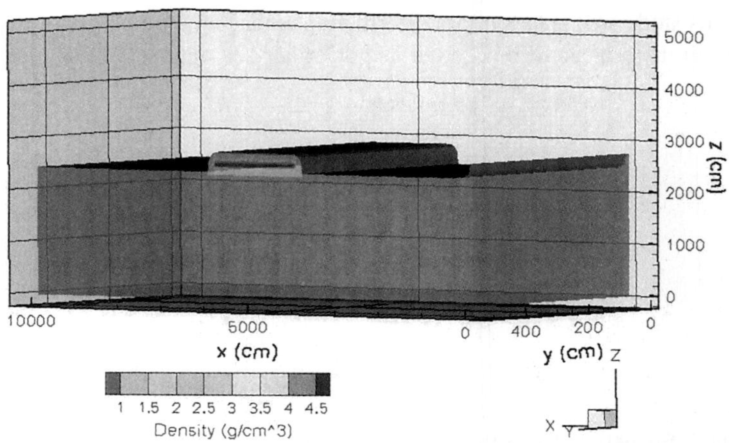

Figure 9. Snapshot of a TRUCHAS simulation for waves hitting a fixed bridge deck.

meter wave impacting a bridge deck of height 5 meters and width 20 meters. We are hopeful that TRUCHAS will give us a better understanding of the wave forcing on bridge decks, and we expect the model to play an integral role in this study.

ACKNOWLEDGMENTS

Funding for this work was provided by the Office of Naval Research as well as the University of Florida. Additional financial assistance was administered by the American Society for Engineering Education in the form of a National Defense Science and Engineering Fellowship.

REFERENCES

Dean, R.G., and R.A. Dalrymple. 1991. *Water wave mechanics for engineers and scientists*, World Scientific, River Edge, New Jersey, 353 pp.

Donelan, M.A., F. Anctil, and J.C. Doering. 1992. A simple method for calculating the velocity field beneath irregular waves, *Coastal Engineering*, 16, 399-424.

IMPROVED SHALLOW-WATER SPECTRAL WAVE MODELING

Jane McKee Smith[1] and Donald T. Resio[1]

The STWAVE spectral wave model has been upgraded to include spatially variable winds, surge, bottom friction, and full-plane propagation and generation. The model has been applied to simulations of Hurricane Katrina, and compared to the limited wave data available in Lake Pontchartrain. A two-scale approximation of the nonlinear four-wave interaction source term is under development. The two-scale approximation is skillful in representing the full wave-wave interactions with a computational time of approximately three times the Discrete Interaction Approximation.

INTRODUCTION

As coastal engineering projects become more complex, nearshore wave generation/transformation models are required to provide more sophisticated results with greater efficiency. In the past, studies of nearshore sediment transport required only the longshore variation of wave height and angle. Today, morphology change models require not only refraction, shoaling, and depth-limited breaking, but also wave-current interaction, wind input, wave-wave interactions, whitecapping, wave-bottom interactions, and interaction with structures. Nearshore transformation models must also be efficient and robust. Model efficiency is especially important as projects grow in scale (regional applications) and resolution. Morphology models are sensitive to the interaction of waves, currents, and bed changes, thus requiring model applications to be closely coupled.

The nearshore wave model STWAVE has been a workhorse for the US Army Corps of Engineers (Smith et al. 2001). The model has evolved over the past 15 years to include wave-current interaction, improved breaking, and enhanced efficiency. Presently, an effort is ongoing to implement substantial improvements in STWAVE: full-plane, time-dependent transformation and generation, diffraction, bottom interaction, and development of improved third-generation source terms. The impetus for this effort is failure of existing models to represent complex environments and the details of spectral evolution under both complex (tropical cyclones) and simple wind forcing. This paper will discuss recent upgrades to STWAVE, simulations of Hurricane Katrina, and on-going improvements to the wave-wave interaction source term.

[1] Engineer Research and Development Center, Coastal and Hydraulics Laboratory, 3909 Halls Ferry Road, Vicksburg, MS, USA 39180-6199; Jane.M.Smith@erdc.usace.army.mil, Donald.T.Resio@erdc.usace.army.mil

STWAVE UPGRADES

STWAVE is phase-averaged, spectral wave model that solves the steady-state conservation of spectral action balance along backward-traced wave rays (Smith et al. 2001). It includes parametric wave growth (wind input, nonlinear wave-wave interactions, and dissipation), refraction and shoaling, wave-current interaction, simplified diffraction, and depth- and steepness-induced wave breaking. Recent upgrades to STWAVE are spatially variable wind and surge, bottom friction, and full-plane implementation.

Spatially Variable Wind and Surge

Generally, STWAVE applications have been on a scale of ~10 km, and over these domains, winds and tides are relatively uniform (from a wave growth and transformation perspective). Finer-scale winds are rarely available. However, modeling waves for Hurricane Katrina (IPET 2006, Smith 2006) required representation of the complex, spatially variable wind and surge fields. Much of southeastern Louisiana is relatively flat and near sea level in elevation. Accurately representing the surges (which were on the order of 5 to 9 m) was critical to modeling the waves because the surge was most or all of the water depth. Interpolation routines were developed to interpolate wind and surge from the finite-element ADCIRC circulation model to the STWAVE grid (IPET 2006). The interpolation functions are pre-computed, so interpolations execute efficiently.

Bottom Friction

The spectral energy loss from bottom friction is formulated as

$$S_{bf} = \frac{-1}{g} c_f \frac{\sigma^2}{\sinh^2 kd} E(f,\alpha) \tag{1}$$

where g is acceleration of gravity, c_f is the bottom friction coefficient, σ is radial frequency, k is wave number, d is water depth, and E is the spectral energy density for frequency f and direction α. The value of c_f can be specified as a constant over the domain or specified for each STWAVE grid cell. For the JONSWAP bottom friction formulation, c_f is specified as Γ/g, where the recommended values of Γ are in the range 0.038 to 0.067 m^2/s^3 (or $c_f = 0.004$ to 0.007) for sand beds based on experiments in the North Sea. Equation 1 has a weak inverse dependence on water depth (related to the increase in bottom wave orbital velocity as the relative depth, kd, decreases).

A Manning formulation for c_f is also available in STWAVE,

$$c_f = \frac{gn^2}{d^{1/3}} \tag{2}$$

where the value of the Manning coefficient, n, is specified for input to STWAVE (either spatially constant or variable). With the Manning formulation, bottom friction dissipation has an addition inverse dependence on water depth. Estimates of Manning coefficients are available in most fluid mechanics reference books (e.g., 0.01 to 0.05 for smooth to rocky/weedy channels). Bottom friction in STWAVE has been used to represent dissipation over coral reefs in Hawaii and wetlands in southern Louisiana.

Full-Plane Implementation

The standard version of STWAVE has included wave generation and propagation over a half plane (±90 deg of the cross-shore axis). The half-plane limitation is generally appropriate for nearshore coastal applications, where the waves of interest are generated in the ocean and propagate towards the shore. The half-plane model has the advantage of substantially lower memory requirements (~ two orders of magnitude) and faster execution (~ an order of magnitude) than full-plane models. Despite clear advantages of a half-plane model, simulation of enclosed or semi-enclosed bays, estuaries, and lakes require a full-plane model that propagates and generates waves in all directions.

Full-plane propagation in STWAVE is calculated using back-traced wave rays, solved in four sweeps. The order of the sweeps is dependent on the wind direction. The half-plane opposite the wind direction is solved first, and then the half-plane coincident with the wind direction. The propagation is solved by alternating sweep directions in each half plane. Wave generation is calculated following the final propagation sweep for each cell. The alternating method reduces the number of iterations required. Sea and swell components are tracked separately in the model.

The full-plane version provides some additional improvements to the model: greater flexibility in cell spacing and directional resolution, a spectral generator for input boundary conditions, simple bathymetry generator, and optional one-dimensional transformation on lateral boundaries to reduce boundary effects (and potentially reduce grid domains).

Both the half- and full-plane versions of STWAVE have a Windows-based graphical user interface within the Surface-Water Modeling System (Smith et al. 2001, Smith and Zundel 2006). Parallel versions of both codes exist for execution on high-performance computers.

HURRICANE KATRINA SIMULATIONS

Both half- and full-plane versions of STWAVE were applied in the simulation of Hurricane Katrina. These simulations took advantage of new capabilities in STWAVE to simulate spatially variable wind, surge, and bottom friction. Storm surge for Hurricane Katrina was simulated using the ADCIRC model (Westerink et al. in prep, IPET 2006). STWAVE and ADCIRC were loosely coupled by passing storm surge fields from ADCIRC to STWAVE and passing radiation stress fields (to calculate wave setup) from STWAVE to ADCIRC.

STWAVE Domains

STWAVE was applied on four grids for southern Louisiana: Lake Pontchartrain, Louisiana Southeast, Louisiana South, and Mississippi/Alabama (Figure 1). Four grids were used to take advantage of the efficient half-plane version of STWAVE for the three outer grids (which must approximately align with the shoreline) and to concentrate grid coverage in the areas of interest. The input for each grid includes the bathymetry (interpolated from the ADCIRC domain), Manning n values (generated from land use maps), surge fields (interpolated from ADCIRC), and wind (interpolated from the ADCIRC wind fields, which apply land effects to the wind fields generated by Ocean Weather, Inc.). The wind applied in STWAVE is spatially and temporally variable for all domains. STWAVE was run at 30-min intervals from 0030 UTC on 28 August 2005 to 0000 UTC on 30 August 2005 (Katrina's landfall was on 29 August 2005).

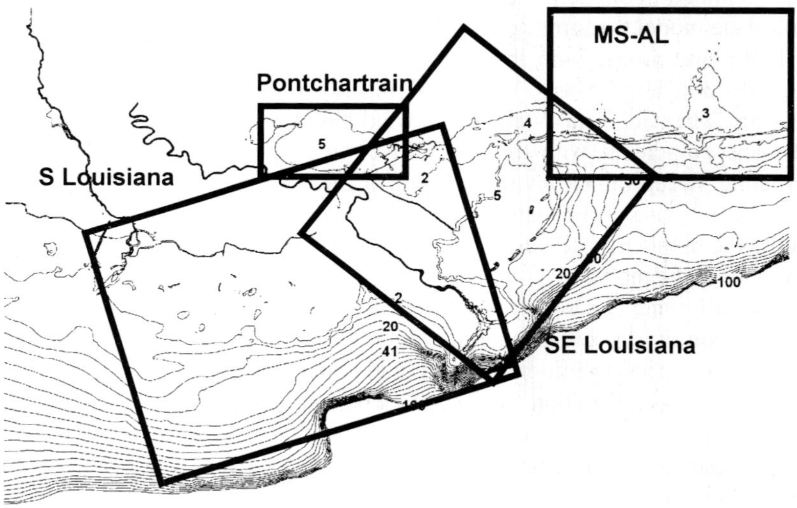

Figure 1. STWAVE modeling domains (contours in meters to 100 m depth)

The total domain coverage area was 60,000 km^2, and each STWAVE grid had a resolution of 200 by 200 m (Δx by Δy). The grids covered the region from east of Mobile Bay in Alabama to the Atchafalaya River in Louisiana. The three offshore grids extended to depths of approximately 30 m. These grids were forced with both the local winds and wave spectra interpolated on the offshore boundary from the regional and basin WAM model simulations (IPET 2006). Lake Pontchartrain is a shallow lake with maximum depths of approximately 5 m, and the Pontchartrain grid encompassed the lake and surrounding wetlands. Lake Pontchartrain was run with the full-plane version of STWAVE and was forced with local winds.

Results

The peak wave conditions on the south shore of Lake Pontchartrain occur at approximately 1330-1430 UTC on 29 August 2005. Figure 2 shows the maximum wave height for each grid cell within the domain for the entire simulation period. Areas contoured in darkest grey with no vectors (zero wave height or period) are wet areas with minimal wave activity. White areas with no contours are land. Figure 3 shows the peak wave period corresponding to the maximum wave height for each cell. The maximum wave heights range from 2.3 to 2.6 m on the New Orleans lakefront and the associated peak periods are 7 sec. These are approximately the design conditions for the lakefront levees (which performed well during the storm).

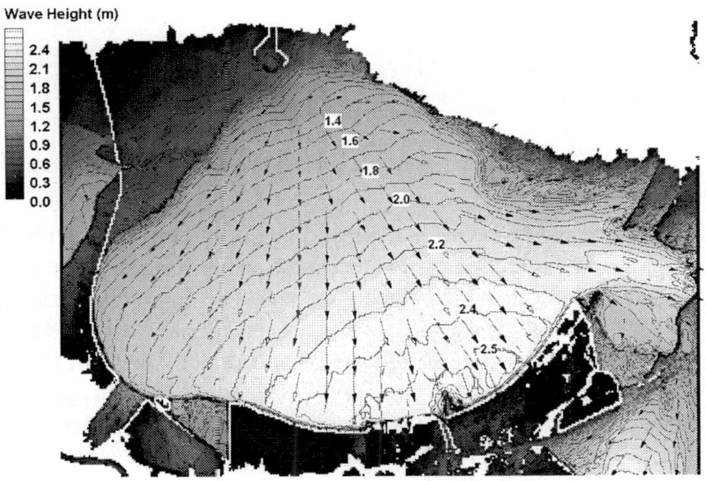

Figure 2. Lake Pontchartrain maximum significant wave height and corresponding mean direction (meters).

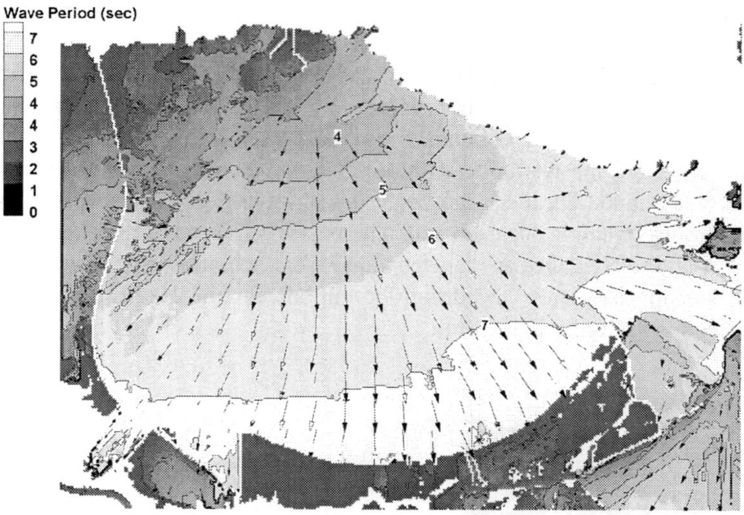

Figure 3. Lake Pontchartrain modeled peak wave period corresponding to the maximum wave (sec).

Three small wave buoys were deployed in Lake Pontchartrain on 27 August 2005 to capture wave conditions in Hurricane Katrina. Two of those gauges were recovered and provide valuable comparison data. The deployment locations were 30 deg 2.053' North, 90 deg 7.358' West for Gauge 22 and 30 deg 1.989' North, 90 deg 7.932' West for Gauge 23. Gauge 22 was directly north of the 17th Street Canal entrance and Gauge 23 was west of Gauge 22. Both gauges were in approximately 4 m water depth. The sampling records were a relatively short 8.5 min, so there is a lot scatter in the data. At the peak of the storm (~29 August 2005 1200 to 1530 UTC), the measured wave heights drop from approximately 2.4 m to 1.5 m. This is the time of maximum wind speed and thus the time when the maximum wave height would be expected. The wave height measurements do not appear to be reliable during the storm peak. The buoys may have experienced excesses tilt due to the extreme winds or may have been submerged or overturned. Figures 4 and 5 show comparisons of significant wave height and peak period, respectively, for the buoy locations. The solid lines with and without symbols are the measurements with the spectra averaged over 3 records (25.5 min), and the dashed line is the modeled parameters (30-min average). The STWAVE results give essentially the same results for the two gauge sites. The modeled wave heights are average of 0.3 m lower than the measurements in the growth stage of the storm (0000-1200 UTC 29 Aug 2005) and 0.15 m lower than the measurements in the decaying stage of the storm (1530-2200 UTC 29 Aug 2005). Comparisons at the storm peak are not meaningful. The modeled peak periods are consistent with the measurements, but 1.0 sec shorter in the decaying stage of the storm.

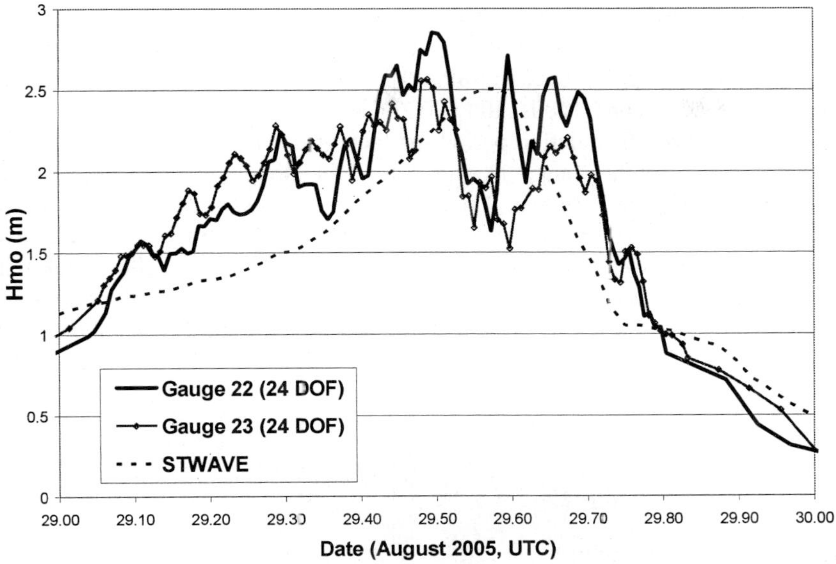

Figure 4. Lake Pontchartrain measured and modeled significant wave height, modeled wind speed, and measured wave height.

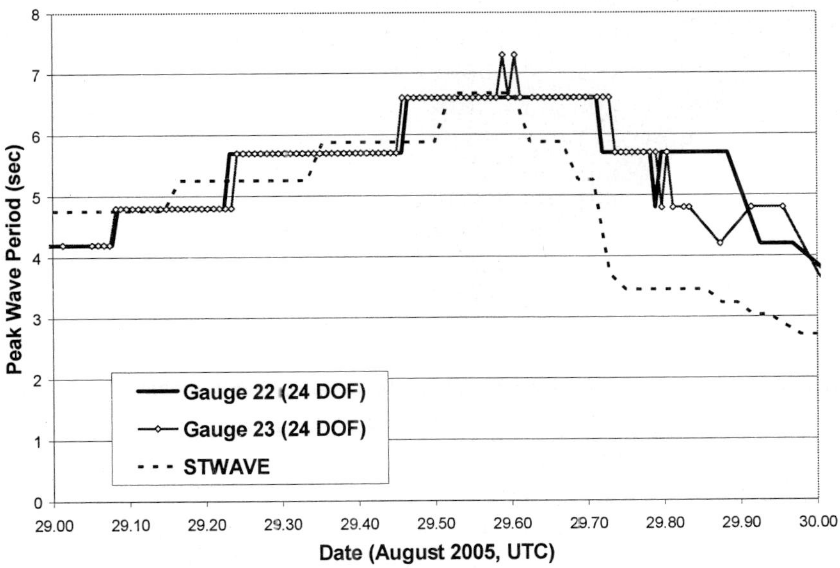

Figure 5. Lake Pontchartrain measured and modeled peak wave period.

The peak wave conditions on the southeast grid occur between approximately 1000 and 1500 UTC on 29 August 2005. The highest waves along the Mississippi River levees occur around 1000-1200 UTC and along the Lake Borgne shoreline around 1400-1500 UTC. Figures 6 shows the maximum wave heights for the entire simulation period for each grid cell within the domain. The maximum wave heights range from 1.2 to 3 m along the shoreline and the associated periods are 7-16 sec. The longer wave periods originate from wave energy traveling between the islands from the Gulf of Mexico. Larger wave heights occur in lower Plaquemines Parish (1.8 - 3 m) and smaller heights in upper Plaquemines and St. Bernard Parishes (1.2 – 1.8 m). The peak periods are relatively large (up to 16 sec) because of wave penetration through the barrier islands.

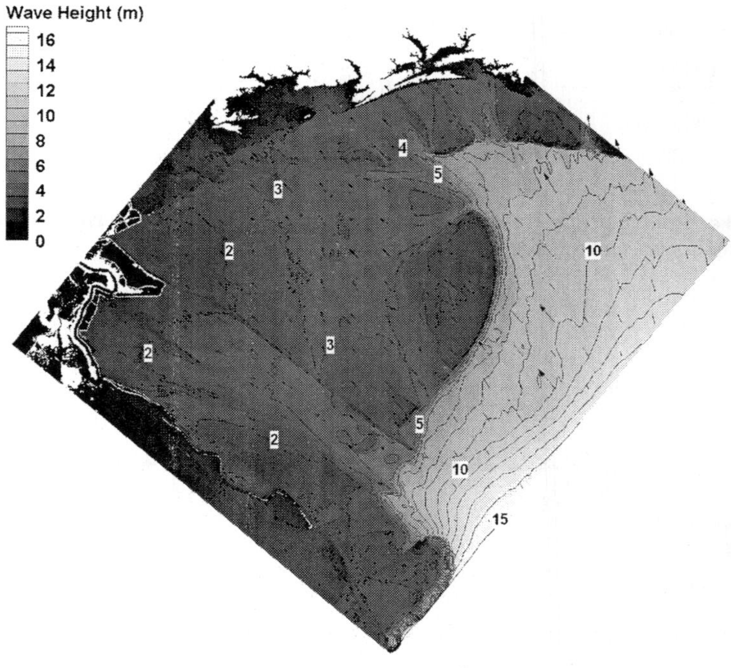

Figure 6. Southeast Louisiana maximum modeled wave height (meters).

The peak wave conditions on the south grid occur between 0800 and 1030 UTC on 29 August 2005. The water level changes due to surge on this grid are generally less than the Southeast Louisiana grid; therefore, wave penetration over the marsh is less severe. Figure 7 shows the maximum wave heights for the entire simulation period for each grid cell within the domain. The maximum wave heights at the barrier islands are approximately 3-4 m (depth limited) and associated periods are 15-16 sec. Wave heights were significantly lower along the west bank Mississippi River levees. The barrier islands dissipated much of the wave energy arriving from the Gulf of Mexico and help protect the interior shorelines. These simulations were made with pre-Katrina bathymetry, so as barriers eroded, this protection may be overstated in the modeling results. The local winds were less important on this grid because the winds generally blow along the shore or offshore in the area. The portion of the south Louisiana grid east of the Mississippi River should be disregarded because the model is not forced along the lateral boundary (that area is modeled with the southeast Louisiana grid).

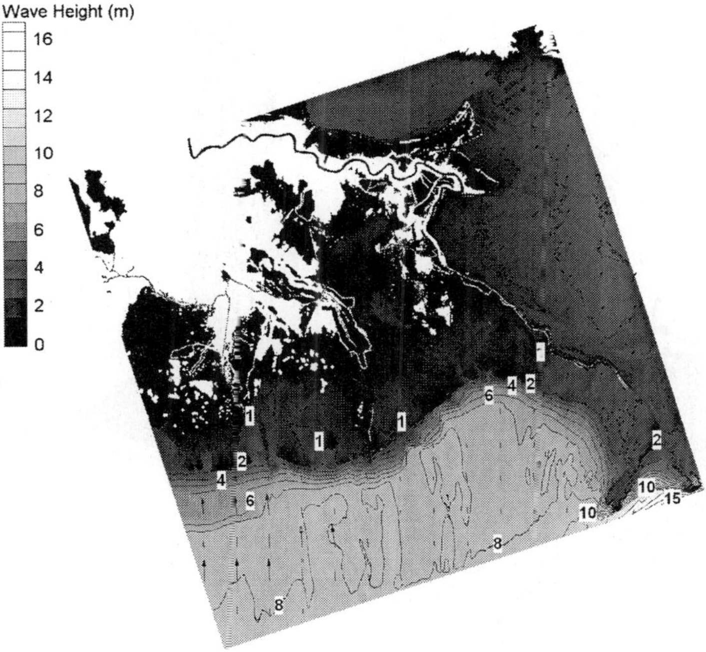

Figure 7. South Louisiana maximum modeled wave (meters).

The peak wave conditions on the Mississippi-Alabama grid occur around 1430 UTC on 29 August 2005, near the time of the hurricane landfall in Mississippi. Figure 8 shows the maximum wave heights for the entire simulation for each grid cell within the domain. The maximum wave heights at the barrier islands are approximately 6 m (depth limited) and associated periods are 15 sec. The barrier islands dissipated much of the wave energy arriving from the Gulf of Mexico and help protect the interior shorelines. These simulations also were made with pre-Katrina bathymetry, so as barriers eroded, this protection may be overstated. Wave heights in Mississippi Sound and Mobile Bay generally range from 1.5 to 3 m, but are 3 to 6 m in the lee of the inlets on the Mississippi coast. Similar to the Louisiana south domain, the barrier islands on the Mississippi and

Alabama coasts dissipated much of the wave energy arriving from the Gulf of Mexico and help protect the interior shorelines. Large wave periods (15 sec) penetrate to the interior shorelines. The depth-limited wave breaking on the Mississippi and Alabama coasts generates wave setup in Mississippi Sound and Lake Borgne, which then forces additional water into Lake Pontchartrain (simulated with ADCIRC).

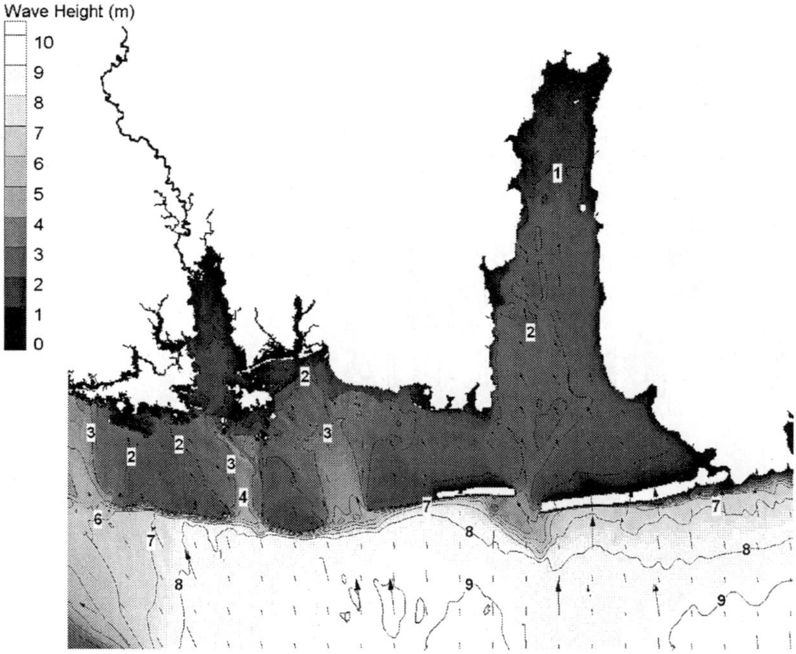

Figure 8. Mississippi-Alabama maximum modeled wave height for 0030 on 28 August 2005 to 0000 UTC on 30 August 2005 (meters).

Simulations of nearshore waves in coastal Louisiana for Katrina are presented. The nearshore wave modeling required relatively high resolution to capture transformation processes over large, shallow areas and radiation stress gradients that were fed back to the circulation model. Wave propagation and generation are significantly altered by surge in shallow wetland areas and Lake Pontchartrain. The best information possible was used as input to the model (bathymetry, winds, bottom friction), these values have a significant degree on uncertainty. A sensitivity analysis was performed to quantify uncertainty in the results. This analysis examined ±5 percent changes to the wind fields, which resulted in ±0.3 m (or less) changes in wave height at the shoreline (up to ±1 m at the barrier islands). Modeling degraded bathymetry of the Chandeleur Islands, which were known to erode during Katrina, resulted in increased wave height of up to 1.8 m in the lee of the island, but near zero difference a the shoreline. Modeling bottom friction to represent pre- and post-Katrina vegetation cover showed that surge and waves were higher in some areas due to the rougher pre-Katrina vegetation cover (larger Manning's *n*) and lower in other areas. Higher waves occurred in areas where surge increased and lower waves where Manning's *n* values were larger pre-Katrina. Follow-on work will continue to evaluate the role of wetlands in attenuating surge and waves.

SOURCE TERMS

Existing spectral wave models are relatively successful in reproducing bulk wave statistics, but often cannot reproduce spectral shape (self-similar equilibrium range (Resio et al. 2001), directional bimodality above the spectral peak, overshoot of the PM limit). The correct distribution of energy with direction is critical in sediment transport modeling (Smith and Gravens 2002), as well as an indicator of the accuracy of the source terms. The dominant source term in the evolution of the spectral shape is the nonlinear four-wave interactions. This term is the mechanism for redistribution of energy input from wind. The nonlinear interactions are described by the Boltzmann equation, but the full solution is too computationally intensive for practical models. Present finite-depth models rely on a scaled version of the Discrete Interaction Approximation (DIA) (Hasslemann et al. 1985) or other parametric relationships. The DIA includes only a subset of the four-wave interactions (those with two wave numbers equal). The DIA was calibrated parametrically for total energy in the low-frequency lobe (energy fluxes at frequencies lower than the peak), but the flux distribution is typically inaccurate at the higher frequencies. The DIA was the critical link that allowed wave generation modeling to advance, but now is a weakness in existing models, particularly when scaled to shallow water and applied to complex wave fields.

A two-scale approximation to the nonlinear interactions has been developed. The two-scale approximation splits the interactions into a broad scale, a local scale, and the cross terms between the two. The broad-scale interactions are calculated using a full Boltzmann model for representation wave spectra and parameterized using spectral peakedness. The local scale and cross-term interactions are reduced to four terms which are solved explicitly. The formulation conserves action, energy, and momentum because each elemental transfer is conservative. Figure 9 shows an example of the nonlinear interaction source term calculated using the full Boltzmann integral, the two-scale approximation (TSA), and the DIA for a shallow-water spectrum ($k_pd = 0.7$). The agreement of the TSA with the full integral is excellent. The TSA is approximately three times slower than the DIA computationally. Revision of the nonlinear interactions also requires rebalancing the wind input and whitecapping source terms within a wave model.

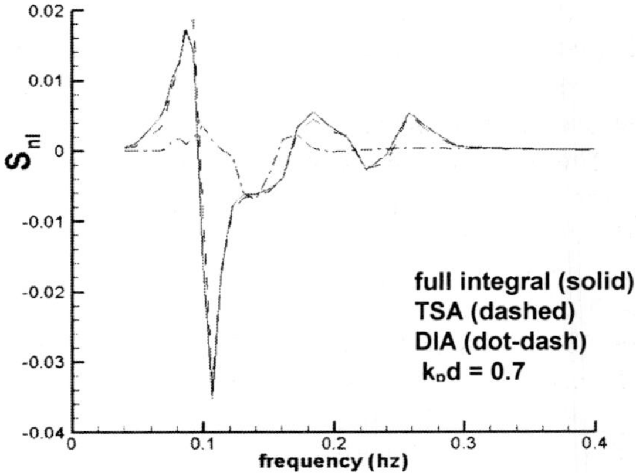

Figure 9. Full integral, two-scale, and DIA nonlinear interaction source terms.

SUMMARY

The Corps workhorse spectral transformation model STWAVE is evolving to include improved fidelity, additional options, and more flexibility. STWAVE has been applied extensively for hurricane applications over the past year, and is being applied to improve our understanding of wave-surge-wetland interaction. A two-scale approximation is under development to efficiently represent nonlinear four-wave interactions. Initial results are excellent, but additional testing is required.

ACKNOWLEDGEMENTS

Permission to publish this paper was granted by the Office, Chief of Engineers, U.S. Army Corps of Engineers. This research was conducted under the Coastal Simulation Focus Area in the System-Wide Water Resources Program of the Coastal and Hydraulics Laboratory, U.S. Army Engineer Research and Development Center and the Interagency Performance Evaluation Task Force (IPET). Vince Cardone and Andy Cox, Ocean Weather, Inc., Mark Powell, National Hurricane Center, Robert Jensen, CHL, Joannes Westerink, Notre Dame, and Mary Cialone, CHL, are acknowledged for providing the input driving conditions through IPET. Ann Sherlock assisted in STWAVE simulations. Harley Winer, New Orleans District USACE, provided the Lake Pontchartrain wave data.

REFERENCES

Hasselmann, S., K. Hasselmann, J.A. Allender and T.P. Barnett. 1985. Computation and parameterizations of the nonlinear energy transfer in grave-wave spectrum. Part II: Parameterization of the nonlinear energy transfer for application in wave models. *J. Phys. Oceanogr.*, 15, 1378-1391.

Interagency Performance Evaluation Task Force (IPET). 2006. Performance evaluation of the New Orleans and southeast Louisiana Hurricane Protection System, Volume IV – The Storm (main text and technical appendices). US Army Corps of Engineers. 1197 pp. https://ipet.wes.army.mil/

Resio, D.T., J.H. Pihl, B.A. Tracy and C.L. Vincent. 2001, Nonlinear energy fluxes and the finite depth equilibrium range in wave spectra, *J. Geophys. Res.*, 106(C4), 6985-7000.

Smith, J.M. 2006. Modeling Nearshore Waves for Hurricane Katrina. 2006. *Proceedings*, 9th International Workshop on Wave Hindcasting and Forecasting, http://www.waveworkshop.org/9thWaves/index.htm

Smith, J.M., and M.B. Gravens (2002), Incident boundary conditions for wave transformation, *Proceedings*, 7th Intl. Wrkshp on Wave Hindcasting and Forecasting, Env. Canada, 373-384.

Smith, J.M., A.R. Sherlock, and D.T. Resio (2001), STWAVE: steady-state spectral wave model; user's manual for STWAVE V 3.0, SR-01-1, US Army ERDC, Vicksburg, MS.

Smith, J.M., and A. Zundel. 2006. STWAVE-FP in SMS: Graphical Interface. Coastal and Hydraulics Engineering Technical Note CHETN-I-69, U.S. Army Engineer Research and Development Center, Vicksburg, MS. http://chl.erdc.usace.army.mil/library/publications/chetn/pdf/chetn-i-71.pdf

Westerink, J. J., R. A. Luettich, J. C. Feyen, J. H. Atkinson, C. Dawson, M. D. Powell, J. P. Dunion, H. J. Roberts, E. J. Kubatko, and H. Pourtaheri. In Preparation. A Basin to Channel Scale Unstructured Grid Hurricane Storm Surge Model for Southern Louisiana, *Monthly Weather Review*.

IMPLEMENTATION AND TEST OF IMPROVED METHODS FOR EVALUATION OF NONLINEAR QUADRUPLET INTERACTIONS IN A THIRD GENERATION WAVE MODEL

Michel Benoit[1]

Nonlinear quadruplets wave-wave interactions play a dominant role in the evolution of ocean and coastal wave spectra. Although the theoretical model equation representing these interactions is available for more than 40 years, most of existing operational and applied wave models rely on a rather crude and very simplified computing scheme to evaluate these interactions, namely the Discrete Interaction Approximation (DIA). Recent research results suggest that improved or quasi-exact techniques can be available to estimate this transfer term quite accurately within a computational time ratio in the order of 20 to 200, compared to DIA (depending on the level of approximation). This leads us to consider the possibility to use such improved computational techniques in wave models applied on real cases (not only academic simplified 1D cases as it used to be up to now). To that end the present study address several issues related to the inclusion of improved techniques for computing the quadruplet interactions term in a real wave model environment. Test simulations are performed to examine and quantify the effects of several parameters and choices, such as physics for wind input and white-capping dissipation, wave growth limiter, prognostic limit of the discrete spectrum. Results are presented for homogeneous wind and wave conditions (one single sea point) in deep water under constant wind (duration-limited case) and turning wind conditions (instantaneous shift of wind direction).

MOTIVATION OF THE STUDY

The third-generation (3G) wave model TOMAWAC was developed about 10 years ago for oceanic and coastal applications (Benoit *et al.* 1996). TOMAWAC solves the wave action density balance equation (e.g. Bretherton and Garret 1969, Komen *et al.* 1994) and computes the evolution (in space and time) of the directional wave spectrum on an unstructured spatial grid, under unsteady wind forcing and possibly unsteady tidal currents and sea levels. It uses unstructured meshes for the maritime domain. This permits to refine the grid over areas of particular interest as well as over areas where the bottom varies rapidly or in the presence of islands or irregular coastline. The early version of the model was mainly applicable to deep water and continental shelves. Subsequently depth-induced breaking and non-linear triad interactions in shallow water were also considered and implemented.

Regarding the physics of wave generation and wind-wave interactions, several state-of-the-art models and parameterizations have been implemented for wind input and white-capping dissipation (e.g. Komen *et al.* 1984, Janssen

[1] Laboratoire National d'Hydraulique et Environnement (LNHE), EDF R&D, 6 quai Watier, BP 49, 78400 Chatou, France

1991, Tolman and Chalikov 1996, Alves and Banner 2003). Beside these processes the nonlinear third-order resonant interactions (the so-called "quadruplet" interactions) play an important role in the evolution of the wave spectrum (Young and Van Vledder 1993). For this term a theoretical model is available since Hasselmann (1962) and Zakharov (1968) and is expressed as a Boltzmann-type integral:

$$\frac{\partial N_1}{\partial t} = \int_{\underline{k}_2}\int_{\underline{k}_3}\int_{\underline{k}_4} G(\underline{k}_1,\underline{k}_2,\underline{k}_3,\underline{k}_4)\{N_3N_4(N_1+N_2)-N_1N_2(N_3+N_4)\}$$
$$\times\,\delta(\underline{k}_1+\underline{k}_2-\underline{k}_3-\underline{k}_4)\,\delta(\sigma_1+\sigma_2-\sigma_3-\sigma_4)\,d\underline{k}_2 d\underline{k}_3 d\underline{k}_4 \qquad (1)$$

where N is the action density, defined as $N(\underline{k}) = F(\underline{k})/\sigma$, $F(\underline{k})$ is the directional variance spectrum of ocean waves, expressed here as a function of the wave-number vector $\underline{k} = (k_x, k_y)$ in the wave-number plane. The notation N_j stands for $N(k_j)$ with $j = 1$ to 4. $G(\underline{k}_1, \underline{k}_2, \underline{k}_3, \underline{k}_4)$ is the coupling coefficient.

From this expression, it is clear that interactions only occur between quadruplets of four wave components which fulfill the two following resonance conditions:

$$\underline{k}_1 + \underline{k}_2 = \underline{k}_3 + \underline{k}_4 \qquad (2.a)$$

$$\sigma_1 + \sigma_2 = \sigma_3 + \sigma_4 \qquad (2.b)$$

BRIEF OVERVIEW OF EXISTING METHODS FOR COMPUTING THE NONLINEAR QUADRUPLET INTERACTIONS

From a numerical point of view, the exact evaluation of the full Boltzmann integral is an awkward and time-consuming task. Although some algorithms were successfully developed for such computations on research applications with a limited number of sea points (e.g. Webb 1978, Masuda 1980, Hasselmann and Hasselmann 1981), they can not yet be employed in operational third-generation wave models due to their prohibitive demand on CPU time.

Approximations have thus to be derived. By far the most popular one and the most widely used is the Discrete Interaction Approximation (DIA) of Hasselmann *et al.* (1985). The DIA has triggered the development of third-generation models, starting from WAM (WAMDIG 1988) and is used in almost all operational wave models nowadays. The main advantage of DIA lies in its computational efficiency. By considering only one particular quadruplet configuration (and its mirror image) the dominant features of the nonlinear transfer are retained with a low computational effort. However DIA is a crude approximation of the actual nonlinear transfer rate and exhibits a number of deficiencies (e.g. Van Vledder *et al.* 2000, Benoit 2005). Therefore research efforts are necessary to improve the modeling of these nonlinear interactions and to develop efficient schemes that could be used on real applications (with typically 1 000 to 10 000 sea points).

Benoit (2005) reviewed and compared several possible improvements or alternative techniques to obtain higher accuracy in the evaluation of the Boltzmann integral while keeping acceptable CPU time (say in the order of 10 to 100 times more than the present DIA). Basically four main directions can be investigated:

1. the **Multiple DIA** (MDIA) approach improves the DIA by adding several configurations, typically 3 to 4 (e.g. Van Vledder 2001, Tolman 2004),
2. **Reduced Exact Methods** start from potentially "exact" methods and use a similar algorithm with either a coarser resolution and/or a reduced integration range (e.g. Lin and Perrie 1998, Lavrenov 2001, Benoit 2005),
3. **Diffusion-like operators** (e.g. Zakharov and Pushkarev 1999), which are very computational efficient. Tests performed in Benoit (2005) however showed that the accuracy of this technique is lower than DIA,
4. **Neural-network technique** (not addressed in this study).

Based on preliminary tests, methods from classes 1 and 2 appear to offer the best balance between accuracy and efficiency for implementation in 3G wave models. Following the conclusions of our recent work on a single spectrum (Benoit 2005), two improved methods for computing the non-linear transfer rate are implemented in a single-point version of TOMAWAC. Our goal is to study the changes (compared to the standard DIA on one hand and to exact evaluation on the other hand) brought in the evolution of the wave spectrum and characteristic wave parameters (significant wave height, mean and peak periods, directional spread, etc.) when these methods are applied in "real model" environment (i.e. used in conjunction with other source terms and with a numerical scheme for integration in time of the wave action balance equation).

ISSUES RELATED TO THE INCLUSION ON IMPROVED ALGORITHMS FOR QNL4 IN 3G WAVE MODELS

Replacing the DIA with an other method for computing the nonlinear interactions in a real numerical wave code is not as straightforward as one could naively imagine. It raises several issues, some of which are addressed here:

1. Interactions with other source/sink terms:

In existing wave codes the formulations used for modeling wind input and/or white-capping dissipation were developed and tuned to work in combination with the DIA scheme, and thus inherently contain "correction" to accommodate for DIA shortcomings and limitations. Replacing DIA with a more exact algorithm will thus require recalibration and/or retuning of the other sink and source terms (and in particular the white-capping dissipation). It is also of interest to examine how different physics for wind-input and white-capping react to the change of method for computing the nonlinear interactions.

2. Numerical integration in time of the source/sink terms:

Numerics of operational wave models may also have some effect on the inclusion of improved algorithms for nonlinear interactions. The main aspects are: (i) the type of numerical scheme for time integration (e.g. explicit, implicit, semi-implicit, etc.), (ii) the value of the time-step (it is taken as large as possible, usually 5 to 20 min) and (iii) the presence of a wave growth limiter.

3. Discrete grid for the wave spectrum:

In wave models the directional spectrum is defined on a discrete grid, typically 20 to 30 frequencies and 12 to 36 directions. The finite frequency range $[f_{min}, f_{max}]$, as well as the frequency and angle resolution may have some effects in the computation of the various source and sink terms.

4. Shape of the diagnostic tail of the spectrum:

Finally the wave model may be sensitive to the characteristics of the parametric (diagnostic) high-frequency tail of the spectrum. Usually a f^n tail is assumed for the spectrum for frequencies above a given frequency f_d. Two aspects are relevant here for their effect on nonlinear transfer computations: (i) the value of the prognostic limit f_d (usually taken as f_{max} or $2.5\, f_{mean}$) and (ii) the value of the high frequency exponent n (usually -4 to -5).

DEFINITION OF TEST-CASES AND NUMERICAL PARAMETERS

Model equation used for the study

In the present work, we limit ourselves to (i) deep water conditions and (ii) spatially homogeneous cases, by considering the evolution in time of a single point in physical space (x, y). The model equation we solve thus reads:

$$\frac{\partial F}{\partial t} = Q_{nl4} + Q_{in} + Q_{wc} \tag{3}$$

where Q_{nl4} represents the nonlinear interactions term, Q_{in} the wind input term and Q_{wc} the white-capping dissipation term.

Initial conditions (t = 0)

The initial spectrum is obtained as a combination of a JONSWAP frequency spectrum $E(f)$ and a frequency-independent angular spreading function $D(\theta)$ as $F(f,\theta) = E(f).D(\theta)$. The JONSWAP frequency spectrum is defined by a significant wave height $H_{mo} = 0.1$ m, peak frequency $f_p = 0.333$ Hz (peak period $T_p = 3$ s), peak enhancement factor $\gamma = 3.3$, $\sigma_a = 0.07$ if $f < f_p$ and $\sigma_b = 0.09$ if $f > f_p$. This corresponds to short wind waves with very small wave height. The angular spreading function is $D(\theta) = \Delta\, cos^6((\theta - \theta_0)/2)$ with $\theta_0 = \pi/2$.

Discretization (frequency-direction grid, time-step)

The discrete frequency-direction grid is composed of 32 frequencies, with logarithmic spacing from $f_1 = 0.04$ Hz to $f_{32} = 0.5$ Hz ($f_{n+1}/f_n = 1.085$), and 36 directions ($\Delta\theta = 10°$). Except otherwise stated the time-step is 900 s (15 min).

Description of test-cases

Two cases for wind forcing are considered in this study in order to assess the effects of various parameters and model's options.

- **Case A**. Duration limited wave growth (constant wind). A constant wind with velocity U_{10} = 20 m/s is blowing in the same direction as the mean direction of the initial spectrum $\theta_w = \theta_0 = 90°$.
- **Case B**. Turning wind. The wind velocity is the same as for case A (U_{10} = 20 m/s), but its direction θ_w is changing during the simulation. From t = 0 to 8 h, the wind direction is $\theta_w = \theta_0 = 90°$ as for case A. At t = 8 h, the wind suddenly turns towards θ_w = 30° ($\Delta\theta_w$ = -60°) and remains constant until t = 18 h. At t = 18 h, the wind suddenly shifts towards θ_w = 120° ($\Delta\theta_w$ = +90°) and remains constant up to t = 36 h.

TEST A1: EFFECT OF THE METHOD FOR COMPUTING THE QNL4 TERM

We begin by a rather "naive" approach which consists in simply replacing the DIA method for computing the nonlinear interaction term by more precise methods, namely the Multiple DIA method (MDIA), and the Gaussian Quadrature Method (GQM) with raw resolution, without changing anything else. This test is done on case A (steady wind) with the physics and numerical options corresponding to WAM cycle 4 (denoted WAM4) (Komen *et al.* 1994).

Time integration is performed with the semi-implicit scheme (Δt = 900 s). For the wave growth limiter we use the original WAM4 expression:

$$|\Delta F|_{\max} = 6.4\ 10^{-7}\, g^2\, f^{-5}\, \Delta t/1200 \qquad (4)$$

The diagnostic tail of the spectrum starts at frequency:

$$f_d = \min[\, f_{max}\ ;\ \max(4\, f_{PM}, 2.5\, f_{mean})\,] \qquad (5)$$

and the high-frequency (HF) tail exponent is n = -5.

The evolution in time of synthetic sea-state parameters are plotted on Fig. 1 (significant wave height H_{mo}, peak period T_p, directional width σ) while 1D (frequency) spectra are plotted on Fig. 2. Differences in H_{mo} are marked during the first stage of wave growth (e.g. at t = 12 h, DIA produces a wave height H_{mo} which is about 2 m higher than the ones given by MDIA and GQM (i.e. +50%). After 5 days DIA still gives larger wave heights (about 1 meter more compared with GQM). Growth curves for GQM and MDIA are more similar between each other, both for H_{mo} and T_p, although the increase in these parameters is a bit more rapid with GQM and the final values are somewhat lower. We also note that there is an increase of T_p for MDIA during the last 12 hours of the run, which does not sound very realistic. Comparison of 1D spectra is also very instructive (see Fig. 2). The faster development of the spectrum with the DIA method is clearly observed. Although the spectra obtained with MDIA and GQM develop in a quite similar fashion, the former exhibit multiple peaks which are obviously spurious. The increase of T_p with MDIA observed on Fig. 1 (medium panel) is a consequence of this multi-peaked shape of the spectrum.

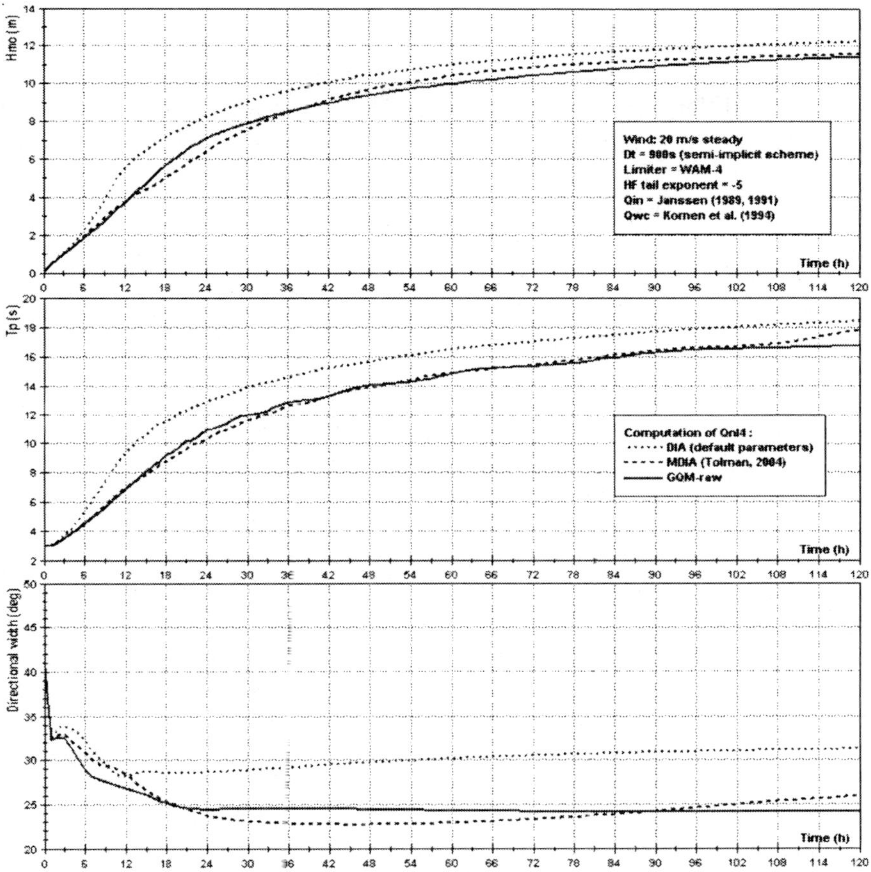

Figure 1. Test A1 - Variation of significant wave height H_{mo} (upper panel), peak period T_p (medium panel) and directional width σ (lower panel) under steady wind conditions (U_{10} = 20 m/s) computed with WAM4 physics for Q_{in} and Q_{wc} in combination with three schemes for Q_{nl4}. These runs use the original WAM4 wave growth limiter (4).

TEST A2: EFFECT OF WAVE GROWTH LIMITER AND PROGNOSTIC LIMIT

The second test is also performed on case A (steady wind) with the physics of WAM4 for wind input (Q_{in}) and white-capping dissipation (Q_{wc}). In order to assess the effect of both the wave growth limiter and the prognostic limit f_d, we only consider here the accurate GQM method for computing (Q_{nl4}). Time integration is performed with the semi-implicit scheme (Δt = 900 s).

For the wave growth limiter we use three options: (i) no limiter, (ii) the original WAM4 expression as given by (4) and (iii) the modified expression in WAM4 by Hersbach and Janssen (1999):

$$|\Delta F|_{\max} = 3.0\ 10^{-7}\ g\ \max(u^*, g\, f_{PM}/f)\ f^{-4}\ f_{max}\ \Delta t \qquad (6)$$

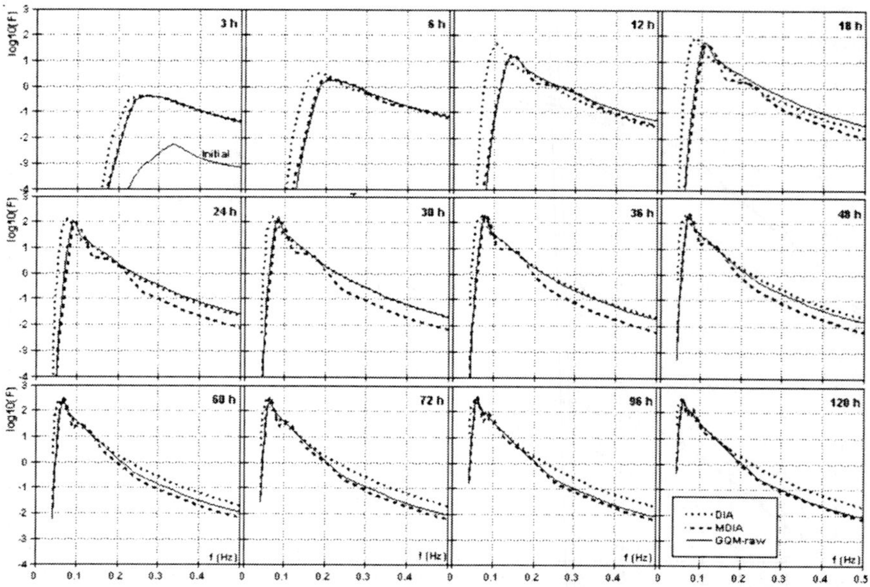

Figure 2. Test A1 - Variation of frequency spectrum (with log-scale on vertical axis).

The diagnostic high-frequency tail of the spectrum starts either at frequency f_d as specified by Eq. (5) or at f_{max} (last discrete frequency of the spectral grid). In both cases the HF tail exponent is $n = -5$.

It is seen on Fig. 3 that the runs done without any limiter are unstable and lead to very large unrealistic wave heights at the beginning of the run. It is however quite surprising that the computation does not blow up. After the first spurious peaks observed on wave height during the first 24 hours the growth curves recover a "normal" shape, although they are significantly lower than the other ones. This however confirms the need to use a growth limiter for practical applications, at least for such large (fixed) time-step. More importantly marked differences appear between the two versions of limiter (Eqs. 4 and 6). The original WAM4 limiter artificially reduces both the wave height and peak period over the first stage of wave development ($t = 0$ to 36 h). The differences in H_{mo} and T_p at $t = 12$ h reach 2 m and 3 s respectively. This highlights the need to use a proper form of wave growth limiter, as pointed out by Hersbach and Janssen (1999) and/or to modify the time-integration scheme (Tolman, 1992). The choice of prognostic limit is of lower effect, at least on the synthetic parameters H_{mo} and T_p. Slightly larger wave heights and periods are obtained with the prognostic frequency limit (6). Analyses of spectra (not shown here) however exhibit significant differences in the high-frequency range. When using f_{max} as prognostic limit the HF tail exponent is significantly lower than expected values (-4 to -5). It is therefore necessary to use a lower prognostic limit, as Eq. (5).

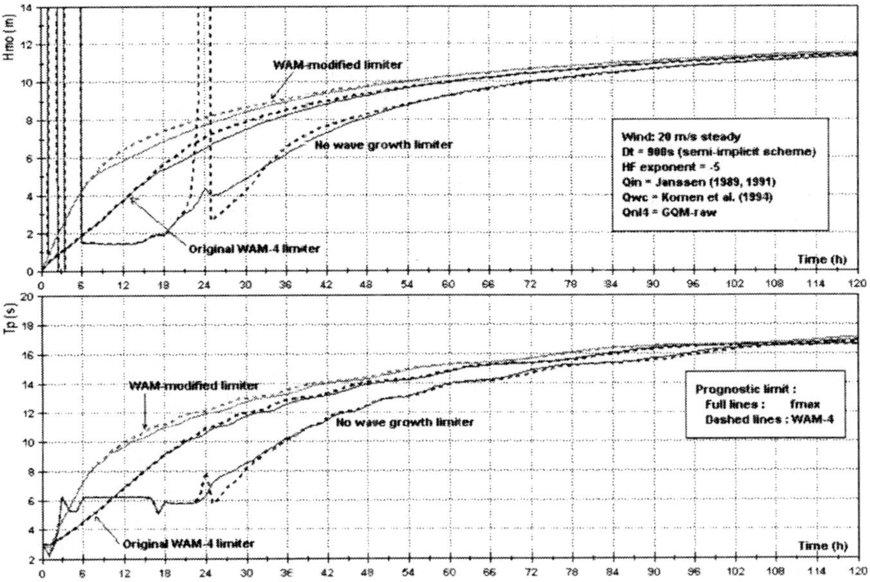

Figure 3. Test A2 - Variation of significant wave height H_{mo} (upper panel) and peak period T_p (lower panel) under steady wind conditions (U_{10} = 20 m/s) computed with WAM4 physics for Q_{in} and Q_{wc} in combination with GQM for Q_{nl4}.

TEST A3: EFFECT OF PHYSICS FOR INPUT AND DISSIPATION

This test is still run on case A (steady wind), but we now turn to a different set of terms for modeling wind input and white-capping dissipation, namely the expressions proposed by Tolman and Chalikov (1996) and used in the WAVEWATCH-III (WW3) code. Time integration is performed with the semi-implicit scheme (Δt = 900 s). We use the modified wave growth limiter (6). The prognostic limit f_d is imposed with a smooth transition between 2.5 f_p and 3.0 f_p as explained in Tolman and Chalikov (1996). The HF tail exponent is n = -5.

Results presented on Figs. 4 and 5 show that there are less differences in synthetic parameters and spectra when we replace the DIA method with MDIA or GQM. On these graphs, two versions of DIA are used: the original one by Hasselmann *et al.* (1985), and the one rescaled by Tolman and Chalikov (1996) (multiplying coefficient C divided by 3). The variations of H_{mo} and T_p computed with GQM follow quite closely the ones computed with the rescaled DIA (see Fig. 4), whereas the original DIA produces different growth curves. This can also be observed on the frequency spectra plotted on Fig. 5. The rescaled DIA and MDIA spectra are closer to the GQM spectra than the original DIA spectra. We however observe once more that the MDIA tends to produce a spurious peak around f = 0.14 Hz (see Fig. 5). In comparison with test A1, the WW3 physics for Q_{in} and Q_{wc} appears less sensitive than WAM4 physics to a replacement of DIA by more precise methods for evaluating Q_{nl4}.

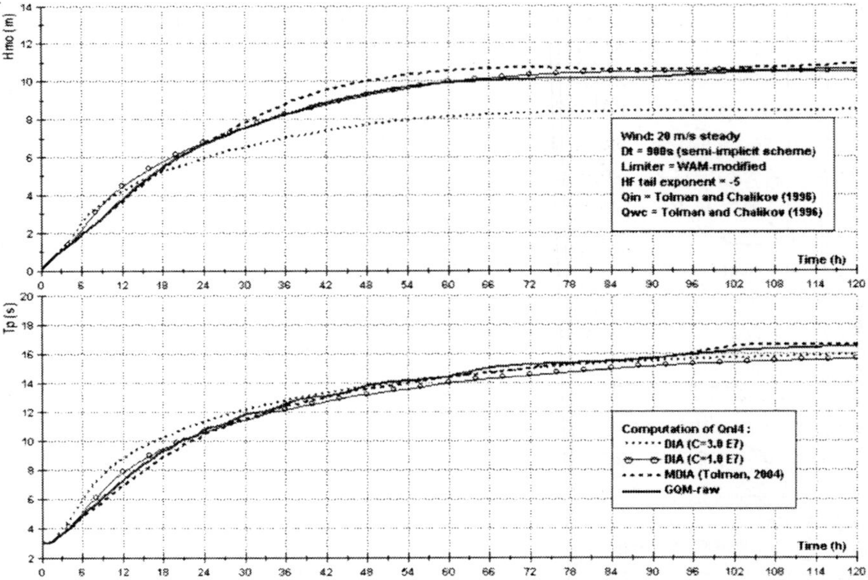

Figure 4. Test A3 - Variation of significant wave height H_{mo} (upper panel) and peak period T_p (lower panel) under steady wind conditions (U_{10} = 20 m/s) computed with WW3 physics for Q_{in} and Q_{wc}, in combination with four options schemes for Q_{nl4}.

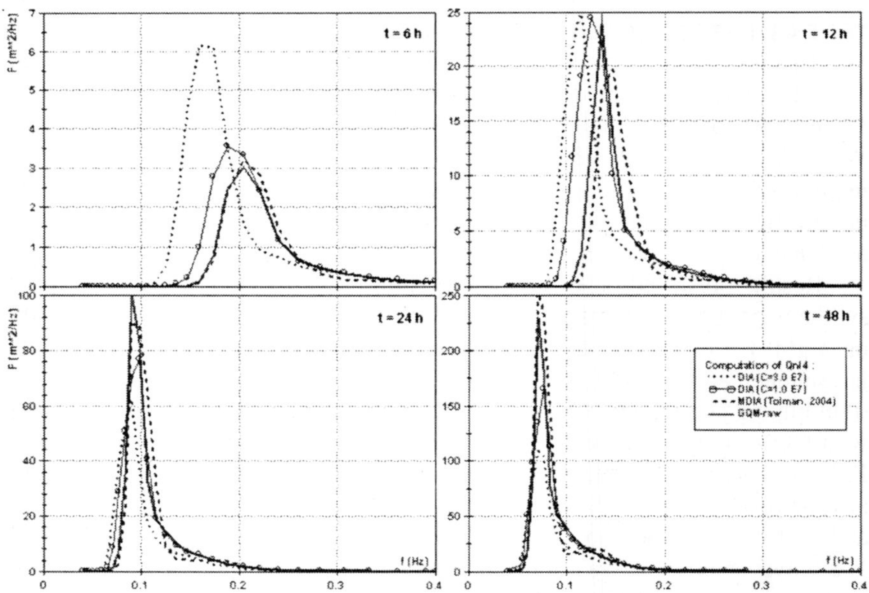

Figure 5. Test A3 - Variation of frequency spectra at 4 times (t = 6, 12, 24 and 36 h) under steady wind conditions (U_{10} = 20 m/s) computed with WW3 physics for Q_{in} and Q_{wc}, in combination with four options schemes for Q_{nl4}.

TEST B: EFFECT OF THE METHOD FOR COMPUTING THE QNL4 TERM

This test is very similar to test A1 (replacement of DIA by MDIA and GQM, without any further changes), but now applied to case B (turning wind). We use the physics of WAM4 for Q_{in} and Q_{wc}. Time integration is performed with the semi-implicit scheme (Δt = 900 s). We use the modified growth limiter given by Eq. (6) and the prognostic limit f_d given by Eq. (5). The HF tail exponent is here taken as n = -4.

This case is the most sensitive to the choice of method for Q_{nl4}. Large differences are observed in the dynamic variations of synthetic parameters (see Figs. 6 and 7). Wave heights and periods obtained with DIA are markedly larger than the ones obtained with GQM and MDIA (see Fig. 6). The mean wave direction θ_m also behaves differently ; its variations are smoother with the DIA (upper panel of Fig. 7). It is also observed (lower panel of Fig. 7) that the angular spreading of energy is much larger with the DIA than the one obtained with more precise methods. The inspection of direction spectra is very instructive (see Fig. 8 for a sample of snapshots of normalized spectra after the second shift of wind direction). Based on DIA the wave spectrum remains unimodal after the shift of wind direction and slowly rotates to adapt to the new wind direction. However the run done with GQM clearly shows the formation of a new peak in the new wind direction. The spectrum is clearly bimodal during several hours. The dynamics of wave spectrum evolution and energy transfer are significantly different between the three schemes for computing Q_{nl4}.

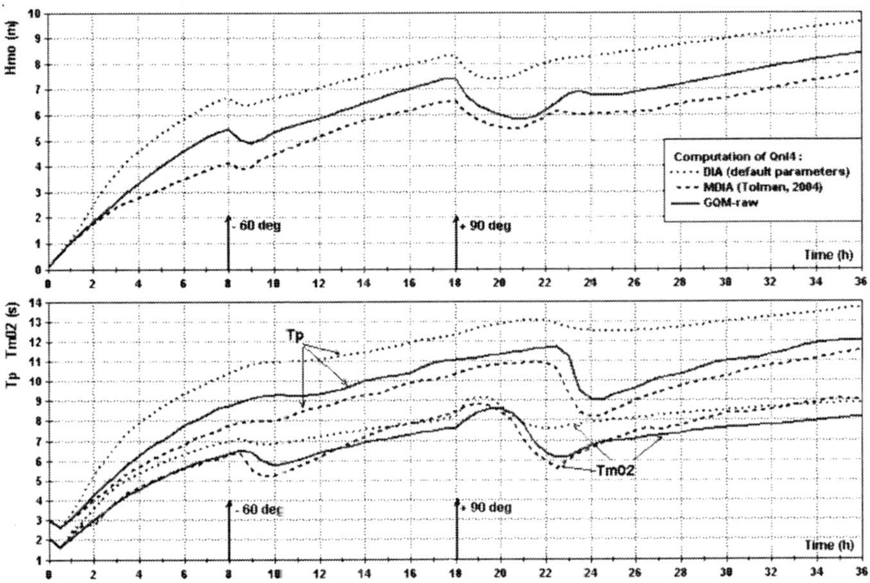

Figure 6. Test B - Variation of significant wave height H_{mo} (upper panel), peak and mean periods T_p, T_{m02} (lower panel) under unsteady (turning) wind conditions computed with WAM4 physics for Q_{in} and Q_{wc}, in combination with 3 schemes for Q_{nl4}

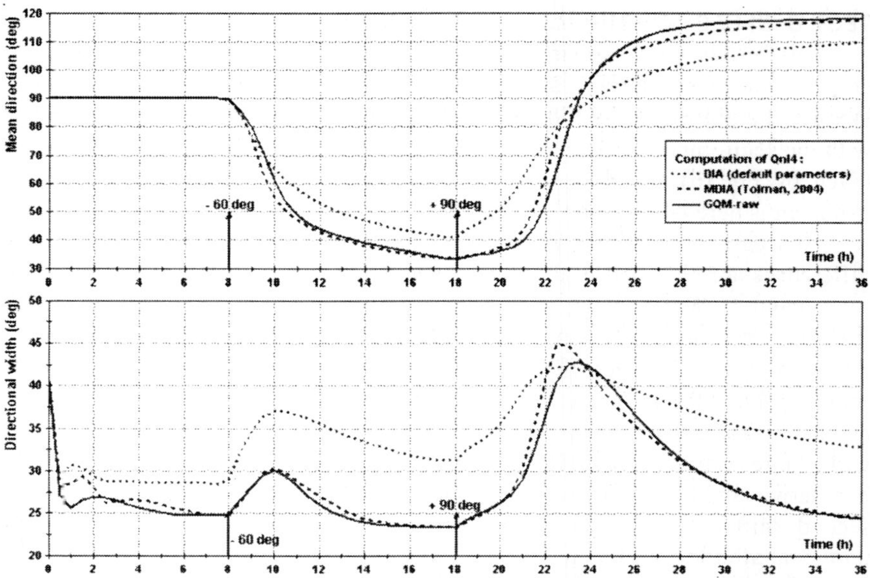

Figure 7. Test B - Variation of mean wave direction θ_m (upper panel) and directional width σ (lower panel) under unsteady (turning) wind conditions computed with WAM4 physics for Q_{in} and Q_{wc}, in combination with 3 schemes for Q_{nl4}.

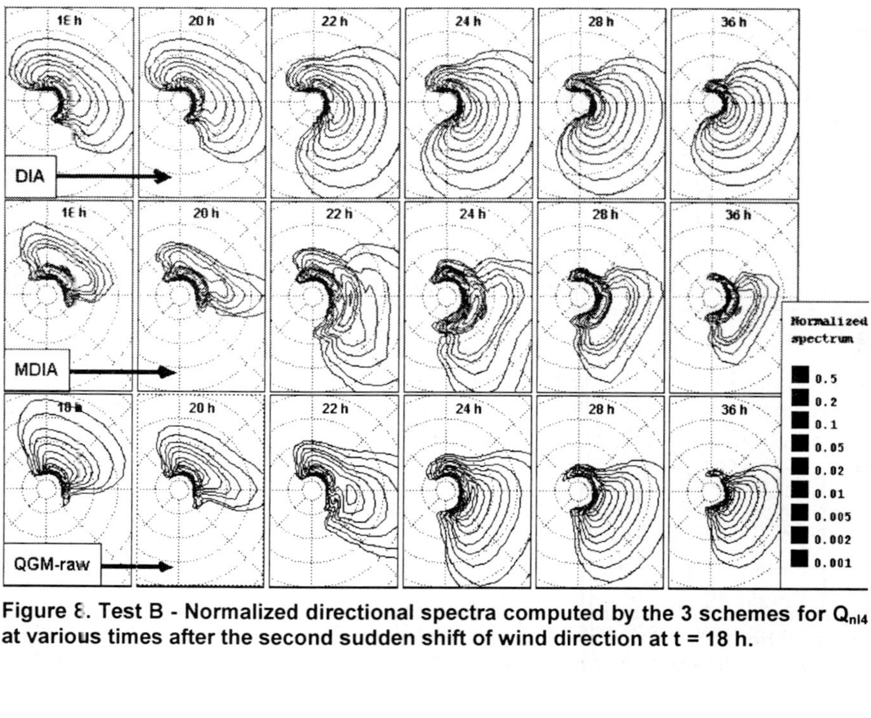

Figure 8. Test B - Normalized directional spectra computed by the 3 schemes for Q_{nl4} at various times after the second sudden shift of wind direction at t = 18 h.

SUMMARY OF CONCLUSIONS AND FUTURE WORK

The numerical experiments presented here confirm that simple substitution of DIA by an improved algorithm for computing the nonlinear wave-wave interactions in the real environment of a 3G wave model is not straightforward. This raises a number of issues, related to the numerical scheme (and time step) for time integration, the effect of a wave growth limiter if present (we have shown that it may have significant effect on spectrum development, at least on the first stage, and thus mask physical effects), the choice of diagnostic frequency limit and prognostic tail shape. Of particular importance is the "compatibility" of Q_{nl4} scheme with formulations used for wind input and white-capping dissipation. Inclusion of improved algorithms for Q_{nl4} will need a recalibration/or retuning of these terms. We noted that WW3 physics seems less sensitive to the replacement of DIA by GQM compared to WAM4 physics.

Regarding the influence of the scheme used to compute Q_{nl4}, significant differences have been noted both in the dynamics of wave spectrum evolution and in the final (fully developed) conditions. It is confirmed that DIA produces too broad spectra and is not appropriate for unsteady cases such as turning winds. MDIA brings significant improvements compared to DIA, but it tends to produce spurious secondary peaks in the spectrum. The "quasi-exact" GQM method appears as a valuable candidate for implementation in 3G models, although there is still a need to optimise the number of configurations and reduce the computational effort for practical applications.

REFERENCES

Alves, J.H.G.M., and M.L. Banner. 2003. Performance of a saturation-based dissipation-rate source term in modeling the fetch-limited evolution of wind waves. *J. Phys. Oceanogr.*, 33, 1274-1299.

Benoit, M., F. Marcos and F. Becq. 1996. Development of a third generation shallow-water wave model with unstructured spatial meshing. *Proc. 25th Int. Conf. on Coastal Eng.*, ASCE, Orlando (Florida, USA), 465-478.

Benoit, M. 2005. Evaluation of methods to compute the non-linear quadruplet interactions for deep-water wave spectra. *Proc. 5th Int. Symp. on Ocean Wave Measurement and Analysis,* 3-7 July 2005, Madrid (Spain).

Bretherton, F.P., and C.J.R. Garret. 1969. Wavetrains in inhomogeneous moving media. *Proc. Roy. Soc. London,* Series A, Vol 302, 529-554.

Hasselmann, K. 1962. On the non-linear energy transfer in gravity-wave spectrum. Part I: General theory. *J. Fluid Mech.,* 12, 481-500.

Hasselmann, S., and K. Hasselmann. 1981. A symmetrical method of computing the nonlinear transfer in a gravity wave spectrum. *Hamburg Geo. Einzelschriften,* Report 1981-52.

Hasselmann, S., K. Hasselmann, J.A. Allender, and T.P. Barnett. 1985. Computations and parameterizations of the nonlinear energy transfer in gravity-wave spectrum. Part II: Parameterizations of the nonlinear energy transfer for application in wave models. *J. Phys. Oceanogr.,* 15, 1378-1391.

Hersbach, H., and P.A.E.M. Janssen. 1999. Improvement of the short-fetch behavior in the wave ocean model (WAM). *J. Atm. Oceanic Tech.*, 16, 884-892.

Janssen, P.A.E.M. 1991. Quasi-linear theory of wind-wave generation applied to wave forecasting. *J. Phys. Oceanogr.*, 21, 1631-1642.

Komen, G.J., S. Hasselmann, and K. Hasselmann. 1984. On the existence of a fully developed wind-sea spectrum. *J. Phys. Oceanogr.,* 14, 1271-1285.

Komen, G.J., L. Cavaleri, M. Donelan, K. Hasselmann, S. Hasselmann and P.A.E.M. Janssen. 1994. *Dynamics and modelling of ocean waves.* Cambridge Univ. Press, 532 p.

Lavrenov, I.V. 2001. Effect of wind wave parameter fluctuation on the nonlinear spectrum evolution. *J. Phys. Oceanogr.,* 31, 861-873.

Lin, R.Q., and W. Perrie. 1998. On the mathematics and approximation of the nonlinear wave-wave interactions. *In "Nonlinear Ocean Waves", Advances in Fluid Mech.,* Perrie Ed., Comp. Mech. Publ., Southampton (UK), 61-88.

Masuda, A. 1980. Nonlinear energy transfer between wind waves. *J. Phys. Oceanogr.,* 10, 2082-2093.

Tolman, H.L. 1992. Effects of numerics on the physics in a third-generation wind-wave model. *J. Phys. Oceanogr.*, 22, 1095-1111.

Tolman, H.L., and D. Chalikov. 1996. Source terms in a third generation wind wave model. *J. Phys. Oceanogr.*, 26, 2497-2518.

Tolman, H.L. 2004. Inverse modeling of Discrete Interaction Approximations for nonlinear interactions in wind waves. *Ocean Modelling,* 6, 405-422.

Van Vledder, G.Ph., T.H.C. Herbers, R.E. Jensen, D.T. Resio and B.A. Tracy. 2000. Modeling of non-linear quadruplet wave-wave interactions in operational wave models. *Proc. 27th Int. Conf. on Coastal Eng.,* Sydney (Australia), 797-811.

Van Vledder, G.Ph. 2001. Extension of the discrete interaction approximation for computing nonlinear quadruplet wave-wave interactions in operational wave prediction models. *Proc. 4th Int. Symp. on Ocean Waves Measurement and Analysis,* (ASCE), San-Francisco (California, USA), 540-549.

WAMDI Group. 1988. The WAM model - A third generation ocean wave prediction model. *J. Phys. Oceanogr.*, 18, 1775-1809.

Webb, D.J. 1978. Non-linear transfers between sea waves. *Deep-Sea Research,* 25, 279-298.

Young, I.R., and G.Ph. van Vledder. 1993. A review of the central role of non-linear interactions in wind-wave evolution. *Phil Trans. R. Soc. London,* A342, 505-524.

Zakharov, V.E. 1968. Stability of periodic waves of finite amplitude on the surface of deep fluid. *PMTF Zh. Prikl. Mekh. Tekh. Fiz.*, 3, 80-94.

Zakharov, V.E., and A.N. Pushkarev. 1999. Diffusion model of interacting gravity waves on the surface of deep fluid. *Nonlinear Processes in Geophysics,* 6, 1-10.

MODELLING AND MEASURING WAVES IN COASTAL WATERS

Judith Wolf[1], Pedro Osuna[1], John Howarth[1] and Alejandro Souza[1]

We investigate some aspects of wave-current interaction in the modelling and observations of waves and currents in the Irish Sea. The interaction of waves and currents through the surface and bottom stress, as well as refraction of waves by currents, is included in a coupled wave-current model. Long-term measurements of currents and waves, starting in November 2002, are being made in the Liverpool Bay Coastal Observatory in the Irish Sea. The large tidal range, relatively shallow depth and waves up to 5m suggest significant wave–current interactions will occur here and the measurements should provide a good test of coupled hydrodynamic-wave models. Some of the problems of getting observational data to validate or challenge these models are discussed.

INTRODUCTION

Coastal development continues to put pressure on wetland habitats, beaches and coastal defences. Concerns about the impacts of climate change exacerbate the problem for coastal managers. In order to properly manage the near-shore coastal zone, we need appropriate models of the physical, biogeochemical and ecological systems. Key challenges in coastal oceanography are the implementation of appropriate local area models and critically testing these with observations. The interactions between waves, currents and turbulence and their impacts on sediment transport become more important, with implications for ecological modelling in terms of water quality and turbidity. We investigate the need for local wind modelling, wave-current interaction, vertical and horizontal current shear, improved representation of bottom friction and turbulence modelling. Various developments in spectral wave modelling have produced a range of possible source terms especially for the wind input and dissipation source terms. Different approximations to the exact nonlinear interaction source term are also available. Our interest is in the practical implementation of the WAM model in shallow water, using the optimal source terms for this environment. In order to test the different options we need adequate observations of waves, currents and reliable wind forcing for the model.

Here we review the mechanisms of interaction which are being incorporated in the POLCOMS-WAM coupled model system and some of the key problems in making observations to test these models. We describe some of this work in progress and recent results. Specifically we look at the surface and bottom stress in coupled models and how to observe and validate these and the implications of the assumption of steady flow in wave models. New data from the POL (Proudman Oceanographic Laboratory) Coastal Observatory are presented.

---

[1] Proudman Oceanographic Laboratory, 6, Brownlow St, Liverpool, L3 5DA, U.K.

ADVANCES IN COUPLED WAVE AND CIRCULATION MODELS

Recently, progress has been made towards coupling atmosphere and wave models (e.g. Janssen 2004) and in developing a consistent treatment of the equations of motion for waves, currents and turbulence (e.g. Mellor 2003, 2005). We know waves are the mechanism for transfer of energy between the atmosphere and ocean and a wave model is an important component of the system. We are interested in testing the implementation of these ideas in pre-operational models such as the POL Coastal Ocean Modelling System (POLCOMS) in which a baroclinic 3D current model is coupled with the 3-G WAM spectral wave model. These are two very different types of model: the former describes explicitly the evolution of sea surface elevation and currents due to tides, density and wind forcing (although with parameterisation of sub-grid processes such as turbulence). The wave model is a phase-averaged model which describes the evolution of the wave spectrum; no information is available on phase of the individual wave components. Procedures to include the two-way interaction between currents and waves in surface and bottom stress and the refraction by changing water depth and currents have been included in the POLCOMS-WAM system (Wolf 2004, Osuna and Wolf 2005). The effect of relative wind is also included.

Some of the problems in modelling the seasonal thermocline (Holt and James 2001) are related to the turbulence and mixing model which may be affected by waves in certain situations (e.g. Langmuir circulation and Stokes' drift) and some alternative formulations have been explored (e.g. Carniel et al. 2004). For the implementation of new terms in the WCI (wave-current interaction) module of POLCOMS, the approaches described in Mellor (2003, 2005), Mellor and Donelan (2006), and Ardhuin and Jenkins (2006) are used. At present, the wave model has been prepared to calculate the following information: Stokes' drift, radiation stress and Doppler velocity for each component of the wave spectrum, to properly account for the vertical variation of current velocity. The Stokes' drift results on the Ekman circulation obtained from the WCI system were qualitatively and quantitatively similar to those reported in Rascle et al. (2006). Results with the enhanced model will be available shortly.

LIVERPOOL BAY COASTAL OBSERVATORY

Long-term measurements of currents and waves, starting in November 2002, are being made in the Liverpool Bay Coastal Observatory in the Irish Sea (see Fig. 1). Directional waves are being measured by wave buoys, ADCPs (acoustic Doppler current profiler) and ADVs (acoustic Doppler velocimeter) within the footprint of a phased-array HF radar system. Simultaneous measurements of currents, waves, bed stress and water level are made (Howarth et al. 2006). In conjunction with this, detailed process studies are being carried out in the mouth of the Dee Estuary measuring wind-stress, currents, waves and turbulence. Liverpool Bay is relatively shallow, with a maximum depth less than 40 m, with tidal currents up to 1 m/s and at the mouth of the Dee, near Hilbre Island, there are extensive sandbanks which are exposed at low water. The Irish Sea is semi-enclosed so waves are mostly locally generated, with significant wave heights up to 5m recorded and mean periods less than 8 seconds. Significant wave–current interactions may occur here and the measurements provide a good test of coupled hydrodynamic-wave models. ADCP and HF radar measure waves using their surface properties and hence are not affected by depth-attenuation like the bottom-mounted ADV.

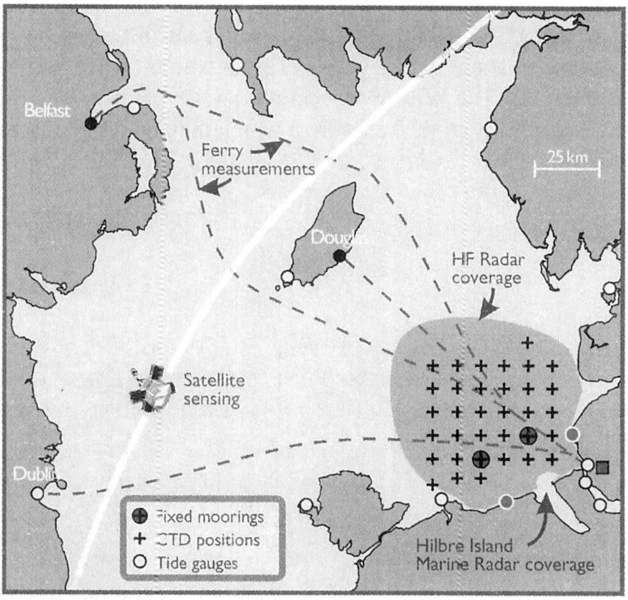

Figure 1. Liverpool Bay Coastal Observatory.

HF radar

A new method of simultaneously observing waves and currents is the WERA HF radar (Gurgel et al. 1999, Wyatt et al. 1999). This measures the directional wave spectrum as well as surface currents at a grid of points over a spatial area. The system (12-16MHz) operates from two shore-based sites, one near Colwyn Bay and the other near Formby Point, giving maximum coverage for resolving currents and wave conditions. The maximum range of the radar is around 100km with a resolution of 2km for sea surface currents and 5km for 2-D wave spectra. The radar was first deployed in March 2004 and real-time data are available on the Coastal Observatory website (http://cobs.pol.ac.uk). At present, only plots are available to view, updated every hour. Although there have been some teething problems we now have some confidence in the wave results since November 2005, see Fig. 2

Fig. 2(a) shows a scatter plot for the HF radar wave height against the WaveNet Waverider buoy (http://www.cefas.co.uk/wavenet/) which is at the same location as the Coastal Observatory primary mooring, using the nearest cell from the HF radar array for comparison. The agreement is reasonable, although further work needs to be done with quality control checks, precluding inversion of the radar Doppler spectrum for waves under certain conditions. A noticeable discrepancy is that the radar often overestimates the wave height in low sea states. It is recognized that the signal to noise ratio can be poor at low wave heights, with the threshold wave height decreasing with increasing radar frequency (Wyatt et al. 2006). Fig. 2(b) shows an instantaneous map of wave height for a time of quite strong winds. The maximum wave height observed by the radar is over 4m. The Waverider recorded 4.46m at the same time. There is a lot of spatial variability in the radar map which may be related to the local water depth and current.

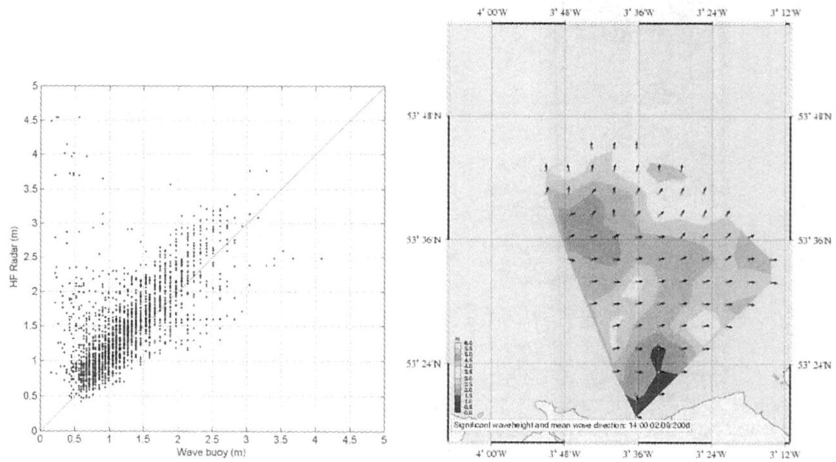

(a) (b)

Figure 2. (a) Comparison between HF radar and buoy significant wave heights (m) at location of Wavenet Waverider buoy (b) Map of radar wave height at 14:00 2 Sep 2006 (wind speed at Hilbre Island = 19.46 m/s)

Fig. 3 shows the distribution of the M_2 tidal component of the surface current measured by the radar. It may be seen that there are substantial gradients of current amplitude and phase in this region with maximum M_2 amplitude of over 1.2m/s.

ADCP measurements

Another instrument which can simultaneously measure waves and the current profile is the ADCP. The Teledyne RDI ADCP Workhorse Waves Array has software which allows the current to be taken into account in the analysis of waves from the near-surface velocities. First results from this for January 2005 showed only occasionally significant differences between the analysis with and without currents. The high frequency measurements allow calculation of the Reynolds' stresses and hence bed stress and bottom drag coefficient, which has been done successfully for current-only (no waves) conditions (Howarth and Souza 2005). The ADCP wave height is in good agreement with the wave buoy but the periods are systematically 0.7s longer. Fig. 4 shows an Hs-Tz plot for the ADCP wave data which is a useful check on data consistency. The solid line indicates the limiting steepness, 1/16 in this case.

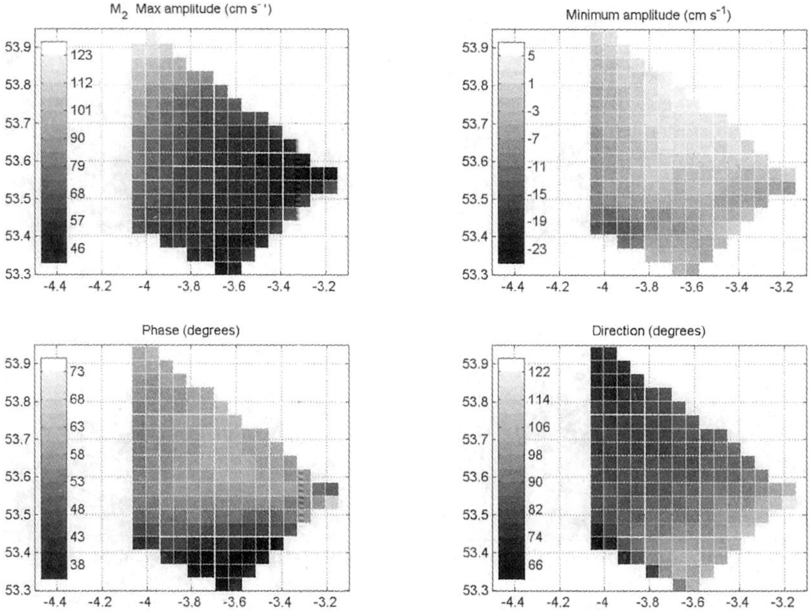

Figure 3. Results of tidal analysis of currents from HF radar. Amplitude of semi-major and semi-minor axis, phase and direction are shown for M_2 tide

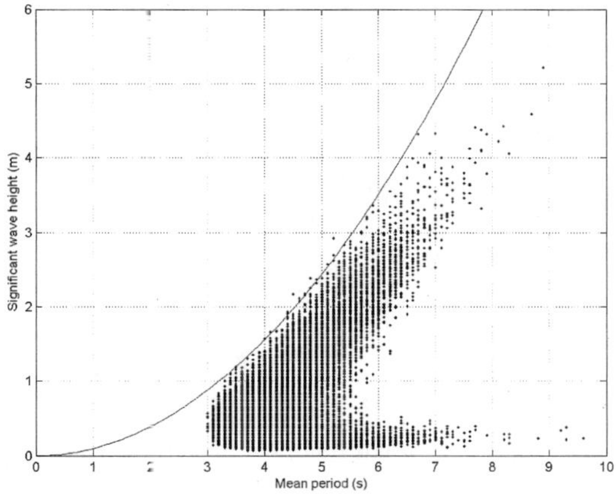

Figure 4. Hs-Tz plot for ADCP wave data December 2002 - October 2006

Wave buoy data

A Triaxys wave buoy has been deployed almost continuously in the Hilbre Channel in the mouth of the Dee Estuary since 2004. This is a relatively sheltered location in about 10m water depth (chart datum), although exposed to the NW there are drying banks (ebb and flood shoals) at the mouth of the estuary. Some results for March 2004 are shown in Fig. 5, along with wind and tide data for the same period. The maximum wave height is between 2 and 2.5m, and can be seen to be closely related to the local wind, which is from the W and NW; the wave period is mostly 3-4 seconds. The tidal modulation is in phase with the water depth where the tidal range varies from 4-8m. Purely due changes in water depth and fetch and using a constant wind of 10m/s the variation in wave height and period from simple fetch-limited growth laws e.g. Hurdle and Stive (1989) the change from low water to high water could be 0.3m to 2.1m in wave height and 1.3 to 4.3 seconds in wave period which is consistent with the order of magnitude observed. This is almost totally due to an assumed change in fetch from 2km to 100km.

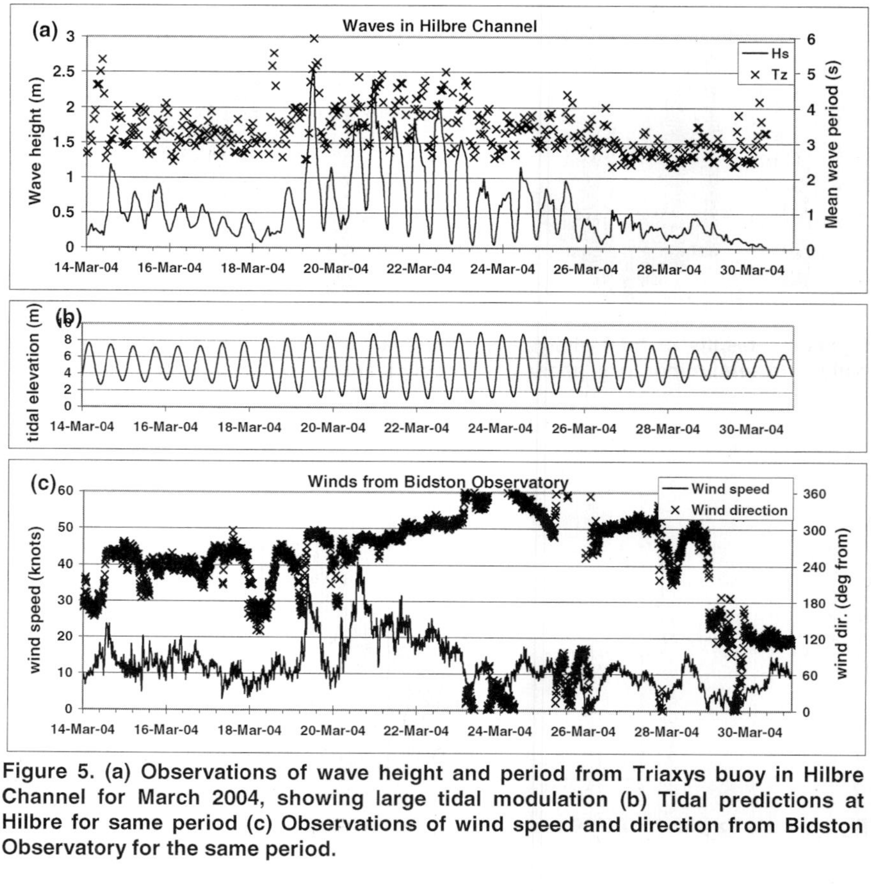

Figure 5. (a) Observations of wave height and period from Triaxys buoy in Hilbre Channel for March 2004, showing large tidal modulation (b) Tidal predictions at Hilbre for same period (c) Observations of wind speed and direction from Bidston Observatory for the same period.

DISCREPANCIES BETWEEN MODEL AND OBSERVATIONS

Although the tidal modulation of wave height in very shallow water has been shown to be consistent with theoretical values we now examine some examples of wave-current interaction in intermediate water depths (~20m) and discuss the related mechanisms. Fig. 6 shows the comparison of a wave model hindcast for 26 January – 7 February 2003, using the coupled POLCOMS-WAM model, forced by Met Office mesoscale model winds, with buoy and ADCP observations for the same period. The general agreement is very good but it may be seen that there are marked tidal modulations, of the wave height especially, and to a lesser extent the wave period, that are not captured by the coupled model. In fact there are only slight differences between the uncoupled (Unc) and coupled (Cou) model runs, although elsewhere in the Irish Sea it has been shown that differences reach 10% in wave height and 20% for mean period (Osuna and Wolf 2005). Although there are also some differences between the buoy and ADCP data the poorer temporal resolution of the buoy data means that it is not clear whether the oscillations are also observed by the wave buoy. The coupled model includes the combined effect of waves and currents in the bottom stress. We now discuss some of the coupling mechanisms and whether they are properly modelled. An illustration of the effects of coupling surface and bottom stress is given in Fig. 7.

For currents, the bottom stress (τ_c) is affected by the waves through the dependency of the drag coefficient on the apparent roughness induced by the waves. The computation of bottom stress for the waves (τ_w) is carried out using Madsen's formulation (Madsen 1994), which takes into account the current at the bottom.

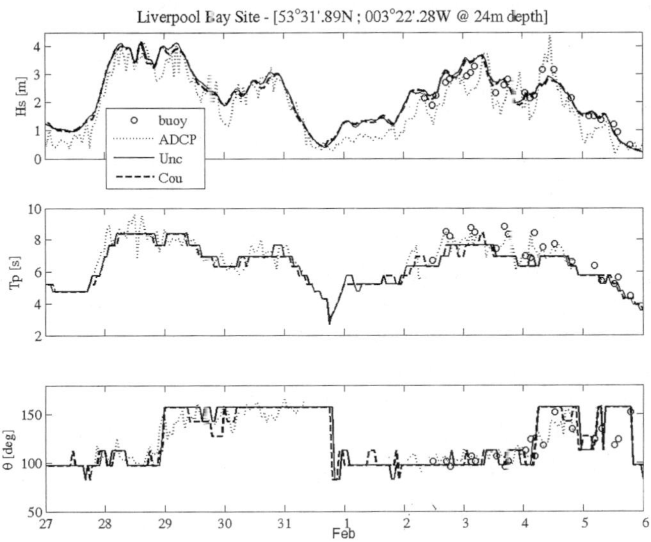

Figure 6. Comparison of model and observations at a location in Liverpool Bay

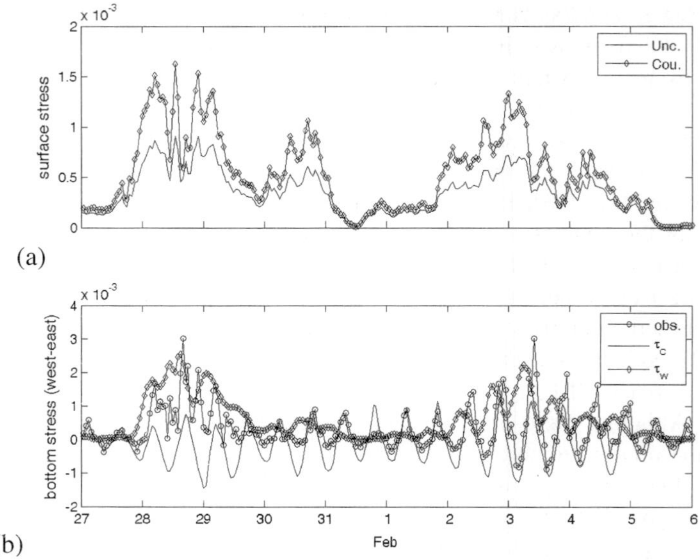

Figure 7. Comparison of (a) surface and (b) bottom stress from model and observations at a location in Liverpool Bay

Surface Stress

Substantial differences in waves and currents due to including wave effects on the surface stress have been found (Osuna and Wolf 2005), using the Janssen (1991) formulation. The effect of the relative wind is included in the coupled model. Some of these differences are related to a different way of calculating the stress, now using the Charnock relationship (Charnock 1955) as opposed to Smith and Banke (1975), hereafter referred to as SB, which was previously employed in POLCOMS, to allow the drag coefficient to be calculated from the surface roughness which comes from the wave model, as may be seen in Fig. 7(a) in which the stress from the coupled model run (Cou) is substantially larger than that from the uncoupled (Unc) run. The Charnock coefficient, $\beta = gz_0 / U_*^2$, is related to the surface drag coefficient, C_D, the 10m wind-speed, U_{10} and the surface roughness length, z_0, (since $U_*^2 = C_D U_{10}^2$) and is often treated as a constant in models with values between 0.01 and 0.03. If the value of β is chosen, it can be used to derive the effective value of the roughness length from a given drag law model e.g. SB. A value of β=0.0275 is now used in the U.K. operational storm surge model (Williams and Flather 2002), having been found to give the optimum results for surges. To compare like with like the equivalent value of the Charnock parameter should be chosen (e.g. Mastenbroek et al. 1993) in our coupled model system which can then be calibrated using the database of surges from the Irish Sea. For β=0.0275 the stress ratio between the new formulation and SB is shown in Fig. 8. The stress is enhanced more at lower wind speeds but in general gives a higher value than SB. This explains part of the increase of wind-stress in Fig. 7(a) even before wave effects are included. A control run using the Charnock formulation with a constant roughness is required.

The European Centre for Medium Range Weather Forecasting coupled atmosphere-wave model allows the 10m winds to be reduced in case of high drag, this feedback will ameliorate the impact of waves on the surface wind stress (Janssen 2004) and introduction of the 2-way coupled model resulted in a substantial improvement of the surface wind field forecast. It may be necessary to use a local wind model for Liverpool Bay to allow modification of the atmospheric boundary layer and also allow for local variations in wind due to the orographic effects, land-sea breezes and the coastal boundary layer.

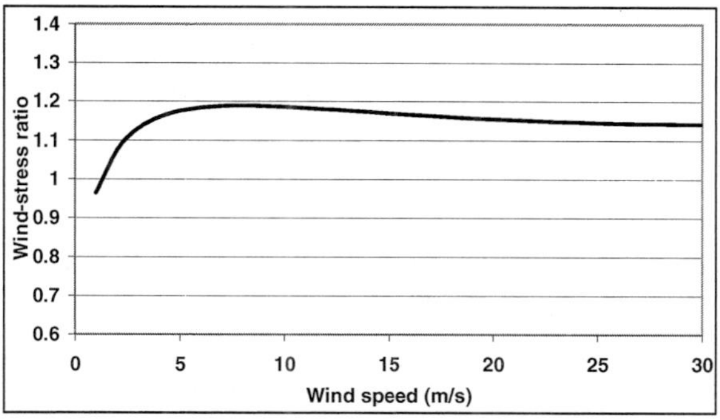

Figure 8. Wind-stress ratio for Charnock parameter β=0.0275, relative to SB

The present method of including wave-induced turbulence in the POLCOMS model, based on Craig and Banner (1994) does not carry turbulence down into the water column. The inclusion of Stokes' drift and potentially Langmuir circulation or other mechanisms in the wave model may enable better modelling of the deepening of the mixed layer without the necessity to infer large eddy viscosity which degrades the sharpness of the thermocline.

Bottom Stress

The use of ADV data for Reynolds' stress calculations is problematical for combined waves and currents Trowbridge (1998) noted that the Reynolds' stress calculated in combined waves and currents is contaminated by waves even though the instrument should be outside the wave boundary layer although within the tidal current boundary layer and although the observations are corrected for tilt (to about 0 1 deg). In theory there should be no contribution from the irrotational wave motion in the covariance method of obtaining the bottom stress but in practice a very small error in tilt can produce an error contribution from the wave motion which can overwhelm the turbulent contribution. This is most likely what we see in Fig. 7(b) where the line marked 'obs' refers to the calculation of Reynolds' stress from the uncorrected turbulent velocities.

It is not straightforward to remove the effect of waves, since waves and turbulence co-exist at wave frequencies. The method used here is to remove the part of the near-bed current fluctuations correlated with bottom pressure in the frequency band 0.05-0.5Hz (Wolf 1999). Fig. 9 shows the original calculation of Reynolds' stress <u′w′> in the dominant current direction compared to that after correction for waves. The predicted quadratic bottom stress due to the mean current is also plotted and the positive part of the corrected stress is now in better agreement. This may not be the whole story however. Babanin (2006) points out that wave-induced turbulence may in fact persist throughout the water column. There has been a persistent tendency for the observed drag coefficient to be about a factor of 2 less than that typically required in numerical circulation models e.g. Howarth and Souza (2005), Wolf (1999) which suggests that the processes of near-bed turbulence and bottom stress are not fully accounted for.

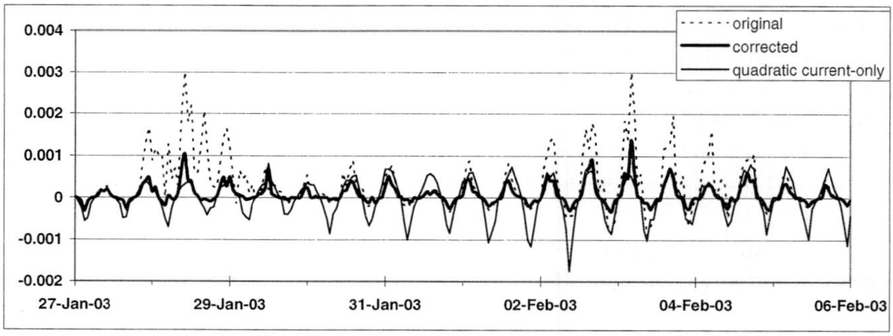

Figure 9. Calculations of bottom stress from observations in combined waves and currents

Wave-current interaction in unsteady flow

Peregrine and Jonsson (1983) and others have reviewed the mechanisms of wave-current interaction and most of these processes are taken account of in our coupled model. Radiation stresses, Stokes' drift and the effect of vertical current shear will soon be included although they are expected to be relatively small or localised. Tidal modulation of wave height and period are often observed in coastal observations e.g. Bradbury et al (2004) and see Figs. 5 and 6 above. Tolman (1991a) discussed tidal modulations of wave parameters as possible observations of wave-current interaction; however he contradicts this in his following paper (1991b) where he attributes most of the modulation to wind effects. However the wind effects discussed above do not seem sufficient to produce the observed modulation of wave height. Modulations of wave period due to Doppler shift in intermediate water depths are modelled, but the observed modulations of wave height are not reproduced. The time-variation of relative frequency is included in the wave model: it can be re-formulated purely in terms of the spatial gradient of current. The time variation of absolute frequency is not included, however:

$\omega = \sigma + \mathbf{k}.\mathbf{U}$ where σ (intrinsic or relative wave frequency) is given by

$$\sigma = \sqrt{gk \tanh(kd)}$$

$$\dot{\omega} = \frac{d\omega}{dt} = \frac{\partial\sigma}{\partial d}\frac{\partial d}{\partial t} + \mathbf{k}.\frac{\partial \mathbf{U}}{\partial t}$$

where ω is absolute frequency, d is the total water depth, **k** is the wavenumber and **U** is the current. Further work is needed to elucidate the implications of this term for the wave model.

SUMMARY

We have discussed various problems of measuring and modelling wave-current interaction. Are discrepancies between model and observations due to errors in the observations or some process not included in the model? What interaction terms may be missing? We are building up new datasets with simultaneous measurements of waves and currents which may be able to critically test the models. There are related problems with separation of waves, currents and turbulence form observations. Key outstanding questions are:

- Are we getting the turbulence right?
- What is the appropriate form for the wave model source terms in shallow water?
- Is Langmuir circulation important and how do we measure and model it?
- What is the effect of unsteady currents on waves?

ACKNOWLEDGMENTS

Some of the wave buoy data were obtained from the Centre for Environment Fisheries and Aquaculture Science (CEFAS) WaveNet system (http://www.cefas.co.uk/wavenet/).

REFERENCES

Ardhuin, F. and A.D. Jenkins. 2006. A practical three dimensional formalism for the ocean, coupling random waves, mean flow, and turbulence. Submitted to *Journal of Fluid Mechanics.*

Babanin, A.V. 2006. On a wave-induced turbulence and a wave-mixed upper ocean layer. *Geophysical Research Letters*, 33, L20605, doi:10.1029/2006GL027308.

Bradbury, A. P., T. E. Mason and M.W. Holt. 2004. Comparison of the performance of the Met Office UK-waters wave model with a network of shallow water moored buoy data, *Proceedings of the 8th International Workshop on Wave Hindcasting and Forecasting, Hawaii, 2004.*

Carniel, S., M. Sclavo, and L.H. Kantha. 2004. The influence of Langmuir cells on the velocity structure in the mixed layer, *Annales Hydrographiques*, 6e Ser., 3, 772, 8-1 – 8-5.

Charnock, H. 1955. Wind stress on a water surface. *Quarterly Journal of the Royal Meteorological Society,* 81, 639-640.

Craig, P.D. and Banner, M.L. 1994 Modelling wave-enhanced turbulence in the ocean surface layer. *Journal of Physical Oceanography*, 24, 2546-2559.

Gurgel, K. -W., G. Antonischki, H. -H. Essen and T. Schlick 1999 Wellen Radar (WERA): a new ground-wave HF radar for ocean remote sensing, *Coastal Engineering*, 37, 219–234.

Holt, J.T. and James, I.D. 2001. An s-coordinate density evolving model of the northwest European continental shelf 1, model description and density structure. *Journal of Geophysical Research*, 106, C7, 14015-14034.

Howarth, M. J. and A. J. Souza. 2005. Reynolds stress observations in continental shelf seas. *Deep-Sea Research II*, 52(9-10): 1075-1086.

Howarth, M.J., R. Proctor, P.J. Knight, M.J. Smithson, D.K. Mills. 2006. The Liverpool Bay Coastal Observatory – towards the goals, *Proceedings of Oceans'06, 18-21 September 2006, Boston*, IEEE, pp6.

Hurdle, D.P and R.J.H. Stive. 1989. Revision of SPM 1984 Wave Hindcast Model to avoid inconsistencies in engineering applications, *Coastal Engineering*, 12, 339-351.

Janssen, P.A.E.M. 2004. *The interaction of ocean waves and wind*, Cambridge University Press. 300pp.

Madsen, O.S. 1994. Spectral wave-current bottom boundary layer flows, *Proceedings of the 24th ICCE*. ASCE, 1:384-398.

Mastenbroek, C., G. Burgers and P.A.E.M. Janssen. 1993. The dynamical coupling of a wave model and a storm surge model through the atmospheric boundary layer. *Journal of Physical Oceanography*, 23, 1856-1866.

Mellor, G. 2003. The three-dimensional current and surface wave equations, *Journal of Physical Oceanography*, 33, 1978-1989.

Mellor, G. 2005. Some consequences of the three-dimensional current and surface wave equations, *Journal of Physical Oceanography*, 35, 2291-2298.

Mellor, G.L. and M.A. Donelan. 2006. A surface wave model for coupling with numerical ocean circulation models. Manuscript in preparation.

Osuna, P. and J.Wolf. 2005. A numerical study on the effect of wave-current interaction process in the hydrodynamics of the Irish Sea, *Proceedings of WAVES 2005*, Madrid, July, 2005.

Peregrine, D.H. and I.G. Jonsson. 1983. Interaction of waves and currents. *U.S. Army Corps of Engineers Miscellaneous Reports,* MR83-6. 88pp.

Rascle, N., F. Ardhuin and E.A. Terray. 2006. Drift and mixing under ocean surface: A coherent one–dimensional description with application to unstratifed conditions. *Journal of Geophysical Research*, 111, C03016, doi:10.1029/2005JC003004.

Smith, S.D. and E.G. Banke. 1975. Variation of the sea surface drag coe±cient with wind speed, *Quarte.ly Journal of the Royal Meteorological Society*, 101, 665-673.

Tolman, H. L. 1991. Effects of tides and surges on North Sea wind waves, *Journal of Physical Oceanography,* **21**, 766-781.

Tolman, H. L. 1991. A third-generation model for wind waves on slowly varying, unsteady and inhomogeneous depths and currents, *Journal of Physical Oceanography,* **21**, 782-797.

Trowbridge, J. H. 1998. On a technique for measurement of turbulent shear stress in the presence of surface waves. *Journal of Atmospheric and Oceanic Technology*, 15, 290-298.

Williams, J. A. and R. A. Flather. 2002. Impact of new atmospheric model dynamics on operational surge model performance, *Proudman Oceanographic Laboratory, Internal Document*, No 145, 14 pp.

Wolf, J. 1999. The estimation of shear stresses from near-bed turbulent velocities for combined wave-current flows, *Coastal Engineering*, 37, 3-4, 529-543.

Wolf, J. 2004. Coupling waves and currents in POLCOMS with measurements from the coastal observatory, *Annales Hydrographiques*, 6th Ser., 3(772): 9-1 - 9-3.

Wyatt, L.R., S.P. Thompson and R.R. Burton 1999 Evaluation of high frequency radar wave measurement, *Coastal Engineering*, 37, 259–282

Wyatt, L.R., J.J. Green, A. Middleditch, M.D. Moorhead, M.J. Howarth, M. Holt, and S. Keogh. 2006. Operational wave, current and wind measurements with the Pisces HF radar, *Journal of Oceanic Engineering* (in press).

WAVE TRANSFORMATION IN FRONT OF THE DUTCH COAST

Jacco Groeneweg[1], Mathijs van Ledden[2] and Marcel Zijlema[3]

The spectral wave model SWAN has been applied to transform a set of offshore wave conditions to nearshore locations along the Dutch coast. The set contains both moderate and severe storm conditions as well as extreme conditions. The resulting transfer function can be applied for design purposes, safety assessment and other purposes for which operational or more extreme wave conditions in the nearshore area are required. The most severe storm in the recent period of five years was chosen to verify the results from the SWAN computations and the assumptions in the boundary condition methodology. Both the predicted significant wave height and the mean wave period agreed very well with the corresponding wave parameters measured at five locations in the nearshore area in front of the Petten sea defense.

INTRODUCTION

A significant part of The Netherlands is below sea level. Dunes and dikes along the Dutch coast protect the mainland from flooding. The safety level of the dikes is expressed in terms of the amount of allowed wave overtopping or damage on the dike revetment under design conditions. Erosion and sedimentation on the dune front have to be evaluated for dunes. The assessment of the safety level under design conditions depends on the water level and wave conditions in front of the construction.

The conventional approach to determine wave conditions for design purposes or evaluation of required height of primary water defenses is to transform wave statistics correlated to wind and water level at one or two offshore stations (out of a total of three in front of the central Holland coast, see Fig. 1) to the location of interest. This transformation needs then to be repeated for every location or study.

Instead of translating these statistics from one location to a limited number of locations, the offshore statistics in the three offshore locations (ELD, YMW and EUR) are transformed to a large set of nearshore locations at once for an entire matrix of wind and water level combinations. These combinations have been chosen such that the resulting nearshore wave conditions can be used for several applications.

The objective of the study is to create a discrete transfer function (of water level, wind speed and wind direction) by means of stationary SWAN computations for the transformation of offshore wave conditions to nearshore

---

[1] WL | Delft Hydraulics, PO Box 177, 2600 MH Delft, The Netherlands; jacco.groeneweg@wldelft.nl

[2] Royal Haskoning, PO Box 151, 6500 AD Nijmegen, The Netherlands;

[3] RIKZ, PO Box 20907, 2500 EX Den Haag, The Netherlands; currently at: Delft University of Technology, Dep. Of Civil Engineering & Geosciences, Stevinweg 1, 2628 CN Delft.

locations along the Dutch coast. These locations consist of output locations in front of primary sea defenses such as dikes and dunes, breakwaters, near the entrance of harbors, and on the 5 m, 10 m and 20 m depth contours. These locations are also indicated in Fig. 1.

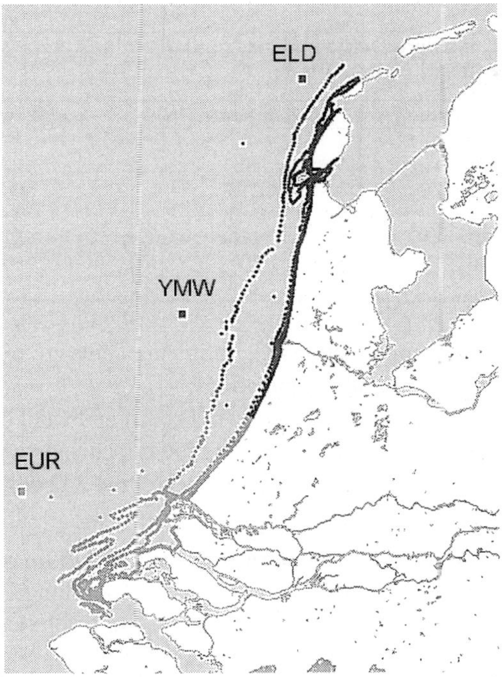

Figure 1. Offshore measuring platforms ELD, YMW and EUR and all nearshore output locations in front of the Dutch coast.

For 252 conditions, given in terms of water level, wind direction and wind speed, the spectral wave model SWAN (version 40.41AB, see Booij et al., 1999, Zijlema and Van der Westhuysen, 2005) has been applied to transform offshore wave conditions to nearshore locations. The set-up of these computations is described in Section 2. In Section 3 some details about the computations are given. In Section 4 some results are shown and the verification with measured wave parameters during a severe storm is presented. Finally, the conclusions are given in Section 5.

SETUP OF SWAN COMPUTATION

The discrete transfer function consists of results from SWAN computations, applied in stationary mode. A matrix of 252 conditions is considered, based on 6 wind speeds, 7 wind directions and 6 water levels (Table 1). The wind speeds are reference values. Based on these values and wind statistics in a number of wind stations spatially varying wind field were constructed. The conditions cover the range of mild storm conditions to extreme conditions, which are representative for North Sea conditions. Correlated with wind is the offshore wave statistics in terms of significant wave height H_{m0}, mean wave period $T_{m-1,0}$ (defined from the 0th and -1st energy moments) and the peak period T_p.

Table 1. Considered wind speeds, wind directions and water levels for SWAN computations.

Wind speed [m/s]	Wind direction [°N]	Surge level [m]
15	0	1
20	30	2
25	210	3
30	240	4
35	270	5
40	300	6
	330	

As input for the SWAN computations the following items are required:

- Grid and bathymetry for all computations;
- Wind fields for each combination of wind speed and wind direction;
- Water level and in principle current field but not considered here;
- Offshore wave boundary conditions for each of the 252 combinations.

Grid and bathymetry

The curvilinear grid covers a coastal strip of 180 km long and 50 km wide and consists of 1980 by 768 grid cells (Fig. 2). The alongshore resolution is approximately 80 m and in cross-shore direction the resolution decreases from 600 m offshore to 30 m nearshore.

In the main part of the bathymetry the most recent depth information is employed, whereas in the surf zone, where the bathymetry shows significant variation due to cyclic morphological behavior, an average over multiple annual surveys is taken which is more realistic for long-term prediction. The offshore water depth varies between 25 m and 35 m.

Wind and water level

For all of the 6x7=42 combinations of wind speed and wind direction (see Table 1) wind fields have been determined, which are based on long-term wind statistics at the three onshore locations Texel (TXH), IJmuiden (YMD) and Hoek van Holland (HVH), indicated in Fig. 4, and on full dependence between the wind at the three locations. For the six wind speeds U_{10} = 15, 20, …, 40 m/s the probability of exceedance in Texel is determined. For the same probability of exceedance the wind speeds at the two other locations have been determined (see Fig. 3) and appear to be lower than in TXH. This is consistent with observations of stronger winds in the North than in the South.

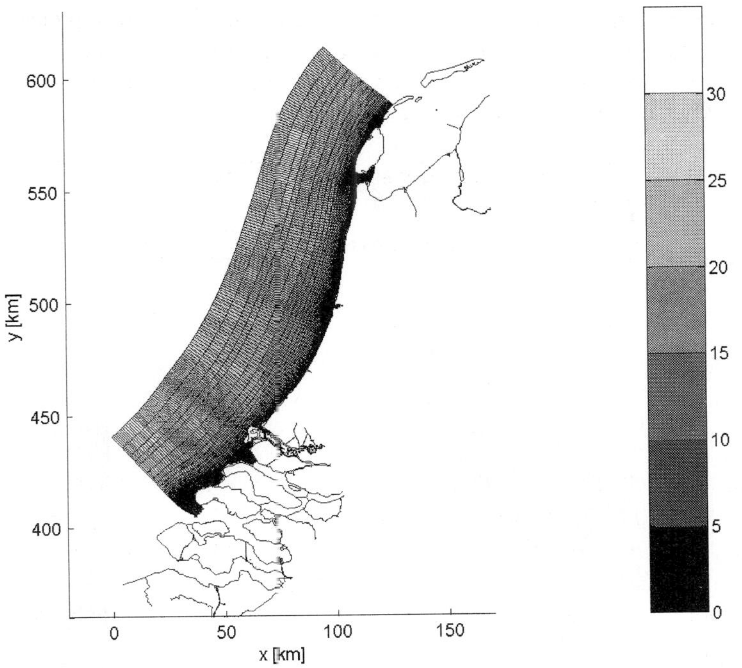

Figure 2. Depth (in m) of computational domain covered by a curvilinear grid. Only every tenth gridline in either direction is indicated.

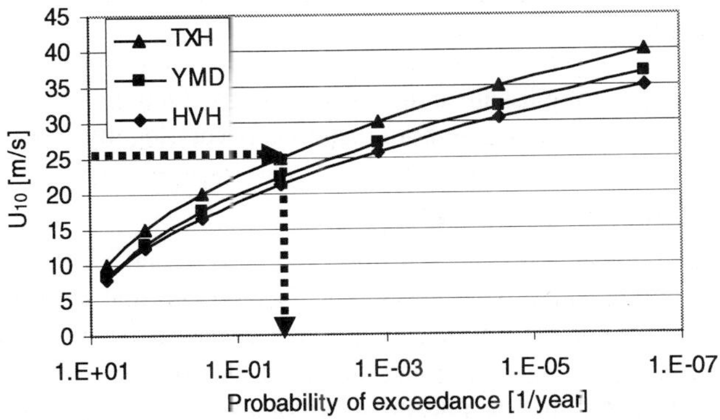

Figure 3. Probability of exceedance of wind speed U_{10} at the three offshore measuring locations.

The wind in these three onshore locations is not representative for the wind at open sea. Therefore Charnock's relation has been applied to account for land-sea transitions. Finally, for each condition a linearly varying wind field has been obtained by taking a 2D plane through the wind speeds at the 3 locations. The wind direction is assumed constant over the entire domain. For a reference wind of 30 m/s and wind direction of 330°N the wind field is given in Fig. 4.

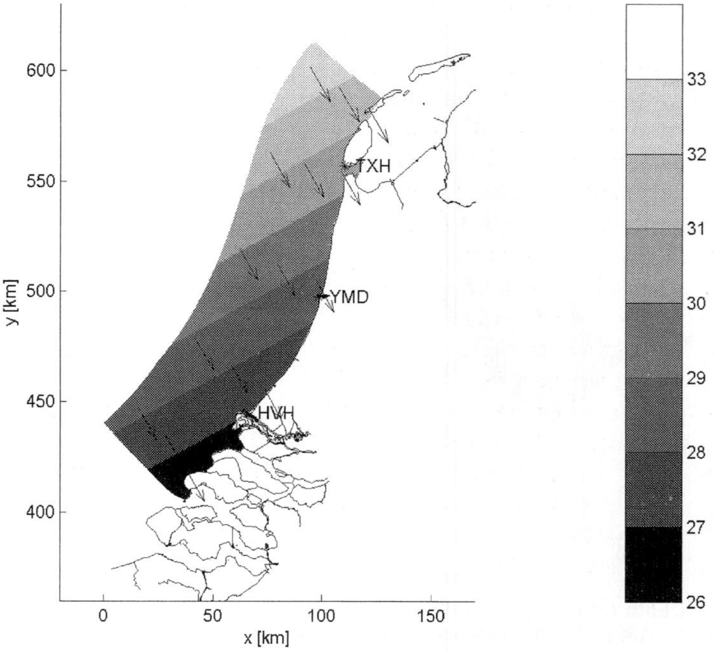

Figure 4. Constructed wind field for a reference wind speed of 30 m/s and wind direction of 330 °N.

The water levels are assumed spatially uniform since the effect of the water level on wave parameters in nearshore locations is very locally.

Offshore wave boundary conditions

At the offshore measuring platforms ELD, YMW and EUR, wave statistics are available for the wave parameters H_{m0}, T_p en $T_{m-1,0}$ that are correlated with wind speed and wind direction. These statistics are used to determine wave boundary conditions along the boundaries of the computational domain, which are further seaward than the platforms. Groeneweg and Van Vledder (2005) showed that simply imposing a JONSWAP spectrum at the boundary led to 'spinning up' effects, especially for extreme situations. The spectral shape of the JONSWAP spectrum is different from the equilibrium spectrum of SWAN, due to the presence of sources and sinks in the model that tend to change the JONSWAP spectrum rapidly within the first couple of grid points near the boundary. As a consequence, imposing a JONSWAP spectrum and tuning the amount of energy and spectral shape to arrive at the required (target)values for

H_{m0}, T_p en $T_{m-1,0}$ at the measurement locations inside the domain, often leads to very unrealistic JONSWAP boundary spectra.

To avoid spinning up effects a boundary spectrum that suits the SWAN physics should be imposed. Groeneweg and Van Vledder (2005) developed a generic method based on wave growth curves. For each wind direction and each offshore measuring location one-dimensional SWAN computations were performed over a sufficiently long trajectory, say 1000 km. The trajectory is taken in the wind direction through the offshore location and intersects the offshore boundary. In Fig. 5 the trajectory through YMW and its boundary location YMWB is given for a wind direction of 330°N. The SWAN-1D computation is performed for the wind speed and water level under consideration over a horizontal bed with the depth of the offshore location.

For a given wind speed and water depth a 2D spectrum is obtained at a certain fetch, which implies a fixed relation between the resulting spectral wave parameters, in particular between H_{m0}, $T_{m-1,0}$ and T_p. Therefore the requirement with respect to the boundary conditions to aim at all three target parameters at the measurement locations has been relaxed and only the computed mean wave period $T_{m-1,0}$ is required to be equal to the statistical imposed value at the three offshore locations. The significant wave height in the area of interest, i.e. the output locations in the nearshore region, is assumed to be dominated by depth-induced breaking and thus the correlation with the offshore significant wave height is minimal.

From the growth curve for $T_{m-1,0}$, resulting from the SWAN-1D computation, the required fetch is determined to arrive at the required value for $T_{m-1,0}$ at the offshore location. The distance between the offshore measuring location (e.g. YMW) and the boundary location (e.g. YMWB) is subtracted from the determined fetch. The 2D spectrum determined with the SWAN-1D computation at this resulting fetch is imposed at the boundary location. This spectrum is in equilibrium, i.e. is a result of the modeled source and sink terms, and does not cause spinning up effects. The procedure is schematically outlined in Fig. 5.

SWAN interpolates the 2D spectra at the grid points in between the three boundary locations. The 2D spectra obtained with additional SWAN-1D computations along the northern and southern boundary of the computational domain have been used as boundary conditions for these parts of the boundary of the computational domain.

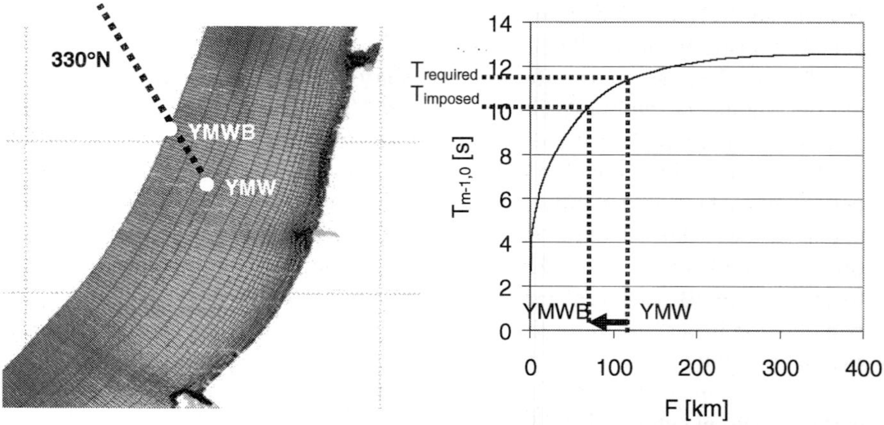

Figure 5. Schematic outline of boundary procedure of Groeneweg and Van Vledder (2005).

Physical and numerical model settings

The physical model settings are mostly the default settings of SWAN, using the third-generation formulations of Komen et al. (1984). The suggestion of Rogers et al. (2002) is used to put more weight on the relative wave number in the whitecapping formulation. This leads to higher wave periods.

Three-wave interactions (triads) have not been taken into account. Several studies showed that the LTA approach, used in SWAN for the modeling of three-wave interactions, leads to significant overestimations of the energy in the higher harmonics on the shallow foreshores in front of the closed Dutch coast (see e.g. Groeneweg et al, 2004).

SWAN determines the solution in an iterative way. Zijlema and Van der Westhuysen (2005) showed that the standard convergence criterion in SWAN is not always strict enough. This so-called 2% criterion means that in less than 2% of all (wet) grid points the relative change in both significant wave height H_{m0} and mean wave period T_{m01} is allowed to be larger than 2%. In the computations not the 2% but the 1% criterion is applied.

SWAN COMPUTATIONS

The SWAN computations have been carried out in batch mode on the Linux-based Hydrax cluster at WL | Delft Hydraulics. The MPI-version of SWAN (Zijlema, 2004) ran on 12 nodes. On average SWAN reached convergence after 22 iterations, taking approximately 45 minutes of CPU time. The entire batch run required 8.5 days. For each condition the offshore wave conditions, which are related to the statistically determined wave conditions at the measurement locations, have been translated to wave conditions at many output locations (see Fig. 1) along the closed part of the Dutch coast. Output has been generated in terms of 1D and 2D spectra and integral wave parameters.

VERIFICATION OF RESULTS

The boundary condition methodology has been verified by comparing the computed wave parameters at the three offshore measuring locations with their target values. In Table 2 the relative mean errors of the three wave parameters, defined as the difference between measured and computed wave parameter averaged over all 252 computations relative to the averaged measured wave parameter, are given for the three measurement lcoations. At all locations the relative mean error for the mean period $T_{m-1,0}$ is small, indicating that the proper boundary conditions were derived. This error is not zero due to a non-uniform bathymetry, non-uniform wind fields and variations of the 2D spectrum along the boundary of the domain.

Table 2. Averaged relative differences between measured (statistics) and computed wave parameters at the offshore measuring locations.

	ELD	YMW	EUR
$\lvert \Delta T_{m-1,0} \rvert$	3.5 %	2.5 %	0.8 %
$\lvert \Delta H_{m0} \rvert$	16.7 %	19.1 %	19.7 %
$\lvert \Delta T_p \rvert$	7.1 %	4.7 %	3.4 %

Since the boundary condition procedure only aimed at getting the mean wave period right, the large difference for the significant wave height is not surprising. It is assumed that the significant wave height decreases significantly in the shallow areas, i.e. the region of interest. That this assumption is valid has been illustrated for an extreme condition (wind speed = 30 m/s, wind direction = 330°N, water level = 4m). The computed spatial distribution of the significant wave height H_{m0} and mean wave period $T_{m-1,0}$ are depicted in Fig. 6 and 7. As expected the significant wave height decreases significantly in shoreward direction due to depth-induced breaking. However the mean wave period remains over 10 s until just outside the surf zone.

This is further illustrated for two trajectories perpendicular to the coast. These trajectories are given in Fig. 6 and 7. The northern trajectory is located in front of the Petten sea defense and consists of a line connecting 5 local measuring devices (directional wave rider 11, radar level meter 33 and pressure sensors 162, 175 and 62). The four most onshore located devices are in the shallow region with migrating sand bars (see Fig. 8 for their location). In Fig. 8 the total water depth, the ratio between significant wave height H_{m0} and water depth, and mean wave period $T_{m-1,0}$ along both trajectories are given. Clearly, the significant wave height is depth-dominated in the region of interest, i.e. in the last kilometers of the rays where it becomes shallow. However, the mean wave period varies only slightly in this region.

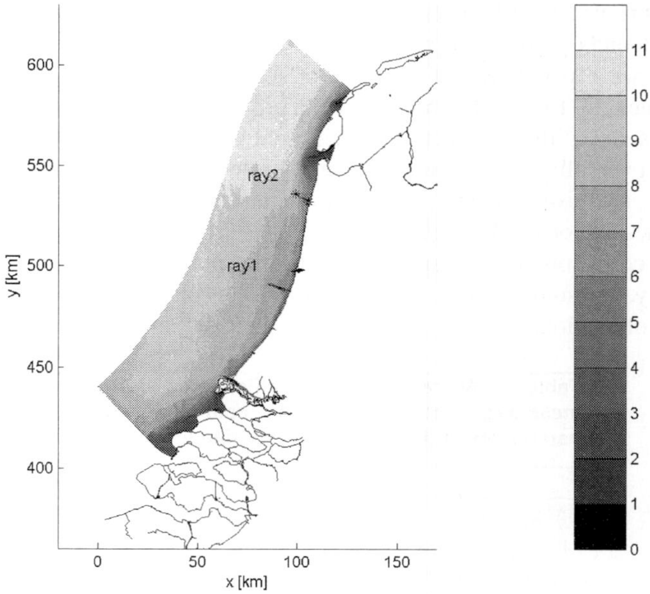

Figure 6. Spatial distribution of significant wave height (in m) for an extreme condition (wind speed = 30 m/s, wind direction = 330°N, water level = 4m).

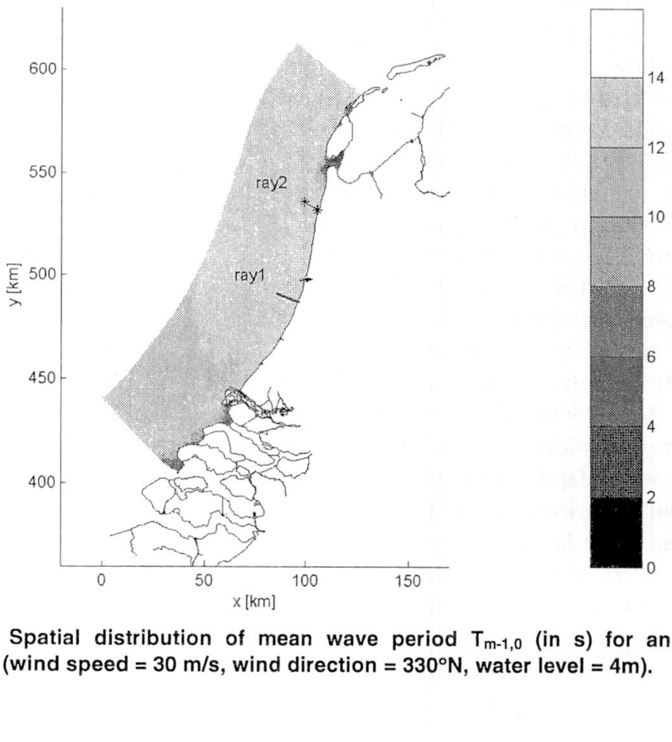

Figure 7. Spatial distribution of mean wave period $T_{m-1,0}$ (in s) for an extreme condition (wind speed = 30 m/s, wind direction = 330°N, water level = 4m).

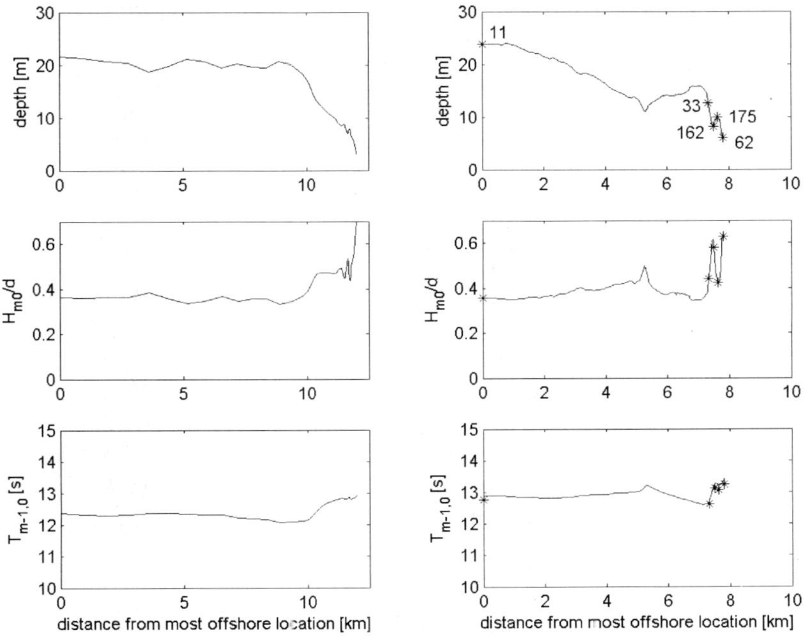

Figure 8. Depth profile and distribution of significant wave height H_{m0} and mean wave period $T_{m-1,0}$ along ray 1 (left) and ray 2 (right, measurement locations indicated with "*") for an extreme condition (wind speed = 30 m/s, wind direction = 330°N, water level = 4m).

The measurements in front of the Petten sea defense are considered for verification purposes. Although extreme conditions have never been measured so far, significant storm events have taken place. Although, the last significant storm was already some years ago, at 26 and 27 October 2002, and the bathymetry considered in the SWAN computations might have changed in the nearshore area in the last three years this storm has been used for verification of the SWAN computations.

At the end of October 2002 a depression above Great Britain caused a severe western storm. During the storm the wind turned from west to south and back to west again. Because southerly winds are outside the range of wind directions considered for the transfer function, storm instants with a southerly wind have not been considered. Four instants have been selected. The wind speed, wind direction and water level at these instants are listed in Table 3.

Table 3. Wind and water level at various instants during the storm of 26-27 October 2002.

	wind speed [m/s]	wind direction [°N]	water level [m]
26 Oct. 2002, 07:00	17	260	1.7
27 Oct. 2002, 11:00	20	230	0.8
27 Oct. 2002, 14:20	24	240	1.0
27 Oct. 2002, 17:00	23	270	1.4

Because none of the combinations of wind and water levels in Table 3 is exactly equal to any of the 252 considered combinations, interpolation is required to obtain the wave conditions in the five measurement locations. For each combination of wind and water level in Table 3 the envelope of (maximum) eight combinations are considered for which SWAN computations have been carried out. Tri-linear interpolation is applied to the significant wave height H_{m0} and mean wave period $T_{m-1,0}$ in each of the five measuring locations. The weights have been determined by the combinations of wind speed, wind direction and water level.

The significant wave height and mean wave period obtained from the transformation matrix is compared to the measured wave parameters in Fig. 10 and 11. On average the computed and measured wave parameters agree very well. Except at location 175 there is a tendency to overpredict the measured significant wave height. The overprediction might be due to changes in the bathymetry, which migrate over a number of years. Most important is that the significant overprediction of H_{m0} at the three offshore measurement locations ELD, YMW and EUR, as shown in Table 2, has completely disappeared.

CONCLUSIONS

The spectral wave model SWAN has been applied to transform a set of offshore wave conditions to nearshore locations along the Dutch coast. These locations are situated in front of primary sea defenses such as dikes and dunes, breakwaters, near the entrance of harbors and on the 5 m, 10 m and 20 m depth contours. The conditions taken into account contain both moderate and severe storm conditions (wind speed 15-25m/s) as well as extreme conditions (wind speed 30-40m/s). Therefore it can be applied for design purpose, safety assessment and other purposes for which operational or more extreme wave conditions in the nearshore area are required.

The SWAN computations require offshore wave boundary conditions, based on measured and statistically extrapolated wave parameters at three offshore measuring locations. To avoid 'spinning up' problems the boundary condition methodology of Groeneweg and Van Vledder (2005) was applied successfully.

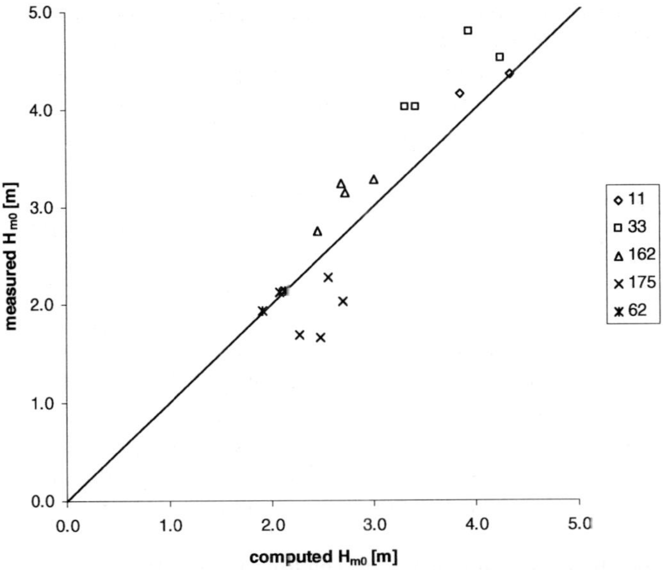

Figure 10. Comparison between measured and computed significant wave height at five measurement locations in front of the Petten sea defense during the storm of 26-27 October 2002.

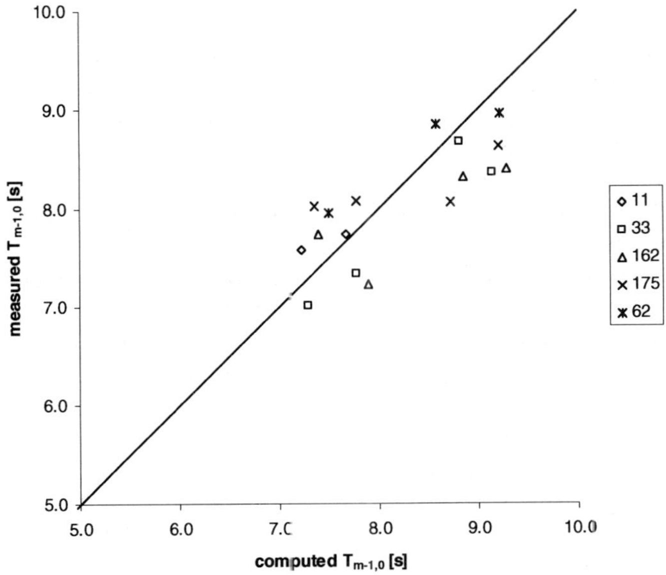

Figure 11. Comparison between measured and computed mean wave period $T_{m-1,0}$ at five measurement locations in front of the Petten sea defense during the storm of 26-27 October 2002.

In the last five years only a few storms hit the Dutch coast. The most severe storm was chosen to verify the results from the SWAN computations and the assumptions in the boundary condition methodology. Both the significant wave height and the mean wave period agreed very well with the corresponding wave parameters measured at five locations in front of the Petten sea defense.

ACKNOWLEDGMENTS

This study was commissioned by the Ministry of Public Works and Transport in the Netherlands, contract RKZ-1488.

REFERENCES

Booij, N., R.C. Ris and L.H. Holthuijsen. 1999. A third generation wave model for coastal regions, Part I, Model description and validation. *J. Geophys. Res.*, 104, C4, 7649-7666.

Groeneweg, J., N. Doorn, M.J.A Borsboom and A.R. van Dongeren. 2004. Nearshore wave modeling with two coupled models: SWAN and Triton. *Proc. of 29th Int. Conf. Coastal Engineering*, Lisbon, 868-880.

Groeneweg, J. and G.Ph. van Vledder. 2005. Tools for improving wave modelling and field measurements in shallow water. *Proc. of WAVES Conference*, Madrid.

Komen, G.J., S. Hasselmann and K. Hasselmann. 1973. On the existence of a fully developed wind-sea spectrum. *J. Phys. Oceanography*, vol. 14, 1271-1285.

Rogers, W.E., P.A. Hwang and D.W. Wang. 2002. Investigation of wave growth and decay in the SWAN model: three regional-scale applications. *J. Phys. Oceanography*, vol. 33, pp 366-389.

Zijlema, M. 2004. Parallelization of a nearshore wind wave model for distributed memory architectures. In Winter, G., Ecer, A., Periaux, J., Satofuka, N., Fox, P. (eds), *Parallel Computational Fluid Dynamics – Multidisciplinary Applications*, Elsevier, pp. 207 – 214.

Zijlema, M. and A.J. van der Westhuijsen. 2005. On convergence behaviour and numerical accuracy in stationary SWAN simulations of nearshore wind wave spectra. *Coastal Engng*, 52, 237-256.

APPLICATION OF ADJOINT WAM TO NOWPHAS WAVE OBSERVATION DATA FOR ESTABLISHMENT OF REASONABLE WAVE OBSERVATION NETWORK

Noriaki Hashimoto[1], Kuniaki Matsuura[2], Toshihiko Nagai[3], and Hiroyasu Kawai[4]

For the purpose of evaluating availability and performance of the data assimilation, we developed two adjoint WAM codes which are capable of correcting the initial condition and/or the boundary condition as the control variables, respectively. Then, we applied them to the NOWPHAS wave observation data. In the paper, as one of the usages of the data assimilation method, we investigate the possibility to evaluate how much the data at each observation station can contribute to the improvement of the estimation accuracy with the model by introducing a few evaluation indexes.

INTRODUCTION

Thirty-six years have passed since the establishment of the Japanese NOWPHAS (Nationwide Ocean Wave Information Network for Ports and Harbors) in 1970. At present, the Port and Airport Research Institute (PARI, Japan) manages the network by collecting and analyzing the wave data measured at about 60 observation stations around Japan. The NOWPHAS has contributed to various purposes such as construction of ports, safety management of coastal facilities, and research work, however, great expense has been invested in the observation. Thus, the establishment of more reasonable and more efficient wave observation network has been required for a long time.

Recently, the third generation wave models, particularly WAM, SWAN and Wave Watch III, are widely used in many countries. These models can certainly describe wave states more accurately than the previous models. However, when we apply these models in practical applications, it is still common that we modify the parameters in the models or the wind field itself by trial and error manner by comparing the numerical results with the observation data.

On the other hand, excellent data assimilation methods recently came to be used in various fields. The methods are vital techniques to connect the model with the reality when we try to reproduce natural phenomena by the model with the data. Applying the method to the model and giving observation data can improve the estimation accuracy with the model.

[1] Kyushu University, 1-10-1, Hakozaki, Higashi-ku, Fukuoka, 812-8581, Japan
[2] Japan Weather Association, 3-1-1 Higashi Ikebukuro Toshima-Ku, Tokyo, 170-6055, Japan
[3] Port and Airport Res. Inst., 3-1-1, Nagase, Yokosuka, 239-0826, Japan
[4] Port and Airport Res. Inst., 3-1-1, Nagase, Yokosuka, 239-0826, Japan

As one of the usages of the data assimilation method, by utilizing this method inversely, we can investigate how much the data at each observation station can contribute to the improvement of the estimation accuracy with the model. Consequently, this makes it possible to verify the importance of each wave observation station on wave observation network.

However, in order to demonstrate the validity of the data assimilation method fully and obtain reliable results, not only the performance of a numerical model but also the performance of the data assimilation method bears the key role. Therefore, we apply the adjoint method as one of the data assimilation methods to WAM with NOWPHAS wave observation data and discuss the applicability and usefulness of the method. And then, we evaluate the importance of each wave observation station on NOWPHAS wave observation network quantitatively.

ADJOINT METHOD

The Adjoint method can deal with the parameters including the nonlinearity in the model and can estimate the optimum values of the control parameters for data assimilation as the maximum likelihood estimate. In the general form, the cost function to correct the parameters such as initial values, boundary values, etc. is usually defined as the sum of background error term and observation error term as follows. The optimum values of the control parameters (model parameters) are estimated by minimizing it.

$$J(\mathbf{x}) = \frac{1}{2}(\mathbf{x}-\mathbf{x}^{\mathrm{b}})^{\mathrm{T}}\mathbf{B}^{-1}(\mathbf{x}-\mathbf{x}^{\mathrm{b}}) + \frac{1}{2}(\mathbf{Hx}-\mathbf{y})^{\mathrm{T}}\mathbf{R}^{-1}(\mathbf{Hx}-\mathbf{y}) \tag{1}$$

where $\mathbf{x}$ is the vector of model parameters, $\mathbf{x}^{\mathrm{b}}$ is the background model parameters (e.g. the first guess), and $\mathbf{y}$ is the vectors of observations. $\mathbf{H}$ is the matrix of the operator that converts the model state $\mathbf{x}$ into $\hat{\mathbf{y}}$. $\mathbf{B}$ is the covariance matrix of the background errors and $\mathbf{R}$ is the covariance matrix of the observation errors, and are expressed as follows, respectively:

$$\mathbf{B} = <\Delta\mathbf{x}^{\mathrm{b}}(\Delta\mathbf{x}^{\mathrm{b}})^{t}>, \quad \mathbf{R}_i = <\Delta\mathbf{y}(\Delta\mathbf{y})^{t}> \tag{2}$$

where $< >$ denotes the ensemble average, $\Delta\mathbf{x}^{\mathrm{b}}$ and $\Delta\mathbf{y}$ are the vector of the background errors and observation errors, respectively.

In order to obtain the optimum value of $\mathbf{x}$, the minimization of the cost function must be performed. Generally, analytical approach is difficult for the minimization. Instead, a method of descent is usually applied, which requires the following descendent value of the cost function.

$$\nabla_{\mathbf{x}} J(\mathbf{x}) = \mathbf{B}^{-1}(\mathbf{x}-\mathbf{x}^{\mathrm{b}}) + \mathbf{H}^{t}\mathbf{R}^{-1}(\mathbf{Hx}-\mathbf{y}) \tag{3}$$

To compute $\nabla_{\mathbf{x}} J(\mathbf{x})$, the transpose of the operator matrix, $\mathbf{H}^{t}$, has to be computed. This matrix corresponds to the adjoint operator $\mathbf{H}^{*}$ of the tangent liner operator of $\mathbf{H}$. In the actual computation of Eq. 3, however, $\nabla_{\mathbf{x}} J(\mathbf{x})$ is directly computed through the adjoint run with the adjoint model code. For constructing the adjoint code of WAM, we used AMC (Adjoint Model Compiler, Giering, 1995). For the minimization of Eq. 1, a descent method with Quasi-Newton Method was used.

EXISTING EXAMPLES OF ADJOINT METHOD APPLIED TO WAM

Hersbach (1998) applied the adjoint method to WAM code and developed ADWAM code with the object to get a better insight into numerical values of several model parameters in the WAM source terms. Kobayashi et al. (2003) applied the adjoint method to WAM code with assuming the open boundary conditions of directional spectra as the control parameters for the data assimilation. However, they took into account only the observation error without considering the background error in the cost function that controls the performance of the data assimilation. For example, in the Hersbach's Adjoint WAM, the minimization of the cost function composed of only the observation errors enables an appropriate correction of the control parameters. The reason why the method goes well is due to the number of the control parameters much fewer than that of the observation data.

On the other hand, however, in this study, we adopted the initial wave condition (directional spectra) and/or boundary condition (directional spectra) as the control parameters, the number of which is much larger than that of the observation data. This is so called an ill-conditional inverse problem, and the minimization of the cost function composed of only the observation errors leads to unstable and unreliable computation. This necessitates the introduction of the term of the background error in the cost function, which imposes the restriction condition that the optimum control parameters should exist in the vicinity of the initial guess of the computation as expressed in Eq. 1.

COVARIANCE MATRIX OF BACKGROUND ERROR

As mentioned above, when we try to estimate parameters inversely under an ill-condition, an appropriate matrix of back ground error should be assumed. We, therefore, referred its setting method in the field of atmospheric research, and then we assumed the following assumptions for the covariance matrixes of both background error and observation error;

a) The structure of the background error reflects the time-spatial scale of target phenomena.
b) The observation errors are independent each other in each observation at each observation station.
c) The background error and the observation error are independent each other.
d) Though the wave parameters are estimated by WAM depending on the directional wave characteristics influenced by geographical conditions, the D-value (the difference between the observed wave parameter and the first guess of wave parameter with WAM) is assumed to be homogeneous in

direction since the D-value shows almost no correlation in direction in our simulation results.

Under these assumptions, we assumed the following covariance matrix for the background error **B** and its element μ_{ij};

$$\mathbf{B} = \left[\mu_{ij}\right], \quad \mu_{ij} = \frac{a}{1+br_{ij}^2} \times c^{\Delta T_{ij}} \tag{4}$$

where μ_{ij} shows time and spatial distribution of correlation coefficient, r_{ij} is the distance between the grid point i and j, ΔT_{ij} is the time lag between the data at the grid point i and j, a, b, and c are parameters to be determined by observation data and numerical simulation results.

Though details are omitted, under the assumptions of a)-d), the variance of the D-value is the sum of the variance of the background error and that of the observation error, which are a and 1-a, respectively. Therefore, the following parameter λ^2 may show relative reliability of the observation data in data assimilation.

$$\lambda^2 = (1-a)/a \tag{5}$$

Fig. 1 and Fig. 2 show examples of correlation function of the D-value of the significant wave height as a function of distance r_{ij} and time lag ΔT_{ij}, respectively. Table 1 shows the parameters estimated using the least square method by applying Eq. 4 to the observation data. In this case, the error ratio is λ^2 =0.406 since the average of a is $\overline{a}$ =0.771. In the adjoint method, the assimilated variables are estimated depending on the error ratio λ^2. That is, in this example, the assimilated variables are estimated under the condition that the variance of the observation error is about 0.4 times of that of the background error.

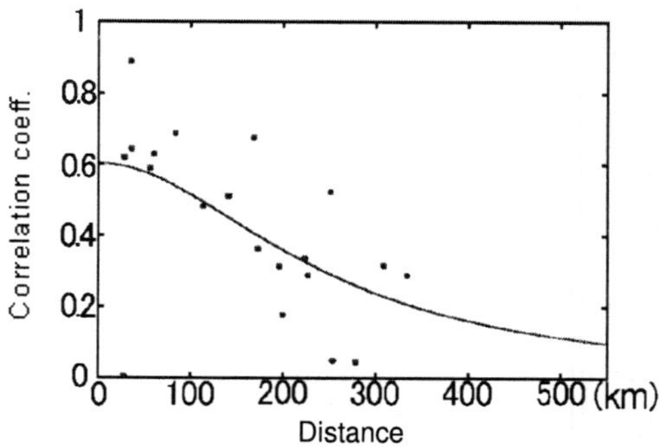

Figure 1 Correlation function of D-value of significant wave height as a function of distance.

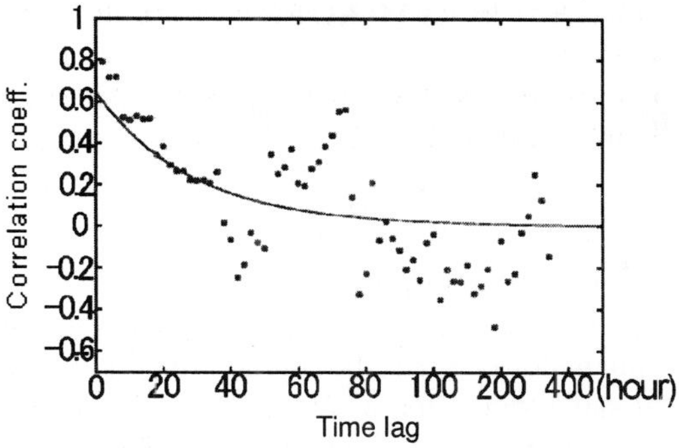

Figure 2 Correlation function of D-value of significant wave height as a function of time lag.

Table 1 parameters in correlation function.

Case	Coefficients		
	a	b	c
1	0.602	1.71×10^{-5}	0.967
2	0.731	1.39×10^{-5}	0.986
3	0.793	1.42×10^{-5}	0.979
4	0.717	1.86×10^{-5}	0.974
Average	0.711	1.60×10^{-5}	0.976

In the following, when we examine the applicability of Eq. 4 for the background error, we use the average value of a, b, and c shown in Table 1. We also examined the case of the unit matrix multiplied by λ^2 with disregarding the off-diagonal components in consideration of the weight of the observation error and the background error.

PROCEDURE OF DATA ASSIMILATION WITH ADJOINT WAM

The procedure of data assimilation with adjoint WAM is almost the same as that of Hersbach (1998). For normal applications of wave hindcasting, the WAM code is used and the energy balance equation is integrated in the forward direction of time t (forward run), while for data assimilation, the ADWAM code is used and the equation is integrated in the reverse direction of time t (adjoint run) to obtain the information with respect to the control parameters to be modified for data assimilation.

Adjoint run is reversely executed with the use of the state variables computed in advance in the forward run. When nonlinearities are included in the forward run, the nonlinearities at each integration time step have to be stored. For realistic applications, this can add up to enormous amount of numbers and may cause the storage problem in the computations. To avoid this problem, a 'check-point' method has been implemented same as WAM Cycle 5 (Hersbach, 1998).

APPLICATION OF ADJOINT WAM TO NOWPHAS WAVE DATA

For the purpose of evaluating availability and performance of the data assimilation, we developed two different adjoint WAM codes which are capable of correcting the initial condition and/or the boundary condition as the control variables, respectively. Then, we applied them to the NOWPHAS wave observation data.

Data assimilation for initial condition

We applied the adjoint method to assimilate the initial condition of wave hindcasting. In computations, first, the initial condition and the boundary condition of directional spectra for a target area were prepared in advance through the computations for a large area with ECMWF wind data. And then, the data assimilation was carried out for a small area shown in Fig. 3 with NOWPHAS data. The reason why the data assimilation was applied for the small areas is that the information of observation data is diffused and fades away with the distance from the observation station and little contribution can be expected to the accuracy improvement in the data assimilation from a far area. The other reason is that the practical application of data assimilation requires huge memories and long computation time when we adopt the initial condition and/or the boundary condition as the control variables.

In computations, the wave field was numerically integrated with the time interval of 15 minutes, and the data assimilation was carried out for 24 hours. The data observed at 6 sites are assimilated. Though observed directional spectra are suitable for data assimilation, the significant wave heights $H_{1/3}$ are used since the directional spectra are not observed at all the sites. In addition, the directional spectra at the open boundary are given by the computation executed in advance for the large computation area.

Fig. 3 (a) shows the initial distribution of significant wave height (the first guess) estimated with the ECMWF wind data. Fig. 3 (b) shows the assimilated

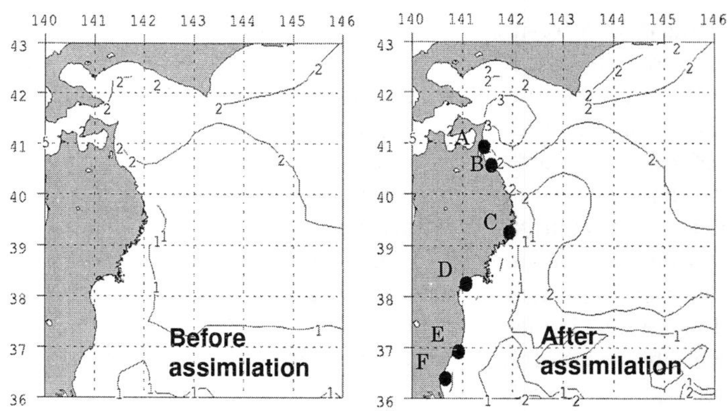

(a) Before assimilation **(b) After assimilation**

Figure 3 An example of data assimilation for initial condition.

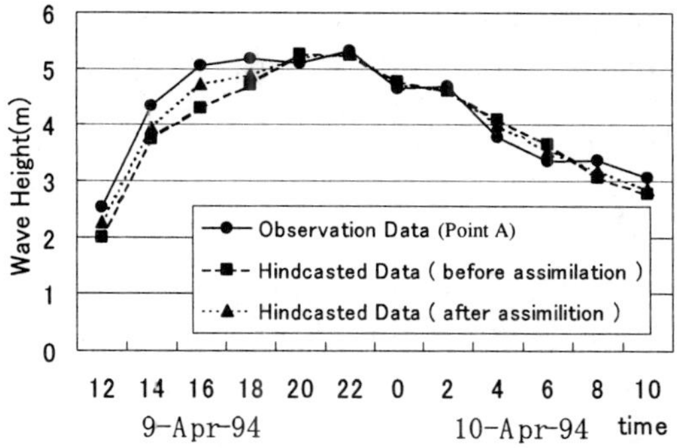

Figure 4 Time series of the assimilated significant wave height.

initial distribution of the significant wave height. Fig. 4 shows the time series of the significant wave height at the site A in Fig.3 (b), where the observation data and the hindcasted data are also shown. Comparing Fig. 3 (a) (before assimilation) and (b) (after assimilation), it is seen that the initial wave height distribution is slightly modified and wave height rises locally in some areas in (b). Fig. 4 shows that the time series of the assimilated significant wave height is corrected to approach the observation data. However, noticeable accuracy improvement seems to be restricted for several hours from the computation beginning. Therefore, we conclude that the initial condition in wave hindcasting is not so crucial compared to that in weather prediction under the condition of sufficient spin up computation.

Incidentally, we discussed the effect of the background error in the cost function through numerical simulations. The result showed that the off diagonal components in the covariance matrix might be ignored for the efficient computations.

Data assimilation for boundary condition

We applied the adjoint method to assimilate the boundary conditions. Fig.5 (a) shows the distribution of the significant wave height at 16:00, 2000/7/8 (first guess). Fig. 5 (b) shows the assimilated distribution of the significant wave height at the same time. Fig. 6 shows the time series of the significant wave height at Ishinomaki, where the observation data and the hindcasted data are also shown. Comparing Fig. 5 (a) (before assimilation) and (b) (after assimilation), it is seen that the wave height distribution is slightly modified and wave height lowers in whole area in (b). Fig. 6 shows that the assimilated significant wave height is successfully corrected to approach the observation data. Comparing the results of Fig. 4 and Fig. 6, we conclude that the data assimilation for boundary condition is more useful than that for the initial condition for 'coastal wave' hindcasting.

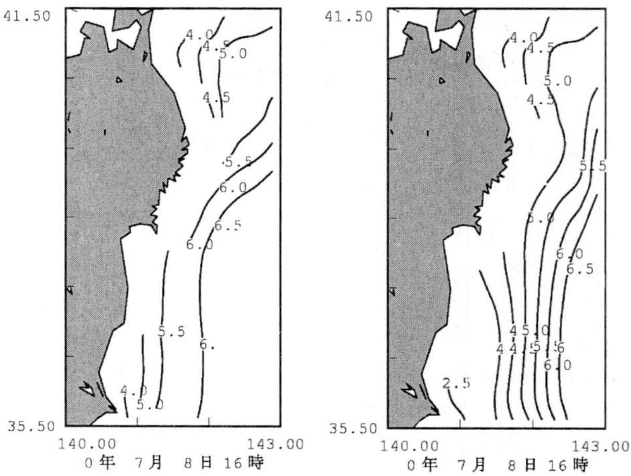

(a) Before assimilation (b)After assimilation

Figure 5 An example of data assimilation for boundary condition.

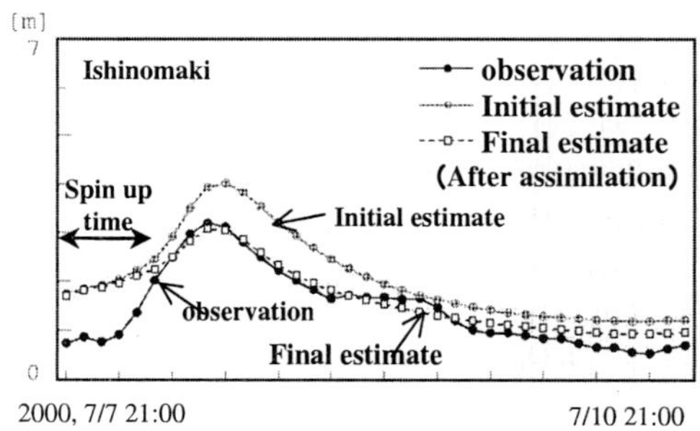

Figure 6 Time series of the assimilated significant wave height.

EVALUATION INDEXES OF WAVE OBSERBATION STATIONS

As explained in the introduction, by utilizing the data assimilation, we can investigate how much the data at each observation station can contribute to the improvement of the estimation accuracy with the model. For examining this, we carried out the wave hindcasting and applied the data assimilation to the wave data in each area in Fig. 7, respectively. Depending on the computation results, we evaluated the contribution of the wave data measured at each observation station to the improvement of wave hindcasting accuracy.

Although the chi-square test is effectual to examine the validity of the data assimilation, it is impossible to use it since the observation errors are unknown in these cases. Instead of this, we, therefore, introduced following three parameters;

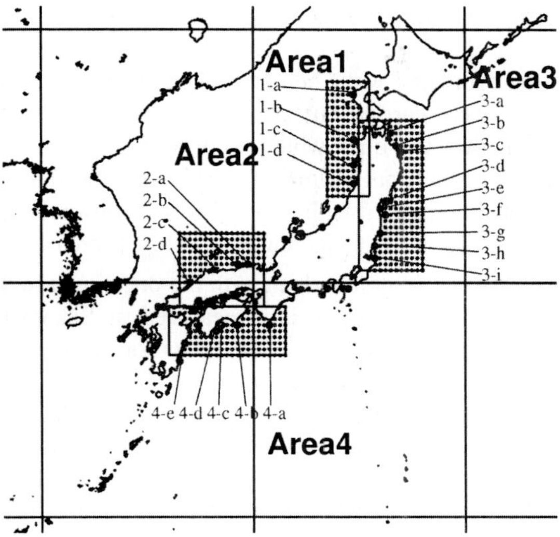

Area	Stn. Code	Stn. Name
Area1	1-a	Setana
	1-b	Fukaura
	1-c	Akita
	1-d	Sakata
Area2	2-a	Shibayama
	2-b	Tottori
	2-c	Hamada
Area3	3-a	Mutsuogawara
	3-b	Hachinohe
	3-c	Kuji
	3-d	Ishinomaki
	3-e	Sendai-Shinko
	3-f	Souma
	3-g	Onahama
	3-h	Hitachi-naka
	3-i	Kashima
Area4	4-a	Shionomisaki
	4-b	Murotsu
	4-c	Kouchi
	4-d	Kamikawaguchi
	4-e	Hososhima

Figure 7 Data assimilation areas and NOWPHAS wave observation stations

$$V_b = \frac{1}{n}\sum_{i=1}^{n}(x_i^b - y_i^0)^2, \quad V_a = \frac{1}{n}\sum_{i=1}^{n}(x_i^a - y_i^0)^2, \quad V_{inc} = \frac{1}{n}\sum_{i=1}^{n}(x_i^a - x_i^b)^2 \qquad (6)$$

where V_b is the variance of background error (before assimilation), V_a is the variance of analysis error (after assimilation), and V_{inc} is the increment (amount of correction by assimilation).

When we evaluate the importance of each wave observation station on the basis of several wave observation data sets in a comprehensive way, the following normalized indexes are used since V_a and V_{inc} show different values depending on each wave data set.

$$\Delta V_{aS} = (V_a^{n-1} - V_a^n)/V_b, \quad \Delta V_{incS} = (V_{inc}^{n-1} - V_{inc}^n)/V_b \qquad (7)$$

where the parameters with suffix n-1 and n indicate that the parameters are estimated in assimilation with the data at n-1 observation stations and the ones with n observation stations, respectively.

EVALUATION OF WAVE OBSERVATION STATIONS

For evaluating the importance of each observation station, first, we apply the data assimilation for the data of all the stations in each area in Fig.7, and next, we apply it again for the data excluding one station. Then, by comparing the difference of the parameters between the two cases, we discussed the importance of the station. That is, if V_a increases and V_{inc} decreases by excluding the data of a station, the contribution from the station to the improvement in the accuracy is significant. This may lead to the conclusion that the excluded station is important on the wave observation network.

Fig.8 shows the examples of the estimated wave height at Akita and Sakata stations, where the data assimilations were carried out using the data of all the stations in assimilation area. Fig.9 shows the ones which were assimilated by excluding Sakata's data. The accuracy of the final estimate (after assimilation) at the both stations in Fig.9 is deteriorated compared with Fig.8. Table 2 shows the parameters V_b, V_a and V_{inc}. As seen in Table 2, V_b increases and V_{inc} decreases in Case 5 (Sakata is excluded), respectively, in comparison with the ones in Case 1 (All stations). This result indicates that the wave data measured at Sakata station is significant in the wave data assimilation, i.e., Sakata station is more important than other wave observation stations around the area.

The above is an example of evaluation using a set of wave observation data. As mentioned previously, when we evaluate the importance of wave observation stations in a comprehensive way using various wave data sets, it is necessary to use the normalized indexes, Eq. 7. In this paper, we evaluated the importance of each station depending on the wave data of five storm cases for each area in

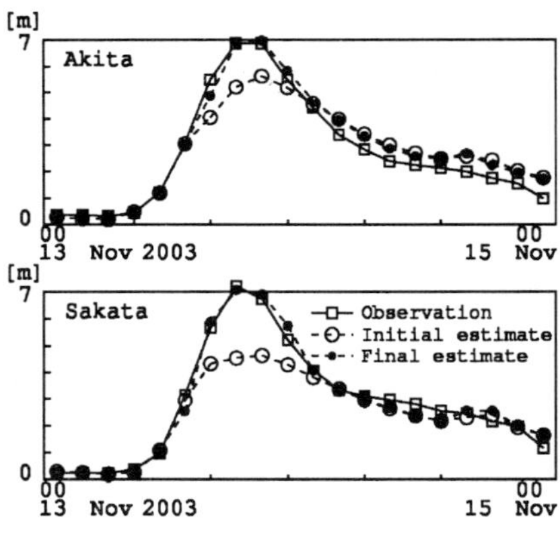

Figure 8 Comparison of time series of wave height (All stations)

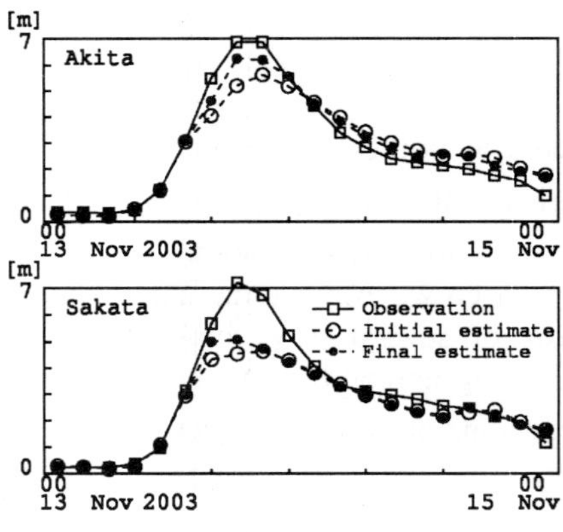

Figure 9 Comparison of time series of wave height (Sakata is excluded)

Table 2 Evaluation indexes V_b, V_a and V_{inc}.

Case	Excluded stn.	Iteration	V_b	V_a	V_{inc}	
1	No stn. is excluded	3	3.70	2.50	0.41	n
2	1-a (Sedana)	3	3.70	2.50	0.41	
3	1-b (Fukaura)	3	3.70	2.56	0.43	n-1
4	1-c (Akita)	3	3.70	2.57	0.44	
5	1-d (Sakata)	3	3.70	2.97	0.26	

Table 3 Average of Normalized Evaluation Indexes.

Area 1	excluded Stn.	ΔV_{aS}	ΔV_{incS}
1	1-a (Setana)	0	0
2	1-b (Fukaura)	0.02	0.01
3	1-c (Akita)	0.04	0.02
4	1-d (Sakata)	0.11	-0.13

Area 2	excluded Stn.	ΔV_{aS}	ΔV_{incS}
1	2-a (Shibayama)	0.01	0.03
2	2-b (Tottori)	0.01	0.03
3	2-c (Hamada)	0.06	-0.05

Area 3	excluded Stn.	ΔV_{aS}	ΔV_{incS}
1	3-a (Mutsuogawara)	0	0
2	3-b (Hachinohe)	0	0
3	3-c (Kuji)	0	0
4	3-d (Ishinomaki)	0	0
5	3-e (Sendai-Shinko)	0.02	-0.01
6	3-f (Souma)	0.03	-0.05
7	3-g (Onahama)	0.01	-0.03
8	3-h (Hitachi-naka)	0.03	-0.05
9	3-I (Kashima)	-0.01	0.02

Area 4	excluded Stn.	ΔV_{aS}	ΔV_{incS}
1	4-a (Shionomisaki)	0	0
2	4-b (Murotsu)	0	0
3	4-c (Kouchi)	0	0
4	4-d (Kamikawaguchi	0	0
5	4-e (Hososhima)	*	*

Fig.7, respectively. Table 3 shows the average values of ΔV_{aS} and ΔV_{incS} calculated for 5 storm cases, respectively. ΔV_{aS} indicates the amount of decrease of error, while ΔV_{incS} indicates the amount of decrease of correction by excluding a wave observation station from n wave observation stations in data assimilation. Therefore, if ΔV_{aS} is positive and larger and at the same time if ΔV_{incS} is negative and smaller, the contribution of the excluded station is considered to be significant in the data assimilation. In Table 3, hatching is covered for the parameters in the cases where ΔV_{aS} is positive and ΔV_{incS} is negative.

Table 4 Top three important wave observation stations in each area.

	1st	2nd	3rd
Area1	1-d (Sakata)	1-c (Akita)	1-b (Fukaura)
Area2	2-c (Hamada)	*	*
Area3	3-f (Souma), 3-h (Hitachinaka)		3-g (Onahama)
Area4	*	*	*

Table 4 shows the top three important wave observation stations evaluated on the basis of Table 3, whose contribution to the data assimilation is considered to be more significant in each area in Fig. 7. In area 1, the contribution of Sakata is the most important since Sakata's ΔV_{aS} is the largest and its ΔV_{incS} is negative and the smallest. Though Akita's and Fukaura's ΔV_{incS} are positive, their ΔV_{aS} are not negative. Therefore, the order of importance is determined by the magnitude of ΔV_{aS}. In area 2, Hamada's ΔV_{aS} is positive and its ΔV_{aS} is negative. Therefore Hamada's contribution is considered to be more important than the other stations in the data assimilation. However, Shibayama's and Tottori's contribution is not so important since ΔV_{aS} is positive but small and ΔV_{incS} is not negative. In area 3, four stations show some contribution since their ΔV_{aS} are positive and ΔV_{incS} are negative. In these stations, Soma's and Hitachinaka's contribution is more important than the other stations. In area 4, the computations were unstable except the case of Hosojima. Moreover, all the station's ΔV_{aS} are small though positive, and their ΔV_{incS} are not negative. This indicates that there are no effective stations in the area. The reason why this area is not appropriately evaluated is due to the computation conditions under deep water condition and relatively coarse grid size which make it difficult to reproduce refraction, shoring, sheltering by the surrounding local geographic conditions. More careful evaluation is required in this case since most of the stations in the area are more or less influenced by the complicated geographic conditions.

CONCLUSIONS

For the purpose of improving the performance of wave model, a data assimilation technique, adjoint method, was applied to a third generation wave model, WAM. Then, we confirmed the usefulness of the adjoint WAM though several numerical computations. We also applied the adjoint WAM to the wave data observed at NOWPHAS wave observation stations in Japan. Then, we evaluated the importance of each wave observation station in accuracy improvement of wave hindcasting quantitatively. However, in the paper, numerical computations were carried out under deep water condition with relatively coarse grid. In fact, the wave data at some wave observation stations include the influences of refraction, diffraction, and shoaling. These make it difficult to evaluate the importance of the wave observation stations adequately. More careful evaluation would be required by taking into account such influences.

ACKNOWLEDGMENTS

The wind data used herein were provided by the European Centre for Medium Range Weather Forecasts (ECMWF). We are grateful to the ECMWF for providing the data.

REFERENCES

Giering, R. 1995. The Adjoint Model Compiler, MPI report.

Hersbach, H. 1998. The adjoint of the WAM model, Tech. Rep. KNMI-WR, 97-01, 28p.

Hersbach, H. 1998. Application of the adjoint of the WAM model to inverse wave modeling, J. Geophys. Res. Vol. 103, No. C5, pp. 10469-10487

Hersbach H. and Janssen P.A.E.M. 1999. Improvement of the Short Fetch Behavior in the WAM model, J. Atmos. Ocean. Techn., 16, pp.884-892.

Holthuijsen, L. H. et al. 2000. SWAN Cycle III version 40.11, USER MANUAL, Delft University of Technology, 124p.

Kobayashi T., Adachi T. and Yasuda T. 2003. Estimation of Wave Fields with Open Boundaries by Ap-plying an Adjoint Model, Asian Pacific Coasts 2003, 8p.

Komen, G. J. et al. 1994. Dynamics and Modelling of Ocean Waves. Cambridge University Press, 532p. 1994.

The WAMDI Group (13 Authors). 1988. The WAM model-A third generation ocean wave prediction model, J. Phys. Oceanogr., 18, pp.1378-1391.

Tolmann, H. L. 1999. User manual and system documentation of WAVEWATCH-III Version 1.18, NOAA/NWS/NCEP/OMB technical note 166, 110p.

WAVE HEIGHT BACKGROUND ERRORS SIMULATION AND FORECASTING VIA STOCHASTIC METHODS IN DEEP AND INTERMEDIATE WATERS

Zoe Theocharis[1] , Constantine Memos[1] and Demetris Koutsoyiannis[1]

Wave forecasting is accomplished today via numerical models. In this work we apply stochastic techniques using actual measurements to improve wave height forecast in real time. Application of these techniques in four locations of the Aegean Sea results in significant improvement of the forecast in the time domain retaining the same pattern of modifications, suggesting, thus, this method for operational use in deep and intermediate waters. The improvement is obtained by four regression models, which take into account the variable of the significant wave height as measured and forecasted by the model. Space-wise extension of the method was also investigated and applied to the Aegean Sea and the Indian Ocean, where its performance was remarkable.

INTRODUCTION

The forecast of the significant wave height is valuable in numerous coastal and offshore investigations and activities. This is currently accomplished numerically via the state of the art third generation deterministic wave models that solve the wave energy balance equation. In this work, we are dealing with the improvement of the wave forecast produced by these models. Waves have a deterministic relation to the wind. Nevertheless, the forecasting error can be important. There are two essential aspects regarding this issue:

- The 3rd generation wave model WAM underestimates the wave height forecast in a global level (Janssen, 1997)
- Wave fields predictability can not exceed wind fields predictability (Young, 1999)

In recent years, data assimilation and artificial neural network techniques have been used in a number of wave height forecast improvement efforts.

Data assimilation technique is a widespread method of satellite data operational exploitation. Its aim is the improvement of the forecast using imported observations of the wind and wave field mainly from satellites. Re-run of the model is taking place at the sequel. Usually, the operation wave models produce forecasts for the next 72-hours (1st guess forecast). Meanwhile, the satellites are overpassing the earth surface collecting information that can operationally being used. Henceforth, depending on the data used for the correction, one can obtain wind and wave data assimilation respectively.

In the last decade there is a tendency of predicting wave parameters via Neural Network Techniques. Tsai et al. (2001) for example present a work to

---

[1] School of Civil Engineering, National Technical University of Athens, Heroon Polytechniou 5, Athens 15780, Greece

forecast the wave height using a set of previous observations. This is achieved with the application of neural networks which use a background algorithm related to the history of wave realizations. Furthermore, in the framework of this research, the wave height forecast was produced with the use of an artificial neural network (ANN), using as input data the wave height observations time series of adjacent buoy stations instead of the wind parameters time series. This is achieved through the application of the ANN's back propagation algorithm learning process, which provides the desirable results. The model computes the weight relations between neighbouring stations, based on short-term data, from which we can produce the forecast wave height time series or the missing data time series of the neighbouring station. The results showed that ANN achieved satisfactory records for both kinds of application. Deo et al. (2001), also, dealt with the wave height forecast in real time. They applied at the same time series two models, in order to compare the results: an ANN and a 2nd order autoregressive model. The ANN resulted in a coefficient of determination near 0.66 for a one-step forecast (three hours) and in 0.61 for a two steps forecast. The corresponding results for their bivariate autoregressive model were 0.61 and 0.50.

The stochastic applications in the sector of wave forecasting are relatively limited and concern in the majority of cases gap filling of measurements time series or re-analysis of wave hindcasts. Contrary to the stochastic models used for long-term forecasts, which produce probabilities of appearance of extreme sea states, the stochastic models related to short term forecasting, or now-casting, like the regression models, are using the memory-history to reproduce the following sea states. This fact prompted initially the researchers to use stochastic methods, and specifically the regression models, so that they described sea surface elevation time series as well as its spectrum (Scheffner and Borgman 1992). On the other hand, the stochastic simulation of DelBalzo et al. (2002) uses a twenty years wave height visual observations time series, provided by the COADS database. The simulated time series were compared with the 4-years observations time series of the National Data Buoy Center. The inter-comparison was satisfactory and the deviations occurred are due to the interaction between the buoys and the passing boats.

Finally, Caires and Sterl (2005), proposed a non parametric method to correct ERA-40 re-analysis. This method predicts the error using a background algorithm produced from a learning dataset.

Unlike Caires and Sterl's method, the methods used here are parametric and customized to the specific application basin. Thus, in this work we present the application of linear and non-linear stochastic techniques to show that WAM background errors can be reasonably predicted by using a limited number of buoy observations and improve thus its forecasting robustness. Re-run of the wave model is not required. A full description of the third-generation ocean wave model WAM can be found in WAMDIG (1988).

THE METHOD - STOCHASTIC MODELS USED

In this investigation four types of multiregression models were used:

- a bivariate linear model **BLM1** whose explanatory variables are the WAM prediction of the current step and the measured height at the previous step:

$$\hat{\Psi}_{t+\Delta t} = aX_{t+\Delta t} + \beta\Psi_t + \gamma \tag{1}$$

where,

$\hat{\Psi}_{t+\Delta t}$: the estimated value of the significant wave height at time step t+Δt

$X_{t+\Delta t}$: the value of the WAM forecast for the significant wave height, at time step t+Δt

Ψ_t : the value of the significant wave height measurement at time step t

α : weight parameter corresponding to $X_{t+\Delta t}$

β : weight parameter corresponding to Ψ_t

γ : steady parameter

- a bivariate linear model **BLM2** whose explanatory variables are the WAM prediction of the current step and the measured height at the step prior to the previous step:

$$\hat{\Psi}_{t+2\Delta t} = aX_{t+2\Delta t} + \beta\Psi_t + \gamma \tag{2}$$

where,

$\hat{\Psi}_{t+2\Delta t}$: the estimated value of the significant wave height at time step t+2Δt

$X_{t+2\Delta t}$: the value of the WAM forecast for the significant wave height, at time step t+2Δt

Ψ_t : the value of the significant wave height measurement at time step t

α : weight parameter corresponding to $X_{t+2\Delta t}$

β : weight parameter corresponding to Ψ_t

γ : steady parameter

- a trivariate linear model **TLM** whose explanatory variables are the WAM prediction at the current step and the measured height at the two previous steps:

$$\hat{\Psi}_{t+\Delta t} = aX_{t+\Delta t} + \beta\Psi_t + \gamma\Psi_{t-\Delta t} + \delta \tag{3}$$

where,

$\hat{\Psi}_{t+\Delta t}$: the estimated value of the significant wave height at time step t+Δt

$X_{t+\Delta t}$: the value of the WAM forecast for the significant wave height at time step t+Δt

Ψ_t : the value of the significant wave height measurement at time step t

$\Psi_{t-\Delta t}$: the value of the significant wave height measurement at time step t-Δt

α : weight parameter corresponding to $X_{t+\Delta t}$

β : weight parameter corresponding to Ψ_t

γ : weight parameter corresponding to $\Psi_{t-\Delta t}$

δ : steady parameter

- a bivariate non-linear model, **BNLMa** and **BNLMb,** with the same explanatory variables as in the first bivariate model:

BNLMa:

$$\hat{\Psi}^{\lambda}_{t+\Delta t} = a_1 X^{\lambda}_{t+\Delta t} + \beta_1 \Psi^{\lambda}_t + \gamma_1, \quad \text{if } \Psi_t \le c \tag{4}$$

$$\hat{\Psi}^{\lambda}_{t+\Delta t} = a_1 X^{\lambda}_{t+\Delta t} + \beta_2 \Psi^{\lambda}_t + \gamma_2, \quad \text{if } \Psi_t \ge c \tag{5}$$

BNLMb:

$$\hat{\Psi}^{\lambda}_{t+\Delta t} = a_1 X^{\lambda}_{t+\Delta t} + \beta_1 \Psi^{\lambda}_t + \gamma_1, \quad \text{if } X_{t+\Delta t} \le c \tag{6}$$

$$\hat{\Psi}^{\lambda}_{t+\Delta t} = a_2 X^{\lambda}_{t+\Delta t} + \beta_1 \Psi^{\lambda}_t + \gamma_2, \quad \text{if } X_{t+\Delta t} \ge c \tag{7}$$

where,

$\hat{\Psi}_{t+\Delta t}$: the estimated value of the significant wave height at time step t+Δt

$X_{t+\Delta t}$: the value of the WAM forecast for the significant wave height at time step t+Δt

Ψ_t : the value of the significant wave height measurement at time step t

α : weight parameter corresponding to $X_{t+\Delta t}$

β : weight parameter corresponding to Ψ_t
γ : steady parameter
c : the wave height defining the applicability range of the model equations
λ : exponential index of the non-linear transformation

The parameters listed above are estimated by a weighted least squares procedure and from the continuity equations.

FIRST ASSESSMENT

Improvement in the time domain

Forecasting of the wave climate in the Aegean Sea is accomplished today via the numerical model WAM, run by the National Centre for Marine Research of Greece in the framework of the Poseidon Operational System. A systematic underestimation of the significant wave height is observed for the region of the Aegean Sea, as seen for example in Figure 1, where the observations and WAM prediction for Athos location ($39^0 58'$ - $24^0 43'$) in the northern part of this sea are depicted for the year 2001.

This work deals with the improvement of the significant wave height forecast in real-time. The results were checked against measurements from four pilot-study monitoring stations of the Aegean Sea. These stations are located in the open sea near the Athos peninsula and offshore the islands of Lesvos, Mykonos and Santorini.

In order to develop an efficient wave forecasting system for the Aegean Sea, taking into account the existing peculiarities such as the complex shore-line and the numerous islets as well as the changeable nature of the wind field that influence WAM's predictive power, a set of stochastic methods was examined: four regression models, three linear and one non-linear, that take into account the measured and forecasted significant wave height.

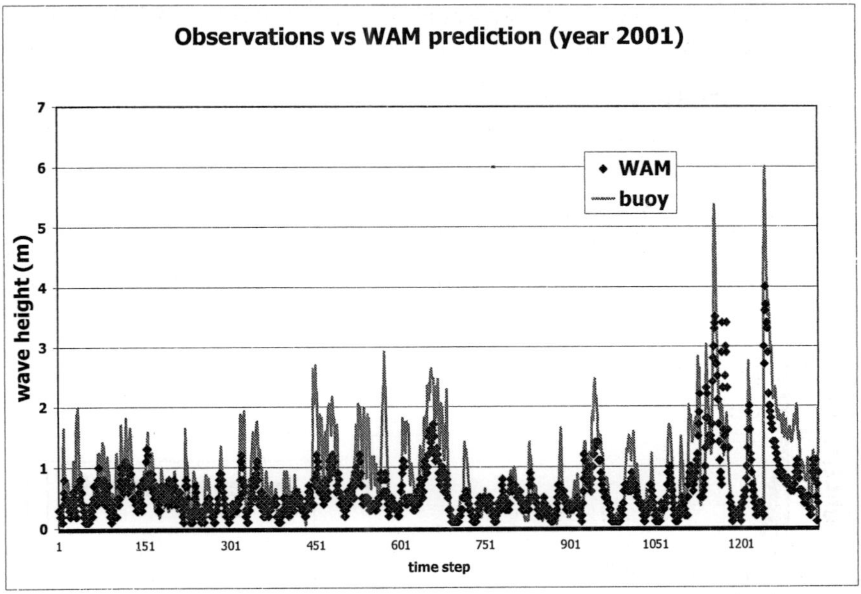

Figure 1: Observed and predicted values of the wave height, Athos station. Severe underestimation observed especially in high sea-states in the year 2001.

The four multiregression models presented in the previous section, resulted in significant improvement of the coefficient of determination, irrespectively of the time period of application. More specifically, the coefficient of determination R^2, that gives the proportion of the variance explained by the model and represents the model efficiency, increased from approximately 0.7 to over 0.9 as shown in Table 1.

Table 1. Coefficients of determination obtained for each station and each regression model.

	R^2 Athos	R^2 Lesvos	R^2 Mykonos	R^2 Santorini
WAM	0.781	0.713	0.722	0.676
BLM1	0.920	0.892	0.927	0.906
BLM2	0.865	0.813	0.860	0.820
TLM	0.924	0.895	0.931	0.911
BNLMa	0.929	0.897	0.936	0.909
BNLMb	0.929	0.898	0.935	0.908

It is shown that the stochastic models result in a significant forecast improvement, irrespectively of the application time period and of the location of the prediction. Better approximation of the measured wave height as compared to WAM's prediction as well as significant decrease of the standard deviation is achieved in all stations. This can be seen in Figure 2, where Athos station results are presented in descending order of wave heights.

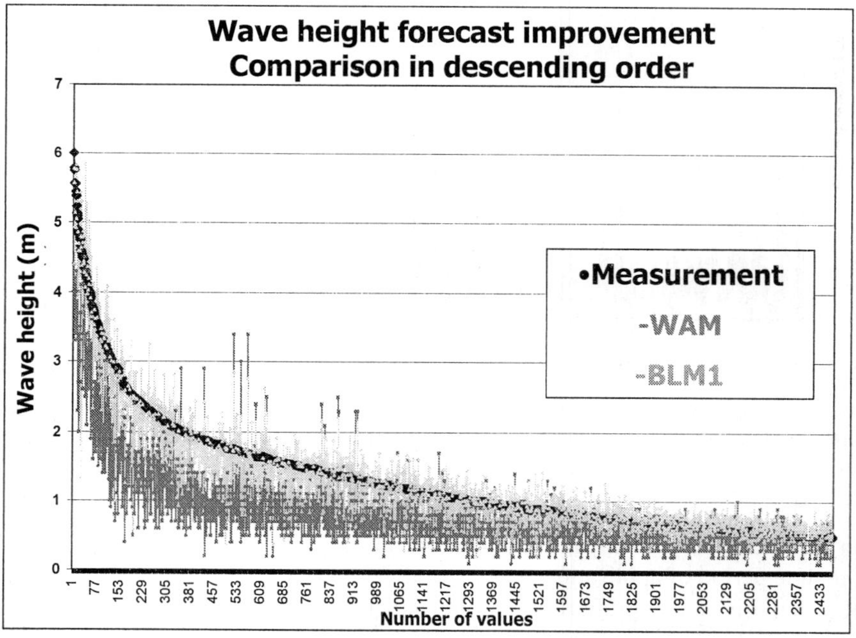

Figure 2: BLM1 results in descending order together with measured and forecasted values of the wave height at Athos station.

By applying the stochastic model BLM1 for lead times between the three hours time step and the 72 hours step, one can observe that the predictive power of the stochastic model reaches the wave model coefficient of determination in all four stations after about 72 hours following the pattern shown in figure 3, suggesting, thus, that the stochastic tools could be valuable for operational use along with the wave model prediction.

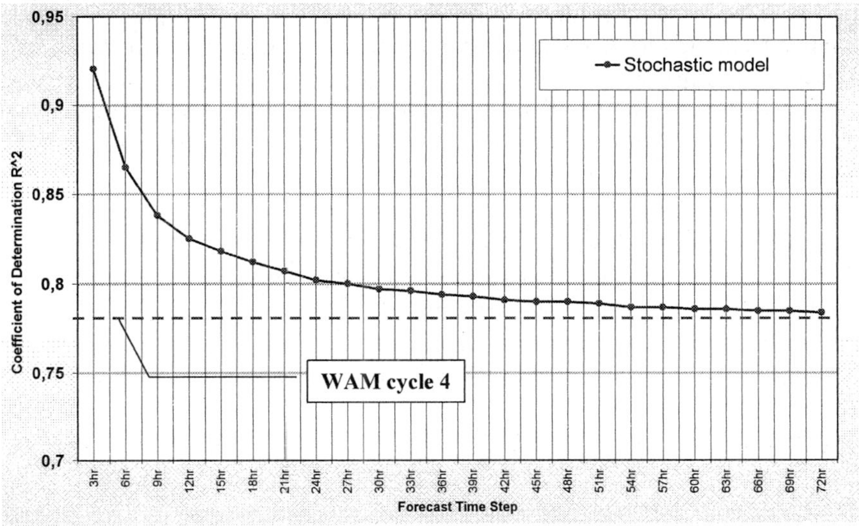

Figure 3: Coefficient of determination decay versus lead time. Model BLM1, Athos station

By observing results of the various applications performed, one can see that by applying the same parameters for all seasons and locations, significant improvement can be obtained. The results for all periods of the year showed that there is no need for individual (depending on the season) implementation of the regression models, since the factors of both linear and non-linear models were found to be almost constant regardless of the station or period analyzed. In Table 2 for example one can see a sample of the parameter values produced for different periods of implementation of BLM1 for Athos station. It is remarkable that by using an overall mean value of α, β and γ parameters produced for different periods the same significant improvement can be achieved.

Table 2. BLM1 parameters for different implementation periods. Athos station.

Imple-mentation period	α	β	γ	R^2 before	R^2 after
All years	0.408	0.879	-0.167	0.781	0.924
1st year	0.394	0.914	-0.192	0.756	0.910
2nd year	0.414	0.869	-0.164	0.801	0.932
5/00 to 9/00	0.402	0.980	-0.254	0.593	0.850
3/01 to 9/01	0.421	0.919	-0.215	0.775	0.914
5/01 to 9/01	0.437	0.915	-0.234	0.603	0.840
10/00 to 2/01	0.412	0.852	-0.152	0.748	0.910
10/00 to 4/01	0.404	0.874	-0.163	0.779	0.922

This fact holds also for all other test stations used, and all other regression models implying that WAM underestimation pattern contains a systematic error source.

The common pattern of the results revealed in all four stations examined, suggested that a space-wise expansion of the present method could be worthwhile.

Spatial modification

This part of the study consists of a space-wise application including spatial stochastic modeling and wave information transfer aiming at extending the improvement described above in space and especially in coastal regions. To accomplish this, the wind speed and direction effects were included in order:

- to cluster the sea states according to their direction and define a reference station whose observations were used to improve the wave height prediction of the whole basin
- to determine the average time lag required for a sea state to be transferred in different locations of the basin which was used in BLM models

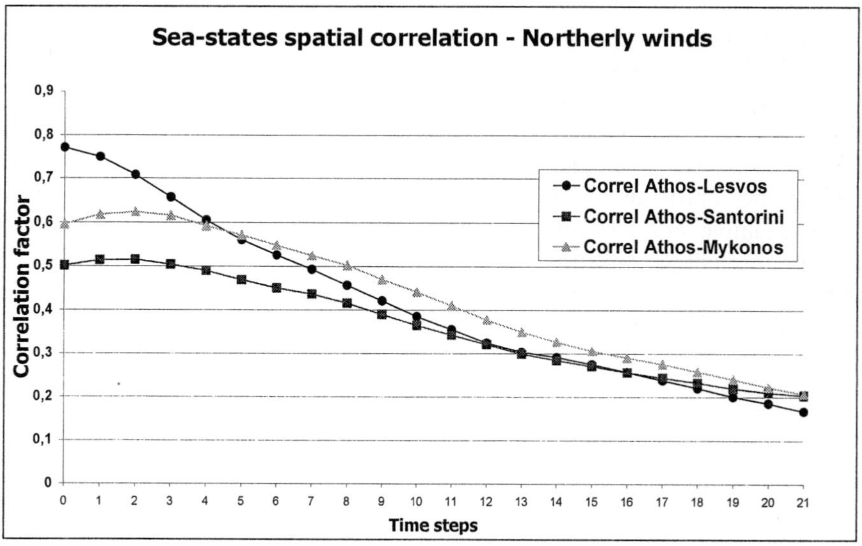

Figure 4: Spatial correlation of sea states for different time lags

For northern winds for example, Athos station was used as a reference station. Then, by calculating the sea states correlation factor, presented in Figure 4, with respect to the stations to the south of Athos, it was found that depending on the distance between the two stations, the value of the correlation factor was changing revealing a temporal delay. This temporal delay was estimated to be equal to the time needed for a sea state to be transferred from one station (Athos) to the other, taking into account the mean wave group velocity. This is the phase lag used to determine the BLM type to be used in each location.

The results of this space–wise extension of the stochastic methods under consideration, show a steady but somehow limited improvement of WAM's predictability power. This is mainly due to the complexity of the Aegean Sea with its numerous islands and complicated shoreline as well as to the irregular wave filed that is frequent over that Sea.

SECOND APPLICATION

To avoid the previously mentioned peculiarities of the Aegean Sea, further examination was conducted regarding the space-wise extension of the method. In this respect two locations of the Indian Ocean were studied by stochastic techniques. WAM cycle 4 without assimilation schemes along with data from two buoys were used. The scope was to improve the wave forecast at the second location which lies in intermediate waters. For this purpose, data of the first location, which lies 900 km offshore the Indian peninsula, along with the prediction of the WAM model were chosen to represent the explanatory variables of the stochastic method.

After examining the correlation factors the reference station was determined and the corresponding time lag was estimated before applying the BNLMa model. The time lag between the two variables was selected 9hr, a value closely associated with the travel time of the waves between the two stations. It is noted that non-linear transformation in the stochastic models is related to the swell content, which can be appreciable in that area. It was found that this technique enhances the improvement of the wave height prediction in intermediate waters by using the offshore measurement. Validation was done by comparing the results with the nearshore buoy data. The improvement of the wave height prediction is remarkable as shown in Figure 5.

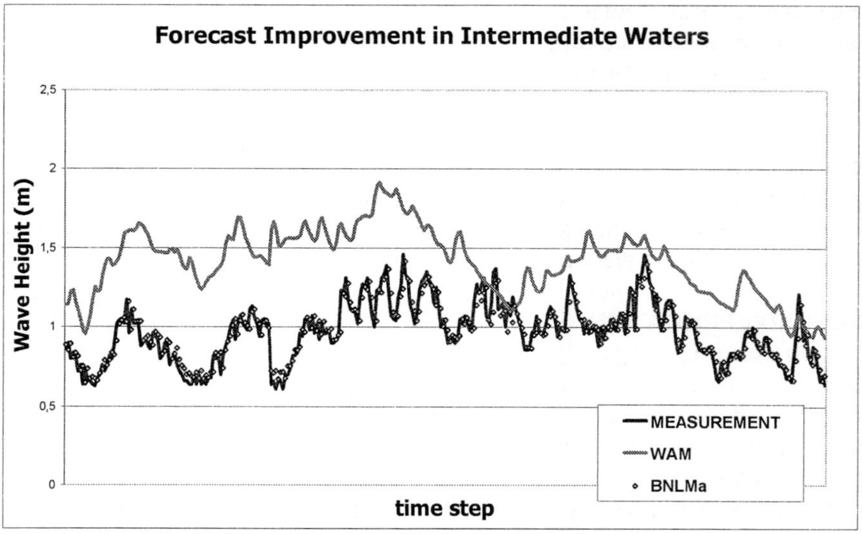

Figure 5: Wave height time series in Indian Ocean intermediate waters. WAM versus buoy and stochastic model

CONCLUSIONS

The results show that:

- A systematic error source exists in the predictions of the wave model tested (WAM), judging by the behavior of the stochastic parameters. This error persisted irrespectively of the value of the wave height and the season of application
- Utilization of measurements in real time in combination with the use of stochastic methods, can significantly improve the wave model forecast in time and space
- Intermediate and shallow waters wave forecast can be improved given that an offshore station provides relevant measurements.

Due to the encouraging results, the method has already been applied to other regions focusing on the possibility of using the stochastic tools for direct exploitation of satellite data in order to improve the wave forecast. Thus direct wave measurements could be abolished. A first evaluation of the method was performed to a 20^0 by 20^0 box in the Southern Indian Ocean, the so called "wave energy storehouse". Results from this case study will be published in the near future.

ACKNOWLEDGEMENTS

The authors would like to thank the National Center for Marine Research and Professor Kallos of the University of Athens for providing the data used in this study.

REFERENCES

Caires, S. and A. Sterl. (2005). "A new non-parametric method to correct model data: Application to significant wave height from the ERA-40 reanalysis", *J. Atmospheric and Oceanic Tech.*, 22(4), pp.443-459.

DelBalzo, D., R., Schultz, J., R., Earle, M., D., 2002, Stochastic time-series simulation of wave parameters using ship observations. Ocean Engineering 30, pp. 1417-1432.

Deo, M.C., Jha, A., Chaphekar, A.S., Ravikant, K., 2001, Neural networks for wave forecasting. Ocean Engineering, 26, pp. 191-203.

Janssen, P.A.E.M., 1991, Quasi-linear theory of wind-wave generation applied to wave forecasting. J. Phys. Oceanogr., 21, pp. 1631-1642.

Scheffner, N., M., Borgman, L., E., 1992, Stochastic time series representation of wave data. Journal of Waterways Port Coastal Ocean Engineering, ASCE, 118(4), pp. 1955-2012.

Theocharis, Z. Memos C. and Koutsoyiannis D., 2006. Wave forecasting improvement in deep and intermediate waters, WISE 2006, Venice.

Theocharis, Z., Koutsoyiannis D., Memos C., and Soukissian T., 2004, Improvement of the wave height real-time forecast in the Aegean Sea using stochastic methods, 1st General Assembly of the European Geosciences Union, Geophysical Research Abstracts, Vol. 6.

Tsai, C., P., Lin, C., Shen, J., N., 2001, Neural network for wave forecasting among multi-stations. J. Ocean Engineering, 29, pp. 1683-1695.

WAMDIG 1988: The WAM model - A third generation ocean wave prediction model. Journal of Physical Oceanography, 18, pp. 1775-1810.

Young, I.R. 1999. *Wind Generated Ocean Waves*. Elsevier Science, Oxford, U.K. 288pp.

A REVISED SYSTEM FOR LONG-TERM WAVE HINDCASTING AND ITS APPLICABILITY

Masataka Yamaguchi[1] and Yoshio Hatada[2]

A long term wave hindcast system is revised by embedding parametric typhoon model-based wind data into NCEP/NCAR (referred to as NCEP) or ECMWF reanalysis/analysis wind data in order to extend the applicability to the estimation of not only wave climate and its trend but also extreme waves through the improvement of the ability for reproducing typhoon-generated waves. Major findings are as follows. 1) Embedding parametric typhoon winds into the original NCEP wind data yields a significant improvement of availability of the system for the estimation of both wave climate conditions and extreme waves, almost without changing the statistical test results for trend analysis in typhoon-dominated sea areas. 2) Embedding parametric typhoon model winds is less effective in the ECMWF winds than in the NCEP winds, because the ECMWF wind data has a higher accuracy than the NCEP wind data due to its finer space resolution.

INTRODUCTION

Hatada and Yamaguchi (1998, 2002) and Yamaguchi and Hatada (2002) have established a long term wave hindcast system for the coastal waters and the open sea areas around Japan, which follows time variation of waves at a single location by use of a backward ray tracing spectral model (Yamaguchi et al. 1987) on a nested grid with a high space resolution under the input condition of 51-year NCEP reanalysis surface wind data or 20-year ECMWF reanalysis/analysis surface wind data. The applicability of each system has been validated by a close agreement with wave climate parameters and their trend obtained from the year-long measurement wave data on the coastal waters and the open sea areas. But both of the systems, particularly the NCEP winds-based system, tend to underestimate waves in cases of strong winds such as typhoons because of a relatively coarse space-time resolution of the wind data.

In this study, the system is revised by embedding parametric typhoon model-based wind data into the NCEP or ECMWF wind data set in order to extend the applicability to the estimation of not only wave climate and its trend but also extreme waves through the improvement of the ability for reproducing typhoon-generated waves.

---

[1] Department of Civil and Environmental Engineering, Faculty of Engineering, Ehime University, Bunkyocho 3, Matsuyama 790-8577, Ehime Pref., Japan

[2] *ditto*

OUTLINE OF A LONG-TERM WAVE HINDCAST SYSTEM

Over-Sea Wind Data

Reanalysis/analysis surface wind data

A 6 hourly NCEP reanalysis wind data set at 10 m height is given on a Gaussian grid with a space resolution of about 1.9 degrees, over a period of 55years ranging from January 1, 1948 to December 31, 2002. Also, a 6 hourly ECMWF reanalysis/analysis wind data set at 10 m height over a period of 24 years from January 1, 1979 to December 31, 2002 is provided on a spherical grid with a space resolution of 1.125 degrees (before September 16, 1991) and 0.5625 degrees (after September 17, 1991). The NCEP and ECMWF wind data sets to be used in the wave hindcasting are remade by bi-linear interpolation of the original wind components data given on a spherical grid onto a Cartesian grid with 80 km grid distance put on the Northwestern Pacific Ocean. The wind data is linearly interpolated every one hour for use in wave hindcasting.

Embedding of parametric typhoon winds

Typhoon-induced wind components are computed with use of a parametric typhoon model under the assumption of an elliptic pressure distribution (Yamaguchi et al. 2004). In this case, the estimation of winds at 10 m height is based on the composition of gradient wind components and wind components related to the movement and the multiplication of a correction factor with magnitude increasing toward the typhoon center from 0.6 in the far field. This correction factor $G(\xi)$, empirically proposed by Fujii and Mitsuta(1986), is expressed as

$$G(\xi)=G(\infty)+\{G(\xi_p)-G(\infty)\}(\xi/\xi_p)^{k-1}\exp[(1-1/k)\{1-(\xi/\xi_p)^k\}]$$
$$\xi=r/r_0\,,\ G(\infty)=0.6\,,\ G(\xi_p)=0.9\,,\ \xi_p=0.5\,,\ k=2.5 \qquad (1)$$

where r is the distance from typhoon center and r_0 the typhoon radius.

The typhoon model wind components (U_{TX}, U_{TY}) on a 80 km grid are calculated for the typhoons with a central pressure lower 980 hPa every one hour and then are embedded into the NCEP or ECMWF wind components (U_{AX}, U_{AY}) on the grid in a way that the distribution of the resulting wind components (U_X, U_Y) smoothly changes from the typhoon center area toward the surrounding area. The expression is given as

$$U_X=C_T U_{TX}+(1-C_T)U_{AX}\,,\ U_Y=C_T U_{TY}+(1-C_T)U_{AY}$$
$$C_T=1\,;r\le r_0\,,\ C_T=\exp\{1-(r/r_0)^{1/2}\};r>r_0 \qquad (2)$$

Wind speed U_{10} and wind direction θ_w are calculated as

$$U_{10}=(U_X^2+U_Y^2)^{1/2}\,,\ \theta_w=\tan^{-1}(U_Y/U_X) \qquad (3)$$

The number of time steps that include the embedded typhoons is 32,196 over the 55-year period in the NCEP wind data case and 16,390 over the 24-year period in the ECMWF wind data case.

Wave Model and Conditions for Wave Hindcast

A shallow water wave model based on backward ray tracing (Yamaguchi et al. 1987) is applied for one-hourly wave hindcasting over a long term of 55 years in the case of the typhoon model winds-embedded NCEP wind data (referred to as TY-NCEP wind data) or of 24 years in the case of the typhoon model winds-embedded ECMWF wind data (referred to as TY-ECMWF wind data) at each of the selected points. The model follows the change of the directional spectrum along a refracted ray of each component wave focusing on a hindcast point. The wind components are bi-linearly interpolated onto a wave ray of each component using either of the above-mentioned wind data on a regular grid. The number of frequencies is 23 from 0.04 to 0.5 Hz and the number of directions is 37 equally divided on the whole circle (10 degrees increment). The directional spectrum parameterized with local wind speed and direction is prescribed at the tip of each ray reaching the open boundary and the condition of zero directional spectrum is imposed at the land boundary.

The model is operated using a nested grid with high space resolution of 5 km and 1 km or 0.5 km grid distances for the Northwestern Pacific Ocean. The wave hindcasting points are each of the 18 wave measurement stations in around 50 m water depth situated along the Pacific Ocean coasts and the East China Sea coasts of Japan. Fig. 1 shows the computation area with small, high resolution sea areas and the measurement stations with water depth in parentheses. The period of the wave measurements is from 13 to 28 years.

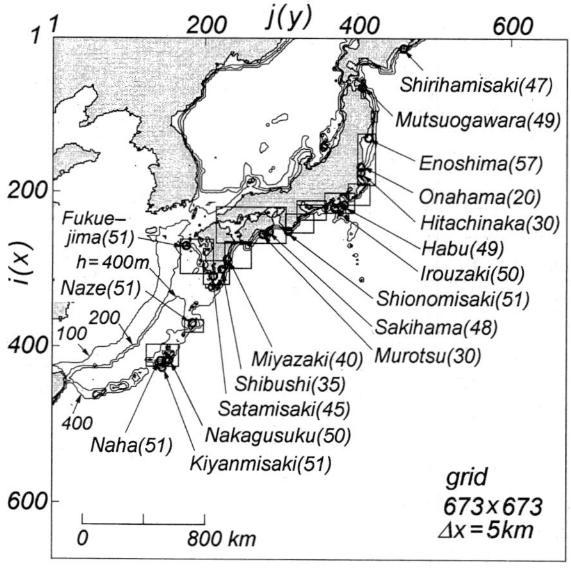

Figure 1. Computation area on the Northwestern Pacific Ocean, rectangular range of nested small sea areas with high space resolution and location of hindcast points.

Method of Wave Data Analysis

Time series analysis

The hindcast data of significant wave height (SWH) H_s or significant wave period (SWT) T_s is individually selected at the time when the measurement data is acquired. Then a correlation diagram between the ascending-ordered SWH data based on the measurement and hindcast is made per season over a whole year period of the measurement. Also, a similar diagram is constructed using the ordered SWT data. Error statistics representing a correspondence between hindcast and measurement data are given by both the correlation coefficient (ρ_H or ρ_T) and the slope value of a correlated straight line with zero intersection value (a_{0H} or a_{0T}).

Wave climate analysis

Using the time series of hindcast and measured SWH data, various kinds of wave height climate parameters and error statistics are calculated separately for months, seasons and years over a whole year period. The parameters are the mean and standard deviation of SWH data H_{sm} and $H_{s\sigma}$, and the highest 1/3 and 1/10 means of SWH data H_{s3} and H_{s10}. Using both data of monthly-separated highest 1/10 mean SWH H_{s10} calculated from hindcast and measurement data, the error statistics are obtained as the correlation coefficient ρ (H_{s10}) and the slope value of a correlated straight line with zero intersection value a_0 (H_{s10}). The error statistics are also estimated for the other type of wave height parameters. Similar parameters and error statistics are calculated using SWT data.

Trend analysis

The trend analyses for yearly-grouped wave height parameters are conducted in two ways. One is to fit a straight line to annual series of the wave height parameters, such that

$$H_{s10} = a(H_{s10}) \cdot t + b(H_{s10}) \tag{4}$$

where t is the year. Slope value a (H_{s10}) and intersection value b (H_{s10}) are determined by the least square method (LSM). A hypothesis of a (H_{s10}) =0 is statistically tested at a 95% upper confidence level using the t-distribution. If the annual series has no trend, the trend parameter is L (H_{s10})=0. Either the trend parameter L (H_{s10})=1 or L (H_{s10})=-1 is given for increasing or decreasing trend of the annual series. The other is to make use of the trend index I_t (H_{s10}) introduced by Suzuki (1975). For the time series with a sample size of more than 20, the criterion at a confidence level of 95% is as follows.

$$\begin{aligned} &|I_t(H_{s10})| < 1.65\text{; no trend} \\ &I_t(H_{s10}) > 1.65\text{; increasing trend} \\ &I_t(H_{s10}) < -1.65\text{; decreasing trend.} \end{aligned} \tag{5}$$

Extreme value analysis

The model (Yamaguchi and Hatada 1997) uses the least square method for the estimation of the scale parameter A and the position parameter B in a candidate distribution such as the Gumbel and shape parameter-fixed Weibull distribution. The shape parameter k is given as any of 27 values ranging from 0.5 to 10. The selection of the optimum distribution is based on a criterion of the largest correlation coefficient between calculated and input data. A jackknife method is applied for the estimation of standard deviation of return wave height. The analysis is conducted for a sample of annual maximum (AM) wave height extracted from either one-hourly hindcast SWH data or one/two/three-hourly measured SWH data over a measurement period from 13 to 28 years.

VALIDATION OF HINDCAST WAVE DATA

Time Series of Waves

Fig. 2 shows time series of 10m-height wind speed U_{10} and SWH H_s at Nakagusuku station, Okinawa Island over 37 days, during a period when 5 typhoons passed through or near that area in succession. Use of the typhoon-embedded TY-NCEP wind data in the wave hindcasting yields a better agreement between hindcast and measurement than use of the NCEP wind data as shown by the increase of the wind speed associated with the passage of a typhoon center.

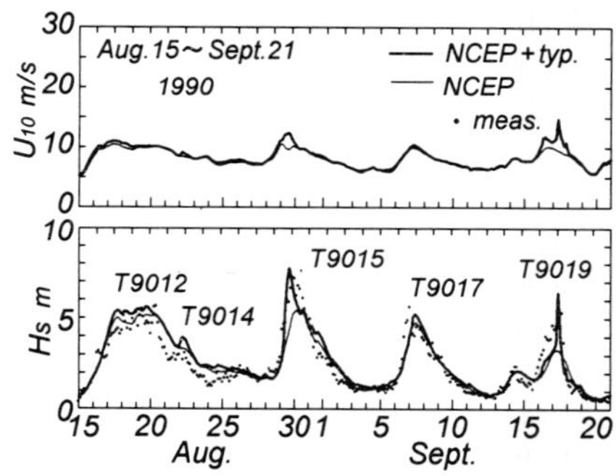

Figure 2. Time series of input wind speed and SWHs based on hindcast and measurement.

In order to have a closer look at the correspondence between hindcast and measurement at higher wave heights or longer wave periods, Fig. 3 illustrates a comparison of the ordered SWH data and the ordered SWT data based on the hindcast and measurement at Nakagusuku station over the July-September period during 26 years in the NCEP wind data cases and over the period during 24 years in the ECMWF wind data cases. The 3-month period is selected owing to a large number of embedding cases of typhoon model winds. Embedding runs of the typhoon model winds give rise to a better correspondence with measurement for higher wave heights or longer wave periods compared to no-embedding runs for both the NCEP wind data and the ECMWF wind data. The accuracy of wave hindcasts may therefore be improved more effectively using the embedding wind data in the NCEP wind data than in the ECMWF wind data.

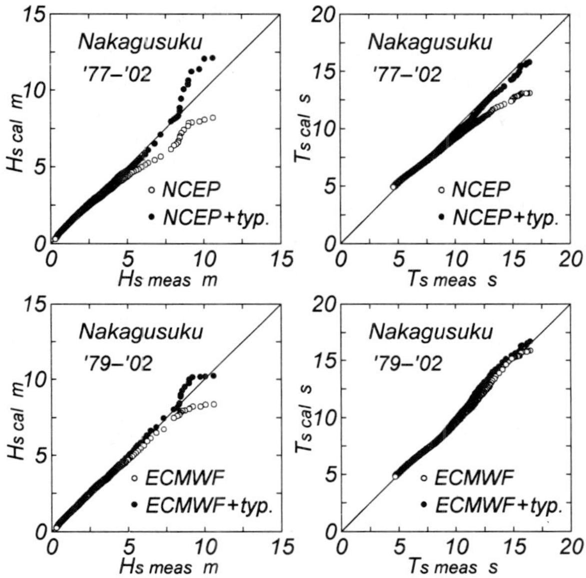

Figure 3. Comparison of ordered SWH or SWT data based on hindcast and measurement.

The correlation coefficient $\rho(H_s)$ or $\rho(T_s)$ has a value rather close to 1 in all cases. This means that this correlation coefficient may not be an effective index for discriminating the degree of correlation between the typhoon winds-embedding case and the non-embedding case. Thus, the correlation coefficient is not given here. Fig. 4 shows the slope value a_0 (H_s) and a_0 (T_s) in the correlation relation between ordered hindcast and measured data in the July-September period (like Fig. 3) at each of the 18 measurement stations along the Pacific Ocean coasts and the East China Sea coasts of Japan, in cases where either the NCEP wind data or the ECMWF wind data are used in the wave hindcasting. The slope value a_0 (H_s) has a consistently greater value in the embedding case than in the non-embedding case. For that reason, the slope value close to 1 in the non-embedding case changes into the slope value a_0 (H_s) greater than 1 in the embedding case at the northern stations of the Pacific Ocean coasts of Japan. As for wave period, the slope value in the embedding case is closer to 1 than that in the non-embedding case. Further improvement of the accuracy of the wave hindcast associated with use of the TY-NCEP or TY-ECMWF wind data is found at the stations located south from Habu station, where the typhoon-generated waves are dominant. Also, the degree of improvement is higher in the NCEP wind data case than in the ECMWF wind data case, because the original

NCEP wind data are produced under the condition of a coarser space resolution than the original ECMWF wind data.

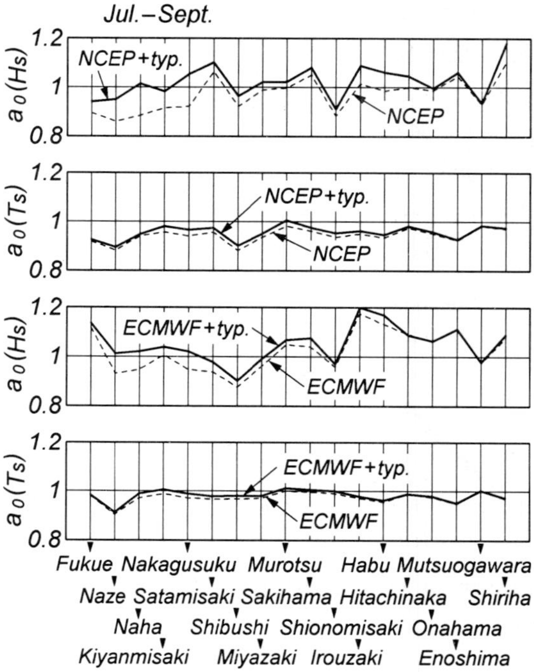

Figure 4. Slope value in the correlation relation of ordered SWH or SWT data based on hindcast and measurement at each of 18 measurement stations.

Wave Climate

Table 1 lists the highest 1/10 mean of SWH H_{s10} and the longest 1/10 mean of SWT T_{s10} on September at each of the 18 measurement stations, in cases where the wave statistics are obtained using measurement data (meas.), hindcast data in the NCEP wind data case (NCP) and hindcast data in the TY-NCEP wind data case (NTY). The month of September (which is accompanied by a frequent passage of strong typhoons) is chosen to elucidate the effect of the typhoon model winds-embedding on the estimated wave statistics. As for wave height, the hindcast results for the embedding case at each of the stations located south of Miyazaki station are closer to the measurement results than those in the non-embedding case, but at each of the stations located north from Miyazaki station, the contrary happens. As for wave period, the hindcast in the embedding case consistently yields results closer to the measurement compared to those in the non-embedding case. That is to say, the improved accuracy of the wave hindcast associated with embedding of typhoon model winds is found more explicitly in the SWT estimates than in the SWH estimates.

Table 1. List of the largest 1/10 means H_{s10}, T_{s10} of SWH or SWT data on September at each of 18 measurement stations.

location	Sept. H_{s10} m			Sept. T_{S10} s		
	meas.	NCP	NTY	meas.	NCP	NTY
Shirihamisaki	2.82	3.39	3.48	12.0	11.3	11.7
Mutsuogawara	2.86	3.00	3.01	11.2	10.8	10.9
Enoshima	3.14	3.46	3.52	11.5	10.8	11.0
Onahama	3.03	2.89	2.98	12.2	11.0	11.3
Hitachinaka	2.79	3.07	3.15	11.4	10.5	10.7
Habu	3.24	3.24	3.41	11.5	10.3	10.7
Irouzaki	2.77	3.02	3.17	11.5	10.3	10.7
Shionomisaki	3.27	3.00	3.10	11.8	11.0	11.8
Sakihama	2.73	3.01	3.14	11.7	11.0	11.6
Murotsu	2.70	2.70	2.75	12.4	11.9	12.3
Miyazaki	3.74	3.60	3.79	12.3	10.9	11.7
Shibushi	2.42	2.35	2.42	11.6	10.0	10.6
Satamisaki	2.40	2.52	2.65	11.7	12.5	11.2
Nakagusuku	3.53	2.98	3.26	11.8	10.2	11.0
Kiyanmisaki	3.59	3.05	3.23	11.6	10.5	11.3
Naha	2.73	2.38	2.71	9.8	9.4	9.7
Naze	2.60	2.40	2.59	9.3	8.6	8.8
Fukuejima	2.48	2.16	2.21	8.4	7.5	7.6

Fig. 5 shows a comparison between the hindcast and measurement both in the non-embedding case and in the embedding case for the largest 1/10 mean of SWH H_{s10} obtained every month over a whole year period of measurement at Nakagusuku station. Similar figures are also given for the largest 1/10 mean of SWT T_{s10}. These figures reveal that the accuracy of the wave hindcast for higher wave height conditions is improved by using the TY- NCEP wind data. This is true of the wave period data. The accuracy is represented by both the correlation coefficient $\rho(H_{s10})$, $\rho(T_{s10})$ and the slope value of a line with zero-intersection $a_0(H_{s10})$, $a_0(T_{s10})$. A better agreement between hindcast in the embedding case and measurement is also obtained for the other wave climate parameters.

The upper two panels of Fig. 6 indicate the correlation coefficient $\rho(H_{s10})$ and the slope value $a_0(H_{s10})$ for monthly-separated H_{s10} data at each of the 18 measurement stations for both the non-embedding NCEP wind data case and the embedding case. For monthly-separated T_{s10} data, the correlation coefficient $\rho(T_{s10})$ and the slope value $a_0(T_{s10})$ are given in the lower two panels of Fig. 6. As for H_{s10}, the correlation coefficient in the embedding case is consistently greater than that in the non-embedding case, whereas the slope value in the embedding case does not always take a value closer to 1 than that in the non-

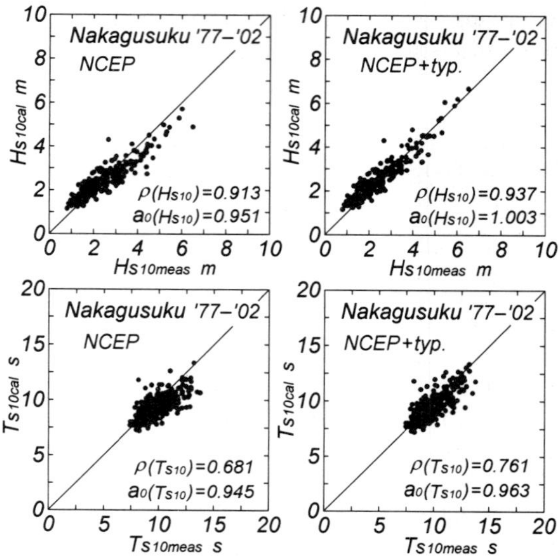

Figure 5. Comparison between hindcast and measurement for monthly-separated H_{s10} or T_{s10}.

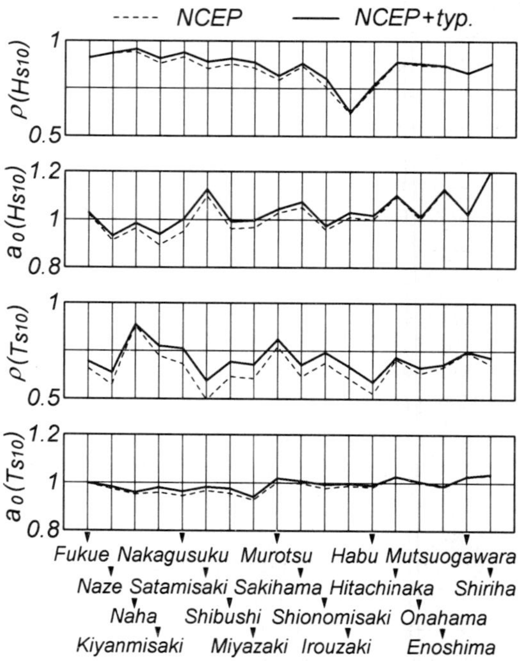

Figure 6. Correlation coefficient and slope value for monthly-separated H_{s10} or T_{s10} data at each of 18 measurement stations.

embedding case, because an increase of wave height hindcasts in the embedding case may enhance a discrepancy from the slope value of 1. As for T_{s10}, both the correlation coefficient $\rho(T_{s10})$ and the slope value $a_0(T_{s10})$ have a value closer to 1 in the embedding case than in the non-embedding case. The above-discussed tendency is more clearly found at each of the stations located south from Habu station along the Pacific coasts, where the typhoon-generated waves are dominant. Use of the TY-NCEP wind data in a long-term wave hindcasting may contribute to a significant improvement in the estimation of waves with higher height and waves with longer period, in particular the latter.

Trend of Wave Height

Fig. 7 demonstrates a 26-year annual series of the highest 1/10 mean of SWH over one year H_{s10} at Nakagusuku station in both the non-embedding NCEP wind data case and the embedding case. The wave hindcasting produces the annual variation of yearly-averaged H_{s10} observed in the measurement irrespective of typhoon winds-embedding. In sequence of the measurement data case, non-embedding data case and embedding data case, the slope value of a straight line fitted to the annual series of H_{s10} data is (0.94, 0.62, 0.93) cm/year, the t-distribution-based trend parameter $L(H_{s10})$ is (0, 0, 0) and the Suzuki trend index $I_t(H_{s10})$ is (0.74, 0.94, 1.09), in cases where the yearly H_{s10} data with measurement rate of a year above 75% are selectively used. These values, which are obtained by these statistical tests for a trend of H_{s10} data, suggest a perfect agreement among the 2 sets of hindcast data and the measurement data in a sense that none of those 3 data sets indicate statistically significant trends.

Table 2 summarizes the Suzuki trend indices $I_t(H_{s10})$ for the measurement-based annual series of yearly-grouped H_{s10} data at each of the 18 measurement stations and the corresponding 2 kinds of hindcast-based annual series. The bold-faced type means the existence of a trend with a statistical significance. The 2 sets of NCEP wind data over 45 years from 1958 to 2002 are used in the wave hindcasting. The trend analyses are conducted for a measurement period indicated by the parenthesized number (in years) on the location column in the table and for a 45-year period. Statistical test results using the trend parameter

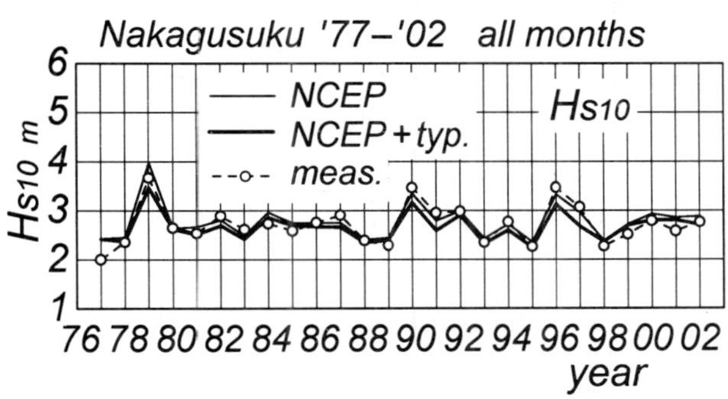

Figure 7. Annual series of yearly-averaged H_{s10} data at Nakagusuku station.

Table 2 Trend index of yearly-grouped H_{s10} data at each of 18 measurement stations.

location (yrs.)	meas. period			'58~'02	
	meas.	NCP	NTY	NCP	NTY
Shirihamisaki(18)	0.27	**1.92**	**2.03**	1.23	1.25
Mutsuogawara(28)	0.35	-0.27	-0.23	0.10	0.23
Enoshima(25)	-0.91	-1.03	-1.15	0.35	0.31
Onahama(21)	0.74	**2.14**	**2.14**	1.17	1.21
Hitachinaka(22)	0.72	-0.06	-0.06	0.08	-0.04
Habu(27)				1.19	1.15
Irouzaki(27)	**1.74**	1.52	1.56	0.47	0.51
Shionomisaki(16)	0.61	1.34	1.22	-0.10	-0.16
Sakihama(26)	**2.05**	**2.23**	**1.87**	**1.82**	1.51
Murotsu(13)				0.37	0.37
Miyazaki(13)				**2.58**	**2.31**
Shibushi(22)	0.42	0.76	0.54	**2.03**	**1.70**
Satamisaki(21)	0.24	**1.99**	**1.85**	**3.19**	**2.74**
Nakagusuku(26)	0.74	0.94	1.09	**2.35**	**1.66**
Kiyanmisaki(22)	-0.38	0.10	0.17	**1.74**	1.59
Naha(28)				0.51	0.74
Naze(26)	0.25	-0.99	-0.82	0.14	-0.43
Fukuejima(23)	**-1.88**	0.19	-0.06	1.60	1.39

$L(H_{s10})$ and the trend index $I_t(H_{s10})$ are consistent with each other in all cases. This leads to a presentation of only the trend index, the value of which changes continuously. As for H_{s10} data over the measurement period, the trend index takes a comparable value between the values in the non-embedding case and the embedding case. The statistical test results show complete consistency between the non-embedding case and the embedding case. The statistical judgments regarding the presence of a trend in the hindcast and the measurement coincide with each other except for the 5 stations, namely Shirihamisaki, Onahama, Irouzaki, Satamisaki and Fukuejima. As for 45-year H_{s10} data, use of the TY-NCEP wind data in the wave hindcasting does not alter the statistical judgment for a trend with the exception of 2 stations like Sakihama and Kiyanmisaki. At these 2 stations, a small increase of the trend index in the embedding case may give rise to exceeding a limiting value of 1.65. Therefore, a weak but statistically significant, increasing trend of wave height statistics like H_{s10} in this recent year period of 45 years may be observed at the southern stations on the Pacific Ocean coasts of Japan.

Return Wave Height

Return wave heights and their standard deviations at each of the 18 measurement stations are estimated by applying the LSM-based model to the samples of AM wave height data extracted from the measurement and the 2 sets of hindcast data over the same period. The upper panel of Fig. 8 shows the results estimated for a 50-year return period at each of the 18 stations for the embedding case and the non-embedding case for the NCEP wind data and for the measurement data case. The results for the ECMWF wind data case are indicated in the lower panel of Fig. 8. The estimated 50-year return wave height H_{50} in the embedding case becomes larger than in the non-embedding case, and moreover the former is closer on the whole to the measurement-based estimate than the latter. The increase is significant at the stations located south from Habu station, where the typhoon-generated waves are dominant, but it is insignificant at the northern stations. Irrespective of either using the hindcast cases or the measurement case, the standard deviation $H_{\sigma 50}$ has a larger value at stations where the largest AM wave height data is much higher than the second largest AM wave height. It results in an increase of the coefficient of variation defined by the ratio of standard deviation to return wave height $H_{\sigma 50}/H_{50}$, which means a lowering of statistical reliability of the estimate for return wave height.

As the space resolution of the ECMWF wind data is rather high, the effect of the embedded typhoon winds on the hindcast waves is not so significant. It turns out that not only the return wave height but also its standard deviation in the embedding case and those in the non-embedding case are comparable respectively at most of the stations.

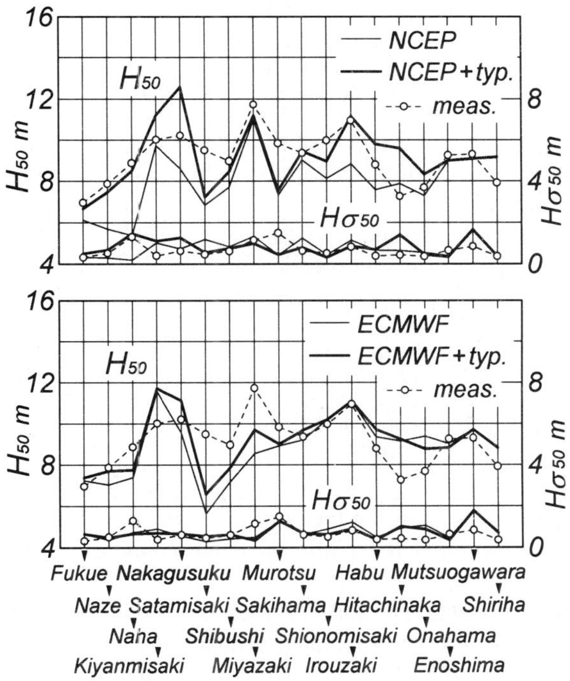

Figure 8. Return wave height and its standard deviation for a 50-year return period at each of 18 measurement stations.

CONCLUSIONS

The main findings in this study are summarized as follows:

1) Embedding parametric typhoon winds into the original NCEP wind data set yields a significant improvement of the applicability of the long term wave hindcast system for the estimation of both wave climate conditions and extreme waves, almost without changing the statistical test results for trend analysis in typhoon-dominated sea areas.

2) Embedding parametric typhoon winds to improve the accuracy of wave estimation is less effective in the ECMWF wind data than in the NCEP wind data, because the ECMWF wind data has a higher accuracy than the NCEP wind data due to its finer space resolution.

ACKNOWLEDGMENTS

Thanks are due to NCAR and ECMWF which kindly provided reanalysis and analysis surface wind data sets.

REFERENCES

Fujii, T. and Y. Mitsuta. 1986. Simulation of winds in typhoons by a stochastic model, *Journal of Wind Engineering*, 28, 1-12(in Japanese).

Hatada, Y. and M. Yamaguchi. 1998. A long term wave hindcast system using ECMWF winds, *Proceedings of the 26th International Conference on Coastal Engineering*, ASCE, 1, 509-522.

Hatada, Y. and M. Yamaguchi. 2002. Estimation of long-term variability of wave climate around the coastal sea areas of Japan, *Proceedings of the 28th International Conference on Coastal Engineering*, ASCE, 1, 61-73.

Suzuki, E. 1975. *Meteorological Statistics*(5th edt.), Chijin Shokan Pub., 314p(in Japanese).

Yamaguchi, M., Hatada, Y. and Y. Utsunomiya. 1987. A shallow water wave prediction model at a single location and its applicability, *Journal of the Japanese Society for Civil Engineers*, 381/II-7, 151-160(in Japanese).

Yamaguchi, M. and Y. Hatada. 1997. An extremal analysis system and its application to the estimation of extremes of meteorological and oceanographic elements around the coasts of Japan, *Proceedings of the 3rd International Symposium on Ocean Wave Measurement and Analysis WAVES97*, ASCE, 2, 932-946.

Yamaguchi, M. and Y. Hatada. 2002. 51-year wave hindcast and analysis of wave height climate trend on the Northwestern Pacific Ocean, *Proceedings of the 7th International Workshop on Wave Hindcasting and Forecasting*, 48-59.

Yamaguchi, M., Nonaka, H. and Y. Hatada. 2004. Estimating the parent distribution of storm type-separated annual maximum wave heights on the Northwestern Pacific Ocean, *Proceedings of the 29th International Conference on Coastal Engineering*, ASCE, 1, 959-971.

WAVE REGIME CHARACTERIZATION ON THE PORTUGUESE COAST USING HINDCAST AND WAVE PROPAGATION MODELS

Rui Capitão[1], Conceição Fortes[1], Fernando Carvalho[2], Alexandre Coli[1], Liliana Pinheiro[1]

To overcome the scarcity or the absence of observed wave data in the vicinity of a site where coastal protection works or maritime structures have to be built, two different methodologies, TRANSFER and MAR3G, are presented, applied and compared in this paper. These methodologies were applied in the definition of the wave regimes, a set of subclimates of the wave climate, at four selected sites along the Portuguese coast. The results are compared in terms of the joint histograms of wave heights, wave periods and wave directions for both methodologies. Statistical values associated with the above variables are also calculated and compared. Finally, the main advantages and disadvantages of using each of the considered methodologies are discussed.

INTRODUCTION

The Portuguese west coast is characterized by a long coastal zone where several beaches, harbours and coastal works can be found. The study of the dynamics and sedimentary equilibrium of the beaches and the design and protection of the harbours, amongst others, depend significantly on the knowledge of the sea wave regime characteristics at the studied location. Unfortunately, along the Portuguese west coast only two buoys are at the moment in operation: at Leixões, on the northwestern coast, and at Sines, on the southwestern coast of Portugal – denoted, respectively, by LXS and SNS in Fig. 1 (shaded rows).

As a consequence, a problem that usually arises in coastal engineering studies is the scarcity or absence of observed wave data during a reasonable number of years at a site (or, at least, at its vicinity) where coastal protection works or maritime structures have to be built. When this occurs, two methodologies are usually used at LNEC to produce significant wave heights, HS, mean wave periods, TZ, and wave directions, Θ, at the coastal site(s) of interest:

1. The methodology TRANSFER, Coli *et al.* (2002), which uses an irregular wave propagation model for back-propagating to offshore a set of observed wave data, normally obtained using directional wave buoys, and then forward--propagating the offshore data to the desired location;
2. The methodology MAR3G which uses a third-generation wind-wave hindcast model, MAR3G, along with a wave ray propagation model, Pires and Carvalho (1996). This methodology uses as input the meteorological (wind) data on the area of interest.

---

[1] National Civil Engineering Laboratory, Avenida do Brasil 101, 1700-066 Lisboa, Portugal.
[2] Meteorological Institute, Rua C ao Aeroporto de Lisboa, 1749-077 Lisboa, Portugal.

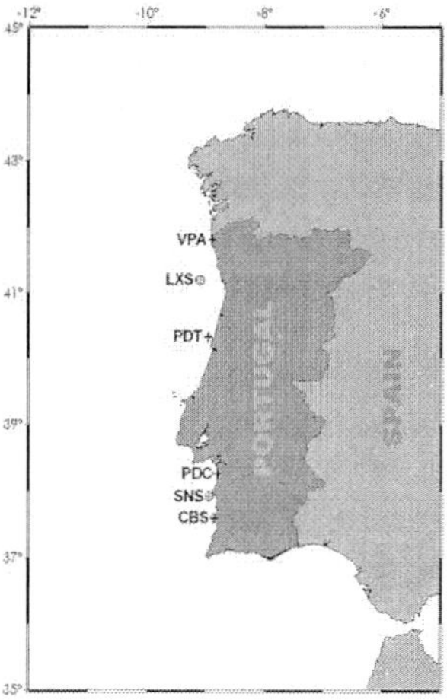

Name location	Code	Latitude (N)	Longitude (W)	Depth (m)
Leixões	LXS	41° 12' 12"	9° 05' 18"	110.0
Vila Praia de Âncora	VPA	41° 48' 31"	8° 52' 39"	10.3
Praia da Tocha	PDT	40° 20' 00"	8° 57' 00"	49.7
Sines	SNS	37° 55' 16"	8° 55' 44"	94.8
Pinheiro da Cruz	PDC	38° 15' 22"	8° 47' 24"	20.1
Cabo Sardão	CBS	37° 35' 46"	8° 50' 31"	47.0

Figure 1. Location of wave buoys and target points.

While the first methodology is more dependent on the quality of the observed wave data, collected using the available wave buoys (although sometimes with many gaps), the second one is based on wind data and so it strongly depends on the precision of the hindcast model used.

In this paper, the two methodologies will be presented, applied and compared for four selected sites on the Portuguese coast where no wave data are available: Vila Praia de Âncora (VPA), Praia da Tocha (PDT), Pinheiro da Cruz (PDC) and Cabo Sardão (CBS), see Fig. 1. These sites were chosen on the basis of their relatively close vicinity to the already mentioned two wave buoys (LXS for VPA and PDT; SNS for PDC and CBS), where wave data exist.

All TRANSFER and MAR3G wave regimes for the above four locations are compared in terms of the joint histograms of wave heights, wave periods and wave directions associated with the different regimes. Also, several statistical parameters based on HS, TZ and Θ obtained values are computed to allow a better comparison of both methodologies. This will make possible the evaluation of the main advantages and disadvantages of each methodology. In this way, a general characterization of wave regimes on the Portuguese coast is attempted.

TRANSFER AND MAR3G METHODOLOGIES

Description of the models in the TRANSFER methodology

The TRANSFER methodology is based on three numerical models: BACKTRACK and REFSPEC, Covas and Pontes (1988) and INTERPOL, Capitão *et al.* (1994). BACKTRACK computes the path of the wave rays reaching one point for a given bottom configuration (fan-type refraction diagrams). This model outputs fan-type refraction diagrams, presented both

graphically and numerically. REFSPEC uses as input the output numerical files from BACKTRACK and the deep-water two-dimensional directional wave spectrum to compute the wave spectrum at the beginning point of the wave rays, by refracting the offshore spectrum. BACKTRACK computes the values, at the point, of the directional wave spectrum ordinates, of the ratio between the significant wave heights at the beginning point and in deep water (H/H0), of the mean zero-upcrossing period (TZ) and of the mean and median wave directions (Θ and Θm). Finally, the INTERPOL program is used to estimate, at a given point, by an interpolation scheme, values of H/H0, TZ and Θo, from values defined on a two-dimensional grid of tabulated values.
The TRANSFER methodology only takes into account the irregular wave refraction due to bottom configuration. It does not take into account other linear phenomena, like diffraction, or nonlinear phenomena, like wave dissipation due to wave breaking and bottom friction. Hence, limits were imposed to the offshore wave directions (less than 350º) and to the offshore peak periods (less than 26 s) for all the regime transfers performed in this study since it was found that over those limits results of the irregular refraction model were compromised.

Description of the models in MAR3G methodology

The spectral numerical model MAR3G, Pires and Carvalho (1996), is a third-generation wind-wave model developed at the Meteorological Institute of Portugal, which is in operational activity since 1994. This deep-water model solves the transport equation where the source terms (generation, growth, propagation and dissipation) have a physical approach similar to the WAM model (WAMDI Group, 1988). The model includes a parameterization for the effect of the wind variability for the wave generation when the wind speed is larger than the wave celerity. The model computes the directional spectra of a sea state in grid points from chronological time series of wind fields provided by the atmospheric model that runs at the European Centre for Medium-range Weather Forecasts (ECMWF) in the same grid. Daily, the model produces one day of analysis (6-hour timestep) and five days of forecasts (12-hour timestep). For points near shore, an inverse tracing ray model (SAIREF) is used which takes into account the following phenomena: shoaling, refraction, shelter by the coastline and dissipation by bottom friction effects.
Several sources of errors are associated with this model: a) Errors in wind direction and intensity due to position, extension and intensity of atmospheric disturbances; b) Errors in the physical and numerical representation of the MAR3G model; c) Errors due to spatial and temporal resolutions adopted in the model configuration; d) Errors in the physical, numerical and bathymetrical representation of the SAIREF model; e) Errors in the observations due to instrumental problems. The MAR3G/SAIREF results have been verified against buoy data for several points near the Portuguese coast, using quality parameters.

APPLICATION OF THE TRANSFER AND MAR3G METHODOLOGIES

Description

The determination of the wave regimes at Vila Praia de Âncora (VPA), Praia da Tocha (PDT), Pinheiro da Cruz (PDC) and Cabo Sardão (CBS), see Fig.1, was made by using the TRANSFER and MAR3G methodologies herein described.

The TRANSFER methodology considers the following steps:

1. Use of wave data from directional wave buoys (Hydrographic Institute, IH) from a position as close as possible to the points of interest. In this case, the following locations were considered:
   - Leixões (LXS) wave buoy, to obtain the Vila Praia de Âncora's (VPA) and the Praia da Tocha's (PDT) wave regimes;
   - Sines (SNS) wave buoy, to obtain the Pinheiro da Cruz's (PDC) and the Cabo Sardão's (CBS) wave regimes.

The above data (at the LXS and SNS locations) consist of wave records of 20-minute duration, taken every 3 hours, which provide the following wave parameters: HS (significant wave height), TZ (mean wave period) and Θ (wave direction). The data refer to the period from 2001-10-1 to 2002-08-31 (11 months) for the LXS wave buoy and from 1996-03-01 to 2002-12-31 (~6 years) for the SNS wave buoy.

2. Use of the BACTRACK-REFSPEC models, to transfer wave data (HS, TZ, Θ) at each buoy location to offshore and then to the four point locations (VPA and PDT are based on LXS data and CBS and PDC are based on SNS data).
3. The wave regimes at each wave buoy, offshore, and at each point of interest are computed. The so-called "IH+TRANSFER" wave regimes are obtained for the selected four locations.

On the other hand, the MAR3G methodology considered the following steps:

1. Use of wind data provided by the European Centre for Medium-range Weather Forecasts (ECMWF) at the four points of interest;
2. Use of the MAR3G hindcast model and of the SAIREF wave propagation model to compute the wave parameters at the points of interest. The results of MAR3G contain the values, taken every 6 hours, of HS0 (significant wave height), TPOT (equivalent to the average wave period) and DMV (equivalent to the mean wave direction);
3. The so-called "MAR3G" wave regimes are obtained at each of four locations.

In this paper, the "IH+TRANSFER" and the "MAR3G" wave regimes, found by using the TRANSFER and MAR3G methodologies, respectively, at the four selected locations - Vila Praia de Âncora (VPA), Praia da Tocha (PDT), Pinheiro da Cruz (PDC) and Cabo Sardão (CBS) - along the Portuguese west coast, will be discussed and compared.

Wave regimes at Vila Praia Âncora (VPA) and Pinheiro da Tocha (PDT)

A first analysis is made by comparing the wave parameters produced by both methodologies at Vila Praia de Âncora (VPA) and Praia da Tocha (PDT). As an example, Figs. 2, 3 and 4 show the comparisons, for about one year (from 2001-10-1 to 2002-08-31), of the time series computed using both methodologies of, respectively: significant wave heights (HS vs. HS0); mean wave periods (TZ vs. TPOT) and mean wave directions (Θ vs. DMV).

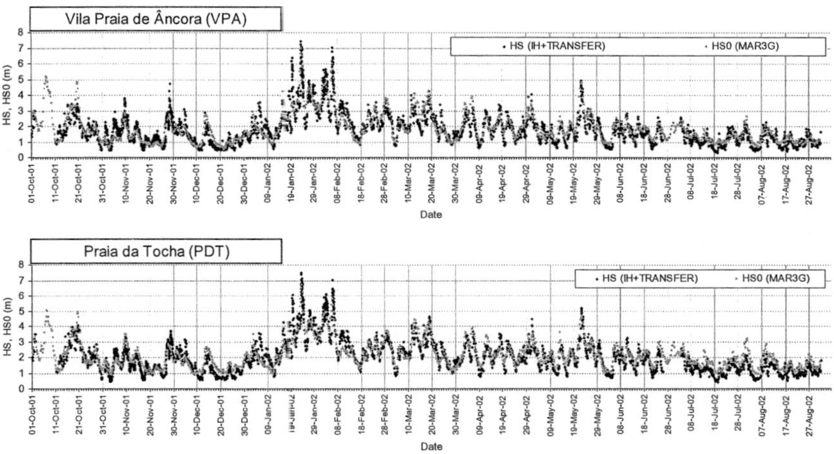

Figure 2 – Comparison of HS and HS0 time series at VPA and PDT.

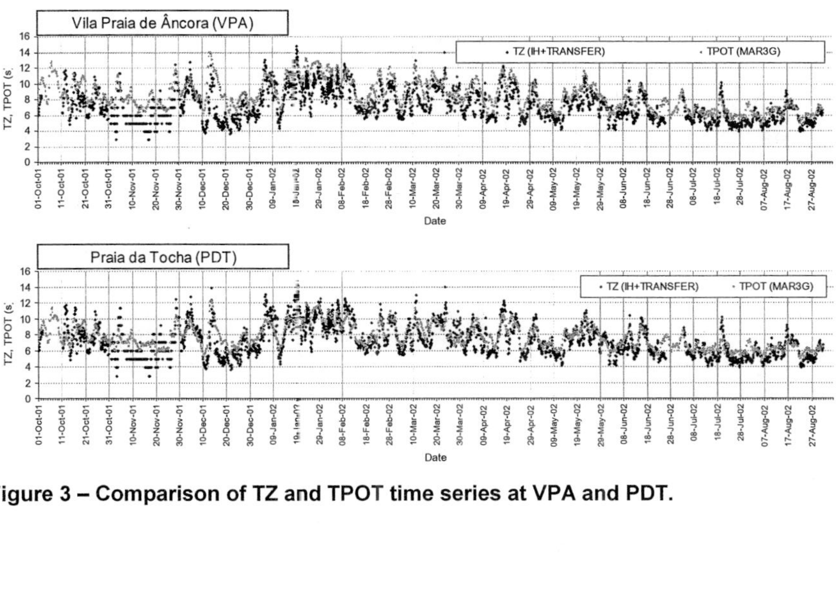

Figure 3 – Comparison of TZ and TPOT time series at VPA and PDT.

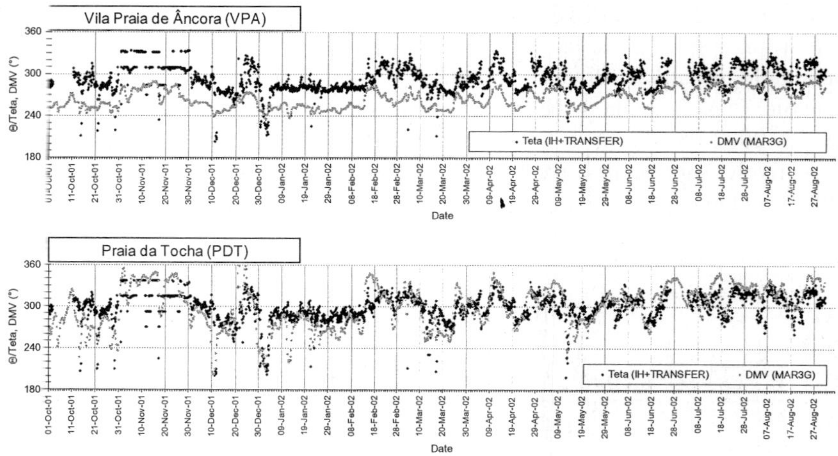

Figure 4 – Comparison of Θ (Theta) and DMV time series at VPA and PDT.

A comparison was also made of: a) the correlations between the wave heights and wave periods using the two methodologies: HS/HS0 vs. TZ/TPOT, HS/HS0 vs. Θ/DMV and TZ/TPOT vs. Θ/DMV, for the whole data that make up the wave regimes "IH+TRANSFER" and "MAR3G" at PDT and VPA, Fig. 5; b) the histograms of HS vs. HS0, TZ vs. TPOT and Θ vs. DMV, for both methodologies, at VPA and PDT sites, also considering all data, Fig. 6.

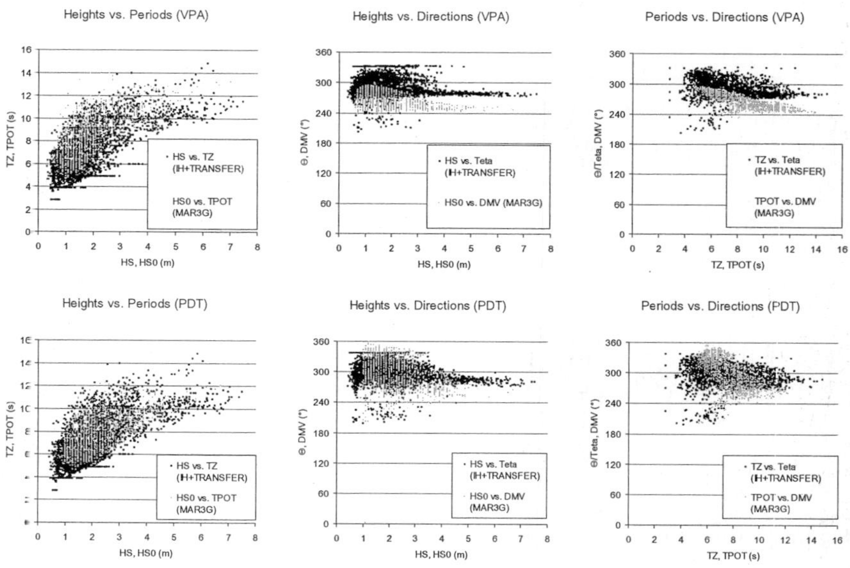

Figure 5 – Correlations HS/HS0 vs. TZ/TPOT, HS/HS0 vs. Θ/DMV and TZ/TPOT vs. Θ/DMV, for the wave regimes IH+TRANSFER and MAR3G at VPA and PDT.

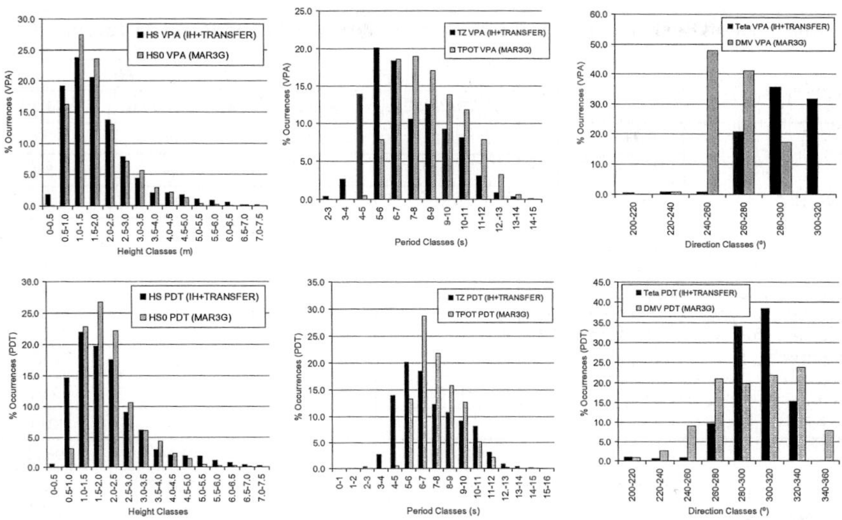

Figure 6 – Histogram comparisons HS vs. HS0, TZ vs. TPOT and Θ vs. DMV, for the wave regimes IH+TRANSFER and MAR3G at VPA and PDT.

The following conclusions can be drawn from the analysis of the above results:

- The parameters (HS vs. HS0, TZ vs. TPOT and Θ vs. DMV), show a similar trend, that is, their variations along the studied period are identical.
- For the wave heights and periods, values have the same order of magnitude. However, the MAR3G methodology is found to produce higher values than TRANSFER, especially for smaller wave height classes.
- For wave directions, larger differences occur between the two methodologies, especially for the VPA site. Note that this point is located in a shallow and sheltered region, where diffraction is significant. None of the wave refraction models used includes diffraction. Moreover, since the MAR3G model operates with wind fields that come from an atmospheric model with a 6-hour time discretization only, its results are no longer sensitive to small local and temporal variations of the wind direction and intensity;
- Differences between the results of both methodologies are found to be smaller for PDT than for VPA. This is due to the fact that VPA is located in a shallower and more sheltered region than PDT, which is located in a deepwater open area. Clearly, the described methodologies are better suited to be applied to the PDT location than to the VPA location.

A more detailed comparison between the two methodologies (TRANSFER and MAR3G) may be produced for the VPA and PDT sites, see Table 1.

Table 1. Statistical analysis at Vila Praia de Âncora (VPA) and Praia da Tocha (PDT)

Parameter		VPA		PDT	
		IH+TRANSFER	MAR3G	IH+TRANSFER	MAR3G
HS / HS0	Mean	1.88	1.90	2.08	2.17
	Standard Deviation	1.150	0.963	1.194	0.886
	Minimum	0.31	0.50	0.41	0.90
	Maximum	7.46	7.00	7.50	6.70
	Count	3127	1340	3129	1340
TZ / TPOT	Mean	7.20	8.51	7.20	7.66
	Standard Deviation	2.074	1.899	2.075	1.507
	Minimum	2.84	4.70	2.80	4.70
	Maximum	14.82	14.00	14.82	12.20
	Count	3127	1340	3129	1340
Θ / DMV	Mean	295.40	264.78	301.36	298.23
	Standard Deviation	19.452	13.615	21.144	32.093
	Minimum	202.80	233.00	198.27	1.00
	Maximum	333.78	294.00	338.60	359.00
	Count	3127	1340	3129	1340

Based on the above results, the following comments can be drawn:

For VPA:

In general, IH+TRANSFER and MAR3G mean values of HS are reasonably similar although MAR3G values are a little higher than IH+TRANSFER ones. For the mean TZ and Θ values, the differences between the two methodologies are higher, especially for the mean wave direction. Moreover, the MAR3G and IH+TRANSFER wave direction ranges are quite different: the MAR3G range is narrower than the IH+TRANSFER. These differences are due to the fact that the VPA location is in a shallow and more sheltered area, where diffraction and other phenomena are important and both numerical models do not take those phenomena into account.

For PDT:

In general, IH+TRANSFER and MAR3G parameters are very similar for this site. The differences between IH+TRANSFER and MAR3G mean values are very small for all parameters, especially for Θ. HS and TZ mean values for MAR3G are a little higher than for IH+TRANSFER. The HS and TZ ranges for IH+TRANSFER are larger than for MAR3G. The wave direction range for IH+TRANSFER is smaller than for MAR3G, since the TRANSFER methodology limits the wave directions to be transferred *a priori*.

Wave regimes at Cabo Sardão (CBS) and Pinheiro da Cruz (PDC)

A similar comparison was made for the Cabo Sardão (CBS) and Pinheiro da Cruz (PDC) sites: a) the correlations between the wave heights and wave periods, HS/HS0 vs. TZ/TPOT, wave heights and wave directions, HS/HS0 vs. Θ/DMV, and wave periods and wave directions, TZ/TPOT vs. Θ/DMV, for all data at CBS and PDC, Fig 7; b) The histogram comparisons of HS vs. HS0, TZ vs. TPOT and Θ vs. DMV, for both methodologies, at the two sites, also considering all data, Fig. 8.

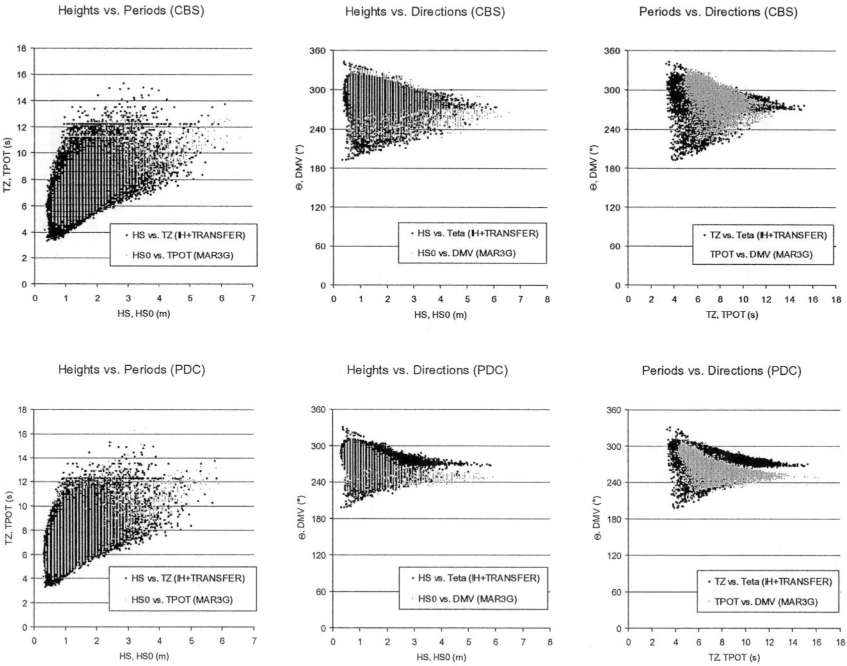

Figure 7 – Correlations HS/HS0 vs. TZ/TPOT, HS/HS0 vs. Θ/DMV and TZ/TPOT vs. Θ/DMV, for the wave regimes IH+TRANSFER and MAR3G at CBS and PDC.

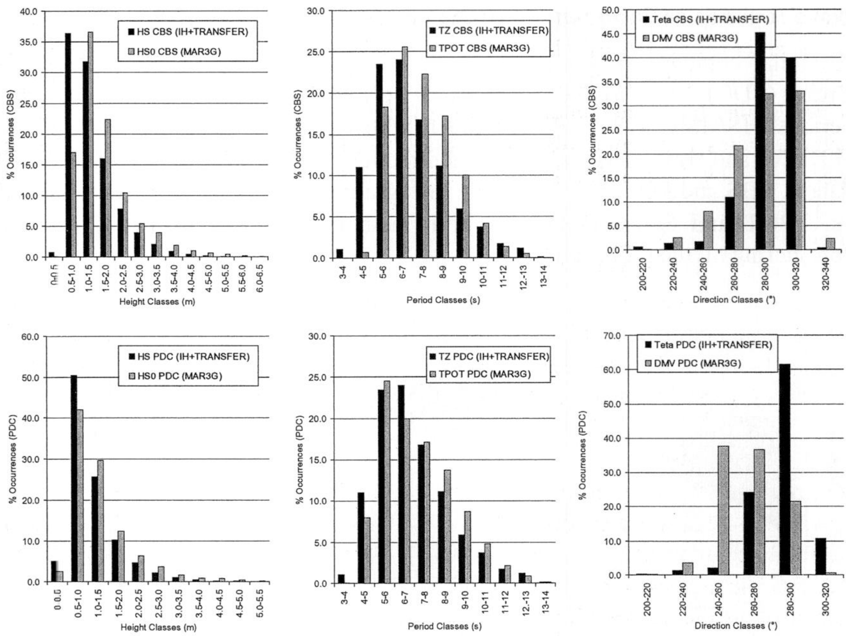

Figure 8 – Histogram comparisons HS vs. HS0, TZ vs. TPOT and Θ vs. DMV, for the wave regimes IH+TRANSFER and MAR3G at CBS and PDC.

A more detailed comparison between the methodologies is found on Table 2.

Table 2. Statistical analysis at Cabo Sardão (CBS) and Pinheiro da Cruz (PDC).

Parameter		CBS		PDC	
		IH+TRANSFER	MAR3G	IH+TRANSFER	MAR3G
HS / HS0	Mean	1.36	1.72	1.11	1.33
	Standard Deviation	0.697	0.838	0.607	0.731
	Minimum	0.33	0.50	0.30	0.40
	Maximum	6.11	6.60	5.82	5.90
	Count	32784	9988	32784	11936
TZ / TPOT	Mean	6.90	7.49	6.90	7.21
	Standard Deviation	1.778	1.501	1.778	1.808
	Minimum	3.32	4.20	3.32	3.90
	Maximum	15.31	13.20	15.31	16.50
	Count	32784	9988	32784	11936
Θ / DMV	Mean	293.81	289.19	286.09	266.26
	Standard Deviation	16.639	21.424	13.374	15.790
	Minimum	192.03	205.00	198.34	215.00
	Maximum	342.56	331.00	331.05	312.00
	Count	32784	9988	32784	11936

From Table 2, one can observe that the ranges of IH+TRANSFER and MAR3G for all parameters (HS vs. HS0, TZ vs. TPOT and Θ vs. DMV), are reasonably similar, especially for CBS. It is also apparent that the MAR3G methodology overestimates the small values of wave heights and wave period values. Both methodologies give good results for the two sites since they are located in an open area. Also from Table 2, one observes that:

For CBS:
In general, IH+TRANSFER and MAR3G parameters are similar. The differences between IH+TRANSFER and MAR3G mean values are very small for all parameters, especially for TZ values. HS and TZ mean values for MAR3G are a little higher than for IH+TRANSFER. The HS, TZ and Θ ranges are larger for IH+TRANSFER than for MAR3G.

For PDC:
For this site, the IH+TRANSFER and MAR3G parameters are quite similar, except for the mean directions. The differences between IH+TRANSFER and MAR3G mean values are very small for all but the direction. However, the MAR3G methodology gives somewhat higher values of HS for the higher classes compared to the values obtained using the TRANSFER methodology. For this site, HS and TZ ranges are larger for MAR3G than for IH+TRANSFER, as opposed to what happens at the CBS site.

CONCLUSIONS

This paper dealt with a problem usually found in coastal engineering studies, which is the scarcity or even absence of observed wave data in the vicinity of a site where coastal protection works or maritime structures have to be built. Two different methodologies were presented, applied and compared in this paper: i) The methodology TRANSFER, Coli *et al.* (2002), which consists in using an irregular wave propagation model for offshore transfer of sea wave data or regimes from a coastal buoy site where available data are sufficient to establish a convincing sea wave regime, followed by a new transfer from offshore to the coastal site(s), not too distant from the original site, where the wave regime(s) is (are) sought; and ii) The numerical model MAR3G, Pires and Carvalho (1996), a third generation wind wave hindcast model plus a refraction model. Both methodologies were applied to four different locations on the Portuguese coast.

The comparison of the computed wave regimes at the different points, using both methodologies, led to the following general conclusions:

- Both methodologies (TRANSFER and MAR3G) show a similar trend for HS, TZ and Θ, but HS, TZ and Θ ranges obtained using TRANSFER methodology are, in general, larger than the MAR3G ones. MAR3G is found to overestimate values for the lower classes of HS.

- For wave heights, although both regimes show somewhat similar behaviours along time, it is apparent that the MAR3G methodology usually produces significantly higher values than the TRANSFER methodology.
- For wave directions, both regimes show identical distributions. Spectral mean wave directions in both regimes are similar. MAR3G ranges are usually narrower than IH-TRANSFER ranges.
- For the wave periods, quite similar ranges were found for all locations.
- For the special case of Vila Praia de Âncora (VPA) one found that although both methodologies show a similar behaviour for HS, significant differences were found for TZ and, especially, for Θ, using the two methodologies. This may be explained by the fact that the VPA point is located in relative shallow waters (10.3 m) and in a fairly sheltered region, where significant effects (other than the ones considered on the used models) become important and therefore are not taken into account in both methodologies.

The significant differences found between both methodologies are due to the different approaches used:

- The TRANSFER methodology uses measured, real, wave data, which is dependent on the quality of the observed wave data, collected using the available wave buoys (sometimes with many gaps), generally available on a distant location. The wave propagation model has its inherent associated errors
- The MAR3G methodology uses wind data, not real wave data, and the results strongly depend on the precision of the hindcast model used. Of course, the hindcast numerical model also has its inherent associated errors.

Even though there are significant errors associated to each methodology and the complexity of the study regions is significant, the results show nevertheless a global satisfactory agreement. When opportunities for other verifications are available, it will surely contribute to the improvement of both methodologies. For now, however, these methodologies are quite important as a first step to understand the wave regime in locations where there are no measured data.
As a final note, the authors of this paper consider that both methodologies are valid and they can be complementary in order to eliminate the disadvantages of each one of them.

ACKNOWLEDGMENTS

The authors are grateful to the financial support of the "Fundação para a Ciência e a Tecnologia" (FCT), through projects no. POCTI/ECM/32446/2000, POCTI/EME/47576/2002 and grant no. SFRH/BD/12488/2003, and of MEDIRES project, funded by the Portuguese Government.

REFERENCES

Capitão, R.; Fortes, C.J.; Carvalho, M.M.; Covas, J.A. and Reis, M.T. (1994). "A Method for obtaining deep water wave characteristics based on the local sea wave climate". "Seminário Luso Francês de Modelação em Hidráulica Marítima", Coimbra.

Coli, A.B.; Santos, J.A.; Fortes, C.J.; Capitão, R.; Carvalho, M.M. (2002). "Metodologia de propagação de regimes de agitação marítima do largo para a costa: análises dos modelos BACKTRACK-REFSPEC e SWAN". 6º Congresso da Água, Porto, Portugal (in Portuguese).

Covas, J.A. and Pontes, M.T. (1988). "Modelo matemático de refracção de ondas irregulares", Joint Report LNEC/LNETI, Lisboa (in Portuguese).

Pires, H.O.; Carvalho, F. (1996). "Fundamento e características do modelo de ondas de terceira geração MAR3G. Technical Note, Instituto de Meteorologia, Lisbon (in Portuguese).

LONG-TERM PREDICTION OF NEARSHORE WAVE CLIMATE WITH AN APPLICATION TO CLIFF EROSION

Peter Stansby[1] , Jianguo Zhou[2], Cuiping Kuang[3], Mike Walkden[4], Jim Hall[4], Mark Dickson[5] , in the Tyndall Centre for Climate Change Research

Nearshore wave climate determines coastal erosion and flooding, along with various other factors. The nearshore wave climate is determined by wave propagation from offshore (where water is 'deep'). Offshore wave conditions are simply defined by the wind speed and direction for this study. The nearshore wave height, period and direction is then parameterised in terms of wind climate to enable efficient long-term prediction. This is applied to various sites along the Norfolk coast in the UK and input to a model for cliff erosion SCAPE (Soft Cliff and Platform Erosion) for comparison with 117 years of historical cliff recession data. Predictions are quantitatively realistic.

INTRODUCTION

Nearshore wave climate is the prominent driver of coastal processes. Computational methods for wave propagation from offshore to nearshore have been developed over the years based on ray tracing. Propagation of wave action, rather than energy, is now generally used as this is conserved in the presence of currents for inviscid flow. For the propagation of random directional waves into shallow coastal regions from offshore, all the following effects are accounted for: wave generation by wind, shoaling, refraction due to depth variation, refraction due to currents, energy loss due to whitecapping, energy loss due to bed friction, depth-induced wave breaking, quadruplet wave-wave energy transfers, triad wave-wave energy transfers. In this study TOMAWAC (Benoit 2002) was used to take advantage of flexible finite-element meshing. Application to a coastal region off East Anglia, similar to the subject of this paper, has previously shown that the choice of broad or narrow spread spectra had negligible effect on nearshore wave climate, defined by the significant wave height H_s, significant period T_s and dominant wave direction θ (Kuang and Stansby 2004). That study also showed that nearshore wave climate was little affected by tidal currents, which were predominantly parallel to the shore, and by wave generation by wind over the field of propagation, which extended about 30km offshore, for wind speeds less than about 20m/s. Also wave-wave interactions had negligible effect. For this region of open coast nearshore wave climate is thus mainly dependent on offshore wave climate (in deep water), bathymetry and tidal level. This will be seen to enable the development of a method suitable for efficient long term nearshore wave climate predictions which

---

1 School of Mechanical, Aerospace and Civil Engineering, University of Manchester, UK
2 now at Department of Engineering, University of Liverpool, UK
3 now at School of Civil Engineering, Tongji University, Shanghai, China
4 School of Civil Engineering and Geosciences, University Newcastle, UK
5 NIWA, Christchurch, New Zealand

will here be used as input into cliff erosion prediction. It should be emphasised that such a limited dependence for wave climate should be assessed for each coastal region of interest. For example it is unlikely to be valid for regions with prominent estuaries.

Long term prediction requires typically hourly offshore wave conditions as input and in the future these should be available from regional climate models (RCMs). However for a long term historical cliff erosion study RCM data is not available. An alternative pragmatic method for defining offshore wave conditions is to use onshore wind data which has been measured by the Meteorological Office for over 100 years. The basic assumption is that wind is blowing offshore with the same speed and direction and the standard formulae for significant wave height $\frac{gH_s}{U^2}=0.283$ and period $\frac{gT_s}{2\pi U}=1.2$ may be applied, assuming a long fetch where H_s is the significant wave height, T_s is the significant wave period, U is wind speed and g is gravitational acceleration (Shore Protection Manual 2001). We still have the problem that long term prediction of say 100 years comprises about 10^6 hourly periods and each wave propagation run takes about 15mins on a modern PC. This is clearly unmanageable even with parallel processing. Here we propose a scheme where combinations of offshore H_s, T_s, θ and tidal level are run covering the ranges of possible values at regularly spaced increments, giving the inshore wave climate at specified sites. This requires about 1000 combinations which took about 10 weeks on a single PC. The inshore H_s, T_s and θ are then determined from offshore values, defined by onshore wind data, by curve fitting and interpolation.

In this paper the procedure is described and the resulting nearshore wave climate for locations on the north East Anglian coast is compared with some measurements from September 2002 to September 2003 which have recently been made available by the UK Environment Agency. Long term onshore wind data for the past 117 years is thus used to provide nearshore wave climate adjacent to Sub Cell 3B which is used as input to the SCAPE model for the prediction of cliff erosion.

DATA ANALYSIS

The aim is to determine, for each nearshore site, significant wave height H, period T (omitting the subscript s to avoid confusion below) and direction θ as a function of offshore values, H_o, T_o, θ_o and tidal level. Wave propagations are run for combinations of H_o, T_o, θ_o and tidal level, covering the possible parameter ranges. A possible approach would then be to obtain nearshore values by interpolation. However a more convenient approach is to represent the variation of H with H_o as a power series with coefficients dependent on T_o, for different θ_o and tidal level. This reduces the interpolation from four variables to three which is easier to manage. This is possible because even a quadratic power series

$$H(\theta_o) = a_0(T_o) + a_1(T_o)H_o + a_2(T_o)H_o^2 \qquad (1)$$

fits the data quite well.

In the same way nearshore T is represented as a quadratic series with coefficients dependent on H_o such that

$$T(\theta_o) = b_o(H_o) + b_1(H_o)T_o + b_2(H_o)T_o^2. \qquad (2)$$

Finally, for each H_o and tidal level, the nearshore direction θ is represented as a quadratic series in θ_o with coefficients dependent on T_o such that

$$\theta(H_o) = c_0(T_o) + c_1(T_o)\theta_o + c_2(T_o)\theta_o^2 \qquad (3)$$

This is chosen because θ is considered to be more dependent on T_o than H_o and it is desirable for interpolation that values show as little variation as possible.

The process is thus as follows. Three tidal levels are chosen: high water spring, mean water level and low water spring. (All levels are relative to chart datum or LAT.) Cromer is the nearest reference point for tidal level. For each level, offshore wave directions are chosen at 45° intervals, with northerly as the origin, $\theta = 0$. Offshore wave height is divided into 1.5m intervals for waves between 2m and 8m and 4m for large waves up to 24m. The '*a*' coefficients for nearshore wave height are determined for each offshore period, divided into 3s intervals for the smaller waves and 4s for the larger waves. The '*b*' and '*c*' coefficients are determined in a similar way.

For a given time and date, the water level everywhere is assumed to be the tidal level at Cromer, available from the BODC (British Oceanographic Data Centre). This defines the two levels to be used for a linear interpolation. This time and date defines the onshore wind speed and direction which in turn defines the offshore significant wave height H_o and period T_o. Data for the two directions just greater than and less than the actual value are to be used for the two water levels. For these two water levels and these two directions the '*a*' coefficients are selected for wave periods just greater than and less than the actual value. '*a*' values are then interpolated linearly for the actual period. Nearshore H is thus obtained from Eq.1 for the two wave directions and two levels. The H values for the actual direction are then obtained by linear interpolation for the two levels. Finally the H value for the actual level is obtained by linear interpolation.

There is a similar procedure for period. Data for the same water levels and direction are chosen and the '*b*' coefficients are selected for wave heights just greater than and less than the actual value. '*b*' values are then interpolated for the actual wave height. Nearshore *T* is thus obtained from Eq.2 for the two wave directions and water levels. The *T* value for the actual direction is then obtained by linear interpolation for the two levels. Finally the *T* value for the actual level is obtained by linear interpolation.

Finally the nearshore direction is obtained following the same procedure as for *H*.

RESULTS

A map of the area of investigation is shown in Fig.1.

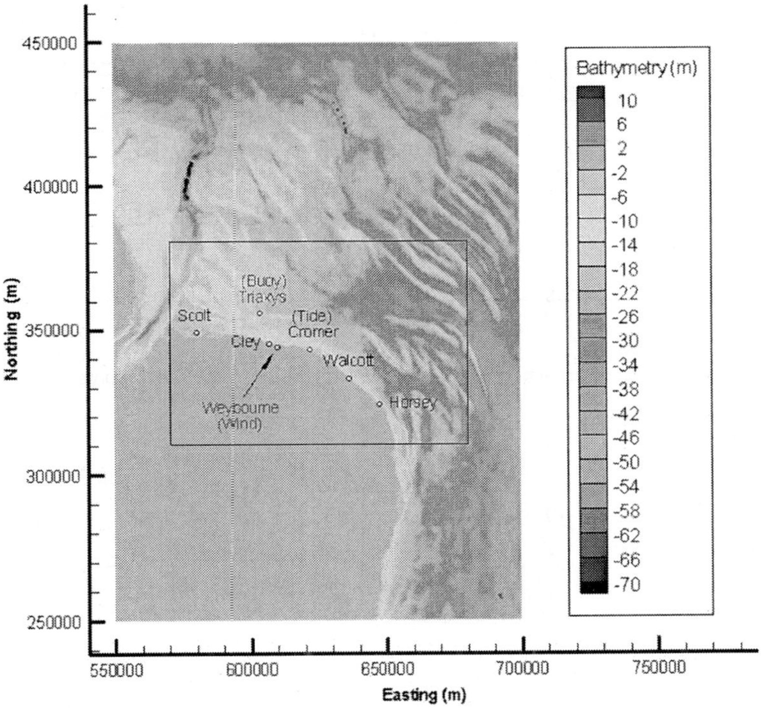

Fig.1 Sketch of region showing computational domain (box) and bathymetry

Hourly data for significant wave height, period and direction are available for the Triaxys wave buoy and for wave height and period only at the nearshore sites of Scolt, Cley, Walcott and Horsey, using bed mounted micro/macrowave sensors. Wind data are available for Weybourne. As stated above tidal levels are available for Cromer and for a given time are assumed to be the horizontal level for the whole domain. The region for wave propagation computation is shown by the rectangle and is covered by about 20000 cells with typical nearshore dimension of about 100m and an offshore dimension of 1000m.

The errors or the difference in H_s between measured and predicted values, as a function of direction, are constructed by splitting up the field into increments of 0.5m in wave height and 30^0 in direction and counting the number of occurrences in each block. This number divided by the total gives the probability of occurrence within that block, leading to contour plots of the probability. The same procedure is followed for period T_s, dividing into increments of 0.5s, and direction θ, only available for the Triaxys buoy.

The Triaxys results for wave height, period and direction are shown in Figs. 2(a), 2(b) and 2(c) respectively.

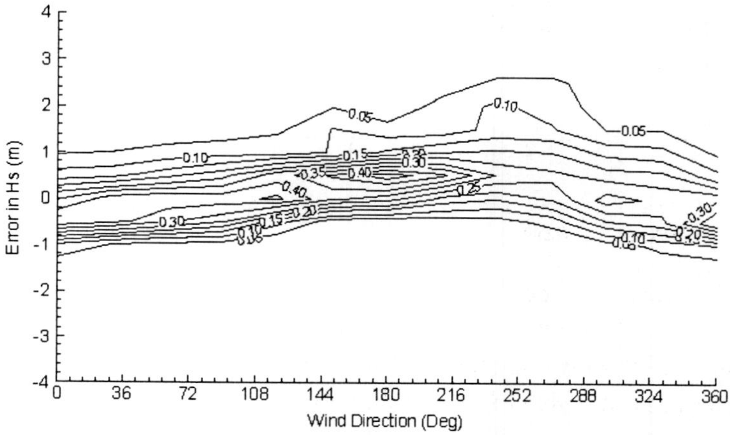

Fig.2(a) Probability of errors in H_s versus wind direction for the Triaxys buoy in 17.5m water depth.

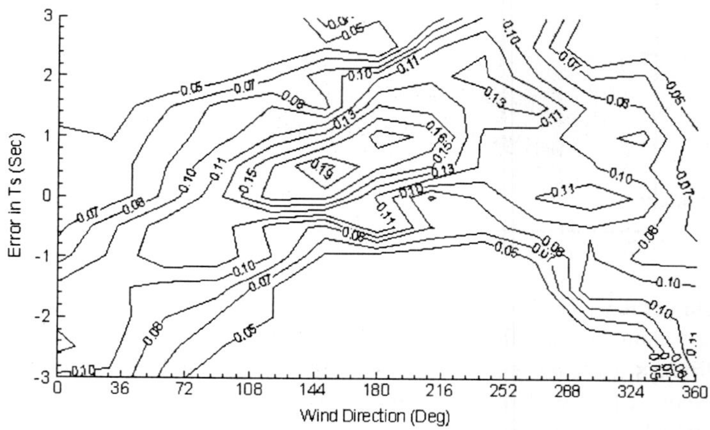

Fig.2(b) Probability of errors in T_s versus wind direction for the Triaxys buoy in 17.5m water depth.

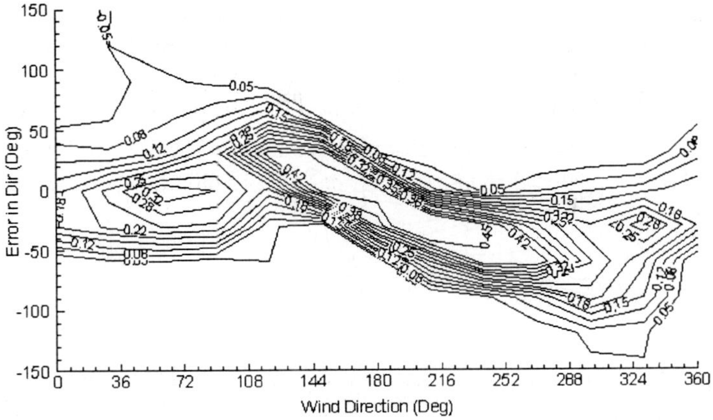

Fig.2(c) Probability of errors in wave direction versus wind direction for the Triaxys buoy in 17.5m water depth.

The mean water depth is 17.55m relative to chart datum. Probabilities of errors in H_s show a little spread about zero while there is more scatter for T_s. For direction there is some skewness in the errors. The errors in H_s and T_s are then shown for Walcott in Figs. 3(a) and (b).

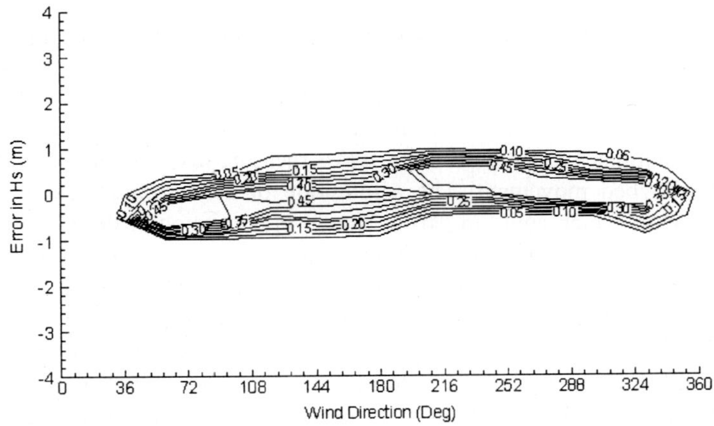

Fig.3(a) Probability of errors in H_s versus wind direction at Walcott in 8.5m water depth.

For Scolt, Cley, and Horsey the plots have quite common features and are not shown. For the nearshore sites the water depth obtained from the water level at Cromer and the bed level is compared with the directly measured values. Our assumption about water depth can be in error by up to 2m in a depth range of about 6m to 11m generally. The errors in H_s are generally quite small while

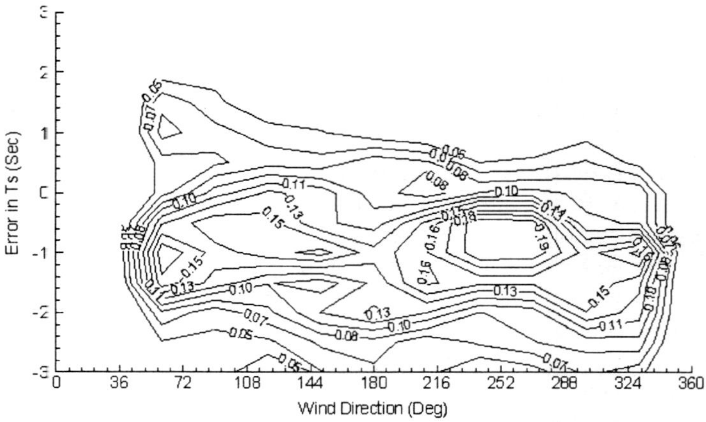

Fig.3(b) Probability of errors in T_s versus wind direction at Walcott in 8.5m water depth.

those for T_s are more scattered with most probable values about 1s below those measured. The ratio of measured $H_{\max}$ to H_s was remarkably close to 1.9 for all cases. The maximum ratio of $H_{\max}$ to depth was about 0.5 and waves were generally not depth limited.

There are many sources of error in this procedure. The basic input data is bathymetry and depth is generally not well predicted. Errors of around 1m are quite common and errors greater than 2m can occur. The bathymetry input was for 2002/3, provided as a digital form of Admiralty charts by METOC Ltd. (now SeaZone Solutions Ltd.), as was the wave measurement but this could have changed with bed movement. The assumption of a horizontal tidal level has uncertain errors which could in part be due to the omission of the effect of tidal currents. Tidal model runs were undertaken in Kuang and Stansby (2004) for a similar region and tidal levels at Lowestoft were predicted within ±6% suggesting that tidal currents do in fact have a significant influence but this would need to be checked for many locations. Storm surge and set up will also have an influence particularly in strong onshore winds which could also be assessed. Another major uncertainty concerns the generation of offshore wave climate by onshore wind data. This may be improved for recent and future times by inputting offshore boundary conditions from a RCM.

CLIFF EROSION

The approach described above allows the representation of wave transformation over very long periods. This capability was developed to generate input to a SCAPE model of the 50km length of coast between Winterton Ness and Blakeney (Fig. 4).

SCAPE, which is described in detail by Walkden & Hall (2005), models a rock shore profile (platform and cliff) with an overlying beach. It can be used to simulate a quasi-3D coast by assembling a series of profiles alongshore and allowing them to interact. SCAPE describes a broad system of processes in simple terms, including shoaling and refraction (linear wave theory), alongshore sediment transport (one-line approach), erosion of the shore platform and lower cliff, cliff slope stability and talus formation. These system components interact and regulate shore evolution in response to the hydrodynamic drivers (wave action, tides and sea-level rise). Two principal feedback pathways govern the shore response: (1) shore platform and cliff recession leads to (→) sediment release → beach building → reduced shore recession; and (2) high rates of cliff recession results in a more elevated shore platform → wider surf zone → reduced cliff recession. In (1) the recession rate is governed by beach denudation through gradients in sediment transport. In (2) the recession rate is determined by the rock strength and is governed by fluctuations in the rock profile shape.

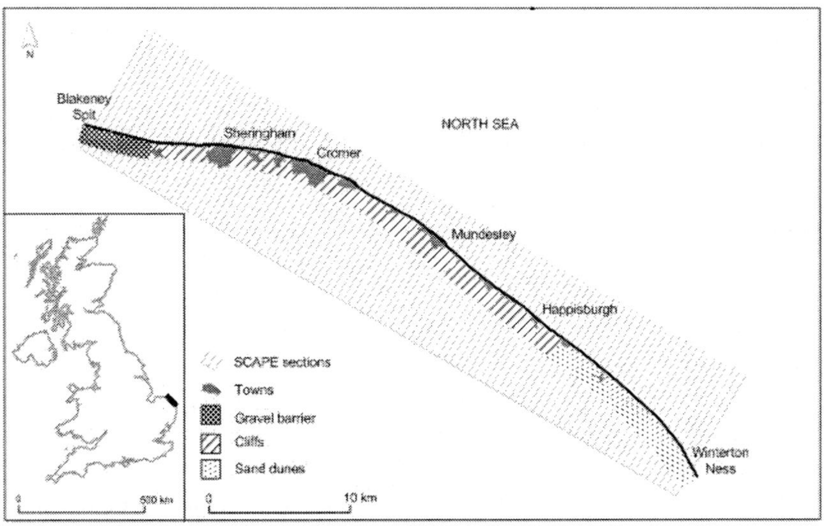

Fig. 4. Map showing the shore model profile locations

This region, and the model assembly, is described in detail by Dickson et al.(2006). Most of this coast has a rock profile (cliff and shore platform) with a perched beach, although the cliffs are replaced by dunes in the south and a coarse-sediment beach barrier in the north. The rock in the centre and south of the study region is composed of soft glacial tills, whilst in the north it is chalk. The transition area is between Weybourne and Cromer, and roughly coincides with the location of a divide in the direction of longshore sediment transport (Vincent, 1979; Clayton, 1989; Chang and Evans, 1992). The drift divide is advantageous in modelling terms as it results in net sediment flux out of the longshore model boundary regions.

SCAPE profiles are not defined by the user, they emerge as dynamic equilibrium forms from the modelling process, typically in several stages over several thousand years (Walkden and Hall, 2005, Dickson et al. 2006). For the Norfolk model this was achieved by first modelling a single profile without a beach, using realistic tides and rate of sea-level rise with offshore wave conditions. Once this profile had achieved dynamic equilibrium it was reproduced one hundred times at 500m intervals to represent the length of the study region, as shown in Fig.4. The inshore wave conditions were then input to the model through transfer functions derived using the method described above. These specified the nearshore wave conditions at four points around the coast each with a water depth of 15m below mean sea-level. The quasi 3D model was then run and a beach was allowed to form. Once the beach had emerged, its volumes had stabilized and the rock profiles had adapted to it, the model was ready for calibration.

In SCAPE in-situ rock strength (R) is treated as a calibration term that is established by comparing model recession to observed cliff recession rates. For this study the variation in rock type near Cromer necessitated two values for R in order to represent the chalk and glacial-till lithologies. The calibration process began with a model of the natural, i.e. undeveloped coast. The resistance values were chosen so that erosion rates approached 1 m yr^{-1} around Cromer. Although R was constant for the glacial till, an alongshore gradient in erosion rate emerged south of Cromer due to changes in factors such as wave conditions, local sediment transport, beach volume, and profile shape. By contrast, lesser erosion rates in the vicinity of Sheringham and Weybourne were largely attributable to the higher resistance value chosen to represent chalk within the cliffs.

The second calibration term is the coefficient of longshore sediment transport in the CERC equation, which was set at 0.77 (the upper limit of the range recommended by Hanson and Kraus, 1989) for all model sections, on the basis of previous studies of longshore sediment transport on this coast (HR Wallingford, 2002, Walkden & Hall, 2006). Additional model parameters included representations of isobath shape and an approximation of the initial shoreline plan shape.

This region has a long history of coastal engineering interventions, which have had a profound effect on the pattern of erosion. Groyne construction began at Mundesley in 1860, now seawalls palisades and groynes protect around 70% of the coast. These are represented in simple terms in the model: groynes are represented as reductions in the CERC coefficient, except when the beach is wide enough to bypass them, palisades reduce the heights of waves passing through them by one-half, and seawalls prevent the erosion of the cliff toe behind them.

Fig.5 shows the recession rates simulated by SCAPE for three periods between 1885 and 2002. These periods were selected to match the years for which good shore location data was available from historic maps, which were analyzed by the British Geological Survey, and a LIDAR survey. In general there is good agreement between historical measured rates and those predicted by the model in the central region where glacial-till cliffs occur (indicated by the solid black line). The alongshore variability in recession rates results largely from the impact of coastal defenses, which is particularly apparent around townships such as Cromer and Sheringham.

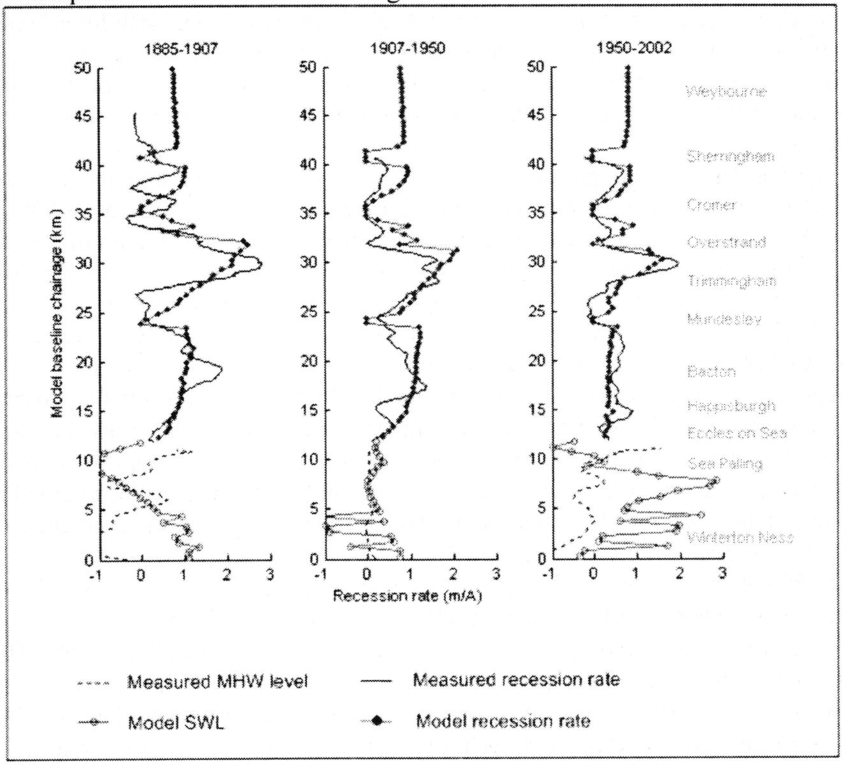

Fig.5. Comparison between measured and modeled shore recession rates between 1885 and 2002 (from Dickson et al, 2006)

CONCLUSIONS

Century scale predictions of the evolution of coastal regions are needed to inform coastal management. Numerical models used to generate such predictions must be driven by similarly long descriptions of wave conditions over large areas. Sophisticated models exist that can describe the hydrodynamic processes well; however the run times involved are prohibitively long. Here a method is presented to enable long term nearshore wave climate to be predicted from onshore wind data. The basic physical assumptions are that wave propagation is decoupled from tidal currents, that wind is blowing offshore with the onshore direction over a very long fetch, and that tidal level is horizontal across the domain of interest. The coastal region is effectively calibrated for the range of possible offshore wave conditions determined by the onshore wind speed and direction. Comparisons with measured data at four nearshore sites show that significant wave height is predicted with little probability of error although the depth assumption can show consistent errors of over 2m. The probability of error in significant wave period is rather more scattered with the most likely values generally underestimated by about 1s. There are many sources of error in such a scheme but it is suggested that sensitivity to errors in depth should be assessed first.

Long-term historical wave climate has thus been input into the SCAPE model of Walkden and Hall (2005) and Dickson et al (2006) for Sub Cell 3b of the East Anglian coast. Good predictions of historical cliff recession rates along the coast, provided by the British Geological Survey, over a 117 year period have been shown.

ACKNOWLEDGEMENTS

Support from the Tyndall Centre for Climate Change Research for the Coastal Simulator is gratefully acknowledged. The Tyndall Centre is funded by the three UK Research Councils: NERC, EPSRC and ESRC. Wave data was made available by the Environment Agency and tide data by the British Oceanographic Data Centre (BODC). Wind data were obtained from the Meteorological Office.

REFERENCES

Benoit, M. 2002. TOMAWAC software for finite element sea state modelling – Release 5.2 theoretical note. HP-75/02/065/A. (available from HR Wallingford)

Chang, S.-C. and Evans, G. 1992, Source of sediment and sediment transport on the east coast of England: Significant or coincidental phenomena? *Marine Geology*, 107, 283-288.

Clayton, K.M. 1989, Sediment input from the Norfolk Cliffs, Eastern England - A century of coast protection and its effect, *Journal of Coastal Research*. 5, 433-442.

Dickson, M.E., Walkden, M.J., Hall, J.W. 2006, Systemic Impacts of Climate Change on an Eroding Coastal Region over the 21st Century, *Climatic Change*, in press.

Hanson, H. and Kraus, N.C. 1989. GENESIS - Generalized model for simulating shoreline change, Vol. 1: Reference Manual and Users Guide. Technical Report CERC-89-19, Coastal Engineering Research Center, p. 247.

HR Wallingford 2002, Overstrand to Mundesley Strategy Study: Littoral Sediment Processes. HR Wallingford Ltd. Report.

Kuang,C.P. and Stansby,P.K. 2004 Efficient Modelling for Directional Random Wave Propagation Inshore, *ICE J. Maritime Engineering*, 157(MA3), 123-131.

Shore Protection Manual, 2001, US Army Coastal Engineering Research Centre.

Vincent, C.E. 1979, Longshore sand transport rates - a simple model for the East Anglian coastline. *Coastal Engineering*, 3, 113-136.

Walkden, M.J.A. and Hall, J.W. 2005 A predictive mesoscale model of soft shore erosion and profile development, *Coastal Engineering*, 52(6), 535-563.

Walkden, M.J.A. and Hall, J.W. 2006 A mesoscale predictive model of the evolution and management of a soft rock coast. under review with *Journal of Coastal Research.*

MEDIUM-TERM WAVE AND CURRENT MODELLING FOR A MESOTIDAL WADDEN SEA COAST

Agnieszka Herman, Ralf Kaiser and Hanz D. Niemeyer[1]

The paper presents an effective and accurate approach towards a medium-term (~decades) modeling of circulation and wind wave processes in a shallow tidal sea. The method is based on a high-resolution modeling with the hydrodynamic model Delft3D and a spectral wave model SWAN, combined with a principal component analysis/neural network technique. The approach enables a very effective simulation of the analyzed phenomena and is shown to give satisfactory results in the whole study area. It enables, among other things, to generate input data for morphodynamic modeling, for studies concerning the concept of a morphologically representative wave or the load on coastal protection structures.

INTRODUCTION

The primary purpose of the study is to produce time series of water levels, currents and mean wave parameters suitable for medium-term (~decades) statistical analysis of the temporal and spatial variability of those quantities in the catchment area of the tidal inlet Norderneyer Seegat (East Frisian Islands, Germany; Fig.1) and its surroundings. The results can be used later e.g. in the analysis of sediment transport processes or in investigations concerning safety of various coastal protection structures. One of the possible applications of the database obtained within the project is related to the concept of a morphologically representative wave climate, analogous to a morphologically representative tide (Latteux 1995) used traditionally as a data-reduction tool in long-term morphodynamic modeling in coastal regions.

Earlier studies concerning the neural network (NN) modeling of coastal processes concentrated on their simulation at *single locations*, see e.g. Lee and Jeng (2002), Tsai *et al.* (2002), Huang *et al.* (2003), Lee (2004), Kalra *et al.* (2005) or Makarynskyy *et al.* (2005). Contrary to these works the approach presented below enables – thanks to the combination of the principal component analysis (PCA) with neural-network modeling – to reconstruct the analyzed parameters in the *whole study area.* Moreover, this can be achieved very effectively, without increasing the complexity of the structure of the applied neural networks in comparison to simulating time series at single locations. As will be demonstrated further, the accuracy of the PCA/NN modeling in the study area is comparable with the one obtained by the above cited authors at single locations. Thus, the approach presented here can be applied as a trustworthy

[1] Coastal Research Station, Lower Saxony Water Management, Coastal Defence and Nature Conservation Agency, An der Mühle 5, 26548 Norderney, Germany

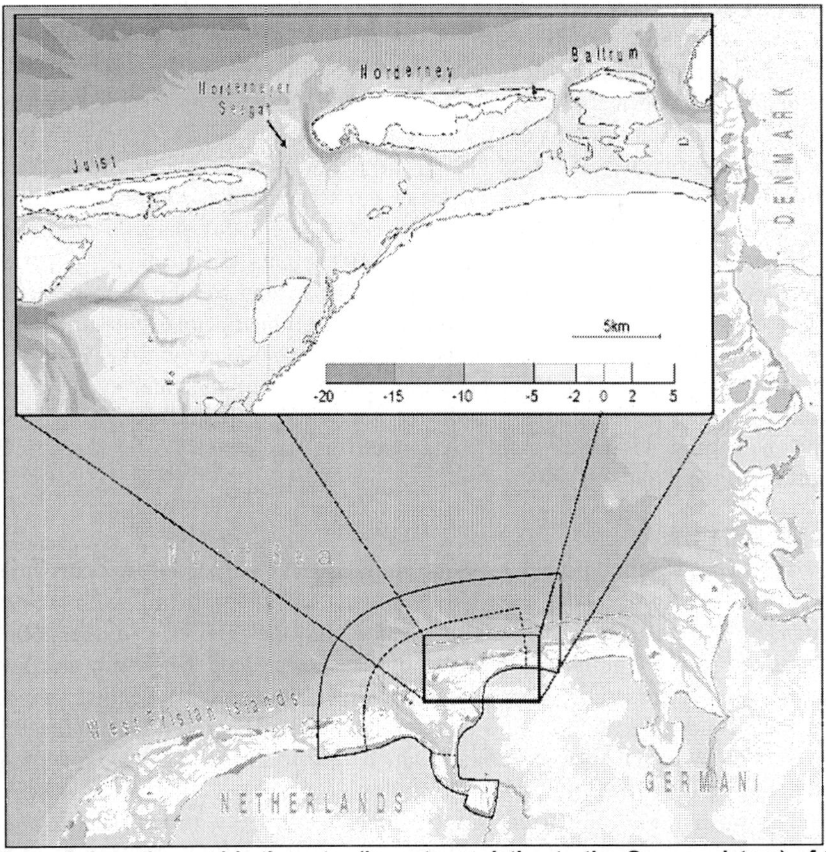

Figure 1. Location and bathymetry (in meters relative to the German datum) of the study area. Continuous and dashed line on the map show the boundaries of the Delft3D and SWAN model grid, respectively.

complementary tool to a computationally extremely time consuming hydrodynamic modeling.

The work presented here is a part of a project MOSES ("Modelling of the medium-term wave climatology at the German North Sea coast"), one of the purposes of which was to produce a database of water levels, currents and mean wave parameters in the German Wadden Sea, covering the period 1962-2002 (time range of the HIPOCAS data used as a source of boundary conditions in MOSES, see below). Because of time constrains only a few years could be simulated with high-resolution hydrodynamic and wave models: years 1962-1963 (used as a source of data for PCA/NN) and 1985, 2002 (used for verification of the PCA/NN modeling). The trained neural networks were then applied to produce the data in the remaining time intervals.

AREA DESCRIPTION

The study area, situated on the German North Sea coast (Fig.1), is marked by all geomorphologic features characteristic for this type of coastline: tidal flats occupying most of the Wadden Sea area and falling dry at low water, deep tidal channels (with water depths over 20 m), inlets and ebb deltas. According to the classical classification of Hayes (1979) the region is subject to upper mesotidal conditions. The amplitudes of the three most energetic tidal components M2, S2 and N2 at the entrance to the Norderneyer Seegat equal about 100 cm, 28 cm and 17 cm, respectively. Mean significant wave heights and mean energy periods to the north of the islands reach about 1.1-1.2 m and 5.5-5.6 s, respectively. Seawards from the islands the conditions are determined primarily by the larger-scale situation in the North Sea. By typical calm to moderate weather conditions the local wind and water level (and thus the phase of the tidal cycle) play a dominating role in the sheltered areas of the Wadden Sea. Only during strong wind events the wind waves from the open sea can penetrate far through the inlets up to the mainland coast.

MODEL SET-UP

The hydrodynamic processes in the study area are simulated on a curvilinear grid (Fig.1; resolution in the Norderneyer Seegat: ~50 m) with a two-dimensional (vertically integrated) version of the Delft3D model (Delft Hydraulics 2003). The model is driven by the interpolated 1-hourly water level time series from the HIPOCAS dataset (Weiße *et al.* 2003) applied as open boundary conditions and by wind fields obtained from a wind atlas produced by the German Weather Service (DWD), which is a structured database for a large number of model runs.

The water level and current velocity fields obtained with Delft3D, together with 3-hourly time series of wave energy spectra from HIPOCAS (results of the North Sea modeling with the WAM model) and wind fields from the wind atlas, are used as input data for the third-generation spectral wave model SWAN (Booij *et al.* 1999). SWAN is used in nonstationary mode with a time step of 10 minutes on a smaller curvilinear grid with the resolution in the Norderneyer Seegat of ~30 m.

Both models have been applied successfully to various modeling problems in a number of projects conducted at the Coastal Research Station in the last years, see e.g. Kaiser and Niemeyer (2001), Niemeyer and Kaiser (2001). The verification studies carried out within this project confirmed the ability of both Delft3D and SWAN to accurately reproduce the most important aspects of the wave and current processes in the study area. An example of modeled and measured time series of the significant wave height and the mean wave period at one of the analyzed measurement stations is shown in Fig.2. For more examples of the verification of Delft3D and SWAN modeling see Herman *et al.* (2006a,b).

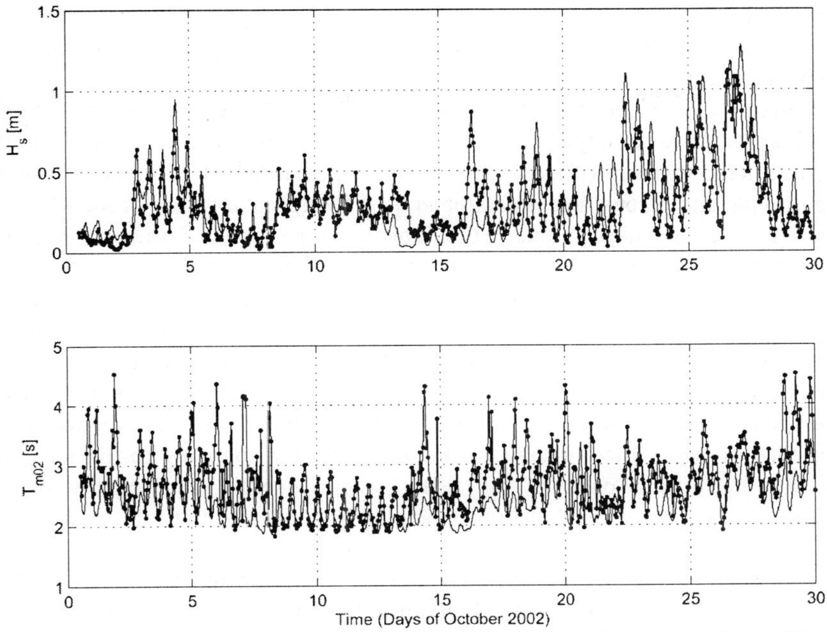

Figure 2. Time series of measured (dots) and modeled with SWAN (lines) significant wave height (above) and mean wave period (below) at the measurement station SGT-NEY in the Norderneyer Seegat

PCA AND NEURAL NETWORK MODELLING

The results of the Delft3D and SWAN simulations from the years 1962-1963, saved hourly, have been analyzed further within an area covering the catchment areas of the three tidal inlets between the islands of Borkum, Juist, Norderney and Baltrum and parts of the open sea region to the north of the islands. To reach a comparable amount of points in both datasets every 5th Delft3D grid point and every 7th SWAN grid point has been chosen for the PCA and neural network modeling. Only points remaining 'wet' during at least 10% of the time have been taken into account, resulting in 3454×17521 matrices of water levels and currents and 3463×17521 matrices of the mean wave parameters.

The distribution of variance of the five analyzed datasets among the ten leading principal components (PCs) is shown in Table 1. As can bee seen, the first PC dominates the other ones: it contains from 70.6% of the variance (mean wave directions) to over 97% (water levels). Based on the analysis of the time variability of the PCs and of the accompanying modes, five leading mode/PC pairs for each parameter have been selected for the neural network (NN) modeling. The inclusion of the further modes made the training of the networks very difficult without improvement of the modeling results. For the discussion of the possible influence of this choice on the quality of the NN modeling see Herman *et al.* (2006a).

As an example Fig. 3 shows the quality of the PCA reconstruction of the significant wave heights in the form of the spatial distribution of the reconstructed variance (δ) and of the standard deviations of differences between the original and reconstructed dataset (σ_{std}). The highest δ values occur seawards from the islands and in the deep tidal channels; the lowest δ values – over the tidal flats and in locations directly at the coast, falling dry at low water. The distribution of σ_{std} is, not surprisingly, reversed: lowest in the open sea region and highest over the tidal flats. The δ and σ_{std} fields for the other analyzed parameters (not shown) have analogous properties related, firstly, to the increasing spatial and temporal variability of the studied phenomena from the open sea region, through the tidal inlets to the Wadden Sea, and secondly, to the presence of long series of zeros in the datasets corresponding to the locations staying dry during parts of the tidal cycle. Generally, the quality of the reconstruction at a given location is positively correlated with the mean water depth at this location. As can be awaited from Table 1, the quality of the PCA reconstruction of the data (measured in terms of δ and σ_{std}) is best for water levels and worst for mean wave directions. However, all five parameters could be reconstructed with high accuracy, without loss of important information contained in the data.

Table 1. Distribution of variance (in %) among the ten leading PCs of water level (ξ), current (V), significant wave height (H_s), energy period (T_e) and mean wave direction (θ_m).

PC	ξ	V	H_s	T_e	θ_m
1.	97.177	84.471	93.956	77.815	70.594
2.	1.803	7.263	3.348	10.917	9.727
3.	0.638	1.966	0.866	1.511	2.553
4.	0.144	1.508	0.475	1.276	1.953
5.	0.052	0.677	0.260	1.053	1.046
Sum 1.-5.	**99.814**	**95.875**	**98.905**	**92.572**	**85.873**
6.	0.034	0.535	0.176	0.700	0.848
7.	0.029	0.403	0.164	0.572	0.690
8.	0.023	0.290	0.091	0.448	0.579
9.	0.017	0.229	0.080	0.407	0.517
10.	0.012	0.202	0.067	0.338	0.468
Sum 1.-10.	**99.929**	**97.534**	**99.483**	**95.037**	**88.975**

In the next stage of the modeling the datasets 1962-1963 have been divided into the training and test data (Jan.1962–Jun.1963 and Jul.–Dec.1963, respectively). For each of the five parameters a number of feed-forward, three-layer neural networks with different number of input and hidden neurons has been set-up and trained. Various combinations of the time series of HIPOCAS water level, wind and wave parameters in a *single* location have been used as

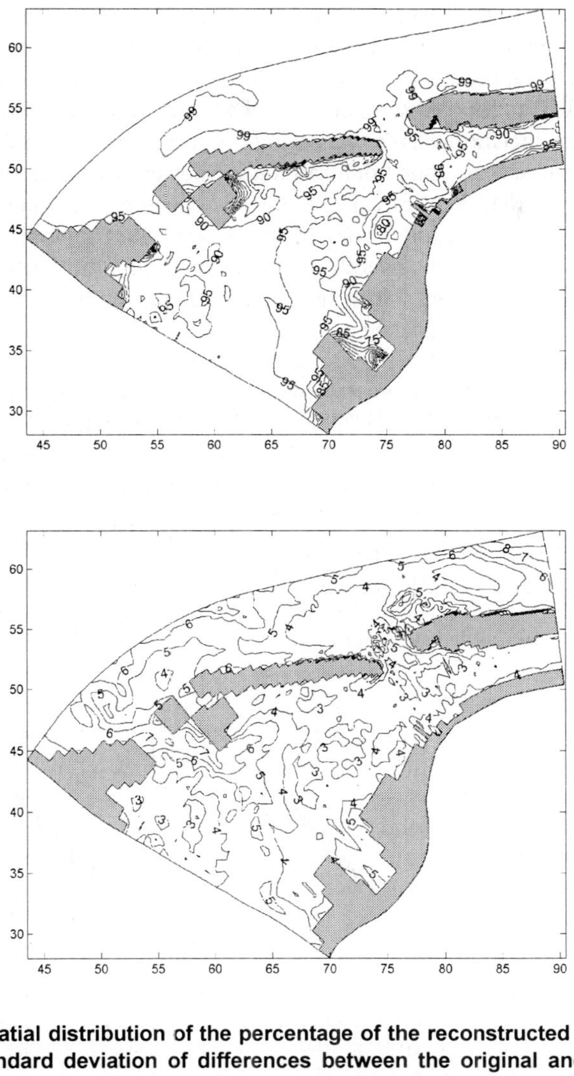

Figure 3. Spatial distribution of the percentage of the reconstructed variance δ (top) and the standard deviation of differences between the original and reconstructed data σ_{std} (bottom; in cm) for significant wave heights from the period 1962-1963.

input for the neural networks. Thus, they could have a simple, easy to train structure (a so-called multi-layer perceptron): at this stage the task has been reduced to finding the relationship between the input (HIPOCAS) and output (PCs of ξ, V, H_s, T_e and θ_m) *time series*, without having to deal with the spatial variability present in the original datasets.

From each ensemble of the trained NNs the network resulting in the lowest standard deviation of differences between the original and NN-modeled PCs has been chosen for the further usage. As an example, the performance of the NN for significant wave heights in the test period July-December 1963 is presented in Table 2. Figure 4 shows the scatterplots of the original and NN-modeled PCs of H_s. Not surprisingly, the quality of the NN modeling is highest for the first PC and decreases for the further PCs. When judging the performance of the NNs it must be remembered that, although all PCs have been normalized to have a unit standard deviation, they are 'responsible' for very different amounts of variance (Table 1). The reconstruction errors of the further PCs have a significantly smaller influence on the overall performance of the modeling. The reconstruction of the first PCs with the highest possible accuracy is from this point of view of crucial importance.

Table 2. Statistical parameters describing the NN performance for the PCs of significant wave heights in the test period Jul.-Dec.1963

	PC1	PC2	PC3	PC4	PC5
Corr.coef. (%)	99.49	98.97	96.37	88.10	95.01
σ_{std} (-)	0.111	0.141	0.266	0.461	0.299
Mean diff. (-)	0.007	0.000	0.001	0.025	0.006

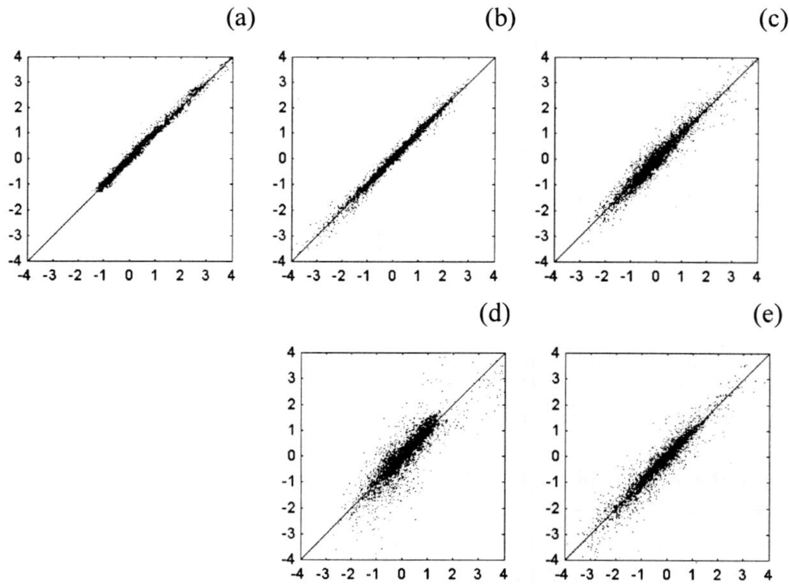

Figure 4. Scatterplots of the original (horizontal axes) and modeled with NN (vertical axes) leading five (a-e) PCs of significant wave heights. All PCs have been normalized to have unit standard deviation.

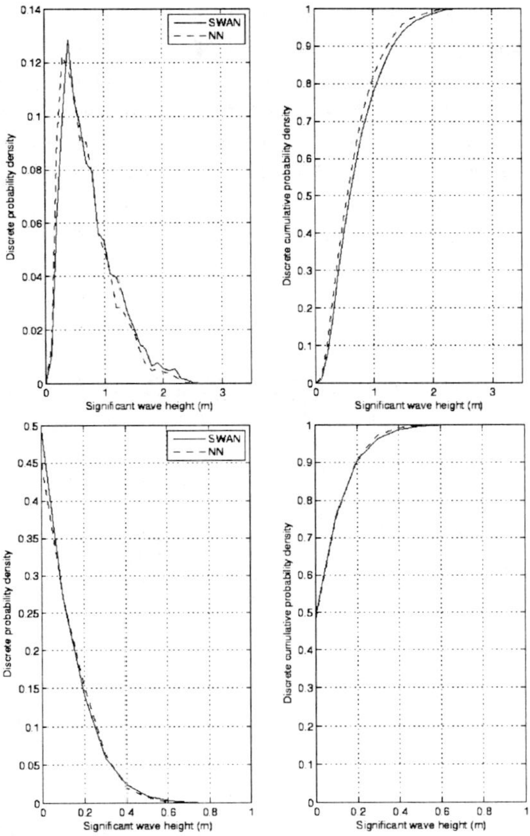

Figure 5. Discrete probability distribution of significant wave heights at a station seawards from the Norderneyer Seegat (above) and in the Wadden Sea (below), modeled with SWAN and with the PCA/NN technique.

The PCs modeled with the NNs, together with the spatial modes resulting from the PCA of the 1962-1963 data, have been then used to reconstruct the Delft3D and SWAN results from the years 1985 and 2002. The quality of the NN modeling can be analyzed – exactly as in the case of the quality of the PCA reconstruction (Fig. 3) – in terms of the δ and σ_{std} parameters. The overall pattern of δ and σ_{std} is of course very similar to those shown in Fig. 3, although δ has slightly lower and σ_{std} slightly higher values. Another measure of the NN-modeling performance is the ability of the networks to reproduce the probability distribution of the analyzed quantities at various locations in the study area. The probability distribution of significant wave heights at two selected locations – to the north of the Norderneyer Seegat and in the Wadden Sea – modeled with

SWAN and with the PCA/NN technique is shown in Fig. 5. At other analyzed locations the agreement between the curves obtained from the SWAN and PCA/NN data was comparably good. The mean wave direction is the only parameter that could not be reproduced with high accuracy at sheltered locations on the tidal flats. The NN performed well only in points with relatively narrow directional distribution (seawards from the islands and in the tidal channels).

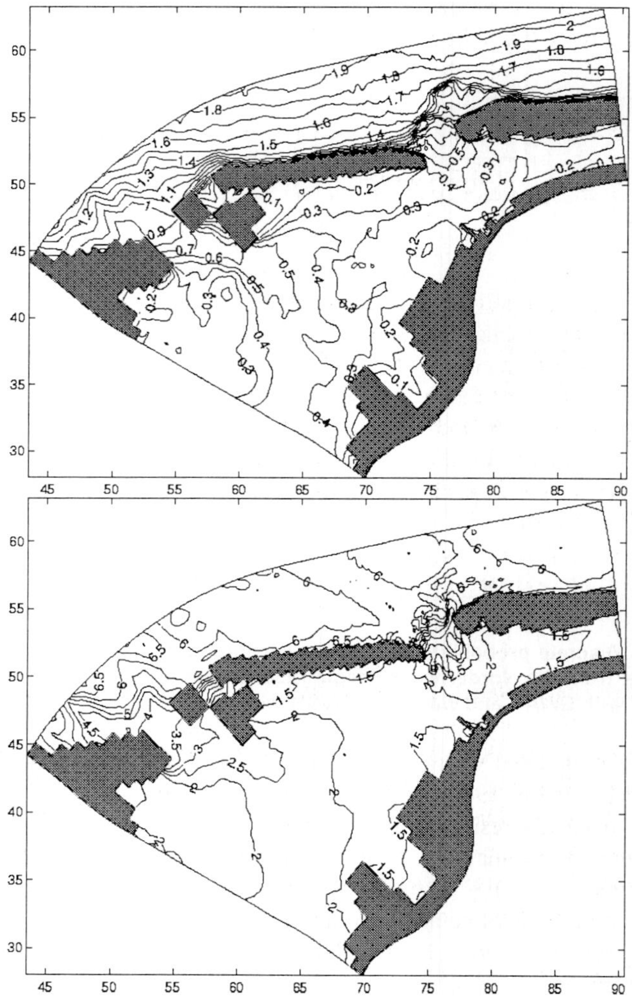

Figure 6. Spatial distribution of the 90th percentile of significant wave heights (above; in meters) and energy periods (below; in seconds) in the study area.

FURTHER APPLICATION OF THE DATA

One of the most straightforward field of application of the database produced within MOSES is the analysis of the medium-term wave and current climate in the study area. It will enable, among other things, to investigate the load on the coastal protection structures at the coast of the islands and the mainland. Estimation of the probability of exceeding certain values of the quantities of interest etc. is particularly important from this point of view. As an example, Fig. 6 shows the spatial distribution of the 90th percentile of significant wave height and energy period in the study area.

The data can be also applied to calculate a number of quantities used as input for morphodynamic modeling, e.g. the morphological impact (CERC 2002), the probability distribution of the wave-height-to-depth ratio H_s/h in the study area (main input parameter to various sediment transport formulae) or the representative wave orbital velocity at the bottom (see e.g. Hirschhäuser and Zanke 1999).

SUMMARY

Medium-term modeling of hydrodynamic processes in shallow coastal areas is a difficult and computationally expensive task. The very effective PCA/NN technique developed in this study has been shown to accurately reproduce the statistical distribution of the analyzed parameters both in the open sea region to the north of the islands and in the Wadden Sea. The main idea behind this approach is the universality of the spatial patterns of the analyzed quantities over time. The very good agreement between the PCA/NN results and the Delft3D/SWAN results in the test years 1985 and 2002 shows that this assumption is justified. Thus, the technique can be used as a practical tool, supplementary to the computationally expensive hydrodynamic/spectral wave modeling.

ACKNOWLEDGMENTS

The work presented here is a part of the research project MOSES ("Modelling of the medium-term wave climatology at the German North Sea coast"; project No. 03 KIS 040) financed by the German Federal Ministry for Education and Research (BMBF) via the carrier BEO, Forschungszentrum Jülich. We thank our colleagues from the Coastal Research Station and our project partners from the German Weather Service, from the GKSS Institute for Coastal Research and from the Coastal Research Laboratory in Kiel for their support and cooperation.

REFERENCES

Booij, N., Ris, R.C., Holthuijsen, L.H. 1999. A third-generation wave model for coastal regions. 1. Model description and validation. *J. Geophys. Res.*, 104, C4, 7649-7666.

CERC (US Army Coastal Engineering Research Center) 2002. *Coastal Engineering Manual. Part III.2: Longshore sediment transport*. Report EM 1110-2-1100, 113 pp.

Delft Hydraulics, 2003. *User manual of Delft3D-FLOW: Simulation of multi-dimensional hydrodynamic flows and transport phenomena, including sediments*, 497 pp.

Hayes, M. 1979. Barrier island morphology as a function of tidal and wave regime. In: Leatherman S.P. (Ed.), *Barrier islands from the Gulf of St., Lawrence to the Gulf of Mexico*, Academic Press, New York, 1-29.

Herman, A., Kaiser, R., Niemeyer, H.D. 2006a. A new approach towards modelling of a medium-term dynamics in a shallow tidal sea, based on combined physical and neural network methods. *Ocean Modelling*, submitted.

Herman, A., Kaiser, R., Niemeyer, H.D. 2006b. Medium-term wave variability in a shallow tidal sea modelled with combined physical and neural network methods. in preparation.

Hirschhäuser, T., Zanke, U. 1999. Long-term morphodynamic modelling of the Hörnum tidal basin near the island of Sylt. *Proc. 3rd German-Chinese Joint Seminar on Sediment Transport*, Taipeh, Taiwan.

Huang, W., Murray, C., Kraus, N., Rosati, J. 2003. Development of a regional neural network for coastal water level predictions. *Ocean Engng*, 30, 2275-2295.

Kaiser, R., Niemeyer, H.D. 2001. Analysis of directional spectra in shallow environment – comparison of field data and results from mathematical modelling. *Proc. Ocean Wave Measurement and Analysis*, Reston/Va., USA.

Kalra, R., Deo, M.C., Kumar, R., Agarwal, V.K. 2005. Artificial neural network to translate offshore satellite wave data to coastal locations. *Ocean Engng*, 32, 1917-1932.

Latteux, B., 1995. Techniques for long-term morphological simulation under tidal action. *Marine Geology*, 126, 129—141.

Lee, T.-L., Jeng, D.S. 2002. Application of artificial neural networks in tide-forecasting. *Ocean Engng*, 29, 1003-1022.

Lee, T.-L. 2004. Back-propagation neural network for long-term tidal predictions. *Ocean Engng*, 31, 225-238.

Makarynskyy, O., Pires-Silva, A., Makarynska, D., Ventura-Soares, C. 2005. Artificial neural networks in wave predictions at the west coast of Portugal. *Computers and Geosciences*, 31, 4:415-424.

Niemeyer, H.D., Kaiser, R. 2001. Design wave evaluation for coastal protection structures in the Wadden Sea. *Proc. Ocean ave Measurement and Analysis*, Reston/Va., USA.

Tsai, C.-P., Lin, C., Shen, J.-N. 2002. Neural network for wave forecasting among multi-stations. *Ocean Engng*, 29, 1683-1695.

Weiße, R., Feser, F., Günther, H., 2003. *A wind and wave climatology for the Southern North Sea based on numerical simulations for the period 1958-2001.* GKSS Forschungszentrum Report 2003/10, 38 pp. (in German).

A MODEL FOR DEFORMATION OF LINEAR WAVES ON A CURRENT WITH ARBITRARY REFLECTION/TRANSMISSION BOUNDARIES

Akira Watanabe, Hidetaka Kobayashi and Toshimitsu Takagi[1]

A numerical model is presented for computing the deformation of highly dispersive waves on a current on the basis of the enhanced Boussinesq-type equation set proposed by Chen et al (1998). Basic properties of the linearized equation set are shown. The present model includes a new technique to handle various fundamental boundary conditions, such as incident, reflective, radiating and transmittal boundaries, in a consistent manner by applying the Levenberg-Marquardt least-squares algorithm. Computational examples are shown for partial reflection with or without a current and for reflection-transmission both for regular and random waves in one-dimensional field and the breaking wave deformation on a current. Two-dimensional wave field around a rip current is also computed.

INTRODUCTION

Wave deformation due to the interaction with ambient currents such as tidal stream and river discharge flow is often substantial as it frequently affects various coastal environments. It is well known that if the river flow or ebb tidal stream is strong, such an opposing current causes rough chopping seas as well as big variations in wave intensity in its vicinity. Such chopping seas often hazard ship navigation, while the intense variations of wave conditions usually generate strong nearshore currents and affect sediment transport processes, bathymetric changes and ecological systems.

It is therefore important to develop a numerical model of waves under the influence of currents, which is not too complicated but gives reasonably accurate solutions even for relatively strong opposing current. In addition, a model should be applicable to random waves and be capable of treating boundaries with arbitrary reflection and/or transmission in the presence of a current. Considering these factors, we have developed a *linear* dispersive wave model, in which a newly-developed flexible boundary condition method is incorporated.

Shown in the following are the properties of the fundamental equation set employed, basic concept of the new boundary condition technique that applies the Levenberg-Marquardt least-squares algorithm, and examples of computations obtained by the present wave-current model.

HIGHLY DISPERSIVE LINEAR WAVE-CURRENT EQUATION SET

Several *linear* wave-current models were so far presented such as mild-slope models by Kirby (1984) and Ohnaka et al. (1988). They are, however, in principle applicable only to monochromatic waves, and besides we need cross-angles between wave-current directions at all points *a priori*.

Rather recently Chen et al. (1998) have presented an enhanced Boussinesq wave-current equation set, which is so highly dispersive (applicable up to deepwater) that even the wave blockage due to a strong opposing current can be treated. In addition, it is solved in time-evolutional manner. Hence to overcome the problems mentioned above, we have decided to apply their equation set in

[1]INA Corporation, 1-44-10, Sekiguchi, Bunkyo-ku, Tokyo, 112-8668, Japan

our new model after its linearization for practical use. Besides, assuming a mean flow field is given, we have set only the wave transformation due to a current as our main target and hence separated the wave-induced orbital velocity components $\boldsymbol{u}_\alpha$ from the mean flow velocity component $\boldsymbol{U}$.

A set of linearized continuity and momentum equations thus obtained is expressed as Eqs. (1) and (2):

$$\begin{aligned}&\eta_t+\nabla\cdot(h\boldsymbol{u}_\alpha)+(\alpha_1+1/2)\nabla\cdot\{h^2\nabla[\nabla\cdot(h\boldsymbol{u}_\alpha)]\}+(\alpha_2-1/6)\nabla\cdot\{h^3\nabla(\nabla\cdot\boldsymbol{u}_\alpha)\}\\&\quad-\nabla\cdot\{\beta_1h^2\nabla[\nabla\cdot(h\boldsymbol{u}_\alpha)]-\beta_2\nabla[h^2\nabla\cdot(h\boldsymbol{u}_\alpha)]\}-\nabla\cdot[\beta_1h^2\nabla\eta_t-\beta_2\nabla(h^2\eta_t)]\\&\quad+\eta\cdot\nabla\boldsymbol{U}+\boldsymbol{U}\cdot\nabla\eta-\boldsymbol{U}\cdot\nabla[\beta_1h^2\nabla^2\eta-\beta_2\nabla\cdot(h^2\nabla\eta)]=0\end{aligned}\tag{1}$$

$$\begin{aligned}&\boldsymbol{u}_{\alpha,t}+g\nabla\eta+h\{(\alpha_1+\gamma_2)\nabla[\nabla\cdot(h\boldsymbol{u}_{\alpha,t})]+(\alpha_2-\gamma_1)h\nabla(\nabla\cdot\boldsymbol{u}_{\alpha,t})\}\\&\quad-gh\{\gamma_1h\nabla(\nabla\cdot\nabla\eta)-\gamma_2\nabla[\nabla\cdot(h\nabla\eta)]\}+(\boldsymbol{u}_\alpha\cdot\nabla)\boldsymbol{U}\\&\quad+(\boldsymbol{U}\cdot\nabla)\boldsymbol{u}_\alpha+(\boldsymbol{U}\cdot\nabla)\boldsymbol{U}+(\boldsymbol{U}\cdot\nabla)\{\alpha_1h\nabla[\nabla\cdot(h\boldsymbol{u}_\alpha)]+\alpha_2h^2\nabla(\nabla\cdot\boldsymbol{u}_\alpha)\}\\&\quad-g\gamma_1h^2(\boldsymbol{U}\cdot\nabla)[\nabla(\nabla\cdot\boldsymbol{u}_\alpha)]+g\gamma_2h(\boldsymbol{U}\cdot\nabla)\{\nabla[\nabla\cdot(h\boldsymbol{u}_\alpha)]\}=0\end{aligned}\tag{2}$$

where η is the free surface elevation, $\boldsymbol{u}_\alpha$ is the wave orbital velocity at the relative level $\alpha_1=z_\alpha/h=-0.541$, $\boldsymbol{U}$ is the depth-uniform mean flow velocity, h is the local depth, and all the coefficients $\alpha_1,\alpha_2,\beta_1,\beta_2,\gamma_1$ and γ_2 are constant (See the original paper for their values).

The equation set is lengthy and seemingly complicated even after the simplifications mentioned above. Chen et al. (1998) have shown the dispersion relation or the Doppler shift corresponding to the above equation set (for one-dimensional and constant depth field) as:

$$\sigma^2=(\sigma_0-Uk)^2=\frac{\left[1+(1/9)k^2h^2+(1/945)k^4h^4\right]k^2gh}{1+(4/9)k^2h^2+(1/63)k^4h^4}\tag{3}$$

where σ_0 and σ are the intrinsic and apparent angular frequencies and k is the wave number. Hence, when we express Eq. (3) as $\sigma^2 = D\,(kh)\,k^2gh$, the phase velocity C and the group velocity C_g are given by Eq. (4).

$$C=\sqrt{ghD(kh)},\quad C_g=\frac{1}{2}\left(2+\frac{kh\cdot dD(kh)/d(kh)}{D(kh)}\right)\tag{4}$$

Using some more basic relations and the conservation of the wave action, we can derive for example change in the wavelength, L/L_0, and in the wave height, H/H_0, due to the presence of a current as functions of the relative depth, k_0h, and of the Froude number, U/C_0, both for the Airy (or the first-order Stokes) theory and for the present equation set, where L_0, $H_0 = 2a_0$, $k_0 = 2\pi/L_0$, and C_0 are the wavelength, the wave height, the wavenumber, and the wave celerity, respectively, at a location where there is no mean flow: $U=0$.

$$\frac{k_0h}{kh}-\sqrt{\frac{D(kh)}{D(k_0h)}}-\frac{U}{C_0}=0,\quad \frac{H}{H_0}=\frac{a}{a_0}=\sqrt{\left(1-\frac{k}{k_0}\cdot\frac{U}{C_0}\right)\cdot\left(\frac{C_{g0}}{C_g}\right)}\tag{5}$$

Figures 1 and 2 show the comparisons of the change in the wavelength and that in the wave height, respectively, between the Airy (the first-order Stokes) theory (line curves) and the present equation set (symbol marks) under conditions of collinear current and uniform depth. It is clear that Eqs. (3) to (5) (or the present equation set) gives remarkably good estimate of the change both in the wavelength and in the wave height nearly up to the limit of wave blocking except for the case of $k_0h = 5.0$. Hence we may conclude that the dispersion property of the present equation set is accurate enough even in the range of deepwater if the relative depth is not too large. (It may be needless to mention that, as is well known, waves on an opposing current decreases the wavelength and increase the height.)

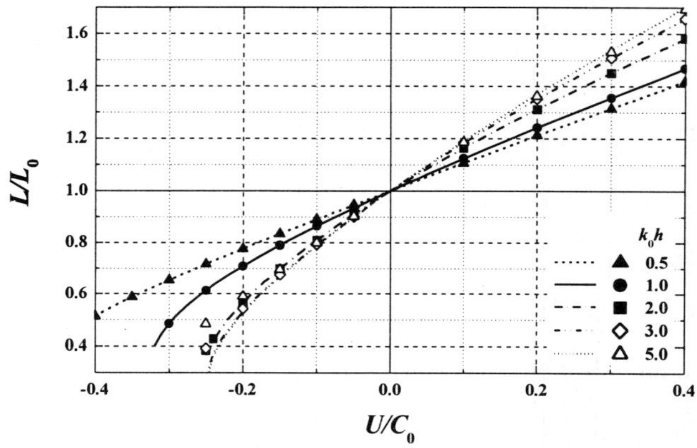

Figure 1. Comparison of the change in wavelength due to collinear currents between the Airy (the first-order Stokes) theory and the present equation set.

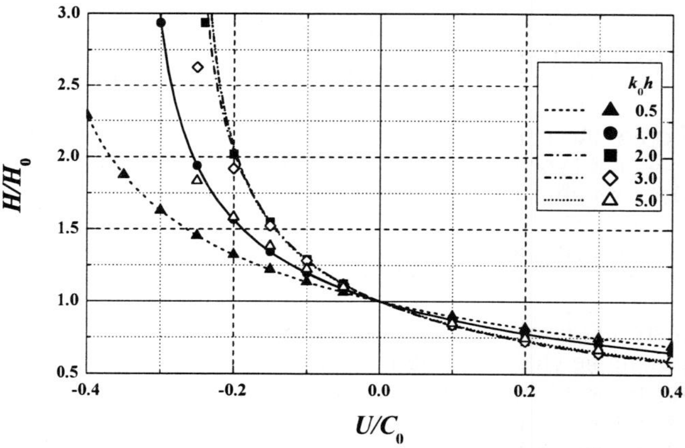

Figure 2. Comparison of the change in wave height due to collinear currents between the Airy (the first-order Stokes) theory and the present equation set.

Hence it has been regarded as reasonable to employ the above equation set in the present linear wave-current model. Although the equations are still lengthy and complicated, the complexity is algebraic and mathematically quite simple *except* for the presence of solution limit corresponding to the wave blocking, which is actually difficult to handle.

The equation set is solved by a finite difference method using staggered grids and a time-evolutional scheme is adopted with consideration of applying to random wave conditions. Specifically, each of variables, η and (u, v), is solved by the Gauss-Seidel method, alternately, applying a new method of treating boundary conditions expressed in terms of η that is stated below.

LEVENBERG-MARQUARDT METHOD TO TREAT ARBITRARY REFLECTION /TRANSMISSION AND NON-REFLECTIVE INCIDENT WAVE BOUNADAIES

Since appropriate boundary conditions are needed for any models to run properly, their treatment is a problem of major interest in the modeling. There are generally four kinds of boundaries to be treated: 1) non-reflective incident wave boundary, 2) reflective boundary, 3) transmittal boundary, and 4) radiating boundary.

In most of previous reflective (or non-reflective) wave models, a sponge layer-type treatment has been employed. It has, however, two kinds of demerits: one is the waste of the computation domain (or memory size and time), and the other is the difficulty in adjusting the dissipation intensity, particularly when we have to include arbitrary transmission boundary at the same time. In the present model, taking advantage of its linearity, we have attempted to build up a new method capable of handling all these four kinds of boundaries in a consistent manner for regular and random waves with or without currents, applying a least-squares parameter estimation method for a nonlinear equation set proposed by Levenberg (1944) and Marquardt (1963).

Arbitrary Reflection and Radiating Boundaries

For convenience of explanation, let's begin with the arbitrary reflection boundary. The newly-proposed method is based on the following four assumptions expressed in terms of the water surface elevation η:

(1) Waves in the immediate vicinity of the boundary can be locally resolved into two *sinusoidal* components, i.e., incident and reflected waves, having the same frequencies and essentially opposite directions of the propagation. It should be noted that the wavenumber is generally different between the two components, if the mean flow exists.

(2) Each component propagates near the boundary without significant deformation at least within very short time intervals.

(3) In case of horizontally two-dimensional wave field, incident and reflected wave directional angles have the same magnitudes, if no mean flow exists in the boundary-normal direction, which deviate according to the change in the wave celerity due to a current (wave refraction).

(4) The wave action of the incident and reflected waves is conserved, respectively, even if the mean flow exists.

For the readers' better understanding, one-dimensional case with a reflective wall boundary and no mean flow is taken as an example. On the assumptions stated above, the water surface elevation $\eta(x, t)$ at points x near the wall at a certain time t is expressed as:

$$\eta(x,t) = a_I \sin(k_I x - \sigma t + \varepsilon) + a_R \sin(-k_R x - \sigma t + \varepsilon) \qquad (6)$$

where a is the wave amplitude, k is the wavenumber, σ is the absolute (or intrinsic) angular frequency (the suffix "0" is omitted for simplicity), and ε is the phase angle. The suffixes "I" and "R" indicate the incident and reflected components, respectively. The wavenumbers k_I and k_R are generally treated as different, though slightly, from each other with the consideration of case of random waves. Hence, for this kind of partially reflective boundary, there are six unknown parameters, a_I, a_R, k_I, k_R, σ and ε, which are generally functions of time.

Now let's assume that the magnitudes of the surface elevation $\eta\,(x, t)$ have already been computed up to a certain time t, namely, values of η are available at $[n_x]$-x-points near the wall at $[n_t]$-t-points. Using these $[n_x \times n_t]$ values of η in (x, t)-plane, if we can separate the computed wave profiles into the two components, in other words, if we can determine values of the six parameters, then we can approximately evaluate a value of η on the boundary at the new time step $t+\Delta t$ by extrapolation and superposition. Then we can proceed the computation to the next time step using this boundary value and so on.

Actually, in addition to the above-mentioned η values, we have to use two dispersion relations, for k_I and k_R, respectively, and a given value of the reflection coefficient K_R that relates a_R to a_I. Hence our problem reduces to that of determining the six unknown parameters from the $[n_x \times n_t]$ + three given equations related *nonlinearly* to each other. To solve this problem, we have adopted the nonlinear least-squares parameter estimation method by Levenberg (1944) and Marquardt (1963). The Levenberg-Marquardt algorithm (LMA) interpolates between the Gauss-Newton algorithm (GNA) and the method of gradient descent. It is stated that the LMA is more robust than the GNA, namely, in many cases it finds a solution even if it starts far off the final minimum. In the present study, a library program of the LMA has been used as a subroutine. Values of n_x and n_t have been selected in the range of three to seven, respectively, depending on the magnitudes of $\Delta x/L$ and $\Delta t/T$.

In the presence of mean flow, or collinear current in the one-dimensional case, we have to take into account both the kinematic and the dynamic influences of the flow velocity U. In case of an opposing current with respect to the incident wave direction, the wavelength, the phase velocity and the group velocity of the incident waves reduce, while the amplitude or the height increases, as shown in Figs. 1 and 2, respectively. The opposite effects appear for the reflected waves. In case of a following current, the inverse influences should take place. As their results, the apparent reflection rate and their dispersion relations will change. The modified dispersion relations can be simply derived from Eq. (3) or Eq. (5) and are used in the model. The apparent reflection rate, or the expected ratio of a_R to a_I, can be obtained theoretically from the law of wave action conservation, Eq. (5), together with the reflection coefficient K_R prescribed for no current condition. Hence it is straightforward to replace the original dispersion relations and the reflection rate by their modified ones.

In case of wave field in horizontal two-dimensions (x, y), the essentially same technique can be employed by assuming the incident and reflected wave directions are symmetric near the boundary with respect to the boundary-normal if there is no mean flow, while the directional wave angles should be added as unknowns. Significant differences from the one-dimensional case take place when the boundary-normal mean flow velocity exists. Then we should incorporate changes in the wave directions due to the flow-caused wave refraction as well as changes in the wave amplitudes for the conservation of wave action. Anyway these differences can be handled rather easily by applying the wave theory.

For a *radiating* or *non-reflective boundary*, the present method explained above is directly applicable by setting a value of the reflection coefficient K_R at zero.

Non-reflective Incident Wave Boundary

On the offshore incident wave boundary, the incoming wave component should be given as prescribed *a priori*, while the outgoing waves due to the reflection inside the computation domain should freely pass through the boundary.

The concept of the present method for arbitrary reflection boundary is also applicable to treatment of such a non-reflective incident wave boundary. For this, we first compute surface elevations η_I (x, y, t) of incident waves on and near the offshore boundary, and then subtract their values from the computed values of η (x, y, t) obtained as solutions of the equation set in the numerical model. The values of η thus obtained by subtraction are regarded as those of reflected components, η_R (x, y, t). Applying the Levenberg-Marquardt algorithm to values of η_R (x, y, t) at several (x, y, t)-points together with the dispersion relations and the amplitude rate relations as well as the non-reflective condition, $K_R = 0$, we can estimate time-dependent values of the amplitude a_R, the angular frequency σ_R, the wavenumber k_R, the phase angle ε_R and the directional angle θ_R of the reflected wave component. Then we compute a value of the surface elevation η_R of the reflected wave on the boundary at the new time step. Adding this value η_R to that of the prescribed incident wave component η_I, we can evaluate η of the composite waves and utilize it as a boundary value at the succeeding time step and so on.

Arbitrary Transmission Boundary

It is generally not so easy to deal with the arbitrary transmission boundary with a conventional sponge layer technique, particularly because prescribed value of the transmission coefficient K_T is to be satisfied simultaneously with that of the reflection coefficient K_R. We often need many trials of adjustment or tuning to determine optimum values of the width and the dissipation intensity of the sponge layer, which are usually very much time-consuming and troublesome.

On contrary, the present boundary treatment technique is very simple and powerful to handle the arbitrary transmission boundary because of its principle. Since the wave profiles are separated to those of the incident and reflected components on the offshore side of an internal boundary by using the arbitrary reflection module mentioned above, it is quite straightforward to obtain the transmitted wave profile on the onshore side of the boundary just by multiplying the incident wave profile by a specified value of K_T, i.e., η_T $(x, y, t) = K_T \times \eta_I$ (x, y, t). As any values of K_T and K_R can be given independently of each other, this technique is very flexible for practical use. Actually even such funny values of $K_T > 1.0$ can be used when $K_R = 1.0$ if one ever wishes!

EXSAMPLES OF NUMERICAL COMPUTATIONS

Application to Partial or Non-Reflection Boundary

Computed results for arbitrary reflection of one-dimensional regular and random waves are shown in Figs. 3 and 4, both for the reflection coefficient $K_R = 0.0$, 0.4 and 0.8. The computations have been conducted with the incident wave condition of $H_I = 1.0$ m and $T = 10$s for regular waves, while $H_{S,\,I} = 1.0$ m

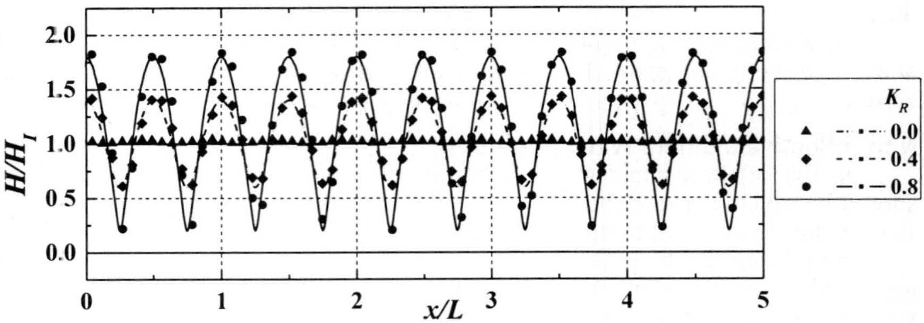

Figure 3. Change in regular wave height distributions for no-reflection (K_R = 0.0) or partial reflection (K_R = 0.4 and 0.8) compared with analytical solutions.

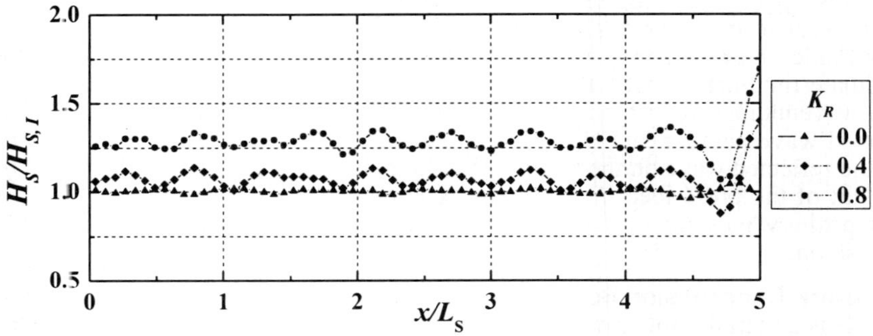

Figure 4. Change in random wave height distributions for no-reflection (K_R = 0.0) or partial reflection (K_R = 0.4 and 0.8).

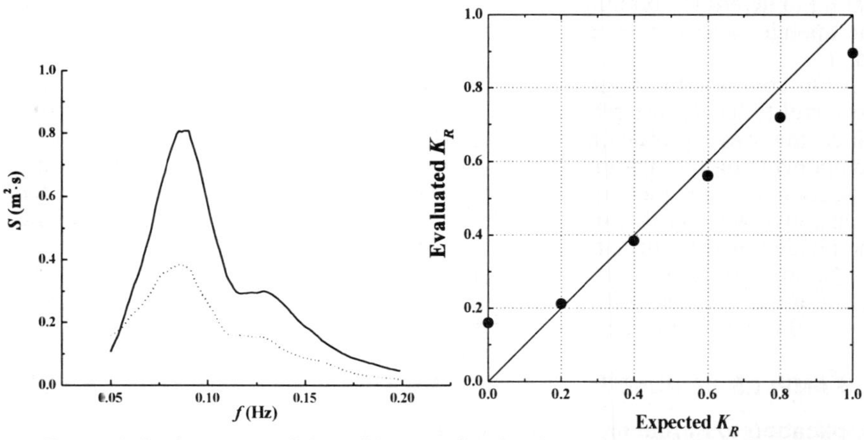

Figure 5. Performance of the arbitrary reflection boundary condition for random waves.

and T_S = 12.0 s in whole constant water depth h = 5 m.

In Fig. 3, analytical solutions are also shown, demonstrating quite a good agreement between the computations and the analytical solutions. It is also shown that the radiating or non-reflective condition, K_R = 0.0, is well satisfied by the present treatment.

In Fig. 4, H_S is the significant wave height obtained from the root-mean-squares of η (t). Their cross-shore distributions seem quite reasonable, featured by the significant wave height variations in front of the wall boundary corresponding to the anti-node and node of standing waves due to the phase relations of various frequency components, which decay rapidly and the wave heights fluctuate only slightly around the expected asymptotic values of $(1+K_R^2)^{1/2}$ at locations beyond about one wavelength away from the boundary, which is the characteristic random wave height distributions.

Figure 5 shows another view on the performance test of the present arbitrary reflection condition technique for random waves. Here values of the reflection coefficient were calculated (Evaluated K_R) with the spectral resolution technique of the incident and reflected waves (Goda and Suzuki, 1976) using computed values of η at two points, and are compared with the expected or given values of the reflection coefficients (Expected K_R). They are well correlated, though not in a perfect agreement but the evaluated values are slightly underestimated. The evaluated K_R corresponding to the non-reflective condition is strangely too large, which might be due to the inevitable error of estimating K_R in the spectral resolution technique, because the radiating condition even for the random waves is found to be well simulated as shown in Fig. 4.

Figure 6 shows the computed wave height distributions in the presence of collinear current, either following (U = 0.5 m/s) or opposing (U = -0.5 m/s) with respect to the incident waves, for K_R = 0.0 and K_R = 0.5, respectively. The incident wave conditions, H_I = 1.0 m and T = 10.0 s, have been imposed in the same constant depth, h = 5 m, where there is no mean flow. It is clearly shown that the following current reduces the wave heights, while the opposing current increases them as expected.

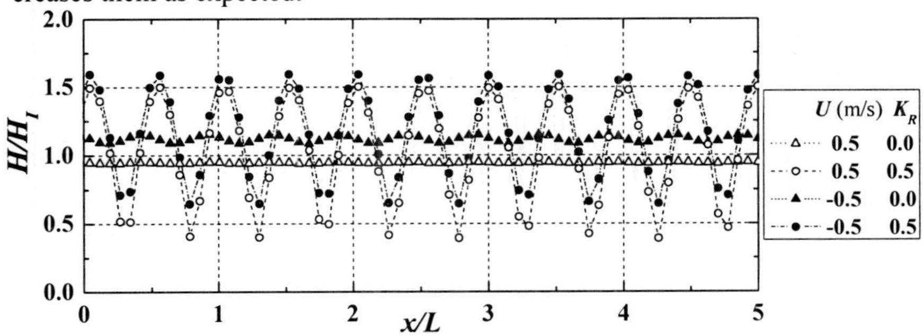

Figure 6. Change in wave height distributions due to following (U = 0.5 m/s) and opposing (U = -0.5 m/s) currents for no-reflection (K_R = 0.0) or partial reflection (K_R = 0.5).

Application to Arbitrary Transmission

Now the present technique of treating the arbitrary transmission is examined. Figure 7 shows one example of the computed surface profiles on the offshore partial reflection side and the onshore transmission side. It is recognized at least qualitatively that the waves are reasonably reflected from and transmitted through the wall boundary without any noticeable disturbance.

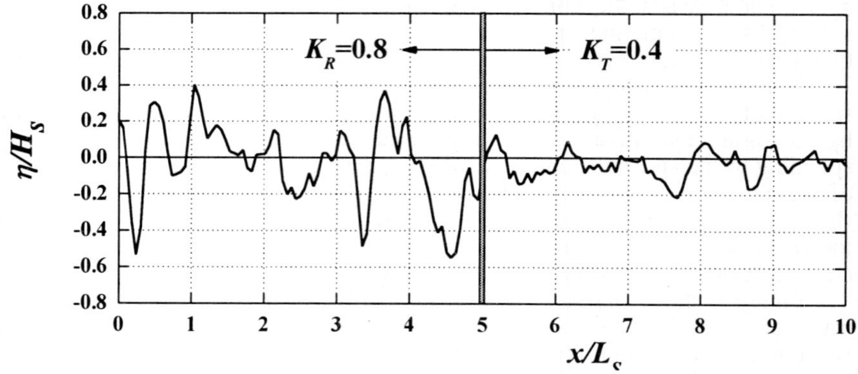

Figure 7. Surface profile of random waves due to the partial reflection and transmission by a vertical wall.

Shown in Fig. 8 are the distributions of computed wave height normalized by the incident wave height imposed at the location of no mean flow. The computations have been made for the partial reflection and transmission; (K_R = 0.5 and K_T =0.4) and (K_R =0.8 and K_T = 0.6), for the latter combination, both no mean flow condition (U = 0) and the opposing current condition (U = -0.5 m/s) are treated. Needless to say, in the transmission region, the wave height ratio agrees with the value of K_T if U = 0, while it becomes slightly larger than K_T as the incoming wave height is amplified by the presence of the opposing current.

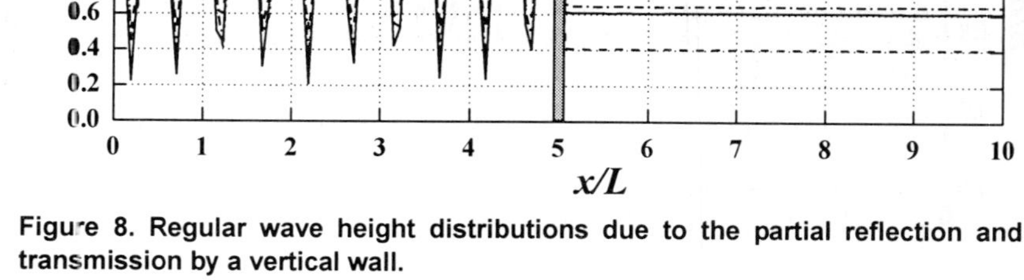

Figure 8. Regular wave height distributions due to the partial reflection and transmission by a vertical wall.

For the partial reflection and transmission of random waves, the computed results are shown in Fig. 9 for the same combinations of (K_R, K_T) values both for no mean flow conditions. As expected, their behavior is quite acceptable.

It has been thus demonstrated that the present method is highly capable of treating the partially reflective boundary, the radiating boundary, and the transmittal boundary as well as the non-reflective incident boundary, though examples are so far shown for the condition of one-dimensional constant depth.

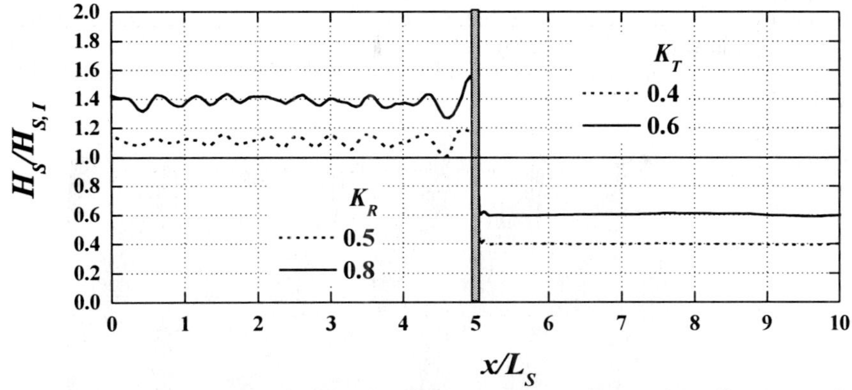

Figure 9. Random wave height distributions due to the partial reflection and transmission by a vertical wall.

Comparison with Measurements: Wave Deformation on Opposing Current on a Slope

To show practical applicability of the present wave-current model, a comparison is made between computations and laboratory measurements in Fig. 10 for wave height distribution on an opposing current with breaking on a slope of 1/20. The flow rate q of the current is -0.0224 m^2/s, or U = -1.12 m/s. The wave breaking was caused mainly by the shoaling, and not due to the blocking by the current, while it accelerated the breaking. In the computation, the breaker-induced momentum diffusion term has been incorporated to treat the wave decay after breaking. We can see a good overall agreement except for slight overestimation of the wave height around the breaking point, which is attributable to the neglect of offshore penetration of the breaker-induced turbulence by the current.

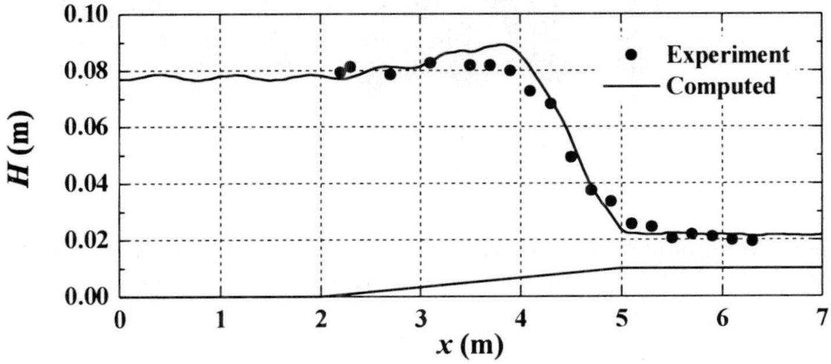

Figure 10. Comparison of wave height distributions on an opposing current with breaking on a slope of 1/20 between the measurement and computation. The flow rate q of the current is -0.0224 m^2/s.

Computations of Two-dimensional Wave Fields

The performance of the newly-proposed technique to treat the partial reflection and non-reflective incident boundaries is examined for two-dimensional wave fields. Figure 11 shows computed surface elevation distributions of the wave field composed of incident waves incoming obliquely at an angle of 30° with respect to the x-axis and partially reflected waves with $K_R = 0.5$. In the computation, the side boundaries have been treated using conventional sponge layers because of the limit of time available for preparing this paper. The expected diamond-shape cross-wave field is clearly observed, demonstrating the capability of the present model.

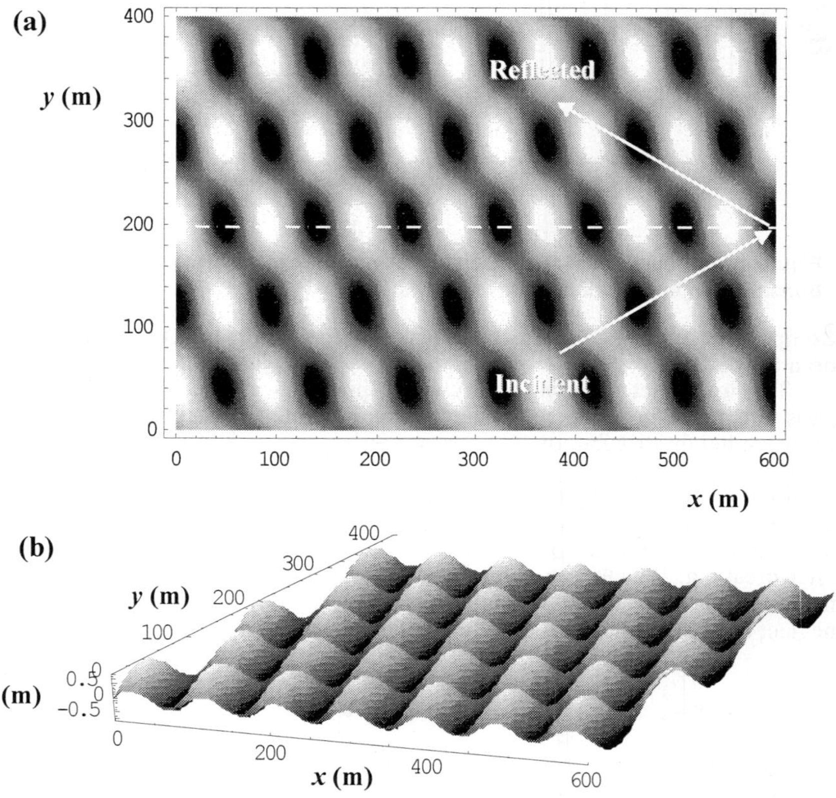

Figure 11. Computed surface elevation distributions of obliquely incident waves combined with partially reflected waves. (a) Two-dimensional plan view; (b) Three-dimensional perspective.

As another example of two-dimensional application, Fig. 12 shows the wave deformation due to a rip current on artificial horizontal bed. In this example, the wave breaking has been caused by the significant increase of wave height due to the presence of opposing current. The breaker-induced momentum diffusion terms are again incorporated in the wave model. The location of wave breaking has been determined from the ratio of the orbital velocity to the local wave celerity, both relative to the mean velocity. Although no measurement data is available, these results seem reasonable at least qualitatively.

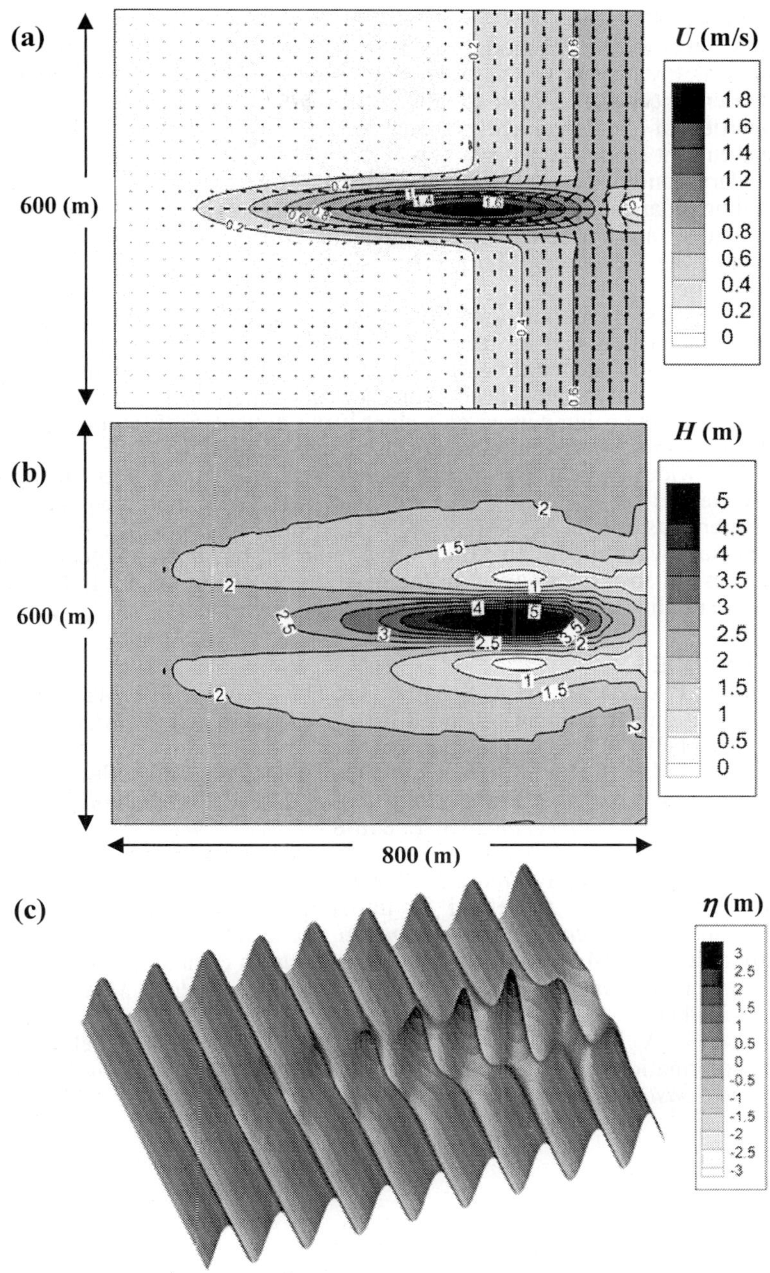

Figure 12. Simulation of a wave field around a rip current. (a) Velocity distribution of the rip current and surrounding; (b) Wave height distribution; (c) Perspective of the water surface elevation.

CONCLUDING REMARKS

A linear numerical model has been presented for computing the deformation of highly dispersive waves on a current on the basis of the linearized form of the enhanced Boussinesq equation set proposed by Chen et al. (1998). Basic properties of the linearized equation set has been discussed. The proposed wave-current model includes a new technique of handling various fundamental boundary conditions, such as incident, partially reflective, radiating and transmittal boundaries, all in a consistent manner by assuming linear superposition of the incident and reflected wave components in the vicinity of boundaries and applying the Levenberg-Marquardt algorithm to determination of parameter values.

Examples of computations have been shown for arbitrary reflection with or without mean flow and for reflection-transmission both for regular and random waves in one-dimensional field. A comparison between computations and laboratory measurements have been made for wave height distribution of the breaking wave on opposing current on a slope. Two-dimensional field of cross-waves and waves around a rip current have also been computed.

The wide applicability of the present technique have been demonstrated fairly well as far as linear waves are concerned. A lot more comparisons with measurements should be performed, if such data are available. Application of the present wave model up to the wave blocking has been attempted but been unsuccessful so far. Extension of the newly-proposed boundary treatment technique to nonlinear wave models should also be attempted.

REFERENCES

Chen, Q., P.A. Madsen, H.A. Schaffer, and D.R. Basco. 1998. Wave current interaction based on an enhanced Boussinesq approach, *Coastal Engineering*, 33, 11-39.

Goda, Y. and Y. Suzuki. 1976. Estimation of incident and reflected waves in random wave experiments, *Proceedings of 15th International Conference on Coastal Engineering*, ASCE, 828-845.

Kirby, J.T. 1984. A note o linear surface wave-current interaction over slowly varying topography, *Journal of Geophysical Research*, 89, 747-747.

Levenberg, K. 1944. A method for the solution of certain problems in least squares, *Quarterly Journal of Applied Mathematics*, 2, 164-168.

Marquardt, D. 1964. An algorithm for least squares estimation of nonlinear parameters, *Journal of Society of Industrial Applied Mathematics*, 11, 431-441.

Ohnaka, S., A. Watanabe, and M. Isobe. 1988. Numerical modeling of wave deformation with a current, *Proceedings of 21st International Conference on Coastal Engineering*, ASCE, 393-407.

NUMERICAL MODELING OF WAVE-CURRENT INTERACTION IN TIDAL AREAS USING AN UNSTRUCTURED FINITE VOLUME TECHNIQUE

Ole R. Sørensen[1], Henrik Kofoed-Hansen[1] and Oliver P. Jones[2]

The present paper describes applications of an integrated numerical modeling system for the study of hydrodynamic conditions in areas with large tidal and surge variations, strong current and very complex bathymetries. The modeling system includes a dynamic coupling between a 2D flow model and a spectral wave model and is based on an unstructured finite volume technique. Two applications are presented: the Bristol Channel, United Kingdom, and Grådyb tidal inlet, Denmark, characterized as macro- and micro-tidal estuaries, respectively. The work demonstrates the advantages of using an unstructured grid approach and that the inclusion of wave-current interaction is mandatory for obtaining accurate results in tidal areas.

INTRODUCTION

Successful prediction of hydrodynamic conditions and morphological changes in tidal areas requires detailed knowledge of water and sediment movement. Many tidal areas are controlled by complex hydrodynamic conditions in which tide, swell, wind waves and wind stress have significant forcing effects on the system. Very often the flow pattern consists of the combined interaction of tidal currents and a wave induced current field. The complexity is further enhanced by the influence of the flow field on wave propagation.

Recently, spectral wave models have been extended to application in coastal areas. Improvements in the physical description of processes and the use of new solution techniques (e.g. flexible mesh) have made it possible to apply coupled flow and spectral wave models to a wide range of complex coastal conditions.

In this paper we shall focus on simulating the interaction between waves and currents using a dynamic coupling between a fully spectral wave model and a 2D flow model. Both models are based on an unstructured cell-centered finite volume method and use an unstructured mesh in geographical space.

Compared to flow modeling, the computational cost of spectral wave modeling is relatively high. Setup of a spectral wave model, therefore, is often a compromise between accuracy and performance: A fine resolution model may improve the results, but may be computationally prohibitive. One way to reduce the computational cost is to solve the wave formulations using a quasi-stationary formulation as an alternative to a full instationary formulation. The quasi-stationary approach performs a series of time-independent solutions throughout

[1] DHI Water & Environment, Agern Allé 5, Hørsholm, DK 2970, Denmark
[2] University College London, Gower Street, London, WC1E6BT, UK

a tidal cycle, which incorporate the effects of dynamic currents and water levels. The quasi-stationary solution is based on the 'stationary assumption' which assumes an aphysical instantaneous travel time of wave energy across the model domain and that waves respond instantaneously to changes in the wind field. Such an approach is usually satisfactory for coastal studies, however in larger domains, the stationary assumption becomes invalid and an instationary (time-dependent) solution must be used.

In this study, we test the validity of using a new, computationally efficient, quasi-stationary approach for combined wave-current modeling in two complex case studies. These include the Bristol Channel, a very large, ebb-dominated, macro-tidal estuary in the Western U.K., and the micro-tidal Grådyb inlet located in the Danish part of the Wadden Sea.

MODELING SYSTEM

The spectral wave model MIKE 21 SW, is part of the MIKE 21/3 Coupled Model FM. It is a truly dynamic modeling system for application within coastal and estuarine environments. The flow and spectral wave models are the basic computational components of the modeling system. The flow model permits both 2D and 3D flows to be simulated and the combined wave-current system simulates the mutual interaction between waves and currents using a dynamic coupling. The same system can also be coupled to a sediment transport model to provide feedback between bed level changes and the hydrodynamics. The numerical approach applied in both the wave and flow models is based on a cell-centered finite volume technique on an unstructured grid.

Flow model

The flow model solves the 2D shallow water equations. The effect of waves on the flow field is taken into account by the radiation stress gradients. The radiation stresses are obtained from the wave model using a spectral formulation (Battjes, 1972).

The shallow water equations are solved using a second-order scheme. An approximate Riemann solver (Roe's scheme, see Roe, 1981) is used to calculate the convective fluxes. Second-order spatial accuracy is achieved by employing a linear gradient-reconstruction technique. The average gradients are estimated using the approach by Jawahar and Kamath, 2000. To avoid numerical oscillations a second-order TVD slope limiter (Van Leer limiter, see Hirch, 1990) is used. A second-order Runge-Kutta method is used for the time integration.

Spectral wave model

The spectral wave model includes two different formulations: a fully spectral formulation and a directionally decoupled parameterized formulation of the wave action balance equation. The fully spectral formulation is used in the present work. In the fully spectral formulation the source functions are based on the WAM Cycle 4 formulations (see Komen et al. 1994). The source term for depth-limited wave breaking is based on the formulation by Battjes et al., 1978. A short description of the source terms can be found in Sørensen et al., 2004. The effect of the time-varying depth and currents is taken into account by the formulation of propagation speeds in the geographical and spectral spaces. The spectral wave model is parallelized using OpenMP technique.

The conservation equation for wave action is solved using a first-order scheme. A first-order upwinding scheme is employed to calculate the convective fluxes both in geographical and spectral space. For the instationary equations the time integration is based on a fractional step approach, where the propagation step is solved using an explicit Euler method. The use of an explicit method can introduce a severe restriction on the time step for a given spatial discretization due to the CFL stability condition. In order to relax the restriction on the time step, a multi-sequence explicit integration scheme is applied. For a more detailed description of the numerical scheme, see Sørensen et al., 2004.

The present paper compares the performance of a new, and more efficient, steady-state or 'quasi-stationary' approach, with that of a full instationary solution. Comparisons have also been made between the performance of a full spectral formulation and that of a parametrically-decoupled approach in which the frequency domain is described in terms of a mean-wave frequency.

Solving the set of non-linear equations

$$\overline{F}(\overline{x})=\overline{0} \tag{1}$$

where $\overline{x}=\{x_i,x_2,\dots,x_N\}$ and $\overline{F}=\{F_1(\overline{x}),F_2(\overline{x}),\dots,F_N(\overline{x})\}$, a modified Newton-Raphson iteration can be expressed as

For k = 0, 1, 2 ... until convergence, do:

$$\overline{\overline{A}}\,\Delta\overline{x}^{k+1}=-\overline{F}(\overline{x}^{k}) \tag{2}$$

$$\overline{x}^{k+1}=\overline{x}^{k}+\alpha\Delta\overline{x}^{k+1} \tag{3}$$

Here $\overline{\overline{A}}$ is an approximation to the Jacobian ($J_{ij}=\partial F_i/\partial x_j$) of the operator $\overline{F}$ with respect to the dependent variables $\overline{x}$ and α is a relaxation factor. The matrix $\overline{\overline{A}}$ is obtained from the convective terms in geographical space and the diagonal part of the functional derivative of the source terms due to bottom friction, wave breaking and non-linear transfer (fully spectral formulation). The relaxation factor must be in the range from 0 to 1. A value of 0.1 has been shown to give a stable and fast convergence in most cases.

The iteration procedure is stopped when the root-mean-square (RMS) norm of a residual vector and the max-norm of the vector containing the increment of the significant wave height between two iteration steps are smaller than two user-specified tolerances

$$\left\|\overline{F}(\overline{x}^{k+1})\right\|_{RMS} \quad < \quad TOL_1 \quad \text{and} \quad \left\|\overline{H}_{m0}^{k+1} - \overline{H}_{m0}^{k}\right\|_{\max} \quad < \quad TOL_2 \tag{4}$$

In the present work $TOL_1=10^{-5}$ and $TOL_2=0.01$ have been used. The linear system in each iteration step is solved using a marching procedure. This is utilized by a simple upwinding approach which discretizes the convective terms and determines a calculation sequence for each discrete direction. This allows the linear system, Eq. (2), to be solved with a single sweep through the domain for each discrete direction.

Dynamic coupling

The time steps for the flow model and the instationary spectral wave model are dynamic and each determined to satisfy separate CFL stability criterion. The two time steps are synchronized at an overall discrete time step at which information between the two models is exchanged. First, flow computations are performed over the overall time step, keeping the radiation stress gradients constant. Then, time integration for the overall time step is performed for the spectral wave model using a linear interpolation of the water level and current conditions. When the spectral wave model is applied in quasi-stationary mode, a steady-state solution is calculated for each overall time step.

BRISTOL CHANNEL

The Bristol Channel is an ebb-dominated, macro-tidal estuary in the Western U.K., characterized by a series of prominent headlands and large sandy embayment's. The main area of interest is the Swansea Bay (see Figures 1 and 2). Maximum tidal range during peak springs varies from 5m at the mouth of the estuary to 12m in some locations in the upper estuary. In the main area of interest the tidal range is 8 m. The region is subject to long period swell and locally generated wind waves and provides one of the most severe test cases for a coupled wave-current model.

Simulations have been carried out for a number of different wind, wave and tidal conditions. In this paper, however, results will be presented for the period of the 12th to the 24th May 2002. This is a period over which suitable directional wave data are available from a U.K. Met. Office 5 year deployment near the Scarweather Sands, labeled herein as 'Scarweather' (U.K. Met. Office National Centre for Ocean Forecasting, 2005) and flow and wave data at two locations in front of and behind the Scarweather Sands, labeled herein as 'site 1' and 'site 2'. (GARDLINE, Titan Surveys, 2002). The locations of the three stations are shown in Figure 2. The water depth at Scarweather, site 1 and site 2 are 29m, 14m and 26m, respectively.

In order to capture the ebb-dominated macro-tidal characteristics of the region, the model mesh (Figure 1) has been extended from the upper Severn Estuary to an open, offshore boundary in the Celtic Sea. The node separation varies from approximately 5km in the outer offshore regions to 500m in the main area of interest (Swansea Bay). A close-up of the mesh covering the Swansea Bay area is presented in Figure 2.

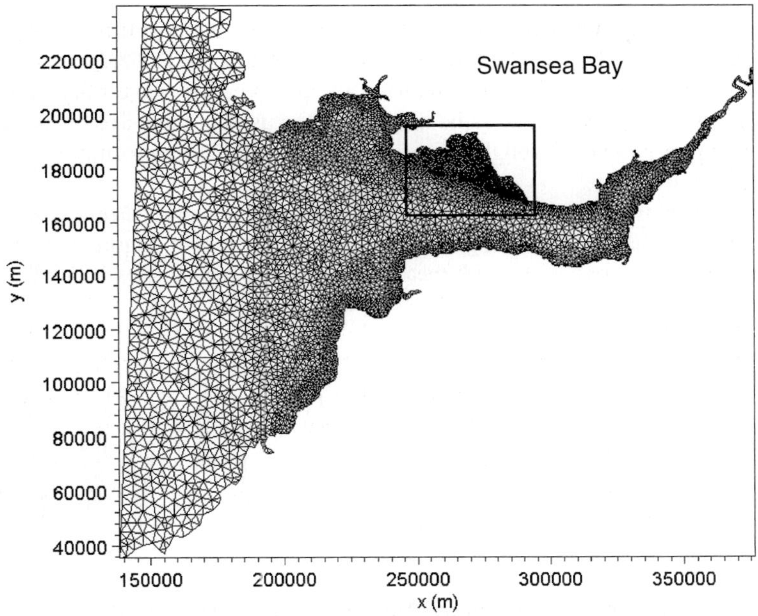

Figure 1. Computational mesh for the Bristol Channel case.

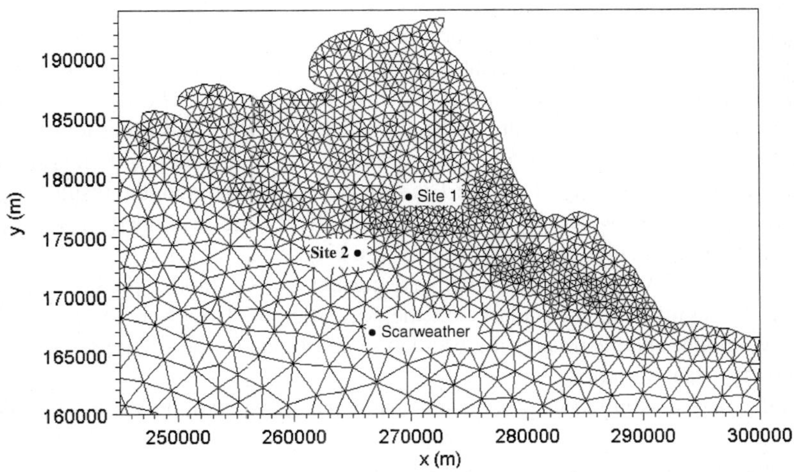

Figure 2. Close-up of the computational mesh for the Bristol Channel case. Swansea bay area.

For the fully spectral experiments, the frequency domain was divided into 30 discrete bins ranging from 0.04 to 0.7 Hz. For the directional domain, 16 discrete directions were selected. For the instationary (time-dependent) experiments, the overall time-step was set to 60 seconds and for the quasi-stationary experiments, stationary computations were performed at intervals of 3600 seconds.

The flow model used tidal elevation data based on global tidal analysis (Andersen, 1995). The spectral wave model used wind sea and swell parameters from the Lundy wave-rider buoy at 51.17N 005.42W (WAVENET, 2002)

Measured wind speeds and directions from the Scarweather wave-rider buoy were used for the wind forcing. A value of $\gamma = 0.8$ was used in the wave breaking formulation, and a Nikuradse roughness of $k_N = 0.05$m applied uniformly across the model domain.

Comparisons between modeled and measured water levels and current speeds around the Bristol Channel show, in general, very good agreement. Figure 3 shows comparisons of model predictions with measured current speeds at sites 1 and 2.

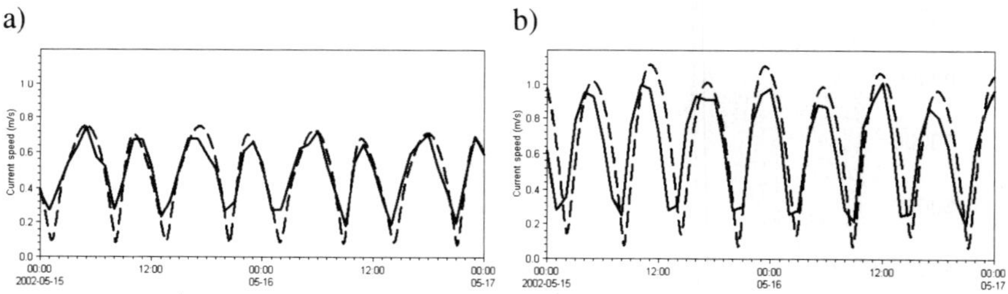

Figure 3. Current speed at site 1 (a) and site 2 (b) for the Bristol Channel case. — Measurements and - - - Calculations.

Time series comparisons of instationary predictions with measured integrated wave parameters at Scarweather are presented in Figure 4. Parameters that have been compared include the significant wave height, H_{m0}, the peak wave period, T_p, the mean wave period, T_{01}, and the peak wave direction, θ_p. The agreement is good although the wave heights are underestimated when the wind is coming from east and south-east. In Figure 5 a time series of the calculated and measured significant wave height, H_{m0}, from site 1 is shown. For comparison, the results of a simulation in which water level and current effects are excluded from the instationary wave calculations are presented (see Figure 5).

The impact of varying water level and current on the model predictions is shown to be significant, particularly on the wave height. However, the coupled flow-wave model underestimates the oscillation amplitude of the wave parameters caused by the tidal variation.

Results of the simulations conducted using the quasi-stationary formulation show excellent agreement with the instationary predictions, particularly at the Scarweather location. Figure 5 shows predictions of significant wave height using the two formulations.

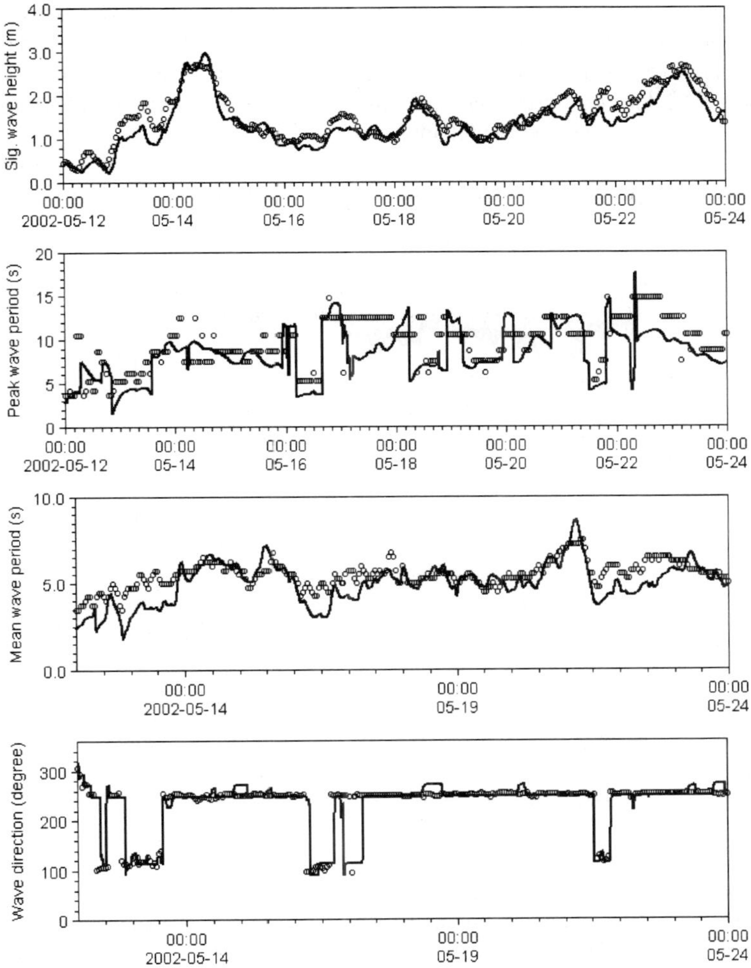

Figure 4. Wave parameters at Scarweather for the Bristol Channel case. O Measurements and — Calculations (instationary formulation).

Generally, the water depths over sandbanks in the Swansea Bay area are very shallow. During low water the depth decreases to less than 2m. The effect of the varying water depth, therefore, is very severe in these areas. Within a tidal period the significant wave height can vary from 0-3m. In the measurement locations, however, the effect of the varying water depth on the waves is limited. Therefore, it is difficult to adequately assess the models performance in these extremely shallow water areas.

Another important factor is the high current speeds that exist over the bank crests. For example, the model predicts maximum current speeds of up to $1.5ms^{-1}$.

Under such conditions it is inevitable that wave blocking will occur. We assume that this is the main reason that the model underestimated the effect of tidal variation on the wave heights and periods measured at sites 1 and 2. This is a result of the difficulty in describing the physics of wave conditions close to blocking using a phase-averaged model approach.

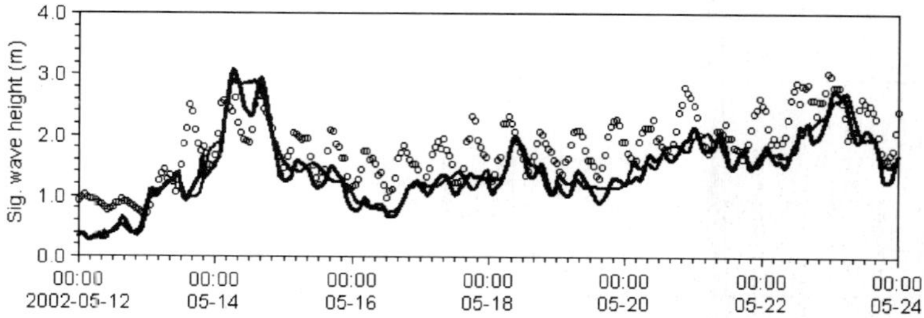

Figure 5. Significant wave height at site 1 for the Bristol Channel case. O Measurements, — Calculations (instationary formulation) and — Calculations (instationary formulation, but excluding level and current effects).

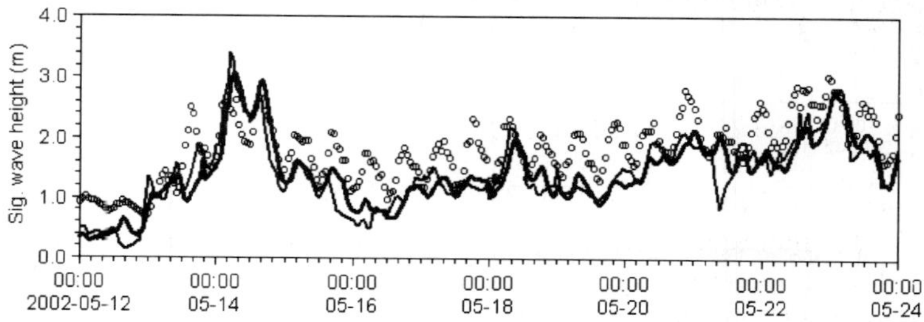

Figure 6. Significant wave height at site 1 for the Bristol Channel case. O Measurements, — Calculations (instationary formulation) and — Calculations (quasi-stationary formulation).

The computations were carried out on a PC (Pentium D 840 Dual Core, 2x3.2GHz/800/2x1MB). The average time step for the flow calculation was 7.7s, and using the instationary formulation, the average time step in the wave calculation was 41.6s. The average number of iterations for the quasi-stationary formulations was 58. For a 12 day simulation, the total CPU time was 51.0 and 25.3 hours for the instationary and quasi-stationary solutions respectively. This demonstrates an impressive increase in efficiency that is achievable due to both the stationary assumption and use of the new stationary solver. The elapsed time for the two simulations was 32.4 hours and 18.8 hours, respectively.

GRÅDYB TIDAL INLET

The second case study is the Skallingen barrier spit and the Grådyb tidal inlet (see Figure 7), located in the Danish part of the Wadden Sea. In this area, the highest tidal range reaches 1.7m at springs, but the storm surge in the area can be as high as 2-4m. The maximum current in the navigation channel, which leads to the harbor of Esbjerg, is in the range of 1-2m/s. The depth in the channel is 10-12m at mean sea level. The dominant wind directions at the location are west and west-southwest.

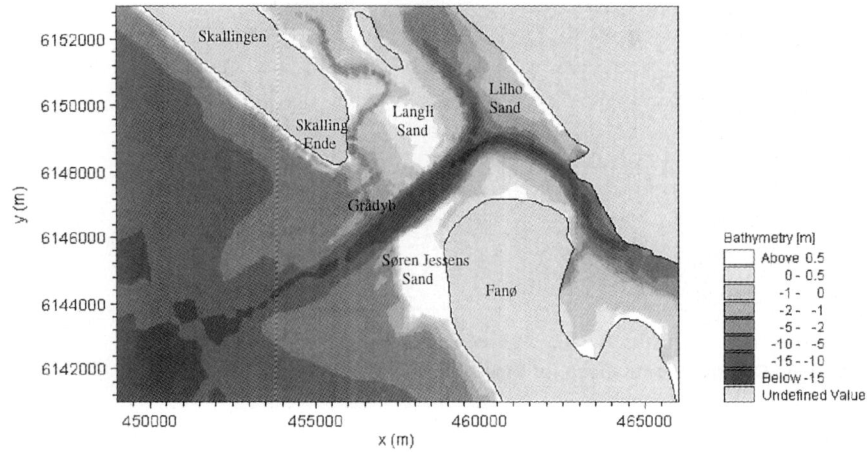

Figure 7. Bathymetry for the Grådyb tidal area.

Simulations have been carried out for a storm event, 2-11 January 1991. During this period the wind direction was in the range 200-270 degree true north and maximum wind speeds of approximately 20 m/s occurred on the 6th and 9th of January. Measurements of wave conditions for this period were carried out at three locations: St. 1 (55° 20' N, 7° 20' E), St. 2. (55° 25,76' N, 8° 12,90' E) and St. 3a (55° 29,05' N, 8° 21,80' E). The water depth at the three stations is 25m, 12m and 10.3m, respectively. The locations of St. 2 and St. 3a are shown in Figure 9.

The mesh contains 20091 elements. The average node separation (edge length) is approximately 1km in the offshore areas and 100m in the shallow areas north and south off the inlet. The mesh is shown in Figure 8 and a close-up of the mesh at the inlet area is shown in Figure 9. For the fully spectral wave formulation, a logarithmic frequency discretization was used including 25 frequencies ranging from 0.05 to 0.85 Hz. The directional space is discretized into 12 directions covering a range from 180 to 360 degrees. For the instationary simulations, the time step was set to 60s and for the quasi-stationary simulations, wave computations were performed at intervals of 3600s.

The boundary conditions for the both the flow and the spectral wave model were obtained from regional simulations covering the North Sea, the Norwegian Sea, the Inner Danish waters, the Baltic Sea and the Gulf of Botnia. A value of γ = 0.8 was used in the wave breaking formulation, and a Nikuradse roughness of k_N = 0.05m, was applied.

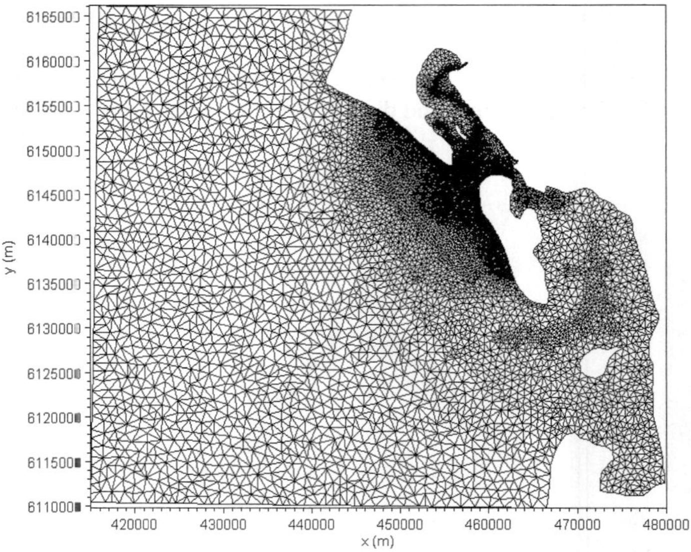

Figure 8. Computational mesh for the Grådyb tidal inlet case.

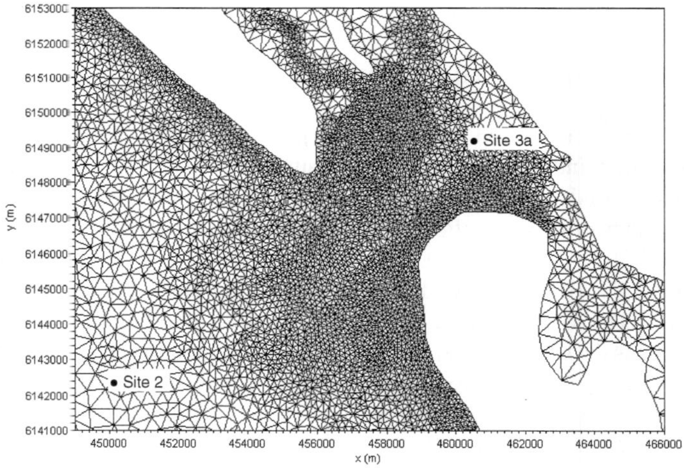

Figure 9. Close-up of the computational mesh for the Grådyb tidal case.

The model shows that shallow water areas including Langli and Søren Jessens Sands to the north of the channel and Skalling Ende to the south of the channel dried at low tide and flooded at high water and during storm surges. The breaking of the waves on the beach on the offshore side of Skallingen generates a south-directed longshore current. The incoming waves propagate partly over the shallow water area to the north and south of the channel and partly through the main channel itself. In the channel, wave refraction over the shallow water areas is observed, which reduces the wave height significantly. In the shallow water areas, wave energy is reduced by wave breaking and bottom friction effects. The effect of the tidal currents is most apparent in the main channel: the model demonstrates that for an opposing current, the wave height increases and that for a following current it decreases.

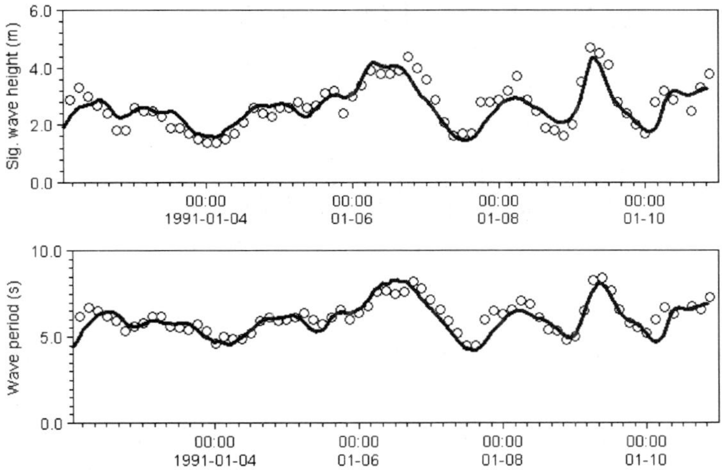

Figure 10. Significant wave height and zero-crossing wave period at St. 2 for the Grådyb tidal inlet case.
O Measurements and — Calculations (instationary formulation).

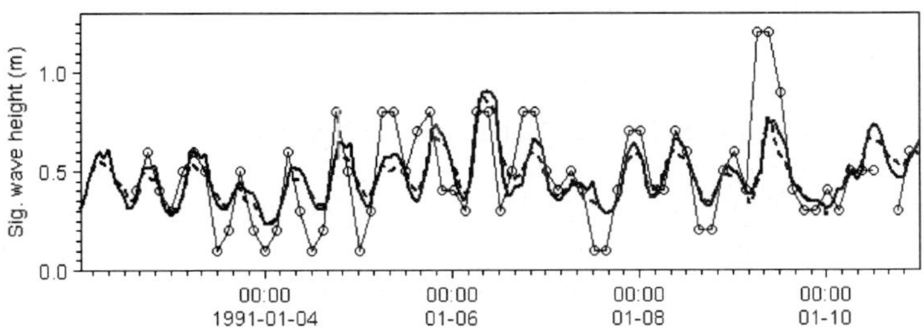

Figure 11. Significant wave height at St. 3 for the Grådyb tidal inlet case.
—O— Measurements, — Calculations (instationary formulation) and - - - Calculations (quasi-stationary formulation).

Time series of calculated and measured significant wave height, H_{m0}, and zero-crossing wave period, T_{02}, are compared in Figure 10 for St. 2. It can be seen that there is no significant effect on the wave height from varying water level and current at this location. Figure 11 shows time series of calculated significant wave height, H_{m0}, at St. 3a compared to measurements. Figure 12 shows the calculated surface elevation and current speed at a location in the center of the channel. At St. 3a there is a significant oscillation of the wave height and the calculated results are in good agreement with the measurements.

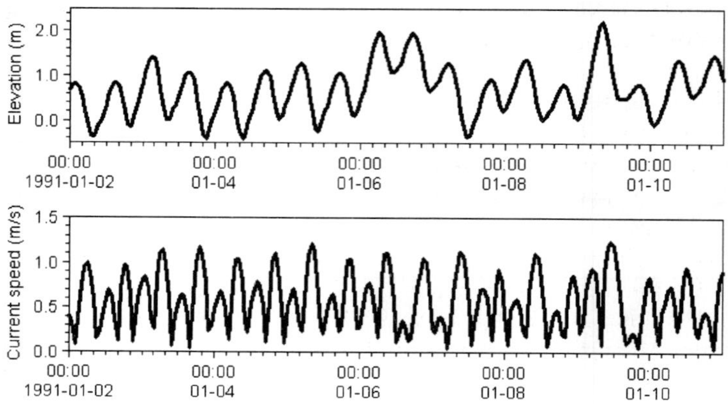

Figure 12. Calculated surface elevation and current speed in the channel in front of St. 3a.

Wave conditions at St. 3a are also influenced by the local generation of wind waves and, therefore, the calculated results are dependent on the air-sea interaction formulation. The use of a coupled formulation (see Komen et al., 1994) for the air-sea interaction gives an overestimation of the wave height. Hence, the results presented here are generated using a decoupled formulation based on a drag law (Smith and Banke 1975). It can be seen (Figures. 11 and 12) that the local maximum of the wave height occurs at maximum opposing current and that the local minimum occurs at maximum following current. The calculated results are very sensitive to the variation in the bathymetry. Hence, the main reason for the deviation between the measured and calculated values is expected to be inaccuracy of the bathymetry.

The coupled simulation was then repeated using the quasi-stationary formulation in the spectral wave model. A time series of significant wave height for St. 3 obtained using this formulation is also included in Figure 12. It can be seen that, even in this case, the difference between instationary and stationary solutions is small.

The average time step used in the instationary wave calculation was 29.6s. The averaged number of iterations required to obtain each quasi-stationary solution was 28. The total CPU time for the instationary and stationary spectral wave calculations was 68.9 and 10.7 hours respectively. In this case study, therefore, an even greater reduction in the CPU time can be achieved using the quasi-stationary solver, without affecting the accuracy of the solution.

CONCLUSION

An integrated numerical modeling system has been presented for applications in areas with large tidal and surge variations and very complex bathymetries. The modeling system contains a dynamic coupling between a 2D flow model and a spectral wave model. The modeling system has been tested in two test areas: The Bristol Channel and the Grådyb tidal inlet. For both cases the calculated results show good agreement with measured data.

The use of a quasi-stationary formulation for the spectral wave model is shown to be a good alternative to the use of an instationary formulation as the computational costs can be reduced significantly.

Simulations using the decoupled parametric formulation in the wave calculations have also been performed as part of this work although not reported in the present paper. These simulations showed that for the Bristol Channel case the prediction of the wave height can be well modeled using the simplified and less computationally demanding approach, but due to the combination of swell and local wind sea the fully spectral approach is needed to accurately predict the wave periods.

REFERENCES

Andersen, O.B. 1995. Global ocean tides from ERS-1 and TOPEX/POSEIDON altimetry, *J. Geophys Res. 100 (C12)*, 25,249-25,259.

Battjes, J.A. 1972. Radiation stresses in short-crested waves, *J. Mar. Res.* 30, 56-64.

Battjes, J.A. and J.P.F.M. Janssen, 1978. Energy loss and set-up due to breaking of random waves, *Proc. 16th Int. Conf. Coastal Eng.*, ASCE, 569-587.

Cefas, 2002. WAVENET directional wave data from 'Lundy wave-rider buoy'.

Hirsch, C. 1990. *Numerical Computation of Internal and External Flows*, volume 2: Computational Methods for Inviscid and Viscous Flows, Wiley.

Jawahar P. and H. Kamath. 2000. A high-resolution procedure for Euler and Navier-Stokes computations on unstructured grids, *Journal Comp. Physics*, 164, 165-203.

Komen, G.J., L. Cavaleri, M. Donelan, K. Hasselmann, S. Hasselmann and P.A.E.M. Janssen. 1994. *Dynamics and Modelling of Ocean Waves*, Cambridge University Press.

Roe, P. 1981. Approximate Riemann solver, parameter vectors, and difference schemes, *Journal of Computational Physics, No. 38, 339-374.*

Smith, S.D. and E.G. Banke. 1975. Variation of the sea surface drag coefficient with wind speed. *Quart. J. Roy. Meteor. Soc*, 101, 665-673.

Sørensen, O.R., H. Kofoed-Hansen, M. Rugbjerg, and L.S. Sørensen. 2004. A third-generation spectral wave model using an unstructured finite volume technique, *Proceedings of the 29th International Conference on Coastal Engineering 2004*, Lisbon, Portugal.

Titan Surveys, 2002. GARDLINE data from the Scarweather Sands Wave Tide and Current Monitoring – Sites 1 and 2. Deployment 3.

U.K. Meteorological Office, 2004. National Centre for Ocean Forecasting. Data from the 'Scarweather wave-rider buoy' 01/06/00 – 31/12/04

A NUMERICAL STUDY OF COUPLING OF WAVES WITH A CURRENT OVER CUT-CELL MESH*

T. Q. LI, P. TROCH AND J. DE ROUCK

Department of Civil Engineering, Ghent University,
Technologiepark 904,
B-9052 Zwijnaarde, Belgium
E-mail: Tingqiu.li@UGent.be, Peter.Troch@UGent.be,
Julien.DeRouck@UGent.be

In this paper, we further study coupling of surface waves with a current over a cut cell, based on our highly efficient Navier-Stokes solver[3,4], named **LVOF**. This solver is constructed by a novel VOF finite volume approach that incorporates surface tension, coupled with a dynamic subgrid-scale (SGS) turbulence model. Additionally, we design two types of numerical wave-current generators: an external and internal generators, while the outgoing waves are dissipated by a breaking-type wave absorber placed at the opposite end. Based on LVOF, we wish that our results can provide valuable base for the analysis of wave-current-body interaction, according to the well-designed wave-current generator.

1. Introduction

When studying the coupling of waves with a current, the periodic tidal current may be simplified as a steady current. On the assumption that waves alone at one end of a numerical wave tank (NWT) is generated, theoretically, methods that model a current basically may be classified into two types of categories: the radiation stresses concept and inflow conditions specified. Of course, most of the studies in this area are based on the former, because of the intrinsic features: the well capture of the dynamics of the wave-induced flows, especially the natural transformation of wave growth and its dissipation. Within the framework of the radiation stress technique, however, the wave motions are distinguished and are in fact separated from the current motions.

*Supported by the Fund for Scientific Research - Flanders (Belgium) within the research project G.0199.06.

On the other hand, inflow conditions (by describing the current and wave profiles) are assigned as one of boundaries, implying that the solution of the Navier-Stokes equations with the free surface directly gives the local current distributions. Importantly, distinction between waves and currents is unnecessary in this theory. It is believed that this type of numerical approach tends to be more promising[5], especially when waves follow a positive current. For the latter, an essential feature may be interpreted as follows:

◇ The positive net flux distributed at the boundary of the domain can be easy to maintain the energy balance in the NWT even over the long-time evolution of wave-current coupling.

But for the particular counter current problem, we found

◇ The fluid leaving through the inflow boundary could drain water out of the tank, leading to a continuous drop in the mean water level.

Definitely, such behaviour eventually terminates the computations, especially in the case of strong opposing currents, based on our experience. To avoid this problem, a net mass flux must be fed into the flow domain through the outlet boundaries for maintaining the mass balance[5]. In this way, the computational domain has to extend to a large distance (due to the use of a damping zone) in order to reduce the effects of the reflected waves. This costs substantial CPU time plus the heavily computational memory, leading to extremely slow convergence in general. Hence, a hopeful alternative is to introduce the internal source function for creating an adverse current and also use a breaking-type wave absorber. Those have benefited greatly, as done in our study[3]. Shortly,

◇ By superimposing a steady current onto waves, our external generator technique provides well-defined wavy inflow conditions, which have been addressed in the study of waves following a current;

◇ For waves propagating against strong currents, the internal source function introduced is used as one type of a net driving force for creation of the desirable opposing current, forming our internal generator.

This paper is organized as follows. We first briefly describe our numerical methods and then discuss our results. Finally, the concluding remarks are given.

2. Numerical Methods

2.1. *A MILES model*

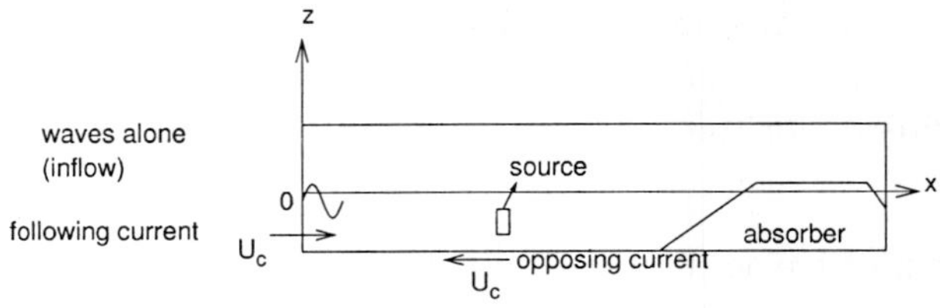

Figure 1. Sketch of the NWT for waves following or opposing a current, defined by U_c.

In a numerical wave tank (see Fig. 1), the mathematical model is governed with the unsteady incompressible Navier-Stokes equations, by introducing MILES (monotonically integrated large-eddy simulation) coupled to a dynamic subgrid-scale model. Namely

$$\frac{\partial \varphi}{\partial t} + \frac{\partial (F_i - F_v + F_a)}{\partial x} + \frac{\partial (G_i - G_v + G_a)}{\partial y} + \frac{\partial (H_i - H_v + H_a)}{\partial z} = Q \quad (1)$$

The variables $\varphi = (0, u, v, w, \alpha)^T$, in which (u, v, w) are the components of the velocity in the (x, y, z)-directions, respectively, and α is the volume fractions in VOF. By defining the inviscid fluxes (F_i, G_i, H_i), the viscous fluxes (F_v, G_v, H_v) and the other fluxes (F_a, G_a, H_a), we have

$$F_i = \begin{pmatrix} 0 \\ u^2 \\ vu \\ wu \\ 0 \end{pmatrix}, G_i = \begin{pmatrix} 0 \\ uv \\ v^2 \\ wv \\ 0 \end{pmatrix}, H_i = \begin{pmatrix} 0 \\ uw \\ vw \\ w^2 \\ 0 \end{pmatrix}. \quad (2)$$

$$F_v = \begin{pmatrix} 0 \\ \nu_{eff}\frac{\partial u}{\partial x} \\ \nu_{eff}\frac{\partial v}{\partial x} \\ \nu_{eff}\frac{\partial w}{\partial x} \\ 0 \end{pmatrix}, G_v = \begin{pmatrix} 0 \\ \nu_{eff}\frac{\partial u}{\partial y} \\ \nu_{eff}\frac{\partial v}{\partial y} \\ \nu_{eff}\frac{\partial w}{\partial y} \\ 0 \end{pmatrix}, H_v = \begin{pmatrix} 0 \\ \nu_{eff}\frac{\partial u}{\partial z} \\ \nu_{eff}\frac{\partial v}{\partial z} \\ \nu_{eff}\frac{\partial w}{\partial z} \\ 0 \end{pmatrix}. \quad (3)$$

$$F_a = \begin{pmatrix} u \\ \frac{1}{\rho}p \\ 0 \\ 0 \\ u\alpha \end{pmatrix}, G_a = \begin{pmatrix} v \\ 0 \\ \frac{1}{\rho}p \\ 0 \\ v\alpha \end{pmatrix}, H_a = \begin{pmatrix} w \\ 0 \\ 0 \\ \frac{1}{\rho}p \\ w\alpha \end{pmatrix}. \tag{4}$$

plus the source terms Q

$$Q = \begin{pmatrix} f \\ \frac{1}{\rho}F_b^x + F_d^x + (1-\theta) fu \\ \frac{1}{\rho}F_b^y + F_d^y + (1-\theta) fv \\ \frac{1}{\rho}F_b^z + F_d^z + (1-\theta) fw - g \\ \alpha\left(\frac{\partial u}{\partial x} + \frac{\partial v}{\partial y} + \frac{\partial w}{\partial z}\right) \end{pmatrix} \tag{5}$$

where g is the gravitational acceleration and p is the total pressure. (F_b^x, F_b^y, F_b^z) and (F_d^x, F_d^y, F_d^z) are three components of a body force $\vec{F_b}$ and an extra dissipative term $\vec{F_d}$ in the (x, y, z)-directions, respectively[3]. $\nu_{eff} = \nu + \nu_t$ is the effective viscous coefficient, in which ν is the molecular viscosity and ν_t is the eddy viscosity to be determined with a dynamic Smagorinsky model. The local density ρ and viscosity ν_{eff} are given as in terms of α in VOF.

$$\rho = \alpha\rho_w + (1-\alpha)\,\rho_a, \quad \nu_{eff} = \alpha\nu_{eff_w} + (1-\alpha)\,\nu_{eff_a} \tag{6}$$

where the subscripts (w, a) denote the water and air, respectively.

A fully implicit VOF finite volume approach that involves surface tension is implemented for solution of unsteady incompressible Navier-Stokes equations with a free surface, in case the boundary conditions are explicitly given. The full description in our solver, LVOF, can be found in our study [4], including a wave-current generator for regular and irregular waves [3].

2.2. *Initial and boundary conditions*

2.2.1. *Initial conditions*

All simulations started from the complete rest via $(u, v, w) = 0$ at $t = 0$, provided that the hydrostatic pressure fields are used as its initial state. Given the flat free-surface (this is measured with respect to the still water line), α are initially assigned by $\alpha = 1$ in the full fluid and $\alpha = 0$ in the air, while $0 < \alpha < 1$ over mixed cells.

2.2.2. *Boundary conditions*

The bottom effects are incorporated by enforcing the slip boundary conditions for cut cells, while the no-slip boundary conditions are applied in the lateral direction (side walls). At interfaces (defined by a free surface) between the water and air, the contribution from the air is only to apply the pressure on the water. Since surface tension is handled as a body force, consequently, the Reynolds stress free-surface boundary condition can be given by Li[6].

$$p = \rho \nu_{eff} \left\{ 2\frac{\partial w}{\partial z} - \frac{\partial \eta}{\partial x}\left(\frac{\partial u}{\partial z} + \frac{\partial w}{\partial x}\right) - \frac{\partial \eta}{\partial y}\left(\frac{\partial v}{\partial z} + \frac{\partial w}{\partial y}\right) \right\} \tag{7}$$

This involves the surface viscous boundary layer on the free surface. A major benefit is for achievement of the accurate prediction in the flows, especially when waves break. At the outlet, a physical wave absorber is introduced, while a wave-current generator fixed at the inlet is employed for inflow conditions by creation of the synthetic motion of waves with currents.

2.3. *The effects of θ*

In our present study, the effects of an opposing current can be considered by adding mass and force terms into the continuity and momentum equations, dependent on the values of θ in Eq. (5). It is the so-called indicator function, that is 0 or 1. When $\theta = 0$, the source function f is incorporated into the continuity and momentum equations as a net driving force, acting on the fluid elements lying within the finite thickness source region, as done in our study[3]. Now, we attempt to investigate behaviour of $\theta = 1$, implying that f is only absorbed to the continuity equation.

Test cases	U_c (ms^{-1})	d (m)	H (m)	T_a (s)
Case 1: open channel flow	−0.11	0.2	0.0394	1.0
Case 2: open channel flow	−0.222	0.45	0.13	2.5
Case 3: vertical wall problems	0.10	0.7	0.16	2.0
Case 4: seadike problems	0.17	0.7	0.16	2.0

To well illustrate such phenomenon, two test cases under consideration are applied (see Table above). Both are in experiments, which involve the fully coupled wave-current motions in the open channel over constant water depth (labelled Case 1 to Case 2). A major difference is Case 1 with small amplitude[2] but Case 2 with large amplitude waves[1].

Figs. 2 and 3 give one typical comparison in terms of the wave train at a certain location and the instantaneous velocity fields, by using $\theta = 0$ and 1, respectively, for Case 1 and Case 2. Broadly, globally similar but distinctly different local features are apparent, especially for Case 1, where significant errors are produced in some transient time at $\theta = 1$. Definitely, such behaviour induces an impact on the accuracy and stability, whereas the reason for that is not clear for small scale motions (Case 1). Consequently, the addition of the mass term seems less efficient. This is our primitive result in the study of waves opposing a turbulent current, by using $\theta = 1$. Further study is, therefore, necessary to clarify these issues.

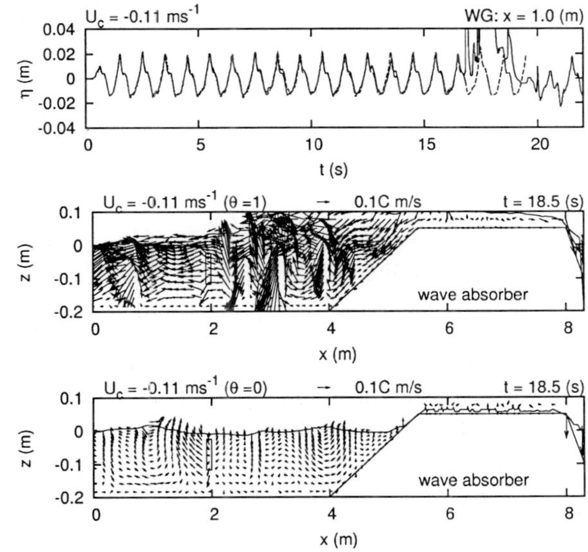

Figure 2. Comparison of adding mass and force terms over grids (291×56) for Case 1 (the regular opposing waves): the wave train at a certain position and the instantaneous velocity fields. —— $\theta = 1$; – – – 0.

3. Results and Discussions

The reminder is that we'll address the numerical aspects as follows:

◇ the history of convergence;

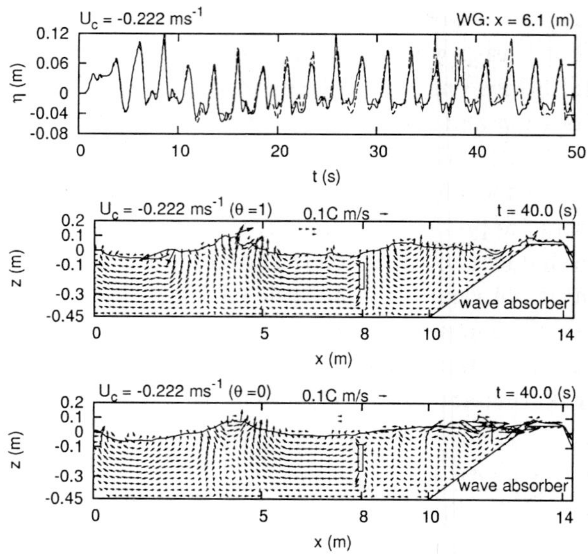

Figure 3. Comparison adding mass and force terms over grids (291 × 56) for Case 2 (the regular opposing waves): the wave train at a certain position and the instantaneous velocity fields. —— $\theta = 1$; – – – 0.

◇ and grid refinement study.

including some solution for a vertical wall problem (Case 3) and a seadike problem (Case 4). These demonstrate the present capability of LVOF solver in modelling of waves opposing or following a current from intermediate to shallow water depths. The calculated results are represented in terms of the instantaneous velocity fields (in which, the symbol, C, stands for the linear shallow wave celerity, unless otherwise stated) and the time history of the surface elevation η (m) at a selected wave gauge (WG).

3.1. *Convergence history*

By examining the L_2 norm of the residuals (Res), we can observe the performance in convergence histories for the pressure (P) and momentum equations ($\bar{U}, W$) in the (x, z)-directions (see Fig. 4).

As expected, the fast convergence rate about the pressure is achieved. This is divided into two regimes according to the trend of curves: first to decay nearly exponentially in the transient phase with a relatively short time, and then reach saturation (see Fig. 4). The results exhibit the well

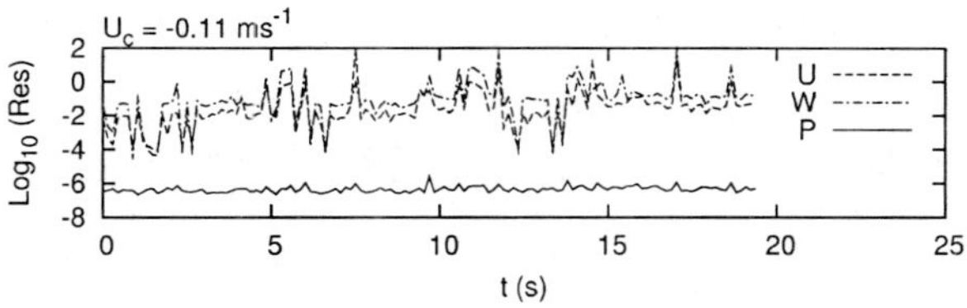

Figure 4. Convergence histories of the L_2 norm of the residuals (U, W, P) over grids (291×56) for Case 1 (the regular opposing waves).

overall performance in computations, based on the fact that solution of the pressure-Poisson equation is essentially the most time-consuming part. This behaviour ensures that our simulations are stable even during lengthy computations, as in this case the iterative number for the pressure is completely dependent on the residual norm, whereas local peaks for the velocities are clearly magnified over their curves.

3.2. *Grid refinement effects*

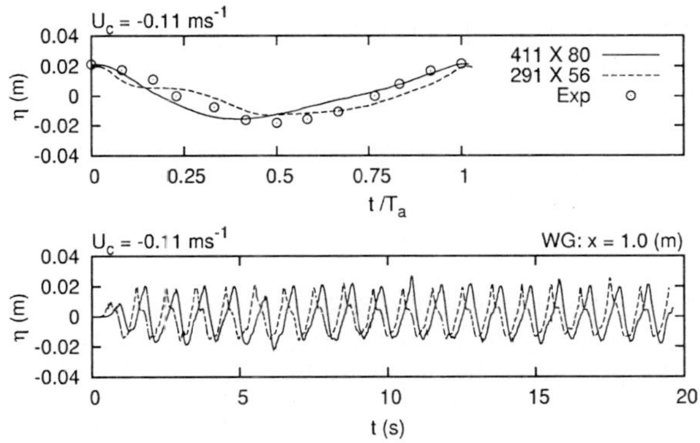

Figure 5. Grid refinement study with two meshes $(411 \times 80$ and $291 \times 56)$ for Case 1 (the regular opposing waves): the mean surface profiles and the wave train at a certain location.

By observation, a common knowledge is that more detailed physical phenomena of motions can be represented with increased mesh refinement (411×80), as illustrated in Fig. 5. It is shown that the shape of the wave profiles may be resolved adequately for both meshes. Consequently, the differences attributed to the grid effects are relatively small for the surface elevation but mainly limited to the capture of fine structures for the velocity fields[3].

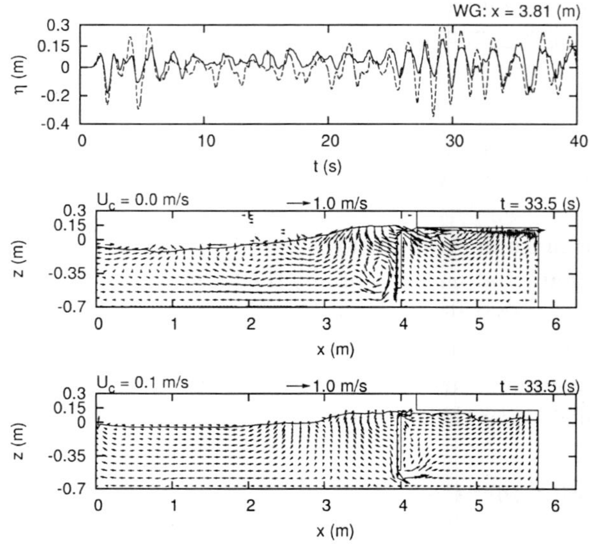

Figure 6. Wave-body coupling with and without a current over grids (251 × 40) for Case 3 (irregular waves): the wave train at a certain location and the instantaneous velocity fields. —— $U_c = 0.1\ ms^{-1}$; – – – $U_c = 0.0$.

3.3. *A verticl wall*

Fig. 6 displays a typical example regarding wave-body coupling problems in the presence of a positive current or not for Case 3. By observation, the waves overtop over a vertical barrier in the front of a pier, until strongly reflected at the seawall, as one case of diffraction of the irregular waves by a barrier sitting above the bottom. By comparison, the wave-current interaction tends to be less severe, especially in the near-barrier region, characterized by a horizontal large-scale eddy shed from the barrier. Additionally, the overall shapes in the wave trains exhibit the typical feature of nonlinearity after perturbing from its rest state.

3.4. *A sloping structure*

For a well-defined problem, Case 4 (a seadike problem) is a more challenging case, because of its impressive natural phenomenon, characterized by various combination of wave propagation with a current, shoaling, reflection, breaking and overtopping. Interestingly, the results in 3D remain very similar to the simulations in 2D in the non-breaking region (see Fig. 7), and as a result of wave dynamics being two-dimensional in this area.

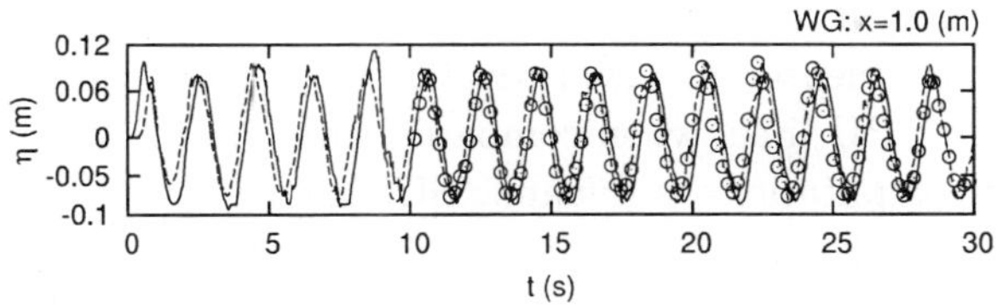

Figure 7. Effects of 3D on 2D for Case 4 (regular waves alone). —— 3D (205 × 31 × 40); – – – 2D (205 × 1 × 40); ◎ Measurement.

Finally, we give the perspective view of the free-surface profiles at two certain time levels (see Fig. 8). This provides a straightforward observation for the wave development in the region of interest. As expected, the water surface is still smooth in the nonbreaking region but becomes corrugated due to disturbances caused by the turbulent current in waves breaking.

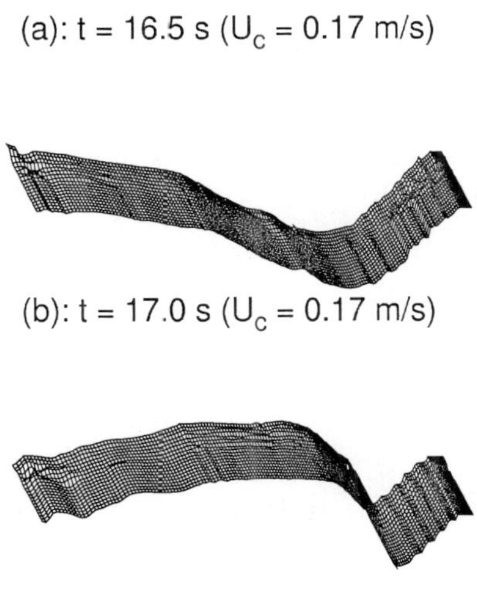

Figure 8. Perspective view of the wave profiles on grids (205 × 21 × 40) for Case 4 (regular waves following the current speed, $U_c = 0.17\ m/s$).

4. Conclusions

For waves propagating in a numerical wave tank, we emphasise a particular situation: the wave dynamics is still in 2D within the nonbreaking region, by comparison between 2D and 3D. Consequently, a breaking-type wave absorber is superior to the other standard conditions without loss of generality. Importantly, it is very valuable in assessing our design wave-current generators (where measurements are available), as computations in 2D cost much more less CPU time, when compared with that of 3D.

In our 2D computations, a dynamic SGS model can be considered as a turbulence model, which helps to damp turbulence, caused in waves breaking, as it is found that turbulence model-based solver can give a better accuracy, especially for the lengthy computations.

References

1. Fredsϕe, J., Andersen, K. H. and Sumer, B. M., Wave plus current over a ripple-covered bed, *Coastal Engineering*, 38 (1999).
2. Kemp, P. H. and Simons, R. R., The interaction between waves and a turbulent current: waves propagating with the current, *J. Fluid Mech.*, 116 (1982).
3. Li, T., Troch, P. and De Rouck, J., Interactions of breaking waves with a current over cut cells, *J.Comp.Phys.*, (accepted, 2006).
4. Li, T., Troch, P. and De Rouck, J., Wave overtopping over a sea dike, *J.Comp.Phys.*, 198 (2004).
5. Li, T., Troch, P. and De Rouck, J., On the study of wave-current-structure interactions in a numerical wave tank, In Proc. 16th International Offshore and Polar Engineering Conference, 3 (2006).
6. Li, T., Computation of turbulent free-surface flows around modern ships, *Int.J.Numer.Meth.Fluids*, 43 (2003).

ESTIMATING FREAK WAVE OCCURRENCE IN THE AGHULHAS CURRENT WITH SWAN

A.J. Lansen[1, 2], L.H. Holthuijsen[1] and A.J. van der Westhuysen[1]

The third-generation wave model SWAN, which accounts for current-induced refraction, frequency shifting, wave-current energy transfer, wind generation, white-capping and non-linear wave-wave interactions, has been experimentally extended to compute the Benjamin-Feir Index as an indicator for the occurrence of freak waves. This index is defined as the ratio of the wave steepness over the width of the frequency spectrum. Computations in the coastal jet and a ring of the Agulhas current under conditions of strong currents and (locally) moderate wind and waves, show an increase of a factor two of the probability of encountering individual waves that are higher than twice the significant wave height and an additional factor three due to wind generation.

INTRODUCTION

Freak waves are sometimes seen as the cause of serious accidents on the open seas. It is generally accepted that a freak wave is an exceptionally high wave that occurs unexpectedly, seemingly without any relation to its immediate surroundings, which may well be quite moderate. In other words: a freak wave stands out from the ambient statistical conditions. The effect is often that such a wave surprises the crew of a ship, sometimes with disastrous effects. Due to the fact that freak waves are often observed in strong non-uniform counter currents such as the Agulhas Current, current-induced refraction has been suggested as one of the underlying mechanisms of generating such waves. The focusing effect of these currents may cause energy to concentrate in caustics, which would then result in large, steep waves. This in itself does not produce freak waves but high, steep waves may induce non-linear effects such as small-scale self-focusing, which may create a short train of waves with unusual characteristics. Freak waves may then occur due to instability in the modulation in such a train (controlled by the ratio of the width of the energy spectrum over the average wave steepness; also known as the Benjamin Feir Index, BFI; see Janssen, 2003 and others). We use the SWAN wave model (Booij et al., 1999) in various wind and wave conditions to investigate whether waves in the Agulhas Current can reach critical values of the BFI, i.e., to what extent the probability of encountering freak waves is enhanced by the presence of the current.

[1] Faculty of Civil Engineering and Geosciences, Delft University of Technology, Stevinweg 1, Delft, 2628 CN, the Netherlands, [2] now at Royal Haskoning, Barbarossastraat 35, Nijmegen, 6522 DK, the Netherlands

REFRACTION IN THE AGULHAS CURRENT

Current-induced refraction is formulated in our wave model SWAN as a separate term for wave propagation in spectral (direction) space; in addition to propagation in physical space (we used the spherical option in SWAN). The spectral balance equation, in which these formulations are used, is expressed in terms of action density. SWAN thus accounts for the energy exchange by radiation stresses between waves and currents. It also accounts (in deep water) for energy-bunching (the shoaling-like effect of currents), frequency-shifting (related to Doppler shifting), generation by wind (using the wind speed at 10 m elevation, relative to the local current; Yan, 1987), quadruplet wave-wave interactions (the DIA approximation of Hasselmann et al., 1985) and white-capping (van der Westhuysen et al., 2006, based on the saturation model of Alves and Banner, 2003).

For the computations, we consider two regions of the Agulhas Current: the coastal jet (along the east coast of South Africa) and a ring (some 2000 km to the south west; see Fig. 1). The actual current fields that we used were numerically simulated by the U.S. Naval Research Laboratory with their NRL Layered Ocean Model (NLOM, see Smedstad, 2003) over a large region (as shown in Fig. 1) with an exceptionally high spatial resolution ($1/32^0$).

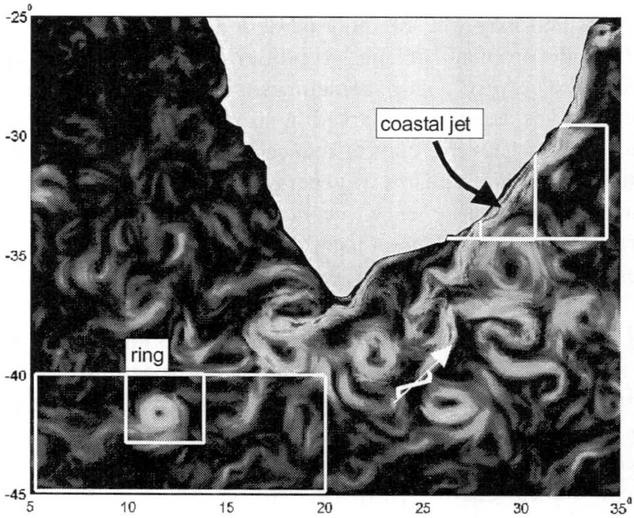

Fig. 1 The Agulhas Current (coastal jet and ring) as used in the computations. The computational regions are indicated with the large boxes and output regions with the smaller boxes (coordinates in longitude, latitude degrees).

To illustrate the effects of refraction, we selected one moment in time with strong currents (August 21st, 2001) and we imposed moderate wave conditions at the boundaries of the computational regions (see Fig. 1) with the wind and the waves opposing the main current along the south-east coast of South Africa (the coastal jet towards NE, wind speed 15 m/s, significant wave height 1 m and mean wave period 3.5 s). A generic illustration for such a situation (without wind) is given by Lavrenov (1992; see Fig. 2), which shows that the current-induced refraction diverts wave energy away from lateral regions (as seen in the direction of wave propagation) to the centre of the jet. The actual current field that we used (magnitude of the velocity), and the results of the computations in terms of the significant wave height, are shown in Fig. 3. The (relative) increase in wave height in this case, i.e., compared to a situation without currents, reaches a maximum of a factor 1.4 in the north-eastern section of the coastal jet. As indicated by the generic picture, the energy required for this seems to be taken from adjacent areas alongside the jet, particularly near the south-western section where the waves enter the jet. The increasing enhancement of the wave height suggests that the wave grow steeper and steeper as they propagate into the Agulhas Current and therefore also more and more nonlinear.

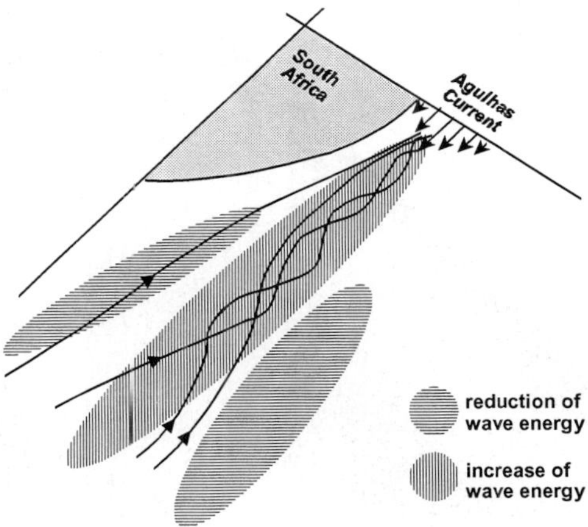

Fig. 2 The diversion of wave energy from lateral regions to the centre of the coastal jet of the Agulhas Current due to current-induced refraction (after Lavrenov, 1992).

ENHANCED FREAK WAVE PROBABILITY IN THE COASTAL JET

The wave heights of random, deep water waves are usually assumed to be Rayleigh distributed (in the sense that the waves are usually assumed to behave as linear waves):

$$p\left(\underline{H} \geq hH_s\right) = \exp\left(-2h^2\right) \tag{1}$$

where H_s is the significant wave height. To quantify the effect of the nonlinear character of the waves on the probability of encountering high (i.e., freak) waves, Janssen (2003) introduced the Benjamin-Feir Index (BFI), which is defined as the ratio of wave steepness over normalized spectral width:

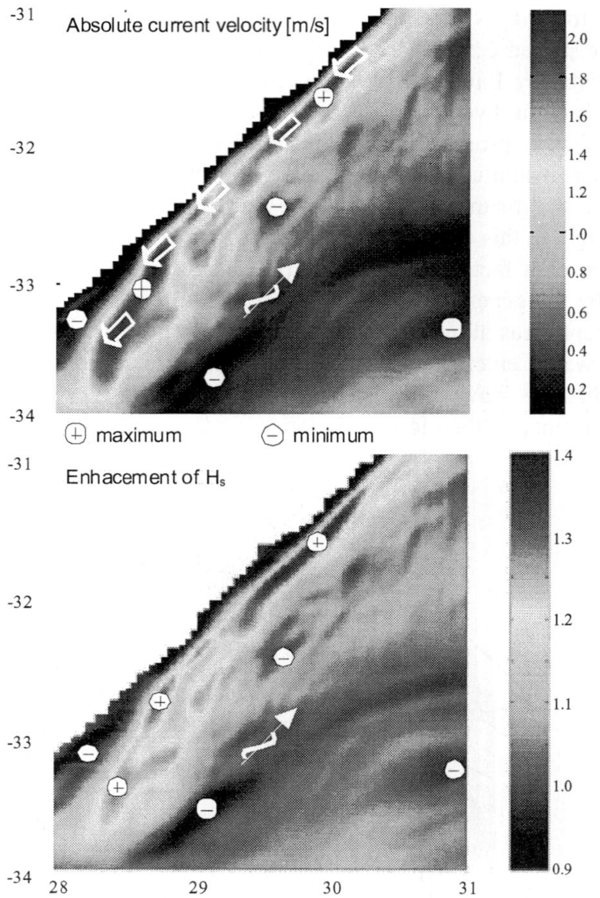

Fig. 3 The absolute current velocity and the enhancement of the significant wave height due to current-induced refraction.

$$BFI = \varepsilon\sqrt{2}/\sigma_{\omega}^{'} \tag{2}$$

in which $\varepsilon = k_0 m_0^{1/2}$ and $\sigma_{\omega}^{'} = \sigma_{\omega}/\omega_0$ (k_0 and ω_0 are a characteristic wave number and frequency respectively, m_0 is the variance of the waves, and σ_{ω} is a measure of the spectral width). The distribution of the wave heights is then the Rayleigh-Edgeworth distribution:

$$p\left(\underline{H} \geq hH_s\right) = \exp\left(-2h^2\right)\left[1 + C_{BFI}\right] \tag{3}$$

in which $C_{BFI} = \frac{\pi}{3\sqrt{3}} 2h^2\left(h^2 - 1\right) BFI^2$. The corresponding enhancement of

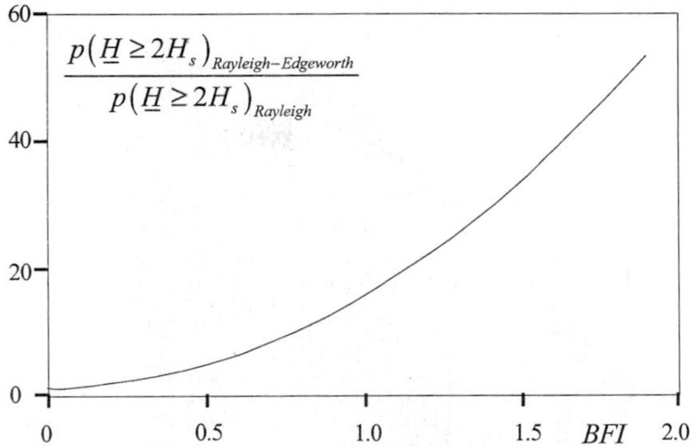

Fig. 4 The enhancement of the probability that an individual wave height is larger than twice the significant wave height due to nonlinear effects.

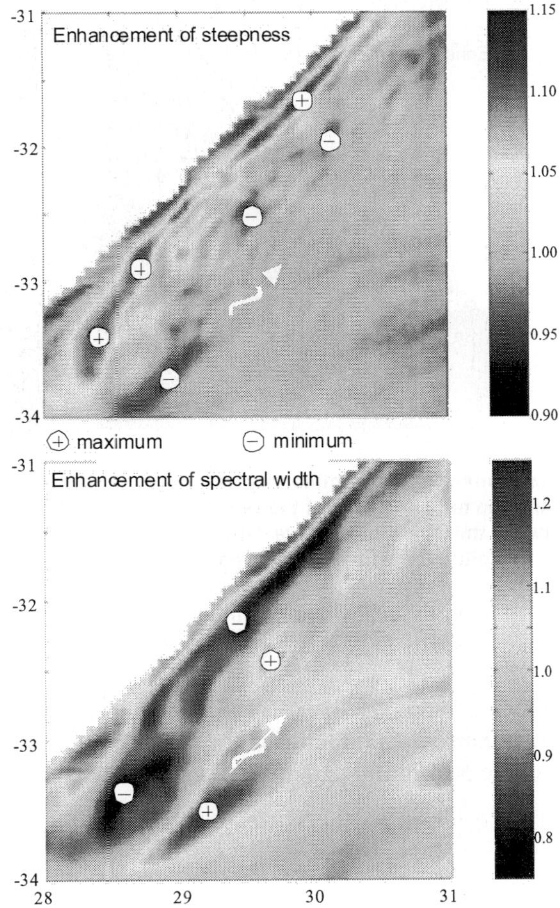

Fig. 5 The enhancement of the wave steepness and spectral width (due to the current) in the coastal jet of the Agulhas Current.

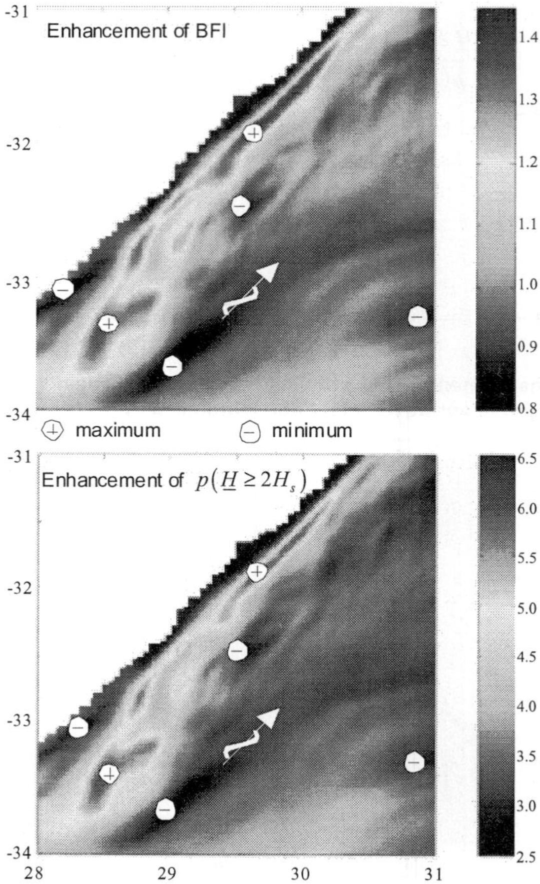

Fig. 6 The enhancement of the nonlinearity of the waves (BFI; due to the presence of the current) and the enhancement of the probability that an individual wave height is larger than twice the significant wave height in the coastal jet of the Agulhas Current (relative to a situation with linear waves; i.e. due to current and wind).

the probability that an individual wave height is larger than twice the significant wave height is shown in Fig. 4.

For the computations, we have chosen the mean wave number as characteristic wave number and, following the suggestion of Janssen (2003), we use the inverse of the peakedness parameter of Goda (1976) as spectral width:

$$\varepsilon = k_{m_{01}} m_0^{1/2} \quad \text{and} \quad \sigma_\omega' = Q_p^{-1} \tag{4}$$

in which the mean wave number $k_{m_{01}}$ is based on the zero-th and first-order moment of the wave spectrum and $Q_p = 2\int_0^\infty \omega E^2(\omega)\, d\omega / (\int_0^\infty E(\omega)\, d\omega)^2$. These parameters and the BFI were added to the output of SWAN and results of the computations in the coastal jet of the Agulhas Current are shown in Figs. 5 and 6. It is obvious that the parameters follow the same general geographic pattern as the current field: where the counter current is strong, the significant wave height and wave steepness are enhanced (and the spectral width is reduced). The effect is to enhance the BFI by a factor as much as 1.45. The increase in the probability of encountering a high wave is illustrated here with the probability that an individual wave is higher than twice the significant wave height. For the actual values of the BFI (not shown here) this probability is up to a factor 6.5 higher (in the upper left-hand quadrant of the coastal jet; shown in the lower panel of Fig. 6) than in a situation with linear waves (when a Rayleigh distribution is used). Additional computations indicate that without currents but with wind, the enhancement of this probability would be a factor 3 (due to the wind steepening the waves).

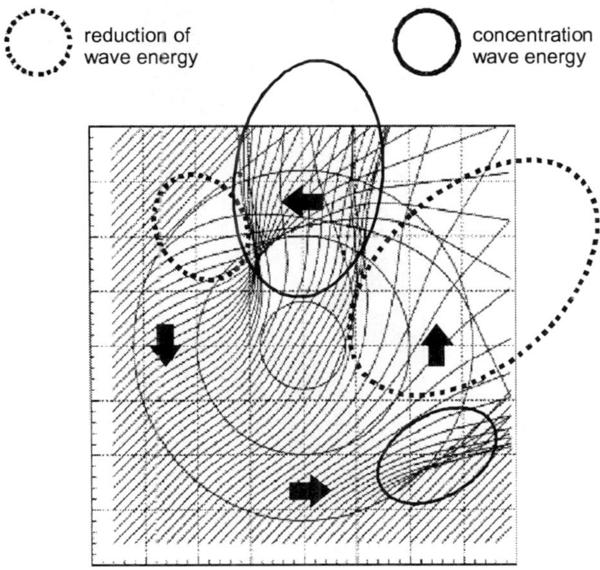

Fig. 7 The diversion of wave energy from the upper right-hand quadrant (as seen in the wave direction) to the upper left-hand quadrant of an Agulhas Ring due to current-induced refraction (after Gerber, 1993 and White and Fornberg , 1998).

REFRACTION AND ENHANCED FREAK WAVE PROBABILITY IN A RING

To illustrate the effects of refraction in an Agulhas Ring, we use the same ambient situation as above (August 21st, 2001; significant wave height = 1 m and mean wave period = 3.5 s; wind and wave directions are towards NE and the wind speed is 15 m/s). A generic illustration for such a situation is given by Gerber (1993) and White and Fornberg (1998; see Fig. 7). It shows that the current-induced refraction diverts wave energy from the upper-right-hand quadrant of the ring to the upper left-hand quadrant (all seen in the direction of the wave propagation). In addition, there is a slight concentration on the right-hand side. The actual current field (magnitude of the velocity) and the results of the computations in terms of the significant wave height are shown in Fig. 8.

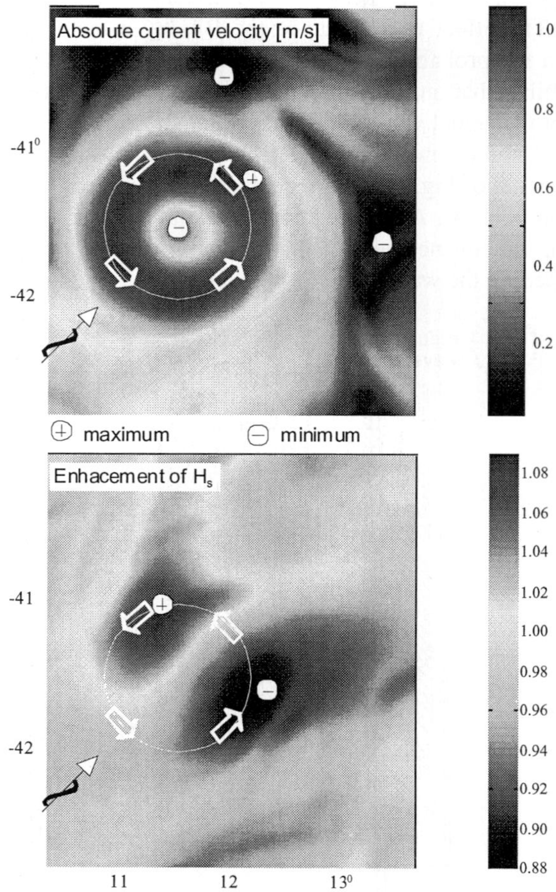

Fig. 8 The absolute current velocity and the enhancement of the significant wave height due to current-induced refraction in an Agulhas Ring.

The increase in significant wave height in this case reaches a maximum of a factor 1.08 in the upper left-hand quadrant of the ring and a minimum of 0.88 in the upper-right-hand quadrant (seen in the direction of incident wave propagation). These enhancements and reductions agree qualitatively with the generic picture above and it suggests that the waves grow steeper and steeper as they propagate into the upper left-hand quadrant and therefore also more and more nonlinear. The enhancement of the wave steepness and spectral width are shown in Fig. 9. The enhanced values of the BFI and the enhancement of the probability that an individual wave height is larger than twice the significant wave height is shown in Fig. 10 (enhancement relative to a situation with linear waves; i.e. due to current and wind).

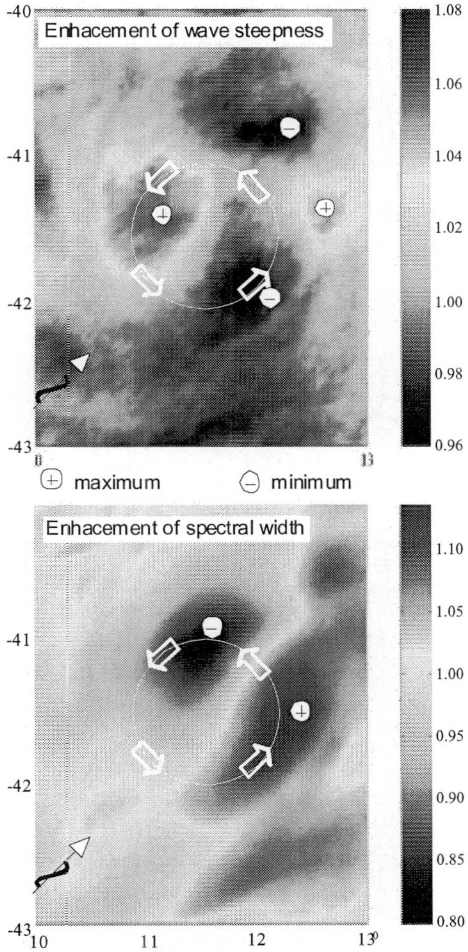

Fig. 9 The enhancement of wave steepness and of the spectral width (due to the current) in the Agulhas Ring.

It is obvious that the parameters react to the focusing effect of the ring: where the significant wave height is enhanced, the wave steepness is enhanced and the spectral width is decreased. The effect is to enhance the BFI by a factor as much as 1.3. The probability of encountering a high wave is illustrated here with the probability that an individual wave is higher than twice the significant wave height is considerable. For the actual values of the BFI (not shown here) this probability (shown in Fig. 10) is, in the upper left-hand quadrant, up to a factor 5.5 higher than for a situation with linearly behaving waves. In the upper-right-hand quadrant the enhancement is less (up to a factor 3). The computations also indicate that without the currents, the enhancement of this probability would be a factor 3 (due to the wind steepening the waves).

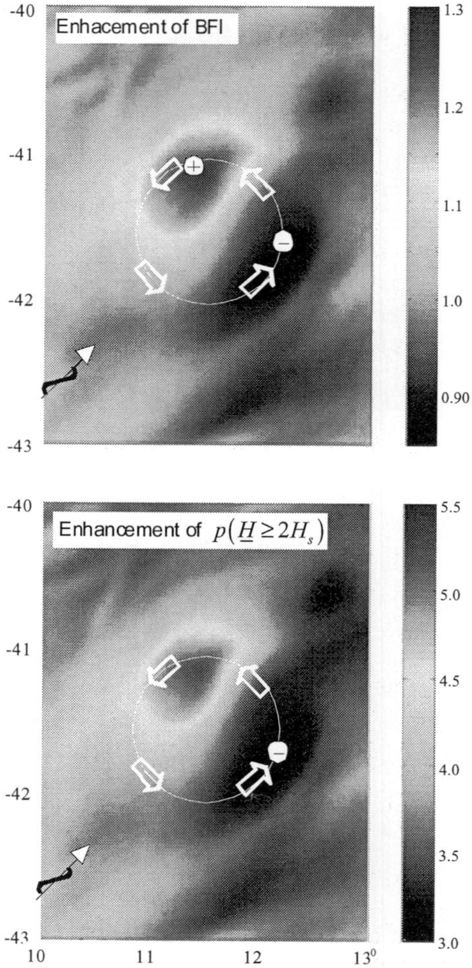

Fig. 10 The enhancement of the nonlinearity of the waves (BFI) and of the probability that an individual wave height is larger than twice the significant wave height (due to the current) in the Agulhas Ring.

CONCLUSIONS

The speculation that the probability of encountering a freak waves is enhanced in strong oceanic counter currents is substantiated with the results of this study for the coastal jet of the Agulhas Current. Such enhancement is partly due to current-induced refraction, which induces higher wave steepness, but it is also due to direct wave-current interactions, which induces narrower spectra. The combination increases the Benjamin-Feir Index (BFI), which acts as a freak wave indicator. Additional computations show that the same combination of enhanced steepness and narrower spectra occurs at the lee side of the counter-current side of a ring of the Agulhas Current, where the ring concentrates wave energy. There too, the spectra are narrower, so that the BFI is higher. The opposite occurs at the other half of the ring (the following-current side), where the ring decreases the wave energy and broadens the spectrum. These results are essentially generic and therefore also applicable to other, similar conditions (western boundary currents). However, in view of the meteorological and oceanographic situation (dominant western wind conditions and easterly currents), the combination of high, steep waves propagating against the current (a jet or a ring) seems to be (far) more likely in the Agulhas Current than in the Gulf Stream or the Kuro Shio (where the wind, the currents and coastal geography do not favor such conditions).

ACKNOWLEDGMENTS

We gratefully acknowledge the gracious permission of U.S. Naval Research Laboratory to use their numerically simulated current fields of the region of the Agulhas Current, which were carried out with exceptionally high geographic resolution (see text). We also gratefully acknowledge the financial support of Royal Haskoning and the Delft University of Technology for the first author (AJL).

REFERENCES

Alves, J. H. G. M., Banner, M. L., 2003. Performance of a saturation-based dissipation-rate source term in modelling the fetch-limited evolution of wind waves. *J. Phys. Oceanogr.*, 33, 1274--1298

Booij, N., R.C. Ris and L.H. Holthuijsen, 1999, A third-generation wave model for coastal regions, Part I, Model description and validation, *J. Geophys. Res.*, 104, C4, 7649-7666

Gerber, M., 1993, The interaction of deep-water gravity waves and an annular current – linear theory, *J. Fluid Mech.*, 248, 153 - 172

Goda, Y, 1976, On wave groups, *Proc. Intl. Conf. Behaviour Offshore Structures*, The Norwegian Institute of Technology, 1, 115 - 128

Hasselmann, S., K. Hasselmann, J.H. Allender and T.P. Barnett, 1985a, Computations and parameterizations of the nonlinear energy transfer in a gravity wave spectrum. Part II: Parameterizations of the nonlinear transfer for application in wave models, *J. Phys. Oceanogr.*, **15**, 11, 1378-1391

Janssen, P.A.E.M., 2003: Nonlinear four-wave interactions and freak waves. *J. Phys. Oceanogr.*, 33, 863–884.

Lavrenov, I.V., 1998, The wave energy concentration in the Agulhas current off South Africa, *Natural Hazards*, 17, 117 – 127

Van der Westhuysen, A. J., M. Zijlema and J. A. Battjes, 2006, Nonlinear saturation-based whitecapping dissipation in SWAN for deep and shallow water, *Coastal Engineering* , in press

White, B.S. and B. Fornberg , 1998, On the change of freak waves at sea, *J. Fluid Mech.*, 355, 113 - 138

Yan, L., 1987. An improved wind input source term for third generation ocean wave modelling, Rep. No. 87-8, Royal Netherlands Meteorological Institute, 20 pp.

TYPHOON EARLY WARNING SYSTEM BASED ON A COUPLED ATMOSPHERE-OCEAN-WAVE MODEL

Jun Yoshino[1], Tomokazu Murakami[1], Masanori Hayashi[1], Takashi Yasuda[1]

A high-resolution coupled atmosphere-ocean-wave model is developed for typhoon early warning, and is applied to the case of Typhoon Chaba (2004). The coupled model is based on three existing models: MM5 as the atmospheric model, CCM as the oceanic model, and SWAN as the wave model. The ocean coupling has an important effect on the storm intensity change, due to the cooling of the sea surface temperature impacted by the storm. It is argued that inclusion of air-sea interaction processes is necessary to reduce the positive intensity bias often observed in mature storms.

INTRODUCTION

Background

Recorded 10 typhoons have destroyed many lives of coastal residents on the Japan islands in the 2004 Northwest Pacific typhoon season. Typhoon- and flood-related claims for the fiscal 2004 were 18 times higher than that of the previous year (5 billion dollars). To reduce future disaster risks, an early warning system, which can provide highly accurate information concerning not only typhoon track (latitude and longitude of storm center) but also typhoon intensity (central pressure, maximum sustained wind speed, and radius of 50-kt winds), must be implemented and refined.

Many meteorological organizations, JTWC Guam, NHC Miami, CPHC Honolulu and RSMC Tokyo routinely provide first-level basic information of future position and intensity of tropical cyclones in each basin. Recently, a dynamic model, rather than a statistical model, is generally used for tropical cyclone prediction. Although rapid development of meteorological, remote sensing and computing technologies makes it possible to simulate the track of tropical cyclone with high accuracy, it would be difficult to accurately evaluate the storm intensity even in short-term forecasts. This difficulty arises from the facts that the current models are: (1) too coarse a spatial grid to correctly capture the sharp discontinuous structure near the eyewall, (2) beyond using the specific parameterizations for key air-sea interaction processes, and (3) computed using the inaccurate and arbitrary initialization using a tropical cyclone bogussing scheme. In particular, the second problem is crucial for realistic intensity simulation since the major energy source of tropical cyclone is the diabatic heating and moistening from the warm ocean surface. It is considered that coupling an ocean dynamic model with a typhoon model can be one possible solution to this problem.

---

[1] Graduate Course of Engineering, Gifu University, 1-1 Yanagido, Gifu, 501-1193, Japan

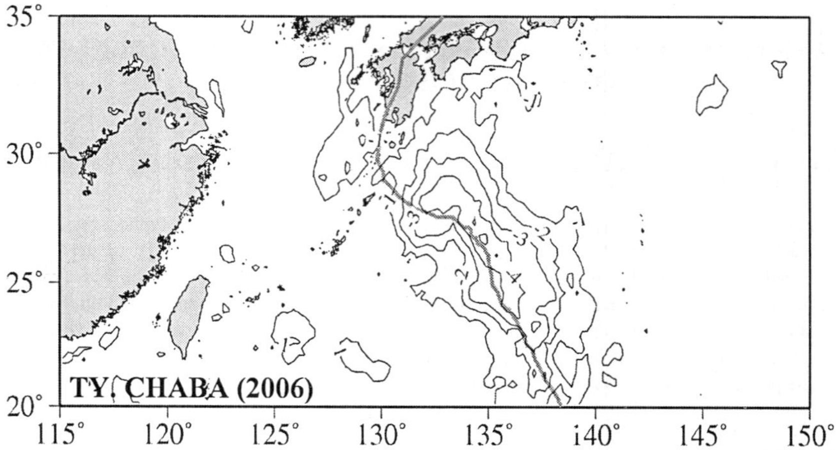

Figure 1. Sea surface temperature difference (26 August – 30 August 2004) between before and after the passage of Typhoon Chaba (2004).

A number of numerical studies (Sutyrin and Khain 1984; Khain and Ginis 1991; Bender et al. 1993; Schade and Emanuel 1999) used their coupled model to investigate that storm intensity change indeed is a very sensitive to air-sea interaction processes. It is clearly shown that using a coupled model can accurately predict the intensity of many tropical cyclones.

The goals of the present study are to develop a high-resolution limited-area atmosphere-ocean-wave coupled model for typhoon intensity forecasts over the Northwest Pacific Ocean, and to quantify the impact of complex sea surface processes on storm intensity using the coupled model. We select Typhoon Chaba (2004), as a case study, which has caused severe damage along the Japan coastal area in August 2004.

Typhoon Chaba (2004)

Chaba first reached the typhoon strength around the Marshall Islands on August 19, 2004. The storm moved slowly northwest, and the central pressure fell to 910hPa, and wind gusts were estimated in the excess of 50 m/s on August 25. Landfall was made in the Kyushu islands on August 30. The typhoon brought disaster to the coastal areas in Japan, and it quickly weakened and was turned sharply northeastward and merged with an extratropical system over the Okhotsk Sea.

Fig. 1 shows the sea surface temperature (SST) differences between before and after the passage of Chaba, observed by the passive microwave satellite sensors TRMM/TMI and Aqua/AMSR-E (downloaded from Remote Sensing Systems website at http://www.ssmi.com/). In the mature stage, Chaba drew its energy from extensive ocean area with a temperature greater than 29 degree C and moved slowly northwestward over the warm ocean (see Fig. 3). However, the strong storm caused a large and rapid SST decrease, finally reaching a maximum decrease of about 4 degree C likely along and just to the right of the storm track. Chaba experienced the gradual intensity decay, from 930 to 940 hPa during the two days. Bender et al. (1993) suggested that such a large SST decrease tends to be produced by a slowly moving storm with a speed of less than 5 m/s. The time period of interest in this study is during this dissipating stage from 00Z 27 to 00Z 29 August 2004.

MODEL DESCRIPTIONS

A high-resolution coupled atmosphere-ocean-wave model (typhoon forecast version) developed in this study is based on three existing models: MM5 as the atmospheric component (Dudhia 1993), CCM as the oceanic component (Murakami et al. 2004), and SWAN as the wave component (Booij et al. 1997). The coupled model simultaneously represents both atmospheric and ocean processes, and interactively exchanges information at the interface among these three model components, at every 10 minutes (See Fig. 2).

MM5 (Dudhia 1993) is a fully compressible, non-hydrostatic, finite-differencing, terrain-following sigma-coordinate, cloud-resolving primitive equation model, which contains a comprehensive physics parameterization packages, developed by the Pennsylvania State University and National Center for Atmospheric Research. This model is designed to simulate complex mesoscale meteorological phenomena on the order of a few to several hundred kilometers. Specific physical parameterizations for this simulation include the Mellor-Yamada level 2.5 PBL scheme (Janjić 1994), the Grell (1993) cumulus parameterization, and an explicit cloud microphysical scheme (including separate treatment of cloud water, rain, cloud ice, and snow) implemented by Reisner et al. (1993).

CCM (Murakami et al. 2004) is a multi-sigma coordinate coastal current

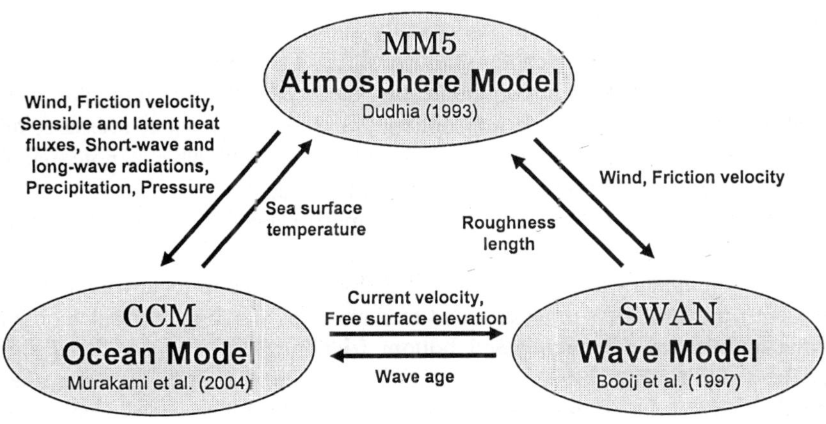

Figure 2. A schematic diagram of the coupled atmosphere-ocean-wave model developed in this study.

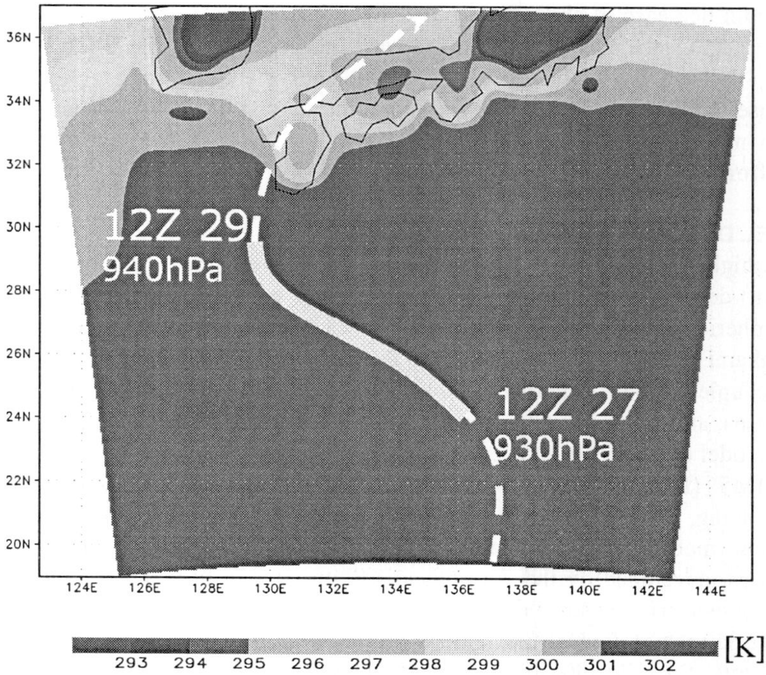

Figure 3. A computational domain for CPLD and SNGL, showing the sea surface temperature distribution (shaded) with the typhoon track.

model originally developed by the authors, and is composed of momentum, mass, temperature, salinity and density equations and is closed mathematically with the Mellor and Yamada level 2.5 turbulent closure scheme. The vertical multi-sigma coordinate approach allows to realistically simulating 3-D coastal ocean flow and water quality from deep to shallow water. The simultaneous coupling is physically achieved through communicating the ocean surface temperature field to MM5 and the near-surface atmospheric variables (e.g. wind, pressure, heat fluxes, radiations, and precipitation) to CCM. It would have a potential for profound impacts on both typhoon and ocean at large.

SWAN (Booij et al. 1997) is a sophisticated third-generation time-dependent spectral wave model designed for the near-shore zone and can effectively simulate wave generation and propagation due to local winds, as well as wave transformation processes due to refraction, shoaling, current blocking, whitecapping, wave breaking and bottom friction. Since wave-induced stress may be a significant component of the total momentum stress in the atmospheric boundary layer, the coupling effect cannot be ignored. The coupling processes are represented by mutual interaction of the wind waves and boundary layer

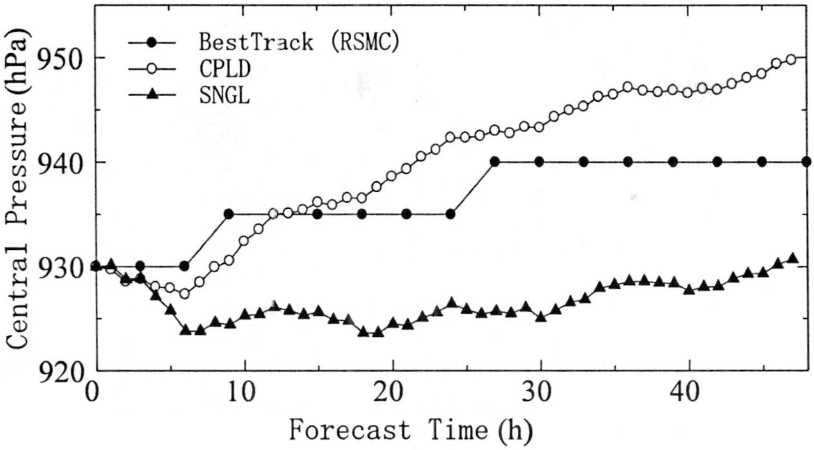

Figure 4. Time series (48 h) of the minimum surface pressure of Typhoon Chaba (2004), observed by RSMC and simulated by CPLD and SNGL.

stress. The high frequency waves (i.e. young ocean waves) have the fastest growth rate due to wind forcing (Janssen 1989), probably resulting in weaker surface winds inside the typhoon boundary layer affected by enhanced roughness length.

The coupled model runs with horizontal grid spacing of 10 km for MM5 and 14 km for CCM and SWAN, which cover a 2000 × 2000 km wide region that contains the nearshore waters of the Japan islands (Fig. 3). The horizontal resolution of MM5 is much higher than that of the current conventional typhoon model in JMA (Japan Meteorological Agency). The forecasting period is 48 hours from 1200 UTC 27 through 1200 UTC 29 August in 2004. Initial and boundary conditions are taken from NCEP global analyses (1-degree resolution) for MM5, JCOPE (Japan Coastal Ocean Predictability Experiment) regional ocean analyses (10-km resolution) for CCM, and steady-state values diagnosed by initial surface winds for SWAN. The JMA best track report is applied for typhoon initialization using a typhoon bogussing scheme. This modeling system closely emulates a real-time numerical prediction in spite of the simulations being performed in a hindcast mode.

To isolate the importance of ocean coupling, two experiments are conducted for this case. The first simulation makes use of the coupled model developed in this study (henceforth CPLD). The second simulation is done by the uncoupled model using a stand-alone atmospheric model, which assumes that the standard Charnock parameter (Charnock 1955) and steady-state SST condition for the energy flux calculations (hereafter SNGL).

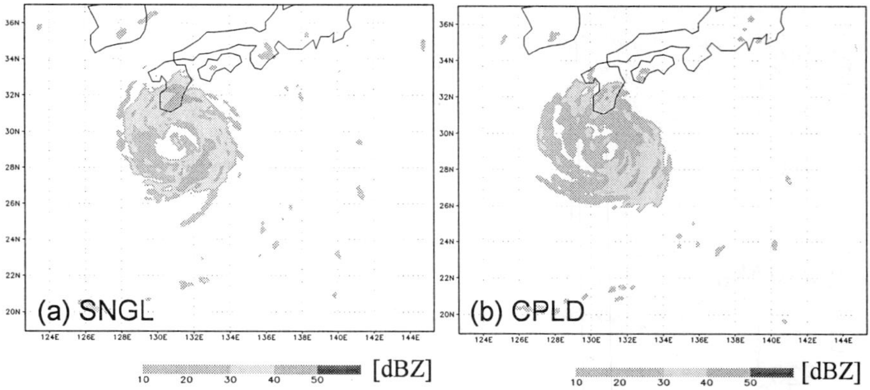

Figure 5. Horizontal distributions of the 800-hPa radar reflectivity (shaded) simulated by (a) SNGL; and (b) CPDL, at 12Z 29 August 2004.

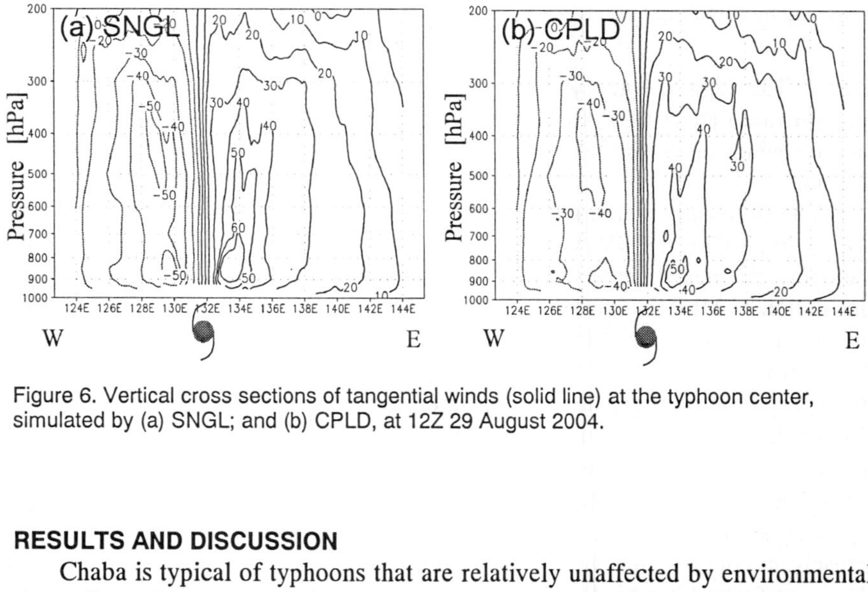

Figure 6. Vertical cross sections of tangential winds (solid line) at the typhoon center, simulated by (a) SNGL; and (b) CPLD, at 12Z 29 August 2004.

RESULTS AND DISCUSSION

Chaba is typical of typhoons that are relatively unaffected by environmental shear flow. The uncoupled and coupled simulations predict the storm track accurately and contain similar small error, implying that track simulation is insensitive to any ocean coupling processes (not shown). However, the typhoon intensity or structure in CPLD is significantly affected by the complex air-sea interaction.

Fig. 4 compares the time evolution of the central pressure among observed and simulated storms. There is a large difference in the typhoon intensity change between these CPLD and SNGL. The intensity in CPLD varies from 930 to 950 hPa during the 48-h integration, in better agreement with the best-track intensity.

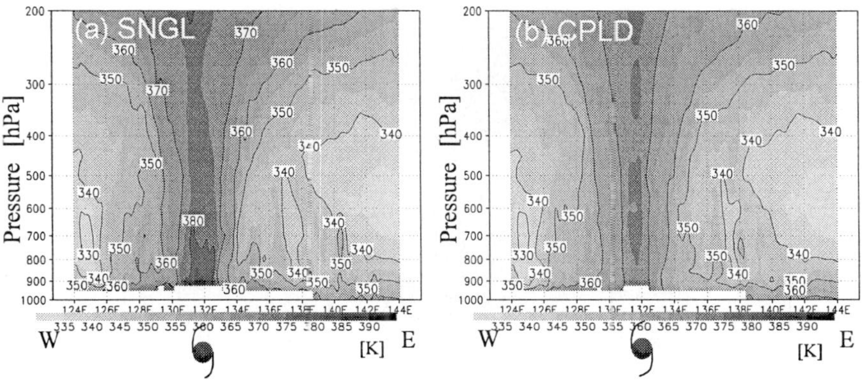

Figure 7. Vertical cross sections of the equivalent potential temperature (shaded) at the typhoon center, simulated by (a) SNGL; and (b) CPLD, at 12Z 29 August 2004.

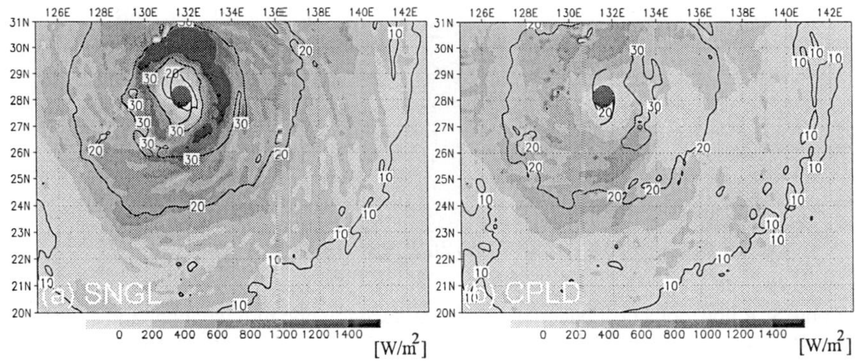

Figure 8. Horizontal distributions of the 10-m wind speed (solid line) and latent heat flux (shaded) simulated by (a) SNGL; and (b) CPLD, at 12Z 29 August 2004.

In contrast, the storm in SNGL keeps central pressure constant about 930 hPa during the forecast period, which is comparatively low (i.e. stronger). The results indicate that the difference in the treatment of ocean surface processes leads to be a crucial error in typhoon intensity forecast, and the inclusion of ocean surface coupling reduces the positive intensity bias (Fig. 4).

We discuss the simulated characteristics of the atmospheric and ocean fields at the final time of the forecasts (48-h simulation time). Figs. 5a and 5b show the cloud structure simulated by SNGL and CPLD, respectively. Both typhoons contain the near-axisymmetric form near the center. Notice that CPLD lacks the typical eye and eyewall cloud structure, compared to SNGL. The 20-hPa difference between both simulations influences on the typhoon cloud structure and even rainfall rate.

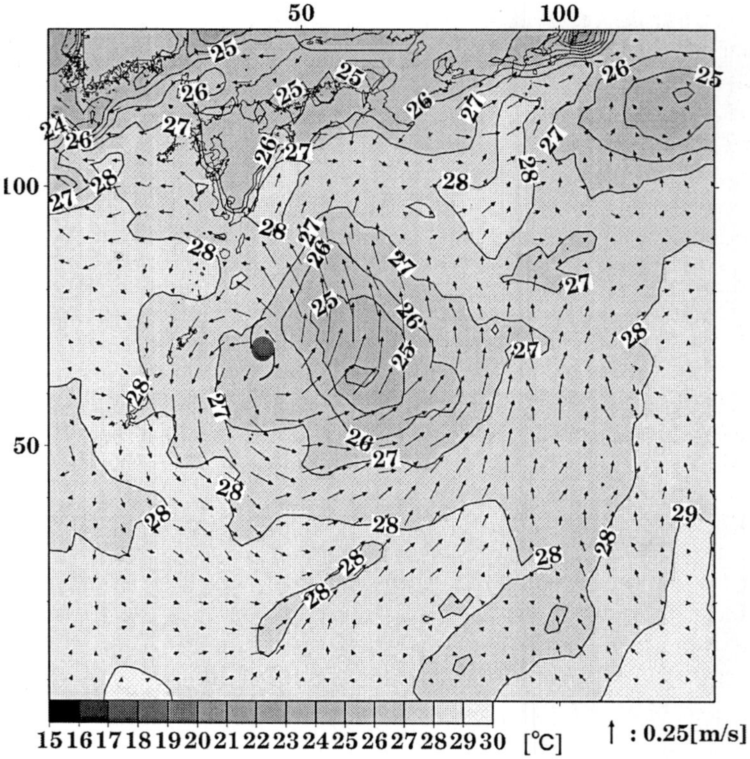

Figure 9. A horizontal distribution of the sea surface temperature (shaded) and surface current (vectors) simulated by the ocean model in CPLD, at 00Z 29 August 2004.

Likewise, the typhoon-associated tangential wind fields, given in Figs. 6a (SNGL) and 6b (CPLD), show a large difference between both simulations. It is evident that the wind field near the eyewall in SNGL becomes more intense, more compact, and more symmetric than that in CPLD. In other words, the ocean coupling effect increases from 50 m s^{-1} in CPLD to 60 m s^{-1} in SNGL.

Such features are also visible in the thermal and moisture structure. Figs. 7a and 7b show the equivalent potential temperature profiles in both simulations, showing the typhoon warm core structure. The low-level equivalent potential temperature in SNGL is approximately 10 degree C warmer than that in CPLD near the typhoon center. The boundary layer in SNGL is warmer and more humid as a consequence of the exclusion of the ocean and wave models. Emanuel (1986) theoretically suggested that the central pressure decrease in tropical cyclone has close relation to the equivalent potential temperature increase in the storm center. Our results are also consistent with the theory.

Thus, while CPLD reasonably reproduces the typhoon structure both in horizontal and vertical, the storm in SNGL seems to be excessively strong and extremely reinforced during a 48-h forecast. It is seen that these differences of simulated typhoon structure arise from the different energy exchanges at the air-sea interface between CPLD and SNGL.

To check whether the ocean coupling effect alters the energy inflow into the typhoon core, we compare the distributions of the latent heat fluxes at the sea surface between SNGL and CPLD, as shown in Figs. 8a and 8b, respectively. In SNGL, the evaporation rate (more than 1400 W m^{-2}) is much larger than that in CPLD (600–800 W m^{-2}), and shows more symmetric distribution at a radial distance of 100–200 km from the typhoon center. The sensible heat flux distributions also show similar patterns, but the effect is secondary importance (not shown). The huge surface energy inflow in SNGL leads to be the excessive development over the warm ocean and causes a positive intensity bias error. We stress that using relevant parameterizations about the air-sea interaction is quite important and essential for typhoon intensity forecast.

Finally, we explore the upper ocean response impacted by the strong typhoon in CPLD (see in Fig. 9 and Fig. 10). A rapid ocean cooling of about –5 degree C occurs at the east of the typhoon center, where a strong southerly current induces strong vertical turbulent mixing (Fig. 9). As shown in Fig. 10, under the strong typhoon impact (0300 UTC 28 August), the ocean mixed layer depth is shallower and the vertical stratification is stronger (see a 25 degree C

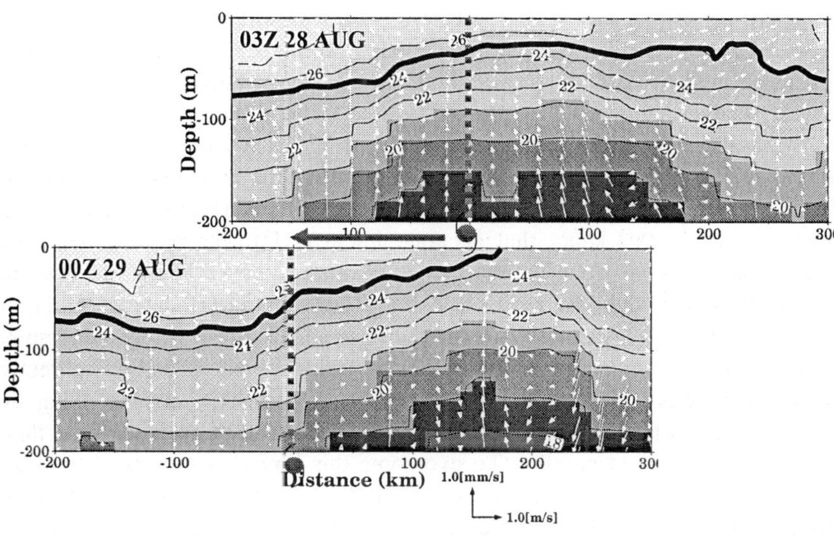

Figure 10. Vertical cross sections of the temperature (shaded) and *x-z* current (vectors) at the typhoon center, simulated by the ocean model in CPLD, at (upper) 03Z 28 and (lower) 00Z 29 August 2004. Temperature at 25 degree C is indicated by the thick line.

line), resulting in the favorable condition for ocean surface cooling (0000 UTC 29 August). This condition is enhanced by the cold water pumped up from the deep ocean due to the large-scale upwelling at the typhoon center, which is likely induced by the slow-moving typhoon. The result implies that in order to predict the realistic typhoon intensity, a full 3-D ocean model, rather than a 1-D mixed layer model generally often used, should be implemented into a coupled model.

SUMMARY AND CONCLUSIONS

In this study, a very sophisticated coupled atmosphere-ocean-wave model was developed for typhoon early warning system. The model consists of a nonhydrostatic atmospheric model, MM5, coupled with an ocean model, CCM, and a wave model, SWAN. The modeling system is capable of enhancing ocean cooling, due to ocean surface mixing and upwelling. It is found that the ocean coupling had an important effect on storm intensity. Inclusion of the ocean surface processes was essential to reduce the positive intensity bias during the forecast when the typhoon was greatly impacted by the reduced, reasonable supply of heat and moisture.

Although the ocean coupling indeed reduces the central pressure bias, CPLD still contains a +10 hPa bias error at the final time of the simulation (48-h simulation time), as shown in see Fig. 3. It must be noted that accurate intensity prediction is still difficult even if taking into account the complex ocean surface interaction leads to better surface fluxes. The changes in the central pressure are rather indirect consequences due to many mutual interactions localized in time and space. The respective parameterizations in the coupled model have compensating errors. Given that our results depend strongly on the choice of parameterization, further research is needed to decide which most reliably represents realistic phenomena, and to extend the validity of algorithms in extreme weather conditions.

In addition, there still remain some big problems to operationally forecast the typhoon intensity change using this model. First one is related to the quality of initialization in the ocean model as well as typhoon model. Although it is clearly shown in this study that accurate intensity forecast requires real-time information of upper ocean thermal structure, in fact, observations are few and limited. However, there is reasonable hope that the recent data assimilation techniques (e.g. 3DVAR, 4DVAR and the ensemble Kalman filter) using some satellite observations available can estimate ocean thermal structure indirectly. The other serious problem, moreover, is the computational cost in using the coupled model, which carries out calculations on such huge grids with the three independent models in parallel. To overcome this limitation, we need to either use a very recent high-performance parallel computer, such as the Earth Simulator in Japan, or largely simplify the basic equations in each component model, similar to the method of the Coupled Hurricane Intensity Prediction System (CHIPS) described in Emanuel et al. (2004).

Finally, if we can resolve these issues, the advanced coupled modeling system can operationally provide useful information not only about future position and intensity but also about the possible hazard of typhoon-related disasters (e.g. wind gusts, high waves and storm surge) along the coastal areas.

ACKNOWLEDGEMENTS

We would like to thank Dr. T. Ohsawa (Kobe University) who provided a tropical cyclone bogussing scheme for typhoon initialization. This research was originally funded by the 2005 Grants-in-Aid program for Scientific Research by the Japan Society for the Promotion of Science.

REFERENCES

Bender, M. A., I. Ginis, and Y. Kurihara. 1993. Numerical simulations of tropical cyclone-ocean interaction with a high-resolution coupled model, *J. Geophys. Res.*, 98, 245–263.

Booij, N., L. H. Holthuijesn, N. Doorn, and A. T. M. M. Kieftenbur. 1997. Diffraction in a spectralwave model, *Proc. Third Int. Symp. Ocean Wave Measurement and Analysis*, 234–255.

Charnock, H. 1955. Wind stress on a water surface, *Quart. J. Roy. Meteor. Soc.*, 81, 639–640.

Dudhia, J. 1993. A nonhydrostatic version of the Penn State-NCAR mesoscale model: Validation test and simulation of an Atlantic cyclone and cold front, *Mon. Wea. Rev.*, 121, 1493–1513.

Emanuel, K. A. 1986. An air-sea interaction theory for tropical cyclones. Part I: Steady-state maintenance, *J. Atmos Sci.*, 43, 585–604.

Emanuel, K. A., C. DesAutels, C. Holloway, and R. Korty. 2004. Environmental Control of Tropical Cyclone Intensity, *J. Atmos. Sci.*, 61, 843–858.

Grell, G. A. 1993. Prognostic evaluation of assumptions used by cumulus parameterizations, *Mon. Wea. Rev.*, 121, 764–787.

Janjic´, Z. 1994. The step-mountain eta coordinate model: Further developments of the convection, viscous sublayer, and turbulence closure schemes, *Mon. Wea. Rev.*, 122, 927–945.

Janssen, P. A. E. M. 1989. Wave-induced stress and the drag of air flow over sea waves, *J. Phys. Oceanogr.*, 19, 745–754.

Khain, A., and I. Ginis. 1991. The mutual response of a moving tropical cyclone and the ocean, *Beitr. Phys. Atmos.*, 64, 125–141.

Murakami, T., Y. Yasuda, and T. Ohsawa. 2004. Development of a multi-sigma coordinate model coupled with an atmospheric model for the calculation of coastal currents, *Ann. J. Coastal Eng.*, 51, 366–370 (in Japanese).

Reisner, J., R. T. Bruintjes, and R. M. Rasmussen. 1993. Preliminary comparisons between MM5 NCAR/Penn State model generated icing forecasts and observations, *Preprints, Fifth Int. Conf. on Aviation Weather Systems*, AMS, 65–59.

Schade L. R., and K. A. Emanuel. 1999. The ocean's effect on the intensity of tropical cyclones: results from a simple coupled atmosphere-ocean model, *J. Atmos. Sci.*, 56, 642–651.

Sutyrin, G. G. and A. P. Khain. 1984. Effect of the ocean atmosphere interaction on the intensity of a moving tropical cyclone, *Izvest. Atmos. Ocean. Phys.*, 20, 697–703.

WAVE HEIGHT LONG TERM PREDICTION BASED ON THE USE OF THE SPREAD PARAMETER

R. Gencarelli[1], G.R. Tomasicchio[2], P.Veltri[1]

A new method for the geographical transposition of extreme wave height is proposed. It extends the use of the spread parameter (Goda 2002, 2004) in order to determine the value of the extreme significant wave height with a long term return period at a sea area where only a few years of wave records are available. The procedure has been verified at the Italian and Spanish sea areas. The application has considered the buoys of the Italian National Sea Wave measurement Network (RON) and of the Calabria Regional Network. In particular, the value of the significant wave height with 50 years return period has been determined offshore Ancona (January 1999-Dicember 2003), along the Adriatic coastline, and at the Calabria Regional buoys of Capo Spartivento and Roccella Ionica (since 2003).

INTRODUCTION

In 1989 the Italian Ministry of Public Works started up the National Sea Wave measurement Network (RON) providing measurements and analysis of wave data in the Italian seas at deep water conditions with excellent results in terms of data acquisition rates, temporal coverage and reliability. The number of RON buoys is continuously increased through the time.

The reliability of the extreme wave height analysis depends on the number of available years of registration. Traditionally, when this number is small, the extreme wave height analysis considers a geographical transposition method based on the ratio between the effective fetches at the considered locations (Steijn, 1989; Shore Protection Manual, 1992; Leenknecht et al., 1992).

The present study proposes and verifies a geographical transposition method allowing to transfer information on the extreme wave climate from one location to another location. The method considers the spread parameter, $\gamma_{50,10}$, defined as (Goda, 2001)

$$\gamma_{50,10} = \frac{H_{50}}{H_{10}} \tag{1}$$

with H_{50} and H_{10} significant wave height, H_s, with 50 and 10 years return periods, respectively.

[1] Soil Conservation Dept., University of Calabria, Ponte Bucci 41 B, Rende, 87036, Italy
[2] Engineering Dept., University of Lecce, Via per Monteroni, Lecce, 73100, Italy

THE SPREAD PARAMETER

The characteristics of the spread parameter have been showed and discussed by Goda (2004) with reference to the extreme wave climate at the Japan seas. The sea around the Japanese island has been divided in three different parts: North Pacific, South Pacific and Japan Sea.

A different number of buoys has been considered at each of these areas. For each buoy in the same sea area, the type of selected probability distribution function results the same, but the values of the scale, A, and location, B, parameters are different (Table 1). A mean value of the spread parameter, $\bar{\gamma}_{50,10}$, at a certain sea area is given. The value of the standard deviation, $\sigma(\gamma_{50,10})$, of $\gamma_{50,10}$ at the considered buoys results negligible. The sea area with these characteristics is defined as *homogeneous* for the extreme wave climate.

Table 1. Mean value and standard deviation of the spread parameter around Japan (Goda, 2004).

Region	Probability distribution	Numbers of buoys	$\bar{\gamma}_{50,10}$	$\sigma(\gamma_{50,10})$
North Pacific	Weibull, k =1	10	1.21	0.0133
SouthPacific	Gumbel	9	1.24	0.0190
Japan Sea	Weibull, k =1.4	11	1.13	0.0038

Goda (2004) compared the results of the extreme wave analysis at the Japan sea with the results from the *Working Group of Section of Maritime Hydraulics* (WG-*IAHR*) for the two locations of Haltenbanken, along the Atlantic coast of Norway, and Kodiac, along the North-East Pacific of Alaska. Table 2 reports the selected probability distribution functions and the values of A, B, H_{10}, H_{50} and $\gamma_{50,10}$ for the two locations. In particular:

1. the value of $\gamma_{50,10}$ at Kodiac is equal to $\bar{\gamma}_{50,10}$ at the Japan Sea area ($\gamma_{50,10}$=1.13);
2. Kodiac and the Japan Sea area have the same selected probability distribution function (Weibull, k =2), but values of A and B are different.

Table 2. Extreme wave characteristics of Haltenbanken and Kodiac (Goda,2004).

Buoy	Probability distribution	A [m]	B [m]	H_{10} [m]	H_{50} [m]	$\gamma_{50,10}$
Haltenbanken	Weibull, k =2	5.03	0.7	13.25	14.78	1.115
Kodiak	Weibull, k =1.4	1.86	5.81	10.51	11.91	1.133

PROPOSED EXTENSION OF THE SPREAD PARAMETER

If the number of wave records at one buoy allows the reliable use of the POT (Peak Over Threshold) and Goda's extreme wave height analysis (Goda, 1988; Goda and Konube, 1990), values of H_{50}, H_{10} and $\gamma_{50,10}$ are obtained. If a value of $\bar{\gamma}_{50,10}$ characterizes an homogeneous sea area, the proposed method allows to determine the value of H_{50} at all buoys that have been recently installed within the sea area (Figure 1).

The method is suitable for a buoy having a number of wave records which is to small for a reliable estimate of H_{50}, but sufficient for a reliable estimate of H_{10} based on the wave height extreme analysis. Specifically, the value of H_{10} at a recently installed buoy is obtained by means of the extreme wave height analysis; the knowledge of $\bar{\gamma}_{50,10}$ allows to estimate H_{50}.

The proposed extension of the use of spread parameter has been verified at the Italian and Spanish seas.

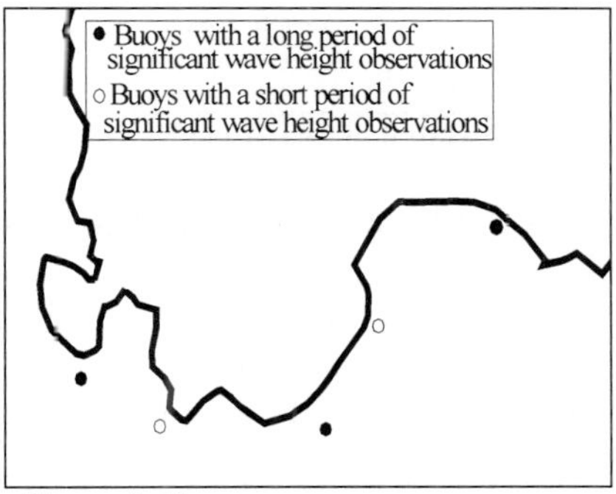

Figure 1. Homogeneous sea area.

VERIFICATION AT THE ITALIAN SEAS

The Italian seas can be grouped in four different areas: South Adriatic, Ionian, Tyrrhenian and Ligure (Figure 2). The buoys of Monopoli, Ortona and Ancona fall in the South Adriatic; La Spezia buoy in the Ligure sea; Ponza and Cetraro buoys in the Tyrrhenian sea; Catania and Crotone buoys in the Ionian sea.

Values of the spread parameter at the RON buoys with a long period of significant wave height records (1989-2003) have been estimated (Table 3). The values of H_{50} and H_{10} have been calculated with the probabilistic extreme wave height analysis (Goda, 1988; Goda and Konube, 1990) method. These values of H_{10} and H_{50} favourably compare with the values in "Italian Seas Wave Atlas" (Corsini et al., 2004) which considered the use of POT method.

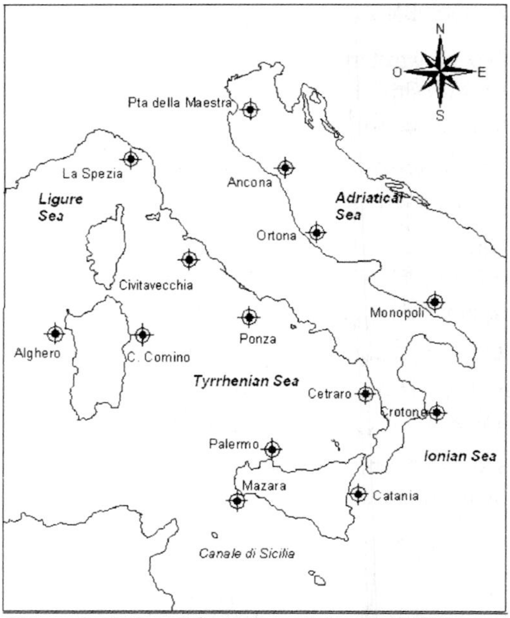

Figure 2. National Sea Wave Network (RON).

Table 3. Values of the spread parameter at RON buoys (July 1989-Dicember 2003).

Buoy	Probability distribution	A [m]	B [m]	H_{10} [m]	H_{50} [m]	$\gamma_{50,10}$
La Spezia	Weibull, k =1.4	1.33	2.13	6.6	7.8	1.18
Alghero	Weibull, k =2	3.79	0.66	9.8	11	1.12
Ponza	Weibull, k =0.75	0.43	2.78	7.1	8.8	1.24
Mazara	Weibull, k =2	2.72	0.015	6.5	7.3	1.12
Catania	Weibull, k =2	2.65	0.2	5.9	6.8	1.15
Crotone	Weibull, k =2	2.27	0.86	6.1	6.8	1.11
Monopoli	Weibull, k =2	1.6	1.24	4.9	5.4	1.10
Ortona	Weibull, k =2	2.11	1.02	5.8	6.5	1.12

With reference to the Monopoli and Ortona buoys, in the Adriatic sea area, it is possible to observe that:

- they present the same selected probability distribution function (Weibull, k =2), but values of A and B at the two buoys are different;
- $\gamma_{50,10}$ is a constant a the two locations ($\bar{\gamma}_{50,10}$ =0.11, $\sigma(\gamma_{50,10})$=0.0065).

The South Adriatic sea area can be assumed as an homogeneous area for the extreme wave climate. Consequently, the proposed geographical transposition method has been applied. In particular, in Ortona, the value of H_{50} calculated for the full set of data (1989-2003), with the use of traditional extreme wave analysis, has been compared with the value of H_{50} calculated with the use of a reduced set of data and $\bar{\gamma}_{50,10}$; it has been supposed that the period of significant wave height records at the Ortona buoy is only 5 years (1989-1993). The probability distribution function has been selected for this reduced set of data and the value of A , B , and H_{10} have been calculated (Figure 3 and Table 4). The value of H_{50} for this reduced data set has been calculated with the use of mean value of spread parameter at the South Adriatic sea. The error between the value of H_{50} for the reduced data set (1989-1993) and the full set of data (1989-2003) is lower then 1% (Table 5). Similar results have been obtained when considering a reduced set of data for a different temporal interval.

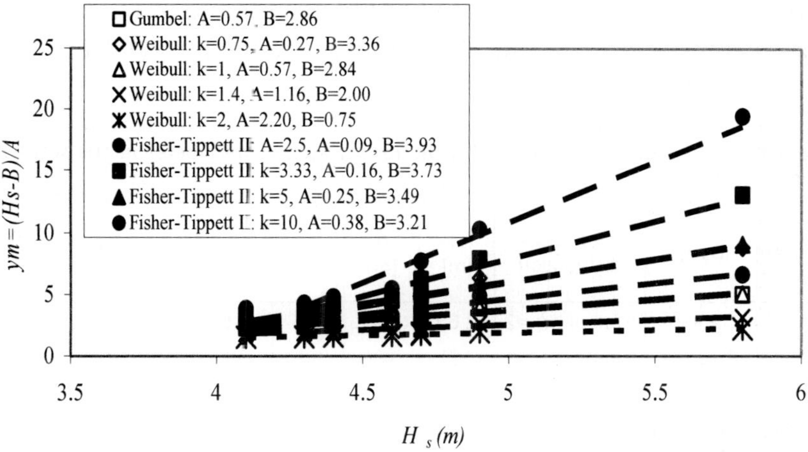

Figure 3. H_s **extreme probability distribution function (Weibull,** k **=2). Ortona buoy (July 1989-December 1993).**

Table 4. Correlation coefficient and standard deviation for the candidate distributions. Ortona buoy (July 1989-December 1993).

Ortona buoy	FT-I	Weibull				Fisher-Tippett II			
		k =0.75	k =1	k =1.4	k =2	k =2.50	k =3.33	k =5	k =10
Correlation Coefficient	0.974	0.978	0.975	0.971	0.982	0.97	0.98	0.98	0.978
Standard deviation	0.09	0.081	0.09	0.11	0.06	0.13	0.07	0.74	0.81

Table 5. Values of H_{50} for the reduced and the entire set of data. Ortona buoy.

Probability Distribution	A	B	y_{10}	H_{10} [m]	$\bar{\gamma}_{50,10}$	H_{50} (reduced) [m]	H_{50} (full) [m]	error [%]
Weibull, k =2	2.20	0.75	2.21	5.7	1.11	6.4	6.5	0.8

The analysis has been conducted considering the Catania and Crotone buoys in the Ionian sea area (Table 6). Similarly to the South Adriatic case, although they present the same selected probability distribution function, the values of A and B at the two buoys are different; in addition $\gamma_{50,10}$ is nearly constant. As a consequence, the Ionian sea is homogeneous for the extreme wave climate and the value of H_{50} for the buoys with a short period of significant wave record can be calculated with the proposed geographical transposition method.
The value of H_{50} at the Catania buoy has been calculated for the full set of data (1989-2003); it has been compared with H_{50} determined with a reduced set of data (1989-1993) and the use of $\bar{\gamma}_{50,10}$. The error is less then 1% (Table 7).

Table 6. Mean value and standard deviation of spread parameter at the Ionian sea area.

Region	Buoy	Probability distribution	A [m]	B [m]	H_{10} [m]	H_{50} [m]	$\gamma_{50,10}$	$\bar{\gamma}_{50,10}$	$\sigma(\gamma_{50,10})$
Ionian Sea	Catania	Weibull, k =2	2.65	0.2	5.9	6.8	1.15	1.135	0.01
	Crotone	Weibull, k =2	2.27	0.86	6.1	6.8	1.12		

Table 7. Values of H_{50} for the reduced set and the full set of data.

Probability distribution	A [m]	B [m]	y_{10}	H_{10} [m]	$\bar{\gamma}_{50,10}$	H_{50} (reduced) [m]	H_{50} (full) [m]	error [%]
Weibull, k =2	2.7	0.36	2.16	6.1	1.135	6.9	6.8	0.7

VERIFICATION AT THE SPANISH SEAS

The verification of the proposed transposition method has been extended at the Spanish seas considering two different areas in the Mediterranean Sea and in the Atlantic Ocean. In particular, the Cap Tortosa (1990-2002) and Llobregat (1984-2002) buoys have been considered in the Mediterranean (Figure 4). The extreme wave analysis for the two locations has considered the Goda (1988) and

Goda and Konube (1990) method. The following is observed (Table 8):

- the two buoys have the same selected probability distribution function (Weibull, k =0.75), but values of A and B at the two buoys are different;
- $\gamma_{50,10}$ is a constant a the two locations ($\bar{\gamma}_{50,10}$=1.245, $\sigma(\gamma_{50,10})$=0.015).

The Mediterranean Spanish sea is homogeneous for the extreme wave climate.

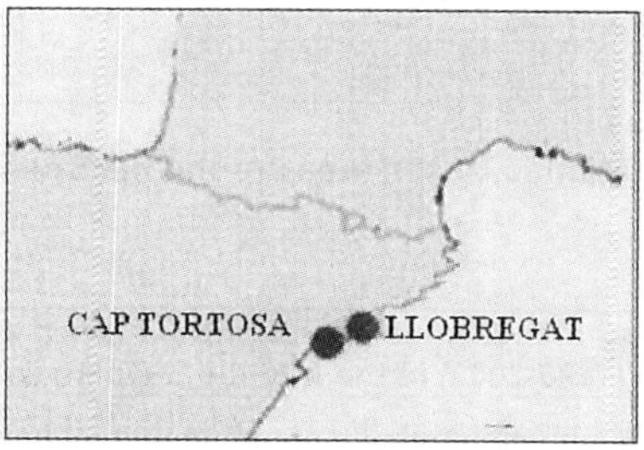

Figure 4. Mediterranean Sea area around the Spanish Sea.

With reference to the Llobregat buoy, Table 9 reports the values of H_{50} from the entire (1984-2002) and a reduced (1984-1988) set of data. The resulting error from the adoption of the proposed method for the reduced set of data is equal to 0.09%.

Table 8. Mean value and standard deviation of spread parameter for Llobragat (1984-2002) and Cap Tortosa (1990-2002) in the Mediterranean Spanish sea area.

Buoy	Probability Distribution	A [m]	B [m]	H_{10} [m]	H_{50} [m]	$\gamma_{50,10}$	$\bar{\gamma}_{50,10}$	$\sigma(\gamma_{50,10})$
Llobregat	Weibull, k =0.75	0.22	2.48	4	4.9	1.23	1.245	0.015
Cap Tortosa	Weibull, k =0.75	0.34	2.34	5	6.3	1.26		

Table 9. Values of H_{50} for the reduced set and the full set of data. Llobregat buoy.

Buoy	Probability distribution	A [m]	B [m]	H_{10} [m]	$\bar{\gamma}_{50,10}$	H_{50} (full)	H_{50} (reduced)	Error [%]
Llobregat	Weibull, k=0.75	0.13	2.66	4.1	1.245	4.9	5.0	0.09

Similarly, La Coruña and Silleiro buoys (Figure 5) define an homogeneous sea area (Table 10) and H_{50} can be determined with a reduced set of data. The error for La Coruña buoy has resulted equal to 0.8 % (Table 11).

Figure 5. Atlantic Ocean area around the Spanish Sea.

Table 10. Mean value and standard deviation of the spread parameter at la Coruna (1984-2002) and Silleiro (1990-2002) in the Atlantic Spanish sea.

Buoy	Probability distribution	A [m]	B [m]	H_{10} [m]	H_{50} [m]	$\gamma_{50,10}$	$\bar{\gamma}_{50,10}$	$\sigma(\gamma_{50,10})$
La Coruna	Weibull, k=2	3.93	0.12	9.1	10.4	1.14	1.15	0.009
Silleiro	Weibull, k=2	5	-0.78	10	12.1	1.16		

Table 11. Values of H_{50} for the reduced set and the full set of data (1984-2002). La Coruna buoy.

Buoy	Probability distribution	A [m]	B [m]	H_{10} [m]	$\bar{\gamma}_{50,10}$	H_{50} (full)	H_{50} (reducet)	Error [%]
La Coruna	Weibull, k =2	4.6	-0.43	9.2	1.15	10.4	10.5	0.8

APPLICATION: ANCONA BUOY

After the verification of the proposed method at the Italian and Spanish seas, the value of $\bar{\gamma}_{50,10}$ has been used to calculate the value of H_{50} for the Ancona buoy having a few years of wave data records (1999-2003). The results are shown in Figure 6 and reported in Tables 12 and 13.

In order to have a deeper verification, the value of H_{50} calculated with the proposed transposition method has been compared with the value of H_{50} estimated with the use of the traditional transposition method based on the ratio of the effective fetches (Steijn,1989; Shore Protection Manual, 1992; Leenknecht et al., 1992). Following the traditional transposition method, the values of the effective fetch along the dominant direction (345°N) have been determined at the Ortona (Figure 7) and Ancona (Figure 8) buoys. Value of H_{50} at Ancona has been determined with the knowledge of H_{50} at Ortona and of the ratio between the effective fetches at Ancona and Ortona: the result is H_{50} = 5.5 m in Ancona and is very close to the result from the use of the proposed method (Table 13).

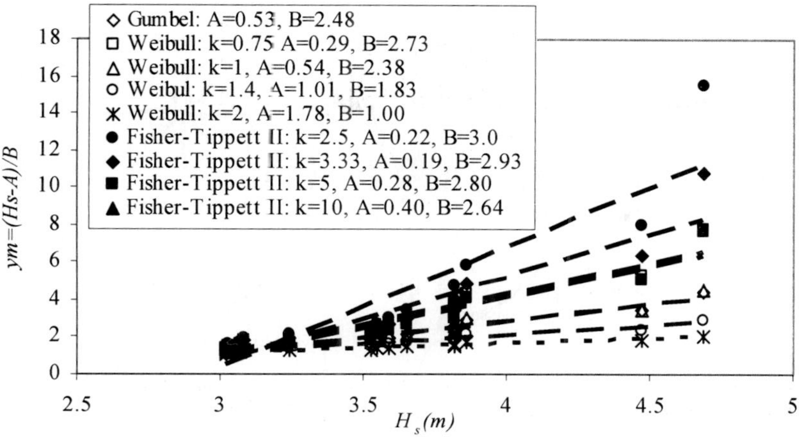

Figure 6. H_s **extreme probability distribution function (Weibull,** k **=2) Ancona buoy (July 1999-December 2003).**

Table 12. Correlation coefficient and standard deviation for the candidate distributions. Ancona buoy (January 1999-December 2003).

Ancona	Gumbel	Weibull				Fisher-Tippett II			
		k =0.75	k =1	k =1.4	k =2	k =2.5	k =3.33	k =5	k =10
Correlation coefficient	0.974	0.956	0.97	0.977	0.979	0.9	0.927	0.948	0.946
Standard deviation	0.2	0.32	0.23	0.16	0.16	3.8	0.52	0.37	0.28

Table 13. Value of H_{50} **at the Ancona buoy calculated with the use of new geographical transposition method (January 1999-December 2003).**

Probability distribution	A [m]	B [m]	y_{10}	H_{10} [m]	$\bar{y}_{50,10}$	H_{50} [m]
Weibull k =2	1.78	1	2.25	5	1.11	5.6

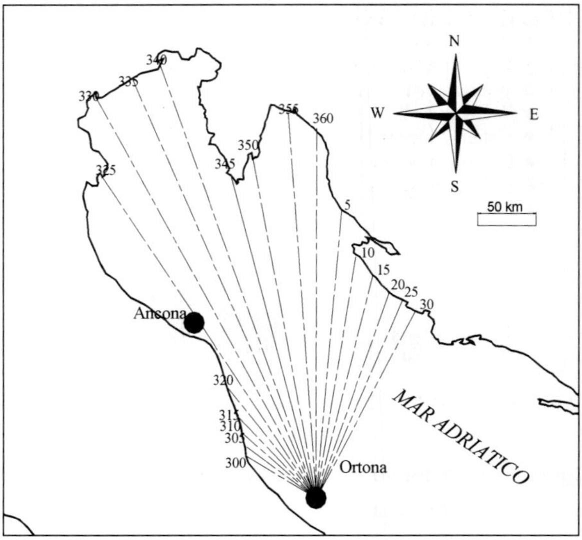

Figure 7. Geographic fetches at Ortona buoy along the dominant wind direction.

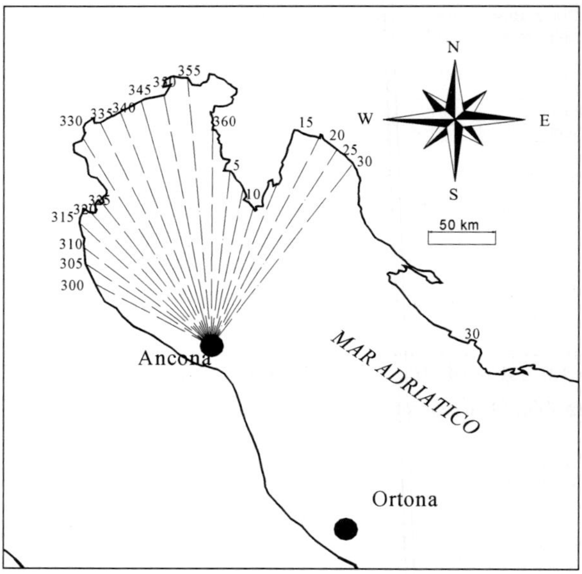

Figure 8. Geographic fetches at Ancona buoy along the dominant wind direction.

REGION CALABRIA NETWORK

In the last few years, in order to obtain more information on the extreme wave climate around the Italian coastline, some Italian Regions have been installing their own network (ROR). In particular, since 2003, Regione Calabria manages 4 deep water wave buoys sampling wave data (Figure 9): three buoys are in the Ionian Sea (Corigliano Calabro, Roccella Ionica and Capo Spartivento) and one is in the Tyrrhenian sea (Tropea).

Extreme wave analysis along the Calabria coastline

The existence of wave height records from the ROR buoys together with the new geographical transposition method give the chance to calculate the value of H_{50} at the Regione Calabria buoys, where only two years (2003-2004) of wave records are available. In particular, the ROR buoys of Roccella Ionica and Capo Spartivento are in the Ionian homogeneous area, where the value of $\bar{\gamma}_{50,10}$ is known.

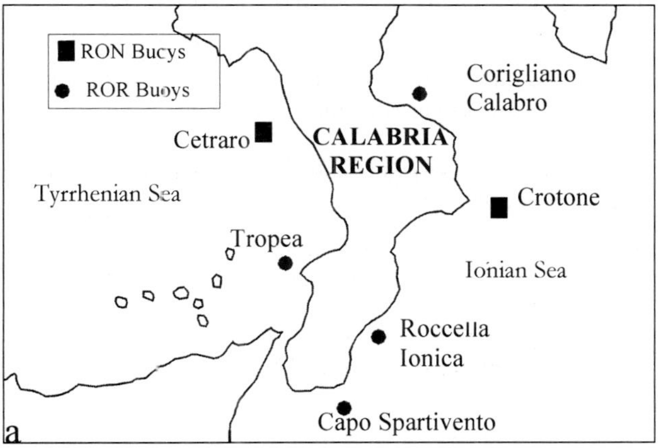

Figure 9. Regione Calabria buoys (ROR) and National buoys (RON).

For all the Calabria ROR buoys the records length does not allow to obtain reliable values of H_{10}. The spread parameter is defined as the ratio between H_{50} and H_5 values

$$\gamma_{50,10} = \frac{H_{50}}{H_5} \tag{2}$$

where H_5 is the significant wave height with 5 years return period which has been calculated by means of the extreme wave height analysis (Goda and Konobe, 1990). The parameter $\gamma_{50,5}$ shows the same characteristics of $\gamma_{50,10}$; in fact, it is constant within an homogeneous area (Table 14).

Table 14. Values of $\gamma_{50,5}$ for RON buoys (1989-2003).						
Buoy	Probability distribution	A [m]	B [m]	H_5 [m]	H_{50} [m]	$\gamma_{50,5}$
La Spezia	Weibull, k =1.4	1.33	2.13	6.3	7.8	1.24
Alghero	Weibull, k =2	3.79	0.66	9.2	11	1.20
Ponza	Weibull, k =0.75	0.43	2.78	6.7	8.8	1.31
Mazara	Weibull, k =2	2.72	0.015	6.1	7.3	1.20
Catania	Weibull, k =2	2.65	0.2	5.4	6.8	1.26
Crotone	Weibull, k =2	2.27	0.86	5.7	6.8	1.19
Monopoli	Weibull, k =2	1.6	1.24	4.6	5.4	1.17
Ortona	Weibull, k =2	2.11	1.02	5.5	6.5	1.18

With the adoption of $\gamma_{50,5}$ in the proposed method, the values of H_{50} at Roccella Ionica and Capo Spartivento have been calculated (Table 15). It has been possible to determine the value of H_{50} at only two of the Calabria ROR buoys: Corigliano Calabro buoy suffered problems during the installation; for Tropea, an homogeneous sea area cannot be defined.

Table 15. Significant wave heights with 50 years return period for Roccella Ionica and Capo Spartivento.							
Homogeneous area	Buoy	Probability distribution	A [m]	B [m]	H_5 [m]	$\bar{\gamma}_{50,5}$	H_{50} [m]
Ionian sea	Roccella	Weibull, k =2	2.7	0.78	5.05	1.22	7.2
	C.Spartivento	Weibull, k =2	2.5	0.63	5.2		8.1

CONCLUSIONS

The use of the spread parameter is extended in order to obtain a new geographical transposition method for the extreme wave heights. The method allows to determine the value of the significant wave height with 50 years return period at a buoy with only a few years of observations.

The proposed transposition method has been verified at the Italian and Spanish seas.

As an application, the value of H_{50} at the Ancona buoy (installed in 1999) has been calculated; for a deeper verification, the result has been compared with the value of H_{50} obtained with the canonical geographical transposition method based on the ratio of the effective fetches.

Application to the Regional Calabria buoys network which was started in 2003 has showed the capability of the method to determine H_{50} even if only two years of observations are available.

The proposed method does not suffer any error compared to the traditional transposition method based on the effective fetches.

AKNOWLEDGMENTS

The authors thank Prof. Yoshimi Goda and Prof. Nobuhisa Kobayashi for the fruitful discussions and useful comments.

REFERENCES

Archetti, R., and Franco, L. 1995. Nuove analisi di dati ondametrici nei mari Italiani, *Italian Section of PIANC Congress* (in Italian), 151-166.

Barbaro, G., and Tomasicchio, G.R. 1996. Applicazione e verifica di una soluzione analitica per i periodi di ritorno delle mareggiate, *Proc XXV Italian Conference on Hydraulics*, Torino, Vol. 2, 347-357 (in Italian).

Corsini, S., Inghilesi, R., Franco, L., and Piscopia, R. 2004. *Italian Wave Atlas*, Final Report – APAT, Roma.

Goda, Y. 1988. On methodology of selecting the design wave, *Proc. 21st Int. conference on Costal Eng.*, 899-913, ASCE, Malaga.

Goda, Y., and Konube, K. 1990. Distribution function fitting for storm wave data, *Proc. 21st Int. Conference on Costal Eng.*, 18-31, ASCE, Delft.

Goda, Y. 2001. Extreme wave statistics for reliability-based of caisson breakwaters, *Advanced design of maritime structures in the 21st century*, Port and harbour research institute, pp. 1-13.

Goda, Y. 2004. Spread parameter of extreme wave height distribution for performance-based design of maritime structures, *J. Waterway, Port, Ocean and Coastal Eng.*, vol. 130, No. 1, pp. 29-38, ASCE.

Leenknecht, D.A., A. Szeuwalski, and Sherlock, A.R. 1992. Automated Coastal Engineering System: User's guide, *U.S. Army Corps of Engineers*, Washington DC.

Steijn, R.C. 1989. Schematization of the natural conditions in multi-dimensional numerical models of costal morphology, *Report on Investigation*, Delft Hydraulics

FREAK WAVE PREDICTION FROM DIRECTIONAL SPECTRA

Nobuhito Mori[1] and Peter A.E.M. Janssen[2]

Based on a weakly non-Gaussian theory the occurrence probability of freak waves is formulated in terms of the number of waves in a time series and the surface elevation kurtosis. Finite kurtosis gives rise to a significant enhancement of freak wave generation in comparison with the linear narrow banded wave theory. For fixed number of waves, the estimated amplification ratio of freak wave occurrence due to the deviation from the Gaussian theory is 50% – 300%. The results of the theory are compared with laboratory and field data.

INTRODUCTION

The last decade freak waves have become an important topic in engineering and science and are sometimes featured by a single and steep crest causing severe damage to offshore structures and ships. Freak wave studies started in the late 80's Dean (1990) and the high-order nonlinear effects on the freak waves were discussed in the early 90's Coastal Engineering Conference (Yasuda et al., 1992; Yasuda and Mori, 1994). Due to the many research efforts, the occurrence of freak waves, their mechanism and detailed dynamic properties are now becoming clear Trulsen and Dysthe (1997); Lavrenov (1998); Osborne et al. (2000); Onorato et al. (2001); Haver (2001); Mori et al. (2002). The state of the art on freak waves was summarized at the *Rogue Wave Conference*, held in 2000 and 2004 Olagnon and Athanassoulis (2000); Olagnon (2004). It was concluded that the third order nonlinear interactions enhance freak wave appearance and are the primary cause of freak wave generation in a general wave field except for the case of strong wave-current interaction or wave diffraction behind the islands.

Numerical and experimental studies have demonstrated that freak-like waves can be generated frequently in a two-dimensional wave flume without current, refraction or diffraction Yasuda et al. (1992); Trulsen and Dysthe (1997); Onorato et al. (2001); Stansberg (1990). Moreover, the numerical studies clearly indicate that a freak wave having a single, steep crest can be generated by the third order nonlinear interactions in deep-water Yasuda et al. (1992). Also, the theoretical background of freak wave generation has become more clear Osborne et al. (2000), but the quantitative occurrence probabilities in the ocean remain uncertain. In addition, it is still questionable how to characterize the dominant statistical properties of the freak wave occurrence in terms of nonlinear parameters, spectral shape, water depth and so on.

Nevertheless, although there is no doubt that the third order nonlinear interactions relate with the steep wave generation in the random wave train, the

[1] Graduate School of Engineering, Osaka City University, 3-3-138 Sugimoto, Sumiyoshi-ku, Osaka 558-8585, JAPAN.

[2] European Centre for Medium-Range Weather Forecasts, Shinfield Park, Reading RG2 9AX, UK

theoretical background of the relationship between the freak wave generation and the third order nonlinear interactions is not well-established. Freak wave generation is sometimes discussed in the context of the Benjamin-Feir instability in deep-water waves because of the similarity of the steep wave profile itself Yasuda et al. (1992); Onorato et al. (2001). The last two decades, the Benjamin-Feir type instability of the deep-water gravity waves has been studied by many researchers using the nonlinear Schrödinger type of equations Yuen and Lake (1982); Caponi et al. (1982); Dysthe (1979), mode-coupling equations Stiassnie and Shemer (1987), pseudo-spectral methods Yasuda and Mori (1997) and experiments Su (1982). However, there is disparity between the periodic wave instabilities and random wave behavior, because the broad banded spectra and random phase approximation are essential describing the ocean waves in nature Yasuda and Mori (1997); Mori (2003). Thus, the energy transfer of random waves due to four-wave interactions has been studied for describing spectral evolutions Hasselmann (1962); Krasitskii (1990). By means of a series of numerical investigations Yuen and Lake Yuen and Ferguson (1978) stated that the instability is confined within an initially unstable range and becomes weak if the spectral bandwidth broadens. Alber Alber (1978) mathematically demonstrated that for a random sea the Benjamin-Feir instability vanishes if the wave spectrum is sufficiently broad. Therefore, there is discrepancy between the nonlinear behavior of periodic waves and random waves.

Janssen Janssen (2003); Mori and Janssen (2006) investigated the freak wave occurrence as a consequence of four-wave interactions including the effects of non-resonant four-wave interactions. He found that the homogeneous nonlinear interactions give rise to deviations from the Gaussian distribution for the surface elevation on the basis of the Monte Carlo simulations of the Zakharov equation. Surprisingly, inhomogeneities only play a minor role in the evolution of the wave spectrum. He also formulated the analytical relationship between spectral shape and the kurtosis of the surface elevation. These results have the potential to unify previous freak wave studies covering nonlinear interactions, spectral profiles to nonlinear statistics and etc.

The purpose of this study is to investigate the relationship between kurtosis and the occurrence probability of freak waves through the nonlinear four-wave interactions. First, for a nonlinear stochastic wave field the relationship between high-order moments including kurtosis of surface elevation and nonlinear transfer function is derived. Second, the wave height and maximum wave height distributions are formulated as a simple function of kurtosis by the non-Gaussian theory. Third, the wave height distribution is compared with laboratory experiments and the occurrence probabilities of freak waves are compared with field observations. Finally, the dependence of the occurrence of freak waves on the number of waves and kurtosis will be analyzed and discussed.

HIGH-ORDER MOMENTS IN THE NONLINEAR STOCHASTIC WAVE FIELD

Our starting point is the Zakharov equation Zakharov (1968), which is a deterministic nonlinear evolution equation for surface gravity waves in deep water. Zakharov obtained from the Hamilton equations an approximate evolution equation for the amplitude of the free surface gravity waves Zakharov (1968), that contained the third order non-resonant and resonant four-wave interactions. In order to eliminate the effects of bound waves he applied on B a canonical transformation of the type

$$B = B(b, b^*), \tag{1}$$

where b is the normal variable of the free gravity waves. The evolution equation for b, called the Zakharov equation, becomes

$$\frac{\partial b_1}{\partial t} + i\omega_1 b_1 = -i \int \mathrm{d}\vec{k}_{2,3,4} T_{1,2,3,4} b_2^* b_3 b_4 \delta_{1+2-3-4}, \tag{2}$$

where, for brevity we have introduced the notation $b_1 = b(\vec{k}_1)$, etc, and the nonlinear transfer function $T_{1,2,3,4}$ as found by KrasitskiiKrasitskii (1990) enjoys a number of symmetries which guarantee that the Zakharov equation is Hamiltonian and conserves wave energy. In Janssen's Paper Janssen (2003) it was shown that in the context of the deep-water version of the Zakharov equation extreme surface gravity waves are generated by nonlinear focusing in a random wave field. This process also causes the Benjamin-Feir instability of a uniform wave train. As a consequence, for deep-water waves a considerable enhancement of the probability for extreme waves is found.

The nonlinear term in Eq.(2) will generate deviations from the normal, Gaussian probability distribution function (PDF) for the surface elevation. It is of interest to determine these deviations because it gives us information on the occurrence of extreme sea states. According to Eq.(29) of Janssen (2003), the fourth moment $\langle \eta^4 \rangle$ and kurtosis μ_4 can be obtained in terms of the action density N and of the nonlinear transfer function $T_{1,2,3,4}$. The result is

$$\kappa_{40} = \frac{\langle \eta^4 \rangle}{m_0^2} - 3 \tag{3}$$

$$= \mu_4 - 3 \tag{4}$$

$$= \frac{12}{g^2 m_0^2} \int \mathrm{d}\vec{k}_{1,2,3,4} T_{1,2,3,4} \sqrt{\omega_1\omega_2\omega_3\omega_4}\, \delta_{1+2-3-4} R_r(\Delta\omega, t) N_1 N_2 N_3 \tag{5}$$

where κ_{40} is the fourth order cumulant of the surface elevation η and is equivalent to $\mu_4 - 3$, where μ_4 is the normalized fourth order moment, kurtosis of the surface elevation. The transfer function $R_r = (1 - \cos(\Delta\omega t))/\Delta\omega \to \mathcal{P}/\Delta\omega$ for large time t, where $\Delta\omega = \omega_1 + \omega_2 - \omega_3 - \omega_4$ and $\mathcal{P}$ denotes the principle value of the integral to avoid singularity in the integral and m_0 is the total energy of the wave field.

Eq.(5) can be converted from the action density N into wave number spectrum F description.

$$C_4 = \frac{4g}{m_0}\mathcal{P}\int d\vec{k}_{1,2,3}T_{1,2,3,4}\sqrt{\frac{\omega_4}{\omega_1\omega_2\omega_3}}\frac{F_1F_2F_3}{\Delta\omega} \tag{6}$$

where $\mu_4 = 3(C_4+1)$, $\omega_n = \omega(k_n)$ the angular frequency defined by wave number k_n through the linear dispersion relation, $F = F(\vec{k})$ the wave number spectrum.

Assuming the Gaussian function for frequency spectrum $E(\omega)$ and unidirectional for angular direction,

$$E(\omega) = \frac{m_0}{\sqrt{2\pi}\sigma_\omega}e^{-\frac{1}{2}\nu^2} \tag{7}$$

$$\nu = \frac{\omega-\omega_0}{\sigma_\omega} \tag{8}$$

where ω_0 is the primary angular wave frequency. Eq.(6) is can be simplified as

$$C_4 = \frac{\pi}{3\sqrt{3}}BFI^2 \tag{9}$$

$$BFI = k_0\sqrt{m_0}Q_p^2 \tag{10}$$

$$Q_p = \frac{2}{m_0}\int d\omega\,\omega E^2(\omega) = \frac{1}{\sqrt{\pi}}\frac{\omega_0}{\sigma_\omega} = \frac{1}{\sqrt{\pi}\Delta_\sigma} \tag{11}$$

where $\Delta = \omega_\sigma/\omega_0$ the relative frequency spectrum band width.

For a narrow-band, unidirectional weakly-nonlinear wave train, the resonant and nonresonant interactions give rise to a much larger contribution to the kurtosis than the bound waves. The narrow band wave train, correct to third order in amplitude,

$$\eta = \alpha\left(1+\frac{1}{8}k_0^2\alpha^2\right)\cos\theta + \frac{1}{2}k_0\alpha^2\cos 2\theta + \frac{3}{8}k_0^2\alpha^3\cos 3\theta, \tag{12}$$

where α is connected to the free wave normal variable b through $\alpha = b\sqrt{\omega/2g}$. The first-order wave train $\eta = \alpha\cos\theta$ obeys Gaussian statistics. Hence, θ is uniformly distributed. The Stokes wave model (Eq.12) predicts wave moments of the form

$$E[\eta^n] = \int_0^\infty\int_0^{2\pi} d\alpha\, d\theta\, \eta^n(\alpha,\theta)A(\alpha,\theta) \tag{13}$$

where $A(\alpha,\theta)$ is the joint probability density of α and θ. For example, in lowest significant order, the skewness becomes

$$\mu_3 = \frac{\langle\eta^3\rangle}{\langle\eta^2\rangle^{3/2}} = 3\epsilon, \tag{14}$$

where $\epsilon = k_0 m_0^{1/2}$ is a measure of spectral steepness, while the kurtosis becomes

$$\kappa_{40} = 24\epsilon^2. \tag{15}$$

Similar results were obtained by Vinje (1989). Comparing Eq.(15) with Eq.(11), the contribution of the (non)-resonant waves dominates the one from the bound waves when the relative width $\Delta = \omega_\sigma/\omega_0$ satisfies the inequality $\Delta^2 < \pi/12\sqrt{3} \simeq 0.15$.

We have different notations of kurtosis as $\mu_4 = 3(C_4+1) = \kappa_{40}+3$ but these are equivalent.

WAVE HEIGHT DISTRIBUTION

Following a central limit theorem, linear, dispersive random waves have a Gaussian PDF for the surface elevation. Finite amplitude effects result, however, in deviations from the Normal distribution, as measured by a finite skewness and kurtosis. We assume that waves to be analyzed here are unidirectional with narrow banded spectra and satisfy the stationary and ergodic hypothesis. Let $\eta(t)$ be the sea surface elevation as a function of time t and $\zeta(t)$ an auxiliary variable such that $\eta(t)$ and $\zeta(t)$ are not correlated. Assuming both $\eta(t)$ and $\zeta(t)$ are real zero-mean function with variance σ, we have

$$Z(t) = \eta(t) + i\zeta(t) = A(t)e^{i\phi(t)}, \tag{16}$$

$$A(t) = \sqrt{\eta^2(t) + \zeta^2(t)}, \tag{17}$$

$$\phi(t) = \tan^{-1}\left(\frac{\zeta(t)}{\eta(t)}\right), \tag{18}$$

where A is the envelope of the wave train and ϕ the phase. Mori and Yasuda Mori and Yasuda (2002) investigated the wave height distribution as a function of kurtosis and skewness using the joint probability density function of $\eta(t)$ and $\zeta(t)$ for a narrow banded weakly nonlinear wave train. We will follow this approach closely. For weakly nonlinear waves deviations from the Normal distribution are small. In those circumstances the PDF of the surface elevation can be described by the Edgeworth distribution. As there is no-correlation between $\eta(t)$ and $\zeta(t)$, the joint probability density function of $\eta(t)$ and $\zeta(t)$ becomes

$$\begin{aligned} p(\eta,\zeta) = {}& \frac{1}{2\pi}\exp\left[-\frac{1}{2}(\eta^2+\zeta^2)\right] \\ & \times\left[1 + \frac{1}{3!}\sum_{n=0}^{3}\frac{3!}{(3-n)!n!}\kappa_{(3-n)n}H_{3-n}(\eta)H_n(\zeta)\right. \\ & \left. + \frac{1}{4!}\sum_{n=0}^{4}\frac{4!}{(4-n)!n!}\kappa_{(4-n)n}H_{4-n}(\eta)H_n(\zeta)\right] \end{aligned} \tag{19}$$

where H_n is the nth order Hermite polynomial. All variables will be normalized by the variance of the surface elevation $\sigma = m_0^{1/2}$ (where m_0 is the zero moment of the wave spectrum) and have zero-mean.

The PDF of the envelope A follows now immediately from an integration of the joint probability distribution

$$p(A,\phi) = A\,p(\eta,\zeta) \tag{20}$$

over ϕ, hence

$$p(A) = \int_0^{2\pi} \mathrm{d}\phi\, p(A,\phi). \tag{21}$$

Performing the integration over ϕ it is found that the first term of (19) gives the usual Rayleigh distribution for wave amplitude. Assuming the narrow band spectra, wave height H equals $2A$ and hence the wave height PDF becomes

$$p(H) = \frac{1}{4} H e^{-\frac{1}{8}H^2} \left[1 + \kappa_{40} A_H(H)\right], \tag{22}$$

where

$$A_H(H) = \frac{1}{384}\left(H^4 - 32H^2 + 128\right). \tag{23}$$

The comparison of exceedance probability of wave heights with experimental data is shown in Fig.1. The filled circles • denote experimental data, the Rayleigh distribution is represented by the dotted line, Eq.(22) corresponds to the solid line, and the wave height distribution including skewness effects proposed by Mori and Yasuda Mori and Yasuda (2002) (denoted as ER, Edgeworth-Rayleigh, in the figure) corresponds to the dashed line. For simplicity we refer to Eq.(22) as Modified ER (MER) hereafter. Due to the nonlinear effects, the exceedance probability obtained from the laboratory data departs for large wave height from the Rayleigh distribution. Both the MER and ER distributions for the exceedance probability of wave heights follow this separation at large amplitude region. Surprisingly, the MER distribution shows better agreement with the laboratory data than the ER distribution, although the corrections to the Rayleigh distribution only stem from effects of finite kurtosis.

MAXIMUM WAVE HEIGHT DISTRIBUTION AND FREAK WAVE OCCURRENCE

The PDF of maximum wave height p_m in wave trains can be obtained once the PDF of wave height $p(H)$ and exceedance probability of wave height $P(H)$ is known Goda (2000), thus

$$p_m(H_{max})dH_{max} = N[1 - P(H_{max})]^{N-1} p(H_{max})dH_{max} \tag{24}$$

with N the number of waves. For sufficiently large N one may use the approximation

$$\lim_{N\to\infty}^{N} [1 - P(H_{max})]^N \simeq \lim_{N\to\infty}^{N} \exp\left[-NP(H_{max})\right], \tag{25}$$

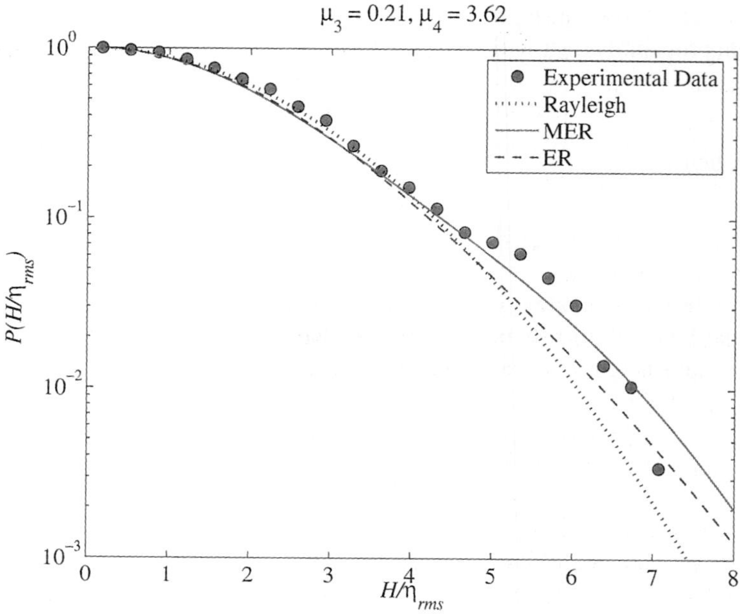

FIG. 1. Comparison of wave height distribution from laboratory data and theory ($\mu_3 = 0.21, \mu_4 = 3.62$; •: laboratory data, solid line: Eq.(22), dashed line: Mori and Yasuda (2002), dotted line: Rayleigh distribution).

Substituting Eq.(22) into Eq.(24), gives the PDF of the maximum wave height, p_m,

$$p_m(H_{max})dH_{max} = \frac{N}{4} H_{max}\, \mathrm{e}^{-\frac{H_{max}^2}{8}} \left[1 + \kappa_{40} A_H(H_{max})\right] \times \exp\left\{-N\mathrm{e}^{-\frac{H_{max}^2}{8}} \left[1 + \kappa_{40} B_H(H_{max})\right]\right\} dH_{max} \quad (26)$$

Equations Eq.(26) are evaluated as a function of N and κ_{40} (or μ_4). For $\kappa_{40} = 0$ results are identical to the ones following from the Rayleigh distribution. For simplicity it will be assumed that $H_{1/3} = 4m_0^{1/2}$, although it is $H_{1/3} = 4.004m_0^{1/2}$ in an exact linear random wave theory. The freak wave condition in this study therefore becomes $H_{max}/m_0^{1/2} > 8$, and we obtain from Eq.(26) the following simple formula to predict the occurrence probability of a freak wave as function of N and κ_{40},

$$P_{freak} = 1 - \exp\left[-\beta N(1 + 8\kappa_{40})\right] \quad (27)$$

where $\beta = e^{-8}$ is constant.

Using Eq.(27) it is seen that the effect of kurtosis becomes already of the same order as linear theory for $\kappa_{40} = 1/8$. This corresponds to $\mu_4 = 3.125$, and is not a strong nonlinear condition. Hence, both the effects of finite kurtosis and the number of waves N are important for determining the probability of maximum wave height in the nonlinear wave train.

The observed data was originally collected using an ultra sonic wave gauge at a depth of 30m, off the coast of the Pacific Ocean. The length of each record was 20min and the data were collected every hour from March 1 to the end of June in 2001. The wave statistics such as H_{max}, $H_{1/3}$, $T_{1/3}$, N, μ_3, and μ_4 were operationally calculated and archived. Note that the water depth of 30m is relatively shallow water. Therefore, to eliminate shallow water effects, the data are excluded if the dimensionless water depth $k_p h$ is less than 2.0 (it corresponds to $T_{1/3} \geq 8$s). The total number of valid data was about 2546.

We discuss the general behavior of the PDF of maximum wave height in the nonlinear wave field, by showing the ensemble averaged $H_{max}/H_{1/3}$ of each bin as a function of μ_4 and N in Fig.2. The brackets <> indicate the ensemble averaged value. Fig.2 (a) is observed data and (b) is the expected value of Eq.(26) through numerical integration. The dependence of $H_{max}/H_{1/3}$ on N is weaker than expected from Eq.(26). This is because the length of observed time series was fixed to 20min, so we cannot discuss the dependence of $H_{max}/H_{1/3}$ on number of waves in detail. On the other hand, the dependence of $\langle H_{max}/H_{1/3}\rangle$ on μ_4 is clear. The theoretically predicted $\langle H_{max}/H_{1/3}\rangle$ is underestimated compared to the observed data but it agrees with the observed data in a qualitative sense. The observed $\langle H_{max}/H_{1/3}\rangle$ monotonically increases for increasing μ_4, but for high values of kurtosis the theoretically estimated value of $\langle H_{max}/H_{1/3}\rangle$ is lower.

Finally, we examine the occurrence probability as function of the period of the waves. For historical reasons, the wave data have been obtained from 20 min records. The number of waves is related to the characteristic wave period T as $N = 1200/T$. Fig.3 shows Eq.(27) with $N = 1200/T$ as a function of T and μ_4. From Fig.3, freak waves in 20 min records are more frequently observed for short waves than long waves. Assuming windsea and following Toba's 3/2 law, the wave period T can be converted to significant wave height H_S. Then, freak waves occur more frequently for conditions of small wave height than for large wave heights.

CONCLUSION

For a narrow-band, random wave train we have shown that the kurtosis of the surface elevation is mainly determined by resonant and non-resonant wave-wave interactions, while bound waves only give a small contribution. Thus, the kurtosis and related high-order cumulants can be evaluated on the basis of Janssen's work Janssen (2003). Second, we have shown that for a narrow-band wave train the wave height and the maximum wave height probability distribution depends to a good approximation on the wave variance and the kurtosis. As a consequence it

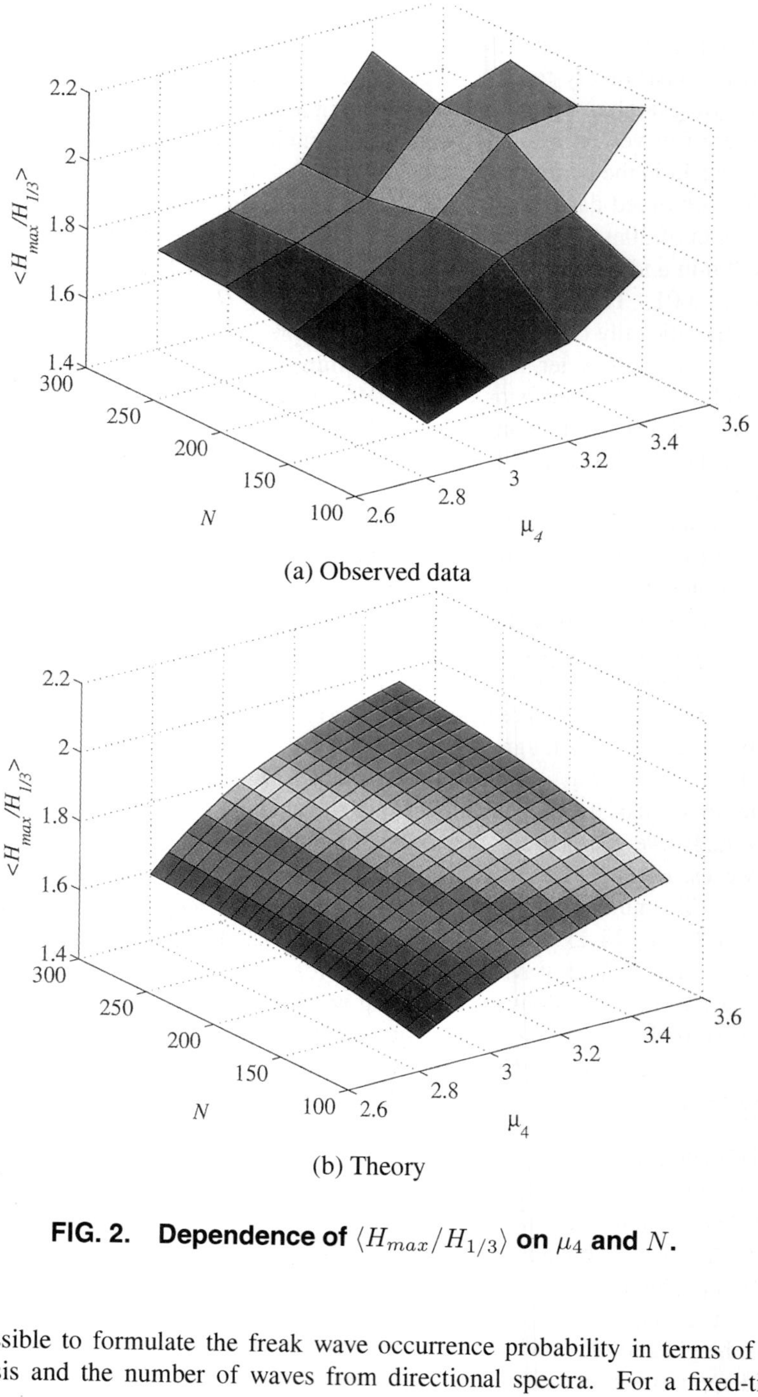

FIG. 2. Dependence of $\langle H_{max}/H_{1/3}\rangle$ on μ_4 and N.

is possible to formulate the freak wave occurrence probability in terms of the kurtosis and the number of waves from directional spectra. For a fixed-time

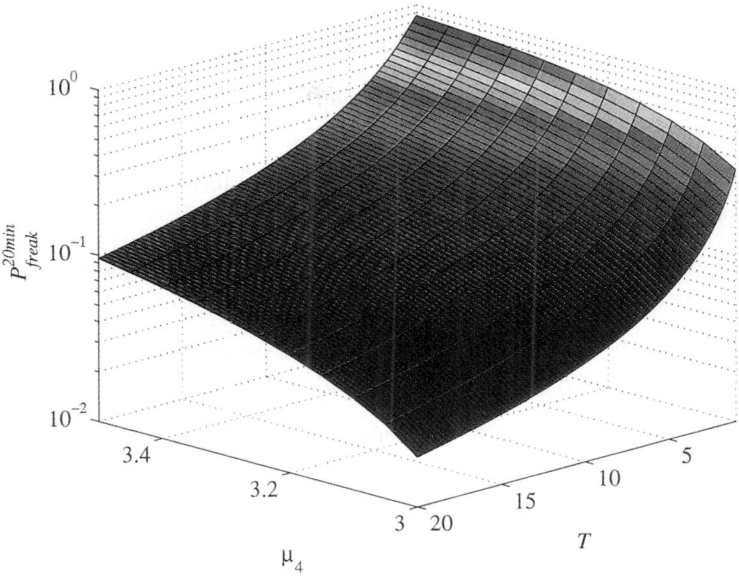

FIG. 3. Predicted occurrence probability of freak wave in 20 min record as a function of wave period T and μ_4.

series of 20 min measurement, the strong dependence of freak wave occurrence on kurtosis is investigated.

REFERENCES

Alber, I.. 1978. The effects of randomness of the stability of two-dimensional surface wavetrains, *Proc. Roy. Soc. Lond.,* A363, 525–546.

Caponi, E., P. Saffman, and H. Yuen. 1982. Instability and confined chaos in a nonlinear dispersive wave system, *Phys. Fluids,* 25(12), 2159–2166.

Dean, R.. 1990. Freak waves: a possible explanation, A. Tørum and O. Gudmestad (Eds.), *Water wave kinematics*, 609–612. Kluwer Academic Pub.

Dysthe, K.. 1979. Note on modification to the nonlinear schrödinger equation for application to deep water waves, *Proc. Roy. Soc. Lond.,* A369, 105–114.

Goda, Y.. 2000. *Random Seas and Design of Maritime Structures* (2nd ed.),, World Scientific.

Hasselmann, K.. 1962. On the nonlinear energy transfer in gravity-wave spectrum. I. Gerenral theory, *J. Fluid Mech.,* 12, 481–500.

Haver, S.. 2001. Evidences of the existence of freak waves, M. Olagnon and G. Atanassoulis (Eds.), *Rogue Waves*, 129–140.

Janssen, P.. 2003. Nonlinear four-wave interactions and freak waves, *J. Phys. Oceanogr.,* 33(4), 863–884.

Krasitskii, V.. 1990. Canonical transformation in a theory of weakly nonlinear waves with a nondecay dispersion law, *Soviet Physics JETP,* 71, 921–927.

Lavrenov, I.. 1998. The wave energy concentration at the Agulhas current off South Africa, *Natural Hazard,* 17, 117–127.

Mori, N.. 2003. Effects of wave breaking on wave statistics for deep-water random wave train, *Ocean Engineering,* 30(2), 205–220.

Mori, N. and P. Janssen. 2006. On kurtosis and occurrence probability of freak waves, *J. Phys. Oceanogr.,* 36(7), 1471–1483.

Mori, N., P. Liu, and T. Yasuda. 2002. Analysis of freak wave measurements in the Sea of Japan, *Ocean Engineering,* 29(11), 1399–1414.

Mori, N. and T. Yasuda. 2002. A weakly non-gaussian model of wave height distribution for random wave train, *Ocean Engineering,* 29(10), 1219–1231.

Olagnon, M. (Ed.). 2004. *Rogue Waves 2004*. IFREMER.

Olagnon, M. and G. Athanassoulis (Eds.). 2000. *Rogue Waves 2000*. IFREMER.

Onorato, M., A. Osborne, M. Serio, and S. Bertone. 2001. Freak waves in random oceanic sea states, *Physical Review Letter,* 86(25), 5831–5834.

Osborne, A., M. Onorato, M. Serio, and S. Bertone. 2000. The nonlinear dynamics of rogue waves and holes in deep water gravity wave trains, *Physical Letter A,* 275(5-6), 386–393.

Stansberg, C.. 1990. Extreme waves in laboratory generated irregular wave trains, A. Tørum and O. Gudmestad (Eds.), *Water wave kinematics*, 573–590. Kluwer Academic Pub.

Stiassnie, M. and L. Shemer. 1987. Energy computations for evolution of class I and II instabilities of Stokes waves, *J. Fluid Mech.,* 174, 299–312.

Su, M.. 1982. Three-dimensional deep-water waves. Part 1. Experimental measurement of skew and symmetric patterns, *J. Fluid Mech.,* 124, 73–108.

Trulsen, K. and K. Dysthe. 1997. Freak waves - A three-dimensional wave simulation, *Proceedings of the 21st Symposium on Naval Hydrodynamics*, 550–558. (National Academy Press.

Vinje, T.. 1989. The statistical distribution of wave heights in a random seaway, *Appl. Ocean Research,*, 11(3), 143–152.

Yasuda, T. and N. Mori. 1994. High order nonlinear effects on deep-water random wave trains, *International Symposium: Waves-Physical and Numerical Modelling*, Volume 2, Vancouver, 823–332.

Yasuda, T. and N. Mori. 1997. Roles of sideband instability and mode coupling in forming a water wave chaos, *Wave Motion,* 26(2), 163–185.

Yasuda, T., N. Mori, and K. Ito. 1992. Freak waves in a unidirectional wave train and their kinematics, *Proc. of the 23th Int. Conf. on Coastal Eng.*, Volume 1, Venice, ASCE, 751–764.

Yuen, H. and W. Ferguson. 1978. Relationship between Benjamin-Feir instability and recurrence in the nonlinear schrödinger equation, *Phys. Fluids,* 21(8), 1275–1278.

Yuen, H. and B. Lake. 1982. Nonlinear dynamics of deep-water gravity waves, *Advances in Applied Mech.,* 22, 67–327.

Zakharov, V.. 1968. Stability of periodic waves of finite amplitude on the surface of a deep fluid, *J. Appl. Tech. Phys.,* 9, 190–194.

A BOUNDARY ELEMENT SOLUTION APPLIED TO THE DESCRIPTION OF EXTREME WAVES IN COASTAL WATERS

Caroline H. Hague[1] and Chris Swan[1]

This paper describes a fully nonlinear three-dimensional wave model based on the boundary element method (BEM). The model is applied to the focusing of wave components whereby a large number of wave crests of varying frequency and propagating in different directions are superimposed at one point in space and time to create an extreme wave event. This process, which is commonly referred to as frequency focusing, is believed to be responsible for the formation of the largest waves and may be further strengthened by bathymetric changes. Sea states are simulated in a number of water depths and are described by realistic (JONSWAP) frequency spectra and directional spreads. To our knowledge no previous boundary element based model has been applied to realistic wave fields of this type.

INTRODUCTION

The description of large waves represents an important input in the design of most coastal structures; providing the basis on which the applied loads, the occurrence of wave slamming and the extent of any over-topping are predicted. In reality the largest waves have three underlying physical characteristics: (i) they are irregular or unsteady, reflecting the distribution of energy across the frequency domain; (ii) they are short-crested, with energy travelling in a range of different directions; (iii) they are nonlinear, with the most extreme waves being at or close to their breaking limit. Although recent advances in wave modelling (see for example Bateman, Swan & Taylor 2001; 2003) have gone a long way towards including these effects, important limitations remain. In particular, Fourier based models are not easily adapted to simulate non-uniform water depths and cannot be applied to the description of limiting and overturning waves. In order to address these limitations, a numerical model based on the boundary element method (BEM) has been developed which encapsulates all of the underlying physical characteristics.

The success of the model is largely based on two features. First, the model utilises multiple fluxes at domain corners and edges. This approach, first suggested for two-dimensional potential problems in a text by Brebbia & Dominguez (1992), is shown to be suitable for domains used for wave propagation studies and is extended to three dimensions. This allows the accurate simulation of fluid/boundary intersections; avoiding the difficulties and inaccuracies associated with the commonly applied split node approach. Second, the model employs an effective semi-Lagrangian to Lagrangian switching

---

[1] Department of Civil and Environmental Engineering, Imperial College London, South Kensington, London, SW7 2AZ, United Kingdom

scheme. This allows the majority of the calculations to be undertaken in a semi-Lagrangian frame of reference, removing the difficulties associated with particle drift and the inaccuracies associated with the need to add additional nodes or re-grid existing nodes. Although Lagrangian simulations are not shown in the present paper, the option of switching to a Lagrangian representation as the largest waves approach their breaking limit allows the model to simulate overturning waves (Hague 2006).

NUMERICAL METHOD

The numerical domain is defined in physical space as shown in Fig.1, consisting of lateral boundaries Γ_l, Γ_{front}, Γ_r and Γ_{back}, an impermeable bed Γ_b and the free surface Γ_s.

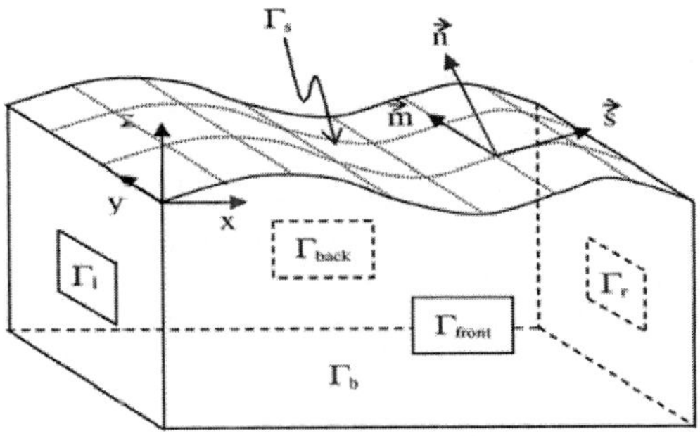

Figure 1. The numerical domain used for computations.

The fluid is assumed to be irrotational and inviscid, thus a velocity potential $\phi(x,y,z,t)$ may be defined such that conservation of mass is described by Laplace's equation

$$\nabla^2 \phi = 0 \tag{1}$$

The vertical coordinate z is taken in a positive sense upwards from the still water level and the velocity field is given by $\mathbf{u} = \nabla\phi = (u,v,w)$. Using Green's second identity with the free space Green's function $G(r) = 1/4\pi r$ as a fundamental solution to Laplace's equation, the boundary integral equation (BIE) is

$$c_p \phi_p = \int_\Gamma G(r)\frac{\partial \phi_q}{\partial n} - \phi_q \frac{\partial G(r)}{\partial n} d\Gamma \tag{2}$$

This relates the values of the potential ϕ and the potential flux $\partial\phi/\partial n$ at each point p to every other point q around the boundary. The distance between the points is $r = |\mathbf{x}_p - \mathbf{x}_q|$, where the position vectors of the points are $\mathbf{x}_p = (x_p, y_p, z_p)$ and $\mathbf{x}_q = (x_q, y_q, z_q)$. The outward normal vector to the domain boundary is $\mathbf{n}$, and c_p is a geometric coefficient. The use of a rigid mode technique avoids the explicit calculation of c_p.

Boundary conditions are specified along all sections of the domain boundary, according to

$$\phi = \phi\big|_{known} \qquad \text{on } \Gamma_s \tag{3}$$

$$\frac{\partial\phi}{\partial n} = \left.\frac{\partial\phi}{\partial n}\right|_{known} \qquad \text{on } \Gamma_l, \Gamma_r, \Gamma_{front}, \Gamma_{back} \tag{4}$$

$$\frac{\partial\phi}{\partial n} = 0 \qquad \text{on } \Gamma_b \tag{5}$$

Every point on the boundary is therefore assigned either a known potential or a known potential flux, allowing the BIE to be solved for the corresponding unknown quantities. In order to achieve this a total of N nodes are placed around the domain, discretising the boundary into M isoparametric serendipity elements, whereby groups of eight nodes describe quadrilateral patches. Each element is mapped onto intrinsic curvilinear coordinates (ξ, ζ) as shown in Fig.(2).

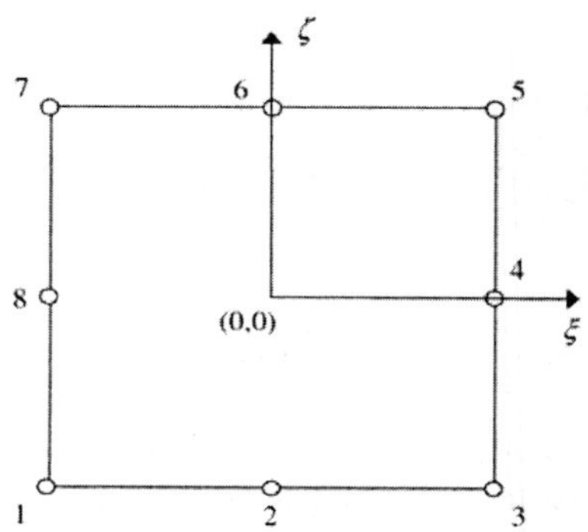

Figure 2. An example of a bi-quadratic serendipity element consisting of eight nodes, locally numbered anticlockwise from 1 to 8.

The variation of both geometry and fundamental variables is assumed to be quadratic along each element side, and is therefore approximated by

$$\varphi(\xi,\zeta) = \sum_{k=1}^{8} N_k(\xi,\zeta)\varphi_k \tag{6}$$

where φ represents any variable, k denotes the local numbering of nodes forming the element and N_k are polynomial shape functions. Derivatives of φ are calculated by

$$\frac{\partial\varphi(\xi,\zeta)}{\partial\xi} = \sum_{k=1}^{8} \frac{\partial N_k(\xi,\zeta)}{\partial\xi}\varphi_k \tag{7}$$

$$\frac{\partial\varphi(\xi,\zeta)}{\partial\zeta} = \sum_{k=1}^{8} \frac{\partial N_k(\xi,\zeta)}{\partial\zeta}\varphi_k \tag{8}$$

Equation (2) can now be rewritten in terms of evaluations across elements as

$$c_p\phi_p + \sum_{j=1}^{M}\sum_{k=1}^{8} \phi_k \int_{\xi_j}\int_{\zeta_j} N_k(\xi,\zeta)\frac{\partial G(r)}{\partial n} J(\xi,\zeta)d\xi d\zeta$$

$$\ldots = \sum_{j=1}^{M}\sum_{k=1}^{8} \int_{\xi_j}\int_{\zeta_j} N_k(\xi,\zeta)G(r)J(\xi,\zeta)d\xi d\zeta \tag{9}$$

with the Jacobian of the transformation given by

$$J(\xi,\zeta) = \left|\frac{\partial\mathbf{x}}{\partial\xi} \times \frac{\partial\mathbf{x}}{\partial\zeta}\right| \tag{10}$$

The evaluation of the integrals in Eq.(9) is carried out numerically using Gauss quadrature. The BIE can then be rearranged in a matrix formulation

$$\mathbf{H\Phi} = \mathbf{G\Phi_n} \tag{11}$$

consisting of coefficient matrices **H** and **G** which multiply both the known and unknown values of ϕ and $\partial\phi/\partial n$ contained in vectors $\mathbf{\Phi}$ and $\mathbf{\Phi_n}$.

At points around the domain boundary where the geometry ceases to be smoothly varying, i.e. at corners or edges, the "corner problem" exists. At these positions the direction of the outward normal vector and hence the outward flux $\partial\phi/\partial n$, is ill-defined. Having said this, the elements which meet at an edge or corner do have a well-defined flux associated with each element individually. Therefore, by keeping each of these fluxes in the solution step and considering them all to be valid at the edge or corner, an accurate representation of the behaviour at these discontinuities can be achieved. This approach is the multiple flux technique. Multiple fluxes are implemented in the model such that they exist at all points where elements meet, one for each element which intersects at this point, each associated with the direction of the outward normal to that element. The contribution to the integration which is associated with each of the fluxes is then treated separately. This results in arrays being created such that, whilst **H** is of size ($N \times N$) since there is only one potential at each position, **G** is rectangular of size ($N \times 8M$) and $\mathbf{\Phi_n}$ is a vector of length $8M$.

In order to find the unknowns of the system, Eq.(11) is transformed into the system matrix formulation

$$\mathbf{AX} = \mathbf{B} \tag{12}$$

where **A** is the system matrix (size $N \times N$), **X** is the vector of unknown potentials and potential fluxes (size N), and **B** is a vector of known quantities (size N). This is achieved by a set of swapping routines which either swap the coefficients in **G** and **H**, absorb them into the system matrix **A**, or retain them to multiply known boundary values to form the vector **B**.

For each element ($j=1..M$), the locally-numbered nodes making up that element ($k=1..8$) are nodes which have an associated global index i. Considering each of the elements and nodes within that element in turn, each locally-numbered node having one of the boundary conditions ($b=1..8M$), columns of **G** and **H** are swapped according to the following rules:

1. If the boundary condition b at k is $\partial\phi/\partial n$, do nothing. In this way, all fluxes which are known are treated separately on the right hand side of the system matrix formulation.
2. If the boundary condition b at k is ϕ
   - If column i of **H** has not yet been swapped, swap column i of **H** with column b of **G**, reversing the sign of the column entries. This ensures coefficients which act on an unknown flux are placed in the system matrix.
   - If column i of **H** has been swapped previously, add column b of **G** to column i of **H**, reversing sign, then set column b of **G** to 0. By doing this, any coefficients which act on the same unknown are absorbed into the system matrix.

This procedure enables the coefficients generated by the numerical integration across elements to act only on the fluxes associated with that element, thus considering these to be separate from those fluxes associated with neighbouring elements, even if they share the same position.

Within a three dimensional domain, edges are defined as a geometric discontinuity formed at the intersection of two elements and corners are defined as geometric discontinuities created at the intersection of three elements. The multiple flux approach is applicable for all corners and edges except those for which more than one potential flux is unknown. By allowing multiple fluxes to exist at an edge or corner, either the potential or one of the fluxes can be solved for. Therefore, if more than one flux is an unknown, the exact nature of the flux that is solved for is undefined. However, if boundary conditions are specified around the boundary according to Eqs.(3-5), only junctions between elements along the smoothly varying free surface could produce instances where more than one flux would be unknown. Since the unknown fluxes would in this case be describing the same flux, the swapping procedure merges the unknown fluxes to form a single unknown flux in the solution step. Because of this, the multiple flux approach is applicable to any three-dimensional numerical problem specified in this way.

By solving Eq.(12) for the vector of unknowns **X** using a GMRES scheme (Saad & Schultz 1986), the normal velocity along the free surface, $\partial\phi / \partial n$, is obtained. By combining this information with the variation of ϕ in the tangential directions, $\partial\phi / \partial s$ and $\partial\phi / \partial m$, the velocity field **u** may be calculated from

$$\frac{\partial\phi}{\partial x} = \frac{\partial\phi}{\partial s}\cos\vartheta_s \cos\alpha_s + \frac{\partial\phi}{\partial m}\cos\vartheta_m \cos\alpha_m + \frac{\partial\phi}{\partial n}\cos\vartheta_n \cos\alpha_n \tag{13}$$

$$\frac{\partial\phi}{\partial y} = \frac{\partial\phi}{\partial s}\cos\vartheta_s \sin\alpha_s + \frac{\partial\phi}{\partial m}\cos\vartheta_m \sin\alpha_m + \frac{\partial\phi}{\partial n}\cos\vartheta_n \sin\alpha_n \tag{14}$$

$$\frac{\partial\phi}{\partial z} = \frac{\partial\phi}{\partial s}\sin\vartheta_s + \frac{\partial\phi}{\partial m}\sin\vartheta_m + \frac{\partial\phi}{\partial n}\sin\vartheta_n \tag{15}$$

where ϑ_s, ϑ_m, ϑ_n are the angles made by the vectors $\vec{s}, \vec{m}, \vec{n}$ to the *x-y* plane, and α_s, α_m, α_n are the azimuthal angles made by $\vec{s}, \vec{m}, \vec{n}$ with respect to the positive *x*-axis.

Now, the motion of the free surface must satisfy the free surface boundary conditions, which are taken in a semi-Lagrangian frame as

$$\frac{\delta\eta}{\delta t} = \frac{\partial\phi}{\partial z} - \frac{\partial\phi}{\partial x}\frac{\partial\eta}{\partial x} - \frac{\partial\phi}{\partial y}\frac{\partial\eta}{\partial y} \tag{16}$$

$$\frac{\delta\phi}{\delta t} = -g\eta - \frac{1}{2}\nabla\phi\cdot\nabla\phi + \frac{\partial\phi}{\partial z}\frac{\delta\eta}{\delta t} \tag{17}$$

where the semi-Lagrangian time derivative is denoted by $\delta/\delta t$ and the extra terms on the right hand side arise from taking the material derivative for a particle following the fluid in the vertical direction only. In a Lagrangian frame these equations are expressed by

$$\frac{D\eta}{Dt} = \frac{\partial \phi}{\partial z} \tag{18}$$

$$\frac{D\phi}{Dt} = -g\eta + \frac{1}{2}\nabla\phi \cdot \nabla\phi \tag{19}$$

Along with the spatial gradients $\partial\eta/\partial x$ and $\partial\eta/\partial y$ obtained directly from geometry if using a semi-Lagrangian frame, the velocities calculated from Eqs. (13-15) are substituted into the free surface boundary conditions. These are then integrated with respect to time in order to follow the evolution of the free surface. Time-marching is carried out within the model by the use of an Adams-Bashforth-Moulton predictor-corrector method, employing fourth-order predictor and fifth-order corrector steps. In order to provide information at the first three time steps, a fourth-order Runge-Kutta scheme is implemented.

RESULTS

The numerical method is applied to the simulation of directionally spread wave groups in a number of water depths. These examples provide an opportunity to assess the viability of the boundary element method as a practical engineering tool for free surface flows.

Computations of the focusing of a JONSWAP spectrum in a number of water depths are considered. The JONSWAP spectrum used is specified by

$$G_{\eta\eta}(\omega) = \frac{\alpha g^2}{\omega^5} \exp\left(-\beta\left(\frac{\omega_p}{\omega}\right)^4 \right) \gamma^{\exp\left(-\frac{(\omega-\omega_p)^2}{2\omega_p^2\sigma^2} \right)} \tag{20}$$

where

$$\begin{aligned} &\sigma = 0.07 \quad \mathit{for} \quad \omega \le \omega_p \\ &\sigma = 0.09 \quad \mathit{for} \quad \omega > \omega_p \\ &\alpha = 0.0081 \\ &\beta = 1.25 \end{aligned} \tag{21}$$

Discrete wave amplitudes a_n are obtained for N_c components according to the *New-Wave* formulation (Tromans, Anaturk & Hagemeijer 1991), such that

$$a_n = \frac{A_{input}}{\sum_{n=1}^{N_c} G_{\eta\eta}(\omega_n)} G_{\eta\eta}(\omega_n) \tag{22}$$

for a given input amplitude A_{input}. For directional cases, each of these frequency components is discretised into M_c directional components, according to (Mitsuyasu 1975):

$$a_{nm} = a_n \frac{\cos^s\left(\frac{\theta_m}{2}\right)}{\sum_{m=1}^{M_c} \cos^s\left(\frac{\theta_m}{2}\right)} \tag{23}$$

where s is the spreading parameter. $s=7$ corresponds closely to a standard deviation of 30° for a normal distribution, and is a typical value recorded in North Sea storms (Jonathan and Taylor 1996).

A three dimensional numerical domain is defined, of length $L=7.08m$, width $W=3m$ and depth $d=0.24m$. The left lateral boundary forms the input boundary and is placed at $x=-4.08m$, whilst the back boundary is chosen to act as a reflective wall and hence is placed along the line of symmetry at $y=0$. This allows only half the domain to be modelled. The front boundary also contributes to the input. For $x \leq 0$, the boundary condition specified along the front boundary is that of the input velocity v, which is then smoothly tapered to zero from $x=0$ to $x=2m$. In order to remove excess wave energy from the domain, a combination of passive absorption and radiation condition is applied at the downstream end of the domain (similar to that of Clement 1996). The passive absorption layer is placed across the width of the domain at the downstream end between $x=2m$ and $x=3m$, within which the dynamic free surface boundary condition is modified by a damping coefficient μ according to

$$\frac{\delta\phi}{\delta t} = -g\eta - \frac{1}{2}\nabla\phi\cdot\nabla\phi + \frac{\partial\phi}{\partial z}\frac{\delta\eta}{\delta t} - \mu\frac{\partial\phi}{\partial n} \tag{24}$$

where μ is given by

$$\mu = 3\beta\left(\frac{x - x_{sponge}}{L_{sponge}}\right)^2 - 2\beta\left(\frac{x - x_{sponge}}{L_{sponge}}\right)^3 \tag{25}$$

and x_{sponge} denotes the start of the damped layer, which is of length L_{sponge}, and the damping strength β is chosen to be 0.3. Along the downstream boundary a Sommerfeld radiation condition is implemented. A zero-flux condition is applied along the bed.

Elements are placed around the entire domain with a spatial discretisation of $\Delta x=\Delta y=\Delta z=0.06m$. A total of 75 frequency components are specified according to Eq.(22), then further discretised into 75 directions according to Eq.(23). The input amplitude is taken to be $A_{input}=0.05m$ and a spreading parameter of $s=7$ is used. Taking the peak of the spectrum as k_p, the inline parameters k_pd and Ak_p are therefore $k_pd=0.92$ and $Ak_p=0.19$. Computations are started at $t_0=-15s$ and time- marched with a time step of $\Delta t=0.02s$ which, in combination with the negative x-position of the input boundary, acts to implicitly phase the components. In this way, linear random wave theory (LRWT) predicts the focal event to be at $x=0$, $t=0$. In order to avoid the generation of spurious long waves due to the initial start up of wave generation, the input velocity is ramped over the first 3s according to a smooth cosine function.

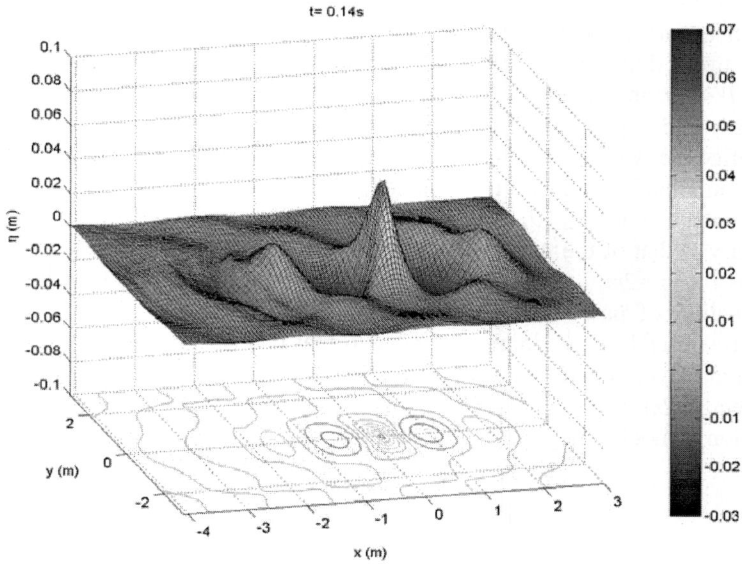

Figure 3. Spatial surface plot at the time of maximum crest elevation. T_p=1.2s, γ=2.5, A_{Input}=0.05m, s=7, d=0.24m. Contours of surface elevation are also shown.

Fig.(3) shows the free surface at the time of maximum crest elevation, showing the ability of the three-dimensional model to simulate directionally spread extreme events. Contour plots show the shape of the focal event in each case. Note that no smoothing or filtering was applied at any stage in the computations. It can be seen from the figure that a single extreme event occurs within the domain, whereby there exists a wave crest of significantly greater magnitude than the surrounding sea state.

For comparison, the same case is simulated in deep water, with the numerical domain specified as before except the depth is now $d=1.05m$ and the boundary is discretised using $\Delta x=\Delta y=0.06m$, $\Delta z=0.075m$. Fig.(4) shows the free surface at the time of maximum crest elevation for this case.

As expected, whilst both cases produce a focused event, highlighted by the symmetric nature of the surface elevation contours, the reduction in water depth leads to a narrower, steeper maximum wave crest with more rapidly varying contours. This is due to the nature of the dispersion relationship, whereby smaller depths results in shorter wavelengths for wave components.

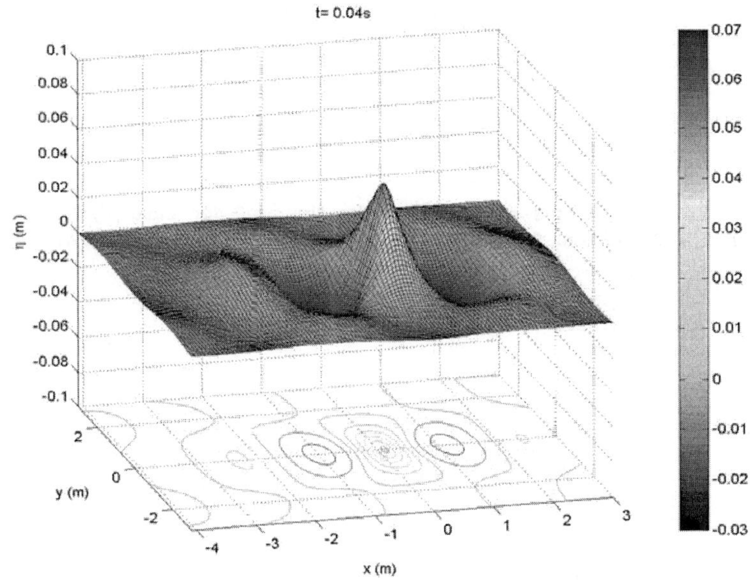

Figure 4. Spatial surface plot for the deep water case at the time of maximum crest elevation. T_p=1.2s, γ=2.5, A_{input}=0.05m, s=7, d=1.05m. Contours of surface elevation are shown, highlighting the focused nature of the extreme event.

By considering the vertical movement in time of a point placed at the focal location, the time history of the focal event can be obtained. Fig.(5) shows the time history obtained in this way for the shallow water case. This temporal profile is compared to the time history predicted by both LRWT and the second-order theory of Sharma & Dean (1981) at the focal position. It is obvious from the figure that the trace obtained by the multiple flux model has a maximum crest elevation which is significantly lower than that predicted by second-order.

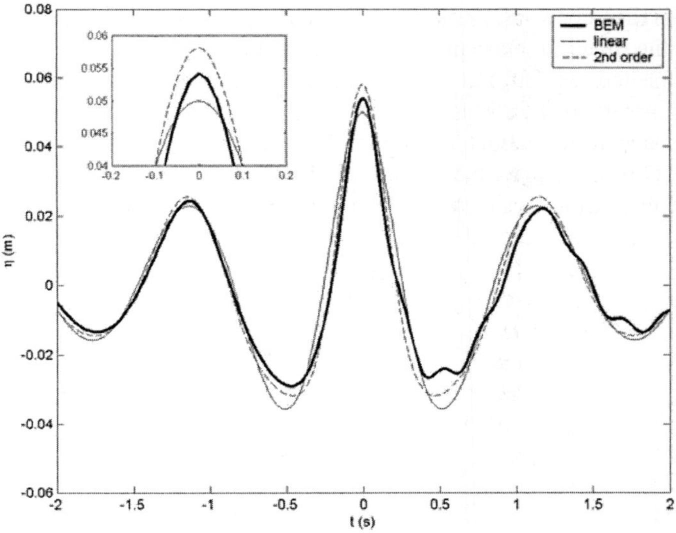

Figure 5. Temporal profile along the centre line *y=0* for case T_p=1.2s, γ=2.5, A_{input}=0.05m, s=7, d=0.24m. A close up view of the crest region is also shown.

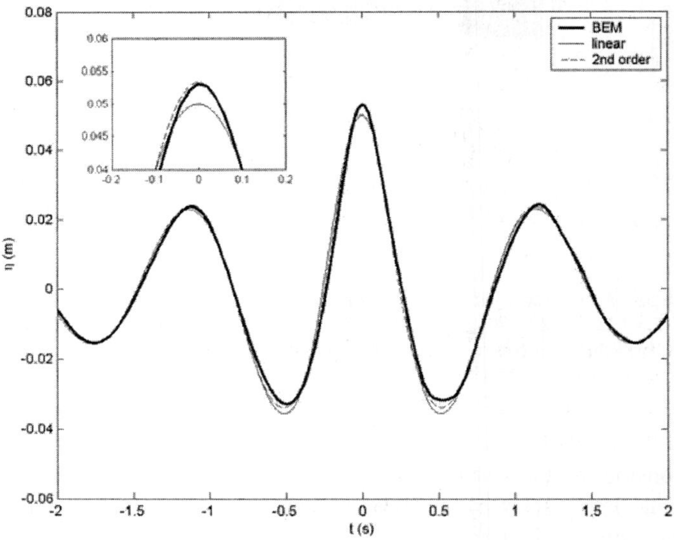

Figure 6. Temporal profile along the centre line *y=0* for deep water case T_p=1.2s, γ=2.5, A_{input}=0.05m, s=7, d=1.05m. A close up view of the crest region is also shown.

This result is interesting in that it provides independent confirmation of some aspects of the numerical results presented by Katsardi and Swan (2005); where the latter calculations were undertaken using a Fourier based method.

Also evident in the figure is a high-frequency wave trailing the largest event. The exact nature of this high frequency wave is, at the present time, unknown. If we consider the time history of the free surface at the focal location for the deep water case, as shown in Fig.(6), the high frequency wave is not present. Given that this case resulted in a higher crest elevation with no evidence of this wave, there is little reason to surmise this is simply numerical error. Indeed, this may represent important local changes in the spectrum describing the wave field (Gibson and Swan 2006). Further work concerning such changes in a wide variety of limiting and near-limiting cases is ongoing. However, at this stage it is important to stress that the occurrence of this high-frequency wave does not appear to be related to the sawtooth instability reported by other authors undertaking BEM-based simulations, for example Longuet-Higgins and Cokelet (1976).

CONCLUSIONS

In this paper, a fully nonlinear numerical model based on the Boundary Element Method (BEM) has been developed in three dimensions for use in investigations of free surface fluid flows. The use of multiple fluxes at points of geometric discontinuity in the domain (corners and edges) has been described and implemented within the method as a technique to avoid the so-called "corner problem".

The focusing in space and time of a large number of directionally spread wave components has been simulated, represented by the *New-Wave* formulation of the JONSWAP spectrum, successfully modelling the differences in the nature of the extreme event resulting from shallow water conditions. The fact that no smoothing or regridding was needed for any of the computations shows the improvement to overall accuracy gained by the multiple flux treatment of domain corners.

Given the success of the model as presented in this paper, the use of a BEM-based model with multiple fluxes can be considered to be a useful numerical tool for the investigation of highly nonlinear extreme wave events. Since the formulation is not restricted in terms of its application to non-periodic, multi-valued surfaces with varying bed geometry, there is significant potential for the multiple flux technique to be applied to the simulation of the most extreme free surface flows in the future.

ACKNOWLEDGEMENTS

This work was undertaken as part of EPSRC grant GR/R64971, and the authors gratefully acknowledge this support.

REFERENCES

Bateman, W.J.D., Swan, C. and Taylor, P.H. 2001. On the Efficient Numerical Simulation of Directionally Spread Surface Water Waves, *Journal of Computational Physics,* Vol 174, No 1, pp 277-305.

Bateman, W.J.D., Swan, C. and Taylor, P.H. 2003. On the Calculation of the Water Particle Kinematics Arising in a Directionally Spread Wavefield, *Journal of Computational Physics,* Vol 186, No 1, pp 70-92.

Brebbia, C.A. and Dominguez, J. 1992. *Boundary Elements: An Introductory Course*. Computational Mechanics Publications (McGraw-Hill).

Clement, A. 1996. Coupling of Two Absorbing Boundary Conditions for 2D Time-Domain Simulations of Free Surface Gravity Waves, *Journal of Computational Physics,* Vol 126, pp 139-151.

Gibson, R. and Swan, C. 2006. The Evolution of Large Ocean Waves: The Role of Local and Rapid Spectral Changes, *Proceedings of the Royal Society of London A (in press).*

Hague, C.H. 2006. Fully Nonlinear Computations of Directional Waves, Including Wave Breaking, *PhD Thesis, Imperial College London*

Jonathan, P. and Taylor, P.H. 1996. Irregular Nonlinear Waves in a Spread Sea, *ASME Trans J Offshore Mech Arctic Eng,* Vol 119, pp 37-41.

Katsardi, V. and Swan, C. 2005. The Evolution of Large Waves in Shallow Water," *Proceedings of 29th International Conference on Coastal Engineering (ICCE)*, pp 69-81.

Longuet-Higgins, M.S. and Cokelet, E.D. 1976. The Deformation of Steep Surface Waves on Water. I. A Numerical Method of Computation, *Proceedings of the Royal Society of London A*, Vol 350, No 1660, pp 1-26.

Mitsuyasu, H. 1975. Observation of Directional Spectrum of Ocean Waves Using a Cloverleaf Buoy, *Journal of Physical Oceanography*, Vol 16, pp 459-482.

Saad, Y. and Schultz, M.H. 1986. GMRES: A Generalized Minimum Residual Algorithm for Solving Nonsymmetric Linear Systems. *SIAM Journal of Statistical and Scientific Computing*. Vol 7, pp 856-869

Sharma, J.N. and Dean, R.G. 1981. Second-Order Directional Seas and Associated Wave Forces, *Society of Petroleum Engineers Journal*, Vol 21, No 1, pp 129-140.

Tromans, P.S., Anaturk, A.R. and Hagemeijer, P. 1991. A New Model of the Kinematics of Large Ocean Waves, *Proceedings of 1st Inernational Offshore and Polar Engineering Conference (ISOPE)*, Vol 3, p 64.

ESTIMATION OF EXTREME WAVE CONDITIONS FROM HINDCAST SIMULATIONS WITH APPLICATION TO THE WAVE CLIMATE ALONG FRENCH COASTS

Florence Lafon[1] and Michel Benoit[1]

A numerical wave data-base has been built by hindcasting sea-state conditions over a period of 23 years and 8 months, from 1979/01/01 to 2002/08/31, using the TOMAWAC spectral wave model, driven by wind fields from the ERA-40 meteorological reanalysis produced by the ECMWF. The resolution of the finest model's grid is about 2 to 3 km along the French coasts of Atlantic ocean, English Channel and North Sea. From this data-base, synthetic charts representing mean and severe wave conditions over the European continental shelf are established. A statistical analysis of extreme wave heights is also carried out assuming that the wave climate is stationary. The so-called peak over threshold (POT) approach is applied and a chart of estimates of 50-year return spectral significant wave height is obtained. The results are compared with those obtained considering a block maxima approach. Finally potential trends in the regional wave climate are investigated over the period 1979-2002.

INTRODUCTION – SCOPE OF THE STUDY

A proper estimation of extreme wave conditions is important in order to identify coastal areas of high potential risks and to improve the design of nearshore structures and coastal defenses. The estimation of extreme wave conditions requires long-term series of data. As only limited amounts of observations and measurements are in general available (through *in situ* buoys and/or satellite sensors), a now common way on getting continuous data-sets over long durations is to run hindcast simulations with an appropriate wave model driven by reanalysed wind fields. The present study relies on this approach and uses nearshore wave data based on numerical hindcast simulations performed at EDF R&D LNHE (Benoit and Lafon 2004).

In this paper the main features of the numerical wave data-base are first described (Section 2). Second the data-base is used to characterize medium and severe wave conditions along the French Atlantic coasts and the English Channel (Section 3). Then "traditional" extreme value theory is applied in order to characterize extreme values of wave height, assuming that the wave climate is stationary over the simulated period of time (Section 4). Finally potential trends in the wave climate are investigated (Section 5).

[1] Laboratoire National d'Hydraulique et Environnement (LNHE), EDF R&D, 6 quai Watier, BP 49, 78400 Chatou, France

NUMERICAL WAVE DATA-BASE

The wave data used in this study is based on numerical simulations performed with the third generation spectral wave model TOMAWAC (Benoit et al. 1996). TOMAWAC solves the wave action density balance equation (e.g. Bretherton and Garret 1969; Komen et al. 1994) and computes the evolution in space and time of the directional wave spectrum on an unstructured spatial grid, under unsteady wind forcing, and possibly unsteady currents and sea levels.

For this project, two nested numerical models have been set up and calibrated. The coarser model, called "OCEANIC" model, covers the Northern part of the Atlantic Ocean, with a grid of variable mesh size, from about 1 degree offshore down to 20 km along the French coastline (see Fig. 1). Some finite depth effect are taken into account in the simulations (refraction, shoaling, dissipation due to bottom friction) but not depth-induced breaking. The finer model, called "COASTAL" model, covers the European continental shelf with a mesh size of about 2 to 3 km along the coasts of France (see Fig. 2). Hourly directional wave spectra provided by the OCEANIC model are imposed on its boundaries. Shallow-water processes including surf-breaking are considered in this model. In this first version of the database the simulations are performed with steady hydrodynamics conditions, that is with a constant water level equal to mean sea level and without any effects related to ambient current.

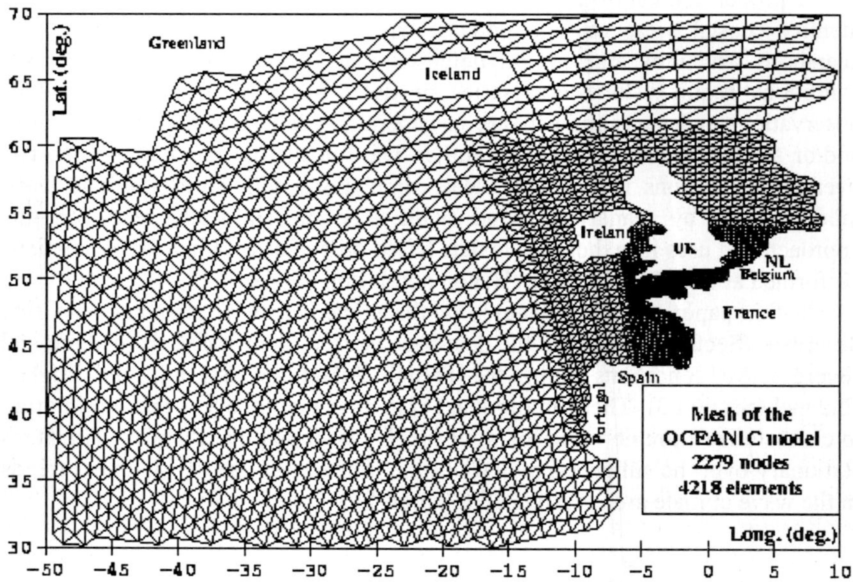

Figure 1. Spatial mesh of the OCEANIC model.

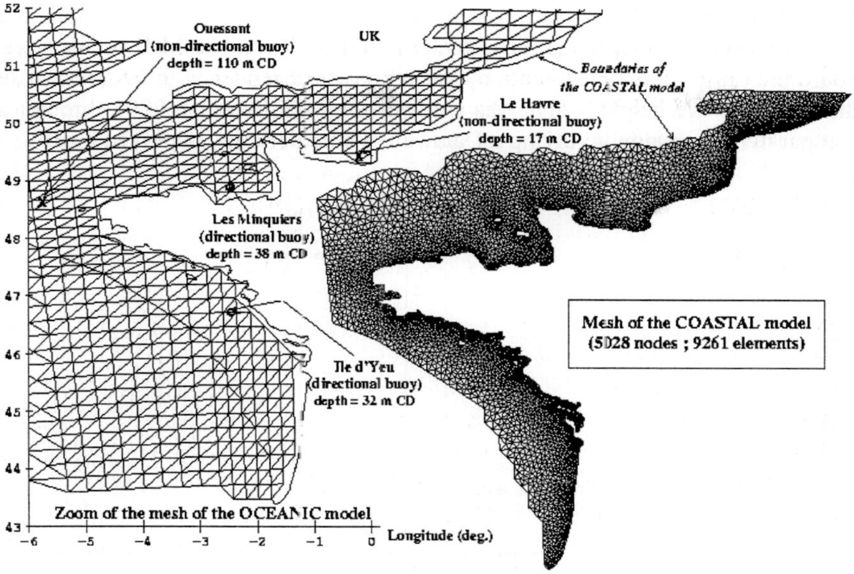

Figure 2. Mesh of the COASTAL model (right panel) compared with the mesh of the OCEANIC model zoomed near the French coasts (left panel), with the locations of wave buoys used in this study.

Two sources of wind data have been considered as input to force the models: ERA-40 re-analysis coming from the European Centre for Medium-range Weather Forecast (ECMWF), provided over a rectangular grid of 0.5 x 0.5 degrees resolution, and version 2 of NOAA/NCEP reanalysis, provided over a Gaussian-type grid, which mesh size is about 1.875 degrees. The influence of the source of wind data was examined by Benoit and Lafon (2004). The comparison revealed that the results obtained with ERA-40 winds were slightly better than those obtained with NCEP2 winds. Therefore the present study is based on the simulations performed with ERA-40 wind fields. Finally the numerical wave data-base considered here consists of hindcast results of the OCEANIC and COASTAL models driven by ERA-40 winds over the period 1979/01/01 to 2002/08/31 (23 years and 8 months).

The model's performances were checked by comparing the numerical results with buoy measurements. Detailed results are presented in Benoit and Lafon (2004). Additional calibration runs led us to slightly modify the amplification factor applied to the 10m wind speed (now varying between 1.08 and 1.20 depending on the wind speed). The correspondence between measured and computed percentiles of spectral significant wave height (H_{mo}) over the years 1999 and 2000 is shown by so called Q-Q plots (i.e. quantile-quantile plots) at the locations of four buoys of the French network operated by the

CETMEF (see Fig. 3). The agreement between the simulations and the measurements is fairly good. It is to be mentioned that the MRU-sensor onboard the (large size) Ouessant buoy has a consistent tendency to underestimate the highest wave heights. This was confirmed by CETMEF in 2004 through a comparative campaign involving a Datawell Waverider buoy deployed nearby.

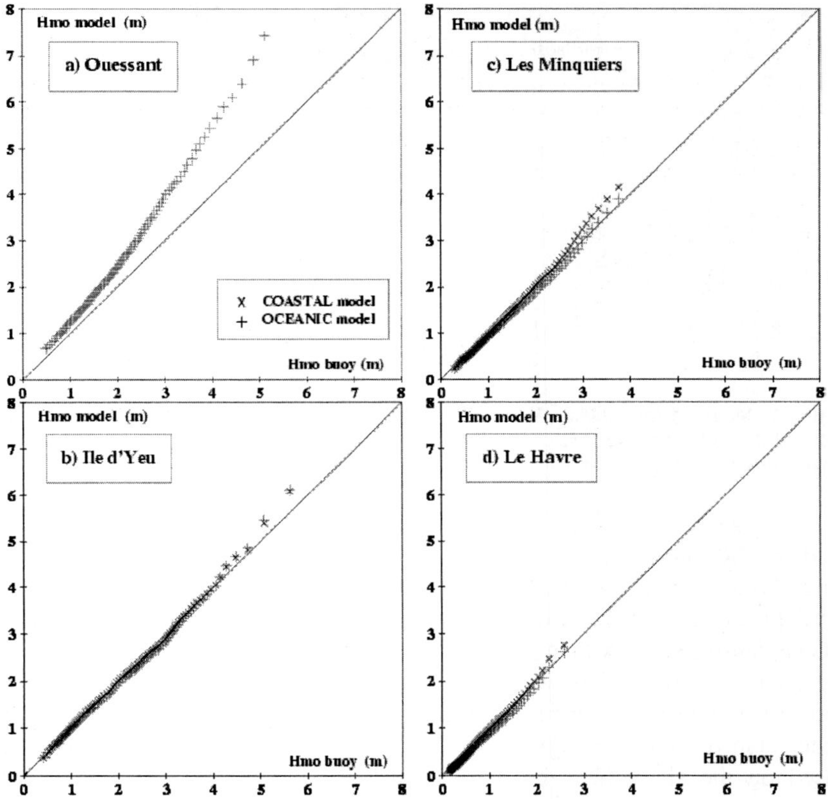

Figure 3. Q-Q plot of significant wave height H_{mo} for the years 1999-2000 at the 4 wave buoys. a) Ouessant, b) Ile d'Yeu, c) Les Minquiers, d) Le Havre.

MEAN CLIMATOLOGY OF SEA-STATES

In this section statistical analyses are performed on the time series of sea-state reduced spectral parameters (H_{mo}, T_m, θ_m, etc.) at each node of the OCEANIC and COASTAL grids over 23 full years, from 1979/09/01 to 2002/08/31. The first 8 months of year 1979 are excluded of the analysis in order to avoid introducing seasonal bias. The mean and quantile values of the wave parameters, as well as their joint distributions are determined. As examples of results, charts of the mean value and charts of 99% quantile value of H_{mo} (value which is exceeded 1% of the time on the average) are presented on Fig. 4 (OCEANIC model) and Fig. 5 (COASTAL model). Both models give consistent results over the European continental shelf, but as expected the results of the COASTAL model are of better resolution in intermediate and shallow water areas. This is particularly noticeable in the vicinity of islands. Over the COASTAL domain (see Fig. 5), the mean value of H_{mo} varies between 1 and 3 m, while the 99% quantile of H_{mo} can reach 8 to 9 m on the Atlantic boundary of the model. Such charts of characteristic wave heights, as well as similar charts for mean or peak period, mean incoming directions, etc. are very useful to get a first description of the local wave climate when a particular site has to be studied.

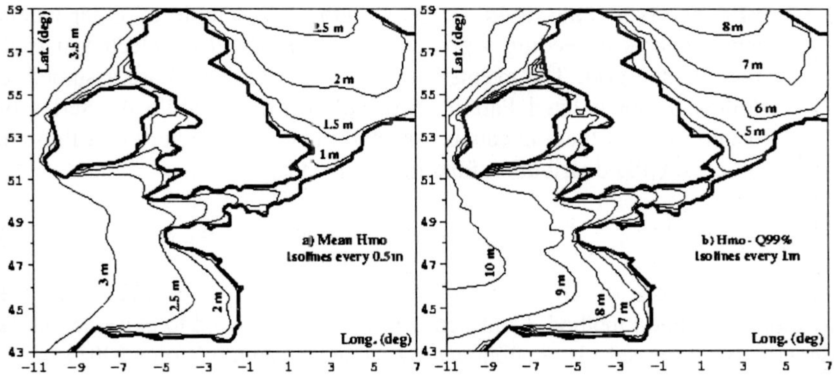

Figure 4. Charts of the mean value (a) and Q99 value (b) of wave height H_{mo} computed over the period 1979-2002 with the OCEANIC model.

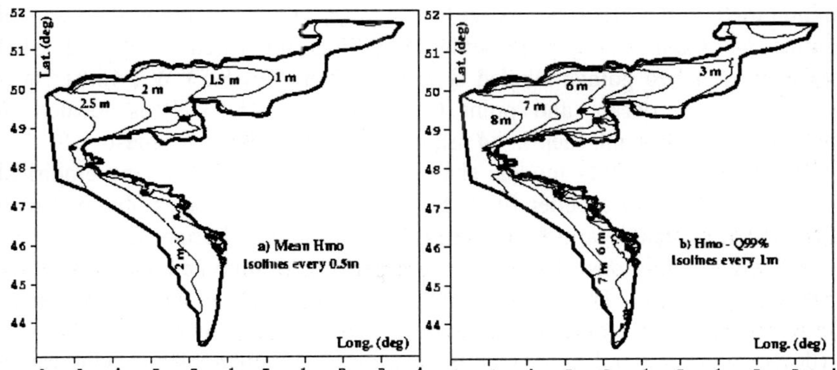

Figure 5. Charts of the mean value (a) and Q99 value (b) of wave height H_{mo} computed over the period 1979-2002 with the COASTAL model.

STATISTICAL ANALYSIS OF EXTREME SEA-STATE CONDITIONS

Methodology

The statistical analysis of extreme wave heights is carried out under the assumption of stationary asymptotic value theory (Coles 2001). Two approaches are applied and compared in this study:

1) **“Block maxima” method**: Time series of H_{mo} are blocked into sequences of equal duration and the block maxima (typically the largest wave height every year) are fitted to a Generalized Extreme Value (GEV) distribution. GEV distribution function is given by:

$$GEV(y) = \exp\left\{ -\left[1+\xi\left(\frac{y-\mu}{\sigma}\right)\right]^{-\frac{1}{\xi}} \right\}, \quad y: 1+\xi\left(\frac{y-\mu}{\sigma}\right) > 0, \mu \in \Re, \sigma > 0, \xi \in \Re \tag{1}$$

2) **“Peaks Over Threshold” (POT) method**: a sample of independent storm events exceeding a predetermined threshold is selected at each node of the grid. The events that excess the threshold are assumed to follow a Generalized Pareto Distribution (GPD), while the series of occurrences of these events is assumed to be the trajectory of a Poisson process. GPD distribution function is given by:

$$GPD_{\text{Threshold u}} = 1-\left(1+\xi\frac{y-u}{\sigma}\right)^{-\frac{1}{\xi}}, \quad y > u: 1+\xi\left(\frac{y-u}{\sigma}\right) > 0, \sigma > 0, \xi \in \Re \tag{2}$$

The parameters of GPD and GEV distributions are estimated by the maximum likelihood method. Confidence intervals associated to extreme values with various return periods are obtained by the delta-method (Coles 2001).

Application to the wave data-base

Both methods are first applied to a limited number of points in order to assess the accuracy of GEV and GPD distribution models. For the block maxima method the block size is set to one year, so that a sample of 23 annual maxima is obtained. Fig. 6 presents Q-Q plots for fitted GEV and GPD distributions at two points, labeled O and C. Point O (-8.53°W ; 46.61°N) is located in deep ocean, and point C (- 2.48°W ; 46.53°N) is located in intermediate water depth (depth: 64 m), near the French Atlantic coast.

Both GEV and GPD models show a good agreement with the empirical distribution. However a better fit to the measured distribution is obtained with GPD model, especially around the upper end of the distribution, that is for the highest wave conditions. The return levels of H_{mo} computed with GPD are represented on Fig. 7. The 70% confidence interval is plotted with dashed lines. The water depth has a clear influence on the shape of the distribution. At point O (deep water), the tail behaviour is close to an exponential one, whereas at point C the tail of the distribution is more curved. As the water depth decreases this trend is more pronounced, showing the effect of depth-induced breaking.

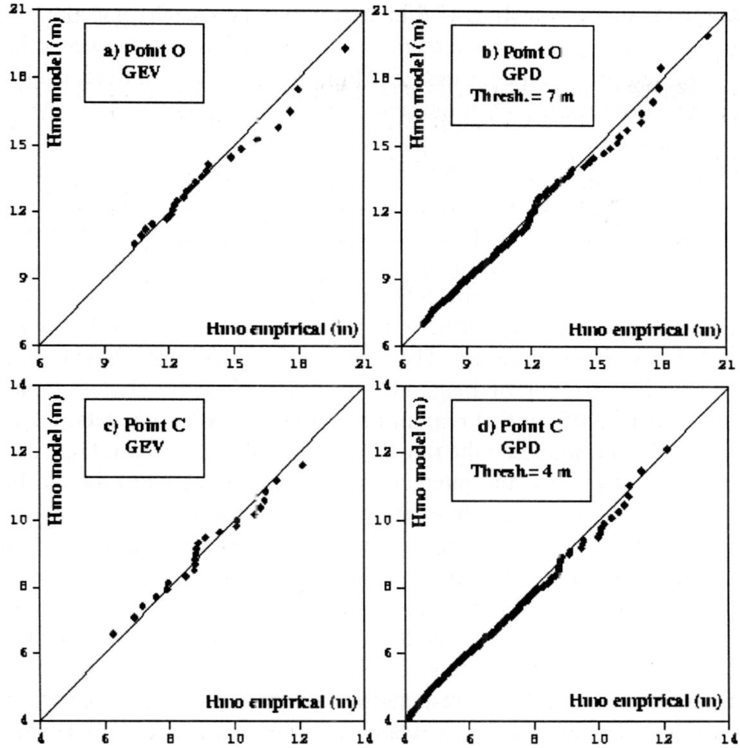

Figure 6. Model diagnostic for GEV and GPD distributions: a) b) point O c) d) point C

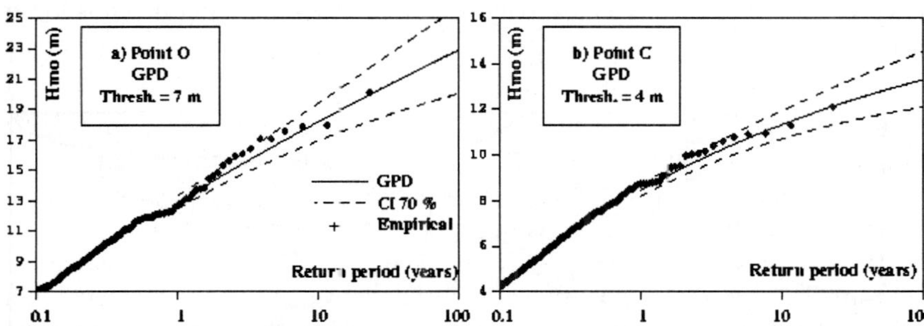

Figure 7. Return levels of H_{mo} computed with GPD: a) point O b) point C

As GEV and GPD appear as suitable models, the statistical processing of simulated wave heights is automatically performed at each node of the OCEANIC and COASTAL grids. For POT method several criterions for threshold selection have been tested and compared. Finally the value of the threshold is chosen as the 96% quantile value of the local distribution of H_{mo}. This choice of threshold leads to the selection of 6 to 12 storm events each year along a large part of the French coasts. This seems a satisfying balance between the total length of the sample and the necessity of limiting the selection to real significant independent storm events. Then the return values of significant wave height and the associated confidence interval are computed for several return periods, from 10 years to 50 years. Charts of 50-year return value of significant wave height are presented on Fig. 8. Isolines of H_{mo} 50-year return value follow quite regularly the bathymetry of the domain. Over the European continental shelf (see Fig. 8b) the best estimate of H_{mo} 50-year return value varies between 6 m and 17 m. Fig. 8a shows that H_{mo} 50-year return value can reach 22 m over the oceanic part of the domain, northern of Ireland. 100-year return H_{mo} were also computed in the same region. The results obtained (not shown here) are consistent with the order of magnitude of 100-year return H_{mo} estimated by Caires and Sterl (2005) in this region, based on ERA-40 wave data, taking into account a linear correction on the raw estimate of 100-year return H_{mo}.

It is well known that the interaction between waves and tide can become significant in shallow water. These effects are not accounted for in the present version of the wave data-base. In particular the constant water level (equal to mean sea level) considered in the simulations intensifies the effect of plateau due to depth-induced breaking because the highest waves tend to always break with almost the same wave height at each (constant) water depth. This introduces a bias in the adjustments of GEV and GPD distributions for shallow water areas. Therefore areas corresponding to depth lower than 20 m are masked in gray on Fig. 8. Moreover areas where the shape parameter ξ is lower than -0.5 are also masked because in this case the maximum likelihood estimators do not have the standard asymptotic properties (Coles 2001). For the POT method however this latter case is almost only occurring when the water depth is lower than 20 m.

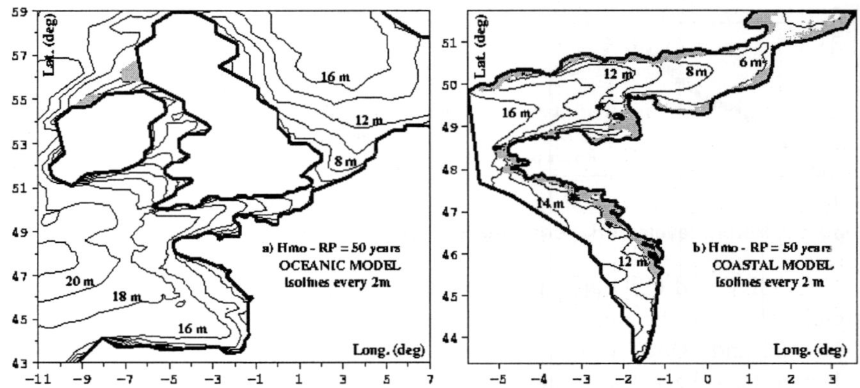

Figure 8. Charts of 50-year return value of H_{mo} (obtained with GPD model)

The shape parameter ξ determines the type of extreme value distribution (e.g. Gumbel, Frechet or Weibull family for GEV), that governs the tail behavior. Because of the duality between GEV and GPD families, the shape parameter should in theory be the same in both distributions. Charts representing the spatial variations of the shape parameter were constructed (not shown here). For water depths greater than 30 m, the estimation of the shape parameter ξ of GPD varies between -0.3 and 0.0 over almost the whole represented domain. In intermediate and shallow waters the value of ξ is clearly negative, indicating bounded distributions.

So as to test the sensitivity of the return levels to the choice of distribution (GEV or GPD) the relative difference between H_{mo} 50-year obtained with GEV and H_{mo} 50-year return value obtained with GPD is computed at each node of the COASTAL grid. The results are presented on Fig. 9. For the block maxima method (GEV) two analyses were carried out: a first one with the block size set to one year (graph on the left hand side), leading to the selection of a sample of 23 values, and a second one with the block size set to half a year (graph on the right hand side), leading to the selection of a sample of 46 values. In the latter case, the period considered begins on 1979/01/01 (instead of 1979/09/01) and ends on 2001/12/31 (instead of 2002/08/31), in order to limit the bias introduced by seasonal effects.

When the block size is set to a year (see Fig. 9a) the relative difference between H_{mo} 50-year return values respectively obtained with GEV and GPD is lower than 15% over a large part of the domain but it may locally reach 40% to 50% over an area located in the southern part of the grid. Moreover the maximum likelihood estimation of the shape parameter is lower than -0.5 over an area located in the northern part of the grid, which is left blank on the chart. When the block size is set to half a year (see Fig. 9b) the relative difference between H_{mo} 50-year return value respectively obtained with GEV and GPD is lower than 15 % over the whole domain, and the areas where the maximum likelihood estimation of the shape parameter is lower than -0.5 correspond to shallow water areas located very close to the coastline.

These results confirm the necessity of considering long series of data when using the block maxima approach, in order to be able to build a rich enough and homogeneous set of block maxima and to limit the variance of the estimation (in the present case, 23 values were not sufficient). Assuming that this condition is satisfied, it is noticed that both methods lead to consistent results over the COASTAL domain (as shown by the rather weak differences depicted on Fig. 9b). This allows to give some confidence in the estimations of extreme wave conditions obtained by the GPD method for e.g. the 50-year return values of H_{mo}.

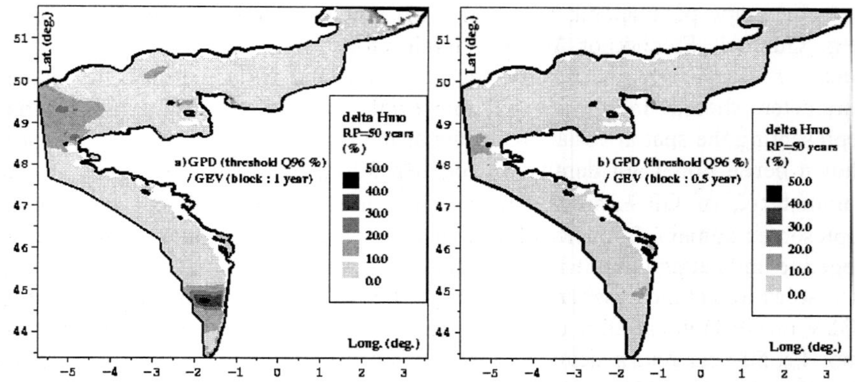

Figure 9. Comparison between 50-year return value of H_{mo} obtained with GPD and GEV over the COASTAL domain. a) the block size for GEV method is set to 1 year ; b) the block size for GEV method is set to 0.5 year

LONG-TERM TRENDS OF EXTREME WAVE CLIMATE

In section 3 the extreme value analysis has been carried out assuming that the wave climate is stationary. Several studies have been undertaken in the last decade in order to investigate the variability of wave conditions and storms in the North Atlantic ocean (e.g. WASA group 1998, Günther et al. 1998, Sterl et al. 1998, Wang and Swail 2002, Caires and Swail 2004). The advent of reanalysis data produced by meteorological centers has indeed brought a major tool for such analyses. The above-mentioned studies are based on large scale models, covering the Northeast Atlantic ocean (Günther et al. 1998), the North Atlantic ocean (Wang and Swail 2002) or even on global models (Sterl et al. 1998, Caires and Swail 2004). Here potential trends in the intensity of storm events is investigated at a more regional scale, using the numerical wave data-base presented in section 2, with a focus on the results of the COASTAL model along the French coasts. However it should be emphasized that a period of 23 years of simulations is rather short regarding the time scales of climate variability, so that the results presented here have to be considered as preliminary. The duration of the data-base will be extended in a future phase of the project, allowing further analysis. At this stage, two basic tests are performed and described below.

The first test consists in dividing the sample of annual maxima selected at each node of the OCEANIC and COASTAL grids in two parts of equal duration, in order to compare their mean values. As it is, the first sample contains the first 11 annual maxima, the second sample contains the last 11 annual maxima. Then an hypothesis test is carried out over the difference between the mean μ_1 of the first sample and the mean μ_2 of the second sample. The null hypothesis of the test is $\mu_1=\mu_2$. The significant level is set to 10%, that is the risk of wrongly rejecting the null hypothesis is 10%. The value of the relative difference δ (in %) between μ_1 and μ_2 is computed. δ is defined by:

$$\delta = 2\frac{\mu_2 - \mu_1}{\mu_1 + \mu_2} \times 100 \qquad (3)$$

The results are presented on Fig. 10. The areas where the hypothesis $\mu_1 = \mu_2$ (corresponding to a stationary storm climate) is accepted are left blank. Over a large part of the COASTAL domain (see Fig. 10b), the results do not put in evidence any significant long-term change in the values of annual maxima of H_{mo}. However, an area where the null hypothesis is rejected is identified West of the French Brittany. This pattern extends over the OCEANIC domain (see Fig. 10a), suggesting a decrease of annual maxima of H_{mo} intensity in a larger area West of the French Atlantic coast.

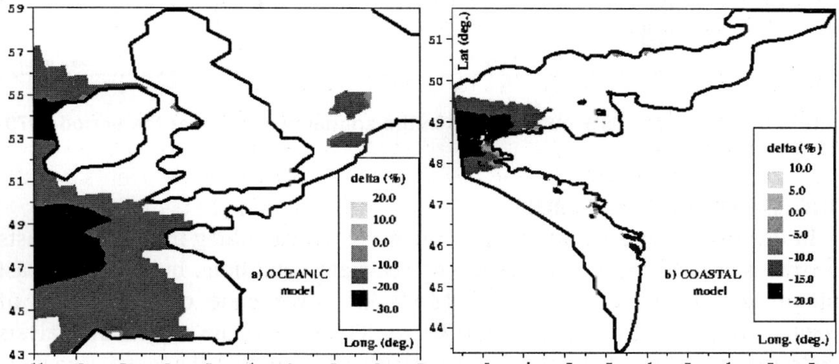

Figure 10. Hypothesis testing of the difference between the mean of annual maxima over the periods sept.1979-aug.1990 and sept.1991-aug.2002. Significant level: 10 %

The second test consists in computing the trend of the time series of respectively annual 90% quantile, annual 99% quantile and annual maxima of H_{mo} at each node of the COASTAL grid, using a linear regression. Then an hypothesis test is performed in order to determine whether the trend value is significant at the 10% level. Fig. 11 shows the results for the annual 99% quantile and for the annual maxima of H_{mo}. Areas where the trend is not significant are again left blank. For the annual 90 % quantile no significant trend has been identified over the COASTAL domain. Fig. 11a reveals two areas of negative trend in the annual 99% quantile of H_{mo}: one is located West of the French Brittany and the other is located in the South of Biscay. The trends in the annual maxima of H_{mo} (see Fig. 11b) exhibit almost the same pattern West of Brittany, with a larger decrease, but nothing significant in the South of Biscay.

It is emphasized that the studied period (1979-2002) is too short to be able to conclude to a significant long-term decrease of storm intensity in these regions. However these two tests made possible to identify areas where potential trends have to be further investigated.

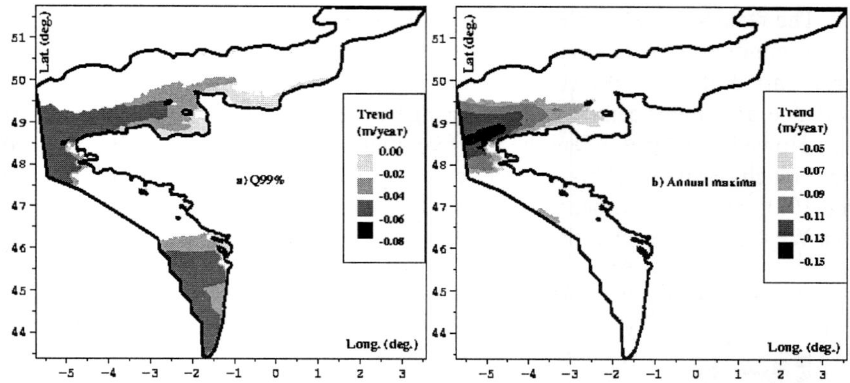

Figure 11. Trend in the 99% quantile and in the annual maxima over the period 1979-2002. Significant level: 10 %.

CONCLUSION AND OUTLOOK

In this paper the average and extreme wave climate along the French coasts has been analyzed using a data-base of sea-state conditions built by hindcast simulations performed with the TOMAWAC wave code over a period of 23 years and 8 months, from 1979/01/01 to 2002/08/31. The statistical analysis of extreme wave heights has been carried out assuming that the wave climate is stationary. The comparison between POT method and block maxima method showed that the two approaches lead to consistent estimates of return values over the COASTAL domain, when the set of block maxima is rich enough. However this condition can be difficult to fulfill because of the necessity of working with an homogenous sample. So the POT method appears (as yet often stated) as a more relevant approach in the present case. It was observed that the water depth has an influence on the shape parameter of GPD and GEV distributions: in deep ocean the tail behaviour is close to an exponential, whereas the distributions are more curved in intermediate water depth, and clearly bounded in shallow water. Finally two basic statistical tests were performed in order to check the hypothesis of stationary wave climate. They revealed that the hypothesis was accepted over a large part of the domain. However an area of potential non-stationarity was identified West of the French Brittany.

The steady hydrodynamical conditions considered in this version of the data-base appear to be a limitation for extreme wave analysis in shallow water, because the effects of varying sea levels and currents (due to tides) on the breaking process are not well represented on the distribution of highest wave conditions. Therefore it is planned to include the effects of tides, currents and storm surges in the next version of the data-base through the coupling with a flow model. Moreover the period of hindcast simulations will also be extended, in order to further investigate the variability of wave climate.

ACKNOWLEDGMENTS

This project is a joint collaboration between EDF R&D LNHE and CETMEF (French Ministry in charge of Coastal and River Works).

REFERENCES

Benoit, M., F. Marcos, and F. Becq. 1996. Development of a third generation shallow-water wave model with unstructured spatial meshing. *Proc. 25th Int. Conf. on Coastal Eng., ASCE, Orlando (Florida, USA),* 465-478.

Benoit, M., and F. Lafon. 2004. A nearshore wave atlas along the coasts of France based on the numerical modelling of wave climate over 25 years. *Proc. 29th Int. Conf. on Coastal Eng., Lisbon (Portugal),* 714-726.

Bretherton, F.P., and C.J.R. Garret. 1969. Wavetrains in inhomogeneous moving media. *Proc. Roy. Soc. London,* Series A, Vol. 302, 529-554.

Caires, S., and V. Swail. 2004. Global wave climate trend and variability analysis. *Proc. 8th Int. Workshop on Wave Hindcasting and Forecasting.*

Caires, S., and A. Sterl. 2005. 100-year return value estimates for ocean wind speed and significant wave height from the ERA-40 data. *Journal of Climate,* 18, 1032-1048.

Coles, S. 2001. *An introduction to statistical modelling of extreme values.* Springer Series in Statistics, Springer Verlag.

Günther, H., W. Rosenthal, M. Stawarz, J-C. Carretero, M. Gomez, I. Lozano, O. Serano, and M. Reistad. 1998. The wave climate of the North Atlantic over the period 1955-1994 : the WASA wave hindcast. *Global Atmosph. Ocean Syst.,* 6, 121-161.

Komen, G.J., L. Cavaleri, M. Donelan, K. Hasselmann, S. Hasselmann, and P.A.E.M. Janssen. 1994. *Dynamics and modelling of ocean waves.* Cambridge Univ. Press, 532 p.

Sterl, A., G.J. Komen, and P.D. Cotton. 1998. Fifteen years of global wave hindcast using winds from the European Center for Medium-Range Weather Forecast analysis. Validating the reanalyzed winds and assessing the wave climate. *Journal of Geophys. Research,* 103 (C3), 5477-5492.

Wang, X.L., and V.R. Swail. 2002. Trends of Atlantic wave extremes as simulated in a 40-Yr wave hindcast using kinematically reanalyzed wind fields. *Journal of Climate,* 15, 1020-1035.

WASA group, 1998. Changing waves and storm in the Northeast Atlantic ? *Bulletin of the American Meteorological Society,* 79, 9704-9706.

GENERATION OF EXTREME WAVE CONDITIONS FROM AN ACCELERATING TROPICAL CYCLONE

David P Callaghan[1], Jeff Callaghan[2], Peter Nielsen[1] and Tom Baldock[1]

It has been well established that tropical cyclones are able to generate large wave conditions. However, the limited fetch dimensions from the moving circular wind field often limits wave growth. Here, we look at an accelerating tropical cyclone which generated extreme wave conditions bigger than expected given its size. The generation process involved the unusual situation in which the growing wave field was trapped within the tropical cyclone's wind field, due to the system movements matching that of the growing waves. Consequently, the trapped waves experienced the cyclonic winds for over one day, resulting in twice the wave heights that would be expected with a stationary cyclone.

INTRODUCTION

Large and intensive tropical cyclones are able to generate extreme wave conditions (e.g., Phinn and Hastings, 1995). However, the limited fetch dimensions from the circular wind fields often limit wave growth. Here, we look at tropical cyclone Heta, which generated extreme wave conditions not usually seen from cyclones of its size, that destroyed buildings located on top of a 20 m high cliff on the South Pacific island of Niue. Three distinctions exist between this cyclone and others of similar size and severity. Firstly, this cyclone accelerated from near stationary to 20 knots while the wave field was growing. Secondly, the wind field and hence the fetch spatial extents were elongated from constraining high pressure systems as the cyclone traveled towards the south-east. Thirdly, the cyclone traveled in approximately the same direction as the growing waves from the first quadrant, which contains the strongest winds. These features provide similar meteorological conditions as study by Dysthe and Harbitz (1987), who showed that very small polar lows where able to generate 16 m wave heights.

This mechanism, called fetch enhancement (Callaghan, 2004), is achieved when the tropical cyclone accelerates from near zero to in excess of 20 knots in a manner that approximates the growth of the group velocity of the growing wave field. Consequently, the waves see high wind velocity over an unlimited fetch length (at least while the waves and cyclone propagate together). This mechanism produces two peculiar outcomes; very large waves with no forewarning from preceding swells and the spatial extents containing these very large waves is small (e.g., for Heta, the extremely large waves were estimated to cover an area of 12,320 km^2 or 3600 nm^2). Hence, the spatial scale of this

---

[1] Civil Engineering, The University of Queensland, Brisbane, QLD, 4072, Australia
[2] Bureau of Meteorology, 295 Ann Street, Brisbane, QLD, 4000, Australia

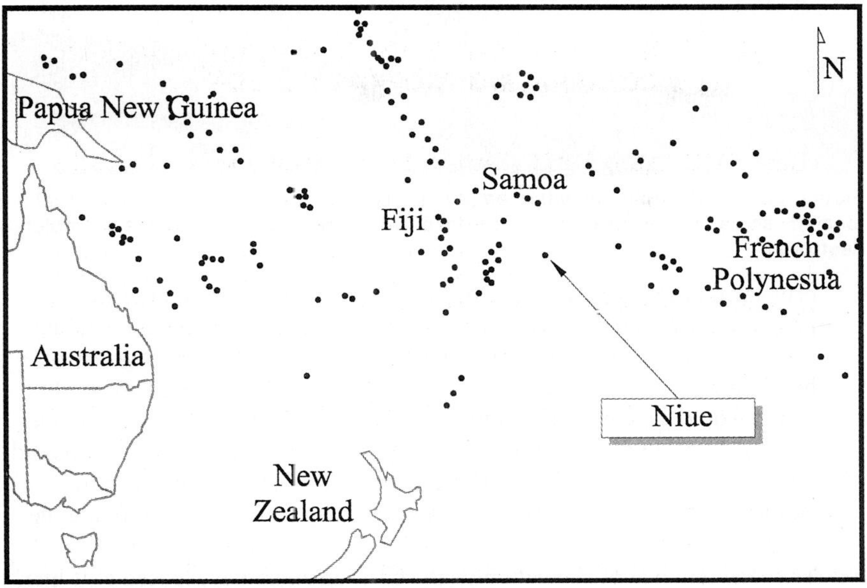

Figure 1. Locality map for Niue.

mechanism is often missed by large scale ocean wave models. This explains why wave height predictions are often less than that measured when fetch enhancement occurs. For example, the Canadian wave forecast model predicted $H_{sig} \approx 8$ m for hurricane Danielle at wave buoy 44141 (43°N 58°W) which, however, measured $H_{sig} \approx 13.4$ m.

FIELD EVIDENCE

Niue (fig. 1) is a small independent state on top of a large dormant volcano rising steeply from a 3000 m to 4000 m deep ocean (i.e., no surrounding shelf). Located between Tonga on the west and the Cook Islands further east, it marks the boundary of the paths taken by large tropical cyclones (Pearce, 2005), which are born in warm tropical seas above Niue. This limestone island measures about 23 by 18 km with a total surface area of 250 km^2. It rises to 68 m above sea level.

The damage at Niue included two significant structures, a motel and shopping centre, reportedly destroyed by waves breaking over the 20 m cliff face. These two structures are located at Alofi and are approximately 25 m above mean sea level, supporting the estimated deep water wave height.

The field inspection by the first and third authors in September 2005 (some 18 months after TC Heta past Niue) identified the following additional support that building damage was principally from waves and not the wind (local inhabitance reported water waves from the ocean):

- The buildings on Desert Road, constructed with concrete foundations and a combination of timber and steel frames and located along the cliff that was

Figure 2. Desert Road showing empty housing blocks except for one dwelling located some 120 m from the cliff face. All other structures were completely destroyed.

orientated perpendicular to cyclone track (and consequently large wave propagation direction), were all completely destroyed except one building located 120 m from the cliff face (see figs. 2 and 3). The vast majority of building debris was also never located. If wind generated this damage, debris would have been found further inland (as it was at other sites on the island). A local witness reported waves propagating across the cliff top road for approximately three hours. Consequently, the lack of debris can be explained by it being removed by water draining from the cliff top into the ocean.

- The total removal of vegetation on the cliff top along the exposed cliff line (fig. 3). Locals reported that vegetation further from the cliff line was damaged from the wind, however, as shown by the aerial image, wind damaged vegetation remained. This indicates again that the cliff-top zone near the ocean was exposed to water waves (consistent with local observations).
- Concrete head stones were sheared off at the base, which is only possible by forces generated by water wave impact.
- A large coral rock (*ca* 2 m diameter, see fig. 4) was moved from approximately half way between the shoreline and its present location during TC Heta. This coral rock was first noticed after TC Oda (1990) and was then not moved by any other wave events or human efforts before TC Heta.
- Buildings located away from the ocean were typically not damaged by the cyclonic winds.

The survey information available from the Niue Government indicates that Desert Road, the area that experienced most damage, is typically between 22 m and 26 m above mean sea level (MSL). The estimated variation in cliff-top height above MSL between Desert Road and the cliff edge is 2 to 4 m (using GPS equipment).

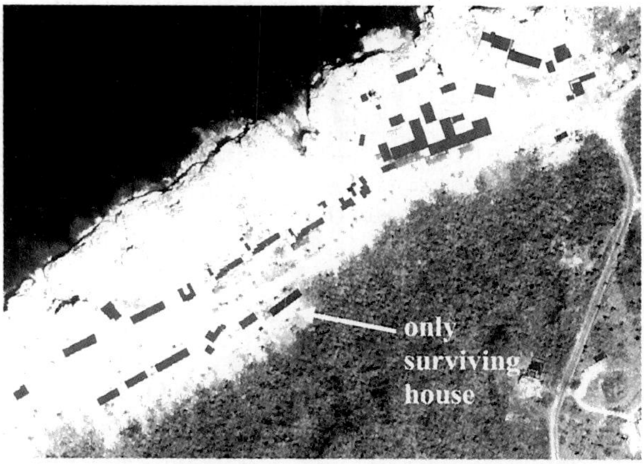

Figure 3. Aerial image of Desert Road, Niue taken several days after TC Heta. The buildings shown with solid black hatching were destroyed during TC Heta. In this image, the gray hatched building was the only one to survive TC Heta (albeit severely damaged).

Figure 4. Large coral rock moved by TC Heta, destroying a residential dwelling in the process. Between TCs Oda and Heta, this coral rock was located halfway between its current location and the ocean.

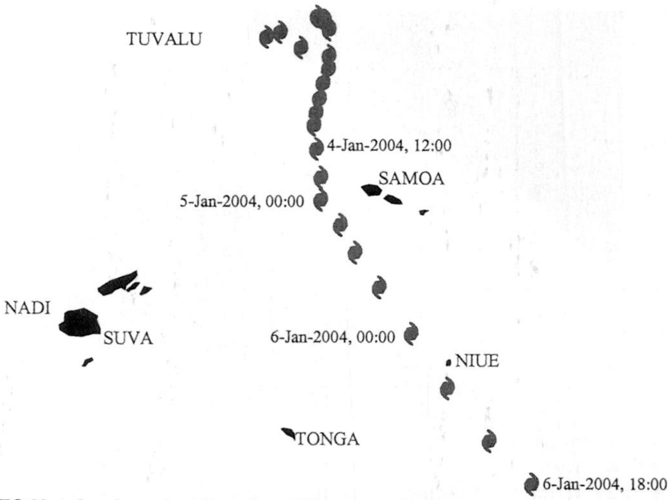

Figure 5. TC Heta's observed location (●) every six hours.

MODELLING

TC Heta started as a tropical depression north of Fiji (December 27, 2003) before moving northeastwards and intensifying into a tropical cyclone and when nearing Atafu (Tokelau Islands, January 1, 2004). On January 3, TC Heta's direction of propagation shifted significantly, towards a polar heading. This movement was initially with constant speed of 10 knots from 176°W 11°S until it past Independent Samoa (fig. 5), from where it irregularly accelerated to 21.5 knots by the time it past Niue.

Wind field estimates from remote sensing was analyzed by Callaghan (2004) as a 222 km (120 nm) square region with 30 m/s wind speeds in the direction of the cyclone propagation. An idealized cyclone model similar to that of Bowyer and Mac Afee (2005) was applied to TC Heta's track (table 1) from when TC Heta past Independent Samoa until it past Niue. This was done to assess if the waves generated within the cyclone fetch were trapped in the fetch. Fig. 6 shows seven snap shots of this analysis, which is summarized as follows: for each time step (¼ hour used)

- determine the wind field location from the observed cyclone track;
- estimate the wave field using the previous distribution and the new group

Table 1. Tropical Cyclone Heta summary track measurements.

Segment Number	Start date/time [UTC]	Finish date/time [UTC]	Direction [°TN]	Speed [knots]
1	4/01/2004 23:25	5/01/2004 4:25	324	9.5
2	5/01/2004 4:25	5/01/2004 9:25	333	12.5
3	5/01/2004 9:25	5/01/2004 15:25	321	11.5
4	5/01/2004 15:25	5/01/2004 20:25	326	14.5
5	5/01/2004 20:25	6/01/2004 1:25	326	20
6	6/01/2004 1:25	6/01/2004 4:25	326	21.5

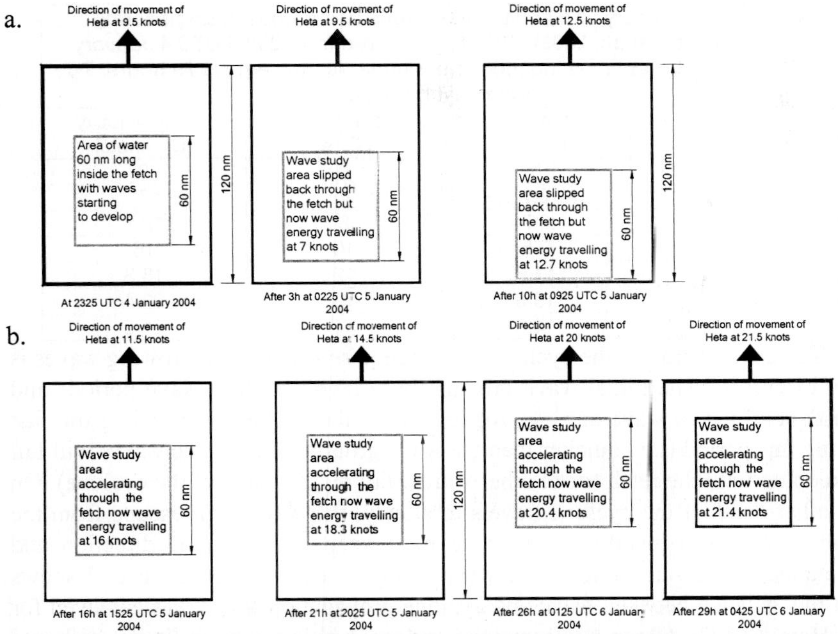

Figure 6. Propagation of tropical cyclone Heta wind field and the growing wave field as the cyclone approached Niue at a. 0, 3, and 10 hours, b. 16, 21, 26 and 29 hours after 23:25 UTC 4 January 2004, with time zero and 29 hours being when TC Heta past Independent Samoa and Niue respectively. (nm = nautical miles)

velocity multiplied by the time step;

- increase time counter; and
- repeat as necessary.

The wave field is considered trapped if some or all of the wave energy transmitted from the cyclone remains in its fetch. Here, we track a 111 km (60 nm) box, initially located towards the rear of the wind field. Fig. 6 shows that initially the waves, having short periods and therefore small group velocities, travel slower than TC Heta and move towards the rear of the fetch (fig. 6a). To keep the waves trapped, the cyclone needs to initially move slowly. However, as the waves grow and their periods increase, the cyclone needs to accelerate in conjunction with the wave group velocity. This is in fact what TC Heta did (see fig. 6b), consequently we have the trapped fetch scenario. Using the Coastal Engineering Manual (Resio et al., 2005) with a 350 nm fetch length and $U_{10} \sim 30$ m/s, the significant wave height, as TC Heta past Niue of the trapped waves, is estimated at 15 m with 14 s peak period. Not considering the trapped fetch mechanism yields 9 m significant wave height. Table 2 provides the summary of wave heights and period for the seven time steps shown in (fig. 6).

Table 2. Summary of manual calculations using Coastal Engineering Manual (Resio et al., 2005). Time is measured from 23:25 UTC 4 January 2004 when TC Heta past Independent Samoa. At time equals 29 hours, TC Heta past by Niue.

time [hours]	Wave height [H_{sig}, m]	Wave period [T_p, s]	Group velocity [C_g, knots]
0	0	0	0
3	2.7	4	7
10	6.7	8	12.7
16	9.5	10	16
21	11.6	12	18.3
26	13.6	13	20.4
29	14.8	14	21.5

The acceleration of the cyclone system required to trap the growing waves is now discussed. The initial wave growth phase involves short wave periods and consequently slow wave energy propagation. If the cyclone is traveling too fast during this period (i.e., quicker than the wave group velocity), the waves will fail to keep up with the fetch (or in other words, fall out the back of the cyclone). On the other hand, if the cyclone travels to slowly, the waves will escape from the wind field. Consequently, the cyclone must keep going in one direction and accelerate at the same rate as the wave group velocity increases. Fig. 7 shows that this balance between wave and cyclone propagation speed was achieved for TC Heta, and the 60 nm box remained entirely within the fetch limits. It should be noted however, that between eight to 13 hours, the wave energy box was on the brink of not remaining within the cyclone fetch, indicating that if TC Heta had accelerated at a higher rate, it would not have produced the large waves. While trapping waves in fetches may seem an unusual situation, it is a common phenomenon for polar lows (Dysthe and Harbitx, 1987).

When the cyclone speed does not match that of the growing waves, two possible outcomes occur. That is, people are either exposed to the tropical cyclone winds without significant waves (cyclone traveling too fast for the waves to keep up) or large waves without the winds (cyclone traveling slower than the wave energy). However, in the trapped scenario, when the wave energy and the tropical cyclone travel together, people are exposed to both cyclonic winds and very large waves.

It is this combination that may explain how the waves were able to overtop the 20 to 25 m cliffs at Niue. The bathymetry at Niue slopes at 1:3 near the expected wave breaking point for a deep water wave height of $H_{sig} \approx 15$ m. The breaking wave height is ca 19.5 m and the break point is located within one wavelength from the cliff face. Two different theories are proposed for how these waves may have propagated onto the top of the cliffs, noting that we have no direct evidence to support either theory. The first theory is that the waves impacted directly on the cliff face after the initial breaking, with some wave heights being high enough for the wave crests to break over the cliff (i.e., crests are sheared off and continue as a solitary units of water propagating across the cliff top). The second theory is similar in that waves impact on the cliff face and generate near vertical water jets that were blown onto the cliff top by the cyclonic winds, again resulting in solitary units of water propagating across the cliff top. In either case, large water volumes are elevated onto the cliff top (as witnessed by locals who sheltered near the back cliff—120 m from the cliff face). Both theories then include debris removal by gravity flow back into the ocean.

SUMMARY

There are many coastlines and islands regularly exposed to tropical cyclone, typhoon or hurricane activity (e.g., mid South Pacific Islands, Australian eastern, northern and western coasts, U.S. east coast, Taiwan, Japan, Philippines). Waves generated from these tropical cyclones are typically fetch limited and do not propose significant threat compared to the large mid-latitude low pressure systems in, for example, Atlantic and Southern oceans. However, accelerating tropical cyclones have the potential to generate very large waves. One example of this mechanism was TC Heta. TC Heta accelerated towards Niue, trapping waves within its fetch, which consequently grew to twice the size they would have reached with a stationary cyclone. These large wave heights arguably caused the extensive building damage during TC Heta.

While it is recognized that this particular trapping of waves under cyclones or hurricanes is a rare event, Niue serves as a warning for those designing important infrastructure in wave climates that include tropical cyclone or

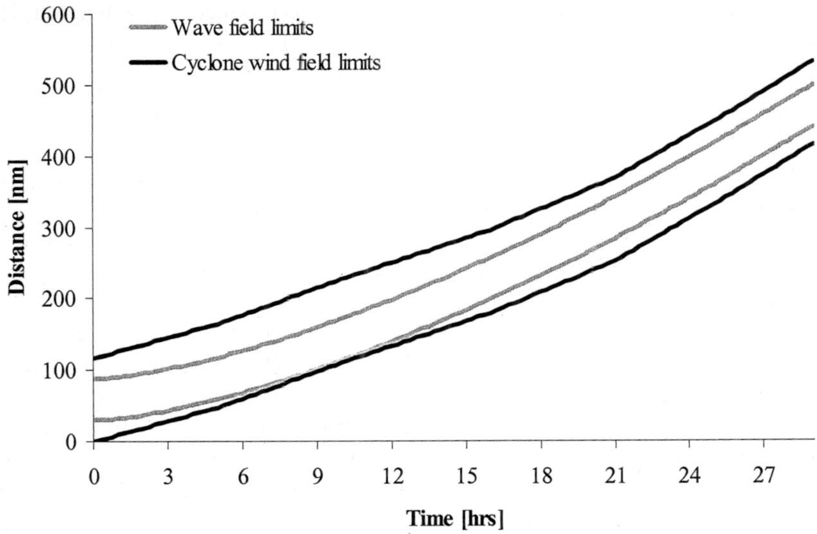

Figure 7. Propagation of tropical cyclone Heta wind field (▬) and the growing wave field (▬) limits as the cyclone approached Niue (nm = nautical miles). Time is measured from 23:25 UTC 4 January 2004 as TC Heta past Independent Samoa.

hurricane activity. Niue has been hit by the phenomenon twice (TCs Oda & Heta)!

ACKNOWLEDGMENTS

The authors gratefully acknowledge support from the CRC Sustainable Tourism Project #52001, Niue Meteorological Service (Mr S. Pulehetoa) and Justice, Land and Survey Department of Niue (Mrs M. Enote).

REFERENCES

Bowyer, P.J. and MacAfee, A.W., 2005. The theory of trapped-fetch waves with tropical cyclones-an operational perspective. *Weather and Forecasting*, 20(3), 229-244.

Callaghan, J., 2004. Forecasting extreme waves. *International Conference on Storms.*

Dysthe, K.B. and Harbitx, A., 1987. Big waves from polar lows. *Tellus*, 39A, 500-508.

Pearce, F., 2005. The gathering storm. *New Scientist*, 188(2528), 36-41.

Phinn, S.R. and Hastings, P.A., 1995. Southern Oscillation influences on the Gold Coast's summer wave climate. *Journal of Coastal Research*, 11(3), 946-958.

Resio, D.T., Bratos, S.M. and Thompson, C.S., 2005. Meteorology and Wave Climate. In: L. Vincent and Z. Demirbilek (Editors), *Coastal Engineering Manual, Part II, Hydrodynamics, Chapter II-2, Engineer Manual 1110-2-1100*. U.S. Army Corps of Engineers, Washington, DC.

AN EXTREME VALUE MODEL FOR WAVE CLIMATE CONSIDERING DURATION AND SEASONALITY

Fernando J. Méndez[1], Melisa Menéndez[1], Alberto Luceño[2], Raúl Medina[1], Nicholas E. Graham[3]

A statistical model to characterize the long-term extreme value distribution of intensity of significant wave height conditioned to the duration of the storm and taking into account seasonality is presented in this paper. The model is based on a time-dependent version of the Peak Over Threshold (POT) approach and is applied to specific reanalysis time series and NOAA buoy records. The model allows consideration of the annual and semiannual cycles which are parameterized in terms of harmonic functions. The inclusion of seasonal variability substantially improves the correlation between the model and the data.

INTRODUCTION

Wave studies at a particular site have usually focused on frequency analysis of the wave intensity alone, typically the significant wave height, *H*. However, in the design of a maritime work, the persistence or duration of exceedance (*D*) of the sea state intensity *H* is of considerable interest, as it makes possible a more adequate description of the storm history from a statistical viewpoint. A number of authors have developed several models trying to obtain the relationship between sea state intensity and the duration of persistence for extreme values (f.i. Teisson, 1990; Sobey and Orloff, 1999).

The seasonal variability of wave conditions is a well known feature. Several authors have analyzed the non-stationary behavior of extreme wave heights by means of extreme value distributions (f.i., Carter and Challenor, 1981; Morton et al., 1997; Méndez et al., 2006), thus modeling the intra-annual variability. To illustrate the importance of seasonality, Figure 1 shows the independent extreme events of significant wave height within a year (crosses) in buoys situated near Alaska (#46001) and San Francisco (#46013). In dashed lines we represent the results for a stationary model (location parameter μ_0 of the Generalized Extreme Value (GEV) distribution and the 20-year return period quantile $H_{20,0}$). The solid lines stand for the results of a time-dependent POT model (Mendez et al. 2006), $\mu(t)$ and $\psi(t)$ are the time-dependent location and

---

[1]Ocean & Coastal Research Group, Universidad de Cantabria, Avda de los Castros s/n, 39005 Santander, SPAIN
[2]Department of Applied Mathematics, Universidad de Cantabria. Avda de los Castros s/n, 39005 Santander, SPAIN
[3]Hydrologic Research Center – Scripps Institution of Oceanography, University of California, San Diego, CA

scale parameters of the GEV distribution – $\psi(t)$ is represented over the threshold, and $H_{20}(t)$ is the estimated 20-year return period quantile). As seen, the time-dependent parameters are able to properly model the data variability along a year, better than the stationary POT model. Moreover, the 20-year return period quantile modulates along the winter season, depending on each particular case.

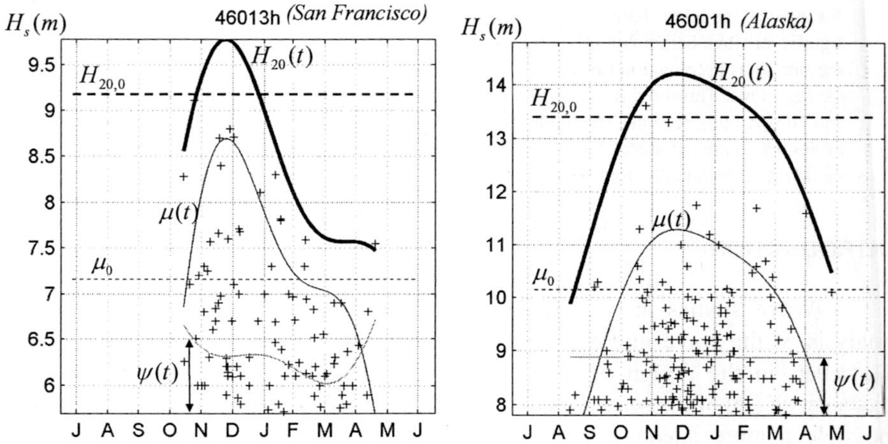

Figure 1. Application of a time-dependent POT model in #46013 buoy (left) #46001 buoy (right)

Therefore, the objective of this work is to statistically characterize the long-term extreme value distribution of intensity of significant wave height conditioned to the duration of the storm and taking into account seasonality. It will be shown that both factors allow a more correct design of the maritime structures.

The paper is organized as follows. First, the time series analysis is presented. Second, a brief description of the generalized extreme value and POT models (the so-called generalized Pareto distribution (GPD) - Poisson model) for time-dependent random variables is shown, focusing on the parameterization employed in terms of covariates and functions of time. Next, some examples of application are shown. Finally, some conclusions are given.

TIME SERIES ANALYSIS

The time series analysis can be shown in Figure 2: it is based on the selection of the peaks of every independent storm (in terms of the significant wave height *H*) that are exceeded or equalized during *D* continuous hours.

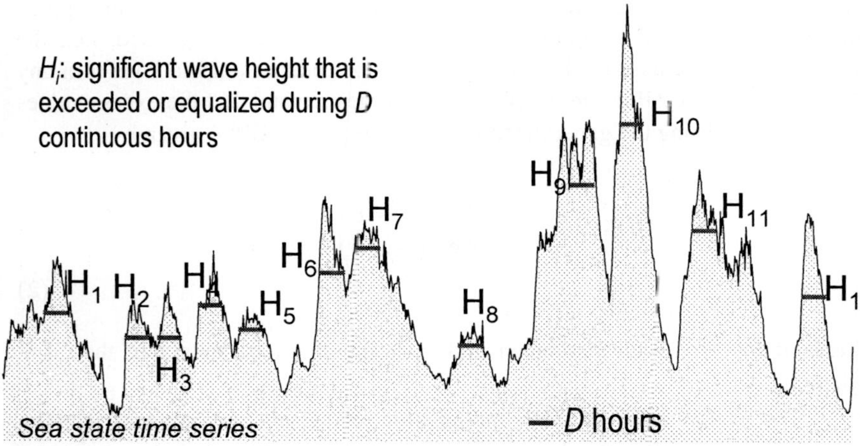

Figure 2. Example of time series analysis

Therefore, the data consists on *N* independent events $x_1, x_2, ..., x_N$ that are observed at days $t_1, t_2, ..., t_N$ over a *T*-year period (in this study the time span that will be required to assume the independence of consecutive storm is $\Delta t = 72$ hours).

THEORY

Statistical model

The simplest model in extreme value theory is the Annual Maxima method, which uses only annual maxima values. This is modeled using the generalized extreme value distribution (GEV) given by

$$H(x;\mu,\psi,\xi) = \exp\left\{-\left(1+\xi(x-\mu)/\psi\right)^{-1/\xi}\right\} \tag{1}$$

where $1+\xi(x-\mu)/\psi>0$, μ is a location parameter, ψ a scale parameter and ξ is the shape parameter which determines the nature of the tail of the distribution. The main problem in extreme value analysis is the scarcity of data for model estimation. To mitigate this problem, the POT method for independent and identically distributed random variables is proposed (Davison and Smith, 1990). The basic idea is to pick up a high threshold *u* and to study all the exceedances over *u*; in particular, the number of exceedances over *u* for a given time of period, and the amounts by which the threshold is exceeded, i.e., the excess values. We assume that the number of exceedances of the level *u* in any given year follows a Poisson distribution with mean ν and that the excess values *y* are modelled using the generalized Pareto distribution (GPD) given by

$$G(y;\sigma,\xi)=\begin{cases}1-\left(1+\xi y/\sigma\right)^{-1/\xi} & \xi\neq 0\\ 1-\exp(-y/\sigma) & \xi=0\end{cases} \tag{2}$$

where $\sigma>0$ is a scale parameter and ξ is the shape parameter of the GEV distribution.

The combination of both models for frequency and intensity can be expressed in terms of the GEV distribution for annual maxima if $\sigma=\psi+\xi(u-\mu)$ and $\nu=\left(1+\xi(u-\mu)/\psi\right)^{-1/\xi}$, Moreover, the hypothesis of homogeneity in the random variable *y* can be relaxed, being able to model non-stationarity as seasonal effects.

Therefore, an extension of the GPD-Poisson model is obtained by allowing its parameters to be time-dependent, so that $\nu(t)>0$, $\sigma(t)>0$ and $-\infty<\xi(t)<\infty$ may vary along the year. Accordingly, the GEV distribution has time-dependent parameters $-\infty<\mu(t)<\infty$, $\psi(t)>0$ and $-\infty<\xi(t)<\infty$. The functions $\mu(t)$, $\psi(t)$ and $\xi(t)$ can contain sine waves representing seasonal effects (see f.i. Katz et al., 2002; Méndez et al., 2006).

Fitting the model

If the *N* exceedances $y_1,y_2,\ldots,y_N$ are observed at days $t_1,t_2,\ldots,t_N$ over a *T*-year period, the log-likelihood function for the time-dependent GPD-Poisson model is expressed as,

$$l_{y,t}(\theta)=\sum_{i}\left\{\eta_{t_i}\log\nu(t_i)-\frac{\nu(t_i)}{T_S}-\eta_{t_i}\log\sigma(t_i)-\eta_{t_i}\left(1+\frac{1}{\xi(t_i)}\right)\log\left(1+\xi(t_i)\frac{y_i}{\sigma(t_i)}\right)\right\} \tag{3}$$

where $\sigma(t)$ and $\nu(t)$ are related to the GEV parameters; η_{t_i} equals 1 if there is an exceedance on day t_i and 0 otherwise; T_S is a time scaling constant so that $T_S = 365.25$ (for a daily time interval); y_i is the threshold excess on day t_i; θ is a vector of regression parameters; and the sum over t is extended to all the days with data available (i.e., days without data are omitted).

Model diagnostics

There are a number of procedures for model-checking when data are assumed to be identically distributed, but they are not directly applicable to time-dependent random variables unless some modifications are performed. One possibility is to standardize the data, conditional on the fitted parameter values. Smith (2001) (see also Mendez et al., 2006) provides a detailed discussion of graphical procedures to examine the goodness of fit of extreme value models for frequency (Z statistic) and for intensity (W statistic). Diagnostic Probability-Probability (PP) and Quantile-Quantile (QQ) plots can be made for both Z and W statistics. If the suggested model is adequate, the PP and QQ plots should consist of points close to the unit diagonal. Note that both plots must be checked simultaneously as they contain the same information but on different scales.

Regression model

Likelihood-based methods allow modeling the effects of temporal non-stationarity. In this study, we propose models for the yearly seasonality using

$$\mu(t) = \beta_o + \beta_1 \sin(2\pi t) + \beta_2 \cos(2\pi t) + \beta_3 \sin(4\pi t) + \beta_4 \cos(4\pi t) \tag{4}$$

for the location parameter and

$$\psi(t) = \alpha_o + \alpha_1 \sin(2\pi t) + \alpha_2 \cos(2\pi t) + \alpha_3 \sin(4\pi t) + \alpha_4 \cos(4\pi t) \tag{5}$$

for the scale parameter. We assume constant values for the shape parameter $\xi(t) = \xi_0$. In these equations t is given in years. A possible model (see example of Alaska #46001 buoy in Figure 1) can be expressed in terms of the vector parameter $\theta = (\beta_0, \beta_1, \beta_2, \beta_3, \beta_4, \alpha_0, \xi_0)$. In this model we consider an annual and semiannual cycle for the location parameter and constant values for the scale and shape parameters. Figure 3 shows how the seasonality affects to the shape of the GEV probability density function.

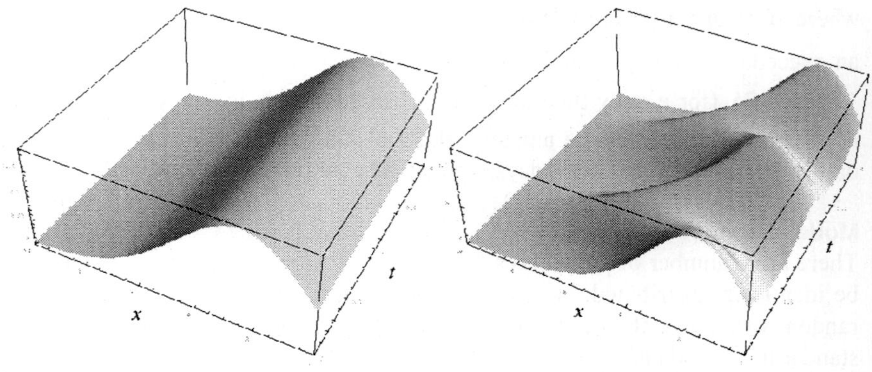

Figure 3. GEV pdf for a stationary random variable (left) and time-dependent random variable (right)

Model selection

One of the main requirements is to obtain (following the principle of parsimony) the simplest model that explains the data variability sufficiently well. Thus, the quality of the models is assessed systematically by minimizing a penalty function such as the Akaike Information Criteria (AIC) (Akaike, 1973),

$$\mathrm{AIC} = -2\hat{l}(p) + 2p\,, \tag{6}$$

where p is the number of parameters and $\hat{l}(p)$ is the maximum of the log-likelihood resulting from each model. Therefore, the best model is selected trying a number of possible parameterizations using the AIC criterion given by eq. (6) as the fitness function to minimize.

Estimation of quantiles

Estimates of the time-dependent extreme quantiles are obtained by inverting eq. (1),

$$\hat{y}_R(t) = \begin{cases} \hat{\mu}(t) - \dfrac{\hat{\psi}(t)}{\hat{\xi}(t)}\left\{1 - \left[-\log(1-1/R)\right]^{-\hat{\xi}(t)}\right\}, & \hat{\xi}(t) \neq 0, \\ \hat{\mu}(t) - \hat{\psi}(t)\log\left[-\log(1-1/R)\right], & \hat{\xi}(t) = 0, \end{cases} \tag{7}$$

where $\hat{y}_R(t)$ is the time-dependent return level associated with the return period R (years) and $R > 1$.

APPLICATION OF THE MODEL

For illustrative purposes, the Poisson-GPD model is applied to an hourly 44-years time series numerically obtained in a wave reanalysis of the Mediterranean in the frame of the HIPOCAS project (Ratsimandresy et al., 2003). The "numerical buoy" is located at the eastern side of the Balearic Islands (Spain). For each selected duration (1, 3, 6, 12 and 24 hours), we obtain $Y_1, Y_2, ..., Y_N$ exceedances over a T-year period.

The application of the model in the stationary mode ($\mu(t) = \beta_0, \psi(t) = \alpha_0, \xi(t) = \xi_0$) is displayed in Figure 4. As seen, the return level plots show a good agreement between data and the model. Besides, a bounded tail behavior (Weibull for maxima family) is detected.

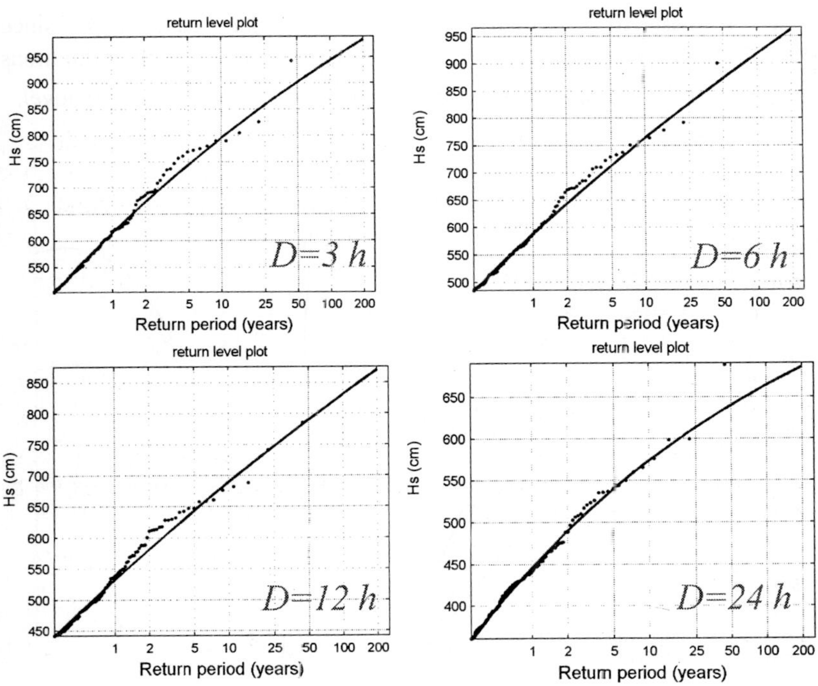

Figure 4. Return level plots for different durations (HIPOCAS Reanalysis data near Minorca)

Figure 5 shows the plots of exceedances for D=1 h versus day within year and the time-dependent 50-year return period quantiles for different durations.

For this particular case, the best parameterization was obtained using $\theta=(\beta_0,\beta_1,\beta_2,\beta_3,\beta_4,\alpha_0,\xi_0)$ .The result shows a clear seasonality, with many fewer exceedances during summer months, reinforcing the proposed regression model.

The inclusion of seasonality in the model (M_1) notably improves the goodness of fit and the significance of the parameters, with respect to the stationary case (dashed lines, M_0). This is measured using the likelihood ratio test: we can assure that model M_1 explains substantially (at the α -level of significance) more variability in the data than M_0 if $2[\ell_1(M_1)-\ell_0(M_0)]>\chi^2_{k,1-\alpha}$, where k is the number of parameter that are introduced in model M_1 with respect to M_0 , $l_1(M_1)$ and $l_0(M_0)$ are the maximized log-likelihood functions under models M_1 and M_0 , respectively, and $\chi^2_{k,1-\alpha}$ is the 1- α quantile of the χ^2 distribution with k degrees of freedom (in this particular case k=4 since $\theta_1=(\beta_0,\beta_1,\beta_2,\beta_3,\beta_4,\alpha_0,\xi_0)$ and $\theta_0=(\beta_0,\alpha_0,\xi_0)$). For all the durations selected, M_1 model was significant with respect to M_0 model at the 0.01-level.

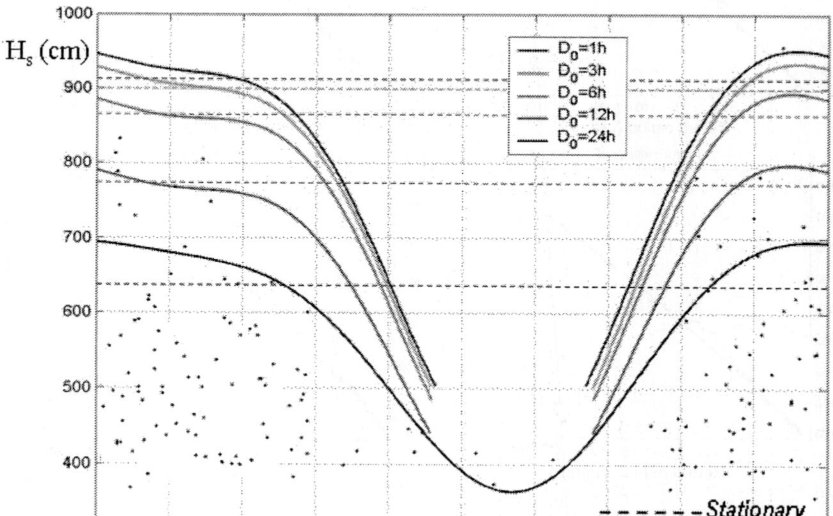

Figure 5. Plot of exceedances over the 99.5% SWH percentile for duration *D*=1 hour and time-dependent quantiles for different durations

We have used the *Z* and *W* statistics described previously to test the goodness of the fit of the model. As an example, Figure 6 shows the results for *D*=12 hours. The fit of the time-dependent GPD-Poisson model is remarkably good (PP and QQ plots for both *W* and *Z* statistics).

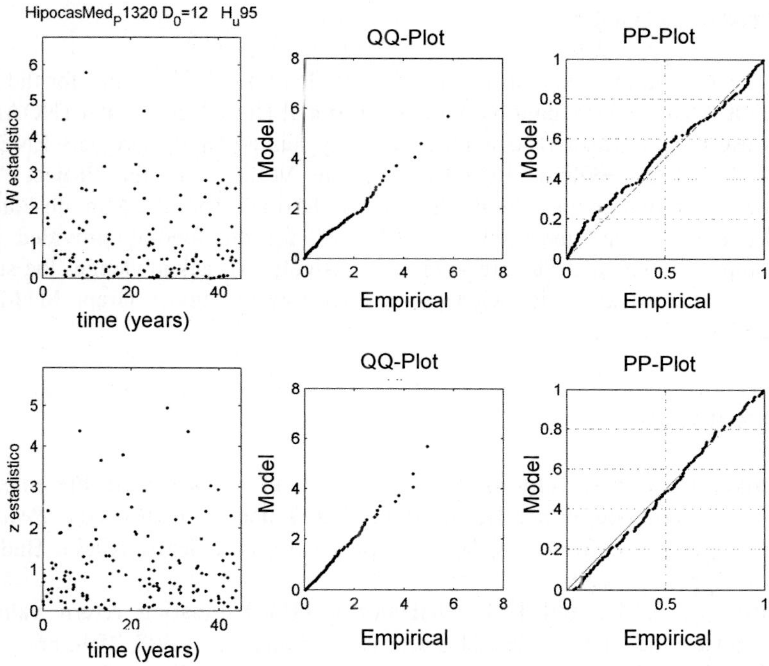

Figure 6. Diagnostic plots for D=12 hours

CONCLUSIONS

A model for the extreme value analysis is developed, considering storm duration and a time-dependent behaviour. The model is based on a time-dependent version of the Peak Over Threshold (POT) model, which is based on the modelling of the intensity using the Generalized Pareto Distribution and the frequency as a Poisson process. The GPD-P allows using a larger amount of extreme values, compared to the classical annual maxima method.

The inclusion of non-stationarity in the model (seasonality) notably improves the goodness of fit and the significance of the parameters. The model has been applied to specific significant wave height time series showing the ability of the proposed method.

The information obtained in this study can be useful in the design of maritime works, due to the fact that (a) the model allows a better explanation of the variability of extreme wave climate along a year and (b) the duration of the storm is a key factor in several formulations of breakwater stability.

ACKNOWLEDGMENTS

The authors thank to Puertos del Estado (Min. de Fomento) for the use of HIPOCAS reanalysis data base and to National Buoy Data Center (NOAA) for the use of buoy data. The work is partially funded by the projects ENE2004-08172 and CGL2005-05365/CLI from the M.E.C. and the Project "Wave climate duration analysis along the Spanish littoral" from the Min. de Fomento. FJM and MM are indebted to the M.E.C. for the funding provided in the "Ramon y Cajal" and FPI programs, respectively. AL acknowledges the support of the Spanish Dirección General de Investigación under Grant MTM2005-00287.

REFERENCES

Akaike, H. (1973) Information theory and an extension of the maximum likelihood principle, in: B.N. Petrov and F. Csáki, eds. *Proc. 2nd Internat. Symp. on Information Theory* (Akadémia Kiadó, Budapest) pp. 267–281.

Carter, D. J. T., and P.G. Challenor (1981) Estimating return values of environmental variables. *Quart. J. Roy. Meteor. Soc., 107*, 259-266.

Coles, S.G., (2001) An introduction to statistical modelling of extreme value. Springer, London.

Davison, A. C., and R.L. Smith (1990) Models for exceedances over high thresholds (with discussion), *J. Roy. Stat. Soc. B, 52*, 393-442.

Katz, R.W., M.B. Parlange, and P. Naveau (2002) Statistics of extremes in hydrology. *Advances in Water Resources, 25*, 1287-1304.

Méndez, F.J., Menéndez, M., Luceño, A., Losada, I.J. (2006) Estimation of the long-term variability of extreme significant wave height using a time-dependent POT model, *Journal of Geophysical Research*, Vol. 111, C07024.

Morton, I.D., J. Bowers, and G. Mould (1997) Estimating return period wave heights and wind speeds using a seasonal point process model, *Coastal Eng., 31*, 305-326.

Ratsimandresy, A. W., Sotillo, M. G., Carretero, J. C., Alvarez Fanjul, E., (2003) The HIPOCAS multi-decadal atmosphere and oceanographic hindcast of the Mediterranean area. EGS - AGU - EUG Joint Assembly, Abstract 1946.

Smith, R. L. (2001). Environmental Statistics, Technical Report, Department of Statistics, University of North Carolina, Chapel Hill, NC, USA.

Sobey, R.J. and Orloff, L.S., (1999). Intensity-duration-frequency summaries for wave climate, *Coastal Engineering*, 36, 37-58.

Teisson, C. (1990). Statistical approach of duration of extreme storms: consequences on breakwater damages, ICCE1990, ASCE, 1851-1860.

NUMERICAL SIMULATION OF EXTREME EVENTS FROM FOCUSED DIRECTIONALLY SPREAD WAVEFIELDS

David R. Fuhrman[1] and Per A. Madsen[1]

So-called extreme, rogue, or freak waves are currently receiving increased attention within the coastal and offshore engineering communities, as they are thought to be important in the design of ships and offshore structures. In this work we simulate extreme three-dimensional wave events from focused, directionally spread wavefields with a fully-nonlinear, high-order Boussinesq-type model. The simulations are based on recently conducted laboratory experiments, involving waves out to (wavenumber times depth) $kh = 13.25$. An excellent match in both free surface elevations and vertical velocity profiles is demonstrated. While the accuracy of free surface computations has previously been established with the present method, this work marks the first validation of velocity kinematics directly against experimental measurements, generally confirming previously claimed depth limitations.

INTRODUCTION

Extreme wave events in the open sea are being increasingly blamed for damage to ships and other offshore structures. (For a recent review of so-called freak or rogue waves see Kharif and Pelinovsky, 2003). Methods for accurately predicting the internal kinematics of such events are therefore important within the coastal and offshore engineering communities, as these form the basis of corresponding force predictions. In the present study we simulate extreme wave events using a finite difference model (see Fuhrman and Bingham, 2004) based on the high-order Boussinesq-type method of Madsen et al. (2002, 2003), which is capable of providing accurate surface descriptions for roughly $kh \leq 25$ (k the wavenumber, h the depth), and accurate velocity kinematics for roughly $kh \leq 12$.

Numerical models based on the present Boussinesq-type approach have proved useful for the study of numerous wave phenomena, including deep-water wave instabilities (see *e.g.* Madsen et al., 2002; Fuhrman et al., 2004, 2006), wave-structure interactions (Fuhrman et al., 2005; Jamois et al., 2006), and the generation of three-dimensional short-crested waves (Fuhrman and Madsen, 2006). From these (and other) studies, the accuracy of free surface computations is now well established. However, while the accuracy of the embedded velocity kinematics has been analyzed previously against linear and nonlinear steady wave solutions, it has not been previously confirmed directly against laboratory measurements. This is therefore among the motivations for the present work.

[1] Department of Mechanical Engineering, Technical University of Denmark, Nils Koppels Allé, DTU - Building 403, DK-2800 Kgs. Lyngby, Denmark

For validation of the model we consider a series of experiments from Johannessen and Swan (2001), involving large three-dimensional focusing events under variable spectral width and directional spreading. Both surface elevations and velocity kinematics were measured, making detailed comparisons possible. We focus on the cases having the largest input amplitude sum for each frequency distribution and directional spreading combination, which were reported to be very near breaking. We note that similar simulations have been carried out previously using a pseudo-spectral approach by Bateman et al. (2001, 2003), and our comparison will resemble their work.

THE HIGH-ORDER BOUSSINESQ MODEL

The numerical model used in the present work is based on the recently derived fully-nonlinear, high-order Boussinesq-type formulation of Madsen et al. (2002, 2003). This method uses exact representations of the kinematic and dynamic free surface conditions expressed in terms of surface quantities as

$$\frac{\partial \eta}{\partial t} = \tilde{w}\left(1 + \nabla\eta \cdot \nabla\eta\right) - \tilde{\mathbf{U}} \cdot \nabla\eta, \tag{1}$$

$$\frac{\partial \tilde{\mathbf{U}}}{\partial t} = -g\nabla\eta - \nabla\left(\frac{\tilde{\mathbf{U}} \cdot \tilde{\mathbf{U}}}{2} - \frac{\tilde{w}^2}{2}(1 + \nabla\eta \cdot \nabla\eta)\right), \tag{2}$$

where $\tilde{\mathbf{U}} = (\tilde{U}, \tilde{V}) = \tilde{\mathbf{u}} + \tilde{w}\nabla\eta$. Here $\tilde{\mathbf{u}} = (\tilde{u}, \tilde{v})$ and $\tilde{w}$ are the horizontal and vertical velocities directly on the free surface $z = \eta$, $g = 9.81$ m/s^2 is the gravitational acceleration, $\nabla = (\partial/\partial x, \partial/\partial y)$ is the horizontal gradient operator, and t is time. The vertical distribution of fluid velocity is approximated via a Padé-enhanced truncated series solution of the Laplace equation, corresponding to

$$\begin{aligned} \mathbf{u}(x,y,z,t) &= (1 - \alpha_2\nabla^2 + \alpha_4\nabla^4)\hat{\mathbf{u}}^*(x,y,t) \\ &\quad + ((z-\hat{z})\nabla - \beta_3\nabla^3 + \beta_5\nabla^5)\hat{w}^*(x,y,t), \end{aligned} \tag{3}$$

$$\begin{aligned} w(x,y,z,t) &= (1 - \alpha_2\nabla^2 + \alpha_4\nabla^4)\hat{w}^*(x,y,t) \\ &\quad - ((z-\hat{z})\nabla - \beta_3\nabla^3 + \beta_5\nabla^5)\hat{\mathbf{u}}^*(x,y,t), \end{aligned} \tag{4}$$

where

$$\alpha_2 = \frac{(z-\hat{z})^2}{2} - \frac{\hat{z}^2}{18}, \tag{5}$$

$$\alpha_4 = \frac{(z-\hat{z})^4}{24} - \frac{\hat{z}^2(z-\hat{z})^2}{36} + \frac{\hat{z}^4}{504}, \tag{6}$$

$$\beta_3 = \frac{(z-\hat{z})^3}{6} - \frac{\hat{z}^2(z-\hat{z})}{18}, \tag{7}$$

$$\beta_5 = \frac{(z-\hat{z})^5}{120} - \frac{\hat{z}^2(z-\hat{z})^3}{108} + \frac{\hat{z}^4(z-\hat{z})}{504}. \qquad (8)$$

Optimal velocity distributions are obtained with the expansion point near $\hat{z} = -h/2$, and we adopt this value throughout. Considering (3) and (4) at the sea bottom $z = -h$, the kinematic bottom condition (neglecting bottom slope) becomes

$$\left(1 - \frac{4}{9}\gamma^2\nabla^2 + \frac{1}{63}\gamma^4\nabla^4\right)\hat{w}^* + \left(\gamma\nabla - \frac{1}{9}\gamma^3\nabla^3 + \frac{1}{945}\gamma^5\nabla^5\right)\hat{\mathbf{u}}^* = 0, \qquad (9)$$

where $\gamma = h/2$. It is straight forward to include variable depth terms, however as the present work is restricted to flat bottoms, they are not presented here. (In fact, the method has recently been extended to allow rapidly varying bathymetry by Madsen et al., 2006). Analysis of this system has, again, shown that it provides excellent linear and nonlinear properties to (wavenumber times depth) $kh \approx 25$, and accurate velocity kinematics to $kh \approx 12$, largely eliminating any shallow water limitations conventionally associated with Boussinesq-type methods. Thus, within a large range of kh, the system may be regarded as a highly accurate approximation to the exact Laplace problem for nonlinear water waves.

The system of PDEs is solved numerically using the high-order (37-point) finite difference spatial discretizations originally presented by Fuhrman and Bingham (2004), combined with an explicit fourth-order, four-stage Runge-Kutta time-stepping scheme. As shown by Fuhrman and Bingham (2004), the dominant computational expense of the model involves the iterative solution of an ill-conditioned sparse matrix problem during each stage evaluation, and the simulations presented here use the matrix-free Fourier space preconditioner and irrotational operators, as discussed therein. Closed transverse boundaries are created by assuming appropriate symmetries about the numerical sidewalls. During simulations a 10th-order (109-point, octagon shaped) Savitzky and Golay (1964) type smoothing filter is applied intermittently, which is necessary to remove high wavenumber disturbances arising from the discretization of the nonlinear terms.

SIMULATION OF FOCUSED DIRECTIONALLY-SPREAD WAVEFIELDS

To perform simulations involving focused directionally-spread wavefields, we will consider the work of Johannessen and Swan (2001), who conducted a series of 84 experiments in a three-dimensional wave basin, where the frequency-amplitude spectrum, the directional-amplitude spectrum, and the input amplitude sum were all systematically varied. They considered three different frequency spectra, where the amplitude a was varied with the frequency f in each case according to

$$a \propto f^{-2}, \qquad (10)$$

Table 1. Tabulated input amplitude sum values (A, in mm) from the numerical simulations of Bateman et al. (2001), as well as the present study. Values in parentheses represent the percentage scaling relative to the reported experimental values of Johannessen and Swan (2001).

Case	Bateman et al. (2001)	Present study
B52ud	57.2 (+10%)	59.0 (+13.5%)
B66ud	70.62 (+7%)	66.0
B78s4	78.0	78.0
D61ud	61.0	61.0
D78s45	79.56 (+2%)	78.0
D93s4	93.0	93.0

scaled such that their linear amplitude sum A is given by

$$A = \sum_{n=1}^{N} a_n, \tag{11}$$

with N the total number of frequency components. We will consider cases from both their most broad- and narrow-banded experiments. Their broad-banded experiments (hereafter denoted by the marker B) cover the frequency range $46/64 \leq f \leq 106/64$ Hz, whereas their narrow banded cases (marker D) cover the frequency range $53/64 \leq f \leq 80/64$ Hz. The experiments were performed on a depth $h = 1.2$ m, which (from the linear dispersion relation) means that these cases respectively cover the range of dimensionless depths $2.53 \leq kh \leq 13.25$ (B) and $3.32 \leq kh \leq 7.55$ (D). We note that the maximum $kh = 13.25$ from the broad-banded experiments in fact slightly exceeds the previously mentioned upper limit (in terms of wave kinematics) $kh = 12$ for the Boussinesq model. As they are reasonably close, however, comparison shall still be made.

Within each experiment, the amplitudes for a given frequency were directionally spread according to

$$b(\theta) = B\cos^s\left(\frac{\theta}{2}\right), \tag{12}$$

where s is a spreading parameter. For experimental reasons the directions were limited to the range $-\pi/4 \leq \theta \leq \pi/4$. Note that $s = \infty$ corresponds to uni-directional cases, while $s = 0$ corresponds to uniform directionally spreading. For full details on the experimental set up, the reader is referred to Johannessen and Swan (2001).

Following Bateman et al. (2001) who made the first successful numerical comparisons against these experiments, we will consider the six cases listed in table 1, where the experimental information is indicated by the case identifier in the first column. For example, case B52ud corresponds to a broad-banded case (B), with linear amplitude sum $A = 52$ mm, and with

Figure 1. Computed free surface near the time of focusing for case D93s4.

$s = \infty$ (*i.e.* uni-directional). Alternatively, *e.g.* case D93s4 corresponds to a narrow-banded case (D), with $A = 93$ mm and $s = 4$. These six cases in fact correspond to those with the largest linear amplitude sum for a given frequency-amplitude spectrum and directional spreading combination, from the experiments of Johannessen and Swan (2001). These were all reported to produce waves very near breaking, and are therefore physically challenging.

It should be noted that, although it is not specifically mentioned in their paper, Bateman et al. (2001) actually used linear amplitude sum values in some of their numerical simulations that deviate slightly from the experimentally reported values (this can be seen *e.g.* by looking closely at the linear free surface peaks at the time of focusing in their figure 5). For the sake of repeatability, we list their values in table 1. In the present work, we have used the exact experimentally reported values in our simulations, with one exception: In case B52ud the linear amplitude sum is increased to $A = 59$ mm, which we find is necessary in order to obtain a reasonable match. This was also the case modified most strongly by Bateman et al. (2001), hence it seems that the experimentally reported value $A = 52$ mm may be underestimated in this case, though additional experiments are necessary before more firm conclusions can be drawn.

For all the simulations we specify an initial condition corresponding to the previous experimental description, where the relative phase of each wave component is determined from linear theory, such that all the wave crests will superimpose at a particular point in space and time. The initial conditions are specified at a time corresponding to 8 s prior to the focusing event. All cases are simulated on a 513×257 grid, where $\Delta x = 0.05$ m, $\Delta y = 0.0488$ m, with time step $\Delta t = 0.01$ s. Due to symmetry about the side-walls, only the lower half (in plan) is modeled. The previously mentioned smoothing filter is applied infrequently, after every 100 time steps. In each case the time is shifted, such that $t = 0$ corresponds to the instant of nonlinear focusing. As an example, a computed free surface near the time of nonlinear focusing for case D93s4 is provided in figure 1, where the results have been mirrored across the upper side-wall a single time for presentation purposes. To give an indication of the computational efficiency of the present method, simulations for 1300 time steps typically require roughly 6 hr on a single 3.2 GHz Pentium 4 processor.

We will begin our comparison against this data set by considering time series of free surface elevations measured at the location of the focusing events. These are presented for the six cases in figure 2. For both broad-banded and narrow-banded cases, the comparison is generally excellent. While there are some local areas of minor disagreement, there is no systematic error from case to case. The present results compare well with those shown previously by Bateman et al. (2001), confirming the accuracy of the high-order Boussinesq model in simulating extreme free surface events from directionally spread irregular waves in quite deep water. This accuracy should in fact be expected, as the kh values from these experiments are well within the previously mentioned range of nonlinear free surface accuracy (again, roughly $kh \leq 25$).

A much more challenging task, at least for the present model (and Boussinesq-type models in general), is the accurate prediction of underlying velocity kinematics. A comparison of the horizontal velocity profiles underneath the computed and measured extreme wave events is similarly provided in figure 3. Also included on these plots for comparison are velocity profiles based on linear wave theory. The results for the three narrow-banded cases (right) are again impressive, giving a generally excellent match against the measurements. These again compare well against the previous results of Bateman et al. (2003). The experimental and numerical values both deviate noticeably from those based on linear theory, with the uni-directional results perhaps giving the strongest deviations, especially near the wave crest. The numerical results for the more difficult (for the present model) broad-banded cases (left) are also in generally good agreement with the experiments. These results are, perhaps expectedly, slightly less accurate than in the narrow-banded cases. For example, there seems to be a slight overestimation of the results *e.g.* for case B78s4 over part of the profile. Minor oscillations in the computed profiles are also apparent near the sea-bed, which is consistent with the fact that these cases are near (actually slightly exceeding) the upper kh limit for the model kinematic accuracy. Regardless, the fact that a reasonable match in the velocity kinematics is achieved in these challenging physical situations (kh up to 13.25, with waves near their breaking limit), is a testament to the

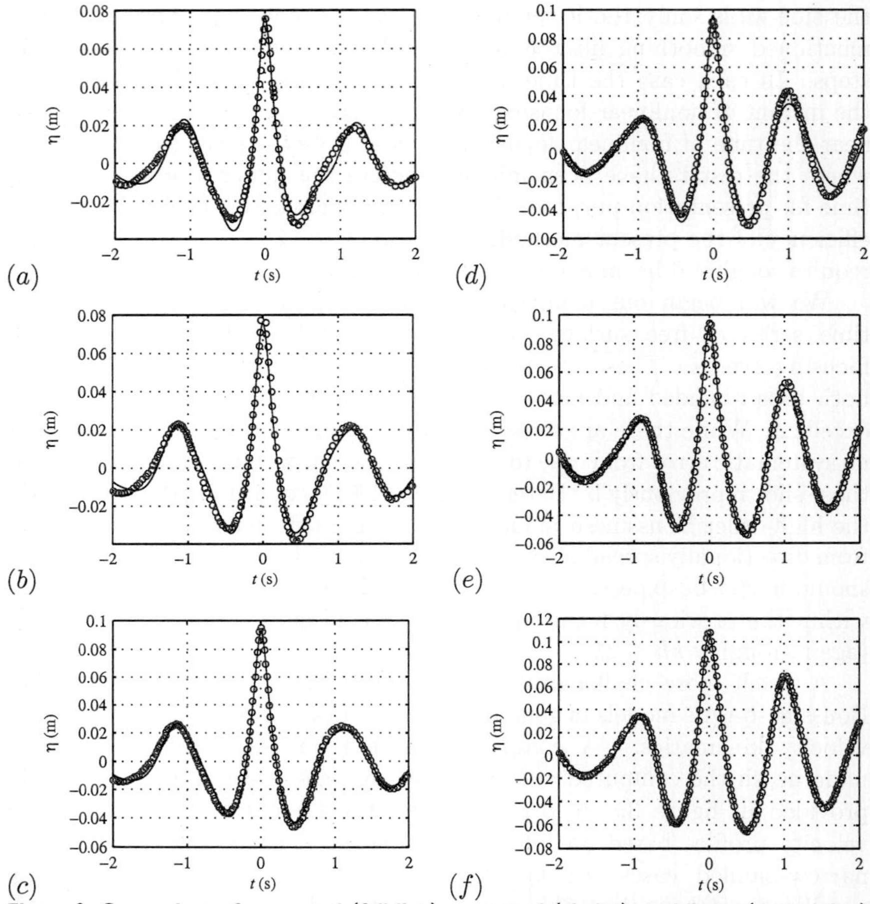

Figure 2. Comparison of computed (full line), measured (circles), and linear (dotted line) free surface time series case (*a*) B52ud, (*b*), B66s45, (*c*) B78s4, (*d*) D61ud, (*e*) D78s45, and (*f*) D93s4.

accuracy of the present high-order Boussinesq method. We generally find the agreement in all cases to be most acceptable.

CONCLUSIONS

In this work we have simulated physically challenging cases involving highly nonlinear extreme wave events caused by the focusing of irregular, directionally spread water waves, based on the experiments of Johannessen and Swan (2001). The numerical model used is based on the high-order Boussinesq-type method of Madsen et al. (2002, 2003, 2006), utilizing the efficient finite difference solution strategies of Fuhrman and Bingham (2004). Following Bateman et al. (2001, 2003), we have focused on

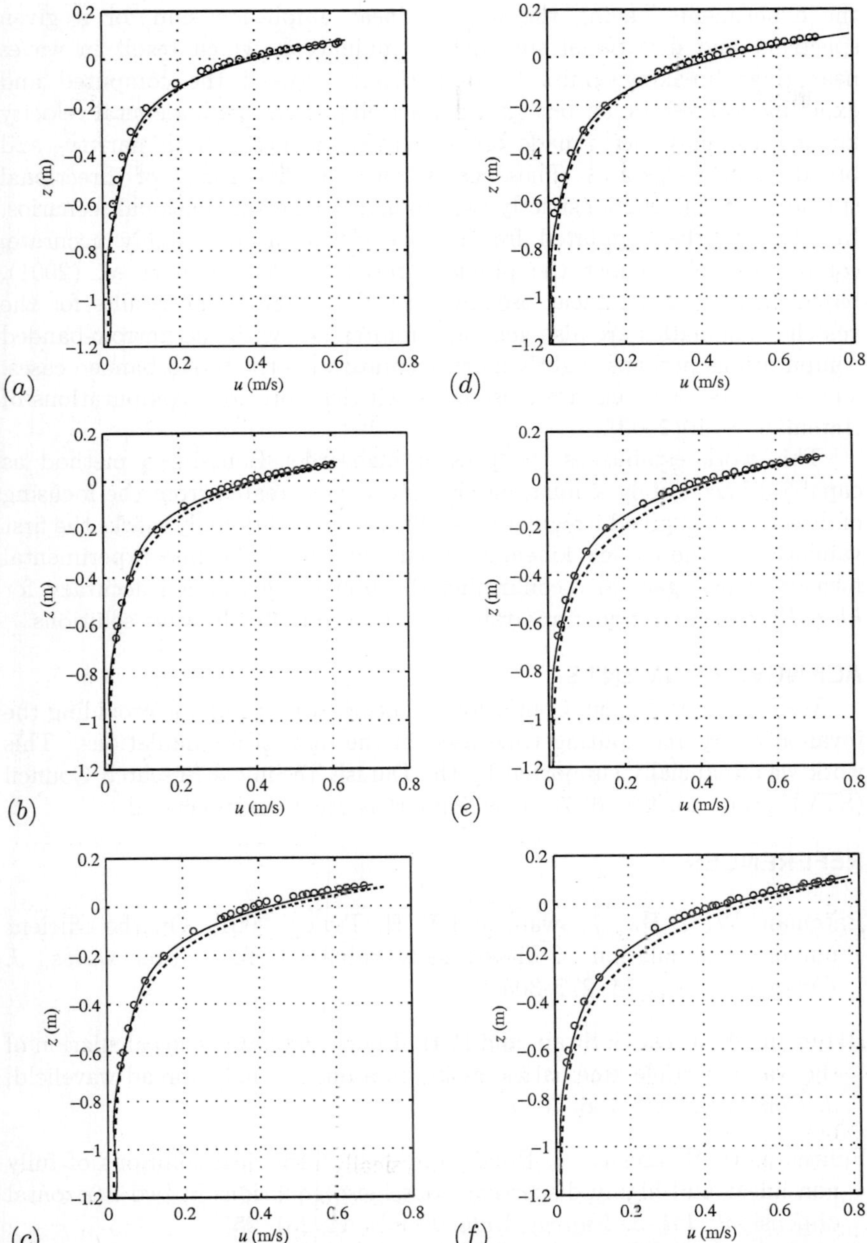

Figure 3. Comparison of computed (full line), measured (circles), and linear (dotted line) horizontal velocity profiles for case (a) B52ud, (b), B66s45, (c) B78s4, (d) D61ud, (e) D78s45, and (f) D93s4.

the experiments having the largest linear amplitude sum for a given spectrum and directional spreading combination, which result in waves near their breaking limits. Comparisons between the computed and experimental results, for both free surface elevations and horizontal velocity kinematics, have been made for six cases, involving both narrow- and broad-banded spectra. These cases cover a wide range of directional spreading, ranging from strictly two- to highly three-dimensional scenarios. In all cases the computed free surface elevations are highly accurate, comparing well against the previous results of Bateman et al. (2001), based on a fully nonlinear pseudo-spectral method. The results for the velocity kinematics are also generally impressive, with the narrow-banded computations perhaps slightly more accurate than the broad-banded cases. These results also compare reasonably with the previous computations of Bateman et al. (2003).

This work establishes the present high-order Boussinesq method as capable of accurately simulating extreme waves created from the focusing of directionally spread irregular seas. This work additionally marks the first validation of the models kinematic accuracy directly against experimental measurements, generally confirming the previously claimed accuracy for $kh \leq 12$, based on earlier analysis against various steady wave solutions.

ACKNOWLEDGMENTS

We thank the Danish Center for Scientific Computing for providing the invaluable supercomputing time used in the numerical simulations. This work was financially supported by the Danish Technical Research Council (STVF grant no. 9801635). This support is greatly appreciated.

REFERENCES

Bateman, W. J. D., C. Swan, and P. H. Taylor. 2001. On the efficient numerical simulation of directionally spread surface water waves, *J. Comput. Phys.*, 174, 277–305.

Bateman, W. J. D., C. Swan, and P. H. Taylor. 2003. On the calculation of the water particle kinematics arising in a directionally spread wavefield, *J. Comput. Phys.*, 186, 70–92.

Fuhrman, D. R. and H. B. Bingham. 2004. Numerical solutions of fully non-linear and highly dispersive Boussinesq equations in two horizontal dimensions, *Int. J. Numer. Meth. Fluids,* 44, 231–255.

Fuhrman, D. R., H. B. Bingham, and P. A. Madsen. 2005. Nonlinear wave-structure interactions with a high-order Boussinesq model, *Coast. Eng.*, 52, 655–672.

Fuhrman, D. R. and P. A. Madsen. 2006. Short-crested waves in deep water: a numerical investigation of recent laboratory experiments, *J. Fluid Mech.*, 559, 391–411.

Fuhrman, D. R., P. A. Madsen, and H. B. Bingham. 2004. A numerical study of crescent waves, *J. Fluid Mech.*, 513, 309–341.

Fuhrman, D. R., P. A. Madsen, and H. B. Bingham. 2006. Numerical simulation of lowest-order short-crested wave instabilities, *J. Fluid Mech.*, 563, 415–441.

Jamois, E., D. R. Fuhrman, H. B. Bingham, and B. Molin. 2006. A numerical study of nonlinear wave run-up on a vertical plate, *Coast. Eng.*, 53, 929–945.

Johannessen, T. B. and C. Swan. 2001. A laboratory study of the focusing of transient and directionally spread surface water waves, *Proc. R. Soc. Lond. A*, 457, 971–1006.

Kharif, C. and E. Pelinovsky. 2003. Physical mechanisms of the rogue wave phenomenon, *Eur. J. Mech. B/Fluids*, 22, 603–634.

Madsen, P. A., H. B. Bingham, and H. Liu. 2002. A new Boussinesq method for fully nonlinear waves from shallow to deep water, *J. Fluid Mech.*, 462, 1–30.

Madsen, P. A., H. B. Bingham, and H. A. Schäffer. 2003. Boussinesq-type formulations for fully nonlinear and extremely dispersive water waves: derivation and analysis, *Proc. R. Soc. Lond. A*, 459, 1075–1104.

Madsen, P. A., D. R. Fuhrman, and B. Wang. 2006. A Boussinesq-type method for fully nonlinear waves interacting with a rapidly varying bathymetry, *Coast. Eng.*, 53, 487–504.

Savitzky, A. and M. J. E. Golay. 1964. Smoothing and differentiation of data by simplified least squares procedures, *Anal. Chem.*, 36, 1627–1639.

NON-GAUSSIAN PROPERTIES OF SURFACE ELEVATION IN CROSSING SEA STATES IN SHALLOW WATER

Toffoli A.[1], Onorato M.[2], Osborne A.R.[2] and Monbaliu J.[1]

The Kadomtsev-Petviashvili equation, an extension of the Korteweg-de Vries equation in two horizontal dimensions, is used to study the statistical properties of shallow water waves in crossing sea states. Numerical simulations indicate that the interaction of two non-collinear wave trains generates steep and high amplitude peaks, thus enhancing the deviation of the surface elevation from the Gaussian statistics.

INTRODUCTION

A recent study by Toffoli et al. (2005), based on data collected from January 1995 to April 1999 by the Lloyd's Marine Information Service (Bitner-Gregersen and Eknes 2001), has revealed that a large percentage of ship accidents, reported as being due to bad weather, might have occurred in crossing seas. The latter represents sea states, which are characterized by two spectral peaks with different mean directions. Aside from the difficulties of keeping a ship under proper control, such conditions are also suspected to enhance the probability of occurrence for extreme wave events (see, for example, Lehner et al., 2005). A theoretical study (Onorato et al., 2006) on the influence of a second spectral peak that propagates in deep water with a mean direction that differs from the principal wave system, showed that waves can become unstable in crossing seas. By using second-order wave theory, Toffoli et al. (2006) have investigated the statistical properties of random crossing seas in deep and intermediate water depths. They concluded that crest amplitudes can be expected to be higher when two wave systems with not well separated directions coexist.

In the case of shallow water, Peterson et al. (2003) and Soomere and Engelbrecht (2005; 2006) have studied the interaction of multi-directional wave trains by considering the two-soliton solutions of the Kadomtsev-Petviashvili equation (Kadomtsev and Petviashvili 1970), which represents an extension of the Korteweg-de Vries equation in the case of propagation in the (x,y) plane (see also Segur and Finkel, 1985). The most interesting feature that arises from the interaction of two non-collinear solitary waves is an unexpectedly steep and high amplitude peak in the intersection region owing to resonance phenomena (Miles 1977). Such a peak can reach values up to four times the amplitude of the incoming waves, which is far beyond the value predicted by linear theory.

---

[1] Hydraulics Laboratory, Katholieke Universiteit Leuven, Kasteelpark Arenberg 40, Heverlee, 3001, Belgium

[2] Dip. Fisica Generale, Universitá di Torino, Via P. Giuria 1, Torino, 10125, Italy

Theoretically, these structures have unlimited life-time and they may cross large sea areas in favorable conditions.

The aim of the present study is to extend the investigation of crossing, shallow water waves to the more general case of random sea states; the influence of a second spectral peak on the statistical properties of the wave field is consequently looked at. In order to accomplish this task, many random simulations have been performed by means of the Kadomtsev-Petviashvili equation. The advantage of this approach is that the numerical solution of the Kadomtsev-Petviashvili equation is computationally fast. This makes it possible to run many numerical experiments considering different random phases, different degrees of nonlinearity, and different angles between the incoming wave systems.

The paper is organized as follows: we first begin with a short description of the numerical experiments. In the sections following that, the statistical properties of simulated wave profiles are investigated; and their variation as a function of the angle between crossing wave trains is also shown. Some concluding remarks are presented in the last section.

THE KADOMTSEV-PETVIASCHVILI EQUATION AND THE NUMERICAL EXPERIMENTS

In order to model the evolution of weakly directional waves in shallow water, a two-dimensional generalization of the Korteweg-de Vries (KdV) equation can be used: the Kadomtsev-Petviashvili (KP) equation (Kadomtsev and Petviashvili 1970). For an irrotational, inviscid fluid in constant depth, the KP equation can be derived from the Euler equations (see, e.g., Infeld and Rowlands 1990), if the following assumptions are satisfied:

1. waves have a small amplitude a, i.e., $\alpha = a / h << 1$;
2. the water is shallow, i.e., $\beta = (kh)^2 << 1$;
3. the waves are weakly two-dimensional, i.e., $\varsigma = (k_y / k_x)^2 << 1$, where k_x and k_y are the components of the wavenumber vector **k**;
4. these three small effects are comparable, i.e., $\varsigma \approx O(\beta) \approx O(\alpha)$;
5. waves propagate only towards positive values of x.

In dimensional form, the KP equation can be written as follows:

$$\frac{\partial}{\partial x}\left(\frac{\partial \eta}{\partial t}+\frac{3}{2}\frac{C_0}{h}\eta\frac{\partial \eta}{\partial x}+\frac{1}{6}h^2 C_0{}^2\frac{\partial^3 \eta}{\partial x^3}\right)+\frac{1}{2}C_0\frac{\partial^2 \eta}{\partial y^2}=0 \tag{1}$$

where η is the surface elevation, $C_0 = (gh)^{0.5}$ is the phase velocity, and h is the water depth. To perform numerical simulation with Eq. (1), a method developed by Fornberg and Whitham (1978) to solve KdV-type wave equations is used.

In order to generate the surface elevation in crossing sea conditions, bimodal directional wave spectra have been used as input. For each spectral peak, the JONSWAP formulation with dominant wavelength $\lambda_p = 76.5$ *m*, peak enhancement factor $\gamma = 3.3$ and Phillips parameter $\alpha = 0.004$ has been used to model the energy in the frequency domain, and a cos-2s function with s = 25 has been used to describe the energy in the directional domain. The two peaks, then, have been oriented along different mean directions; the input, bimodal, wave spectra can therefore be expressed as $\widetilde{E}(\omega,\vartheta)=E(\omega,\vartheta-\vartheta_1)+E(\omega,\vartheta-\vartheta_2)$, where ω is the angular frequency, ϑ is the wave direction, and ϑ_1 and ϑ_2 are the mean wave directions of the two spectral peaks; the total significant wave height H_s is 2.4 *m*. By using the linear dispersion relation, the energy spectrum in the frequency and directional domain can be conveniently expressed in terms of wavenumber **k** (see Tanaka 2000). The input spectra $\widetilde{E}(\mathbf{k})$ have been defined by using 128 equally spaced wavenumbers in the x-direction and 64 equally spaced wavenumbers in y-direction, i.e., Δk_x and Δk_y are constants; this automatically generates periodic boundaries for the simulated surface. A schematic representation of an input spectrum $\widetilde{E}(\mathbf{k})$ is presented in Fig. 1; note that the two peaks are symmetric with respect to the k_x axis, and hence $\vartheta_1=\vartheta_2=\vartheta^*$.

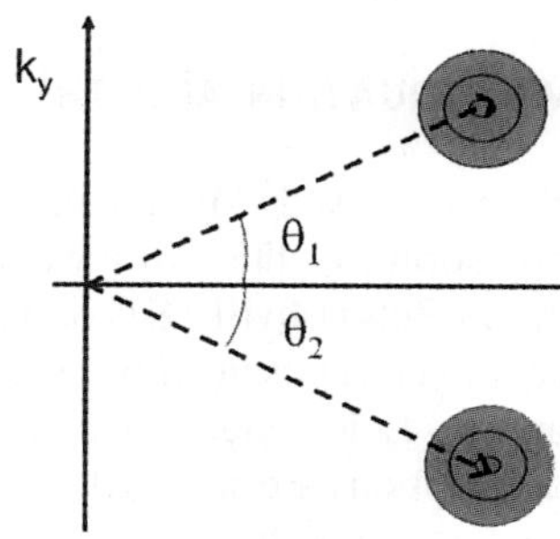

Figure 1. Schematic representation of an input spectrum

Sets of numerical simulations have been performed by considering the energy as constant but with different angles between the crossing wave systems ($0^\circ \le \vartheta^* \le 45^\circ$); note that when $\vartheta^*=0^\circ$ the energy remains concentrated in one single peak, i.e. unimodal sea state. To obtain statistically stable results, 500 repetitions of each experimental case have been made by using different random phases.

The experiments were also repeated for three different water depths h = 7, 6, and 5 *m*, which automatically define different degrees of nonlinearity as measured by the Ursell number (Ursell 1952)

$$Ur=\frac{3}{4}\frac{ka}{k^3h^3} \tag{2}$$

where the wave number k is here related to the dominant wavelength λ_p, and the amplitude a has been taken as half of the significant wave height ($a = H_s/2$). For the selected depths, the Ursell number values are 0.39, 0.62 and 1.10 respectively.

SKEWNESS AND KURTOSIS OF THE SURFACE ELEVATION

For each time step, the model produces a two-dimensional surface as output, which measures 400 *m* in the x-direction and 200 *m* in the y-direction. The skewness, λ_3, and kurtosis, λ_4, which represent the third and fourth order moments of the probability density function respectively (see, for example, Bendat and Piersol, 2000), have been computed from the space series associated with each instant in time. Whereas λ_3 describes the vertical asymmetry of the wave profile, λ_4 is an indication of extreme events. For waves in shallow water, i.e. $kh \rightarrow 0$, it is expected that $\lambda_3 > 0$ and $\lambda_4 > 3$, where 0 and 3 are the values expected for a Gaussian (linear) process. Considering the 500 random repetitions, an ensemble average of the moments is used. Note furthermore that values of the skewness and kurtosis have been obtained after a sufficiently large number of time steps in the numerical solution of Eq. 1 such that the statistical properties of the simulated surfaces have stabilized.

We now investigate how the statistical properties of the simulated surface elevation can change when the wave energy is distributed over two spectral peaks with different mean directions. In Figs. 2 and 3, we show the skewness and kurtosis as a function of the angle ϑ^*. The existence of a second spectral peak results in an increase of the skewness and kurtosis as the angle between the peaks increases. This means that the wave field becomes vertically more asymmetric and extreme values become more likely, when two non-collinear wave systems coexist. This enhancement, however, is almost nil when the systems are almost collinear ($\vartheta^* \leq 5^\circ$) and becomes more relevant as the wave trains assume well separated directions. In this respect, it is important to remember that the energy remains constant and hence this variation is only due to the nonlinear interaction between non-collinear wave trains.

The maximum magnitudes of the third and fourth order moments are obtained for $\vartheta^* \approx 30^\circ$. This is approximately the same angle, at which two interacting solitons with amplitude equal to $H_s/2$ and in water depth of 6 *m* would produce the highest amplitude peak (cf. Peterson et al., 2003). For $\vartheta^* > 30^\circ$, and in the case of low and moderate nonlinearity, the value of for the kurtosis remains approximately constant; the value for the skewness, in contrast, slowly decreases. For high degrees of nonlinearity, both skewness and kurtosis clearly show a decreasing trend as $\vartheta^* \rightarrow 45^\circ$. Note, however, that the results for $\vartheta^* > 30^\circ$ are rather uncertain, since the hypothesis of weakly directionality is no longer respected. Further investigations using e.g. the Boussinesq equation are therefore needed.

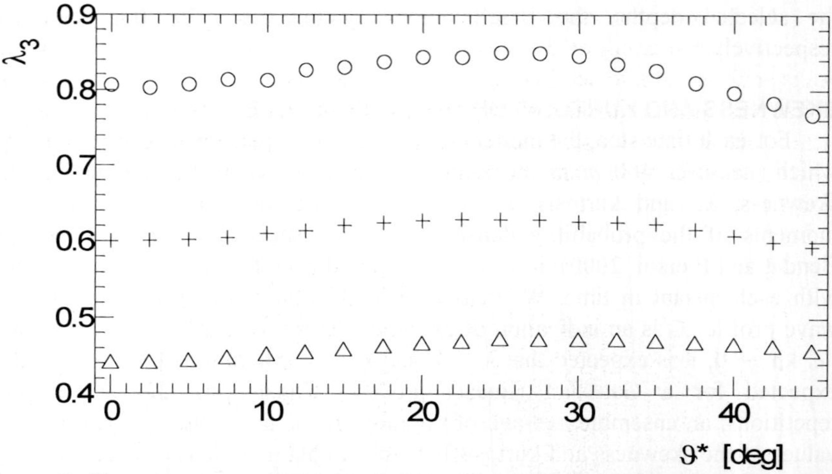

Figure 2. Skewness as a function of the mean direction of the spectral peaks ($\vartheta_1 = \vartheta_2 = \vartheta^*$) from numerical simulations: Ur = 0.39 (Δ); Ur = 0.62 (+); Ur = 1.10 (o).

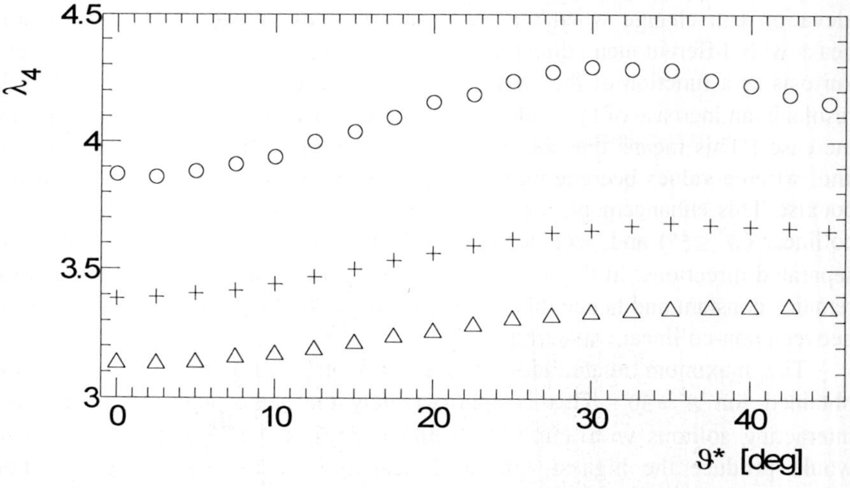

Figure 3. Kurtosis as a function of the mean direction of the spectral peaks ($\vartheta_1 = \vartheta_2 = \vartheta^*$) from numerical simulations: Ur = 0.39 (Δ); Ur = 0.62 (+); Ur = 1.10 (o).

PROBABILITY DENSITY FUNCTION OF THE SURFACE ELEVATION

As waves propagate into shallow water, a definite excess of steep crests and shallow troughs can be observed due to the increasing of nonlinearity (see, e.g., Ochi 1998). In Figs. 4 and 5, we show, as an example, two-dimensional shallow water wave fields simulated by using unimodal ($\vartheta^* = 0^\circ$) and bimodal ($\vartheta^* = 30^\circ$) spectral conditions.

When two wave systems coexist, not only there is a sharpening of the wave crests and a flattening of the wave troughs, but also steep and large amplitude peaks arise in the intersection region of crossing wave trains. In the case of two, interacting, non-collinear solitary waves, these peaks can propagate for an infinite distance (see, e.g., Peterson et al. 2003). In a dispersive system, however, their amplitude arises and decreases within a short space frame of a few wavelengths.

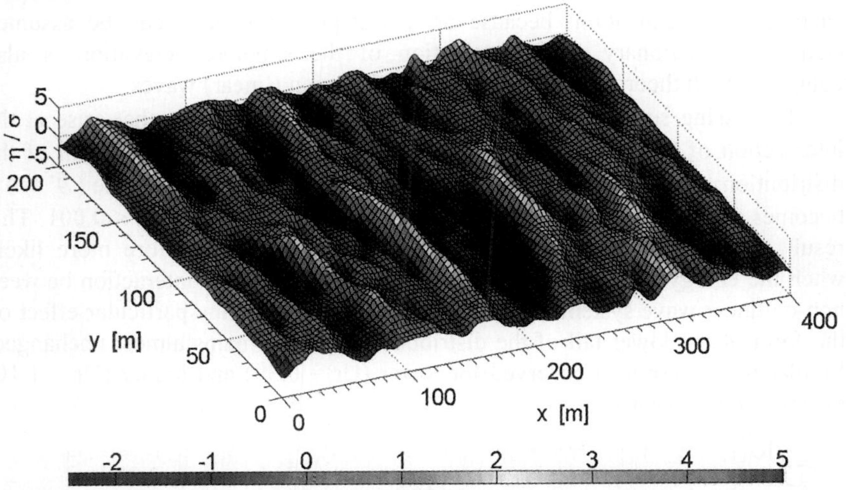

Figure 4. Nonlinear sea surface (η/σ) from numerical simulations: $\vartheta^* = 0°$ (unimodal sea).

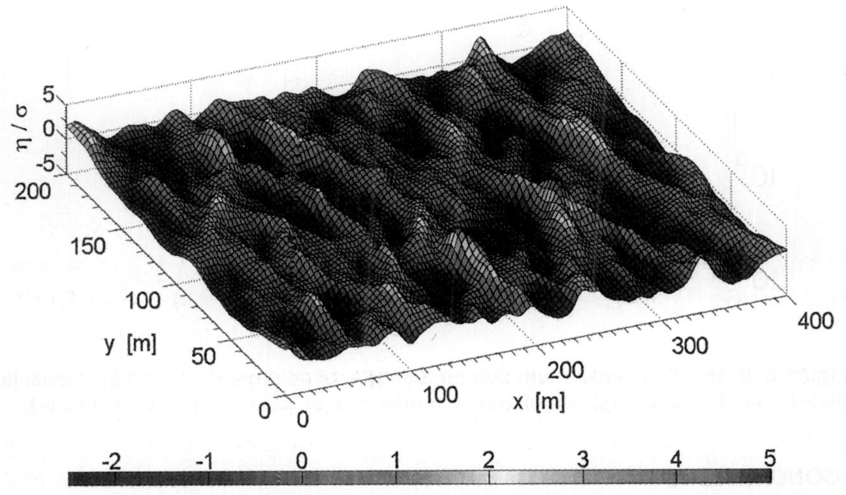

Figure 5. Nonlinear sea surface (η/σ) from numerical simulations: $\vartheta^* = 30°$ (bimodal sea).

It is now instructive to investigate how these interaction peaks modify the probability density function for the surface elevation. In Fig. 6, we present, as an example, the probability density function in unimodal ($\vartheta^* = 0^\circ$) and bimodal ($\vartheta^* = 30^\circ$) wave fields for a degree of nonlinearity corresponding to Ur = 0.62. The wave elevation is expressed in normalized form by using the standard deviation (σ) of the surface displacements. Note that probability density function is calculated by only using the surface elevation η(x,y) at the final time step of each random realization, because at this stage the surface can be assumed statistically stationary. The distribution of the simulated elevation is also compared with theoretical distributions for Gaussian (linear) waves.

If crossing seas (e.g. $\vartheta^* = 30^\circ$) are considered, the peaks that arise at the intersection of the wave fronts produce a modification of the upper tail of the distribution. As appear in Fig. 6, the deviation from the unimodal case ($\vartheta^* = 0^\circ$) becomes statistically significant at low probability levels, i.e., $p(\eta) < 0.001$. This result clearly shows that extreme wave amplitudes are therefore more likely when the energy is distributed over two spectral peaks. The interaction between non-collinear wave systems, however, seems not to have any particular effect on the form of the lower tail of the distribution, which remains almost unchanged. Similar results are also observed for lower (Ur = 0.39) and higher (Ur = 1.10) degrees of nonlinearity.

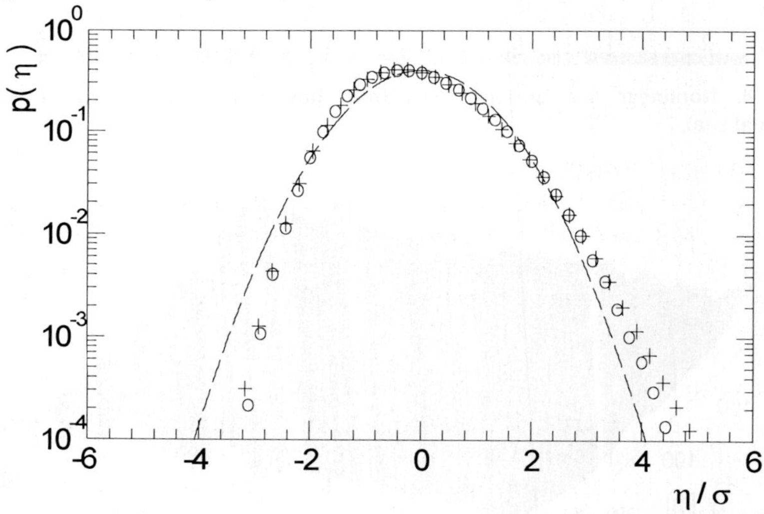

Figure 6. Probability density function for a degree of nonlinearity Ur = 0.62: Gaussian distribution (dashed line); simulated unimodal sea (o); simulated bimodal sea (+).

CONCLUSIONS

The statistical properties of bimodal, shallow water wave fields have been discussed. A generalization of the Korteweg-de Vries equation in the (x,y) plane, the Kadomtsev-Petviashvili equation, has been used to model two-dimensional waves in shallow water. The input spectra have been generated as a summation of two equivalent, single-peaked spectra with different mean directions; we have used a JONSWAP spectral formulation to model the energy distribution in the frequency domain and a cos-2s function to model the directional domain. Considering several spectral cases with constant energy and different angles between the two peaks, many realizations have been simulated with different random phases. Different degrees of nonlinearity, as measured by the Ursell number, have been investigated.

When two non-collinear wave trains are considered, their nonlinear interaction produces steep and high amplitude peaks. These peaks yield a modification of the upper tail of the probability density function for the surface elevation $\eta(x,y)$, which significantly deviates from the distribution of wave elevation in the unimodal condition. The coexistence of two spectral peaks, therefore, enhances the nonlinearity of the wave field, which results in an increase of the skewness and kurtosis. Whereas this enhancement is negligible for nearly collinear waves, the skewness and kurtosis reaches high values when the two spectral peaks have well separated directions. In particular, it has been observed that for $\vartheta^* = 30^\circ$ their magnitudes maximize. Beyond the threshold $\vartheta^* > 30^\circ$, the increasing trend of the skewness and kurtosis vanishes; for high degrees of nonlinearity, furthermore, skewness and kurtosis reduce in magnitude. Note, however, that the use of the KP equation for $\vartheta^* > 30^\circ$ might no longer be justified since the system can no longer be considered weakly directional. Therefore additional verification should be made with e.g. the Boussinesq equation to confirm this finding.

ACKNOWLEDGEMENTS

This work was carried out in the framework of the F.W.O. project G.0228.02 and G.0477.04., and the E.U. project SEAMOCS (contract MRTN-CT-2005-019374). The numerical simulations were performed by using the K.U. Leuven's High Performance Computing (HPC) facilities. M.O. was supported by MIUR (PRIN project).

REFERENCES

Bendat J.B., A.G. Piersol, 2000. *Random Data: analysis and measurement procedures,* John Wiley & Sons Inc., New York, 594 pp.

Bitner-Gregersen E., M. Eknes, 2001. Ship accidents due to bad weather, Tech. report 2001-1330, Det Norske Veritas.

Fornberg B., G. Whitham, 1978. A numerical and theoretical study of certain nonlinear wave phenomena, *Phil. Trans. Royal Soc. London*, Series A 289, 373-404.

Kadomtsev B.B., V.I. Petviashvili, 1970. On the stability of solitary waves in weakly dispersive media, *Sov. Phys. Dokl.,* 15, 539-541.

Infeld E., G. Rowlands, 1990. *Nonlinear waves, solitons and chaos*, Cambridge University Press, Cambridge, 423 pp.

Lehner S., H. Günther, W. Rosenthal, 2005. Extreme wave observations from radar data sets. *In: Ocean Waves Measurements and Analysis, Fifth International Symposium WAVES 2005*, paper 69.

Miles J.W., 1977. Diffraction of solitary waves. *Z. Ang. Math.*, 28, 889-902.

Ochi M.K., 1998. *Ocean Waves: The Stochastic Approach,* Cambridge University Press, Cambridge, 319 pp.

Onorato M., A. Osborne, M. Serio, 2006. Imodulation instability in crossing sea states: A possible machanism for the formation of freak waves, *Phys. Rev. Lett.,* 96.014503.

Peterson P., T. Soomere, J. Engelbrecht, E. van Groesen, 2003. Soliton interaction as a possible model for extreme waves in shallow water, *Nonlin. Proc. Geophys.,* 10, 503-510.

Segur H., A. Finkel, 1985. An analytical model of periodic waves in shallow water, *Studies in Applied Mathematics,* 73, 183-220.

Soomere T., J. Engelbrecht, 2005. Extreme elevation and slopes of interacting solitons in shallow water, *Wave Motion*, 41, 179-192.

Soomere T., J. Engelbrecht, 2006: Weakly two-dimensional interaction of solitons in shallow water, *Europ. J. Mech. B/Fluids*, 25, 636-648.

Tanaka M., 2001. A method of studying nonlinear random field of surface gravity waves by direct numerical simulations, *Fluid Dynamics Research*, 28, 41-60.

Toffoli, A., J. Lefèvre, E. Bitner-Gregersen, J. Monbaliu, J., 2005. Towards the identification of warning criteria: analysis of a ship accident database, *Applied Ocean Research*, 27, 281-291.

Toffoli A., M. Onorato, J. Monbaliu, 2006. Wave statistics in unimodal and bimodal seas from a second-order model, *Europ. J. Mech. B/Fluids*, 25, 649-661.

Ursell F., 1953. The long wave paradox in the theory of gravity waves, *Proc. Cambridge Phil. Soc.*, 49, 685-694.

PROBABILISTIC ANALYSIS OF TYPHOON INDUCED HYDRAULIC BOUNDARY CONDITIONS FOR SUO-NADA BAY

E.N. Klaver[1], J.K.Vrijling[1], S.N. Jonkman[1,2], P.H.A.J.M. van Gelder[1], L.H. Holthuijsen[1], S. Takahashi[3], H. Kawai[3]

For risk-based design of flood defence systems it is necessary to know the probability of failure of the system. This can only be derived if the probability of occurrence of the loads on the flood defence is known. Hydraulic loads caused by typhoons such as water levels and wave characteristics are mutually dependent and this has to be included in their joint probability density functions. This paper describes a method to derive the dependent joint probability density functions of hydraulic loads resulting of typhoons for a bay in Japan. An example is given on how to apply the derived probability function of loads to the design and failure analysis of a flood defence.

INTRODUCTION

The design of coastal structures, sea defences or breakwaters is especially difficult at coasts threatened by typhoons, sometimes also indicated as hurricanes or cyclones. The problem is that, although the coast may be hit by a typhoon one or more times per year, it is not unlikely that the exact location of the project has never experienced a hurricane. Further, observations of hydraulic characteristics sometimes fail due to the extreme loads during typhoon events. This may mean that no representative observations of wind velocities, wind set up, wave height and period are available for the project location. In the best case a few observations are available because the project location is hit on average every N-th year by a typhoon. So for a ten year observation period one might expect 10/N independent observations, which is insufficient to make a reliable statistical inference about say the 1/50 or 1/100 year design condition by extrapolation of available data. However not only statistical considerations might deem the number of observations insufficient, also a complicated geometry of the coast with shielded bays or protruding headlands may lead to this conclusion. In case of a complicated coastal geometry the project location may very well have been shielded from the full force of the typhoon in the few occurrences, but there is no reason why it will not be hit fully by a future typhoon with a slightly different (unfavourable) path.

[1] Delft University of Technology, Faculty of Civil Engineering, Hydraulic Engineering Section, PO Box 5048, 2600 GA, Delft, The Netherlands

[2] Rijkswaterstaat, Road and Hydraulic Engineering Institute, PO Box 5044, 2600 GA, Delft, The Netherlands

[3] Port and Airport Research Institute, Nagase 3-1-1, Yokosuka 239-0826, Japan

Such a more or less complicated coastal geometry makes the hydraulic data gathered at other locations useless for direct use in the estimation of the say 1/100 year conditions at the project location. Only in case of a long straight coast with parallel depth contours data of other sites might support the estimation of the project's design conditions.

Katrina, the hurricane that flooded New Orleans on 29-th of August in 2005 provides a powerful illustration. Although the coast of the Mexican Gulf is hit by several hurricanes every year, New Orleans was only seriously threatened thirty years before. So observations of surge levels and wave heights along the flood defence were not sufficiently available. In Katrina's case the hurricane was only of category 3 strength, but the path was the most threatening for the city. First the southern storm pushed the water into the waterways, Lake Borgne and Lake Pontchartrain loading the defences with extreme water levels and waves. Some dikes withstood the overtopping, but the dike along the Mississippi River Gulf Outlet and the floodwall protecting the Ninth Ward failed. Then the hurricane moved on the northern storm on the left side swept the increased water mass of Lake Pontchartrain against the lake coast of the city and into the canals. The flood walls along the canals were unable to withstand the pressure and gave way. Not the extreme force of the storm but the combination with the effectiveness of its path caused the extreme and fatal loading of the NO flood defence system.

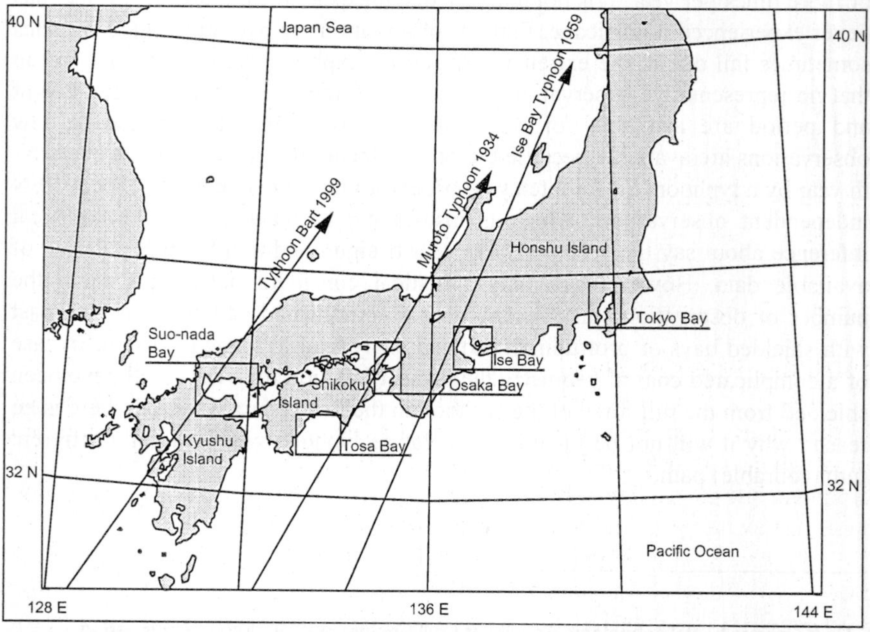

Figure 1: Japanese coastline with relatively shielded bays (www.agora.ex.nii.ac.jp)

Similar situations and conditions are found along the Japanese coast (Fig. 1) where there are many bays that provide shelter against most storms. However a typhoon with the right path may cause an extreme threat for a coastal structure. For coastal design it is important to relate the threat to the probability of exceedance. In this paper it is proposed to solve the question by means of a simulation model of physical phenomena in combinations with probabilistic methods. As an illustration the hydraulic boundary conditions for a location in the Suo-nada Bay in Japan (Fig. 2) are calculated.

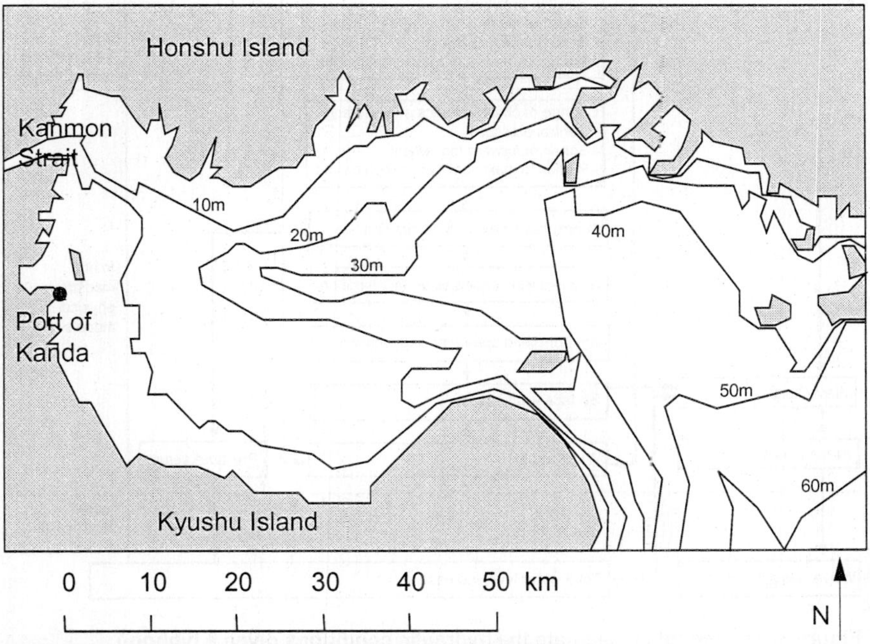

Figure 2: Geometry of Suo-nada Bay (Kawai et al. 2004)

PHYSICAL PHENOMENA

It is clear that a structure on the coast is threatened by the joint occurrence of tide, storm surge and waves. However for reasons explained above only a few observations of such combined occurrences are available due to the rarity of typhoons hitting the bay and the complicated geometry of the bay coast which limits the exposure of the structure to severe loading.

The objective is to determine the joint probability density function of the water level dh resulting from tide level, pressure set up and wind set up, the significant wave height H_s and the peak period T_p. Due to the scarcity of data and the dependence structure between the variables dh, H_s, T_p, the extrapolation to extreme values cannot be based on purely statistical methods. Moreover a statistical extrapolation neglects physical bounds that may be met in extrapolation such as the depth limitation to wave height.

Therefore a model (Fig. 3) is formulated that derives from the typhoon parameters, the tide level and the geometry/bathymetry of the bay, via the pressure/windfield model, a wave growth model and a wind/pressure set up model the required combination of dh, H_s and T_p at the structures location.

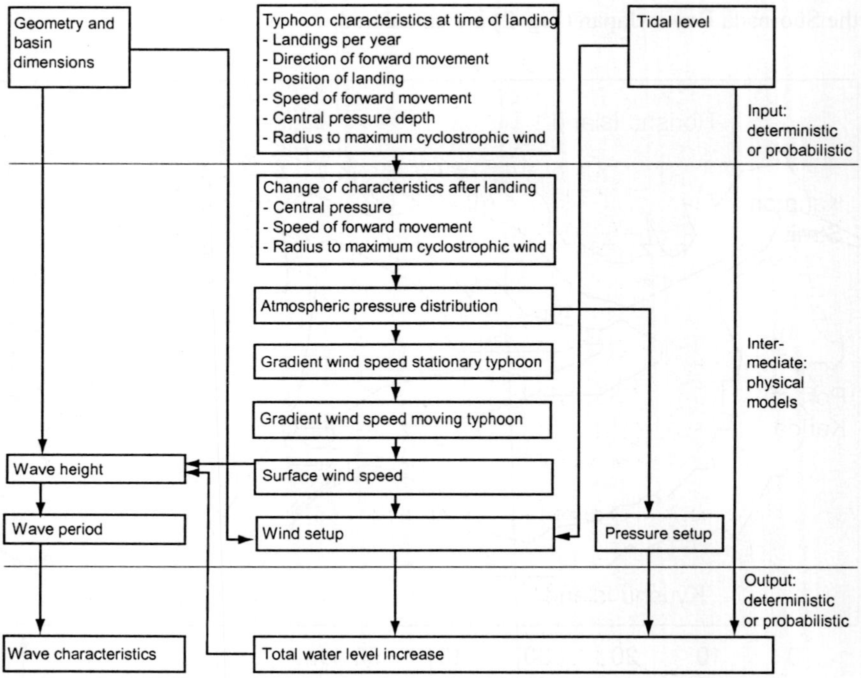

Figure 3: The model to calculate the hydraulic conditions given a typhoon

Although more refined numerical models are available, here simple analytical models are applied to illustrate the principles and to minimise computation time. For the wind/pressure field of the typhoon the model of Schloemer (1954) is used.

Kato and Torii (2002) performed an analysis with an analytical model proposed by the Japan Meteorological Agency (1999) for several Japanese bays. This simple model combines a linear pressure set up model with a quadratic wind set up model. A similar model, with different parameters is applied here.

For the wave height and the period the well known Sverdrup-Munk-Bretschneider model is chosen. The significant wave period is derived from the wave height using the wave steepness.

This combined model was calibrated by hindcasting the available observations with historical typhoons. Typhoon parameters are used from studies of Mitsuta and Fujii (1986) and Fujii (1998). In only four cases the observed water levels were available overlapping these studies. Wave observations started in 1991; in this respect also four storms have been observed. The result of the calibration is shown in Fig. 4 and Fig. 5.

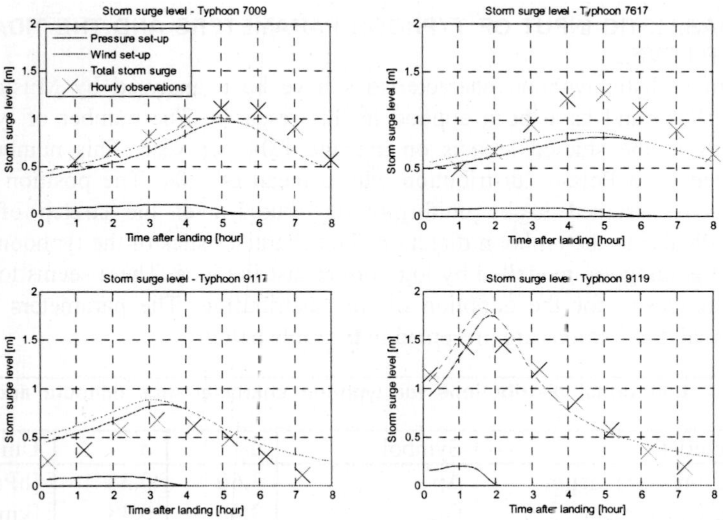

Figure 4: Comparison of observed and calculated water levels during typhoons

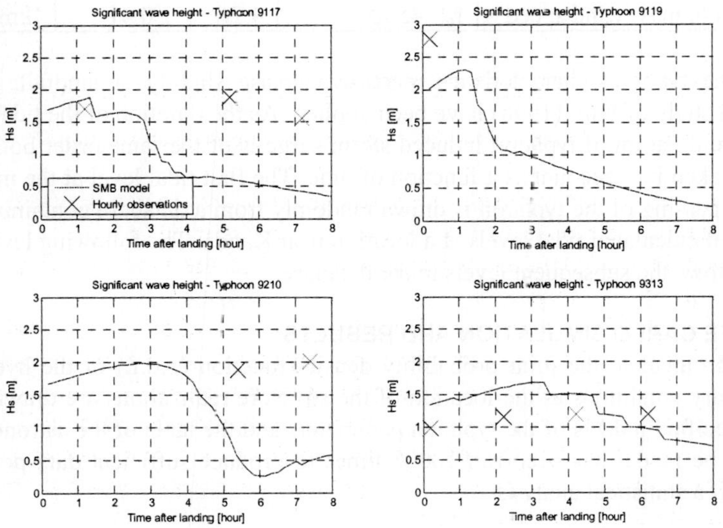

Figure 5: Comparison of observed and calculated wave heights during typhoons

The fit of the models is not perfect, but reasonably in view of the purpose of the project; especially the maxima of water level and wave height are predicted reasonably well and these are most relevant for the design. It is expected that more advanced numerical models will predict the phenomena better and in more detail.

PROBABILISTIC INPUT OF TYPHOON PARAMETERS AND THE TIDAL WATER LEVEL

Probabilistic typhoon characteristics have been analysed by Mitsuta and Fujii (1986) and have been applied in this research. The number of typhoon landings in the studied area is on average 1.04 per year. This number was modelled by a Poisson distribution with a mean of 1.04. The position of the landing is assumed to be uniformly distributed over the stretch of coast perpendicular to the typhoon direction. The characteristics of the typhoon at the time of landing are modelled by log-normal distributions. There seems to be no physical reason for the adoption of this distribution. The parameters of the lognormal distributions are presented in the table below.

Table 1: Lognormal distributions for typhoon characteristics (Mitsuta and Fujii 1986)

Characteristic	Symbol	μ	σ	Units
Central pressure depth	Δp	3.60	0.38	[hPa]
Forward movement	C_{fm}	3.40	0.38	[km/h]
Direction of movement	γ	4.20	0.45	[deg]
Radius to max. wind speed	r_m $0<\Delta p<30$	4.61	0.31	[km]
Radius to max. wind speed	r_m $30<\Delta p<45$	4.39	0.59	[km]
Radius to max. wind speed	r_m $45<\Delta p$	4.26	0.41	[km]

The typhoon characteristics increase/decrease slightly at landfall; values from a study of Fujii (1998) have been applied. As the duration of the tidal cycle and the duration of typhoon induced storm surge is of the same order both have to be taken into account as a function of time. The first tidal level at the moment of the passing of the typhoon is drawn randomly from a database containing two years of calculated tidal levels at a location near Kanda. The following levels are taken from the subsequent levels in the database.

MONTE CARLO SIMULATION AND RESULTS

To calculate the joint probability density function (p.d.f.) of the hydraulic boundary conditions at the location of the structure realizations are drawn from the specified p.d.f.'s of the typhoon parameters and the table of the astronomical tide. The model has been run 10000 times to produce sufficient data points to support a statistical analysis.

The results presented in Fig. 6 are the simulated data points and the fitted marginal probability distribution of the raised still water level during typhoons i.e. the sum of astronomical tide and wind/pressure set up caused by the typhoons. It appears that a lognormal distribution fits the simulated data well. The same is done for the wave characteristics.

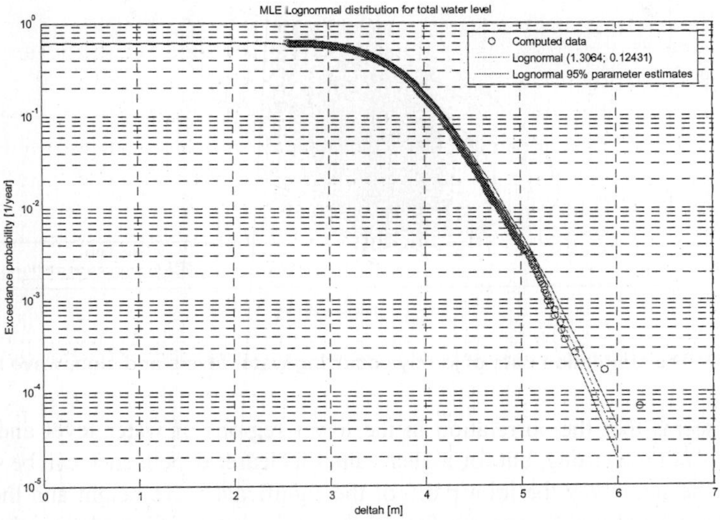

Figure 6: Approximation of the simulated water levels by a lognormal distribution

In reliability analyses the joint p.d.f.'s of the hydraulic loads have to be known as well as these joint loads are of importance for the failure of the structure. The joint distribution can be obtained via independent variables that are related to the hydraulic variables that have to be described with the joint p.d.f. for the reliability function (Vrijling and Van Gelder 2002). The derived independent marginal p.d.f.'s have been multiplied by the Jacobian to determine the dependent joint p.d.f.'s. The results of the water level and the significant wave height are plotted in Fig. 7.

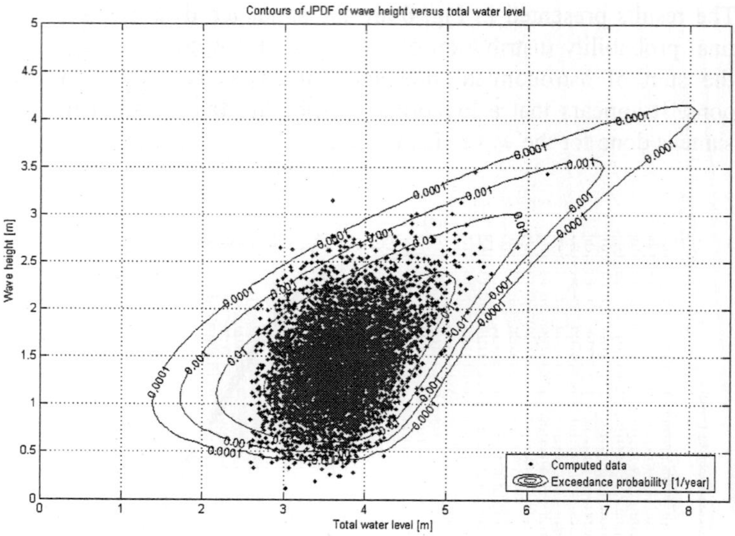

Figure 7: The simulation data of jointly occurring water levels and sign. wave height

It seems that the correlation in the lower regions of water level and wave height is not too strong, but for higher values a clearer dependence can be seen.

In the same way the joint p.d.f. of the significant wave height and the peak period can be found. So the aim formulated at the outset of establishing the joint p.d.f. of water level, significant wave height and peak period is reached. Some observations have been used to validate the joint p.d.f.'s. The results were not exact but the model showed similarity with the observations.

APPLICATION IN A DESIGN

Now the joint p.d.f. of water level, significant wave height and peak period is applied in the design of coastal structures. As an illustration the design of the crest level of a simple dike is shown here (Fig. 8). The crest of the dike is threatened by overtopping. The load, the wave run up is the sum of the water level and the actual wave run up. The actual wave run up is a function of the outer slope of the dike, the significant wave height and the peak period. In Fig. 9 the line gives the combination of all water levels and wave heights that produce a wave run up that just reaches the crest of the dike (the reliability function) according to formulae provided by the TAW (2002).

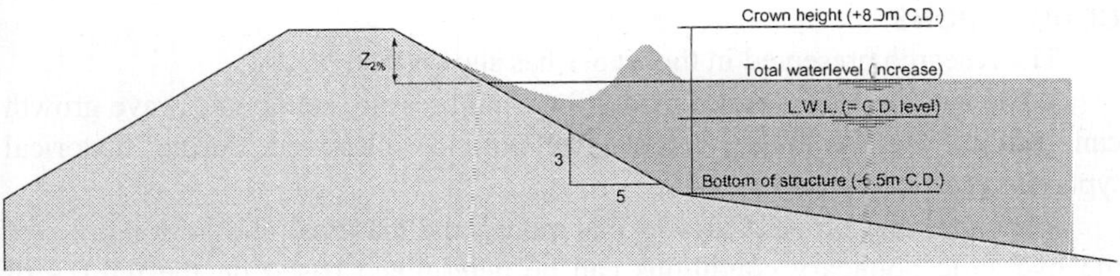

Figure 8: Cross section of a dike threatened by wave overtopping

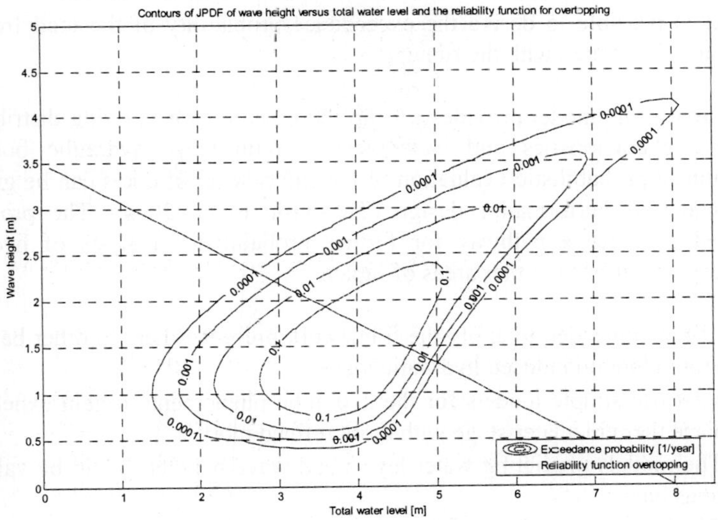

Figure 9: The joint p.d.f. of water level and wave height with the reliability function

The failure probability of this failure mode is the integral of the p.d.f. over the upper right "unsafe" part above the reliability function. This failure probability serves to estimate the risk of flooding. This risk is defined as the expected value of the loss i.e. the product of the probability of failure and the loss in economic and societal terms. When the risk is known a decision can be made about the optimal investment in a higher dike that will reduce the probability of a failure.

CONCLUSIONS

The research presented in this paper has shown that

- The simple deterministic model for wind/pressure set up and wave growth can predict the hydraulic boundary conditions observed during historical typhoons reasonably well.
- By applying Monte Carlo to this model, the marginal and joint p.d.f.'s of the hydraulic boundary conditions can be determined based on the p.d.f.'s of typhoon parameters and the astronomical tide.
- The observations of water levels and wave heights of historical typhoons are reasonably in line with the results of the Monte Carlo simulation.
- It is possible to derive the exceedance probability of the load from the p.d.f., here illustrated with the run up.

Overall, this model provides a simplified method for deriving distributions of wave characteristics and water levels. With these hydraulic boundary conditions a probabilistic evaluation of the safety level of dikes can be given in the context of performance design (Takahashi et al. 2004). The procedure followed provides a pathway for further probabilistic analysis of hydraulic boundary conditions in other areas of Japan.

It is recommended to apply the framework presented here to other bays that suffer from typhoon induced hydraulic loads.

- First use simple models for the hydraulic phenomena to gain experience, but replace them in a later stage with more refined models.
- The models predicting water levels and wave heights should be validated for various bays.
- The knowledge of the statistical properties of typhoon parameters should be extended and generalized.

Finally the good cooperation between Delft University of Technology, the Port and Airport Research Institute of Japan and the Dutch Ministry of Public Works is acknowledged.

REFERENCES

Fujii, T. 1998. Statistical analysis of the Characteristics of Severe Typhoons Hitting the Japanese Main Islands. *Monthly weather review*, April 1998, Vol 126, no. 4

Japan Meteorological Agency. 1999. Tide table for the year 2000, pp. 283

Kato, F., Torii, K. 2002. Risk assessment on storm surge floods. *Proceedings of the 2nd Workshop on the Development of Integrated Disaster Reduction Systems on Compound Urban Floodings*, 2003, pp. 20-31

Kawai, H., Kawaguchi, K., Hashimoto, N. 2004. Development of storm surge model coupled with wave model and hindcasting of storm wave and surges caused by Typhoon 9918. *Proceedings of 14th ISOPE*, conference paper No 2004-NM-9, p.8

Mitsuta, Y., Fujii, T., 1986. Analysis of typhoon pressure patterns over Japanese islands (II). *Journal of Natural Disaster Science*, Vol 8, No. 2, 1986, pp. 19-28

Schloemer, R.W. 1954. Analysis and synthesis of hurricane wind patterns over Lake Okeechobee, Florida. *Hydrometeorological Report*, USWB, No. 31, pp. 1-49

Takahashi, S., Kawai, H., Takayama, T., Tomita, T. 2004. Performance design concept for storm surge defenses, *Proc. of ICCE 2004*

TAW. 2002. Technical Report Wave Run-up and Wave Overtopping at Dike. Technical Advisory Committee for Water Defences, The Hague, May 2002

Vrijling, J.K., Bruinsma, J. 1980. Hydraulic Boundary Conditions. *Symposium on Hydraulic Aspects of Coastal Structures*, pp. 109-132

Vrijling, J.K., Van Gelder, P.H.A.J.M. 2002. Probabilistic design in hydraulic engineering. Delft University of Technology

EXTREME WAVES REVISITED

Kenneth Willems[1], Agustin Sanchez-Arcilla[2], Jesus Gomez Aguar[2], Julien De Rouck[3], Peter Troch[3], R. Bolaños[2]

An important difficulty in the field of maritime engineering is the probabilistic nature of wave drivers. To quantify the magnitude of forces that a structure will be exposed to in its lifetime, a maritime engineer needs to have a good understanding of the extreme climate of the region. By conducting extreme wave analyses on previously gathered wave data one can estimate design values, but there will always be uncertainty present. The objective of this paper is evaluating that uncertainty. More specifically, we will formulate conclusions about the uncertainty that lies in the model as well as the method that is used to conduct the analyses. But the latter is also influenced by regional particularities. In order to have a better view on those regional particularities, we conducted our investigation with data from two distinct maritime environments: the NW Mediterranean and the Southern North Sea. A last point of interest is the effect of the amount of the recorded data on the uncertainty within extreme wave analyses. We will do this by conducting extreme wave analyses on well-chosen datasets: developed with different selection methods, representing different time periods, having different sizes. It will therefore be necessary to establish storm databases for both regions and with different methods. Besides conducting non-directional and directional wave analyses, we also look into other storm characteristics. The results presented in this paper do not only give additional insights in the way to conduct extreme wave analyses, but will also provide more insight in the storm behaviour of both regions, the evolution of storm characteristics in the past two decades and its implications on the uncertainty in extreme wave analyses.

INTRODUCTION

This work deals with the uncertainty and its effects on the extreme wave analysis, its results and the implications for maritime design. To obtain generic conclusions which are not site-dependent, we will review this for two distinct cases: the Southern North Sea (Westhinder, 1982-2004) and the NW Mediterranean (Cap Tortosa Waverider buoy, 1990-2004). Both locations are shelf sea environments and the depth of the studied deep-sea buoys is in the range of 20-60m.

Due to its location, the Southern North Sea is highly influenced by the Atlantic Ocean. This explains that N and SW are the most important storm directions. The Westhinder buoy is the most suitable buoy due to the fact it has the deepest buoy location and has a long recording period.

[1] Dredging International, D.E.M.E., Scheldedijk 30, Zwijndrecht, Belgium, kennethgwillems@hotmail.com
[2] LIM/UPC, Universidad Politécnica de Cataluña, Spain, info.lim@upc.edu
[3] Department of Civil Engineering, Ghent University, Belgium, Julien.DeRouck@Ugent.be

Cap Tortosa is located at the Ebro Delta, 200km southwest of Barcelona. The Catalan coastline can be seen as a highly complex situation of geographical and climatic characteristics, with highly localised wave climate conditions. There are three mayor storm directions: E, NW and S, which is less important.

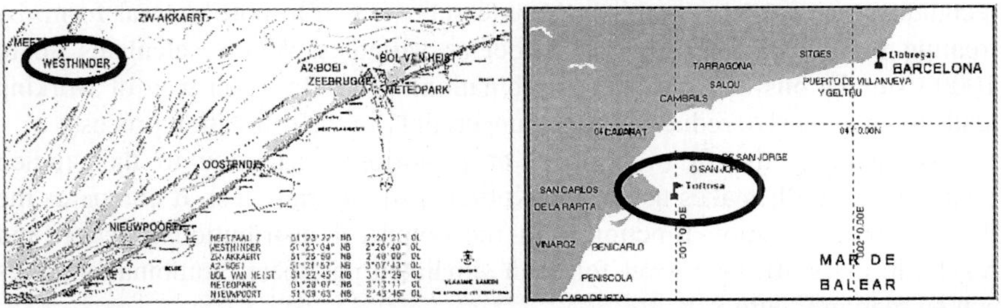

Figure 1. The buoy locations along the Belgian coast (left) and the XIOM buoys along the Catalan coast (right).

Four main types of uncertainty will be examined:

1. Uncertainty related to dataset sampling and size.
2. The uncertainty in the postulated model which among other things state that the extreme wave behaviour is stationary on a long-term basis.
3. The uncertainty due to the chosen methodology to conduct the extreme wave analysis.
4. The uncertainty caused by regional differences (and all other types of "errors").

BACKGROUND

BGPE: Bayesian Generalised Pareto Estimation

The Generalised Pareto Distribution is widely used for modelling exceeding of a random variable over a high threshold. The main reason that assures the appropriateness of the GPD as a model for excesses lies in the fact that Peak Over Threshold is the most useful approach to establish a dataset for the specific risk analysis in this work. The conditional distribution of excesses y over a high enough threshold u converges to a generalised Pareto distribution, with shape parameter ξ and scaling parameter β (Pickands theorem, 1975) for independent observations:

$$G_{\xi,\beta}(y) = 1 - \left(1 + \frac{\xi}{\beta} \cdot y\right)^{-\frac{1}{\xi}} \qquad \text{for} \neq 0 \qquad (1)$$

$$G_{\xi,\beta}(y) = 1 - \exp\left(-\frac{y}{\beta}\right) \qquad \text{for} = 0 \qquad (2)$$

The above described GPD gives a mathematical model for the sizes of the events. To determine the time-occurrence of these events, the hypothesis that extreme events occur approximately as a Poisson process is of common use in hazard estimation. The Poisson process has one parameter: $\lambda(u_0)$, the occurrence rate of events over the absolute threshold.

This means that the model has four parameters: ξ, β, u_0 and $\lambda(u_0)$. Bayesian techniques are used to establish the uncertainty with a more natural form, by treating these parameters of the model as variables. BGPE calculates up to 10000 simulations of parameter combinations. This Bayesian way of working leads to an extensive reduction of the uncertainty in the estimation process.

To quantify the fit of a priori or posterior distribution to the studied population, BGPE makes use of the Kolmogorov-Smirnov test. It compares the theoretical distribution function with the empirical distribution function and results in a significance level for each simulated parameter combination. This can be seen in Figure 2.

BGPE: Methodology

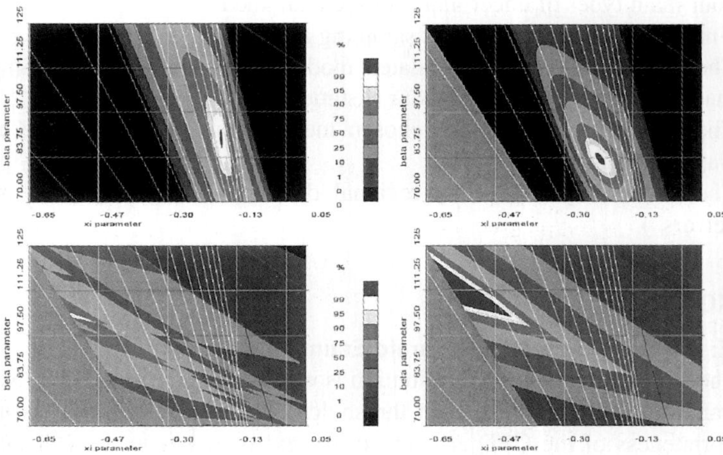

Figure 2. BGPE-generated plots: Prior density (top left), Posterior density (top right), multinomial Goodness of Fit (bottom left), KS-test (bottom right)

First we have to determine the absolute threshold u_0, which determines the extreme values that are appropriate for establishing the Generalised Pareto Distribution. This is done by analysing the excesses plot of the extreme values population.

Once the absolute threshold is established, we need to focus on the priori input and the simulations of the parameter combinations. The priori input consists of some general estimated values, such as "characteristic storm height" (exceeding probability: ¼) or "possible but unobserved storm height". By defining small applicable intervals for the priori input, we are able to fine-tune the prior density plot to the posterior density plot and the results of the KS-test. The objective is to have the core of all plots (see Figure 2) as similar as possible. There lies the most appropriate parameter combination, which represents the best fitting Generalised Pareto Distribution for the studied extreme wave height population. From this distribution, the significant wave height with a return period of 100 years is derived as well as its confidence intervals.

SAMPLING UNCERTAINTY

Implications of dataset's size

We have noticed a correlation between the results of the non-directional extreme wave analyses and the size of the used Westhinder datasets. The $H_{s,design}$ of the 22-year dataset is respectively 25cm and 42cm smaller than the average $H_{s,design}$ of the 10-year and 5-year datasets. We believe this discrepancy is caused by difficulties related to the choice of priori input for datasets with distinct observation periods.

Implications of anomalous years in small datasets

In situations with short buoy recording periods, the occurrences of abnormal events has a significant influence on the results of extreme wave analyses. This was observed for 5-year datasets for the two studied field cases. The Westhinder buoy recorded a very small amount of storms during 2003. When replacing this abnormal year from the 5-year dataset, both the storm ratio and the $H_{s,design}$ change considerably. The same effects were observed for Cap Tortosa when excluding a very extreme storm in November 2001. In the case of Westhinder, the anomaly leads to an excessively mild assessment of the wave climate, while the anomaly in Cap Tortosa leads to a more energetic assessment.

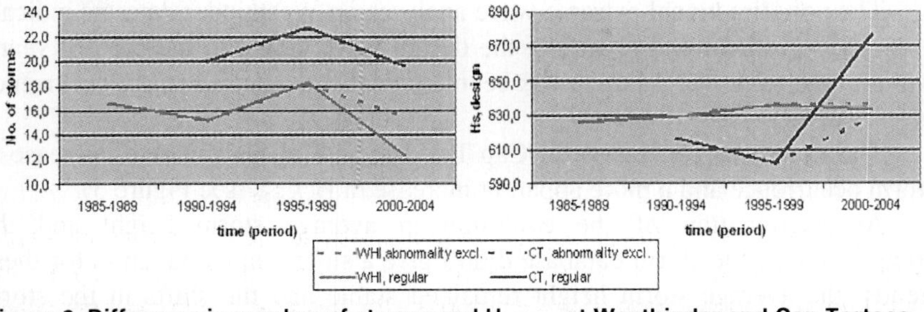

Figure 3. Difference in number of storms and $H_{s,design}$ at Westhinder and Cap Tortosa

Small datasets with unfortunate observation periods

When only a very short recording period is available, one could doubt the representativeness of the derived design wave height on account of the presumption of "an unfortunate observation period". This means that it would be possible that the small amount of registered years is a succession of very mild/energetic years. We examined this by developing datasets with the storms of the most/least energetic years and comparing resulting design wave heights.

We took into account the invalidity of years with a low buoy recording activity (see Figure 4). For Westhinder, the design wave height derived from 5 energetic years is only 0.6% larger than the design wave height from 5 mild years. The confidence interval for 5 energetic years is 2.9% larger. We concluded that the results are very similar and thus that there is no uncertainty on account of "an unfortunate observation period" when encountered with a very small recording period. For Cap Tortosa, similar results were obtained but it was more difficult to draw conclusions because of the small observation period and the small fluctuations in the annual storminess. In conclusion, no extra uncertainty has to be accounted for due to "an unfortunate observation period". In contrast, the anomalous year is of the greater relevance for observational periods below ten years.

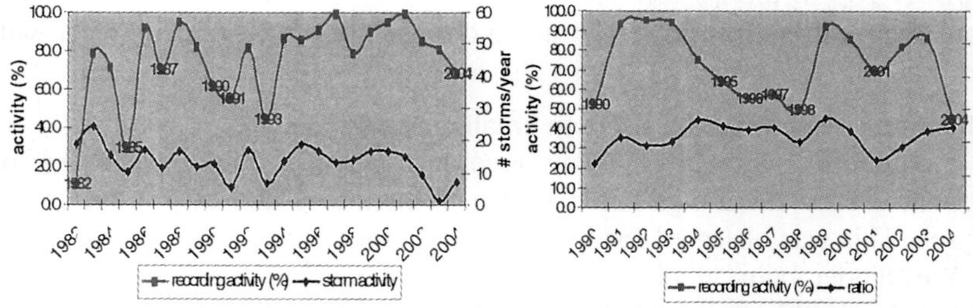

Figure 4. Evolution of storm ratio and recording activity in Westhinder (left) and Cap Tortosa (right).

MODELLING UNCERTAINTY: LONG-TERM TRENDS

Non-directional Extreme Wave Analyses

The non-directional extreme wave analyses on Westhinder datasets indicate a decadal upward trend regarding the design wave height in the past 20 years, while in the same period there has been a downward trend in storm occurrence (see Figure 3).

The same was concluded for Cap Tortosa, with a less apparent decrease in storm occurrence and a more apparent increase in $H_{s,design}$ (see Figure 3).

An examination of the evolution in average storm height and the composition of the storm database didn't give a sufficient explanation for these trends: the average storm height remained stable and the shifts in the storm composition are too small to be substantial. Nor are the evolutions caused by a few arbitrary but extreme storm surges: the anomalous year in Westhinder does not significantly affect these trends. However the severe storm in Cap Tortosa greatly enhances the upward trend in $H_{s,design}$, when dismissing this outlier, the trend still remains significant. This last observation also allows concluding that such outliers play a relatively more important role in middle wave climates such as the NW Mediterranean.

Directional Extreme Wave Analyses

The design wave heights of the directionally divided datasets of Westhinder on the other hand show for some directions a downward trend. This seems contradictory with the conclusions above, but the downward trend is caused by the very substantial difference in observation time between the directionally divided 10-year datasets of Westhinder. This is caused by the necessary elimination of some additional recorded years during the establishment of the storm database.

The design wave heights of the datasets of the two most important directions at Cap Tortosa show an upward trend and are thus in agreement with the results of the non-directional analyses.

Other Storm Characteristics

Not only has the evolution of the design wave height been investigated, we also put focus on two other relevant characteristics regarding storm behaviour: the storm energy and the relative storm growth rate.

We quantify the storm energy with the surface between the H_s-line and the storm threshold in a H_s – time plot:

Observation time interval $\Delta t = x$ h: $$Surface = \sum_{i=1}^{N} x \cdot H_{s,i} \quad (3)$$

The relative storm growth rate is defined as follows:

$$pd1_{relative} = \frac{1}{\Delta T} \quad (4)$$

With ΔT as the time difference between the start of the storm and the moment the storm reaches its peak.

We concluded that there is a big increase in average storm energy during the past 15 years in Cap Tortosa. A clear-cut upward evolution couldn't be observed for Westhinder. We also noticed a steadily decreasing storm growth rate at Westhinder. In Cap Tortosa on the other hand, there has been an increase in storm growth rate of 25%. Nevertheless we observed a parallel trend for both regions regarding the relative storm growth rate: an upward trend in the past 15 years, but this was in Westhinder preceded by a big downfall. The comparison also showed that storms reach their peaks faster in Cap Tortosa than in Westhinder.

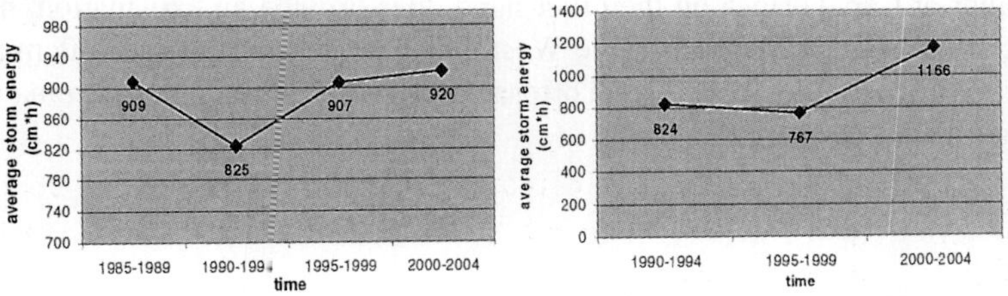

Figure 5. The evolution of the storm energy for 5-year datasets at Westhinder (left) and at Cap Tortosa (right)

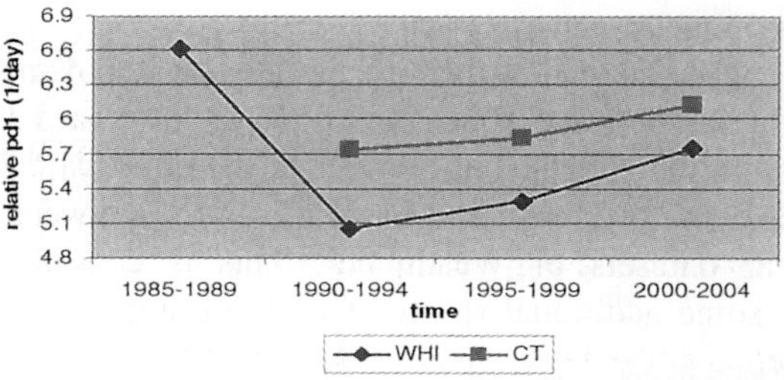

Figure 6. The evolution of the relative storm growth rate for 5-year datasets at Westhinder and Cap Tortosa

REGIONAL UNCERTAINTY

Because the storm behaviour is region-specific in this paper, the storm database is sampled in different ways. Different methods have been used in the preliminary investigation, e.g. to determine the total observation period of a certain database, or determine the validity of the storms in a storm database. But most importantly, the implications of the different directional storm behaviour between both regions on the sampling methodology have been investigated.

Regional Confidence Intervals

A first mayor indication for less apparent differences in the storm behaviour of both regions is the size of the confidence intervals of the design wave heights. The average confidence interval of the non-directional analyses of the Cap Tortosa storms is 181cm and thus up to 2.5 times as big as the average confidence interval of Westhinder. Both were derived with a numerous amount of datasets and the same sampling procedures and the same analysing techniques were used, so the differences can't be attributed to methodological aspects. Therefore we need to search the cause in the differences in storm behaviour. The fact that Cap Tortosa possesses more mayor storm directions was first thought to be the cause. This would mean that the non-directional storm database of Cap Tortosa consists of more subpopulations than at Westhinder and thus will lead to a bigger uncertainty. The directional investigation showed however that these subpopulations result individually in a very high confidence interval as well. The fact that the storm threshold is a lot smaller at Cap Tortosa on the other hand does provide an explanation, but only to certain extend: when using the Westhinder storm threshold of 230cm as well, the confidence interval of Cap Tortosa reduces to 135cm, but still remains a lot bigger than at Westhinder.

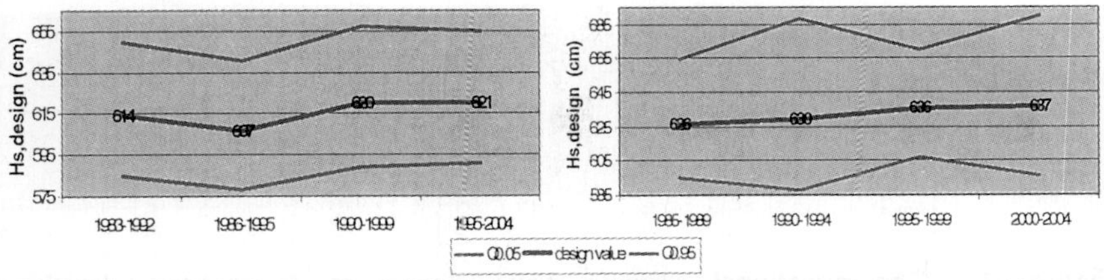

Figure 7. Evolution of the design wave height at Westhinder for decennial (left) and 5-year (right) datasets

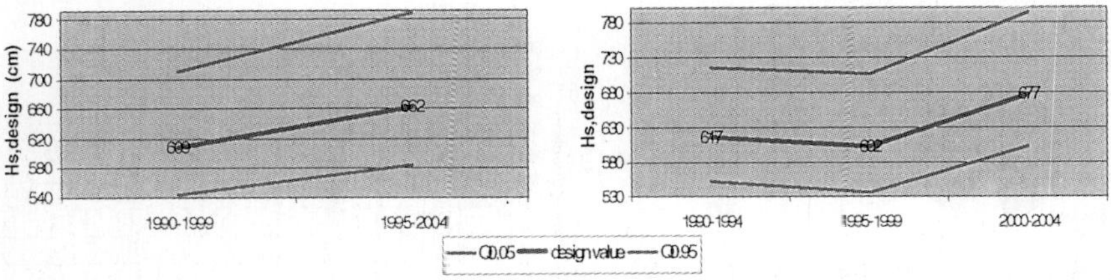

Figure 8. Evolution of the design wave height at Cap Tortosa for decennial (left) and 5-year (right) datasets

Directional Distribution of Storms

A big difference in the directional behaviour was observed for storms of Cap Tortosa and Westhinder. In Westhinder there are storms of which the direction gradually changes a great deal (see Figure 10). This never occurs in Cap Tortosa. In Cap Tortosa on the other hand it often happens that the direction changes a great deal very suddenly during the storm. Such a "bi-directional scattered storm" can be seen in Figure 9 and 11. This directional behaviour was never encountered for all recorded Westhinder storms.

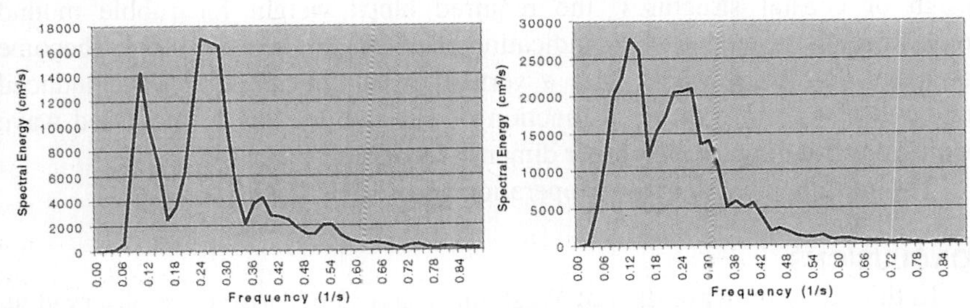

Figure 9. Evolution of spectra during bi-directional scattered storms, in which the absolute peak changes: from the small-period domain to the long-period domain

We believe this could be caused by the different orientation of the sea with respect to the propagation of low-pressure areas. Low-pressure areas have a west-east propagation and are therefore positioned above the sea a long time before reaching the North Sea, while at the NW Mediterranean the low-pressure areas are coming from the Iberian Peninsula.

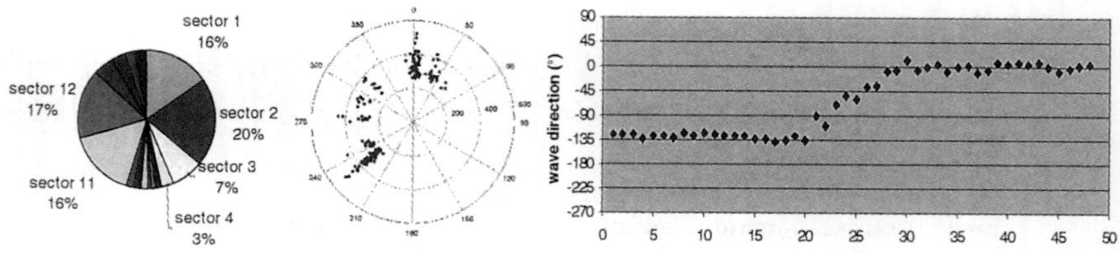

Figure 10. The directional extreme wave climate at Westhinder: directional distribution of all wave observations (left), polar plot of the storm database (middle), direction-time characteristic of a storm (right)

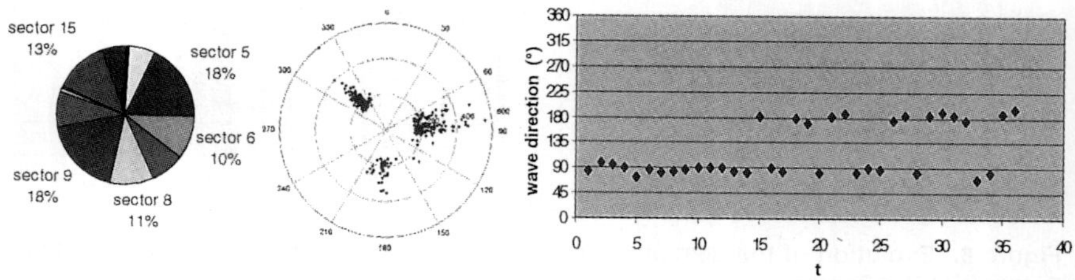

Figure 11. The directional extreme wave climate at Cap Tortosa: directional distribution of all wave observations (left), polar plot of the storm database (middle), direction-time characteristic of a storm (right)

As a consequence there are several methodologies to establish the directional storm population, which has implications for the methodology of the analysis.

IMPLICATIONS FOR MARITIME RISKS

A resulting upward trend in $H_{s,design}$ could have serious implications for the design of coastal structures: the required block weight for rubble mound breakwaters is related to H_s^3, indicating that (much) heavier blocks become necessary. For breakwaters with a vertical wall, typically the wave induced pressures along the wall are a function of H_s, also leading to increased wave loading on the structure and larger dimensions of the breakwater.

The same applies to harbour operation and coastal risks.

CONCLUSIONS

In this study on wave data of Westhinder and Cap Tortosa, we observed an upward trend regarding the design wave height in the past 20 years and in the same period a downward trend in storm occurrence.

Between both regions, there are differences regarding the importance of anomalous events in small observation periods. Moreover, the dataset's size doesn't augment the uncertainty inherent to the extreme wave analysis as much as the presence of anomalous events in the dataset.

APPENDIX

In table A-1 some frequently used H_s values have been presented along with the confidence intervals, derived with the non-directional analyses on both Westhinder and Cap Tortosa datasets.

Table A-1. $H_{s,100y}$ and Confidence Intervals of important datasets from the Westhinder and Cap Tortosa wave database

	$H_{s,100y}$ (cm)	$Q_{0.05}$ (cm)	$Q_{0.95}$ (cm)	Confid. Interval (cm)
WHI, 1982-2004	590.0	567.2	614.5	47.3
WHI, 1995-1999	636.3	606.9	670.4	63.5
WHI, 2000-2004	637.0	596.2	690.0	93.8
WHI, 1990-1999	620.1	589.3	657.0	67.7
WHI, 1995-2004	620.6	590.8	655.2	64.4
CTO, 1990-2004	656.6	580.9	779.9	199.0
CTO, 1995-1999	601.6	536.6	706.6	170.0
CTO, 2000-2004	677.5	603.7	792.5	188.8
CTO, 1990-1999	608.8	544.8	709.6	164.8
CTO, 1995-2004	662.2	585.3	787.9	202.6

REFERENCES

Shore Protection Manual. 1984. Coastal Engineering Research Center – US Army Corps of Engineers. Vicksburg-Mississippi. USA

M.J. Tucker. 1991. Waves in Ocean Engineering. Ellis Hardwood Ltd. UK.

Guide to Wave Analysis and Forecasting. 1998. World Meteorological Organization. Switzerland.

Y. Goda. 2000. Random Seas and Design of Maritime Structures. World Scientific Publishing. Singapore.

Z. Liu & H. Burcharth. 1999. Port Engineering. Aalborg Universitet. Denmark.

X. Gironella, J.P. Sierra, V. Gràcia, J. Sospedra. Analisis Extremal del Oleaje para el Diseño de Obras Maritimas. UPC. Spain

Pickands J. III. 1975. Statistical Inference Using Extreme Order Statistics. *The Annals of Statistics 3*. 119-131.

L. Herreras Martinez. 2002. Estudio de los Parametros que Caracterizan una Tormenta. LIM-UPC. Spain

Hydro Meteo Atlas, Meetnet Vlaamse Banken. 1999. Afdeling Waterwegen Kust. Ministerie van de Vlaamse Gemeenschap.

J.J. Egozcue & R. Tolosana-Delgado. 2002. Program BGPE: Bayesian Generalised Pareto Estimation Help Document. UPC. Spain.

GENERATION OF REFLECTION–FREE WAVES IN A NUMERICAL FLUME USING PID FEEDBACK CONTROL

Naresh Subramaniam[1], F.F. Li[2], D.M. Ingram[3] and C.G. Mingham[4]

This paper proposes an approach to active absorption of reflected waves in a numerical wave flume that makes use of Proportional, Integral and Derivative (PID) control strategy. Numerical simulations using the shallow water equations are presented to support the use of PID control methods. Standard PID controllers allow three adjustable parameters. Manual tuning is applied to obtain appropriate values for the three parameters. Simulations showing the feasibility of using PID control strategies are conducted with sinusoidal left boundary conditions with more than one frequency component along with a fully reflective right boundary.

INTRODUCTION

In numerical simulations of wave structure interactions, the problem of reflected waves corrupting the incident waves generated by a numerical algorithm is well-known (Hughes, 1993). In such situations the measurement of the pure incident component at or near a simulated coastal structure is complicated by the presence of reflected waves from the structure. Further, the reflected waves travelling towards the wave generator must be handled using absorping boundary conditions at the left boundary to maintain the stability of the system.

There have been several attempts to minimise the effects of reflected waves on measurement and on the wave generator using offline reflection analysis methods (Goda and Suzuki, 1976; Mansard and Funke, 1980) and also using a bank of filters to actively absorb reflected waves (Schaffer and Hyllested, 1999).

Early approaches to the wave separation problem were made by Thornton and Calhoun, 1972 and also by Kajima, 1969 and further developed by Morden et al., 1976 and Goda and Suzuki, 1976. Thornton and Calhoun, 1972 were chiefly concerned with extending the resolving methods based on the monochromatic waves to more complex wave forms occurring in nature. Their method consisted of measuring the combined temporal profile of incident and reflected components simultaneously at two spatial locations in a line parallel to the direction of propagation of the wave. The distance of sensors from the wave generator and spacing between the sensors are known. By initially considering a monochromatic

[1] Centre for Mathematical Modelling and Flow Analysis, Manchester Metropolitan University, Chester Street, Manchester, M1 5GD, UK

[2] School of Computing, Science and Engineering, University of Salford, Salford, M5 4WT, UK

[3] School of Engineering and Electronics, University of Edinburgh, Kings Buildings, Mayfield Road Edinburgh, EH9 3JL, UK

[4] Centre for Mathematical Modelling and Flow Analysis, Manchester Metropolitan University, Chester Street, Manchester, M1 5GD, UK

wave and applying linear wave theory (Dean and Dalrymple, 1991) the amplitude and phase angle of the incident and the reflected wave components is obtained as a function of amplitude and the relative phase difference of the measured temporal profiles at the two sensor locations. The method is then naturally extended to random waves by again invoking linear wave theory and considering the sea surface as a superposition of an infinite number of sinusoids. Spectral analysis is used to decompose the measured temporal profiles into constituent frequency components. The method developed for a monochromatic wave is then individually applied to each frequency component to obtain the reflected wave of that frequency component. Since the accuracy of the two-sensor method greatly depends on the spectral estimate of the individual frequency components of the measured temporal profiles, the further developments proposed by Morden et al., 1976 and Goda and Suzuki, 1976 were mostly concerned with better ways of estimating the combined wave spectra. Frigaard and Brorsen, 1995, however, demonstrated a time-domain implementation of a two-sensor method using two linear digital filters. In terms of accuracy, this method is demonstrated to be on a par with the two-sensor method as derived and popularised by Goda and Suzuki, 1976.

The two-sensor method, described above, is only applicable to a limited frequency range because the resolution of incident and reflected waves is not possible at singularity points, limiting the range of frequency components that can be decomposed. However, this can be circumvented by measuring the velocity profile at fixed locations instead of surface elevation (Troch and Rouck, 1999).

Mansard and Funke, 1980 proposed a three-sensor method by utilising least squares optimisation. By using a third sensor this method was able to include the combined effects of nonlinear interactions and unwanted signal noise at the sensor locations into a third single unknown. Their method is similar to the two-sensor method in using measured temporal profiles in the resolution procedure but relies on least squares optimisation to estimate incident and reflected component amplitudes and phase angles when fitting the measured temporal profiles to the model given by linear wave theory. Parameters of the model are estimated by minimising an error function equal to unwanted signal noise at the sensors. The notable advantage of this is, using three sensors they were able to widen the available range of frequencies that could be resolved. However, the method is still sensitive to critical sensor spacing due to singularity points. Lin and Huang, 2004 extended the least square optimisation method using four sensors to separate higher harmonic waves. Their model derived from linear wave theory considers the amplitudes and phase angles of bound waves and free waves as separate unknowns thus requring the extra sensor to estimate the additional unknown variable.

Medina, 2001 took a different approach to the separation problem by using simulated annealing. This method is claimed to be able to separate nonlinear wave components. The basic idea is to use local approximations instead of trying to separate the whole measured wave field by seeking a global solution. Accordinly, the surface of the flume is divided into several local segments to formulate a parametric model. The problem is then posed as an optimisation problem and simulated annealing is used to find the optimal parameters in the mean squared sense. When compared with the three-sensor method the difference is in using several local segments and in using simulated annealing for solving the optimisation problem.

SYSTEM MODEL

One possible way to generate numerically predefined type of waves without undesirable reflections is to use feedback control coupled with the wave generator (Schaffer and Hyllested, 1999). From a control theory perspective, a wave flume driven by a predefined input can be modelled as a feedback system as illustrated in Figure 1. This is a typical feedback regime used in many industrial process control problems. For process control, the forward transfer function, the wave flume in this case, is typically known or can be assumed. But in the wave generation problem, although the propagation of waves is modelled using the equations of fluid flow, it is not possible to model them as a linear system. Hence, a transfer function is not known for the wave flume. Further, in Figure 1 the controller is shown in the forward path but it could also be inserted in the feedback loop and both representations are valid.

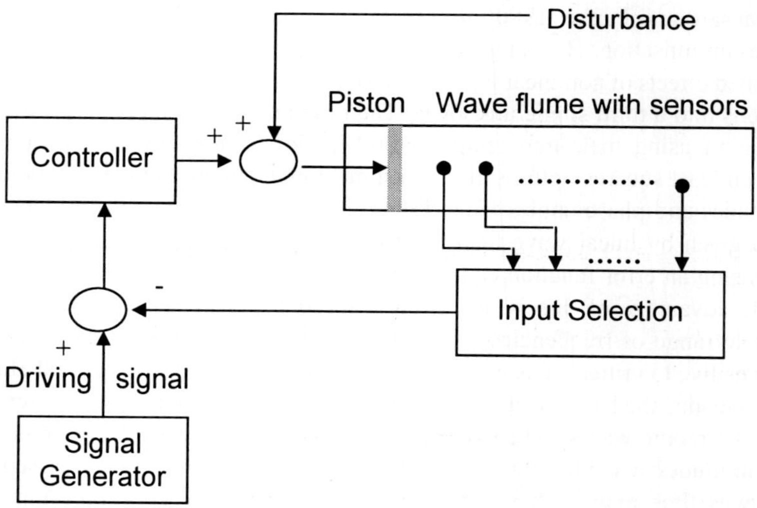

FIG. 1. **Schematic of propsed control system**

For the wave generation problem, the input of the feedback system is the piston driving signal and the outputs are observations obtained by a number of sensors placed in the wave field near the piston. The feedback loop ensures that the output from the system follows the input regardless of external interference. Here, the reflected waves travelling towards the generator are viewed as interference. The feedback loop functions by minimising the error signal, which is the difference between the signal generator output and the output of the wave flume measured closer to the piston. Numerically, the actuating signal - i.e., the output of the controller, is the left or the seaward boundary condition specified in the solver. Hence, the controller dynamically modifies the boundary condition during simulation to compensate for the reflected waves.

PID ADVANTAGE

PID controllers (Tewari, 2002) are extensively used in industrial process control applications due to their robustness over a wide range of frequencies and processes, capability to deal with weak non-linearity and independently adjustable parameters (see Figure. 2).

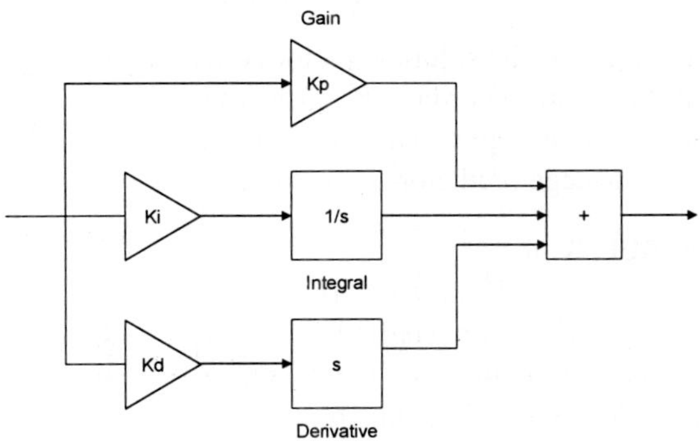

FIG. 2. **PID controller with three adjustable parameters, K_p, K_i and K_d**

The transfer function, $G(s)$ of the PID controller block in the Laplace domain is given by

$$G(s) = K_p + \frac{K_i}{s} + sK_d \tag{1}$$

where K_p, K_i and K_d are the three adjustable PID parameters. Existing control systems, based on filter banks and hydrodynamic feedback adjust the output proportional to the computed error only. In contrast, the PID controller takes into account the accumulated error over time by adjusting the error in proportion to instantaneous (K_p) and accumulated (K_i) errors . Furthermore, a third parameter (K_d) proportional to the difference between present and previous output of the system is used to control the dynamics of the system. This setup provides three independently adjustable parameters that can be attuned to optimise the active wave absorption control system. Therefore, a PID controller is used as an alternative to existing control strategies to test its effectiveness in solving the wave separation problem.

NUMERICAL MODEL

The shallow water equations over a flat bed of the following form are used in the numerical simulations.

$$\frac{\partial}{\partial t}\mathbf{U} + \frac{\partial}{\partial x}\mathbf{F}(\mathbf{U}) = 0 \tag{2}$$

where

$$\mathbf{U} = \begin{pmatrix} \phi \\ \phi u \end{pmatrix}$$

$$\mathbf{F}(\mathbf{U}) = \begin{pmatrix} \phi u \\ \phi u^2 + \frac{1}{2}\phi^2 \end{pmatrix}$$

In the above form U is the solution vector containing the conserved variables and $F(U)$ is the flux vector. Further, $\phi = gh$ where g is the acceleration due to gravity and h is the water depth. The MUSCL-Hancock scheme (Shiach et al., 2004), which is a two stage predictor-corrector method is used to solve Eq. 2.

SIMULATION RESULTS

Table 1 presents the PID parameters, K_p, K_i and K_d for two separate numerical simulations with a sinusoidal left boundary and a fully reflective right boundary. The first simulation was conducted with a simple one component driving signal at the left boundary with a frequency of 0.67Hz and an amplitude of 0.1m. For the second simulation the sinusoidal left boundary was chosen with two frequency components with frequencies 0.67Hz, 0.33Hz and both components with an amplitude of 0.1m. Both simulations use a fully reflective right boundary with the water depth set to satisfy the shallow water condition.

TABLE 1. PID parameters

Simulation	K_p	K_i	K_d
1	0.01	0	0
2	0.09	0	0.015

The results are given in Figure 3 and Figure 4 for simulation numbers 1 and 2 respectively. The appropriate values for K_p, K_i and K_d were obtained using

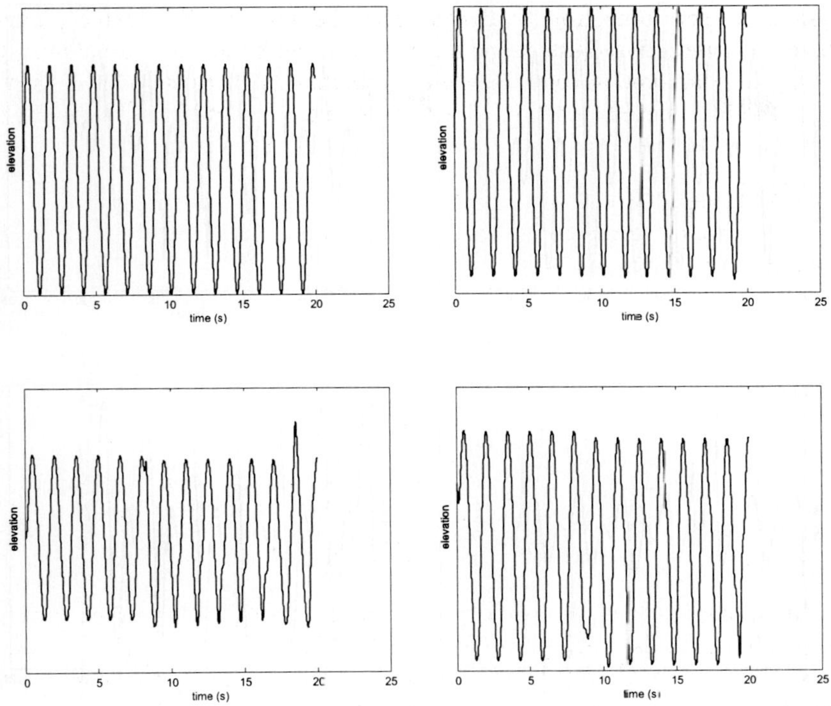

FIG. 3. Simulation 1 results: input signal (top left), actuating signal (top right), output signal without PID (bottom left), output signal with PID (bottom right)

manual tuning. The input signal is the sinusoidal left boundary condition, the output measured with and without the PID controller, in both cases is the wave height profile measured at a grid point very close to the left boundary. The actuating signal is the output of the PID controller, which in turn is the input to the solver block. The actuating signal in effect drives the solver by dynamically modifying the left boundary condition during simulation to minimise the effects of the reflected waves.

DISCUSSION

Computer simulations were conducted using the shallow water equations with the PID controller arranged in the feedback control loop. Pilot results indicate that the waveforms from the output can be maintained nearly reflection-free even if reflective boundaries exist. This reconfirms the underpinning mechanism of the proposed reflection cancellation method: close-loop feedback control with a PID controller having three adjustable parameters, K_p, K_i and K_d can be attuned to

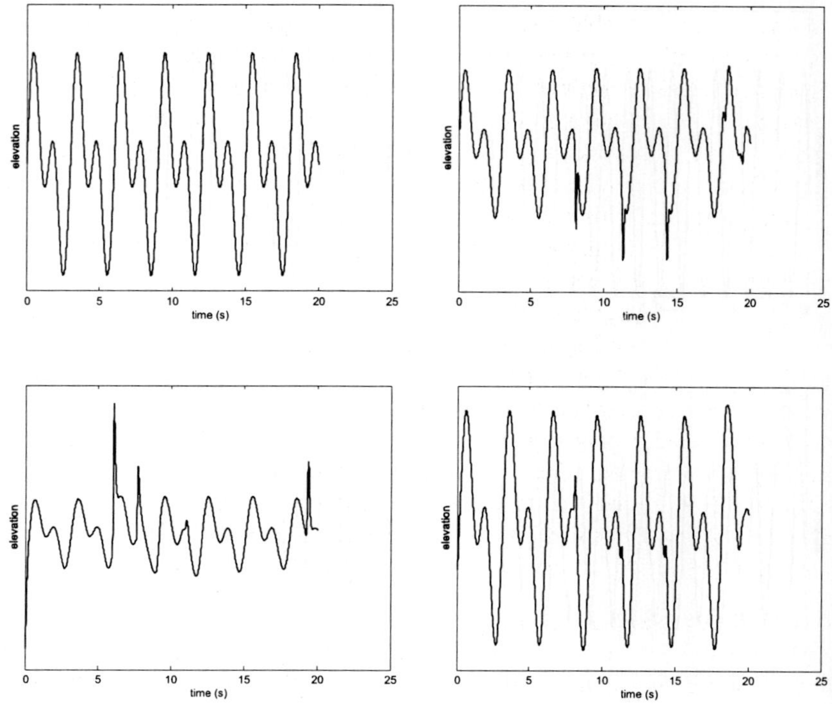

FIG. 4. **Simulation 2 results: input signal (top left), actuating signal (top right), output signal without PID (bottom left), output signal with PID (bottom right)**

attenuate the effects of reflection in a numerical wave flume.

There are, however, two caveats to be noted in using the PID control based reflection cancellation method. First, the shallow water equation solver used in this study has stability constraints imposed on the step-length and each time-step of the solver output has to satisfy the CFL condition to give stable numerical performance. Therefore, outputs are only available at non-uniformly spaced time instants. However, the controller block typically adopts uniform time-step. To circumvent time-step synchronization problems in this study, the shallow water solver is simulated once to obtain the minimum stable time-step satisfying the CFL condition. This in turn is used in a second run of simulation to enable numerical implementation of the PID controller. To eliminate the computational overhead from this two-sweep approach, interpolation and re-sampling might be used to resolve the time-step paradox of both the wave equation solver and the PID controller. Second, the scale of the output when the PID control mechanism is applied is smaller then that of open loop operation of the numerical solver. This is not surprising as stable feedback systems typically have lower gain than its open-loop operation. For this particular application the required scale (i.e wave height) can be easily rectified by increasing the system gain. A straightforward approach is to insert and amplification stage between the function generator and the input of the feedback loop. Since the solver block cannot be characterised using a linear transfer function, the system gain is difficult to quantify using traditional methods. This complexity is aggregated by considering different types of solvers. Empirical approaches might be a practical solution. But this has not been considered in this paper.

Regulating the three parameters of a PID controller has always been an art in the design of control systems. For linear system control there exist a set of effective methods to determine the PID parameters to specifications. For systems with non-linearity heuristic methods can be used. In the context of numerical simulation of ocean waves, the characteristics of this system is given by the wave equation solvers. For each individual type of solver, a suitable set of PID parameters are needed. It is therefore a significant advantage to develop a self-tuning or self adaptive PID regime to accommodate the complexity of wave equation solvers. This will be a major future work to the current pilot study.

ACKNOWLEDGMENTS

The authors would like to thank Mr. Jon Shiach for providing the solver for the shallow water equations in one dimension. The first author would like to acknowledge the UK Overseas Research Students (ORS) awards committee for part-funding his PhD studies.

REFERENCES

Dean, R. G. and R. A. Dalrymple. 1991. *Water Wave Mechanics for Engineers and Scientists*, World Scientific, Singapore.

Frigaard, P. and M. Brorsen. 1995. A time-domain method for separating incident and reflected irregular waves, *Coastal Engineering,*, 205–215.

Goda, Y. and Y. Suzuki. 1976. Estimation of incident and reflected waves in random wave experiments, *Proceedings of the 15th coastal engineering conference*, 828–845.

Hughes, S. A.. 1993. *Physical Models and Laboratory Techniques in Coastal Engineering*, World Scientific, Singapore.

Kajima, R.. 1969. Estimation of incident wave spectrum in the sea area influenced by reflection, *Coastal Engineering in Japan,*, 10–16.

Lin, C.-Y. and C.-J. Huang. 2004. Decomposition of incident and reflected higher harmonic waves using four wave gauges, *Coastal Engineering,*, 395–406.

Mansard, E. P. D. and E. R. Funke. 1980. The measurement of incident and reflected spectra using a least squares method, *Journal of Waterways, Habour and Coastal Engineering,*, 443–460.

Medina, J.. 2001. Estimation of incident and reflected waves using simulated annealing, *Journal of Waterways, Port, Coastal, and Ocean Engineering,*, 443–460.

Morden, D. B., E. P. Richey, and D. R. Christensen. 1976. Decomposition of co-existing random wave energy, *Proceedings of the 15th coastal engineering conference*, 846–865.

Schaffer, H. A. and P. Hyllested. 1999. Wave measure and analysis in laboratory flumes, *Proc. Coast. Struct. '99*, 93–99.

Shiach, J., C. Mingham, D. ingram, D. Causon, and T. Bruce. 2004. The applicability of the shallow water equations for modelling violent wave overtopping, *Coastal Engineering,*, 1–15.

Tewari, A.. 2002. *Modern Control Design*, John Wiley & Sons, England.

Thornton, E. B. and R. J. Calhoun. 1972. Spectral resolution of breakwater reflected waves, *Journal of Waterways, Habour and Coastal Engineering,*, 443–460.

Troch, P. and J. D. Rouck. 1999. An active wave generating-absorbing boundary condition for vof type numerical model, *Coastal Engineering,*, 223–247.

ACTIVE ABSORPTION OF WAVES IN A 3D BASIN

Oscar Cruz-Castro[1], Billy L. Edge[2], Lambert N.G. Romijnders[3] and Aimee Thurlow[4]

This paper aims at describing theoretical and practical aspects of the implementation of 3D active wave absorption (ARA) in a real basin. Implementation of the 3D ARA depends on the combining algorithm and on proper frequency response control through tuned filtering of the system response. The 48-paddle 3D wave generator with Active Reflection Absorption system, recently installed at the Haynes Coastal Engineering Laboratory, uses two signals for waveboard steering with simultaneous reflection absorption. The file with both signals is supplied by a purpose-built GEDAP package. The ARA control loop is closed in an industrial PC running Bosch Rexroth' HyPCoS real-time control software; no special hardware is used.

INTRODUCTION

In the last decade, wave makers have been routinely used together with control systems for simultaneous generation and absorption of waves, both on flume wave makers and on multidirectional wave makers around the world. These wave makers are known as absorbing wave makers or active wave absorption systems, because they act as a controlled moving boundary to absorb waves (Schaffer, 2001). Absorption of reflected waves is especially important in hydraulic testing as there are waves reflected from physical models and structures inside the basin, as well as from the basin's boundaries. As a consequence, waves gradually reflected back to the wave generator increase disturbances and reduce the accuracy of experiments and its duration.

Modern laboratories with wave makers are typically equipped with some form of mechanism for absorbing wave energy during a test. Previously, this was done with passive absorption filters and L-shaped tanks; now it is more effectively done with absorption as an active part of the wave generation mechanism. The theory of 2D and 3D active absorption of waves has been presented by several authors including (Schaffer, 2001) and (Schaffer and Klopman, 2000). Most of the efforts to implement 2D active wave absorption

---

[1] PhD student and Research Assistant, Haynes Coastal Engineering Laboratory, Zachry Department of Civil Engineering, Texas A&M University, Ocean Engineering Program, 600 Discovery Drive, 3136 TAMU, College Station, Texas, 77843-3136, USA.

[2] Professor and Director Haynes Coastal Engineering Laboratory, Zachry Department of Civil Engineering, Texas A&M University, Ocean Engineering Program, CE/TTI 808-I, 3136 TAMU, College Station, Texas, 77843-3136, USA.

[3] Senior Design Engineer, Wave Generator Systems, Bosch Rexroth, P.O. Box 32, NL 5280 AA Boxtel, The Netherlands.

[4] Intec Engineering, Marine Survey and Geohazards, 15600 JKK Blvd 9th Floor, Houston, Texas, 77032, USA.

have been successful. For 3D active absorption, many of the references describe success with absorption in 3D basins simulated by numerical wave tanks. (Hiraishi et al. 1998) demonstrated in the laboratory the effectiveness of the directional active absorption.

Importance of an Active Reflection Absorption system becomes evident when a wave maker is placed in a line in a relatively wide basin for multidirectional wave tests, because an experiment area with a uniform wave field (including oblique waves) is limited and reflected waves cannot be absorbed properly. On the other hand, the typical 3D control of the system itself (ARA) presents a problem that is related to the way in which multi-directional wave makers generate multi-directional waves. Generally speaking reduction of the system reaction is required to effectively absorb oblique-reflected waves (ITO, et al. 1996). If not, the system overreacts for oblique waves, creating spurious waves, which greatly reduce the wave maker's absorptivity.

The next sections describe the 3D ARA system available for testing; the 2D setup used during experiments with 3D ARA and its results, the initial work during experiments for the 3D setup with 3D ARA, conclusions and future work.

SYSTEM OVERVIEW

Active reflection absorption systems can be used to generate waves, and at the same time avoid re-reflections of waves. Such systems can be used also as active wave absorbers instead of passive absorbers (Schaffer, 2000). This is the case of the 3D multidirectional wave generator with Active Reflection Absorption (ARA) system, recently installed by Bosch Rexroth at the Haynes Coastal Engineering Laboratory at Texas A&M University. The system uses two signals for waveboard steering with simultaneous reflection absorption, see Figure 1. This wave maker has the ability of absorbing orthogonal and oblique waves by using a hydrodynamic feedback provided by water surface elevation gages integrated in the paddle front.

The data gathered for three paddles is used to calculate a corrected displacement to compensate reflected waves traveling to the paddles. The file with the required signals is supplied by a purpose-built GEDAP package. The ARA control loop is closed in an industrial PC running HyPCoS real-time control software developed by Bosch Rexroth; no special hardware is used. As shown in Figure 1 error signals from the adjacent paddles are combined with the filtered error signal for the center paddle in an algorithm designed to incorporate the net error to be used in modifying the action of the center paddle. Implementation of the 3D ARA depends on the combining algorithm and on proper frequency response control through tuned filtering of the signals. Response of the system is demonstrated in Figure 2.

The wave generator is located on the west wall of the three-dimensional basin. It has 48 paddles or wave boards; each is driven by a ball spindle and nut which are in turn driven by a digital AC-servomotor, or actuator. By converting the rotary motion of the actuator, the ball spindle is able to move in a linear motion guided by two ball guide rails. Each actuator contains feedback of the position and velocity of the paddle and is controlled by a Motor Controller which is mounted inside a nearby Motor Control Cabinet (MCC).

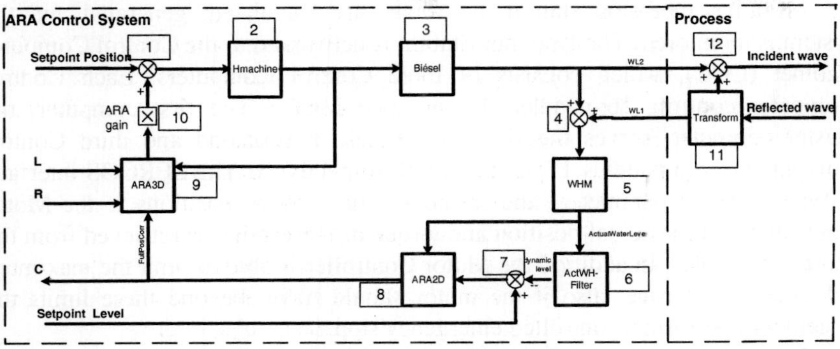

Figure 1. Schematic of Control System for Individual Paddle.

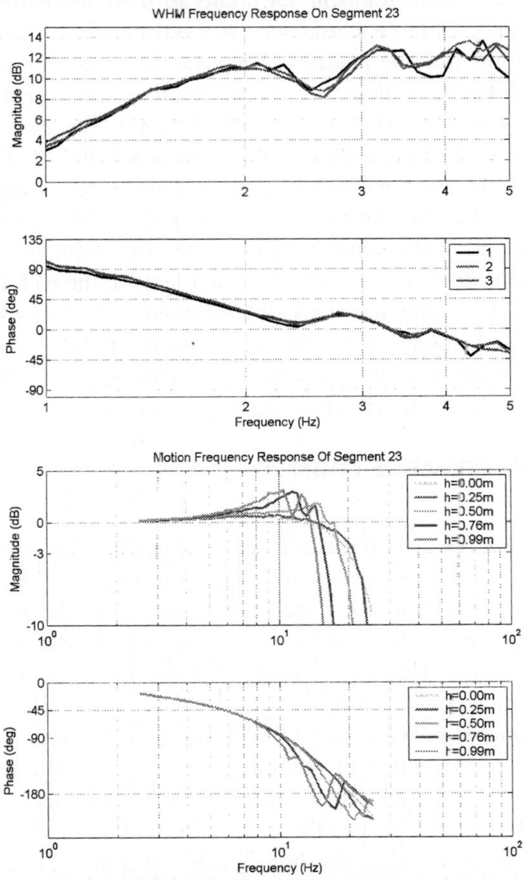

Figure 2. Frequency Response Analyses for Typical Individual Paddle.

Each MCC serves eight actuators and is controlled by a Serial Real-time Communication System (SERCOS) interface via fiber-optic cable from its respective Control Computer.

From the Operator Station the waves are calculated, generated and the system is monitored. The Operator Station is networked to the Control Computer Cabinet (CCC), which consists of three Control Computers. Each Control Computer controls 16 paddles through two MCCs. The first computer, the master computer, serves the first 16 paddles. The second and third Control Computers serve paddles 17-32 and 33-48 respectively. The SERCOS interface provides control parameters and set points for segment positions to the Motor Controller and the actual position and torque of the paddles is retrieved from the Motor Controller. In addition the Motor Controller is able to limit the maximum velocities and torque; also if the motor should travel beyond these limits the system will perform a controlled emergency stop.

The Control Computers receive data and commands from the Operator Station and then distribute segment set points to their respective MCC via the SERCOS interface. All three computers are synchronized, and allow smooth start-up and shutdown of the paddles. They also monitor the wave generator's behavior. The software utilized to compute the wave signal is GEDAP, developed by the Canadian Hydraulics Centre (CHC). The user enters the wave properties into GEDAP, which then creates a set point file. LIMITING software is then used to ensure it is not exceeding the system's operational limits. The set point file is then sent to the master Control Computer, which then distributes the set points locally and to the slave computers. The set point file contains a time series for both the paddle position and the water level at the paddle.

Finally, this wave generator is capable of generating various waves such as regular long-crested waves, irregular long-crested waves, and short-crested waves in perpendicular and oblique directions. The next section focuses on the practical performance of the ARA by executing some experiments.

PRACTICAL PERFORMANCE FROM 2D

Model Tests

In order to test ARA it would be ideal to have another wave generator to direct its waves towards the paddles of the generator with ARA activated. Since this is not available, a series of experiments were considered using reflected waves to test the three dimensional Active Reflection Absorption system, for the wave generator at the Haynes Coastal Engineering Laboratory. The setup consisted in placing a concrete wall on the south side of the 3D basin to create a closed area of about 9 m wide and 27 m long, that covered 19 paddles of the wave maker, with the intention of guiding waves to the end wall of this closed area and then having reflected waves coming back to the generator. The end wall was placed parallel with respect to the intended direction of wave travel to form an incident wave angle of 0 degrees (non-directional action, 2D). A general view of the wave tank and the position of the wireless gages can be seen in Figure 3, where the setup for these tests is shown in detail.

The first part of the tests consisted of using the configuration mentioned above for one wave height H= 0.07 m, three wave periods T= 1.0, 1.5 and 2.0 sec, and water level d= 0.40 m, with the basin closed. The second phase repeated the same wave conditions but this time with the basin opened, leaving the rock beach exposed to absorb the waves. For all of these tests the waves were monochromatic with a wave incident angle of 0 degrees, the 3D ARA was activated with local and overall ARA gain factors of "1" (100%), and also without ARA to make comparisons. The filters used in the HyPCoS control software for the 3D ARA, were designed earlier based on initial tests of the frequency response analysis of the paddles. The 3D ARA signal distribution for the experiments was with Left= 25%, Center= 50%, and Right= 25%.

A burst mode generation of waves for the closed cases for each of the three wave periods was used in order to get an easy separation of the incident, reflected and re-reflected waves, from the wave time series, although here generation and absorption of waves is not performed at the same time. Therefore, a continuous mode generation of waves was also employed to be able to see how the generation and absorption of waves was performing simultaneously.

A total of eight wireless capacitance type wave gages were employed to measure the wave field inside the basin as shown in Figure 3. Along the center line, an array of three wave gages were placed to measure the water surface elevation and the wave reflection coming from the end wall, for the continuous wave mode. The array was setup to calculate the incident and reflected wave traces at the position of gage 1, which is located 13.90 m away from the wall. The separation of the wave gages is defined according to the recommendations of (Mansard and Funke, 1980) for the array located in the middle of the basin. The next section shows the results obtained from the experiments.

Experimental Results-burst mode

A total of 12 tests were made for the conditions indicated in Table 1. Tests 1 to 3 were made with ARA, and tests 4 to 6 without ARA, in order to make a comparison of the effectiveness of the ARA absorption in the closed basin condition with wave burst mode. Tests 7 to 12 for the opened and closed basin cases were made to asses the behavior of the ARA for continuous wave generation and absorption. Figure 4 shows the wave traces for experiments 1 to 6 at the position of gage 1. Figure 4a) shows the comparison for tests 1 and 4 (T=1.0 sec). It shows in detail the reflected wave, originated by the reflector wall, traveling towards the wave maker, as well as the re-reflected wave from the wave maker, for both conditions ARA and NO ARA. In a similar fashion Figures 4b) and 4c) show the comparisons for experiments 2 and 5 (T=1.5 sec), and for 3 and 6 (T=2.0 sec), respectively.

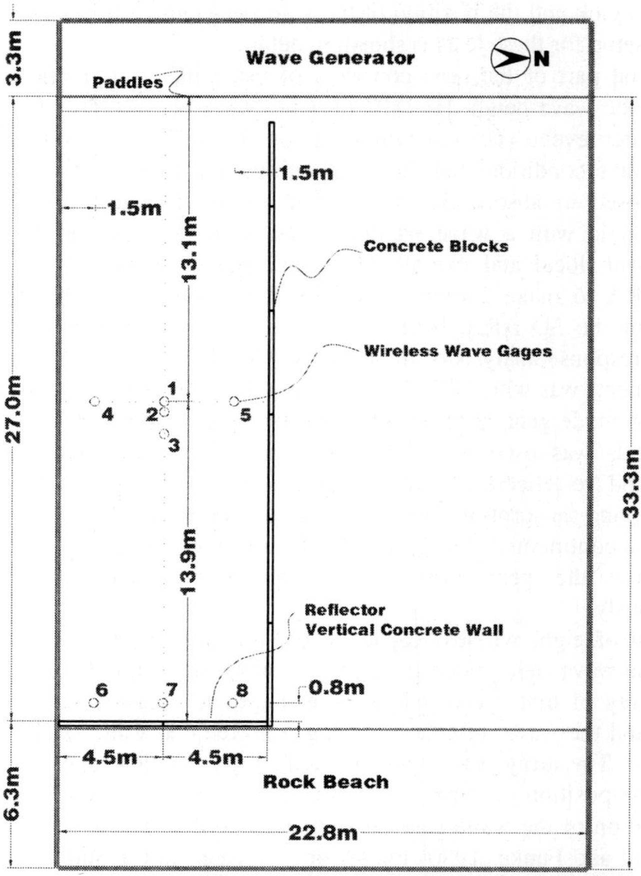

Figure 3. General View of the ARA Test, 2D Setup, conducted at the Haynes Coastal Engineering Laboratory.

Table 1. Test Matrix configuration for ARA experiments.

H (m)	d (m)	Test	T (sec)	ARA condition	Wave Generation Mode	Basin Condition
0.07	0.40	1	1.00	100% 3D ARA	burst	closed
		2	1.50			
		3	2.00			
0.07	0.40	4	1.00	NO ARA	burst	closed
		5	1.50			
		6	2.00			
0.07	0.40	7	1.00	100% 3D ARA	continuous	closed
		8	1.50			
		9	2.00			
0.07	0.40	10	1.00	100% 3D ARA	continuous	opened
		11	1.50			
		12	2.00			

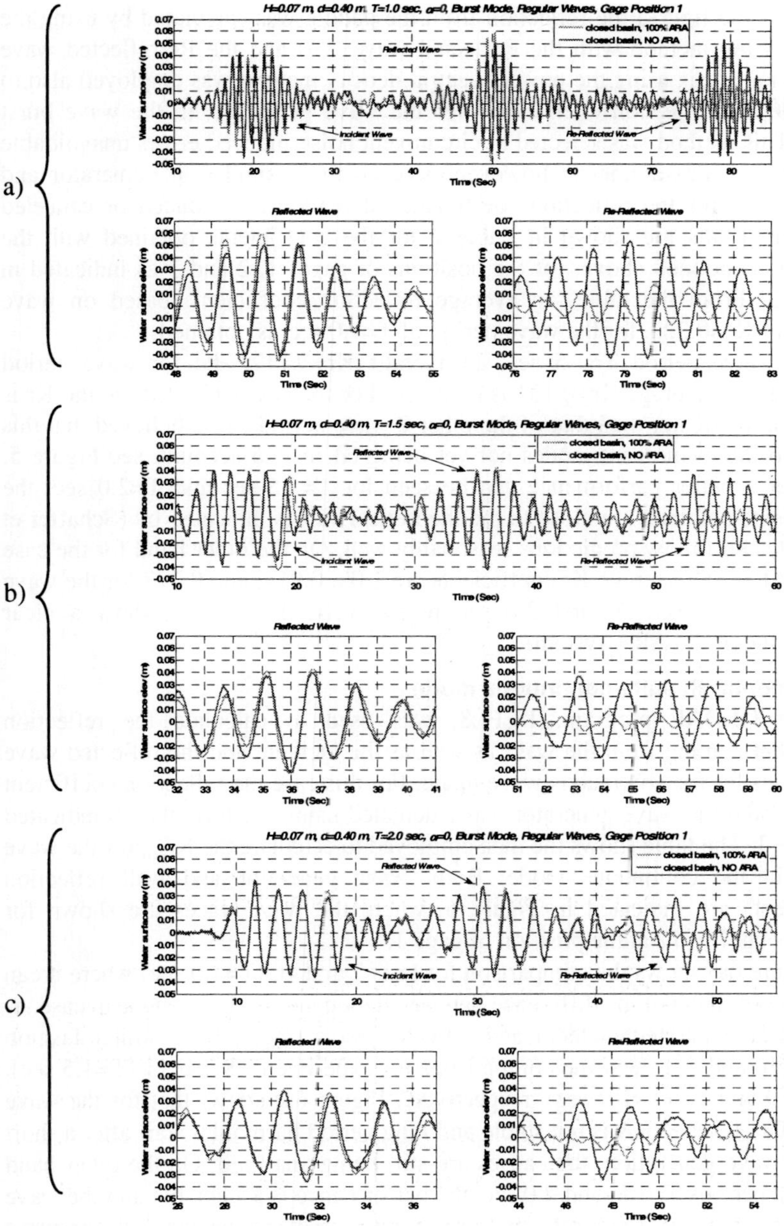

Figure 4. Wave traces for experiments 1 to 6 at the position of gage 1, Burst Mode. 4a) T= 1.0 sec, 4b) T= 1.5 sec, and 4c) T= 2.0 sec, for closed basin, with and without ARA.

From the figures the reflection from the paddle was calculated by using the zero-crossing method with the Reflected wave (Hi) and the Re-reflected wave (Hr). An analysis using the power spectral density method was employed also to calculate the reflection coefficients for the wave generator, in the wave burst mode. This method was selected to circumvent noise-induced errors unavoidable in the wave measurements; however these errors if similar in numerator and denominator for the reflection coefficient calculation are reduced or canceled out. The results are shown in Table 2 for the coefficients obtained with the methods mentioned above, for the positions of gages 1, 2 and 3 as indicated in Figure 3. From the table, an average reflection coefficient (based on wave height) of Kr=0.323, for the wave period of T=1.0 sec, is obtained.

For wave period T=1.5 sec, an average Kr=0.117, and for wave period T=2.0 sec, an average Kr=0.131 is obtained. For the case of T=1.0 sec the Kr is significantly greater than that of the other 2 wave periods. It is believed that this is due to the evanescent modes not yet being taken into account, see Figure 5. However a better performance can be seen for T=1.5 sec and T=2.0 sec; the coefficients obtained are close to the theoretical values indicated by (Schaffer et al. 1994) and follow predictions reasonably well. On the other hand for the case of NO ARA, the average Kr coefficients are 1.09, 0.914, and 0.777 for the wave periods of T=1.0, 1.5, and 2.0 sec respectively. The results show a clear functioning of the ARA system.

Experimental Results-continuous mode

The GEDAP program, REFLS, was used to calculate the reflection coefficient coming from the wall, as well as the incident and the reflected wave time series for the continuous wave mode. For this case the reflection coefficient coming from the wave generator was calculated using the formulation indicated in Table 3. The table shows the data employed to calculate the Kr_{gen} for the wave maker in the continuous mode. Both, wave generator and wall reflection coefficients are indicated in Figure 6, where the wave traces are shown for experiments 7 to 12 at the position of gage 1.

Figure 6a) shows the comparison for tests 7 and 10 (T=1.0 sec) where it can be observed in detail the difference between the opened and closed basin case, as well as the separated incident and reflected wave traces. In a similar fashion Figure 6b) and 6c) show the comparisons for experiments 8 and 11 (T=1.5 sec), and for 9 and 12 (T=2.0 sec), respectively. Figure 6a) shows that for the wave period T=1.0 sec, ARA is not stable and long period harmonics start after a short time of wave generation; the causes are still being analyzed. On the other hand for wave periods T=1.5, and 2.0 sec, the performance is acceptable and the wave traces of the incident and reflected waves are in good agreement with the wave conditions used as input.

Table 2. Summary of the results obtained for ARA experiments burst mode.

H (m)	T (sec)	d (m)	ARA	Gage pos.	Zero-crossing: Reflected wave Hi (m)	Zero-crossing: Re-Reflected wave Hr (m)	Power spectral density: Total energy Hi (m/Hz)	Power spectral density: Total energy Hr (m/Hz)	Kr Zero-cross	Kr using PSD.	Kr Avg.
0.07	1.00	0.40	100%	1	0.0821	0.0253	0.0466	0.0043	0.308	0.304	0.323
				2	0.0836	0.0281	0.0458	0.0049	0.336	0.327	
				3	0.0814	0.0273	0.0397	0.0043	0.335	0.330	
0.07	1.00	0.40	NO	1	0.0691	0.0797	0.0416	0.0472	1.153	1.065	1.090
				2	0.0692	0.0786	0.0414	0.0471	1.136	1.066	
				3	0.0770	0.0818	0.0382	0.0427	1.062	1.057	
0.07	1.50	0.40	100%	1	0.0597	0.0077	0.0259	0.0006	0.129	0.150	0.117
				2	0.0617	0.0064	0.0297	0.0004	0.104	0.109	
				3	0.0596	0.0067	0.0315	0.0003	0.112	0.095	
0.07	1.50	0.40	NO	1	0.0618	0.0570	0.0291	0.0218	0.922	0.865	0.914
				2	0.0594	0.0638	0.0301	0.0227	1.074	0.868	
				3	0.0599	0.0581	0.0326	0.0199	0.970	0.782	
0.07	2.00	0.40	100%	1	0.0624	0.0097	0.0368	0.0001	0.155	0.059	0.131
				2	0.0584	0.0096	0.0385	0.0004	0.164	0.106	
				3	0.0606	0.0107	0.0371	0.0006	0.177	0.123	
0.07	2.00	0.40	NO	1	0.0545	0.0510	0.0384	0.0217	0.936	0.751	0.777
				2	0.0565	0.0490	0.0418	0.0204	0.867	0.699	
				3	0.0580	0.0449	0.0402	0.0161	0.774	0.634	

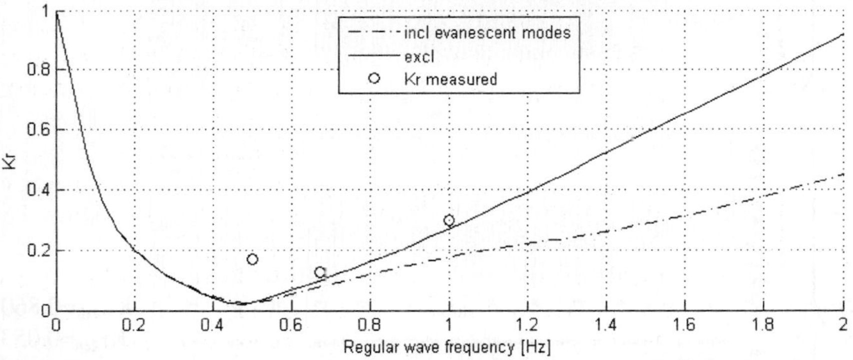

Figure 5. Predicted 2D ARA performance, with burst mode measurements. Plotted measured data are the net ARA contribution.

Table 3. Summary of the results for ARA experiments continuous mode 2D setup.

H (m)	T (sec)	d (m)	ARA	Hi_{closed} (m)	Hr_{closed} (m)	Hi_{open} (m)	Kr_{gen} formulation	Kr_{gen}
0.07	1.00	0.40	100%	0.0741	0.0613	0.0651	$Kr_{gen} = \dfrac{H_{I closed} - H_{I open}}{H_{r closed}}$	0.147
0.07	1.50	0.40	100%	0.0646	0.0530	0.0601		0.085
0.07	2.00	0.40	100%	0.0714	0.0664	0.0679		0.053

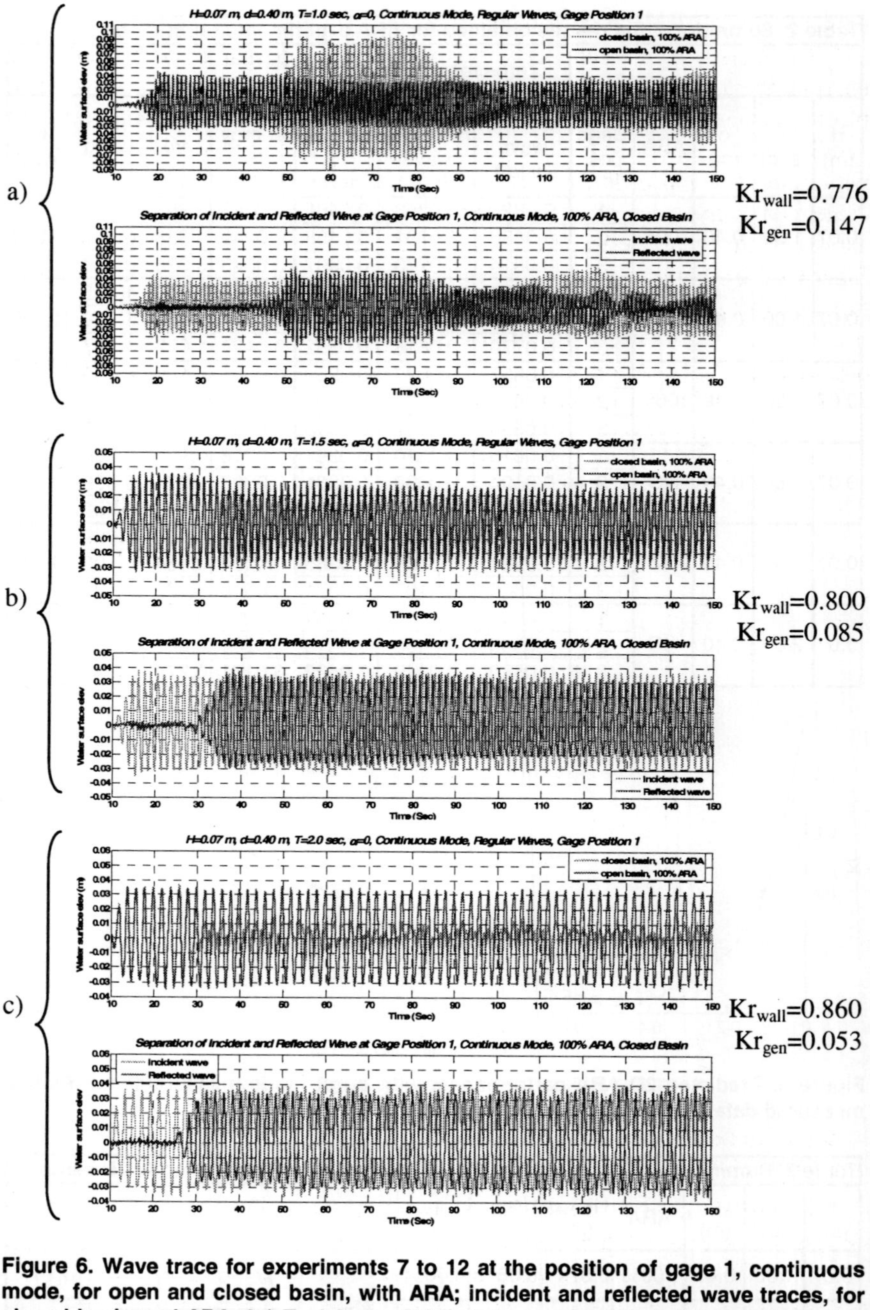

Figure 6. Wave trace for experiments 7 to 12 at the position of gage 1, continuous mode, for open and closed basin, with ARA; incident and reflected wave traces, for closed basin and ARA. 6a) T= 1.0 sec, 6b) T= 1.5 sec, and 6c) T= 2.0 sec. Reflection coefficients Kr are calculated for the reflector wall as well as for the wave generator.

PRACTICAL PERFORMANCE FROM 3D

Three tests for the continuous wave mode were developed to preliminarily measure the performance of the 3D ARA with a 3D setup. The setup consisted in placing concrete blocks in the basin to create a closed area that will cover 32 paddles of the wave maker. The end wall formed an angle of about 30 degrees with respect to the paddles of the wave generator. Within this area, along the center line, an array of wave gages was placed to measure water surface elevation and wave reflection. A detailed view of the setup is illustrated in Figure 7. This configuration was used for a combination of wave height H=0.06 m, T= 1.00, 1.50 and 2.00 sec and d= 0.30 m, with the basin closed and later opened. The waves were monochromatic with a wave incident angle of 30 degrees, the 3D ARA was activated with local and overall ARA gain factors of "1" (100%), and also without ARA (0%).

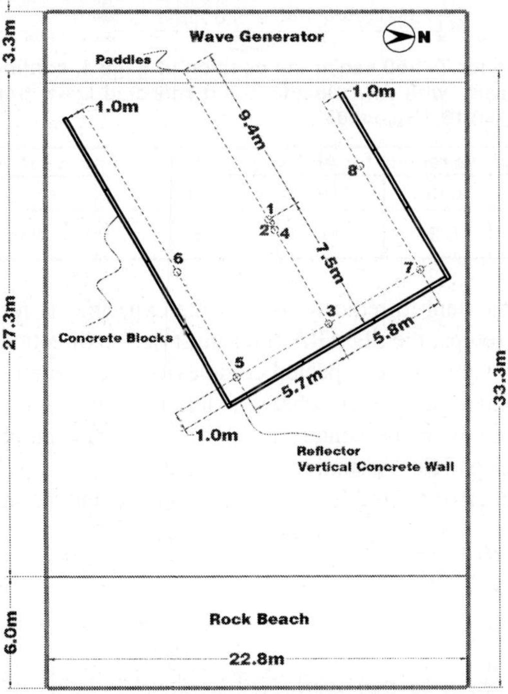

Figure 7. General View of the ARA Test, 3D Setup, conducted at the Haynes Coastal Engineering Laboratory.

The filters used in the HyPCoS control software for the 3D ARA, were designed based on initial tests of the frequency response analysis of the paddles. The 3D ARA signal distribution for the experiments was with Left= 20%, Center= 60%, and Right= 20%. The results are shown for wave period T= 1.50 sec, in Figure 8, and the analysis is presented in Table 4. From the data a reflection coefficient is estimated for the wall of Kr_{wall}=0.83, and a Kr_{gen}=0.10 for the wave generator.

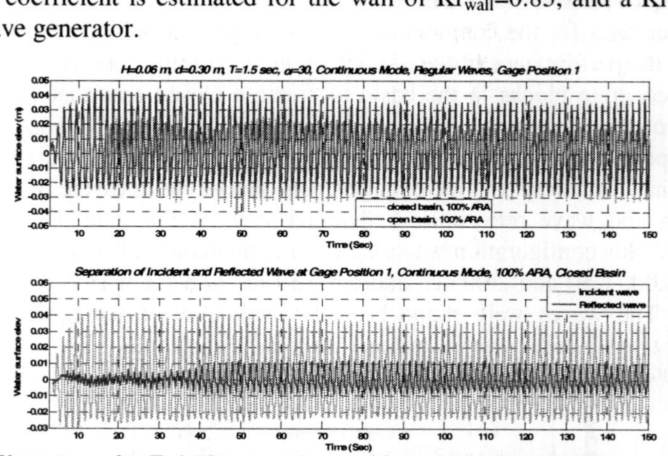

Figure 8. Wave trace for T=1.50 sec at the position of gage 1, continuous mode, for open and closed basin, with ARA; incident and reflected wave traces, for closed basin and ARA. Kr_{gen}=0.10, Kr_{wall}=0.83

Table 4. Summary of the results for ARA experiments continuous mode 3D setup.							
H (m)	T (sec)	d (m)	ARA	Hi_{closed} (m)	Hr_{closed} (m)	Hi_{open} (m)	Kr_{gen}
0.06	1.50	0.30	100%	0.0676	0.0560	0.0620	0.100

Finally it is important to indicate that for the experiments mentioned above for the 2D and 3D setups, the 3D ARA filter must derive directional information from the signals. In view of the expected complexity of this filter, it was initially simplified into a spatial gain mixer, where the left and right adjacent wave height signals are combined with the center signal into a representative wave height signal for the center paddle. Assuming small paddle motion, a relation between wave frequency and error caused by wave angle was developed and is presented in Figure 9.

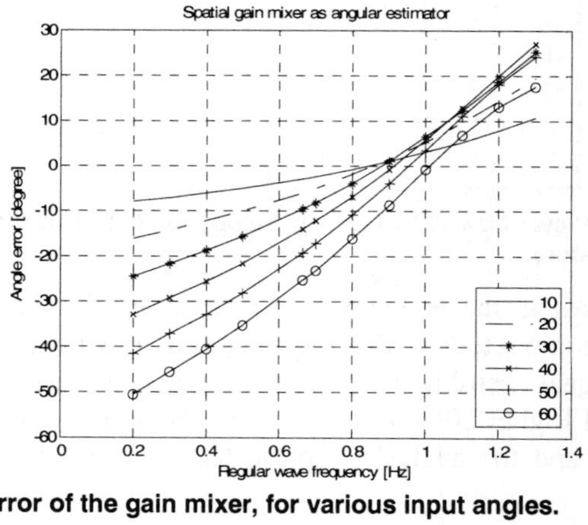

Figure 9. Angle error of the gain mixer, for various input angles.

From Figure 9 it is observed that only at around 0.9 Hz it is somewhat accurate for all directions. For the lower frequencies, the angle is underestimated, therefore the motion reduction will be too small, and absorption performance will suffer. Similarly, for higher frequencies the angle is overestimated, and motion reduction is excessive; absorption performance will again be reduced. Nevertheless, qualitative experience shows that this approach is fruitful (see Figure 8).

CONCLUSIONS AND FUTURE WORK

A series of tests were developed to measure the performance of the 3D ARA for the wave maker located at the Haynes Coastal Engineering Laboratory at Texas A&M University. The results obtained are in reasonable agreement with the theory proposed by Schaffer et al. (1994). However further improvement is needed based on the higher reflection coefficients obtained for the wave period of T=1.0 sec for the 2D setup. Inclusion of evanescent mode activity, better stability and refinement of the 3D filter will be the main topics of future work. Future work is also related to testing a similar 3D setup but with different incident wave angles, allowing better tuning for the ARA for this basin.

REFERENCES

Hiraishi, Tetsuya, Katsuya Hirayama and Haruhiro Maruyama. 1998. Applicability of Dual Face Serpent-type Wave Generator, *Report of the Port and Harbour Research Institute Japan*, Vol 37, No. 4.

Ito, K., Katsui, H., Mochizuki, M. and Isobe, M. 1996. Non-reflected multidirectional wave maker theory and experiments of verification, *Proceedings of 25th International Conference on Coastal Engineering*, ASCE, 443-456.

Mansard, E. & Funke E. R. 1980. The Measurement of Incident and Reflected Spectra Using a Least Squares Method, *Proceedings of the 17th Coastal Engineering Conference*, Sydney, ASCE. 3, 154-172.

Miles, M.D. 1997. GEDAP User's Guide for Windows NT. Canadian Hydraulics Centre, Ottawa, Ontario.

Schaffer, H. A., and Klopman, G. 2000. Review of Multidirectional Active Wave Absorption Methods, *Journal of Waterways, Port, Coastal and Ocean Engeneering,* ASCE, 126(2), 88-97.

Schaffer, H. A., and Stolborg, T., and Hyllested, P. 1994. Simultaneous generation and active absorption of waves in flumes, *Proceedings of the Waves-Physical and Numerical Modeling*, Department of Civil Engineering, University of British Columbia, Vancouver, 90-99.

Schaffer, Hemming A. 2001. Active Wave Absorption in Flumes and 3D Basins, *Proceedings of the Fourth International Symposium Waves: Ocean Wave Measurement and Analysis*, ASCE, 1200-1208.

RISK-BASED PREDICTIONS FOR SHIP UNDERKEEL CLEARANCE

Michael J. Briggs[1], Andrew L. Silver[2], and Leon E. Borgman[3]

This paper describes two new risk-based models for predicting ship underkeel clearance (UKC) in deep-draft entrance channels. In the first model, recurrence intervals are estimated for the number of years between accidents or groundings using Poison and Bernouilli probability distributions for ship arrivals and groundings. The second model predicts channel accessibility based on an acceptable level of risk for different wave, ship, and channel combinations by modeling the uncertainty in these parameters using Gaussian and Rayleigh distributions. Comparisons between both methods are presented

INTRODUCTION

Ports in the United States are currently handling $700B in cargo each year. Annual maintenance dredging amounts to several hundred million cubic yards per year at a cost of over $1B. It is estimated that the present world navigation requirements will exceed channel capacity by the year 2020. In the last decade, containerships and tankers have become much larger and wider. Thus, there is plenty of economic incentive to achieve an optimum design of entrance channels to insure safe navigation while minimizing downtime.

Historically, port designers relied on deterministic approaches with large safety factors for channel design (McCartney 1993, PIANC 1997). Recent trends are to incorporate a risk-based methodology in the design process to minimize dredging and define a useful lifetime with an acceptable level of risk of accidents or groundings (McBride et al. 1998, ROM 1999). The PIANC Maritime Commission is undertaking a new working group (WG49) to update the 1997 Guidelines on vertical and horizontal dimensions of fairways.

This paper describes two risk-based models for entrance channel design. The first model is based on Poisson and Bernoulli probability laws for ship arrivals and groundings using laboratory data for a limited dataset of channel configurations and environmental conditions. The second model is a channel analysis and design tool (CADET) that is based on Gaussian and Rayleigh distributions and an Ochi extremal analysis prediction of underkeel clearance (UKC) from ship motions allowances. The paper describes the project site at Barbers Point Harbor, HI. The laboratory setup including harbor model, C9 containership, environmental conditions, and navigation study procedures are described in the next section. The probability assessment method (PAM) model is described

---

[1] U.S. Army Engineer Research and Development Center, Coastal and Hydraulics Laboratory, 3909 Halls Ferry Rd., Vicksburg, MS 39180-6199
[2] Naval Surface Warfare Center, Carderock Division, Seakeeping Department, 9500 MacArthur Blvd., West Bethesda, MD 20817-5700
[3] Borgman Inc., 2526 Park Avenue, Laramie, WY

in the next section and includes descriptions of the Poisson and Bernouilli probability laws used for the number of ship arrivals and groundings, the combined laws, and the recurrence interval calculations. The CADET motions analysis, I/O parameters, and an example are presented in the next section. Finally, a summary and conclusions compares the two different models at the project site.

PROJECT SITE

Fig. 1 shows Barbers Point Harbor (BPH) on the southwest coastline of Oahu, HI. It consists of a deep-draft harbor, barge basin, resort marina, and entrance channel. The deep-draft harbor has a depth of 11.6 m. The entrance channel is approximately 1,006 m long, 137.2 m wide, and 12.8 m deep. It is aligned in a northeast/southwest (harbor to offshore) direction. The 1-way channel was designed to accommodate containerships, tankers, and bulk carriers.

Harbor modifications were proposed in seven different plans. However, only two plans are relevant for this paper: the existing plan 1a and the recommended plan 4c (Fig. 2). Plan 4c has (a) a deeper harbor of 13.7 m, (b) a 335 by 335 m harbor expansion, (c) a deeper channel of 15 m, (d) a wider approach channel in the outer 305 m, with a flare from the existing channel width of 137.2 to 228.6 m, and (e) a 137-m-long, rubble-mound, shore-connected jetty. The channel depth for plan 4c was selected based on existing PIANC and Corps guidance.

LABORATORY SETUP

Harbor Model

Fig. 2 the 1:75 scale laboratory model of Barbers Point Harbor for plan 4c. It included offshore contours to 30 m and wave absorption along the perimeter. Waves were generated with the directional spectral wave generator (DSWG). The 27.4-m-long electromechanical DSWG consisted of 60 piston-mode paddles, each 46 cm wide and 76 cm high. Twenty capacitance wave gages were used to calibrate incident directional wave spectra and measure harbor response. A pump, PVC piping, and multiport diffusers were calibrated to generate tidally-driven, longshore currents.

C9 Design Vessel

The design vessel was the APL's C9 *President Lincoln* containership. It has an overall length of 262 m, a beam of 32.2 m, and a draft of 10.7 m. The 1:75 scale model C9 had a length of 3.5 m, beam of 43 cm, and a draft of 14.2 cm. Remote control features included forward and reverse speeds, rudder angle, and bow-thruster direction and speed. It was statically and dynamically balanced.

Environmental Conditions

Environmental conditions selected for the model tests included waves, longshore currents, and winds. Each was divided into three categories, A, B, or C (reflecting frequency of occurrence and severity), to define the different combinations of waves, currents, and winds that might be present when the ship is transiting the entrance channel. Category (Cat.) A was composed of "extreme"

Figure 1. Project site at Barbers Point Harbor, HI.

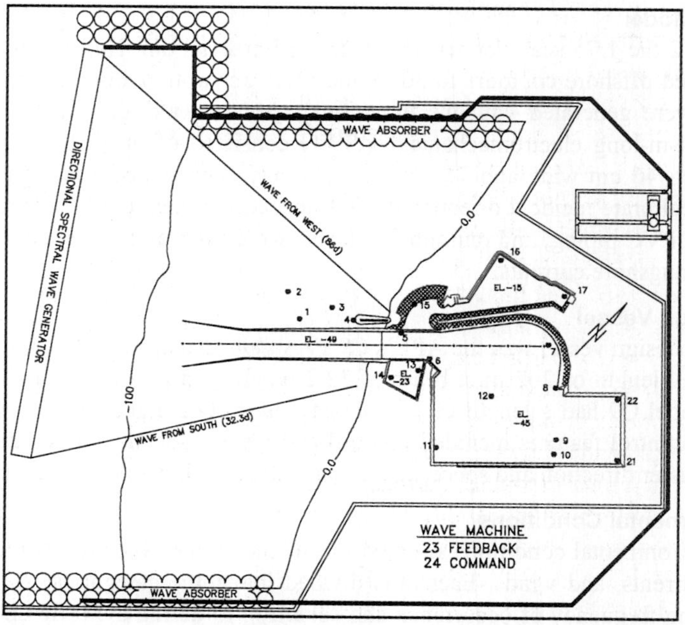

Figure 2. Model of Barbers Point Harbor for recommended plan 4c.

environmental conditions that have a low frequency of occurrence and occur about 3 to 5% of the time. Cat. B ("normal operational" conditions) was characterized by frequently occurring environmental conditions that would influence navigation on a weekly basis. Cat. C was the "gentle" category that occurs daily and has a negligible influence on navigation.

Waves. Wave data were collected over a 4 year period from 1986 to 1990 by Scripps Institution of Oceanography (SIO) using a four-gage S_{xy} array. Primary wave directions are from the NW and SW. The largest waves, around 3 m, occur during the winter months and are caused by a north swell generated in the Northwest Pacific. Nine model waves were selected from measured prototype wave conditions in Cat. A and B. Cat. A waves were characterized by wave heights greater than 2 m and wave periods greater than 11 sec. Cat. B waves began with wave heights greater than 1 m and wave periods greater than 11 sec.

Longshore currents. Longshore current data were collected over a 65-day period in 1988 at two channel locations (Sea Engineering 1988). Samples were taken at the surface and at three mid-depths to quantify flows that could affect the ship. The measured field data indicated a bi-directional, tidally-driven, longshore current flowing parallel to the bottom contours. Cat. A conditions (0.65 to 0.80 kts) occur less than once a month with a level of exceedance less than 0.1% of the time. Cat. B conditions represents a current (0.50 kts) that could occur a little more than once a day and could be exceeded only 5% of the time. Cat. C (0.35 kts) corresponds to conditions frequently occurring in the prototype, which a ship must be able to handle on a daily basis. Model currents were calibrated using dye, surface floats, and ultrasonic, triaxial current meters

Winds. Wind information was obtained from a 19-year dataset collected at the Barbers Point Naval Air Station from 1949 to 1967. Wind speed is defined by the Beaufort scale from 1 to 9 (B1 to B9). Winds greater than B5 were designated as Cat. A, between B3 and B5 as Cat. B, and less than B3 as Cat. C. Cat. A, B, and C winds for all directions occur about 4, 31, and 17 weeks a year, respectively. For this study, only Cat. A winds were tested.

Next available transit environment. The environmental wave, longshore current, and wind conditions were separated into Cat. A, B, or C for extreme, normal, and gentle conditions, respectively. Each of these environmental conditions had unique environmental probabilities for each category, based on the measured data. The next step was to define what mixture of environmental conditions is included during a ship transit at time t. Extreme (Env=1), normal (Env=2), and gentle (Env=3), "next available transit environments" were defined for this unique mix of waves, currents, and winds in the model tests. The final overall environmental probability P[Env = j] for each of the three transit environments was based on some simplifying assumptions and engineering judgment.

Initially, all possible combinations were considered. However, some were discarded since they are physically unlikely or impossible. In Env=1, waves in Cat. A were selected as dominant. They were combined with Cat. A and B currents and winds. Any Cat. C currents and winds were considered insignificant in

Env=1. Thus, the probability for transit environment 1 (Env=1) was equal to that for waves in Cat. A, or P[Env=1]=P[Wav_A]=0.021.

In transit environment Env=2, waves were still dominant, but longshore currents were judged to have a significant impact. In the model tests, Cat. B waves and currents were combined. Thus, the probability for Env=2 was equal to the probability of Cat. B waves (P[Wav_B]) times the conditional probability of currents and winds in Env=2, given that waves in Cat. B were present. Thus, P[Env=2] = P[Wav_B] $*$ P[Currents and Winds in Env=2 | Wav_B]= 0.284 (0.7) = 0.199. The derivation of this conditional probability was based on linear programming assuming estimates of the probabilities for waves and currents.

Finally, the transit environment 3 (Env=3) probability is equal to the remainder given by 1 − P[Env = 1] − P[Env = 2] = 0.780. A summary of the environmental probabilities in the three transit environments Env=1,2 or 3 is:

$$\begin{aligned} P[Env = 1] &= 0.021 \\ P[Env = 2] &= 0.199 \\ P[Env = 3] &= 0.780 \end{aligned} \qquad (1)$$

Laboratory Navigation Study

Test procedures. The different combinations of waves, longshore currents, and winds from the three Cat. A, B, and C were combined in Env=1, 2, or 3 and run in both plans 1a and 4c for inbound and outbound transits of the C9 containership model. A total of 414 inbound and 75 outbound transits were completed. More inbound runs were completed since they were more difficult than outbound runs. In general, the same person from our staff conducted all the model runs and local pilots from Hawaii confirmed the accuracy (Fig. 3).

Test criteria. An acceptable run occurred if the ship did not hit bottom or get within half a beam width of the channel sides. Acceptable and unacceptable (groundings) runs were grouped for both plans for inbound and outbound runs for the different combinations of waves, longshore currents, and winds. An ideal ship track followed the channel centerline. Observations from the ship operator and video recordings were used to evaluate success or failure of each run. The video recordings permitted a more objective evaluation.

Model results. For plan 1a, a total of 77 inbound tests were conducted. Video analysis showed that 42% of these runs were unacceptable. For plan 4c, 90 inbound tests indicated that 93% of the runs were acceptable in Cat. A.

PROBABILITY ASSESSMENT MODEL (PAM)

The first risk-based model is the probability assessment method (PAM) proposed by Borgman (Briggs, et al. 2003). The unacceptable runs were converted into a statistical representation of the expected number of groundings based on frequency of occurrence and environmental severity during a transit. This four-component probability model included (a) a Poisson probability law for number of ship arrivals, (b) a Bernouilli probability law for grounding in a single random ship arrival, with an estimation of the probability parameter in the Bernouilli law from laboratory data, (c) a compound Poisson and Bernouilli probability law for total number of ship groundings in a particular time interval, and (d) a determination of recurrence intervals or return periods between groundings.

Figure 3. Model testing with C9 containership during staff series.

Poisson Probability Law for Number of Ship Arrivals

The first step in the PAM model is to define a distribution for the number of ship arrivals S=n in an interval of t years (0 ≤time ≤t) for a fixed expected rate of arrival λ. The Poisson distribution is a good candidate since it is used in queuing problems that involve discrete events in a continuous interval of time.

$$P[S=n]=\frac{e^{-\lambda t}(\lambda t)^{n}}{n!} \tag{2}$$

BPH has a short, 1-way channel with limited berths in the harbor, so it experiences only limited transits. These transits can be considered independent random processes as long as the time between them is much longer than the duration of environmental conditions (i.e., no change during the passage of a ship).

Plans 1a and 4c have berths for two and four C9 containerships, respectively. Realistic estimates of normal and full utilization for plan 1a are 100 and

200 ship calls/year, respectively. Similarly for plan 4c, normal, medium, full, and maximum utilization rates are 200, 300, 400, and 500 ships/year, respectively.

Bernouilli Probability Law for Ship Groundings

A Bernouilli random variable G was defined such that G=1 if the ship experiences a grounding and G=0 otherwise.

$$P[G=1]=p_{g,j}$$
$$P[G=0]=1-p_{g,j} \tag{3}$$

where $p_{g,j}$=probability of a grounding for a single random ship arrival (SRSA) and Env=j. The SRSA requirement was relaxed and simplified since a weighted average of the probabilities for each environment for the random arrival time was assumed. Thus, $p_{g,j}$ is defined as the product of the joint probability P[G=1 | Env=j] that there is a grounding given Env=j is present times the probability P[Env=j] from Eq. 1 that Env=j is present during the SRSA:

$$p_{g,j}=P[G=1|Env=j]P[Env=j] \tag{4}$$

Because of the limited number of environmental cases and inbound and outbound runs in the model tests, engineering judgment was used to simplify this joint probability:

$$P[G=1|Env=j]=\sum_{k=1}^{K} Wgt(k)\{Grd\,|\,Env=j\} \tag{5}$$

where Wgt(k)=weighting function based on the maximum fraction of a year a ship could be exposed to grounding during a single transit=2.85×10^{-5}; k=number of selected environmental conditions representing both inbound and outbound runs; and {Grd | Env=j}=grounding fraction from the model tests in Env=j. The grounding fractions from the model tests were used to calculate a "best," "average," and "worst" case scenario within each transit environment. The "best" case was the minimum fraction of groundings recorded in the model tests. Similarly, the "worst" case was the maximum fraction recorded. The "average" was the average of these two values. Finally, the net P[G = 1 | Env = j] for a round trip in a transit environment was the sum of inbound and outbound probabilities.

Compound Poisson/Bernoulli Probability Law for Total Groundings

The next step was to define the probability for the number of ship groundings $P[N_{g,j}=n_g]$ occurring in time interval t when Env=j as a compound Poisson/Bernoulli probability law given by:

$$P[N_{g,j}=n_g]=\frac{e^{-\lambda t p_{g,j}}(\lambda t p_{g,j})^{n_g}}{n_g!} \tag{6}$$

where λt=mean number of ship arrivals in interval t, and n_g=number of groundings. Thus, the expected number of groundings $E[N_{g,j}]$ in Env=j is given by:

$$E[N_{g,j}] = \lambda t p_{g,j} = (\lambda t)P[G=1 | Env=j]P[Env=j] \quad (7)$$

Finally, the expected total number of ship groundings for all three transit environments $E[N_g]$is the sum of the individual $E[N_{g,j}]$:

$$E[N_g] = \lambda t p_g = \sum_{j=1}^{3} E[N_{g,j}] \quad (8)$$

Recurrence Intervals

The last step in this model is to estimate the recurrence interval or return period between groundings. Return periods are defined as the expected number of years between groundings, and are the inverse of the expected number of groundings in t-years. Fig. 4 shows the number of years between groundings for plans 1a and 4c for the different levels of port utilization.

An example for the "best" case scenario for existing plan 1a with 100 ship calls/year will help clarify this procedure. Using Eq. 7 for inbound ships in Env=1, the $E[N_{g,1}]_{In}=100*(1.20\times10^{-5})*0.021=2.5\times10^{-5}$. For outbound ships, $E[N_{g,1}]_{Out}=100*(0.09\times10^{-5})*0.021=0.2\times10^{-5}$. Thus, the combined number of groundings for both inbound and outbound transits in Env=1, $E[N_{g1}]=2.7\times10^{-5}$. Similarly, for inbound ships in Env=2, $E[N_{g,2}]_{In}=100*(0.31\times10^{-5})*0.199 = 6.0\times10^{-5}$. For outbound ships in Env=2, $E[N_{g,2}]_{Out}=100*(0.03\times10^{-5})*0.199= 0.4\times10^{-5}$. The combined number of groundings in Env=2, $E[N_{g,2}]=6.4\times10^{-5}$. The combined number of groundings in Env=3 is defined to be $E[N_{g,3}]=6.0\times10^{-5}$, a constant for both inbound and outbound ship runs. Unacceptable navigation in Env=3 is assumed to be due to circumstances other than environmental conditions. This value is based on worldwide experience and represents a ship accident 6 times every 100,000 years or once every 16,667 ship calls (Bratteland 1988). Thus, using Eq. 8 the total number of ship groundings $E[N_g]$ for all three transit environments is 15.1×10^{-5}. Inverting this expected number, one obtains a total of 6,631 ship calls between groundings. This corresponds to one grounding every 66 years for 100 ships calling the port every year.

The "best" $E[N_g]$ for recommended plan 4c for 100 ship calls/year is $7.4\text{x}10^{-5}$, or 13,451 ship calls between groundings. Again, this corresponds to one grounding every 135 years. The plan 1a grounding values are much higher because the existing harbor and entrance channel was never designed for the larger C9 containerships, and their deeper channel depths.

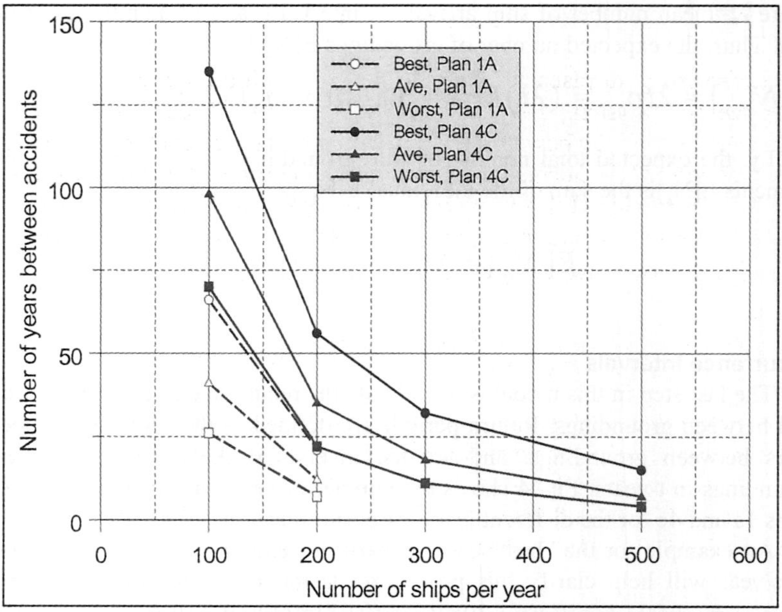

Figure 4. C9 containership recurrence intervals using the PAM model.

CHANNEL ANALYSIS AND DESIGN EVALUATION TOOL MODEL

The second risk-based model is the Channel Analysis and Design Evaluation Tool (CADET) that provides accessibility of deep-draft ships for a range of channel depths. From this information, a realistic depth for the channel can be selected. The theory behind CADET is based on the Navy's Environmental Monitoring and Operator Guidance System (EMOGS). EMOGS operationally assists deep draft US Navy ships that are required to transit shallow entrance channels at their home ports. EMOGS runs in near-real-time, and calculates the effective clearance of the deep-draft ship based on values of the estimated channel depth at mean lower low water (MLLW), predicted change in water level from MLLW due to astronomic tides and meteorological conditions, estimated static draft during transit, predicted sinkage and trim at speed, and a motions allowance. The details of EMOGS can be found in Silver and Dalzell (1997) and the adaptation of the theory in CADET is explained in Kopp and Silver (2004).

Motions Allowance

The motions allowance is defined within EMOGS and CADET as the extreme, wave-induced vertical displacement of the ship during a given transit of the channel where the risk of exceeding this value is 1 in 100. This allowance is calculated by first considering the vertical ship motions at the bow and stern, which are the coupled sum of heave and pitch motions represented by:

$$H_j = z + X_j\theta \tag{9}$$

where H_j=vertical displacement transfer function with the subscript j identifying either the bow or stern, z=vertical heave motion, X_j=distance from the ship's center of gravity to the bow or stern, and θ=pitch transfer function. The variance of the motions σ_j^2 is calculated by integrating the product of the square of the modulus of the vertical displacement transfer function $|H_j(\omega,\mu)|^2$, referred to as the ship's response amplitude operator (RAO), with the directional wave spectrum $S(\omega,\mu)$ over frequency ω and direction μ:

$$\sigma_j^2 = \int_0^{2\pi}\int_0^{\infty} S(\omega,\mu)\left|H_j(\omega,\mu)\right|^2 d\omega d\mu \tag{10}$$

Because the directional wave spectrum used in the variance does not necessarily represent the actual extreme waves the ship will experience during a transit of the channel, motions allowance is used in the calculation to represent an expected extreme motion during a given transit. This motions allowance A_j is calculated using Ochi's (1973) theory of extreme values:

$$A_j = \sigma_j \sqrt{2\ln\left[\frac{L_r \sigma_{v,j}}{2\pi\alpha U \sigma_j}\right]} \tag{11}$$

where σ_j, =rms motion at the bow or stern, $\sigma_{v,j}$ =rms time derivative of the motion, or velocity, at the bow or stern, L_r =length of the reach of the channel, U=ship speed, and α=risk parameter, taken as 100 in EMOGS and CADET.

CADET I/O Parameters

Several critical parameters in CADET must be computed and prepared as lookup tables before running the program. They include underway sinkage at the bow and stern, heave, pitch and roll RAOs, and the local wave climatology. The values of other parameters, including the ship's static draft and trim, the range of ship speeds in the channels, the channel orientation, length, and the range of channel depths, are supplied as input to CADET.

The output from CADET provides a channel accessibility estimate on a yearly basis for a range of channel depths. The channel accessibility is based on both the annual occurrence of waves with significant height, period and direction from the local wave climatology and the risk of touching project depth for each of these wave conditions. The risk of touching project depth is defined as the proportion of all transits, under statistically constant conditions, where the minimum instantaneous UKC is less than zero. The project depth definition assumes that the channel bottom is perfectly flat. The risk model accounts for the uncertainty in each of the parameters by assuming a Gaussian distribution for static ship draft, and underway sinkage and trim, and a Rayleigh distribution for the vertical motion and velocity variances. The Rayleigh distribution reflects the most likely probability distribution of the waves. Using these distributions, the probability density of the largest motion excursion or the minimum UKC, is determined and its area up to a minimum clearance of zero is calculated.

Barbers Point Harbor Example

The CADET was used to compare predicted channel depth to the PAM model results for the recommended plan 4c. The same BPH channel configuration and *President Lincoln* containership were used. Instead of one channel depth, a range of channel depths from 12.2 to 15.2 m was specified.

As previously stated, CADET requires tables of pitch, heave, and roll RAOs and underway sinkage and trim. The RAOs for the *President Lincoln* in shallow water were calculated using the finite depth version of SCORES II (Kaplan, 1996). The sinkage values at the bow and trim of the *President Lincoln* were generated for 3, 5, and 7 knots and the range of channel depths in 0.3 m increments using a program based on Beck, Newman and Tuck (1975) algorithms.

The same SIO measured wave data at BPH that was used for the PAM model was used for this comparison. However, the entire 5-year duration was used since CADET can easily process 1 year or 20 years of measured or hindcast wave data. On the SIO website (http://cdip.ucsd.edu/) data are listed in joint probability matrices by wave period and height, wave period and direction, and wave direction and height. These three matrices were combined into one matrix covering the joint distributions of wave height, period, and direction and the number of occurrences of each combination over the dataset. The wave height and period matrix was then divided into seven separate matrices, each representing a total band of directions covering 245°. This procedure yielded 130 different wave conditions with probability of occurrence within the dataset. The spectral energy of each of these wave conditions was spread over frequency and direction to represent a directional JONSWAP spectrum for limited fetch nearshore conditions.

The CADET was then run for the *President Lincoln* over the range of channel depths, ship speeds, and inbound and outbound transits for the specified wave conditions. The output showed the days/year the ship can safely transit the channel for a given project depth and ship speed. This is based on the percent occurrence of wave conditions that yielded a less than 1% risk of the ship touching the specified project depth. As shown in Table 1, the optimum depth is between 13.7 and 14.0 m. This is significantly shallower than the 15 m used in the PAM model, and results in lower dredging costs.

SUMMARY AND CONCLUSIONS

This paper has presented two risk-based models for deep-draft entrance channel design. The first model consists of a probability assessment method (PAM) for assessing the expected value of a ship grounding in an entrance channel for different recurrence intervals using physical model data. This model

Table 1. CADET Output Channel Accessibility in days per year.

Project Depth (m)	3 knots	5 knots	7 knots
12.2	185	222	31
12.5	269	295	156
12.8	320	311	294
13.1	346	350	346
13.4	359	361	362
13.7	363	364	365
14.0	365	365	365
14.3	365	365	365
14.6	365	365	365
14.9	365	365	365
15.2	365	365	365

included a Poisson probability law for number of ship arrivals, a Bernouilli probability law for grounding in a single random ship arrival, a compound Poisson and Bernouilli probability law for total number of ship groundings in a particular time interval, and a determination of return periods between groundings.

The second model was the CADET channel design tool that predicts the accessibility of deep-draft ships for a range of entrance channel depths. It is based on probabilistic risk analysis techniques to evaluate the accessibility of a series of channel reaches for multiple vessel geometries and loading conditions.

In the PAM model, a remote-controlled C9 containership was used to evaluate the navigability of the existing (plan 1a) and recommended (plan 4c) configurations for Barbers Point Harbor (BPH), Hawaii. For 100 ship calls per year, plan 4c increased the years between groundings from 66 to 135 years by increasing the channel depth from 12.8 to 15.0. The CADET predicted good accessibility for a significantly reduced depth of 13.7 to 14.0 m. The PAM depth was based on existing PIANC and Corps guidance. The CADET has the advantage that it allows the designer to evaluate more depth options than the PAM model. However, it is only based on vertical depth and not horizontal width considerations that were included in the PAM results.

The PAM model was based on physical model data that is necessarily limited in the number of runs that can realistically be included. Even so, it provided a rational criterion for evaluating the various channel configurations (both channel depth and width) for a range of realistic environmental and annual number of ship calls. Of course, numerical or ship simulator data could be used in the PAM model. The CADET has the advantage that it allows the planner to determine the "optimum" channel depth, given acceptance of a user-determined level of risk of grounding and reduced accessibility. The PAM model might be more useful for insurance rating purposes where one needs to know the number of groundings for a particular channel configuration over its lifetime; whereas, the CADET model could be used while the channel is being designed to optimize its usefulness or cost benefit ratio.

ACKNOWLEDGMENTS

The authors wish to acknowledge the Headquarters, U.S. Army Corps of Engineers, and the Naval Surface Warfare Center, Carderock Division for authorizing publication of this paper.

REFERENCES

Beck, R.F., J.N. Newman, and E.O. Tuck, 1975, Hydrodynamic forces on ships in dredged channels, *Journal of Ship Research*, Vol. 19, No. 3.

Bratteland, E. (Ed.)1988. *Advances in berthing and mooring of ships and offshore structures*, NATO ASI Series, Kluwer Academic Publishers, Dordrecht.

Briggs, M.J., L.S. Lillycrop, G.S. Harkins, E.F. Thompson, and , D.R. Green. 1994. Physical and numerical model studies of Barbers Point Harbor, Oahu, Hawaii, Technical Report CERC–94-14, U.S. Army Engineer Waterways Experiment Station, Vicksburg, MS, Sep.

Briggs, M. J., L E. Borgman, and E. Bratteland. 2003. Probability assessment for deep draft navigation channel design, *Elsevier Coastal Engineering Journal*, 48, 29-50.

Briggs, M.J., A.L Silver,. and P.J. Kopp. 2004. CADET: A tool for predicting underkeel clearance in deep-draft entrance channels, *Proceedings of the CEO6 Conference*, Baltimore, MD, Oct.

Kaplan, P. 1996. Sample calculations and verification of SCORES II – finite depth program, Hydrodynamics, Inc. Report No. 96-101B.

Kopp, P.J. and A.L. Silver. 2004. Program documentation for the Channel Analysis and Design Evaluation Tool (CADET), Naval Surface Warfare Center, Carderock Division, Hydromechanics Directorate Report, NSWCCD-50-TR-2004/004, May.

McCartney, B. 1993. Report on ship channel design, ASCE Manual and Report No. 80.

McBride, M.W., J.V. Smallman, and S.W. Huntington. 1998. Guidelines for design of approach channels, Ports 98, Long Beach, CA, 1315-1324.

Ochi, M.K. 1973. On prediction of extreme values, *Journal of Ship Research*, Vol 17, No. 1.

PIANC. 1997. *Approach channels: a guide for design*, Supplement No. 95, II-30(2).

ROM. 1999. Recommendations for maritime works – ROM 3.1-99, Puertos del Estado, Spain.

Sea Engineering, Inc. 1988. Barbers Point entrance channel current study, Waimanalo, HI, Technical Report No.8, prepared for the Estate of James Campbell, Waimanalo, HI, Oct.

Silver, A.L. and J.F. Dalzell. 1997. Risk based decisions for entrance channel operation and design, *Proceedings of the 7th ISOPE Conference*, 815-822, May.

Swash, Nearshore Currents, and Long Waves

San Diego and Mission Bay, California,
(Photo courtesy of US Army Corps of Engineers, Los Angeles District)

SWASH ZONE BOUNDARY CONDITION: THE LONGSHORE FLOW.

M.Antuono[1] , M.Brocchini[2] and G.Grosso[3]

The aim of the present work is to analyze in detail flow motions within the swash zone and define suitable Shoreline Boundary Conditions for the longshore flow for wave-averaged circulation models. An analytical solution for the longshore motion representing the drift velocity of the whole swash zone water mass is found. A complete set of Shoreline Boundary Conditions is obtained on the basis of fully-numerical solutions of the Nonlinear Shallow Water Equations. The main focus of the work is to clarify the structure of the Shoreline Boundary Conditions for the longshore flow but care has also been placed on their derivation and assessment from the numerical solutions. The latter have been obtained on the basis of a fairly broad range of input wave conditions.

Introduction

Swash zone dynamics deeply influence the surf zone hydro- and morphodynamics . The role played by such dynamics on the longshore drift and sediment transport is crucial. Moreover swash zone flows are of fundamental importance not only because of their local effects, but also because they influence the surf zone dynamics as a whole (e.g. Brocchini, 2006). For this reason, any models reproducing flow motions in the nearshore should not neglect swash zone dynamics. However, problems arise because wave-averaged circulation models (averaging over the typical period of wind waves) cannot deal with swash zone flows directly. Very often a pragmatic perspective is taken for which a "wall boundary", i.e. a perfectly reflecting vertical wall positioned at a given cross-shore location (usually the still-water shoreline) is used. This approach, however, significantly alters the natural flow pattern of seaward-reflected waves (e.g. Brocchini, 2006). Hence, the need of suitable Shoreline Boundary Conditions (SBCs) which take into account the presence of the swash zone. If the wet/dry boundary is taken as the intersection of the mean water level with the beach face, both theoretical and practical problems arise: flow integration would occur also during periods of local dry conditions and wave-wave interactions cannot be properly accounted for. Analysis of different definitions for the mean shoreline made by Brocchini and Peregrine (1996), has revealed that such an interface cannot be uniquely defined; but problems can be overcome if the boundary between wet and dry is taken as the envelope of the rundown positions, since flow properties can be unambiguously defined in the

[1] D.I.Am, Università di Genova, via Montallegro 1, Genova, 16145, Italy

[2] Istituto di Idraulica e Infrastrutture Viarie, Università Politecnica delle Marche, via Brecce Bianche 12, Ancona, 60131, Italy

[3] D.I.Am, Università di Genova, via Montallegro 1, Genova, 16145, Italy

wet region. The main aim of our work is to define, along such envelope, suitable SBCs for wave-averaged circulation models which take into account the proper dynamics of the swash zone. The main challenge in describing SBCs for the longshore flow is represented by (i) making use of an appropriate definition of the swash zone drift velocity and (ii) the correct modelling of the continuous forcing of the longshore velocity component caused by nonlinearities and wave breaking.

The integral model

The model we consider makes use of the Nonlinear Shallow Water Equations (NSWE) obtained assuming that vertical accelerations of the water or those normal to the beach are negligible compared with gravity. We introduce basic definitions, choosing the still water level to be z=0 (z pointing upward), the onshore coordinate x to point landward with x=0 at the still water shoreline and the total water depth to be

$$d(x,y,t)=h(x,y)+\eta(x,y,t) \tag{1}$$

where $z=-h(x,y)$ is the seabed location, $z=\eta(x,y,t)$ is the position of the free surface and y is the longshore coordinate oriented to make with x and z a right-handed cartesian reference frame. The onshore and longshore velocity components $u(x,y,t)$ and $v(x,y,t)$ are the depth-independent horizontal velocities. We assume motions to occur over an uniformly-plane beach with mild longshore variations. Then we can suppose $h_{,x} = \alpha$ and $h_{,y} = 0$ (the notation $(.)_{,i}$ is used for partial differentiation with respect to the variable i). The dimensional, general set of equations in conservation form is (see Brocchini and Peregrine, 1996):

$$d_{,t} +(ud)_{,x} +(vd)_{,y} = 0, \tag{2,a}$$
$$(ud)_{,t} + (u^2d + g/2\, d^2)_{,x} +(uvd)_{,y} +\alpha g d + \tau_1 = 0, \tag{2,b}$$
$$(vd)_{,t} +(uvd)_{,x} + (v^2d +g/2\, d^2)_{,y} +\tau_2 = 0, \tag{2,c}$$

where g is gravity acceleration and $\tau=(\tau_1, \tau_2)=(C_f|u|u, C_f|u|v)$ is the bed friction evaluated, in terms of the friction coefficient C_f, on the basis of a steady flow approach. In our model, the swash zone limits are x_l (seaward or lower limit) and x_s (instantaneous shoreline i.e. shoreward or higher limit for integration). The seaward boundary is taken as the lowest limit of the shoreline (see figure 1). The above equations are integrated over the swash zone with the assumption that wave motion evolves on a "fast time scale" T_s (short-period waves), while the lowest swash zone boundary, x_l (run-down), is supposed to vary on a "slow time scale" $T_s^{(2)}$ (long-period waves). We identify the "fast time scale" T_s with the typical period of each single incident wave and the "slow time scale" $T_s^{(2)}$ with the period of modulation of incident short waves. In accordance with the assumptions used in Brocchini and Peregrine (1996) to formulate the integral swash zone model we take that swash motions are almost entirely governed by short-wave contributions and that the only long-wave contribution comes from parameterising the longshore drift due to wave-breaking by a longshore current velocity W=W(y,t). Hence, we use the following decomposition:

$$u = \frac{\partial x_l}{\partial t} + \hat{u}, \tag{3}$$

where û is the cross-shore velocity inside the swash zone. Since we aim atdefining a longshore drift velocity W for the entire swash zone a slightly different decomposition, which makes use of the spatial average over the swash width, is employed.

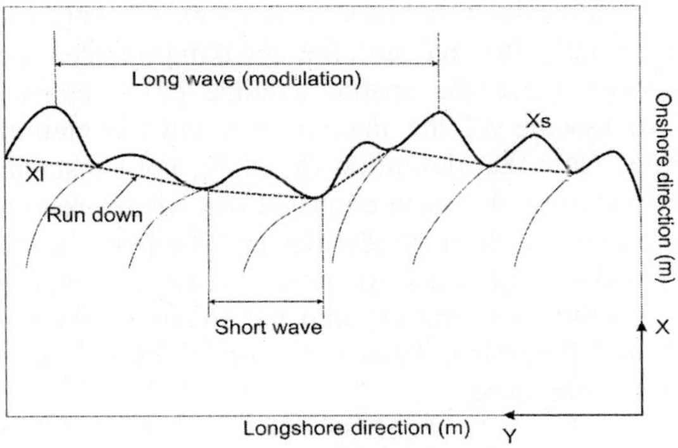

Figure 1: Sketch of the flow evolution of the swash zone at a given time. Bimodal waves evolve over a uniformly-sloping beach. The instantaneous, local shoreline (x_s), the envelope of the run-down positions (x_l) and an indication of the bore paths (thick dashed lines) are illustrated.

We thus introduce the swash-averaging operator M{·} which applied to the generic function f , leads to:

$$M\{f\} \equiv \frac{1}{x_s - x_l} \int_{x_l}^{x_s} f \, dx \, . \tag{4}$$

Hence, with a definition similar to that of a Stokes drift-type velocity (e.g. Brocchini, 1997), we put:

$$v = W + \acute{v} \qquad \text{in which} \qquad W \equiv \langle M\{v\} \rangle \quad \text{and} \quad \langle M\{\acute{v}\} \rangle = 0\,, \tag{5}$$

in which < > indicates the average over the short-period waves. Note that, differently from (3), the decomposition in short and long waves of the longshore momentum equation is made after the spatial average over the swash is performed. Hence, the velocity fluctuation $\acute{v}$ has to obey an integral constraint as given in (5). Moreover we assume that only W can take into account the action of bores eventually forcing the swash dynamics.

Defining t_k the k-th instant for which $x_s=x_l$, if $|t-t_k| << 1$, we can write $x_s=x_l+\varepsilon$ with $\varepsilon<<1$ and, consequently, we have:

$$\int_{x_l}^{x_s} f\,dx = \int_{x_l}^{x_l+\varepsilon} f\,dx \equiv F(\varepsilon) = F(0) + \frac{dF}{dx}(0)\varepsilon + \mathcal{O}(\varepsilon^2) \simeq f\Big|_{x_l}(x_s - x_l). \tag{6}$$

Finally, using (6), it follows:

$$\lim_{t\to t_k} W = \lim_{t\to t_k}\langle M\{v\}\rangle = \Big\langle v\Big(x_s(t_k), y_s(t_k), t_k\Big)\Big\rangle. \tag{7}$$

in which y_s indicates the longshore coordinate of the shoreline.

Since a bore eventually occurs between two subsequent swashes, from (7) we obtain $W \rightarrow \langle v_k\rangle^-$ as $t \rightarrow (t_k)^-$ and $W \rightarrow \langle v_k\rangle^+$ as $t \rightarrow (t_k)^+$, so that the jump in the long-period component of the longshore velocity is $[[\langle v\rangle]] \equiv \langle v_k\rangle^+ - \langle v_k\rangle^- = [[W]]$. Finally, since we can assume $[[\langle v\rangle]] \approx [[v]]$, such result confirms that W can take into due account the breaking-induced jump in longshore velocity. Moreover, since the spatial average of $\acute{v}$ represents a short-wave contribution, we assume all the integral quantities containing $\acute{v}$ to represent short-wave terms. Note that definition (5) of W, notwithstanding its simplicity, cannot explicitly express the main contributions we are searching for: the local instantaneous forcing due to nonlinearities and the periodic momentum transfer due to wave breaking. In order to have a clear decomposition of the flow dynamics, we substitute (3) and (5) into the equation of longshore momentum (see Brocchini and Peregrine, 1996 for more details on the derivation of the integral equations), obtaining:

$$\frac{\partial\langle WV\rangle}{\partial t} + \frac{\partial\langle\check{P}_2\rangle}{\partial t} + \frac{\partial\langle W^2V\rangle}{\partial y} + 2\frac{\partial\langle W\check{P}_2\rangle}{\partial y} + \frac{\partial\langle\check{M}_{22}\rangle}{\partial y} + \langle\Upsilon_2\rangle = \left[\left\langle\left(u - \frac{\partial x_l}{\partial t}\right)vd\right\rangle\right]_{x_l} =$$
$$= \left(\langle u\rangle - \frac{\partial x_l}{\partial t}\right)\langle v\rangle\langle d\rangle + \langle\tilde{u}\tilde{v}\rangle\langle d\rangle + \langle\tilde{u}\tilde{d}\rangle\langle v\rangle + \langle\tilde{v}\tilde{d}\rangle\left(\langle u\rangle - \frac{\partial x_l}{\partial t}\right) + \langle\tilde{u}\tilde{v}\tilde{d}\rangle \tag{8}$$

where

$$V \equiv \int_{x_l}^{x_h} d\,dx\,, \qquad \Upsilon_2 \equiv \int_{x_l}^{x_h} \tau_2\,dx\,, \tag{9}$$

$$\check{P}_2 \equiv \int_{x_l}^{x_h} (\check{v}\,d)\,dx\,, \qquad \check{M}_{22} \equiv \int_{x_l}^{x_h}\left(\check{v}^2 d + \frac{g}{2}d^2\right)dx\,, \tag{10}$$

and the following short/long wave decomposition is used seaward of x_l i.e. in the inner surf zone:

$$u = \langle u\rangle + \tilde{u}, \qquad v = \langle v\rangle + \tilde{v}, \qquad d = \langle d\rangle + \tilde{d}, \qquad \langle\tilde{u}\rangle = \langle\tilde{v}\rangle = \langle\tilde{d}\rangle = 0. \tag{11}$$

Evaluation of the integral model (see also Brocchini and Bellotti, 2002) is made on the basis of fully-numerical solutions of the 2DH NSWE (2), i.e. inclusive of the longshore flow propagation, computed with the shock-capturing solver of Brocchini et al. (2001).

Table 1. Input parameters used to obtain the fully--numerical NSWE solutions used to assess the value of the model equations. T_S and H_S are the short-wave period and height θ is the angle of incidence to the beach normal of the waves evaluated at the seaward boundary of the sloping region C_f is the frictional coefficient. The period indicated by * is used only for JONSWAP spectra, while that indicated by ◊ only for bimodal waves

C_f	T_S (s)	H_S (m)	Θ (°)
0.01	3.2*	0.10	4
0.03	4.8◊	0.14	6
0.05	6.4	0.17	8
0.07		0.20	10
		0.23	

To clearly assess the value of the above model equation, we use, like in Brocchini and Bellotti (2002), either irregular waves obtained through a JONSWAP spectrum and bimodal waves to check the balance between the left-hand side (LHS) and the right-hand side (RHS) of (8). For such initial analysis we used an uniformly-sloping beach of slope $\alpha = 0.1$ and a Froude-type scaling was considered resulting in the input data of table 1.

In the case of JONSWAP spectra, T_s and H_s are the short-wave significant period (about 0.78 of the peak period) and height, for regular waves $T_s \equiv T$ and $H_s \equiv H$ are the short-wave period and height while for bimodal waves, modulations were obtained by the interaction of the short wave with a secondary signal $(H_s^{(2)}, T_s^{(2)}) = (0.25H_s, 5T_s)$. Longshore flow components are forced by varying the incidence angle θ of unidirectional waves at the offshore boundary over the range reported in table 1. Incident waves evolve over a very wide computational domain in the longshore direction and open lateral boundary conditions are used. Also note that in the cases forced by regular waves (see analysis of section 4), periodicity is established after 3-4 short-wavelengths at most. In summary, a total of 80 numerical solutions for regular waves, 20 for JONSWAP spectra and 5 for bimodal waves have been used for the comparative analysis which follows and based on simulations run for a time sufficiently long to remove any influence of the initial transient (i.e. approximately $15T_s$). With use of the mentioned conditions a detailed evaluation of the model equation balance has been made. More details of such analysis can be found in Antuono, Brocchini and Grosso (2006). Since average imbalances are in the order of a few percent it seems clear that (8) provides a good starting point to define the longshore SBCs which follow.

An analytical solution for the longshore motion.

Analyzing in detail equation (8) we extend the results of Brocchini and Bellotti (2002) to the case of weakly-2DH motion. We introduce Ryrie's (1983) approximation by considering waves incident at the seaward boundary of the sloping region with a small angle θ to the beach normal. We, thus, introduce a pseudotime t' and a small parameter ε such that

$$t' = t - \varepsilon y \tag{12}$$

and assume that this is the only y-dependence for the flow. For example, this occurs for a regular train of waves approaching the beach at an angle θ with the shore normal, with an offshore velocity c, where ε=sin(θ)/c (constant by Snell's law) and c is the shallow water wave velocity √(gd). This kind of approach can be extended to any beach that has a weakly-varying topography in the longshore direction. Use of (12) into (8) leads to:

$$\frac{d}{dt'}\left[W\langle V\rangle + \langle\check{P}_2\rangle - \epsilon W^2\langle V\rangle - 2\epsilon W\langle\check{P}_2\rangle - \epsilon\langle\check{M}_{22}\rangle\right] = \left\langle\left(u - \frac{\partial x_l}{\partial t}\right)vd\Big|_{x_l}\right\rangle - \langle\Upsilon_2\rangle. \tag{13}$$

For the sake of brevity, we rewrite (13) in the following compact form:

$$\frac{d}{dt'}P_W(t') = \mathfrak{B}(t') \quad \Rightarrow \quad P_W(t') - P_W(t'_0) = \int_{t'_0}^{t'} \mathfrak{B}(\tau')\,d\tau' \tag{14}$$

in which:

$$P_W(t') \equiv \left[W\langle V\rangle + \langle\check{P}_2\rangle - \epsilon W^2\langle V\rangle - 2\epsilon W\langle\check{P}_2\rangle - \epsilon\langle\check{M}_{22}\rangle\right], \tag{15}$$

$$\mathfrak{B}(t') \equiv \left\langle\left(u - \frac{\partial x_l}{\partial t}\right)vd\Big|_{x_l}\right\rangle - \langle\Upsilon_2\rangle. \tag{16}$$

We denote the time of beginning of the first run-up with t_0' and the time at which the k-th swash ends with t_k'. At such instants it is $x_s = x_l$ and so all the integral quantities vanish. In particular:

$$P_W(t'_k) = 0 \quad \forall k \in \mathbb{N} \quad \Rightarrow \quad P_W(t'_0) = -\int_{t'_0}^{t'_k} \mathfrak{B}(\tau')\,d\tau' \quad \forall k \in \mathbb{N}. \tag{17}$$

Substituting this result in (14) we obtain:

$$P_W(t') = \int_{t'_k}^{t'} \mathfrak{B}(\tau')\,d\tau' \qquad \forall k \in \mathbb{N}, \tag{18}$$

and with the usual independent variables (t,y), this can be rewritten in the form:

$$P_W(t,y) = \int_{t_k(y)}^{t} \mathfrak{B}(\tau,y)\,d\tau \qquad \forall k \in \mathbb{N} \quad \text{with} \quad t_k(y) = t'_k + \epsilon y. \tag{19}$$

The general solution of (13), which, as shown in figure 2, well compares with definition (5) directly evaluated from the numerical data, is:

$$W = \frac{b - \sqrt{b^2 - 4aC}}{2a} \tag{20}$$

in which:

$$a \equiv \epsilon\langle V\rangle, \qquad b \equiv \langle V\rangle - 2\epsilon\langle\check{P}_2\rangle, \qquad C \equiv \int_{t_k(y)}^{t} \mathfrak{B}(\tau,y)\,d\tau + \epsilon\langle\check{M}_{22}\rangle - \langle\check{P}_2\rangle. \tag{21}$$

The other root is not taken into account as physically meaningless. Since $\varepsilon \ll 1$, we can use the approximate solution $W \approx C/b$. Analysis of the fully-numerical solutions described in the second section shows that the term $\langle M_{22} \rangle$ is negligible (less that 5% of C), i.e.

$$W \approx \frac{\int_{t_k(y)}^{t} \mathfrak{B}(\tau, y)\, d\tau \; - \; \langle \breve{P}_2 \rangle}{\langle V \rangle}. \tag{22}$$

As shown in figure 2, this approximated solution agrees very well with the exact solution (20) and, consequently, with the numerically-evaluated definition (5). Moreover (22) confirms our initial inference on the physical meaning of W. In fact, the numerator is an "effective water mass flux" within the swash zone, while the denominator represents the whole water mass of the swash zone itself. Hence, such a ratio is clearly the drift velocity of the whole water mass (volume) in the swash zone and W can be correctly regarded as the longshore current which represents it. Definition (5) and the analytical exact (20) and approximated (22) solutions to the longshore problem have all an important role in the following analyses which lead to the SBCs for the longshore flow. More details can be found in Antuono, Brocchini and Grosso (2006).

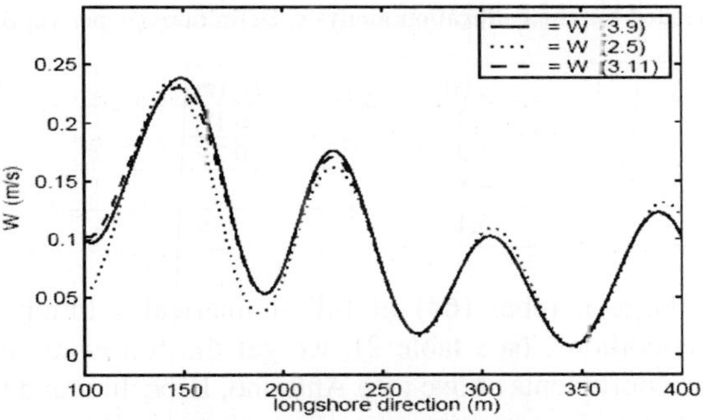

Figure 2. Typical example of drift velocity comparisons. The numerical solution (5) is illustrated by the dotted line while the exact solution (20) and the approximate one (22) are respectively given by the solid and dashed lines. It is evident the growth of W due to wave breaking.

Closures and shoreline boundary conditions.

In this section we further elaborate the approximate solution (22) to find suitable SBCs for the longshore motion. We want to express the integral quantities defined in the swash zone as functions of long-wave properties calculated at x_l. Through the relations obtained, we are able to assign the appropriate SBCs from knowledge of the integral quantities in the swash and local wave properties. We introduce standard variables, as follows:

$$\bar{d} \equiv \langle d \rangle, \qquad \bar{u} \equiv \langle u \rangle + \frac{\langle \tilde{u}\tilde{d} \rangle}{\langle d \rangle}, \qquad \bar{v} \equiv \langle v \rangle + \frac{\langle \tilde{v}\tilde{d} \rangle}{\langle d \rangle}. \tag{24}$$

Finally, from Part 2, we have that, once the incoming characteristic variable $R^+ = \bar{u} + 2\sqrt{d}$ is known, the SBCs for 1DH flows are:

$$\bar{d}(x_l) \approx \frac{H_l}{2}\,, \qquad \bar{u}(x_l) \approx \mathscr{R}_+ - \sqrt{2gH_l}\,, \qquad \frac{dx_l}{dt} \approx \left(\bar{u} - \frac{2}{H_l}\frac{d\langle V\rangle}{dt}\right)\bigg|_{x_l}. \tag{25}$$

By analogy with Part 2 and using Buckingham's Theorem, we assume:

$$\begin{aligned}
|\langle \breve{P}_2\rangle| \equiv \left|\left\langle \int_{x_l}^{x_h} (\breve{v}\, d)\, dx \right\rangle\right| &\approx C_{P_2}\sqrt{g}/\alpha\, \sin(\theta)\, H_l^{5/2}\,,\\
\langle \breve{M}_{22}\rangle \equiv \left\langle \int_{x_l}^{x_h} \left(\breve{v}^2 d + \frac{g}{2} d^2\right) dx \right\rangle &\approx C_{M_{22}}\, g/\alpha\, H_l^3\,,\\
\langle \tilde{Q}_2\rangle \equiv \langle \tilde{v}\tilde{d}\rangle &\approx C_{Q_2}\sqrt{g}\, \sin(\theta)\, H_l^{3/2}\,,\\
\langle \tilde{S}_{12}\rangle \equiv \langle \tilde{u}\tilde{v}\tilde{d}\rangle &\approx C_{S_{12}}\, g\, \sin(\theta)\, H_l^2\,,\\
\langle \Upsilon_2\rangle \equiv \left\langle \int_{x_l}^{x_h} \tau_2\, dx \right\rangle &\approx C_{\Upsilon_2}\, g\, \sin(\theta)\, H_l^2\,,
\end{aligned} \tag{26}$$

Defining the time scale $t_0 = 1/\alpha\sqrt{(d_0/g)}$ and the dimensionless period $T_s^* = T_s/t_0$, we assume the coefficients of (26) to depend both on C_f and T_s^*.

Table 2. Input parameters used to obtain fully-numerical periodic NSWE solutions needed for the calibration analysis. Definitions as per caption of table 1.

C_f	T_s (s)	H_s (m)	Θ (°)
0.01	3.2	0.17	6
0.03	4.0	0.23	10
0.05	4.8		
0.07	6.4		

Fitting a fairly large number (64) of fully-numerical solutions obtained with periodic input conditions, (see table 2), we get the following structure for the regression of the coefficients above (see Antuono, Brocchini and Grosso, 2006):

$$C_{(\dots)} = \frac{a_2 T_s^{*2} + a_1 T_s^* + a_0}{C_f} + \left(b_2 T_s^{*2} + b_1 T_s^* + b_0\right) + C_f\left(c_2 T_s^{*2} + c_1 T_s^* + c_0\right). \tag{27}$$

Note that the dependence on $1/C_f$ is foreseen by Ryrie's work (1983) on periodic breaking waves. We also underline that this formulation is only valid when the longshore flow has reached a quasi-steady state (e.g. Ryrie, 1983).

Full expressions for coefficients $C_{(\dots)}$ can be found in Antuono, Brocchini and Grosso (2006). Now using (24), we can expand to its main contributions the longshore momentum flux appearing in B(t) [see also equation (16)] :

$$\left\langle \left(u - \frac{\partial x_l}{\partial t}\right) vd\bigg|_{x_l} \right\rangle = \left[\left(\overline{u} - \frac{\partial x_l}{\partial t}\right)\overline{v}\,\overline{d} - \frac{\langle \tilde{Q}_1\rangle\langle \tilde{Q}_2\rangle}{\overline{d}} + \langle \tilde{S}_{12}\rangle + \langle \tilde{u}\tilde{v}\rangle\,\overline{d}\right]\bigg|_{x_l}. \tag{28}$$

So, defining the time scale $1/\alpha\sqrt{(H_l/g)}$, using (25) and the regressions in (26), we assume that (Antuono, Brocchini and Grosso 2006):

$$\int_{t_k(y)}^{t} \mathfrak{B}(\tau, y)\, d\tau \approx \frac{H_l^2}{\alpha}\left[C_1\,\bar{\bar{v}} + C_2 \sin(\theta)\sqrt{g H_l}\right] \tag{29}$$

where the first contribution models the mean momentum flux across x_l while the second all the short-wave interactions and the frictional forces in the swash zone. Here C_1 and C_2 are constants while v double overbar is the mean value of v overbar at x_l for the single swash at hand:

$$\bar{\bar{v}} \equiv \frac{1}{t'_{k+1} - t'_k}\int_{t'_k}^{t'_{k+1}} \bar{v}\Big|_{x_l}\, d\tau'\,. \tag{30}$$

Definition (30) is introduced, in analogy with Ryrie (1983), to use a single variable for each swash which can take into account breaking events. Since a bore implies a growth of v from each swash to the following one, v double overbar can, implicitly, account for the jump in longshore velocity [[v]] and also allows to clarify the dependence of the drift W on such forcing. In fact substituting (29) in (22) and using the previous regressions, we obtain the following theoretical form of dependence for W:

$$W \approx \left[\widetilde{C}_1\,\bar{\bar{v}} + \widetilde{C}_2 \sin(\theta)\sqrt{g H_l}\right]. \tag{31}$$

The final step is to make the dependence of v double overbar on H_l explicit. Since bore generation is most frequent in shallow waters, we want to isolate the contribution due to breaking. Following the approach used by Ryrie (1983) to analyze periodic breaking waves at the shoreline, we assume the following relation for v double overbar:

$$\bar{\bar{v}} \approx \left[D_1 + \frac{D_2}{C_f}\frac{T_{swash}}{T_s}\right]\sin(\theta)\sqrt{g H_l}. \tag{32}$$

Here D_1 models the purely non-breaking, nonlinear contributions to W while the second term represents the contribution due to breaking. V double overbar is unbounded as C_f goes to zero which is in good agreement with Ryrie's (1983) results. Substituting (32) into (31) we get the final expression for W:

$$W \approx \left[A + \frac{B}{C_f}\frac{T_{swash}}{T_s}\right]\sin(\theta)\sqrt{gH_l}\,. \tag{33}$$

The best-fitting coefficients are:

$$A = 187.9467\,C_f^2 - 23.4849\,C_f - 0.7939\,, \qquad B = 4.9438\,C_f^2 - 0.1512\,C_f + 0.0239 \tag{34}$$

and have been derived from the fitting curves reported on figure 3 where the data scatter reduces for increasing C_f. Equation (33) clearly shows the dependence of the drift velocity W on the expected fraction of the local shallow-water velocity (i.e. $\sin(\theta)\sqrt{(gH)}$) through: a first term (A) which accounts for short-wave interactions, frictional swash zone forces, the purely non-breaking, nonlinear contributions and a second term (depending on B) which models breaking forcings. Evaluation of the performance of the proposed SBCs is given on the basis of further numerical solutions of the NSWE. These have been run on purpose with input values different from those used during both the preliminary assessment of the integral model performances (table 1) and the calibration of the model (table 2). Table 3 summarizes the input values used for the final analysis of the SBCs' assessment. Figure 4 shows that equation (33) predicts well the numerically-computed longshore velocity for both monochromatic waves (panel a) and bimodal waves (panel b).

Table 3. Input parameters used to obtain the fully--numerical NSWE solutions used to assess the value of the proposed SBCs. Definitions as per caption of table 1.

C_f	T_s (s)	H_s (m)	Θ (°)
0.02	4.8	0.10	6
0.04	$4.8^{◊}$	0.14	
		0.17	
		0.20	
		0.23	

In summary a simple "recipe" is suggested to prescribe the SBCs. The motion of the mean shoreline is given by the third of equations (25) in which the rate of change of the volume in the swash zone, related to H_l through the first of equations (23), acts to decelerate the shoreline motion during the run-up and accelerate it during the run-down. At the mean shoreline the mean water depth is computed to be about half of H_l (first of equations 25). The onshore velocity depends on both local (H_l) and global information (R_+ coming from the interior of the domain brings to the shoreline knowledge of the deeper water dynamics). Finally, the local longshore velocity is a function of the incident wave angle θ and period T_s beyond the local wave height ($T_{swash}=\sqrt{(gH_l)}/(g\alpha)$).

Conclusions

Complete SBCs for wave-averaged 2DH circulation models have been derived on the basis of an integral version of the NSWE which naturally accounts for the interactions of long and short waves near the shoreline. The integral model can be used to provide a SBCs for the longshore drift velocity W only if short-wave contributions are suitably closed. To this purpose and to clarify the dependence of W on the short-wave properties we have:

1. obtained an analytical solution for the longshore drift velocity of the swash zone water masses;
2. computed short-wave contributions by wave-averaging fully-numerical solutions of the wave-resolved NSWEs.

Contributions to the drift velocity, coming from both nonlinear wave-wave interaction and wave breaking, are written as functions of known parameters.

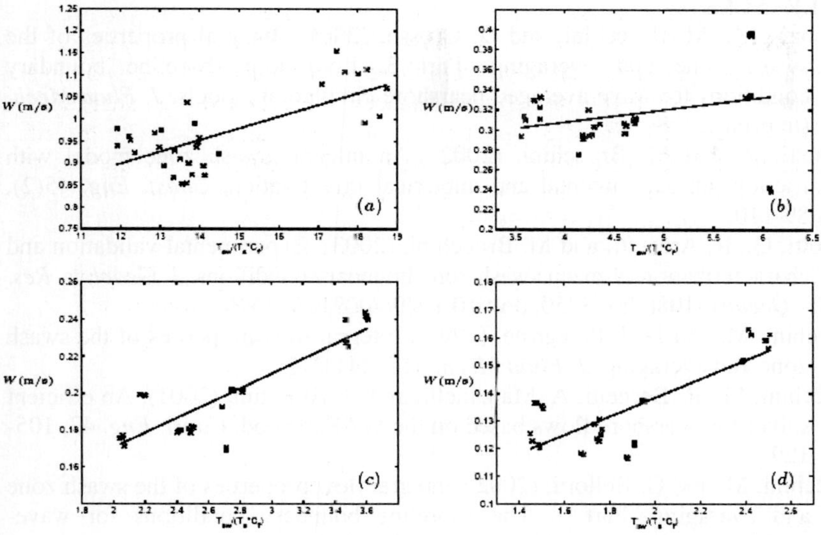

Figure 3. Fitting with (33) of the numerically-computed values of W (see table 2) as function of $T_{swash}/(T_sC_f)$ with: a) C_f=0.01, b) C_f=0.03, c) C_f=0.05, d) C_f=0.07.

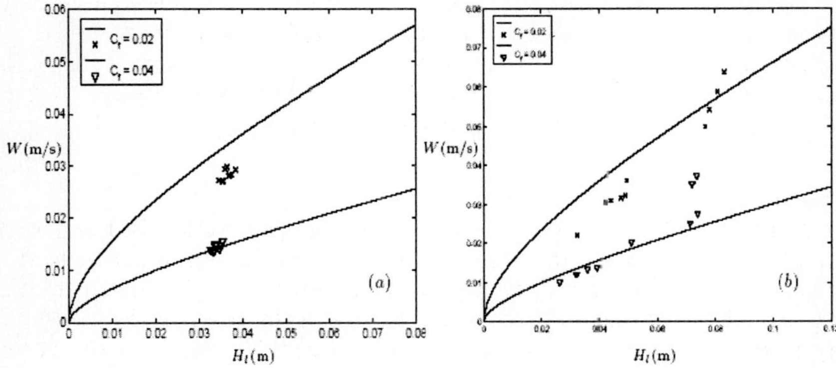

Figure 4. Assessment of the predictive capacity of solution (33) through a comparison with the numerically-computed values of W for the waves of table 3: a) monochromatic waves, b) bimodal waves.

REFERENCES

Antuono, M., M. Brocchini, and G. Grosso, (2006). Integral properties of the swash zone and averaging. Part 3. Longshore shoreline boundary conditions for wave-averaged nearshore circulation models. *J. Fluid Mech.* (in print).

Archetti, R. and M. Brocchini, (2002). An integral swash zone model with friction: an experimental and numerical investigation. *Coast. Eng.* 45(2), 89-110.

Bellotti, G., R. Archetti, and M. Brocchini, (2003). Experimental validation and characterization of mean swash zone boundary conditions. *J. Geophys. Res. – Oceans.* 108(C8), 3250, doi: 10.1029/2002JC001510.

Brocchini, M. and D.H. Peregrine, (1996). Integral flow properties of the swash zone and averaging. *J. Fluid Mech.* 317, 241-273.

Brocchini, M., R. Bernetti, A. Mancinelli, and G. Albertini, (2001). An efficient solver for nearshore flows based on the WAF method. *Coast. Eng.* 43, 105-129.

Brocchini, M. and G. Bellotti, (2002). Integral flow properties of the swash zone and averaging. Part 2. The shoreline boundary conditions for wave-averaged models. *J. Fluid Mech.* 458, 269-281.

Ryrie, S.C. (1983). Longshore motion generated on beaches by obliquely incident bores. *J. Fluid Mech.* 129, 193-212.

THE INFLUENCE OF THE NEARSHORE BATHYMETRY ON THE EVOLUTION OF BREAKING WAVE-INDUCED MACROVORTICES

Luciano Soldini[1], Carlo Lorenzoni[1], Maurizio Brocchini[1] and Andrew B. Kennedy[2]

The horizontal mixing on nearshore flows characterized by non-uniform topographies is studied with specific focus on the interpretation of the statistics of passive tracers. Two different data sets have been used and the discrepancies/similarities between the same statistics derived from the two experiments have been explained in relation to the specific experimental conditions. The main sources of discrepancy seem to be the different seeding procedure (leads to changes in the ballistic regime), the varying confinement effects due to the basin boundaries and to the different still-water free board in use (both lead to differences in the asymptotic value of the eddy diffusivity). Notwithstanding these differences use of both data sets leads to an improved formulation for the eddy viscosity (equation 6)

INTRODUCTION

The mixing of nearshore flows is characterized by shallow waters and by turbulent structures of different sizes. This aspect is very important for the analysis of the water quality and the modeling of nutrients and the transport of pollutants for which use is most often made of convection-diffusion equations like:

$$\frac{\partial C}{\partial t}+u_i\frac{\partial C}{\partial x_i}=-\frac{\partial q_i^T}{\partial x_i}+K_m\frac{\partial^2 C}{\partial x_i^2} \qquad (1)$$

in which C is the turbulence-averaged concentration of particles in motion, q_i^T is the turbulent mass flux and K_m is the molecular diffusivity.

For times long enough we can use for the q_i^T, a "Fick-type closure":

$$q_i^T=-K_{ij}\frac{\partial C}{\partial x_j} \qquad (2)$$

and equation (1) can be rewritten as follow:

[1] Istituto di Idraulica e Infrastrutture Viarie, Università Politecnica delle Marche, Via Brecce Bianche 12, Ancona, 60131, Italy

[2] Department of Civil & Coastal Engineering, University of Florida, Gainesville, FL, 32611-6590, USA

$$\frac{\partial C}{\partial t}+u_i\frac{\partial C}{\partial x_i}=-\frac{\partial}{\partial x_i}\left(K_{ij}\frac{\partial C}{\partial x_j}\right)+K_m\frac{\partial^2 C}{\partial x_i^2} \tag{3}$$

in which the K_m can be omitted, because negligible with respect to the turbulent diffusivity K_{ij}. Equation (3) can only be solved for C once a "closure" for the unknown turbulent diffusivity is made. Various techniques are available to estimate K_{ij}. One of the goals of the present and other related ongoing studies is to assess the possibility of making such a "closure" on the basis of laboratory and numerical experiments in alternative to field experiments (e.g. La Casce and Bower, 2000; Takewaka et al., 2003).

In this paper we have used experimental data to describe the motions of passive tracers and to quantify the mass and the momentum mixing induced by the flow turbulence.

THEORETICAL BACKGROUND

Nearshore flows mixing is strongly promoted by the presence of large-scale horizontal vortices, called "macrovortices". According to Peregrine (1998), macrovortices arise in the presence of breaking waves of finite length in the longshore direction, like the uniform fronts of breaking waves of finite length overpassing a submerged obstacle, for example a submerged breakwater (see Figure 1).

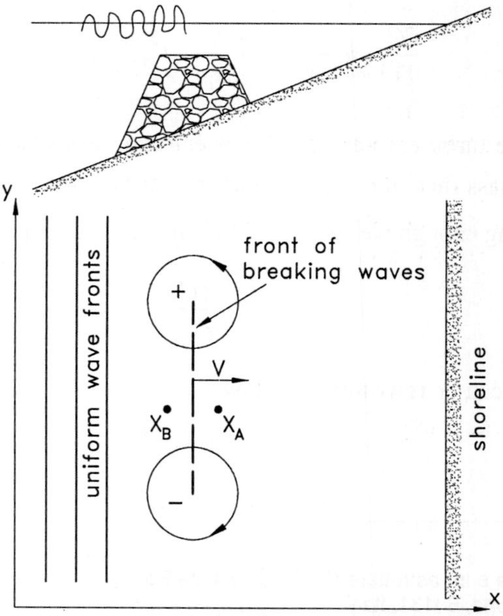

Figure 1. Example of macrovortices generation in nearshore region

Brocchini et al. (2004) and Kennedy et al. (2006), analysed experimentally, computationally and analytically the behaviour of breaking-wave-induced macrovortices during startup conditions. Both studies provide insight into the fundamental deterministic features of macrovortex evolution.

Turbulence Characteristics

The turbulence properties of shallow-water motions are quite similar to those of the two-dimensional turbulence, i.e. the spectral properties of this turbulence are characterized by a direct enstrophy cascade at large wavenumbers and an inverse energy cascade at small wavenumbers, with decay rates typical of two-dimensional turbulence i.e. a "−5/3 law" and a "−3 law" respectively for energy and enstrophy cascading (Kraichnan, 1967).

Statistical Analysis of Turbulent Motions

The study of turbulent motions can also be done on the basis of statistics of passive tracers.

Two different types of statistical properties can be defined:

- absolute properties, referred to each single particle in motion;
- relative properties, referred to groups of two or more particles.

The main absolute properties are defined as follows:

- the absolute dispersion X^2, which is the mean of squared displacements of the various particles from their initial positions;
- the absolute diffusivity $K^{(1)}$, which is the time derivative of the absolute dispersion.

According to Taylor's theory (1921), these variables have a well-defined behavior:

- for times much smaller than T_L the absolute diffusivity is linearly proportional to the time t,
- while for times much greater than T_L it is simply a constant.

The first regime is called "ballistic" and the second is the "brownian" regime. T_L is the Lagrangian time scale which can be calculated from the autocorrelation coefficient of the velocity field

$$T_L = \int_0^{\infty} R(t)dt \tag{4}$$

in which $R(t)$ is equal to

$$R(t-t_0) = \frac{\left\langle \overrightarrow{u_i}(t_0+t)\cdot\overrightarrow{u_i}(t_0)\right\rangle}{\left(\left\langle \overrightarrow{u_i}(t_0+t)\right\rangle^2 \cdot \left\langle \overrightarrow{u_i}(t_0)\right\rangle^2\right)^{1/2}} \tag{5}$$

and $\overrightarrow{u_i}$ is the Lagrangian velocity of the *i*-th particle while the angular brackets indicate ensemble averaging. With the above definitions, in a brownian regime, "particles have forgotten knowledge of their initial velocity", i.e. the autocorrelation coefficient of their velocity is close to zero. Although T_L is, in principle, estimated by computing the autocorrelation coefficient of the velocity field, we prefer here to estimate such time scale directly from the diagrams of the absolute diffusivity $K^{(1)}$.

For relatively small times the motion is dominated by the small-scale turbulence, whose effect is to deform the cloud of passive tracers. In this case it is more useful to consider relative statistical properties, than absolute ones, because more sensitive to small-scale changes. These are:

- the relative dispersion D^2, which is the mean of squared distances between particles of the various pairs (n-plets);
- the relative diffusivity $K^{(2)}$, defined as the time derivative of the relative dispersion.

In a two-dimensional turbulent field these properties, typically, obey the following laws:

- for times smaller than T_p $(t < T_p)$ the relative diffusivity (and the absolute) is linearly proportional to the time t (where T_p represents the "time for which pairs of particles have forgotten their initial separation" and it is much smaller than T_L);
- for intermediate times $(T_p < t < T_L)$ various behaviors are possible in dependence of the flow field. For our flows the most common are:
  - <u>enstrophy cascading</u>, which dominates in the absence of intense external shear and for which the diffusivity is proportional to the relative dispersion D^2 (which itself depends on time);
  - <u>external field shearing/wave advection</u> for which the turbulent relative diffusivity is proportional to $D^{4/3}$.
- for times greater than T_L $(t > T_L)$ and large distances, the relative diffusivity $K^{(2)}$ is constant and is simply equal to twice the absolute diffusivity $K^{(1)}$.

EXPERIMENTAL SETUP

Two different data set have been available to study nearshore mixing induced by macrovortices.

The first data set we have analyzed comes from the experimental tests executed at the large wave basin of the Laboratory of Coastal Engineering of the Politecnico of Bari (Lorenzoni et al., 2004, 2005) (Bari experiments hereinafter). The studied model, with a movable sandy bed, was characterized by an almost-isolated submerged breakwater, called "single breakwater configuration" and by an array of three submerged breakwaters, called "rip-current configuration". The breakwaters used into the model were rubble-mound structures. Analysis of surface hydrodynamic circulation was made by tracking the trajectories of tracers released around the breakwaters before the beginning of each test. The tests have been performed releasing some wooden spheres in the gaps between breakwaters and around the heads of the isolated breakwater and recording the tracers movements with a fixed videocamera. Figure 2 shows two examples of tracers measurements. The picture on the top shows that, in the case of an isolated breakwater, particles were dispersed both offshore and onshore while, the picture on the right, with an array of breakwaters, called rip-current configuration, shows that almost all floaters were dispersed towards the offshore by the rip current.

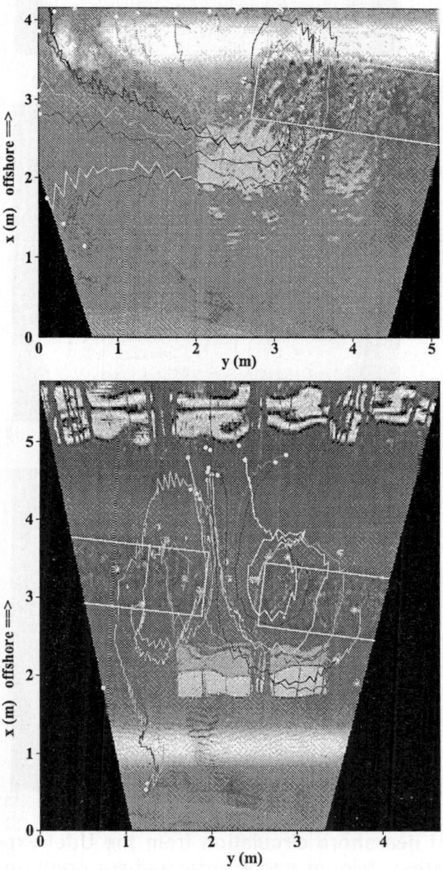

Figure 2. Example of passive tracers trajectories from the Bari experiments. Top panel: isolated breakwater configuration. Bottom panel: rip current configuration.

The second group of experiments, described in detail in Kennedy and Thomas (2004), was performed at the wave basin of the Laboratory of Coastal Engineering of the University of Delaware, USA (UDel experiments hereinafter).

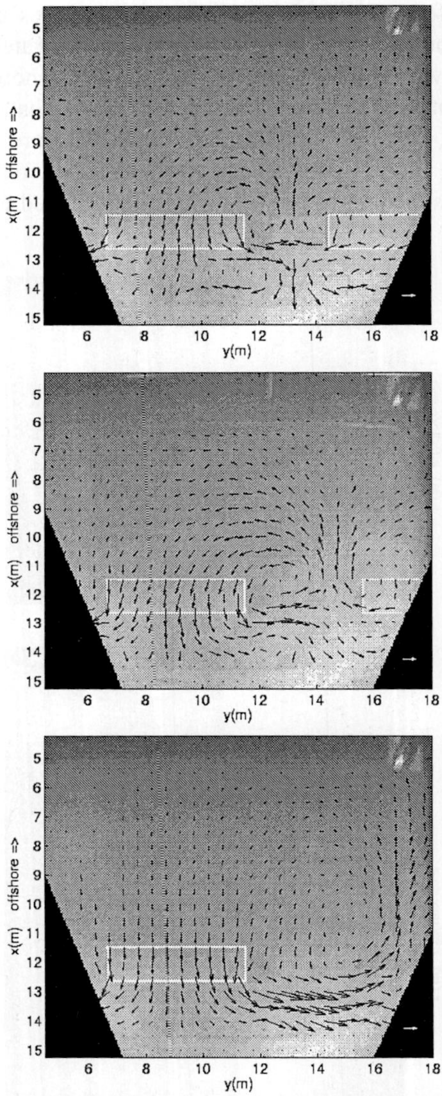

Figure 3. Example of nearshore circulation from the Udel experiments. Top panel: rip current configuration. Middle panel: intermediate configuration. Bottom panel: isolated breakwater configuration.

The model consists of submerged obstacles similar in shape to natural bars. Also in this case different configurations were analyzed: an almost-isolated bar and various similar obstacles placed nearby in an array of bars. The bottom and the bars of the model were made of concrete. Analysis of surface hydrodynamic circulation was made. The tests have been performed by continuously releasing plastic, mildly buoyant, drifters at the gaps and around the heads of the bars during each test and recording the tracers movements with a fixed videocamera. The results, obtained by tracking the trajectories of floaters continuously released around the bars, show very different mean current fields for the three analyzed conditions (see Figure 3).

ANALYSIS OF EXPERIMENTAL DATA

Flow mixing is quantified here in terms of absolute and relative statistics (i.e. dispersion and diffusivity) of experimental floaters' motion applying the approach by Piattella et al. (2006). Due to the different characteristics of the two models (breakwater vs. natural bars and movable vs. fixed bed) the results are illustrated in two subsections, one for each data set.

Bari experiments

The analysis of all tests shows that the absolute diffusivity grows in time similarly to what predicted by the theory. More details are found in Piattella et al. (2006).

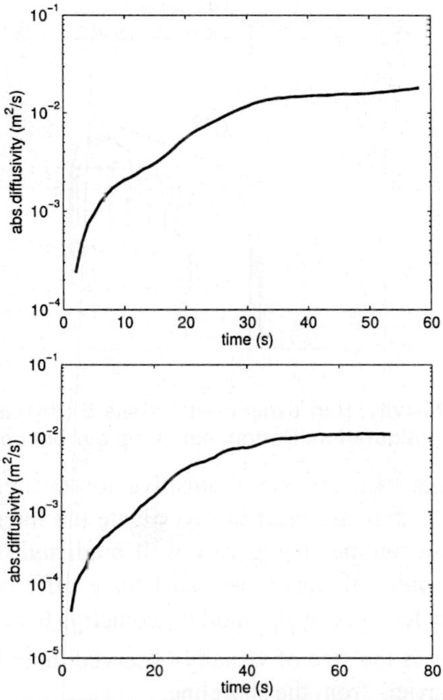

Figure 4. Absolute diffusivity. Bari experiments. Case E (H=5cm; T=0.9s). Top panel: isolated breakwater configuration. Bottom panel: rip current configuration.

Figure 4 shows typical results for the eddy diffusivity. For times much smaller than the Lagrangian time scale T_L $(t \ll T_L)$ the dispersion linearly increases in time, while for times larger than T_L $(t \gg T_L)$ it reaches an almost constant value. This suggests the presence of both a ballistic regime at small times and a Brownian regime for large times. The Lagrangian time scale, here determined as the time after which the diffusivity is almost constant, falls between $35s$ and $55s$ for all the tests. The asymptotic values of absolute diffusivity are larger for the isolated breakwater than for the rip current configuration.

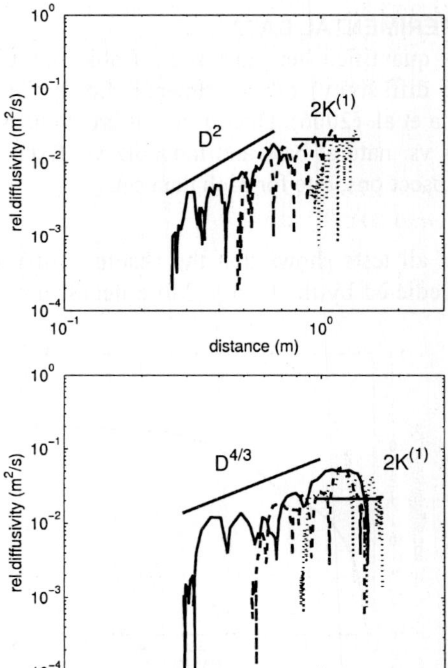

Figure 5. Relative diffusivity. Bari experiments. Case E (H=5cm; T=0.9s). Top panel: isolated breakwater configuration. Bottom panel: rip current configuration.

Since relative statistics are more sensitive to the floaters evolution with small-scale turbulence they are used to investigate the intermediate regime, the ballistic and brownian regimes being very well predicted by absolute statistics. To select groups of pairs of tracers we used three different initial separations D_0, which represent features of the model geometry: $0.4m$ is representative of the berm width; $0.8m$ is the size of visually-observed macrovortices and $1.5m$ is the distance of the barriers from the shoreline.

Figure 5 shows the relative diffusivity plotted as a function of the mean separation of pairs of particles and it suggests different behaviors for the two configurations: for the isolated breakwater (top panel), the relative diffusivity is proportional to the square of the mean separation (D^2), while for the array of breakwaters (bottom panel) it is proportional to $D^{4/3}$. The first law is typical of turbulent regimes dominated by a direct cascade of enstrophy. The second one, and the absence of merging of macrovortices, suggests dominance of a shear advection rather than a possible inverse cascade of energy. Both configurations reach an asymptotic value of relative diffusivity, characteristic of the situation in which the pairs of passive tracers have forgotten their initial separation and the asymptotic value of the relative diffusivity is estimated as twice the absolute one. The scale for a constant diffusivity, called Lagrangian spatial scale (L_L), is about (0.7-1.0)m; this, confirms the visual observation on the typical size of macrovortices.

UDel experiments

The same analysis of statistical variables has been used to interpret the flow evolution of the Delaware experiments.

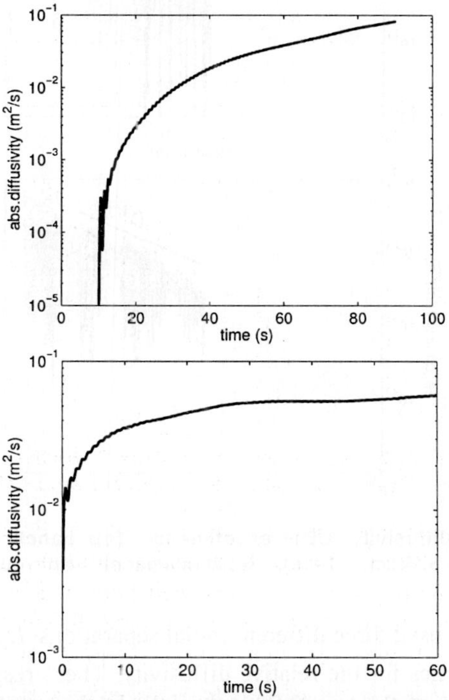

Figure 6. Absolute diffusivity. UDel experiments. Top panel: isolated breakwater configuration (H= 5.14cm; T=1s). Bottom panel: rip current configuration (H=5.107cm; T=1s).

The plots of the absolute diffusivity show both similarities and differences with the corresponding ones of the Bari experiments (see Figure 6). First of all the absolute diffusivity linearly grows in time over a very short period. We doubt a ballistic regime can be clearly assessed from these analyses only. Moreover, it is much harder to detect asymptotic conditions for the case of an isolated bar. These may be argued to exist for $t \geq 80s$, i.e. after a much longer transient than that characterizing the case of an array of bars (the asymptotic conditions of a Brownian regime can be observed for $t \geq 30s$).

Also note that the asymptotic absolute diffusivity typical of the array of bars is larger than that of the isolated bar. This result, in apparent contrast with the results coming from the Bari experiments, can be explained in view of the still growing diffusivity for the case of an isolated breakwater.

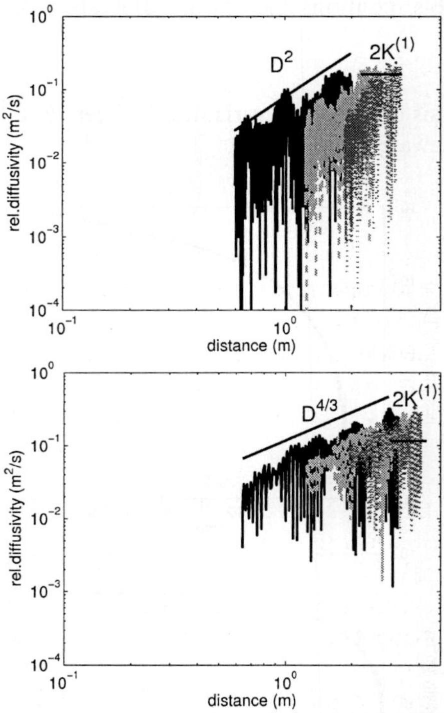

Figure 7. Relative diffusivity. UDel experiments. Top panel: isolated breakwater configuration (H= 5.14cm; T=1s). Bottom panel: rip current configuration (H=5.107cm; T=1s).

Again we have used three different initial separations D_0 to build groups of homogeneous statistics for the relative diffusivity. These respectively represent: the crosshore width of the bar (0.9m); the estimated size of the macrovortices (1.8m) and the distance of the bars from the shoreline (2.9m). The results are plotted in Figure 7.

For the isolated bar the relative diffusivity is proportional to the square of the mean distance (D^2), while for the array of bars it is proportional to $D^{4/3}$. So, trends are the same found for the data of Bari and it seems that the flow induced by an isolated bar is dominated by a direct cascade of enstrophy, while that induced by an array of bars is dominated by a shear advection. Both configurations reach an asymptotic value of relative diffusivity, characteristic of the situation in which the pairs of passive tracers have forgotten their initial distance. In this situation the asymptotic value of the relative diffusivity is estimated as twice the absolute one.

The scale for a constant diffusivity is about (2.2-2.8)m, so the Lagrangian length scale (L_L) is slightly larger than the visually-estimated size of the macrovortices, equal to $1.8m$, and two-to-three times larger than that observed during the Bari experiments.

PARAMETRIC ANALYSIS AND DISCUSSION

The discrepancies found between the results obtained from the two data sets may be explained by the various differences characterizing the two experiments. In increasing order of importance these are:

- the different bottom configurations: movable for Bari versus rigid for the UDel tests;
- the different types of the submerged obstacles: rubble mound breakwaters for the Bari test and impermeable bars for the UDel tests;
- the different methods for seeding the flow with tracers: a complete seeding before running the waves in the case of Bari versus a continuous seeding of almost steady flow conditions in the case of the UDel experiments: this might explain the very short period of its ballistic regime;
- the different confinement effects of the basin boundaries, which is much more intense in the case of the UDel test (smaller basin and relatively large macrovortices) than in the case of the Bari experiments (larger basin and relatively small macrovortices). It is known that confinement effect may influence the type/existence of a Brownian regime;
- the large differences in the still-water free-boards (4 times larger for the UDel tests) is thought to influence the size of the macrovortices and the resulting diffusivities.

Hence, to account for the influence of the free-board, we have proposed a parametric formula, based on geometric and kinematic parameters, which modifies the predictive law of Piattella et al. (2006) to estimate the absolute diffusivity. The equation is:

$$K_{calc}^{(1)} = 5 \cdot \left(\frac{F}{H_0} \right)^{1/5} \cdot \sqrt{I_r \frac{L_t}{L_c}} \cdot \sqrt{g h_b} \cdot H_0 \tag{6}$$

in which F is the still-water free-board, H_0 is the offshore wave height, I_r is the Irribarren number, h_b is the water depth at the offshore toe of the structure and L_c and L_t are the structure and the gap lengths respectively. We have calculated the absolute diffusivities and compared them to the measured ones to find a fairly good agreement, especially for the lower values obtained from the data of the Bari experiments.

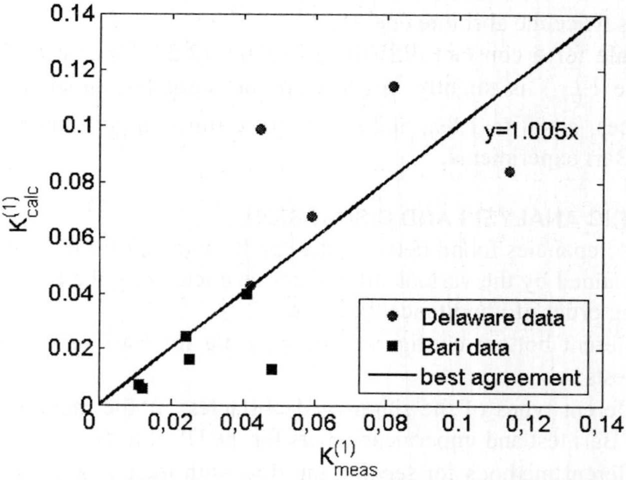

Figure 8. Comparison between the absolute diffusivity determined from the experimental measures and by the equation (6).

REFERENCES

Brocchini M., A.B. Kennedy, L. Soldini and A. Mancinelli. 2004. Topographically-controlled, breaking wave-induced macrovortices. Part 1. Widely separated breakwaters, *J. Fluid Mechanics*, 507, 289-307.

Kennedy A.B., and D. Thomas. 2004. Drifter measurements in a laboratory rip current, *J. Geophysical Research-Oceans*, 109, doi:10.1029/2003JC001982.

Kennedy A.B., M. Brocchini, L. Soldini and E. Gutierrez. 2006. Topographically-controlled, breaking wave-induced macrovortices. Part 2. Changing geometries, *J. Fluid Mechanics*, 559, 57-80.

Kraichnan, R. H. 1967. Inertial ranges in two-dimensional turbulence, *Physics of Fluids*, 10, 1417–1423.

La Casce J.H., and A. Bower. 2000. Relative dispersion in the subsurface North Atlantic, *J. Marine Research.*, 58, 863-894.

Lorenzoni, C., L. Soldini, A. Mancinelli, A. Piattella and M. Brocchini. 2004. La circolazione idrodinamica in presenza di barriere sommerse: un'analisi sperimentale. *Proceedings of XXIX Convegno di Idraulica e Costruzioni Idrauliche*, 3, 573–580 (in Italian).

Lorenzoni C., A. Piattella, L. Soldini and A. Mancinelli, M. Brocchini. 2005. An experimental investigation of the hydrodynamic circulation in the presence of submerged breakwaters, *Proceedings of 5th International Symposium on Ocean Wave Measurement and Analysis*, paper 125, pp. 10, CD-ROM.

Peregrine D. H. 1998. Surf zone currents. *Theoretical and Computational Fluid Dynamics*, 10, 295–309.

Piattella A., M. Brocchini and A. Mancinelli. 2006. Topographically controlled, breaking-wave-induced macrovortices. Part 3. The mixing features, *J. Fluid Mechanics*, 559, 81-106.

Soldini L., C. Lorenzoni, A. Piattella, A. Mancinelli and M. Brocchini. 2004. Nearshore macrovortices generated at a submerged breakwater: experimental investigation and statistical modeling, *Proceedings of 29th International Conference on Coastal Engineering*, 2, 1380-1392.

Takewaka S., S. Misaki and T. Nakamura. 2003. Dye diffusion experiment in a longshore current field, *Coastal Engineering*, 45, 471-487.

Taylor, G. I. 1921. Diffusion by continuous movements, *Proceedings London Mathematic Society*, 20, 196–212.

FLUID ACCELERATION IN THE SWASH ZONE

Ali FARHADZADEH[1], Jack A. PULEO[2]and Nobuhisa KOBAYASHI[3]

A Navier-Stokes solver (NEWFLUME) is used to evaluate velocity, local, convective and total accelerations in the swash zone for three different types of breaking, surging, plunging and nearly spilling waves which are obtained by altering the beach slope holding the forcing conditions constant. The hydrodynamic results have been presented for a level right above the bed which is more important for sediment transport issues regarding the three breakers. Also, depth-averaged velocity and acceleration parameters are shown. The results show that regardless of the breaker types, moving from the inner surf to swash zone, the flow velocities magnitude increases and the largest onshore-directed velocity occurs at the beginning of uprush, also the greatest offshore velocities take place at the end of backwash phase. It has been found that the convective accelerations are mainly onshore-directed or nearly zero and reach their highest values at the early uprush and at the end of backwash phases in the onshore direction. We also find that in the surging case the local accelerations are offshore-directed during a swash event except at the beginning of swash cycle when the local acceleration value decreases during the time and reaches its maximum offshore-directed value at the mid-swash time. Just outside and at the seaward edge of the swash zone, local accelerations are shoreward-directed for the plunging case and near zero but shoreward-directed for the spilling wave.

1. Introduction

The swash zone is the area of the nearshore that is intermittently covered and uncovered by wave run-up. Since swash hydrodynamics are the main contributor to the evolution of beach morphology, understanding these motions is of paramount importance. Many studies have investigated basic swash zone flow characteristics and statistics on reflective and dissipative beaches (e.g. Masselink et al.,2006) generally using either electro-magnetic or acoustic Doppler current meters. A potentially important fluid parameter that has received attention more recently is the fluid acceleration. Initial work suggested that the collapse of the bore provides a push of water up the beach face that causes a short-lived burst of fluid acceleration (Puleo et al., 2003). Other

---

[1] Ph.D. Candidate, Center for Applied Coastal Research, University of Delaware, Newark, DE 19716:, office: 302-831-4171, email: alif@coastal.udel.edu

[2] Assistant Professor, Center for Applied Coastal Research, University of Delaware, Newark, DE 19716, office: 302-831-2440, fax: 302-831-1228, email: jpuleo@coastal.udel.edu (corresponding author)

[3] Professor, Center for Applied Coastal Research, University of Delaware, Newark, DE 19716, office: (302) 831-6531, fax: 302-831-1228, email: nk@coastal.udel.edu

studies, suggest that the fluid acceleration away from this collapse area should always be offshore-directed (Hughes and Baldock, 2004). One difficulty in inter-comparing past work is based on the location within the swash zone where measurements were collected. Indeed, it is probably correct to say that some of the previously collected measurements were actually obtained in the inner surf zone. The point of this study is not to delve into the nuances of the exact description of the swash zone, but rather to investigate the existence and persistence of accelerations, and indeed, decelerations in swash zone flows.

2. Numerical model

2.1 Model Description

A 2D numerical model is used in this study to investigate the cross-shore structure of fluid acceleration in the swash zone. The model, NEWFLUME, developed by Lin and Xu (2005) is an extended version of two-dimensional Navier-Stokes (eq.1,2) solver, RIPPLE, in which $k-\varepsilon$ turbulence closure has been added.

$$\frac{\partial u_i}{\partial x_i}=0 \tag{1}$$

$$\frac{\partial u_i}{\partial t}+u_j\frac{\partial u_i}{\partial x_j}=-\frac{1}{\rho}\frac{\partial p}{\partial x_i}+g_i+\frac{1}{\rho}\frac{\partial}{\partial x_j}\left(\mu\frac{\partial (u_i)}{\partial x_j}-\rho\langle u_i'u_j'\rangle\right) \tag{2}$$

where u_i is the velocity vector, u_i' is the velocity fluctuation, p is the pressure, μ is dynamic viscosity, g is gravity, ρ is fluid density and t is time, and $\rho\langle u_i'u_j'\rangle$ is the Reynolds stress

This model has been validated for wave overtopping and wave forces on a submarine pipeline [Lin and Xu, 2005], while other versions of the RIPPLE model have been validated against breaking waves in the surf zone [Lin and Liu, 1998; Zhao and Tanimoto, 1998], solitary wave runup [Lin, et al., 1999] and swash zone flows driven by regular and irregular waves [Puleo 2004].

The model domain is discretized into individual control volumes, which may be empty, full or partially filled with water. This volume of fluid (VOF) approach enables the capture of breaking waves and other swash hydrodynamics by calculating the appropriate force balances in each control volume and the flux of water across each of the control volume surfaces. A two step projection method has been used for the mean flow solution, aided by the incomplete Cholesky conjugate gradient technique solving the Poisson equation for the mean pressure field.

2.2 Model Domain and Forcing

To evaluate swash flow acceleration, the model is forced with sinusoidal waves over an impermeable and smooth slope (Figure 1). Surf similarity yields spilling, surging and plunging breakers, however in order to have more continuous data being used to calculate accelerations, the surging case is mostly discussed. A steep slope was chosen to avoid the excessive computational time required for a larger domains associated with a gentler slope. Results from nearly spilling and plunging breakers will also be discussed, however splashing associated with these breaker types increases discontinuity in the data making it more difficult to quantify accelerations.

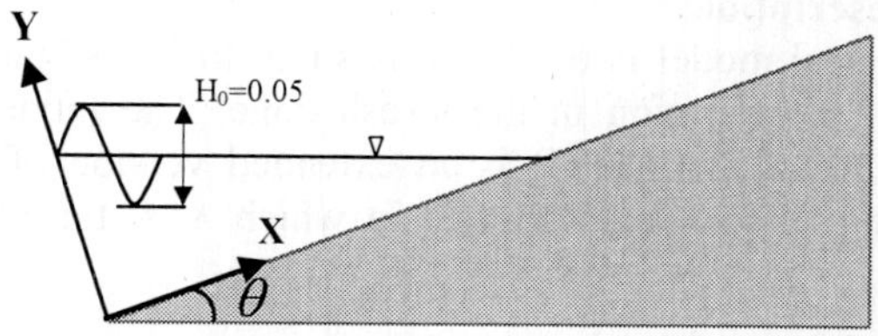

Figure 1. The numerical modeling domain and coordinate system

Three cases of breakers have been produced only by changing the beach slope based on surf similarity number (ξ_0). Table 1 shows the characteristics of waves and beach slope.

Table1. Parameters and conditions for the three wave cases

Breaking Type	T(s)	H_0(m)	ξ_0	Slope (θ)
Surging	1.5	0.05	3.64	23.5
Plunging	1.5	0.05	2.0	13.42
Nearly Spilling	1.5	0.05	0.67	4.57

In the model, the offshore (left side) boundary has a specified inflow condition (sinusoidal waves), the upper boundary is free surface, the onshore (right side) boundary has a continuative outflow condition so that the flow can run over the beach up to the location where swash ends, and a turbulent boundary layer is assumed near the bed .

Table 2 shows the length, height, the mesh numbers and sizes for each particular breaker type. To simplify use of the numerical results, the x direction has been considered along the beach slope and y direction perpendicular to the bed as shown in Figure 1.

Table 2. The computed domains characteristics for the each breaker type

Breaking Type	x ,y length (m)	Cell number (x, y)	Swash mesh size (x, y (m))
Surging	6 , 2	1000, 195	0.005, 0.005
Plunging	9.5 , 2	1000, 195	0.005, 0.005
Nearly Spilling	23 , 2	1000, 195	0.005, 0.005

Bed-parallel fluid acceleration, a_x, in a 2D system has three components (eq.3).

$$a_x = \frac{\partial u}{\partial t} + u\frac{\partial u}{\partial x} + v\frac{\partial u}{\partial y} \quad (3)$$

The first part is local and the second and the third parts represent the convective acceleration. The coordinate system has been chosen with the bed-normal vertical velocity component, v and its variation in that direction is small. The third part is neglected so that only the velocity component along the beach, u is considered when calculating the acceleration in the swash zone.

Bed-parallel and local acceleration have been calculated using centered differences when possible and total acceleration is simply given by algebraic summation of local and convective components. Centered differences could not be used when a location on the foreshore was unwetted. During these times, only one sided differences using four adjacent cells values are used to approximate a derivative.

3. Computed Results

Figure 2 shows the bed-parallel velocity flow field during a single swash event for the surging, plunging and nearly spilling cases respectively. At the surging case, the swash starts at t=8.1s and ends at t=9.6s. The swash cycle has been divided in 6 panels so that the Figure 2a-A is the beginning of uprush and Figure 2a-D is the time at which the uprush phase ends and the backwash starts. Finally Figure 2a-F is the end of backwash phase. Taking a look at these figures, it can be seen that the velocity during the uprush phase decreases in magnitude during time and along the slope. Similar gross patterns can also be seen in Figures 2b and 2c which are related to plunging and spilling breakers. However, for the plunging wave there is considerable splashing and more variability with respect to the bed-normal direction. In the spilling case, the dominant feature is the incoming bore which, to be exact, is not considered swash. Once this bore reaches the shoreline the uprush begins and is seen to span only a short distance in this case.

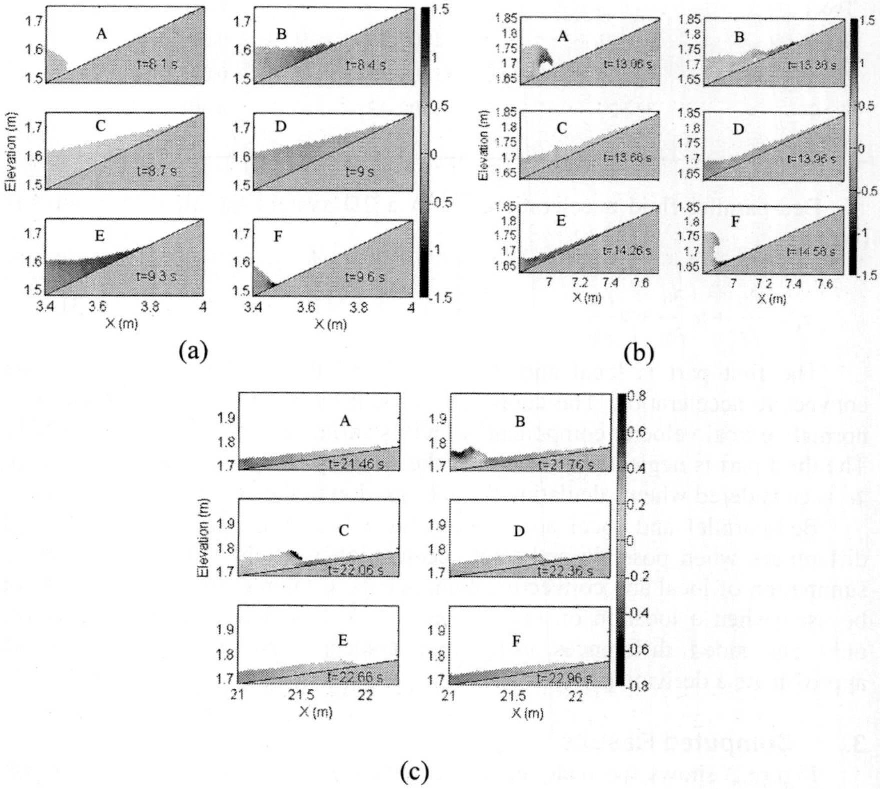

Figure 2. Bed-parallel velocity snapshots of surging (a) plunging (b) and nearly spilling breakers (c) for a single swash event. Colorscale represents velocity with positive being shoreward-directed.

To assess the velocities and accelerations in the swash zone for the various wave types, these quantities have been normalized in time and space (Figure 3). For a single swash cycle, the duration of swash event, D, and the maximum swash length along the slope, L have been used to normalize the velocity and acceleration as follows:

$$x' = (x - x_{start}) / L$$
$$t' = (t - t_{start}) / D \quad (4)$$

where x_{start} and t_{start} are respectively the location and the time at which a single swash event initiates. It should be mentioned that a full swash cycle starts at the region of greatest seaward rundown (t_{start} and x_{start}). However, in this paper we will also investigate the variability in swash hydrodynamics in regions landward of the initial uprush location.

Figure 4 shows the normalized bed parallel velocity, convective, local and total acceleration variation during a single swash event for the nearly spilling, plunging and surging breakers just above the bed in the inner surf zone, beginning of swash zone and in a distance about one third of swash length onshore. As it can be seen, moving onshore, regardless of the breaker type, the flow velocity magnitude increases for this portion of the swash cycle. Among the breakers, the surging wave has a larger velocity followed by the plunging and then spilling.

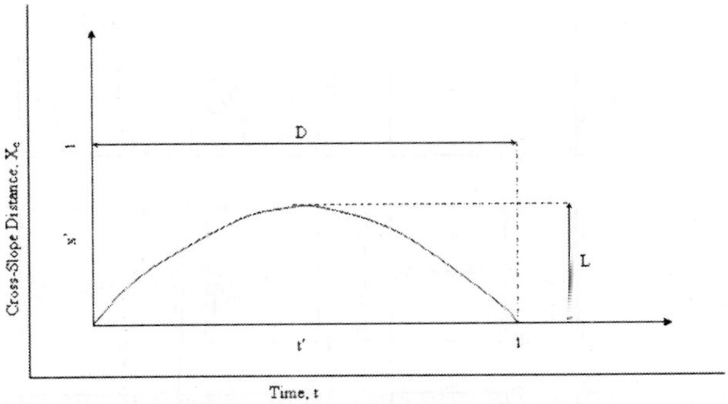

Figure 3. Schematic demonstrating swash-normalization based on excursion distance (L) and swash duration (D)

It is also seen that at the beginning of uprush phase, the largest onshore-directed and at the end of backwash, the greatest offshore-directed velocities take place. The convective accelerations are mostly onshore-directed or nearly zero especially in the nearly spilling breaker case. The convective accelerations reach their highest values at the early uprush and at the end of backwash phases in onshore direction. The same ranking of magnitudes for the breakers which exist for the velocities is repeated here for the convective acceleration, the surging, the plunging and eventually the nearly spilling. The local accelerations are negative during a swash cycle except at the beginning of swash cycle for the surging wave and near the end of the swash cycle at which they are onshore-directed. Onshore-directed local accelerations are often observed at the rundown position during backwash. If onshore-directed local accelerations occur at the beginning of the swash cycle, the values decrease drastically with the time and their direction changes suddenly. The spilling and plunging breakers also show the same pattern, in fact both breaker have positive local acceleration at the beginning of the swash cycle however the local acceleration of the spilling breaker has very small values. Again the biggest values of acceleration belong to the surging wave and after that the plunging and finally the nearly spilling which will be seen later (Figure 7).

Total acceleration which is the summation of convective and local acceleration has been shown in the 4th column of Figure 4. As it has been already mentioned, the convective acceleration is generally small with respect to local acceleration. Therefore, the total acceleration and the local acceleration are similar in pattern and behavior. However, the convective acceleration has large positive quantities at the beginning and end of swash cycle with a small time lag with respect to the maximum local acceleration resulting in the amplification and the extension of the onshore-directed total acceleration.

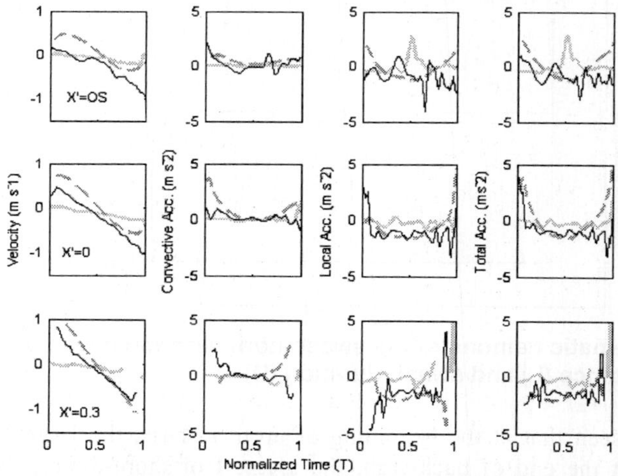

Figure 4. Normalized time series of fluid parameters for swash driven by surging (dashed line), plunging (thin dark line) and nearly spilling (gray line) waves just above the bed and at locations within the normalized swash excursion (OS denotes just outside the swash zone).

For better understanding of the accelerations in the swash zone, Figure 5 shows the timestack of the velocity, convective, local and the total acceleration fields just above the bed for the surging breaker. In Figure 5A, in every swash event's velocity field, two regions can be recognized which are nearly symmetric. At the left side of a single swash velocity field, the onshore direction and at the right side, the offshore direction velocity field can be seen which are related to the uprush and the backwash phases respectively.

These two regions meet each other at the center which is the time at which the flow direction changes, so the velocity values go to zero at the mid-swash cycle. The maximum bed-parallel velocity magnitudes for all breakers studied was on the order of 1.5 m s^{-1}. The maximum onshore-directed velocity happens at the beginning of uprush phase and also at the location near to beginning of swash zone, while the offshore-directed velocity occurs at the end of backwash phase and almost at the same location in which the maximum onshore-directed velocity occurred.

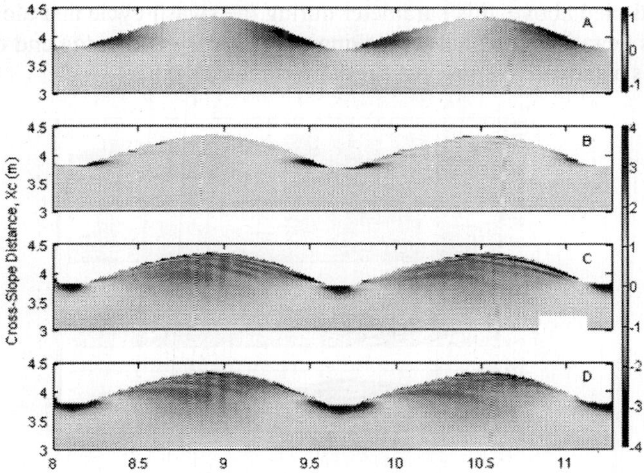

Figure 5. Timestack of bed-parallel velocity (A; m s^{-1}), bed-parallel convective acceleration (B; m s^{-2}), bed-parallel local acceleration (C; m s^{-2}) and bed-parallel total acceleration (D; m s^{-2}) for the surging wave case. (Values are denoted by the respective colorscales with positive being shoreward-directed)

Figure 5B shows that the convective accelerations in the swash zone have small and almost positive (onshore-directied) or zero values in time and space, except at the beginning and end of swash cycle which have larger values. As Figure 5C shows, the local acceleration field in the swash zone has negative (offshore-directed) quantities at any time and space except at the time between two successive swash events. Comparing the local and convective acceleration shows that the occurrence of the maximum onshore convective and local acceleration are not coincident and they do not take place at the same location as well, this leads to the wider time-space zone of positive (onshore) total acceleration between two successive swash events which can be seen in Figure 5D.

Figure 6 shows the velocity, convective, local and total acceleration for plunging case. Due to the breaking nature of the wave and air entrapment, it is hard to delineate the actual swash and calculate accelerations. However, the accelerations appear to be negative during a swash event except at the very beginning and end of swash cycle.

Figure 7 shows the local acceleration field during a single swash event for the surging, plunging and nearly spilling cases respectively. As it can be seen and mentioned above, this parameter during the swash cycle and along the slope is always negative except at the beginning of uprush and at the end of backwash which has a short-lived positive (onshore-directed) value.

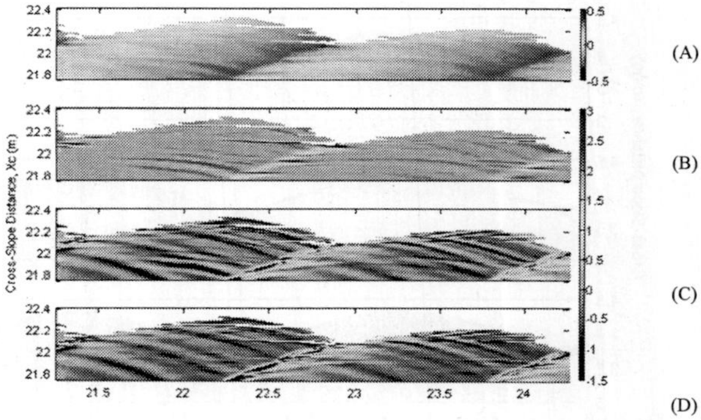

Figure 6. Timestack of bed-parallel velocity (A; m s^{-1}), bed-parallel convective acceleration (B; m s^{-2}), bed-parallel local acceleration (C; m s^{-2}) and bed-parallel total acceleration (D; m s^{-2}) for the plunging wave case.

Since the depth-averaged velocity, convective, local and total accelerations quantities are fairly smoother than their values at a given elevation which may include some discontinuities due to air entrainment and splashing, the depth-averaged quantities are shown in Figure 9. The depth averaged velocity, convective, local and total acceleration at the location offshore (x=3.63m), at the beginning of the swash zone (x=3.75m) and inside the swash zone (x=3.86) for the surging case.

As is seen, the depth averaged flow velocity in the swash zone is nearly linear during much of its duration and experiences the largest magnitude right at the seaward swash boundary. The depth-averaged convective acceleration is found to be small in the swash zone except at wave rundown where it reaches its maximum magnitude. The depth-averaged convective acceleration has two positive peaks one right after flow starts the uprush phase and second at the end of backwash phase. In contrary, the depth-averaged local acceleration is negative except at the beginning of swash cycle which reaches its maximum onshore-directed magnitude.

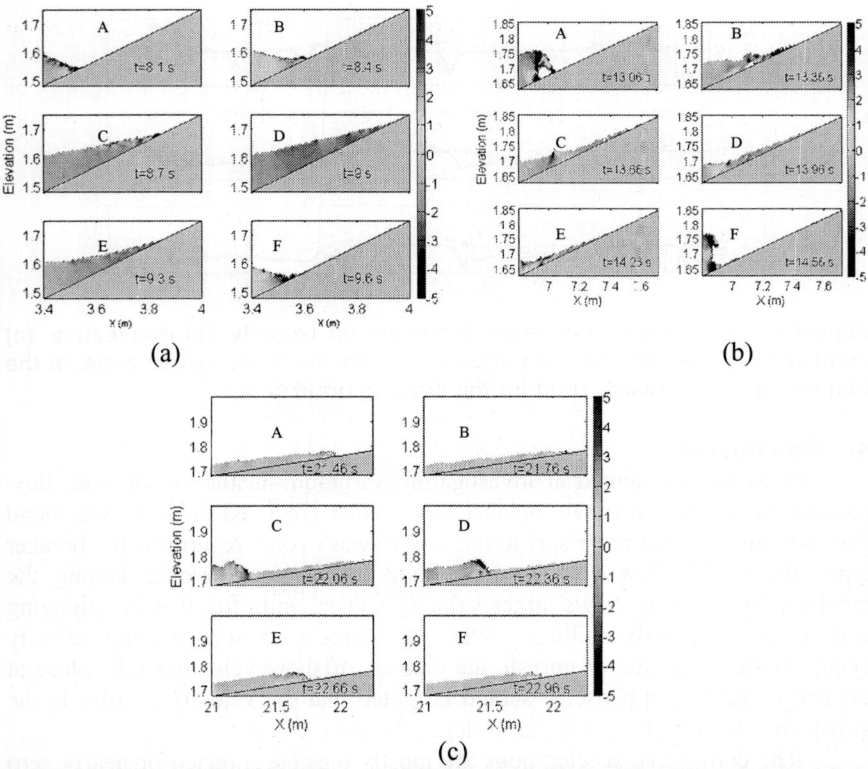

Figure 7. Bed-parallel local acceleration snapshots of surging (a) plunging (b) and nearly spilling breakers for a single swash event. Colorscale represents velocity with positive being shoreward-directed.

Here, the interesting point is that while the depth-averaged convective acceleration has two maxima during a single swash event, the depth averaged local acceleration has only one peak at the beginning of the swash cycle. This will result the amplification of the depth-averaged total acceleration at the time between two successive swash events and also broaden the time scale of onshore-directed total acceleration at this time.

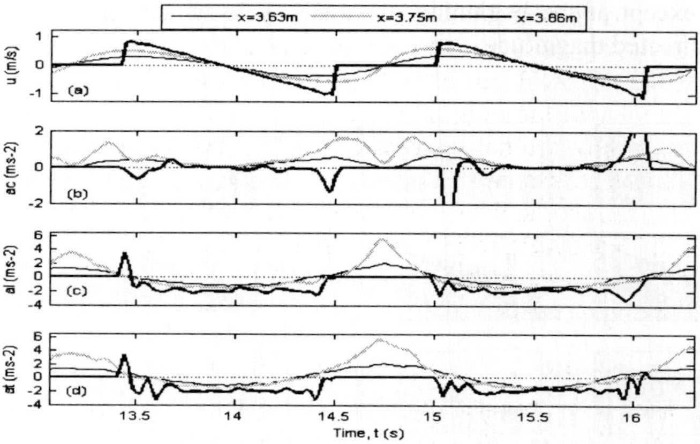

Figure 9. The variation of depth averaged (a) velocity, (b) convective, (c) local and (d) total acceleration at three locations, before swash zone, at the beginning and in swash zone for the surging breaker

4. Conclusion

This study was aimed at investigating variations in the swash zone flow parameters for three different breaker types under idealized forcing. We found that moving from the inner surf to the lower swash zone, regardless the breaker types the swash flow velocity magnitude increases. Moreover among the breakers, the surging shows larger velocity values, and after that the plunging and finally the nearly spilling. While the largest onshore-directed velocity occurs at the beginning of uprush, the greatest offshore velocities take place at the end of backwash phase. It should be noted that the velocity profiles in the swash zone tends to be nearly linear during the swash cycle.

The convective accelerations are mostly onshore-directed or nearly zero especially in the nearly spilling breaker case. The convective accelerations reach their highest values at the early uprush and at the end of backwash phases in onshore direction. The largest convective acceleration values were found for the surging wave case followed by the plunging and nearly spilling wave cases.

For surging waves, the local accelerations are negative during a swash cycle except at the beginning of swash cycle at which they have large positive values. Furthermore the onshore-directed local acceleration only occurs near the wave rundown limit.

The local accelerations for swash driven by spilling and plunging breakers tend to be negative for most of the swash duration. The spilling wave breaker has a short-lived positive value of local acceleration at the beginning of swash cycle on the leading edge of flow. Also, the plunging wave experiences positive acceleration at the transition between the surf and swash zones where the wave plunges on the beach. Based on the computed results, it can also be concluded that, the time and location in which the maximum convective and local acceleration happen are not the same. There appears to be a time lag between these two accelerations, while two high velocity and convective acceleration areas are recognizable in the velocity and acceleration fields (in space and time), just one high local acceleration zone tends to be identifiable. Therefore, the total acceleration time-space field has a wider high total acceleration zone due to the phasing of its components.

Acknowledgments

Supported was provided by Tarbiat Modarres University as a visiting scientist to CACR (AF), the University of Delaware (JAP) and the US Army Corps of Engineer's MORPHOS project (NK).

References

1. Hughes, M.G. and Baldock, T.E., 2004. Eulerian flow velocities in the swash zone: Field data and model predictions. J. Geophys. Res., 109: 11.
2. Kobayashi, N., DeSilvia, G.S. and Watson, K.D., 1989. Wave transformation and swash oscillation on gentle and steep slopes. J. Geophys. Res., 94(C1): 951-966.
3. Kothe, D.B., Mjolsness, R.C. and Torrey, M.D., 1991. RIPPLE: A computer program for incompressible flows with free surfaces. LA-12007-MS, Los Alamos National Laboratory.
4. Lin, P., and P. L.-F. Liu (1998), A numerical study of breaking wave in the surf zone, J. Fluid Mech., 359, 239-264.
5. Lin, P. Z., et al. (1999), Runup and rundown of solitary waves on sloping beaches, J. Waterw. Port Coast. Ocean Eng.-ASCE, 125, 247-255.
6. Lin, P., and W. Xu (2005), NEWFLUME: A numerical water flume for two-dimensional turbulent free surface flows, J. Hydraul. Res., 1-15.
7. Masselink, G., Evans, D., Hughes, M.G. and Russell, P., in press. Suspended sediment transport in the swash zone of a dissipative beach. Mar. Geol(2006).

8. Puleo, J.A. 2004. Hydrodynamics and sediment transport in the inner surf and swash zones. Ph.D. Dissertation, University of Florida.
9. Puleo, J.A., Beach, R.A., Holman, R.A. and Allen, J.S., 2000. Swash zone sediment suspension and transport and the importance of bore-generated turbulence. J. of Geophys. Res., 105(C7): 17021-17044.
10. Puleo, J.A., Holland, K.T., Plant, N., Slinn, D.N. and Hanes, D.M., 2003. Fluid acceleration effects on suspended sediment transport in the swash zone. J. Geophys. Res., 108.
11. Puleo, J.A., Slinn, D.N., Holland, K.T. and Smith, E., Submitted. Numerical simulation of inner surf and swash zone hydrodynamics using the volume of fluid technique. Coastal Engineering Journal.
12. Raubenheimer, B., Guza, R.T., Elgar, S. and Kobayashi, N., 1995. Swash on a gently sloping beach. J. Geophys. Res., 100(C5): 8751-8760.

OPTICAL MEASUREMENTS OF SWASH KINEMATICS

Margaret Palmsten[1] and Robert Holman[1]

Foreshore bathymetry is known to respond to variations in offshore wave forcing, despite the fact that incident wave heights are typically depth limited in the surf zone, so do not directly represent those wave height variations. In an effort to investigate the nature and dynamics of swash velocity moments under varying wave conditions we have developed an algorithm for optical measurement of cross-shore swash velocity by exploiting foam traces on waves. The resulting velocities were compared with in situ measurements of swash velocity in the frequency domain. Spectral comparisons of optically derived velocity time series with in-situ velocity measurements showed similar spectral shapes and energies. Coherence between the two times series was high, especially at low frequencies. A time lag identified between velocity records increased towards shore.

INTRODUCTION

Beaches erode during storms and accrete during fair weather. Paradoxically, incident band wave heights in the inner surf zone do not vary between these conditions, since wave height is depth-limited in the surf zone and thus is insensitive to offshore wave conditions. However, previous work has shown that low frequency, infragravity motions become energetic during high-energy periods and these may affect storm-driven sediment transport in the swash zone.

We hypothesize that offshore transport during high energy periods may be the result of offshore-directed swash velocity moments of infragravity motions. Our hypothesis is supported by the in situ measurements of Butt and Russell (1999), where high-energy conditions were associated with negative infragravity skewness and backwash exceeded the sediment transport threshold. During low energy conditions Butt and Russell (1999) observed rapid, turbulent transitions between offshore and onshore flow and advection of sediment onshore.

Our motivation in this work is to develop an understanding of swash velocity moments in an effort to understand swash zone sediment transport. To avoid the complications of traditional swash zone sampling, we have developed a technique for remotely measuring swash velocity, which provide a spatially dense, logistically simple, and non-invasive alternative to in situ measurements of the thin lens of water composing swash. The algorithm will be described, then applied to optical measurements obtained during infragravity-dominated conditions during the Nearshore Canyon Experiment (NCEX) at La Jolla, CA, USA in Fall 2003. The optically-derived surface cross-shore velocities will be compared with in situ measurements of swash made close to the seabed.

[1] College of Oceanic and Atmospheric Sciences, Oregon State University, 104 COAS Admin Building, Corvallis, Oregon, 97311, USA

ALGORITHM DEVELOPMENT

Method

Cross-shore swash velocity was estimated from the movement of foam patterns on breaking waves (Fig. 1). The bright foam patches act as a tracer, representative of surface velocity on the breaking wave. The analysis was based on optical image intensity data, collected along a single cross-shore transect of pixels through the swash zone over many video frames. The resulting time stack records the time evolution of cross-shore foam trajectories across the swash zone. The cross-shore transect used in this analysis is shown by the black line through the surf and swash zone in Fig. 1. The propagation of waves across the cross-shore transect is shown in Fig. 2. Pixel resolution in the transect is O(10 cm).

Figure 1. Oblique view of a video frame showing the cross-shore transect used to generate a time stack.

The dominant features in the time stack are the high intensity bore fronts (e.g. Fig. 2, solid arrow), which progress from the offshore, through the surf zone, until the bore becomes runup on the beach. The slope of these bore front features in the time stack, equivalent to the inverse of the bore front velocity, changes with time as the bore decelerates toward the beach. The other predominant features in the time stack are the trajectories of foam left on the water surface following passage of the bore (Fig. 1, dashed arrow). Determining the slopes of these foam streaks, equivalent to the inverse of the water velocity, is the central goal of the algorithm described below.

Analysis involves a number of steps. To isolate the foam patch trajectories and reduce the signatures of bores, a band pass filter was applied in the x-direction, passing foam patches with widths of 1 m to 4 m. The cross-shore velocity of any region of foam patches was then calculated from the shift in the cross-shore direction of the foam patches between consecutive samples.

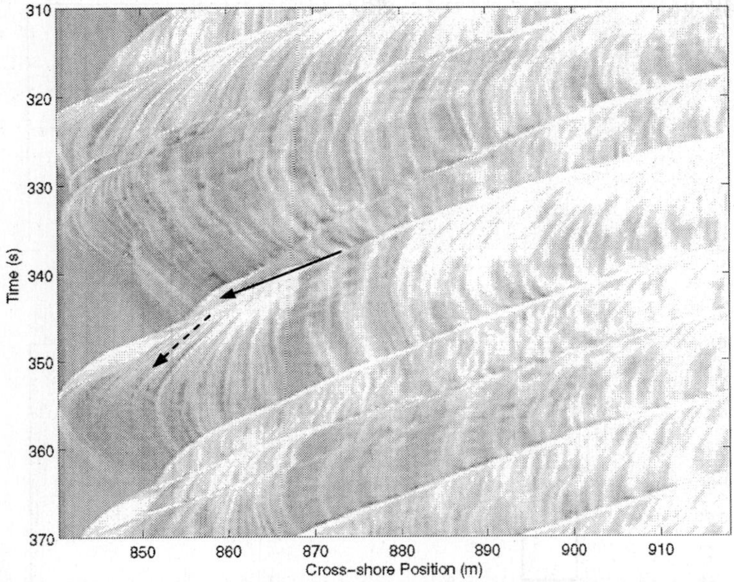

Figure 2. A time stack generated from a cross-shore transect of pixels shows the progression of foam streaks in the cross-shore direction. Waves approach the beach from the right side of the time stack, move through the surf and swash zone (solid arrow), then become runup at the left side of the time stack. The slope of the foamy bore front is the inverse of cross-shore celerity. The dashed line identifies an example foam streak trajectory after the bore has passed. The slope of these trajectories correspond to the inverse of the cross-shore water velocity.

A dynamic time warping algorithm was applied to calculate the movement of foam streaks. This class of algorithm has previously been applied to voice recognition and stereo matching problems (Munich and Perona 1999) and generates a robust estimate of surface velocity, not influenced by potentially noisy statistics of individual foam patches. A running window, 6.1 m in length, is applied to two consecutive time samples of band-passed image intensity (Fig. 3, middle panel). Lag correlations between the two consecutive samples are calculated within the window, up to a maximum lag of 0.5 m, which, for the 6 Hz sample rate, is equivalent to allowing a maximum velocity of 3 m/s. The window is shifted to the next pixel in the cross-shore transect and lag correlation for the window at that location is calculated. In this way, lag correlation, R, is calculated for each pixel in the cross-shore transect (Fig. 3, lower panel). The lag of maximum correlation, $D(x,t)$, is found through the cross-shore transect using the method described by Munich and Perona (1999); white line in the lower panel of Fig. 3). Cross-shore velocity is then calculated as,

$$u(x,t) = \frac{D(x,t)}{\Delta t} \tag{1}$$

where Δt is the time step between samples.

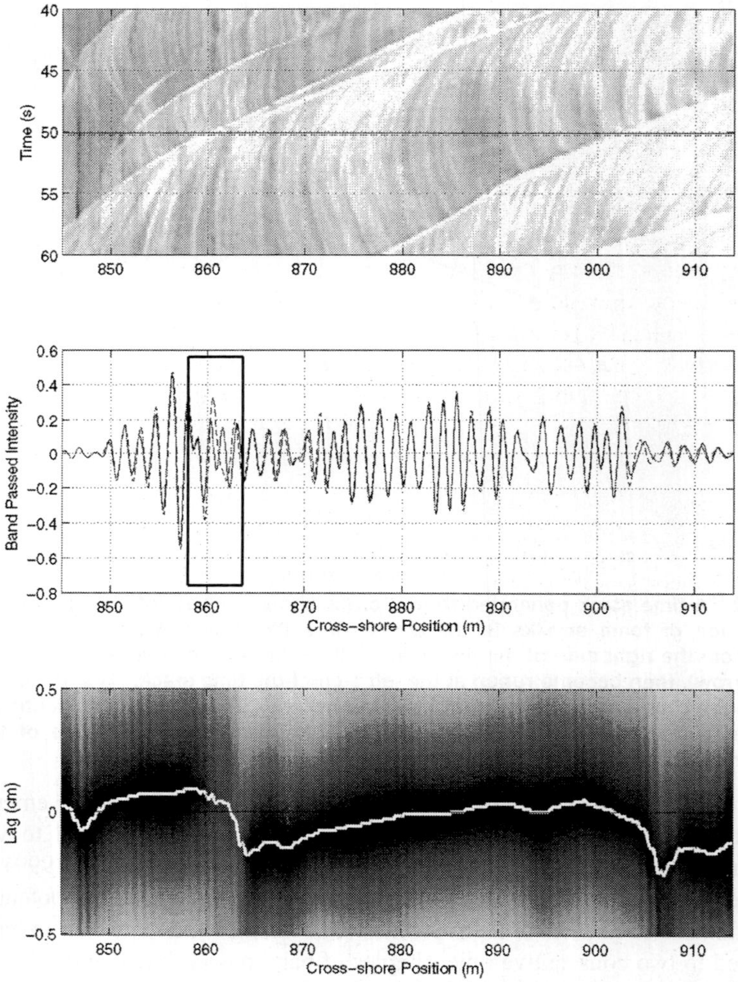

Figure 3. (Upper panel) A time stack with foam traces, identifiable as white steaks, moving onshore and offshore in time. Two consecutive time samples are highlighted in solid and dashed lines. (Middle panel) Band passed intensity is plotted for the two time samples shown above. An example of the running window is plotted in the black box. Onshore movement of foam occurred where the dashed line is onshore of the solid line and offshore movement occurred where the solid line is onshore of the dashed line. (Lower panel) Lag correlation is computed for the 6 m running window at each pixel location in the cross-shore transect. Darker colors indicate higher correlation. The path of maximum correlation through the transect is plotted in white.

Results

The dynamic time warping algorithm yields a spatially dense measure of cross-shore velocities for the swash and inner surf zone results (Fig. 4). Cross-shore velocity vectors in Fig. 4 overlie the time stack of foam traces. High negative velocities generally align with passage of the bore front. Onshore movement of foam streaks after bore passage matches with negative velocity measurements and positive velocities match with offshore movement of foam tracers. The transition of the foam streaks from onshore movement to offshore movement is associated with 0 m/s velocity.

Velocities near a passing bore are usually biased toward the bore front velocity because the strong bore front signal dominates the correlation window. In Fig. 4, the width of the bore front is exaggerated from approximately 1 m width in the time stack to the width of the correlation window. This cross-shore shift due to the correlation window causes the bore front to appear to arrive earlier in the velocity record than in the time stack. In addition, time derivatives of foam movements, used to calculate velocity, were computed over 1/6 second frame rates, a process that introduced some high frequency velocity fluctuations, visible in Fig. 4.

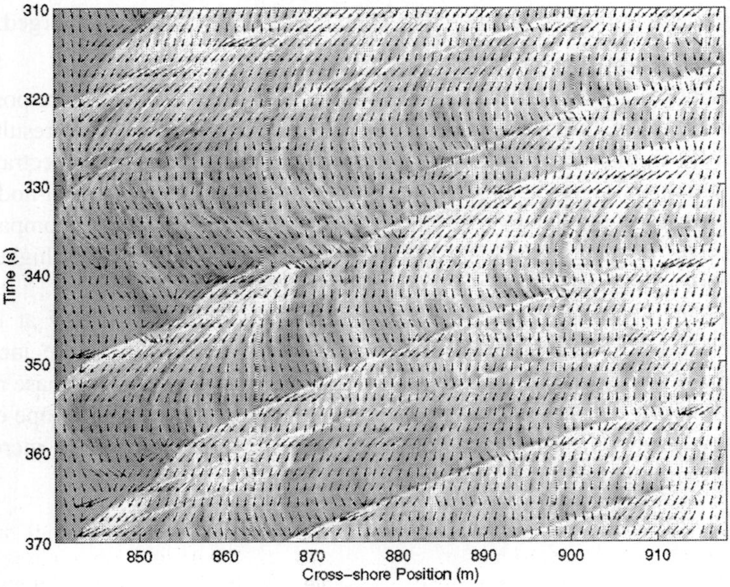

Figure 4. Cross-shore velocity vectors are overlain on the image intensity time stack.

COMPARISON WITH IN SITU DATA

Field Data

Velocities resulting from the dynamic time warping method were compared with in situ measurements in an effort to validate the algorithm. Video images and in situ measurements were collected October 31, 2006 during the Nearshore Canyon Experiment (NCEX) in La Jolla, CA, USA. The video camera was located on a cliff above the beach overlooking a cross-shore array of 5 quadpods equipped with Acoustic Doppler Velocimeters (ADVs) in the inner-surf and swash zones. Video images were collected at a rate of 6 Hz and pixel resolution in the vicinity of the ADVs was approximately 0.1 m. The cross-shore pixel transect (Fig. 1) used to generate a time stack (Fig. 2) was located 3 m north (alongshore) from the ADVs to avoid influence of the stationary quadpod legs on the optical measurements of velocity. Assuming shallow water waves and an angle of incidence of less than 20 ^{o}, the time lag associated with alongshore nonuniformity will be less than 0.5 s. The most offshore quadpod (QP5) was located in 0.5 m mean water depth. The two most onshore quadpods (QP1 and QP2) were frequently subaerially exposed during the backwash. The in situ velocity time series were collected at 16 Hz, then deglitched to remove data with low signal to noise ratio and smoothed using 0.5 second moving window. Velocity was set to 0 m/s during times when the ADVs were not submerged.

Results

The optical and in situ velocity time series data at each cross-shore position were compared by computing spectra and cross-spectra (Fig. 5). The results are qualitatively similar at all cross-shore locations. In situ and optical spectra have similar structures (Fig. 5, upper panel), with spectral nodes and anti-nodes at similar locations, particularly for lower frequencies. Variance is comparable between 0.02 Hz and 0.04 Hz. At higher frequencies, variance at is higher for optically measured velocity.

Coherence–squared values (Fig. 5, middle panel) are above 0.9 at lower frequencies and remain significant, although lower, across the entire incident wave band. Cross-spectral phase values (Fig. 5, lower panel) show a phase ramp, consistent with a constant time lag between the two time series. The slope of the phase ramp, hence the corresponding time lag between the two series, increases significantly toward shore (Fig. 6 and Table 1).

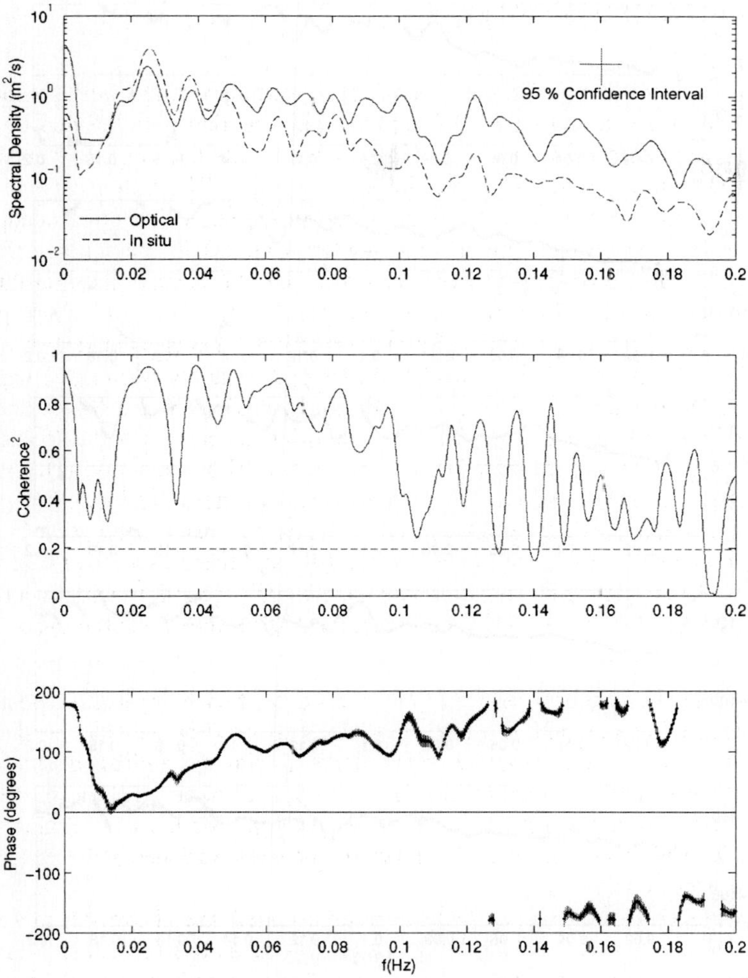

Figure 5. (Upper panel) Spectral structures of the optical and in situ velocity time series are similar at the mid-swash position. Variance of the time series is similar at lower frequencies but variance of the optical velocity time series is greater at higher frequencies. A phase ramp indicates a constant time lag between the optical and in situ velocity time series. Phase is plotted with the 95 % confidence interval for frequencies with significant coherence squared.

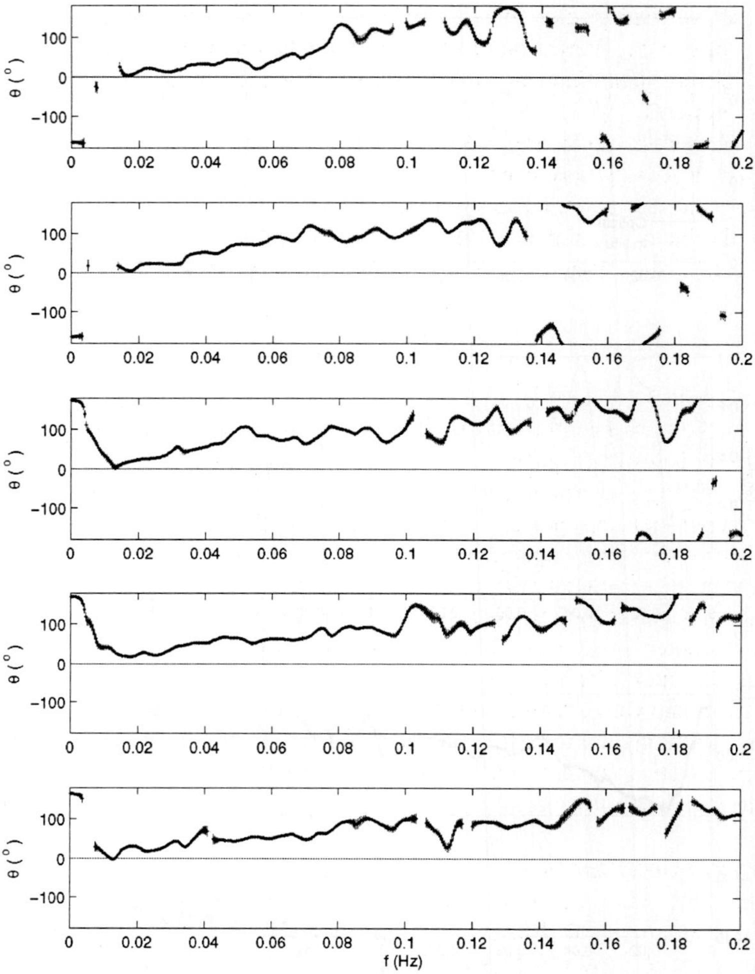

Figure 5. Cross-spectral phase plots for each of the five locations of the in-situ instruments, from landward-most (upper panel) to seaward-most (lower panel). The slope of the phase ramp decreases off shore.

Table 1. Time lag associated with phase ramp at each Quapod.					
Quadpod	1	2	3	4	5
Lag (s)	5.3	6.1	4.1	2.3	2

DISCUSSION

Observation of a cross-shore variable time lag between the in situ and optically-derived time series has interesting physical consequences. However, three possible causes of lags have been identified. These are due to:

- A vertical gradient in velocity between surface- and near-bed velocities, due to the interaction of an uprush with the previous backwash.
- A constant time lag offset between optical and in situ data logging systems.
- Variable-length drop-outs in the ADV signal associated with the foam in a passing bore front.

To isolate the first potential source of lag, which is of scientific interest, we must resolve the latter two potential sources of lag. The methods and results of this identification are issues of current research and are beyond the scope of this paper.

CONCLUSIONS

A method has been developed for determining cross-shore swash velocities from the statistics of foam trajectories in video imagery. The algorithm relies on the dynamic time warping approach, previously applied stereo matching and voice recognition problems. The algorithm produced cross-shore velocity estimates with high spatial density but velocity estimates had some contamination associated with bore front passage. Cross-spectral comparisons indicated that the technique works well at low frequencies when compared with in situ measurements. However, a cross-shore variable time lag was identified between the optical and in situ velocity records. The source of the time lag is presently unknown but may be due to vertical structure in the flow resulting from uprush overtaking backwash. Possible contamination due to a lag between measuring systems or drop outs in the ADV record due to passage of the bore front must also be ruled out.

ACKNOWLEDGEMENTS

This work was funded by the Office of Naval Research Coastal Geosciences program, grant N00014-02-1-0154 and under the SECNAV/CNO Chair program, grant N00014-03-1-0973. The authors are grateful to Britt Raubenhiemer and Todd Holland for providing data from NCEX and valuable advice.

REFERENCES

Munich, M.E., and P. Perona. 1999. Continuous Dynamic Time Warping for translation-invariant curve alignment with applications to signature verification, *Proceedings of 7th International Conference on Computer Vision*.

Butt, T. and P. Russell. 1999. Suspended sediment transport mechanisms in high-energy swash, *Marine Geology*, 161, 361-375.

MODELING INNER SURF HYDRODYNAMICS DURING STORM SURGES

J.S.M. van Thiel de Vries[1], J. van de Graaff[1], B. Raubenheimer[2], A.J.H.M. Reniers[1], M.J.F. Stive[1]

Delft3D surfbeat model predictions of surf and swash zone water level and fluid velocity fluctuations are evaluated using field data (including dissipative and reflective conditions) and a large scale, laboratory dune erosion test. The simulated hydrodynamics compare well with measurements for the dissipative and dune erosion conditions.

INTRODUCTION

During large wave events, storm surge may cause big waves to impact the sandy dune face. As a consequence, sand episodically slumps from the dunes onto the active beach, where it is transported offshore by inner surf and swash zone flows.

A good understanding of the near shore hydrodynamics is essential to improve and model dune erosion. In this paper a model is presented that solves the nearshore hydrodynamics on wave group time scales. It is expected that inner surf and swash zone hydrodynamics are dominated by long (infragravity) waves that are not dissipated strongly, whereas short wave energy is small owing to depth-induced breaking.

In deep water, short wave energy, and therefore also radiation stresses, vary on a wave group time scale. These temporally and spatially fluctuating radiation stresses are compensated by water level gradients that form a bound long wave that is 180° degrees out of phase with the short wave envelope (Longuet-Higgins and Stewart 1961). The bound long waves propagate with the wave group velocity and are released in the surf zone where the groupiness of the short waves disappears due to depth-induced breaking (Figure 1). In addition, the presence of wave groups causes the break point, defined as the location with most intensive wave breaking, to vary in time. This moving break point works like a wave paddle generating free long waves (Symonds et al. 1982).

Shoreward from the break point, released long waves propagate at the shallow water wave celerity, which results in a phase shift relative to the short wave envelope (Roelvink and Stive 1989). As the water depth decreases, the long-wave fronts become steeper and may break (Van Dongeren et al. 2006). At the shore line, remaining long waves reflect.

---

[1] Section of Hydraulic Engineering, Delft University of Technology, Stevinweg 1 2628 CN, Delft, The Netherlands

[2] Group, Woods Hole Oceanographic Institution

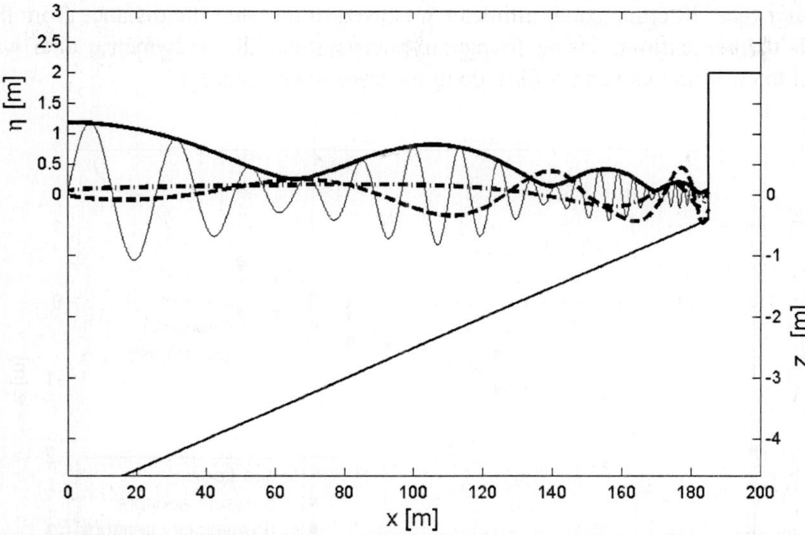

Figure 1: Waves normally incident on a sandy dune coast (thin solid line) with the corresponding short wave envelope (thick solid line), incoming bound (and released) long waves (dashed line) and outgoing shoreline-reflected free long waves (dashed dotted line).

Here, the measurements used for model validation and analysis are discussed first, followed by a brief description of the model and nine simulations. The simulation results are used to validate the model and to obtain insight into the physical processes that drive the nearshore hydrodynamics. Finally there are some conclusions.

MEASUREMENTS

La Jolla field experiment

A field experiment focused on measuring inner surf and swash zone hydrodynamics was conducted on a mildly sloped (about 0.02), fine grained (mean diameter about 0.2 mm) beach near La Jolla (California, USA) during September and October 2000 (see Raubenheimer 2002).

Eight buried pressure sensors and acoustic Doppler velocity meters (ADV) located 5 and 20 centimeters above the bed were deployed on a cross-shore transect (Figure 2). Data were sampled at 16 Hz for 3072 second (51.2 min) bursts starting every hour. The position of the instruments relative to the surf and swash zones varied with the tide, which had a vertical range of about two meters. Measured wave angles at the most seaward sensor were within ten degrees of normal incidence.

The beach and foreshore profiles were measured several times during the experiment with a differential global positioning system (GPS). In water depths greater than 50 cm, a sonar altimeter was used to measure the distance from the GPS to the seafloor. Using triangular interpolation, the bathymetric data was used to construct eleven profiles along the cross shore transect.

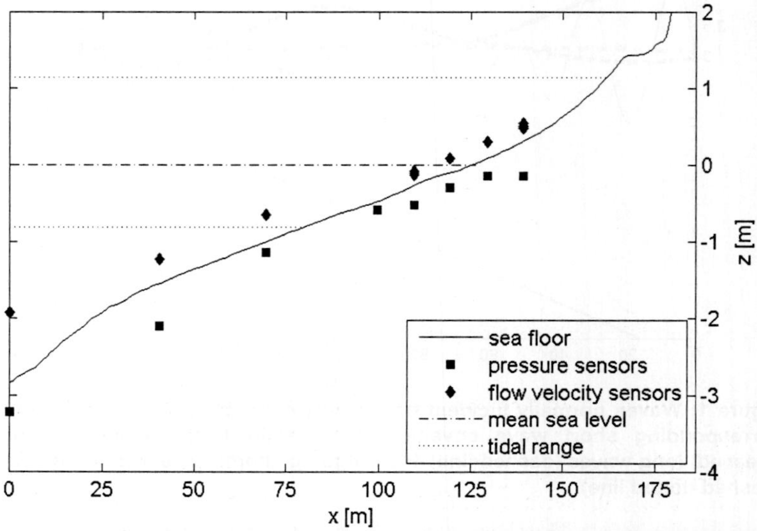

Figure 2: Cross shore beach profile and instruments deployed during the field experiment.

Large scale dune erosion tests

Large scale dune erosion tests were conducted in the Deltaflume, a WL|Delft Hydraulics research facility, from November 2005 till February 2006 (Van Gent et al. 2006). Eight tests with different hydrodynamic boundary conditions were produced.

In this paper the last hour of the sixth test (T06) is simulated (Table 1). Measurements to evaluate the model performance are all from a cluster of sensors that was fixed to the flume wall. It consisted of ten pressure sensors to measure wave transformation over the foreshore, three resistance wires and a co-located pressure sensor and electromagnetic velocity sensor to separate incoming and -reflected waves and two electromagnetic velocity meters to measure inner surf and swash zones hydrodynamics (Figure 3).

The initial profile during test T06 is representative of the Dutch Holland coast, with a depth scale of six, a contraction coefficient of two in the cross shore direction, and fine sand (mean diameter 0.200 mm) (Vellinga 1986) (Figure 3). The profile was measured at 0.0 0.1 0.3 1.0 2.0 and 6.0 hours.

Table 1: Hydrodynamic boundary conditions test T06

Test	Hm0 [m]	Tp [s]	Tm-1,0 [s]	Still water level [m]
T06	1.50	4.90	4.45	0.0

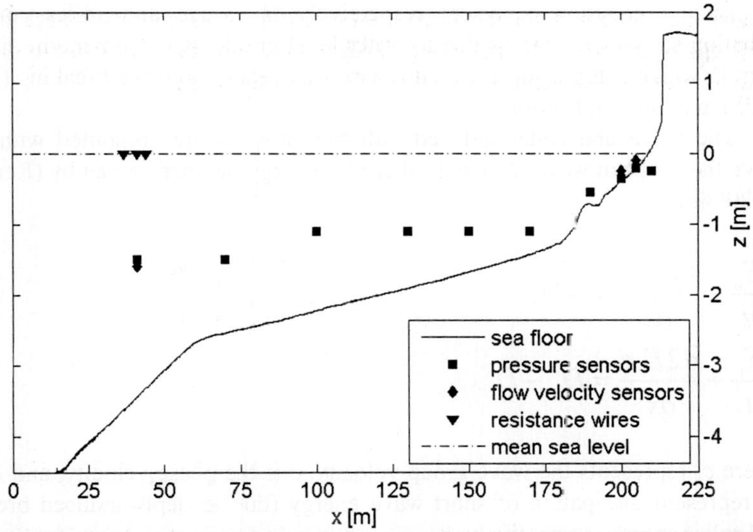

Figure 3: Cross shore transect deployed on the flume wall during the Deltaflume experiments:

MODEL

The Delft 3D surfbeat model solves the nearshore hydrodynamics on wave group time scales (Reniers et al. 2004a). In this paper, the model is depth averaged and does not include the longshore direction.

The numerical scheme was modified to switch between a momentum conservative scheme in flow expansions and an energy conservative scheme in flow contractions to simulate the rapidly varying flows in hydraulic jumps and bores in the inner surf and swash zones (Stelling and Duinmeijer 2003).

Model equations

The hydrodynamics are solved using the nonlinear shallow water equations for continuity and conservation of momentum (NSWE):

$$\frac{\partial \eta}{\partial t}+\frac{\partial (hu)}{\partial x}=0$$
$$\frac{\partial u}{\partial t}+u\frac{\partial u}{\partial x}=-F_x-g\frac{\partial \eta}{\partial x}+\upsilon_t\frac{\partial^2 u}{\partial x^2}-\frac{\tau_x}{h} \qquad (1)$$

where η is the water surface elevation, h is the water depth, u is the Generalized Lagrangian Mean (GLM) velocity, g is gravitational acceleration, t is time, and x the cross shore position (positive onshore). The terms on the right hand side of the second equation represent respectively the wave and roller computed radiation stresses F_x, forces due to water level gradients, a momentum diffusion term in which υ_t is the turbulent eddy viscosity related to wave breaking (Battjes 1975), and bottom friction.

The wave and roller induced radiation stresses are computed with linear wave theory from wave (E_w) and roller (E_r) energy balances given by (Reniers et al. 2004a):

$$\frac{\partial E_w}{\partial t}+\frac{\partial E_w c_g}{\partial x}=-D_w$$
$$\frac{\partial E_r}{\partial t}+\frac{\partial 2E_r c}{\partial x}=D_w-D_r \tag{2}$$

where c_g represents the wave group velocity, c is the phase velocity, and D_w and D_r represent dissipation of short wave energy (due to depth-induced breaking) and roller energy, respectively.

Dissipation due to depth-induced breaking is computed as (Roelvink 1993):

$$D_w=\frac{2\alpha E_w}{T_{m-1,0}}\left[1-\exp\left[\left(-\frac{E_w}{\gamma^2 E_{ref}}\right)^{\frac{n_d}{2}}\right]\right] \text{,where } E_{ref}=\frac{1}{8}\rho g h_{bd}^2 \tag{3}$$

In these equations α is a coefficient $O(1)$, γ accounts for wave-height saturation, n_d relates the dissipation to the randomness of the incoming waves, and ρ is water density. The term h_{bd} is a water depth weighted over about one wave length seaward from the cross shore position of interest (Roelvink et al. 1995). Dissipation of roller energy is modeled with a constant β ($\beta=O(0.1)$) (Reniers and Battjes 1997):

$$D_r=\frac{2g\sin(\beta)E_r}{c} \tag{4}$$

Finally, computed Lagrangian velocities are corrected for the short wave Stokes drift to obtain Eulerian velocities (Walstra et al. 2000):

$$u=u^E+u^S \tag{5}$$

where u is the GLM velocity, u^E is the Eulerian velocity, and u^S is the Stokes drift computed from (Phillips 1977):

$$u^S = \frac{M_x}{\rho h}, \text{ where } M_x = \frac{(E_w + 2E_r)}{c} \quad (6)$$

Model Boundaries

The two lateral boundaries are modeled as frictionless, impermeable walls. At the seafloor, which is assumed fixed in space, a shear stress due to the combined effect of waves and currents is computed (Soulsby et al. 1993). Pressure is assumed to be zero at the water surface. An absorbing-generating boundary condition is applied at the seaward model boundary (Van Dongeren and Svendsen 1997), which allows outgoing waves to leave the model domain with minimum reflection while incoming long waves can still be specified.

To calculate the incoming long wave time series, the measured water surface elevation and velocity time series are low-pass filtered with a high frequency cut-off computed from a characteristic wave period $T_{m-1,0}$: $f_{split} = 0.5/T_{m-1,0}$. The low-passed time series are then used to separate incoming and reflected waves (Guza et al. 1984):

$$\eta_{in} = \frac{\eta\sqrt{gh} + Q}{c_g + \sqrt{gh}} \text{ and } \eta_{out} = \frac{\eta\sqrt{gh} - Q}{c_g + \sqrt{gh}} \quad (7)$$

where η_{in} and η_{out} are the water surface elevation time series associated with the incoming and reflected long waves, respectively, and Q is a discharge expressed as $Q = u^*h$.

The incoming long waves are assumed to be bound and to propagate with the wave group velocity, while the outgoing waves are assumed to be free and to propagate with the shallow water wave velocity. In reality, however, it is expected that both free and bound incoming long waves are present, which introduces small errors in equation (7).

Within the model domain, incoming long waves can be free or bound depending on the phase coupling between simulated short wave energy and water surface elevation. Incoming bound long waves that do not correspond to the wave group forcing are released at the model boundary.

Finally, the high frequency wave components from the measured variance spectrum at the model boundary are utilized to construct a short wave envelope, which is related to the short wave energy as:

$$E_w(x,t) = \frac{1}{2}\rho g \,|\, A(x,t) \,|^2 \quad (8)$$

where A is the short wave envelope. The resulting wave energy time series varies on wave group time scales and is input into the short wave energy balance. The roller energy is set to zero at the model offshore boundary.

SIMULATIONS

Nine simulations, including dissipative and reflective conditions during the La Jolla field experiment and a dune erosion test, are used to evaluate the predictions and to obtain more insight in the processes that drive the surf and swash zone hydrodynamics.

Detailed model performance for the dissipative case

Simulated and measured sea-surface elevation variance spectra agree well for frequencies below the split frequency (Figure 4). The measured distribution of variance spectra shifts toward lower frequencies with decreasing water depth. Low frequency energy (below the split frequency) dominates the measured and modeled spectra in the inner surf and swash zones. The model generation of water surface variance in frequencies larger than the split frequency could be owing to nonlinear interactions between long waves or to the presence of wave groups with envelope frequencies larger than the split frequency.

For water depths smaller than 0.7 m, measured and simulated water surface elevation time series are well correlated (Figure 5). Steep wave fronts observed in both measurements and simulations suggest that long waves may break in shallow water, as suggested by (Van Dongeren et al. 2006). Measured and simulated velocity time series also compare well in the inner surf and swash zones (Figure 6). Strong (de-) accelerations are observed in both measurements and simulations again indicating depth induced breaking of the long waves.

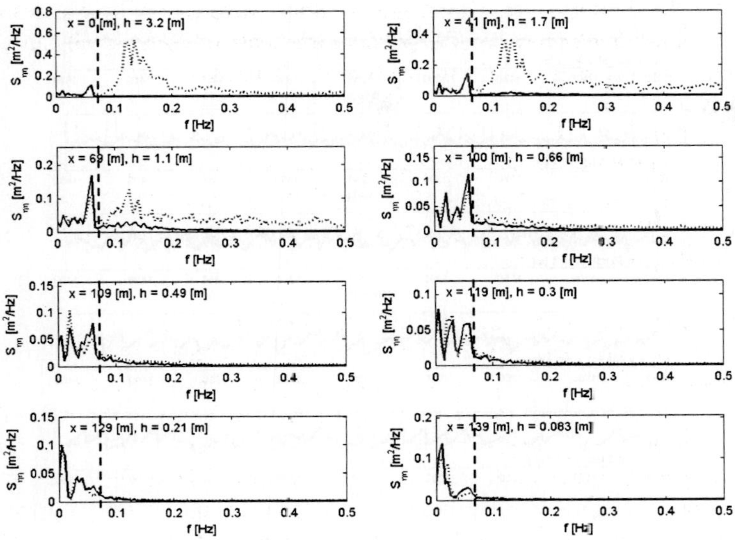

Figure 4: Measured (dotted curve) and simulated (solid curve) variance spectra (solid line) at different cross shore positions and water depths as indicated in the upper left corner of each subplot. The split frequency is indicated by the black dashed line.

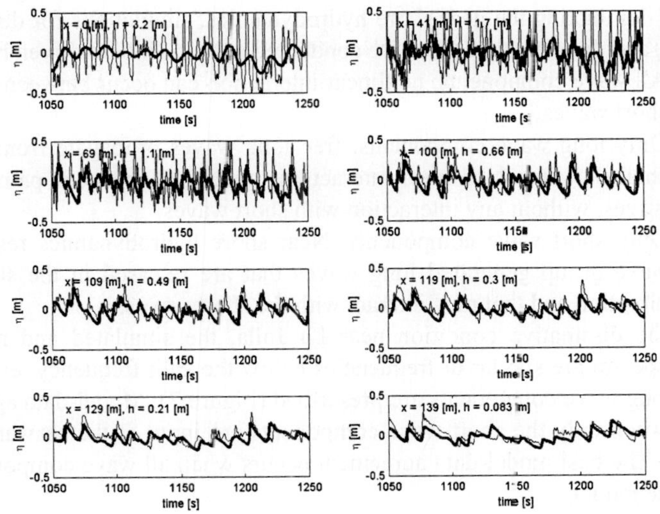

Figure 5: Measured (thin solid curve) and simulated (thick solid curve) water surface elevations at different cross shore positions and water depths indicated in the upper left corner of each subplot.

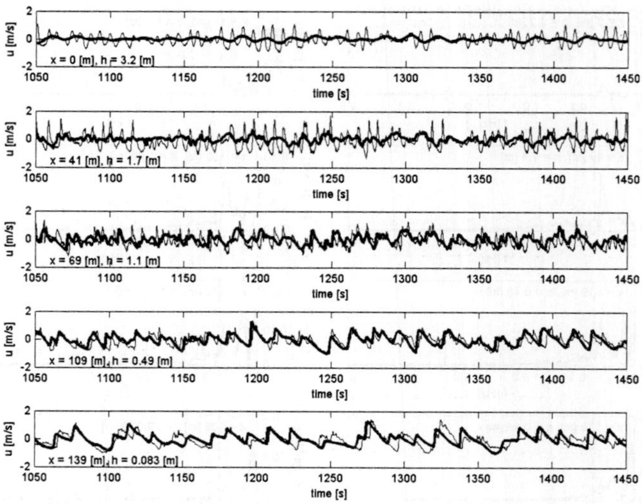

Figure 6: Measured (thin line, thin dashed line, thin dashed dotted line) and simulated (thick line) current velocities at different cross shore positions and water depths indicated in the lower left corner of each subplot.

Model sensitivity

To study the model performance further and to obtain more insight into the processes dominating the near shore hydrodynamics, simulations for dissipative, reflective, and dune erosion cases are conducted with the model driven by:

1. All wave components: nonlinear interaction can occur between long and short waves.
2. Only long wave components: free long waves propagate from deep to shallow water. Nonlinear interactions occur only between pairs of long waves, without any interaction with short waves.
3. Only short wave components: Near shore hydrodynamics result from wave group generated long waves that are released in the shoreward direction and that may interact with the short waves.

For the dissipative condition near La Jolla, the simulated and measured variance spectra are similar at frequencies below the split frequency, even when only the long wave components are prescribed (Figure 7). Model-data agreement is worse when only the short wave components are input at the seaward model boundary. The best model-data agreement occurs when all wave components are input to the model.

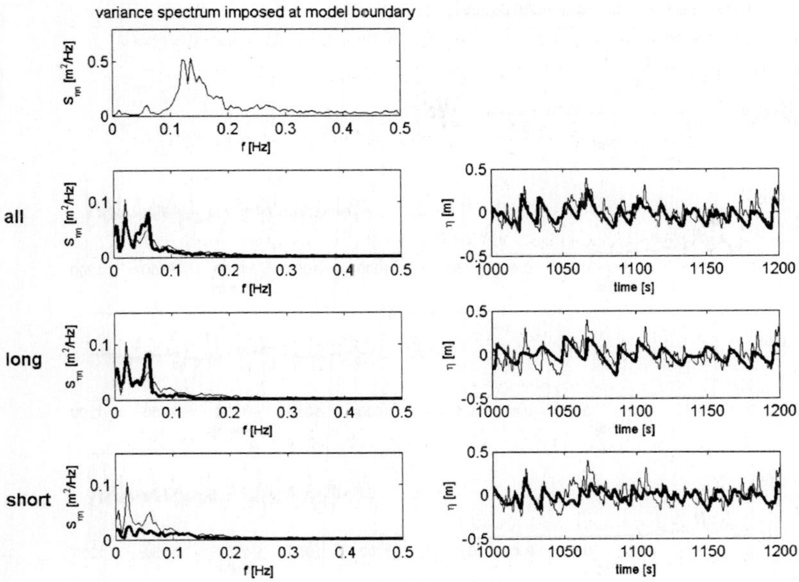

Figure 7: Measured (thin lines) and simulated (thick lines) variance spectra (left column) and water surface elevation fluctuations (right column) at 0.49 m water depth for the dissipative case. The variance spectrum imposed at the model boundary is shown as a reference (upper plot). Results are shown for imposing all, only long and only short wave components.

Similar to the dissipative conditions the measured distribution of variance spectra for the reflective conditions is observed to shift towards lower frequencies with decreasing water depth (even though the shallowest sensor is in quite deep water), (Figure 8). A significant fraction of the measured wave variance is simulated when the model is driven with either all wave components or only the short wave components. Model-data agreement is worse when the model is driven with only the long wave components. However, in all input conditions, the correlation between measured and simulated water surface elevation fluctuations is lower for the reflective case than for the dissipative case.

Like the dissipative and reflective cases, the measured distribution of variance spectra for the dune erosion case shifts to lower frequencies with decreasing water depth (Figure 9). The predicted and measured variance spectra compare well for frequencies below the split frequency when the model is driven with either all wave components or short wave components. The model overestimates the measured variance if only the long wave components are input. The measured and simulated water surface elevation time series are well correlated in 0.21 m water depth. However, when only the short wave

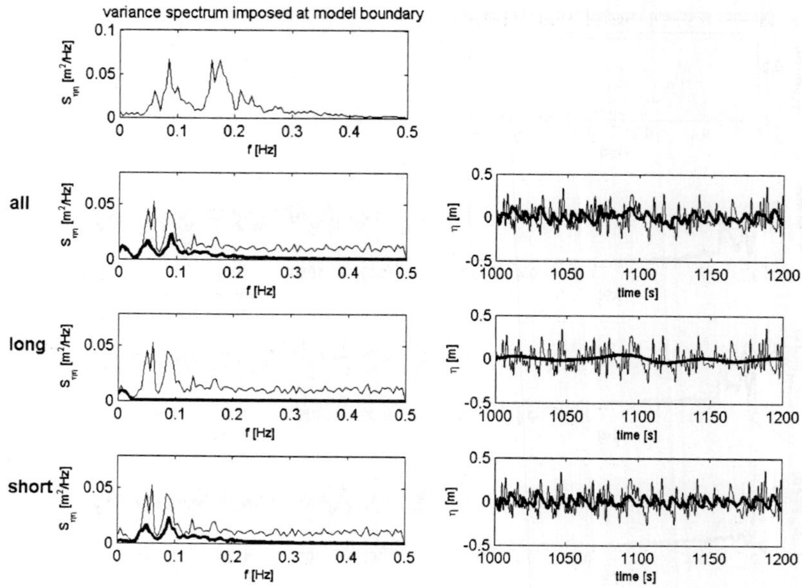

Figure 8: Measured (thin lines) and simulated (thick lines) variance spectra (left column) and water surface elevation fluctuations (right column) at 0.69 m water depth for the reflective case. The variance spectrum imposed at the model boundary is shown as a reference (upper plot). Results are shown for imposing all, only long and only short wave components.

components are input to the model, a small phase shift is visible between measurements and simulations.

DISCUSSION: Importance of nonlinear interactions

It is unclear whether the near shore dominance of long waves is owing solely to depth-induced breaking of short waves, or to nonlinear transfer of energy from short to long waves.

The best model-data agreement occurs when the model is driven with all wave components, and thus when nonlinear interactions are included in the simulations. However, imposing only long wave components at the model boundary results in reasonable predictions of near shore variance for both the dissipative and dune erosion conditions, suggesting that nonlinear interactions have little affect on the near shore hydrodynamics.

However, for both the dissipative and dune erosion studies, the offshore model boundary is within the surf zone, onshore of the region in which significant interactions are expected. Many of the long waves may have already been released owing to depth-induced short wave breaking.

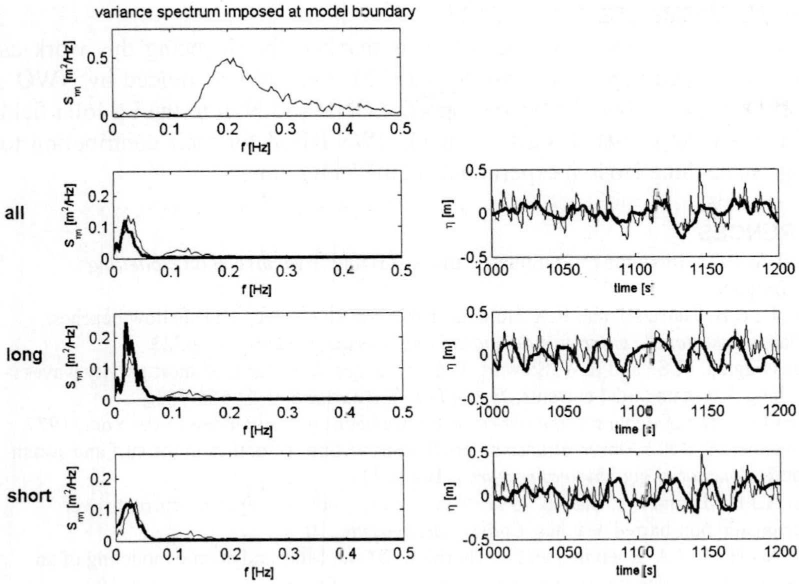

Figure 9: Measured (thin lines) and simulated (thick lines) variance spectra (left column) and water surface elevation fluctuations (right column) at 0.21 m water depth for the dune erosion case. The variance spectrum imposed at the model boundary is shown as a reference (upper plot). Results are shown for imposing all, only long and only short wave components.

CONCLUSIONS

- For the dissipative and dune erosion conditions, the model simulates near shore hydrodynamics accurately.
- For the reflective condition, predictions are less accurate. Further research will be necessary to find out more about the model performance during reflective conditions.
- Forcing the model with all wave components (prescribing a slowly varying water surface elevation signal to the non linear shallow water equations and a wave group varying energy signal to the wave energy balance) results in the best model performance.
- Forcing the model with only long waves results in good predictions of the near shore hydrodynamics for the dissipative and dune erosion conditions. For the reflective condition, only a small part of the measured near shore variance is explained.
- Forcing the model with only short wave components results in accurate hydrodynamic predictions for the dune erosion test. A significant part of the measured variance is explained for the reflective scenario. Poor results are obtained for the dissipative condition.

ACKNOWLEDGMENTS

The Dr.ir Cornelis Lely Foundation is thanked for financing this work as part of a PhD study on dune erosion. A.J.H.M. Reniers was funded by NWO`, contract DCB.5856. The financial support of ONR and NSF to the La Jolla field experiments is appreciated. Last we thank RWS RIKZ for their contribution to the large scale dune erosion experiments in the Deltaflume.

REFERENCES

Battjes, J.A. 1975. Modelling of turbulence in the surfzone, *Symposium on modelling techniques*

Guza, R.T., E.B. Thornton, and R.A. Holman. 1984. Swash on steep and shallow beaches., *19th International conference on coastal engineering*, ASCE 1, 708-723.

Longuet-Higgins, M.S., and R.W. Stewart. 1961. Changes in the form of short gravity waves on long waves and tidal currents, *Journal of fluid mechanics*, 8, 565-583.

Phillips, O.M., *The dynamics of the upper ocean*, Camebridge Univ. Press, New York, 1977.

Raubenheimer, B. 2002. Observations and predictions of fluid velocities in the surf and swash zones, *Journal of geophysical research*, 107 (C11)

Reniers, A.J.H.M., and J.A. Battjes. 1997. A laboratory study of longshore currents over barred and non-barred beaches, *Coastal engineering*, 30

Reniers, A.J.H.M., J.A. Roelvink, and E. Thornton. 2004a. Morphodynamic modeling of an embayed beach under wave group forcing, *Journal of geophysical research*, 109

Roelvink, J.A., Surf beat and its effect on cross-shore profiles, Ph.D. Thesis thesis, Delft University of Technology, Delft, Netherlands, 1993.

Roelvink, J.A., T.J.G.P. Meijer, K. Houwman, R. Bakker, and R. Spanhoff. 1995. Field validation and application of coastal profile model, *Coastal dynamics*, 818-828.

Roelvink, J.A., and M.J.F. Stive. 1989. Bar-generating cross-shore flow mechanisms on a beach, *Journal of geophysical research*, 94, 4785-4800.

Soulsby, R.L., L. Hamm, G. Klopman, D. Myrhaug, R.R. Simons, and G.P. Thomas. 1993. Wave-current interaction within and outside the bottom boundary layer, *Coastal engineering*, 21, 41-69.

Stelling, G.S., and S.P.A. Duinmeijer. 2003. A staggered conservative scheme for every Froude number in rapidly varied shallow water flows, *International journal for numerical methods in fluids*, 43, 1329-1354.

Symonds, G., D.A. Huntley, and A.J. Bowen. 1982. Two dimensional surf-beat: Long wave generation by a time varying breakpoint, *Journal of geophysical research*, 87, 492-498.

Van Dongeren, A., J.A. Battjes, T.T. Janssen, K. Steenhauer, and A.J.H.M. Reiniers, Shoaling and shoreline dissipation of low-frequency waves, 2006.

Van Dongeren, A., and I.A. Svendsen. 1997. Absorbing-generating boundary conditions for shallow water models, *Journal of waterway, port, coastal and ocean engineering*, 123, 303-313.

Van Gent, M.R.A., E.M. Coeveld, D.J. Walstra, J. Van de Graaff, H. Steetzel, and M. Boers. 2006. Dune erosion tests to study the influence of wave periods on dune erosion, *ICCE*

Vellinga, P., Beach and dune erosion during storm surges, PhD thesis, Delft University of Technology, 1986.

Walstra, D.J.R., J.A. Roelvink, and J. Groeneweg. 2000. Calculation of wave-driven currents in a 3D mean flow model, *Coastal Engineering ASCE*, vol. 2, 1050-1063.

SWASH ZONE CHARACTERISTICS AT OCEAN BEACH, SAN FRANCISCO, CA

L. H. Erikson[1], D. M. Hanes[2], P. L. Barnard[2], and A. E. Gibbs[2]

Runup data collected during the summer of 2005 at Ocean Beach, San Francisco, CA are analyzed and considered to be typical summer swash characteristics at this site. Analysis shows that the beach was dissipative with Iribarren numbers between 0.05 and 0.4 and that infragravity energy dominated. Foreshore slopes were mild between 0.01 and 0.05 with swash periods on the order of a minute. Predicted runup heights obtained with six previously developed analytical runup formulae were compared to measured extreme runup statistics. Formulations dependent on offshore wave height, foreshore slope and deep water wavelength gave reasonable results.

INTRODUCTION

The swash zone, defined as that part of the beach extending from a nearshore shallow depth to the limit of maximum inundation, is a relatively narrow region of great importance for the exchange of sediment between land and sea. Morphological processes such as storm-induced erosion, post-storm recovery, seasonal variation in foreshore shape, and evolution of rhythmic shoreline features are all driven by the inter-relationship between swash zone hydro- and sediment dynamics.

Much research has been done to understand and predict swash zone hydrodynamics and sediment transport (e.g., CSR 2006). Numerous experiments and datasets heave been collected from both the laboratory and field, and various numerical and analytical models have been developed with many significant improvements in recent years. It is now fairly well established that swash motion is driven by a combination of low frequency infra-gravity motions and incident wave bores which propagate up the beach face. The two mechanisms do not appear to be exclusive, but rather, one dominates over the other depending on the incident waves and foreshore slope.

One of the most important parameters in dealing with swash and morphological evolution is the runup length or height, defined as the limit of landward inundation. Figure 1 is a definition sketch showing a time-series of fluctuations of water elevation about its time-mean (set-up), and discrete water level elevation maxima, (R), referenced to the height above still water line (SWL). In this study, we present and analyze runup data measured during the

[1] G&G Environmental, 2601 Hambleton Lane, Santa Cruz, CA 95065 contracting to USGS

[2] Coastal and Marine Geology Program, U.S. Geological Survey, 400 Natural Bridges Drive, Santa Cruz, CA, 95060, USA

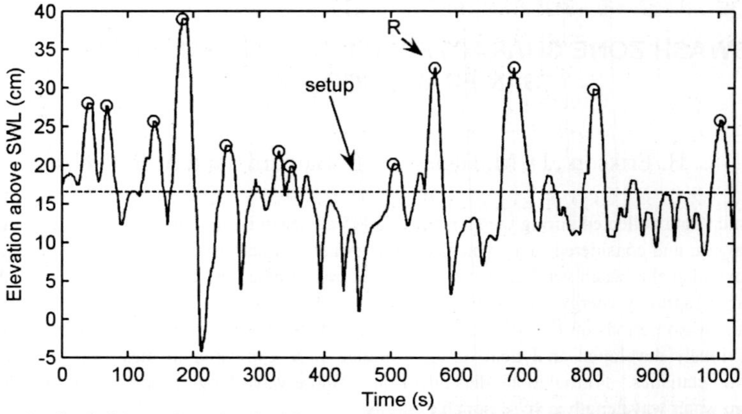

Figure 1. Definition sketch of a time-series showing swash oscillations, setup, and runup.

summer of 2005 at Ocean Beach, San Francisco, CA (Fig 2). Ocean Beach is a high-use recreation area located on the west side of San Francisco and within the Golden Gate National Recreation Area. The field effort in 2005 is part of an ongoing study that began in April 2004 to document, analyze, and simulate the processes that control sand transport and sedimentation patterns along Ocean Beach and the mouth of San Francisco Bay.

In the following sections, measurement methods and runup results are presented, followed by an assessment of the summer wave conditions and its affect on the runup limits. The paper then compares some previously developed analytical models to observed runup statistics.

Figure 2. Ocean Beach, San Francisco, CA study site (source: Google maps).

METHODS AND MEASUREMENT RESULTS

Time-series of runup elevations were determined by employing the method devised by Aagaard and Holm (1989) and Holland and Holman (1993 and 1999), whereby the leading edge of the swash is extracted from digital time-stacked images of a fixed cross-shore transect. A video camera, encased in a protective housing and mounted on top of the Cliff House Restaurant (http://www.evsboca.com/usgs/default.htm) at the north end of Ocean Beach, San Francisco, CA, collected four hours of video from which pixels along five predefined cross-shore transects (T1 through T5, approximately 50 meters apart) were extracted at 1 Hz (Fig. 3). A section of one of the time-stacks along transect 3 is shown in Fig. 3c. The location of the digitized leading edge was

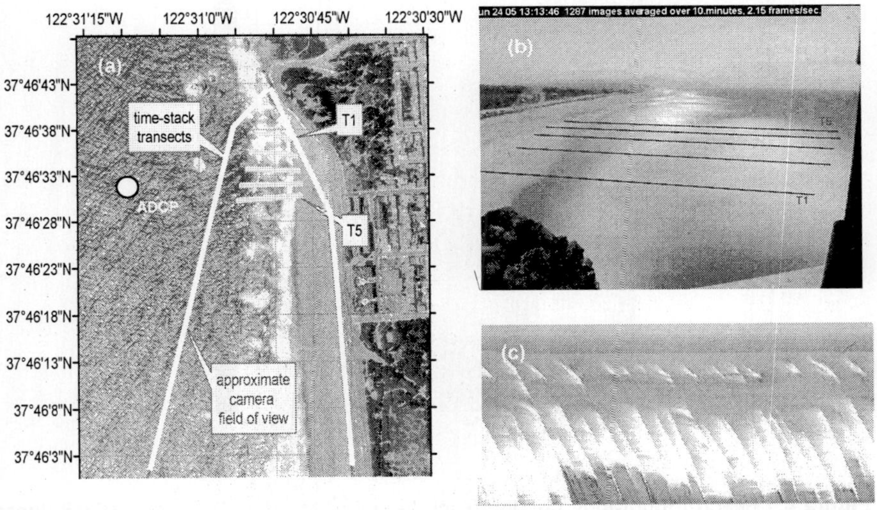

Figure 3. Overview of field measurements setup: (a) bird's-eye-view of camera, sampling transects, and location of offshore ADCP; (b) oblique camera view; (c) typical time stack image.

converted to real world coordinates with a transformation matrix determined from known ground control points, associated pixels, and assuming a plane beach (Holland et al. 1997). The resulting time-series from the time-stack images are leading edge swash motions obtained in a moving (LaGrangian) reference frame. The time-stack pixel resolution depends on the camera's position, field of view, and distance to target. For this particular setup it was 85 cm per pixel or better which, due to the mild beach slopes, translates to a vertical resolution of approximately 2 cm. Runup lengths were converted to runup heights, *R*, using a 'look-up table' of 1 meter-gridded beach topography measured with Differential Global Positioning Systems (DGPS) mounted on all terrain vehicles. Beach topography data at Ocean Beach were collected bi-monthly by the USGS as part of the Coastal Evolution and Processes project; the dataset from June 27th 2005 was used for this study. It is recognized that runup height is dependent on beach slope, particularly for reflective beaches, and the use of beach topography data obtained three days *after* the runup data may introduce some error. However, due to the mild wave climate of the summer, the three day differential is not expected to affect the results greatly as exemplified with the topographical change measured 14 days prior to and 3 days following the runup experiment (mean absolute error MAE=0.30m, Fig. 4).

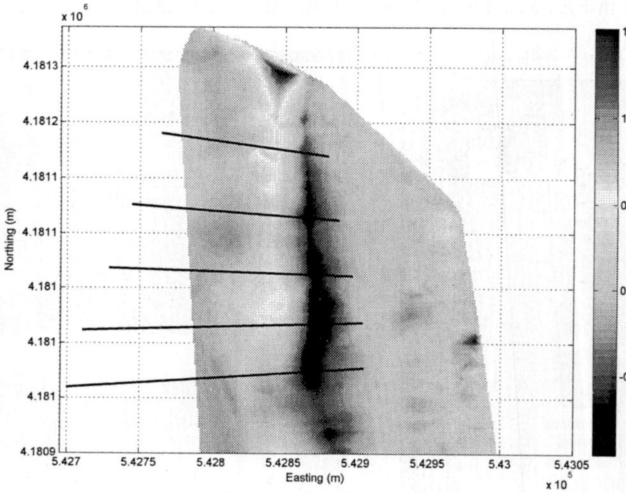

Figure 4. Beach topography change from June 12th to June 27th 2005. Black lines are cross-shore transect locations. Elevation in meters.

In order to minimize the effect of the rising tide on the location of wave breaking and area of active foreshore, and to provide consistency between the datasets, each four hour time-series at each transect was partitioned into ten 17-minute segments (1024 points at 1 Hz) for a total of $5 \times 10 = 50$ time-series. A best fit line through that part of the cross-shore profile corresponding to the minimum and maximum runup lengths during each 17-minute segment was used to define the active beach slope at each transect. Profiles of the five cross-shore transects are shown in Fig. 5. The beach slopes are mild (β<0.1) throughout but with an increase further up the coast at the two northern most transects T1 and T2 (max β=0.05).

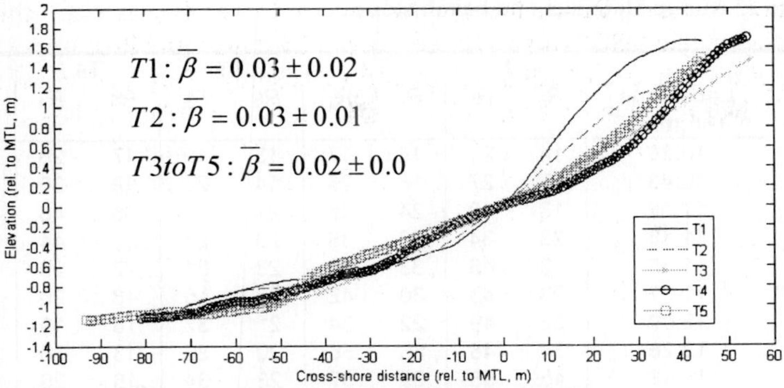

Figure 5. Cross-shore profiles along sampled time-stack transects.

Table 1. Measured and calculated offshore wave parameters.

	Measured		Back-calc deep water H_s (m)				Wave period (s)			
	H_s	h	H_o (T_{p1})	H_o (T_{p2})	H_o (T_{mo1})	H_o (T_{mo2})	T_{p1}	T_{p2}	T_{mo1}	T_{mo2}
min	1.01	7.69	0.90	1.02	1.08	1.10	13.29	9.31	6.71	6.10
max	1.35	8.49	1.18	1.31	1.46	1.47	13.47	10.20	7.42	6.81
stdev	0.12	0.29	0.10	0.10	0.13	0.13	0.06	0.30	0.26	0.25

Two runup statistics were focused upon in this study. The significant runup, R_s, was calculated as $R_s = 1.46 R_{rms}$ where R_{rms} is the root-mean-square runup height of discrete runup elevations determined from zero-down crossing in the time-domain. The significant runup is statistically equivalent to the mean of the one-third highest runup elevations assuming a Rayleigh distribution. The second runup statistic is R_2, representing the runup height that is exceeded 2% of the time. The 2% runup was found by picking off the 98th percentile from the cumulative probability density function of the discrete runup measurements. Runup statistics listed in Table 2 were calculated following removal of the tide measured at the offshore ADCP (Fig. 3), are referenced to the still water level, and include setup and both infragravity and incident wave components. Distinctions between the infragravity and incident components are discussed later.

Wave parameters utilized in this study were measured with an RDI Workhorse Sentinel ADCP deployed outside the surf zone at 8m water depth (Fig. 3). Although the instrument collected data for over a week, it was quickly buried in the sand limiting further measurements, and hence, only data from the first day and coincident with the four-hour runup data is analyzed herein. Waves were measured at two hour intervals during 68 minutes of acoustic bursts at 2Hz. Bulk wave statistics were calculated from the resulting velocity spectra shown in

Table 2. Runup statistics and beach slope.

	Local time (hh:mm)	T1 R_s	T1 R_2	T2 R_s	T2 R_2	T3 Rs	T3 R_2	T4 Rs	T4 R_2	T5 Rs	T5 R_2
1	**10:26**	14	21	13	17	13	19	17	23	25	36
2	**10:43**	17	27	18	26	14	23	18	30	24	34
3	**11:00**	13	19	24	35	21	24	38	48	21	36
4	**11:18**	23	34	25	36	20	23	31	44	20	27
5	**11:35**	33	46	33	46	23	31	22	29	24	31
6	**11:52**	28	43	30	42	25	45	18	20	18	27
7	**12:09**	32	49	22	34	21	32	16	19	21	32
8	**12:26**	34	46	36	56	20	26	13	20	18	26
9	**12:43**	45	56	39	57	23	34	16	26	22	31
10	**13:00**	40	50	42	51	21	29	19	28	25	39
	mean	28	39	28	40	20	29	21	29	22	32
	min	13	19	13	17	13	19	13	19	18	26
	max	45	56	42	57	25	45	38	48	25	39
	stdev	11	13	9	13	4	7	8	10	3	4

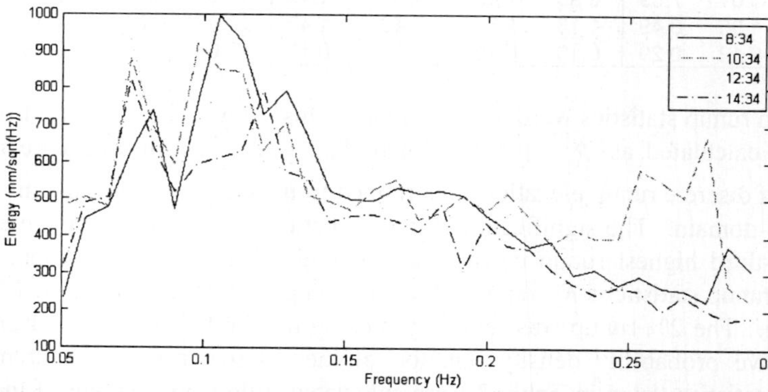

Figure 6. Velocity spectra measured with the ADCP (Site 1) concurrent with runup measurements on June 24th, 2005.

Fig. 6 with frequency cut-offs at 0.05 Hz and 0.30 Hz. The spectra show that the incident wave climate is bi-modal and hence two peak periods ($T_{p1} = 1/f_{p1}$ and $T_{p2} = 1/f_{p2}$, where f_p is the frequency at maximum energy) and mean periods ($T_{mo1} = m_0/m_1$ and $T_{mo2} = m_0/m_2$, where m_0, m_1, and m_2 are the zero'th, first, and second moments of the velocity PSD, and PSD is the power spectral density) were calculated in addition to the significant wave height ($H_s = 4\sqrt{\sum(PSD)df}$). Bulk parameters were calculated for each of the four measured time-periods in Fig. 6 and linearly interpolated to time-windows of the runup data (Table 2).

Because the instrument was positioned in transitional water, deep water significant wave heights were back-calculated employing the conservation of wave energy flux from linear wave theory via Eq. (1) and using the various measured wave periods:

$$H_o = H_h\left(\left(\frac{C}{2}\left(1+\frac{2kh}{\sinh(2kh)}\right)\right)\frac{4\pi}{gT}\right)^{0.5} \tag{1}$$

where H_h is the measured wave height at depth h, g is gravitational acceleration, C the phase speed (L/T), and k the wave number ($2\pi/L$). During the four-hour experiment, the significant wave height increased by about 35 cm while the incident wave period remained fairly constant at 13s, 10s, 7s, and 6s, for T_{p1}, T_{p2}, T_{mo1}, and T_{mo2}, respectively. The maximum peak energy switched over from a slightly shorter peak period, $T_{p2} \cong 10s$, to $T_{p1} \cong 13s$ about two-thirds of the way through the field study.

RUN-UP ANALYSES AND CHARACTERISTICS

Depending on beach slope and wave conditions, total runup is often described by two or three primary components (e.g., Komar 1998): (1) set-up of the mean water line due to radiation stresses induced by breaking waves (Longuet-Higgins and Stewart 1962); (2) fluctuations about that mean, due to the swash of incident waves producing runup and run-down (backwash); and (3) oscillations about the mean at infragravity frequencies, such that swash periods are greater than 20s, beyond the usual range of incident wave periods.

The runup time-series were band-pass filtered to determine the relative contributions of sea and swell (incident band, R_{inc}) $(0.05Hz < f < 0.2Hz)$ to infragravity (R_{ig}) $(f < 0.05Hz)$ components (Fig. 7). On average, the infragravity band contained 93% of the total runup variance, ranging from 84% (T1) to a maximum at T3 of 98%. The relative contribution of runup in the infragravity band to total runup (R_{ig}/R) was 0.83, similar to what Ruessink et al. (1998) found $(R_{ig}/R = 0.85)$, but less than the Oregon data set presented by Ruggiero et al. (2001 and 2004).

Runup energy density spectra were calculated using Welch's averaging method and a bandwidth of 9E-4 Hz for each of the 17-minute segments. Averages of the ten time-series segments are shown for each transect in Fig. 8. The dashed vertical line depicts the cutoff between infragravity and incident runup frequencies. It is clear that the peak energy lies within the infragravity

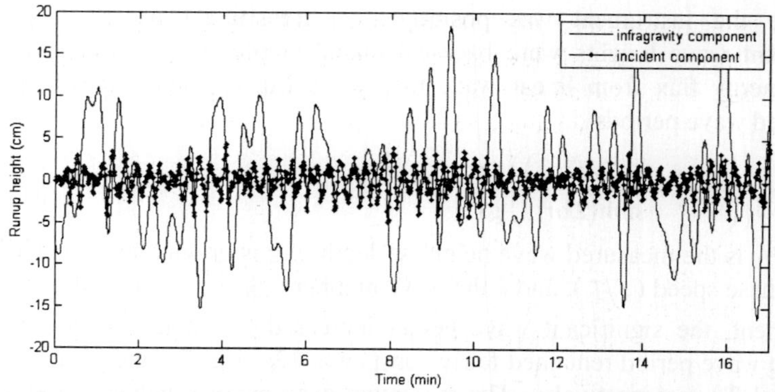

Figure 7. Typical 17-minute detrended time-series illustrating the dominant infragravity component (transect T3 on June 24, 2005, 10:09 to 10:23 am).

band and at incident band frequencies there is a steep drop-off in the energy. The peak periods ($1/f_p$) range from 41s to 68s with the longer periods (56s to 68s) observed at the three middle transects.

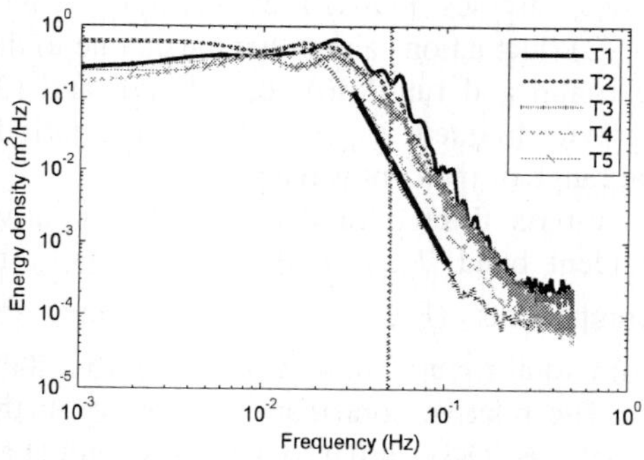

Figure 8. Total energy density (averaged over 10 subsequent 17-minute time-periods) measured at each cross-shore transect. Heavy solid line is best fit through T3 – rest are not shown. Dashed vertical line is cutoff between infragravity and sea swell frequencies.

The steep drop-offs in Fig. 8 suggest saturated surf and swash zones. Miche (1951) theorized that monochromatic incident waves consist of both a progressive and a standing wave component and that runup is proportional to the amount of shoreline reflection from the standing wave amplitude. The amplitude of the standing wave is depth limited as an increase in incident wave heights causes the breaking to occur further offshore, allowing energy dissipation through wave breaking, and less energy for swash. As such, saturation implies that the incident band runup height does not increase with increased offshore wave height.

Saturated runup energy density spectra have shown to decay as $E(f) = \alpha f^{-C}$, where α is a dimensionless constant (Huntley et al., 1977). Based on various field studies, the coefficient c, describing the degree of roll-off, is typically on the order of 3 to 4 (e.g., Guza and Thornton 1982; Ruessink et al. 1998;

Raubenheimer and Guza 1996; Kobayashi et al. 1989; Ruggiero, et al. 2004). Best fit lines (Marquardt least-squares minimization) were fit through the roll-offs in Fig. 8 and found to range from c = 4.0 to as much as c = 4.8 at T3 (coefficients of determination were >0.95 for all five transects). At this time it is uncertain why the roll-off at T3 is larger than obtained at other sites.

Run-up dependencies

Although numerous factors influence the extent of swash oscillations (e.g., grain size, friction, turbulence, swash interaction, and infiltration to name a few), three parameters are generally considered to be strongly linked to observed runup statistics: offshore wave height (H_o), deepwater wave length (L_o=1.56T^2) and beach slope (β; $\tan(\beta) \cong \beta$ for mild slopes). Various combinations of these three parameters have been proposed, and in particular, there have been some recent developments in the prediction of runup on dissipative beaches. In this section, some of these dependencies and proposed analytical models are compared to the data obtained for this study.

In total, six different semi-empirically-based analytical models were used for the comparison: 1) Hunt, 1959; 2) Larson and Kraus 1989; 3) Mayer and Kriebel 1994; 4) Miche 1951, Guza et al. 1984, and Raubenheimer and Guza 1996; 5) Ruggiero et al. 2001, and 6) Stockdon et al. 2006. The last three were specifically developed with data from both reflective and dissipative beaches. Both Ruggiero et al. (2001) and Stockdon, et al. (2006) provide separate equations for strongly dissipative beaches, and although the data of this study shows that the site was dissipative under the measured summer conditions, there was an incident component that should not be ignored. In addition, employing equations that encompass both the incident and infragravity swash on dissipative beaches provides a broader spectrum of applicable scenarios.

Correlation coefficients (CODs) and root-mean-square-error (rmse) between measured and predicted runup elevations are summarized in Table 3. In order to assess the sensitivity of the models to the choice of bulk parameters obtained from the bimodal velocity spectra measured with the offshore ADCP, runup statistics were compared using the two peak periods (T_{p1} and T_{p2}) and mean representations (T_{mo1} and T_{mo2}). The choice of wave period affects both the wavelength and back-calculated deepwater wave height (H_o) whether reverse shoaling or conservation of energy flux is employed. In all cases, the significant runup height, R_s, is better correlated than the extreme runup statistic, R_2. This may be because the deepwater wave height used is the H_s, based on the same statistic as R_s ($R_s = 1.46 R_{rms}$; $H_s = 1.46 H_{rms}$). With respect to both the highest COD and smallest rmse, Hunt's original formula (1959) with either T_{mo1} or T_{mo2} appears to perform best overall (COD=0.70 and rmse=5cm, Fig. 9a). The relationship presented by Guza (1984) and others does not do as well even though it specifies a separate equation for saturated conditions which this dataset includes (see previous section). The relationship is independent of offshore wave height whereas all the others depend on H_o.

Table 3. Matrix of coefficients of determination (COD) and root-mean-square-error (rmse) between measured and predicted runup statistics.

COD/Rmse	T_{p1}		T_{p2}		T_{mo1}		T_{mo2}	
	R_s	R_2	R_s	R_2	R_s	R_2	R_s	R_2
1)	0.71/18	0.66/10	0.72/9	0.67/8	0.70/5	0.66/13	0.70/5	0.65/15
3)	0.72/48	0.67/37	0.72/36	0.67/25	0.71/25	0.67/14	0.71/22	0.67/13
4)	0.71/22	0.66/13	0.69/12	0.64/9	0.63/6	0.61/13	0.62/6	0.60/14
5)	0.66/18	0.59/29	0.67/21	0.61/32	0.65/22	0.58/34	0.65/23	0.58/33
6)	0.71/45	0.67/35	0.70/30	0.65/19	0.70/17	0.66/9	0.70/14	0.66/7
7)	0.70/54	0.62/43	0.56/36	0.50/26	0.72/22	0.66/12	0.71/18	0.66/11

Because there was not much variation in the offshore wave height during the field experiment, a plot of runup versus H_o does not reveal any clear trend; rather, the dependency is inferred from the various runup models tested. The Mayer and Kriebel formulation does surprisingly well as it was not directly intended for dissipative conditions. However, their formulation addresses the physics behind the saturated surf and swash by incorporating the location of the break point and thus accounting for surf zone decay and consequently runup limited by depth-induced breaking. The location of the breakpoint, X_b, was estimated by assuming an equilibrium beach profile (A=0.11, based on D_{50}=0.30mm) and with a breaker index based on the non-dimensional Iribarren number, ε_o, (Iribarren and Nogales 1949; Battjes 1974; Kaminsky and Kraus 1993).

The formulations presented by Ruggiero et al. (2001) and Stockdon et al. (2006) perform about equally well except that the former appears to be somewhat less sensitive to the choice of wave period. Ruggiero et al.'s (2001) formulation was developed from both US west coast (Oregon) and East coast datasets, while Stockdon et al's is mostly from the East coast (but includes all of Ruggerio et al.'s data) with possibly less bimodal conditions. Ruggiero et al.'s formulation is quite similar to Hunt's original relationship except that Hunt has a stronger dependence on the beach slope (β vs $\sqrt{\beta}$). A least-squares fit to Ruggiero et al.'s governing parameter, $\sqrt{\beta H_o L}$, yields a coefficient of 0.27, equivalent to their empirical coefficient (Fig. 9b). It is expected that the empirical coefficient is dependent on some variables such as breaker type or friction. The median grain size at Ocean Beach is about D_{50}=0.27 mm (Barnard et al. under review), similar to the grain sizes of the datasets used by Ruggiero et al. (2001).

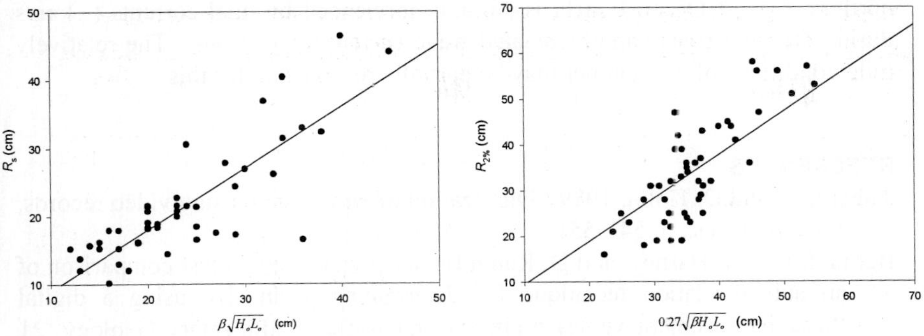

Figure 9. Measured runup statistics versus (a) Hunt's formula (1959), and (b) Ruggiero et al.'s (b) formulation. Both are shown with $T = T_{mo2}$.

SUMMARY AND CONCLUSIONS

Swash measurements were obtained along five cross-shore transects during a 4-hour sampling period on June 24th, 2005 at the north end of Ocean Beach, San Francisco. The data was band-partitioned into incident ($0.05Hz < f < 0.2Hz$) and infragravity ($f < 0.05Hz$) components and runup statistics were calculated separately and jointly. Beach slopes were mild ranging from 0.01 to 0.05, and analysis showed that the swash was dissipative with Iribarren numbers between 0.05 and 0.4 (using T_{mo2}). Infragravity components were dominant and made up approximately 80% of the total swash oscillation. Energy spectra of all measured runup data suggest that the surf and swash zones were saturated so that an increase in offshore wave height did not increase the runup as a significant portion of the incident waves were depth-limited breaking offshore. The spectra also indicate that the peak frequency of the swash ranged from 0.014Hz to 0.025Hz (i.e., swash periods between 40s and 70s).

An ADCP located just outside the surf zone of the cross-shore transects measured wave heights coincident with the runup data. Wave heights were on the order of 0.9m to 1.5m. Velocity spectra from the ADCP clearly show bi-modal conditions with incident peak wave periods ranging from 9s to 13.5s. Measured wave heights and periods were back-calculated to deep water conditions and used to predict the 2% and significant runup heights (R_2 and R_s, respectively). Predicted runup statistics using six previously published relationships were compared to field observations. It was found that Hunt's 1959 formula performed best when comparing to R_s, and Ruggiero et al.'s formulation (2001), which is similar to Hunt's but with less dependence on the foreshore worked well for predicting R_2. Using peak wave periods increased the root-mean-square-error (rmse) in all cases, while the mean period, in particular T_{mo2}, which is theoretically similar to the wave period obtained through zero-down crossing, gave results with the smallest rmse. In all, the analytical runup formulations did quite well considering that Ocean Beach, and in particular the northern part of Ocean Beach, is strongly influenced by tidal currents (~1 m/s during the field study) and associated wave-current interactions. The relatively mild conditions of the summer may be partially responsible for this.

REFERENCES

Aagard, T. and D. Holm, 1989. Digitization of wave runup using video records, J. Coastal Res., 5, 547-551.

Barnard, P.L., J. Harney, and D. Rubin (under review). Field test comparison of an autocorrelation technique for determining grain size using a digital 'beachball' camera versus traditional methods. Sedimentary Geology, 21 pp.

Continental Shelf Research (CSR). 2006. Volume 26, Issue 5, Swash-Zone Processes, April 2006,

Battjes, J.A. 1974. Surf similarity, Proceedings of the 14th International Coastal Engineering conference, ASCE, Copenhagen, Denmark.

Guza,R.T. and E.B. Thornton. 1982. Swash oscillations on a natural beach. Journal of Geophysical Research, Vol. 87, pp. 438-491.

Holland, K.T. and R.T. Holman. 1999. Wavenumber frequency structure of infragravity swash motions, J. Geophys. Res. 104, 13,479-13,488.

Holland, K.T. and R.T. Holman. 1993. The statistical distribution of swash maxima on natural beaches, J. Geophys. Res. 87, 10,271-10,278.

Holland, K.T., R.T. Holman, and T. Lippman. 1997. Practical use of video imagery in nearshore oceanographic field studies. J. Oceanic Engr., 22(1), 81-92.

Holland, K.T., B. Raubenheimer, R.T. Guza, and R.A. Holman. 1995. Runup kinematics on a natural beach. J. Geoph. Res. 100 (C3), 4985-4993.

Holman, R.A. and R,T, Guza. 1984. Measuring runup on a natural beach. Coastal Engineering, 8, 129-140.

Holman, R.A. and A.H. Sallenger. 1985. Setup and swash on a natural beach. Journal of Geophysical Research, 90(C1), 945-953.

Hunt, I.A. 1959. Design of seawalls and breakwaters, J. Waterway Harbors Coastal Eng. ASCE, 85, 123-152.

Huntley D., R.T. Guza, and A.J. Bowen. 1977. A universal form for shoreline runup spectra? J. Geophys. Res. 82, 2577-2581.

Iribarren, R.R. and C. Nogales. 1949. Protection des Ports, proceedings from the XVIIth International Navigation Congress, Lisbon, Portugal.

Kaminsky, G.M. and N.C. Kraus. 1993. Evaluation of depth-limited wave breaking criteria, Proceedings of the Second International Symposium on Ocean Wave Measurements and Analysis. ASCE

Kobayashi, N., G.S. De Silva, and K.D., Watson. 1989. Wave transformation and swash oscillation on gentle and steep slopes, J. Geophys. Res., 94, 951-966.

Komar, P.D. 1998. Beach Processes and Sedimentation. Second Edition. Prentice-Hall, Inc.

Larson, M., and N.C. Kraus. 1989. "SBEACH: Numerical Model for Simulating Storm-Induced Beach Change. Report 1: Empirical Foundation and Model Development." Technical Report CERC-89-9. Coastal Engineering Research Center, U.S. Army Engineer Waterways Experiment Station, Vicksburg, MS.

Longuet-Higgins, M.S. and R.W. Stewart. 1962. Radiation stress and mass transport in gravity waves, with application to surfbeats, J. Fluid Mech., 13, 481-504.

Mayer, R.J. and D.L. Kriebel. 1994. Wave runup on composite-slope and concave beaches." 24th International Conference on Coastal Engineering, ASCE.

Miche, R. 1951. Le Pouvoir reflechissant des ouvrages maritimes exposes a l'action de la houle, Ann. Ponts Chaussees, 121,285-319.

Raubenheimer, B. and R.T. Guza. 1996. Observations and predictions of runup. J. Geophys. Res., 101, 25,575-25,587.

Ruessink, B.K., M.G. Kleinhans, and P.G.L. van den Beukel, 1998. Observations of swash under highly dissipative conditions, J. Geophysical Research, 103, 3111-3118.

Ruggiero, P., R.A. Holman, and R.A. Beach. 2004. Wave runup on a high energy dissipative beach, Journal of Geophysical Research, 109, 1-12.

Ruggiero, P., P.D. Komar, W.G. McDougal, J.J. Marra, and R.A. Beach, 2001. Wave runup, extreme water levels and the erosion of properties backing beaches, Journal of Coastal Research, 17(2), 407-419.

Stockdon, H.F., R.A. Holman, P.A. Howd, and A.H. Sallenger, Jr. 2006. Empirical parameterization of setup, swash and runup. Coastal Engineering, 53, 573-588.

ACKNOWLEDGEMENTS

The study was funded by the USGS Coastal Evolution Process-Based, Multi-Scale Modeling project. Many thanks are extended to Peter Ruggiero and Jodi Eshleman for insightful reviews and comments and Jodi Eshleman for rigorous wave data processing.

BORE-DRIVEN SWASH ON BEACHES: NUMERICAL MODELING AND LARGE-SCALE LABORATORY EXPERIMENTS

TOM O'DONOGHUE[1], LUKE J. HONDEBRINK[2], DUBRAVKA POKRAJAC[3]

Novel laboratory experiments have been carried out to measure swash hydrodynamics on smooth and rough impermeable beach slopes. The experiments use a swash rig which produces highly repeatable, full-scale, single swash events, enabling detailed measurements of velocities and depths across the swash zone. Example results are presented, showing comparisons between smooth and rough beach hydrodynamics. The effects of roughness are most evident in the maximum run-up, in depths at the top end of the beach and in the backwash velocities. Experimental results have been compared with predictions of depth and depth-averaged velocity from a 1D NLSWE-based numerical model with a simple quadratic stress law for the bed shear stress. Comparison of experimental and model results shows reasonably good agreement for depth and velocities for much of the swash cycle. However, differences in the backwash velocities point to poor bed shear stress modeling. The results imply that good agreement between measured and predicted depths and velocities for swash uprush is not an indication that the bed shear stress is well predicted by the model.

INTRODUCTION

Bore-driven swash is associated with reflective beaches, which tend to be coarse-grained and relatively steep, with a narrow surf zone and surging and collapsing breakers. The swash is generated by the collapse of bores onto the beach at the period of the incident waves, producing a highly dynamic environment in terms of flow and sediment transport rates. Swash flow is highly unsteady and non-uniform and is complicated further by the poorly understood effects of beach roughness and beach permeability. A number of field studies of swash dynamics have been conducted on coarse-grained beaches (e.g. Holmes *et al.*, 2002; Horn *et al.*, 2003), but a lack of control over the variables involved, the impossibility of isolating particular processes and difficulties in obtaining detailed and accurate measurements mean that fundamental swash processes remain poorly understood. Previous laboratory experiments (e.g. Petit and Longo, 2001; Cowen et al., 2003) have tended to be at small scale with small swash depths, low velocities and short run-up and may suffer from scale effects relating to beach roughness and beach permeability. The present paper reports on a study of bore-driven swash hydrodynamics based on large-scale laboratory experiments in which detailed measurements have been made of depths and

---

[1] Reader, University of Aberdeen, Department of Engineering, King's College, Aberdeen AB24 3UE, Scotland.
[2] Research Student, University of Aberdeen, Department of Engineering.
[3] Senior Lecturer, University of Aberdeen, Department of Engineering.

velocities across the swash zone. The experiments were conducted on smooth and rough impermeable beaches in order to evaluate the effect of roughness on the depths and velocities. The experimental results are compared with predictions of depth and depth-averaged velocity from a numerical model based on the non-linear shallow water equations, with a simple quadratic stress law for the bed shear stress.

LARGE-SCALE LABORATORY EXPERIMENTS

Experimental Facility

The experiments were conducted using a laboratory swash rig built into a 20 m long, 0.9 m high and 0.45 m wide, glass-sided flume (Fig. 1). The rig consists of a water reservoir placed at one end of the flume, fronted by a gate designed to be raised at high speed while at the same time providing good seal when closed. The sudden opening of the gate produces a bore, which generates a large swash event on a beach located downstream. Swash repeatability is excellent, which means that a detailed description of the flow hydrodynamics can be built up using measurements from many repeats of the swash event. Bores with speeds of ~3m/s are produced with a reservoir depth of ~800mm and these have produced swash excursions of ~7m on a 1:10 impermeable, marine plywood beach and ~4m on a 1:10 highly permeable beach consisting of very large shingle. Swash of this magnitude is similar to large swash in the field and is at least an order of magnitude greater than swash produced in typical laboratory wave flumes.

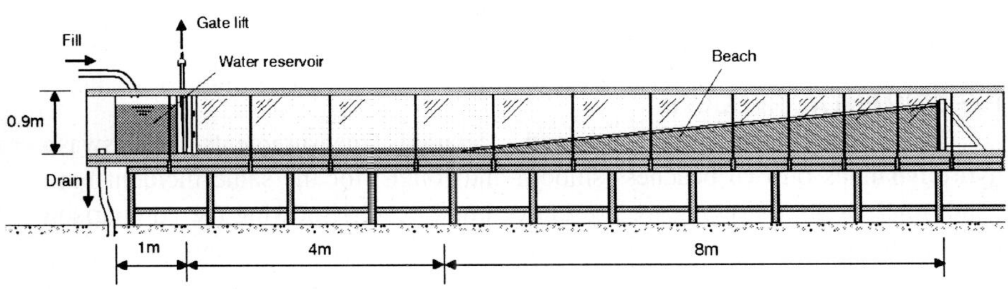

Figure 1. Laboratory rig for large-scale swash experiments.

Experimental Conditions and Measurements

For the present experiments the beach toe was located 4 m from the reservoir. Two beach types were used: a 1:10 "smooth", impermeable perspex beach and a 1:10 rough impermeable beach, made by gluing 5-6 mm pebbles uniformly over the perspex beach panels. The experiments involved a 600 mm deep reservoir with a 60 mm water depth in front of the reservoir. Ensemble-averaged depths, velocities and turbulent quantities were obtained from 50 repeats of the swash event in each case.

Measurements were made of the time-dependent depths and velocities at a number of locations across the swash zone for the two beach types. Swash depths were measured at increments of 0.2 m cross-shore using standard double-wire capacitance-type gauges. For run-up measurement, a capacitance gauge was placed parallel to the beach slope at a height of 3 mm above the beach face. Velocities were measured at five cross-shore locations (x = 0.5, 2, 2.35, 3.0 and 3.5 m, where x = 0 corresponds to the initial shoreline position) and offshore at x = -1.8 m. Velocities were measured using a cross-correlation particle image velocimetry (PIV) system, comprising a 1k x 1k cross-correlation camera and a double pulse Yag laser with a pulse separation of 2 ms; the camera and laser are synchronised and image pairs were recorded at a rate of ~12.5 Hz. The 2-d PIV laser light sheet entered the flow via narrow windows in the underside of the swash flume and the underside of the 1:10 beach. The flow was seeded with 15 μm hollow glass spheres. Most measurements were made with a measuring area of 100 x 100 mm and the analysis was carried out using two interrogation areas: a larger interrogation area of 64 x 64 pixels and a smaller interrogation area of 32 x 32 pixels; the measurement area is therefore 6.4 mm and 3.2 mm for the 64 x 64 and 32 x 32 interrogation areas respectively. Analysis was carried out with a 75% overlap of interrogation areas to yield velocity vectors every 1.6 mm above the bed in the case of the 64 x 64 area and every 0.8 mm above the bed in the case of the 32 x 32 area. As for other measurement methods, the high degree of air entrained in the bore means that good PIV velocity measurements are not possible at the time of bore arrival. Difficulties were also encountered with velocity measurements during the later stage of backwash when water depths were very shallow.

EXPERIMENTAL RESULTS

The experiments have yielded detailed measurements of swash hydrodynamics on two beaches (smooth and rough) for the same incident bore conditions. In the following we present example results showing the comparison between the rough and smooth beach measured depths and velocities.

Swash Depths

Figure 2 presents measured depth time-series at 12 cross-shore locations for the two beaches. Differences in depths are seen mainly at the lower end of the beach and near maximum run-up at the top end of the beach and. Measured maximum run-up, defined as the highest point on the beach where the depth reaches 3 mm, was 5.8 m and 4.7 m for the smooth and rough beaches respectively. The higher run-up in the case of the smooth beach means depths at locations high up the beach are greater for the smooth than for the rough beach; this is clearly seen in the results for x = 4 m and 4.2 m. The greater depths high up the beach are balanced by lower depths at the low end of the beach. This is seen in the results at $0.8 \leq x \leq 2.0$ m where the rough beach depths are greater; in this region the biggest difference in depth occurs at $x \approx 1.1$ m. Rough and smooth beach depths are very similar between the lower and upper beach, i.e. at $2.4 < x < 3.6$ m. Depths are also very similar very low down the beach ($x = 0.4$ m in Fig. 2), except towards the end of the backwash when the water is slower to drain completely off the rough beach

The bore travels slightly more slowly across the rough beach compared to the smooth beach. With the exception of location $x = 0.4$ m, bore arrival at each x in Figure 2 is earlier in the case of the smooth beach. The difference is small low down the beach, approximately 0.02 s at $x = 1.2$ m, reaching approximately 0.6 s by $x = 4.2$ m.

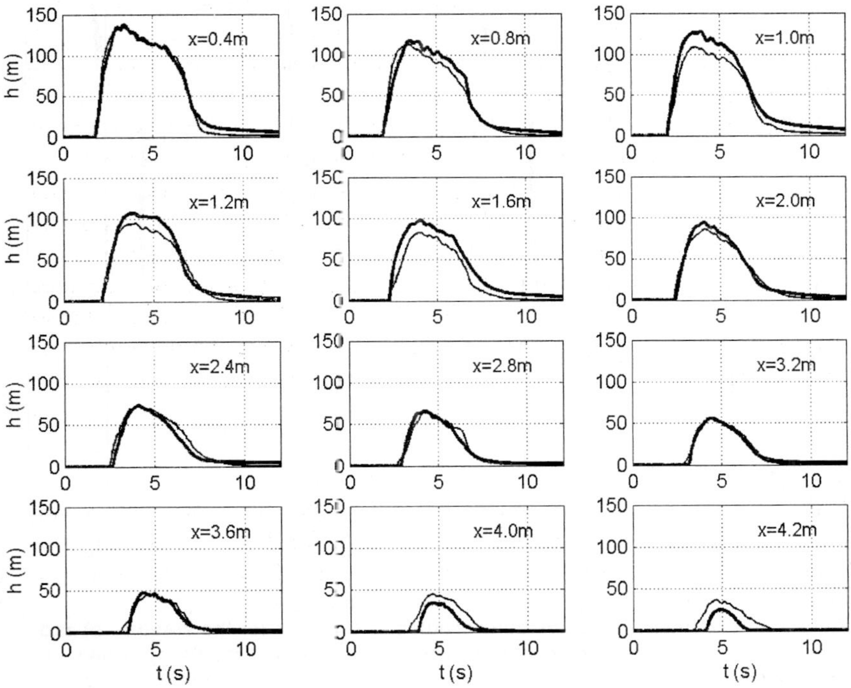

Figure 2. Depth time-series at 12 cross-shore locations on the smooth (thin line) and rough (thick line) beaches. $x = 0$ corresponds to the initial shoreline position.

Velocity Profiles

Ensemble-averaged velocity profiles have been obtained by averaging the measured velocity profiles from the 50 repeats of each experiment. The process is complicated somewhat by the fact that not all runs produced valid velocity data over the same range of z, mainly because of cross-tank oscillations of the water mass. Example ensemble-averaged velocity profiles are presented in Figures 3 and 4, showing bed-parallel velocity profiles at 12 phases of the swash cycle for the rough and smooth beaches at cross-shore locations x = 0.5 m and x = 2.35 m respectively. All of the results shown in Figures 3 and 4 are from the PIV analysis with a 64 x 64 pixel interrogation area. z = 0 corresponds to the surface of the Perspex, i.e. ~5 mm below the top of the roughness elements in the case of the rough bed. The last panel in each of Figure 3 and Figure 4 shows velocities measured near the end of the backwash when depths are very shallow and it is difficult to get good measurements of the velocity profile.

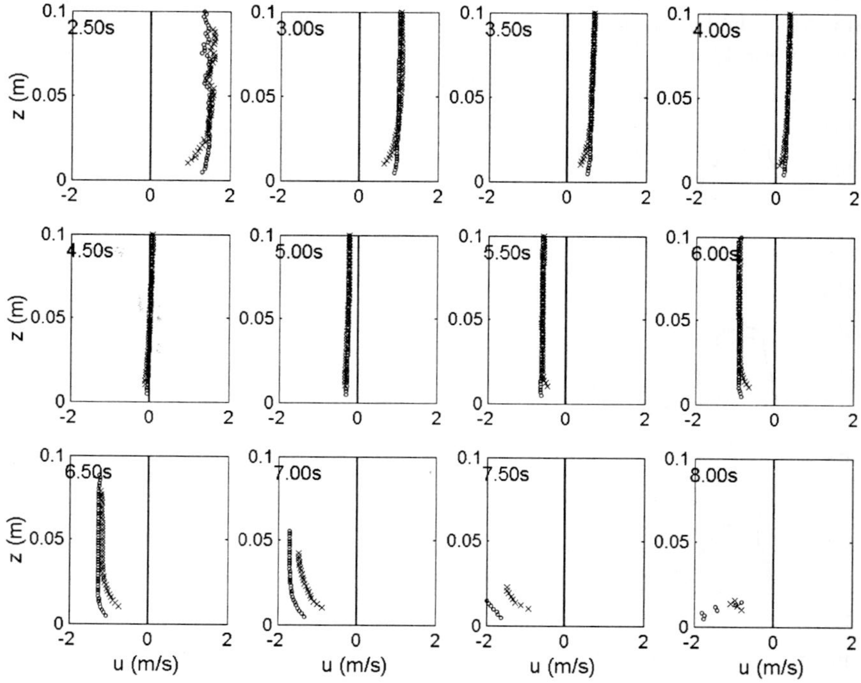

Figure 3. Ensemble-averaged velocity profiles at x = 0.5 m for the smooth (o) and rough (x) beach experiments. (z=0 corresponds to the top of the Perspex bed, i.e. ~5mm below the top of the roughness elements in the case of the rough bed.)

Results at x = 0.5 m (Figure 3) and at x = 2.35 m (Figure 4) show very similar behaviour throughout the swash cycle. During uprush and for a significant part of the backwash ($2.5 \leq t \leq 6.0$ s in Figure 3 and $3.2 \leq t \leq 5.75$ s in Figure 4) there is no great difference between the rough and smooth beach velocity profiles. During these times the velocity magnitudes are nearly the same for the rough and smooth beaches and the profiles are mostly depth uniform. The bottom boundary layer is confined to a relatively small region close to the bed and, as expected, is more pronounced in the case of the rough beach. The most noticeable difference between the rough and smooth beach results in Figures 3 and 4 is seen during the second part of the backwash when velocities become increasingly greater on the smooth beach compared to the rough beach. Also, as the water depth becomes shallower the velocity profile becomes less uniform over the depth.

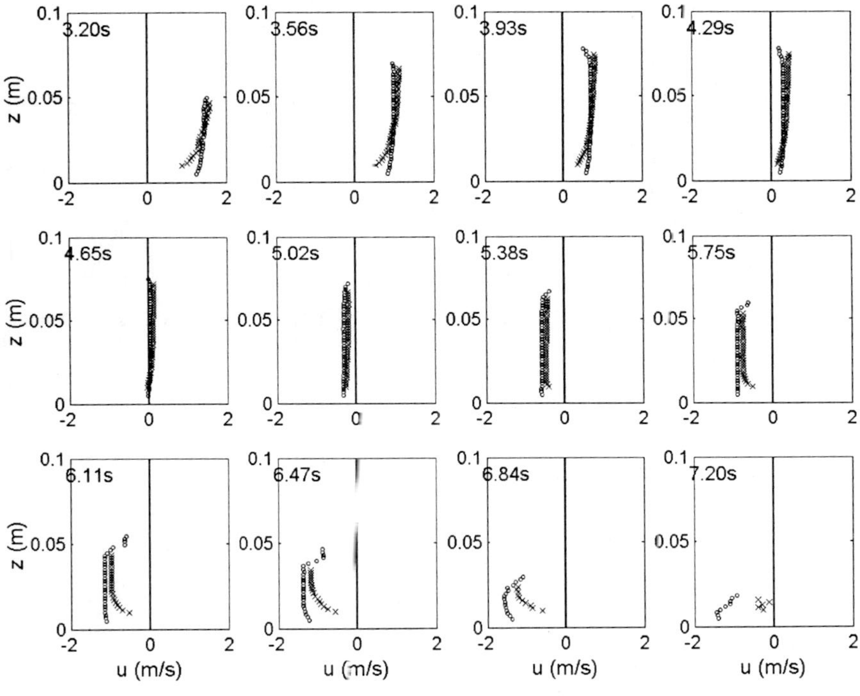

Figure 4. Ensemble-averaged velocity profiles at x = 2.35 m for the smooth (o) and rough (x) beach experiments.

Depth-averaged Velocities

The ensemble-averaged velocity profiles have been depth-averaged to produce time series of depth-averaged velocity at the 5 cross-shore locations on the beach where velocities were measured. Figure 5 presents the results at x = 0.5, 2.35 and 3.0 m for the rough and smooth beach experiments. The corresponding ensemble-averaged depth time-series are also shown. As already mentioned, considerable amounts of entrained air at the time of bore arrival mean that good PIV measurements at the moment of bore arrival are not possible. The first valid PIV velocity measurements appear a little time later when the entrained air volume is greatly diminished.

The results in Figure 5 show that the peak positive velocity is slightly lower on the rough beach and occurs slightly later in time, consistent with the lower bore speed on the rough beach. After the peak, uprush velocity decays rapidly on both beaches, but the deceleration seems slightly lower on the rough beach and the difference in deceleration seems to increase with distance up the beach. Flow reversal occurs at the same time on both beaches at x = 0.5 m but occurs somewhat later on the rough beach at x = 3 m because of the difference in decelerations.

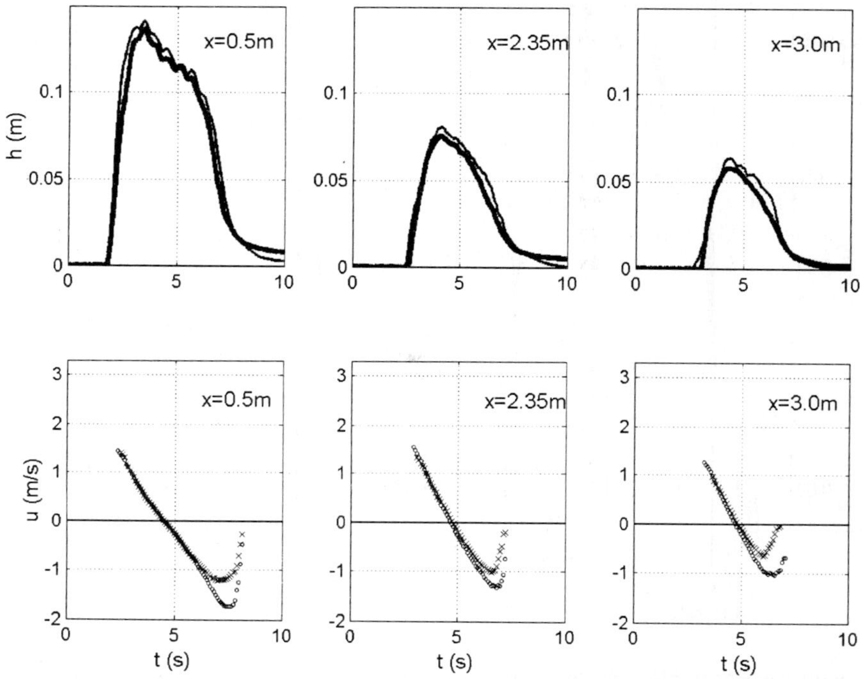

Figure 5. Depths (thin line = smooth beach, thick line = rough beach) and depth-averaged velocities (o = smooth, x = rough) at 3 cross-shore locations.

The biggest differences between the rough and smooth beach depth-averaged velocities are seen in the backwash, echoing the earlier observations from the velocity profiles. Depth-averaged velocities during backwash show that water accelerates off the beach during backwash until it reaches a maximum (negative) velocity at low depth and subsequently decelerates towards zero velocity during the final stage of the backwash. Higher accelerations and higher peak negative velocities are reached in the case of the smooth beach and the change from accelerating to decelerating flow tends to occur a little later in time. Backwash flow is driven by the weight of water acting down the slope and is resisted by the effects of bottom friction. Bed friction increases as backwash velocity increases and at the same time the weight of water on the slope decreases. There comes a point towards the end of the backwash when the retarding force due to friction becomes greater than the driving force due to gravity and the flow starts to decelerate. The results in Figure 5 show that bed roughness is important in this process: bed friction increases more rapidly in the case of the rougher beach with the result that backwash velocities are lower and the arrest of flow acceleration occurs sooner.

Bed roughness and friction affect uprush flow as well as backwash, as has been seen, but uprush flow is less sensitive to bed effects. This is because uprush flow is driven by incident bore momentum and is resisted by the combined effects of gravity and bed friction and the gravity force far exceeds bed friction for much of the uprush.

SWASH NUMERICAL MODEL

The swash numerical model used for the present study is based on the flux-conservative, nonlinear shallow water wave equations (NLSWE). For an impermeable bed the governing pair of mass and momentum conservation equations are

$$\frac{\partial h}{\partial t}+\frac{\partial (hu)}{\partial x}=0 \qquad (1)$$

$$\frac{\partial (hu)}{\partial t}+\frac{\partial \left(hu^2+\frac{1}{2}gh^2\right)}{\partial x}+\frac{\tau_o}{\rho}+gh\tan\theta=0 \qquad (2)$$

where h is flow depth, u is depth-averaged velocity, τ_o is bed shear stress, ρ is water density, g is acceleration due to gravity, θ is beach angle, t is time and the x-axis is along the beach face and is positive shoreward. A number of swash models are based on the Lax-Wendroff method for solving the NLSWE, described by Hibberd and Peregrine (1979). However, the Lax-Wendroff approach leads to numerical oscillations at the bore front and requires special treatment of the moving shoreline. The model used for the present study uses a Godunov-type, flux-limited, finite-volume method as described by Toro (2001). The shock-capturing properties of the method prevent numerical oscillations developing behind the bore and allow the moving shoreline to be modeled straightforwardly. The swash model described by Clarke et al. (2004), which includes exchange between the main swash flow and a permeable bed, uses a similar approach.

One of the difficulties associated with NLSWE-modeling of swash is proper treatment of the bed shear stress. In the absence of quantitative understanding, many swash models simply use $0.5\rho C_f u|u|$, with the friction factor C_f spatially and temporally constant. Besides the question of what C_f value should be chosen for a given beach type (C_f values range 0.005 – 0.22 in the literature), the simple quadratic stress law is unlikely to be a good model of bed shear stress in a flow which is highly unsteady and highly non-uniform.

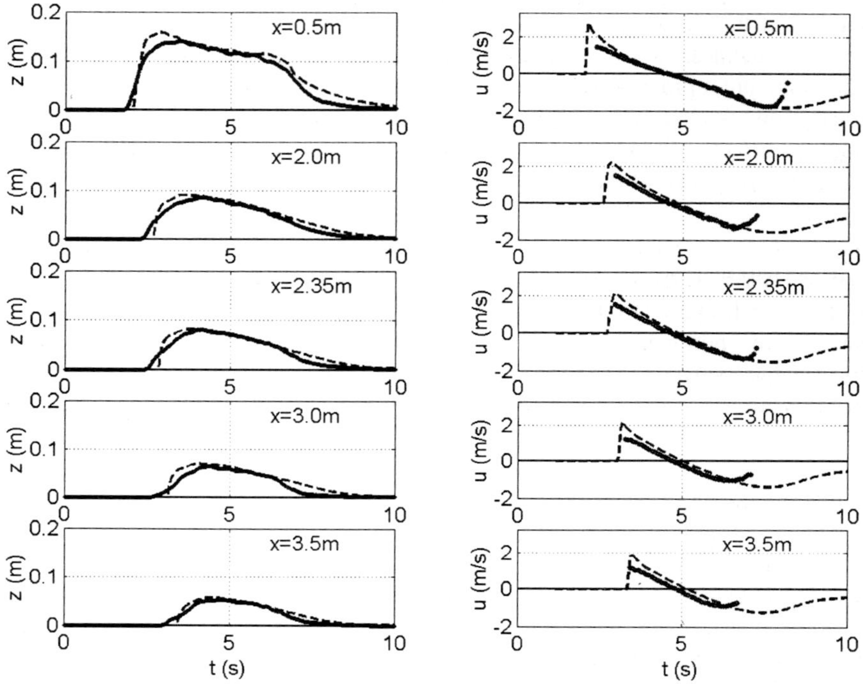

Figure 6. Comparison of measured depths and depth-averaged velocities with NLSWE model predictions (broken line). Smooth beach case with C_f = 0.012 in the model.

In the following, we compare NLSWE model predictions of depth and depth-averaged velocities across the swash zone with the experimental measurements. For the model we use the quadratic stress law with C_f constant in time and space and with value selected to give the same maximum run-up as measured in the experiments: C_f = 0.012 for the smooth beach experiments and C_f = 0.035 for the rough beach experiments. The offshore boundary for the model was set at x = -1.8 m and the experimentally-measured time-series of h and u at x = -1.8 m were used to specify the boundary conditions.

The comparison of model and measured results for the smooth beach case is presented in Figure 6, showing results for h and u at x = 0.5, 2.0, 2.35, 3.0 and 3.5 m. The comparison shows quite good agreement between the model and measured results. With the exception of the end of the backwash, the depth time-series are in good agreement in terms of time-series shape and depth magnitudes, although the bore front tends to be steeper in the model. Another difference is the arrival time, which tends to be slightly later in the case of the model results. This is also seen in the velocity time series, with the difference in arrival time increasing with distance up the beach. This difference aside, the measured and modeled velocities are in quite good agreement: a small time shift of either the model or experimental results would result in good overlap of the results for much of the swash cycle. There is however a substantial difference in the results during the second part of the backwash. In the case of the experimental data, the arrest of the backwash acceleration occurs much sooner, with the result that the measured peak backwash velocities and the velocities near the end of the backwash when the depths are very shallow are much less than the corresponding velocities predicted by the model. Similar behaviour is seen in the comparison of experimental and modeled results for the rough beach. Given, as discussed earlier, that the effects of bed roughness and bed friction are most felt during the latter part of the backwash, these comparisons suggest that bed shear stress is not well modeled using the simple quadratic stress law. This is not surprising as one would expect swash bed shear stress to depend at least on flow acceleration and flow depth in addition to flow velocity.

CONCLUSION

Novel laboratory experiments have been carried out to measure swash hydrodynamics on smooth and rough impermeable beach slopes. The experiments use a swash rig which produces highly repeatable, full-scale, single swash events, enabling detailed measurements of velocities and depths across the swash zone. Example results have been presented, showing comparisons between the smooth and rough beach hydrodynamics. With reference to swash depths, the effects of roughness are most evident in the maximum run-up and in depths at the top end of the beach, but they are also evident in depths measured on the lower part of the beach. The greatest effect of bed roughness on swash velocities is seen during the backwash. Bed shear stress increases more rapidly in the case of the rougher beach as the flow accelerates down the slope. The result is that backwash velocities are lower on the rougher beach and the change from accelerating to decelerating flow occurs sooner.

The experimental results have been compared with predictions of depth and depth-averaged velocity from a 1D NLSWE-based swash numerical model. Bed shear stress was modeled using a simple quadratic stress law with constant friction coefficient tuned to give the same maximum run-up as in the corresponding experiment. The comparison between experimental and model results shows good agreement for depth and velocities for much of the swash cycle (allowing a small shift in time). However, significant differences are seen during the later part of the backwash when bed shear stress effects are most felt. The results indicate that the simple quadratic stress law is an over-simplification for swash. They imply that good agreement between measured and predicted depths and velocities for swash uprush is no indication of good bed shear stress modeling and this result has very important implications for modeling swash sediment transport. Analysis of the experimental data is on-going and includes evaluation of bed shear stress via log-law fitting to the near-bed velocity profiles.

The experiments have produced a valuable dataset on full-scale, bore-driven swash hydrodynamics on impermeable beaches. Further laboratory experiments are planned, extending the work first to permeable, immobile beaches and later to mobile beaches.

Acknowledgments

Hondebrink acknowledges receipt of a PhD studentship from the University of Aberdeen. Our on-going swash research is supported by the UK's Engineering and Physical Sciences Research Council (grant EP/E011330/1).

References

Battjes, J.A., and J.P.F.M. Janssen. 1978. Energy loss and set-up due to breaking of random waves, *Proceedings of 14th International Conference on Coastal Engineering*, ASCE, 466-480.

De Vriend, H.J., J. Zyserman, J. Nicholson, J.A. Roelvink, P. Pechon, and H.N. Southgate. 1993. Medium-term 2DH coastal area modeling, *Coastal Engineering*, 21, 193-224.

Wiegel, R.L. 1965. *Oceanographical Engineering*, Prentice-Hall, Englewood Cliffs, New Jersey, 531 pp.

Clarke, S., Dodd, N. and Damgaard, J.S. 2004. Modeling flow in and above a porous beach, ASCE, *J. Waterway, Port, Coastal and Ocean Engineering*, 130(5): 223-233.

Cowen, E.A., Sonu, I.M., Liu, P.L.-F., Raubenheimer, B. 2003. PIV measurements within a laboratory generated swash zone. ASCE, *J. Engineering Mechanics*, 129: 1119-1129.

Hibberd, S. and Peregrine, D.H. 1979. Surf and run-up on a beach: a uniform bore, *J. Fluid Mechanics*, 95, 323–345.

Holmes, P., Horn, D., Blewett, J., Blanco, B., Peel-Yates, T. and Shanehsaz-zadeh, A. 2002. Hydraulic gradients and bed level changes in the swash zone on sand and gravel beaches, *Proceedings of 28th International Conference on Coastal Engineering*, ASCE, 1016–1027.

Horn, D., Li, L. and Holmes, P. 2003. Measurement and modelling of gravel beach groundwater response to wave runup, *Proceedings of Coastal Sediments '03*, ASCE, CD-ROM World Scientific Publishing Corp. and East Meets West Productions, Corpus Christi, Texas, USA, ISBN 981-238-422-7, 11pp.

Petti, M. and Longo, S. 2001. Turbulence experiments in the swash zone, *Coastal Engineering*, 43: 1-24.

Toro, E.F. 2001. *Shock Capturing Methods for Free-Surface Shallow Flows*. Wiley and Sons Ltd., 309 pp.

AIR-BUBBLE MEASUREMENTS OF THE SURF ZONE BREAKING WAVES BY BUBBLE TRACKING VELOCIMETRY

Hiroaki Kashima[1], Nobuhito Mori[1] and Shohachi Kakuno[2]

The two-phase flow measurement using imaging technique, called Bubble Tracking Velocimetry: BTV, is developed to measure instantaneous elliptic bubble shape and its moving velocity from the two dimensional project plain by the shadow graph method. The BTV measurements is applied to the surf zone waves. The characteristics of void fractions, air bubble distributions, and bubble size spectra are measured for surf zone breaking waves in the laboratory. The two unique power-law scaling of bubble diameter spectra is investigated by the BTV measurements.

INTRODUCTION

Ocean surface gravity waves propagates from offshore to shoreline and are often regarded as a single phase flow using potential flow theories or the Navier-Stokes equation. Generally, the single phase flow approach to the ocean wave is successful for simulating wave transformation in the coastal area. However, waves steepen and break owing to bottom bathymetric effects in the nearshore. The wave breaking creates dense plumes of bubbles, and dissipates the energy and momentum. An accurate estimation of bubble size and population distributions in the surf zone is important for understanding two-phase flow characteristics, solving engineering problems and environmental mechanisms of the coastal area (?). Recent photographic studies have illustrated the disintegration of entrapped air cavities divided into bubbles (Deane and Stokes, 1999, 2002). However, there are unexplained aspects of the phenomena, such as enhanced bubble populations in salt rather than freshwater (Thorpe, 1982), scale effects of void and bubble size distribution in the laboratory experiments, and the relation between void fraction and turbulence.

Measurements of bubble size and population distributions by in-situ instruments (Chanson et al., 2002), video or photographic instruments (Leifer and de Leeuw, 2002), laser instruments (Mori, 2002) and acoustic measurements (Deane, 1997) have been conducted for several purpose. The bubble size measurements using lasers show high accuracy but are impossible to use in the presence of high void fractions owing to technical limitations. Acoustic measurements of bubble are useful in deep-water but have limitation for very shallow water region owing to the multi-reflection of sound beams between the sea surface and the bottom. Therefore, conventional optical or resistivity type void probe are useful for surf zone waves (Chanson et al., 2002). Vagle and

[1] Graduate School of Engineering, Osaka City University, 3-3-138 Sugimoto, Sumiyoshi-ku, Osaka 558-8585, JAPAN.

[2] Vice President, Osaka City University, 3-3-138 Sugimoto, Sumiyoshi-ku, Osaka 558-8585, JAPAN.

Farmer (1998) compared four methods to measure bubble size and void fraction and concluded that electrical conductivity probes should be used for the high void fractions.

Understanding air entrainment and bubble distribution induced by wind wave breaking has long history of the research and is advanced compared to the surf zone. The air bubble distributions under the wind wave breaking in the ocean surface layer is summarized by Thorpe (1982), theoretically. Baldy (1993) discussed bubble formation and dependence on the turbulent energy dissipation rate of fluids and proposed a -2 power-law scaling with bubble diameter. Garrett et al. (2000) proposed $-10/3$ power-law scaling based on the discussion of bubble fragmentation and bubble spectrum owing to strong turbulent shear flow. Mori (2003) also proposed a $-5/2$ power-law scaling based on the dimensional analysis of wind wave breaking.

Despite the fruitful knowledge of air entrainment for wind-wave breaking, a few studies of air entrainment and of surf zone wave breaking exist. Peregrine (2003) reported that the entrapped air of breaking wave gives influence to the wave impact owing to their greater compressibility compared with pure water. Loewen et al. (1996) found a little difference in the bubble populations beneath mechanically generated surface waves in saltwater and freshwater. The compressibility owing to air-water mixture decreases the velocity of sound and is being used to estimate large-scale prototype impacts, since the usual Froude scaling is unlikely to be corrected for engineering problems. Therefore, the connection between the air-mixture, the bubble distribution and the wave breaking induced turbulence is essential to understand the gas-liquid interaction in the surf zone. Cox and Shin (2003) investigated the dependence of void fraction on the turbulent intensity in the bore region of surf zone waves. The wave breaking induced bubbles in the surf zone are split by the strong local turbulent shear induced by breaking waves at the bubble scale (*i.e.* Deane and Stokes, 1999). The authors demonstrated on the dependence of void fraction on wave scale using different scales of wave flume following the Froude similarity law (Mori et al., 2007). However, the bubble characteristics and connections between bubble characteristics and wave breaking are not well known owing to the lack of understanding of phenomena in the surf zone. The detail information of the two-phase flow characteristics is required for mathematical modeling.

The purpose of this study is to investigate the characteristics of void fractions, bubble distributions and turbulent properties in the surf zone waves using visualization technique. First, the automatic bubble shape recognition and moving velocity technique is developed. Second, the analysis of the experimental data of void fractions, bubble spectra of the surf zone waves, the power scaling law of bubble population spectra will be discussed.

METHOD OF VISUALIZATION

The visualization method is developed to measure both bubble shape and its moving velocity in a gas phase simultaneously. The method is basically based on

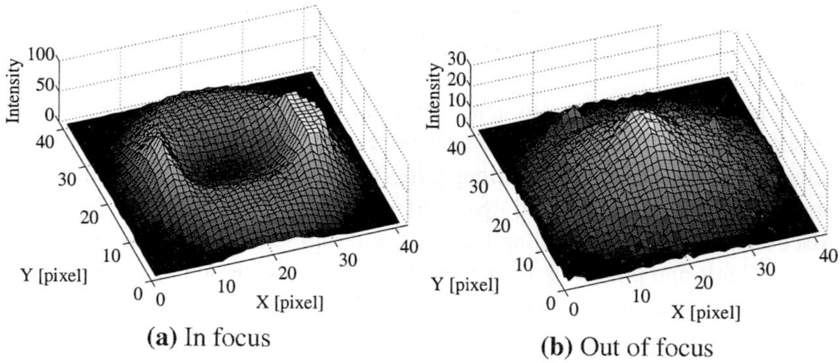

FIG. 1. Example of captured bubble image by SGM

the shadow graph method (SGM, hereafter for simplicity) and particle tracking method. Therefore, we denotes this method as Bubble Tracking Velocimetry or BTV for simplicity here after, First, the bubble is projected into the vertical two dimensional domain by the shadow graph method using a flat back light. The back light gives clear shadow image of the bubble or bulk of air, although it is a two dimensional projection from a three dimensional bubble shape. Fig. 1 shows examples of captured bubble image both in and out focus by SGM. The in focus captured bubble image as shown in Fig. 1 (a) shows clear edge of the bubble. The projected bubble shape can be calculated by tracing the edge. However, the out of focus bubble image as shown in Fig. 1 (b) has no clear boarder on the edge and is hard to analyze. Although the actual edge is slightly larger than observed edge owing to the refraction of the light, the refraction of the light through the bubble is neglected in here. The captured bubble image has exact location within the thickness of focus range that is 30cm approximately in this setup. Secondary, the traced edge bubbles are approximated by elliptic curves. The major and minor axes, and angles of the axis are estimated by the least-square method. The estimated error depends on the number of pixels of bubble image and is 2% approximately confirmed by the numerical simulation. Finally, the bubble moving velocity is estimated by the conventional three time particle tracking method. This is an extension of the particle tracking velocimetry using the bubble as a tracer. Thus, this method denotes bubble tracking velocimetry, BTV, hereafter for simplicity.

The image processing procedure of BTV can be summarized as follows (see Fig. 2): The background of the images was removed, first. Then original bubble shape is extracted by the edge detection. After that, the elliptic approximation of bubble shape is calculated and the major, minor and angle of the elliptic shape bubble are estimated to avoid sub-pixel error. Finally the bubble velocity is

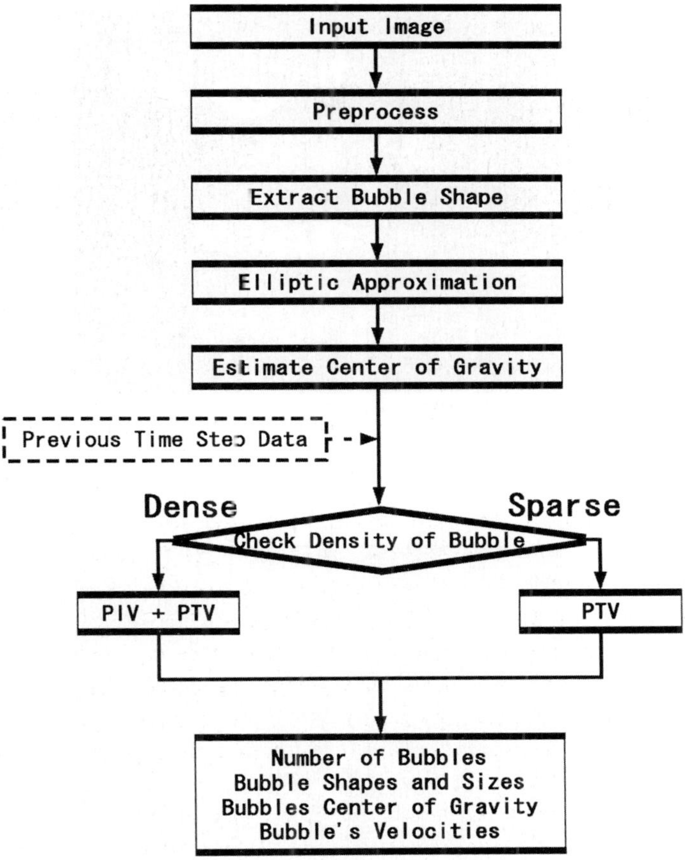

FIG. 2. Flowchart of image processing

estimated by the particle tracking method (PTV). Basically, the PTV is applied for the bubble velocity estimation but the particle image velocimetry (PIV) without bubble shape estimation is applied for the highly dense entrapped air area owing to the overlapped bubble image.

Fig. 3 shows example of BTV method. Fig. 3 (a) is the original image and (b) is the BTV analyzed image. The BTV analyzed image shows bubble shape of in focus bubble and velocity. The BTV can be estimated 2D projected bubble shape and velocity within 10cm×10cm area simultaneously. This is impossible using a standard in-situ probe.

EXPERIMENTAL SETUP

A flat LED light (10cm×10cm) was used for the shadow graph to avoid flicker of the back light. A continuous diode laser operated at a power of 5W and a

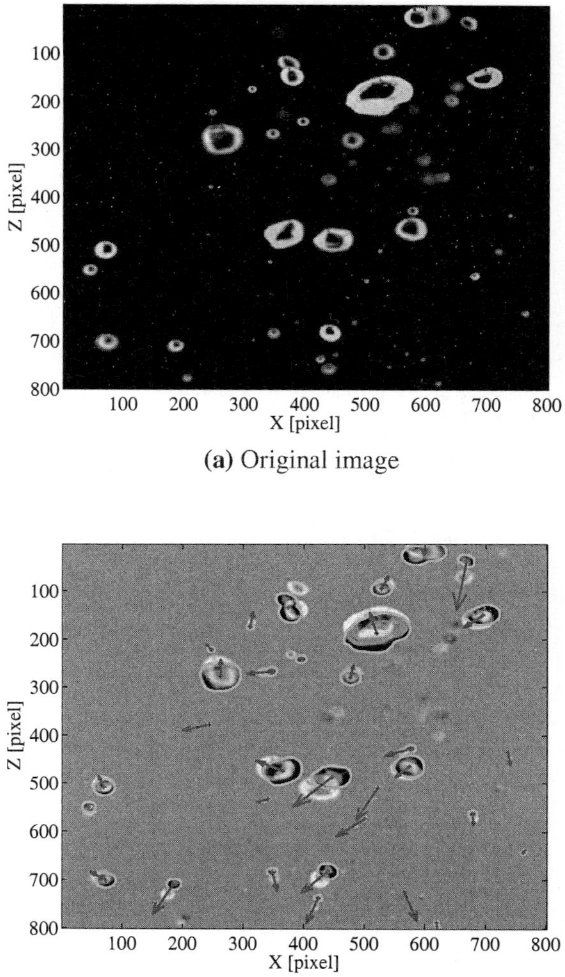

(a) Original image

(b) BTV analyzed image

FIG. 3. Example of image processing

wavelength of 532 nm was used as the light source to adjust the exact location of bubble in transverse direction of bubble between the camera and the back light. A cylindrical lens mounted in front of the laser was used to create the light sheet. An 8-bit high speed CMOS camera with a resolution of 1280×1024 pixels was used to capture the images. The images were captured at a framing rate of 125-250 frames per second. Each captured image was stored in a PC and later analyzed using the

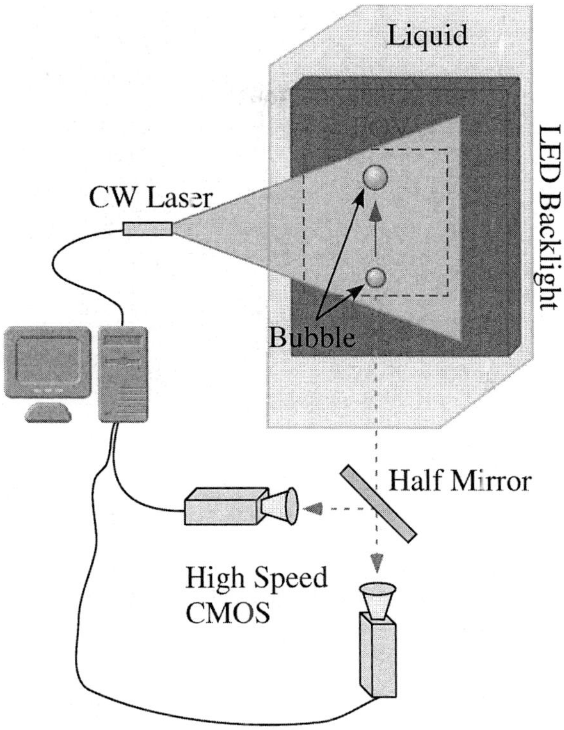

FIG. 4. Setup of BTV measurements

BTV image processing technique with results similar to those shown in Fig. 2. The uncertainty owing to image distortion was also checked using fixed markers in the tank and found that the maximum error was 1.5 mm (about 5 pixels). Fig. 4 shows the sample of BTV experimental setup using two cameras. The half mirror was used to divide one image into two cameras. The two cameras can be installed different lens to enhance the resolution of field of view (FOV). However, only one camera was used in this study.

The breaking wave experiments were conducted in a wave tank that is 50.0 m long, 1.0 m wide, and 1.5 m deep located in Osaka City University. The one side wall of the tank was constructed using glass for optical access. The BTV imaging system shown in Fig. 4 was set up in the middle of the tank. A fixed, impermeable 1/30 slope was installed with the toe 20 from the wavemaker. The water depth was 1.0 m, and regular wave trains were generated by a computer-controlled piston-type wavemaker with active absorber A detail of the experimental setup is shown in Fig. 5

The three different locations were chosen as field of view (FOV) of the BTV

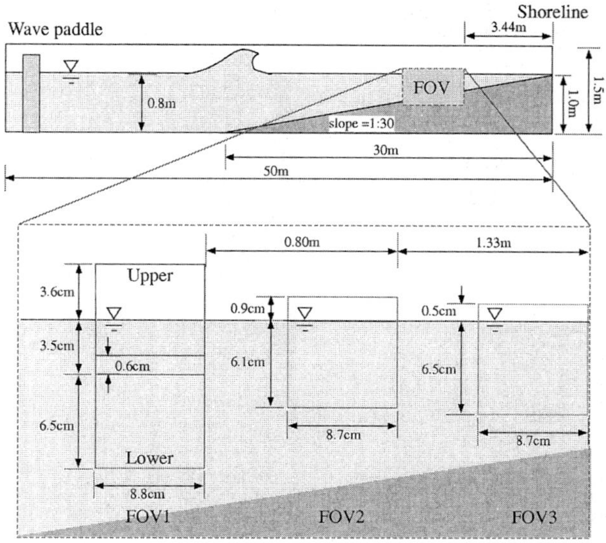

FIG. 5. Illustration of experimental setup

measurements. The FOV 1-3 were located at $3h_b$, $7h_b$ and $15h_b$, respectively. Here h_b denotes the water depth at the breaking point (B.P.). The void fraction in this study is defined by the spatial ratio to bulk of air and water below the free surface. Therefore, the location of the free surface becomes important to define the void fraction. The edge detection method as same as bubble shape detection was used to identify the free surface. The threshold value of edge detection for the free surface was selected through the trial and error through the checking process.

Fig. 6 shows an example of the instantaneous time series of void fraction measured by the BTV. The solid and dashed line denote void fraction, and area of liquid in the image. The spatial-averaged void fractions do not decreased monotonically as Cox and Shin (2003) investigated. This is because the difference between the instantaneous (BTV measured) and phase-averaged void fraction (void fraction measured). The measured time series of void fraction have several interesting characteristics, although measurement locations were limited to discuss in detail. First, the maximum values of void fraction were decreased as breaking wave propagates to shoreline (FOV1→ FOV3). The void fraction at FOV1 ($x = 3h_b$) exceeds 0.2 but is about 0.1 at FOV3 ($x = 15h_b$). Second, the rising up of void fraction from 0 to the maximum at FOV1 is faster than the others and is decreased as FOV1→ FOV3. In addition, the decreasing rates of void fraction after reached maximum value show the similar tendency. Thus, the bulk of air was massively injected into the water near the B.P. and was diffused as wave propagated to shoreline. The BTV can be measured these characteristics

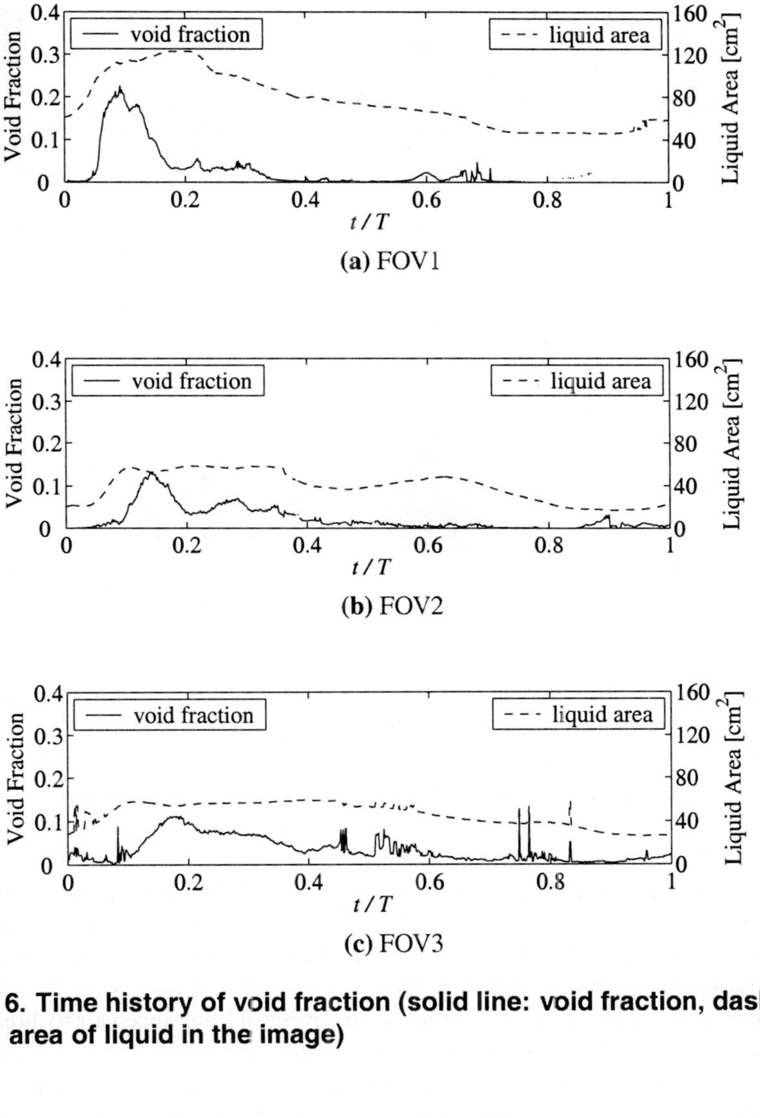

(a) FOV1

(b) FOV2

(c) FOV3

FIG. 6. Time history of void fraction (solid line: void fraction, dashed line: area of liquid in the image)

from the spatial-averaged (depth-averaged) point of view.

Fig. 7 shows the time series of bubble diameter distribution. The bubble diameter denotes mean value of major and minor diameter of elliptic approximation of 2D projected plane. The horizontal axes are time normalized by the incident wave period and the characteristics diameter. The vertical axis is number of bubbles per unit area. The bubble number density usually has a dimension of [Number of bubbles/m^{-4}] but the BTV is 2D measurement therefore

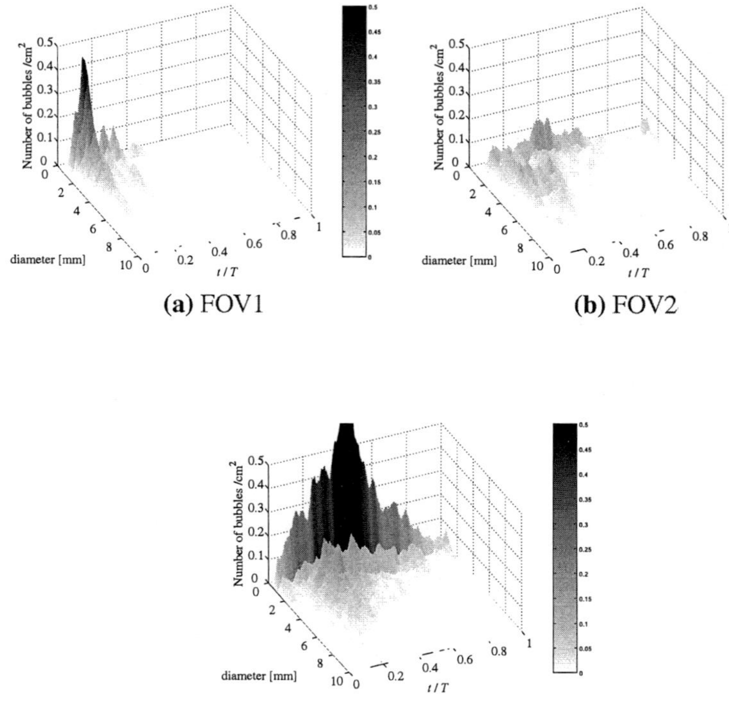

(a) FOV1 (b) FOV2

(c) FOV3

FIG. 7. Time series of bubble size distribution

the bubble number density is normalized by unit area. As already shown in Fig. 6, the bulk of air injected into the free surface rapidly and decreased near the breaking point (FOV1) rather than far field (FOV3). The time series of the bubble number density distribution, Fig. 7, shows similar tendency but the number of small scale bubble has different behavior. Generally, entrapped air by breaking waves is split into small bubbles owing to the shear and turbulence of surrounding flows. The early work of this problem for a general condition discussed by Kolmogorov (1949) and Hinze (1955) in the middle of the last century and a large bibliography has been generated (*i.e.* Martínez-Bazán et al., 1999). The bubble size spectra in FOV1 and 3 show two clear peaks, but the appeared time of peaks is different as shown in Fig. 7 (a) and (c). The number of small size bubble smaller than 1mm increased significantly. The difference of the bubble size spectra between FOV1 and 3 can be explained by the bubble splitting owing to strong shear flow generated by the breaking waves. The detail of small scale bubble characteristics related with the power-law scaling of bubble size spectra is

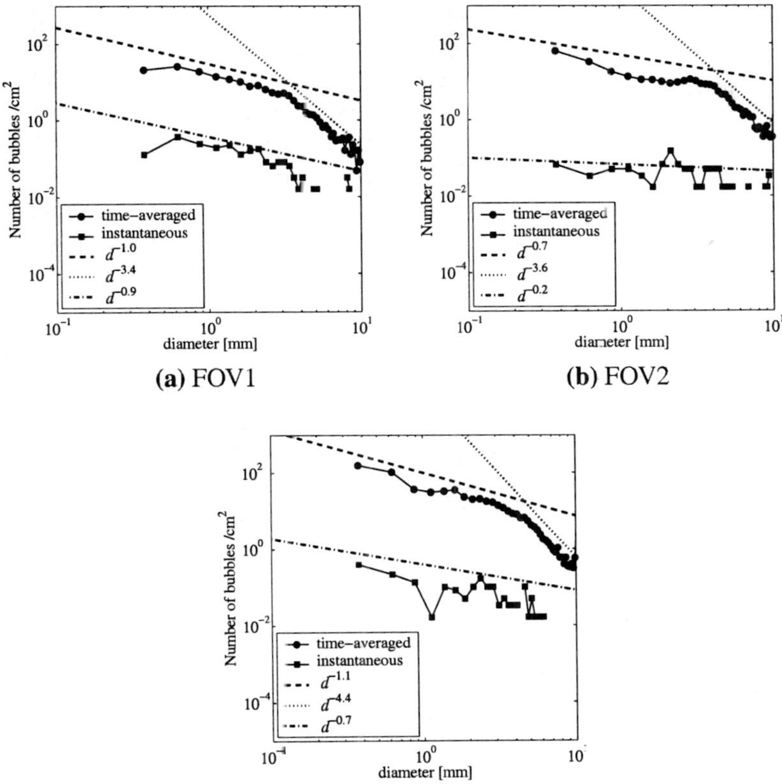

(a) FOV1 (b) FOV2

(c) FOV3

FIG. 8. Power law of bubble size distribution

discussed next.

To check the power-law scaling of the bubble size spectra in the surf zone, Fig. 8 shows the instantaneous and time-averaged bubble size spectra with the estimated power-law by the least-square method. The instantaneous bubble size spectra corresponds at the time of the maximum void fraction. There are two scaling laws can be seen at all FOVs. The small and large size bubble power scaling laws are $d^{-0.7\sim-1.1}$ and $d^{-3.4\sim-4.4}$, respectively. There is no significant difference between instantaneous and time-averaged spectra. The critical point in between two power scaling corresponds to the Hinze scale of bubble splitting theory (Hinze, 1955). Garrett et al. (2000) proposed semi-empirical $-10/3$ power-law scaling for the bubble which is larger than the Hinze scale based on the discussion of bubble fragmentation in the strong turbulence flow below the trough level. The measured power-law scaling of large size bubble size spectra is similar to Garrett et al. (2000). On the other hand, Deane and Stokes (2002) measured a $-3/2$ power-law scaling, smaller than the Hinze scale (1mm) of the acoustically active phase near the crest. The power-law scaling of small size bubble shown in Fig. 8 is close to $-3/2$ power-law predicted by Deane and Stokes (2002). The Hinze scale of time-averaged spectra at FOV1 is smaller than FOV3. This is related with the turbulent intensity and energy dissipation induced by the breaking waves. The detail mechanisms and relationship between the Hinze scale and energy dissipation require liquid phase measurements and should keep for future study.

CONCLUSIONS

The two-phase flow measurement using imaging technique, called Bubble Tracking Velocimetry: BTV, has been developed to measure instantaneous elliptic bubble shape and its moving velocity simultaneously. The characteristics of void fractions, air bubble distributions, and bubble size spectra were measured for surf zone breaking waves in the laboratory. The unsteady characteristics of two-phase flow in the surf zone is important. The further detail measurements using the BTV for surf zone breaking waves will be investigated.

ACKNOWLEDGMENTS

This research was funded by the Ministry of Education, Science, Sports and Culture through Grant-in-Aid for No.17760409 and also acknowledges the support of the Sumitomo Foundation No.053493.

REFERENCES

Baldy, S.. 1993. A generation-dispersion model of ambient and transient bubbles in the close vicinity of breaking waves, *J. Geophysical Research,* C10, 18277–18293.

Chanson, H., S. Aoki, and M. Maruyama. 2002. Unsteady air bubble entrainment and detrainment at plunging breaker: Dominant time scales and similarity of water level variation, *Coastal Engineering,* 46, 139–157.

Cox, D. and S. Shin. 2003. Laboratory measurements of void fraction and turbulence in the bore region of surf zone waves, *Journal of Engineering Mechanics, ASCE,* 129(10), 1197–1205.

Deane, G. B.. 1997. Sound generation and air entrainment by breaking wave in the surf zone, *J. Acoustic Soc. America,* 102, 2671–2689.

Deane, G. B. and M. D. Stokes. 1999. Air entrainment processes and bubble size distributions in the surf zone, *J. Phys. Oceanogr.,* 29, 1393–1403.

Deane, G. B. and M. D. Stokes. 2002. Scale dependence of bubble creation mechanisms in breaking waves, *Nature,* 418, 839–844.

Donelan, M., W. Drennan, E. Saltzman, and R. Wanninkhof. 2002. *Gas Transfer at Water Surfaces*, AGU, Washington DC.

Garrett, C., M. Li, and D. Farmer. 2000. The connection betweeen bubble size spectra and energy dissipation rates in the upper ocean, *J. Phys. Oceanogr.,* 30, 2163–2171.

Hinze, J.. 1955. Fundamentals of the hydrodynamic mechanism of splitting in dispersion processees, *Amer. Inst. Chem. Eng. Jour.,* 1, 289–295.

Kolmogorov, A.. 1949. On the breakage of drops in a turbulent flow, *Dokl. Akad. Navk. SSSR,* 66, 825–828.

Leifer, I. and G. de Leeuw. 2002. Bubble measurements in breaking-wave generated bubble plumes during the LUMINY wind-wave experiment, M. Donelan, W. Drennan, E. Saltzman, and R. Wanninkhof (Eds.), *Gas Transfer at Water Surfaces*, 303–310. Washington DC, AGU.

Loewen, M., M. O'Dor, and M. Skafel. 1996. Bubbles entrained by mechanically generated breaking waves, *J. Geophysical Research,* 101, 20759–20769.

Martínez-Bazán, C., J. Montanés, and J. Lasheras. 1999. On the breakup of air bubble injected into a fully developed turbulent flow. Part I: Breakup frequency, *J. Fluid Mech.,* 401, 157–182.

Mori, N.. 2002. Air bubble distribution induced by wind wave breaking, *Proc. Annual Conference of Japcnese Society for Multiphase Flow*, Japanese Society for Multiphase Flow, 193–194.

Mori, N.. 2003. Experimental study of air bubble distributions induced by wind wave breaking, *Proc. Asia Pasicifc Coast*, Makuhari, JSCE, CD–ROM.

Mori, N., T. Suzuki, and S. Kakuno. 2007. Experimental study of air bubbles and turbulence characteristics in the surf zone, *J. Geo. Res.,*, in press.

Peregrine, D.. 2003. Water wave impact on walls, *Annual Review of Fluid Mechanics,* 35, 23–43.

Thorpe, S.. 1982. On the clouds of bubbles formed by breaking wind waves in deep water, and their role in air-sea gas transfer, *Phil. Trans. R. Soc. London,* A304, 155–210.

Vagle, S. and D. Farmer. 1998. A comparison of four methods of bubble measurement, *IEEE Oceanic Engineering,* 23(3), 211–222.

A NOTE ON SETUP SENSITIVITY AND PREDICTION ACCURACY

Alex Apotsos[1], Britt Raubenheimer[1], Steve Elgar[1], R. T. Guza[2], and Jerry Smith[2]

The sensitivity of numerical model predictions of the cross-shore profile of wave-driven setup, the increase in the mean sea level associated with breaking waves, to the accuracy of the observational inputs and to different physical processes is explored using data from three field experiments. Using different parametric wave models to estimate the cross-shore distribution of wave heights and energy used to drive the setup model can result in up to 90% changes in the predicted setup. Approximating the surfzone beach profile used in the models as planar results in increased setup on concave beaches and reduced setup on convex beaches. Using a planar beach profile and a constant mean water level rather than the measured bathymetry and tidal fluctuations increases the mean and root-mean square errors between the predictions and observations by up to a factor of 5. Including bottom stress and wave rollers significantly improves setup predictions in water depths less than 1 m.

1. INTRODUCTION

The importance of breaking-wave-driven setup to flooding and storm damage was recognized in the early 1900's, when it was observed that storm-driven increases in sea level sometimes were more than a meter higher along coastlines exposed to breaking ocean waves than along protected shores. Furthermore, setup is a dominant forcing mechanism for the mean offshore-directed surf zone flows (undertow) [*Garcez-Faria et al.,* 2000] that drive offshore motion of sand during storms [*Thornton et al.,* 1996; *Gallagher et al.*, 1998].

Assuming alongshore uniform waves and bathymetry and negligible wind and bottom stress, the cross-shore pressure gradient associated with the time-averaged wave setup, $\overline{\eta}$, theoretically balances the cross-shore gradient of the time- and depth-averaged cross-shore wave momentum flux (i.e., the wave radiation stress, S_{xx}) [*Longuet-Higgins and Stewart,* 1962, 1964]

$$\frac{\partial}{\partial x}S_{xx} + \rho g(\overline{\eta} + d)\frac{\partial}{\partial x}\overline{\eta} = 0, \tag{1}$$

where x is the cross-shore coordinate (positive onshore), d is the time-averaged still water depth, ρ is the water density, and g is the gravitational acceleration. The wave radiation stress can be estimated as

$$S_{xx} = E_w\{[\cos^2(\theta) + 1]\frac{c_g}{c} - \frac{1}{2}\}, \tag{2}$$

[1] Woods Hole Oceanographic Institution, Woods Hole, MA, 02540, USA
[2] Scripps Institution of Oceanography, La Jolla, CA 92093, USA

where θ is the mean wave direction relative to beach normal, c_g is the group speed, and c is the phase speed. The wave energy, E_w, can be estimated from linear theory as

$$E_w = \frac{1}{8}\rho g H_{rms}^2, \tag{3}$$

where H_{rms} is the root-mean-square wave height (defined as $2\sqrt{2}$ times the standard deviation of the sea-surface elevation fluctuations).

The accuracy of the data used to drive models based on (1), (2), and (3) may be important when predicting setup. Here, the sensitivity of setup predictions to errors in the offshore wave height, angle, and period, the local bathymetry, tidal fluctuations, the wave theory used to estimate E_w, and the technique used to estimate the cross-shore distribution of wave heights is examined. Within this paper, these terms, which concern the accuracy of the model inputs, are referred to as "Accuracy" terms.

The model based on (1) and (2) is a simplification of the cross-shore momentum balance. Here, the effects of wind stress, wave rollers, bottom stress, convective acceleration of the cross-shore current, directional spread of the incident waves, alongshore variations in the wave radiation stress, wave nonlinearities, infragravity wave reflection from the beach, wave skewness, and the Coriolis force on the cross-shore, surfzone setup profile are examined. Within this paper, these terms, which represent additional physical processes not included in the simple model, are referred to as "Additional" terms.

The observations and methodology are described first (Sections 2 and 3), followed by the sensitivity analyses (Section 4), model data comparisons, (Section 5), and conclusions (Section 6).

2. OBSERVATIONS

Wave-induced pressures and mean water levels were measured for 90 days during Sep to Nov 1997 (SandyDuck) on a barred beach near Duck, NC, and for 14 days during Sep and Oct 2000 (SwashX) and 23 days during Oct and Nov 2003 (NCEX) on near planar beaches near La Jolla, CA. The 3-hr-long (SandyDuck) and 1-hr-long (SwashX and NCEX) data records were subdivided into 8.5-min-long sections for processing to ensure stationarity in the presence of tidally induced depth changes. Wave heights were measured at 11 (SandyDuck) and 8 (SwashX and NCEX) cross-shore locations (Figure 1A circles and 1B & C diamonds, respectively). Mean water levels (i.e., setup) were measured at 13 (SandyDuck) and 8 (SwashX and NCEX) cross-shore locations (Figure 1A-C diamonds, respectively).

Offshore root-mean-square wave heights ranged from 0.20 to 2.10 m (SandyDuck), 0.19 to 1.05 m (SwashX), and 0.20 to 1.00 m (NCEX). Setup ranged from -0.06 to 0.48 m (SandyDuck), -0.01 to 0.18 m (SwashX), and -0.01 to 0.21 m (NCEX). Wave angles during SandyDuck ranged between ±45° with respect to beach normal, whereas waves were near-normally incident (ranging from about ±5°) during both SwashX and NCEX. Centroidal wave frequencies ranged from 0.09 to 0.24 Hz (SandyDuck), 0.09 to 0.20 Hz (SwashX), and 0.08 to 0.21 Hz (NCEX).

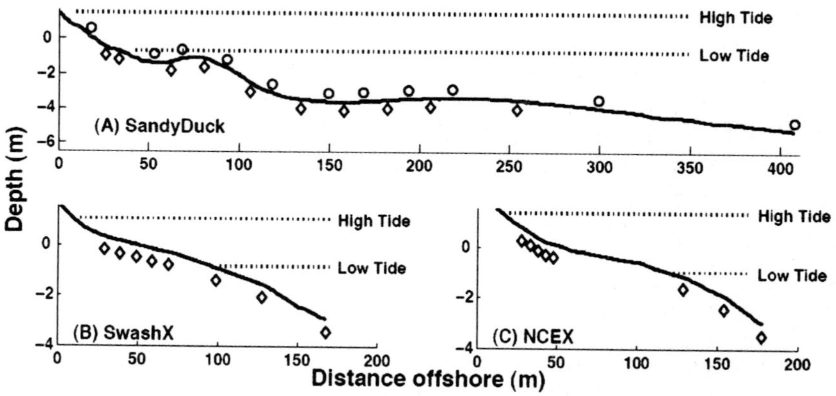

Figure 1: Observed bathymetry (solid curves), unburied pressure gages (circles), and buried pressure gages (diamonds) vs. distance offshore for SandyDuck (A), SwashX (B), and NCEX (C).

Cross-shore bathymetric profiles were measured relative to mean sea level (MSL) approximately every other day using an amphibious vehicle (SandyDuck) or a sonar and GPS mounted on a waverunner (SwashX and NCEX). The time-varying still water levels are obtained by adding the mean water level above MSL owing to tidal fluctuations measured at the most offshore setup sensor to the surveyed bathymetric profiles. The tidal ranges relative to MSL were -0.67 to 1.50 m (SandyDuck), -0.87 to 1.04 m (SwashX), and -0.96 to 1.40 m (NCEX).

3. THEORY AND METHODOLOGY

The sensitivity of setup predictions to the Accuracy and Additional terms is examined with respect to predictions of a model based on (1), (2), and (3) and driven with the measured offshore wave conditions and the tidally adjusted bathymetric profiles. In this "base" model, the cross-shore distribution of H_{rms} is estimated with the parametric wave model developed by *Thornton and Guza* [1983] using fixed constants of $B = 1$ and $\gamma = 0.42$. Snell's Law is used to refract the input offshore wave angle into shallow water for all models. Each Accuracy or Additional term is varied or added to the base model one at a time.

3.1 Accuracy Terms

The sensitivity of setup predictions based on (1), (2), and (3) to errors in the offshore wave conditions is estimated by varying the input offshore wave height by ±20%, angle by ±10°, and period by ±2 s, as well as by using constant offshore wave angles (0 to 45°), the experiment-averaged wave angle, and the experiment-averaged wave period for all data records. The sensitivity of the predictions to errors in the bathymetric profile is examined by comparing predictions based on the bathymetry measured nearest in time to the data record with predictions based on a single, randomly selected profile measured during an experiment and to an artificially planar profile. At all three experiments, five different profiles are selected and used for all data records to estimate the effects on setup predictions of morphological change during the experiment. To obtain planar profiles, a constant linear slope is fit from the shoreline to the most offshore measured depth (SwashX and NCEX) or to the local depth maximum in the most offshore bar trough (SandyDuck, e.g., x = 140 m in Figure 1A). Additionally, the sensitivity to the depth profile is examined by comparing predictions based on the tidally adjusted profiles with predictions for tidal levels ±0.20 m, for a tidal level of zero, and for a linear model in which the contribution of setup to the water depth is ignored [e.g., (1) is linearized by assuming that $\overline{\eta} << d$].

The sensitivity to errors in the cross-shore distribution of H_{rms} is examined by comparing setup predictions based on the cross-shore distribution of wave heights estimated with a wave model [*Thornton and Guza*, 1983] with predictions based on fitting a cubic spline to the measured cross-shore distribution of wave heights, and with predictions based on 7 different parametric wave models [*Lippmann et al.*, 1996; *Ruessink et al.*, 2003; *Whitford*, 1988; *Baldock et al.*, 1998, *Battjes and Stive*, 1985; *Nairn*, 1990; *Battjes and Janssen*, 1978]. The effect of using linear theory to determine E_w [i.e., (3)] is examined by estimating E_w from a weakly nonlinear cnodial theory.

3.2 Additional Terms

The momentum balance equation (1) can be extended to include bottom stress τ_B, wind stress τ_w, the effects of the Coriolis force, the convective acceleration of the time and depth-averaged cross-shore current U_r, and alongshore gradients of the alongshore radiation stress S_{xy} such that

$$\frac{\partial}{\partial x}S_{xx} + \rho g(\overline{\eta}+d)\frac{\partial}{\partial x}\overline{\eta} + \tau_B - \tau_w + fv\rho(\overline{\eta}+d) - \frac{\partial \rho U_r^2(\overline{\eta}+d)}{\partial x} + \frac{\partial S_{xy}}{\partial y} = 0 \cdot \quad (4)$$

Bottom stress is estimated following *Apotsos et al.* [in press]. Wind stress is found as

$$\tau_w = \rho_a C_{10}\left|V_w\right|V_w, \quad (5)$$

where ρ_a is the density of air, V_w is the wind speed measured approximately 10 m above the water surface, and C_{10} is given by [*Wu*, 1980]

$$C_{10} = C_{fw}\left(0.8 + 0.065V_w\right), \quad (6)$$

in which C_{fw} is a friction coefficient equal to 10^{-3}. While (6) is not dimensionally consistent, it is a useful parameterization based on observations. The Coriolis term is found from the latitude dependant Coriolis parameter, f, and the time- and depth-averaged alongshore current estimated from measurements, v. The time- and depth-averaged cross-shore current, U_r, is estimated as

$$U_r = \frac{M_w}{\rho d}, \tag{7}$$

where the mass flux of the wave $M_w = E_w/c$. The alongshore radiation stress is estimated as

$$S_{xy} = E_w \cos(\theta)\frac{c_g}{c}\sin(\theta). \tag{8}$$

To estimate the alongshore gradient of the radiation stress, the wave model [*Thornton and Guza,* 1983] and Snell's Law were used to calculate E_w and θ along cross-shore profiles measured roughly ±20 m alongshore of the instrument transect.

The wave radiation stress equation (2) can be extended to account for a wave skewness parameter C_s, the directional spread of the incoming waves σ_θ [*Feddersen*, 2004], and wave roller energy E_r such that

$$S_{xx} = \left[E_w\{[\cos^2(\theta)+1]\frac{c_g}{c} - \frac{1}{2} + C_s\} + 2E_r\left(\cos^2\theta\right)\right]\left(1-\frac{2}{3}\sigma_\theta^2\right). \tag{9}$$

The wave skewness parameter is given by [*Johnson and Kobayashi*, 1998]

$$C_s = \frac{\sigma}{d}s - \left(\frac{\sigma}{d}\right)^2, \tag{10}$$

where σ is the standard deviation of the sea surface elevation and s is the wave skewness. The directional spread of the incoming waves is estimated both using Snell's Law to refract the observed offshore directionally spread waves into shallow water, and by fitting a cubic spline to the observed cross-shore distribution of directional spreads, which may increase in the surfzone [*Hendersen et al.,* 2006]. The wave roller energy is estimated using four different formulations for E_r [*Lippmann et al.*, 1996; *Ruessink et al.*, 2001; *Tajima*, 2003; *Smith*, unpublished].

Using the linear theory approximation of equipartition of energy [e.g., (3)] may underestimate the wave energy E_w by as much as 50% in 0.30 m water depth [*Raubenheimer et al.*, 2004]. Consequently, wave nonlinearities in shallow water are examined by multiplying E_w calculated with (3) by a factor that increases linearly from 1.0 (in 1.30 m water depth) to 1.5 (in 0.30 m water depth).

The effect of infragravity wave reflection is examined by assuming that [*Longuet-Higgins and Stewart*, 1964]

$$\overline{\eta}_r = \int_f \left[2A^2(f)k(f)\coth 2k(f)h\cos 2k(f)x\right], \qquad (11)$$

where $A^2(f)$ is determined from the spectral amplitude of the reflected wave estimated from observations and $k(f)$, the frequency dependent wave number, is estimated from the linear dispersion relation. It is assumed that $\overline{\eta}_r$ can be linearly superimposed on $\overline{\eta}$.

3.3 Methodology

Setup in deep water ($h = \overline{\eta}+d > 1.0$ m) can be smaller than the measurement error, and the shoreline setup ($h = 0.1$ m) is sensitive to small changes in the local bathymetry. Thus, the focus here is on setup in shallow (0.1 to 0.5 m) and mid-depth (0.5 to 1.0 m) water.

The models were run for all data records, and each term was evaluated using four separate indicators: biases, 50% spreads, best fit linear slopes, and squared correlation coefficients (r^2). Biases and spreads are found from a histogram of the percent differences between the base and modified models for all data records at each experiment (Figure 2). The bias is defined as the median of the percent differences (e.g., 50% of the records have a smaller (or larger) percent difference). The 50% spread is defined as the difference between the values for which 25% and 75% of the records have smaller differences. Best-fit linear slopes and squared correlations are calculated between the predictions for each term and the predictions of the base model. Slopes less than 1 and negative biases signify underprediction.

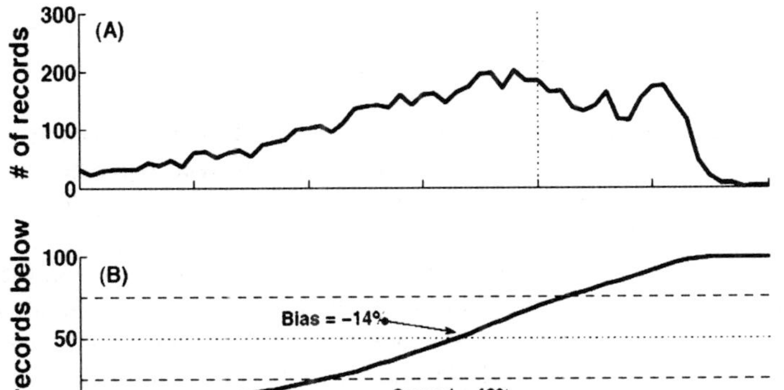

Figure 2: (A) Histogram of the differences between the base model and the model modified by adding a roller following *Ruessink et al.*, [2001], and (B) the cumulative sum of the histogram with the bias (dotted line) and the limits of the spread (dashed lines). The dotted line in (A) indicates a bias of 0.

4. RESULTS

4.1 Accuracy Terms

4.1.1 Summary. For the ranges of parameters considered here, variations in the wave height and water depth have the largest effect on the setup predictions, whereas variations in the offshore wave period and angle, and the theory used to estimate E_w are of secondary importance (Table 1). Inclusion of $\overline{\eta}$ in the calculation of the local water depth has a negligible effect on the setup predictions.

Table 1: Ranges of biases and spreads in shallow and mid-depth water for the primary and secondary Accuracy Terms.

Accuracy Term	Shallow (0.1 to 0.5 m)		Mid-depth (0.5 to 1.0 m)	
	Bias	Spread	Bias	Spread
Offshore Wave Height ±20%	-20% to 26%	6 to 12%	-48% to 50%	34% to 50%
Wave Model	-12% to 36%	8% to 90%	-18% to 26%	4% to 92%
Linear Profile	-64% to -12%	40% to 54%	-68% to -46%	64% to 86%
Single Profile	-38 to 0%	10% to 64%	-38% to 18%	28% to 94%
Tide ±0.2 m	-34 to 44%	6 to 50%	-44% to 54%	22% to 72%
No Tidal Fluctuations	-34% to -18%	60% to 72%	-14% to 44%	72% to 120%
Offshore Wave Angle = 45°	-30% to -24%	6% to 8%	-34% to 42%	8% to 20%
Offshore Peak Period ±2 s	-14% to 8%	2 %to 12%	-16% to 8%	6% to 18%
Cnodial Theory	-24% to -20%	10% to 22%	0% to 2%	30% to 46%

4.1.2 Primary Terms. Percent changes in the offshore wave height result in roughly equal percent changes in shallow water setup (i.e., a 20% increase in $H_{rms,o}$ results in approximately a 20% increase in shallow water setup) at all three experiments (Table 1). The change in setup owing to increasing or decreasing the offshore wave height typically occurs in $h > 1$ m, thus larger percent changes in setup occur in mid (and deep) water depths because the absolute setup decreases (e.g., the denominator decreases). Using different parametric wave models or a cubic spline of the observed wave heights results in biases as large as 30%, and spreads up to 90% for both shallow and mid-depth water (Table 1).

Changes in the local water depth can affect setup predictions significantly. On average, using a linear profile reduces the predicted setup in shallow (mid-depth) water by 64% (54%), 24% (46%), and 12% (68%) at SandyDuck, SwashX, and NCEX, respectively (Table 1). However, limiting the data records from NCEX to high tides during which the surfzone bathymetry is concave results in a positive bias in shallow and mid-depth setup (Figure 3), suggesting the negative biases calculated using all data records are owing to the surfzone beach profiles being typically convex (e.g., Figure 1A-C).

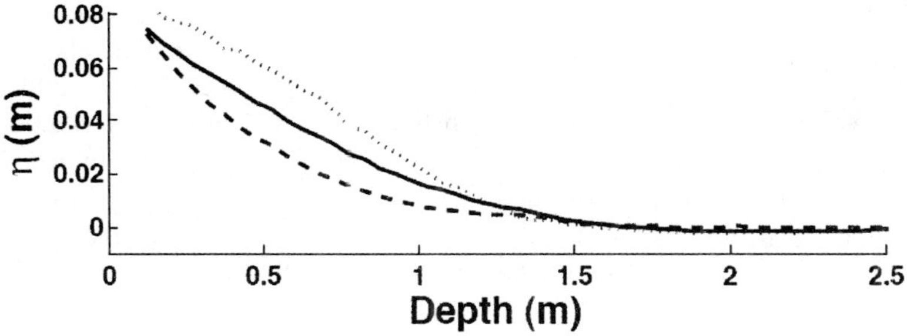

Figure 3: Average predicted setup at NCEX for all data records and the most recent bathymetry (solid curve), and the average setup predicted approximating the profile as planar for data records when the measured surfzone beach profile was concave (dotted curve) or convex (dashed curve) vs. water depth.

Using a single bathymetric profile for the duration of the experiment can result in a positive, negative, or near zero bias depending on the profile used. At all three experiments, the calculated spread for predictions based on a single profile can be large (Table 1), indicating that the accuracy of setup predictions depends partially on the accuracy of the cross-shore depth profile. Thus, accurate and possibly repeated measurement of the beach profile is important for predicting setup, especially during times of significant morphological change.

Model predictions are most sensitive (largest spreads and smallest r^2 values, Table 1) to accurate estimates of the tidal fluctuations. Varying the tidal level by ±0.2 m causes approximately a ±40% bias in the predicted setup in both shallow and mid-depth water (Table 1). As the tide increases (decreases), the predicted setup decreases (increases) at a given cross-shore location because the water depth is larger (smaller). Neglecting the tidal fluctuations results in negative bias (-34 to -18%) in shallow water for all three experiments, possibly owing to the positively skewed tidal ranges.

4.1.3 Secondary Terms. Using an experiment-averaged offshore wave angle or period, varying the measured offshore wave angle by ±10°, or using an offshore wave angle < 20° for all data records results in less than a 5% bias and small spreads (< 15%) at all three experiments. However, using an offshore wave angle > 25° for all data records causes increasingly negative biases up to about -30% for $\theta = 45°$ (Table 1).

Varying the wave period by ±2 s results in small biases (<16%) and small spreads (< 20%) (Table 1). As the period increases (decreases) the predicted setup increases (decreases) owing partly to changes in the cross-shore distribution of E_w estimated from the wave models and partly to changes in the calculation of c and c_g in (2).

Using cnodial theory instead of linear theory to estimate E_w in (2) results in a maximum bias of -24% in shallow water (Table 1), and has a larger effect for larger waves. However, it is unclear if a nonlinear wave theory should be used in conjunction with (2), which is based on linear theory, or if a relationship between linear and nonlinear waves should be developed [e.g., *Tajima*, 2003].

4.1.4 Negligible Terms. The bias owing to neglecting the contribution of $\overline{\eta}$ to the water depth is less than 6%, and is reduced to almost 0% by using $\overline{\eta}$ calculated at the previous cross-shore step. However, it is likely that for waves larger than those observed during these three experiments (i.e., for $H_{rms,o} > 2$ m), $\overline{\eta}$ will become more important to the local water depth for $h < 1$ m.

4.2 Additional Terms

4.2.1 Summary. Including bottom stress and wave rollers in the model significantly affects setup predictions [*Apotsos et al.,* in press], whereas infragravity wave reflection and wave nonlinearities may be of secondary importance (Table 2). The effects of the Coriolis force, wind stress, convective acceleration of the cross-shore current, wave directional spread, alongshore variations in the wave radiation stress, and wave skewness are negligible for the three experiments considered.

Table 2: Ranges of biases and spreads in shallow and mid-depth water for the primary and secondary Additional Terms.

Additional Term	Shallow (0.1 to 0.5 m)		Mid-depth (0.5 to 1.0 m)	
	Bias	Spread	Bias	Spread
Bottom Stress	64% to 68%	8% to 16%	72% to 82%	20% to 38%
Wave Rollers	-38% to 28%	4% to 74%	-56% to 10%	10% to 110%
Wave Reflection	8% to 24%	6% to 18%	8% to 16%	8% to 20%
Wave Nonlinearities	-2% to 6%	8% to 12%	-14% to -8%	16% to 32%

4.2.2 Primary Terms. Including bottom stress results in a positive bias of about 65% in shallow water and 77% in mid-depth water at all three experiments (Table 2). Small spreads and r^2 values close to 1 indicate a consistent bias with little scatter over a wide range of incident wave and bathymetric conditions.

Addition of a roller can cause significant negative or positive biases, along with spreads up to about 100% (Table 2), depending on the water depth and formulation used. Rollers have a bigger effect when offshore waves are large. Furthermore, the effects of rollers depend on the surfzone bathymetry. For example, although all roller formulations decrease setup in mid-depth water and increase setup in shallow water on the near-planar SwashX and NCEX profiles, predicted setup can increase or decrease on the nonplanar profiles observed during SandyDuck depending on the roller formulation used, presumably owing to the barred bathymetry.

4.2.3 Secondary Terms. Including either infragravity wave reflection or wave nonlinearities in the model can cause a 10-25% bias in shallow and mid-depth water (Table 2). The increase in setup owing to reflection was larger at SandyDuck (24%) than at SwashX and NCEX (10%), possibly owing to the steeper foreshore at SandyDuck. However, the corrections used here for infragravity wave reflection (which neglects the sloping bed) and linear-theory underprediction of E_w (which is based empirically on limited observations) are crude, and these terms may be more important than estimated.

4.2.4 Negligible Terms. Including the Coriolis term, wind stress, convective acceleration of the cross-shore current, wave directional spreads, alongshore variations in the wave radiation stress, and wave skewness results in biases smaller than 5%. Furthermore, including a wind stress based on 15 m/s winds throughout the experiments results in biases smaller than 10%.

5. COMPARISONS OF PREDICTIONS WITH OBSERVATIONS

5.1 Extended Model

The base model was extended to include bottom stress and wave rollers. For each data record the cross-shore distribution of E_w in both the base and extended models was estimated using the wave transformation model with the smallest root-mean-square (rms) percent error between the observed and predicted wave heights. Only data records for which all observational inputs were measured are used.

The base model underpredicts setup for all water depths (i.e., negative mean errors and slopes less than 1, solid curves in Figures 4A-C and 4G-I, respectively). Addition of the roller and bottom stress in the extended model increases predicted setup, and typically results in smaller mean errors and similar r^2 values relative to the base model. In water depth ranges $0.3 < h < 0.5$ m, $0.5 < h < 1.0$ m, and, $1.0 < h < 2.0$ m, the best fit slopes between the extended model predictions and observations are 0.66, 0.74, and 1.06 (SandyDuck), 1.13, 1.13, and 1.42 (SwashX), and 0.90, 0.74, 0.84 (NCEX), respectively. Setup and setdown in the deepest water at SwashX and NCEX often were smaller than the measurement error, and thus the slope of the linear regression may be inaccurate (e.g., Figure 4H & I, see values in 2.25 m water depth). For $h < 2$ m, the r^2 values between model predictions and observations are greater than about 0.59 (SandyDuck), 0.67 (SwashX), and 0.47 (NCEX).

At SandyDuck and NCEX the extended model reduces the mean and rms errors relative to the base model, and results in slopes closer to 1 for $h < 1.5$ m. However, the extended model overpredicts setup in deeper water at SandyDuck. At SwashX, the extended model results in larger rms errors in mid-depth water, and overpredicts setup in most water depths, possibly owing to poor representation of the bottom stress in deeper water. Specifically, a model that includes bottom stress only when $h < 1$ m has smaller rms errors than the base model in all depths, and results in no overprediction of the observed setup (not shown).

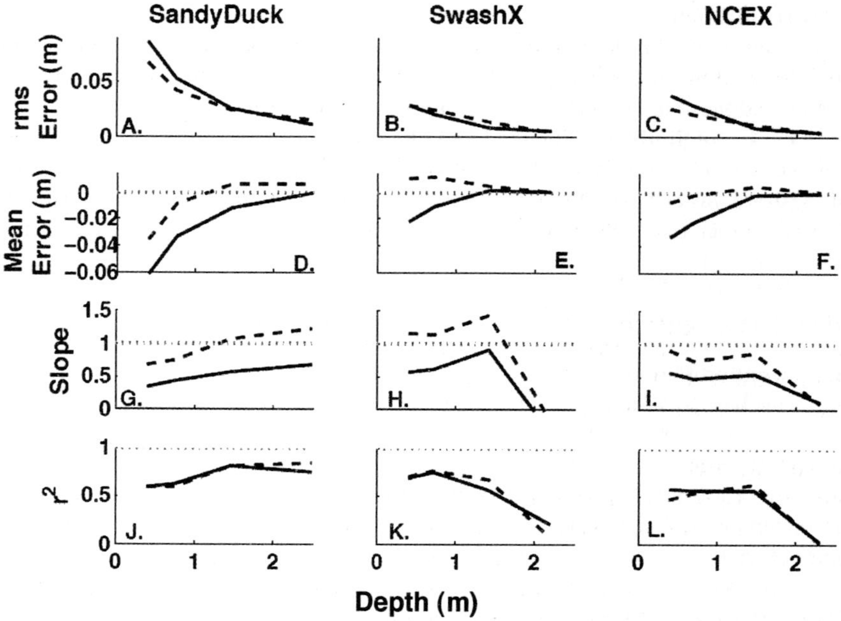

Figure 4: (A-C) Root-mean square (rms) and (D-F) mean errors, and (G-I) best-fit slopes and (J-L) correlations (r^2) between the observations and predictions for the base model (solid curve) and the extended model (dashed curve) for (A, D, G, and J) SandyDuck, (B, E, H, and K) SwashX, and (C, F, I, and L) NCEX vs. water depth. The horizontal dotted lines represent either zero (mean errors) or 1 (slope and r^2 values).

5.2 Modeling Setup without Observational Inputs

The accuracy of setup predictions based on limited knowledge of the input data is examined by assuming only the offshore wave height and water depth is known. The model is driven using a planar beach profile, an experiment-averaged wave period, normally incident waves, and a constant mean sea level (i.e., no tidal fluctuations). The resulting setup predictions from this "simulation" model are significantly less accurate than the predictions of the base model driven with all observational inputs.

For all three experiments, the r^2 values between the observations and the simulation model predictions are small (e.g., $r^2 < 0.25$ in $h < 1$ m), and the mean and rms errors are approximately 5 times larger than those between the observations and the base model. For the data considered here, almost all of the reduction in model skill is owing to the assumptions of a planar profile and a constant mean sea level. At SwashX and NCEX, where the profile is nearly planar, addition of tidally fluctuating water levels greatly increases the skill of the model. In contrast, at SandyDuck accurate representation of the barred profile is of equal importance to accurate measurement of the tidal levels. For all experiments, using an experiment-averaged wave period or assuming normally incident waves does not affect the setup predictions significantly.

6. CONCULSIONS

Accurate measurement of the cross-shore water-depth and wave-height profiles is important when predicting setup. Tidal water-level fluctuations can affect setup significantly, and accurate measurement of the nearshore bathymetry is particularly important on barred beaches. Including bottom stress and wave rollers results in improved model predictions of the observed setup. Infragravity wave reflection may have a smaller, yet important, effect on setup in shallow water, especially on a beach with a steep foreshore.

AKNOWLDEGEMENTS

Funding was provided by NSF and ONR. Staff from the CCS at Scripps Institution of Oceanography, PVLAB at Woods Hole Oceanographic Institution, and the US Army Corps of Engineers FRF are thanked for their heroic efforts in obtaining the observations.

REFERENCES

Apotsos, A., B. Raubenheimer, S. Elgar, R.T. Guza, and J. Smith. In Press. The effects of wave rollers and bottom stress on wave setup, *J. Geophys. Res.*

Baldock, T.E., P. Holmes, S. Bunker, and P. Van Weert. 1998. Cross-shore hydrodynamics within an unsaturated surf zone, *Coastal Eng., 34*, 173-196.

Battjes, J.A. and J.P.F.M. Janssen. 1978. Energy loss and set-up due to breaking of random waves, *Proceedings of the 14th International Conference on Coastal Engineering*, ACSE, 466-480.

Battjes, J.A. and M.J.F. Stive. 1985. Calibration and verification of a dissipation model for random breaking waves, *J. Geophys. Res., 90*, 9159-9167.

Feddersen, F. 2004. Effect of wave directional spread on the radiation stress: Comparing theory and observations, *Coastal Eng., 51*, 473-481.

Gallagher, E.L., S. Elgar, and R.T. Guza. 1998. Observations of sand bar evolution on a natural beach, *J. Geophys. Res., 103*, 3203-3215.

Garcez-Faria, A.F., E.B. Thornton, T.C. Lippmann, and T.P. Stanton. 2000. Undertow over a barred beach, *J. Geophys. Res., 105*, 16999-17010.

Hendersen, S.M., R.T. Guza, S. Elgar, and T.H.C. Herbers. 2006. Refraction of surface gravity waves by shear waves, *J. Phys. Oceanogr., 36,* 629-635.

Johnson, B.D. and N. Kobayashi. 1998. Nonlinear time-averaged model in surf and swash zones, *Proceedings of the 26th International Conference on Coastal Engineering*, ACSE, 2785-2798.

Lippmann, T.C., A.H. Brookins, and E.B. Thornton. 1996. Wave energy transformation on natural profiles, *Coastal Eng., 27*, 1-20.

Longuet-Higgins, M.S. and R.W. Stewart. 1962. Radiation stress and mass transport in gravity waves, with application to 'surf-beats,' *J. Fluid Mech., 13*, 481-504.

Longuet-Higgins, M.S. and R.W. Stewart. 1964. Radiation stresses in water waves: A physical discussion with application, *Deep Sea Res., 11*, 529-562.

Nairn, R.B. 1990. Prediction of cross-shore sediment transport and beach profile evolution. PhD Thesis, Imperial College.

Raubenheimer, B., S. Elgar, and R.T. Guza. 2004. Observations of swash zone velocities: A note on friction coefficients, *J. Geophys. Res.* 109, C01027, doi:10.1029/2003JC001877.

Ruessink, B.G., J.R. Miles, F. Feddersen, R.T. Guza, and S. Elgar. 2001. Modeling the alongshore current on barred beaches, *J. Geophys. Res., 106*, 22451-22463.

Ruessink, B.G., D.J.R. Walstra, and H.N. Southgate. 2003. Calibration and verification of a parametric wave model on barred beaches, *Coastal Eng., 48*, 139-149.

Tajima, Y. 2003. Surf-zone hydrodynamics, PhD Thesis, MIT/WHOI, Cambridge.

Thornton, E.B. and R.T. Guza. 1983. Transformation of wave height distribution, *J. Geophys. Res., 88*, 5925-5938.

Thornton, E.B., R.T. Humiston, and W. Birkemeier. 1996. Bar-trough generation on a natural beach, *J. Geophys. Res., 101*, 10297-12110.

Whitford, D.J., 1988. Wind and wave forcing of longshore currents across a barred beach, Ph.D Thesis, Naval Postgraduate School, Monterey.

Wu, J. 1980. Wind-stress coefficients over sea surface near neutral conditions - A revisit, *J. Phys. Oceanogr., 10*, 727-740.

THREE DIMENSIONAL NUMERICAL SIMULATIONS OF WETTING AND DRYING BED DUE TO TIDAL CURRENTS USING FRACTIONAL STEP METHOD

Mazen Abualtayef[1], Masamitsu Kuroiwa[1], Kentarou Tanaka[1], Yuhei Matsubara[1] and Junichi Nakahira[2]

This paper describes the development and application of a three dimensional (3D) multi-layer hydrodynamic model to tidal motions in Ariake Sea. The governing equations are discretized into finite difference horizontally and finite element vertically in a uniform staggered grid system. The velocity components in the x, y, and z directions are directly obtained. Our study introduces a new 3D, time-dependent, non-hydrostatic, tidal current model that can compute wetting and drying bed (WAD) in tidal flats due to tidal motion. The 3D model is tested for artificial tidal flat surrounded by breakwaters in order to investigate the performance of the presented model. The model results for the velocities and the free surface water elevations coincide with observation for Saigo Port (Abualtayef et al., 2006). Then, the model is applied to the Ariake Sea. In the first simulation, the tidal current behavior including current circulation and time history of sea levels is predicted. The numerically predicted results show good agreement with field observations.

INTRODUCTION

The use of numerical models has become standard procedure for simulating tidal processes in estuaries and for predicting variations of the tidal properties. Tidal inlets form complex and dynamic morphological elements of the coastal system. Therefore, a prediction model for simulating a tidal motion, which is consider WAD in the inter-tidal zone, is needed in order to exactly predict changes of hydro- and morpho-dynamic in a region with tidal flats. Conventionally, 2D models (e.g. Andrea Balzano, 1998) or 3D multilevel models (e.g. Zhang et al., 2000), which are layer integrated models, are mainly used to predict the tidal motion in the tidal flats. Although the multi-level model can express vertical structure at deep water in the inter-tidal zone, but the number of layers of vertical direction decreases, and it is difficult to predict the vertical distribution of currents. Therefore, a new 3D tidal current model based

---

[1] Civil Eng. Department, Tottori University, 4-101, Koyama, Tottori, 680-8552, Japan

[2] Yachiyo Consulting Eng., Tokyo Shinjukuku Nishiotiai, 161-8575, Japan

on fractional step method (FSM), suggested by Koutitas et al. (1980) and used by Kuroiwa et al. (1998), is proposed in this study. By using this approach, it is possible to divide the water depth into equal layers and near sea-bottom layer can be divided finely. Therefore, it is possible to calculate the vertical flow distribution even in extremely shallow sea areas.

The main goal of this study is to formulate and develop a 3D numerical model to clarify the morphodynamic processes on the estuary due to the tidal current, siltation and mixing processes of temperature and salinity. As a second phase, a 3D tidal current model, which can express the WDB in the tidal flats and salinity circulation, was developed.

MODEL DESCRIPTIONS

Governing Differential Equations

The model presented in this study consists of two modules: (1) the tidal current module, and (2) the mixing process module including a dynamic pressure. Governing equations for the tidal current velocities, with the Bussinesq turbulence approximation, are derived from the continuity equation and 3D Navier-Stokes equations. The tidal current computations include the effect of the mixing processes due to the changes of salinity and water temperature were considered. The 3D distribution of the salinity was determined using 3D diffusion equations. The eddy viscosity due to turbulent components was approximated from Bussinesq assumption. Fig.1 shows the definition of the coordinate system. The governing equations of continuity and momentum for incompressible fluid flow in a Cartesian coordinate system with the z-axis directed vertically upwards are:

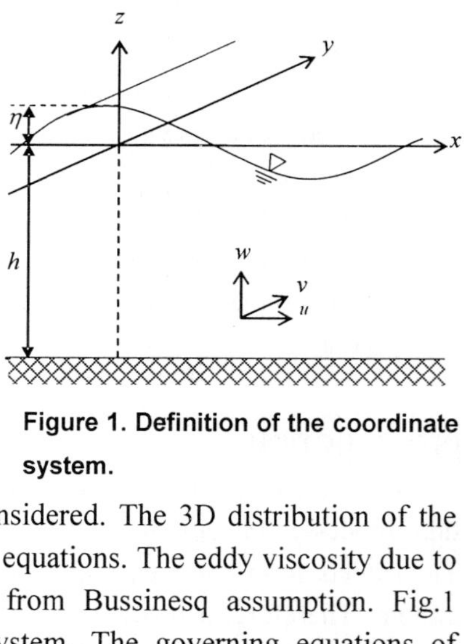

Figure 1. Definition of the coordinate system.

1. Equation of continuity:

$$\frac{\partial u}{\partial x}+\frac{\partial v}{\partial y}+\frac{\partial w}{\partial z}=0 \tag{1}$$

2. Equations of motion:

$$\frac{\partial u}{\partial t}=fv-u\frac{\partial u}{\partial x}-v\frac{\partial u}{\partial y}-w\frac{\partial u}{\partial z}-g\frac{\partial \eta}{\partial x}+\frac{\partial}{\partial x}\left(\varepsilon_h\frac{\partial u}{\partial x}\right)+\frac{\partial}{\partial y}\left(\varepsilon_h\frac{\partial u}{\partial y}\right)+\frac{\partial}{\partial z}\left(\varepsilon_v\frac{\partial u}{\partial z}\right) \tag{2}$$

$$\frac{\partial v}{\partial t}=-fu-u\frac{\partial v}{\partial x}-v\frac{\partial v}{\partial y}-w\frac{\partial v}{\partial z}-g\frac{\partial \eta}{\partial y}+\frac{\partial}{\partial x}\left(\varepsilon_h\frac{\partial v}{\partial x}\right)+\frac{\partial}{\partial y}\left(\varepsilon_h\frac{\partial v}{\partial y}\right)+\frac{\partial}{\partial z}\left(\varepsilon_v\frac{\partial v}{\partial z}\right) \quad (3)$$

$$g+\frac{1}{\rho}\frac{\partial p}{\partial z}=0 \quad (4)$$

$$w=w_{z=-h}-\frac{\partial}{\partial x}\int_{-h}^{z}udz-\frac{\partial}{\partial y}\int_{-h}^{z}vdz \quad (5)$$

3. The condition at the free surface is given by:

$$\frac{\partial \eta}{\partial t}=-\frac{\partial}{\partial x}\int_{-h}^{\eta}udz-\frac{\partial}{\partial y}\int_{-h}^{\eta}vdz \quad (6)$$

4. The equation of salinity conservation is:

$$\frac{\partial S}{\partial t}=-u\frac{\partial S}{\partial x}-v\frac{\partial S}{\partial y}-w\frac{\partial S}{\partial z}+A_C\left(\frac{\partial^2 S}{\partial x^2}+\frac{\partial^2 S}{\partial y^2}\right)+\frac{\partial}{\partial z}\left(K_C\frac{\partial S}{\partial z}\right) \quad (7)$$

5. The sate equation of seawater is defined by Knudsen formula:

$$S=0.0305+1.805Cl \quad (8)$$

$$\rho=\sigma_t+1000 \quad (9)$$

$$\sigma_0=-0.069+1.4708Cl-0.001570Cl^2+0.0000389Cl^3 \quad (10)$$

$$\sigma_t=-\frac{(T-3.98)^2}{503.570}\times\frac{T+283.0}{T+67.26}+(\sigma_0+0.1344)\{1-A_t+B_t(\sigma_0-0.1324)\} \quad (11)$$

$$A_t=T\left(4.7869-0.098185T+0.0010843T^2\right)\times10^{-3} \quad (12)$$

$$B_t=T\left(18.030-0.8164T+0.01667T^2\right)\times10^{-6} \quad (13)$$

where u, v, w are Cartesian velocity components along axis x, y, z respectively, f is the Coriolis force, η is the water surface displacement, g is acceleration of gravity, p is the water pressure, ρ is the density of the fluid, S is the salinity (‰), A_c and K_c are the horizontal and vertical diffusion coefficients, Cl is chloride (‰), T is the water temperature (°C), and ε_h and ε_v are the horizontal and vertical eddy viscosity coefficients, which are estimated using Smagorinsky model as follows:

$$\varepsilon_h = C_h \Delta x \Delta y \left[\left(\frac{\partial u}{\partial x} \right)^2 + \frac{1}{2} \left(\frac{\partial v}{\partial x} + \frac{\partial u}{\partial y} \right)^2 + \left(\frac{\partial v}{\partial y} \right)^2 \right]^{0.5} \tag{14}$$

$$\varepsilon_v = C_v \Delta z^2 \left[\left(\frac{\partial u}{\partial z} \right)^2 + \left(\frac{\partial v}{\partial z} \right)^2 \right]^2 \tag{15}$$

where C_h and C_v are dimensionless coefficients in horizontal and vertical directions, respectively.

Boundary Conditions

Free surface

- The shear stresses due to wind and wave breaking are not considered, namely the fluxes across the water surface are set to zero,

$$\frac{\partial u}{\partial z} = 0, \quad \frac{\partial v}{\partial z} = 0 \qquad (z = \eta) \tag{16}$$

- The chloride condition is given by

$$K_c \frac{\partial Cl}{\partial z} = Q_{Cl} \qquad (z = \eta) \tag{17}$$

where, Q_{Cl} is chloride flux at the sea surface.

Sea bed

- The shear stresses are estimated as follows

$$\varepsilon_z \frac{\partial u}{\partial z} = \frac{\tau_{bx}}{\rho}, \quad \varepsilon_z \frac{\partial v}{\partial z} = \frac{\tau_{by}}{\rho} \qquad (z = -h) \tag{18}$$

in which τ_{bx} and τ_{by} are given by

$$\left. \begin{aligned} \tau_{bx} &= \rho g u \sqrt{u^2 + v^2} / C^2 \\ \tau_{by} &= \rho g v \sqrt{u^2 + v^2} / C^2 \end{aligned} \right\} \tag{19}$$

where C is Chezy coefficient (Kakinoki et al., 2000), which is given by

$$C = \frac{1}{n}(h + \eta - 1.0)^{1/6} \qquad : \qquad h + \eta \geq 1.1\text{m} \tag{20}$$

$$C = \frac{0.683}{n}(h+\eta) \qquad : \qquad h+\eta \le 1.1\text{m} \tag{21}$$

- Vertical velocity is given by

$$w = -u\frac{\partial h}{\partial x} - v\frac{\partial h}{\partial y} \qquad (z=-h) \tag{22}$$

- The salinity is given by

$$K_c\left(\frac{\partial S}{\partial z}\right) = 0 \qquad (z=-h) \tag{23}$$

Open sea boundary

For closed sea areas, the water level change of the tidal wave is given at the open part. On the other hand, when the alongshore current is parallel to coastline, it is necessary to give the cross-shore and alongshore distribution of water level or current velocity at open seaside according to the conditions of the field site. However in this model, it is supposed that the water level on the side or the velocity distribution becomes complex from the consideration of WAD. The change of water level due to M_2 was set only for open sea boundary at offshore region. The change of water level by sinusoidal function, which has the phase in the alongshore direction, was given and both open sea sides in the alongshore direction were made as a periodic boundary.

NUMERICAL METHODOLOGY

The governing equations for the tidal currents and 3D diffusion equations for the salinity and water temperature are solved using the FSM, originally suggested by Koutitas et al. (1980), combining the finite difference method in the horizontal plane, and a Galerkin finite element method (FEM) in the vertical direction. This hybrid technique

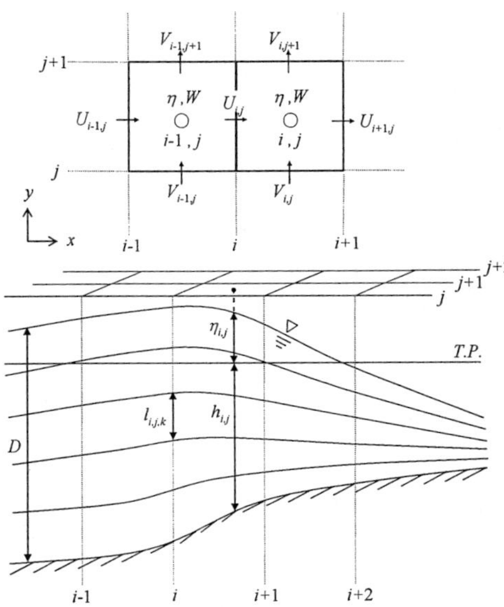

Figure 2. Definition of space-staggered grid system.

is used for solving the equation of motion by dividing it into two differential sections (in horizontal and vertical directions), and integrating at time separately for two stages. As the FEM was used in the vertical direction, the vertical direction was divided into 5 equal level thicknesses, l_k, with 6 nodes, and near sea-bottom layer is possible to divide finely. Therefore, it is possible to calculate the vertical flow distribution even in extremely shallow sea areas. In this computation, the position of variables was defined using the space-staggered grid system as shown in Fig. 2. Kuroiwa et al. (1998) has applied it to the Quasi-3D analysis of the nearshore current.

The governing equations are discretized by using the Galerkin FEM. Firstly, the variables are located on each element according to the staggered grid system shown in Fig. 2. Secondly, the vertical integration of Eqs. (1)~(3) over a discrete level of thickness, l_k, were taken. The computational domain in the x-y plane is divided into a rectangular grid ($\Delta x=\Delta y$). The vertical component of current velocity w and mean water level η are computed at the center of each grid cell, whereas the horizontal velocity components u and v are corrected midway along the cell faces. The mean water surface level η for the next time step ($n+1$) is evaluated using that at the previous time step (n-1). The computational process is depicted in Fig. 3.

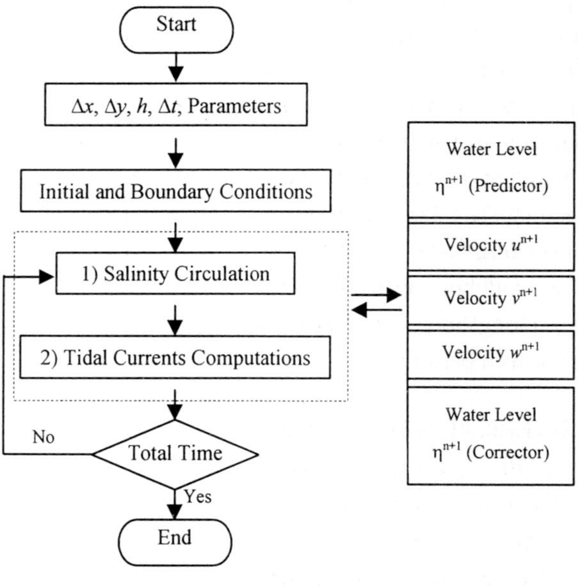

Figure 3. Flowchart of numerical computations.

Wetting and Drying Scheme

Many schemes of WAD have been implemented in coastal models. The readers are referred to Flather et al. (1990), Casulli et al. (1992), and Ji et al. (2001) for comprehensive summaries. WAD implementations have been routine in hydraulic modeling for quite some time (Stelling et al., 1986; and Balzano, 1998). A common theme in all the implementations is to invoke blocking and

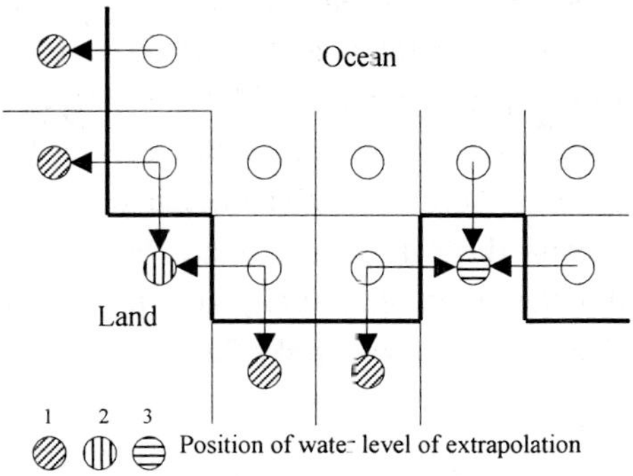

Figure 4. Extrapolation method of water level.

de-blocking conditions at cells' interfaces, for example setting flow across the interface to zero when the real water depth, $d(=h+\eta)$, becomes shallower than a pre-assigned depth, d_{dry}. We give here a WAD scheme proposed by Kato et al. (1979). The model determines the coastline position at each time-step and extrapolates the 3D predictive variables onto the tidal flat only when the water depth exceeds that value, assuring that the extrapolation is consistent with physics as well as with the numerical scheme involved. At the end of every time-step, the WAD subroutine was used to scan those cells where the number of wet layers was one.

In this model, a 2D mask (IP) is used such that IP=0 for water cell, and IP=1 for land cell at the cell's face. This model uses IP to facilitate the zero-flux conditions across solid boundaries. To implement WAD, we first define an absolute land boundary over which water can never spill. This is accomplished by making the modeled region sufficiently large to include on its coastal side the highest elevated land points that are always dry (IP=10).

The treatment of moving boundary is as follows:

1. at the flood tide: the water level is calculated at time $t=t_1$, where the position of the boundary at i is exist, then the boundary moves to $i+1$ at time $t=t_2$, and its set as a sea region if d became above d_{dry}.
2. at the ebb tide: the local water depth is calculated at time $t=t_1$, where the position of the boundary at $i+1$ is exist, then the boundary moves to i at time $t=t_2$, and its set as a land region if d drops below d_{dry}.

a. The method of extrapolation of water level is shown in Fig. 4, where 17 patterns were used to check every cell to determine its condition after the new water level.

As a result of trial and error method, d_{dry} was found 0.2m. It is necessary also for the velocity to be extrapolated when the water level is extrapolated, and the velocity is set to zero if d drops below d_{dry}.

VERIFICATION OF NUMERICAL SCHEME

The full validation of the model results is carried out through comparison with field measurements for Saigo port, which is located at Isahaya Bay in the Ariake Sea (Abualtayef et al., 2006), and the Northern part of the Ariake Sea.

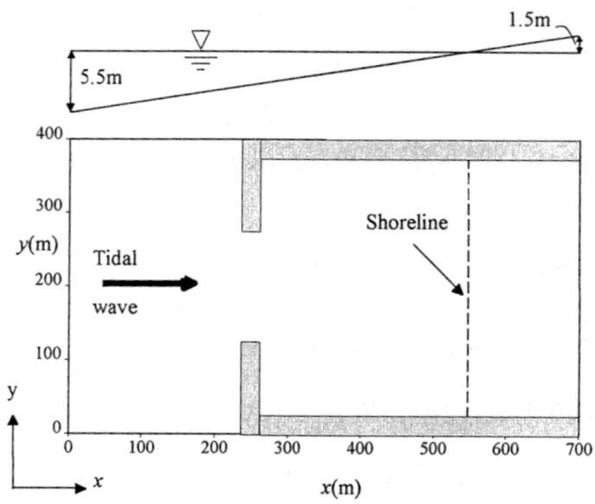

Figure 5. Study area of tidal flat model

Model Setup

The current calculation was carried out for the artificial tidal flat that is surrounded with the structure, in order to investigate the performance of the model as shown in Fig. 5. The computational conditions are shown in Table 1. The water depth is divided into 5 equal layers. The change of water level due to M_2 constituent was set only for open sea boundary at offshore region. The change of water level by sinusoidal function, which has the phase in the alongshore direction, was given and both open seasides in the logshore direction were made as a periodic boundary.

Table 1. Computational conditions of the hydrodynamic model.

Water level and velocity of incident boundary				Velocity of side boundary	
$\varsigma = a\cos\alpha\sin(kx\cos\alpha + ky\sin\alpha - \sigma t)$ $\partial u/\partial x = 0$				$\partial v/\partial y = 0$	
Amplitude *a* [m]	α [deg.]	Cycle *t* [hours]	Δx [m]	Δy [m]	Δt [sec]
0.5	0	12	10	10	0.5

Results and Considerations

The time change in the depth-averaged current vectors at the flood tide are shown in Figs. 6(a) ~ (c). The velocity distribution in vertical section at the position of y=200m where the flood tide was the strongest is shown in Fig. 6(d). Figs. 6(e) ~ (h) show similar results at the ebb tide. From these figures, the circumstances of the wetting and drying bed were calculated, the velocity became fast at open part, and it was found that the vortex due to flood tide was generated behind the breakwater.

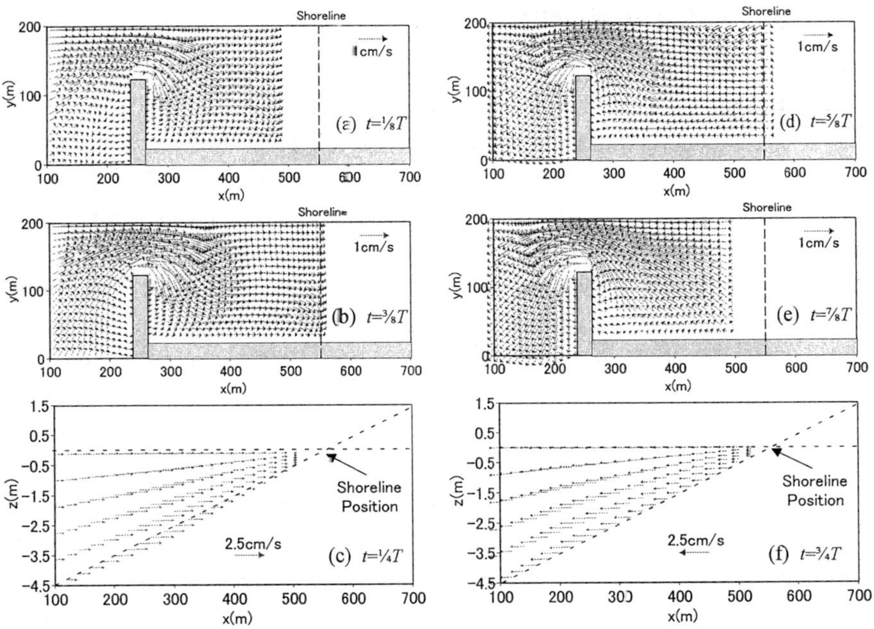

Figure 6. (a) and (b) the time change in the depth-averaged current vectors at the flood tide; (c) the vertical current profile at strong flood; and (d)~(f) show similar results at ebb tide.

MODEL APPLICATION

Study Area

The Ariake Sea is the long inner bay on the west of Kyushu Island, on the west part of Japan, which has an area of 1,700km^2, the extended 96km of the bay axis, 18km of the average width, and 20m in average depth. The amplitude of tidal wave increases gradually from the mouth to the head of bay areas. At the flood tide, the mean tidal range is about 3.0m in the bay mouth area. The model was applied to the prediction of the tidal current in the North Bay of

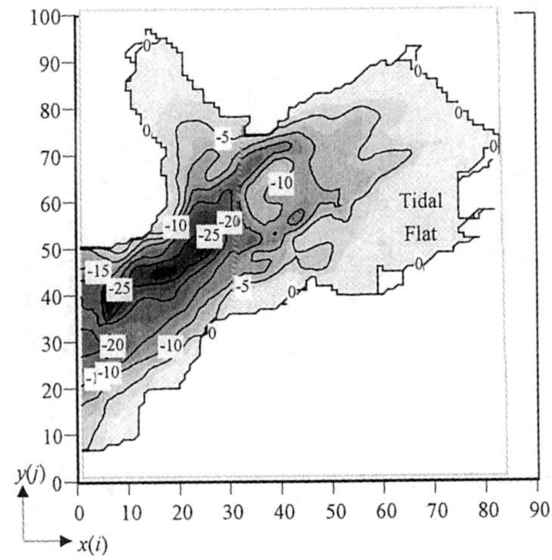

Figure 7. Depth contour map of North Bay of Ariake Sea.

Ariake Sea in order to be verified and calibrated. Fig. 7 shows the water depth contour map of the study area.

Model Setup

The horizontal computational domain considered in this study covers an area of 50.5km [NS] and 43.0km [EW], number of horizontal meshes was 101 [NS] and 86 [EW] with grid interval of 500m, horizontal and vertical eddy viscosity coefficients were 10 and 0.1 respectively, period of 12 hours, the amplitude at open boundary was set 1.4m, and time step was 1.0sec. The water depth was divided into 5 equal layer thicknesses. At the water level fluctuation type is $\eta = -a \cdot \sin(ky \cdot \sin\alpha - \sigma t)$, $\alpha = 0$, in which α is an incidence angle of the tidal wave in an open seaside boundary, k is the wave number, which was computed using of long wave theory. x and y have displayed the distance from origin of coordinates.

Since we assume that there is neither current nor elevation in the sea initially, the calculated tide is a type of progressive wave in the early stage. Therefore, at least computation of 4 tidal cycles are required to get periodically stationary solution, and only the computed results for the fourth tidal cycle is taken and analyzed.

Results and Considerations

The data acquisition was difficult in the study area. Therefore, the results were compared with data observation in previous studies. The tidal amplitude results compared with observation provided by Isozaki et al. (1977) and Kyozuka et al. (1993), and the tidal current results were compared with the observation provided by Sasaki et al. (2005).

Tidal amplitudes

The distribution of the computed amplitude of M_2 constituents, relative to the amplitude at the open boundary, is shown in Fig. 8. In Table 2 a comparison between the computed and observed amplitudes during M_2 constituent at various tidal stations is presented. The results show good

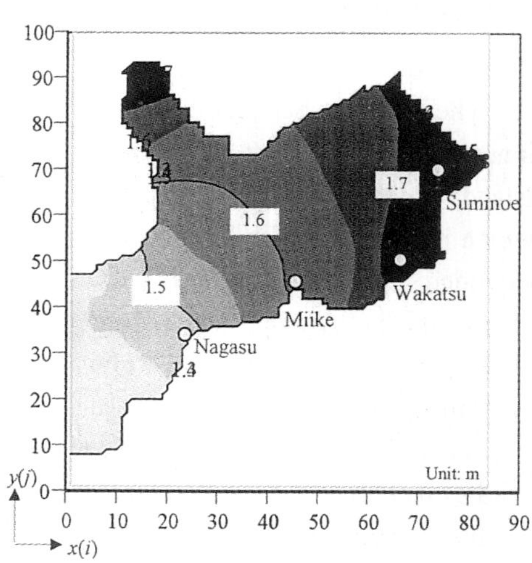

Figure 8. Computed amplification of M_2 in the North Bay of Ariake Sea.

agreement with observation except at Wakatsu station, where the computed amplitudes are slightly higher than the observation by 0.031m. However, little discrepancies are remarkable at the head of the bay. Whereas, these discrepancies are due to that the model is not considering the inflow of rivers to the Ariake Sea, and some of water depths at Wakatsu tidal flat region are not accurate.

Table 2. Computed and observed amplitudes at various tidal stations

Station name	Computed amplitude [m]	Observed amplitude [m]
Nagasu	1.478	1.475
Miike	1.598	1.590
Wakatsu	1.641	1.610
Suminoe	1.724	1.721

Tidal currents

In the Ariake Sea, the fastest currents are seen near Nagasu tidal station, which exceed a speed of 1.0 m/sec. The currents in the sea decrease their speed toward the bay head. Fig. 9(a) is a representative example of the computed tidal currents at the flood tide, and Fig. 9(b) shows similar result at the ebb tide. Currents presented here are given as vertically averaged values.

From these figures, the circumstances of WAD were demonstrated clearly. Furthermore, as a result of using this model, the tidal currents tend almost similar to the observations as shown in Fig. 10, and the velocity was excessively appraised at the ebb tide. Therefore, only water level distribution was given at the boundary of open seaside, and it was not possible to consider tidal lag.

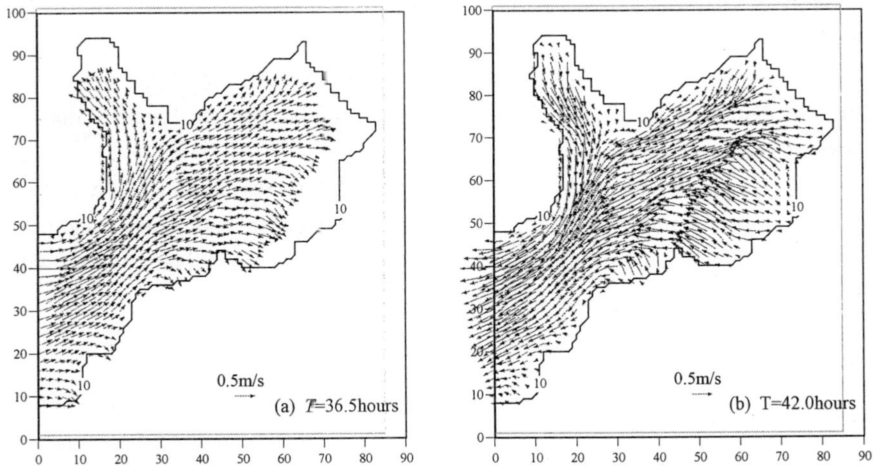

Figure 9. Computed depth-averaged tidal currents, (a) at flood tide, (b) at ebb tide.

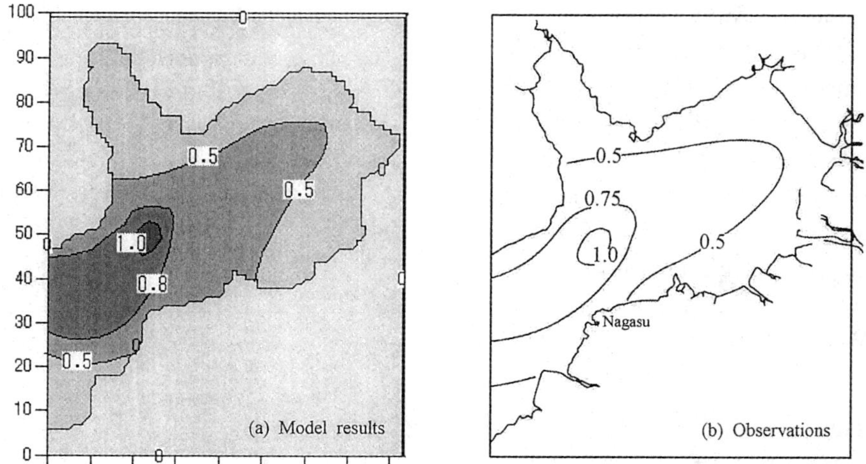

Source: Sasaki, K., *et. al.* (2005)

Figure 10. Tidal currents contour of (a) computed results; and (b) observation results.

CONCLUSIONS

A new 3D numerical model was developed to simulate the tidal motion in Ariake Sea, which is able to consider vertical distribution of the flow for the WAD in the tidal flat region based on FSM that is used the finite difference method together with the finite element method.

The applicability of the model was demonstrated on artificial tidal flat. The model was then applied to Ariake Sea. The results indicate that the predicted tidal elevations and current velocities match well with the field measurements. An appearance of the WAD of the tidal flat region, with the tidal wave and vertical distribution of the flow result, was demonstrated clearly and a comparatively an easy technique.

The 3D numerical model developed in this study was found to be simple and efficient. As future problem, method of assuming the boundary condition will be reexamined for siltation, and fresh water inflow from rivers, which will affect significantly the feature of tide over a shallow region.

REFERENCES

Abualtayef, M., M. Kuroiwa, Y. Yamashita, K. Tanaka, Y. Matsubara, and J. Nakahira. 2006. Three dimensional numerical modeling of tidal currents in inter-tidal zone, *Int. J. Offshore and Polar Eng.*, ISOPE, 16, 592-599.

Balzano, A. 1998. Evaluation of methods for numerical simulation of wetting and drying in shallow water flow models, *Journal of Coastal Engineering*, Elsevier Science, 34, 83-107.

Casulli, V., and R. Cheng. 1992. Semi-implicit difference methods for three-dimensional shallow water flow, *Int. J. Numer. Methods Fluids*, 15, 629-648.

Flather, R.A., and K.P. Hubbert. 1990. Tide and surge models for shallow water-Morecambe Bay revisited, In: Davies, A.M. (Ed.), *Modeling Marine Systems*, 1, CRC Press, 135-166.

Isozaki, I., and E. Kitahara. (1977). Tides in the bays of Ariake and Yatsushiro, *The Oceanographical Magazine*, 28, 1-32.

Ji, Z.G., M.R. Morton, and J.M. Hamrick. 2001. Wetting and drying simulation of estuaries processes, *Estuarine Coastal Shelf Science*, 53, 683-700.

Kakinoki, T., K. Takigawa, and F. Yamada. 2000. Tidal currents, water waves and topography changes on the tideland of Ariake Sea, *Proceedings of Coastal Engineering*, JSCE, 47, 636-640.

Kato, K., N. Tanaka, and K. Nadaoka. 1979. Tidal simulation on tidal marsh and numerical forecasting of its topographic deformation, *Port and Harbor Technical Research Institute*, 18, 3-75.

Komatsu, T., T. Adachi, S. Kinnou, S. Yano, Kohashi, and K. Fujita. 2003. Field observation of current and mass transport in Ariake Sea, *Proceedings of Coastal Eng.*, JSCE, 50, 936-940.

Koutitas, C., and B. O'Connor. 1980. Modeling 3D wind-induced flows, *Journal of the Hydraulic Division*, ASCE, 11, 1843-1865.

Kuchiishi, T., K. Kato, M. Kuroiwa, Y. Matsubara, and H. Noda. 2004. Applicability of a 3D morphodynamic model with shoreline change using a quasi-3D nearshore current model, *International Journal of Offshore and Polar Engineering*, ISOPE, 15, 715-722.

Kuroiwa, M., H. Noda, and Y. Matsubara. 1998. Applicability of a quasi-3D numerical model to nearshore currents, *Proceedings of 26th International Conference on Coastal Engineering*, ASCE, 2815-827.

Kyozuka, Y., and K. Kawanami. 1993. Tidal simulation on tidal marsh in Ariake bay, *The West-Japan Society of Naval Architects*, 86, 131-146.

Sasaki, K., Y.Matsukawa, H. Tsutsumi and M. Azuma. 2005. *The ecosystem regeneration of Ariake Sea*, the Oceanographic Society of Japan, Koseisha Koseikaku Co. Ltd., 9-21 pp.

Stelling, G.S., A.K. Wiersma, and J.B.T.M. Willemse. 1986. Practical aspects of accurate tidal computations. *Journal of Hydraulic Eng.*, ASCE, 9, 802-817.

Zhang, Q.Y., and K.Y.H. Gin. 2000. 3D numerical simulation for tidal motion in Singapore's coastal waters, *Coastal Engineering*, Elsevier Science, 39, 71-92.

ESTIMATION OF TIME-DPENDENT WAVE SET-UP HEIGHT IN A RIVER ENTRANCE

Hyun-seok Lee[1] and Hitoshi Tanaka[2]

Precise prediction of water level in a river mouth is of practical importance for flood inundation modeling in the vicinity of a river mouth. The water level in a river entrance is influenced by several external forces, such as river discharge, tidal motion and wave action, though, very little attention has been paid to the effect of waves. In recent years, several wave set-up observations at a river mouth have been reported. From those results, it has been concluded that the height of wave set-up shows distinct dependence on the geomorphology of river entrance. However, quantitative relationship between the magnitude of wave set-up height and wave condition has not thoroughly been clarified yet. In this study, a method is proposed for estimating time-dependent wave set-up height at a river entrance, and the result is compared with laboratory data by the present authors.

INTRODUCTION

An extreme river mouth water level is very important in connection with river planning as a boundary condition at the downstream end. In order to determine possible highest water level at the downstream of a river with a certain return period, a probabilistic approach can be used based on long-term measured data including the combined effect of flood, tide and storm surge. Although the effect of wave set-up is also included in part in the measured data, it has not been taken into consideration in more direct manner. It is well-known that wave set-up height attains to from 10% to 20% of deep water wave height. Therefore, depending on wave conditions, the height of wave set-up can be around 1m or so. Although this magnitude is almost comparable to the water level rise due to flood, very limited number of investigations have been made on water level rise due to waves in a river entrance.

Yamamoto (1978) has performed experimental study on wave set-up height in a river mouth using a laboratory flume. Moreover, one of authors investigated the relationship between wave condition and the amount of water level rises based on field measurement in three river mouths (Tanaka et al., 1998). According to this research, there are rivers with and without distinct water level rise by wave set-up. Haslow and Nielsen (1992) studied the water level variation in the Brunswick River, Australia which has two training jetties.

[1] Korean Water Resources Corporation, San 6-2, Yeonchuk-dong, Kaedeok-gu, Daejeon 306-711, Korea.

[2] Department of Civil Engineering, Tohoku University, 6-6-06 Aoba, Aoba, Sendai 780-7985, Japan.

In spite of remarkable wave set-up on the neighboring coat, it was hardly seen in this river mouth. As described above, there are various types of river mouth with respect to occurrence of wave set-up in a river mouth. This result implies that further investigations are required for various river mouths, and that quantitative evaluation is necessary by integrating knowledge at these river mouths.

In this study, research on a river mouth water level rise was done for the Shiribetsu River, Japan. First, the field data sets are extracted in which remarkable wave set-up is observed. Moreover, a method is proposed for evaluating quantitatively the wave set-up height in a mouth of a river from a measured water level data, and the relation with wave condition is discussed. Although it is expected that the wave set-up is closely connected with river mouth geographical feature, the geographical data at a river mouth is insufficient in general for correlating it with measured wave set-up height. In the study area at the mouth of Shiribetsu River, weekly oblique photograph is available from the neighboring summit of a hill for these 21 years. Thus, the present study area is quite suitable for studying interaction between river mouth morphology and wave set-up height.

FIELD SITE AND DATA COLLECTION

The Shiribetsu River pours into the Sea of Japan in the western area of Hokkaido, Japan. Its catchment area is 1,640km^2, and the length of the river is 126km. The outline of the river mouth geographical feature is shown in Photo 1.

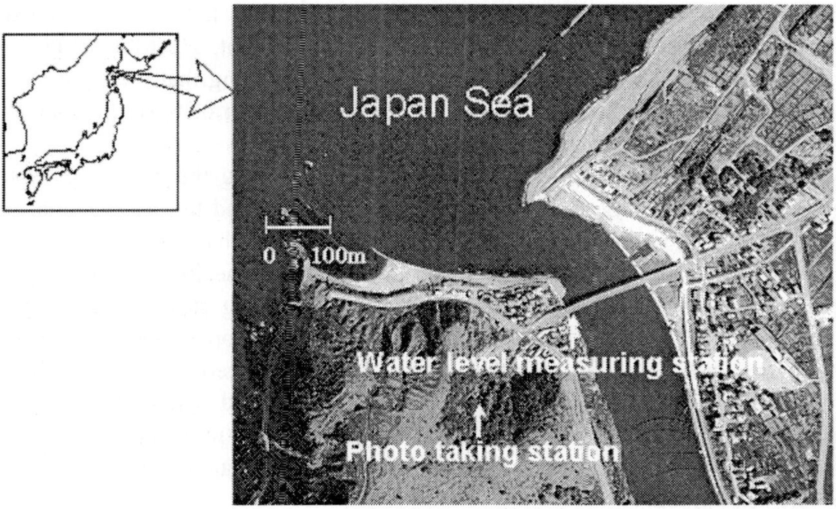

Photo 1. Study site.

River mouth water level is measured at the observation site shown in Photo 1, while the tidal data is obtained at the Setana Port. By taking the difference between the river mouth water level and tidal level with one hour interval from 1991 to 1998, the amount of water level rise in the river mouth is calculated. Furthermore, by using the river flow discharge at 14km upstream and wave data off Setana Port by 3.6km was used for classifying into flood-originated and wave-originated water level rise.

Sand spit length and the river mouth width are obtained from oblique photographs taken over 21 years weekly from the hill near the river mouth, and are used for a numerical analysis described below. Moreover, laboratory experiment was also carried out in order to compare with the measured data in the Shiribetsu River mouth.

It is here noted that, at the Shiribetsu River mouth, a training jetty on the right bank has been completed in 1999 (Otaru Development and Construction Department, 2000). The analysis of the water level rise in the present study is restricted to the period before the completion of the jetty construction. After the construction of the training jetty, the height of the water level rise has remarkably reduced as reported by the authors (Tanaka and Lee, 2002).

RESULT AND DISCUSSIONS

Characteristics of water level variation

Figure 1(a) shows an example of typical wave set-up observed in December, 1996, in which η_R and η_T denote the water level in the river mouth and the tidal level, respectively. It is clearly observed the water level rises (η_R-η_T) fluctuates in response to deep water wave height, H_0. Meanwhile, the river discharge Q is constant during this period, indicating that the water level rise is induced by waves, not by flood. A similar behavior of water level can be seen in Fig. 1(b) in February, 1991.

Figure 2(a) and 2(b) show the relationship among the maximum water level rises in a day, the corresponding river discharge and the wave height for 1995 and 1996, respectively. It is seen in each figure that the closed circles, corresponds to higher water level rise, are located in the lower right and the upper left, respectively induced by high waves and big floods. The number shown near the closed circles in the figure denotes the corresponding height of water level rise. It is seen that the water level rise induced by waves with 7m wave height is almost equal to that induced by flood with the maximum discharge of 900m^3/s. Thus, it should here noted that the maximum water level rises by wave set-up has the approximately similar magnitude to that by a middle-scale flood.

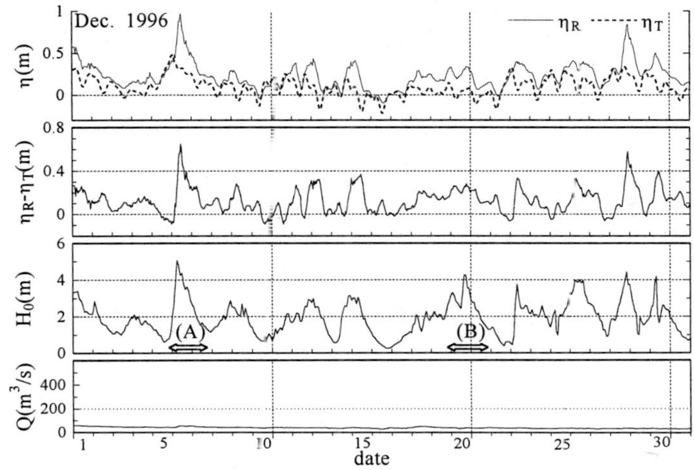

(a) December, 1996.

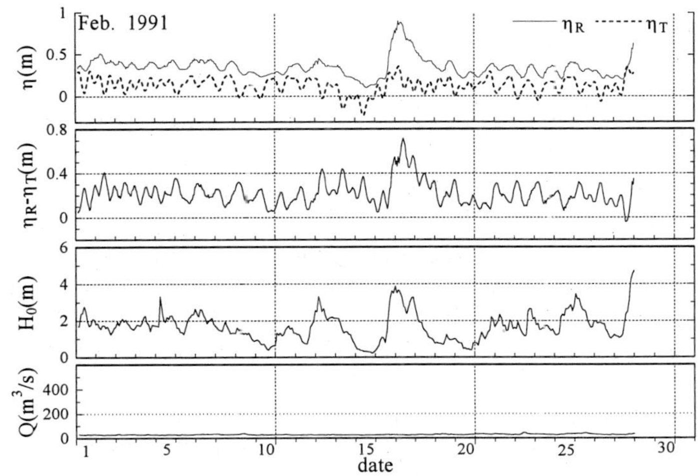

(b) February, 1991.

Figure 1. Change of water level in the river mouth, a tidal level, wave height, and river discharge.

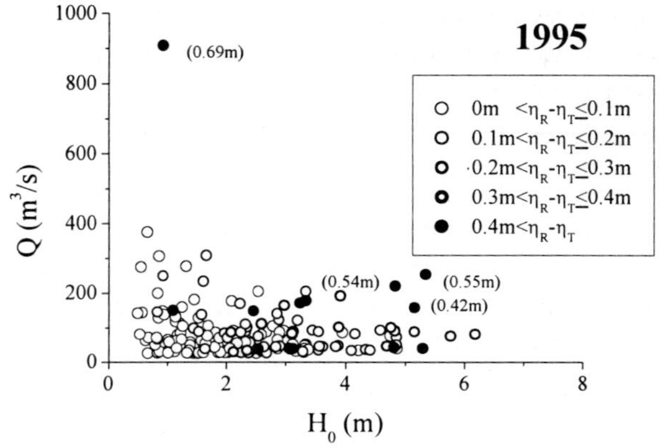

(a) 1995

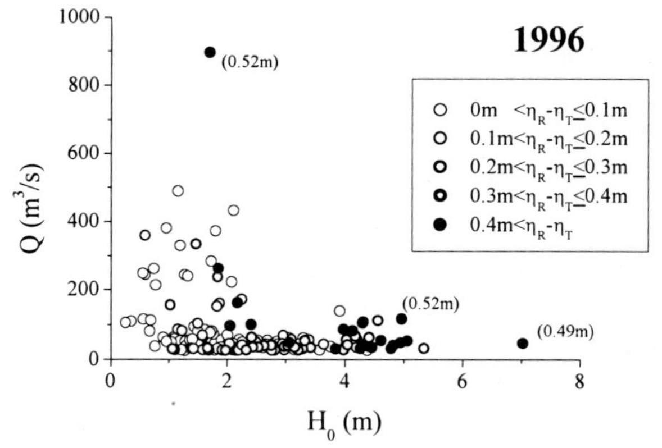

(b) 1996

Figure 2. Relationship among of water level rises, fresh water discharge, and wave height.

Figure 3 shows the relationship between hourly wave set-up height in the river mouth and the corresponding deep water wave height H_0. Figure 3(a) corresponding to the period (A) in Fig.1 (a). It is clearly seen that the relationship between wave set-up and wave height is not linear, but represented by a counterclockwise loop, and similar behavior can be formed in many cases in which distinct wave set-up height was observed. This denotes that the river mouth water level does not immediately respond to the water level in the sea, but has a marked phase lag behind the sea water level. The river mouth in winter is almost closed as shown in Photo 2, which causes phase difference in the water level response in the river mouth behind tidal variation, due to distinct energy loss the gorge section.

On the contrary to the case shown in Fig. 3(a), the data obtained in period (B) in Fig.1(a) shows a linear relationship as seen in Fig. 3(b). In case with such a gradual increase in wave height, the rapid water level response can be seen in the river mouth. It was found, however, that such cases are very rarely found in the data sets at the Shiribetsu River mouth.

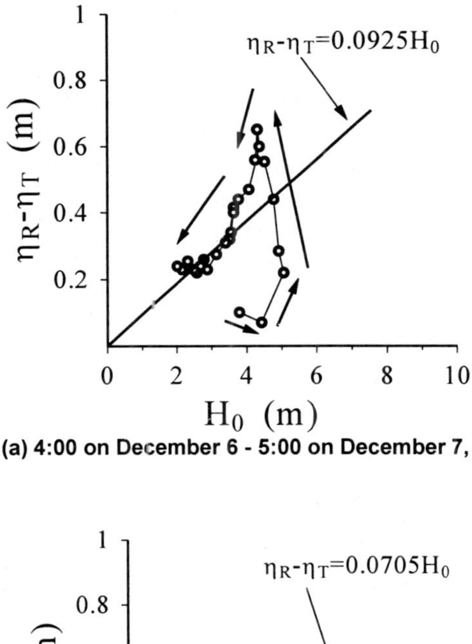

(a) 4:00 on December 6 - 5:00 on December 7, 1996.

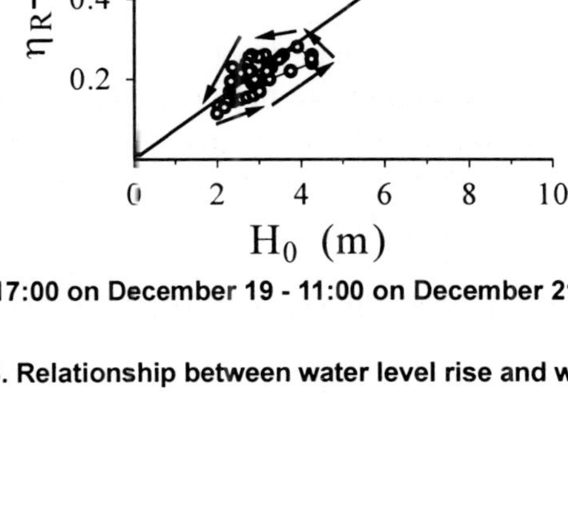

(b) 17:00 on December 19 - 11:00 on December 21, 1996.

Figure 3. Relationship between water level rise and wave height.

Photo 2. Sand spit development at the Shiribetsu River mouth (December 10, 1996).

Formulation of wave set-up height

As shown in the previous section, the response of the river mouth water level to tidal variation is accompanied by the phase lag due to constriction at the narrow section. For this reason, the instantaneous water level rise in the river mouth and the corresponding wave characteristics measured at the same time cannot necessarily be correlated. Then, the relationship between these two is investigated using a simple hydrodynamic model, in which the time lag caused by energy loss at the gorge section is considered.

The following governing equation, commonly used for an analysis of tidal inlet-bay system, is employed in the present study (Mehta and Ozsoy, 1978; Tanaka, 1995).

$$\frac{d^2\eta_R}{dt^2}+\left\{\left(\frac{K_{en}+K_{ex}}{2L_C}+\frac{gn^2}{R_C^{4/3}}\right)\frac{A_R}{A_C}\left|\frac{d\eta_R}{dt}-\frac{Q}{A_R}\right|-\frac{B}{A_C}\frac{d\eta_R}{dt}\right\}\left(\frac{d\eta_R}{dt}-\frac{Q}{A_R}\right)+\frac{gA_C}{L_C A_R}\{\eta_R-(\eta_O+\Delta\eta)\}=0 \qquad (1)$$

where $\Delta\eta$: the wave set-up height above the tidal level, L_C, R_C, A_C, B: the length, the hydraulic radius, the cross-sectional area, and the width of a channel which connects a river to the sea, g: the acceleration due to gravity, t: the time, K_{en}, K_{ex}: the loss coefficients of entrance and exit, n: Manning's roughness coefficient, and A_R: the area of tidal compartment. In the following computation for the Shiribetsu River mouth, the energy loss coefficient are assumed K_{en}=0.5 and K_{ex}=1.0, and also n=0.026. The length of the channel and the river mouth width are determined to be 27m and 98m, respectively, from oblique photographs taken from the hill. In evaluation of the cross-sectional area and the hydraulic radius, the river mouth section was assumed to be parabolic, as done by Tanaka (1995). For the integration of Eq.(1), the Runge-Kutta method was used in the numerical analysis.

Considering the previous studies on wave set-up height on a sloping beach (e.g., Goda, 1975), the following expression is assumed for the wave set-up height in a river entrance.

$$\frac{\Delta\eta}{H_0} = \alpha\left(\frac{H_0}{L_0}\right)^{\beta} \tag{2}$$

where α and β are the unknown quantities. For an arbitrary combination of α and β, the water level in the river mouth can be calculated from Eq. (1) numerically. Next, the error of the computed result is evaluated using the following equation.

$$Error = \frac{1}{n_t}\sum_{i=1}^{n}\left|\eta_{Rmeas} - \eta_{Rcal}\right| \tag{3}$$

where n_t is the total number of the water level data in the calculation. An estimation of α and β can be done by minimizing the error between measurement and computation defined by Eq. (3). The above calculation procedure was applied to four cases listed in Table 1, in which Case 2 corresponds to measured data shown in Fig. 3(b).

Figure 4 is an example of a contour lines of error obtained for various combinations of α and β for Case 1. It is observed that the minimum error, 6.4cm is achieved by the combination of α and β indicated by the open circle in the figure. Optimum α and β obtained for all cases are summarized in Table 1. Moreover, Eq. (2) obtained by the present calculation method is shown in Fig.5. Although there is a slight difference among the coefficients obtained in each case, it is seen in Fig.5 that the difference is substantially negligible. The plotted point in the figure is an experimental value which will be explained in the next session.

Table 1. Cases of calculation and estimated result.

Case	Duration	Estimated result	
		α	β
Case 1	Feb.12--Feb.19, 1991 (8days)	0.0080	-0.80
Case 2	Dec.8--Dec.11, 1991 (4days)	0.0040	-0.96
Case 3	May 17--May 20, 1993 (4days)	0.0060	-0.84
Case 4	Dec.18--Dec.21, 1996 (4days)	0.0220	-0.36

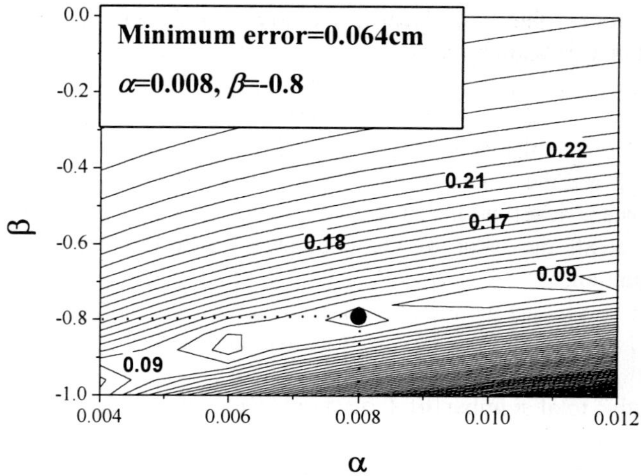

Figure 4. Contour lines of error (Case 1).

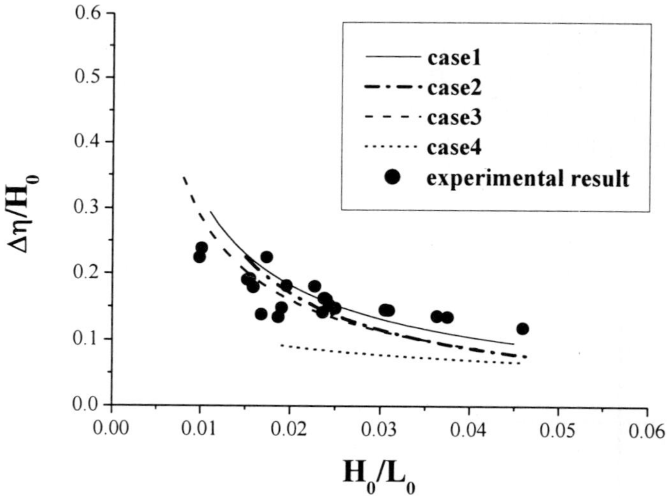

Figure 5. Relation between wave set-up height and wave steepness.

In Fig. 6, the water level in the river mouth from estimation (η_{Rcal}) and that from the measurement (η_{Rmeas}) are compared for Case 1 and Case 4, in which measured tidal variation is also illustrated. Furthermore, the water level variation estimated using a linear regression equation, such as those shown in Figs. 3(a) and 3(b) is also shown (η_{Rcc}) in Fig. 6.

During the Period (A) in Fig.6, the water level rise by wave set-up is most predominant. The water level rise attained to about 60cm due to high waves with the height of 4m. The difference between these two estimations is negligibly small during the Period (A). On the contrary, during Period (B) in Fig. 6(a) when the water level shows gradual descend, it is seen that the simple linear regression yields underestimation. Meanwhile, the estimation method based on Eq. (1) gives better result, although the measured water level does not show regular oscillation in response to tidal motion.

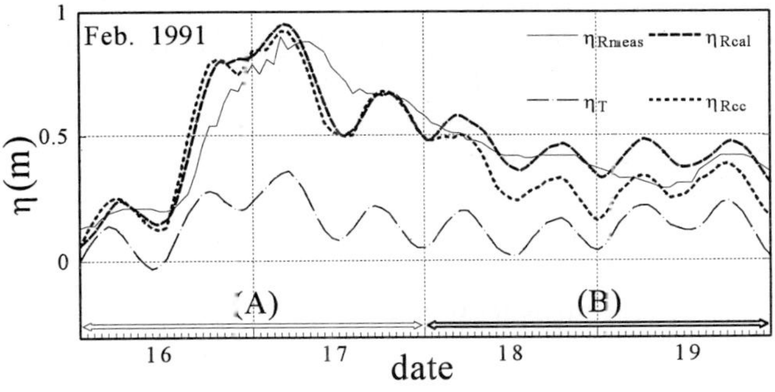

(a) Case 1

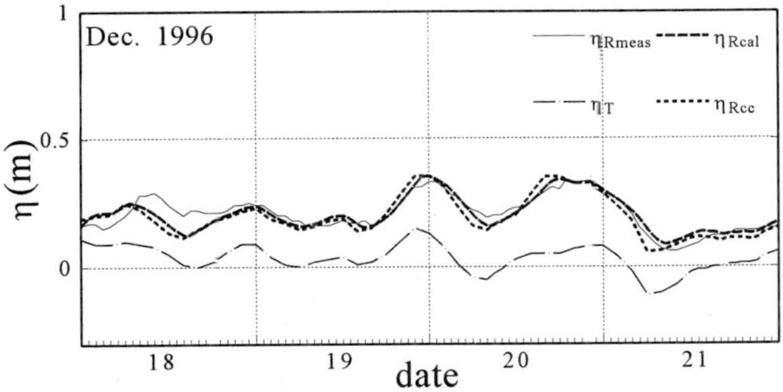

(b) Case 4

Figure 6. River mouth water level comparison of calculation and survey.

Comparison with laboratory experiment

In order to reproduce the water level rise observed in the Shirinetsu River mouth, a laboratory experiment was conducted. The experiment was carried out in a two dimensional wave flume shown in Fig. 7, with 16m length, 0.8m width, and 0.6m height. A wave generator was installed at the offshore end, whereas a model of river mouth deployed at the middle of the flume was connected with offshore area by a sloping bottom with 1/20 slope. The river mouth part consists of a channel with a constant width of 20cm. The uniformity of water level in the transverse direction was achieved by training jetties at the river entrance. The experiments have been made without supply of river discharge from the upstream end.

Immediately after the start of wave generation in the experiment, a unidirectional flow into a river mouth occurred due to combined effect of mass transport and water level rise in the surf zone. After some time, the mean water level reached a stationary state under the action of waves. Then, the mean water level was measured from the offshore end to the river mouth at several measuring points as depicted in Fig. 7. Moreover, wave height was measured at several points in cross-shore direction. The experiment has been carried out for 22 cases with different river mouth depth and wave conditions.

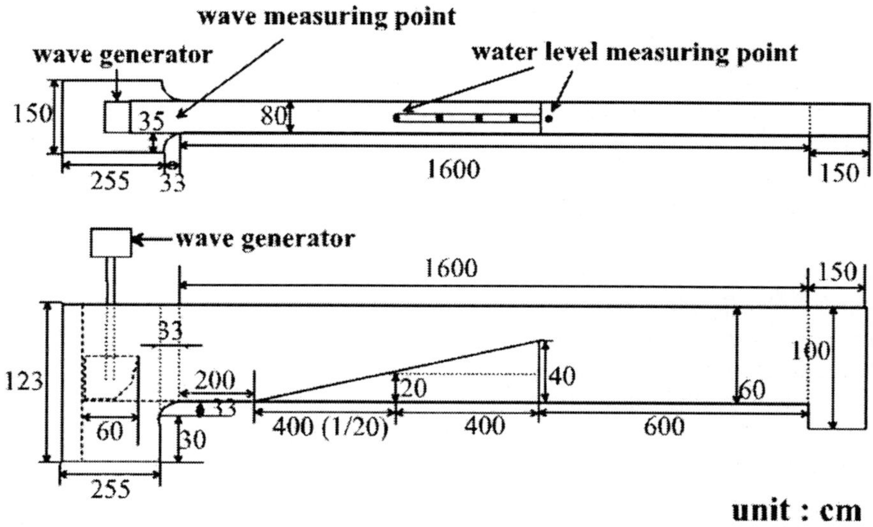

Figure.7 Wave flume.

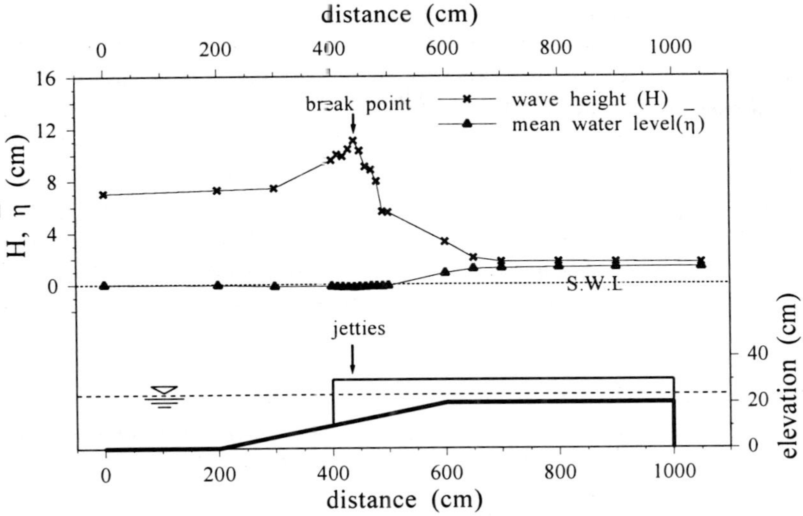

Figure 8. Cross-shore distribution of wave height and mean water level.

Figure 8 shows cross-shore distribution of wave height and mean water level for Case 1 (H_0=8.4cm, H_0/L_0=0.024, h_r (water depth in the river mouth) = 3.6cm). Wave set-down and wave set-up are clearly seen, which is quite similar to experimental result of Bowen et al. (1968) on a sloping beach without a river mouth. In a river channel, wave height is kept almost constant, resulting in constant mean water level in the river mouth. It is well-known that wave set-up height on a sloping beach shows gradual increase after wave breaking up to the shoreline position. In a river mouth, on the contrary, the wave set-up height reaches maximum value at a river entrance, which is somewhat lower than that observed on a sloping beach.

Figure 5 shows the relationship between dimensionless wave set-up height and wave steepness for whole of the experimental data in the present study. The figure shows that the wave set-up height in a river mouth shows distinct dependence on wave steepness. Moreover, it is seen that the experimental results shown herein are in good agreement with the estimated curve from the Shiribetsu River mouth.

CONCLUSIONS

The main conclusions drawn in this study are summarized as follows.

(1) In the Shiribetsu River mouth, distinct water level rise induced by wave set-up was observed. The maximum amount of the water level rises due to wave set-up is almost comparable to that caused by a middle-scale flood.

(2) Using an estimation method proposed herein, instantaneous wave set-up

height in the river mouth is formulated in dimensionless form as a function of wave steepness.

(3) The equation thus formulated from the field data shows good agreement with laboratory data in a wave flume.

ACKNOWLEDGEMENT

This research has been carried out as a joint research by the Committee of Hydro-science and Hydraulic Engineering, Japan Society of Civil Engineers. Furthermore, this research was partially supported by Grant-in-Aid for Scientific Research (B) from JSPS (No.17360230), Japan. The authors would like to express their gratitude to these supports.

REFERENCES

Bowen, A.J., D.L., Inman, and V.P. Simmons. 1968. Wave 'set-down' and set-up. *Journal of Geophysical Research*, 73, 2569-2577.

Goda, Y. 1975. Irregular wave deformation in the surf zone. *Coastal Engineering in Japan*, 18, 13-25.

Hanslow, D. J. and P. Nielsen. 1992. Wave setup on beaches and in river entrances. *Proceedings of 23rd International Conference on Coastal Engineering*, 240-252.

Mehta, A.J. and E. Ozsoy. 1978. *Inlet Hydraulics, in Stability of Tidal Inlets - Theory and Engineering* (Bruun, P. ed.). Elsevier, 83-161.

Otaru Development and Construction Department. 2000. Field observation of hydraulic characteristics at the Shiribetsu River mouth in 2000. *Technical Report, Hokkaido Development Bureau*, 234pp. (in Japanese)

Tanaka, H. and N. Shuto. 1992. Field investigation at a mouth of small river. *Proceedings of 23rd International Conference on Coastal Engineering*, 2486-2499.

Tanaka, H., H. Nagabayashi, and K. Yamauchi. 2000. Observation of wave set-up height in a river mouth. *Proceedings of 27th International Conference on Coastal Engineering*, 3458-3471.

Tanaka, H. and H.S. Lee. 2003. Influence of jetty construction on morphology and wave set-up at a river mouth. *Coastal Engineering Journal*, 45, 4, 659-683.

Tanaka, H. and H.S. Lee. 2004. Effectiveness of jetty construction at Shiribetsu River mouth, Japan. *Proceedings of 9th International Symposium on River Sedimentation*, 1875-1881.

Yamamoto, K. 1978. River Mouth Improvement [I] –for river mouths with sand bar-. *Technical Report, Public Works Research Institute*, 1394, 223pp. (in Japanese)

A 2DH IMPLICIT MODEL TO NUMERICALLY STUDY HYDRODYNAMICS AND SEDIMENT TRANSPORT TRENDS OF A REAL SHALLOW TIDAL EMBAYMENT: THE ARCACHON LAGOON (FRANCE)

Aurélie Le Dissez[1], Aldo Sottolichio[2] and Jean-Paul Caltagirone[1]

In the present study, a new numerical model is proposed, devoted to the simulation of the shallow coastal flows in tidal bays, capable to tackle with difficulties arising from varying bottom, wetting and drying processes as well as large time step simulation with reasonable computing time cost. For numerical treatment of dry zones emerging during wetting and drying processes, a Darcy type term is addeding in the classical SWE and a penalty method is employed. Governing equations are discretized using an implicit Finite Volume approach on a staggered grid. An adaptation of the augmented Lagrangian method is used for resolution of water depth and current fields. The present numerical modeling is validated by comparison with several literature test cases from the literature before being applied to a real tidal embayment. The application of the model to the Arcachon lagoon allows to highlight some trends on tidal circulation and possible sediment transport in the lagoon, where the alternative dominance of the study of the hydrodynamics and the consecutive sediment transport trends that occur in a real tidal bay: the Arcachon lagoon.

INTRODUCTION

Protected from the ocean and the wave conditions by a sand spit or a bar system, the tidal bays are connected to the ocean by an inlet made of channels dug through the bars. This configuration prevents the entrance of the oceanic waves inside the lagoon whose dynamics is consequently governed by the tide.

In these shallow environments dominated by the tide, bed elevations in some areas are located between the low water and the high water of a spring tide which means alternatively wetting and drying over a tidal cycle. Such zones are called intertidal flats and they experience highly complicated flows over strongly irregular bottom, generating sediment transport processes that we still did not accurately simulate, not even understooand.

Tidal bays arouse numerous interests such as ecological, economical or scientific. These highly attractive sites are appealing intensive tourism and sailing, their rich ecosystems are experiencing intensive fishing, oyster-farming, but all these activities generate high anthropogenic pressure which causes damage like erosion, growth, pollution. This leads to a growing demand for engineering studies concerning environmental impact evaluation, simulation of management proposals, and so on.

---

[1] TREFLE, UMR 8508, 16 av. Pey Berland, Pessac, 33607, France
[2] EPOC, UMR CNRS 5805, av. des facultés, Talence, 33400, France

Consequently, it appears a growing demand for numerical models and for research devoted to always improve numerical hydrodynamics tools. The need forof robust, accurate and efficient models to forecast and reproduce currents in complex sites, with complicated hydrodynamics and consecutive sediment transport trends, leads to numerical developments, new formulation or characterization of physical processes, like sediment transport over intertidal flats depending on sediment properties (LeHir et al., 2000) for instance, or tidal wave distortions when propagating in shallow bays occupied by large intertidal flats (Greenberg 2005), or the phenomenon of predominance of the ebbing in presence of large intertidal flats (Wang et al. 1995).

In the present study, a new numerical model is proposed, devoted to the simulation of the shallow coastal flows in tidal bays, capable to tackle with difficulties arising from varying bottom, drying and wetting processes as well as large time step simulation with reasonable computing time cost. For numerical treatment of dry zones emerging during wetting and drying processes, a Darcy type term is adding in the classical SWE and a penalty method is employed. Governing equations are discretized using an implicit Finite Volume approach on a staggered grid. An adaptation of the augmented Lagrangian method has been used for resolution of the flow field. The present numerical modeling is validated by comparison with several literature test cases before being applied to a real site. the studySimulations are then analysed in order to discuss trends of circulation and possible implications for of the hydrodynamics and the consecutive sedimentary evolutiontransport trends that occur in a real tidal bay: the Arcachon lagoon..

STRATEGY OF MODELING

The fisrt stage of modeling consists in implementing a hydrodynamics module in a pre-existent numerical tool globally devoted to numerical fluid mechanics, Aquilon (Glockner 2001). Its purpose is to simulate tidal flows in shallow coastal areas, handling with the presence of large intertidal flats and strongly irregular bottom simultaneously.

Accounting, on the one hand, for the objectives inherent to tidal flow modeling, on the other hand for the current library of numerical methods already implemented in our numerical tool, we base our quite original hydrodynamics module on a finite volume method, using an implicit time resolution, a spatial discretization on a fixed staggered mesh, and a penalty method for handling with the Wetting And Drying (WAD) processes.

Here are presented the objectives, the consecutive mathematical model and the numerical methods used for its resolution

Context and objectives

The behavior of waves and currents in coastal environments is often controlled by the water depth that is generally small compared to the characteristical horizontal scale of the flow (Pedlosky 1987). Therefore the Shallow Water Equations (SWE) are commonly used to describe nearshore hydrodynamics and investigate the two main types of coastal phenomena: on the one hand, surf zone hydrodynamics (small time and space scales) is studied with explicit shock capturing schemes (Vincent et al. 2001), on the other hand, larger scale coastal flows like tidal currents are dealt with implicit discretizations (Ip et al. 98).

In the framework of this last class of problems, a two-dimensional horizontal (2DH) model for coastal hydrodynamics modeling, based on SWE, must satisfy the four following essential requirements:

- Robust numerical schemes are necessary to describe the propagation of tidal flows. Indeed, the hyperbolic nonlinear character of the equations (SWE) induces the appearance of discontinuous solutions. Moreover, the propagation of waves, such as tide, bore or breaking waves, in the very shallow water areas, implies the presence of strong velocity gradients at the interface between the immersed and emerged zones (Carrier et al. 1957, Ip et al. 98). Here is the first mathematical difficulty to take into account by providing the model with spatial and time discretization capable to ensure stability and accurate calculation.
- The model has to be able to work with large time steps. Hydrodynamics simulations commonly last over considerable periods, from over few tidal cycles to few months, even few years, which is widely longer than encountered in other fields of numerical fluids dynamics. As coastal hydrodynamics is concerned, time scale are about the month or the year, whereas spatial scale is about few kilometers. These characteristical scales consequently require large time steps in order to reduce calculation cost and duration and to increase the efficiency of the model.
- It has to cope with bathymetries strongly irregular due to the net of channels dug, sometimes deeply, in the mudflats.
- The modeling should achieve an accurate reproduction of the covering/uncovering of tidal flats which are subjected to alternating wetting and drying due to the tidal oscillation of the free surface. As said previously, the presence of intertidal flats influence flows in the lagoon and generate specific hydraulic phenomena. Yet, as mentioned in (Greenberg 2005), numerically accounting for these zones is still awkward and imprecise for many numerical approaches.

The previous four points have to be taken into account simultaneously by the model.

Mathematical model

The modeling of flooding and ebbing of shallow embayments presents special numerical difficulties, as the domain of simulation must evolve in response to the computed solution itself (Ip et al. 1998). Numerous research works have been devoted to this issue, which propose various answers

(Leendertse 1970, Flather and Heaps 1975, Falconer and Owens 1987, Stelling et al. 1986, Casulli 1990, Thomas 1990, Balzano 1998, Heniche et al. 2000, Jiang et al. 2005), using structured or unstructured finite-volume methods, unstructured finite elements or finite element sigma coordinate model, semi-implicit finite difference method... Nevertheless, Pluss et Schüttrumpf (Pluss et al. 2004) have compared four different choices of methods and none of them has finally appeared to be clearly the best regarding the others when modeling a real site.

Accounting for the numerical methods available in Aquilon, the authors' aim is to build a physical model that intrinsically takes Wetting And Drying (WAD) process into account by introducing a Darcy term into the momentum equation of the classical three-dimensional Navier-Stokes equations (3D-NS). This term will enable us to characterize the dry coastal zones.

Interaction between incompressible flow and porous media

The model is based on the introduction of a Darcy term into the classical three-dimensional Navier-Stokes equations (3D-NS). Initially used to represent the interaction of a fluid with a porous media, this term will enable us to characterize the dry coastal zones. The system of equations is then integrated over the water depth by assuming the hydrostatic distribution of the pressure. This leads to a new formulation of the SWE model, and allows taking into account the dry zones in an original way.

The 3D-Navier-Stokes equations including a Darcy term are written as:

$$\begin{aligned} &\frac{\partial \rho}{\partial t} + \nabla \cdot (\rho \mathbf{V}) = 0 \\ &\frac{\partial (\rho \mathbf{V})}{\partial t} + \nabla \cdot (\rho \mathbf{V} \otimes \mathbf{V}) = -\nabla p + \rho \mathbf{g} + \mathbf{f} + \nabla \cdot \left(\mu \left(\nabla \mathbf{V} + \nabla^t \mathbf{V}\right)\right) - \frac{\mu}{\kappa} \mathbf{V} \end{aligned} \qquad (1\text{-}2)$$

Where V is the velocity vector, p the pressure, t the time, ρ the density, g the acceleration of gravity, μ the dynamic viscosity, $\eta=\mu/\rho$ the kinematics' viscosity and κ the permeability tensor.

The Darcy term or volume drag term $-(\eta/\kappa)\,\rho\mathbf{V}$ allows to take into account the flow of a fluid in mixed fluid-porous system (Angot et al. 1999, Khadra et al. 2000). In fact, these motion equations of Navier-Stokes-Brinkman type represent the flow of a fictitious fluid made of a Newtonian fluid including fixed solid particles. Depending on the value of κ, the Eq. 2 allows to describe an intermediate situation between two extreme behaviours: Newtonian fluid or solid.

New shallow water model including a dry zone term

Assuming the pressure field to be near to the hydrostatic equilibrium, the density constant and the shallow water approximation, the three-dimensional conservative Navier-Stokes-Brinkman (Eq 1-2) are integrated over the water depth. This leads to a novel two dimension horizontal (2DH) shallow water model with a smaller number of unknowns, written as:

$$\frac{\partial h}{\partial t}+\nabla\cdot(h\mathbf{V})=0$$
$$\frac{\partial(h\mathbf{V})}{\partial t}+\nabla\cdot(h\mathbf{V}\otimes\mathbf{V})=-gh\nabla(h+h_b)+\nabla\cdot\left(\mu h\left(\nabla\mathbf{V}+\nabla^t\mathbf{V}\right)\right)-C_f h\|\mathbf{V}\|\mathbf{V}-\frac{\nu}{\kappa}(h\mathbf{V}) \tag{3-4}$$

u and *v* are the components of the velocity vector **V** in the horizontal plane (x, y), *h* is the water depth, h_b the solid bottom elevation and Cf the friction coefficient. The new term $-(\eta/\kappa)(h\mathbf{V})$ added to SWE takes into account the permeability of the medium and will be helpful for numerical treatment of dry zones, with no fluid flow, as impermeable porous zones characterized by $\kappa=0$.

NUMERICAL METHODS

Discretization schemes

AQUILON is a library of numerical models, discretization schemes and solvers devoted to the resolution of fluid mechanics and thermics problems. The equations are discretized by implicit Finite Volumes on fixed staggered Cartesian grids. This method is particularly well suited when accounting for conservative balance equations since it consists in integrating motion equations in each control volume of the mesh.

Moreover the implicit scheme allows the use of large time steps.

In a first approach of our case, the advection terms in Eq. 4 are approximated by first order upwind schemes and the viscous terms by centered schemes, but higher order are also possible.

Resolution of the coupling velocity - hydraulic discharge, V/hV

The major difficulty when solving our system (Eq. 3-4) is the coupling between V and hV. This problem is analogous to the coupling between p and V in Navier-Stokes (NS) equations written for incompressible flows or to the coupling ρ/ρV in NS equations written for compressible flows.

For incompressible flows, several resolution methods exist, among which is the method called augmented Lagrangian (Fortin and Glowinski 1982), and an Uzawa minimization algorithm implemented in our numerical library.

Inspired by this minimization algorithm devoted to incompressible flows, a method for the resolution of the coupling $\rho/\rho\mathbf{V}$ has been developed concerning compressible flows (Figué 1996). Its adaptation to the SWE is reported here.

In order to simplify the notation, friction, Brinkman and viscous terms do not appear during the following proof. Here, $\mathbf{q} = h\mathbf{V}$ is the hydraulic discharge and the coupling $h/h\mathbf{V}$ is analogous to the coupling $\rho/\rho\mathbf{V}$ in cases of NS equations for compressible flows.

The SWE system is written as:

$$\frac{\partial \mathbf{q}}{\partial t} + \nabla \cdot (\mathbf{q}\mathbf{V}) = -gh\nabla(h + h_b)$$
$$\frac{\partial h}{\partial t} + \nabla \cdot \mathbf{q} = 0 \tag{5-6}$$

The system (Eq. 5-6) is discretized in time by a fist order scheme, but it would still be validate for higher order.

$$\frac{\mathbf{q}^{n+1} - \mathbf{q}^n}{\Delta t} + \nabla \cdot \left(\mathbf{q}^{n+1}\mathbf{V}^n\right) = -gh^n\nabla\left(h^{n+1} + h_b\right)$$
$$\frac{h^{n+1} - h^n}{\Delta t} + \nabla \cdot \left(\mathbf{q}^{n+1}\right) = 0 \tag{7-8}$$

Water depth gradient Δh^{n+1} can be expressed by applying the gradient operator to the discretized mass balance equation Eq. 8. It leads to:

$$\nabla h^{n+1} = \nabla h^n - \Delta t \nabla\left(\nabla \cdot \mathbf{q}^{n+1}\right) \tag{9}$$

When rewriting:

$$-gh^n\nabla\left(h^{n+1} + h_b\right) = -gh^n\nabla\left(h^n + h_b\right) + gh^n\Delta t\nabla\left(\nabla \cdot \mathbf{q}^{n+1}\right) \tag{10}$$

Substituting the second member of the momentum equation Eq. 7 by its expression given by Eq. 10 allows highlighting two terms. The first one stay in the second member $-gh^n\nabla\left(h^n + h_b\right)$, the other one $+gh^n\Delta t\nabla\left(\nabla \cdot \mathbf{q}^{n+1}\right)$ is transposed in the first member.

The discretized system (Eq 7-8) becomes:

$$\frac{\mathbf{q}^{n+1} - \mathbf{q}^n}{\Delta t} + \nabla \cdot \left(\mathbf{q}^{n+1}\mathbf{V}^n\right) - gh^n\Delta t\nabla\left(\nabla \cdot \mathbf{q}^{n+1}\right) = -gh^n\nabla\left(h^n + h_b\right)$$
$$h^{n+1} = h^n - \Delta t\nabla \cdot \mathbf{q}^{n+1} \tag{11-12}$$

The number of variables is reduced to only one $\mathbf{q}^{n+1}$ in Eq. 11: this is the only equation that is solved. Velocity field is deduced by the relation $\mathbf{V}^{n+1} = \mathbf{q}^{n+1}/h^n$. Last, Eq. 12 allows to update h^{n+1}.

VALIDATION ON THEORETICAL TEST CASES

The novel 2DH model for shallow flows is first validated on several theoretical test cases (see Le Dissez, 2006). The results of these tests demonstrate the robustness and the accuracy of the model, which does not require any 'stabilizing' terms or artificial viscosity to handle correctly the shallow water processes. The following major points are particularly tested: the use of large time steps, the flow over irregular bathymetry and intertidal zones as well as the sensibility of the numerical model to the bottom friction and horizontal turbulent diffusion. In (LeDissez et al. 2005) is presented the validation of the treatments of the strong bottom irregularities and the wetting/drying process.

APPLICATION TO A REAL SITE: THE ARCACHON LAGOON

Presentation of the site

The Arcachon Lagoon is a triangular shaped coastal lagoon situated on the south-western Atlantic coast of France (Fig. 1a). Tidal range varies between 0.9 and 4.9 m at the mouth. This 180-km2 lagoon opens to the Atlantic Ocean by two narrow channels (2 to 3 km wide and about 12 km long) separated by a complex system of sandy banks. The inner lagoon extends for about 156 km2. Tidal channels represent 41 km2, and penetrate between large intertidal areas (115 km2). These tidal flats are essentially muddy and are covered by Zostera noltii meadows, while channels are roughly sandy (Thauront 2004).

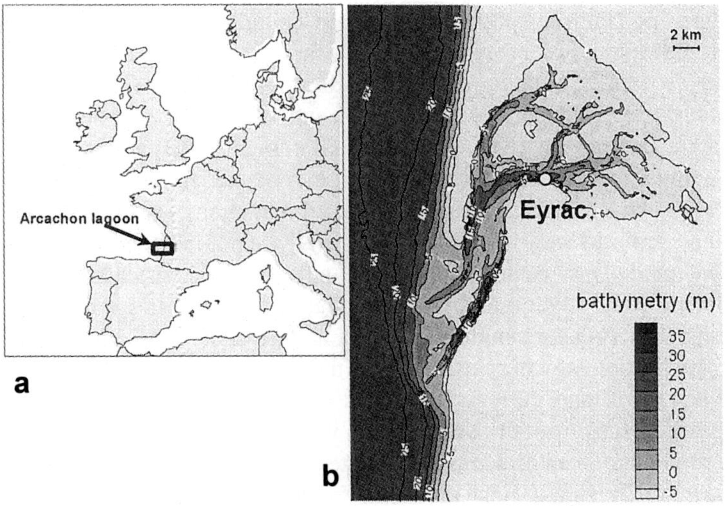

Figure 1: (a) Localization (a) and bathymetry (b) of the Arcachon lagoon. Eyrac is the station considered in figure 2

The lagoon receives freshwater inputs from its northeastern and southern parts and mostly by the river Leyre situated in the south-eastern part of the bay. Despite the development of some vertical salinity gradients, the lagoon appears as a well mixed environment, where circulation is essentially tidally-induced. A depth-integrated model is then enough to reproduce barotropic currents in the bay. The embayment presents a strongly irregular bathymetry and an active morphology (Thauront 1994), in both the tidal inlet and the inner zone of the baysin (Ffig. 1b). Understanding sediment transport and predicting the evolution of the bathymetry are of a great interest for the Arcachon Lagoon users (sailors, fishers, oyster breeders, tourism...) and scientists. As morphodynamics is mainly controlled by hydrodynamics, modeling sediment transport requires accurate hydrodynamics computation, especially for the intertidal zones, subjected to alternating wetting and drying due to the tidal oscillation of the free surface. In such areas, experiencing strong water elevation fluctuations and covering 70% of the total surface of the lagoon (Plus, 2004), sediment transport processes are strongly dependent on the tidal flow (LeHir at al. 2000). Consequently, a good treatment of dry zones as well as of varying bathymetry and the use of large time steps are particularly essential.

Model settings and validation

A regular Cartesian grid of the Arcachon lagoon has been constructed. Dimensions of the computational domain are 31km x 36km, discretized into a meshgrid of 200m x 200 m. At the open boundary, the tide is imposed. Tidal water elevations are provided by the Mars model, developed by the French Center for the Exploration of Oceans (Ifremer, Plus 2004) and able to simulate the tide propagation in the entire shelf and in the bay, considering four main harmonic components. The time step chosen for calculations is 20 s. In the simulations, fresh water input from the Leyre is neglected. Also, meteorological forcing is not considered in the present computations, as the terms of pressure gradients and surface friction due to the wind are not included in the equations.

The model has been calibrated by adjusting the Strickler bottom friction coeffcient. Runs cover one fortnightly tidal cycle, beginning most of the time by a spring tide. As an example, figures 2a and 2b show results after calibration at the Eyrac station (southern boundary of the lagoon, see Figure 1b).

Water level and currents velocities computed by the Aquilon model were compared to runs from the Mars model, which is a relevant comparison since the Mars model has been previously calibrated and validated (Plus 2004).

Figure 2a shows tidal elevation for a sequence of 11 consecutive tides. There is a good correspondance between the two models in terms of amplitude and phase, which demonstrates the capability of our model to correctly simulate the tidal propagation in the Arcachon lagoon. Figure 2b shows current velocities for the same run. In general, intensities are well reproduced, also the alternative dominance of flood velocities during sprigs and ebb velocities during neaps is clearly well simulated. There are some differences in terms of maximum velocity between the two models. Our model seems to underestimate systematically the peak velocity. This can be due to the differences on horizontal resolution. The Mars model has a grid resolution of 64 m, while our model has a resolution of 200 m. Thus, a roughly representation of the channel bathymetry can smooth velocities in our model, as currents are more sensitive to bottom topography than tidal elevation, which depends on the global channel section.

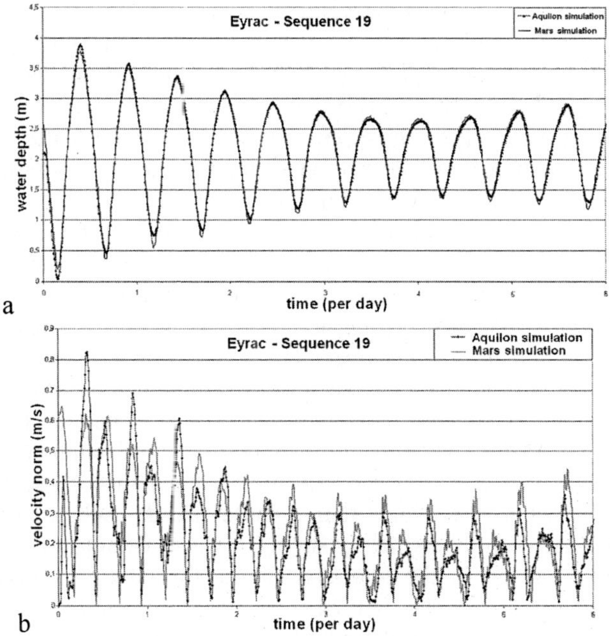

Figure 2: Numerical simulations of water elevation (a) and current velocities (b) at Eyrac

Tide-induced dynamics and possible implications for sediment transport

In this part, tidal hydrodynamics simulated by the model are analyzed in order to assess trends occurring in the lagoon and their possible implications on sediment transport. In terms of sediment dynamics, in the absence of any sediment transport model, we can consider the horizontal distribution of two hydrodynamic variables:

- First, the friction velocity, which is directly calculated from the depth averaged velocity by

$$\overline{u} = \frac{u^*}{\kappa} Log\left(\frac{h}{ez_0}\right) \tag{13}$$

where u is the depth-averaged velocity, k the von Karman constant, h the water depth, z0 the rugosity length and u* the friction velocity. Maximum friction velocity over a tidal cycle indicates zones where resuspension is expected to be relatively important.

- Secondly, the Eulerian tide-average velocity, which gives the residual horizontal transport of water over a tidal cycle, and consequently the residual transport of a suspended particle.

Figure 3 show the distribution of the maximum u* and the residual velocity over a spring tidal cycle. A distinction is made between flood and ebb phase, as they are supposed to have opposite directions of transport. Fig. 3-left points out areas subject to strong erosive capacity of the flow. It appears large ebb dominance in the channels and, inversely, flood dominance over the intertidal flats, with higher intensities in the southern part of the lagoon. Inversely, linking channels and the north-eastern part of the lagoon experience the lowest friction velocities, which suggest that these areas are favorable for sediment deposition.

Residual velocities (fig. 3-right) confirm ebb dominance in the channels and flood dominance in the tidal flats. These trends substantiate previous observations based on bathymetric surveys which demonstrated that a net sedimentation-accretion in the intertidal flats (especially in the north-eastern part) and relative erosion in the subtidal channels (L'Yavanc 1995).

It has to be pointed out that these trends result from a first approach analysis that does not take into account for sediment availability and/or erodibility, which are supposed constant in time and space. The actual sedimentary functioning of the lagoon must be investigated by a sediment transport model which considers a consistent set of sedimentary processes. The 2DH flow model has been recently coupled with a local sediment transport formula and is being developed to compute separately the bed-load and the suspended load contributions.

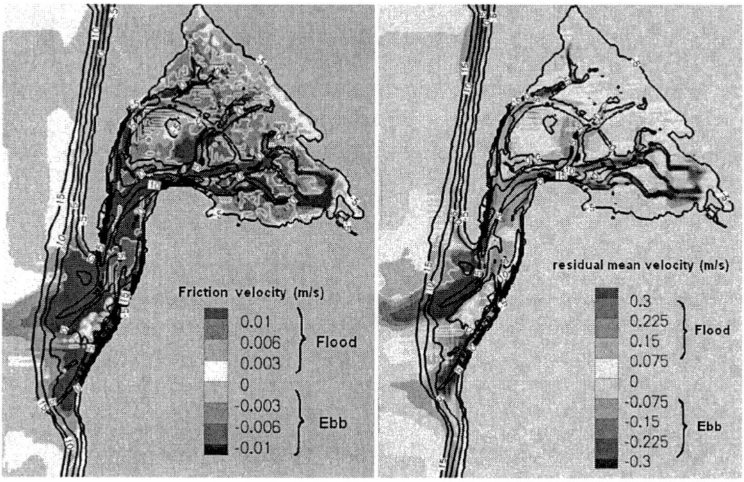

Figure 3: Maximum friction velocity (left) and tide-averaged velocity (right) over one spring semi-diurnal tide

CONCLUSIONS

Simulation of hydrodynamics and - and mmorpho-dynamics processes of coastal areas requires robust numerical methods, able to deal with the particular features of shallow water flows. Numerical models have to handle large time steps required by sediment transport phenomena, to accurately represent the processes of wetting and drying in intertidal flats and to cope with strongly irregular bathymetry. In the present nNote, a novel numerical approach is presented, based on the introduction of a Darcy term into the classical 3D Navier–Stokes equations. An original Shallow Water Equation system is solved thanks toby an implicit resolution procedure provided by an adaptation of the Augmented Lagrangian algorithm and a penalty method. The model accuracy and robustness have been tested on several academic test cases. A real application is finally carried out to show the model accuracy in representing the tidal propagation in a complex coastal embayment: the Arcachon lagoon. Our results are compared with Ifremer simulations from a reference model, which hasve been previously validated with fields measurements. Comparison of water levels and current velocities shows a very good agreement between both models.

These results do encourage us to carry out more detailed studies on the Arcachon baybasin. Studying Investigation on sediment processes in its inner zone would locally require a finer meshgrid in order to better describe this area. An adaptative mesh refinement technique will be used to increase the local mesh accuracy according to bathymetry, velocity or water level gradients. The new treatment of the dry zone will then provide more accurate velocity fields on the intertidal flats and, consequently, a better transport computation.

ACKNOWLEDGMENTS

We would like to thank Martin Plus, from IFREMER, for fruitful discussions and for having accepted to provide us with water level and velocity data in different stations of the Arcachon lagoon.

REFERENCES

Angot P., C.H. Bruneau and P. Fabrie. 1999. A penalization method to take into account obstacles in viscous flows. *Num. Math.*, 81, 497-520.

Balzano A. 1998. Evaluation of methods for numerical simulation of wetting and drying in shallow water flow models. *Coast. Eng.*, 34, 83-107.

Carrier G. F. and H.P. Greenspan. 1957. Water waves of finite amplitude on a slopping beach. *J. Fluid. Mech*, 4, 97-109.

Casulli V. 1990. Semi-implicit Finite-Difference Methods for the Twodimensional Shallow Water Equations. *J. Comput. Phys.*, 86, 56-74.

Cayocca F. 2001. Long-term morphological modeling of a tidal inlet: The Arcachon Basin, France. *Coast. Eng.*, 42, 115-142.

Falconer R. A. and P.H. Owens. 1987. Numerical simulation of flooding and drying in a depth-averaged tidal flow model. *Proc. Instn. Civ. Engrs*, Part 2, Mar. 83, 161-180.

Figué J. F. 1996. *Modélisation numérique des écoulements compressibles en milieu poreux. Application à la détente de Joule-Thomson.* Thèse de doctorat, Univ. Bordeaux 1, France.

Flather R. A. and N. S. Heaps. 1975. Tidal computations for Morecambe Bay. *Geophys. J. R. Astr. Society*, 42, 489-517.

Fortin M. and R. Glowinski, 1982. *Méthode de Lagrangien augmenté. Application à la résolution numérique des problèmes aux limites*, Collection Méthodes Mathématiques de l'Informatique, Dunod.

Glockner S. 2001. *Aquilon, Notice théorique et numérique*. Technical report.

Greenberg D. A., J. A. Shore, F. H. Page and M. Dowd. 2005.A finite element circulation model for embayments with drying intertidal areas andits application to the Quoddy region of the Bay of Fundy. *Ocean Modelling*, 10, 211-231.

Heniche M., Y. Secretan, P. Boudreau and M. Leclerc. 2000. A two-dimensional finite element drying-wetting shallow water model for rivers and estuaries. *Adv. Water Resources*, 23, 359-372.

Jiang Y.W. and W.H. Wai Onyx. 2005. Drying-wetting approach for 3D finite element sigma coordinate model for estuaries with large tidal flats, *Adv. in Water Resources*, 28, 779-792.

Khadra K., S. Parneix, P. Angot and J.P. Caltagirone. 2000. Fictitious domain approach for numerical modelling of Navier-Stokes equations. *Int. J. Num. Meth. in Fluids*, 341, 651-684.

Le Dissez A., D. Sous, S. Vincent, J.P. Caltagirone, A. Sottolichio. 2005. A novel implicit method for coastal hydrodynamics modeling: application to the Arcachon lagoon, *C. R. Mecanique*, 333 (11), 796-803.

Le Dissez A. 2006. *Modélisation numérique des écoulements tidaux en milieux peu profonds. Application à l'étude de l'hydrodynamique du bassin d'Arcachon.* Thèse de doctorat, Univ. Bordeaux 1, France, 233 p.

Le Hir P., W. Roberts, O. Cazaillet, M. Christie, P. Bassoulet and C. Bacher. 2000. Characterization of intertidal flat hydrodynamics. *Continental Shelf Res.*, 20, 1433-1459.

Leendertse J.J. February 1970. A Water-quality Simulation Model for Wellmixed Estuaries and Coastal Seas. *Principles of Computation*, RM-6230-RC, Santa Monica, USA.

L'Yavanc J. 1995. Evolutions bathymétriques et morphologiques du Bassin d'Arcachon. *Rap. Ifremer DEL/95-12*, 82 p. + annexes.

Pedlosky J. 1987. *Geophysical Fluid Dynamic,* Springer Editions

Plus M. and N. Durand. 2004. Le modèle hydrodynamique du Bassin d'Arcachon. Description générale et validation. *Rapport interne, Ifremer DEL/AR/04-04*

Pluss A. D. and H. Schüttrumph. 2004.Comparison of numerical tidal models for practical applications. *Proceedings of 29th International Conference on Coastal Engineering*, ASCE.

Stelling G.S., A.K. Wiersma and J.B.T.M. Willemse. 1986. Practical aspects of accurate tidal computations. *J. Hydraulic Eng.*, 112 (9), 802-817.

Thauront F. 1994. Les transits sédimentaires subtidaux dans les passes internes du Bassin d'Arcachon. Thèse de doctorat, Univ. Bordeaux 1, France

Vincent S., P. Bonneton and J.P. Caltagirone. 2001.Numerical modelling of bore propagation and run-up on sloping beaches using a Mac-Cormack TVD scheme. *J. Hyd. Res.*, 39, 41-49.

Wang Z.B., T. Louters and H.J. de Vriend. 1995. Morphodynamic modeling for a tidal inlet in the Wadden Sea. *Marine Geology*, 126, 289-300.

A SIMPLE AND EFFICIENT WELL-BALANCED MODEL FOR 2DH BORE PROPAGATION AND RUN-UP OVER A SLOPING BEACH

Fabien Marche[1] and Philippe Bonneton[2]

We consider numerical solutions of the 2DH Non-Linear Shallow Water equations with a bed slope source term. To accurately describe the evolution of the shoreline over a strongly varying topography, we investigate a new model, called $SURF\_WB$. It relies on a reconstruction method for the treatment of the bed-slope and can handle strong variations of topography. The use of the recent VFRoe-ncv Riemann solver leads to a robust treatment of wetting and drying phenomena, which preserves the water depth positivity. An adapted "second order" reconstruction generates accurate bore-capturing abilities. This model is validated against analytical solutions involving a time-dependent moving shoreline and a study of mean-currents over a ridge and runnel system is investigated. This model should have an impact in time-varying numerical simulations in the surf zone.

INTRODUCTION

Several studies (Hibberd and Peregrine 1979), (Kobayashi et al. 1989), (Bonneton 2006) and (Bonneton et al. 2004) have demonstrated the ability of the 1D non-linear shallow water equations to describe cross-shore wave transformation in the surf and swash zones. However, recent studies (Brocchini et al. 2002) and (Peregrine et al. 2000) have shown that 2D bore propagation over a complex topography is a very difficult task from a numerical point of view. Dissipation in 2D bores crossing the surf zone leads to the generation of large scale vertical vorticity. Numerical investigations (Brocchini et al. 2002) revealed that study of vorticity demands higher accuracy than those of the classical shock-capturing methods. In particular, the treatment of the bed slope source term requires an accurate well-balanced scheme. In this paper, we present a new model which gives good results for 2D bore propagation and run-up over complex topography. The first difficulty which appears in the simulation of run-up phenomena is the treatment of the shoreline. Flow properties change very rapidly near the shoreline, as both water depth and velocity vanish and the position of the shoreline is always evolving. Various methods have been proposed to cope with these problems and there are two main approaches.

The first one relies on coordinate transformations (which generate a map between the time-varying physical domain and a time-invariant computational domain) and uses a Lagrangian description of the flow (Prasad and Svendsen 2003). The second approach is based on the use of conservative high resolution bore-capturing schemes on fixed grids and the direct computation of flow properties is used to evaluate the shoreline position, without any tracking method (Hu et al. 1998),

[1] Mathématiques Appliquées de Bordeaux, Université Bordeaux 1, 351 cours de la libération, 33405 Talence, France.

[2] Département Géologie Océanographie, Université Bordeaux 1, Avenue des Lucioles, 33405 Talence, France

(Brocchini et al. 2002) and (Hubbard and Dodd 2002) for a 2D model. Our investigations lie in this context.
The second difficulty which arises in the study of wave transformation on a sloping beach is the treatment of the bed slope source term. Most of the models introduced above rely on fractional step methods (FSM). This is well known today that FSM are not designed to preserve discrete steady states and are unable to compute the convergence toward a steady state or at least the evolution of small perturbations around these states. Thus, we need to discretize the flux gradient and the source term as a whole, achieving some particular properties, as the preservation of a large class of steady states. Since some early works (Greenberg and Leroux 1996), discretisations of the source term which preserve steady states and avoid the drawbacks of FSM have been pointed out. These methods are called well-balanced methods since they preserve balancing properties for steady states. In this context we investigate in (Marche et al. 2006) the use of a new "second order" well-balanced model, called SURF-SVWB, for the study of moving shoreline problems. This model relies on a robust bore-capturing and positivity preserving VFRoe-ncv homogeneous Riemann solver (Gallouet et al. 2003a). This property of positivity is rarely clearly exposed and enables the solver to deal with dry areas. The use of the recent "hydrostatic" reconstruction method (Audusse et al. 2004) for the treatment of the source term ensures the required well-balanced properties. Combined with a MUSCL reconstruction, it raises a stable and simple "second order" bore-capturing well-balanced scheme. This model is able to accurately simulate run-up and back-wash phenomena on various 2D topography and to compute the preservation and the convergence toward steady states. In this paper we will show the ability of the scheme to compute rip currents induced by 2D bores propagating over a complex topography.

NUMERICAL MODEL

Basic equations

The two-dimensional shallow water equations are written as follows :

$$\mathbf{U}_{,t} + F(\mathbf{U})_{,x} + G(\mathbf{U})_{,y} = S(\mathbf{U}), \tag{1}$$

$$\mathbf{U} = \begin{pmatrix} h \\ hu \\ hv \end{pmatrix}, \quad F(\mathbf{U}) = \begin{pmatrix} hu \\ hu^2 + \frac{g}{2}h^2 \\ huv \end{pmatrix}, \quad G(\mathbf{U}) = \begin{pmatrix} hv \\ huv \\ hv^2 + \frac{g}{2}h^2 \end{pmatrix},$$

and

$$S(\mathbf{U}) = \begin{pmatrix} 0 \\ -g\,h\,d_x \\ -g\,h\,d_y \end{pmatrix},$$

where $()_{,x}$ (resp. $()_{,y}$) stands for the derivate along the x direction (resp. the y direction), $\mathbf{u} = {}^t(u,\, v)$ is the depth-averaged velocity with u and v the scalar components in the horizontal x, y directions and h is the local water depth. $\mathbf{U}$

is the vector for the conservative variables, $F(\mathbf{U})$ and $G(\mathbf{U})$ stand for the flux functions respectively along the x and y directions and $S(\mathbf{U})$ represents the bed slope source term with the bed slope $\nabla d = {}^t(d_{,x},\, d_{,y})$.

A two-dimensional semi-discrete finite volume formulation of system (1) is given by :

$$\frac{d}{dt}\mathbf{U}_{ij}(t) + \frac{1}{\Delta x}(\mathbf{F}^*_{i+\frac{1}{2},j} - \mathbf{F}^*_{i-\frac{1}{2},j}) + \frac{1}{\Delta y}(\mathbf{G}^*_{i,j+\frac{1}{2}} - \mathbf{G}^*_{i,j-\frac{1}{2}}) = \mathbf{S}_{ij} \quad (2)$$

where the cell-centered vector of conservative discrete variables is $\mathbf{U}_{ij} = {}^t(h_{i,j}, h_{i,j}u_{i,j}, h_{i,j}v_{i,j})$, $\mathbf{F}^*_{i\pm\frac{1}{2},j}$ and $\mathbf{G}^*_{i,j\pm\frac{1}{2}}$ stands respectively for the numerical flux functions through interfaces in each horizontal direction and $\mathbf{S}_{ij}$ is a discretization of the source term. For the sake of simplicity, we directly give here the expressions of these numerical flux functions and the reader is referred to (Marche et al. 2006) for further theoretical details :

$$\mathbf{F}^{n,-}_{i+\frac{1}{2},j} = \mathbf{F}^n_{i+\frac{1}{2},j} + \mathbf{S}^n_{i+\frac{1}{2}-,j} \quad , \quad \mathbf{F}^{n,+}_{i+\frac{1}{2},j} = \mathbf{F}^n_{i+\frac{1}{2},j} + \mathbf{S}^n_{i+\frac{1}{2}+,j}, \quad (3)$$

$$\mathbf{F}^n_{i+\frac{1}{2},j} = F(\mathbf{U}_{i+\frac{1}{2},j}(0,\, \mathbf{U}^n_{i+\frac{1}{2}-,j}, \mathbf{U}^n_{i+\frac{1}{2}+,j})). \quad (4)$$

and

$$\mathbf{G}^{n,-}_{i,j+\frac{1}{2}} = \mathbf{G}^n_{i,j+\frac{1}{2}} + \mathbf{S}^n_{i,j+\frac{1}{2}-} \quad , \quad \mathbf{G}^{n,+}_{i,j+\frac{1}{2}} = \mathbf{G}^n_{i,j+\frac{1}{2}} + \mathbf{S}^n_{i,j+\frac{1}{2}+}, \quad (5)$$

$$\mathbf{G}^n_{i,j+\frac{1}{2}} = G(\mathbf{U}_{i,j+\frac{1}{2}}(0,\, \mathbf{U}^n_{i,j+\frac{1}{2}-}, \mathbf{U}^n_{i,j+\frac{1}{2}+})). \quad (6)$$

In the following, we only detail the one-dimensional case for the sake of simplicity.

Interface values

The interface value $\mathbf{U}_{i+\frac{1}{2}}(0,\, \mathbf{U}_{i+\frac{1}{2}-},\, \mathbf{U}_{i+\frac{1}{2}+})$ between cell i and $i+1$ is obtained from a VFRoe-ncv linearized Riemann solver. This solver relies on the symmetrizing change of variable $\mathbf{W}(\mathbf{U}) = {}^t(2c,\, u)$ which preserves the water depth positivity at least for interface values (Gallouet et al. 2003b). The formalism of this solver is not recalled in details here. It relies on the exact resolution of a linearized problem around an averaged state $\tilde{\mathbf{W}}$. The eigenvalues of the linearized convection matrix are

$$\tilde{\lambda}_1 = \tilde{u} - \tilde{c}, \quad \tilde{\lambda}_2 = \tilde{u} + \tilde{c},$$

where $c = \sqrt{gh}$. Then, the exact solution of the linearized Riemann problem is given by the sign of the eigenvalues $\tilde{\lambda}_k$:

- If $\tilde{\lambda}_1 > 0$ or $\tilde{\lambda}_2 < 0$, then the flow is super-critical and we recover a classical upwinding. The interface value is defined as follows :

$$\mathbf{W}^*_{i+\frac{1}{2},j} = \begin{cases} \mathbf{W}_i \text{ if } \tilde{\lambda}_k > 0\ \forall k \\ \mathbf{W}_{i+1} \text{ if } \tilde{\lambda}_k < 0\ \forall k. \end{cases} \quad (7)$$

- If $\tilde{\lambda}_1 < 0$ and $\tilde{\lambda}_2 > 0$ then the flow is subcritical and we obtain :

$$c^*_{i+\frac{1}{2},j} = \tilde{c} - \frac{1}{4}(u_{i-1,j} - u_{i,j}) \text{ and } u^*_{i+\frac{1}{2},j} = \tilde{u} - (c_{i+1,j} - c_{i,j}). \quad (8)$$

We can finally recover conservative variables, using the inverse change of variable. The key point is that this interface value is computed from values $\mathbf{U}_{i+\frac{1}{2}-}$ and $\mathbf{U}_{i+\frac{1}{2}+}$ issued from a combined limited linear/hydrostatic reconstruction.

Variable reconstruction

More precisely, considering the cell i, we compute first linear reconstructions $\mathbf{U}_{i,r}$ and $\mathbf{U}_{i,l}$ respectively at $i+\frac{1}{2}-$ and $i-\frac{1}{2}+$, using a *minmod* limiter. Values of $H_{i,l}$ and $H_{i,r}$, where $H = h + d$, are also reconstructed, and we deduce reconstructions of $d_{i,l}$ and $d_{i,r}$. Then, interface topography values $d_{i+\frac{1}{2},j}$ are defined with $d_{i+\frac{1}{2}} = max(d_{i,r}, d_{i+1,l})$.

The positivity preserving hydrostatic reconstruction of the water height is thus defined as follows :

$$h_{i+\frac{1}{2}-} = max(0, h_{i,r} + d_{i,r} - d_{i+\frac{1}{2}}), \quad h_{i+\frac{1}{2}+} = max(0, h_{i+1,l} + d_{i+1,l} - d_{i+\frac{1}{2}}),$$

and we deduce reconstructed values on each side of the interface :

$$\mathbf{U}_{i+\frac{1}{2}-} = \begin{pmatrix} h_{i+\frac{1}{2}-} \\ h_{i+\frac{1}{2}-}\, u_{i,r} \end{pmatrix}, \quad \mathbf{U}_{i+\frac{1}{2}+} = \begin{pmatrix} h_{i+\frac{1}{2}+} \\ h_{i+\frac{1}{2}+}\, u_{i+1,l} \end{pmatrix}. \quad (9)$$

Finally we introduce :

$$\mathbf{S}^n_{i+\frac{1}{2}-} = \begin{pmatrix} 0 \\ \frac{g}{2} h^2_{i+\frac{1}{2}-} - \frac{g}{2} h^2_{i,r} \end{pmatrix}, \quad \mathbf{S}_{i-\frac{1}{2}+} = \begin{pmatrix} 0 \\ \frac{g}{2} h^2_{i,l} - \frac{g}{2} h^2_{i-\frac{1}{2}+} \end{pmatrix}, \quad (10)$$

and a centered source term discretization is added to achieve second order well-balancing :

$$\mathbf{S}_{c,i} = \begin{pmatrix} 0 \\ g\frac{h_{i,l} + h_{i,r}}{2}(d_{i,l} - d_{i,r}) \end{pmatrix}. \quad (11)$$

The extension to the two-dimensional framework on Cartesian meshes is straightforward.

Wetting and drying procedure

No special tracking procedure is used in this model. A distinction is made between wet and dry cells with the introduction of an artificial threshold value $h_{tol} = 10^{-10}\, m$ for the definition of a dry cell. The detailed wetting/drying procedure is simple. At the beginning of each time step, we search for dry cells, i.e. cells for which the water depth is less than h_{tol}, and set them to $h = h_{tol}$. The keypoint is to apply this "wetting procedure" to the quantity which are issued

from the hydrostatic reconstruction instead of the natural quantities, leading to a better accuracy at the shoreline and a better stability in practice.
Thereafter, the shoreline is constructed from the mesh interface between wet and dry cells, without any interpolation. We have three possible types of inernal edge, which are *wet/wet*, *wet/dry* and *dry/dry*, the case *wet/dry* corresponding to the flooding or the drying of a specific area. Then, thanks to the robustness of the VFRoe-ncv solver, the computation is directly performed, without any distinction between these possible configurations.

NUMERICAL ASSESSMENTS

We propose here some validations relying on analytical solutions.

The periodic Carrier and Greenspan solution

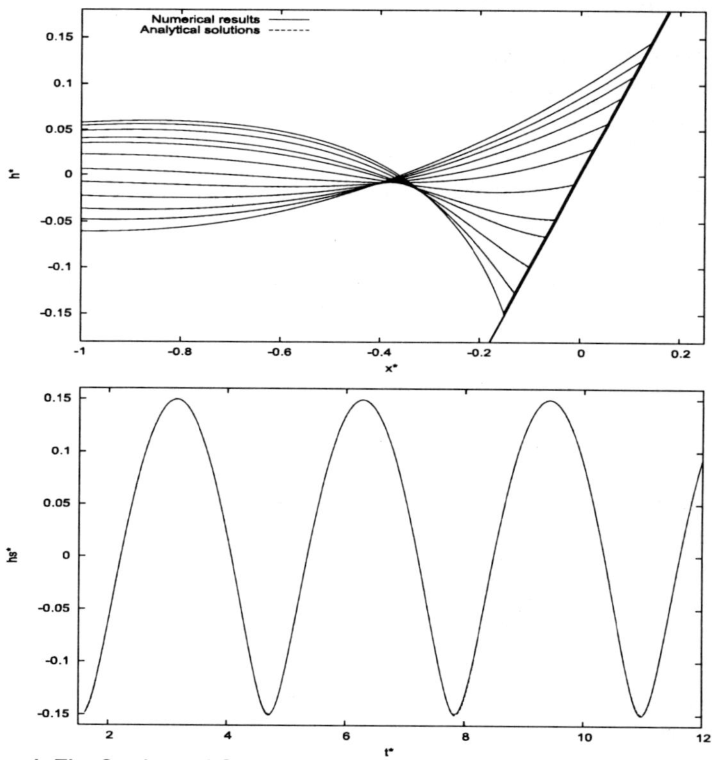

Figure 1. The Carrier and Greenspan solution. Numerical and analytical results.

We compare here numerical results (in solid lines) with the periodic Carrier and Greenspan 1D (Carrier and Greenspan 1958) analytical solutions (in dotted lines). This solution represents the motion of a periodic wave traveling shoreward and being reflected out to sea, generating a standing wave which is let run-up and run-down on a plane beach. It has been observed that even though waves may break just before they reach the shoreline, the tip of the resulting swash motion has many similarities with the nonbreaking run-up of waves and this test case is generally acknoledged to be a suitable test case for the assessment of shoreline motions computation. We show on Fig. 1, a comparison between numerical and analytical solution, highlighting the ability of this new model in the computation of moving shoreline. There is a very good agreement between the two curves, both for the surface elevation profiles and time series of the shoreline motions. The analytical solution is used as a left inlet boundary condition to force the motion and we use $\Delta x = 0.01$ and $CFL = 0.7$. The shoreline oscillations are accurately computed even after a large number of period. Evaluation of the L^2-error for this test can be found in (Marche et al. 2006). We stress out that many other well-balanced methods, highly validated in hydraulic situations, are unable to provide such results (Marche et al. 2006).

The two-dimensional Thacker solution

We perform a comparison with the analytical solution introduced in (Thacker 1981), describing free boundary and free surface oscillations in a 2D parabolic basin (see Fig. 2). This case is perhaps one of the more difficult case to handle for a numerical model, since it involves a two-dimensional periodic moving shoreline with a no radially symmetric configuration. The shoreline is a circle in the (x, y) plane and the motion is such that the center of the circle orbits the center of the basin, while the surface remains planar with constant gradient at any given instant. We use here $\Delta x = \Delta y = 0.01\,m$ and $CFL = 0.7$.

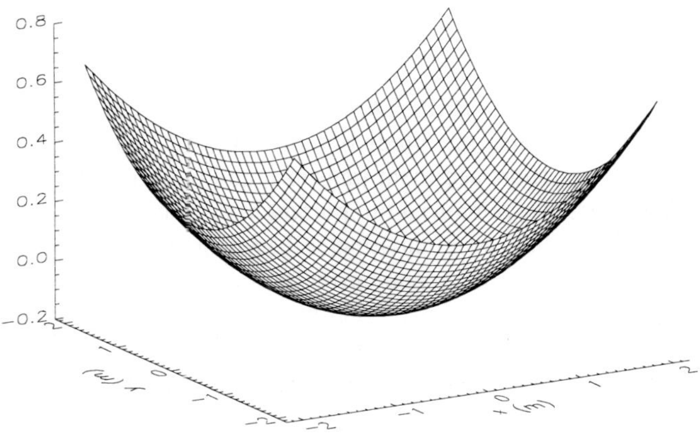

Figure 2. Topography of the basin in the Thacker case.

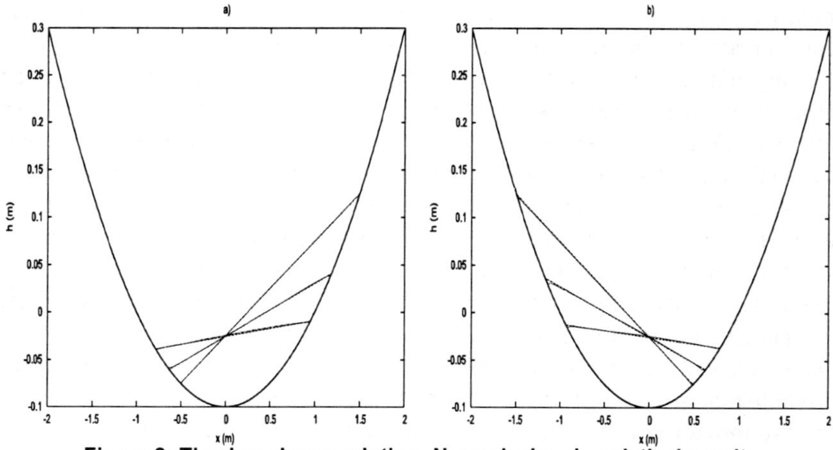

Figure 3. Thacker planar solution. Numerical and analytical results

We show on Fig. 3 a comparison between numerical (in solid lines) and analytical solution (in dotted lines) for free surface profiles during the fourth period. Tiny distortions can be observed but the planar aspect of the surface is preserved even after several periods. Moreover, we are able to compute a large number of period without any large instability (almost 10 periods have been effectively computed).

Note that these results are almost as accurate as those presented in the literature, relying on more complex or expensive algorithms, like extrapolation and filtering (Lynett et al. 2002) or even adaptive refinement method (Hubbard and Dodd 2002).

MEAN CURRENTS OVER A RIDGE AND RUNNEL SYSTEM

On natural beaches, accurate prediction of wave-induced current circulations which control the nearshore sediment transport, and therefore the morphological changes in the nearshore, remains an important topic. The classical approach for modelling ISZ wave-induced circulation is based on time averaged equations. 2DH numerical studies of ISZ mean currents using time-varying models is still not well developped. Few studies relying on Boussinesq-like models have been reported, like (Sorensen et al. 1998) or (Chen et al. 1999). These time-varying approaches may be very fruitful, providing informations not available with time-averaged models, like velocity oscillations, instantaneous vorticity fields or swash zone motions.

The Aquitanian coast beaches exhibit longshore non-uniform multiple bars. We mainly focus here on the rigde and runnel system, in which ridges are regularly interrupted by down-currents oriented runnels, with a mean wave

length of about $400\,m$. Observations have shown that topographic feed-back may generate rip circulation cells. On the Aquitanian coast, these rip currents occur at each runnel outlet and their intensity depends on the length, the shape of the system, the offshore wave conditions and the tide level. The mechanism leading to the generation of a rip current is summarized in (Castelle et al. 2006). The main idea is that longshore variations of the topography are associated with similar variations in the pressure, due to mean water level changes and hence, lead to longshore pressure gradients. The longshore pressure gradients induce an intensification of the longshore current in the runnel. On the upper part of the beach, the longshore pressure gradients force the feeder currents. Water convergence inshore the runnel outlet results in the formation of a seaward oriented current, a rip current. Observations show that this seaward flow tends to be narrow and relatively strong but rapidly desintegrates outside the surf zone.
For our analysis, we have neglected first the tidal variations, since it has been shown that tidal currents modulations can be neglected in a first approach, when compared to the wave-induced currents (Castelle et al. 2006). However, we have performed several simulations for various mean water levels to highlight tidal variations impact on the circulation. Fig. 4 shows the idealized ridge and runnel topography used for our simulations.

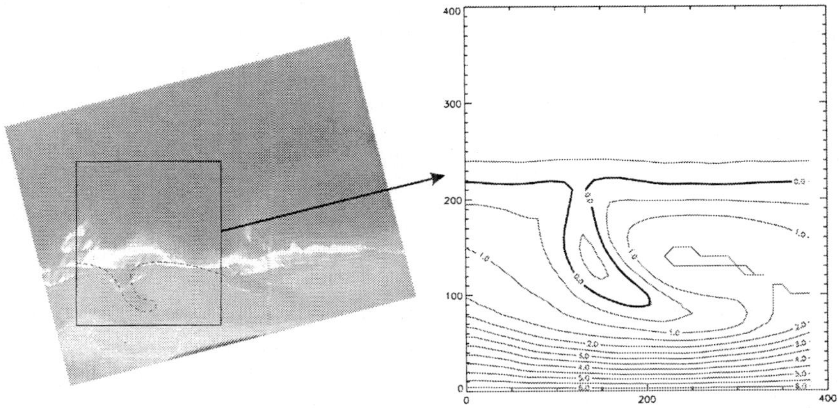

Figure 4. Contour plot of a ridge and runnel system

At this level of analysis, the study will only be qualitative. It is well-known that NSW equations lead to an anticipated steepening of the solution too far offshore. However, we have chosen to generate a periodic train of wave offshore of the surf zone. When this train of waves reaches the ISZ, the wave shape has change to the characteristic sawtooth shape. Hence, the simulation may be realistic in the ISZ enough. Such qualitative studies in the ISZ have been performed for instance in (Brocchini et al. 2002) for the study of macro-vortices induced by single breakwaters.

We set the wave period at $T_w = 12\,s$. The offshore wave amplitude is the main characteristic controlling the rip velocities, as changes in the breaking patterns control the rip current characteristics. Our aim is to obtain a large enough bore-like wave amplitude in the area corresponding to the ISZ, taking into account the energy dissipation which occurs during all the propagation. For this purpose, we set the offshore wave height at $H_w = 1\,m$. We use a variable mesh in the cross-shore direction. The largest cell size is $\Delta x = 2\,m$ near the offshore boundary, while the finest cell size is $\Delta x = 0.05\,m$ near the shoreline. In the longshore direction, the mesh is regular and the discretization step is $\Delta y = 1\,m$. The time step is constant and fixed at $\Delta t = 0.05\,s$.

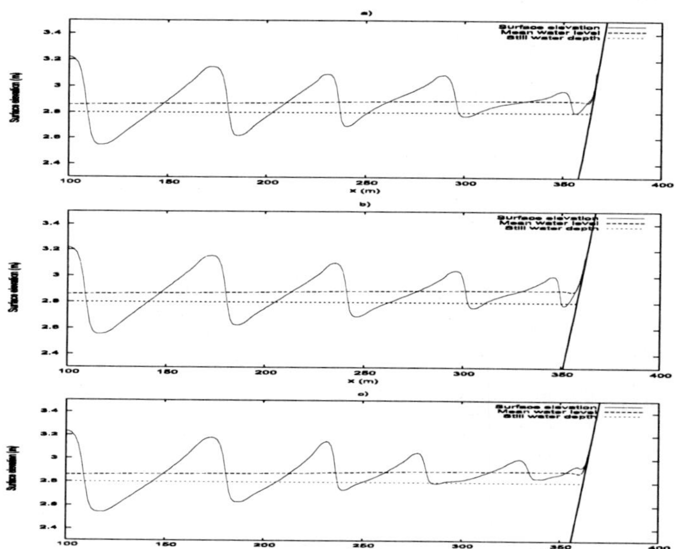

Figure 5. Section views of the the free surface elevation and set-up

Wave refraction and set-up

The refraction can be observed on Fig. 5 where profiles of the free water surface are plotted at three different locations in the longshore direction : a) y = 95.25 m, b) y = 190.5 m, c) y = 285.75 m. We have also plotted on the same Figure the three corresponding set-up profiles.

Wave-induced mean currents

In the sequel h_m stands for the value of water depth over the zero isobath. Fig. 6 and Fig. 7 show the two dimensional mean current patterns simulated during a normally incident swell, for respective values of water depth $h_m = 1\,m$,

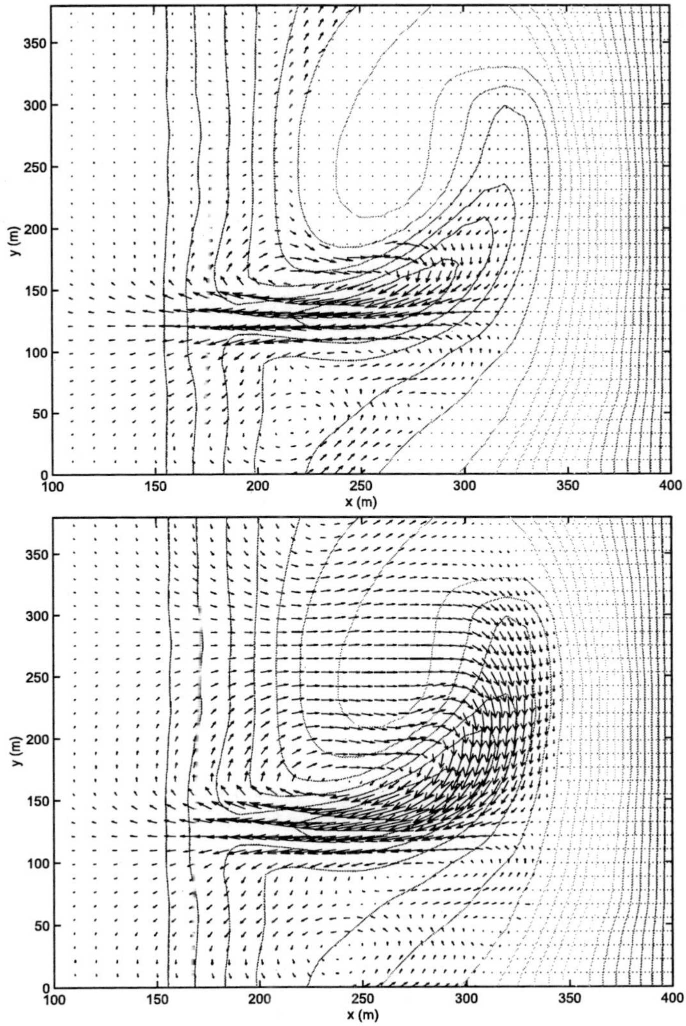

Figure 6. Two dimensional view of the computed mean currents in the system

$h_m = 1.5\,m$ (Fig. 6), $h_m = 2.0\,m$ and $h_m = 2.5\,m$ (Fig. 7). Note that $h_m = 2.0\,m$ corresponds to mid-tide and high-tide occurs around $h_m = 4.0\,m$. The steady state is reached after approximatively $800\,s$ and the instantaneous velocity field is averaged over a wave period. A narrow seaward oriented current can be observed at the runnel outlet, associated with a large circulation cell, even when the whole system is not entirely submerged. For $h_m = 1\,m$, the maximum value of this current is about $0.14\,m/s$. For $h_m = 2\,m$, it is about $0.2\,m/s$. As expected, this current rapidly vanishes offshore of the surf zone.

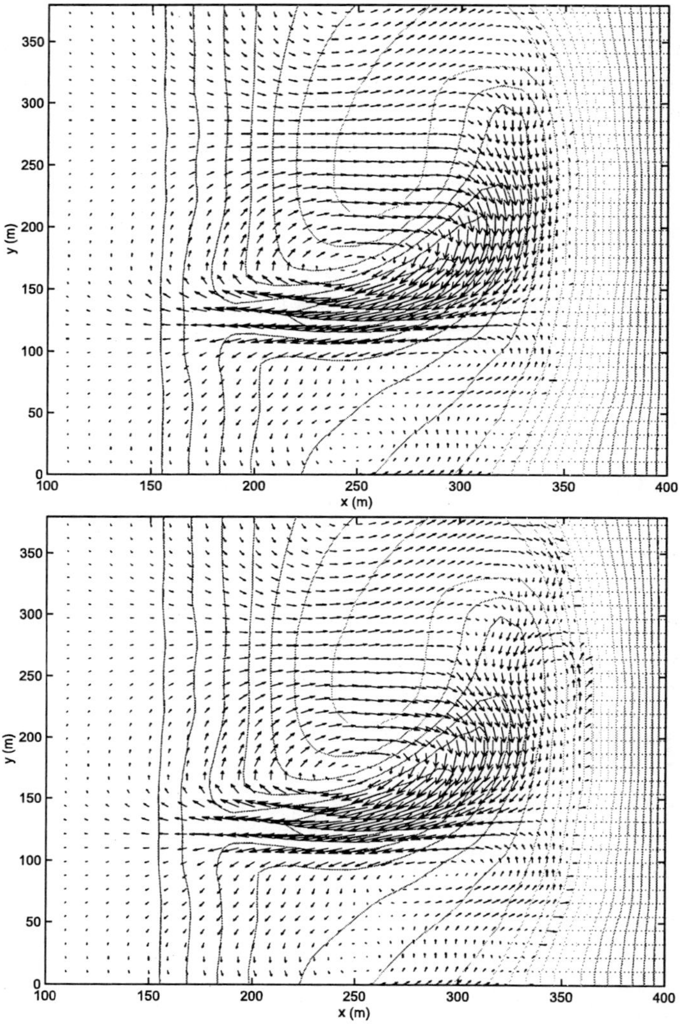

Figure 7. Two dimensional view of the computed mean currents in the system

As suggested in (Castelle et al. 2006), the maximum of rip velocity occurs at mid tide. The intensity of this maximum rip velocity then decreases with respect to the increasing mean water level. Still comparing with (Castelle et al. 2006), the second circulation cell near the lateral boundary is slightly shifted in the cross-shore direction, towards offshore. All these results are qualitatively similar to those shown in (Castelle et al. 2006) with only slight differences.

CONCLUSION

From these validations and applications, this new model $SURF\_WB$ appears as a promising and efficient tool for the study of many coastal hydrodynamic processes, including wave breaking, run-up and two-dimensional wave-induced mean-currents over complex topography. The two-dimensional physical processes involved in the last application can be accurately reproduced and the Finite-Volume framework, with high order shock-capturing abilities, clearly enables to handle the propagation and run-up of breaking waves, which was not possible with Finite-Difference approaches. Moreover, the shoreline tracking method proposed here is far more simpler to implement and less computationnaly expensive than many methods found in the literature. However, we need to validate the model with laboratory and field experiment data to assess the real improvements brought up by the well-balanced approach to simulate 2DH bore propagation and run-up.

REFERENCES

Audusse, E., F. Bouchut, M. Bristeau, R. Klein, and B. Perthame. 2004. A fast and stable well-balanced scheme with hydrostatic reconstruction for shallow water flows, *SIAM J.Sci.Comp.,* 25(6), 2050–2065.

Bonneton, P.. 2006. Modelling of periodic wave transformation in the inner surf zone, *Ocean Engineering,*. in press.

Bonneton, P., V. Marieu, H. Dupuis, N. Sénéchal, and B. Castelle. 2004. Wave transformation and energy dissipation in the surf zone: comparison between a non-linear model and field data, *J. Coast. Res., 39.*

Brocchini, M., A. Mancinelli, L. Soldini, and R. Bernetti. 2002. Structure-generated macrovortices and their evolution in very shallow depths, *Proceedings of 28th International Conference on Coastal Engineering,* ASCE, 772–783.

Brocchini, M., I. A. Svendsen, R. S. Prasad, and G. Belloti. 2002. A comparison of two different types of shoreline boundary conditions, *Comput. Methods Appl. Mech. Engrg, 191*, 4475–4496.

Carrier, G. F. and H. P. Greenspan. 1958. Water waves of finite amplitude on a sloping beach, *J. Fluid Mech, 4*, 97–109.

Castelle, B., P. Bonneton, N. Sénéchal, H. Dupuis, R. Butel, and D. Michel. 2006. Dynamics of wave-induced currents over an alongshore non-uniform multiple-barred sandy beach on the aquitanian coast, france, *Continental shelf research, 26*, 113–131.

Chen, Q., J. T. Kirby, R. A. Dalrymple, A. B. Kennedy, and M. Haller. 1999. Boussinesq modelling of a rip current system, *J. Geophys. Res.,* 109(9), 20617–20637.

Gallouet, T., J. M. Herard, and N. Seguin. 2003a. On the use of some symetrizing variables to deal with vacuum, *Calcolo, 40*, 163–194.

Gallouet, T., J. M. Herard, and N. Seguin. 2003b. Some approximate godunov schemes to compute shallow-water equations with topography, *Computers and Fluids, 32*, 479–513.

Greenberg, J. M. and A. Y. Leroux. 1996. A well balanced scheme for the numerical processing of source terms in hyperbolic equations, *SIAM J.Numer.Anal.,* 33(1), 1–16.

Hibberd, S. and D. H. Peregrine. 1979. Surf and run-up on a beach: a uniform bore, *J. Fluid. Mech.,* 95(2), 323–345.

Hu, K., C. G. Mingham, and D. M. Causon. 1998. A bore-capturing finite volume method for open-channel flows, *Int.J.Numer.Meth.Fluids, 28*, 1241–1261.

Hubbard, M. E. and N. Dodd. 2002. A 2d numerical model of wave run-up and overtopping, *Coastal Engineering, 47*, 1–26.

Kobayashi, N., G. S. Desilva, and K. D. Watson. 1989. Wave transformation and swash oscillation on gentle and steep slopes, *J. Geophys. Res., 94*, 951–966.

Lynett, P. J., T. S. Wu, and P. L. F. Liu. 2002. Modeling wave runup with depth-integrated equations, *Coastal Engineering, 46*, 89–107.

Marche, F., P. Bonneton, P. Fabrie, and N. Seguin. 2006. Evaluation of well-balanced bore-capturing schemes for 2d wetting and drying procedure, *Int. J. Num. Meth. Fluids,*. in press, article online in advance of print.

Peregrine, D. H., M. D. Patterson, and O. Bokhove. 2000. Large eddies and vorticity in the surf and swash zone, *Book of Abstract*, SASME.

Prasad, R. S. and I. A. Svendsen. 2003. Moving shoreline boundary condition for nearshore models, *Coastal Engineering, 49*, 239–261.

Sorensen, O. R., H. A. Schaffer, and P. A. Madsen. 1998. Surf zone dynamics simulated by a boussinesq type model. part iii. wave-induced horizontal nearshore circulation, *Coastal Engineering., 33*, 155–176.

Thacker, W. C.. 1981. Some exact solutions to the nonlinear shallow-water wave equations, *J. Fluid Mech., 107*, 499–508.

CALCULATION OF INFRAGRAVITY EDGE WAVE ALONG A LONG AND STRAIGHT COAST

Guan-Yu Chen[1]

Hua-Lien Harbor is located in a long and straight coast and hence a complete study on the influence of infragravity edge wave to the harbor should include the coast. This study starts with surveys on theoretical computation and both the applicability and accuracy of the theory are discussed. It is concluded that the exact resonance between the varying breaking points and the edge wave phase has to be satisfied, or the theoretical prediction will one-order over predict the growth rate. For the bathymetry in Hua-Lien, Stokes edge waves of 100-200 second period are shown to dominate and this observation will significantly simplify the calculation. Finally, we analyze new field data and some evidences of infragravity edge wave are shown.

Introduction

The motivation of the present study is based on our previous studies where edge wave was shown to play an important role in the harbor resonance of Hua-Lien Harbor, Taiwan. In the south of the Harbor, there is a long and straight coast. Edge waves are generated not only locally, but also along the whole long coast. Therefore, to calculate the edge wave that induces harbor resonance, the computational domain has to include the whole coast.

Hua-Lien Harbor is the largest commercial harbor in eastern Taiwan. When the center of the typhoon is beside the Philippine islands, even if it is one thousand kilometers away, its swell is known to induce resonance inside the harbor. The ships inside the harbor have to flee before the wave conditions become dangerous.

In treating the resonance problem, engineers have usually assumed that the harbor entrance is facing the open sea and the forcing is a plane infragravity wave incident shoreward. This is the most common layout; however, instead of facing the sea, the Hua-Lien harbor is along the long straight east coast of Taiwan. This specific layout of Hua-Lien harbor makes the basin vulnerable to the influence of edge waves.

An interesting evidence for the importance of edge wave in Hua-Lien Harbor resonance was also reported. A series of offshore breakwaters constructed outside the harbor results in a signifgicant reduction of the oscillation and hence delays the resonance inside for more than one hour. This phenomena can be expalined if edge wave is the forcing for basin oscillation: In the existence of offshore breakwaters, two edge wave modes of the same frequency are generated: one is reflected at the breakwaters and then

---

[1] Institute of Physical Oceanography, National Sun Yat-Sen University, 70 Lien-hai Rd., Kaohsiung 80424, Taiwan

continuously refracted by the coastal topography and is associated with the breakwater. The other mode is the original edge wave without the breakwater. The former is shorter in longshore wavenumber while the later is longer. After passing the offshore breakwater zone, these two edge wave modes are out of phase and hence make the oscillation minimized.

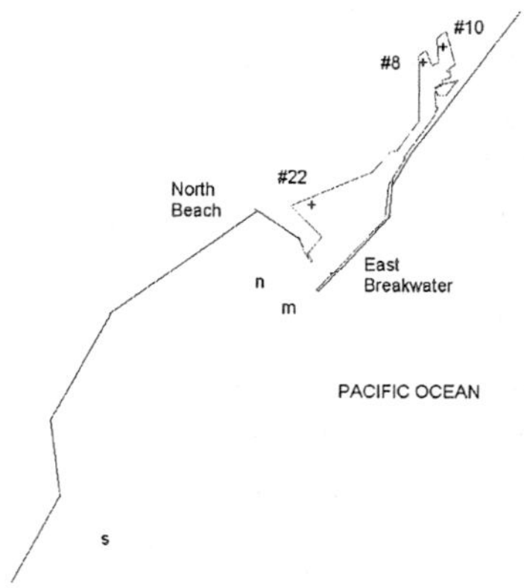

Figure 1. The layout of Hua-Lien Harbor and the locations of three measurement stations m, n and s.

The bad news for harbor resonance study in Hua-Lien is that infragravity edge wave has never been observed in the east coast of Taiwan. This means all countermeasures designed to reduce edge wave height cannot be directly applied.

How to solve this problem? The cheaper way is to theoretically or numerically compute edge waves in a large domain and compare the results with the field measurement near Hua-Lien Harbor. The other method, which is much more expensive, is to find field evidences for the existence of infragravity edge wave by deploying more wave gages and related instruments. In the following, we will first discuss the theoretical approach.

Review of existing theories in edge wave generation

We surveyed the existing theories and find the theory of Lippmann et al. (1997) is applicable. The accuracy of this theory is also discussed and the difference between this theory and a laboratorial experiment is explained in the next section. In the third section, we find that this theory can be further simplified in applying it to Hua-Lien area.

There are two ways to compute infragravity edge wave. One theoretically estimates the edge waves and theories of Gallagher (1971), Foda and Mei (1981), Schäffer (1994), and Lippmann et al. (1997) are developped. The other numerically computes the boundary value problem and gives infragravity components mixed with both free and bound waves. Reniers et al. (2002) and Van Dongeren et al. (2003) are a few recent examples.

In simulating harbor resonances, the biggest problem for the later approach is that the bound waves have to be removed. For shorter waves, their phase changes significantly due to irregular basin geometry. The resulting destructive interferences will make resonance unlikely to happen. These phase effects will not be shown if we directly apply the bound wave forced by the short wave groups. Therefore, only free infragravity wave can stimulate harbor resonance. Unless free waves are separated from the bound wave, the accumulation of edge wave along the coast cannot be calculated and the forcing for the harbor resonance cannot be specified.

The theory of Lippmann et al. (1997) focuses on edge waves that are obtained by taking inner product between the forcing and the edge wave eigenfunction. Therefore, the results are edge waves that propagate freely. We do not have to worry if they are contaminated by bound wave.

Another reason that the theory of Lippmann et al. (1997) is chosen is its bathymetry assumption. Some researchers, for example Schäffer (1994), assumed a flat offshore sea-bottom while Lippmann et al. (1997) assumed a constant slope. In practice, Hua-Lien Harbor and the east coast of Taiwan is next to the deep Pacific Ocean. The sloping bottom extends more than 40 kilometers with no continental shelves. Comparing with the wavelength of an infragravity edge wave which is only a few kilometers long, a constant slope is a better representation for Hua-Lien coast. Therefore, we choose the theory of Lippmann et al. (1997).

The first attempt to predict infragravity wave is Symonds et al. (1982) who let the breaking point vary and serves as a wavemaker. This approach is extended to three dimensions by Lippmann et al. (1997). In these two theories, the surf zone forcing is obtained by assuming saturated wave condition and this is justified only when the Iribarren number, $s/\sqrt{H_0/L_0}$, is less than 0.5. Therefore, we have to check the surf similarity parameter or Iribarren number first.

Hua-Lien coast has 1/25 bottom slope near the shoreline and typical offshore wave height is 3m or higher in typhoon period. Typical offshore wave period is 12 sec; hence, deep water wavelength is approximately 200 m. Iribarren number is 0.3 or less, the 0.5 criterion is satisfied, and spilling breakers and saturated surf zone typically occur. The wave amplitude inside the surf zone then can be asasumed to be proportional to the water depth only:

Combining continuity and momentum equations will result in a single equation for the infragravity wave surface elevation (Mei and Benmoussa, 1984):

$$\frac{\partial^2 \eta}{\partial t^2} - g\frac{\partial}{\partial x}\left(h\frac{\partial \eta}{\partial x}\right) - g\frac{\partial}{\partial y}\left(h\frac{\partial \eta}{\partial y}\right)$$

$$= -\frac{1}{\rho}\left[\frac{\partial}{\partial x}\left(\frac{\partial}{\partial x}S_{xx} + \frac{\partial}{\partial y}S_{yx}\right) + \frac{\partial}{\partial y}\left(\frac{\partial}{\partial x}S_{xy} + \frac{\partial}{\partial y}S_{yy}\right)\right]$$

Following Bowen and Guza (1978) and Holman (1981), Lippmann et al. (1997) consider the temporal growth of infragravity edge waves and assume the amplitude a function of time. Exact resonance is also assumed so that the break point varies at the same temporal and spatial periods as an infragravity edge wave. To use the orthogonality of the cross-shore structure, the inner product is taken to explicitly derive the surf zone forcing by assuming saturated wave condition.

However, comparing the theoretical growth rate with the experiment by Bowen and Guza (1978), Lippmann et al. (1997) concluded that this theoretical value is an order larger than the experimental value, and proposed a damping to explain this difference. In the next section, this difference will be explained based on the assumption of exact resonance.

Exact resonance assumption

Lippmann et al. (1997) assumed exact resonance and the break point varies at the same temporal and spatial periods as an infragravity edge wave. In the experiment of Bowen and Guza (1978), the alongshore velocity of the incident wave is exactly the group velocity of first mode edge wave. Lippmann et al. (1997) thus concluded the exact resonance condition satisfied.

However, we note that for edge waves, the group velocity is just half of the phase velocity. The phase relative to an edge wave group is always changing. Therefore, if the forcing moves at the group velocity, it will reinforce edge wave at some time but reduce the edge wave height when the phase is changes to be opposite to the forcing.

Another way to see this phase difference is to consider the alongshore periodicity. The break point varies at the same temporal period as some infragravity edge wave, but its spatial period is just half of edge wave's wavelength. That is why the experiment of Bowen and Guza (1978) gives a growth rate an order less than the theory

The above discussion demonstrates that if there is no exact resonance, the growth rate will be much less than the prediction of the theory. However, to most engineering purposes, an order-of-magnitude accuracy is good enough. For example, in the case of Hua-Lien harbor, the result of this theory is used to compare with the infragravity component of field measurement which has only order-of-magnitude accuracy since the reflective and bound waves cannot be accurately estimated.

Edge wave mode generated

In laboratorial and numerical experiments, plane infragravity wave can easily generate edge waves of the same period by oblique incidence in Hua-Lien area. However, only the lowest Stokes mode tends to exist for infragravity edge waves of 140-160 seconds: As Figure 2 shows, some energy is trapped by the coast and seperated from the original plane wave. The flow-elevation relation shows that the flow at the wave crest is similar to the wave velocity, the flow at the trough is opposite to the wave velocity.

Furthermore, onshore flow occurs behind the crest while offshore flow is behind the trough. Theses features are exactly similar to Stokes edge wave.

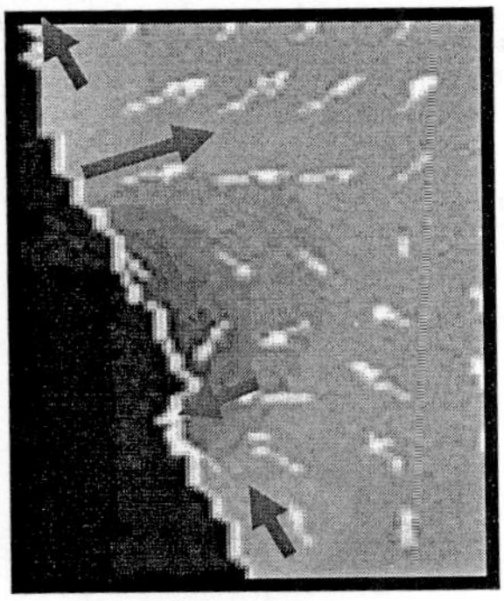

Figure 2. The relations between the flow and the surface elevation suggest that Stokes edge wave dominates the infragravity wave nearshore. The wave propagates along the shore to the upper side in the figure: the arrow indicates the instantaneous flow direction, while the surface elevation is represented by the shading.

We can also identify a Stokes edge wave by the dispersion curve based on bottom slopes 1/30 and 1/20. These slope variations are consistent with the sea chart.

The reason for this single mode dominacy can be explained by the local bathymetry. The nearshore bottom slope is only 1/25 which can support various edge wave modes for the wave period of interest. However, the offshore slope is 1/10 and the influence of this slope change on various edge wave modes depends on the crossshore decay rate of each mode.

If an edge wave mode decays so quickly that its amplitude is important only in the onshore region, then the idelized 1/25 slope can support this mode fairly well. On the other hand, if the amplitude predicted by the constant slope theory is not negligible in the offshore area, then the real bathymetry with larger offshore slope cannot support this mode.

Since Stokes mode decays fastest, it is not surprising that its amplitude beyond the slope transmission is negligible. Higher edge wave modes, however, decays less quick and cannot be supported by the real sea bottom. Although the theory of Lippmann considers every edge wave mode, in practice this should be modified because most modes are not supported by the bathymetry. Therefore, in the case of Hua-Lien, we only have to consider the lowest mode (Stokes mode) for edge waves of infragravity period.

Measurement

In 2005, a few new measurements are done by the Center of Harbor and marine Technology and have been analyzed. In the following, we try to use these observations to verify the existence of edge waves.

Unfortunately, based on the pressure record, the spatial structure cannot be clearly identified: wavenumber cannot be calculated due to poor spatial resolution. However, some infragravity energy can be resolved. For example, as Figure 3 shows, peaks of infragravity components of period 130-150 seconds in the year 2005 are consistent with the occurrence time of typhoon. Besides, the nearshore location (n) has larger amplitude than the offshore (m), a feature of edge wave. This increasing should be the result of shoaling effect and cannot be explained by edge wave theory because the Stokes edge wave amplitude at the m station is negligible.

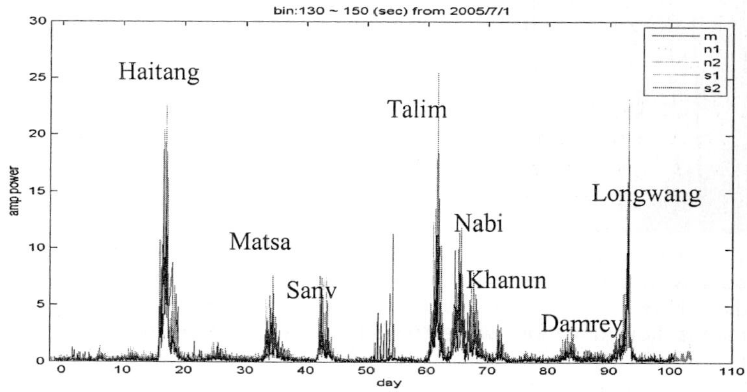

Figure 3. Peaks of infragravity components of period 130-150 seconds in the year 2005 are consistent with the occurrence time of typhoon.

To identify a Stokes edge wave from the flow field is possible because Stokes edge wave has circular trajectory in the horizontal plane, while a leaky infragravity wave only induces flows in the propagation direction. An example taken in the period of typhoon Haitang, based on the phase and amplitude of flow spectrum, is given in Figure 4.

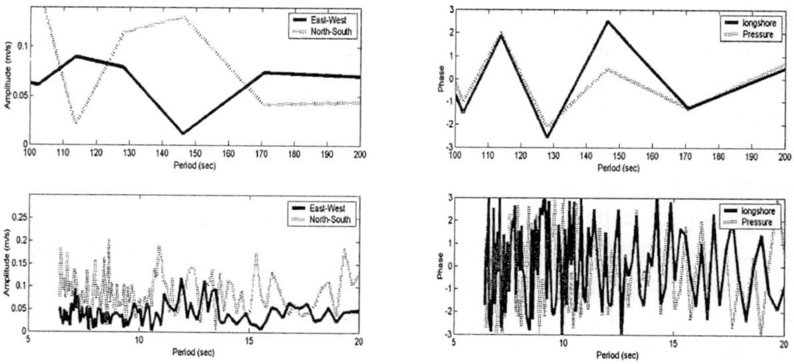

(a) (b)

Figure 4. (a) is the amplitude of the east-west and north-south flow spectra, (b) is the measured phase of the spectrum of alongshore current and pressure. The upper diagrams represent infragravity components, while the lower diagrams are short wind-wave components. These spectra are based on the 19 minute measurement that starts from 10:30, 7/18/2005 taken at station s.

From the lower diagram of Figure 4(a), the north-south flow is significantly higher than the east-west flow for wind waves. However, for infragravity waves, the east-west flow is comparable to the north-south flow. This suggests the existence of Stokes edge wave which induces the same flow speed in all directions. The locked or plane long waves should have similar directionalities as the wind wave and hence cannot explain the infragravity flow. Figure 4(b) shows that most alongshore flow components are in phase with the measured pressure.

Acknowledgments

This research was funded by the National Science Council of Taiwan, the Republic of China, under grants NSC- NSC94-2611-E-110-001. The field measurement is done by Mr. H.M. Tseng. Mr. J. J. Liu helped in analyzing the field data.

REFERENCES

Battjes, J.A., 1988. Surf zone dynamics. *Annu. Rev. Fluid Mech.* 20, 257– 293.

Eckart, C., *Surface waves on water of variable depth*, Wave Rep. 100, Scripps Inst. of Oceanogr., Univ. of Calif., La Jolla, 1951.

Elfrink, B. and Tom Baldock, Hydrodynamics and sediment transport in the swash zone: a review and perspectives, *Coastal Engineering* 45 (2002) 149–167

Foda, M. A., and C. C. Mei, Nonlinear excitation of long trapped waves by a group of short swells, *J. Fluid Mech.*, 111, 319–345, 1981.

Gallagher, B., Generation of surf beat by non-linear wave interactions, *J. Fluid Mech.*, 49, 1 –20, 1971.

Iribarren, C.R., Nogales, C., 1949. Protection des ports. Sect. 2, comm. 4, *17th Int. Nav. Cong.*, Lisbon, p. 31– 80.

Komar, P.D., 1998. *Beach Processes and Sedimentation*, Prentice Hall, New York.

Lippmann, T. C., R. A. Holman, and A. J. Bowen, Generation of edge waves in shallow water, *J. Geophys. Res.*, 102, 8663–8679, 1997.

Longuet-Higgins, M. S., and R. W. Stewart, Radiation stresses in water waves: A physical discussion, with applications, *Deep Sea Res.*, 11, 529–562, 1964.

Mei, C. C. 1983 *The Applied Dynamics of Ocean Surface Waves.* J. Wiley & Sons.

Mei, C. C., and C. Benmoussa, Long waves induced by short wave groups over an uneven bottom, *J. Fluid Mech.*, 139, 219– 235, 1984.

Munk, W. H., Surf beats, *Eos Trans.*, AGU, 30, 849– 854, 1949.

Schäffer, H. A., Edge waves forced by short-wave groups, *J. Fluid Mech.*, 259, 125–148, 1994.

Symonds, G., D. A. Huntley, and A. J. Bowen, Two-dimensional surf beat: Long wave generation by a time varying breakpoint, *J. Geophys. Res.*, 87, 492–498, 1982.

A NEW LOOK AT WAVE GROUP FORCED NEARSHORE CIRCULATION

Joseph W. Long[1] and H. Tuba Özkan-Haller[1]

A series of idealized model simulations are performed to investigate the dynamics of depth-integrated vortical motions generated by spatially and temporally varying wave groups. In particular, the effect that subsequent wave groups of varying magnitude have on an existing circulation cell is examined. The primary diagnostic tool used in this study is the time history of the enstrophy (mean-squared vorticity) computed over a control area, defined by a particular vortex. We find that variations in the enstrophy occur at time scales associated with the individual wave groups. Hence, the vortical motions appear to be a directly forced response to the alongshore variations in wave group momentum flux and do not exhibit any free very low frequency oscillations or decay as postulated in previous literature.

INTRODUCTION

Shorelines are continuously exposed to complex hydrodynamic processes that occur at a wide range of spatial and temporal scales. These processes induce a change in the beach configuration that is influenced by the incident wave conditions, beach type (i.e. steepness), sediment grain size distribution, and the subsequent positive or negative feedback imposed on the water column by the underlying changes in bathymetry. Despite the inherent randomness of the ocean waves and the wide variability in bathymetric conditions, it is not uncommon to observe rhythmic morphologic patterns in the form of beach cusps, transverse and oblique bars attached to the shoreline, crescentic sandbar systems, and periodic rip channels with adjacent shoals (e.g. Komar 1976; Sonu 1972; Lippmann and Holman 1990). Results from some field and laboratory experiments however, have shown that identifying the dominant hydrodynamic forcing responsible for the initial formation of these morphological features is far from straightforward (e.g. Holland and Holman 1996). In fact, two of the most reasonable and exploited theories differ mainly in whether these features are formed by response to a direct hydrodynamic forcing or through a "self-organization" process stimulated by hydrodynamic or morphodynamic instabilities.

Recently, a numerical modeling study by Reniers et al. (2004) hypothesized that circulation cells generated by temporally and spatially varying wave groups may exhibit a very low frequency decay, providing sufficient time for a bathymetric perturbation to develop. Fundamentally, this theory is based on the work of Ryrie (1983) which theoretically showed that a single wave group approaching the shoreline at an oblique angle could have two effects:

[1] College of Oceanic and Atmospheric Sciences, Oregon State University, 104 COAS Admin Bldg., Corvallis, OR, 97333, USA

the generation of transient infragravity motions that propagate alongshore and eventually disperse (i.e. edge waves), and a residual quasi-steady state circulation cell that remains at the wave group breaking position until it decays due to bottom friction. Extending on these studies, the excitation of an instability response could be a corollary to the perturbations caused by these quasi-steady circulation cells eventually leading to common morphologic features. These pockets of residual velocity, rather than the highly studied transient effects, have not yet been observed in field data and the details of their formation and dynamics under realistic wave group conditions remains in question, however their potential in shaping nearshore morphology is significant.

In this study we aim to further evaluate, through systematic numerical testing of the circulation system, the hypothesis that wave groups induce quasi-steady circulation patterns which decay at a timescale related only to frictional dissipation. Our work is concentrated on understanding the development of these quasi-steady circulation cells, and more importantly the rate at which they decay in the presence of subsequent wave groups. It is essential to first investigate a series of simple idealized cases in order to gain intuition into the vortex dynamics that are observed in more complex situations. This paper focuses on four simulations that test how a quasi-steady circulation cell (that develops from the first incoming wave group) changes with subsequent groups, particularly those of different magnitude. Our primary research question is whether we can expect vorticity signals generated by spatially varying wave groups to pulse on the wave group time scale or exhibit a much longer, uninterrupted frictional dissipation time variation.

MODEL DESCRIPTION

A phase-averaged wave-circulation model series is used to simulate the set of idealized conditions described in the following sections. The wave propagation model is based on the conservation of energy flux and accounts for wave rollers and depth-induced wave breaking (Thornton and Guza 1983). Wave forcing, in the form of radiation stress gradients is provided to the circulation model of Özkan-Haller and Kirby (1997) and (1999) to obtain two-dimensional horizontal velocity information in the nearshore region. The circulation model solves the continuity equation (1) and the depth-integrated time-averaged Navier Stokes equations (2 and 3) to simulate the temporally and spatially varying circulation patterns.

$$\frac{\partial \eta}{\partial t} + \frac{\partial}{\partial x}[u(h+\eta)] + \frac{\partial}{\partial y}[v(h+\eta)] = 0 \qquad (1)$$

$$\frac{\partial u}{\partial t} + u\frac{\partial u}{\partial x} + v\frac{\partial u}{\partial y} = -g\frac{\partial \eta}{\partial x} + \tilde{\tau}_x + \tau'_x - \tau_{bx} \qquad (2)$$

$$\frac{\partial v}{\partial t} + u\frac{\partial v}{\partial x} + v\frac{\partial v}{\partial y} = -g\frac{\partial \eta}{\partial y} + \tilde{\tau}_y + \tau'_y - \tau_{by}. \qquad (3)$$

The incident wave forcing provided from the wave model is represented by $\tilde{\tau}$ as given by Longuet-Higgins and Stewart (1964). For the set of governing equations (1-3), $\vec{u}$ is the depth-integrated current velocity with cross-shore and alongshore components, u and v, respectively. The still water depth is given by h, while the mean water surface elevation is represented by η. The lateral momentum mixing, τ', is parameterized using an eddy viscosity formulation where the eddy viscosity is given by $\nu_t = Md\,(\epsilon_b/\rho)^{1/3}$. The breaking wave dissipation is denoted by ϵ_b, ρ is the water density, $d = h + \eta$, and M is the lateral mixing coefficient (Battjes 1975). We use a constant bottom friction represented by, $\vec{\tau_b} = 2/(\pi d)c_f u_0\ \vec{u}$, where u_0 is the amplitude of the horizontal orbital velocity and c_f is the bottom friction coefficient. There are two free parameters in the model; the bottom friction coefficient (c_f) and the lateral mixing coefficient (M). Their specific values were not the focus of this research and were assigned constant values throughout this study with c_f = 0.01 and M = 0.25. The spatial derivatives in the solution are approximated using a finite difference approach in the alongshore direction and a Chebyshev collocation method in the cross-shore direction. Temporal derivatives are treated using a high order explicit time-integration scheme. A wall boundary was defined at the onshore boundary and wave run-up (shoreline inundation) has been neglected. Periodic boundary conditions were used at the lateral boundaries. Finally, the offshore boundary conditions consist of the time series of the incident wave groups which are spatially and temporally variable and an open boundary condition that allows seaward propagating reflected wave energy to exit the domain.

MODEL SIMULATIONS

A series of four idealized case studies, increasing in complexity, were performed during this study. For all simulations a planar beach configuration was used with a beach slope of 0.038 m/m and offshore water depth of 38 m. The model domain extends 1000m in both the cross-shore and alongshore directions with a constant alongshore grid spacing of 7.75 m and a variable cross-shore grid spacing ranging from 1.6 m to 6.4 m. A time step of 0.05 s was used to compute the time-varying circulation field and the properties of the flow field were output every 5 s. In each of the case studies a control area was defined between y = 331 m $<$ y $<$ 655 m that extends the entire length of the cross-shore domain. This region was chosen because it encompasses one individual circulation cell. As a quantitative measure, the enstrophy, defined as the mean-squared vorticity in the control area ($\zeta = \frac{1}{2}\int q^2 dA$), was computed at each output time step. The enstrophy, often used in the study of two-dimensional turbulence, provides a method of comparing the signal of the response to the wave forcing characteristics.

Impulsive Wave Group

The first idealized case study follows the work of Ryrie (1983) who showed that an alongshore short-crested impulse of momentum in the surf zone would result in two motions: transient motions such as edge waves, and quasi-steady

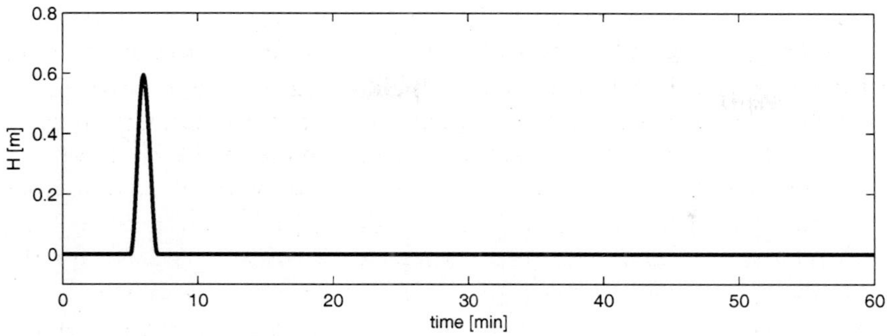

Figure. 1. **Time series of offshore wave height at y = 340 m.**

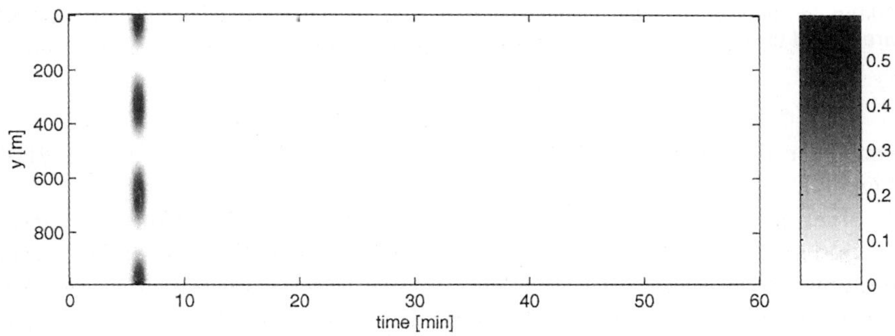

Figure. 2. **Time stack of wave height, in meters, at the offshore boundary.**

state motions such as a circulation cell that would eventually decay due to bottom friction. This slow decay of the quasi-steady circulation cell would then appear in a frequency spectrum in one of the lowermost frequency bins. While this case is admittedly unrealistic it serves as an important basis in the evaluation of the subsequent cases that investigate the role that additional wave groups have on these quasi-steady circulation patterns. A time series of the offshore wave height at a single point on the boundary is shown in Figure 1 and the alongshore variations of the wave group are illustrated by the time stack of wave height at the offshore boundary in Figure 2. The period of the wave group impulse is 2 minutes which encompasses 12 carrier waves with a period of 10 seconds.

The circulation patterns that develop in this scenario consist of three cells that are located in the alongshore minima of the wave groups. The circulation cells are set up initially by the impulsive wave group and then slowly dissipate because they are influenced by only bottom friction. A decay of the circulation cell can be observed in the snapshots of the vorticity taken at five time intervals shown in Figure 3. The control area previously described is outlined by the solid black lines

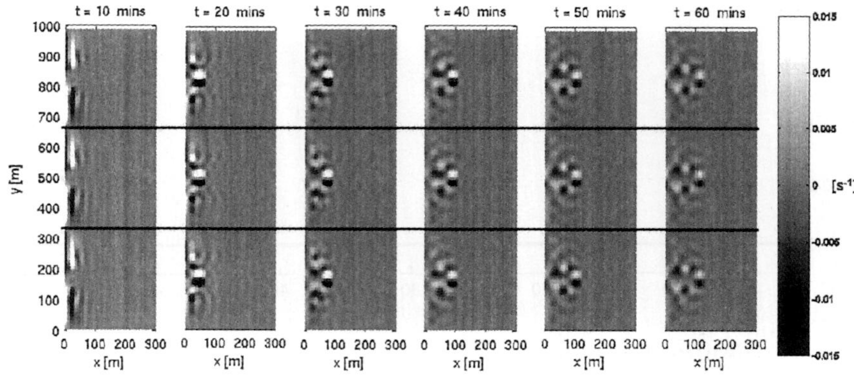

Figure. 3. Snapshots of vorticity at 10-minute time intervals. From left to right: t = 10min, 20min, 30min, 40min, 50min, 60min. The solid black lines denote the control area used for enstrophy calculations (see Figure 4).

and the enstrophy computed during the simulation is given in the middle panel of Figure 4.

The enstrophy signal consists of a slow exponential decay due to bottom friction superimposed with high frequency oscillations. The slow decay has an e-folding decay time of approximately 350 seconds and investigation revealed that the high frequency oscillations were due to edge waves that were present in the domain. This was confirmed by comparing a frequency spectrum of the high-passed enstrophy signal (Figure 4, bottom panel) and a frequency analysis of the shoreline position at a single point which resulted in identical peak frequencies. It is currently unclear why the edge waves contribute to the vorticity signal, however additional research is being conducted and we postulate it may be due to nonlinear effects. Using the method described in this case, three additional case studies were performed to examine the effect of subsequent wave groups of varying magnitude.

Sawtooth Wave Group

One of the questions surrounding the hypothesis of Reniers et al. (2004) is what effect subsequent smaller wave groups would have on a circulation cell formed by a previous larger wave group. In this case we chose a linearly decaying sawtooth function to represent the longer period amplitude variations of the individual wave groups. Similar to the previous example, a time series of the wave height just before breaking is shown in the top panel of Figure 5. The individual wave groups again have a period of 2 minutes, however we have now included a longer period amplitude modulation with a period of 10 minutes. Hence, each sawtooth has a period of 10 minutes and includes five separate wave group impulses of different magnitude. Choosing the same control area, the enstrophy

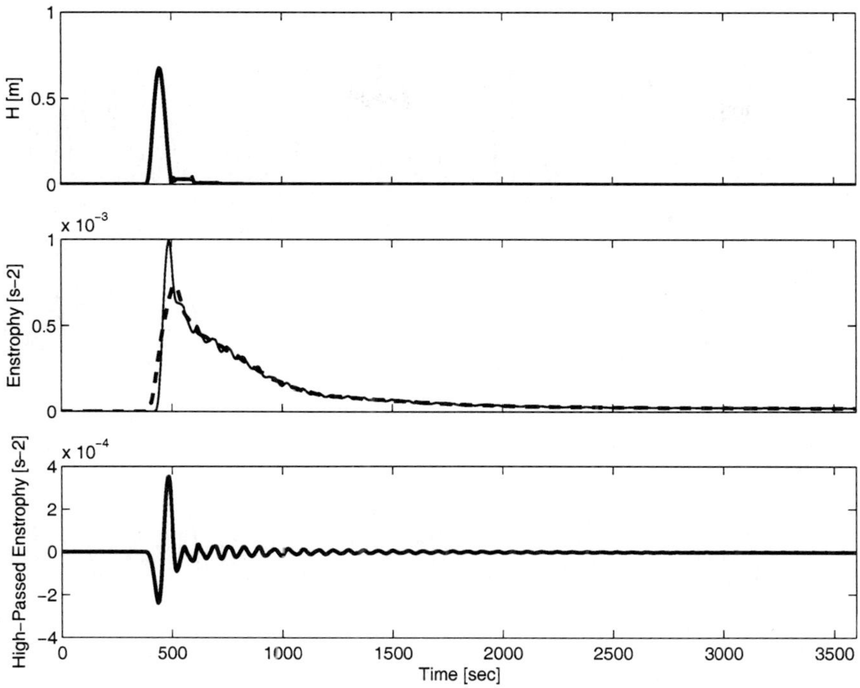

Figure. 4. Top: Time series of offshore wave height at y = 340 m. Middle: Enstrophy (mean squared vorticity) time series computed over the control area (solid line). Low-passed (f $<$ 0.008) enstrophy time series (dashed line). Bottom: High passed enstrophy signal.

(middle panel) contains oscillations at two primary frequencies including both the individual wave groups (f = 0.00833) and the longer period amplitude modulation of the wave group (f = 0.00167) which is represented by a sawtooth function in this case. There is also energy at additional higher frequencies (bottom panel) that again correspond to edge wave motions present in the domain.

Exponentially Decaying Sawtooth Wave Group

In the previous case the change in amplitude between two wave groups was only significant between the tail and face of the sawtooth function. To create wave group amplitude variations that were more defined we used an exponentially decaying sawtooth function as shown by the wave height just outside of the surfzone in the top panel of Figure 6. Similar to the enstrophy signal in the traditional sawtooth function (Figure 5), we also observe high frequency energy in the enstrophy response (Figure 6, middle and bottom panels). This energy is consistent with the wave group time scale (T = 2 min) as well as edge wave contributions. Nonetheless, the most pronounced feature is that associated with

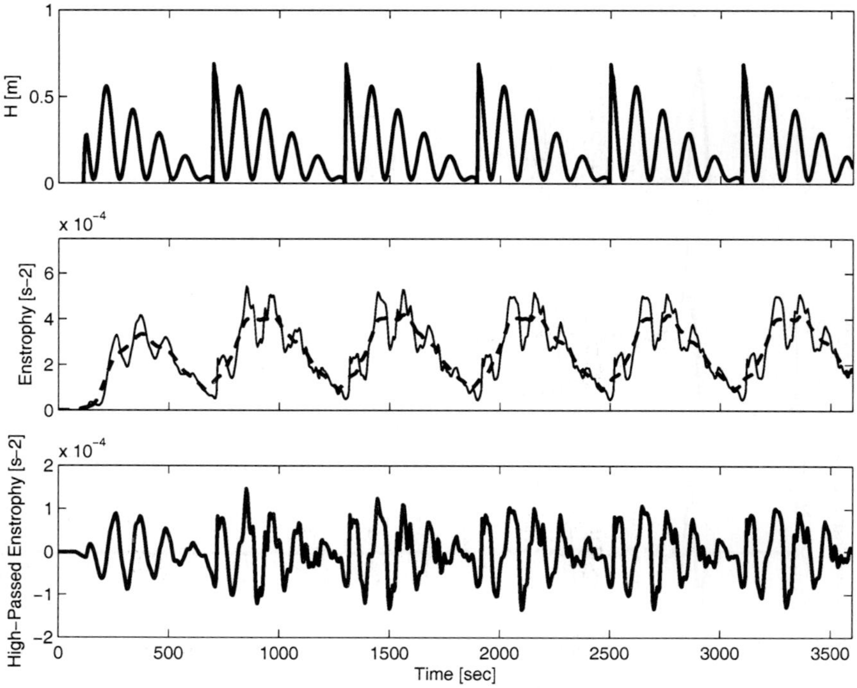

Figure. 5. Top: Time series of offshore wave height at y = 340 m. Middle: Enstrophy (mean squared vorticity) time series computed over the control area (solid line). Low-passed ($f < 0.008$) enstrophy time series (dashed line). Bottom: High passed enstrophy signal.

the longer period amplitude variations of wave signal (T = 10 min).

Sinusoidal Wave Group

The final case consists of what may be viewed as more realistic, sinusoidal wave group amplitude modulations. It is comprised of three pairs of wave components that have the same frequency and equal but opposite incident angles. This is an extension of the work of Dalrymple (1975) that showed stationary rip currents generated by the summation of two incident wave components of identical frequency but equal and opposite angles of incidence. The time series of wave group amplitude prior to breaking and the corresponding enstrophy are shown in Figure 7 for this case. The results mirror the conclusions discussed in the previous cases with an enstrophy response that appears directly forced by the individual wave groups and the lower frequency wave group amplitude modulations. The edge wave contribution to the enstrophy signal is again evident.

CONCLUSIONS

Idealized case studies were used to evaluate the circulation patterns that

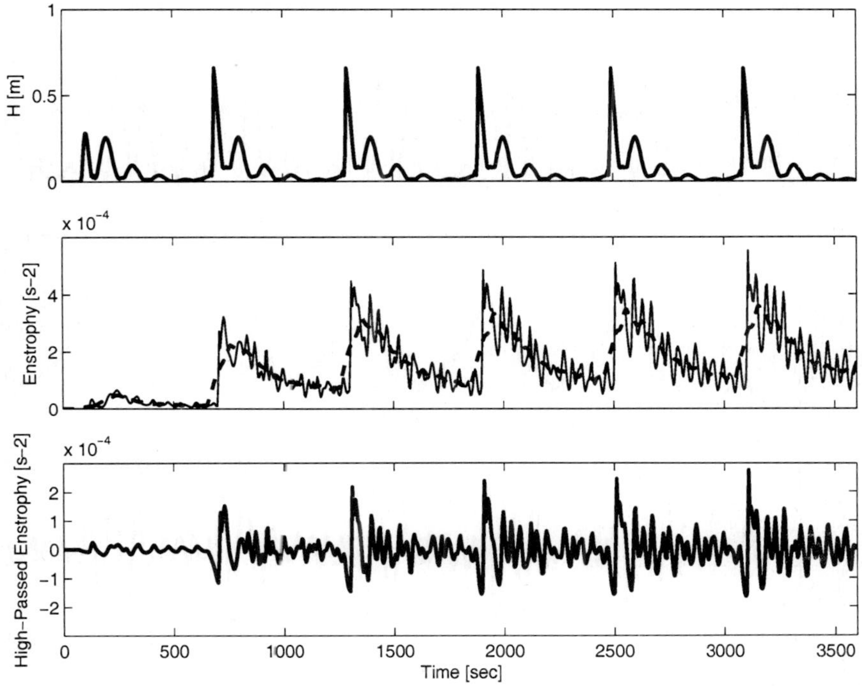

Figure. 6. **Top: Time series of offshore wave height at y = 340 m. Middle: Enstrophy (mean squared vorticity) time series computed over the control area (solid line). Low-passed (f < 0.008) enstrophy time series (dashed line). Bottom: High passed enstrophy signal.**

arise from spatially varying wave groups. The primary effort was to investigate whether vortical structures exposed to repeated forcing can be expected to exhibit a very low frequency signal associated with frictional dissipation. Results from the four cases shown are consistent with a directly forced system where the circulation responds at the individual wave group frequency (f = 0.00833 Hz) and the frequency of any longer period wave group amplitude modulations (f = 0.00167 Hz). It is noted that these amplitude modulations would not appear in a Fourier analysis of the offshore wave forcing, therefore comparing frequency-wavenumber spectra from the offshore wave groups and the nearshore current velocities would show a very low frequency signal in the response not observed in the forcing. This is similar to a Fourier analysis of the water surface elevation where modulations of wave groups would not appear. In each of the cases involving subsequent wave groups, no free very low frequency decay was observed. Rather, the enstrophy signal represents a circulation response directly forced by the varying wave groups.

The aim of this study was to begin evaluating nearshore vortex dynamics

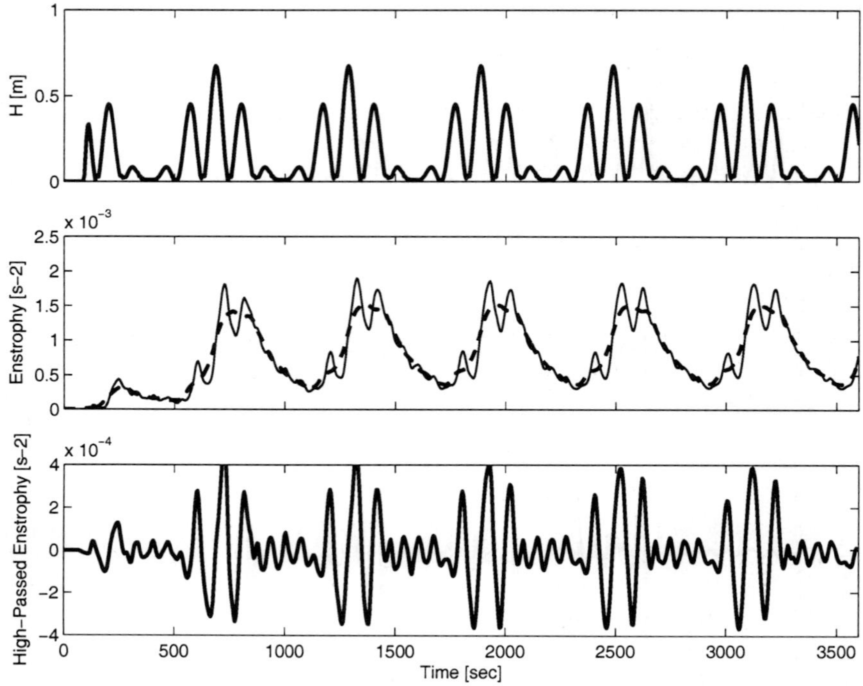

Figure. 7. **Top: Time series of offshore wave height at y = 340 m. Middle: Enstrophy (mean squared vorticity) time series computed over the control area (solid line). Low-passed ($f < 0.008$) enstrophy time series (dashed line). Bottom: High passed enstrophy signal.**

at a fundamental base. Armed with these results we can build some intuition of what may be expected to occur in more complex situations. This is on-going research and, using similar methodology, we have begun evaluating the mobility and strength of vortices forced by spatially and temporally varying wave groups generated from a realistic frequency-direction energy density spectrum. Preliminary results, on planar and barred beach profiles confirm the expectations derived from this set of idealized cases.

ACKNOWLEDGMENTS

The authors would like to thank Rob Holman for the insightful discussions throughout this continuing research. This research is sponsored by the Office of Naval Research, Coastal Studies Program under Grant N00014-02-1-0198.

REFERENCES

Bowen, A. J. and R. A. Holman. 1989. Shear instabilities of the mean longshore current 1. theory, *Journal of Geophysical Research,* 94, 18,023–18,030.

Dalrymple, R. A.. 1975. A mechanism for rip current generation on the open coast, *Journal of Geophysical Research,* 24, 3485–3487.

Dodd, N. J., J. Oltman-Shay, and E. B. Thornton. 1992. Shear instabilities in the longshore current: A comparison of observations and theory, *Journal of Physical Oceanography,* 22, 62–82.

Fowler, R. E. and R. A. Dalrymple. 1990. Wave group forced nearshore circulation, *Proceedings of 14th International Conference on Coastal Engineering*, ASCE, 729–742.

Holland, K. T. and R. A. Holman. 1996. Field observations of beach cusps and swash motions, *Marine Geology,* 134, 77–93.

Komar, P. D.. 1998. *Beach Processes and Sedimentation* (2nd Ed. ed.),, Prentice-Hall, New Jersey, 336-350 pp.

Lippmann, T. C. and R. A. Holman. 1990. The spatial and temporal variability of sand bar morphology, *Journal of Geophysical Research,* 95(C7), 11575–11590.

Oltman-Shay, J., P. A. Howd, and W. A. Birkemeier. 1989. Shear instabilities of the mean longshore current 2. field observations, *Journal of Geophysical Research,* 94, 18,031–18,042.

Özkan Haller, H. T. and J. T. Kirby. 1997. A fourier-chebyshev collocation method for the shallow water equations including shoreline runup, *Applied Ocean Research,* 19, 21–34.

Özkan Haller, H. T. and J. T. Kirby. 1999. Nonlinear evolution of shear instabilities of the longshore current: A comparison of observations and computations, *Journal of Geophysical Research,* 104, 25,953–25,984.

Reniers, A. J. H. M., J. A. Roelvink, and E. B. Thornton. 2004. Morphodynamic modeling of an embayed beach under wave group forcing, *Journal of Geophysical Research,* 109(C01030), 21–34. doi:10.1029/2002JC001586.

Ryrie, S. C.. 1983. Longshore motion due to obliquely incident wave group, *Journal of Fluid Mechanics,* 137, 273–284.

Sonu, C. J.. 1972. Field observation of nearshore circulation and meandering currents, *Journal of Geophysical Research,* 77, 3232–3247.

Tang, E. C.-S. and R. A. Dalrymple. 1989. Rip currents and wave groups, R. J. Seymour (Ed.), *Nearshore Sediment Transport Study*, 205–230. Philadelphia, Pa., Plenum Press.

Thornton, E. B. and R. T. Guza. 1983. Transformation of wave height distribution, *Journal of Geophysical Research,* 88(C10), 5925–5938.

Quasi-3D Nearshore Circulation Equations: a CL-Vortex Force Formulation

Fengyan Shi[1], James T. Kirby[1] and Kevin Haas[2]

We formulate a CL-vortex form of surface wave force for a quasi-3D nearshore circulation model. The CL-vortex force formulation is obtained by applying surface wave equations to depth-integrated and wave-averaged momentum equations. A new splitting algorithm of current velocity is used to facilitate the application of wave equations to the wave-averaged equations. The derivation shows that the CL-vortex term arises from both the wave refraction by current shear and the wave-current interaction shown in the radiation stress type momentum equations. In the vertical direction, the equation governing the vertical structure of current velocity is driven by a surface stress related to wave dissipation. Numerical tests show the CL-vortex formulation performs identically with the radiation stress formulation in modeling of rip currents. However, without fully coupling of wave and current models, the CL-vortex formulation gives a more reasonable result than the radiation stress formulation does.

INTRODUCTION

Recently much attention has been paid to different formulations of surface wave force in wave-driven ocean and coastal circulations (e.g., McWilliams et al., 2004, Mellor, 2003, Smith, 2006 and others). Basically, the analytical expressions for surface wave force and wave-current interaction can be classified into two types. One is the classical wave 'radiation stress' concept presented by Longuet-Higgins and Stewart (1962, 1964) and many others in depth-integrated and short wave-averaged equations. Mellor (2003) recently used the same concept to derive short wave-averaged 3-D equations with a depth-dependent wave-induced force. A direct application of this kind of depth-dependent wave-induced force was conducted by Xia et al. (2004) who related the vertical variation of current to the vertical structure of radiation stresses.

Another type of wave force is the surface wave force initially derived by Garrett (1976) in the study of Langmuir circulation generation. The wave driving forces include the wave dissipation term and the wave-averaged vortex forcing term which has been identified later by Leibovich (1980) and Smith (1980) as the vertically integrated form of the 'CL vortex-force' derived by Craik and Leibovich (1976). Dingemans et al. (1987) also presented a similar formulation of this type of wave driving force, though the current refraction, that may result in the vortex-force term, was recognized to give insignificant contributions under the conditions of slowly varying wave fields. Smith (2006) extended the formulation of Garrett (1976) to include finite-depth effects and provided some insight into physical interpretation of each forcing term in depth-integrated equations. A

[1] Center for Applied Coastal Research, University of Delaware, Newark, DE 19716, USA
[2] Department of Civil and Environmental Engineering, Georgia Tech, Savannah, GA 31407, USA

depth-dependent wave force of this type can be found in McWilliams et al. (2004) who showed a series of equations in different time scales for surface waves, infra-gravity waves and low-frequency currents in a coupled system.

Most recently, Newberger and Allen (2006) applied a similar form of the CL-vortex formulation to forcing a three-dimensional hydrostatic primitive-equation model in the surf zone. Their analysis was focused on shallow water dynamics involving interactions of linear waves with wave-averaged mean currents. The short wave forcing in their approach consists of a surface stress and a body force. The surface stress is proportional to the wave energy dissipation which is basically caused by wave breaking in the surf zone. The body force arises from the so-called local radiation stresses. Under an assumption of shallow water currents with linear waves, the body force includes one term that is related to the vortex force and a second term that is related to gradients of part of the radiation stress tensor. The vortex force, which is a product of the mean wave momentum and the vertical component of the depth-averaged mean vorticity vector, is similar to the "refraction force" in Smith (2006) except that Smith's refraction force is evaluated using the mean vorticity at the mean surface. The body force was uniformly applied to the vertical water column in their three-dimensional model.

Although the consistency in the two types of theoretical formulations can be found without difficulties, the numerical models based the different wave forcing formulations may perform differently. Dingemens et al. (1987) pointed out that the formulation in terms of the wave dissipation yields more trustworthy results as the radiation stress tensor can be a rapidly varying function of the spatial coordinates, numerical differentiation can lead to poor results.

Among two- or three-dimensional models of wave-induced nearshore circulation, a quasi-3D nearshore circulation model developed by Svendsen et al. (2004) is a simple approach to 3-D modeling of wave-induced nearshore circulation. The quasi-3D equations reveal three-dimensional dispersion of momentum in wave-induced nearshore currents, wave-current interaction, and the contribution of short wave forcing to a solution of vertical current profile. The objective of this study is to describe the CL-vortex type short wave force formulation in the quasi-3D model frame.

In the present paper, we re-formulate the quasi-3D nearshore circulation equations (SHORECIRC equations, Putrevu and Svendsen, 1999) using a new splitting algorithm of current velocity proposed by Haas et al. (2003). A new type of wave forcing is derived for both the depth-integrated and wave-averaged momentum equations and the equation governing the vertical structure of current velocity. Numerical consistency between two different wave force formulations is discussed in idealized rip current simulations.

DERIVATION

Split of wave-averaged current velocity

Slightly different from the splitting method used by Putrevu and Svendsen (1999), the short wave-averaged current velocity $V_\alpha(z)$ is split into a 'undisturbed' depth-averaged mean current $V_{m\alpha}$ and vertical variation of mean current $V_{d\alpha}(z)$:

$$V_\alpha(z) = V_{m\alpha} + V_{d\alpha}(z) \tag{1}$$

where

$$V_{m\alpha} = \frac{1}{h}\int_{-h_0}^{\bar{\zeta}} \bar{u}_\alpha dz \tag{2}$$

or

$$V_{m\alpha} = \frac{1}{h}\left(\overline{\int_{-h_0}^{\zeta} u_\alpha dz} - \overline{\int_{\bar{\zeta}}^{\zeta} u_\alpha dz}\right) \tag{3}$$

and

$$\int_{-h_0}^{\bar{\zeta}} V_{d\alpha} dz = 0 \tag{4}$$

u_α represents the instantaneous horizontal velocity including the wave component. The first term on the right of (3) is the total volume flux and the second term is the net wave volume flux:

$$Q_{w\alpha} = \overline{\int_{\bar{\zeta}}^{\zeta} u_\alpha dz} \tag{5}$$

Notice that Putrevu and Svendsen (1999) used the total volume flux as a main variable in their split. This splitting method was also used by Haas et al. (2003) who pointed out that this split is more physical and has advantages in the simplification of the 3D dispersive mixing terms.

Equations of depth-integrated and wave-averaged current

Following the derivation of Putrevu and Svendsen (1999), the depth-integrated, short-wave-averaged momentum equations read

$$\frac{\partial \bar{\zeta}}{\partial t} + \frac{\partial}{\partial x_\alpha}(V_{m\alpha}h + Q_{w\alpha}) = 0 \tag{6}$$

and

$$\begin{aligned} &\frac{\partial}{\partial t}(V_{m\alpha}h) + \frac{\partial Q_{w\alpha}}{\partial t} + \frac{\partial}{\partial x_\beta}[V_{m\alpha}V_{m\beta}h + Q_{w\alpha}V_{m\beta} + V_{m\alpha}Q_{w\beta} \\ &+ \overline{\int_{-h_0}^{\zeta} V_{d\alpha}V_{d\beta} dz} + \overline{\int_{\zeta_t}^{\zeta}(u_{w\alpha}V_{d\beta} + V_{d\alpha}u_{w\beta})dz}\ ] \\ &+ \frac{1}{\rho}\frac{\partial T_{\alpha\beta}}{\partial x_\beta} + gh\frac{\partial \bar{\zeta}}{\partial x_\alpha} - \frac{\tau_\alpha^s}{\rho} + \frac{\tau_\alpha^B}{\rho} + \frac{1}{\rho}\frac{\partial S_{\alpha\beta}}{\partial x_\beta} = 0 \end{aligned} \tag{7}$$

where τ_α^s and τ_α^B represent the surface stress and bottom stress, respectively. In (7) the radiation stress is defined by

$$S_{\alpha\beta} = \overline{\int_{-h_0}^{\zeta} (\rho u_{w\alpha}u_{w\beta} + p\delta_{\alpha\beta})dz} - \frac{1}{2}\rho g h^2 \delta_{\alpha\beta} \tag{8}$$

where p is the total pressure and $\delta_{\alpha\beta}$ is the Kronecker delta function. Similarly to Putrevu and Svendsen (1999), we assume that $V_{d\alpha}$ is approximately constant in the interval $\bar{\zeta}$ to ζ and thus (4) may be simplified as follows

$$\overline{\int_{-h_0}^{\zeta} V_{d\alpha} dz} \approx \int_{-h_0}^{\bar{\zeta}} V_{d\alpha} dz = 0 \tag{9}$$

With the same approximation, the integrals in (7) may be written as

$$\overline{\int_{-h_0}^{\zeta} V_{d\alpha} V_{d\beta} dz} + \overline{\int_{\zeta_t}^{\zeta} (u_{w\alpha} V_{d\beta} - V_{d\alpha} u_{w\beta}) dz} \approx \int_{-h_0}^{\bar{\zeta}} V_{d\alpha} V_{d\beta} dz + V_{d\beta}(\bar{\zeta}) Q_{w\alpha} + V_{d\alpha}(\bar{\zeta}) Q_{w\beta} \tag{10}$$

Using (10), (7) is reorganized as

$$\begin{aligned} &\frac{\partial}{\partial t}(V_{m\alpha} h) + \frac{\partial}{\partial x_\beta}\left[V_{m\alpha} V_{m\beta} h + \int_{-h_0}^{\bar{\zeta}} V_{d\alpha} V_{d\beta} dz\right] \\ &+ \frac{1}{\rho}\frac{\partial T_{\alpha\beta}}{\partial x_\beta} + gh\frac{\partial \bar{\zeta}}{\partial x_\alpha} - \frac{\tau_\alpha^s}{\rho} + \frac{\tau_\alpha^B}{\rho} \\ &+ \frac{\partial Q_{w\alpha}}{\partial t} + \frac{\partial}{\partial x_\beta}\left[\left(V_{m\beta} + V_{d\beta}(\bar{\zeta})\right) Q_{w\alpha} + \left(V_{m\alpha} + V_{d\alpha}(\bar{\zeta})\right) Q_{w\beta}\right] \\ &+ \frac{1}{\rho}\frac{\partial S_{\alpha\beta}}{\partial x_\beta} = 0 \end{aligned} \tag{11}$$

Notice that the wave-current interaction terms in the third line of (11) include the mean current value at the mean surface, i.e., $(V_{m\alpha} + V_{d\alpha}(\bar{\zeta}))$ or $V_\alpha(\bar{\zeta})$, rather than the deviation of the mean current at the surface, as in the equations of Putrevu and Svendsen (1999).

To connect (11) to wave evolution equations, the radiation stresses may be expressed using the form evaluated by Longuet-Higgins and Stewart (1962, 1964), i.e.,

$$S_{\alpha\beta} = \rho Q_{w\alpha} C_\beta^g + \rho h J \delta_{\alpha\beta} = E^w\left(\frac{k_\alpha C_\beta^g}{\sigma}\right) + \rho h J \delta_{\alpha\beta} \tag{12}$$

where E^w is the wave energy, k_α presents the wave number, C_α^g the wave group velocity, σ the intrinsic radian frequency and

$$\rho h J = \frac{1}{2}\rho h(\bar{u_w^2} - \bar{w^2}) = \rho Q_{w\alpha}(C_\alpha^g - \frac{1}{2}C_\alpha) = E^w\left(k\frac{C^g}{\sigma} - \frac{1}{2}\right) \tag{13}$$

Note that the definition of J is slightly different from that in Smith (2006) by a factor of ρ.

Following Smith (2006), we employ the conservation of wave action and wave crests, i.e.,

$$\frac{\partial A}{\partial t} + \frac{\partial}{\partial x_\alpha}\left(A(C_\alpha^g + V_\alpha(\bar{\zeta}))\right) = -D_w \tag{14}$$

and

$$\frac{\partial k_\alpha}{\partial t} + \frac{\partial}{\partial x_\alpha}\left(\sigma + k_\beta V_\beta(\bar{\zeta})\right) = 0 \tag{15}$$

to get a wave momentum evolution equation:

$$\begin{aligned}\frac{\partial Q_{w\alpha}}{\partial t} &+ \frac{\partial}{\partial x_\beta}\left[Q_{w\alpha}(C^g_\beta + V_\beta(\bar{\zeta}))\right] = -\frac{1}{\rho}k_\alpha(D_w) \\ &-Q_{w\beta}\frac{\partial V_\beta(\bar{\zeta})}{\partial x_\alpha} - J\frac{\partial h}{\partial x_\alpha}\end{aligned} \tag{16}$$

where D_w represents wave dissipation caused by wave breaking. In derivation of (16), the relation between Q_w and wave action A, i.e.,

$$Q_{w\alpha} = Ak_\alpha/\rho \tag{17}$$

is used. Using (16) to replace the wave-induced terms shown in the third and fourth lines of (11), we get the mean current equations:

$$\begin{aligned}&\frac{\partial}{\partial t}(V_{m\alpha}h) + \frac{\partial}{\partial x_\beta}\left[V_{m\alpha}V_{m\beta}h + \int_{-h_0}^{\bar{\zeta}} V_{d\alpha}V_{d\beta}dz\right] \\ &+\frac{1}{\rho}\frac{\partial T_{\alpha\beta}}{\partial x_\beta} + gh\frac{\partial\bar{\zeta}}{\partial x_\alpha} - \frac{\tau^s_\alpha}{\rho} + \frac{\tau^B_\alpha}{\rho} - \frac{F_{w\alpha}}{\rho} = 0\end{aligned} \tag{18}$$

where $F_{w\alpha}$ is the new form of wave force:

$$\begin{aligned}F_{w\alpha} &= k_\alpha D_w + \rho Q_{w\beta}\left(\frac{\partial V_\beta(\bar{\zeta})}{\partial x_\alpha} - \frac{\partial V_\alpha(\bar{\zeta})}{\partial x_\beta}\right) \\ &-\rho V_\alpha(\bar{\zeta})\frac{\partial Q_{w\beta}}{\partial x_\beta} - \rho h\frac{\partial J}{\partial x_\alpha}\end{aligned} \tag{19}$$

Or its vector form

$$\begin{aligned}\vec{F}_w &= \vec{k}D_w + \rho\vec{Q}_w \times (\nabla \times \vec{V}(\bar{\zeta})) \\ &-\rho\vec{V}_\alpha(\bar{\zeta})(\nabla \cdot \vec{Q}_w) - \rho h\nabla(J)\end{aligned} \tag{20}$$

It can be seen from the derivation that the vortex term $\vec{Q}_w \times (\nabla \times \vec{V}(\bar{\zeta}))$ arises from both the wave-current interaction term in (11) and wave refraction by current shear in (16). The equivalent wave forcing formulation in the radiation stress form is

$$\vec{F}_w = -\rho\frac{\partial\vec{Q}_w}{\partial t} - \rho\nabla\cdot\left(\vec{V}(\bar{\zeta})\vec{Q}_w + \vec{Q}_w\vec{V}(\bar{\zeta})\right) - \nabla\cdot\vec{S} \tag{21}$$

Without including the wave refraction by current shear (generally shown in wave equations), the radiation stress formulation (21) would not explicitly show the vortex term.

Equation for vertical variation of mean current

Using the same strategy as in Putrevu and Svendsen (1999), we derive the equation governing the vertical structure of $V_{d\alpha}$. By subtracting the mean current momentum from the 3D wave-averaged equations, the lowest order of the equation for vertical variation of mean current can be written as

$$\frac{\partial V_{d\alpha}}{\partial t} - \frac{\partial}{\partial z}(\nu \frac{\partial V_{d\alpha}}{\partial z}) = -\frac{1}{\rho h}F_{w\alpha} - \frac{1}{\rho}f_\alpha - \frac{1}{h}V_{m\alpha}\frac{\partial Q_{w\beta}}{\partial x_\beta} - \frac{1}{\rho h}(\tau_\alpha^s - \tau_\alpha^B) \quad (22)$$

where f_α is local radiation stress defined by

$$f_\alpha = \rho \overline{\frac{\partial}{\partial x_\beta}(u_{w\alpha}u_{w\beta})} + \rho \overline{\frac{\partial w_w u_{w\alpha}}{\partial z}} - \rho \overline{\frac{\partial w_w^2}{\partial x_\alpha}} \quad (23)$$

In shallow water, f_α may be evaluated as (Newberger and Allen, 2006)

$$f_\alpha = -\frac{\rho}{h}Q_{w\beta}\left(\frac{\partial V_\beta(\bar{\zeta})}{\partial x_\alpha} - \frac{\partial V_\alpha(\bar{\zeta})}{\partial x_\beta}\right) + \rho\frac{\partial}{\partial x_\alpha}\left(\frac{J}{h}\right) \quad (24)$$

Applying (24) to (22) yields the equation governing $V_{d\alpha}$ in shallow water:

$$\frac{\partial V_{d\alpha}}{\partial t} - \frac{\partial}{\partial z}(\nu \frac{V_{d\alpha}}{\partial z}) = \frac{1}{\rho h}(-k_\alpha D_w - \tau_\alpha^s + \tau_\alpha^B) \quad (25)$$

(25) can be solved analytically or numerically by applying the surface and bottom boundary conditions given by Putrevu and Svendsen (1999) and the condition (4). It should be mentioned that only leading terms are retained in the derivation of (22) in order to maintain consistency with the SHORECIRC equations.

NUMERICAL TEST

To make a simple test on the numerical consistency between two different wave force formulations, we implemented (20) in the 2D mode of SHORECIRC equations (Svendsen et al., 2004). The lateral mixing caused by vertical variation of current velocity was not taken into account by neglecting (25) and switching off the 3D dispersion terms in (18). An unsteady wave-driver developed by Kennedy and Kirby (2004) was used to provide the circulation model with non-stationary wave forcing. The model coupling system which can switch to two different wave force formulations was used in rip current simulations.

We used an idealized bathymetry (Yu and Slinn, 2003) as shown in Figure 1. A normally incident wave, with a wave period of 10s and a wave height of 1.8 m, was simulated in the wave model. Wave-current interaction can be modeled by coupling the wave model and the circulation model each time step. Similar to the results shown in Yu and Slinn (2003), our results indicate that wave-current interaction is important in the rip current system. The interaction reduces the strength of the currents and restricts their offshore extent. When the wave

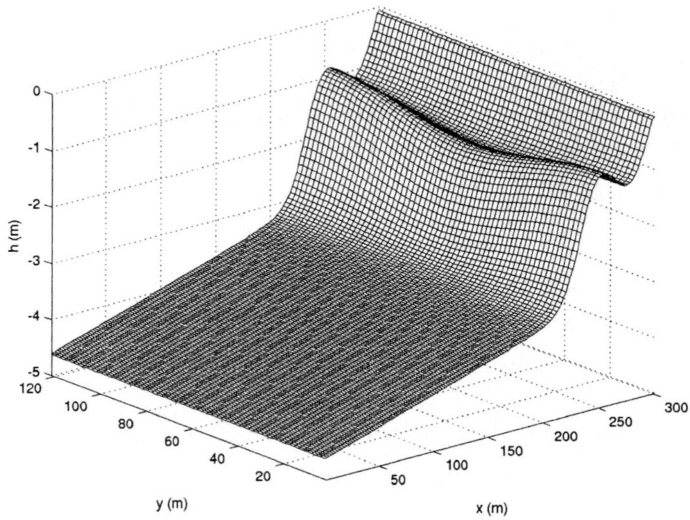

Figure 1. Idealized bathymetry in Yu and Slinn (2003)

model and the circulation model are fully coupled, the two different wave force formulations basically give very similar results as shown in Figure 2. Figure 3 shows the comparison of time-averaged rip current profiles between the radiation stress formulation (dashed line) and the CL-vortex formulation. The term-by-term comparisons of the two wave force formulations also indicate that the two different formulations are basically equivalent in the rip current simulations. In x-direction, the dominant term in (21) is $-\nabla \cdot \vec{S}$ which is shown in Figure 4 (a). The dominant terms in (20) are $\frac{1}{\rho}\vec{k}D_w$ and $-h\nabla(J)$. The sum total of the two terms is shown in Figure 4(b) which is similar to that in Figure 4 (a). In y-direction, the dominant terms in (21) are $-\rho\nabla \cdot \left(\vec{V}(\bar{\zeta})\vec{Q}_w + \vec{Q}_w\vec{V}(\bar{\zeta})\right)$ and $-\nabla \cdot \vec{S}$. The sum total of the two terms is shown in Figure 4 (c). The dominant term in (20) is $\rho\vec{Q}_w \times (\nabla \times \vec{V}(\bar{\zeta}))$ and is shown in Figure 4 (d). The comparison between Figures 4 (c) and 4 (d) clearly shows that the effect of the vortex type force seems to occur in the integration of the two dominant terms in (21), that is similar to the CL-vortex force in (20).

Without wave-current interaction, however, the two wave force formulations perform very differently in the rip current simulations. The effect of vortex type force does not show up in the radiation stress formulation (21). Figure 5 (a) shows the sum value of the two dominant terms in (21) in y-direction. It differs from the value calculated from the CL-vortex force and shown in Figure 5 (b). Figure 6 shows a comparison between the time-averaged rip current profiles from the

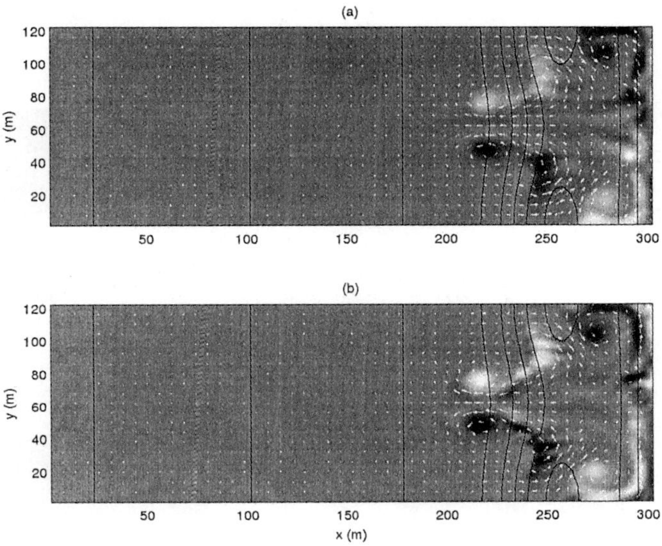

Figure 2. Snapshot of current and vorticity (color) field: (a)radiation stress formulation (b) CL-vortex formulation.

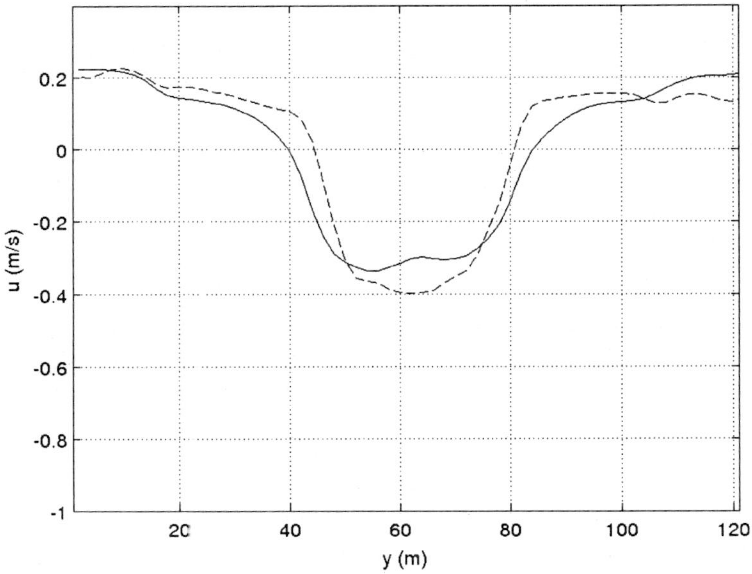

Figure 3. Comparison of time-averaged rip current profiles at x = 220 m with wave-current interaction (dashed line: radiation stress formulation, solid line: CL-vortex formulation).

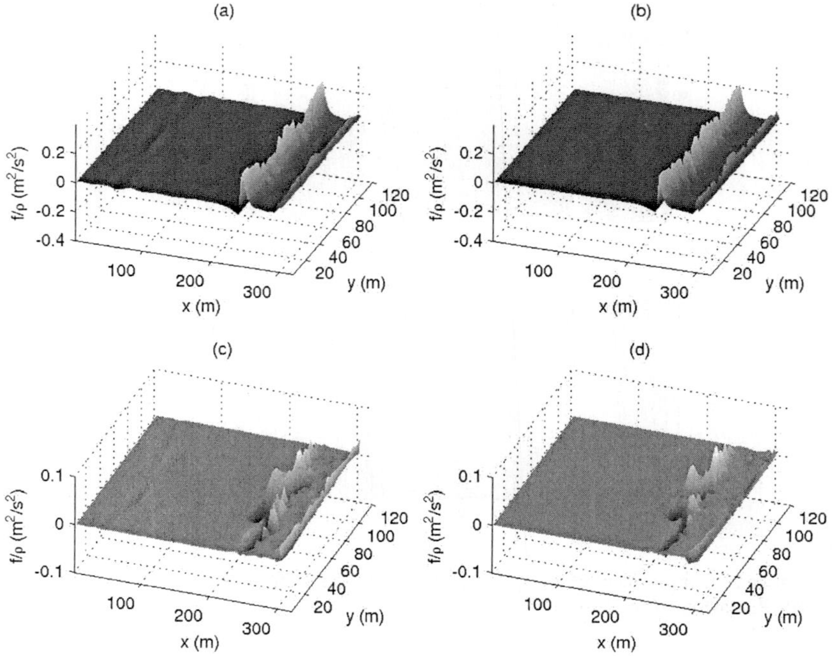

Figure 4. Comparisons of wave forcing when wave and current are fully coupled (a) radiation stress formulation in x direction (b) CL-vortex formulation in x direction (c) radiation stress formulation in y direction (d) CL-vortex formulation in y direction

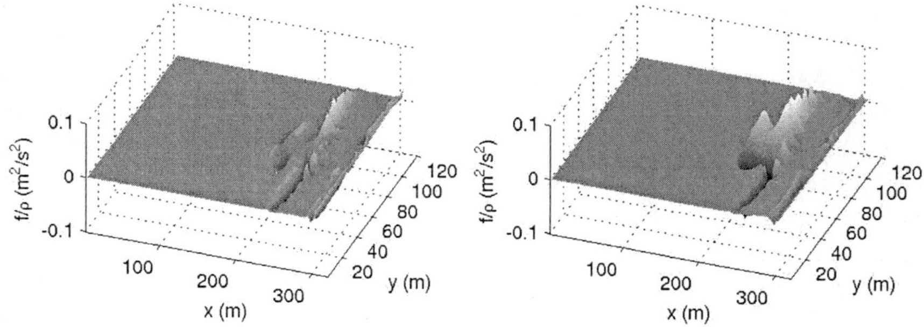

Figure 5. Comparisons of wave forcing when wave and current are not coupled (a) radiation stress formulation in y direction (b) CL-vortex formulation in y direction

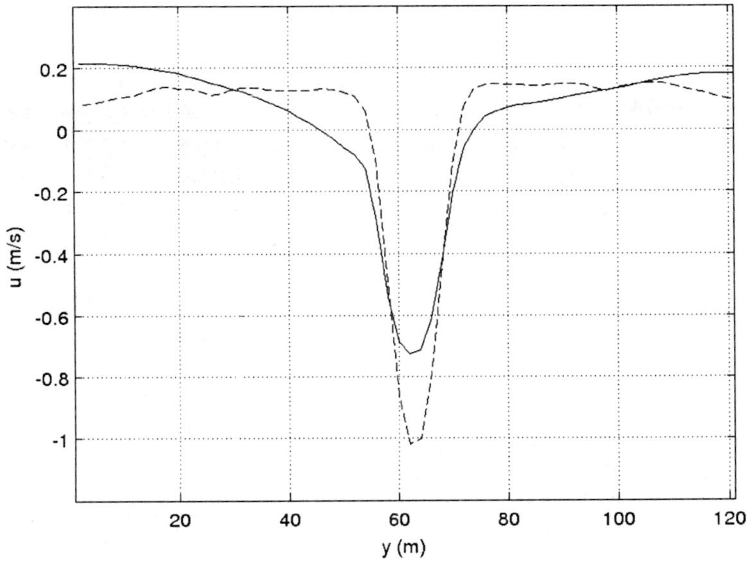

Figure 6. Comparison of time-averaged rip current profiles at x = 220 m without wave-current interaction (dashed line: radiation stress formulation, solid line: CL-vortex formulation)

radiation stress formulation (dashed line) and the CL-vortex formulation. The CL-vortex formulation predicts a weaker rip current than the radiation stress formulation because the CL-vortex forcing plays an role in widening the rip current neck and thus weakening the strength of the rip current.

CONCLUSIONS

In this study, we described a CL-vortex wave force for a quasi-3D nearshore circulation model. A new splitting algorithm of current velocity presented in Haas et al. (2003) is applied to keep aligned with Smith's (2006) derivation of the CL-vortex force in general application of wave-induced circulations. Our derivation shows that the wave force driving the depth-integrated and wave-averaged momentum equations is the same as in Smith's derivation with finite-depth effects included. The new wave formulation includes a wave dissipation term, a CL-vortex term, a correction term due to mass-flux lost on non-zero current velocity at surface, and a term described as a hydrostatic pressure gradient in Smith (2006). In the equation governing the vertical structure $V_{d\alpha}$, the force from the short wave contribution only consists of a wave dissipation term.

It is interesting to compare our formulation to the short wave force derived by Newberger and Allen (2006) in application of a three-dimensional circulation model in the surf zone. The short wave force in their approach includes a surface stress caused by wave dissipation and a body force which consists of the

CL-vortex force and an extra force related to gradients of part of the radiation stresses. In our application to Quasi-3D equations, both the surface stress and the body stress (integrated) are included in the depth-integrated and wave-averaged momentum equations. In the equation governing the vertical structure of current velocity, the surface stress is the only force from the short wave contribution. Our results are theoretically consistent with Newberger and Allen (2006).

The numerical consistency in using the different types of wave forces was discussed in the paper. Simulations of rip current with a idealized bathymetry show that the two formulations perform identically when wave-current interaction is included in model coupling. However, if leaving out wave-current interaction, the CL-vortex formulation gives a result closer to the result with the wave-current interaction than the radiation stress formulation does. The vortex forcing is significant in the rip current case with a fast-varying wave field. It plays an important role in widening the rip current neck and thus weakening the strength of rip current.

Our further study may include several developments and applications of the new wave formulations, such as, 1) implementation of the full set of equations, i.e., (18) and (25), to take into account 3D dispersion effects; 2) quantitative comparison between models with different wave force formulations; 3) evaluation of model results using measurement data; 4) optimization of numerical time step used in wave-current model coupling; 5) model performance in predictions of various nearshore phenomena such as longshore currents, infra-gravity waves, and shear waves.

ACKNOWLEDGMENTS

This study was supported by Office of Naval Research, Coastal Geoscience Program under grant N00014-05-1-0423.

REFERENCES

Craik, A. D. D. and S. Leibovich, 1976, A rational model for Langmuir circulations, *J. Fluid Mech.*, 73, 401-426.

Dingemans, M. W., A. C. Radder, and H. J. de Vriend, 1987, Computation of the driving forces of wave induced currents, *Coastal Engineering*, 11, 539-563.

Garrett, C., 1976, Generation of Langmuir circulations by surface waves - A feedback mechanism, *J. Mar. Res.*, 34, 117-130.

Haas, K. A., I. A. Svendsen, M. C. Haller, and Q. Zhao, 2003, Quasi-three-dimensional modeling of rip current systems, *J. Geophys. Res.*, 108, C7, 3216, doi:10.1029/2001JC001313.

Hasselmann, K., 1971, On the mass and momentum transfer between short gravity waves and larger-scale motions, *J. Fluid Mech.*, 50, 189-205.

Kennedy, A. and T. J. Kirby, 2003, An unsteady wave driver for narrow-banded waves: modeling nearshore circulation driven by wave groups, *Coastal Engrng*, 48, 257-275.

Leibovich, S., 1980, On wave-current interaction theories of Langmuir circulations, *J. Fluid Mech.,* 99, 715-724.

Longuet-Higgins, M. S. and R. W. Stewart, 1962, Radiation stress and mass transport in gravity waves, with application to 'surf-beats', *J. Fluid Mech.*, 13, 481-504.

Longuet-Higgins, M. S. and R. W. Stewart, 1964, Radiation stress in water waves; a physical discussion, with applications, *Deep Sea Res.*, 11, 529-562.

McWilliams, J. C., J. M. Restrepo, and E. M. Lane, 2004, An asymptotic theory for the interaction of waves and currents in coastal waters, *J. Fluid Mech.*, 511, 135-178.

Mellor, G., 2003, The three-dimensional current and surface wave equations, *J. Phys. Oceanogr.*, 33, 1978 - 1989.

Newberger P. A. and J. S. Allen, 2006, Forcing a three-dimensional, hydrostatic primitive-equation model for application in the surf zone, Part 1: Formulation, submitted to *J. Geophys. Res.*

Phillips, O. M., 1977, The Dynamics of the Upper Ocean. 2nd ed. Vol. 1, Cambridge Monographs on Mechanics and Applied Mathematics, Cambridge University Press, 336 pp.

Putrevu, U. and I. A. Svendsen, 1999, Three-dimensional dispersion of momentum in wave-induced nearshore currents, *Eur. J. Mech. B/Fluids*, 83-101.

Smith, J. A., 1980, Wave, currents, and Langmuir circulation, Dalhousie University.

Smith J. A., 2006, Wave-current interactions in finite-depth, *J. Phys. Oceanogr.*, in press.

Svendsen I. A., K. A. Haas, and Q. Zhao, 2004, Quasi-3D Nearshore Circulation Model SHORECIRC: Version 2.0, Research Report, Center for Applied Coastal Research, University of Delaware.

Xia, H., Z. Xia, and L. Zhu, 2004, Vertical variation in radiation stress and wave-induced current, *Coastal Engrng.*, 51, 309-321.

Yu, J. and D. N. Slinn, 2003, Effects of wave-current interaction on rip currents, *J. Geophys. Res.*, Vol. 108, C3, 3088, doi: 10.1029/2001JC001105.

NUMERICAL MODELING OF ALONGSHORE SURF AND SWASH ZONE CURRENTS

Qin Chen[1] and Riccardo Briganti[2]

The paper presents a numerical investigation on alongshore surf and swash currents. The study has been carried out in the framework of wave-resolving Boussinesq-type models. Two moving shoreline treatment techniques, namely slotted-beach and wetting-drying methods introduced for this class of models have been implemented into an enhanced version of FUNWAVE2D. Both methods are compared in the simulation of longshore currents with simple, realistic bathymetry. The numerical model, tested under the SandyDuck field conditions, has produced fairly good results. An insight into the capability of the model to predict the alongshore drift velocity of the water mass in the swash zone is also given.

INTRODUCTION

The swash zone on a beach forms the mobile interface of the seawater and the land. The swash motion is characterized by the turbulent bore on the front face of a broken wave, the infiltration and exfiltration on the permeable beach, and the moving shoreline boundary. Previous studies (e.g. Bodge and Dean 1987 and Kamphuis 1991) have revealed bimodal distributions of the longshore sediment transport rate across the surf zone with a peak shoreward of the breaking point and the second peak in the swash zone. In the case of a large surf similarity parameter, experimental data suggested that up to 60% of the total longshore sediment transport can occur in the swash zone.

Although considerable effort has been devoted to the study of swash motions in the literature, wave-resolving, numerical modeling of the swash zone dynamics needs further improvements. In particular, an accurate model for swash motions over complex topography is needed whenever we want to simulate both gravity waves and currents on the beach under field conditions.

We aim at investigating the swash motions with longshore currents in the framework of Boussinesq-type wave models. Recent advances in Boussinesq modeling of surf zone hydrodynamics (e.g. Madsen et al. 1997, Chen et al. 1999, 2003) have offered a promising approach to the study of swash motions, including wave runup as well as cross-shore and longshore velocities in the swash zone. Several different approaches that resolve the shoreline movement

[1] Associate Professor, Department of Civil and Environmental Engineering, Louisiana State University, 3418D CEBA Building, Baton Rouge, LA 70803, USA. Email: qchen@lsu.edu

[2] Postdoctoral Research Fellow, Department of Civil and Environmental Engineering, Louisiana State University, 3418 CEBA, Baton Rouge, LA 70803, USA. Email: ricbrig@lsu.edu

have been proposed for Boussinesq-type wave models, but their performances in the presence of both waves and currents have not been fully tested.

As a matter of fact, in comparison with the understanding of the swash motion in the cross-shore direction, little is known about the magnitude and structure of alongshore currents in the swash zone generated by obliquely incident waves. The field experiment SWASHX conducted at Scripps Beach, California in 2000 has shown that there exist strong alongshore currents in the swash zone and further field measurements were carried out in the Nearshore Canyon Experiment (NCEX) in 2003. These datasets are useful for model development and verification. Moreover it is well known that an accurate prediction of swash motions is crucial for sediment transport modeling.

In this paper, we test the performances of an improved version of FUNWAVE2D in modeling alongshore surf and swash zone currents, using laboratory and field data. Emphasis is given to the interaction of swash motion with unstable longshore currents in the surf zone.

ENHANCED BOUSSINESQ WAVE MODEL

The wave-resolving numerical model utilized in this study is based on the fully nonlinear Boussinesq-type equations for nearshore waves and currents with the enhancement of the conservation property for potential vorticity (Chen 2006). The equations in a simple form may be expressed as

$$\frac{\partial \eta}{\partial t} + \nabla \cdot M = O(\mu^4 \sqrt{g h_0}) \tag{1}$$

$$\begin{aligned} &\frac{\partial u_\alpha}{\partial t} + (u_\alpha \cdot \nabla) u_\alpha + g \nabla \eta + \mu^2 (V_1 + V_2 + V_3) \\ &- R_b - R_s + R_f = O(\mu^4 \frac{g h_0}{l_0}) \end{aligned} \tag{2}$$

Equation (1) results from depth integration of the continuity equation, in which η is the free surface elevation relative to the still water level; M is the Boussinesq volume rate of flow; ∇ is the horizontal gradient operator; and t denotes time. Equation (2) stems from depth integration of the Euler equation of motion, where u_α is the horizontal velocity vector at the reference elevation z_α (= -0.54122h) in the water column; h is the still water depth; and g is the gravitational acceleration. Both continuity and momentum equations are truncated at $O(\mu^4)$ in which $\mu = h_0/l_0$ is the measure of wave dispersion, and h_0 is the characteristic water depth and l_0 is the characteristic wave length. The second-order Boussinesq dispersion terms in the momentum equations are denoted by V_1, V_2 and V_3, which can be found in details in Chen (2006). The additional terms, R_b, R_s and R_f, represent the parameterizations of the effects of wave breaking, subgrid lateral turbulent mixing and seabed shear stress, respectively. Detailed descriptions can be found in Chen et al. (1999).

In comparison with the pre-existing Boussinesq-type equations for wave and current interaction, this model has the advantage of conserving the potential vorticity with the accuracy consistent with the level of approximation for the pure wave motion (Chen 2006). Taking the curl of Equation (2) and invoking the continuity equation yield the equation for the potential vorticity ζ

$$\frac{D\varsigma}{Dt} = \varsigma_t + \overline{u} \cdot \nabla\varsigma = \frac{i_z}{H}(\nabla \times R_b + \nabla \times R_s - \nabla \times R_f) + O(\mu^4 \frac{gh_0}{l_0^3}) \qquad (3)$$

where $\varsigma = \omega/(h+\eta)$, ω = vertical vorticity component, $\overline{u}$ = depth-averaged horizontal velocity vector, and i_z= unit vector in the vertical direction. If there are no sources of vorticity, Equation (3) returns to the conventional form of conservation of potential vorticity for depth-uniform flows in shallow water.

TREATMENTS OF MOVING SHORELINE

To simulate swash motions and longshore currents, it is necessary to include an appropriate treatment of the moving shoreline boundary condition in a Boussinesq wave model. Various techniques have been proposed in the literature (e.g. Madsen et al. 1997, Kennedy et al. 2000, Bellotti and Brocchini 2001, and Lynett et al. 2002). In the present study, we focus on two relatively simple methods, namely the slotted-beach technique and the wetting-drying method, for handling the shoreline movement with breaking-induced currents.

Instead of tracking the wet and dry cells during wave run-up/run-down on the beach, the slotted-beach technique treats the entire computational domain as an active fluid domain. The basic idea behind this technique is to replace the solid bottom, where there is very little or no water covering the land, by a porous seabed, or to assume that the solid bottom contains narrow slots. This allows the water level to fall below the beach elevation. The replacement of the solid bottom by narrow slots results in a modification of the mass equation. Three parameters influence the extent of such a modification. They are the relative width of slot with respect to a unit width of beach (δ), the shape parameter (λ) for the smooth transition from unity to a narrow slot, and the depth of a slot, or the offshore still water depth (h_0) where a slot begins.

Madsen et al. (1997) showed that, even though a very narrow slot width is used, there is still about a 10% error in the computed maximum run-up in comparison with the analytical solution of Carrier and Greenspan (1958). This is attributed to the additional cross-sectional area introduced by the narrow slot, because the maximum run-up is very sensitive to the total volume of mass at the run-up tip. In an attempt to improve the conservation property for mass in the presence of a slot, Kennedy et al. (2000) retained an equivalent cross-sectional area of a unit width of beach by slightly elevating the physical seabed elevation, leading to the improvement in the simulation of wave run-up. The optimal values of δ and λ are found to be 0.002 and 80, respectively, which give the best agreement with the analytical solution obtained by Carrier and Greenspan (1958). Although the slotted-beach technique implemented in the Boussinesq model FUNWAVE2D is simple for the complex problem of moving boundaries, the accuracy of the technique is limited by the need for a larger slot width and numerical filtering near the shoreline to suppress the noise generated by numerical instabilities associated with the moving shoreline for wave run-up on a steep slope and under field conditions. The accuracy of the velocity prediction near the shoreline also is not as good as the run-up prediction.

The second technique we have employed in the present study is the wetting-drying method of Lynett et al. (2002) who used a conceptually simple method of linear extrapolation to extend the dependent variables in the wet cells to the dry beach without alternating the seabed. In contrast to the slotted-beach method, the computational domain is explicitly divided into wet and dry cells, and the equations of motion are solved in the wet region only. The distinction between wet and dry regions is performed on the basis of a threshold value for the local water depth. This threshold is set to be $a_0/50$, where a_0 is the offshore wave amplitude. If the local total depth is smaller than or equal to the threshold value, it is considered as a dry cell.

To treat the boundary condition at the interface between the wet and dry regions, a linear extrapolation method is used to assure the existence of the shoreline between two grid points and to provide the boundary conditions at the shoreline. This approach is simple and straightforward in one-dimension (1D) but becomes quite involved for complex shoreline geometry in two horizontal dimensions (2HD). In order to model alongshore surf and swash zone currents, periodic lateral boundary conditions are desirable. We therefore implement and extend Lynett et al.'s (2002) run-up scheme to periodic cross-shore boundaries in FUNWAVE2D. In the case of 2HD, an extrapolated variable in a dry grid point is obtained from an average of the values resulting from 1D linear extrapolation of the surrounding wet points. At each of the lateral boundary, the extrapolation and average have to involve the counterparts at the opposite lateral boundary to assure the alongshore periodicity of the domain.

MODEL TESTS

A number of model tests have been carried out to examine the skills of the extended wetting-drying scheme implemented into FUNWAVE2D with an emphasis on the velocity and current prediction. The tests include 1) Carrier and Greenspan's (1958) analytical solution of nonlinear long wave run-up on a sloping bottom, 2) Thacker's (1981) analytical solution of wave resonance in a circular parabolic basin, 3) Briggs et al's (1995) laboratory experiment on solitary wave run-up on a circular island, and 4) field observations of longshore currents at the U.S. Army Corps of Engineers Field Research Facility in Duck, North Carolina. We shall focus on the comparison with the analytical solutions and the dataset collected during the SandyDuck field experiments in 1997.

The analytical solution of Carrier and Greenspan (1958) has served as the classic benchmark test for many wave run-up schemes. We consider a 10s wave train with an initial wave height of 0.006m in a wave flume containing a 1:25

slope. The still water depth in the flat portion of the flume is 0.5m. Figure 1 shows the comparison of the numerical (dashed lines) and exact (solid lines) solutions of the free surface profiles and velocity profiles. It is seen in the left panel that the wetting-drying scheme implemented into FUNWAVE2D is able to accurately reproduce Carrier and Greenspan's analytical solution, including the maximum swash excursion, and the nodes and anti-nodes of the standing waves. More importantly, the modeled swash velocities are in excellent agreement with the exact solution, as seen in the right panel.

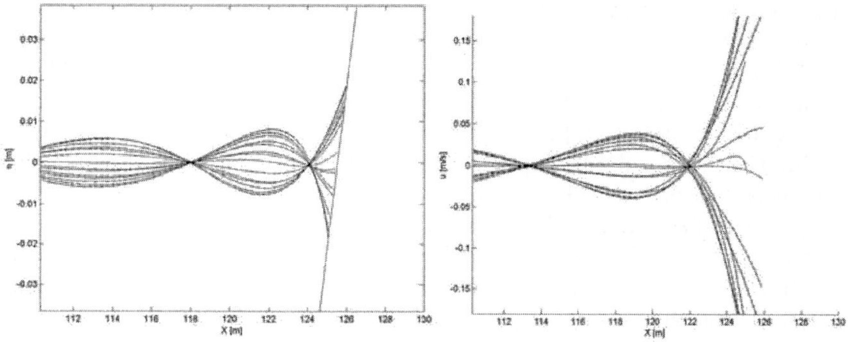

Figure 1. Comparison of numerical (dashed lines) and analytical (solid lines) solutions for wave run-up and rundown on a sloping bottom using the wetting-drying scheme. Left: surface profiles; Right: velocity profiles.

Another benchmark test often utilized in the literature for run-up schemes is Thacker's (1981) analytical solution of wave resonance in a circular basin with a parabolic bottom profile under the framework of nonlinear shallow water equations. Given a parabolic bottom profile in a circular basin of radius R at the still water level and the maximum still-water depth h_0, the radial velocity u is

$$u = \frac{\varphi\, rA\sin(\varphi\, t)}{2(1-A\cos(\varphi\, t))} \tag{4}$$

$$A = \frac{R^4 - r^4}{R^4 + r^4} \tag{5}$$

$$\varphi = \frac{\sqrt{8gh_0}}{R} \tag{6}$$

where φ = resonance frequency, r_o = the initial radius of the shoreline

Using the same input conditions as those in Lynett et al. (2002) with r_o= 2000 m and R = 2500 m, we obtained both numerical and analytical velocity and surface profiles as shown in Figure 2, where the full lines are model result while the dotted lines represent the exact solutions. It is seen from the figure that the numerical result of FUNWAVE2D with the extended wetting-drying scheme matches very well the analytical solution in both elevation and velocity. Notice that the Boussinesq dispersive terms were switched off in order to be consistent with the nonlinear shallow water equations used in Thacker's (1981) analytical solution.

Apparently, the wetting-drying scheme of Lynett et al. (2002) implemented into FUNWAVE2D predicts the swash velocities very well under non-breaking, long wave conditions with simple bathymetry.

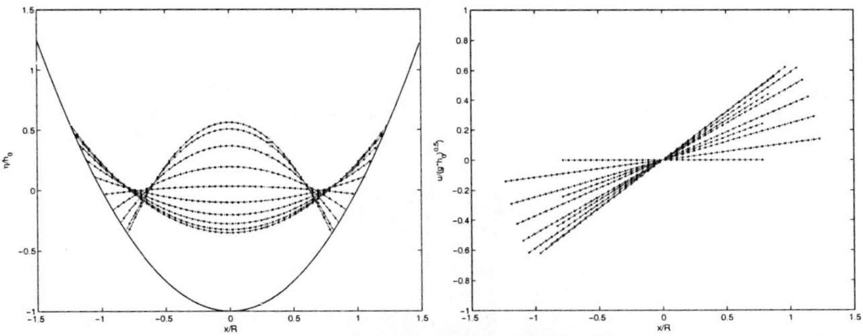

Figure 2. Comparison of numerical (full lines) and analytical (dots) solutions of long wave resonance in a parabolic bowl using the wetting-drying scheme. Left: surface profiles; Right: velocity profiles.

ALONGSHORE DRIFT VELOCITY IN THE SWASH ZONE

After validating the model using the simple, non-linear shallow water solutions with an emphasis on the cross-shore swash velocity, it is important to explore the capabilities of the model in predicting the main features of the swash zone with broken waves. An interesting aspect of swash motions is the drift velocity of the longshore flow in the swash zone. The importance of this component was portrayed in Brocchini (2006) and Antuono et al. (2006), who have suggested that swash motions can influence the entire surf zone dynamics.

We compute the alongshore drift velocity of the water mass in the swash zone by integrating the instantaneous longshore velocity component over the cross-shore excursion of the swash zone and averaging it over two short-wave periods. The drift velocity serves as a measure of the swash influence at the seaward limit of the swash zone on the alongshore surf zone current. For simplicity, we consider an obliquely incident, monochromatic wave train, modulated by a longer wave and propagating on a longshore uniform beach with a constant slope of 1:30. The offshore water depth is 0.4 m, and the incoming short waves have an amplitude $a_0 = 0.024$ m and a period of 1 s. This simple simulation is also useful for testing the periodic cross-shore boundary conditions in the extended Boussinesq wave model with respect to surface waves and longshore flows in both surf and swash zones.

Figures 3 & 4 show a snapshot of the surface elevation and the drift velocity after 40s from the beginning of the simulation, respectively. First, it is seen that the periodic cross-shore boundary conditions work well. Second, we notice that the alongshore drift velocity averaged over two wave periods and over the cross-shore excursion of the swash zone varies in the longshore direction and grows with time in the early stages of development. Further improvement in the algorithm for tracking the seaward limit of the swash motion and the velocity in the swash zone is needed to remove the kinks in Figure 4. Obviously, the magnitude of the alongshore drift velocity in the swash zone is sensitive to the bottom friction, similar to the longshore current in the surf zone. Detailed comparisons with the analytical solution in Antuono et al. (2006) are currently undergoing. It is also desirable to further test the model under field conditions with respect to breaking-induced longshore currents.

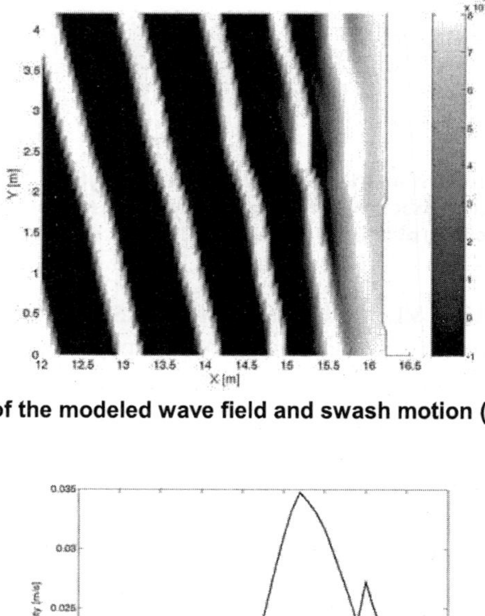

Figure 3. Snapshot of the modeled wave field and swash motion (40s after cold start).

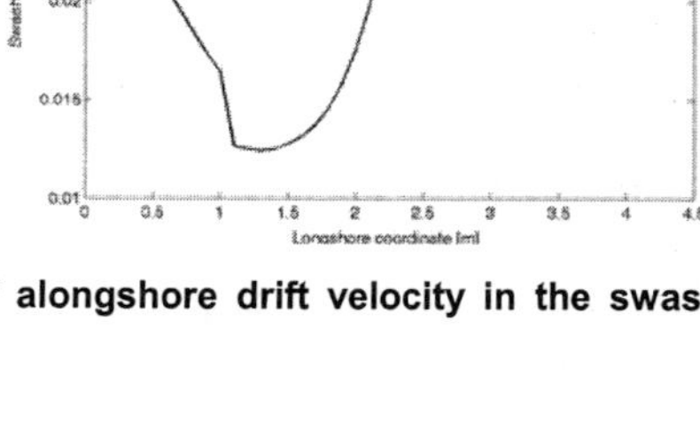

Figure 4. Snapshot of alongshore drift velocity in the swash zone (40s after cold start).

SIMULATION OF SANDYDUCK FIELD EXPERIMENT

The SandyDuck coastal field experiment was conducted in October 1997 at the U.S. Army Corps of Engineers Field Research Facility (FRF) at Duck, North Carolina. Figure 5 shows the bottom topography at the FRF close to October 2, 1997 when the field dataset for the time period 0400-0700 EST is used to test the model. The model bathymetry was constructed from a composite of high resolution mini-grid data and an interpolation between the two larger-area surveys that bracket the study time (Kirby and Chen 2003). It is seen that the bottom depression under the FRF research pier significantly alters the barred-beach, and consequently influences the surface waves and wave-driven longshore currents.

Input wave conditions are derived from published frequency-direction spectra which are available from the FRF. The random waves are mapped onto a finite number of components that satisfy the requirement of longshore periodicity (Chen et al. 2003). Realizations of 3.5 hour duration are constructed using linear superposition of directional components and random phases, which has a root-mean-square wave height of 1.06 m and a peak wave period of 6.6 s. The peak angle of the directional spectrum is 26°, which means the waves came from the northeast. The bottom friction coefficient is set to be 0.0015.

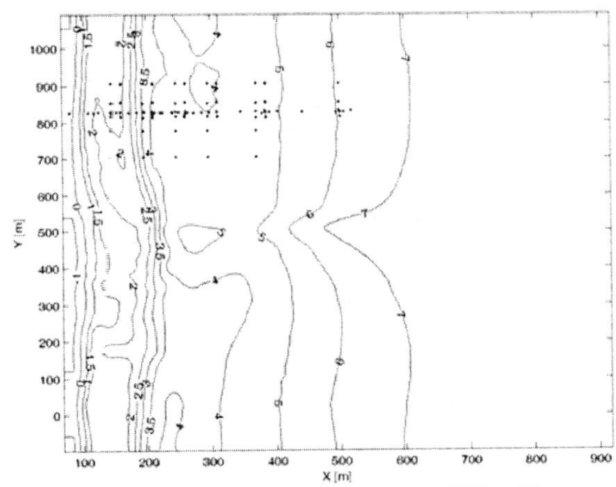

Figure 5. Instrument array (dots) and bathymetry at the FRF with scour holes under the research pier.

Figure 6 illustrates the time sequence of the velocity and vertical vorticity fields predicted by the enhanced Boussinesq wave model under the SandyDuck field conditions. The solid lines are the contours of water depth and the shoreline is on the left. We notice that the wetting-drying scheme extended to periodic cross-shore boundary conditions handles the moving shoreline and swash motion fairly well although filtering is required to stabilize the model.

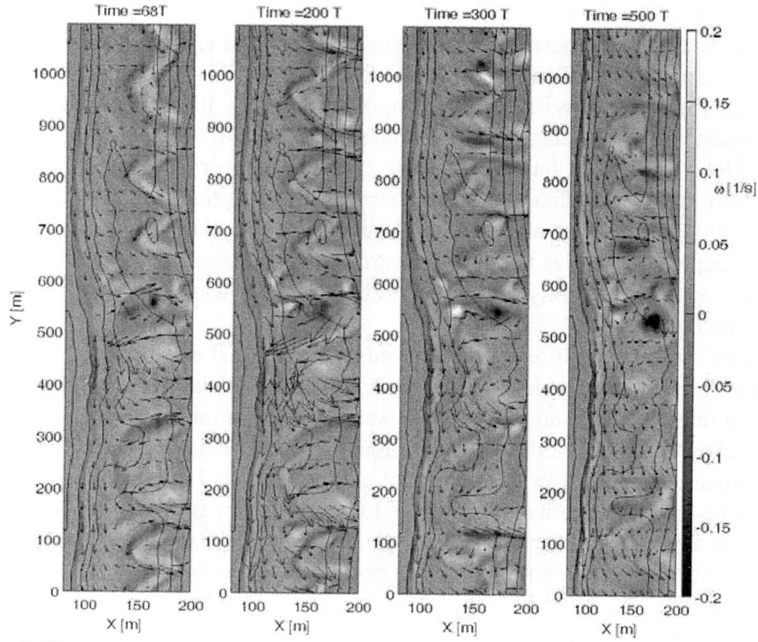

Figure 6. Time sequence of the velocity (arrows) and vertical vorticity fields predicted by the enhanced Boussinesq model under the SandyDuck field conditions. The solid lines are contours of water depth. The shoreline is on the left.

The longshore current shown in Figure 6 was obtained by averaging the computed fluid particle velocity over 5 peak wave periods to filter out the short wave motion. It is seen that the longshore current meandered and the flow pattern changed with time. Strong vortices accompanied by transient seaward currents were triggered by shear instabilities and bathymetric variations, in particular at the seabed depression under the research pier. The predicted flow field differs from previous studies using Boussinesq wave models, as the vertical vorticity is now transported by the depth-average velocity, similar to phase-averaged models based on the nonlinear shallow water equations. A comparison of the modeled and measured mean longshore currents averaged over two hours is shown in Figure 7. Fairly good agreement is found.

EFFECT OF SHORELINE BOUNDARY ON SURF ZONE CURRENTS

To investigate the effect of the two different treatments of shoreline boundary conditions on the prediction of longshore surf zone currents, a series of numerical experiments using a longshore uniform beach is carried out. We construct the model bathymetry using the cross-shore beach profile along the instrumentation array at 827.5 m (see the bottom panel in Figure 7) as the representative beach profile from the field experiment. The wave input uses a monochromatic wave train to represent the wave conditions in the field.

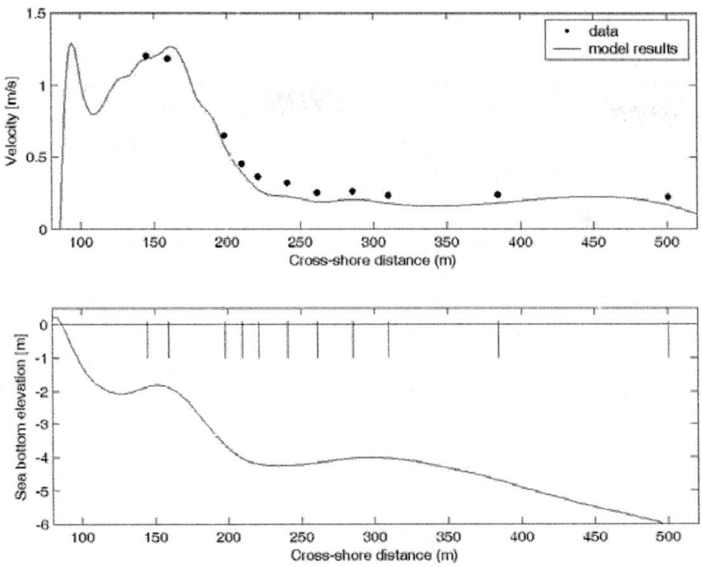

Figure 7. Comparison of the modeled and measured mean longshore currents under the SandyDuck field conditions using the enhanced Boussinesq wave model.

Comparisons are made between the flow fields predicted by the Boussinesq model with the slotted-beach and wetting-drying techniques for a treatment of the moving shoreline. We focus our attention on the mean longshore current velocity and vertical vorticity component as well as the relative standard deviation that is a measure of the variability of the flow field owing to shear instabilities. Figure 8 shows these comparisons. Notice that the sand bar crest is located approximately at 160 m in the FRF coordinate system. Albeit a strong similarity between the predicted maximum longshore currents by the two shoreline treatment techniques, it is evident that several differences exist.

First, it is seen that the velocity of the mean longshore current is sustained close to the shoreline in the wetting-drying case while its counterpart reduces more rapidly to a very small value in using the slotted-beach technique. There is also a slight difference in the predicted mean water levels at the shoreline. In the trough of the bar, the wetting-drying technique leads to a slightly weaker current. Nevertheless, both shoreline treatment techniques predict very similar *mean* longshore currents inside the surf zone, including the maxima near the shoreline and shoreward of the bar crest.

Outside the surf zone, it is surprising to see that the slotted-beach technique results in a larger mean longshore current than does the wetting-drying method. This is caused by the use of a larger and deeper slot to avoid numerical instabilities in the simulation. The slot parameters were set to be $\delta = 0.01$ and λ

= 20, which considerably deviate from the optimal values. In addition, the slots begin further offshore in the present numerical experiment, which is different from the starting point at the edge of the surf zone in Chen et al. (2003). The significant alternation of the seabed by the slots can increase the wave celerity and the mean longshore current outside the surf zone. Therefore, care must be taken when choosing the three slot parameters.

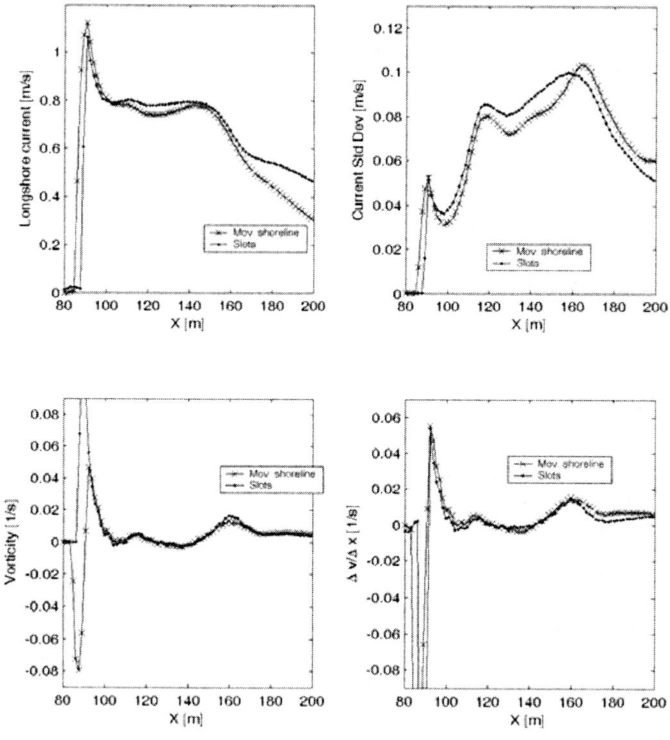

Figure 8. Comparison of the longshore currents predicted by the slotted-beach (dotted lines) and wetting-drying techniques.

Secondly, we compute the standard deviation of the "instantaneous" longshore current averaged over two wave periods as a measure of the flow variability owing to shear instabilities. It is seen from the right top panel in Figure 8 that the wetting-drying technique yields lower standard deviation (SD) of velocity in the trough and higher SD at the seaward edge of the surf zone than the prediction of its counterpart. This is a possible indication of more energetic shear waves outside the surf zone in the wetting-drying model in comparison to the slotted-beach model. The average vorticity profiles shown in the lower left panel are quite similar except for those near the shoreline.

It is seen that both the shoreline treatment techniques give much larger vertical vorticity with opposite signs near the shoreline compared to the rest of the domain. Differences in the velocity shears of the longshore currents are also found near the shoreline and outside the surf zone, as seen in the lower right panel. Inside the surf zone, there are no remarkable differences in the mean values of vorticity and velocity shear predicted by the two methods. Nevertheless, the considerable differences in the vertical vorticity and velocity shear near the shoreline suggest that more in-depth investigation on the alongshore drift velocity produced by the two shoreline treatment techniques is indeed needed. Such a study requires three things. First, much higher resolution of the flow in the swash zone is necessary in comparison to the typical resolution of the surf zone in a Boussinesq model with rectangular grids. Second, a consistent and accurate algorithm for defining and tracking the seaward limit of the swash zone should be used for both moving shoreline schemes in order to properly partition the swash and inner surf zones where the mean alongshore drifts are defined differently. Third, a better-devised filter is desirable in order to suppress noises near the shoreline associated with both moving shoreline treatment techniques because none of them is free from numerical instabilities under breaking wave and realistic seabed conditions.

CONCLUSIONS

The paper presented a numerical investigation on longshore flows in the surf and swash zones. The study was carried out in the framework of wave-resolving Boussinesq-type models. Two moving shoreline treatment techniques, namely slotted-beach and wetting-drying methods introduced for this class of models have been implemented into an enhanced version of FUNWAVE2D. They were compared in the simulation of longshore currents with simple, realistic bathymetry. The numerical model, tested under the SandyDuck field conditions, has produced fairly good results. An insight into the capability of the model to predict the alongshore drift velocity of the water mass in the swash zone was given.

Recent research carried out by Otta and Pedrosa-Acuna (2004) focused on the cross-shore swash motion using these two approaches and they found evidence of a discrepancy between the two models in predicting the cross-shore velocities. In this work, we compared the performances of these two approaches in 2HD wave and current simulations. The two moving shoreline treatments were implemented and tested in the same solver in order to exclude any cause of difference induced by the differences in the numerics of two different solvers. The result of benchmark tests for the wetting-drying scheme has confirmed that the model is capable of simulating the shoreline movement of long, non-breaking waves. The result of longshore current simulations has shown considerable differences in the modeled vertical vorticity and velocity shear near the shoreline by the two methods. Further studies are obviously needed.

ACKNOWLEDGMENTS

This study was supported by the Office of Naval Research Coastal Geosciences Program through research Grant No. N00014-04-1-0310.

REFERENCES

Antuono, M., Brocchini, M., and Grosso, G. (2006). Integral properties of the swash zone and averaging. Part 3: Longshore shoreline boundary conditions for wave-averaged nearshore circulation models. *J. Fluid Mech.*, in press.

Bellotti, G. and Brocchini, M. (2001). On the shoreline boundary condition for Boussinesq type models. Int. J. for Num. Meth. In Fluids. 37: 479-500.

Bodge, K. R., and R. G. Dean, (1987) Short-term impoundment of longshore transport, *Proc. Coastal Sediments '87*, 468-483,

Briggs, M.J., Synolakis, C.E., Harkins, G.S., Green, D.R. (1995). Laboratory experiments of tsunami runup on a circular island, *Pure and Applied Geophysics*, 144: 3/4, 569-593.

Brocchini, M. (2006). Integral swash zone models. *Con. Shelf Res.* 26: 653-660

Carrier, C. F. and Greenspan, H. P. (1958). Water waves of finite amplitude on a sloping beach. *J. Fluid Mech.*, 4: 97-109.

Chen, Q. (2006). Fully nonlinear Boussinesq-type equations for waves and currents over porous beds. *J. Eng. Mech.* 132: 220-230.

Chen, Q., Dalrymple, R. A., Kirby, K. T., Kennedy, A. B. and Haller, M. C. (1999). Boussinesq modeling of a rip current system. *J. Geophys. Res.*, 104 (C9): 20,617-20,637.

Chen, Q., Kirby, K. T., Dalrymple, R. A., Shi, F. and Thornton, E. B. (2003). Boussinesq modeling of longshore currents. *J. Geophys. Res.*, 108 (C11): 3362-3380, doi: 10.1029/2002JC001308.

Kamphuis, J.W. (1991), Alongshore sediment transport rate, *J. Waterw., Port Coastal and Ocean Eng.*, 117: 624-640,

Kennedy, A. B., Chen, Q., Kirby, J. T., and Dalrymple, R. A. (2000). Boussinesq modeling of wave transformation, breaking and runup. I: 1D *J. Waterw., Port, Coastal and Ocean Eng.*, 126: 39-47

Otta, A. K. and Pedrosa-Acuna, A. (2004). Swash boundary and cross-shore variation of horizontal velocity on a slope. *Proc. ICCE*, 1616-1628.

Lynett, P. J., Wu, T. and Liu, P. L.-F. (2002). Modeling wave runup with depth-integrated equations. *Coastal Eng.*, 46: 89-107.

Madsen, P. A., Sorensen, O., and Schaffer, H. A, 1997. Surf zone dynamics simulated by a Boussinesq type model. Part I: model description and cross-shore motion of regular waves. *Coastal Eng.,* 32: 255-287.

Thacker, W. C. (1981). Some exact solutions to the nonlinear shallow water wave equations. *J. Fluid Mech.*, 107: 499-508.

THE MIDDLE WAY OF SURF MODELING

Daniel C. Conley[1] and Eugen Rusu[2]

Motivated by the assumption that variations in the wave forcing field are the dominant factor in longshore variations in the nearshore circulation, a hybrid surf zone modeling approach has been developed. This approach combines high resolution 2-D wave generation and shoaling modeling with a simple surf zone circulation model based on 1-D momentum balances. Model-data comparisons are presented from two different field experiments performed in locations where the 1-D beach paradigm is not valid. As a rule of thumb, locations for model evaluation were chosen where the shoreline and shallow water isobaths, shows observable curvature over spatial scales on the order of the incident wavelength. The results indicate that the proposed methodology can be used to extend the range of applicability of the 1-D momentum balances but that it is not a general solution.

BACKGROUND

The surf zone environment is of central importance in many military and civilian applications and the ability to be able to predict environmental characteristics such as breaking wave height and incident wave direction, breaker type, surf zone width and longshore current strength is widely desired but dedicated oceanographic centers for the development of such predictions are not always available. For this reason, a surf-zone modeling capability has been developed which is designed to be run by operators with limited training. Thus, in order to minimize the model complexity and maximize the range of applicability of such a model, it was decided to apply a hybrid approach in which a restrictive one-line surf zone circulation model is matched with a more generalized fully 2-D wave transformation and generation model. It was sought to expand the conditions for which the one line model could be applied by initializing it with the wave transformation model outputs adjacent to the surf-zone.

MODEL DESCRIPTION

The Interface for SWAN and Surf Model (ISSM) has been created by combing the wave model SWAN (Simulating WAves Nearshore) and the surf forecasting model SURF3.2 which is the most recent release of the U.S. Navy's "Navy Standard Surf Model" (NSSM) an operational surf model developed by the Naval Research Lab, Stennis Space Center [Earle, 1989]. SWAN was

---

[1] NATO Undersea Research Centre, Viale San Bartolomeo 400, 19128, La Spezia, Italy; conley@nurc.nato.int
[2] Technical University of Lisbon, Instituto Superior Tecnico Av. Rovisco Pais, 1049-001 Lisboa, Portugal; eugen.rusu@mar.ist.utl.pt

selected based on its relative sophistication, its wide application, and the high level of calibration and validation already achieved. SURF3.2 was chosen primarily because of the level of physics included in the model as well as its focus on predicting surf zone parameters of military operational importance.

SWAN

SWAN (http://fluidmechanics.tudelft.nl/swan/) is a freely available 3rd generation model that was developed by at the Delft University of Technology (NL). The model was designed to simulate the generation and propagation of random short-crested waves in shallow, coastal waters. The model [*Booij, Ris, and Holthuijsen*, 2001] calculates waves on a phase averaged basis using the action balance equation and explicitly accounts for wave refraction over currents and bathymetry as well as white capping, quadruplet and triad wave-wave interactions, bottom dissipation and depth-induced wave breaking. The implicit numerical scheme used in SWAN is the feature which renders it particularly applicable for shallow water. The model is in wide use both in research as well as in industry and is even incorporated in one widely distributed coastal engineering modeling package. The model has been widely tested and validated both by its authors as well as by other objective parties [e.g. *Jin and Ji*, 2001;*Ou et al*, 2002] and is considered validated for almost any environment where wave diffraction does not play a large role.

Navy Standard Surf Model

The profile based surf zone circulation model used is SURF3.2. Inputs to the model include, profile depths, wave conditions at the seaward end of the profile, and local wind conditions. The model estimates the maximum and significant wave height, the strength and direction of the maximum longshore current, the width of the surf zone and the number of lines of breaking surf as well a cross-shore distribution of both the wave height and currents.

As a 1-line model or profile model, the calculations in SURF3.2 are based on the following assumptions:

- •Bottom contours are straight and parallel to the coast
- •Currents are uniform throughout the depth-of the water column
- •Linear wave theory is valid
- •Wave spectra are narrow-banded in both frequency and direction
- •Wave heights are Rayleigh distributed

Based on these SURF3.2 solves the following equation for the longshore current,

$$\tau_y^r + \rho \frac{\partial}{\partial x}\left[\mu h \frac{\partial V}{\partial x}\right] - \left\langle \tau_y^b \right\rangle + \tau_y^w = 0 \qquad (1)$$

Where the first term represents the longshore directed radiation stress due to the incident waves, the second term represents the horizontal mixing term due to cross-shore gradients in the longshore current V, the third term is the wave averaged bottom stress and the last term represents the longshore wind stress. The model includes a parametric relation for cross-shore growth and dissipation of waves due to breaking and additional relations are included for estimating percent breaking, the number of lines of breakers and breaker type. This model has also been extensively tested [*Mettlach and May*, 1997;*Hsu, Mettlach, and Earle*, 2002].in environments where the "infinitely long" beach paradigm is a reasonable approximation and in general it has been found to predict the breaking wave height and the strength of the longshore current when it is provided with high quality estimates of the wave conditions at the boundary.

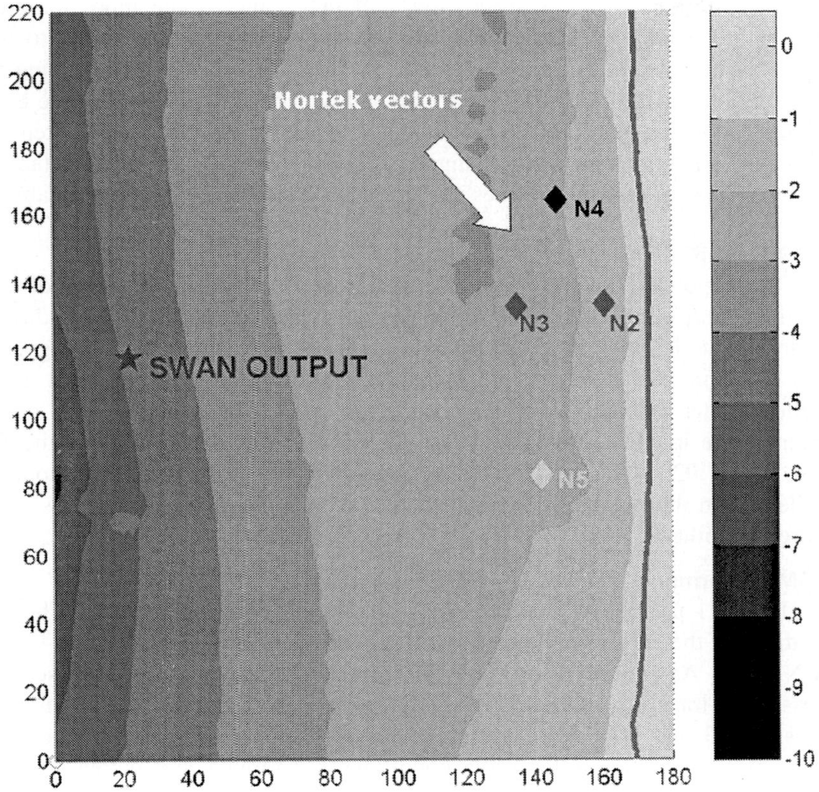

Figure 1: Bathymetry developed for the SURF'02 field site. The figure indicates the location for the nearshore pressure-current meters (Nortek Vectors N2-N5) and shows an approximate location for the 3 sites for SWAN output.

ISSM

The ISSM [*Rusu et al, 2006*] exists as a MATLAB based Graphical User Interface (GUI) which was designed to facilitate the combined operation of SWAN and SURF3.2. Not only does the model provide a framework for running both models and passing the relevant data between them but it is also a comprehensive package for much of the potential data pre-formatting which is necessary to prepare the model inputs. One of the goals of the development was to maximize both simplicity but also reliability of the system. The package is distributed as a MATLAB toolbox and, in addition to the SWAN & SURF3.2 executables, also relies on portions of other public domain MATLAB toolboxes such as M_MAP [*Pawlowicz*, 2004] and DIWASP [*Johnson,* 2002]. There is considerable flexibility built into the ISSM and users are permitted to operate using input data of varying levels of quality and complexity. Inputs may represent either hindcasts, nowcasts, or forecasts. A default set of model options has been designed for the basic user although some of the more important options are readily selectable. However an expert user has the ability to fully adjust all parameters as seen necessary. In either case the data processing and presentation capabilities of ISSM are still fully accessible. The package easily allows simulating multiple scenarios and does not require repeating steps for which the conditions have not changed and, in the absence of one executable or the other, the package capabilities can still be utilized for the other executable.

MODEL VERIFICATION

Two field experiments were performed in order to generate a record of wave shoaling, breaking and longshore current generation in non-traditional environments [*Conley et al,* 2006]. These trials were specifically designed to take place in pocket beach like locations where the shoreline as well as the shallow water isobaths show observable curvature over spatial scales on the order of the incident wavelength (i.e. O[100m]). The bathymetry in the first trial, SURF'02, was characterized by smooth, shore parallel bottom contours while that in the second, SURF'04, included strong longshore as well as cross-shore variability.

ISSM Performance on a Mildly 3-D Shoreline

Figure 1 provides a close up view of the SURF'02 bathymetry. The four locations of the various pressure current measurements are identified and labeled as N2-N5. As can be seen, these four sites are distributed on three separate cross-shore transects, one each for site N4, and N5 respectively, and one transect for sites N3 and N2. SWAN simulations were performed for the study area using measured deep water wave conditions for forcing and output for each transect was collected at the intersection of the profile and the 4 m isobath. This output was then used as input for SURF3.2 run along each profile. Model data comparisons of wave parameters for two of the test sites are presented in Figure 2.

The top panels in each figure shows comparison between the observed significant wave height (H_{sig}) and the SWAN output height. As can be seen (Table 2), the H_{sig} comparisons at all 4 locations are extremely good with a maximum rms error of 9 cm and a correlation of greater than 0.95 at all stations. While the surf model itself outputs a wave height distribution across the surf-zone, it has been observed that the maximum wave heights at the instruments are under predicted because the model mis-predicts the onset of wave breaking and over-predicts the wave dissipation which will have occurred at the instrument stations. For this reason it was decided that the SWAN model output provided a more robust estimate of wave height for instrument model comparisons. The middle panel in these figures shows the model-data comparison for wave period. Discrepancies occur only during periods of low wave conditions when the determination of wave period from data is particularly problematic.

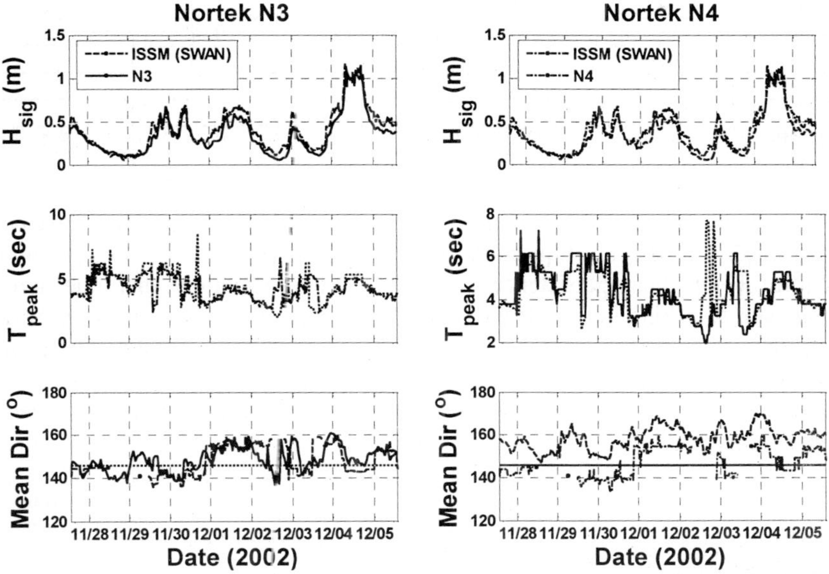

Figure 2: SWAN (H_{sig}, T_{peak}) and SURF3.2 (mean direction) modeling results versus wave parameters derived from current and pressure measurements collected at N3 and N4. The top panels present significant wave height (H_{sig}), the middle panels peak wave period (T_{peak}) and the bottom panels show mean wave direction (direction of wave propagation).

The bottom panel provides a comparison of peak wave direction between that calculated by the model and that determined from the measurement based directional spectra estimates. Unfortunately the comparisons are quite poor for all stations except for N3. However close inspection of the behavior of the model data mismatch suggests that the presence of a strong reflected wave signal in the recorded pressure-velocity data is leading to deterioration in the estimate of wave direction from the PUV measurements [*Huntley & Davidson*, 1998]. If this is the case, the most reliable estimation of incident wave angle would be the measurements which are furthest from the shoreline where the coherence between the incident and reflected waves is minimal. For that reason we will use only the results from N3 for comparisons between "measured" and simulated wave angles. At this location the bias between predicted and observed directions is only 1.1 degrees and the correlation is 0.59, a value which is considered satisfactory considering that the observations are partially in doubt.

Table 2: Water depths and various wave prediction statistics at the 4 instrument locations. Significant wave height statistics are calculated using the SWAN forcing wave height.

Instrument Location	Mean Water Depth (m)	H_{sig} Correlation	H_{sig} RMS error (m)	T_{peak} Correlation	Incidence Angle Bias (Degrees)
N2	1.56	0.97	0.09	0.50	15.8
N3	2.92	0.97	0.09	0.68	1.1
N4	2.46	0.97	0.09	0.58	12.0
N5	2.50	0.96	0.08	0.74	9.4

While the simulations appear to do a reasonable job in predicting nearshore wave heights, a base level question in any modeling endeavor is, does the model perform better than persistence? In the current situation, that question is equivalent to asking how well the model reproduces the change in wave height from the offshore boundary to the surf-zone. Figure 3 gives the results of just such an exercise where we see in general that the model does a very good job of modeling the changes in wave height. In fact, the correlation between the two is 0.96.

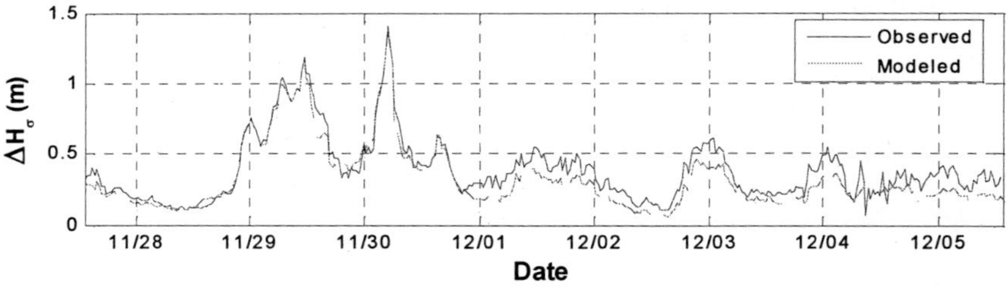

Figure 3: Comparison of modeled and observed differences in wave height from offshore (100 m isobath) to surf-zone at N03.

A comparison between modeled and measured longshore currents is presented in Figure 4. For the model-data comparisons presented here we use the value of SURF3.2 predicted longshore current which corresponds to the cross-shore position of the instrument. As can be readily seen from Figure 4, the only station which is predicted to exhibit longshore currents at any point other than the storm on 4 December is at N2. In examination of the results at N2, a general observation can be made in that the model predictions represent a worst-case scenario. In other words, when large currents are observed in the data, they are also predicted in the model. However there are cases where reasonably large currents are predicted by the model and not seen in the data. This can be due to relatively minor errors in the prediction of the cross-shore distribution of the longshore current or it could be due to a true suppression of longshore currents due to 3-D bathymetry in the study area. Examples of apparent 3-D behavior are observed in the figures such as the current oscillation which is seen in all current meters at approximately midnight on 30 November, a period of particularly low wave forcing at the shoreline. While the correlations at N2-N4 are somewhat low (r=0.38-0.62) the majority of the disagreement occurs for the case of observed currents when waves are low and correlations increase significantly (r=0.54-0.75) when only conditions with wave heights greater than 0.5 m are considered.

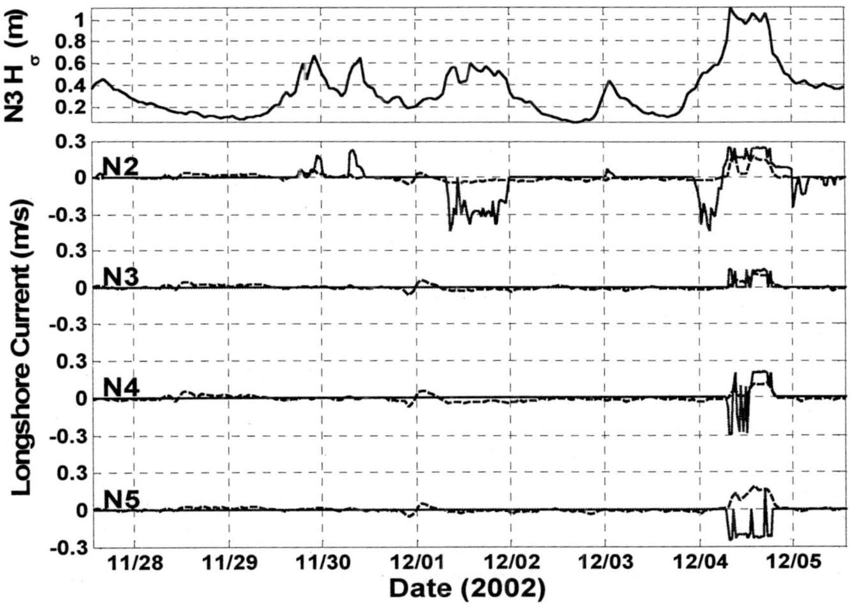

Figure 4: Comparison of Nortek measured currents and SURF3.2 predicted currents for all four instrument stations. Top panel shows significant wave height as measured at N3 for reference. Model derived velocities are solid and measured velocities are dashed. The model velocities represent those predicted by the model at sensor location, not the peak longshore current parameters.

The comparisons between longshore current observations and predictions at N5 however indicate a different story. At this location the surf model results clearly predict a dominantly negative longshore current while the observations measured a positive current. This result is due either to the fact that the wave model is performing poorly in modeling wave directions, or the local currents at N5 are driven more by longshore gradients than by local momentum balances. As the model has so far given us no reason to believe the former, we suspect the latter. Thus it is worth noticing that, even in this locally 2-D stretch of shoreline, for the beach profile which is closest to topographical curvature, the hybrid system of 2-D wave simulations and 1-D surf modeling is not capable of properly simulating the longshore currents.

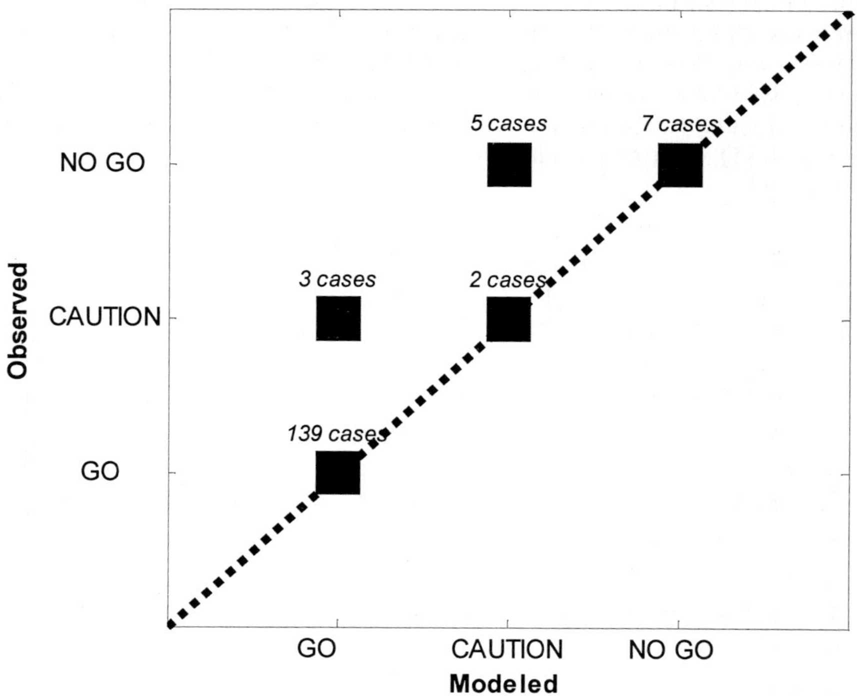

Figure 5: A comparison of amphibious operation TDAs created using modeled (horizontal axis) and measured (vertical axis) data. Results provide an operational description of model accuracy and show that, operationally, the model results are the same as measurements 95% of the time. These results are general for all the stations at SURF'02.

While providing statistical measures of model-data comparisons does provide quantitative measures with which to compare other models or other conditions, it may be difficult for an end user to understand exactly the relevance of these numbers. In order to arrive at a more user relevant measure for comparison, we have used the model results and observations to create value added product which is of direct interest to the end user. In the current application, those products are called Tactical Decision Aids which provide a traffic light assessment (green-"go", yellow-"caution" or red-"no go") of the feasibility of performing an operation. The method of Grasso and Giannecchini [*2006*] has been applied to create model TDAs which are based upon significant and maximum breaker heights, magnitude of the longshore current and Modified Surf Index (MSI) [*Commander Naval Surface Force Pacific and Commander Naval Surface Force Atlantic*, 1987]. In operational use, these TDAs would be based on maximum longshore current and the wave heights at breaking but in order to permit model data comparisons, we have used local values of both quantities in creating the TDAs. Maximum wave heights are approximated as 155% of the significant wave height. A plot of the results of this exercise is presented in Figure 5 where it is seen that in the great majority of the cases (≥95%), the TDAs derived from model results were identical to those calculated from observations. When discrepancies did occur, at all stations, they were only on a basis of 1 level of pessimism or optimism without an apparent preference to one type or another.

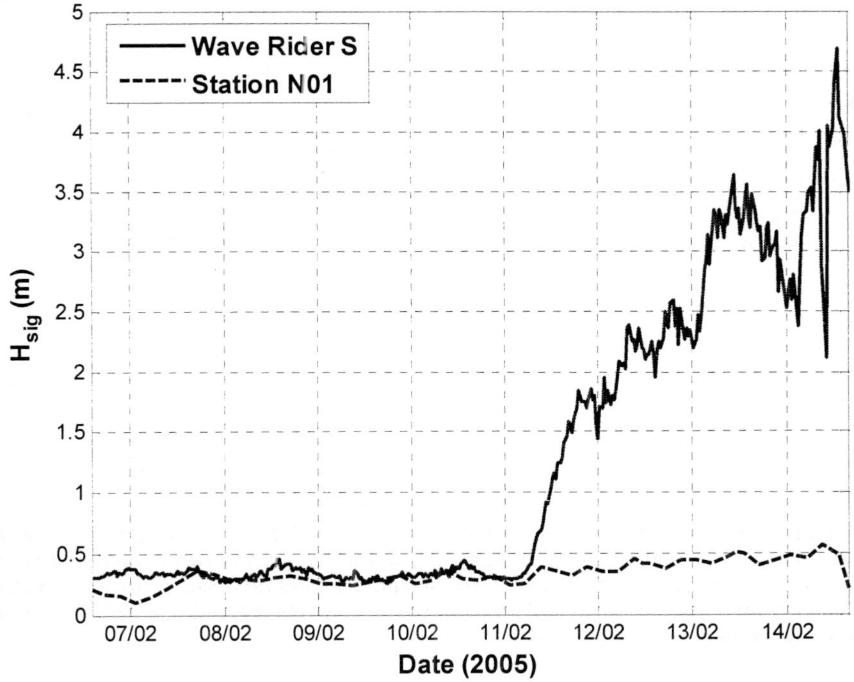

Figure 6: Plot of offshore wave conditions as measured at the southern wave rider (solid line) and surf-zone wave heights measured at station N01 (dotted line). Wave heights in this location are relatively insensitive to the offshore wave height.

ISSM Performance on a Strongly 3-D Shoreline

The ISSM procedure was applied to the SURF'04 field data which was collected at a site with strongly 3-D characteristics in which assumption of smooth bottom contours parallel to the coast is strongly violated. Qualitatively, the wave results from three separate cross-shore profiles vary little so we shall concentrate here on the N1 profile. Figure 6 shows a plot comparing offshore and surf-zone wave parameters as derived from measurements. The key point to observe from this figure is that while the range in offshore wave heights is greater than an order of magnitude (0.2 m to 5 m), the surf zone heights vary only between 0.1 and 0.5 m. This result tells us that the surf zone measurements were collected in a location where waves have already undergone significant energy dissipation, particularly whenever the offshore waves are high. In other words, the location of the surf zone measurements is often well inside the surf zone and this fact must be considered when interpreting the following results.

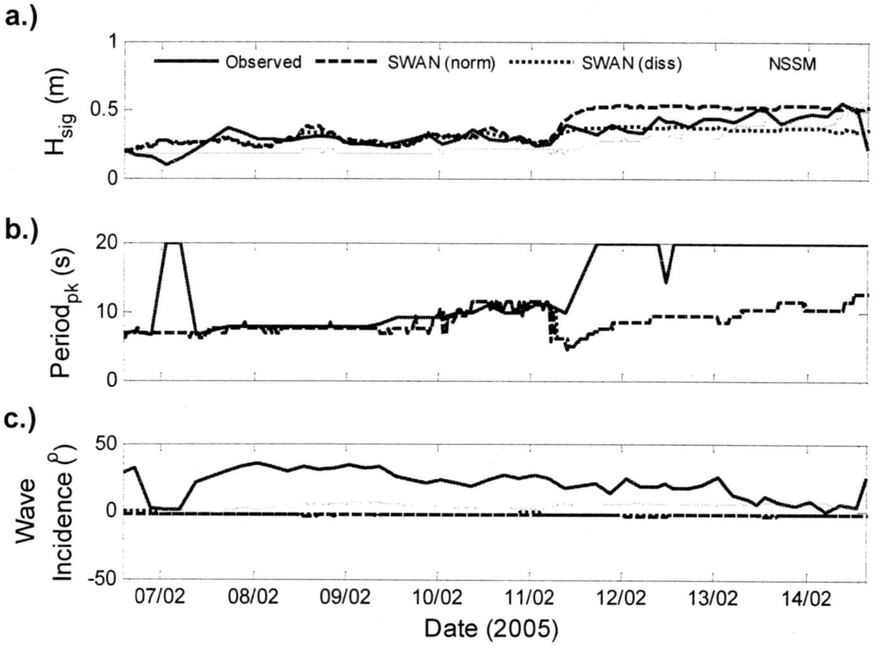

Figure 7: Plot of observed and modeled values of significant wave height (a.), peak wave period (b.), and angle of incidence (c.), at instrument site N01. The heavy solid line represents observed values, the dashed and dotted lines represent SWAN results with default and enhanced dissipation respectively and the thin solid line represents model output from SURF3.2.

Figure 7a provides the comparison between measured wave heights at N01 and those simulated directly from SWAN. It is observed that while there is an excellent match between the two for mild conditions offshore (Hs < 1 m), SWAN appears to consistently overestimate the surf zone wave heights for energetic offshore conditions. Recalling that this location is deep in the surf zone, it was found that this result could be significantly improved by increasing the wave breaker dissipation factor (1.0-1.5) and by reducing the breaking ratio from 0.73 to 0.65 [*Raubenheimer, Guza, and Elgar*, 1996]. Comparisons between observed and predicted wave periods (Figure 7b) indicate good correlation between predicted and observed values except for the conditions where bound long period energy that has been freed by shallow water wave dissipation becomes the dominant process. This process appears not to be properly simulated in SWAN but it is also not a feature which is of general operational relevance. A comparison between modeled wave directions and those derived from measurements is presented Figure 7c. Contrary to the results from SURF'02, the wave directional estimates improve as the shoreline is approached. However to be consistent, the results from N01 are presented here and there is seen to be little relation between observations and simulations. Once again, this is felt to represent a weakness in the observational estimates rather than the simulations.

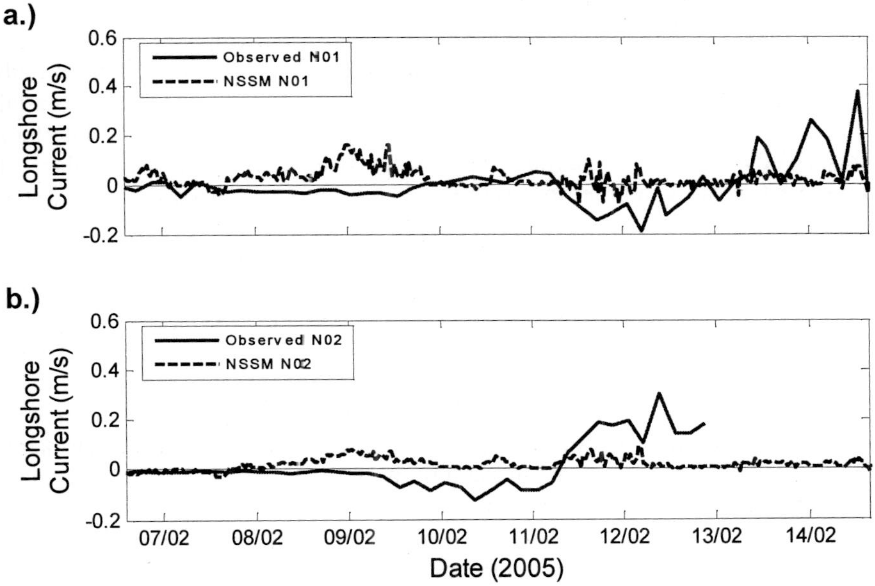

Figure 8: Plot of observed and modeled values of longshore current velocity at instrument sites N01 (a.) and N02 (b.). N02 is on the same cross-shore profile as N01 but 40 m closer to the shoreline. The solid line represents observed values and the dashed line represents model output from SURF3.2.

Taking into consideration the offshore wave heights observed during SURF'04, it was decided to utilize the SWAN output at 6.5 m depth in order to force the SURF3.2 simulations. The nowcast vs. observation comparisons for longshore currents at N1 are presented in Figure 8. From this figure it is immediately obvious that SURF3.2 is essentially incapable of predicting the longshore currents at this location. Indeed the model data correlation is 0.01 indicating a total lack of correlation. Examination of the results indicates this lack of correlation is due to the fact that longshore current conditions observed at Porto Ferro are not predominantly driven by the local cross-shore momentum balance but are instead driven by gradients in the longshore momentum balance and by infra-gravity driven low frequency currents. This inference is highly supported by the presence of opposing cross-shore currents along the N1 profile which is evident at times in Figure 8.

Despite the apparent relatively poor job of modeling the surf zone conditions, an operational assessment of the error similar to that in Figure 5 indicates no error. It can be shown that this is due completely to the calculation of TDAs from inappropriate data and when the TDAs are correctly determined, 100% error is observed during energetic wave conditions. It is critical to note that in this case, the issue is not the reliability of simulations but the reliability of decision aids that are inappropriately calculated.

OBSERVATIONS AND CONCLUSIONS

A hybrid surf-zone modeling methodology has been proposed which has the potential to provide a convenient framework for making surf predictions in both traditional (e.g. long straight beaches) and non-traditional (e.g. coastal embayments) scenarios. An example of this modeling framework, the Interface for SWAN and Surf Modeling (ISSM) has been developed and applied to multiple test cases in order to assess its effectiveness. In the cases studied, the framework has been shown to be very effective in modeling offshore to onshore wave height changes where the correlations between predictions and observations are greater than 0.9. The framework has been seen to be less effective in predicting longshore currents where correlations between observations and predictions have a mean value around 0.6. Despite these results, when performing an operational evaluation of the framework effectiveness, it is seen that over 90% of the time operational decisions determined from model predictions are identical to those that would be made based on in-situ observations. In the few cases where there were discrepancies, the discrepancy was always only one level of operational consideration. In one of the test case locations, it has been demonstrated that the observed errors due to limitations in the modeling approach can be much lower than errors due to faulty application of the analysis procedure.

ACKNOWLEDGMENTS

The authors would like to thank, the Naval Oceanographic Office for permission to use the NSSM, Rick Allard, Larry Hsu and Erick Rogers of the Naval Research Laboratory, Stennis Space Center, for their guidance and advice on the use of SURF and SWAN.

REFERENCES

Booij, N., R. C. Ris, and L. H. Holthuijsen, 2001, A third-generation wave model for coastal regions. 1. Model description and validation., *J.Geophys.Res.*, 104, 7649-7666.

Commander Naval Surface Force Pacific and Commander Naval Surface Force Atlantic. 1987, *Joint Surf Manual.* 3840.1B, 1-36. 1-2-1987. COMNAVSURFPAC/COMNAVSURFLANTINST.

Conley, Daniel C., Rusu, E., Ferreira-Coelho, R. Allard 2006, Hybrid surf modelling: final report, NURC-FR-2006-016, 1-49, La Spezia, NATO Undersea Research Centre.

Earle, M.D. 1989. *Surf Forecasting Software Scientific Reference Manual*, NORDA Report, Naval Ocean Research and Development Activity, 261 p.

Report, Naval Ocean Research and Development Activity, 261 p.Grasso, Raffaele and Giannecchini, Simone, 2006, Geo-spatial tactical decision aid systems: fuzzy logic for supporting decision making. 2006. Florence, International Society of Information Fusion. *International conference on information fusion, 9th.*

Hsu, Y. Larry, Mettlach, Theodore R., and Earle, Marshall D., 2002, *Validation for the Navy Standard Surf Model.* NRL/FR/7322-02-10008, 1-28. 6-21-2002. Stennis Space Center, MS, Naval Research Laboratory.

Huntley, D. A. and M. A. Davidson, 1998, Estimating the Directional Spectrum of Waves near a Reflector, *Journal of Waterway, Port, Coastal and Ocean Engineering*, 124, 312-319.

Jin, K. R. and Z. G. Ji, 2001, Calibration and verification of a spectral wind-wave model for Lake Okeechobee, *Ocean.Engineering*, 28, 571-584.

Johnson, David. 2002, *DIWASP, a directional wave spectra toolbox for MATLAB®: User Manual.* WP 1601 DJ (V1.1), i-18. 2002. Centre for Water Research, University of Western Australia.

Mettlach, Theodore R. and May, Douglas A., 1997, *The accuracy of the navy-standard surf model-derived modified surf index and its sensitivity to nearshore bathymetric profile errors*. NRL/FR/7240--97-9665, 1-29. 10-24-1997. Stennis Space Center, MS, Naval Research Center.

Ou, S. H., J. M. Liau, T. W. Hsu, and S. Y. Tzang, 2002, Simulating typhoon waves by SWAN wave model in coastal waters of Taiwan, *Ocean.Engineering*, 29, 947-971, 2002.

Pawlowicz, Rich. 2005, *M_Map Matlab toolbox. (1.4b).* http://www.eos.ubc.ca/~rich/map.html.

Raubenheimer, B., R. T. Guza, and S. Elgar, 1996, Wave transformation across the inner surf zone, *J.Geophys.Res.*, 101, 25589-25598.

Rusu, Eugen, Daniel C. Conley, and Emanuel Ferreira-Coelho, 2006, *A NATO tool for predictions of waves and longshore currents in the surf zone.* NURC-FR-2006-014. 1-35, La Spezia, NATO Undersea Research Centre.

Wave-averaged and Wave-resolving Numerical Modeling of Vorticity Transport in the Nearshore Region: the SANDYDUCK Case Study

Riccardo Briganti[1], James T. Kirby[2], Fengyan Shi[2], Maurizio Brocchini[3], Qin Chen[1] and Bruce L. Lipphardt[4]

The paper analyzes the mixing features in numerically generated shear waves by simulating the circulation during SANDYDUCK experiments. Different wave resolving Boussinesq type models and a wave averaged model have been compared. They produce different velocity/vorticity fields and in turn different mixing properties. Lagrangian statistics have been used to study this aspect. Differences in absolute and relative dispersion and diffusion have been pointed out. All the models seem to follow a Richardson-like intermediate regime for the longshore relative dispersion even though some results seem to suggest a dependency from particles separation weaker than the classical $4/3$ law.

INTRODUCTION

Numerical modeling of unstable nearshore flows has become quite popular in the last decade. A class of models, usually adopted for this purpose, relies on averaging the equations over the time scales of the short waves (hence wave averaged models). The wave characteristics, and in turn the forcing terms are provided by another model (wave driver). Examples of longshore current studies using these models are Özkan Haller and Kirby (1999), Zhao et al. (2003), among others. On the other hand in wave resolving models, all the time scales of the hydrodynamics are explicitly solved. Boussinesq-type models (BTMs hereinafter) belong to this class of models and have been used, for instance, in Chen et al. (2003), Gobbi et al. (2000) and Johnson and Pattiaratchi (2006) among others. When applied in similar conditions, the two approaches may produce different results in terms of flow description due to different forcing mechanisms. Differences may arise within the same class of models that share a similar set of equations. In particular it has been shown in Gobbi et al. (2000) that different transport equations for vertical vorticity may be written starting from different BTMs, depending on the order of magnitude of the retained vertical vorticity. The modeled flows in wave averaged models are characterized by less energetic fluctuations with respect to those predicted by BTMs. In turn the vorticity transport is radically different,

[1] Dept Civil Engineering, Louisiana State University, Baton Rouge, LA 70803-USA, email: ricbrig@lsu.edu

[2] Center for Applied Coastal Research, University of Delaware, Newark, DE 19716 - USA.

[3] Dipartimento di Idraulica e Ingegneria Ambientale, Universita' di Genova, IT.

[4] College of Marine Studies, University of Delaware, Newark, DE 19716 - USA.

producing different mixing features. An interesting approach which allows to investigate this latter aspect, is the study of the trajectories of particles (tracers) released in the fluid, as done, for example in oceanic field experiments (e.g. LaCasce and Bower, 2000). Following the theory of turbulence, emphasis is given to Lagrangian statistics rather than on the description of the single drifter trajectory. The statistics allow to investigate the regimes of the absolute and relative dispersion of the floaters. The outcome of both Johnson and Pattiaratchi (2004) and Piattella et al. (2006) is that relative dispersion depends from particles separation with a $4/3$ power law both within and outside the surf zone. The present study aims at identifying the characteristics of the dispersion induced by the unstable longshore currents as modeled by both wave resolving BTMs and a wave averaged model and investigate the reasons of these differences.

NUMERICAL MODELS USED

Wave resolving models

The wave resolving model used in this study is based on the Boussinesq type equations described also in Gobbi et al. (2000). The equations are defined in a cartesian coordinate system x, y, z, with z directed upward. We adopt the classical non-dimensionalization of the governing equations choosing the offshore short waves amplitude a_0, water depth h_0 and inverse of the wave number $1/k_0$ as characteristic length scales for wave amplitude, water depth and wave length, leading to dimensionless parameters $\delta = a_0/h_0$ and $\mu = k_0 h_0$. The equations are obtained without any limitation on δ. Terms only up to $O(\mu^2)$ are retained. The velocity is defined introducing $\mathbf{u}_\alpha = (u_\alpha, v_\alpha)$ as the horizontal velocity vector at an arbitrary elevation z_α. The integration over depth of the Euler equation leads to a dimensionless set of governing equations in which the mass conservation equation is given by:

$$\eta_t + \nabla \cdot \mathbf{M} = O(\mu^4) \tag{1}$$

where M is the horizontal volume flux vector, η is the free surface. The momentum conservation equation is given by:

$$\frac{D\mathbf{u}_\alpha}{Dt} + \nabla\eta + \mu^2 \left\{\mathbf{V}_1 + \mathbf{V}_2 + \mathbf{V}_3(z)\right\} + \mathbf{R} = 0 \tag{2}$$

$$\mathbf{R} = \mathbf{R}_f - \mathbf{R}_b - \mathbf{R}_l \tag{3}$$

The wave breaking terms are expressed by $\mathbf{R}_b$, while $\mathbf{R}_f$ model the bottom friction and $\mathbf{R}_l$ the subgrid mixing terms. The term $\mathbf{V}_3(z)$ introduces dependence of the equations from z, closure of the system depends on choice of method for evaluating this term. We here compare results from three different closures:

1. Wei et al. (1995): Neglects $\mathbf{V}_3$.
2. Chen et al. (2003): Evaluates $\mathbf{V}_3$ at $z = z_\alpha$
3. Chen (2006): In this model the momentum equations are integrated over depth.

The models differ in terms of the advection properties for vertical vorticity. The eulerian description of the flow field in rip and longshore currents case has been discussed in Gobbi et al. (2006). The differences in the predicted vorticity transport and flow organization, suggested to investigate the results in terms of the mixing features.

Wave-averaged model

It is well known that wave-averaged circulation models rely on a decomposition of the velocity $\mathbf{u}(x, y, z, t) = (u, v)$ in components corresponding to different time scales, following the approach in (Putrevu and Svendsen, 1999). We report here only the momentum equation in non conservative form for sake of brevity:

$$\tilde{\mathbf{U}}_{,t} + (\tilde{\mathbf{U}} \cdot \nabla)\tilde{\mathbf{U}} =$$

$$= -g\nabla\bar{\eta} - \frac{1}{gD}\nabla \cdot (\mathbf{S} - \mathbf{T} + \mathbf{L}) - \frac{1}{\rho D}\tau_B + \frac{1}{\rho D}\tau_S \quad (4)$$

where $\bar{\eta}$ is the mean surface elevation with and h_0 the still water depth with. The local total water depth h, hence is $D = h_0 + \bar{\eta}$. $\tilde{\mathbf{U}}$ is the depth-uniform current velocity vector, $\mathbf{S}$, $\mathbf{L}$ and $\mathbf{T}$ are, respectively, the radiation stress tensor, the dispersive mixing tensor and the turbulent stress tensor, while τ_B and τ_S are the bottom and surface shear stress vectors respectively. Since Boussinesq type models do not retain a description of the vertical structure of the mean current the terms describing the vertical structure of the current are neglected also in the wave averaged model, being the comparison between the two the objective of the paper. Hence the original equations solved with SHORECIRC, coincide with non-linear Shallow water equations (NSWE).

NUMERICAL EXPERIMENTS

The study carried out here involved numerical tests using both a real bathymetry and an idealized one. The SANDYDUCK experiment was selected as it is suitable for studying particles dispersion in a longshore current over a relatively simple bottom configuration. However, before considering the real bathymetry, an idealized one will be taken into account to reduce the complexity of the problem introduced by a real sea bottom. This bathymetry is generated considering a cross-shore section of the surveyed SANDYDUCK one (the model bathymetry obtained from the survey is shown in Fig. 1). Also wave conditions have been simplified with respect to the ones relative to the time interval between $4:00am$ and $7:00am$ on Oct, 2nd 1997.

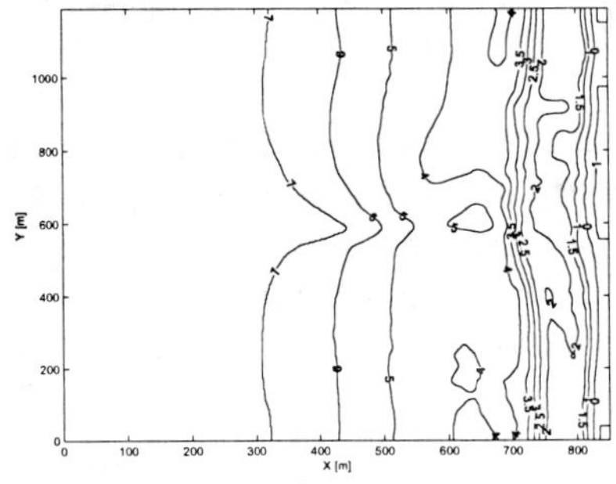

Figure 1. **Model bathymetry for SANDYDUCK**

The bathymetry is uniform along the y coordinate, repeating the cross-shore section located at $y = 267m$ in the model coordinate system shown in Fig. 1 (we here use model coordinates and not the FRF ones). The numerical mesh size is 567×298 with $\Delta x = 1.5m$ and $\Delta y = 4.0m$ and it is periodic in y. The wave attack has been assumed to be monochromatic with 1.06 m amplitude, and 6.6 s period and incidence angle at the generation point is $26°$. The turbulent mixing and bottom friction coefficients are 0.15 and 1.5×10^{-3}, respectively. All the simulations start from cold start and cover a 3 hours time interval. In order to study the mixing properties described by the models it is necessary to compute the trajectories of virtual particles displaced in the domain. For each model, the velocity field at each time step is stored and processed Particles trajectories are obtained by numerical integration of the Eulerian velocity components using a ODE solver (Runge Kutta forth order) implemented in MATLAB. The OdEs are simply: $Dx/dt = u$ and $Dy/dt = v$, where u and v are defined by (u, v)= $(u_\alpha$,$v_\alpha)$ for the BTMs and by considering both waves and currents in NSWE. Linear interpolation in space has been used. No interpolation in time is used, dt being the same as the one used for the numerical models time marching. The virtual particles are launched in randomly filled circular patches. In order to avoid the interaction with the lateral boundaries the cluster is released always at aa distance (usually $100m$) from one of the upstream lateral boundary. As in LaCasce and Bower (2000) the attention will be focused on the dispersion and diffusion of the particles. Once the trajectories $(x_i(x_0, t), x_j(x_0, t))$, $i, j = 1, 2$ (i,j=1, cross-shore direction; i,j=2, longshore direction) are known, it is possible to compute the elements of the absolute dispersion tensor:

$$\langle X_{ij}^2(t)\rangle = \frac{1}{M} \sum_{particles} x_i x_j \tag{5}$$

where x_i is the particle displacement from its initial position in the ith-direction and in the jth-direction and M is the number of particles in the cluster. The trace of this matrix is the total absolute dispersion, i.e. $\langle X_{tot}^2 \rangle = \langle X_{11}^2 \rangle + \langle X_{22}^2 \rangle$. The derivative in time of $\langle X_{tot}^2 \rangle$ is defined as the total absolute dispersion of the cluster$K$. If only the cross-shore and longshore elements are considered cross-shore and longshore diffusivity are obtained. In analogy with the absolute dispersion we define a relative dispersion. This is computed on pairs of particles located within a given distance at the time of the tracers release. The elements of the relative dispersion tensor are defined as:

$$\langle D_{ij}^2(t) \rangle = \frac{1}{N} \sum_{pairs} y_i y_j \tag{6}$$

where y_i is the separation between two particles in the same pair in the ith-direction and in the jth-direction and N is the number of pairs. As done for the absolute statistics, the trace of the tensor represents the total relative dispersion $\langle D_{tot}^2 \rangle = \langle D_{11}^2 \rangle + \langle D_{22}^2 \rangle$, hence a relative diffusion μ may be defined as the derivative of D_{tot}. The behavior of the relative dispersion and diffusivity in time and the relationship between these properties and particles/drifters separation (S), is often used as an indicator of the mixing regime of the flow. Of particular interest is the regime reached and before the relative and absolute diffusion both reach a constant value (browinian regime). In the presence of a background shear the turbulent relative diffusivity follows a $4/3$-law in this intermediate regime. Similar behaviors can also occur in the presence of waves. Johnson and Pattiaratchi (2004) found evidence of this law inside and outside the surf zone studying pulsating rip currents.

Longshore uniform barred beach

We here discuss the numerical experiments carried out using the uniform barred beach. In this case a cluster of 80 particles has been released on the bar, inside the surf zone in all the models considered here. In all the models the cluster is released after 2 hours and 15 minutes from the beginning of the simulation. Fig. 3 shows the trajectories of the particles as computed starting from the models results. Even though the trajectories displayed here come from a particular launch, other attempts have been made, all resulting in similar statistics for each of the model, even though the details of the trajectories may differ.

The virtual particles have been released at the beginning of the third hour of each simulation. The Eulerian field at that time is different for each model, details may be found in Gobbi et al. (2006) and will not been shown here. Inspection of the figure reveals several important differences.Results coming from Wei et al. (1995) show that all the particles are transported

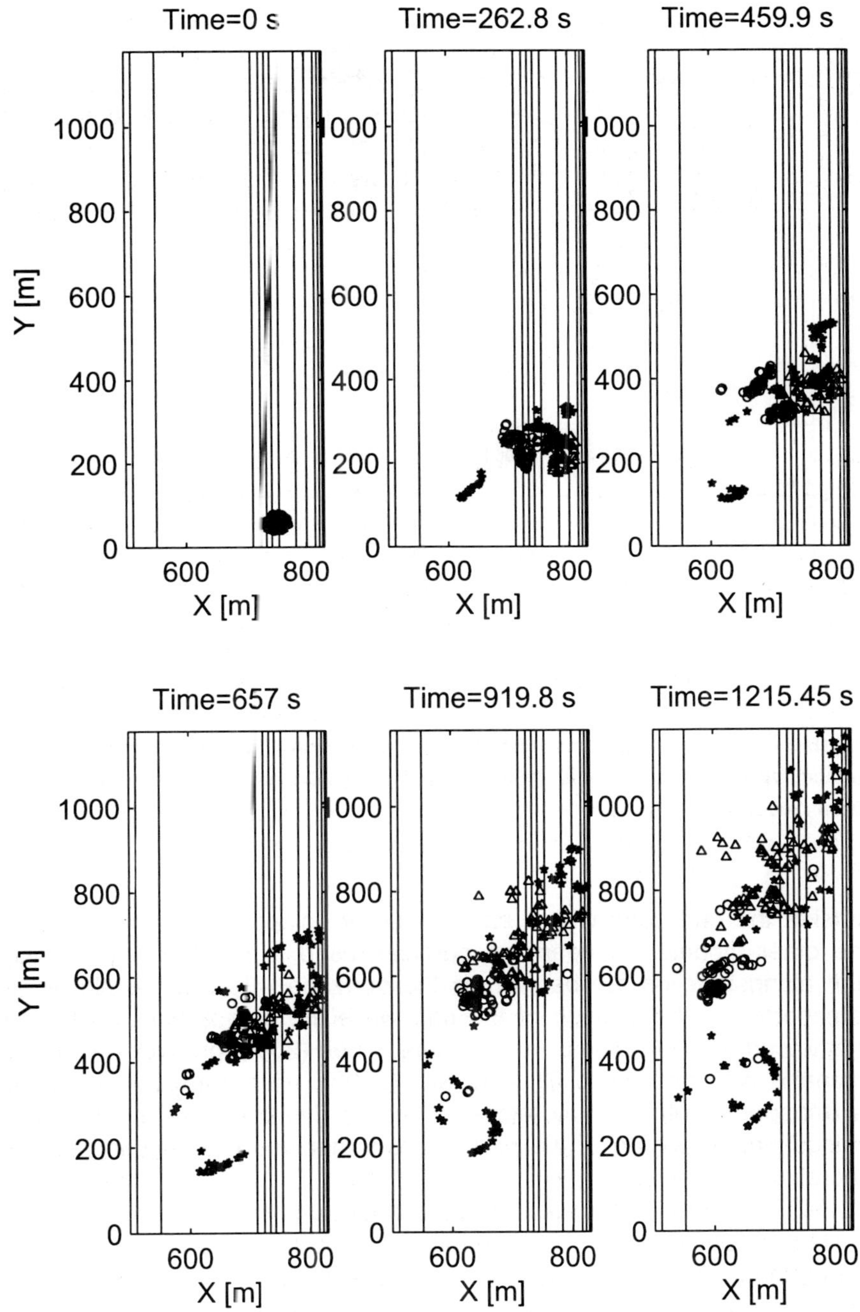

Figure 2. Particles trajectories on a longshore uniform barred breach. Circles Wei et al. (1995), triangles Chen et al. (2003), stars Chen (2006).

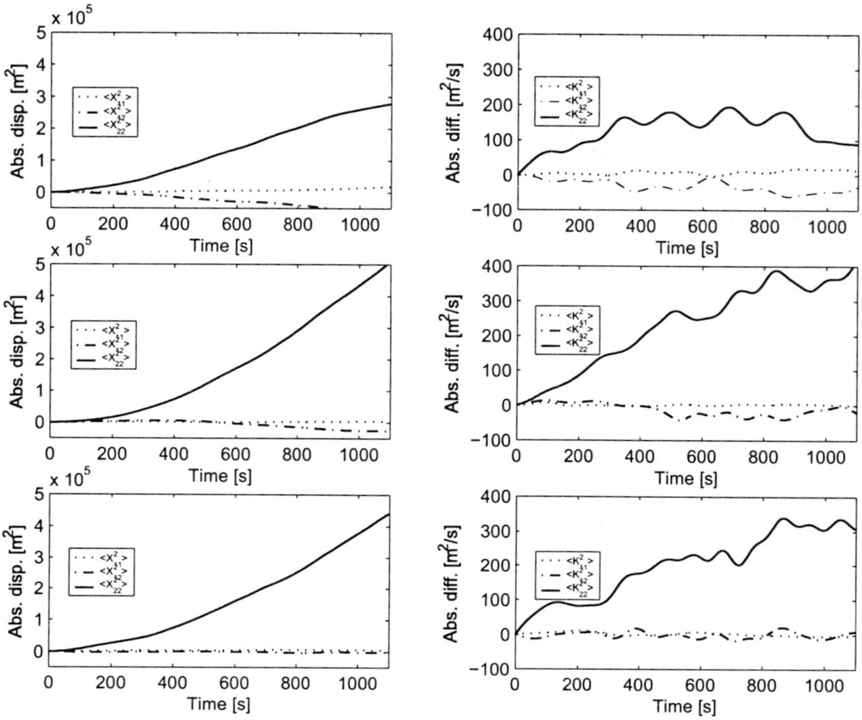

Figure 3. Absolute dispersion (left) and diffusion (right), longshore uniform barred breach. Upper panels Wei et al. (1995), central panels Chen et al. (2003), lower panels Chen (2006).

outside the surf zone with a smaller relative dispersion if compared to the other models. Instead the particles used in Chen et al. (2003) drift alongshore on the bar across the surfzone. Part of the cluster is transported in the through of the bar, while the part is drifted offshore by the current instabilities. Furthermore, Chen (2006) produces a similar scenario with respect to Chen et al. (2003), while part of the cluster is transported offshore by jets and part in the trough of the bar, the alongshore spreading of the particles is larger and the transport to the shoreline is stronger.

As it is evident, after about $1200s$ a large number of particles reaches the lateral boundary, hence, in order to avoid the influence of this region on the boundary, the results will include only the first $1100s$ of the simulations. Fig. 3 shows the elements of the absolute dispersion tensor ($\langle X_{ij}^2 \rangle$) and absolute diffusivity ($\langle K_{ij} \rangle$) Absolute statistics (see Fig. 3) allow some consideration:

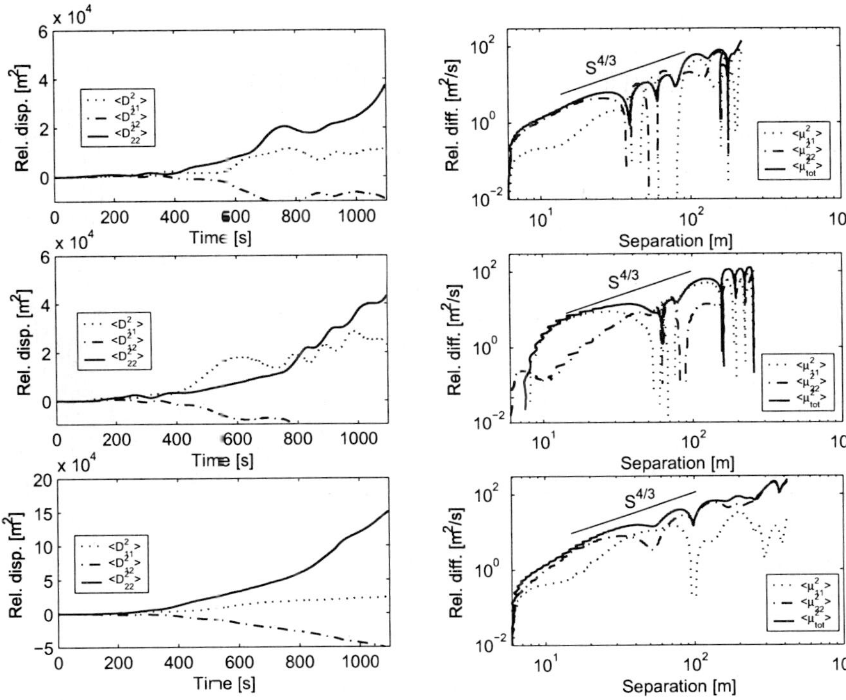

Figure 4. **Relative dispersion (left) and diffusion (right), longshore uniform barred breach. Upper panels Wei et al. (1995), central panels Chen et al. (2003), lower panels Chen (2006).**

1. Wei et al. (1995) model shows the lower longshore absolute dispersion, and the larger cross-shore one.
2. The cross-shore dispersion is from one to two orders of magnitude lower than the longshore one for each model.
3. Within the time interval in which the trajectories are studied, a constant value of longshore diffusivity seems to be reached for Chen et al. (2003) and Chen (2006) after $800s$, even though oscillations persist. For Wei et al. (1995) diffusivity oscillates around the value of $150m^2/s$ after $300s$ and decreases after $1000s$. However the existence of a plateau for the diffusivity is not clear in the interval studied as it is influenced by the mean longshore current which causes a growth in diffusivity.

More important indications are given by the relative (two particles) statistics shown in Fig.3. The relative statistics shown here are computed considering an initial separation smaller than $4m$. The results suggest

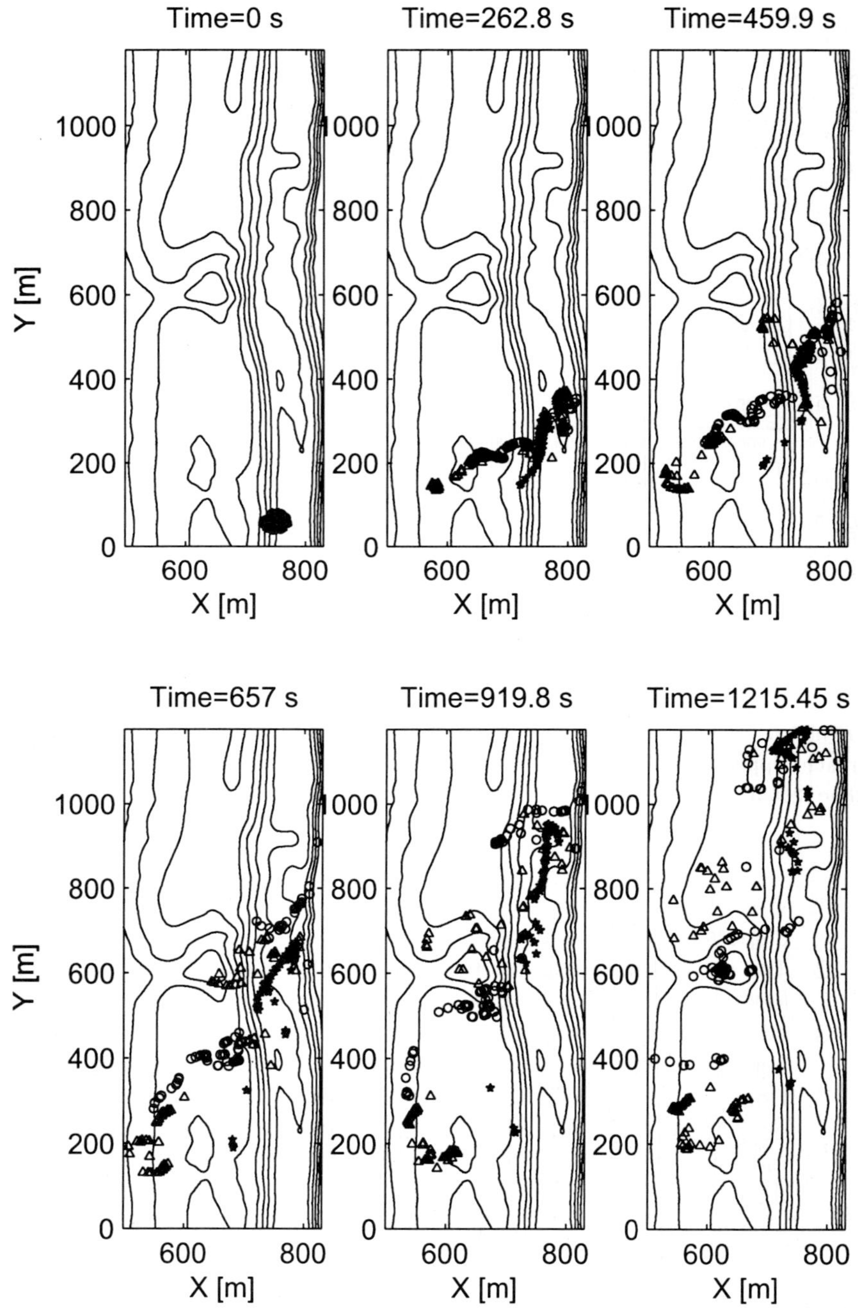

Figure 5. **Particles trajectories on SANDYDUCK bathymetry. Circles Chen et al. (2003), triangles Chen (2006), stars NSWE.**

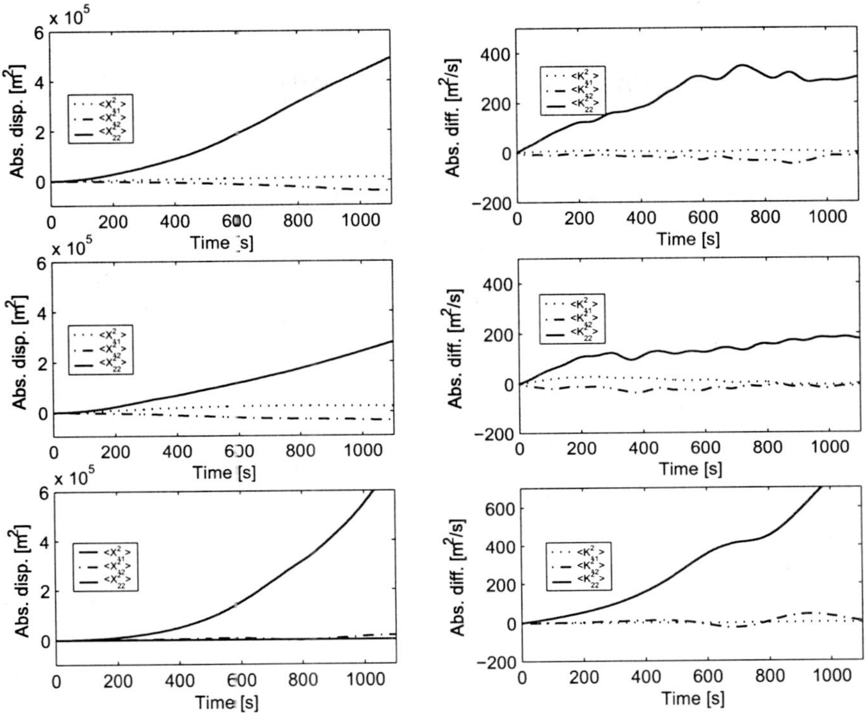

Figure 6. **Absolute dispersion for Sandyduck case. Upper panels Chen et al. (2003), central panels Chen (2006), lower panels NSWE.**

the existence of an intermediate regime close to a Richardson like $4/3$ law, for all the models, even though some results seem to support a weaker dependency. At separation smaller than $100m$, Chen et al. (2003) and Chen (2006) show a different relative importance of cross-shore and longshore dispersion/diffusivity. Longshore relative diffusion and dispersion dominates in Chen (2006), the opposite in Chen et al. (2003). Though this similarity exists from a statistical point of view between Wei et al. (1995) and Chen (2006) at small scales, it has to be reminded that the scenario described by the particles trajectories is totally different between the two models. Chen et al. (2003) and Chen (2006) trajectories describe a similar scenario, however, a large difference between the two models exists in term of relative statistics and in the magnitude of the longshore dispersion which is larger in Chen (2006). This may be explained by considering that the absolute statistics is influenced by the presence of particles that drift offshore in a zone where the longshore current is not effective. Besides, in Chen (2006) for separations larger than

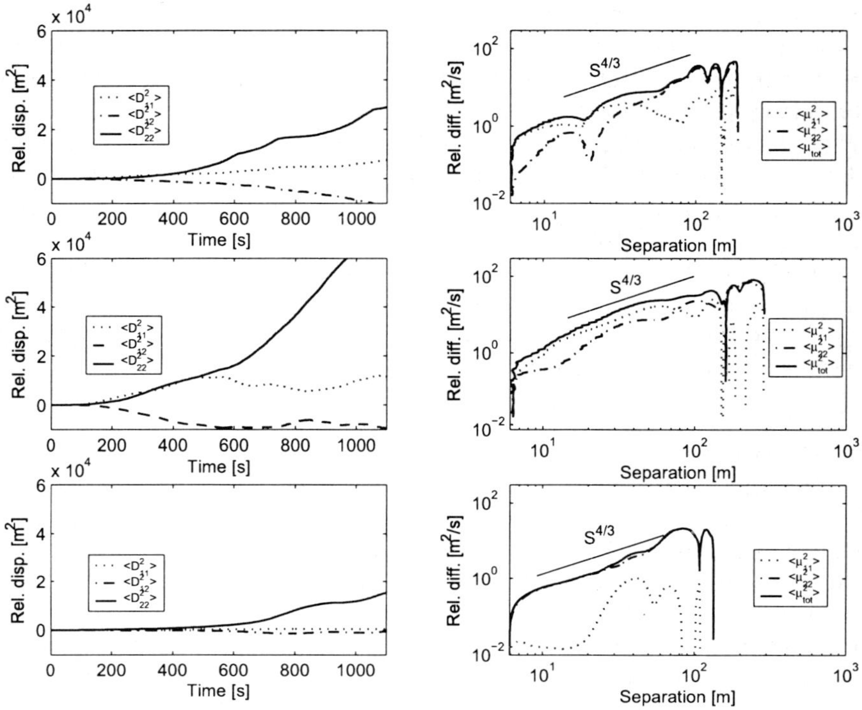

Figure 7. **Relative dispersion for Sandyduck case. Upper panels Chen et al. (2003), central panels Chen (2006), lower panels NSWE.**

about $100m$ a plateau characterizes cross-shore diffusivity while longshore one still grows. The figure also shows a comparison with a $4/3$ power law, which is well followed by Wei et al. (1995) and (Chen et al., 2003)

The Sandyduck case

In this section we discuss the simulation of the the propagation of regular waves over a real bathymetry coming from the SANDYDUCK experiments. The bathymetry is illustrated in Fig. 1. In this case only two BTEs were tested and also the results coming from the NSWE approach are shown. Fig. 3 shows the different trajectories coming from Chen et al. (2003), Chen (2006) and the NSWE. Most of the features found in the previous tests are confirmed here, the transport offshore is generally larger considering the real bathymetry and field wave conditions and in both BTMs is not localized in the rip channel only. The main differences arise from a weaker longshore current offshore the bar in Chen (2006). This cause the particles to drift offshore without a large longshore transport as in Chen et al. (2003). Totally different from these results are those coming from the NSWE. In this case, due to weaker instabilities, no vortex detachment from the longshore flow is detected and the transport is almost exclusively directed alongshore. As a result both the absolute dispersion (Fig.3) and the relative one (Fig.3) are dominated by the longshore component. Interesting results are given by the relative statistics. The total diffusivity in Chen (2006) seems to follow a power law with an exponent lower than $4/3$ for separations larger than $30 - 40m$ because of the contribution of the cross-shore diffusivity. The longshore diffusivity follows the aforementioned law. Also in Chen et al. (2003)the cross-shore diffusivity is important at separations up to $30 - 40m$, where the longshore diffusivity does not follow a $4/3$ law. The longshore diffusivity seems to reach a plateau for scales larger than $100m$. The $4/3$ law is well followed by the particles coming from the NSWE model.

CONCLUSIONS

The study has analyzed the trajectories of virtual particles released in the surf zone in numerically simulated shear waves. Three wave resolving Boussinesq type models with different vorticity transport properties have been compared using an idealized bathymetry and a real bathymetry coming from the SANDYDUCK field experiment. In this latter case, together with two among the selected BTMs, a wave averaged model based on NSWE (i.e. SHORECIRC without the vertical dispersion description) has been used. Results coming from a representative launch have been shown. The analysis of the trajectories and the Lagrangian statistics pointed out several differences among the models. Wei et al. (1995) predicts a very low transport inshore. On the other hand Chen et al. (2003) and Chen (2006) trajectories are quite close with each other. The main difference between the two models is the role played by the cross-shore transport directed offshore that is relatively more important in Chen (2006) case for the considered launch at small separations. The Richardson $4/3$ law is found reasonably well supported by the longshore relative diffusivity in the BTMs under study and in the wave averaged model even though some results suggest a lower exponent at large ($>100m$) separations expecially for Chen (2006). This occurs in both the considered bathymetries. Results obtained with the wave-averaged model can only give few indications, the flow dynamics at time scales smaller than that of the wave being ignored by definition. However, given the weaker longshore current instabilities generated the cross-shore transport is very weak and the longshore dispersion dominates the process. Analysis is underway in order to overcome the limitation of the present version of the ODE solver at the lateral boundary and to explain the role of the shoreline boundary conditions in influencing the shear dispersion that is observed close to shoreline.

ACKNOWLEDGEMENTS

R.B. and Q.C. have been supported by the ONR Coastal Geosciences Program research Grant No. N00014-04-1-0310. J.T.K and F.S. have been supported by ONR Coastal Geosciences Program research Grant No. N00014-05-1-0069. M.B. has been supported by Italian M.I.U.R. through the grant INTERLINK-II04C02L8E.

REFERENCES

Chen, Q.. 2006. Fully nonlinear boussinesq-type equations for waves and currents over porous beds, *Journal of Engineering Mechanics,* 132, 220–230.

Chen, Q., J. T. Kirby, R. A. Dalrymple, F. Shi, and E. B. Thorton. 2003. Boussinesq modeling of longshore currents, *Journal of Geophysical Research,* 108 (c11), 3362, doi:10.1029/2002JC001308.

Gobbi, M. F., J. T. Kirby, R. Briganti, and Q. Chen. 2006. Boussinesq wave models and vertical vorticity fields, *In preparation,*.

Gobbi, M. F., T. J. Kirby, and A. B. Kennedy. 2000. On the consistency of boussinesq models and their ability to predict vertical vorticity fields, *Proc., XXVIIth Intl Conf. Coastal Engrng.*, Sidney, Aus., ASCE, 380–388.

Johnson, D. and C. Pattiaratchi. 2004. Transient rip currents and nearshore circulation on a swell dominated beach., *Journal of Geophysical Research,* 109, C02026, doi: 10.1029/2003JC001789.

Johnson, D. and C. Pattiaratchi. 2006. Boussinesq modelling of transient rip currents, *Coastal Engineering,* 53, 419439.

LaCasce, J. H. and A. Bower. 2000. Relative dispesion in the subsurface north atlantic, *Journal of Marine Research,* 58, 863–894.

Özkan Haller, T. and J. Kirby. 1999. Nonliner evolution of shear instabilities of the longshore current: A comparison of observations and computations, *Journal of Geophysical Research-Oceans,* 104(C11), 25953–25984.

Piattella, A., M. Brocchini, and M. A.. 2006. Topographically-controlled, breaking wave-induced macrovortices. part 3. the mixing features., *Journal of Fluid Mechanics,* 58, 863–894.

Putrevu, U. and I. Svendsen. 1999. Three-dimensional dispersion of momentum in wave-induced nearshore currents, *Eur. J. Mech. B/Fluids,* 18(3), 409–427.

Wei, G., J. T. Kirby, S. T. Grilli, and R. Subramanya. 1995. A fully nonlinear boussinesq model for surface waves. i. highly nonlinear, unsteady waves., *J. Fluid Mech.,* 294, 71–92.

Zhao, Q., I. Svendsen, and K. Haas. 2003. Three-dimensional effects in shear waves, *J.Geophys. Res.,* 108(C8), 3270.

VIDEO OBSERVATIONS OF LONGSHORE CURRENTS, MYRTLE BEACH, SOUTH CAROLINA

Kevin A Haas[1] and M. Kemal Cambazoglu[2]

Field observations of longshore currents near Myrtle Beach, South Carolina have been made using both the optical current meter technique and in situ instrumentation. The procedure for providing quantified estimates of the longshore current from the video imagery is briefly described. Time series of the longshore current from both the in situ instruments and the video estimates are compared showing good agreement at least for the very low frequency variations. In addition, reasonable cross-shore profiles of the longshore current are obtained from the video.

INTRODUCTION

Longshore currents on beaches play an important role for many processes in the nearshore environment. For example, longshore sediment transport is primarily driven by the longshore current. Video imagery, in particular, has been used by many researchers to quantify a number of naturally occurring nearshore physical processes. Optically based methods not only give the advantage of low-cost and logistic-ease studies, but also provide wide coverage in nearshore estimations. In this study, video observations are used in order to apply optical techniques to determine the characteristics and temporal variations of longshore currents during a field experiment. The basic idea is to use the foam left behind by breaking waves as a tracer for the longshore current using the optical current meter (OCM) technique developed by Chickadel et al. (2003).

EXPERIMENTAL SETUP

As part of the South Carolina Coastal Erosion Study (SCCES) a nearshore field experiment was carried out from December 10th- 15th, 2003 near Myrtle Beach, South Carolina, providing measurements of the waves, currents and morphological evolution (Haas et al. 2004 and Obley et al. 2004). The experiment consisted of in situ measurements of the longshore currents using two Sontek Triton downward-looking Acoustic Doppler Velocimeters, and two Nortek AquaDopp profilers arranged in cross-shore line from inner surfzone to several surf zone widths past the breakers. In addition, a video camera was located on a balcony of a nearby hotel 45 m above the beach providing an excellent field of view of the entire experimental area. The digital video was captured directly onto a computer via a firewire connection during daylight

[1] Georgia Tech Savannah, 210 Technology Circle, Savannah GA 31407, USA, kevin.haas@gtsav.gatech.edu.
[2] Georgia Tech Savannah, 210 Technology Circle, Savannah GA 31407, USA, mkc@gatech.edu.

hours at 30 Hz (down sampled to 2 Hz in this study) and many control points were surveyed each day to facilitate rectification of the imagery.

A variety of conditions were encountered during the experiment (see the wind data shown in Figure 1) including two fronts which passed through the region, generating wind speeds up to 15 m/s. This observed weather system during the experiment is typical of the weather systems passing through coastal South Carolina during the winter months. The first front (December 11) generated waves from the south driving a longshore current towards the north as seen in Figure 2. After several relatively calm days with nearly normal incident waves the second front passed through the area (December 13-14) with strong wind and waves approaching the shore with a large angle of incidence from the north. This drove an extremely strong longshore current toward the south which was strong enough to knock over and bury the tripods holding the Triton ADV's.

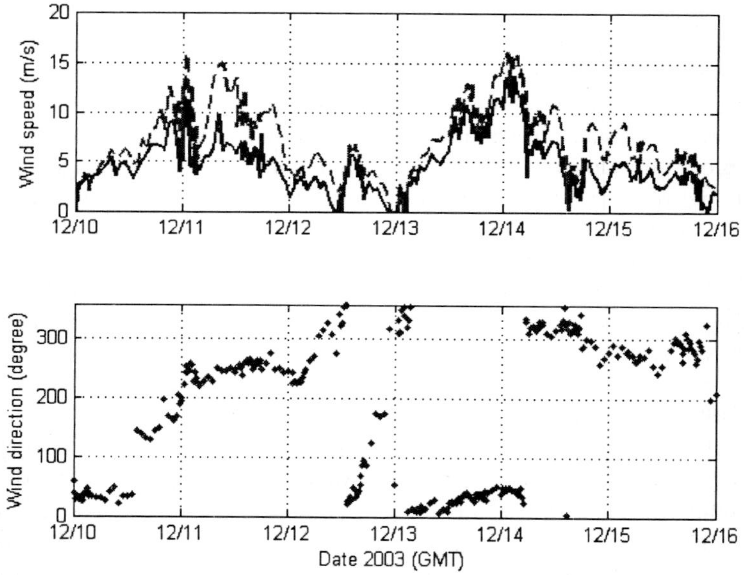

Figure 1 Mean (solid) and maximum (dashed) wind speed (top) and direction (bottom) during the 2003 field experiment.

IMAGE PROCESSING

The procedure used for processing the video imagery into quantified information about the longshore currents closely follows the methodology described by Chickadel et al. (2003). Therefore, only a brief description outlining the steps is presented here.

The first step is to identify a longshore array of points centered about the location where the longshore current is to be measured. An array of 64 points with a spacing of 0.5 m is selected as shown in Figure 3. The pixel coordinates (with subpixel accuracy) are then calculated for each of the points in the array. An intensity value is calculated from the individual frames for each array point by utilizing a 2D spatial linear interpolation scheme. These intensity values from 128 frames are stacked on top of each other forming a so-called "time stack" as shown in Figure 4.

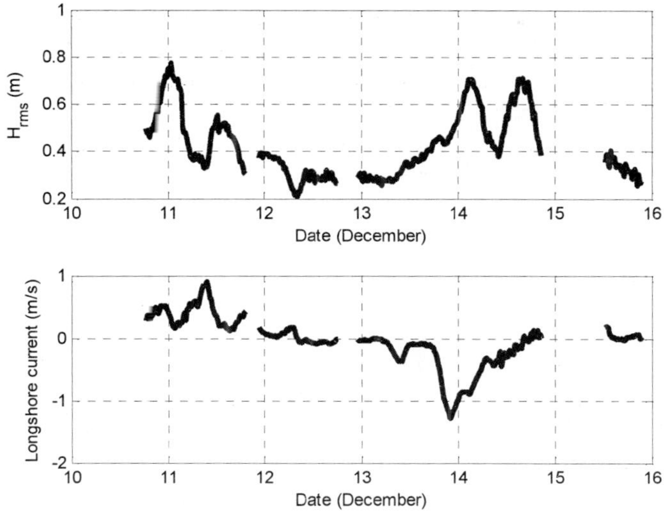

Figure 2 Wave height (top) and Longshore current (bottom) time series from the inshore AquaDopp.

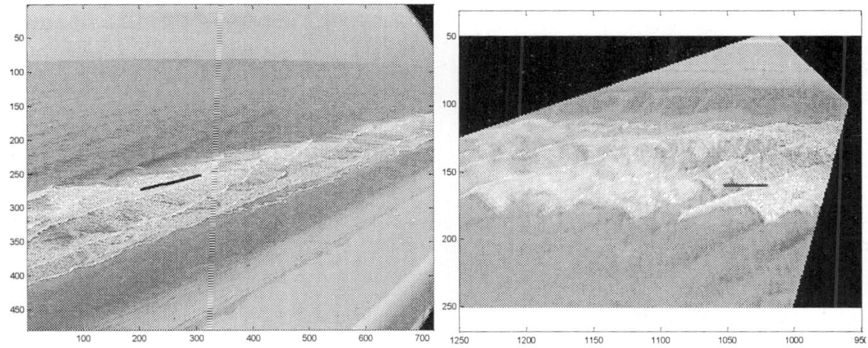

Figure 3 Raw frame (left) and rectified image (right).

A time stack is a mapping of the array intensities in longshore space on the horizontal axis and time on the vertical axis. The bright points in the time stack are from the patches of foam. The bright nearly horizontal lines are traces of the wave crest passing through the longshore array. The steeper stripes of bright and dark patches are from the foam being advected down the array with the slope being inversely proportional to the velocity. The goal is to use some sort of a reliable measure of this velocity and to automate that process.

Several filters which are applied to the time stack, including a median filter, have been found to increase the percentage of acceptable longshore current calculations. A 2 dimensional FFT is applied to generate a foam intensity spectrum in wave number and frequency space. Next the spectrum is transformed to velocity and wave number space and integrated across all wave numbers to produce a one dimensional spectrum in velocity space as shown in Figure 5.

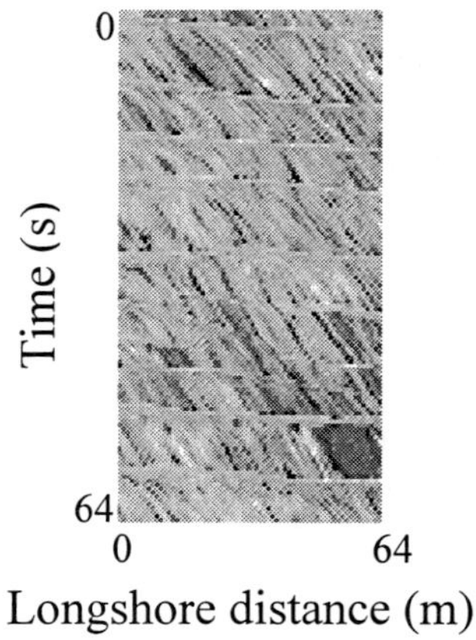

Figure 4 Example of a time stack.

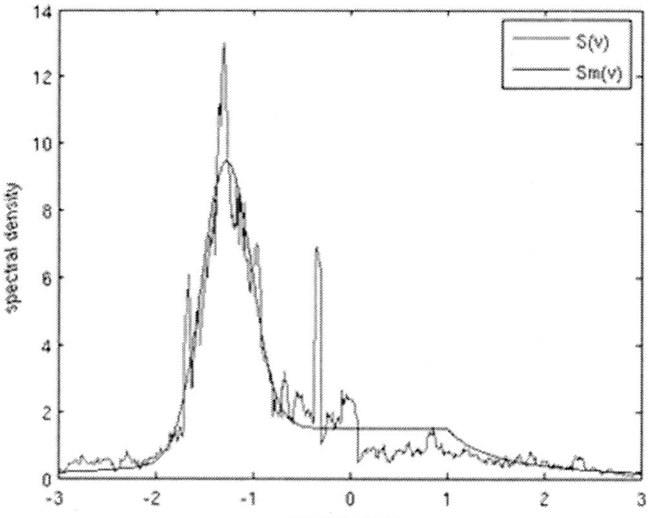

Figure 5 Computed foam spectrum, S(v), with modeled spectrum, Sm(v).

This spectrum is generally fairly noisy; therefore a model accounting for noise in the image and assuming a Gaussian distribution of foam wave lengths is applied. This model contains several free parameters which are determined by applying a least squares fit of the model to the computed spectrum. The velocity corresponding to the peak of the modeled spectrum is taken to be the estimate of the longshore current. There are time stacks for which the model is incapable of fitting the computed spectrum well. Therefore the next step is to assess the quality of the data using a χ^2 statistic and 95% confidence range and to throw out any results that do not satisfy these requirements.

Several corrections to the estimates of the longshore current must be applied as outlined by Chickadel et al. (2003). One of the most critical was found to be the correction due to the viewing angle bias. Because of the oblique field of view, any vertical motion of the foam produces an apparent horizontal motion in the video imagery. This is corrected by using the in situ measurements of the wave height and angles and computing the apparent horizontal velocity and subtracting it from the estimated mean longshore velocity.

RESULTS

To generate time series of the longshore current from the video imagery, time stacks are created for 64 second periods. This produces a single mean value of the longshore current for that period. The next time stack is created with a 16 second increment; therefore the time stacks do have overlap in time. The resulting time series are put through various low-pass filters depending on the feature of interest.

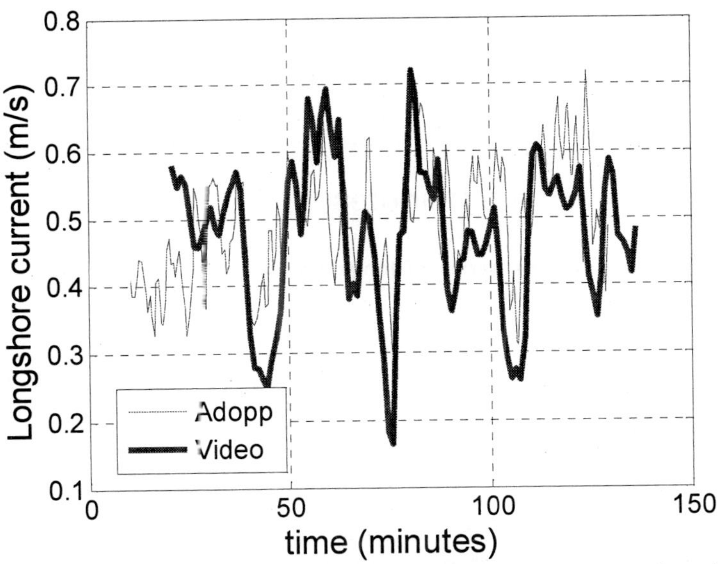

Figure 6 Low-pass filtered time series of longshore current from December 11, 2003.

Figure 6 shows a time series from a two hour period as the first front passed through the region on December 11. This figure shows the longshore current from the top bin measured by the AquaDopp as well as the estimated longshore current from the OCM. Both of these time series have been low-pass filtered with a cutoff frequency corresponding to a period of 2 minutes. It is readily apparent that the measurements from the in situ instrument and the OCM are very similar in both overall magnitudes as well as in temporal variability. The low frequency variability on the order of 20 minutes (the time scale of shear waves) is particularly well resolved by the OCM with excellent representation of the magnitude of the variability. On the other hand, while the variations on the scale of a few minutes (within the infragravity band) are present in both measurements, the magnitudes of these are not nearly as similar.

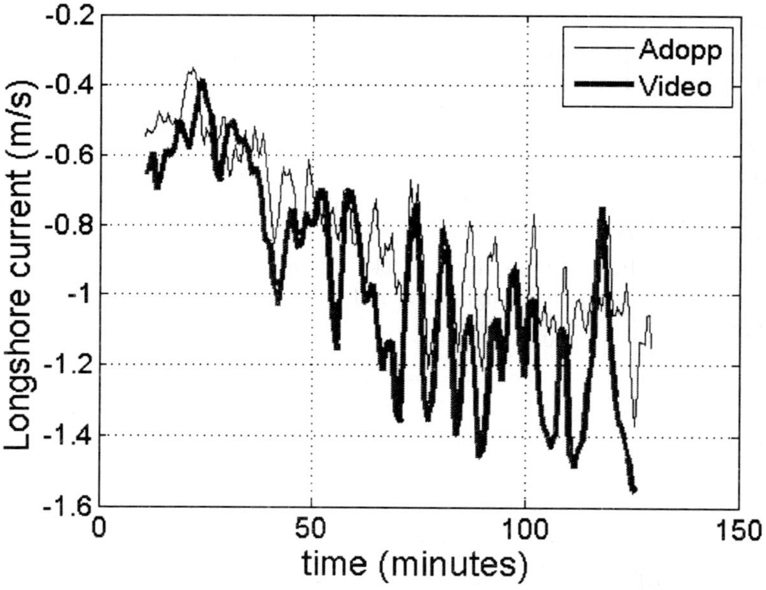

Figure 7 Low-pass filtered time series of longshore current from December 13, 2003.

The time series of the longshore current measured during the passage of the second front is are shown in Figure 7. Once again the overall magnitude of the longshore current is quite similar; however the infragravity variations are not. Instead of presence of shear waves, this time series indicates a slow strengthening of the longshore current.

The full advantage of the OCM technique is demonstrated by computing the cross-shore variation of the 60 minute mean longshore as shown in Figure 8. This figure shows the longshore currents from December 13 spanning a time period of nine hours. The resulting cross-shore profiles have the typical parabolic distribution and correspond to a period of increasing longshore current due to the passage of the front. As the tide falls from hour 5 to hour 8 the peak of the longshore current increases and moves offshore. The wave height during this period remains fairly constant; the large increase in longshore current is directly attributed to the significant increase in the breaking wave angle.

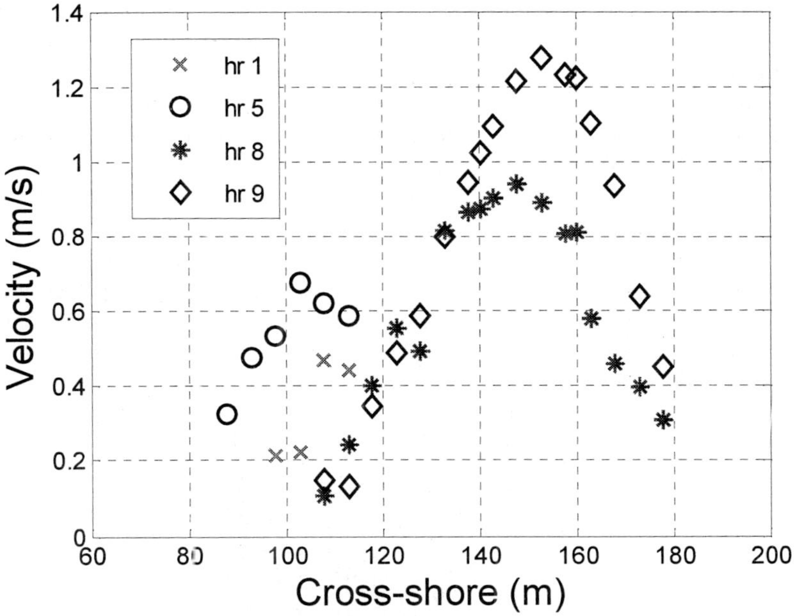

Figure 8 Cross-sections of the 60 minute mean longshore current showing the increase in longshore current from December 13, 2003.

CONCLUSIONS

In this paper the optical current meter technique developed by Chickadel et al. (2003) has been applied to a data set collected near Myrtle Beach, South Carolina. This technique successfully estimated the longshore currents, especially the very low frequency variations. The technique also estimates the infragravity portion of the longshore current, however this did not agree with the in situ measured longshore current nearly as well. The beauty of the technique has been demonstrated by showing the cross-shore distribution of the longshore current estimated by the OCM.

Future work includes using this method for the entire time period of the experiment and to quantify the differences between all of the in situ measurements and the video estimates of the longshore current. This technique will also be applied to a second field experiment which used multiple cameras to expand the field of view. The goal is to use the video imagery to observe shear waves which based on the preliminary results presented within this paper, has a high probability of success. In addition, perhaps the infragravity band can also be observed in the longshore currents with further work.

ACKNOWLEDGMENTS

This work was supported by the National Geospatial-Intelligence Agency, the United States Geological Survey and South Carolina Sea Grant. Their support is gratefully acknowledged.

REFERENCES

Chickadel, C.C., R. A. Holman and M.H. Freilich (2003), An optical technique for the measurements of longshore currents. *J Geophys. Res.*, 108(C11), 3364, doi:10.1029/2003JC001774.

Haas, K., H. Demir, P. Work, G. Voulgaris, and S. Obley, 2004. Myrtle Beach Nearshore Experiment, Dec 10 to Dec 15th, 2003. Part II: Morphodynamic & Remote Imagery Measurements. Technical Report, School of Civil and Environmental Engineering, Georgia Tech – Savannah, Savannah, Georgia.

Obley, S., G. Voulgaris, K. Haas, H. Demir and P. Work, 2004. Myrtle Beach Nearshore Experiment, Dec 10 to Dec 15th, 2003. Part I: Hydrodynamic Measurements. Technical Report CPSD/04/01. Dept. Geological Sciences, University of South Carolina, Columbia, SC. 31p plus Appendices.

VERY LOW FREQUENCY RIP CURRENT PULSATIONS DURING HIGH-ENERGY WAVE CONDITIONS ON A MESO-MACRO TIDAL BEACH

N. Bonneton[1], P. Bonneton[1], N. Sénéchal[1] and B. Castelle[2]

Very Low Frequency (VLF) rip current pulsation investigations are still poorly documented in the literature. In this paper, we analyse data acquired during an in-situ field experiment on Truc Vert beach, on the Aquitanian coast. This beach was exposed to high energy wave conditions during a spring tide cycle and has high tidal range and gentle slope. VLF motions were observed within the surf zone, with a characteristic frequency of 6 10^{-4} Hz which corresponds to a 27 minute period. These VLF motions were associated with high velocity. This paper shows that the VLF velocity pulsations were larger in the inner surf zone and increased with increasing offshore wave energy. We found that the significant wave height and the velocity evolved in-phase. These results are in good agreement with Callaghan et al. 2004. This paper also suggests that wave groups might have generated these VLF pulsations.

INTRODUCTION

This paper presents in-situ field experiment results about Very Low Frequency (VLF) rip current pulsations. Many authors have been interested in rip currents over the past 3 decades (MacMahan et al., 2006), but to our knowledge, the VLF contribution has not been so well documented.

Two recent papers in 2004 investigated these VLF motions, both of them involving in-situ field experiment. MacMahan et al. (2004) made some observations on a beach exhibiting quasi-periodically spaced rip current channels at Sand City, in the southern Monterey Bay (California, USA). These observations took place during the RIPEX field experiment from April to May 2001. The beach face was relatively steep (slope of about 1:10). They showed that the VLF motions were located outside the gravity restoring region, and are believed to be related to surf zone eddies. They also showed that the VLF root mean squared velocity was constant within the surf zone. Callaghan et al. (2004) also investigated these current pulsations in another type of environment. The experiment took place on a beach exhibiting a Transverse Bar and Rip (TBR) system on the eastern coast of Moreton Island (Queensland, Australia). Moreton Island is a sand barrier east of Brisbane and exposed to Pacific Ocean high wave energy. The bathymetry is conformed to the classic rip-feeder system, with two feeder channels inshore of an inner bar converging to the rip channel. All of

[1] UMR EPOC, Bordeaux, France

[2] Griffith Centre for Coastal Management, Australia

these authors agreed with the fact that VLF velocity pulsations are increasing with increasing wave energy. But Callaghan et al (2004) pointed out that, in contrast with MacMahan et al. (2004), the VLF wave energy pulsations were synchronized with both the cross-shore and the alongshore velocity pulsations.
We get interested in that question and we analyse data acquired during an in situ field experiment on the Aquitanian coast, in particular on Truc Vert Beach. The specificities of this site are high wave energy, high tidal range with a gentle sloping beach. Figure 1 shows an aerial photograph of a typical intertidal Transverse Bar and Rip (TBR) system on the Aquitanian coast, with the presence of a wave-induced rip current. This rip current plays a key role in nearshore morphodynamics and is also responsible for beach safety issues during the summer months. This rip current, with an approximately 300 m alongshore wavelength (Fig. 1), is associated with complex circulation cells.
This complexity is well shown in the frequency domain. A typical cross-shore velocity spectra of the rip system of the Aquitanian coast is represented on Figure 2. One can notice the different energetic contribution. The swell contribution is in the frequency band 0.04-035 Hz and the infragravity contribution (IG) is in the frequency band 0.004-0.04 Hz (see Bonneton et al (2004)), the very low frequency (VLF) is in the frequency band 0.0005-0.004 Hz.
On figure 3, you can see the time evolution of the five minute averaged cross-shore velocity.

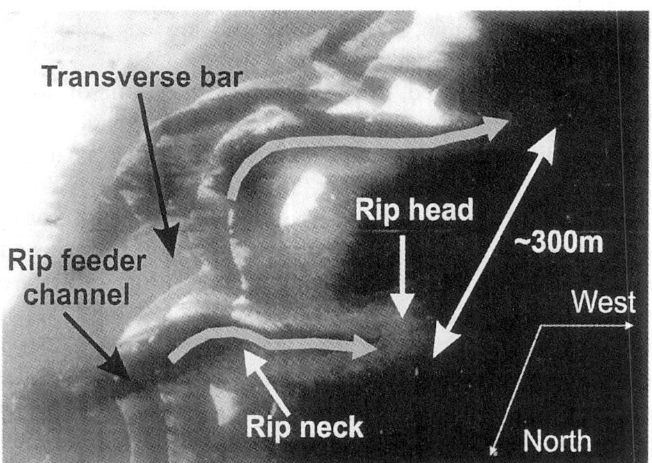

Figure 1. Aerial photograph of a typical Transverse Bar and Rip (TBR) morphology on the Aquitanian coast (Castelle and Bonneton (2006)).

We can notice that the cross-shore velocity is high with a maximum of about 1 m/s which is quite strong. Figure 3 shows cross-shore velocity oscillations of large amplitude. For example, at the time t=200 minutes, the cross-shore velocity oscillations, that is to say **ΔUx** is about 90 cm/s. We can clearly observe, with the naked eye, a period of this velocity which is roughly estimated at 25 minutes, and which corresponds to very low frequency motions.

In this paper, we will discuss the physical processes associated to VLF motions, observed during the Truc Vert beach field experiment.

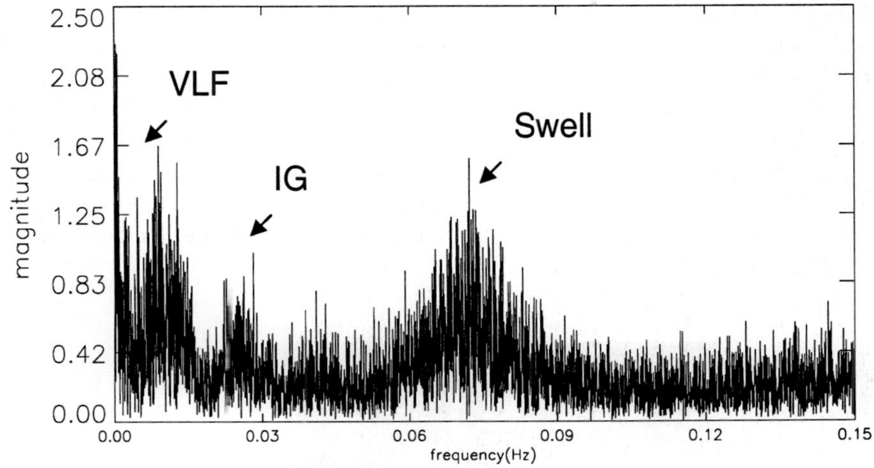

Figure 2. Cross-shore velocity spectra.

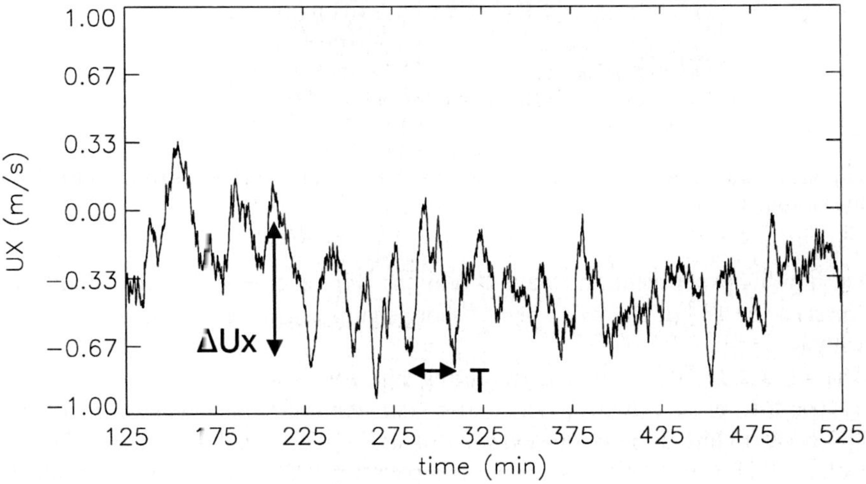

Figure 3. Five minute averaged cross-shore velocity.

FIELD SITE

The experiment took place from the 14th to the 19th of October 2001. The nearshore area was characterized by the presence of two distinct sandbar systems. The outer bar exhibited crescentic patterns and a TBR system connected to an inner crescentic bar system (see Fig. 4). The present study focuses on hydrodynamics over the TBR system. Three transect lines were positioned along which pressure sensors were deployed to provide the alongshore variability of the cross-shore distribution of wave heights. Three bottom-mounted directional wave current meters were deployed. Two S4 positioned at the end of each transect and one ADV (see Figure 4). Such instrument deployment was undertaken in order to assess the 2-dimensional variability of both waves and wave-induced currents. A Triaxys directional wave rider, moored approximately 15 km offshore, provided the incident wave conditions.

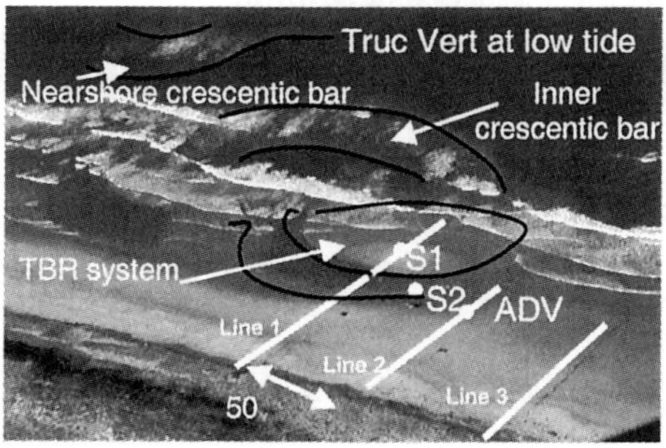

Figure 4. Instrument deployment at Truc Vert Beach during PNEC 2001 field measurement.

During this experiment, the wave conditions were energetic. For this study, we focus on the 18th of October when offshore significant wave heights reached 3 m with a shore-normal incidence.

The wave-induced current patterns during this field experiment were investigated by Castelle et al. (2006) with a time and depth averaged (2DH) modeling approach. Figure 5 shows an example of undertaken simulation on the 18th of October, showing the presence of complex circulation cells over the three-dimensional bedform features. In particular, a strong narrow seaward oriented current is observed with a maximum mean flow velocity of about 1 m/s. Castelle et al. (2006) showed that rip current was favoured by the presence of shore-normal wave incidence and that the maximum rip current flow velocity occurred at mid-tide for low-energy to moderate wave-energy conditions

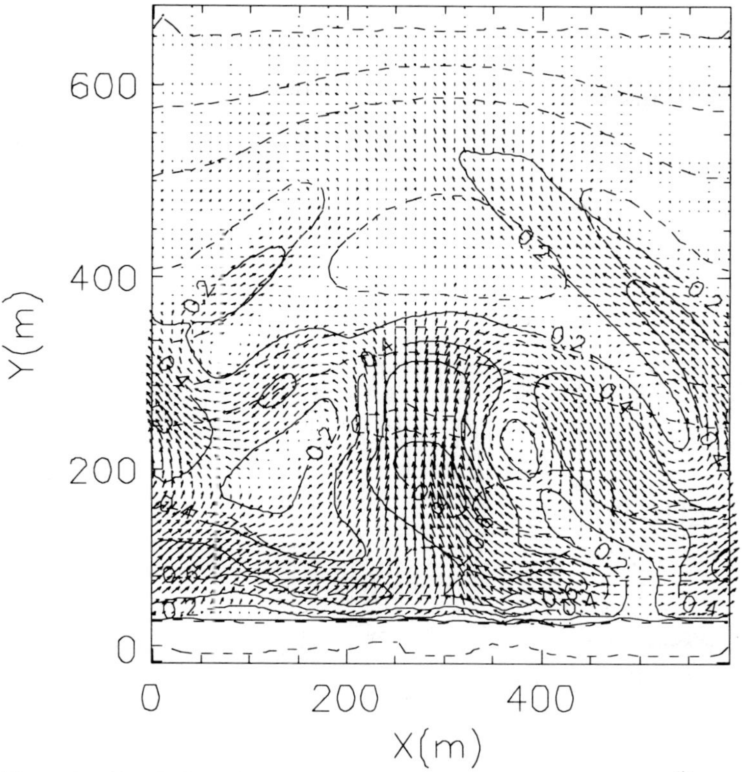

Figure 5. Simulation of mean wave-induced current during the 18th of October 2001, the solid lines represent the iso-values of mean current velocities in m/s, the dashed lines correspond to the bathymetry (Castelle et al., 2006).

RESULTS

You can observe on Figure 6 the time evolution of both the cross-shore (Ux) and alongshore (Uy) velocity during the first tide cycle on the 18th October 2001, when the significant wave height (Hso) was in order of 3 m. The cross-shore velocity intensity is negative if the current goes offshore, and the alongshore velocity intensity is positive in the north direction.

For this very high wave intensity conditions, the cross-shore current can reach the value of 1 m/s, which is quite large.

We observe very low frequency motions with a period in order of 25 minutes and notice quasi-periodic oscillations with very large amplitude for both the cross-shore and alongshore velocity components (see figure 6). To study this VLF motions more precisely, a spectral analysis of the cross-shore velocity has been done. The Ux spectra are computed from the autocorrelation function where the white noise has been filtered (see figure 7). This analysis showed clearly a very energetic peak at f = 6 10^{-4} Hz which corresponds to a period equal to 27 minutes

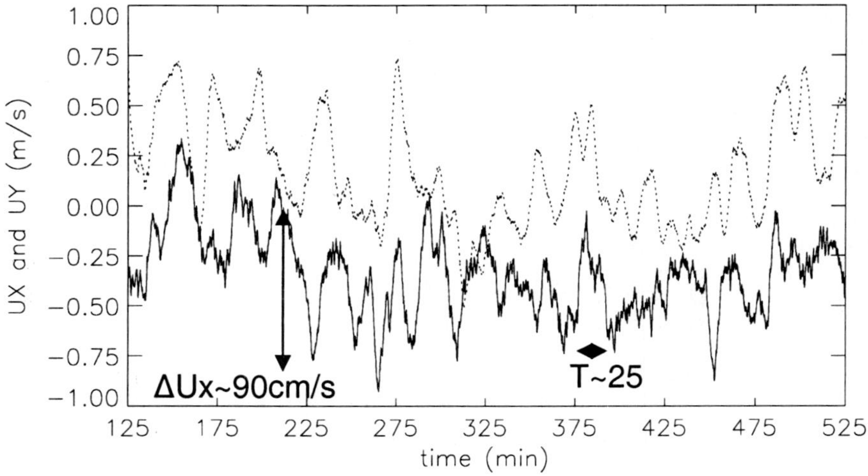

Figure 6. Time evolution of the 5 minute averaged cross-shore velocity (Solid line) and alongshore velocity (dashed line), during the 18th of October 2001.

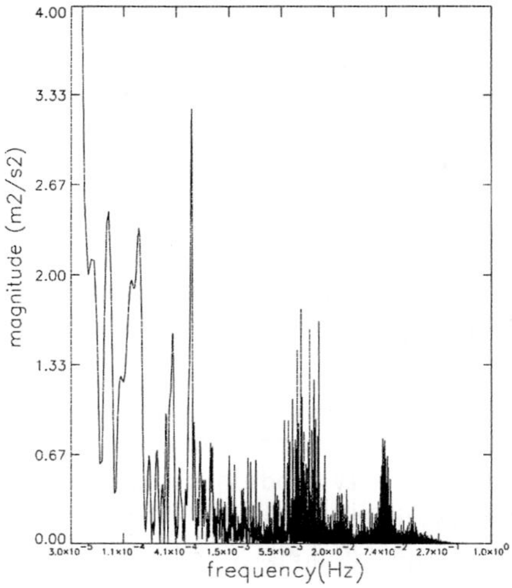

Figure 7. Cross shore velocity spectra.

The five minute averaged cross-shore and alongshore velocity are represented on figure 8. Figure 8 (a) and (b) show respectively their time evolution for the first tide and the second tide of the 18th of October. For both tides, the wave were

frontal, the significant wave height was high, up to 3m (Hso=3m) for the first tide and more moderate for the second tide (Hso=2.3m).
We notice that the intensity of these oscillations is larger when the offshore wave conditions are more energetic.
Sénéchal et al. (2004) have shown that the minimum value of γ = Hs/hm (where Hs is the significant wave height and hm the mean water depth) is equal to 0.5 within the inner surf zone. When γ is smaller than 0.5, the S4 is located outside the inner surf zone. For the first tide (see figure 8 (c)), the instrument was always in the inner surf zone, and for the second tide (see figure 8 (d)), the instrument was outside the inner surf zone from 950 minutes to 1150 minutes. We remark that both the Ux and Uy oscillations intensity are larger in the inner surf zone.

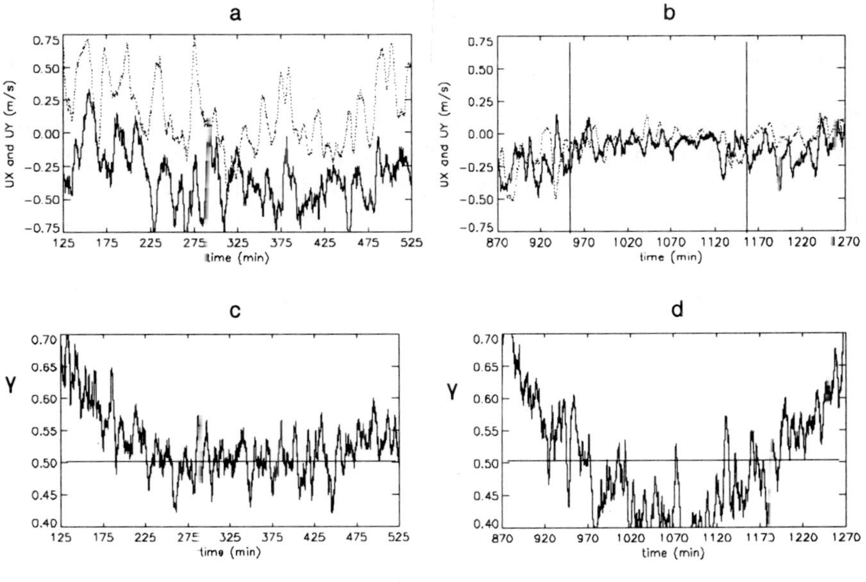

Figure 8. Five minute averaged Ux in solid line and Uy in dashed line during the first tide (a) and the second tide (b), and γ during the first tide (c) and the second tide (d).

On Figure 9 is plotted in solid line, the time evolution of the five minute averaged water depth and in dashed line, the time evolution of the significant wave height for the second tide of the 18th of October. The five minute averaged water depth presents no oscillation. On the contrary, the significant wave height oscillates, with a period in order of 25 minutes, the same as we already pointed out for the cross-shore velocity.

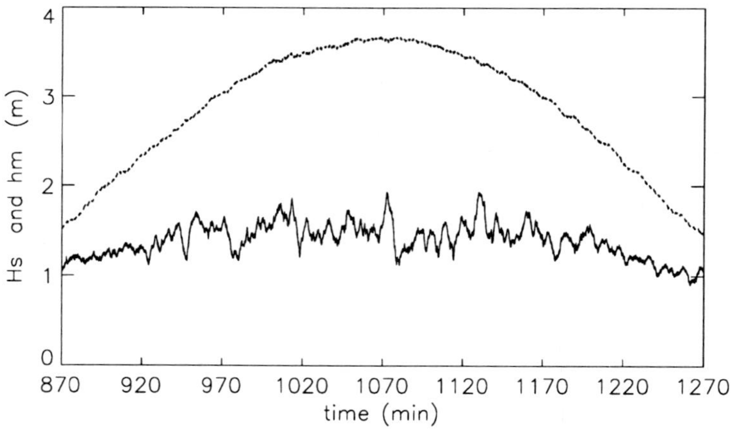

Figure 9. Significant wave height (solid line) and five minute averaged water depth (dashed line).

If we superimpose the significant wave height and the velocity intensity during the second tide of the 18th of October (Figure 10), we find that the significant wave height and the velocity intensity are evolving in-phase. This result is in agreement with Callaghan et al. (2004) observations and indicates that the VLF motions are probably generated by wave groups.

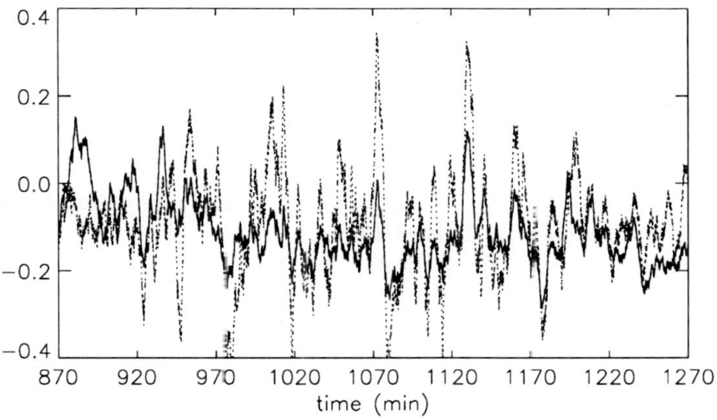

Figure 10. Significant wave height (dashed line) and Velocity intensity (solid line) during the 18th of October 2001.

CONCLUSION

A field experiment was undertaken at Truc Vert Beach during a spring tide cycle for high energy wave conditions. VLF motions are observed in the surf zone with a peak energy frequency of about 6 10^{-4}Hz. These VLF motions are characterized by high velocity (in order of 1m/s) and very large amplitude velocity pulsations. We have shown that the VLF velocity pulsations are larger in the inner surf zone, and increase when the offshore significant wave height is increasing. We found that the significant wave height and the velocity are well correlated. These results are in good agreement with Callaghan et al. (2004). The mean water depth has no oscillations. It seems that wave groups might have generated very low frequency currents.

A new international field experiment on Truc Vert beach in 2008 will be organized to have a better understanding in wave induced hydrodynamic and sand bar morphodynamic. One of the components of that experiment will try to answer to the role of the wave groups on the VLF motions and to confirm or invalidate the relation between large horizontal vortices and VLF motions proposed by Mac Mahan et al (2004). For that, accurate offshore conditions will be accessible.

REFERENCES

Bonneton, P., Marieu,V., Dupuis, H., Sénéchal, N. and Castelle, B. 2004 Wave transformation and energy dissipation in the surf zone : comparison between a non-linear model and field data., *J. Coast. Res,* SI 39, 329-333.

Callaghan, D.P., Baldock, T.E., Nielsen, P., Hanes, D.M., Haas, K and MacMahan, J.H 2004 Pulsing and circulation in rip current system. *Proc 29th Int. Conf. on Coastal Eng.,* 2, 1493-1505.

Castelle, B., Bonneton, P., Sénéchal, N., Dupuis, H., Butel, R. and Michel, D. 2006 Dynamics of wave-induced currents over a multi-barred beach on the Aquitanian coast. *Cont. Shelf Res.,* 26(1), 113-131.

Castelle, B. and Bonneton, P. 2006. Modélisation du courant sagittal induit par les vagues au-dessus des systemes barre/baine de la cote Aquitaine (France), *C.R. Geoscience*, 338(10), 711-717.

MacMahan, J.H., Reniers, A.J.H.M., Thornton, E.B. and Stanton, T.P. 2004 Surf zone eddies coupled with rip current morphology. *J. Geophys. Res,* 109,C07004.

MacMahan, J.H., Thornton, E.B. and Reniers, A.J.H.M. 2006. Rip current review, *Coastal Eng.*, 53(2-3), 191-208.

Sénéchal, N., Dupuis, H. and Bonneton, P. 2004 Preliminary hydrodynamic results of a field experiment on a barred beach, Truc Vert beach on october 2001. *Ocean Dynamics*, 54, 408-414.

AUTHOR INDEX

AUTHOR INDEX

AUTHOR INDEX

AUTHOR INDEX

AUTHOR INDEX

AUTHOR INDEX

AUTHOR INDEX

AUTHOR INDEX

AUTHOR INDEX

AUTHOR INDEX

AUTHOR INDEX

AUTHOR INDEX

AUTHOR INDEX

SUBJECT INDEX

SUBJECT INDEX

SUBJECT INDEX

SUBJECT INDEX

SUBJECT INDEX

SUBJECT INDEX

SUBJECT INDEX

SUBJECT INDEX

SUBJECT INDEX

SUBJECT INDEX

SUBJECT INDEX

SUBJECT INDEX

SUBJECT INDEX

SUBJECT INDEX

SUBJECT INDEX

SUBJECT INDEX

SUBJECT INDEX

SUBJECT INDEX

SUBJECT INDEX

SUBJECT INDEX